A Course in Probability and Statistics

Charles J. Stone

University of California at Berkeley

Duxbury Press
An Imprint of Wadsworth Publishing Company
I(T)P™ An International Thomson Publishing Company

Belmont • Albany • Bonn • Boston • Cincinnati • Detroit • London • Madrid • Melbourne
Mexico City • New York • Paris • San Francisco • Singapore • Tokyo • Toronto • Washington

Editorial Assistant   *Janis Brown*
Production   *Ruth Cottrell*
Print Buyer   *Karen Hunt*
Permissions Editor   *Peggy Meehan*
Copy Editor   *Betty Duncan*
Cover Designer   *Craig Hanson*
Compositor   *SuperScript*
Printer   *Quebecor Printing Book Group/Fairfield*

Copyright © 1996 by Wadsworth Publishing Company
A Division of International Thomson Publishing Inc.
I(T)P The ITP logo is a trademark under license.
Duxbury Press and the leaf logo are trademarks used under license.

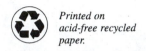 *Printed on acid-free recycled paper.*

Printed in the United States of America
1 2 3 4 5 6 7 8 9 10—02 01 00 99 98 97 96

For more information, contact Duxbury Press at Wadsworth Publishing Company.

Wadsworth Publishing Company
10 Davis Drive
Belmont, California 94002, USA

International Thomson Publishing Europe
Berkshire House 168-173
High Holborn
London, WC1V 7AA, England

Thomas Nelson Australia
102 Dodds Street
South Melbourne 3205
Victoria, Australia

Nelson Canada
1120 Birchmount Road
Scarborough, Ontario
Canada M1K 5G4

International Thomson Editores
Campos Eliseos 385, Piso 7
Col. Polanco
11560 México D.F. México

International Thomson Publishing GmbH
Königswinterer Strasse 418
53227 Bon, Germany

International Thomson Publishing Asia
221 Henderson Road
#05-10 Henderson Building
Singapore 0315

International Thomson Publishing Japan
Hirakawacho Kyowa Building, 3F
2-2-1 Hirakawacho
Chiyoda-ku, Tokyo 102, Japan

Library of Congress Cataloging-in-Publication Data

Stone, Charles J.
   A course in probability and statistics / Charles J. Stone
    p.   cm.
   Includes index.
   ISBN 0-534-23328-7
   1. Probabilities. 2. Mathematical statistics. I. Title.
QA273.S7545 1995
519.5—dc20
                                      95-22403
                                        CIP

Contents

Preface

The writing of this textbook began a decade ago with my realization that, in this age of computers, the core of a first course in theoretical statistics should no longer consist of the classical material on mathematical statistics involving sufficiency, unbiasedness, efficiency, optimality, and decision theory.

At that time I decided to write a textbook for a one-year upper division or graduate level course on probability and statistics, in which the first half would serve as a textbook for a one-semester course in probability and the second half would serve as a textbook for a one-semester course in statistics having probability as a prerequisite. The probability portion would provide the proper background for the treatment of statistics, and the statistics portion in turn would apply and reinforce most of the probability material. The level of the book would be compatible with having two years of calculus, including a modicum of linear algebra (properties of vectors and matrices that are summarized in Appendix A), as its prerequisite.

The decision was made to avoid the convention of requiring random variables to be real-valued. In the more general treatment of random variables given here, it is not necessary to introduce "ω" or to distinguish between probability spaces and distributions of random variables. By this means and by making use of vectors and matrices (with which the students have been surprisingly comfortable), we get a more thorough and systematic treatment of mean vectors, variance–covariance matrices, multivariate transformations, and the multivariate normal distribution in Chapter 5 than is usual for a book at this level. Moreover, by treating independence directly and thereby delaying the introduction of conditional probability, we get a more thorough treatment of conditioning in Chapter 6.

Rather than having the statistics portion of the book deal with a wide variety of models, each on a superficial basis, I decided to concentrate on two closely related major topics: (i) normal models and the corresponding linear models; and (ii) binomial and Poisson models and the corresponding generalized linear models—logistic regression and Poisson regression. The first of these two major topics, treated in Chapters 7–11, covers confidence intervals based on the t distribution, t tests, F tests, the least-squares method, and the design and analysis of experiments. The second topic, treated in Chapters 12 and 13, covers inference for binomial and Poisson distributions as exponential families, the maximum-likelihood method, iteratively-reweighted least squares, normal and multivariate normal approximation to the distribution of maximum-likelihood estimates, Wald tests, and likelihood-ratio tests. Although the emphasis is on developing a thorough conceptual understanding of the material, both the text and supplementary projects introduce the students to various aspects of the analysis of real data.

While writing this textbook, my research has involved the functional approach to statistical modeling—specifically, the use of polynomial splines to model main effects and their tensor products to model interactions. This approach has been successfully applied to model regression functions and the logarithms of density and conditional density functions, hazard and conditional hazard functions (in survival analysis), and spectral density functions (in the analysis of stationary time series). The model fitting involves least-squares and maximum-likelihood estimation. The theory involves rates of convergence, and the corresponding practical methodology involves stepwise addition of basis functions using Rao statistics, stepwise deletion using Wald statistics, and a variety of approaches to final model selection.

After a few years of working on the textbook, I decided to switch from the purely parametric approach to linear and generalized linear modeling to one motivated by my research interests. Thus the treatment presented here involves an unknown function that is assumed to be a member of a specified finite-dimensional linear space of functions. Once we choose a basis of the space, we get the usual formulation in terms of unknown regression coefficients. Nevertheless, this innovative approach has a number of important advantages:

- It is compatible with the modern approach to numerical approximation and other aspects of numerical analysis.

- It strengthens the connections among statistical theory, methodology, and applications.

- It emphasizes conceptual understanding over obscure matrix manipulations in the treatment of linear and generalized linear models.

- It facilitates the proper interpretation of regression coefficients and their linear combinations, especially in the context of logistic regression and models that are not linear in the covariates.

- It provides a smooth transition from the usual ANOVA-type analysis of fractional factorial experiments and other orthogonal arrays to the use of quadratic polynomials in response-surface exploration.

- It provides a common framework for using linear and quadratic functions of factor-level combinations in analyzing small-scale experiments and using polynomial splines or other flexible function spaces in analyzing large observational studies (a topic for a more advanced text).

The book has been developed through the experience of using preliminary versions in teaching upper-division and graduate-level one-year introductory courses in probability and statistics in the Statistics Department at the University of California at Berkeley. The students have been drawn from a variety of majors—including statistics, mathematics, engineering, economics, business administration, and biostatistics. The course format has invariably involved three one-hour (50-minute) lectures and one 1–2 hour discussion section conducted by teaching assistants per week. The discussion sections have mainly been devoted to going over problems and exams, preparations for exams, and preparation for the four computer projects per semester. [Current versions of these projects, which are interfaced to the statistical package S, along with the relevant data sets for the statistics projects, can be obtained from the author (stone@stat.berkeley.edu) or publisher.]

The sections in the book are in one-to-one correspondence with individual lectures. There are 72 such sections: 36 sections in the first six chapters—on probability; and 36 in the last seven chapters—on statistics. In particular, Section 1 in Chapter 1 and Section 1 in Chapter 7 are designed to complement introductory lectures at the beginning of the two semesters. Typically there have been 43 class meetings per semester, with 36 being devoted to primary lectures on the various sections, 5 to reviewing this material, 2 to midterm exams.

This textbook can be used for first or second courses for students at various levels, from a first course at the upper-division level for relatively strong students to a course at the graduate level. For a first course at the undergraduate level, the lectures and discussion sections should focus on the motivation, intuition, basic concepts and applications. Let the students read the derivations on their own and do not hold them responsible for this material on exams. Also, it would be appropriate to have open book exams, containing tasks similar to those in the problems in the text. For a second course for graduate-level students, on the other hand, the lectures and discussion sections should put more emphasis on the derivations, and the students should be held responsible for some such derivations on the exams, which should be at least partly closed book. Naturally, in either case, it would be helpful to make hard copies of the instructor's lectures available to the students as a cohesive, but individualized, treatment intermediate between the full treatment in the book and the summary at the end of the book.

It is my pleasure to thank my students and teaching assistants over the years for their numerous corrections, comments, and suggestions, which have led to substantial improvements in the quality of this textbook.

1

Random Variables and Their Distributions

1.1
Introduction

We are all familiar with a variety of games such as coin tossing, dice, cards, roulette, and darts, in which the outcomes vary in a seemingly haphazard manner. We will refer to such outcomes as random variables. Randomness occurs in more complicated ways in virtually all of our activities: Think of taking a test, getting into an automobile accident, catching a disease, growing crops, predicting the weather, conducting an opinion poll, buying and selling commodities, designing a safe nuclear reactor. The branch of mathematics that treats random phenomena is known as probability theory.

To introduce some of the basic concepts, terminology, and notation of this theory, we first consider the simple experiment of rolling a fair die in an honest manner. It is self-evident that each of the six sides should have the same chance of showing (being the top side) as any other side. We say that each side has one chance in six of showing or that each side has probability $1/6 \doteq .167$ ($\doteq$ means "is approximately equal to" when the error in approximation is due entirely to rounding and other such minor errors in numerical calculations).

Consider instead an honest roll of an unfair die. Perhaps it has been loaded by placing a hidden lead weight near one of the sides to make that side more likely to land on the bottom and hence less likely to show; perhaps one of the corners has been shaved off to make the adjacent sides less likely to land on the bottom and hence more likely to show. It is now no longer self-evident that each side has the same chance of showing as any other side. How, then, do we determine the probability that a given side will show?

The obvious answer is to conduct an experiment having many trials. On each trial we roll the die in an honest manner and note whether or not the given side shows. The total number of occurrences of that side is referred to as its frequency, and the proportion of occurrences is referred to as its relative frequency. We think of this relative frequency as an estimate of the probability that the side shows on a single trial.

Suppose we roll the die 1000 times and that the frequency of the given side is 193. Then the relative frequency of the side is .193, which we regard as an estimate of the probability that the side shows on a single trial.

FIGURE   1.1

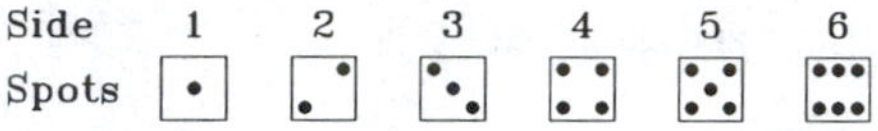

Let the six sides of the die be labeled from 1 to 6 according to the number of spots appearing on these sides (Figure 1.1). Table 1.1 shows possible relative frequencies for an experiment involving 1000 rolls of a die. Table 1.2 shows (hypothetical) probabilities for the six sides. Presumably, if the experiment involved 1 million trials instead of only 1000, the relative frequencies would be closer to the corresponding probabilities.

In general, the *relative frequency interpretation* of probabilities is to think of them as limits of relative frequencies as the number of trials tends to infinity. There

TABLE   1.1

Side	1	2	3	4	5	6
Relative Frequency	.193	.169	.149	.166	.158	.165

TABLE   1.2

Side	1	2	3	4	5	6
Probability	.200	.160	.160	.160	.160	.160

is substantial empirical evidence for the reasonableness of this interpretation in the context of rolling a die and in performing similar experiments involving repeatable trials. It is this interpretation that will be used later on to motivate our development of probability theory.

Probabilities can also be interpreted in a subjective manner. Suppose we examine a die and note that one corner has been shaved but that the die otherwise appears to be quite normal. Before having rolled this unfair die, we might think that the probability of showing for a given side that is not adjacent to the shaved corner would be .15. Such a probability is referred to as a *subjective probability*; it resides in the mind of the subject rather than in the object (the die).

The subject could arrive at his (or her) subjective probability by making a series of thought experiments involving hypothetical bets. Suppose he has the choice of betting on the given side—say, side 1—of the unfair die or on a given side of a fair die. Upon winning the bet, he wins \$1; otherwise, he loses \$1. If the subject prefers to bet on the fair die, he must think that the probability that the unfair die shows side 1 is less than the probability $1/6$ that the fair die shows a given side; that is, his subjective probability that the unfair die shows side 1 must be less than $1/6$.

Upon making this preference, the subject could consider a second hypothetical experiment involving a probability less than $1/6$. For example, he could think about an experiment involving a box and seven tickets, labeled from 1 to 7 but otherwise identical. He puts these tickets into the box and then selects a single ticket in an honest manner (without peeking). We summarize this description by saying that he selects a ticket at random from the box. Evidently, each ticket has probability $1/7 \doteq .143$ of being selected. If the subject would prefer to bet on side 1 of the unfair die rather than on selecting a specified ticket from the box, then his subjective probability that the unfair die shows side 1 is greater than $1/7$.

Upon making this preference, he could consider a third hypothetical experiment involving a box and 100 tickets, labeled from 1 to 100, of which 15 are red and 85 are green. He puts these tickets into the box and then selects a ticket at random from the box. Evidently, each ticket has probability $1/100$ of being selected. Thus, the probability is $15/100 = .15$ that a red ticket will be selected. If the subject is indifferent about betting on side 1 of the unfair die or betting on selecting a red ticket, then his subjective probability that the unfair die shows side 1 is .15.

The theory of probability that is developed in this book can be applied to subjective probabilities, but this will be done only in an incidental manner.

Given a set B, we use $\#(B)$ to denote the number of members of B, which is also referred to as its *cardinality* or *size*. The cardinality of the set $\{y\}$ consisting of the single member y is given by $\#(\{y\}) = 1$. Given integers a and b with $a \leq b$, we use $\{a, \ldots, b\}$ to denote the set of all integers y with $a \leq y \leq b$. Observe that $\#(\{a, \ldots, b\}) = b - a + 1$. In particular, $\#(\{1, \ldots, b\}) = b$ for b a positive integer, and $\#(\{0, \ldots, b\}) = b + 1$ for b a nonnegative integer. Recall that $y \in B$ means that y is a member of B and that $A \subset B$ means that A is a subset of B.

Consider on its own merits the experiment of selecting a ticket at random from a box containing N tickets, labeled from 1 to N. Let $\mathcal{Y}$ denote the set of possible labels, so that $\mathcal{Y} = \{1, \ldots, N\}$, and let $P(B)$ denote the probability that the label of the ticket selected is in the subset B of $\mathcal{Y}$. Then

$$P(\{y\}) = \frac{1}{\#(\mathcal{Y})} = \frac{1}{N}, \qquad y \in \mathcal{Y},$$

and

$$P(B) = \frac{\#(B)}{\#(\mathcal{Y})} = \frac{\#(B)}{N}, \qquad B \subset \mathcal{Y}. \tag{1}$$

Suppose $\mathcal{Y} = \{1, \ldots, 100\}$, so that $N = \#(\mathcal{Y}) = 100$. Then

$$P(\{y\}) = \frac{1}{100}, \qquad y = 1, \ldots, 100.$$

In particular, $P(\{17\}) = 1/100$. Set $B = \{1, \ldots, 15\}$, so that $\#(B) = 15$. Then $P(B) = \#(B)/\#(\mathcal{Y}) = 15/100$.

A function $P(B)$, $B \subset \mathcal{Y}$, that assigns a real number to subsets B of $\mathcal{Y}$ is referred to as a *set function*. The set functions that arise in this book in connection with probability theory are referred to as distributions (or probability distributions or probability measures). The particular distribution given by (1) is referred to as the uniform distribution on $\mathcal{Y}$.

In the context of rolling a fair die in an honest manner or selecting a labeled ticket at random from a box, it is reasonable to assign a uniform distribution to the corresponding random outcome without actually performing any experiments. On the other hand, in the context of rolling a loaded die and in most experimental contexts of practical importance involving random phenomena, it is not obvious how to assign a distribution to the random variable, but we can conduct a compound experiment involving many trials of the simple experiment and use observed relative frequencies to estimate the true but unknown distribution.

In the next section, we will obtain properties of relative frequencies and use them together with the relative frequency interpretation of probabilities to motivate the basic properties of distributions.

Problems

1.1   A fair die is rolled in an honest manner. Determine the probability that the number of spots that shows is more than two but less than five.

1.2   Consider two fair dice that are distinguishable; we label them as die 1 and die 2. Let these dice be rolled in an honest manner. Assume that all $6^2 = 36$ possible outcome pairs have probability $1/36$. Determine the probability that the total number of spots showing on the two dice **(a)** equals 7; **(b)** is an even number.

1.3   (Continued.) Determine the probability that the numbers of spots showing on the two dice coincide.

1.4   A box has 100 tickets, labeled from 1 to 100. A ticket is selected at random from the box. Determine the probability that the label of the ticket is a perfect square $(1, 4, 9, 16, \ldots)$.

1.5   A box has three tickets, labeled from 1 to 3. A ticket is selected at random from the box and then returned to the box, after which a second ticket is selected at random from the box. Assume that all $3^2 = 9$ possible outcome pairs have probability 1/9. Determine the probability that the ticket having label 1 is not selected on either trial.

1.6 A box has three tickets, labeled from 1 to 3. A ticket is selected at random from the box, after which a ticket is selected at random from the two remaining tickets in the box. Assume that all $3 \cdot 2 = 6$ possible outcome pairs have probability 1/6. Determine the probability that the ticket having label 1 is not selected on either trial.

1.7 Let the digits 1, 2, and 3 be written in random order. Assume that each of the $3 \cdot 2 \cdot 1 = 6$ possible orderings has probability 1/6. Determine the probability that at least one digit occupies its proper place.

1.8 An integer is selected at random from the numbers $1, \ldots, 100$. Determine the probability that the selected number is divisible by 2 or 3 (or both).

1.9 Two married couples are seated at random at a round table having four chairs. Determine the probability that no husband is seated next to his wife.

1.10 A drawer has six socks, of which four are tan and two are white. A boy selects two socks at random from the drawer and puts them on. Determine the probability that the two socks on his feet are of the same color.

1.11 Determine your subjective probability that you will receive a grade of A− or better in the course you are taking based on this book. (If you are not taking a course for a letter grade based on this book, modify the exercise appropriately.) In obtaining your subjective probability, do you find it helpful to think about hypothetical experiments involving the selection of a ticket at random from a box?

1.2
Sample Distributions

Notation and Terminology of Set Theory

Before proceeding further, we will summarize the notation and terminology from set theory that will be used in this book. Consider a fixed set $\mathcal{Y}$. Let B be a *subset* of $\mathcal{Y}$—that is, a set every member (or element or point) of which is a member of $\mathcal{Y}$. In general, we use the symbol $\mathbf{y}$ (boldfaced, but not italicized) to denote a particular or possible member of $\mathcal{Y}$. Thus, $\mathbf{y} \in B$ means that $\mathbf{y}$ is a member of B, and $\mathbf{y} \notin B$ means that $\mathbf{y}$ is not a member of B or, equivalently, that $\mathbf{y}$ is a member of the *complement* B^c of B (relative to $\mathcal{Y}$), which consists of all members of $\mathcal{Y}$ that are not in B. Observe that $(B^c)^c = B$. The *empty set* (that is, the set having no members) is denoted by $\emptyset$. Observe that $\emptyset^c = \mathcal{Y}$ and $\mathcal{Y}^c = \emptyset$.

Let B_1 and B_2 be subsets of $\mathcal{Y}$. Then $B_1 \subset B_2$ means that B_1 is a subset of B_2. Alternatively, this relationship can be written as $B_2 \supset B_1$ (B_2 contains B_1). We say that B_1 is a *proper subset* of B_2 if B_1 is a subset of B_2, but there is at least one member of B_2 that is not a member of B_1. Also, $B_1 \cup B_2$ denotes the *union* of B_1 and B_2—that is, the set of points that are in one or both of these sets. Further, $B_1 \cap B_2$ denotes the *intersection* of B_1 and B_2—that is, the set of points that are in both of these sets. Observe that $B_1 \cap B_2 = (B_1^c \cup B_2^c)^c$—that is, that a member of $\mathcal{Y}$ belongs to both of the sets B_1 and B_2 if and only if it is not in the complement of either of these sets. This property is known as *de Morgan's law*. The sets B_1 and B_2 are said to be *disjoint* if they have no members in common—that is, if $B_1 \cap B_2 = \emptyset$. If B is a subset of $\mathcal{Y}$, then B and B^c are disjoint sets whose union is $\mathcal{Y}$.

Let $B_1, B_2, \ldots$ denote a finite or countably infinite collection of subsets of $\mathcal{Y}$. Then $B_1 \cup B_2 \cup \cdots$ is the union of these sets—that is, the set of points that are in one or more of the individual sets. The sets $B_1, B_2, \ldots$ are said to be disjoint if $B_i \cap B_j = \emptyset$ for $i \neq j$; if so, then the cardinality of their union is the sum of their individual cardinalities; that is, $\#(B_1 \cup B_2 \cup \cdots) = \#(B_1) + \#(B_2) + \cdots$. (If the cardinality of the union of an infinite sequence of disjoint sets is finite, then all but finitely many of the individual sets must be empty.) If the sets $B_1, B_2, \ldots$ are disjoint and have union B, then they are said to form a *partition* of B. Similarly, $B_1 \cap B_2 \cap \cdots$ is the intersection of the sets $B_1, B_2, \ldots$; that is, the set of points that are in all of the individual sets. Observe that de Morgan's law

$$B_1 \cap B_2 \cap \cdots = (B_1^c \cup B_2^c \cup \cdots)^c$$

is valid; that is, a point is in all of the sets $B_1, B_2, \ldots$ if and only if it is not in the complement of any of these sets.

The union operation is commutative. In particular, $B_1 \cup B_2 = B_2 \cup B_1$. This operation is also associative. In particular,

$$B_1 \cup B_2 \cup B_3 = (B_1 \cup B_2) \cup B_3 = B_1 \cup (B_2 \cup B_3).$$

Similarly, the intersection operation is commutative and associative. In particular, $B_1 \cap B_2 = B_2 \cap B_1$, and

$$B_1 \cap B_2 \cap B_3 = (B_1 \cap B_2) \cap B_3 = B_1 \cap (B_2 \cap B_3).$$

Moreover, the distributive property is satisfied. In particular,

$$(B_1 \cup B_2) \cap B_3 = (B_1 \cap B_3) \cup (B_2 \cap B_3),$$

and

$$(B_1 \cap B_2) \cup B_3 = (B_1 \cup B_3) \cap (B_2 \cup B_3).$$

Given the sets B_1 and B_2, we let $B_2 \backslash B_1$ denote the set of points that are in B_2 but not in B_1; that is, $B_2 \backslash B_1 = B_2 \cap B_1^c$. As the *Venn diagram* (Figure 1.2) illustrates, $B_1 \cap B_2$ and $B_1 \backslash B_2$ form a partition of B_1; $B_1 \cap B_2$ and $B_2 \backslash B_1$ form a partition of B_2; B_1 and $B_2 \backslash B_1$ form a partition of $B_1 \cup B_2$; B_2 and $B_1 \backslash B_2$ form a partition of $B_1 \cup B_2$; and $B_1 \cap B_2$, $B_1 \backslash B_2$ and $B_2 \backslash B_1$ also form a partition of $B_1 \cup B_2$.

FIGURE   1.2
Venn diagram

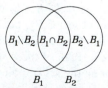

Given points $\mathbf{y}_1, \mathbf{y}_2, \ldots$, the set $\{\mathbf{y}_1, \mathbf{y}_2, \ldots\}$ consists of the indicated points and no other points. In particular, $\{\mathbf{y}\}$ is the set having $\mathbf{y}$ as its only member. We use $\{\mathbf{y} :$ some property involving $\mathbf{y}$ is satisfied $\}$ to indicate the set of points $\mathbf{y}$ satisfying the indicated property. For example, $B^c = \{\mathbf{y} : \mathbf{y} \in \mathcal{Y}$ and $\mathbf{y} \notin B\}$.

The *Cartesian product* $B_1 \times B_2$ of B_1 and B_2 is defined as the set of ordered pairs $(\mathbf{y}_1, \mathbf{y}_2)$ as $\mathbf{y}_1$ ranges over B_1 and $\mathbf{y}_2$ ranges over B_2; that is,

$$B_1 \times B_2 = \{(\mathbf{y}_1, \mathbf{y}_2) : \mathbf{y}_1 \in B_1 \text{ and } \mathbf{y}_2 \in B_2\}.$$

Observe that $\#(B_1 \times B_2) = \#(B_1)\#(B_2)$. More generally, the Cartesian product $B_1 \times \cdots \times B_n$ of sets $B_1, \ldots, B_n$ is defined as the set of ordered n-tuples $(\mathbf{y}_1, \ldots, \mathbf{y}_n)$ as $\mathbf{y}_i$ ranges over B_i for $1 \leq i \leq n$; that is,

$$B_1 \times \cdots \times B_n = \{(\mathbf{y}_1, \ldots, \mathbf{y}_n) : \mathbf{y}_1 \in B_1, \ldots, \mathbf{y}_n \in B_n\}.$$

Observe that

$$\#(B_1 \times \cdots \times B_n) = \#(B_1) \cdots \#(B_n).$$

The set of real numbers is denoted by $\mathbb{R}$. Correspondingly, $\mathbb{R}^2 = \mathbb{R} \times \mathbb{R}$ is the plane, and $\mathbb{R}^3 = \mathbb{R} \times \mathbb{R} \times \mathbb{R}$ is three-space. More, generally, for any positive integer n, Euclidean n-dimensional space $\mathbb{R}^n$ is the Cartesian product of $\mathbb{R}$ with itself n times. When $\mathcal{Y}$ is known to be a set of real numbers (that is, a subset of $\mathbb{R}$), we use the symbol y (italicized, but not boldfaced) instead of $\mathbf{y}$ to denote a particular or possible member of $\mathcal{Y}$.

Let B be a subset of $\mathbb{R}^n$. Then the n-dimensional volume of B is defined by

$$\text{vol}_n(B) = \int_B dy_1 \cdots dy_n.$$

In particular,

$$\text{vol}_1(B) = \int_B dy = \text{length}(B), \qquad B \subset \mathbb{R},$$

$$\text{vol}_2(B) = \int_B dy_1 \, dy_2 = \text{area}(B), \qquad B \subset \mathbb{R}^2,$$

and

$$\text{vol}_3(B) = \int_B dy_1 \, dy_2 \, dy_3 = \text{volume}(B), \qquad B \subset \mathbb{R}^3.$$

Let $B_1, \ldots, B_n$ be subsets of $\mathbb{R}^n$ each having finite length. Then

$$\text{vol}_n(B_1 \times \cdots \times B_n) = \text{length}(B_1) \cdots \text{length}(B_n).$$

In particular,

$$\text{area}(B_1 \times B_2) = \text{length}(B_1) \, \text{length}(B_2)$$

(see Figure 1.3) and

$$\text{volume}(B_1 \times B_2 \times B_3) = \text{length}(B_1) \, \text{length}(B_2) \, \text{length}(B_3).$$

FIGURE   **1.3**
Cartesian product

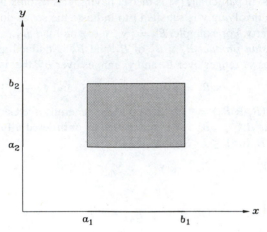

Let a and b be real numbers with $a \leq b$. Recall that the intervals $[a, b]$, $[a, b)$, $(a, b]$, and (a, b) are defined as follows:

$$[a, b] = \{y : a \leq y \leq b\},$$

$$[a, b) = \{y : a \leq y < b\},$$

$$(a, b] = \{y : a < y \leq b\},$$

and

$$(a, b) = \{y : a < y < b\}.$$

Each interval has length $b - a$. Let $a_1 < b_1$ and $a_2 < b_2$. Then $[a_1, b_1] \times [a_2, b_2]$ is the rectangle in the plane having vertices at (a_1, a_2), (b_1, a_2), (a_1, b_2), and (b_1, b_2). For an illustration of the Cartesian product of finite sets, let $B_1 = \{1, 2, 3\}$ and $B_2 = \{1, 2\}$. Then $B_1 \times B_2$ is the set $\{(1, 1), (2,1), (3,1), (1,2), (2,2), (3, 2)\}$ of size $\#(B_1 \times B_2) = \#(B_1)\#(B_2) = 3 \cdot 2 = 6$ (Figure 1.4).

FIGURE   **1.4**
Cartesian product of discrete sets

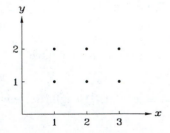

Simple Experiments

Consider the simple experiment of rolling a fair die once in an honest manner and let the outcome be defined as the number y of spots that show. Then the set $\mathcal{Y}$ of possible outcomes is given by $\mathcal{Y} = \{1, \ldots, 6\}$.

Consider instead a simple experiment involving the throwing of a dart at a dartboard (or, more precisely, repeatedly throwing the dart at the dartboard until it sticks). The outcome of a single trial could be defined in terms of the landing point of the dart or the angle or depth of penetration. If, say, the outcome is defined as the ordered pair $\mathbf{y} = (y_1, y_2)$ of real numbers corresponding to the Cartesian coordinates (in inches) of the landing point and if the dartboard is circular, then $\mathcal{Y}$ is a disk in the plane.

Compound Experiments

Consider now a compound experiment consisting of n trials of a simple experiment, where n is a positive integer. Let $\mathbf{y}_i$ denote the outcome on the ith trial for $1 \leq i \leq n$. Then $\mathbf{y}_1, \ldots, \mathbf{y}_n$ are members of $\mathcal{Y}$. Let B be a subset of $\mathcal{Y}$. The number of outcomes $\mathbf{y}_1, \ldots, \mathbf{y}_n$ that are in B is referred to as the *frequency* of B among $\mathbf{y}_1, \ldots, \mathbf{y}_n$ and denoted by $N(B)$; thus, $N(B) = \#(\{i : 1 \leq i \leq n \text{ and } \mathbf{y}_i \in B\})$. Similarly, $N(\{\mathbf{y}\}) = \#(\{i : 1 \leq i \leq n \text{ and } \mathbf{y}_i = \mathbf{y}\})$ is referred to as the frequency of $\mathbf{y}$ among these points.

Observe that

$$N(B) = \sum_{\mathbf{y} \in B} N(\{\mathbf{y}\}). \tag{2}$$

Clearly,

$$N(B) \geq 0, \qquad B \subset \mathcal{Y}, \tag{3}$$

and

$$N(\mathcal{Y}) = n. \tag{4}$$

Also,

$$N(B_1 \cup B_2) = N(B_1) + N(B_2), \qquad \text{if } B_1 \text{ and } B_2 \text{ are disjoint.} \tag{5}$$

To verify (5), we deduce from (2) and a standard property of summation that if B_1 and B_2 are disjoint, then

$$N(B_1 \cup B_2) = \sum_{\mathbf{y} \in B_1 \cup B_2} N(\{\mathbf{y}\})$$

$$= \sum_{\mathbf{y} \in B_1} N(\{\mathbf{y}\}) + \sum_{\mathbf{y} \in B_2} N(\{\mathbf{y}\})$$

$$= N(B_1) + N(B_2).$$

The *relative frequency* (or proportion) $\hat{P}(B)$ of B among $y_1, \ldots, y_n$ is defined by

$$\hat{P}(B) = \frac{N(B)}{n} = \frac{1}{n}\#(\{\,i : 1 \le i \le n \text{ and } y_i \in B\,\}).$$

The set function $\hat{P}(B)$, $B \subset \mathcal{Y}$, is referred to as the *sample distribution* of $y_1, \ldots, y_n$. It follows from (3)—(5), respectively, that

$$\hat{P}(B) \ge 0, \tag{6}$$

$$\hat{P}(\mathcal{Y}) = 1, \tag{7}$$

and

$$\hat{P}(B_1 \cup B_2) = \hat{P}(B_1) + \hat{P}(B_2), \quad \text{if } B_1 \text{ and } B_2 \text{ are disjoint.} \tag{8}$$

We refer to (8) as the additivity of the sample distribution.

EXAMPLE 1.1   Suppose 20 rolls of a die result in the following sequence of values for $y_1, \ldots, y_{20}$: 1, 1, 4, 3, 2, 6, 3, 1, 4, 5, 1, 2, 5, 4, 3, 2, 3, 6, 3, 3. Determine

(a) $N(\{y\})$ for $y = 1, \ldots, 6$;

(b) $N(\{1, 2, 3\})$, $N(\{4, 5\})$, and $N(\{1, 2, 3, 4, 5\})$;

(c) $\hat{P}(B)$ for each of the sets B in part (b).

Solution   **(a)** By inspection, we obtain the following frequency counts:

y	1	2	3	4	5	6
$N(\{y\})$	4	3	6	3	2	2

(b) It follows by inspection of the original data or by using Equation (2) and the frequency counts determined in the solution to part (a) that

$$N(\{1, 2, 3\}) = 13,$$

$$N(\{4, 5\}) = 5,$$

and

$$N(\{1, 2, 3, 4, 5\}) = 18.$$

[Since $\{1, 2, 3\}$ and $\{4, 5\}$ are disjoint sets whose union is $\{1, 2, 3, 4, 5\}$, these results illustrate Equation (5).]

(c) According to the solution to part (b),

$$\hat{P}(\{1, 2, 3\}) = \frac{13}{20} = .65,$$

$$\hat{P}(\{4, 5\}) = \frac{5}{20} = .25,$$

and

$$\hat{P}(\{1, 2, 3, 4, 5\}) = \frac{18}{20} = .9.$$

[These results illustrate Equation (8).] ■

EXAMPLE **1.2** (Continued.) Determine

(a) $N(\{3, 5\})$;   (b) $N([\,3, 5\,])$;   (c) $N([\,3, 5\,))$;

(d) $N((3, 5\,])$;   (e) $N((3, 5))$;   (f) $N([\,3.5, 4.5\,])$.

Solution In the present context, for any set B, $N(B) = N(B \cap \{1, 2, 3, 4, 5, 6\})$.

(a) We have that $N(\{3, 5\}) = N(\{3\}) + N(\{5\}) = 6 + 2 = 8$.

(b) Now $[\,3, 5\,] \cap \{1, 2, 3, 4, 5, 6\} = \{3, 4, 5\}$, so

$$N([\,3, 5\,]) = N(\{3, 4, 5\}) = N(\{3\}) + N(\{4\}) + N(\{5\}) = 6 + 3 + 2 = 11.$$

(c) Now $[\,3, 5) \cap \{1, 2, 3, 4, 5, 6\} = \{3, 4\}$, so

$$N([\,3, 5)) = N(\{3, 4\}) = N(\{3\}) + N(\{4\}) = 6 + 3 = 9.$$

(d) Now $(3, 5\,] \cap \{1, 2, 3, 4, 5, 6\} = \{4, 5\}$, so

$$N((3, 5\,]) = N(\{4, 5\}) = N(\{4\}) + N(\{5\}) = 3 + 2 = 5.$$

(e) Now $(3, 5) \cap \{1, 2, 3, 4, 5, 6\} = \{4\}$, so $N((3, 5)) = N(\{4\}) = 3$.

(f) Now $[\,3.5, 4.5\,] \cap \{1, 2, 3, 4, 5, 6\} = \{4\}$, so $N([\,3.5, 4.5\,]) = N(\{4\}) = 3$. ∎

EXAMPLE **1.3** Consider a dart-throwing experiment in which $\mathcal{Y}$ is the disk in the plane centered at the origin and having radius 12 (inches). The distance d from a possible landing point $\mathbf{y} = (y_1, y_2)$ to the origin is given by $d = \sqrt{y_1^2 + y_2^2}$. Suppose ten throws of the dart at the dartboard yield the data shown in Table 1.3 and Figure 1.5). Determine the frequency and relative frequency of each of the following subsets of $\mathcal{Y}$:

(a) The collection B_1 of points in the plane whose distance from the origin is at most 2;

(b) The collection B_2 of points in the plane whose distance from the origin is greater than 2 but at most 6;

(c) The collection B_3 of points in the plane whose distance from the origin is at most 6.

Solution (a) The dart lands in B_1 on trials 1 and 6, so $N(B_1) = 2$ and $\hat{P}(B_1) = .2$.

(b) The dart lands in B_2 on trials 2, 4, 5, 7, and 10, so $N(B_2) = 5$ and $\hat{P}(B_2) = .5$.

(c) The dart lands in B_3 on trials 1, 2, 4, 5, 6, 7, and 10, so $N(B_3) = 7$ and $\hat{P}(B_3) = .7$.

TABLE **1.3**

Trial	1	2	3	4	5	6	7	8	9	10
y_1	0.66	2.05	4.46	1.29	4.35	−0.69	−5.54	6.75	6.14	−2.23
y_2	−0.21	−5.24	9.23	−1.86	−1.19	0.17	1.08	0.26	4.33	−3.01
d	0.69	5.63	10.25	2.26	4.51	0.71	5.64	6.76	7.51	3.75

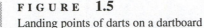

FIGURE   1.5
Landing points of darts on a dartboard

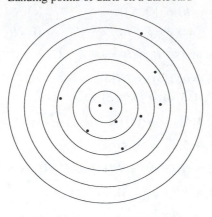

[These results illustrate Equations (5) and (8) since the sets B_1 and B_2 are disjoint and their union is B_3. It should be noted here that B_1, B_2, and B_3 are the collections of all points in the plane satisfying the indicated properties, not just the collections of landing points of darts satisfying these conditions.]   ∎

Law of Relative Frequencies

When experiments involving a large number of trials ($n \gg 1$) have been performed more than once in real-life situations, the relative frequency $\hat{P}(B)$ of a given set B has typically varied only slightly from experiment to experiment. This is compatible with the existence of a number $P(B)$, which is nonrandom and does not depend on n, such that

$$\hat{P}(B) \approx P(B), \qquad \text{when } n \gg 1. \tag{9}$$

(We use $\approx$ to denote "is approximately equal to" when the error in approximation involves more than just rounding and other such minor errors in numerical calculations.) We interpret (9) as meaning that $\hat{P}(B) - P(B)$ can be made arbitrarily small by making n sufficiently large. We refer to (9), which is admittedly imprecise, as the *law of relative frequencies*.

The number $P(B)$ is referred to as the *probability* of B—that is, the probability that the outcome of a single trial will be in B. The interpretation of this probability suggested by the previous paragraph is that when the number of trials is large, the relative frequency $\hat{P}(B)$ of B should be approximately equal to the probability of B; in other words, the probability of B should be the limit of the relative frequency of B as the number of trials tends to infinity.

Suppose the law of relative frequencies is applicable. Since $\hat{P}(B) \geq 0$ by Property (6), we conclude from (9) that

$$P(B) \geq 0, \qquad B \subset \mathcal{Y}. \tag{10}$$

Also, $1 = \hat{P}(\mathcal{Y}) \approx P(\mathcal{Y})$ when $n \gg 1$ by Equation (7) and Property (9), so

$$P(\mathcal{Y}) = 1. \tag{11}$$

Next we will derive the formula

$$P(B_1 \cup B_2) = P(B_1) + P(B_2), \qquad \text{if } B_1 \text{ and } B_2 \text{ are disjoint subsets of } \mathcal{Y}. \tag{12}$$

To this end we observe that, by (9),

$$\hat{P}(B_1) \approx P(B_1), \quad \hat{P}(B_2) \approx P(B_2), \quad \text{and} \quad \hat{P}(B_1 \cup B_2) \approx P(B_1 \cup B_2),$$

when $n \gg 1$, and hence

$$\hat{P}(B_1 \cup B_2) - \hat{P}(B_1) - \hat{P}(B_2)$$
$$\approx P(B_1 \cup B_2) - P(B_1) - P(B_2), \qquad \text{when } n \gg 1. \tag{13}$$

If B_1 and B_2 are disjoint, then $\hat{P}(B_1 \cup B_2) - \hat{P}(B_1) - \hat{P}(B_2) = 0$ for all n by Equation (8), so we conclude from (13) that $P(B_1 \cup B_2) - P(B_1) - P(B_2) = 0$ and hence that (12) holds.

In summary, if a distribution $P(B)$, $B \subset \mathcal{Y}$, arises in a context in which the law of relative frequencies is applicable, then this distribution must satisfy Properties (10)—(12). Starting with the next section, we will in effect regard these three properties as axioms of a distribution.

Problems

1.12 Suppose ten tosses of a coin result in the following sequence of values of H (heads) and T (tails): H, T, H, T, T, H, T, H, H, H. Determine **(a)** the frequencies of heads and tails; **(b)** the relative frequencies of heads and tails.

1.13 Consider the setup of Example 1.1. Determine the frequency and relative frequency of each of the following sets: **(a)** $B_1 = \{2, 3, 4\}$; **(b)** $B_2 = \{5\}$; **(c)** $B_1 \cup B_2 = \{2, 3, 4, 5\}$. [Note that Equations (5) and (8) hold here.]

1.14 Consider the setup of Example 1.1. Determine the frequency and relative frequency of each of the following sets: **(a)** $\{3, 5\}$; **(b)** $[3.5, 5.0)$; **(c)** $(-7, 5)$; **(d)** $[3, 3]$.

1.15 Consider the setup of Example 1.3. Let B_1 denote the collection of points in the plane whose distance from the origin is greater than 6 but at most 10 and let B_2 denote the collection of points in the plane whose distance from the origin is greater than 8. **(a)** Determine the relative frequencies of B_1, B_2, and $B_1 \cup B_2$ and verify that $\hat{P}(B_1 \cup B_2) = \hat{P}(B_1) + \hat{P}(B_2)$. **(b)** Determine the intersection $B_1 \cap B_2$ of B_1 and B_2, that is, the points in the plane that lie in both of these sets. **(c)** According to the answer to part (b), B_1 and B_2 are not disjoint. Why, then, does it turn out that the relative frequency of their union is equal to the sum of their individual relative frequencies?

1.16 Let $\hat{P}$ be a sample distribution on $\mathcal{Y}$ and let B_1 and B_2 be (not necessarily disjoint) subsets of $\mathcal{Y}$. Verify that **(a)** $\hat{P}(B_1) = \hat{P}(B_1 \cap B_2) + \hat{P}(B_1 \backslash B_2)$; **(b)** $\hat{P}(B_2) =$

$\hat{P}(B_1 \cap B_2) + \hat{P}(B_2 \backslash B_1)$;   **(c)** $\hat{P}(B_1 \cup B_2) = \hat{P}(B_1 \cap B_2) + \hat{P}(B_1 \backslash B_2) + \hat{P}(B_2 \backslash B_1)$;
(d) $\hat{P}(B_1 \cup B_2) = \hat{P}(B_1) + \hat{P}(B_2) - \hat{P}(B_1 \cap B_2)$.

1.17   Use the result of Problem 1.16(d) to show that $\hat{P}(B_1 \cup B_2) \leq \hat{P}(B_1) + \hat{P}(B_2)$.

1.18   Use the result of Problem 1.17 to show that $\hat{P}(B_1 \cup B_2 \cup B_3) \leq \hat{P}(B_1) + \hat{P}(B_2) + \hat{P}(B_3)$. [*Hint:* Write $B_1 \cup B_2 \cup B_3 = (B_1 \cup B_2) \cup B_3$.]

1.19   Check the validity of the result of Problem 1.17 in the context of Example 1.1 with $B_1 = \{2, 3, 4\}$ and $B_2 = \{3, 4, 5\}$.

1.20   Check the validity of the result of Problem 1.17 in the context of Example 1.3 with B_1 being the collection of points whose distance from the origin is greater than 2 but at most 6 and B_2 being the collection of points whose distance from the origin is greater than 4 but at most 8.

1.21   A dart is thrown at a dartboard, and the distance from the landing point to the center of the board is reported as 7.28 inches. Which one of the following statements is most reasonable? **(a)** The true distance from the landing point to the center of the board is exactly equal to 7.28 inches. **(b)** The true distance from the landing point to the center of the board is approximately equal to 7.28 inches. **(c)** The true distance from the landing point to the center of the board is not precisely defined.

1.22   **(a)** A possibly loaded die is rolled once and shows three spots. Based on this information, determine a reasonable estimate of the probability that the die shows three spots when rolled again. **(b)** A possibly loaded die is rolled 1000 times; it shows three spots on 250 of these rolls. Based on this information, determine a reasonable estimate of the probability that the die shows three spots if rolled again.

1.23   Let B_1 and B_2 be subsets of $\mathcal{Y}$. Show that **(a)** $(B_1 \cup B_2)^c = B_1^c \cap B_2^c$; **(b)** $B_1 \cup B_2 = (B_1^c \cap B_2^c)^c$; **(c)** $B_1 \cup B_2^c = (B_1^c \cap B_2)^c$; **(d)** $B_1 \cap B_2 = B_2 \cap (B_1 \cup B_2^c)$; **(e)** $B_1 \cap B_2 = B_2 \cap (B_1^c \cap B_2)^c$; **(f)** $B_1 \cap B_2 = B_2 \backslash (B_1^c \cap B_2)$.

1.3
Distributions

In the previous section, we used the law of relative frequencies to motivate three basic properties of a distribution $P(B), B \subset \mathcal{Y}$:

$$P(B) \geq 0, \qquad B \subset \mathcal{Y}, \tag{14}$$

$$P(\mathcal{Y}) = 1, \tag{15}$$

and

$$P(B_1 \cup B_2) = P(B_1) + P(B_2), \qquad \text{if } B_1 \text{ and } B_2 \text{ are disjoint subsets of } \mathcal{Y}. \tag{16}$$

We refer to $P(B)$ as the *probability* of B and to (14) as the nonnegativity of the distribution.

Uniform Distributions

To give some elementary examples of distributions, we introduce the notion of the *size* of a subset B of $\mathcal{Y}$, which we denote by $\text{size}(B)$. This set function is assumed to satisfy three properties:

$$\text{size}(B) \geq 0, \qquad B \subset \mathcal{Y}, \tag{17}$$

$$0 < \text{size}(\mathcal{Y}) < \infty, \tag{18}$$

and

$$\text{size}(B_1 \cup B_2) = \text{size}(B_1) + \text{size}(B_2),$$

$$\text{if } B_1 \text{ and } B_2 \text{ are disjoint subsets of } \mathcal{Y}. \tag{19}$$

For example, let $\mathcal{Y}$ be a finite, nonempty set. Then $\text{size}(B) = \#(B)$ satisfies (17)–(19). For another example, let $\mathcal{Y}$ be an interval in the line $\mathbb{R}$ having finite, positive length. Then $\text{size}(B) = \text{length}(B)$ satisfies (17)–(19). Next, let $\mathcal{Y}$ be a region in the plane $\mathbb{R}^2$ having finite, positive area. Then $\text{size}(B) = \text{area}(B)$ satisfies (17)–(19). Finally, let $\mathcal{Y}$ be a region in space $\mathbb{R}^3$ having finite, positive volume. Then $\text{size}(B) = \text{volume}(B)$ satisfies (17)–(19).

Let $\text{size}(B), B \subset \mathcal{Y}$, satisfy (17)–(19) and set

$$P(B) = \frac{\text{size}(B)}{\text{size}(\mathcal{Y})}, \qquad B \subset \mathcal{Y}.$$

Then $P(B), B \subset \mathcal{Y}$, satisfies Properties (14)–(16), the verification being left as a problem. We refer to this distribution as the *uniform distribution* on $\mathcal{Y}$ and note that it depends on the definition of the size of subsets of $\mathcal{Y}$. In elementary probability problems, the size of a set is implicitly determined by simplicity or familiarity. In such problems, a point selected from $\mathcal{Y}$ according to the uniform distribution on this set is said to be selected *at random* from the set. In this book, the only examples of size (and of uniform distributions and being selected at random) that will be considered are cardinality, length, area, and volume. It is also possible, however, to use other examples of size, such as arc length and surface area, to make up interesting probability problems.

EXAMPLE **1.4**   Let a point be chosen at random from the square $\mathcal{Y} = [1, 6] \times [1, 6]$ shown in Figure 1.6. Determine the probability that the point lies in the subsquare $B = [1, 4] \times [1, 4]$.

Solution   Here it is implicitly understood that $\text{size}(B) = \text{area}(B)$ for $B \subset \mathcal{Y}$. Since $\mathcal{Y}$ has area $5^2 = 25$ and the specified subsquare B has area $3^2 = 9$, the desired probability is given by $P(B) = \text{area}(B)/\text{area}(\mathcal{Y}) = 9/25$. ∎

EXAMPLE **1.5**   Let a point be selected at random from the "lattice square" $\mathcal{Y} = \{1, \ldots, 6\} \times \{1, \ldots, 6\}$ (Figure 1.7) consisting of the points in the plane, each of whose coordinates belongs to the set $\{1, \ldots, 6\}$. Determine the probability that the point lies in

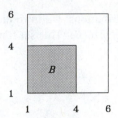

FIGURE **1.6**

the lattice subsquare

$$B = \{1, \ldots, 4\} \times \{1, \ldots, 4\}$$

consisting of the points in the plane each of whose coordinates lies in the set $\{1, 2, 3, 4\}$.

FIGURE **1.7**

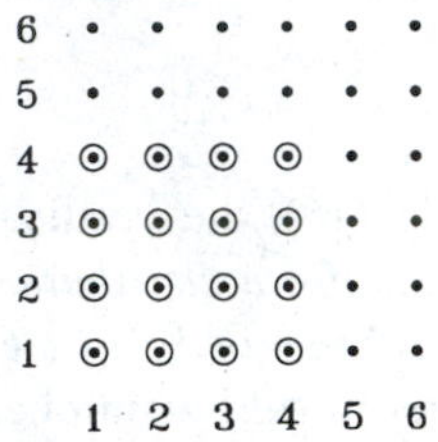

Solution   Here it is understood that size$(B) = \#(B)$ for $B \subset \mathcal{Y}$. Since $\mathcal{Y}$ has $6^2 = 36$ points and the specified lattice subsquare B has $4^2 = 16$ points, the desired probability is given by

$$P(B) = \frac{\#(B)}{\#(\mathcal{Y})} = \frac{16}{36} = \frac{4}{9}. \quad \blacksquare$$

EXAMPLE 1.6   Two fair dice are rolled in an honest manner. Determine the probability that neither die shows more than four spots.

Solution   Thinking of the dice as being distinguishable, we let y_1 denote the number of spots showing on the first die and y_2 the number showing on the second die. Then the ordered pair (y_1, y_2) is a member of the set $\mathcal{Y} = \{1, \ldots, 6\} \times \{1, \ldots, 6\}$. Assuming that the uniform distribution on $\mathcal{Y}$ is applicable in the present context, we see from the solution to Example 1.5 that the desired probability is $4/9$. $\quad \blacksquare$

EXAMPLE 1.7   Two boxes each have four red balls and two white balls. A ball is selected at random from each box. Determine the probability that both of the selected balls are red.

Solution　Think of the balls in each box as being labeled from 1 to 6, with the red balls being labeled from 1 to 4. Let y_1 denote the label of the ball selected from the first box and let y_2 denote the label of the ball selected from the second box. As in the solution to Examples 1.5 and 1.6 (under the assumption of a uniform distribution on $\mathcal{Y}$), we conclude that the probability that both of the selected balls are red is 4/9.　∎

EXAMPLE 1.8　A box has six balls, of which four are red and two are white. A ball is selected at random from the box, its color is noted, and the ball is returned to the box. Then a second ball is selected at random from the box. Determine the probability that a red ball is selected each time.

Solution　As in the solution to Examples 1.5–1.7, we conclude that the desired probability equals 4/9.　∎

EXAMPLE 1.9　A box has six balls, of which four are red and two are white. A ball is selected at random from the box and set aside. Then a ball is selected at random from the five balls remaining in the box. Determine the probability that both of the selected balls are red.

Solution　Think of the balls in the box as being labeled from 1 to 6, with the red balls being labeled from 1 to 4. Let y_1 denote the label of the ball selected on the first trial (that is, the first ball selected) and let y_2 denote the label of the ball selected on the second trial. Since the ball selected on the first trial was not returned to the box before the second trial, we see that $y_1 \neq y_2$. Thus, the ordered pair (y_1, y_2) is a member of the set

$$\mathcal{Y} = \{\, (y_1, y_2) : y_1, y_2 = 1, \ldots, 6 \text{ and } y_1 \neq y_2 \,\},$$

indicated in Figure 1.8. Now both balls are red if and only if (y_1, y_2) is a member of

FIGURE 1.8

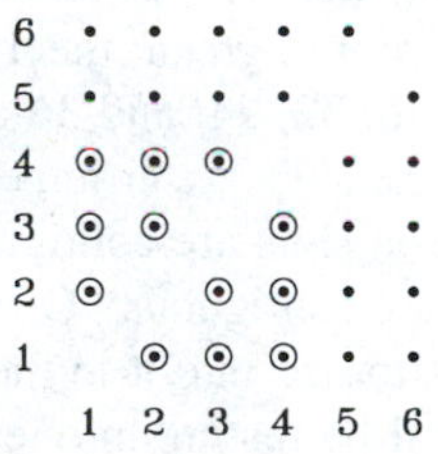

the set

$$B = \{\, (y_1, y_2) : y_1, y_2 = 1, \ldots, 4 \text{ and } y_1 \neq y_2 \,\},$$

also indicated in Figure 1.8. Observe that $\#(\mathcal{Y}) = 30$ and $\#(B) = 12$. Under the reasonable assumption of the uniform distribution P on $\mathcal{Y}$, the desired probability is given by $P(B) = \#(B)/\#(\mathcal{Y}) = 12/30 = 2/5$. ∎

The method of selection described in Example 1.8 is an illustration of sampling with replacement, which is handled by the material in Section 3.3. The method of selection described in Example 1.9 is an illustration of sampling without replacement, which will be discussed in detail in Sections 6.2 and 6.3.

EXAMPLE **1.10**   Determine the probability that eight spots show when two fair dice are rolled in an honest manner.

Solution   Let y_1 and y_2 be as in Example 1.6 and assume again that the ordered pair (y_1, y_2) has the uniform distribution P on $\mathcal{Y} = \{1, \ldots, 6\} \times \{1, \ldots, 6\}$. Then eight spots show if and only if the random outcome lies in the set

$$B = \{(2, 6), (3, 5), (4, 4), (5, 3), (6, 2)\},$$

which has probability $P(B) = \#(B)/\#(\mathcal{Y}) = 5/36$. ∎

EXAMPLE **1.11**   Determine the probability that two heads show when three fair coins are tossed in an honest manner.

Solution   It is reasonable to assume that the random outcome of the experiment has the uniform distribution P on the set $\mathcal{Y} = \{$TTT, TTH, THT, THH, HTT, HTH, HHT, HHH$\}$; here T denotes tails showing and H denotes heads showing. Observe that two heads show if and only if the outcome is in the set $B = \{$THH, HTH, HHT$\}$. Thus, the desired probability is given by $P(B) = \#(B)/\#(\mathcal{Y}) = 3/8$. ∎

Roulette

Consider an American-style roulette wheel, which has 38 slots labeled $1, 2, \ldots, 36$, 0, and 00. The slots labeled 0 and 00 are colored green; the 18 slots labeled 1, 3, 5, 7, 9, 12, 14, 16, 18, 19, 21, 23, 25, 27, 30, 32, 34, and 36 are colored red; and the remaining 18 slots are colored black. (The slots are arranged in a such an order along the rim of the wheel that the two green slots are contiguous, but no two red slots are contiguous and no two black slots are contiguous.) On each trial, the wheel is spun in one direction, and a small ball is tossed into it in the opposite direction. When the wheel stops spinning, the ball will be resting in one of the slots. Before then, each player can bet a positive amount b that the ball will land in one of a number of specified subsets A of $\mathcal{L} = \{1, \ldots, 36, 0, 00\}$ having size 1, 2, 3, 4, 6, 12, or 18. If the ball lands in a slot whose label is in A, the player wins the amount $b[36 - \#(A)]/\#(A)$, which equals $35b$, $17b$, $11b$, $8b$, $5b$, $2b$, or b according as $\#(A)$ equals 1, 2, 3, 4, 6, 12, or 18, respectively. If the ball lands in a slot whose label is not in A, the player loses the amount b of the bet (equivalently, the player wins the amount $-b$). A player can place several such bets on a single play (spin). We make the reasonable assumption that the distribution P of the landing slot of the ball is uniform on $\mathcal{L}$; that is, $P(A) = \#(A)/38$ for $A \subset \mathcal{L}$.

Suppose a roulette player bets \$1 on red. Then he has probability $18/38 = 9/19$ of winning the bet and probability $10/19$ of losing it. If he wins the bet, he wins \$1; otherwise, he loses \$1. Suppose, instead, the player bets \$1 on green. Then he has probability $2/38 = 1/19$ of winning the bet and probability $18/19$ of losing it. If he wins the bet, he wins \$17; otherwise, he loses \$1. Here is a more complicated problem.

EXAMPLE **1.12**   A roulette player bets \$2 on red and \$1 on $\{1, \ldots, 12\}$ on the same play. Determine the probability that the player

(a) wins \$4;   (b) wins \$1;   (c) breaks even;   (d) loses \$3.

Solution   Let $A_1 = \{1, 3, 5, 7, 9, 12, \ldots\}$ denote the set of labels of the red slots and set $A_2 = \{1, \ldots, 12\}$. Then $\#(A_1) = 18$ and $\#(A_2) = 12$. Also, $A_1 \cap A_2 = \{1, 3, 5, 7, 9, 12\}$, so $\#(A_1 \cap A_2) = 6$. Thus, $\#(A_1 \backslash A_2) = 12$ and $\#(A_2 \backslash A_1) = 6$. If the landing slot is in A_1, the player wins \$2 from his bet on red; otherwise, he loses the \$2 he bet on red. If the landing slot is in A_2, the player wins \$2 from his bet on $\{1, \ldots, 12\}$; otherwise, he loses the \$1 that he bet on these numbers. Thus, if the landing slot is in $A_1 \cap A_2$, the player wins \$4; if the landing slot is in $A_1 \backslash A_2$, he wins \$1; if the landing slot is in $A_2 \backslash A_1$, he breaks even; otherwise, he loses \$3.

(a) The probability that the player wins \$4 is given by
$$P(A_1 \cap A_2) = \frac{\#(A_1 \cap A_2)}{38} = \frac{6}{38} = \frac{3}{19}.$$

(b) The probability that the player wins \$1 is given by
$$P(A_1 \backslash A_2) = \frac{\#(A_1 \backslash A_2)}{38} = \frac{12}{38} = \frac{6}{19}.$$

(c) The probability that the player breaks even is given by
$$P(A_2 \backslash A_1) = \frac{\#(A_2 \backslash A_1)}{38} = \frac{6}{38} = \frac{3}{19}.$$

(d) The probability that the player loses \$3 is given by
$$1 - \frac{3}{19} - \frac{6}{19} - \frac{3}{19} = \frac{7}{19}. \quad \blacksquare$$

Properties of Distributions

Suppose $P(B)$, $B \subset \mathcal{Y}$, satisfies Properties (14)–(16). Let B_1, B_2, and B_3 be disjoint subsets of $\mathcal{Y}$; that is, no point belongs to more than one of these sets. Then B_1 and B_2 are disjoint, so
$$P(B_1 \cup B_2) = P(B_1) + P(B_2).$$

Also, $B_1 \cup B_2$ and B_3 are disjoint, so

$$P(B_1 \cup B_2 \cup B_3) = P((B_1 \cup B_2) \cup B_3) = P(B_1 \cup B_2) + P(B_3).$$

Consequently,

$$P(B_1 \cup B_2 \cup B_3) = P(B_1) + P(B_2) + P(B_3).$$

More generally, for any positive integer m,

$$P(B_1 \cup \cdots \cup B_m) = P(B_1) + \cdots + P(B_m),$$

$$\text{if } B_1, \ldots, B_m \text{ are disjoint subsets of } \mathcal{Y}. \qquad \textbf{(20)}$$

In the precise mathematical treatment of distributions in advanced probability theory, it is necessary to assume that (20) holds when m is infinite as well as when it is finite; that is,

$$P(B_1 \cup B_2 \cup \cdots) = P(B_1) + P(B_2) + \cdots,$$

$$\text{if } B_1, B_2, \ldots \text{ are disjoint subsets of } \mathcal{Y}. \qquad \textbf{(21)}$$

Here $B_1, B_2, \ldots$ is either a finite sequence or an infinite sequence of sets, and $B_1 \cup B_2 \cup \cdots$ denotes the collection of points that are in one or more of the sets $B_1, B_2, \ldots$. We refer to (20) as the *additivity* of the distribution and to (21) as its *countable additivity*.

In this book, we define a *distribution P on $\mathcal{Y}$* to be a set function $P(B)$, $B \subset \mathcal{Y}$, satisfying Properties (14), (15), and (21). This definition is technically incomplete. In the complete definition in advanced probability theory (in which a distribution is also referred to as a *probability measure*), $P(B)$ may be undefined for some subsets of $\mathcal{Y}$, which we can think of as being pathological. Fortunately, no such pathological sets arise explicitly in this book. Thus, in the interest of simplicity, we will ignore the possibility that $P(B)$ may be undefined for some B's. Some elementary consequences of the definition of a distribution are summarized in the next result.

THEOREM 1.1   Let P be a distribution on $\mathcal{Y}$. Then

$$P(B^c) = 1 - P(B), \qquad B \subset \mathcal{Y}, \qquad \textbf{(22)}$$

$$P(B) \leq 1, \qquad B \subset \mathcal{Y}, \qquad \textbf{(23)}$$

$$P(\emptyset) = 0, \qquad \textbf{(24)}$$

$$P(B_2 \backslash B_1) = P(B_2) - P(B_1), \qquad \text{if } B_1 \subset B_2 \subset \mathcal{Y}, \qquad \textbf{(25)}$$

and

$$P(B_1) \leq P(B_2), \qquad \text{if } B_1 \subset B_2 \subset \mathcal{Y}. \qquad \textbf{(26)}$$

Proof   To verify (22), observe that, by Properties (15) and (16),

$$1 = P(\mathcal{Y}) = P(B \cup B^c) = P(B) + P(B^c),$$

and hence $P(B^c) = 1 - P(B)$. To verify (23), observe that $P(B^c) \geq 0$. Thus, it follows from (22) that $P(B) = 1 - P(B^c) \leq 1$. To verify (24), observe that, by Properties (15) and (22),

$$P(\emptyset) = P(\mathcal{Y}^c) = 1 - P(\mathcal{Y}) = 1 - 1 = 0.$$

To verify (25), observe that if $B_1 \subset B_2$, then B_1 and $B_2 \backslash B_1$ are disjoint sets whose union is B_2; thus, by Property (16), $P(B_2) = P(B_1) + P(B_2 \backslash B_1)$, and hence (25) holds. To verify (26), note that $P(B_2 \backslash B_1) \geq 0$. Thus, we conclude from (25) that if $B_1 \subset B_2$, then

$$P(B_1) = P(B_2) - P(B_2 \backslash B_1) \leq P(B_2). \quad \blacksquare$$

THEOREM 1.2   Let P be a distribution on $\mathcal{Y}$ and let B_1 and B_2 be subsets of $\mathcal{Y}$. Then

$$P(B_1 \cup B_2) = P(B_1) + P(B_2) - P(B_1 \cap B_2). \tag{27}$$

Proof   Since $B_1 \cap B_2$ and $B_1 \backslash B_2$ are disjoint subsets of $\mathcal{Y}$ whose union is B_1, we conclude from the additivity of P that

$$P(B_1) = P(B_1 \cap B_2) + P(B_1 \backslash B_2). \tag{28}$$

Similarly,

$$P(B_2) = P(B_1 \cap B_2) + P(B_2 \backslash B_1). \tag{29}$$

Since $B_1 \cap B_2$, $B_1 \backslash B_2$ and $B_2 \backslash B_1$ are disjoint subsets of $\mathcal{Y}$ whose union is $B_1 \cup B_2$, we conclude that

$$P(B_1 \cup B_2) = P(B_1 \cap B_2) + P(B_1 \backslash B_2) + P(B_2 \backslash B_1). \tag{30}$$

Equation (27) follows from (28)–(30). $\blacksquare$

COROLLARY 1.1   Let P be a distribution on $\mathcal{Y}$ and let B_1 and B_2 be subsets of $\mathcal{Y}$. Then

$$P(B_1 \cup B_2) \leq P(B_1) + P(B_2).$$

EXAMPLE 1.13   Let P be a distribution on $\mathcal{Y}$ and let B_1 and B_2 be subsets of $\mathcal{Y}$. Show that

(a) if $P(B_1) = 0$ and $P(B_2) = 0$, then $P(B_1 \cup B_2) = 0$;

(b) if $P(B_1) = 1$ and $P(B_2) = 1$, then $P(B_1 \cap B_2) = 1$.

Solution   (a) Suppose $P(B_1) = 0$ and $P(B_2) = 0$. Then $P(B_1 \cup B_2) \leq 0$ by Corollary 1.1, and hence $P(B_1 \cup B_2) = 0$.

(b) Suppose $P(B_1) = 1$ and $P(B_2) = 1$. Then $P(B_1^c) = 0$ and $P(B_2^c) = 0$, so we conclude from (a) that $P(B_1^c \cup B_2^c) = 0$. Thus, by de Morgan's rule,

$$P(B_1 \cap B_2) = 1 - P((B_1 \cap B_2)^c) = 1 - P(B_1^c \cup B_2^c) = 1. \quad \blacksquare$$

Problems

1.24 A roulette player bets $2 on $\{1, \ldots, 6\}$. **(a)** Determine the probability of winning the bet and the probability of losing it. **(b)** How much will the player make if he wins the bet, and how much will he lose otherwise?

1.25 Verify that if size(B), $B \subset \mathcal{Y}$, satisfies Properties (17)–(19), then

$$P(B) = \frac{\text{size}(B)}{\text{size}(\mathcal{Y})}, \qquad B \subset \mathcal{Y},$$

satisfies Properties (14)–(16).

1.26 Let $\mathcal{Y}$ denote the disk in the plane having radius 4 and centered at the origin and let P be the uniform distribution on $\mathcal{Y}$. Determine $P(B)$ for the sets B consisting of the points in $\mathcal{Y}$ satisfying each of the following conditions (recall that the area of a disk of radius r is πr^2): **(a)** The distance from the origin is at most 2; **(b)** The distance from the origin is at most 3; **(c)** The distance from the origin is greater than 2 and at most 3; **(d)** The distance from the origin is at least 2 and at most 3.

1.27 Let $\mathcal{Y}$ denote the set of points in the plane that have integer coordinates and whose distance from the origin is at most 4. Let P be the uniform distribution on $\mathcal{Y}$. Determine $P(B)$ for the sets B consisting of the points in $\mathcal{Y}$ satisfying each of the conditions in Problem 1.26.

1.28 Let P be uniformly distributed on $[\,0, 1\,]$. Determine $P([\,0.025, 0.975\,])$.

1.29 A box has six tickets, numbered consecutively from 1 to 6. Two tickets are selected at random from the box. Determine the probability that the sum of the numbers on the two tickets is an even number **(a)** under sampling with replacement; **(b)** under sampling without replacement.

1.30 A box has three red balls, two white balls, and one blue ball. Two balls are selected at random from the box. Determine the probability that the balls have the same color **(a)** under sampling with replacement; **(b)** under sampling without replacement.

1.31 Two fair dice are rolled in an honest manner. Determine the probability that the number of spots that show equals **(a)** 7; **(b)** 11; **(c)** 7 or 11.

1.32 Determine the probability that at least three heads show when **(a)** four coins are tossed in an honest manner; **(b)** one coin is tossed four times in an honest manner.

Problems 1.33 and 1.34 involve a box containing ten chips, of which four are red, three are white, two are blue, and one is yellow. Red chips are worth $1, white chips are worth $5, blue chips are worth $10, and yellow chips are worth $50.

1.33 A chip is selected at random from the box. Determine the probability that the selected chip is **(a)** white; **(b)** worth $5; **(c)** worth at most $5.

1.34 Two chips are selected from the box. Determine the probability that the sum of the values of the selected chips is at least $15 **(a)** under sampling with replacement; **(b)** under sampling without replacement.

1.35 A roulette player bets $9 on red, $9 on black, and $1 on green, all on the same spin. Determine the probability that the player loses $1.

1.36 A roulette player bets $2 on $\{1, \ldots, 12\}$ and $1 on $\{10, \ldots, 15\}$ on the same play. Determine the probability that the combined amount won on the two bets equals **(a)** 9; **(b)** 3; **(c)** -3; **(d)** some other value.

1.37 Let P be a distribution on $\mathcal{Y}$ and let $B \subset \mathcal{Y}$. Show that $P(B \cap B) = P(B)P(B)$ if and only if $P(B)$ equals 0 or 1.

1.38 Let P be a distribution on $\mathcal{Y}$ and let B_1, B_2, and B_3 be subsets of $\mathcal{Y}$. Show that

$$P(B_1 \cup B_2 \cup B_3) \leq P(B_1) + P(B_2) + P(B_3).$$

[*Hint:* Write $B_1 \cup B_2 \cup B_3$ as $(B_1 \cup B_2) \cup B_3$ and apply Corollary 1.1 twice.]

1.39 Let $B_1, \ldots, B_n$ be subsets of $\mathcal{Y}$. Use Corollary 1.1 and induction to show that $P(B_1 \cup \cdots \cup B_n) \leq P(B_1) + \cdots + P(B_n)$.

1.40 Let $B_1, \ldots, B_n$ be subsets of $\mathcal{Y}$ such that $P(B_1) = \cdots = P(B_n) = 0$. Use the result of Problem 1.39 to show that $P(B_1 \cup \cdots \cup B_n) = 0$.

1.41 Let $B_1, \ldots, B_n$ be subsets of $\mathcal{Y}$ such that $P(B_1) = \cdots = P(B_n) = 1$. Use de Morgan's law and the result of Problem 1.40 to show that $P(B_1 \cap \cdots \cap B_n) = 1$.

1.42 Let B_1 and B_2 be subsets of $\mathcal{Y}$. Use the result of Problem 1.23(f) in Section 1.2 to show that $P(B_1 \cap B_2) = P(B_2) - P(B_1^c \cap B_2)$.

1.4
Random Variables

Consider an experiment in which a fair die is rolled once in an honest manner. Here it is reasonable to assume that the distribution P of the number of spots that show is the uniform distribution on $\mathcal{Y} = \{1, 2, 3, 4, 5, 6\}$. Thus, $P(B) = \#(B)/6$ for $B \subset \mathcal{Y}$. In particular, $P(\{1\}) = \cdots = P(\{6\}) = 1/6$. We use the capital letter Y to denote the random number of spots that show and write $P(B)$ as $P(Y \in B)$ and $P(B^c)$ as $P(Y \notin B)$. Similarly, we write $P(\{1\})$ as $P(Y = 1)$ and $P(\{1\}^c) = P(\{2, 3, 4, 5, 6\})$ as $P(Y \neq 1)$. Thus,

$$P(Y \neq 1) = P(\{2, 3, 4, 5, 6\}) = \frac{\#(\{2, 3, 4, 5, 6\})}{6} = \frac{5}{6};$$

alternatively,

$$P(Y \neq 1) = P(\{1\}^c) = 1 - P(\{1\}) = 1 - \frac{1}{6} = \frac{5}{6}.$$

Consider, in general, an experiment having a random outcome with distribution P on $\mathcal{Y}$. We denote the random outcome by $\mathbf{Y}$, and we refer to $\mathbf{Y}$ as a $\mathcal{Y}$-valued *random variable*, to P as the distribution of $\mathbf{Y}$, and to $P(\mathbf{Y} \in B)$ as the probability of the *event* that $\mathbf{Y}$ is in B. In the expression $P(\mathbf{Y} \in B)$, we can replace $\mathbf{Y} \in B$ by any logically equivalent expression. Thus,

$$P(\mathbf{Y} \in B^c) = P(\mathbf{Y} \notin B),$$

$$P(\mathbf{Y} \in B_1 \cap B_2) = P(\mathbf{Y} \in B_1 \text{ and } \mathbf{Y} \in B_2),$$

$$P(\mathbf{Y} \in B_1 \cup B_2) = P(\mathbf{Y} \in B_1 \text{ or } \mathbf{Y} \in B_2),$$

and

$$P(\mathbf{Y} \in B_2 \backslash B_1) = P(\mathbf{Y} \in B_2 \text{ and } \mathbf{Y} \notin B_1).$$

It is convenient to restate various properties of the distribution of $\mathbf{Y}$ in terms of $\mathbf{Y}$ itself:

$$0 \le P(\mathbf{Y} \in B) \le 1,$$

$$P(\mathbf{Y} \in \mathcal{Y}) = 1,$$

$$P(\mathbf{Y} \in B_1 \cup B_2 \cup \cdots) = P(\mathbf{Y} \in B_1) + P(\mathbf{Y} \in B_2) + \cdots,$$

$$\text{if } B_1, B_2, \ldots \text{ are disjoint,}$$

$$P(\mathbf{Y} \notin B) = 1 - P(\mathbf{Y} \in B),$$

$$P(\mathbf{Y} \in \emptyset) = 0,$$

$$P(\mathbf{Y} \in B_2 \text{ and } \mathbf{Y} \notin B_1) = P(\mathbf{Y} \in B_2) - P(\mathbf{Y} \in B_1), \qquad \text{if } B_1 \subset B_2,$$

and

$$P(\mathbf{Y} \in B_1) \le P(\mathbf{Y} \in B_2), \qquad \text{if } B_1 \subset B_2.$$

Observe that $P(\mathbf{Y} = \mathbf{y}) = P(\mathbf{Y} \in \{\mathbf{y}\}) = P(\{\mathbf{y}\})$ for $\mathbf{y} \in \mathcal{Y}$; here $\mathbf{y}$ is used to denote a possible value of the random variable $\mathbf{Y}$.

EXAMPLE **1.14**   Let $\mathbf{Y}$ be uniformly distributed on a finite nonempty set $\mathcal{Y}$ having N members. Determine $P(\mathbf{Y} = \mathbf{y})$ for $\mathbf{y} \in \mathcal{Y}$.

Solution   Let P denote the distribution of $\mathbf{Y}$. By assumption, P is the uniform distribution on $\mathcal{Y}$. Thus, $P(\mathbf{Y} = \mathbf{y}) = P(\{\mathbf{y}\}) = \#(\{\mathbf{y}\})/\#(\mathcal{Y}) = 1/N$ for $\mathbf{y} \in \mathcal{Y}$.   ∎

EXAMPLE **1.15**   Let $\mathbf{Y}$ be uniformly distributed on a set $\mathcal{Y}$ in the plane having finite positive area. Determine $P(\mathbf{Y} = \mathbf{y})$ for $\mathbf{y} \in \mathcal{Y}$.

Solution   Let P denote the distribution of $\mathbf{Y}$. By assumption, P is the uniform distribution on $\mathcal{Y}$. Thus, $P(\mathbf{Y} = \mathbf{y}) = P(\{\mathbf{y}\}) = \text{area}(\{\mathbf{y}\})/\text{area}(\mathcal{Y}) = 0$ for $\mathbf{y} \in \mathcal{Y}$ since the area of a set consisting of a single point is zero.   ∎

When $\mathcal{Y}$ is understood to be a subset of $\mathbb{R}$, a $\mathcal{Y}$-valued random variable is referred to as a real-valued random variable and denoted by Y. Such random variables play an especially important role in probability theory and its applications.

Let Y be a (real-valued) random variable. If $P(Y \ge 0) = 1$, we refer to Y as a nonnegative random variable; if $P(Y > 0) = 1$, we refer to it as a positive random variable; and if $P(Y \in \{0, \pm 1, \pm 2, \ldots\}) = 1$, we refer to it as an integer-valued random variable.

EXAMPLE 1.16   Let Y be a (real-valued) random variable. Show that

$$P(3 \leq Y \leq 17) = P(Y = 3) + P(3 < Y < 17) + P(Y = 17).$$

Solution   Observe that $\{3\}$, $(3, 17)$ and $\{17\}$ are disjoint sets whose union is $[\,3, 17\,]$. Thus,

$$P(3 \leq Y \leq 17) = P(Y \in [\,3, 17\,])$$
$$= P(Y \in \{3\} \cup (3, 17) \cup \{17\})$$
$$= P(Y \in \{3\}) + P(Y \in (3, 17)) + P(Y \in \{17\})$$
$$= P(Y = 3) + P(3 < Y < 17) + P(Y = 17). \quad \blacksquare$$

EXAMPLE 1.17   Let Y be uniformly distributed on the interval $[\,0, 6\,]$. Determine

(a) $P(0 \leq Y \leq 3)$; **(b)** $P(4 \leq Y \leq 6)$;
(c) $P(0 \leq Y \leq 3 \text{ or } 4 \leq Y \leq 6)$; **(d)** $P(0 \leq Y \leq 3 \text{ and } 4 \leq Y \leq 6)$.

Solution   Since $\text{length}([\,0, 6\,]) = 6$, we see that if $B \in [\,0, 6\,]$, then

$$P(Y \in B) = \frac{\text{length}(B)}{\text{length}([\,0, 6\,])} = \frac{\text{length}(B)}{6}.$$

(a) We have that

$$P(0 \leq Y \leq 3) = P(Y \in [\,0, 3\,]) = \frac{\text{length}([\,0, 3\,])}{6} = \frac{1}{2}.$$

(b) Similarly,

$$P(4 \leq Y \leq 6) = P(Y \in [\,4, 6\,]) = \frac{\text{length}([\,4, 6\,])}{6} = \frac{1}{3}.$$

(c) Since $[\,0, 3\,]$ and $[\,4, 6\,]$ are disjoint intervals, it follows from the solutions to part (a) and (b) that

$$P(0 \leq Y \leq 3 \text{ or } 4 \leq Y \leq 6) = P(Y \in [\,0, 3\,] \text{ or } Y \in [\,4, 6\,])$$
$$= P(Y \in [\,0, 3\,] \cup [\,4, 6\,])$$
$$= P(Y \in [\,0, 3\,]) + P(Y \in [\,4, 6\,])$$
$$= \frac{1}{2} + \frac{1}{3}$$
$$= \frac{5}{6}.$$

(d) Since $[\,0, 3\,]$ and $[\,4, 6\,]$ are disjoint,

$$P(0 \leq Y \leq 3 \text{ and } 4 \leq Y \leq 6) = P(Y \in [\,0, 3\,] \text{ and } Y \in [\,4, 6\,])$$
$$= P(Y \in [\,0, 3\,] \cap [\,4, 6\,])$$
$$= P(Y \in \emptyset)$$
$$= 0. \quad \blacksquare$$

EXAMPLE 1.18   Let $\mathcal{W}$ denote the square $\{(x, y) : 0 \le x, y \le 6\}$ and let the point $\mathbf{W} = (X, Y)$ be chosen at random from $\mathcal{W}$. Determine

(a) $P(0 \le X \le 3)$; **(b)** $P(4 \le Y \le 6)$;

(c) $P(0 \le X \le 3 \text{ or } 4 \le Y \le 6)$; **(d)** $P(0 \le X \le 3 \text{ and } 4 \le Y \le 6)$;

(e) $P(X/3 \le Y \le 2X)$.

Solution   Since area$(\mathcal{W}) = 36$, we see that if $A \subset \mathcal{W}$, then

$$P((X, Y) \in A) = \frac{\text{area}(A)}{\text{area}(\mathcal{W})} = \frac{\text{area}(A)}{36}.$$

(a) Set $A_1 = \{(x, y) : 0 \le x \le 3 \text{ and } 0 \le y \le 6\}$, as shown in Figure 1.9. Then

$$P(0 \le X \le 3) = P(0 \le X \le 3 \text{ and } 0 \le Y \le 6)$$
$$= P((X, Y) \in A_1)$$
$$= \frac{\text{area}(A_1)}{36}$$
$$= \frac{18}{36}$$
$$= \frac{1}{2}.$$

FIGURE 1.9

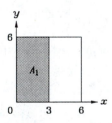

(b) Set $A_2 = \{(x, y) : 0 \le x \le 6 \text{ and } 4 \le y \le 6\}$, as shown in Figure 1.10. Then

$$P(4 \le Y \le 6) = P(0 \le X \le 6 \text{ and } 4 \le Y \le 6)$$
$$= P((X, Y) \in A_2)$$
$$= \frac{\text{area}(A_2)}{36}$$
$$= \frac{12}{36}$$
$$= \frac{1}{3}.$$

FIGURE **1.10**

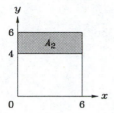

(c) It may be tempting to observe that $[0, 3]$ and $[4, 6]$ are disjoint intervals and conclude that

$$P(0 \leq X \leq 3 \text{ or } 4 \leq Y \leq 6) = P(0 \leq X \leq 3) + P(4 \leq Y \leq 6) = \frac{1}{2} + \frac{1}{3} = \frac{5}{6},$$

but this reasoning would be *invalid* since X and Y are different random variables. The correct answer is given (see Figure 1.11) by

$$P(0 \leq X \leq 3 \text{ or } 4 \leq Y \leq 6) = P((X, Y) \in A_1 \text{ or } (X, Y) \in A_2)$$

$$= P((X, Y) \in A_1 \cup A_2)$$

$$= \frac{\text{area}(A_1 \cup A_2)}{36}$$

$$= \frac{24}{36}$$

$$= \frac{2}{3}.$$

FIGURE **1.11**

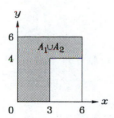

(d) It may be tempting to observe that $[0, 3]$ and $[4, 6]$ are disjoint intervals and conclude that $P(0 \leq X \leq 3 \text{ and } 4 \leq Y \leq 6) = 0$, but this reasoning would be *invalid* since X and Y are different random variables. The correct answer is given (see Figure 1.12) by

$$P(0 \leq X \leq 3 \text{ and } 4 \leq Y \leq 6) = P((X, Y) \in A_1 \text{ and } (X, Y) \in A_2)$$

$$= P((X, Y) \in A_1 \cap A_2)$$

$$= \frac{\text{area}(A_1 \cap A_2)}{36}$$

$$= \frac{6}{36}$$

$$= \frac{1}{6}.$$

FIGURE 1.12

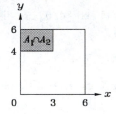

(e) Set $A_3 = \{(x, y) \in \mathcal{W} : x/3 \le y \le 2x\}$, $A_4 = \{(x, y) \in \mathcal{W} : y < x/3\}$, and $A_5 = \{(x, y) \in \mathcal{W} : y > 2x\}$. Then (see Figure 1.13)

$$P\left(\frac{X}{3} \le Y \le 2X\right) = P((X, Y) \in A_3)$$

$$= \frac{\text{area}(A_3)}{36}$$

$$= \frac{36 - \text{area}(A_4) - \text{area}(A_5)}{36}$$

$$= \frac{36 - 6 - 9}{36}$$

$$= \frac{7}{12}. \quad \blacksquare$$

FIGURE 1.13

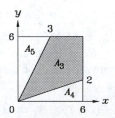

EXAMPLE **1.19**   Let **X** be an $\mathcal{X}$-valued random variable, let **Y** be a $\mathcal{Y}$-valued random variable, let B_1 be a subset of $\mathcal{X}$, and let B_2 be a subset of $\mathcal{Y}$. Show that

$$P(\mathbf{X} \in B_1 \text{ or } \mathbf{Y} \in B_2) = P(\mathbf{X} \in B_1) + P(\mathbf{Y} \in B_2) - P(\mathbf{X} \in B_1 \text{ and } \mathbf{Y} \in B_2).$$

Solution   Set $A_1 = B_1 \times \mathcal{Y}$ and $A_2 = \mathcal{X} \times B_2$. Then **X** is a member of B_1 if and only if the random ordered pair $(\mathbf{X}, \mathbf{Y})$ is a member of A_1, and **Y** is a member of B_2 if and only if $(\mathbf{X}, \mathbf{Y})$ is a member of A_2. Thus, by Theorem 1.2 in Section 1.3,

$$\begin{aligned}
P(\mathbf{X} \in B_1 \text{ or } \mathbf{Y} \in B_2) &= P((\mathbf{X}, \mathbf{Y}) \in A_1 \text{ or } (\mathbf{X}, \mathbf{Y}) \in A_2) \\
&= P((\mathbf{X}, \mathbf{Y}) \in A_1 \cup A_2) \\
&= P((\mathbf{X}, \mathbf{Y}) \in A_1) + P((\mathbf{X}, \mathbf{Y}) \in A_2) \\
&\quad - P((\mathbf{X}, \mathbf{Y}) \in A_1 \cap A_2) \\
&= P((\mathbf{X}, \mathbf{Y}) \in A_1) + P((\mathbf{X}, \mathbf{Y}) \in A_2) \\
&\quad - P((\mathbf{X}, \mathbf{Y}) \in A_1 \text{ and } (\mathbf{X}, \mathbf{Y}) \in A_2) \\
&= P(\mathbf{X} \in B_1) + P(\mathbf{Y} \in B_2) - P(\mathbf{X} \in B_1 \text{ and } \mathbf{Y} \in B_2). \quad \blacksquare
\end{aligned}$$

Let **X** be an $\mathcal{X}$-valued random variable and let **Y** be a $\mathcal{Y}$-valued random variable. When such random variables are simultaneously under consideration, as in Example 1.19, it is implicitly assumed that the random ordered pair $(\mathbf{X}, \mathbf{Y})$ is also a random variable; that is, it has a distribution on $\mathcal{X} \times \mathcal{Y}$, which is referred to as the *joint distribution* of **X** and **Y**. More generally, consider a $\mathcal{Y}_i$-valued random variable $\mathbf{Y}_i$ for $1 \leq i \leq n$. When such random variables $\mathbf{Y}_1, \ldots, \mathbf{Y}_n$ are simultaneously under consideration, it is implicitly assumed that the random ordered n-tuple $(\mathbf{Y}_1, \ldots, \mathbf{Y}_n)$ is also a random variable; that is, it has a distribution on $\mathcal{Y}_1 \times \cdots \times \mathcal{Y}_n$, which is referred to as the joint distribution of $\mathbf{Y}_1, \ldots, \mathbf{Y}_n$.

Transformations

It is common in probability theory for one random variable to be defined in terms of another by means of a transformation. An important task is to determine the distribution of the new random variable from that of the old one.

EXAMPLE **1.20**   Let **W** be uniformly distributed on a disk $\mathcal{W}$ having radius 12 and let Y denote the distance from **W** to the center of the disk. Then Y is a $\mathcal{Y}$-valued random variable, where $\mathcal{Y} = [0, 12]$. Determine $P(2 \leq Y \leq 6)$.

Solution   Set $B = [2, 6]$ and let A be the annulus in $\mathcal{W}$ having inner radius 2, outer radius 6, and the same center as $\mathcal{W}$. Then Y is in B if and only if **W** is in A. Thus,

$$P(2 \leq Y \leq 6) = P(Y \in B) = P(\mathbf{W} \in A) = \frac{\text{area}(A)}{\text{area}(\mathcal{W})} = \frac{\pi(6^2 - 2^2)}{\pi(12^2)} = \frac{32}{144} = \frac{2}{9}. \quad \blacksquare$$

Let $\mathcal{W}$ and $\mathcal{Y}$ be arbitrary sets and let $\mathbf{g}$ be a $\mathcal{Y}$-valued function on $\mathcal{W}$—that is, such that $\mathbf{g}(\mathbf{w}) \in \mathcal{Y}$ for $\mathbf{w} \in \mathcal{W}$. Given a subset B of $\mathcal{Y}$, we refer to the subset A of $\mathcal{W}$ consisting of all members $\mathbf{w} \in \mathcal{W}$ such that $\mathbf{g}(\mathbf{w}) \in B$ as the *inverse image* of B under $\mathbf{g}$. This inverse image is denoted by $\mathbf{g}^{-1}(B)$; thus, $A = \mathbf{g}^{-1}(B) = \{\, \mathbf{w} \in \mathcal{W} : \mathbf{g}(\mathbf{w}) \in B \,\}$.

In Example 1.20, we can write $\mathbf{W} = (W_1, W_2)$, where W_1 and W_2 are the Cartesian coordinates of $\mathbf{W}$. Correspondingly,

$$\mathcal{W} = \left\{ (w_1, w_2) \in \mathbb{R}^2 : \sqrt{w_1^2 + w_2^2} \leq 12 \right\},$$

and

$$A = \left\{ (w_1, w_2) \in \mathcal{W} : 2 \leq \sqrt{w_1^2 + w_2^2} \leq 6 \right\}.$$

The random variable Y can be written explicitly as

$$Y = \sqrt{W_1^2 + W_2^2}.$$

Equivalently, we can write $Y = g(\mathbf{W})$, where g is the function from $\mathcal{W}$ onto $\mathcal{Y}$ given by

$$g(\mathbf{w}) = \sqrt{w_1^2 + w_2^2}, \qquad \mathbf{w} = (w_1, w_2) \in \mathcal{W}.$$

Figure 1.14 shows the relationship between $B = [\,2, 6\,]$ and A.

F I G U R E   1.14
$A = g^{-1}(B) = \{\, \mathbf{w} \in \mathcal{W} : g(\mathbf{w}) \in (B) \,\}$

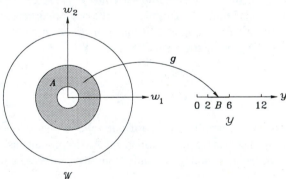

Let $\mathbf{W}$ be a $\mathcal{W}$-valued random variable. Then $\mathbf{Y} = \mathbf{g}(\mathbf{W})$ is a $\mathcal{Y}$-valued random variable, which is referred to as a *transform* of $\mathbf{W}$. The distribution of $\mathbf{Y}$ is uniquely determined by the distribution of $\mathbf{W}$ and the function $\mathbf{g}$. Specifically,

$$P(\mathbf{Y} \in B) = P(\mathbf{W} \in A), \qquad \text{where } B \subset \mathcal{Y} \text{ and } A = \mathbf{g}^{-1}(B).$$

In applications of this formula, of course, we can use different symbols for the various random variables and sets under consideration.

EXAMPLE **1.21**   Let a point $\mathbf{W} = (X, Y)$ be chosen at random from the square

$$\mathcal{W} = [-1, 1] \times [-1, 1]$$

and let $R = \sqrt{X^2 + Y^2}$ denote the distance from the chosen point to the center of the square. Determine

(a) $P(R \le r)$ for $0 \le r \le 1$;   **(b)** $P(R \le r)$ for $1 < r \le \sqrt{2}$;

(c) $P(R \le 2/\sqrt{3})$.

Solution   Observe that area$(\mathcal{W}) = 4$. Given r with $0 \le r \le \sqrt{2}$, let B denote the interval $[0, r]$ and set $A = \{ (x, y) \in \mathcal{W} : \sqrt{x^2 + y^2} \le r \}$.

(a) Suppose $0 \le r \le 1$. Then A is a disk having radius r and hence area πr^2, so

$$P(R \le r) = P(R \in B) = P(\mathbf{W} \in A) = \frac{\text{area}(A)}{\text{area}(\mathcal{W})} = \frac{\pi r^2}{4}.$$

(b) Suppose $1 < r \le \sqrt{2}$ and let $\theta = \text{arcsec } r$ and $\varphi = \frac{\pi}{2} - 2\theta$ be the angles in radians as shown in Figure 1.15. Then

$$\text{area}(A) = 4\sqrt{r^2 - 1} + 4\left(\frac{\varphi}{2\pi}\right)\pi r^2 = 4\sqrt{r^2 - 1} + (\pi - 4\,\text{arcsec } r)r^2$$

(recall that the area of a sector of a disk is proportional to its central angle), so

$$P(R \le r) = P(R \in B)$$

$$= P(\mathbf{W} \in A)$$

$$= \frac{\text{area}(A)}{\text{area}(\mathcal{W})}$$

$$= \sqrt{r^2 - 1} + \left(\frac{\pi}{4} - \text{arcsec } r\right)r^2.$$

(c) According to the answer to part (b),

$$P\left(R \le \frac{2}{\sqrt{3}}\right) = \sqrt{\frac{4}{3} - 1} + \left(\frac{\pi}{4} - \text{arcsec } \frac{2}{\sqrt{3}}\right)\frac{4}{3}$$

$$= \sqrt{\frac{1}{3}} + \left(\frac{\pi}{4} - \frac{\pi}{6}\right)\frac{4}{3}$$

$$= \frac{\sqrt{3}}{3} + \frac{\pi}{9}. \quad \blacksquare$$

FIGURE **1.15**

In our treatment of probability, we have introduced random variables in an informal, mathematically imprecise manner as the random outcome of an experiment. In this way they play a role similar to that of "points" in Euclidean geometry, which also are not precisely defined. A fundamental property of random variables is that they have distributions, which are precisely defined. A second such property is that an ordered n-tuple of random variables is a random variable, and a third fundamental property is that the transform of a random variable is a random variable whose distribution is given explicitly in terms of the distribution of the original random variable and the function used to specify the transform.

Problems

1.43 A box has eight balls, which are labeled from 1 to 8. Let Y denote the label of a ball selected at random from the box. Determine $P(2 \leq Y \leq 5)$.

1.44 Let Y be uniformly distributed on $[-2, 3]$. Determine **(a)** $P(|Y| \leq 1)$; **(b)** $P(Y = 0)$; **(c)** $P(-2 < Y < 3)$.

1.45 Let Y be uniformly distributed on the interval $[0, 1]$. Determine **(a)** $P(Y = .4)$; **(b)** $P(.3 \leq Y \leq .4)$; **(c)** $P(.3 < Y < .4)$; **(d)** $P(Y - .475 \leq .5 \leq Y + .475)$.

1.46 Let $\mathbf{Y}$ be uniformly distributed on a disk $\mathcal{Y}$ in the plane and let B be a square in $\mathcal{Y}$ whose vertices all lie on the boundary of $\mathcal{Y}$. Determine $P(\mathbf{Y} \in B)$.

1.47 Let Y denote the number of spots that show when a fair die is rolled in an honest manner. Determine **(a)** $P(Y = 3)$; **(b)** $P(Y < 3)$; **(c)** $P(3 \leq Y \leq 4)$; **(d)** $P(1 < Y < 6)$.

1.48 Let Y denote the number of spots that show when two fair die are rolled in an honest manner. Determine **(a)** $P(Y = 3)$; **(b)** $P(Y < 3)$; **(c)** $P(3 \leq Y \leq 4)$; **(d)** $P(1 < Y < 6)$.

1.49 A box has 10 tickets, labeled consecutively from 1 to 10. Suppose the value of each ticket is equal to the square of its label. Let Y denote the value of a ticket selected at random from the box. Determine **(a)** $P(Y = 36)$; **(b)** $P(Y = 40)$; **(c)** $P(20 \leq Y \leq 50)$.

1.50 A box has 1000 tickets, labeled consecutively from 1 to 1000. Let L denote the label of a ticket selected at random from the box and set $Y = L/1000$. Determine **(a)** $P(Y = .4)$; **(b)** $P(.3 \leq Y \leq .4)$; **(c)** $P(.3 < Y < .4)$.

1.51 Let X and Y be random variables such that $P(X = 0) = 0$ and $P(Y = 0) = 0$. Show that $P(X = 0 \text{ or } Y = 0) = 0$.

1.52 Let $\mathbf{W} = (X, Y)$ be uniformly distributed on the square having vertices at $(-1, 0)$, $(1, 0)$, $(1, 2)$, and $(-1, 2)$ and let $R = \sqrt{X^2 + Y^2}$ denote the distance from $\mathbf{W}$ to the origin. Determine $P(R \leq 2)$.

1.53 Let $\mathbf{W} = (X, Y)$ be chosen at random from the square with vertices at $(1, 0)$, $(0, 1)$, $(-1, 0)$, and $(0, -1)$. Determine **(a)** $P(X \geq \frac{1}{2})$; **(b)** $P(Y \geq \frac{1}{2})$; **(c)** $P(X \geq \frac{1}{2} \text{ and } Y \geq \frac{1}{2})$; **(d)** $P(X \geq \frac{1}{2} \text{ or } Y \geq \frac{1}{2})$; **(e)** $P(X \geq 0 \text{ and } Y \geq 0)$; **(f)** $P(X^2 + Y^2 \leq \frac{1}{2})$.

1.54   Let $\mathbf{X}$ be an $\mathcal{X}$-valued random variable, let $\mathbf{Y}$ be a $\mathcal{Y}$-valued random variable, and let B_1 and B_2 be subsets of $\mathcal{X}$ and $\mathcal{Y}$, respectively. Show that

$$P(\mathbf{X} \in B_1, \mathbf{Y} \in B_2) = P(\mathbf{Y} \in B_2) - P(\mathbf{X} \notin B_1, \mathbf{Y} \in B_2).$$

[*Hint:* Use the result of Problem 1.42 in Section 1.3, with B_1 and B_2 replaced by suitable sets A_1 and A_2.]

1.5
Probability Functions and Density Functions
Probability Functions

A function f on a set $\mathcal{Y}$ is said to be a *probability function* on $\mathcal{Y}$ if f is nonnegative on $\mathcal{Y}$ and $\sum_{\mathbf{y} \in \mathcal{Y}} f(\mathbf{y}) = 1$. Let f be a probability function on $\mathcal{Y}$. Then

$$P(B) = \sum_{\mathbf{y} \in B} f(\mathbf{y}), \qquad B \subset \mathcal{Y},$$

determines a distribution P on $\mathcal{Y}$, which is said to have probability function f. Every distribution on a finite or countably infinite set $\mathcal{Y}$ has a probability function. Conversely, a distribution on $\mathcal{Y}$ that has a probability function f can be thought of as a distribution on the finite or countably infinite subset $\mathcal{Y}_0 = \{\mathbf{y} \in \mathcal{Y} : f(\mathbf{y}) > 0\}$ of $\mathcal{Y}$. A distribution that has a probability function is referred to as a *discrete distribution*.

Let $\mathbf{Y}$ be a $\mathcal{Y}$-valued random variable having probability function $f = f_{\mathbf{Y}}$; that is, its distribution has probability function f. Then $P(\mathbf{Y} = \mathbf{y}) = f(\mathbf{y})$ for $\mathbf{y} \in \mathcal{Y}$, and $P(\mathbf{Y} \in B) = \sum_{\mathbf{y} \in B} f(\mathbf{y})$ for $B \subset \mathcal{Y}$. A random variable that has a probability function is referred to as a *discrete random variable*.

A $\mathcal{Y}$-valued random variable $\mathbf{Y}$ is said to be a *constant random variable* if $P(\mathbf{Y} = \mathbf{a}) = 1$ for some $\mathbf{a} \in \mathcal{Y}$. Such a random variable has the probability function f given by $f(\mathbf{a}) = 1$ and $f(\mathbf{y}) = 0$ for $\mathbf{y} \neq \mathbf{a}$.

THEOREM 1.3   Let $\mathcal{Y}$ be a finite set, let $\mathbf{Y}$ be a $\mathcal{Y}$-valued random variable, and let f denote the probability function of $\mathbf{Y}$. Then $\mathbf{Y}$ is uniformly distributed on $\mathcal{Y}$ if and only if

$$f(\mathbf{y}) = \frac{1}{\#(\mathcal{Y})}, \qquad \mathbf{y} \in \mathcal{Y}.$$

Proof   Suppose $\mathbf{Y}$ is uniformly distributed on $\mathcal{Y}$. Then

$$f(\mathbf{y}) = P(\mathbf{Y} = \mathbf{y}) = P(\mathbf{Y} \in \{\mathbf{y}\}) = \frac{\#(\{\mathbf{y}\})}{\#(\mathcal{Y})} = \frac{1}{\#(\mathcal{Y})}, \qquad \mathbf{y} \in \mathcal{Y}.$$

Suppose, conversely, that $f(\mathbf{y}) = 1/\#(\mathcal{Y})$ for $\mathbf{y} \in \mathcal{Y}$. Then

$$P(\mathbf{Y} \in B) = \sum_{\mathbf{y} \in B} f(y) = \sum_{\mathbf{y} \in B} \frac{1}{\#(\mathcal{Y})} = \frac{\#(B)}{\#(\mathcal{Y})}, \qquad B \subset \mathcal{Y},$$

so $\mathbf{Y}$ is uniformly distributed on $\mathcal{Y}$.   ∎

EXAMPLE 1.22   Let Y denote the number of spots that show when a fair die is rolled in an honest manner. Determine the probability function of Y and show a plot of this function.

Solution   Since Y is uniformly distributed on $\{1, \ldots, 6\}$, its probability function is given by $f(y) = 1/6$ for $y \in \{1, \ldots, 6\}$, which is shown in Figure 1.16.   ∎

FIGURE 1.16
Uniform probability function

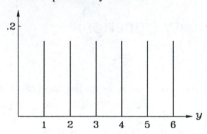

EXAMPLE 1.23   Let two fair dice be rolled in an honest manner, let W_1 denote the number of spots that show on the first die, and let W_2 denote the number that show on the second die. Determine the probability function of $\mathbf{W} = (W_1, W_2)$.

Solution   We interpret the description of the problem as implying that $\mathbf{W}$ is uniformly distributed on $\mathcal{W} = \{ (w_1, w_2) : w_1, w_2 \in \{1, \ldots, 6\} \}$. Since $\#(\mathcal{W}) = 36$, we see that $f_{\mathbf{W}}(\mathbf{w}) = 1/36$ for $\mathbf{w} \in \mathcal{W}$.   ∎

Let $\mathbf{W}$ be a $\mathcal{W}$-valued random variable having probability function $f_{\mathbf{W}}$ and let $\mathbf{Y} = \mathbf{g}(\mathbf{W})$ be a $\mathcal{Y}$-valued transform of $\mathbf{W}$. Then $\mathbf{Y}$ has probability function $f_{\mathbf{Y}}$, which is determined as follows: For $\mathbf{y} \in \mathcal{Y}$, $f_{\mathbf{Y}}(\mathbf{y})$ is the sum of $f_{\mathbf{W}}(\mathbf{w})$ over all $\mathbf{w} \in \mathcal{W}$ such that $\mathbf{g}(\mathbf{w}) = \mathbf{y}$. In particular, if $\mathcal{W}$ is a finite set and $\mathbf{W}$ is uniformly distributed on $\mathcal{W}$, then

$$f_{\mathbf{Y}}(\mathbf{y}) = \frac{1}{\#(\mathcal{W})} \#(\{\mathbf{w} \in \mathcal{W} : \mathbf{g}(\mathbf{w}) = \mathbf{y}\}), \qquad \mathbf{y} \in \mathcal{Y}.$$

EXAMPLE 1.24   (Continued.) Determine the probability function of the number of spots $Y = W_1 + W_2$ that show on the two dice.

Solution   The random variable Y is $\mathcal{Y}$-valued, where $\mathcal{Y} = \{2, 3, \ldots, 12\}$. Observe that

$$\{ (w_1, w_2) \in \mathcal{W} : w_1 + w_2 = 2 \} = \{(1, 1)\}$$

and hence that $f_Y(2) = 1/36$. Also,

$$\{ (w_1, w_2) \in \mathcal{W} : w_1 + w_2 = 3 \} = \{(1, 2), (2, 1)\},$$

so $f_Y(3) = 2/36$. Similarly, $f_Y(4) = 3/36$, $f_Y(5) = 4/36$, $f_Y(6) = 5/36$, $f_Y(7) = 6/36$, $f_Y(8) = 5/36$, $f_Y(9) = 4/36$, $f_Y(10) = 3/36$, $f_Y(11) = 2/36$, and $f_Y(12) =$

1/36. In general, $f_Y(y) = (6 - |7 - y|)/36$ for $y \in \mathcal{Y}$. Figure 1.17 shows the probability function of Y. ∎

FIGURE 1.17

Probability function of the number of spots that show on two dice

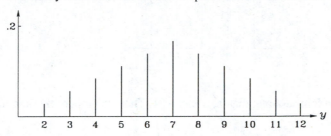

EXAMPLE 1.25   A box has six tickets, labeled from 1 to 6. Two tickets are selected from the box by sampling without replacement. Let L_1 and L_2, respectively, denote the labels of the first and second ticket so selected. Determine the probability function of $L = (L_1, L_2)$.

Solution   We interpret the description of the problem as implying that L is uniformly distributed on $\mathcal{L} = \{ (l_1, l_2) : l_1, l_2 \in \{1, \ldots, 6\}$ and $l_1 \neq l_2 \}$. Since $\#(\mathcal{L}) = 30$, we get that

$$f_{\mathbf{L}}(\mathbf{l}) = \frac{1}{30}, \qquad \mathbf{l} \in \mathcal{L},$$

and $f_{\mathbf{L}}(\mathbf{l}) = 0$ elsewhere. ∎

EXAMPLE 1.26   (Continued.) Determine the probability function of $Y = L_1 + L_2$ and show the plot of this function.

Solution   The random variable Y is $\mathcal{Y}$-valued, where $\mathcal{Y} = \{3, \ldots, 11\}$. Observe that

$$\{ (l_1, l_2) \in \mathcal{L} : l_1 + l_2 = 3\} = \{(1, 2), (2, 1)\},$$

so $f_Y(3) = 2/30$. Also,

$$\{ (l_1, l_2) \in \mathcal{L} : l_1 + l_2 = 4\} = \{(1, 3), (3, 1)\},$$

so $f_Y(4) = 2/30$. Similarly, $f_Y(5) = f_Y(6) = 4/30$, $f_Y(7) = 6/30$, $f_Y(8) = f_Y(9) = 4/30$, and $f_Y(10) = f_Y(11) = 2/30$. Figure 1.18 shows the plot of this probability function. ∎

A $\{0, 1\}$-valued random variable is commonly referred to as an *indicator random variable*. Let Y be such a random variable and set $\pi = P(Y = 1)$. Then the probability function f of Y is given by $f(0) = 1 - \pi$ and $f(1) = \pi$. Observe that Y is a constant random variable if and only if π equals 0 or 1. Suppose Y is a nonconstant

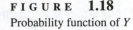

FIGURE   1.18

Probability function of Y

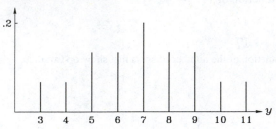

random variable; that is, $0 < \pi < 1$. Then its probability function can be written as

$$f(y) = \pi^y(1-\pi)^{1-y}, \qquad y \in \{0, 1\}. \tag{31}$$

The distribution having the probability function given by (31) is referred to as the *Bernoulli distribution* with parameter π. Observe that the Bernoulli distribution with parameter $\pi = 1/2$ coincides with the uniform distribution on $\{0, 1\}$. A random variable having a Bernoulli distribution is sometimes referred to as a Bernoulli random variable, and similar terminology is used to refer to random variables having other named distributions.

In this book, the symbol π is sometimes used as above to denote a number in $(0, 1)$, which is thought of as a probability; otherwise, it is used to denote the number given by $\pi \doteq 3.14159$. The proper interpretation will be clear from the context. More generally, it is common, especially in statistics, to use Greek letters to denote parameters of distributions.

Let A be a subset of a set $\mathcal{W}$. The corresponding *indicator function* $g(\mathbf{w}) = \mathrm{ind}(\mathbf{w} \in A)$, $\mathbf{w} \in \mathcal{W}$, is the function on $\mathcal{W}$ defined by $g(\mathbf{w}) = 1$ for $\mathbf{w} \in A$ and $g(\mathbf{w}) = 0$ for $\mathbf{w} \notin A$. Let $\mathbf{W}$ be a $\mathcal{W}$-valued random variable. The indicator random variable $I = g(\mathbf{W}) = \mathrm{ind}(\mathbf{W} \in A)$ corresponding to the event $\mathbf{W} \in A$ is defined by

$$\mathrm{ind}(\mathbf{W} \in A) = \begin{cases} 1, & \mathbf{W} \in A, \\ 0, & \mathbf{W} \notin A. \end{cases}$$

Thus, $I = 1$ indicates that the event $\mathbf{W} \in A$ has occurred, while $I = 0$ indicates that this event has not occurred. The indicator random variable for the event $\mathbf{W} \in A^c$ or, equivalently, that $\mathbf{W} \notin A$ is given by $\mathrm{ind}(\mathbf{W} \notin A) = 1 - \mathrm{ind}(\mathbf{W} \in A) = 1 - I$. Observe that $P(I = 1) = P(\mathbf{W} \in A)$ and $P(I = 0) = P(\mathbf{W} \notin A)$. Moreover, I is a constant random variable if and only if $P(\mathbf{W} \in A)$ equals 0 or 1; otherwise, I has the Bernoulli distribution with parameter $\pi = P(\mathbf{W} \in A)$. Observe also that if W is a $\{0, 1\}$-valued random variable and $I = \mathrm{ind}(W = 1)$, then $I = W$.

EXAMPLE 1.27   Consider the sampling without replacement setup of Example 1.25, and let $I = \mathrm{ind}(L_1 + L_2 = 7)$ denote the indicator random variable for the event that the sum of the labels on the two tickets selected equals 7. Identify the distribution of I.

Solution   Let $Y = L_1 + L_2$ as in Example 1.26. Then $I = \text{ind}(Y = 7)$. We conclude from the solution to Example 1.26 that I has the Bernoulli distribution with parameter $\pi = P(Y = 7) = f_Y(7) = 1/5$.   ∎

Let g be a nonnegative function on a set $\mathcal{Y}$ such that $0 < \sum_\mathbf{y} g(\mathbf{y}) < \infty$. Then there is a unique constant a such that the function f on $\mathcal{Y}$ given by $f(\mathbf{y}) = (1/a)g(\mathbf{y})$ for $\mathbf{y} \in \mathcal{Y}$ is a probability function on $\mathcal{Y}$, which is given by $a = \sum_\mathbf{y} g(\mathbf{y})$.

Recall that the sum of a finite geometric series is given by

$$1 + \cdots + t^n = \frac{1 - t^{n+1}}{1 - t}, \qquad t \neq 1,$$

and that the sum of an infinite geometric series is given by

$$1 + t + t^2 + \cdots = \frac{1}{1 - t}, \qquad -1 < t < 1.$$

EXAMPLE **1.28**   Let $0 < \pi < 1$. Determine a probability function f on $\mathcal{Y} = \{0, 1, 2, \ldots\}$ of the form $f(y) = (1/a)\pi^y, y \in \mathcal{Y}$.

Solution   According to the formula for the sum of an infinite geometric series,

$$\sum_{y=0}^{\infty} \pi^y = \frac{1}{1 - \pi}.$$

Thus, the desired probability function is given by

$$f(y) = (1 - \pi)\pi^y, \qquad y \in \mathcal{Y}. \tag{32}$$

In particular, $f(0) = 1 - \pi, f(1) = (1 - \pi)\pi, f(2) = (1 - \pi)\pi^2$, and so forth.   ∎

Let $0 < \pi < 1$. The distribution having the probability function given by (32) is referred to as the *geometric distribution* with parameter π. Let Y have this distribution. Then, by (32),

$$P(Y \geq y) = (1 - \pi) \sum_{z=y}^{\infty} \pi^z = \pi^y(1 - \pi) \sum_{z=0}^{\infty} \pi^z.$$

Thus, by the formula for the sum of an infinite geometric series,

$$P(Y \geq y) = \pi^y, \qquad y = 0, 1, \ldots. \tag{33}$$

The geometric distribution arises naturally as the distribution of the number of successes before the first failure when π is the probability of success on a single trial (see Example 1.66 in Section 1.8).

EXAMPLE **1.29**   Let Y denote the number of successful performances of a certain piece of equipment before its first failure. Suppose Y has the geometric distribution with parameter

$\pi = .99$. Determine the probability that the first failure occurs after the 50th attempt at using the piece of equipment, but before the 200th attempt.

Solution   Observe that the first failure occurs after the 50th attempt if and only if the number of successes before the first failure is at least 50 and that the first failure occurs before the 200th attempt if and only if the number of successes before the first failure is at most 198. Then the desired probability is given by

$$P(50 \le Y \le 198) = \sum_{y=50}^{198} (.01)(.99)^y$$

$$= (.01)(.99)^{50} \sum_{y=0}^{148} (.99)^y$$

$$= (.01)(.99)^{50} \frac{1 - (.99)^{149}}{.01}$$

$$= .99^{50} - .99^{199}$$

$$\doteq .605 - .135$$

$$= .470.$$

Alternatively, by Equation (33),

$$P(50 \le Y \le 198) = P(Y \ge 50) - P(Y \ge 199)$$

$$= .99^{50} - .99^{199} \doteq .605 - .135 = .470. \quad \blacksquare$$

Density Functions

Let P be a distribution on $\mathbb{R}$. Suppose there is a nonnegative function f on $\mathbb{R}$ such that

$$P(B) = \int_B f(y) \, dy, \qquad B \subset \mathbb{R}. \tag{34}$$

It follows by setting $B = \mathbb{R}$ in (34) that

$$\int_{-\infty}^{\infty} f(y) \, dy = 1. \tag{35}$$

The function f is referred to as the *density function* of P or of a random variable having distribution P. A distribution having a density function is referred to as a *continuous distribution*, and a random variable having a density function is referred to as a *continuous random variable*. (Here "continuous" refers to "continuum" rather than to "continuous function" as defined in calculus.)

A function f on $\mathbb{R}$ is called a density function if it is nonnegative and satisfies (35). Let f be a density function on $\mathbb{R}$. Then (34) determines a distribution on $\mathbb{R}$.

Let Y be a random variable having density function f—that is, whose distribution P has density function f. Then, by (34),

$$P(Y \in B) = \int_B f(y)\,dy, \qquad B \subset \mathbb{R}. \tag{36}$$

In particular,

$$P(a \le Y \le b) = \int_a^b f(y)\,dy, \qquad -\infty \le a \le b \le \infty. \tag{37}$$

Conversely, it can be shown that (36) follows from (37). Equation (37) is illustrated in Figure 1.19.

FIGURE 1.19

$P(a \le Y \le b)$ is the area of the shaded region under the graph of the density function of Y.

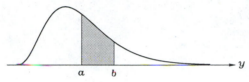

As a special case of (37),

$$P(Y = b) = P(b \le Y \le b) = \int_b^b f(y)\,dy = 0, \qquad b \in \mathbb{R};$$

that is, $P(Y = y) = 0$ for $y \in \mathbb{R}$. Consequently, the formula $P(Y = y) = f(y)$ for $y \in \mathbb{R}$, which is valid for random variables having a probability function, is *invalid* for random variables having a density function. (It is convenient in applications to statistics to use a common symbol, usually f, to denote both probability functions and density functions.)

Density functions are not completely uniquely determined. In particular, let Y be a random variable having a density function f and let f_1 be a nonnegative function on $\mathbb{R}$ that differs from f at only finitely many points. Then f_1 is also a density function of Y. Suppose $P(c < Y < d) = 1$. Then Y has at most one density function that is continuous on the open interval (c, d) and equals zero outside this interval. When possible, we will generally define density functions on $\mathbb{R}$ to have this form.

Let g be a nonnegative function on $\mathbb{R}$ such that $0 < \int_{\mathbb{R}} g(y)\,dy < \infty$. Then there is a unique constant a such that the function f defined by $f(y) = a^{-1}g(y)$ for $y \in \mathbb{R}$ is a density function, which is given by $a = \int_{\mathbb{R}} g(y)\,dy$.

EXAMPLE **1.30**   Let $-\infty < c < d < \infty$.

(a) Determine a density function of the form

$$f(y) = \begin{cases} 1/a, & c < y < d, \\ 0 & \text{elsewhere.} \end{cases}$$

(b) Draw the graph of this density function.

(c) Identify the corresponding distribution.

Solution   (a) Let g be the nonnegative function on $\mathbb{R}$ defined by $g(y) = 1$ for $c < y < d$ and $g(y) = 0$ elsewhere. Then

$$\int_{\mathbb{R}} g(y)\, dy = \int_{c}^{d} 1\, dy = d - c.$$

Thus, the function f defined by

$$f(y) = \begin{cases} 1/(d-c), & c < y < d, \\ 0 & \text{elsewhere} \end{cases} \tag{38}$$

is a density function.

(b) Figure 1.20 shows the graph of the density function.

FIGURE   1.20
Uniform density function

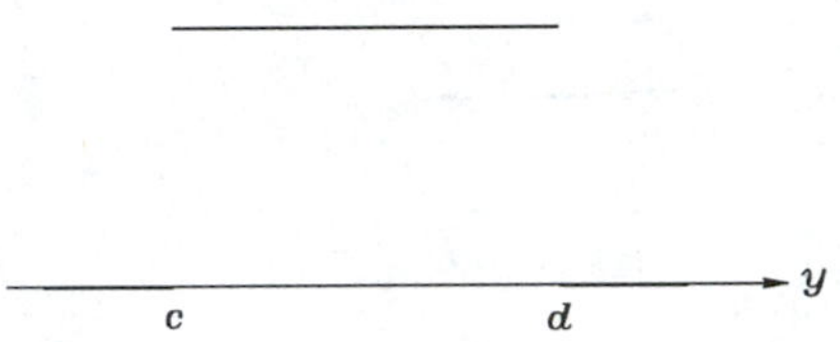

(c) Let P be the distribution having the density function given by (38) and let $B \subset [c, d]$. Then

$$P(B) = \int_{B} \frac{1}{d-c}\, dy = \frac{\text{length}(B)}{d-c} = \frac{\text{length}(B)}{\text{length}([c, d])},$$

so P is the uniform distribution on $[c, d]$.   ∎

EXAMPLE 1.31   Determine $P(2 \le Y \le 3)$ when

(a) Y is uniformly distributed on $\{1, 2, 3, 4, 5, 6\}$;

(b) Y is uniformly distributed on $[1, 6]$.

Solution   (a) The desired probability is given by

$$P(2 \le Y \le 3) = P(Y \in \{2, 3\}) = \frac{\#(\{2, 3\})}{\#(\{1, \ldots, 6\})} = \frac{2}{6} = \frac{1}{3}.$$

(b) The desired probability is given by

$$P(2 \le Y \le 3) = P(Y \in [2, 3]) = \frac{\text{length}([2, 3])}{\text{length}([1, 6])} = \frac{1}{5}.   ∎$$

EXAMPLE **1.32**   Let U be uniformly distributed on $[0, 1]$ and let $0 < \pi < 1$. Determine a nonde-creasing function g on $[0, 1]$ such that $Y = g(U)$ has the Bernoulli distribution with parameter π.

Solution   Let g be defined on $[0, 1]$ by $g(u) = 0$ for $0 \leq u < 1 - \pi$ and $g(u) = 1$ for $1 - \pi \leq u \leq 1$ and set $Y = g(U)$. Then

$$P(Y = 0) = P(0 < U < 1 - \pi) = 1 - \pi$$

and

$$P(Y = 1) = P(1 - \pi \leq U \leq 1) = \pi,$$

so Y has the Bernoulli distribution with parameter π.   ∎

Recall that

$$\frac{d}{dy} e^y = e^y, \qquad -\infty < y < \infty,$$

and hence, by the chain rule, that

$$\frac{d}{dy} e^{-cy} = -ce^{-cy}, \qquad -\infty < y < \infty.$$

Consequently, for $-\infty < a < b < \infty$,

$$\int_a^b e^{-cy}\, dy = -\frac{1}{c} e^{-cy} \Big|_a^b = \frac{1}{c}(e^{-ac} - e^{-bc}).$$

If $c > 0$, we can let $b \to \infty$ in this formula and get that

$$\int_a^\infty e^{-cy}\, dy = -\frac{1}{c} e^{-cy} \Big|_a^\infty = \frac{1}{c} e^{-ca}$$

and, in particular, that

$$\int_0^\infty e^{-cy}\, dy = \frac{1}{c}.$$

EXAMPLE **1.33**   Let $\beta > 0$.

(a) Determine a density function of the form

$$f(y) = \begin{cases} (1/a)e^{-y/\beta}, & y > 0, \\ 0, & y \leq 0. \end{cases}$$

(b) Draw the graph of this density function.

Solution   **(a)** Let g be the nonnegative function on $\mathbb{R}$ defined by $g(y) = e^{-y/\beta}$ for $y > 0$ and $g(y) = 0$ for $y < 0$. Then

$$\int_{-\infty}^\infty g(y)\, dy = \int_0^\infty e^{-y/\beta}\, dy = \beta.$$

Thus, the function f defined by

$$f(y) = \begin{cases} (1/\beta)e^{-y/\beta}, & y > 0, \\ 0, & y \leq 0 \end{cases} \tag{39}$$

is a density function.

(b) Figure 1.21 shows the graph of the density function.   ∎

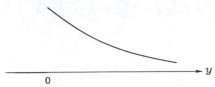

FIGURE   1.21
Exponential density function

The distribution having the density function given by (39) is referred to as the *exponential distribution* with scale parameter β. Decay times of radioactive particles are known to have exponential distributions. In practice, exponential distributions are commonly used to approximate the distributions of other random waiting times as well, such as the time until the next call into a telephone exchange or the time to failure of a piece of equipment.

EXAMPLE 1.34   Suppose the time to failure (in hours of actual use) of a newly installed light bulb is exponential with scale parameter $\beta = 100$ hours. Determine the probability that the time to failure is between 50 hours and 200 hours.

Solution   The desired probability is given by

$$\int_{50}^{200} \frac{1}{100} \exp(-y/100)\, dy = -e^{-y/100}\Big|_{50}^{200} = e^{-1/2} - e^{-2} \doteq .471. \quad ∎$$

Let y be a real number. The *greatest integer in y*, denoted by $[\,y\,]$, is defined as the unique integer i such that $i \leq y < i + 1$. As examples, $[-10.4] = -11$, $[-10] = -10$, $[10] = 10$, and $[10.4] = 10$.

EXAMPLE 1.35   Let W have the exponential distribution with scale parameter β. Identify the distribution of the transform $Y = [\,W\,]$ of W.

Solution   Observe that Y is a nonnegative, integer-valued random variable. Let y be a nonnegative integer. Then

$$P(Y = y) = P([\,W\,] = y)$$

$$= P(y \leq W < y + 1)$$

$$= \int_y^{y+1} \frac{1}{\beta} e^{-w/\beta} \, dw$$

$$= -e^{-w/\beta} \Big|_y^{y+1}$$

$$= e^{-y/\beta} - e^{-(y+1)/\beta}$$

$$= (1 - e^{-1/\beta}) e^{-y/\beta}$$

$$= (1 - \pi)\pi^y,$$

where $\pi = \exp(-1/\beta) \in (0, 1)$. Therefore, Y has the geometric distribution with parameter $\pi = \exp(-1/\beta)$. ∎

Let Y have the density function f and let $a \in \mathbb{R}$ and $\Delta > 0$. Then

$$P(a \leq Y \leq a + \Delta) = \int_a^{a+\Delta} f(y) \, dy.$$

Suppose f is continuous and positive at a. Then

$$\lim_{\Delta \to 0} \frac{1}{\Delta} P(a \leq Y \leq a + \Delta) = \lim_{\Delta \to 0} \frac{1}{\Delta} \int_a^{a+\Delta} f(y) \, dy = f(a).$$

In other words,

$$P(a \leq Y \leq a + \Delta) \approx f(a)\Delta \qquad \text{for } \Delta \approx 0. \tag{40}$$

Typically,

$$P(a \leq Y \leq a + \Delta) \approx f(a + \Delta/2)\Delta \qquad \text{for } \Delta \approx 0 \tag{41}$$

gives a much more accurate approximation than (40). We can include (40) and (41) in the more general approximation

$$P(a \leq Y \leq a + \Delta) \approx f(y)\Delta \qquad \text{for } y \approx a \text{ and } \Delta \approx 0 \tag{42}$$

(see Figure 1.22).

FIGURE 1.22
The approximation given by (42)

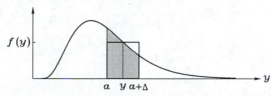

EXAMPLE **1.36** Let Y have the exponential distribution with scale parameter 1. Compare the accuracy of (40) and (41) in approximating $P(1 \leq Y \leq 1.1)$.

Solution   The density function of Y is given by $f(y) = e^{-y}$ for $y > 0$. Thus,

$$P(1 \le Y \le 1.1) = \int_1^{1.1} e^{-y}\,dy = e^{-1} - e^{-1.1} \doteq .035008.$$

According to (40),

$$P(1 \le Y \le 1.1) \approx e^{-1}(0.1) \doteq .0368;$$

according to (41),

$$P(1 \le Y \le 1.1) \approx e^{-1.05}(0.1) \doteq .034994.$$

Therefore, (41) is substantially more accurate than (40) in the present context.   ■

Let P_1 and P_2 be distributions on $\mathbb{R}$ such that P_1 is discrete (that is, it has a probability function) and P_2 is continuous (that is, it has a density function) and let $0 < c < 1$. Then $P = cP_1 + (1 - c)P_2$ is a distribution on $\mathbb{R}$, which is neither discrete nor continuous; it is said to be a *mixture* of P_1 and P_2. There are distributions on $\mathbb{R}$ that are neither discrete, nor continuous, nor mixtures thereof, but they will not arise in this text.

Problems

1.55   A box has eight balls, which are labeled from 1 to 8. Let Y denote the label of a ball selected at random from the box. Determine the probability function of Y.

1.56   Three fair coins are tossed in an honest manner. Determine **(a)** the probability function for the number of heads that show; **(b)** the distribution of the indicator random variable that equals 1 if heads and tails both show at least once and equals 0 if the same side shows all three times.

1.57   Two fair dice are rolled in an honest manner. Determine **(a)** the distribution of the indicator random variable that equals 1 if the number of spots showing is 7 or 11 and equals 0 otherwise; **(b)** the probability function of the absolute value of the difference of the numbers of spots showing on the two dice; **(c)** the probability function of the maximum number of spots showing on the two dice.

1.58   Consider the function f defined on the positive integers by

$$f(y) = \frac{1}{y(y + 1)}, \qquad y = 1, 2, \ldots .$$

(a) Show that f is a probability function. **(b)** Let Y be a random variable having probability function f. Determine $P(4 \le Y \le 7)$.

1.59   Let Y have the density function f given by $f(y) = 1/y^2$ for $y \ge 1$ and $f(y) = 0$ otherwise. Determine $P(Y \le 5)$.

1.60   **(a)** Determine a density function of the form

$$f(y) = \begin{cases} y^2(1 - y)/a, & 0 < y < 1, \\ 0 & \text{elsewhere} \end{cases}$$

and sketch its graph. **(b)** Determine $P(Y \leq 1/2)$, where Y has the density function in part (a).

1.61   Let α be a positive constant. **(a)** Determine the density function f of the form

$$f(y) = \begin{cases} (1/a)y^{\alpha-1}, & 0 < y < 1, \\ 0 & \text{elsewhere.} \end{cases}$$

(b) Sketch the graph of the density function when $\alpha = 0.5, 1, 2$.

1.62   Let Y have the exponential distribution with scale parameter $\beta = 1$. Determine

$$P\left(\frac{Y}{\log 40} \leq 1 \leq \frac{Y}{\log(40/39)}\right).$$

1.63   Let Y have the exponential distribution with scale parameter β. Determine

$$P\left(\frac{Y}{\log 40} \leq \beta \leq \frac{Y}{\log(40/39)}\right).$$

1.64   Determine $P(Y = 1)$ when **(a)** Y is uniformly distributed on $\{0, 1, 2\}$; **(b)** Y is uniformly distributed on $[\,0, 2\,]$.

1.65   Let U be uniformly distributed on $[\,0, 1\,]$. Determine a nondecreasing function g on $[\,0, 1\,]$ such that $Y = g(U)$ is a $\{0, 1, 2\}$-valued random variable having the probability function f given by $f(0) = \frac{1}{4}, f(1) = \frac{1}{2}$, and $f(2) = \frac{1}{4}$.

1.66   Is there a function g such that $Y = g(U)$ has the uniform distribution on $\{1, 2, 3, 4\}$ **(a)** if U is uniformly distributed on $\{1, \ldots, 8\}$; **(b)** if U is uniformly distributed on $\{1, \ldots, 6\}$?

1.67   **(a)** Let U be uniformly distributed on $[\,0, 1\,]$. Is there a function g such that $Y = g(U)$ is uniformly distributed on $\{1, \ldots, 10\}$? **(b)** Let U be uniformly distributed on $\{1, \ldots, 10\}$. Is there a function g such that $Y = g(U)$ is uniformly distributed on $[\,0, 1\,]$?

1.68   Let P_1 be the exponential distribution with scale parameter $\beta = 1$, let P_2 be the geometric distribution with parameter $\pi = 1/2$, and set $P = (P_1 + P_2)/2$. Let Y_1, Y_2, and Y have distributions P_1, P_2, and P, respectively. Determine **(a)** $P(Y_1 \geq 3)$; **(b)** $P(Y_2 \geq 3)$; **(c)** $P(Y \geq 3)$.

1.69   Let Y have the density function f given by $f(y) = y^{-2}$ for $y \geq 1$ and $f(y) = 0$ otherwise. Compare the accuracy of Properties (40) and (41) in approximating $P(1.95 \leq Y \leq 2.05)$.

1.6
Distribution Functions and Quantiles

The *distribution function* $F = F_Y$ of a (real-valued) random variable Y is the function on $\mathbb{R}$ defined by $F(y) = P(Y \leq y)$ for $y \in \mathbb{R}$. It follows from the properties of a distribution that $F(y)$ is a nondecreasing function of y and that $F(y) \to 0$ as $y \to -\infty$ and $F(y) \to 1$ as $y \to \infty$. Observe that $P(Y > y) = 1 - F(y)$ for $y \in \mathbb{R}$ and that

$$P(a < Y \leq b) = F(b) - F(a), \qquad -\infty < a \leq b < \infty.$$

A distribution on $\mathbb{R}$ is uniquely determined by its distribution function; that is, if P_1 and P_2 are distributions on $\mathbb{R}$ having the same distribution function, then $P_1 = P_2$.

Let $p \in (0, 1)$. Suppose there is a real number y_p such that $F(y_p) = p$, $F(y) < p$ for $y < y_p$, and $F(y) > p$ for $y > y_p$. Then y_p is referred to as the *pth quantile* of Y or of its distribution. In particular, if F is a continuous and strictly increasing function on $\mathbb{R}$, then $y_p = F^{-1}(p)$; here F^{-1} is the inverse function to F, which is referred to as the *quantile function* corresponding to Y or to its distribution (see Figure 1.23).

F I G U R E   1.23

$y_p = F^{-1}(p)$

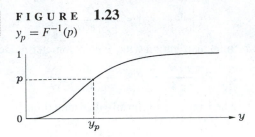

By means of a somewhat more complicated definition, a unique *pth* quantile is assigned to every random variable and its distribution. The *pth* quantile of Y is also referred to as its *100pth percentile*. The 50th percentile $y_{.5}$ is called the *median*. The quantiles $y_{.25}$, $y_{.5}$, and $y_{.75}$ are referred to as *quartiles*. In particular, the 25th percentile $y_{.25}$ is referred to as the *lower quartile*, and the 75th percentile $y_{.75}$ is referred to as the *upper quartile*. Similarly, $y_{.1}, \ldots, y_{.9}$ are referred to as *deciles*; in particular, $y_{.1}$ and $y_{.9}$ are referred to, respectively, as the lower and upper deciles.

In this section, we will discuss distribution functions and quantiles separately for continuous random variables and for nonnegative, integer-valued random variables.

Continuous Random Variables

Let f be a density function on $\mathbb{R}$. The corresponding distribution function is given by

$$F(y) = \int_{-\infty}^{y} f(z)\, dz, \qquad y \in \mathbb{R}, \tag{43}$$

which is a continuous function on $\mathbb{R}$. Moreover, $F'(y) = f(y)$ at all continuity points y of f, where $F'(y)$ is the derivative of F at y.

Observe that

$$1 = \int_{-\infty}^{\infty} f(z)\, dz = \int_{-\infty}^{y} f(z)\, dz + \int_{y}^{\infty} f(z)\, dz = F(y) + \int_{y}^{\infty} f(z)\, dz$$

and hence that

$$1 - F(y) = \int_{y}^{\infty} f(z)\, dz.$$

Since Y has a density function, $P(Y = y) = 0$ for $y \in \mathbb{R}$. Consequently,

$$P(Y < y) = P(Y \le y) = F(y), \qquad y \in \mathbb{R},$$

and

$$P(Y \ge y) = P(Y > y) = 1 - F(y), \qquad y \in \mathbb{R}.$$

Also, for $a, b \in \mathbb{R}$ with $a \le b$,

$$P(a \le Y \le b) = P(a < Y \le b) = P(a \le Y < b) = P(a < Y < b) = F(b) - F(a).$$

Suppose f is positive on an interval (c, d) and equals zero elsewhere. (We allow c to equal $-\infty$ and d to equal ∞.) Then F is a continuous function, $F(y) = 0$ for $y \le c$, F is strictly increasing on (c, d), and $F(y) = 1$ for $y \ge d$. Let $0 < p < 1$. Then the pth quantile of Y (or f) is the unique number $y_p \in \mathbb{R}$ such that $F(y_p) = p$ (see Figure 1.23). Necessarily, $c < y_p < d$. The number y_p is given explicitly by $y_p = F^{-1}(p)$, where F^{-1} is the inverse function to F on (c, d). Observe that

$$P(Y \le y_p) = p \quad \text{and} \quad P(Y \ge y_p) = 1 - p; \tag{44}$$

equivalently,

$$\int_{-\infty}^{y_p} f(z)\, dz = p \quad \text{and} \quad \int_{y_p}^{\infty} f(z)\, dz = 1 - p. \tag{45}$$

In fact, y_p is the unique number satisfying either or both of the equations in (44); equivalently, it is the unique number satisfying either or both of the equations in (45) (see Figure 1.24). Observe that

$$F(y) < F(y_p) = p \text{ for } y < y_p \quad \text{and} \quad F(y) > F(y_p) = p \text{ for } y > y_p. \tag{46}$$

FIGURE 1.24
$P(Y \le y_p) = p$, and $P(Y \ge y_p) = 1 - p$

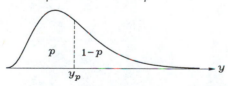

Consider an experiment the distribution of whose outcome has a density function on $\mathbb{R}$. According to the law of relative frequencies, in a large number of trials of the experiment, about one-fourth of the outcomes will be less than the lower quartile of the distribution, and about three-fourths of them will be greater than the lower quartile.

EXAMPLE **1.37** Determine the distribution function and quantiles of the uniform distribution on $[c, d]$.

Solution   The density function is given by $f(y) = 1/(d - c)$ for $c < y < d$ and $f(y) = 0$ elsewhere. Thus, the distribution function is given by $F(y) = 0$ for $y \leq c$,

$$F(y) = \int_c^y \frac{1}{d-c}\, dy = \frac{y-c}{d-c}, \qquad c < y < d,$$

and $F(y) = 1$ for $y \geq d$. Figure 1.25 shows the graph of this function. The pth quantile y_p of the distribution is the unique solution to the equation

$$\frac{y_p - c}{d - c} = p,$$

which is given by $y_p = c + p(d - c)$.   ∎

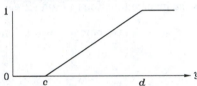

FIGURE   1.25

Uniform distribution function

EXAMPLE 1.38   Determine the distribution function and quantiles of the exponential distribution with scale parameter β.

Solution   The density function is given by $f(y) = (1/\beta)\exp(-y/\beta)$ for $y > 0$ and $f(y) = 0$ for $y \leq 0$. Thus, the distribution function is given by $F(y) = 0$ for $y \leq 0$ and

$$F(y) = \int_0^y \frac{1}{\beta} e^{-z/\beta}\, dz = -e^{-z/\beta}\Big|_0^y = 1 - e^{-y/\beta}, \qquad y > 0.$$

Figure 1.26 shows the graph of F. The pth quantile y_p of the distribution is the unique solution to the equation $1 - \exp(-y_p/\beta) = p$, which is given by

$$y_p = \beta \log \frac{1}{1 - p}.   ∎$$

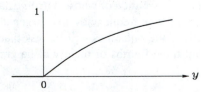

FIGURE   1.26

Exponential distribution function

In the previous solution and in the remainder of the book, *log* means logarithm to the base e.

EXAMPLE 1.39   Suppose the time to failure of a newly installed light bulb is exponential with scale parameter $\beta = 100$ hours. Determine

(a) the probability that the time to failure is between 50 hours and 200 hours;

(b) the quartiles of the distribution.

Solution   (a) The distribution function of the time to failure Y is given by

$$F(y) = 1 - e^{-y/100}, \qquad y \geq 0.$$

Thus,

$$\begin{aligned}
P(50 \leq Y \leq 200) &= F(200) - F(50) \\
&= [1 - e^{-2}] - [1 - e^{-1/2}] \\
&= e^{-1/2} - e^{-2} \\
&\doteq .471,
\end{aligned}$$

which agrees with the solution to Example 1.34 in Section 1.5.

(b) The quantiles of the distribution are given by $y_p = 100 \log(1/(1 - p))$. In particular, the lower quartile is given by

$$y_{.25} = 100 \log \frac{4}{3} \doteq 28.8,$$

the median is given by

$$y_{.5} = 100 \log 2 \doteq 69.3,$$

and the upper quartile is given by

$$y_{.75} = 100 \log 4 \doteq 138.6. \quad \blacksquare$$

EXAMPLE 1.40   Let W have the exponential distribution with scale parameter β and let $0 < \pi < 1$. Determine a nondecreasing function g on $[0, \infty)$ such that $Y = g(W)$ has the Bernoulli distribution with parameter π.

Solution   Let $c = \beta \log 1/\pi$ denote the $(1 - \pi)$th quantile of the distribution of W. Then $P(W < c) = 1 - \pi$ and $P(W \geq c) = \pi$. Define the function g on $[0, \infty)$ by $g(w) = 0$ for $w < c$ and $g(w) = 1$ for $w \geq c$ and set $Y = g(W)$. Then

$$P(Y = 0) = P(W < c) = 1 - \pi \quad \text{and} \quad P(Y = 1) = P(W \geq c) = \pi,$$

so Y has the Bernoulli distribution with parameter π. $\quad \blacksquare$

A density function f on $\mathbb{R}$ is said to be *symmetric about zero* if $f(-y) = f(y)$ for $y \in \mathbb{R}$. Let F be a distribution function having a density function f that is symmetric about zero. It follows from the symmetry of f (see Figure 1.27) that

$$\int_{-\infty}^{-y} f(z) \, dz = \int_{y}^{\infty} f(z) \, dz, \qquad y \in \mathbb{R}, \tag{47}$$

and hence that

$$F(-y) = 1 - F(y), \qquad y \in \mathbb{R}. \tag{48}$$

FIGURE 1.27
If the density function is symmetric about zero, the areas of the shaded regions are equal.

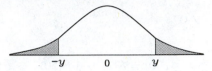

If f is positive on $\mathbb{R}$ and symmetric about zero, then

$$y_{1-p} = -y_p, \qquad 0 < p < 1. \tag{49}$$

To verify (49), observe that, by Equations (43) and (47),

$$F(-y_p) = \int_{-\infty}^{-y_p} f(y)\,dy = \int_{y_p}^{\infty} f(w)\,dw = 1 - p.$$

For $c > 0$, the density function of the uniform distribution on $[-c, c]$ is symmetric about zero. Here is another such example.

EXAMPLE 1.41 **(a)** Determine the constant a such that the function

$$f(y) = \frac{1}{a(1 + y^2)}, \qquad y \in \mathbb{R},$$

is a density function.

(b) Determine the corresponding distribution function F and quantiles.

Solution **(a)** Recall from calculus that

$$\frac{d}{dy} \arctan y = \frac{1}{1 + y^2}, \qquad -\infty < y < \infty,$$

and that

$$\lim_{y \to -\infty} \arctan y = -\frac{\pi}{2} \quad \text{and} \quad \lim_{y \to \infty} \arctan y = \frac{\pi}{2}.$$

(Here $\pi \doteq 3.14159$.) Thus,

$$\int_{-\infty}^{\infty} \frac{1}{1 + y^2}\,dy = \arctan y \Big|_{-\infty}^{\infty} = \frac{\pi}{2} - \left(-\frac{\pi}{2}\right) = \pi.$$

Hence, f is a density function if and only if $a = \pi$. With this choice of a,

$$f(y) = \frac{1}{\pi(1 + y^2)}, \qquad y \in \mathbb{R}. \tag{50}$$

(b) Observe that for $y \in \mathbb{R}$,

$$\int_{-\infty}^{y} \frac{1}{\pi(1+z^2)} \, dz = \frac{1}{\pi} \arctan z \Big|_{-\infty}^{y}$$

$$= \frac{1}{\pi} \left[\arctan y - \left(-\frac{\pi}{2} \right) \right]$$

$$= \frac{1}{\pi} \arctan y + \frac{1}{2}.$$

Thus, the distribution function is given by

$$F(y) = \frac{1}{\pi} \arctan y + \frac{1}{2}, \qquad y \in \mathbb{R}.$$

The pth quantile is the solution to the equation

$$\frac{1}{\pi} \arctan y_p + \frac{1}{2} = p,$$

which is given by $y_p = \tan(\pi(p - 1/2))$. ∎

The distribution having the density function in (50) is referred to as the *standard Cauchy distribution*. (It is a special case of a two-parameter family of Cauchy distributions, which will be introduced in Problem 1.89 in Section 1.7; it is also a special case of the t distribution, which will be introduced in Section 7.2.) The density function and distribution function of the standard Cauchy distribution are shown in Figures 1.28 and 1.29, respectively.

FIGURE 1.28

Cauchy density function

FIGURE 1.29

Cauchy distribution function

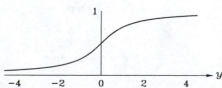

EXAMPLE **1.42**   Show that

$$f(y) = \frac{e^y}{(1 + e^y)^2}, \qquad y \in \mathbb{R},$$

is a symmetric density function and determine the corresponding distribution function and quantiles.

Solution   The function f is clearly nonnegative. Since

$$\int_{-\infty}^{\infty} f(y)\, dy = \int_{-\infty}^{\infty} \frac{e^y}{(1 + e^y)^2}\, dy = -\frac{1}{1 + e^y} \Big|_{-\infty}^{\infty} = 1,$$

f is a density function. Now

$$f(-y) = \frac{e^{-y}}{(1 + e^{-y})^2} = \frac{e^y}{(e^y + 1)^2} = f(y), \qquad y \in \mathbb{R},$$

so f is symmetric about zero. Figure 1.30 shows the graph of this density function.

FIGURE   1.30
Logistic density function

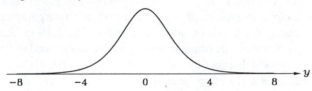

The corresponding distribution function F is given by

$$F(y) = \int_{-\infty}^{y} f(z)\, dz = \int_{-\infty}^{y} \frac{e^z}{(1 + e^z)^2}\, dz = -\frac{1}{1 + e^z} \Big|_{-\infty}^{y} = 1 - \frac{1}{1 + e^y}, \qquad y \in \mathbb{R},$$

and hence by

$$F(y) = \frac{e^y}{1 + e^y}, \qquad y \in \mathbb{R}.$$

The pth quantile y_p of the distribution is the unique solution to the equation

$$\frac{\exp y_p}{1 + \exp y_p} = p$$

or, equivalently, to the equation

$$(1 - p) \exp y_p = p.$$

Thus,

$$y_p = \log \frac{p}{1 - p}. \qquad \blacksquare$$

The distribution having the density function in Example 1.42 is known as the *standard logistic distribution*. Let $0 < p < 1$. When p is thought of as a probability, the quantity $p/(1 - p)$ is referred to as the *odds* corresponding to p and $\log p/(1 - p)$ as the *log-odds*. We also refer to $\log p/(1 - p)$ as the *logit* of p. Similarly, the function on $(0, 1)$ taking on the value $\log p/(1 - p)$ at p is known as the *logit function*. Thus,

$$\mathrm{logit}\, p = \log \frac{p}{1 - p}, \qquad 0 < p < 1.$$

The quantiles of the standard logistic distribution can be written in terms of the logit function as $y_p = \mathrm{logit}\, p$. The logit function is useful in connection with a family of statistical models known as logistic regression models (see Section 13.1).

Consider a random variable Y having a continuous distribution function F. If F is continuously differentiable on $\mathbb{R}$, then $f = F'$ is a density function of Y. Suppose, more generally, that F is continuously differentiable except possibly at finitely many exceptional points. Define the function f on $\mathbb{R}$ by $f(y) = F'(y)$ if y is a nonexceptional point and $f(y) = 0$ if y is an exceptional point. Then f is a density function of Y.

EXAMPLE 1.43 Let the random point (X, Y) be chosen at random from a disk in the plane centered at the origin and having radius c and let $R = \sqrt{X^2 + Y^2}$ denote the distance from the random point to the center of the disk. Determine

(a) the distribution function of R; (b) the density function of R;

(c) the quantiles of R.

Solution (a) Let F_R denote the distribution function of R. Then $F_R(r) = 0$ for $r \leq 0$ and $F_R(r) = 1$ for $r \geq c$. Let $0 < r < c$. Then

$$F_R(r) = P(R \leq r) = P\left(\sqrt{X^2 + Y^2} \leq r \right) = \frac{\pi r^2}{\pi c^2} = \left(\frac{r}{c} \right)^2.$$

(b) Observe that $F_R(r)$ is a continuous function of r and that

$$F_R'(r) = 0, \qquad r \in (-\infty, 0) \cup (c, \infty).$$

Moreover,

$$F_R'(r) = \frac{2r}{c^2}, \qquad 0 \leq r < c.$$

Thus, R has the density function f_R given by $f_R(r) = 2r/c^2$ for $0 < r < c$ and $f_R(r) = 0$ otherwise.

(c) The pth quantile of R is the unique solution to the equation $(r_p/c)^2 = p$, which is given by $r_p = c\sqrt{p}$. ∎

EXAMPLE 1.44 Let a point (X, Y) be chosen at random from the square $[-1, 1] \times [-1, 1]$ and let $R = \sqrt{X^2 + Y^2}$ denote the distance from the chosen point to the center of the square.

(a) Determine the distribution function of R.

(b) Determine the density function of R.

(c) Discuss the determination of the quantiles of R.

Solution   (a) According to the solution to Example 1.21 in Section 1.4, the distribution function of R is given by $F_R(r) = 0$ for $r \leq 0$, $F_R(r) = \pi r^2/4$ for $0 \leq r \leq 1$,

$$F_R(r) = \sqrt{r^2 - 1} + \left(\frac{\pi}{4} - \text{arcsec } r\right) r^2, \qquad 1 < r < \sqrt{2},$$

and $F_R(r) = 1$ for $r \geq \sqrt{2}$. Figure 1.31 shows the graph of this distribution function.

FIGURE 1.31

Distribution function of R

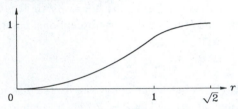

(b) Observe that $F_R'(r) = 0$ for $r \in (-\infty, 0) \cup (\sqrt{2}, \infty)$. Recall from calculus that

$$\frac{d}{dr} \text{arcsec } r = \frac{1}{r\sqrt{r^2 - 1}}, \qquad r > 1.$$

Thus, for $1 < r < \sqrt{2}$,

$$F_R'(r) = \frac{r}{\sqrt{r^2 - 1}} - \frac{r}{\sqrt{r^2 - 1}} + \left(\frac{\pi}{4} - \text{arcsec } r\right) 2r$$

$$= \left(\frac{\pi}{2} - 2\,\text{arcsec } r\right) r.$$

Consequently, R has the density function f_R given by $f_R(r) = \pi r/2$ for $0 < r \leq 1$,

$$f_R(r) = \left(\frac{\pi}{2} - 2\,\text{arcsec } r\right) r, \qquad 1 < r < \sqrt{2},$$

and $f_R(r) = 0$ elsewhere. Figure 1.32 shows the graph of this density function.

(c) Suppose $0 < p \leq \pi/4$. Then the pth quantile of R is the unique solution to the equation $\pi r_p^2/4 = p$, which is given by $r_p = \sqrt{4p/\pi}$. Suppose, instead, $\pi/4 < p < 1$. Then the pth quantile of R is the unique solution to the equation

$$\sqrt{r_p^2 - 1} + \left(\frac{\pi}{4} - \text{arcsec } r_p\right) r_p^2 = p.$$

(Both cases are illustrated in Figure 1.33.) This solution presumably cannot be obtained explicitly, but it is readily obtained numerically for any particular value of p (for example, by successive linear interpolations). Figure 1.34 shows the

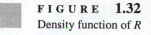

FIGURE 1.32
Density function of R

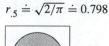

FIGURE 1.33
$r_{.5} \doteq \sqrt{2/\pi} \doteq 0.798$ $r_{.9} \doteq 1.115$

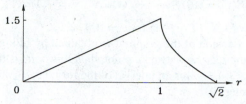

FIGURE 1.34
Quantile function of R

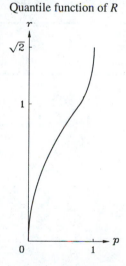

graph of the quantile function $F^{-1}(p)$, $0 < p < 1$, which is easily obtained by plotting the points $(F_R(r), r)$ for the same values of r that were used to plot the above distribution function. ■

Nonnegative, Integer-Valued Random Variables

Let Y now be a nonnegative, integer-valued random variable having probability function f and distribution function F. Then $F(y) = 0$ for $y < 0$, and it is mainly the values of $F(y)$ when y is a nonnegative integer that are of interest. These values can be obtained as a partial sum of the probability function:

$$F(y) = f(0) + \cdots + f(y), \qquad y = 0, 1, 2, \ldots.$$

Conversely, $f(y) = F(y) - F(y - 1)$ for $y = 0, 1, 2, \ldots$.

Consider the extension of the graph of F obtained by filling in its jumps. If the extended graph jumps across p at some point, then this point is the pth quantile y_p of Y. Otherwise, $\{y : F(y) = p\}$ is an interval of the form $[c, d)$, and y_p is the midpoint $(c + d)/2$ of this interval.

EXAMPLE 1.45   Let Y be the number of spots that show when a fair die is rolled in an honest manner. Determine

(a) the distribution function of Y;   **(b)** the quartiles of Y.

Solution   **(a)** The probability function of Y is given by $f(y) = 1/6$ for $y = 1, \ldots, 6$ and $f(y) = 0$ elsewhere. The values of the distribution function at the points $1, \ldots, 6$ are given by $F(y) = y/6$ for $y = 1, \ldots, 6$. The numerical values of f and F are conveniently shown in the form of a table:

y	1	2	3	4	5	6
$f(y)$	.167	.167	.167	.167	.167	.167
$F(y)$	.167	.333	.500	.667	.833	1.000

Figure 1.35 shows the graph of the distribution function.

FIGURE   1.35
Distribution function of Y

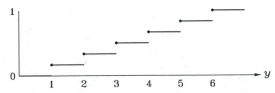

(b) Upon filling in the jumps of the distribution function, we get the graph shown in Figure 1.36. The lower quartile of the distribution is given by $y_{.25} = 2$, the median is given by $y_{.5} = (3 + 4)/2 = 3.5$, and the upper quartile is given by $y_{.75} = 5$.   ∎

EXAMPLE 1.46   Determine the distribution function F of the geometric distribution with parameter π.

Solution   Let y be a nonnegative integer. Then, by the formula for the sum of a finite geometric series,

$$F(y) = \sum_{z=0}^{y}(1 - \pi)\pi^z = 1 - \pi^{y+1}.   ∎$$

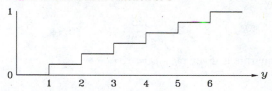

FIGURE **1.36**

Filled-in distribution function of Y

EXAMPLE **1.47**   Determine the quartiles of the geometric distribution with parameter $\pi = .5$.

Solution   Let f denote the probability function of the indicated geometric distribution and let F denote its distribution function. Then $f(y) = .5^{y+1}$ for $y = 0, 1, 2, \ldots$ and, by the solution to Example 1.46,

$$F(y) = 1 - .5^{y+1}, \qquad y = 0, 1, 2, \ldots .$$

The numerical values of $f(y)$ and $F(y)$ for $y = 0, \ldots, 10$ are as follows:

y	0	1	2	3	4	5	6	7	8	9	10
$f(y)$	.500	.250	.125	.062	.031	.016	.008	.004	.002	.001	.000
$F(y)$	.500	.750	.875	.938	.979	.984	.992	.996	.998	.999	1.000

Figure 1.37 shows the extended graph of the distribution function, with its jumps filled in. The lower quartile of the distribution is given by $y_{.25} = 0$, the median is given by $y_{.5} = (0 + 1)/2 = 0.5$, and the upper quartile is given by $y_{.75} = (1 + 2)/2 = 1.5$. ■

FIGURE **1.37**

Filled-in distribution function

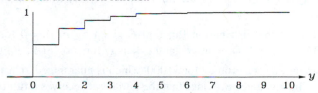

EXAMPLE **1.48**   Determine the quartiles of the geometric distribution with parameter $\pi = .99$.

Solution   Let F be the corresponding distribution. Then $F(y) = 1 - .99^{y+1}$ for y a nonnegative integer. The (real-valued) solution to the equation $1 - .99^{y+1} = .25$ is given by

$$y = \frac{\log .75}{\log .99} - 1 \doteq 27.62.$$

Thus, $F(27) < .25$ and $F(28) > .25$, so the lower quartile is given by $y_{.25} = 28$. Similarly, the solution to the equation $1 - .99^{y+1} = .5$ is given by

$$y = \frac{\log .5}{\log .99} - 1 \doteq 67.97,$$

so the median is given by $y_{.5} = 68$. Finally, the solution to the equation $1 - .99^{y+1} = .75$ is given by

$$y = \frac{\log .25}{\log .99} - 1 \doteq 136.94,$$

so the upper quartile is given by $y_{.75} = 137$.   ∎

Problems

1.70 Let Y have the density function given by $f(y) = 1/y^2$ for $y < -1$ and $f(y) = 0$ for $y \geq -1$. Determine **(a)** the distribution function of Y; **(b)** the median of Y.

1.71 Determine c and d such that the uniform distribution on $[c, d]$ has median 100 and upper decile 130.

1.72 Let $\beta > 0$. Consider the density function f given by $f(y) = \beta y^{\beta-1}$ for $0 < y < 1$ and $f(y) = 0$ elsewhere. Determine **(a)** the distribution function having this density function; **(b)** the corresponding quantiles.

1.73 Let Y have the exponential distribution with scale parameter $\beta = 100$. Determine a and b such that $P(a \leq Y \leq b) = .95$ and $P(Y < a) = P(Y > b)$.

1.74 Given $\beta > 0$, consider the density function f on $\mathbb{R}$ given by

$$f(y) = \frac{1}{2\beta} e^{-|y|/\beta}, \qquad y \in \mathbb{R}.$$

(a) Determine the corresponding distribution function. **(b)** Determine the corresponding quantiles. **(c)** Check that the answer to part **(b)** is compatible with Equation (49). **(d)** Sketch the density function.

1.75 Let Y have the distribution function F given by $F(y) = 0$ for $y \leq 0$ and $F(y) = 1 - \exp(-(y/\beta)^2)$ for $y > 0$. Determine **(a)** the density function of Y; **(b)** the quantiles of Y.

1.76 Let Y have the distribution function F given by $F(y) = 0$ for $y \leq 1$ and $F(y) = 1 - 1/y$ for $y > 1$. Determine **(a)** the density function of Y; **(b)** the quantiles of Y.

1.77 A company sells a product for \$1000, which is guaranteed to last for more than 1000 days. The guarantee provides that the purchaser receives a full refund if the product fails within 500 days of the time of purchase; otherwise, the purchaser receives a 50% refund if the product fails within 1000 days of the time of purchase and no refund if the product lasts for more than 1000 days. Under the assumption that the lifetime of the product has the exponential distribution with scale parameter 1000 days, determine the probability function of the amount of refund.

1.78 A box has eight balls, which are labeled from 1 to 8. Let Y denote the label of a ball selected at random from the box. Determine **(a)** the distribution function of Y; **(b)** the quartiles of Y; **(c)** the deciles of Y.

1.79 Let Y denote the random number of heads that show when three fair coins are tossed in an honest manner and let F denote the distribution function of Y. Determine **(a)** $F(y)$ for $y = 0, \ldots, 3$; **(b)** the corresponding quartiles.

1.80 A fair coin is repeatedly tossed in an honest manner. Let Y denote the number of tails before the first occurrence of heads. Then, under reasonable assumptions, it can be shown that Y has the geometric distribution with parameter $\pi = .5$. Use this fact to determine the probability that the first occurrence of heads **(a)** occurs on the fifth toss; **(b)** occurs after the fifth toss; **(c)** occurs after the second toss but before the fifth toss.

1.81 Two fair dice are rolled in an honest manner. Let F denote the distribution function of the absolute value of the difference of the number of spots showing on the two dice. Determine **(a)** $F(y)$ for $y = 0, \ldots, 5$; **(b)** the corresponding quartiles.

1.82 Two fair dice are rolled in an honest manner. Let F denote the distribution function of the maximum number of spots showing on the two dice. Determine **(a)** $F(y)$ for $y = 1, \ldots, 6$; **(b)** the corresponding quartiles.

1.83 Let W be exponentially distributed with scale parameter β, let M be a positive constant, and set $Y = \min(W, M)$. (For an interpretation, think of W as the failure time of a piece of equipment. Suppose we run it for M units of time. Then Y is the time to failure if the piece of equipment fails during the time it is running and $Y = M$ otherwise.) **(a)** Determine $\{ w \in [0, \infty) : \min(w, M) = M \}$. **(b)** Determine $P(Y = M)$. **(c)** Determine $\{ w \in [0, \infty) : \min(w, M) \leq y \}$ for $0 < y < M$. **(d)** Determine the distribution function of Y. **(e)** Does Y have a density function? **(f)** Does Y have a probability function?

1.84 Let f be the standard logistic density function. Show that $\log f$ is strictly concave. (It suffices to show that the second derivative of $\log f$ is negative.)

1.85 Let T be a positive random variable having a density function f that is continuous and positive on $(0, \infty)$ and equals zero elsewhere and let F denote the distribution function of T. (Think of T as the failure time of a piece of equipment or the survival time of an individual.) The functions $S(t)$, $\lambda(t)$, and $\Lambda(t)$ defined on $(0, \infty)$ by $S(t) = 1 - F(t) = P(T \geq t)$, $\lambda(t) = f(t)/S(t)$, and $\Lambda(t) = \int_0^t \lambda(u)\, du$ are known, respectively, as the *survival function*, *hazard function*, and *cumulative hazard function*. These functions play important roles in reliability theory (in engineering) and biostatistics. Problem 6.15 in Section 6.1 involves a probabilistic interpretation of the hazard function. Show that **(a)** $\frac{d}{dt} \log S(t) = -\lambda(t)$ for $t > 0$; **(b)** $S(t) = \exp(-\Lambda(t))$ for $t > 0$; **(c)** $f(t) = \lambda(t) \exp(-\Lambda(t))$ for $t > 0$.

1.86 Let α and β be positive numbers. The distribution having the distribution function F given by $F(t) = 1 - \exp(-(t/\beta)^\alpha)$ for $t > 0$ and $F(t) = 0$ for $t \leq 0$ is referred to as the *Weibull* distribution with shape parameter α and scale parameter β. Observe that the Weibull distribution with shape parameter one and scale parameter β coincides with the exponential distribution with scale parameter β. **(a)** Determine the density function of the distribution. **(b)** Determine its pth quantile. **(c)** Determine its survival function, hazard function and cumulative hazard function (see Problem 1.85). **(d)** Show that the hazard function is strictly increasing in t if $\alpha > 1$, constant if $\alpha = 1$, and strictly decreasing if $0 < \alpha < 1$.

1.87 Let T_1 have the Weibull distribution with shape parameter α and scale parameter β as defined in Problem 1.86 and let T_2 be a positive random variable whose hazard function is twice that of T_1. Determine the distribution of T_2.

1.7
Univariate Transformations

In this section, we obtain formulas for the distribution function, density function, and quantiles of a real-valued random variable Y that arises as a tranform of another such random variable W. Throughout this section, W is a random variable that ranges over an open interval (c, d) and has a density function f_W that is continuous and positive on this interval and equals zero otherwise. Let F_W denote the distribution function of W. Then $F_W(w)$ is a continuous function of w, $F_W(w) = 0$ for $w \leq c$, $F_W(w) = 1$ for $w \geq d$, and $F'_W(w) = f_W(w)$, except possibly when w equals c or d. Thus, $F_W(w)$ is continuously differentiable in w, except possibly at these two values. The pth quantile of W is given by $w_p = F_W^{-1}(p)$.

Linear Transformations

We start with two simple examples.

EXAMPLE 1.49   Let W have the exponential distribution with scale parameter one, let $\beta > 0$, and set $Y = \beta W$. Determine

(a) the distribution function F_Y of Y;   **(b)** the density function f_Y of Y;

(c) the distribution of Y.

Solution   Here $c = 0$ and $d = \infty$. Recall that the density function of W is given by $f_W(w) = 0$ for $w \leq 0$ and $f_W(w) = e^{-w}$ for $w > 0$ and that its distribution function is given by $F_W(w) = 0$ for $w \leq 0$ and $F_W(w) = 1 - e^{-w}$ for $w > 0$.

(a) Since W is a positive random variable, so is Y. Thus, $F_Y(y) = 0$ for $y \leq 0$. Let $y > 0$. Since β is positive, we see that $\beta W \leq y$ if and only if $W \leq y/\beta$. Thus,

$$F_Y(y) = P(Y \leq y) = P(\beta W \leq y) = P(W \leq y/\beta) = F_W(y/\beta) = 1 - e^{-y/\beta}.$$

(b) Since $F_Y(y)$ is a continuous function of Y and it is continuously differentiable except at $y = 0$, we conclude that Y has the density function f_Y given by

$$f_Y(y) = F'_Y(y) = \frac{d}{dy}\left(1 - e^{-y/\beta}\right) = \frac{1}{\beta}e^{-y/\beta}, \qquad y > 0,$$

and $f_Y(y) = 0$ for $y \leq 0$.

(c) It follows from the solution to part (a) or (b) that Y has the exponential distribution with scale parameter β.   ∎

EXAMPLE 1.50   Let W have the standard logistic distribution, let $\alpha \in \mathbb{R}$ and $\beta > 0$, and set $Y = \alpha + \beta W$. Determine

(a) the distribution function F_Y of Y;　　(b) the density function f_Y of Y;

(c) the pth quantile y_p of Y.

Solution　Here $c = -\infty$ and $d = \infty$. Recall that the density function of W is given by

$$f_W(w) = \frac{e^w}{(1 + e^w)^2}, \qquad w \in \mathbb{R};$$

its distribution function is given by

$$F_W(w) = \frac{e^w}{1 + e^w}, \qquad w \in \mathbb{R};$$

and its pth quantile is given by $w_p = \text{logit}\, p$.

(a) Let $y \in \mathbb{R}$. Since β is positive, we see that $\alpha + \beta W \leq y$ if and only if $W \leq (y - \alpha)/\beta$. Thus,

$$F_Y(y) = P(Y \leq y)$$

$$= P(\alpha + \beta W \leq y)$$

$$= P\left(W \leq \frac{y - \alpha}{\beta}\right)$$

$$= F_W\left(\frac{y - \alpha}{\beta}\right)$$

$$= \frac{\exp((y - \alpha)/\beta)}{1 + \exp((y - \alpha)/\beta)}.$$

(b) Since the distribution function of Y is continuously differentiable, it has the density function f_Y given by

$$f_Y(y) = F'_Y(y)$$

$$= \frac{1}{\beta} F'_W\left(\frac{y - \alpha}{\beta}\right)$$

$$= \frac{1}{\beta} f_W\left(\frac{y - \alpha}{\beta}\right)$$

$$= \frac{1}{\beta} \frac{\exp((y - \alpha)/\beta)}{[1 + \exp((y - \alpha)/\beta)]^2}$$

for $y \in \mathbb{R}$.

(c) The pth quantile of Y is the unique solution to the equation $p = F_Y(y_p) = F_W((y_p - \alpha)/\beta)$. Since there is a unique w_p that satisfies the equation $F_W(w_p) = p$, we conclude that $(y_p - \alpha)/\beta = w_p$ and hence that $y_p = \alpha + \beta w_p$.　∎

More generally, we have the following result.

THEOREM 1.4   Set $Y = \alpha + \beta W$, where $\alpha \in \mathbb{R}$ and $\beta > 0$. Then the distribution function of Y is given by

$$F_Y(y) = F_W\left(\frac{y - \alpha}{\beta}\right), \qquad y \in \mathbb{R};$$

its density function is given by

$$f_Y(y) = \frac{1}{\beta} f_W\left(\frac{y - \alpha}{\beta}\right), \qquad \alpha + \beta c < y < \alpha + \beta d,$$

and $f_Y(y) = 0$ otherwise; and its pth quantile is given by $y_p = \alpha + \beta w_p$.

Proof   Let $y \in \mathbb{R}$. Since β is positive, we see that $\alpha + \beta W \leq y$ if and only if $W \leq (y - \alpha)/\beta$. Thus,

$$F_Y(y) = P(Y \leq y) = P(\alpha + \beta W \leq y) = P\left(W \leq \frac{y - \alpha}{\beta}\right) = F_W\left(\frac{y - \alpha}{\beta}\right),$$

so $F_Y(y)$ is a continuous function of y. Observe also that

$$F_Y'(y) = \frac{1}{\beta} F_W'\left(\frac{y - \alpha}{\beta}\right) = \frac{1}{\beta} f_W\left(\frac{y - \alpha}{\beta}\right),$$

except possibly if y equals $\alpha + \beta c$ or $\alpha + \beta d$ and hence that $F_Y(y)$ is continuously differentiable in y, except possibly at these two values. Thus, Y has the density function f_Y given in the statement of the theorem. The pth quantile y_p of Y is the unique solution to $p = F_Y(y_p) = F_W((y_p - \alpha)/\beta)$. Since there is a unique w_p that satisfies the equation $F_W(w_p) = p$, we conclude that $(y_p - \alpha)/\beta = w_p$ and hence that $y_p = \alpha + \beta w_p$.   ∎

COROLLARY 1.2   Set $Y = \alpha + W$, where $\alpha \in \mathbb{R}$. Then the distribution function of Y is given by

$$F_Y(y) = F_W(y - \alpha), \qquad y \in \mathbb{R};$$

its density function is given by

$$f_Y(y) = f_W(y - \alpha), \qquad \alpha + c < y < \alpha + d,$$

and $f_Y(y) = 0$ otherwise; and its pth quantile is given by $y_p = \alpha + w_p$.

COROLLARY 1.3   Set $Y = \beta W$, where $\beta > 0$. Then the distribution function of Y is given by

$$F_Y(y) = F_W\left(\frac{y}{\beta}\right), \qquad y \in \mathbb{R};$$

its density function is given by

$$f_Y(y) = \frac{1}{\beta} f_W\left(\frac{y}{\beta}\right), \qquad \beta c < y < \beta d,$$

and $f_Y(y) = 0$ otherwise; and its pth quantile is given by $y_p = \beta w_p$.

Think of the distribution of the random variable W as being fixed and suppose $Y = \alpha + W$, where $\alpha \in \mathbb{R}$. Then the distribution function, density function, and quantiles of Y are given by Corollary 1.2. Observe that the distribution of Y depends on the parameter α, which we refer to as the *location parameter* of the distribution. Let $Z = a + Y$, where $a \in \mathbb{R}$. Then $Z = a + \alpha + W$, so Z has location parameter $a + \alpha$.

Suppose, instead, that $Y = \beta W$, where $\beta > 0$. Then the distribution function, density function, and quantiles of Y are given by Corollary 1.3. Here the distribution of Y depends on the parameter β, which we refer to as the *scale parameter* of the distribution. Let $Z = bY$, where $b > 0$. Then $Z = b\beta W$, so Z has scale parameter $b\beta$.

Suppose now $Y = \alpha + \beta W$, where $\alpha \in \mathbb{R}$ and $\beta > 0$. Then the distribution function and quantiles of Y are given by Theorem 1.4. In this context, the distribution of Y depends on the two parameters α and β, which we refer to as the location parameter and scale parameter, respectively, of the distribution. Let $Z = a + bY$, where $a \in \mathbb{R}$ and $b > 0$. Then $Z = a + b(\alpha + \beta W) = a + b\alpha + b\beta W$, so Z has scale parameter $b\beta$ and location parameter $a + b\alpha$.

It follows from the solution to Example 1.49 that the use of the phrase "scale parameter" in Section 1.5 to refer to the parameter β of the exponential distribution is compatible with the general definition of this phrase given above. Thus, we have the following result.

COROLLARY 1.4   Let W have the exponential distribution with scale parameter β and let $b > 0$. Then bW has the exponential distribution with scale parameter $b\beta$.

Let W have the standard logistic distribution and let $\alpha \in \mathbb{R}$ and $\beta > 0$. The distribution of $Y = \alpha + \beta W$ is referred to as the *logistic distribution* with location parameter $\alpha \in \mathbb{R}$ and scale parameter $\beta > 0$. Its distribution function, density function, and quantiles are given in the solution to Example 1.50. Observe that the logistic distribution with location parameter zero and scale parameter one equals the standard logistic distribution. The logistic distribution is commonly used in practice to model the distribution of the tolerance of a random insect in a given species to the log-dose of an insecticide (see Section 13.1).

EXAMPLE 1.51   Assume that the height of adult males in a certain population has the logistic distribution with location parameter 70 inches and scale parameter 2 inches.

(a) Draw the graph of the distribution function of adult height.

(b) Determine the probability that a randomly selected adult male in the population is less than 66 inches tall, less than 60 inches, less than 40 inches, less than 20 inches, or less than 0 inches.

(c) Discuss the reasonableness of the stated assumption.

Solution   (a) Figure 1.38 shows the graph of the distribution function of adult height.

FIGURE   **1.38**

Distribution function of adult height

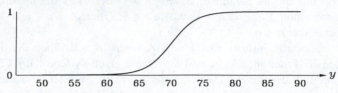

(b) Let Y denote the height of a randomly selected adult male in the population. The distribution function of Y is given by

$$F(y) = \frac{\exp((y - 70)/2)}{1 + \exp((y - 70)/2)}, \qquad y \in \mathbb{R}.$$

Thus,

$$P(Y < 66) = F(66) = \frac{\exp(-2)}{1 + \exp(-2)} \doteq .119.$$

Similarly,

$$P(Y < 60) \doteq .0067,$$

$$P(Y < 40) \doteq 3 \times 10^{-7},$$

$$P(Y < 20) \doteq 10^{-11},$$

and

$$P(Y < 0) \doteq 10^{-15}.$$

(c) One could argue that the stated assumption is unreasonable since it implies that there is a positive probability that an adult male has negative height, which is impossible. One could also argue that the assumption is unreasonable because the logistic distribution is continuous, whereas the distribution of height of members of a finite population must be discrete. In these two counterarguments, the stated assumption is interpreted in a very literal manner; that is, the height of adult males in the population has *exactly* the indicated logistic distribution. A more sensible interpretation would be that the height of adult males in the population has *approximately* the indicated logistic distribution. Under the latter interpretation, the two counterarguments lose their force, and the assumption is not unreasonable. The actual accuracy of the logistic approximation in a particular application is an empirical matter. (There is not a uniquely reasonable way to quantify the accuracy of approximation of one distribution by another. Also, an approximation could be good for some purpose, such as approximating the central portion of the distribution, and poor for another purpose, such as approximating the tail portions).   ∎

EXAMPLE **1.52**   Let W be uniformly distributed on (c, d). Identify the distribution of $Y = a + bW$, where $a \in \mathbb{R}$ and $b > 0$.

Solution   The density function of W is given by

$$f_W(w) = \frac{1}{d-c}, \qquad c < w < d,$$

and $f_W(w) = 0$ elsewhere. Thus, by Theorem 1.4 with $\alpha = a$ and $\beta = b$, the density function of Y is given by

$$f_Y(y) = \frac{1}{b(d-c)}, \qquad c < \frac{y-a}{b} < d,$$

and $f_Y(y) = 0$ elsewhere or, equivalently, by

$$f_Y(y) = \frac{1}{bd-bc}, \qquad a+bc < y < a+bd,$$

and $f_Y(y) = 0$ elsewhere. Therefore, Y is uniformly distributed on $(a+bc, a+bd)$. ∎

Power Transformations

THEOREM 1.5   Let W be a positive random variable. Then $Y = \sqrt{W}$ has the distribution function given by $F_Y(y) = 0$ for $y \le 0$ and $F_Y(y) = F_W(y^2)$ for $y > 0$; its density function is given by $f_Y(y) = 0$ for $y \le 0$ and $f_Y(y) = 2yf_W(y^2)$ for $y > 0$; and its pth quantile is given by $y_p = \sqrt{w_p}$.

Proof   Since W is a positive random variable, Y is a positive random variable and an increasing function of W. In particular, $F_Y(y) = 0$ for $y \le 0$. Let $y > 0$. Then $\sqrt{W} \le y$ if and only if $W \le y^2$, so

$$F_Y(y) = P(Y \le y) = P(\sqrt{W} \le y) = P(W \le y^2) = F_W(y^2).$$

By differentiation, Y has the indicated density function. The pth quantile y_p of Y is the unique positive solution to $p = F_Y(y_p) = F_W(y_p^2)$. Since there is a unique number w_p that satisfies the equation $F_W(w_p) = p$, we conclude that $y_p^2 = w_p$ and hence that $y_p = \sqrt{w_p}$. ∎

EXAMPLE 1.53   Let W have the exponential distribution with scale parameter β. Determine the density function, distribution function, and quantiles of $Y = \sqrt{W}$.

Solution   Recall that $f_W(w) = (1/\beta)\exp(-w/\beta)$ for $w > 0$, $F_W(w) = 1 - \exp(-w/\beta)$ for $w > 0$, and $w_p = \beta \log(1/(1-p))$. Thus, by Theorem 1.5,

$$F_Y(y) = 1 - \exp(-y^2/\beta), \qquad y > 0,$$

$$f_Y(y) = \frac{2y}{\beta}\exp(-y^2/\beta), \qquad y > 0,$$

and

$$y_p = \left(\beta \log \frac{1}{1-p}\right)^{1/2}. \qquad ∎$$

THEOREM 1.6   Let W be a positive random variable. Then $Y = 1/W$ has the distribution function given by $F_Y(y) = 0$ for $y \leq 0$ and

$$F_Y(y) = 1 - F_W\left(\frac{1}{y}\right), \qquad y > 0;$$

its density function is given by $f_Y(y) = 0$ for $y \leq 0$ and

$$f_Y(y) = \frac{1}{y^2} f_W\left(\frac{1}{y}\right), \qquad y > 0;$$

and its pth quantile is given by $y_p = 1/w_{1-p}$.

Proof   Since W is a positive random variable, Y is a positive random variable and a decreasing function of W. In particular, $F_Y(y) = 0$ for $y \leq 0$. Let $y > 0$. Then $1/W \leq y$ if and only if $W \geq 1/y$. Thus,

$$F_Y(y) = P(Y \leq y)$$

$$= P\left(\frac{1}{W} \leq y\right)$$

$$= P\left(W \geq \frac{1}{y}\right)$$

$$= P\left(W > \frac{1}{y}\right)$$

$$= 1 - P\left(W \leq \frac{1}{y}\right)$$

$$= 1 - F_W\left(\frac{1}{y}\right).$$

By differentiation, Y has the indicated density function. The pth quantile y_p of Y is the unique solution to $p = F_Y(y_p) = 1 - F_W(1/y_p)$ and hence the unique solution to the equation $F_W(1/y_p) = 1 - p$. Since there is a unique number w_{1-p} that satisfies the equation $F_W(w_{1-p}) = 1 - p$, we conclude that $1/y_p = w_{1-p}$ and hence that $y_p = 1/w_{1-p}$.   ∎

EXAMPLE 1.54   Let W be a positive random variable having the density function f_W given by $f_W(w) = 0$ for $w \leq 0$ and

$$f_W(w) = \frac{2}{\pi(1 + w^2)}, \qquad w > 0.$$

Show that $Y = 1/W$ has the same distribution as W.

Solution   According to Theorem 1.6, Y has the density function given by $f_Y(y) = 0$ for $y \leq 0$ and

$$f_Y(y) = \frac{1}{y^2} f_W\left(\frac{1}{y}\right) = \frac{2}{y^2 \pi [1 + (1/y)^2]} = \frac{2}{\pi(1 + y^2)}, \qquad y > 0,$$

so Y has the same density function as W and hence the same distribution.   ∎

In Theorems 1.4–1.6, we have that $Y = g(W)$. Observe that $y_p = g(w_p)$ in Theorems 1.4 and 1.5, while $y_p = g(w_{1-p})$ in Theorem 1.6, the reason being that g is increasing in the contexts of Theorems 1.4 and 1.5, while g is decreasing in the context of Theorem 1.6. Theorem 1.7 involves a transform $Y = g(W)$, where g is neither increasing nor decreasing.

THEOREM 1.7 Let W have a density function that is continuous and positive on $\mathbb{R}$ and symmetric about zero. Then $Y = W^2$ has the distribution function given by $F_Y(y) = 0$ for $y \leq 0$ and

$$F_Y(y) = 2F_W\left(\sqrt{y}\right) - 1, \qquad y > 0;$$

its density function is given by $f_Y(y) = 0$ for $y \leq 0$ and

$$f_Y(y) = \frac{1}{\sqrt{y}} f_W\left(\sqrt{y}\right), \qquad y > 0;$$

and its pth quantile is given by $y_p = w_{(1+p)/2}^2$.

Proof Since Y is a positive random variable, $F_Y(y) = 0$ for $y \leq 0$. Observe that w^2 is a continuously differentiable function of w on $\mathbb{R}$; it is a decreasing function on $(-\infty, 0]$ and an increasing function on $[0, \infty)$, so it is neither decreasing on $\mathbb{R}$ nor increasing on $\mathbb{R}$. Let $y > 0$ (see Figure 1.39). Then $W^2 \leq y$ if and only if $-\sqrt{y} \leq W \leq \sqrt{y}$. Thus,

$$F(y) = P(Y \leq y) = P(W^2 \leq y) = P(-\sqrt{y} \leq W \leq \sqrt{y}) = F_W(\sqrt{y}) - F_W(-\sqrt{y}).$$

Since $F_W(-\sqrt{y}) = 1 - F_W(\sqrt{y})$ by Equation (48) in Section 1.6, we conclude that

$$F_Y(y) = 2F_W(\sqrt{y}) - 1.$$

By differentiation, f_Y has the indicated form. The pth quantile y_p of Y is the unique solution to the equation $p = F_Y(y_p) = 2F_W(\sqrt{y_p}) - 1$ and hence the unique solution to the equation $F_W(\sqrt{y_p}) = (1+p)/2$. Since there is a unique number $w_{(1+p)/2}$ that satisfies the equation $F_W(w_{(1+p)/2}) = (1+p)/2$, we conclude that $\sqrt{y_p} = w_{(1+p)/2}$ and hence that $y_p = w_{(1+p)/2}^2$. ∎

EXAMPLE 1.55 Let W have the standard Cauchy distribution. Determine the distribution function, density function, and quantiles of $Y = W^2$.

FIGURE 1.39

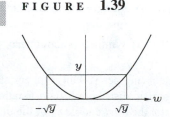

Solution   Recall from Section 1.6 that the density function of the standard Cauchy distribution is given by

$$f(w) = \frac{1}{\pi(1 + w^2)}, \qquad w \in \mathbb{R};$$

its distribution function is given by

$$F(w) = \frac{1}{\pi} \arctan w + \frac{1}{2}, \qquad w \in \mathbb{R};$$

and its pth quantile is given by $w_p = \tan(\pi(p - 1/2))$. Thus, by Theorem 1.7, the distribution function of Y is given by $F_Y(y) = 0$ for $y \leq 0$ and $F_Y(y) = (2/\pi) \arctan \sqrt{y}$ for $y > 0$; its density function is given by $f_Y(y) = 0$ for $y \leq 0$ and

$$f_Y(y) = \frac{1}{\pi \sqrt{y}(1 + y)}, \qquad y > 0;$$

and its pth quantile is given by $y_p = \tan^2 \pi p/2$. (The corresponding distribution is a special case of the F distribution, which will be discussed in Section 7.2.)   ∎

Problems

1.88   Let W have the density function f_W given by $f_W(w) = 1/(w + 1)^2$ for $w > 0$ and $f_W(w) = 0$ for $w \leq 0$. Determine the distribution function, density function, and quantiles of the random variable $Y = 2W$.

1.89   Let W have the standard Cauchy distribution. Determine the distribution function, density function, and quantiles of $Y = \alpha + \beta W$, where $\beta > 0$. (The distribution of Y is referred to as the *Cauchy distribution* with location parameter α and scale parameter β.)

1.90   Set $Y = \alpha + \beta W$, where $\alpha \in \mathbb{R}$, $\beta < 0$, and W has a density function as described in the beginning of this section. Verify that **(a)** $F_Y(y) = 1 - F_W((y - \alpha)/\beta)$ for $y \in \mathbb{R}$; **(b)** $f_Y(y) = (1/|\beta|)f_W((y - \alpha)/\beta)$ for $\alpha + \beta d < y < \alpha + \beta c$ and $f_Y(y) = 0$ otherwise; **(c)** $y_p = \alpha + \beta w_{1-p}$ for $0 < p < 1$.

1.91   Set $X = a_1 + b_1 W$ and $Y = a_2 + b_2 W$, where $b_1, b_2 > 0$. Express the pth quantile y_p of Y in terms of the pth quantile x_p of X.

1.92   Let $Y = \alpha + \beta W$, where $\beta > 0$, and let $0 < p_1 < p_2 < 1$. Show that

$$\alpha = \frac{w_{p_2} y_{p_1} - w_{p_1} y_{p_2}}{w_{p_2} - w_{p_1}} \quad \text{and} \quad \beta = \frac{y_{p_2} - y_{p_1}}{w_{p_2} - w_{p_1}}.$$

1.93   Let W be uniformly distributed on (c, d). Identify the distribution of $Y = a + bW$, where $a \in \mathbb{R}$ and $b < 0$.

1.94   Show that if the density function of W is symmetric about zero, then $-W$ has the same distribution as W.

1.95   Suppose the distribution of the amount (mg per 100 ml) of a certain chemical in the blood of a random adult has the logistic distribution with location parameter 100 and scale parameter 10. Determine **(a)** the pth quantile y_p of the distribution; **(b)** the "normal" range $(y_{.025}, y_{.975})$ (which includes the middle 95% of the distribution and which does not necessarily coincide with the "healthy" range).

1.96 Let W have the exponential distribution with scale parameter 1. Determine the distribution function, density function, and quantiles of $Y = W^2$.

1.97 Let W have the exponential distribution with scale parameter 1. Determine the distribution function, density function, and quantiles of $Y = e^W$.

1.98 Let U be uniformly distributed on $(0, 1)$. Determine the distribution function, density function, and quantiles of $Y = U/(1 - U)$.

1.99 Let U be uniformly distributed on $(0, 1)$. Identify the distribution of $\log(U/(1 - U))$.

1.100 Let W be a positive random variable whose density function is given by $f_W(w) = 1/(1 + w)^2$ for $w > 0$. Identify the distribution of $Y = W/(1 + W)$.

1.101 Let U be uniformly distributed on $[-1, 1]$. Determine the distribution function, density function, and quantiles of $Y = U^2$.

1.102 Let W have the exponential distribution with scale parameter one and let α and β be positive numbers. Show that $\beta W^{1/\alpha}$ has the Weibull distribution with shape parameter α and scale parameter β as defined in Problem 1.86 in Section 1.6.

1.8
Independence

A fair coin is tossed in an honest manner. Let Y_1 be the random variable defined by $Y_1 = 1$ if the coin shows heads and $Y_1 = 2$ if the coin shows tails. It is reasonable to assume that Y_1 is uniformly distributed on $\{1, 2\}$. A fair die is rolled in an honest manner. Let Y_2 be the random variable that denotes the number of spots that show. It is reasonable to assume that Y_2 is uniformly distributed on $\{1, \ldots, 6\}$. Suppose, additionally, that the tossing of the coin and the rolling of the die are completely unrelated to each other. It is then reasonable to assume that the random ordered pair (Y_1, Y_2) is uniformly distributed on $\{1, 2\} \times \{1, \ldots, 6\}$.

Consider, more generally, two experiments, the first of which has finitely many equally likely outcomes that together constitute the set $\mathcal{Y}_1$ and the second of which has finitely many equally likely outcomes that together constitute the set $\mathcal{Y}_2$. Let $\mathbf{Y}_1$ denote the random outcome of the first experiment and let $\mathbf{Y}_2$ denote the random outcome of the second experiment. Then $\mathbf{Y}_1$ is uniformly distributed on $\mathcal{Y}_1$, and $\mathbf{Y}_2$ is uniformly distributed on $\mathcal{Y}_2$. If the two experiments are unrelated to each other, it is reasonable to assume that the random ordered pair $(\mathbf{Y}_1, \mathbf{Y}_2)$ is uniformly distributed on $\mathcal{Y}_1 \times \mathcal{Y}_2$.

Let B_1 and B_2 be subsets of $\mathcal{Y}_1$ and $\mathcal{Y}_2$, respectively. Under the last assumption,

$$P(\mathbf{Y}_1 \in B_1, \mathbf{Y}_2 \in B_2) = P((\mathbf{Y}_1, \mathbf{Y}_2) \in B_1 \times B_2)$$
$$= \frac{\#(B_1 \times B_2)}{\#(\mathcal{Y}_1 \times \mathcal{Y}_2)}$$
$$= \frac{\#(B_1)\#(B_2)}{\#(\mathcal{Y}_1)\#(\mathcal{Y}_2)}$$

and hence

$$P(\mathbf{Y}_1 \in B_1, \mathbf{Y}_2 \in B_2) = \frac{\#(B_1)}{\#(\mathcal{Y}_1)} \frac{\#(B_2)}{\#(\mathcal{Y}_2)}. \tag{51}$$

In particular,

$$P(\mathbf{Y}_1 \in B_1) = P(\mathbf{Y}_1 \in B_1, \mathbf{Y}_2 \in \mathcal{Y}_2) = \frac{\#(B_1)}{\#(\mathcal{Y}_1)}, \tag{52}$$

and

$$P(\mathbf{Y}_2 \in B_2) = P(\mathbf{Y}_1 \in \mathcal{Y}_1, \mathbf{Y}_2 \in B_2) = \frac{\#(B_2)}{\#(\mathcal{Y}_2)}. \tag{53}$$

Thus, $\mathbf{Y}_1$ is uniformly distributed on $\mathcal{Y}_1$, and $\mathbf{Y}_2$ is uniformly distributed on $\mathcal{Y}_2$. It follows from (51)–(53) that

$$P(\mathbf{Y}_1 \in B_1, \mathbf{Y}_2 \in B_2) = P(\mathbf{Y}_1 \in B_1)P(\mathbf{Y}_2 \in B_2), \quad B_1 \subset \mathcal{Y}_1 \text{ and } B_2 \subset \mathcal{Y}_2. \tag{54}$$

Let $\mathcal{Y}_1$ and $\mathcal{Y}_2$ now be arbitrary sets, let $\mathbf{Y}_1$ be a $\mathcal{Y}_1$-valued random variable, and let $\mathbf{Y}_2$ be a $\mathcal{Y}_2$-valued random variable. These random variables are said to be *independent* if they satisfy (54). More generally, let $\mathcal{Y}_1, \ldots, \mathcal{Y}_n$ be arbitrary sets and let $\mathbf{Y}_i$ be a $\mathcal{Y}_i$-valued random variable for $1 \le i \le n$. These random variables are said to be independent if

$$P(\mathbf{Y}_1 \in B_1, \ldots, \mathbf{Y}_n \in B_n) = P(\mathbf{Y}_1 \in B_1) \cdots P(\mathbf{Y}_n \in B_n)$$

for $B_1 \subset \mathcal{Y}_1, \ldots, B_n \subset \mathcal{Y}_n$, and they are said to be *dependent* otherwise. If $\mathbf{Y}_1, \ldots, \mathbf{Y}_n$ are discrete random variables, then they are independent if and only if

$$P(\mathbf{Y}_1 = \mathbf{y}_1, \ldots, \mathbf{Y}_n = \mathbf{y}_n) = P(\mathbf{Y}_1 = \mathbf{y}_1) \cdots P(\mathbf{Y}_n = \mathbf{y}_n)$$

$$\mathbf{y}_1 \in \mathcal{Y}_1, \ldots, \mathbf{y}_n \in \mathcal{Y}_n. \tag{55}$$

It is reasonable to assume that the random outcomes of unrelated experiments form independent random variables. Although random variables defined in terms of the same experimental outcome are typically dependent, sometimes they can be shown to be independent.

EXAMPLE 1.56   Let $\mathbf{W}$ be a $\mathcal{W}$-valued random variable, let A_1 and A_2 be subsets of $\mathcal{W}$, and let I_1 and I_2 be the corresponding indicator random variables, which are defined by $I_1 = \text{ind}(\mathbf{W} \in A_1)$ and $I_2 = \text{ind}(\mathbf{W} \in A_2)$. Show that I_1 and I_2 are independent random variables if and only if

$$P(\mathbf{W} \in A_1 \cap A_2) = P(\mathbf{W} \in A_1)P(\mathbf{W} \in A_2). \tag{56}$$

Solution   Now $P(I_1 = 1) = P(\mathbf{W} \in A_1)$, and $P(I_2 = 1) = P(\mathbf{W} \in A_2)$. Also, $\mathbf{W}$ is a member of $A_1 \cap A_2$ if and only if it is a member of A_1, and it is a member of A_2 and hence if and only if $I_1 = 1$ and $I_2 = 1$. Consequently, $P(I_1 = 1, I_2 = 1) = P(\mathbf{W} \in A_1 \cap A_2)$. Therefore, (56) can be rewritten as

$$P(I_1 = 1, I_2 = 1) = P(I_1 = 1)P(I_2 = 1). \tag{57}$$

It follows directly from the definition of independence that if I_1 and I_2 are independent, then (57) holds. Suppose, conversely, that (57) holds. Then

$$P(I_1 = 1, I_2 = 0) = P(I_1 = 1) - P(I_1 = 1, I_2 = 1)$$
$$= P(I_1 = 1) - P(I_1 = 1)P(I_2 = 1)$$
$$= P(I_1 = 1)[\, 1 - P(I_2 = 1)\,]$$
$$= P(I_1 = 1)P(I_2 = 0)$$

(in the first step, we have used Equation (25) in Section 1.3). Similarly,

$$P(I_1 = 0, I_2 = 1) = P(I_1 = 0)P(I_2 = 1),$$

and

$$P(I_1 = 0, I_2 = 0) = P(I_1 = 0)P(I_2 = 0).$$

Therefore, if (57) holds, then I_1 and I_2 are independent. ∎

EXAMPLE 1.57 Two fair dice are rolled in an honest manner. Let I_1 be the indicator random variable for the event that the number of spots showing on the two dice is 4, 5, 6, or 7 and let I_2 be the indicator random variable for the event that the number of spots is 5, 6, 8, or 9. Show that I_1 and I_2 are independent random variables.

Solution　Let W denote the number of spots showing on the two dice. Then $I_1 = \text{ind}(W \in \{4, 5, 6, 7\})$ and $I_2 = \text{ind}(W \in \{5, 6, 8, 9\})$. According to the solution to Example 1.24 in Section 1.5,

$$P(W \in \{4, 5, 6, 7\}) = \frac{3}{36} + \frac{4}{36} + \frac{5}{36} + \frac{6}{36} = \frac{1}{2},$$
$$P(W \in \{5, 6, 8, 9\}) = \frac{4}{36} + \frac{5}{36} + \frac{5}{36} + \frac{4}{36} = \frac{1}{2},$$

and

$$P(W \in \{5, 6\}) = \frac{4}{36} + \frac{5}{36} = \frac{1}{4}.$$

Thus,

$$P(W \in \{4, 5, 6, 7\} \cap \{5, 6, 8, 9\}) = P(W \in \{5, 6\})$$
$$= P(W \in \{4, 5, 6, 7\})P(W \in \{5, 6, 8, 9\}),$$

so we conclude from Example 1.56 that I_1 and I_2 are independent random variables. ∎

THEOREM 1.8 Let $\mathcal{Y} = \mathcal{Y}_1 \times \cdots \times \mathcal{Y}_n$, where $\mathcal{Y}_1, \ldots, \mathcal{Y}_n$ are nonempty, finite sets; let $\mathbf{Y}_i$ be a $\mathcal{Y}_i$-valued random variable for $1 \le i \le n$; and set $\mathbf{Y} = (\mathbf{Y}_1, \ldots, \mathbf{Y}_n)$. Then $\mathbf{Y}$ is uniformly distributed on $\mathcal{Y}$ if and only if $\mathbf{Y}_i$ is uniformly distributed on $\mathcal{Y}_i$ for $1 \le i \le n$, and $\mathbf{Y}_1, \ldots, \mathbf{Y}_n$ are independent.

Proof   For simplicity, we prove this result when $n = 2$. Thus, let $\mathcal{Y}_1$ and $\mathcal{Y}_2$ be nonempty, finite sets. It has effectively been shown above that if $\mathbf{Y} = (\mathbf{Y}_1, \mathbf{Y}_2)$ is uniformly distributed on $\mathcal{Y} = \mathcal{Y}_1 \times \mathcal{Y}_2$, then $\mathbf{Y}_1$ is uniformly distributed on $\mathcal{Y}_1$, $\mathbf{Y}_2$ is uniformly distributed on $\mathcal{Y}_2$, and $\mathbf{Y}_1$ and $\mathbf{Y}_2$ are independent. Suppose, conversely, that $\mathbf{Y}_1$ is uniformly distributed on $\mathcal{Y}_1$, $\mathbf{Y}_2$ is uniformly distributed on $\mathcal{Y}_2$, and $\mathbf{Y}_1$ and $\mathbf{Y}_2$ are independent. Then

$$P(\mathbf{Y}_1 = \mathbf{y}_1, \mathbf{Y}_2 = \mathbf{y}_2) = P(\mathbf{Y}_1 = \mathbf{y}_1)P(\mathbf{Y}_2 = \mathbf{y}_2)$$

$$= \frac{1}{\#(\mathcal{Y}_1)} \frac{1}{\#(\mathcal{Y}_2)}$$

$$= \frac{1}{\#(\mathcal{Y}_1 \times \mathcal{Y}_2)}$$

for $y_1 \in \mathcal{Y}_1$ and $y_2 \in \mathcal{Y}_2$, so $Y = (Y_1, Y_2)$ is uniformly distributed on $\mathcal{Y} = \mathcal{Y}_1 \times \mathcal{Y}_2$. ∎

EXAMPLE 1.58   Let $0 \le a \le 1/2$ and let X_1 and X_2 be $\{-1, 1\}$-valued random variables such that

$$P(X_1 = -1, X_2 = -1) = P(X_1 = 1, X_2 = 1) = a$$

and

$$P(X_1 = -1, X_2 = 1) = P(X_1 = 1, X_2 = -1) = \frac{1}{2} - a.$$

Determine

(a) the distributions of X_1 and X_2;

(b) the value(s) of a such that X_1 and X_2 are independent.

Solution   **(a)** Now

$$P(X_1 = -1) = P(X_1 = -1, X_2 = -1) + P(X_1 = -1, X_2 = 1)$$

$$= a + \frac{1}{2} - a$$

$$= \frac{1}{2},$$

so $P(X_1 = 1) = 1/2$ and hence X_1 is uniformly distributed on $\{-1, 1\}$. Similarly, X_2 is uniformly distributed on $\{-1, 1\}$.

(b) If X_1 and X_2 are independent, then

$$a = P(X_1 = 1, X_2 = 1) = P(X_1 = 1)P(X_2 = 1) = \frac{1}{2} \cdot \frac{1}{2} = \frac{1}{4},$$

and hence $a = 1/4$. Conversely, if $a = 1/4$, then the random ordered pair (X_1, X_2) is uniformly distributed on $\{-1, 1\} \times \{-1, 1\}$, so we conclude from Theorem 1.8 that X_1 and X_2 are independent. Therefore, X_1 and X_2 are independent if and only if $a = 1/4$. ∎

Consider a population of N objects (for example, tickets in a box) that are labeled consecutively from 1 to N. An object is selected at random from the population, its label is noted, and the object is returned to the population. We refer to this procedure as a single trial of *sampling with replacement*. Let this procedure be repeated until n such trials have been performed and let L_i denote the label of the object selected on the ith trial. In the context of sampling with replacement, it is reasonable to assume that $L_1, \ldots, L_n$ are independent and that each of these random variables is uniformly distributed on $\{1, \ldots, N\}$. According to Theorem 1.8, this is equivalent to the assumption that $(L_1, \ldots, L_n)$ is uniformly distributed on the n-fold Cartesian product of $\{1, \ldots, N\}$ with itself.

Suppose, instead, that on the first trial an object is selected at random from the population of N objects, its label is noted, and the object is set aside. On the second trial, an object is selected at random from the remaining $N - 1$ objects in the population, its label is noted, and the object is set aside. Let this procedure be repeated until n objects in the population have been selected (necessarily, $n \leq N$). We refer to this as *sampling without replacement*. Let L_i denote the label of the object selected on the ith trial. In the context of sampling without replacement, $L_1, \ldots, L_n$ are necessarily distinct. Here it is reasonable to assume that $(L_1, \ldots, L_n)$ is uniformly distributed on the collection of all ordered n-tuples of distinct members of $\{1, \ldots, N\}$, which has size $N \cdots (N - n + 1)$ (see Section 4.1). It follows from the indicated assumption that $L_1, \ldots, L_n$ are each uniformly distributed on $\{1, \ldots, N\}$ (see Section 6.2). When $n \geq 2$, however, these random variables are dependent. To see this, note that $P(L_1 = 1, \ldots, L_n = 1) = 0$, while $P(L_1 = 1) \cdots P(L_n = 1) = (1/N)^n > 0$, so

$$P(L_1 = 1, \ldots, L_n = 1) \neq P(L_1 = 1) \cdots P(L_n = 1).$$

EXAMPLE 1.59 Let n fair dice be rolled in an honest manner. Determine the probability that

(a) each die shows one spot;

(b) the same number of spots shows on every die.

Solution We think of the dice as being distinguishable and labeled from 1 to n. Let Y_i denote the number of spots showing on the ith die for $i = 1, \ldots, n$. We interpret rolling n dice in an honest manner as meaning that $Y_1, \ldots, Y_n$ are independent random variables, each uniformly distributed on $\{1, \ldots, 6\}$ or, equivalently, that $(Y_1, \ldots, Y_n)$ is uniformly distributed on the n-fold Cartesian product of $\{1, \ldots, 6\}$ with itself. (The same assumptions would be applicable to n honest rolls of a single die.)

(a) Under the stated assumptions

$$P(Y_1 = 1, \ldots, Y_n = 1) = P(Y_1 = 1) \cdots P(Y_n = 1) = \left(\frac{1}{6}\right)^n.$$

(b) It follows as in part (a) that

$$P(Y_1 = y, \ldots, Y_n = y) = P(Y_1 = y) \cdots P(Y_n = y) = \left(\frac{1}{6}\right)^n$$

for $y = 1, \ldots, 6$. Thus, the probability that the same number of spots shows on every die is given by

$$P(Y_1 = \cdots = Y_n) = \sum_{y=1}^{6} P(Y_1 = y, \ldots, Y_n = y)$$

$$= \sum_{y=1}^{6} \left(\frac{1}{6}\right)^n = 6 \left(\frac{1}{6}\right)^n$$

$$= \left(\frac{1}{6}\right)^{n-1}. \quad \blacksquare$$

Let $\min(w_1, \ldots, w_n)$ denote the minimum value of the numbers $w_1, \ldots, w_n$ and let $\max(w_1, \ldots, w_n)$ denote the maximum value of these numbers. Observe that

$$\min(w_1, \ldots, w_n) > y \quad \text{if and only if} \quad w_1 > y, \ldots, w_n > y, \qquad y \in \mathbb{R}, \quad \textbf{(58)}$$

and

$$\max(w_1, \ldots, w_n) \le y \quad \text{if and only if} \quad w_1 \le y, \ldots, w_n \le y, \qquad y \in \mathbb{R}. \quad \textbf{(59)}$$

EXAMPLE 1.60   A system consisting of n components *in series* (see Figure 1.40) works if and only if all of the components work. Let W_i denote the failure time of the ith component for $1 \le i \le n$. Then the failure time Y of the system is given by $Y = \min(W_1, \ldots, W_n)$. Derive the distribution of the failure time of the system under the assumption that $W_1, \ldots, W_n$ are independent random variables, each having the exponential distribution with scale parameter β.

F I G U R E   1.40

Solution   Now $F_Y(y) = 0$ for $y \le 0$ since Y is a positive random variable. Let $y > 0$. Then $P(W_i > y) = \exp(-y/\beta)$ for $1 \le i \le n$. According to Property (58),

$$P(Y > y) = P(W_1 > y, \ldots, W_n > y) = P(W_1 > y) \cdots P(W_n > y)$$

$$= (e^{-y/\beta})^n = e^{-ny/\beta},$$

so

$$F_Y(y) = P(Y \le y) = 1 - P(Y > y) = 1 - e^{-ny/\beta}.$$

Since F_Y is continuous at zero, it is continuous on $\mathbb{R}$. By differentiation, $f_Y(y) = 0$ for $y < 0$ and $f_Y(y) = (n/\beta)\exp(-ny/\beta)$ for $y > 0$. Therefore, Y has the exponential distribution with scale parameter β/n. ∎

EXAMPLE 1.61 A system consisting of n components *in parallel* (see Figure 1.41) works if and only if at least one of the components works. Let W_i denote the failure time of the ith component for $1 \le i \le n$. Then the failure time of the system is given by $Y = \max(W_1, \ldots, W_n)$. Determine the distribution function, density function, and quantiles of Y under the assumption that $W_1, \ldots, W_n$ are independent random variables, each having the exponential distribution with scale parameter β.

FIGURE 1.41

Solution Again, $F_Y(y) = 0$ for $y \le 0$ since Y is a positive random variable. Let $y > 0$. Then $P(W_i \le y) = 1 - \exp(-y/\beta)$ for $1 \le i \le n$. According to Property (59),

$$
\begin{aligned}
F_Y(y) &= P(Y \le y) \\
&= P(W_1 \le y, \ldots, W_n \le y) \\
&= P(W_1 \le y) \cdots P(W_n \le y) \\
&= (1 - e^{-y/\beta})^n,
\end{aligned}
$$

so $F_Y(y) = (1 - \exp(-y/\beta))^n$ for $y > 0$. Since F_Y is continuous at zero, it is continuous on $\mathbb{R}$. By differentiation, $f_Y(y) = 0$ for $y < 0$ and

$$
f_Y(y) = \frac{n}{\beta} e^{-y/\beta} (1 - e^{-y/\beta})^{n-1}, \qquad y > 0.
$$

The pth quantile y_p of Y is the unique solution to the equation $[\, 1 - \exp(-y_p/\beta)\,]^n = p$, which is given by

$$
y_p = \beta \log \frac{1}{1 - p^{1/n}}. \qquad \blacksquare
$$

EXAMPLE 1.62 Let $\mathbf{Y}_1$ be a constant random variable and let $\mathbf{Y}_2$ be any random variable. Show that $\mathbf{Y}_1$ and $\mathbf{Y}_2$ are independent.

Solution Let $\mathbf{Y}_1$ be $\mathcal{Y}_1$-valued, let $\mathbf{b}_1 \in \mathcal{Y}_1$ be such that $P(\mathbf{Y}_1 = \mathbf{b}_1) = 1$, let $\mathbf{Y}_2$ be a $\mathcal{Y}_2$-valued random variable, and let B_1 and B_2 be subsets of $\mathcal{Y}_1$ and $\mathcal{Y}_2$, respectively.

Suppose first that $\mathbf{b}_1 \notin B_1$. Then $P(\mathbf{Y}_1 \in B_1) \leq P(\mathbf{Y}_1 \neq \mathbf{b}_1) = 1 - P(\mathbf{Y}_1 = \mathbf{b}_1) = 0$, so $P(\mathbf{Y}_1 \in B_1) = 0$ and hence

$$P(\mathbf{Y}_1 \in B_1, \mathbf{Y}_2 \in B_2) = 0 = P(\mathbf{Y}_1 \in B_1)P(\mathbf{Y}_2 \in B_2).$$

Suppose next that $\mathbf{b}_1 \in B_1$. Then $P(\mathbf{Y}_1 \in B_1) = 1$. Moreover, $\mathbf{b}_1 \notin B_1^c$, and hence $P(\mathbf{Y}_1 \notin B_1, \mathbf{Y}_2 \in B_2) = P(\mathbf{Y}_1 \in B_1^c, \mathbf{Y}_2 \in B_2) = 0$ by what has already been shown. Thus (see Problem 1.54 in Section 1.4),

$$\begin{aligned} P(\mathbf{Y}_1 \in B_1, \mathbf{Y}_2 \in B_2) &= P(\mathbf{Y}_2 \in B_2) - P(\mathbf{Y}_1 \notin B_1, \mathbf{Y}_2 \in B_2) \\ &= P(\mathbf{Y}_2 \in B_2) \\ &= P(\mathbf{Y}_1 \in B_1)P(\mathbf{Y}_2 \in B_2). \end{aligned}$$

Consequently,

$$P(\mathbf{Y}_1 \in B_1, \mathbf{Y}_2 \in B_2) = P(\mathbf{Y}_1 \in B_1)P(\mathbf{Y}_2 \in B_2), \qquad B_1 \subset \mathcal{Y}_1 \text{ and } B_2 \subset \mathcal{Y}_2,$$

so $\mathbf{Y}_1$ and $\mathbf{Y}_2$ are independent.   ∎

Let $\mathbf{Y}_1, \ldots, \mathbf{Y}_n$ be independent random variables. If we permute them, transform them individually, delete one or more of them, or combine two or more of them, we obtain a new sequence of independent random variables. Suppose $Y_1, \ldots, Y_4$ are independent, real-valued random variables. Then Y_1, Y_3, Y_2, Y_4 are independent; Y_1, Y_2^2, $\sin Y_3$, $\exp Y_4$ are independent; Y_1, Y_2, Y_4 are independent; and $Y_1, (Y_2, Y_3)$, Y_4 are independent. Another useful result involving independence is that if $\mathbf{Y}_1$, $\ldots, \mathbf{Y}_n, (\mathbf{Y}_{n+1}, \mathbf{Y}_{n+2})$ are independent and $\mathbf{Y}_{n+1}$ and $\mathbf{Y}_{n+2}$ are independent, then $\mathbf{Y}_1, \ldots, \mathbf{Y}_{n+2}$ are independent.

EXAMPLE 1.63   Let Y_1, Y_2, Y_3, Y_4, and Y_5 be independent random variables. Show that $Y_1 + Y_2 + Y_3$ and $Y_4 Y_5$ are independent.

Solution   Observe first that $(Y_1, Y_2, Y_3), Y_4$, and Y_5 are independent and hence that (Y_1, Y_2, Y_3) and (Y_4, Y_5) are independent. Since $Y_1 + Y_2 + Y_3$ is a transform of (Y_1, Y_2, Y_3) and $Y_4 Y_5$ is a transform of (Y_4, Y_5), we conclude that $Y_1 + Y_2 + Y_3$ and $Y_4 Y_5$ are independent.   ∎

The joint distribution of independent random variables is determined by their individual distributions. In detail, let $\mathbf{X}_1, \ldots, \mathbf{X}_n$ be independent random variables and let $\mathbf{Y}_1, \ldots, \mathbf{Y}_n$ also be independent random variables. If $\mathbf{Y}_i$ has the same distribution as $\mathbf{X}_i$ for $1 \leq i \leq n$, then $\mathbf{Y}_1, \ldots, \mathbf{Y}_n$ has the same joint distribution as $\mathbf{X}_1, \ldots, \mathbf{X}_n$.

EXAMPLE 1.64   Let X and Y be independent random variables having the same distribution. Show that $P(X < Y) = P(X > Y)$.

Solution   Set $A = \{(x, y) \in \mathbb{R}^2 : x < y\}$. Now Y and X have the same joint distribution as X

and Y. Thus,

$$P(X < Y) = P((X, Y) \in A) = P((Y, X) \in A) = P(Y < X) = P(X > Y). \quad \blacksquare$$

Consider again a population of N objects, labeled consecutively from 1 to N. Suppose the lth object has value $g(l)$ for $1 \leq l \leq N$. Let $L_1, \ldots, L_n$ denote the labels of the objects selected on trials $1, \ldots, n$, respectively, under sampling with replacement. Then $L_1, \ldots, L_n$ are independent random variables. The value of the object selected on the ith trial is given by $Y_i = g(L_i)$. Since transforms of independent random variables are independent, $Y_1, \ldots, Y_n$ are independent random variables. On the other hand, if the objects are selected by sampling without replacement, if $n \geq 2$, and if the objects in the population do not all have the same value, then $Y_1, \ldots, Y_n$ are dependent random variables (see Section 6.2).

Sequences of Independent Random Variables

Let $\mathbf{Y}_1, \mathbf{Y}_2, \ldots$ be an infinite sequence of random variables. These random variables are said to be independent if $\mathbf{Y}_1, \ldots, \mathbf{Y}_n$ are independent for $n \geq 2$. The properties of finite sequences of independent random variables extend in an obvious manner to infinite sequences. In particular, if $\mathbf{W}_1, \mathbf{W}_2, \ldots$ are independent random variables and $\mathbf{Y}_i$ is a transform of $\mathbf{W}_i$ for $i \geq 1$, then $\mathbf{Y}_1, \mathbf{Y}_2, \ldots$ are independent.

EXAMPLE 1.65 A fair die is repeatedly rolled in an honest manner until one spot shows for the first time. Determine the probability that ten rolls are required.

Solution Let Y_i denote the number of spots that show on the ith trial. We assume that $Y_1, Y_2, \ldots$ are independent random variables each uniformly distributed on $\{1, \ldots, 6\}$. (We can think of the rolling of the die as continuing even after one spot shows for the first time.) Observe that ten rolls are required if and only if $Y_1 \neq 1, \ldots, Y_9 \neq 1$, and $Y_{10} = 1$. Thus, the probability that ten rolls are required is given by

$$P(Y_1 \neq 1, \ldots, Y_9 \neq 1, Y_{10} = 1) = P(Y_1 \neq 1) \cdots P(Y_9 \neq 1)P(Y_{10} = 1)$$

$$= (\tfrac{5}{6})^9 \tfrac{1}{6}$$

$$\doteq .0323. \quad \blacksquare$$

EXAMPLE 1.66 Consider an experiment having a $\mathcal{Y}$-valued random outcome $\mathbf{Y}$ and let B be a subset of $\mathcal{Y}$. Let us refer to any outcome in B as a "success" and any outcome in B^c as a "failure." Then $\pi = P(\mathbf{Y} \in B)$ is the probability of success, and $1 - \pi = P(\mathbf{Y} \in B^c)$ is the probability of failure. Suppose $0 < \pi < 1$. Consider now an infinite sequence of such experiments; that is, let $\mathbf{Y}_1, \mathbf{Y}_2, \ldots$ be independent random variables each having the same distribution as $\mathbf{Y}$. Determine the distribution of the number of successes before the first failure.

Solution   Let N denote the number of successes before the first failure and let n be a nonnegative integer. Then $N = n$ if and only if there are successes on the first n trials of the experiment and there is a failure on the $(n + 1)$th trial—that is, if and only if $\mathbf{Y}_1 \in B, \ldots, \mathbf{Y}_n \in B$, and $\mathbf{Y}_{n+1} \in B^c$. Consequently,

$$P(N = n) = P(\mathbf{Y}_1 \in B, \ldots, \mathbf{Y}_n \in B, \mathbf{Y}_{n+1} \in B^c)$$
$$= P(\mathbf{Y}_1 \in B) \cdots P(\mathbf{Y}_n \in B) P(\mathbf{Y}_{n+1} \notin B)$$
$$= \pi^n (1 - \pi),$$

so N has the geometric distribution with parameter π.   ∎

EXAMPLE 1.67   Let N denote the number of trials until one spot shows when a fair die is repeatedly rolled in an honest manner. Determine the distribution of $N - 1$.

Solution   Think of one spot as a failure and of any other number of spots as a success. Then the probability of success on each trial is $5/6$. Observe that $N - 1$ is the number of successes before the first failure. Thus, $N - 1$ has the geometric distribution with parameter $5/6$.   ∎

EXAMPLE 1.68   Let $T_1, T_2, \ldots$ be independent random variables each having the exponential distribution with scale parameter β, let $t > 0$, and let N be the minimum value of i such that $T_i \le t$. Determine the distribution of $N - 1$.

Solution   Think of T_i as the lifetime of the ith machine manufactured by a company and of that machine as being a success if $T_i > t$ and a failure if $T_i \le t$. Then the probability of success of a given machine is given by $\pi = P(T_i > t) = e^{-t/\beta}$. Observe that $N - 1$ is the number of successes before the first failure. Thus, by the solution to Example 1.66, $N - 1$ has the geometric distribution with parameter π.   ∎

Problems

1.103   Let Y_1 denote the number of spots that show when a fair die is rolled in an honest manner, let Y_2 be uniformly distributed on the interval $[0, 6]$, and suppose Y_1 and Y_2 are independent. Determine **(a)** $P(Y_1 = 1$ and $Y_2 > 1)$; **(b)** $P(Y_1 < Y_2)$.

1.104   Two fair dice are rolled in an honest manner. Let I_1 be the indicator random variable for the event that the number of spots showing on the two dice is even and let I_2 be the indicator random variable for the event that the number of spots is 5, 6, 7, or 8. Show that I_1 and I_2 are independent random variables.

1.105   A fair coin is tossed twice in an honest manner. Let I_1 denote the indicator of heads on the first toss—that is, the random variable equals 1 if the coin lands heads up on the first toss and equals 0 otherwise. Similarly, let I_2 denote the indicator of heads on the second choice. By assumption, (I_1, I_2) is uniformly distributed on $\{0, 1\} \times \{0, 1\}$; equivalently, I_1 and I_2 are independent random variables, each of which is uniformly distributed on $\{0, 1\}$. Set $S = I_1 + I_2$. Show that **(a)** the ordered pair (I_1, S) is uniformly distributed on some finite set; **(b)** I_1 and S are dependent random variables.

1.106 Let $\mathbf{Y} = (Y_1, Y_2)$ be uniformly distributed on the triangle having vertices at $(0, 0)$, $(2, 0)$, and $(0, 1)$. Determine **(a)** the distribution function and density function of Y_1; **(b)** the distribution function and density function of Y_2; **(c)** $P(Y_1 \geq 1)$, $P(Y_2 \geq 1/2)$, and $P(Y_1 \geq 1, Y_2 \geq 1/2)$ and use the results to show that Y_1 and Y_2 are dependent random variables.

1.107 Verify Theorem 1.8 for $n = 3$.

1.108 Let $0 \leq a \leq 1/4$ and let X_1, X_2, and X_3 be $\{-1, 1\}$-valued random variables such that

$$P(X_1 = -1, X_2 = -1, X_3 = 1) = P(X_1 = -1, X_2 = 1, X_3 = -1)$$
$$= P(X_1 = 1, X_2 = -1, X_3 = -1)$$
$$= P(X_1 = 1, X_2 = 1, X_3 = 1)$$
$$= a$$

and

$$P(X_1 = -1, X_2 = -1, X_3 = -1) = P(X_1 = -1, X_2 = 1, X_3 = 1)$$
$$= P(X_1 = 1, X_2 = -1, X_3 = 1)$$
$$= P(X_1 = 1, X_2 = 1, X_3 = -1)$$
$$= \frac{1}{4} - a.$$

Show that **(a)** X_1, X_2, and X_3 are each uniformly distributed on $\{-1, 1\}$; **(b)** X_1, X_2, and X_3 are pairwise independent; that is, X_1 and X_2 are independent, X_1 and X_3 are independent, and X_2 and X_3 are independent; **(c)** X_1, X_2, and X_3 are independent if and only if $a = 1/8$.

1.109 Let $\mathbf{W}_1$ be a $\mathcal{W}_1$-valued random variable and let $\mathbf{W}_2$ be a $\mathcal{W}_2$-valued random variable. Set $\mathbf{Y}_1 = \mathbf{g}_1(\mathbf{W}_1)$ and $\mathbf{Y}_2 = \mathbf{g}_2(\mathbf{W}_2)$, where $\mathbf{g}_1$ is a $\mathcal{Y}_1$-valued function on $\mathcal{W}_1$ and $\mathbf{g}_2$ is a $\mathcal{Y}_2$-valued function on $\mathcal{W}_2$. Show (directly from the definition of independence and the formula in Section 1.4 for the distribution of a transform) that if $\mathbf{W}_1$ and $\mathbf{W}_2$ are independent, then $\mathbf{Y}_1$ and $\mathbf{Y}_2$ are independent.

1.110 Let $\mathbf{W}_1, \mathbf{W}_2, \mathbf{W}_3$ be independent random variables and set $\mathbf{Y}_1 = \mathbf{W}_2, \mathbf{Y}_2 = \mathbf{W}_3$, and $\mathbf{Y}_3 = \mathbf{W}_1$. Verify directly from the definition of independence that $\mathbf{Y}_1, \mathbf{Y}_2$, and $\mathbf{Y}_3$ are independent.

1.111 Let $\mathbf{Y}_1, \mathbf{Y}_2, \mathbf{Y}_3$ be independent random variables. Verify directly from the definition of independence that $\mathbf{Y}_1$ and $\mathbf{Y}_3$ are independent.

1.112 Let $\mathbf{Y}_1, \mathbf{Y}_2, \mathbf{Y}_3$ be random variables such that $\mathbf{Y}_1$ and $(\mathbf{Y}_2, \mathbf{Y}_3)$ are independent and $\mathbf{Y}_2$ and $\mathbf{Y}_3$ are independent. Verify directly from the definition of independence that $\mathbf{Y}_1, \mathbf{Y}_2, \mathbf{Y}_3$ are independent.

1.113 Let $\mathbf{Y}_1, \ldots, \mathbf{Y}_n$ be independent random variables, each having a probability function. Verify that Equation (51) holds.

1.114   Two fair dice are rolled in an honest manner. Let Y_1 denote the minimum number of spots showing on the two dice and let Y_2 denote the maximum number. **(a)** Determine the joint probability function of Y_1 and Y_2. **(b)** Are Y_1 and Y_2 independent?

1.115   Let X and Y be independent, positive random variables having the same distribution, and let $c \geq 1$. Show that

$$P\left(\frac{X}{c} \leq Y \leq cX\right) = 1 - 2P\left(Y < \frac{X}{c}\right).$$

1.116   Two fair dice are repeatedly rolled in an honest manner until 7 or 11 occurs (as the total number of spots on the two dice) for the first time. Determine the probability that five such rolls of the two dice are required.

1.117   Two fair dice are repeatedly rolled in an honest manner. Let N denote the number of trials until 7 or 11 occurs (as the total number of spots on the two dice). Determine the distribution of $N - 1$.

1.118   Let N denote the number of trials until one spot shows when a fair die is repeatedly rolled in an honest manner. Determine **(a)** $P(N - 1 \geq n)$ for $n = 1, 2, \ldots$; **(b)** $P(N \leq n)$ for $n = 1, 2, \ldots$; **(c)** the median of N.

1.119   Let a dartboard have the form of a disk of radius 12 inches, let the bull's-eye be a concentric disk of radius 1 inch, let darts be thrown randomly at the dart board (so that the successive landing points are independently and uniformly distributed on the dartboard), and let N be the number of throws up to and including the first bull's-eye. Determine the distribution of $N - 1$.

1.120   Let $\mathcal{Y} = \mathcal{Y}_1 \times \mathcal{Y}_2 \times \mathcal{Y}_3$, where $\mathcal{Y}_1$, $\mathcal{Y}_2$, and $\mathcal{Y}_3$ are subsets of $\mathbb{R}$ each having finite, positive length; let $\mathbf{Y}_i$ be a $\mathcal{Y}_i$-valued random variable for $1 \leq i \leq 3$; and suppose $\mathbf{Y} = (\mathbf{Y}_1, \mathbf{Y}_2, \mathbf{Y}_3)$ is uniformly distributed on $\mathcal{Y}$. Show that $\mathbf{Y}_i$ is uniformly distributed on $\mathcal{Y}_i$ for $1 \leq i \leq 3$ and that $\mathbf{Y}_1, \mathbf{Y}_2$, and $\mathbf{Y}_3$ are independent.

2

Expectation

2.1

Introduction

Discrete Random Variables

Consider the roulette player who bets \$2 on red and \$1 on $\{1, \ldots, 12\}$ on the same trial. Let Y denote the amount won by the player on the trial. According to the solution to Example 1.12 in Section 1.3, the probability function f of Y is given by $f(-3) = 7/19, f(0) = 3/19, f(1) = 6/19, f(4) = 3/19$, and $f(y) = 0$ for other values of y.

Suppose now the player makes the same combination of bets on each of n consecutive trials and let Y_i denote the amount won on the ith trial for $i = 1, \ldots, n$. Then $Y_1 + \cdots + Y_n$ is the overall amount won on the first n trials, and $\bar{Y} = (Y_1 + \cdots + Y_n)/n$ is the corresponding average amount won per trial.

Let $N_n(y)$ denote the frequency of y among $Y_1, \ldots, Y_n$ and let $\hat{f}_n(y) = N_n(y)/n$ denote the corresponding relative frequency. Then $Y_1 + \cdots + Y_n = \sum_y y N_n(y)$, and hence

$$\bar{Y} = \frac{Y_1 + \cdots + Y_n}{n} = \sum_y y \frac{N_n(y)}{n} = \sum_y y \hat{f}_n(y).$$

According to the law of relative frequencies, $\hat{f}_n(y)$ should converge to $f(y)$ as $n \to \infty$. If so, then $\bar{Y}$ converges to $\sum_y y f(y)$ as $n \to \infty$. We refer to

$$\sum_y y f(y) = (-3) \cdot \frac{7}{19} + 0 \cdot \frac{3}{19} + 1 \cdot \frac{6}{19} + 4 \cdot \frac{3}{19} = -\frac{3}{19}$$

as the *expected value*, *expectation*, *mean*, or *first moment* of Y, which we denote by EY.

Consider, more generally, a discrete (real-valued) random variable Y having probability function f_Y. Its expected value EY is defined by

$$EY = \sum_y y f_Y(y) = \sum_y y P(Y = y). \tag{1}$$

EXAMPLE 2.1   Suppose $P(Y = a) = 1$ for some $a \in \mathbb{R}$. Determine EY.

Solution   The probability function of Y is given by $f(a) = 1$ and $f(y) = 0$ for $y \neq a$. Thus, $EY = \sum_y y f(y) = a.$ ∎

EXAMPLE 2.2   Let Y have the Bernoulli distribution with parameter π. Determine EY.

Solution   The probability function of Y is given by $f(0) = 1 - \pi$ and $f(1) = \pi$. Thus, $EY = \pi.$ ∎

EXAMPLE 2.3   Let Y be uniformly distributed on $\{c, \dots, d\}$, where c and d are integers with $c \leq d$. Determine EY.

Solution   The probability function of Y is given by $f(y) = 1/(d - c + 1)$ for $y \in \{c, \dots, d\}$ and $f(y) = 0$ elsewhere. Thus,

$$EY = \frac{1}{d - c + 1} \sum_{y=c}^{d} y = c + \frac{1}{d - c + 1} \sum_{y=c}^{d} (y - c) = c + \frac{1}{d - c + 1} \sum_{y=0}^{d-c} y.$$

It is known from algebra (and easily verified by induction) that $0 + \cdots + n = n(n + 1)/2$ for n a nonnegative integer. Thus,

$$EY = c + \frac{(d - c)(d - c + 1)}{2(d - c + 1)} = \frac{c + d}{2}.$$ ∎

EXAMPLE 2.4   Determine the expected number of spots Y that show when a fair die is rolled in an honest manner.

Solution   Since Y is uniformly distributed on $\{1, \dots, 6\}$, we conclude from the solution to Example 2.3 that $EY = (1 + 6)/2 = 7/2.$ ∎

EXAMPLE 2.5   Let Y have the geometric distribution with parameter π. Determine EY.

Solution The probability function of Y is given by $f(y) = (1 - \pi)\pi^y$, for y a nonnegative integer, and $f(y) = 0$ otherwise. Thus,

$$EY = (1 - \pi) \sum_{y=0}^{\infty} y\pi^y$$

$$= \pi(1 - \pi) \sum_{y=0}^{\infty} y\pi^{y-1}$$

$$= \pi(1 - \pi) \sum_{y=0}^{\infty} \frac{d}{d\pi}\pi^y$$

$$= \pi(1 - \pi) \frac{d}{d\pi} \sum_{y=0}^{\infty} \pi^y.$$

Consequently, by the formula for the sum of an infinite geometric series,

$$EY = \pi(1 - \pi)\frac{d}{d\pi}\frac{1}{1 - \pi} = \frac{\pi(1 - \pi)}{(1 - \pi)^2} = \frac{\pi}{1 - \pi}. \quad \blacksquare$$

EXAMPLE 2.6 Let Y have the probability function f given by

$$f(y) = \frac{1}{y(y + 1)}, \qquad y = 1, 2, \ldots,$$

and $f(y) = 0$ elsewhere. Determine EY.

Solution Observe that

$$EY = \sum_{y=1}^{\infty} y\frac{1}{y(y + 1)} = \sum_{y=1}^{\infty} \frac{1}{y + 1} = \frac{1}{2} + \frac{1}{3} + \frac{1}{4} + \cdots = \infty,$$

as is known from calculus. $\blacksquare$

EXAMPLE 2.7 Let Y have the probability function f given by

$$f(y) = \frac{1}{2|y|(|y| + 1)}, \qquad y = \pm 1, \pm 2, \ldots,$$

and $f(y) = 0$ elsewhere. Determine EY.

Solution Now

$$EY = \sum_{y=-\infty}^{-1} y\frac{1}{2|y|(|y| + 1)} + \sum_{y=1}^{\infty} y\frac{1}{2y(y + 1)} = -\frac{1}{2}\sum_{y=-\infty}^{-1}\frac{1}{1 - y} + \frac{1}{2}\sum_{y=1}^{\infty}\frac{1}{y + 1}.$$

Since

$$\sum_{y=-\infty}^{-1}\frac{1}{1 - y} = \sum_{y=1}^{\infty}\frac{1}{y + 1} = \infty,$$

we conclude that EY has the indeterminate form $\infty - \infty$. $\blacksquare$

The expected value of a random variable Y can be finite, it can equal $-\infty$ or ∞, and, as Example 2.7 illustrates, it can be of the indeterminate form $\infty - \infty$.

Let $\mathbf{Y}$ be a $\mathcal{Y}$-valued random variable having a probability function $f_{\mathbf{Y}}$ and let g be a real-valued function on $\mathcal{Y}$. Then the probability function of the random variable $Z = g(\mathbf{Y})$ is given by

$$f_Z(z) = P(Z = z) = P(\mathbf{Y} \in A_z) = \sum_{\mathbf{y} \in A_z} f_{\mathbf{Y}}(\mathbf{y}),$$

where $A_z = \{\mathbf{y} \in \mathcal{Y} : g(\mathbf{y}) = z\}$. Thus, the expected value of Z is given by

$$
\begin{aligned}
EZ &= \sum_z z f_Z(z) \\
&= \sum_z z \sum_{\mathbf{y} \in A_z} f_{\mathbf{Y}}(\mathbf{y}) \\
&= \sum_z \sum_{\mathbf{y} \in A_z} z f_{\mathbf{Y}}(\mathbf{y}) \\
&= \sum_z \sum_{\mathbf{y} \in A_z} g(\mathbf{y}) f_{\mathbf{Y}}(\mathbf{y}) \\
&= \sum_{\mathbf{y}} g(\mathbf{y}) f_{\mathbf{Y}}(\mathbf{y}) .
\end{aligned}
$$

This result can be rewritten as

$$E[\, g(\mathbf{Y})\,] = \sum_{\mathbf{y}} g(\mathbf{y}) f_{\mathbf{Y}}(\mathbf{y}). \tag{2}$$

EXAMPLE 2.8   A box contains six tickets, labeled from 1 to 6. Tickets 1, 2, and 3 each have value 1, tickets 4 and 5 each have value 2, and ticket 6 has value 5. Determine the mean of the value Y of a ticket selected at random from the box.

Solution   Let L be the label of the randomly selected ticket, which is uniformly distributed on $\{1, 2, 3, 4, 5, 6\}$. Then $Y = g(L)$, where g is the function on $\{1, 2, 3, 4, 5, 6\}$ given by $g(1) = g(2) = g(3) = 1$, $g(4) = g(5) = 2$, and $g(6) = 5$. Thus, by Equation (2),

$$EY = E[\, g(L)\,] = \sum_l g(l) f_L(l) = 1 \cdot \frac{1}{6} + 1 \cdot \frac{1}{6} + 1 \cdot \frac{1}{6} + 2 \cdot \frac{1}{6} + 2 \cdot \frac{1}{6} + 5 \cdot \frac{1}{6} = 2.$$

Alternatively, let f_Y denote the probability function of Y, which is given by $f_Y(1) = \frac{1}{2}, f_Y(2) = \frac{1}{3}$ and $f_Y(5) = \frac{1}{6}$. Then $EY = \sum_y y f_Y(y) = 1 \cdot \frac{1}{2} + 2 \cdot \frac{1}{3} + 5 \cdot \frac{1}{6} = 2.$ ∎

EXAMPLE 2.9   A box contains N tickets, labeled from 1 to N. Suppose ticket l has value v_l for $1 \le l \le N$. Determine the mean of the value Y of a ticket selected at random from the box.

Solution   Let L be the label of the randomly selected ticket, which is uniformly distributed on

$\{1, \ldots, N\}$. Then $Y = g(L)$, where $g(l) = v_l$ for $1 \le l \le N$. Thus, by Equation (2),

$$EY = E[g(L)] = \sum_l g(l)f_L(l) = \frac{1}{N}\sum_l v_l. \quad \blacksquare$$

EXAMPLE 2.10   A fair coin is repeatedly tossed in an honest manner. Johnny bets \$1 that it will show heads on the first toss. If it does show heads, he wins \$1. If not, he bets \$2 that it will show heads on the second toss. If it does show heads, he wins \$2 − \$1 = \$1. If not, he bets \$4 that it will show heads on the third toss. If it does show heads, he wins \$4 − \$1 − \$2 = \$1. If not, he loses \$1 + \$2 + \$4 = \$7. Determine the amount Johnny expects to win.

Solution   Let f be the probability function of the amount Y won by Johnny. Then $f(-7)$ is the probability of getting tails of each of the first three tosses, which equals $\frac{1}{8}$, and hence $f(1) = \frac{7}{8}$. Consequently, $EY = (-7) \cdot \frac{1}{8} + 1 \cdot \frac{7}{8} = 0.$ ∎

Let Y have finite mean and let $b \in \mathbb{R}$. It follows from Equation (2), with $g(y) = by$ for $y \in \mathcal{Y}$, that

$$E(bY) = \sum_y byf_Y(y) = b\sum_y yf_Y(y).$$

Thus,

$$E(bY) = bEY, \qquad b \in \mathbb{R}. \tag{3}$$

Consider now a random variable $\mathbf{Y} = (Y_1, Y_2)$ having a probability function $f_{\mathbf{Y}}$, which we write as f_{Y_1, Y_2}, and consider a random variable $g(\mathbf{Y}) = g(Y_1, Y_2)$. Then

$$E[g(\mathbf{Y})] = E[g(Y_1, Y_2)] = \sum_{y_1}\sum_{y_2} g(y_1, y_2)f_{Y_1, Y_2}(y_1, y_2).$$

Suppose, in particular, that Y_1 and Y_2 have finite expectation. Then

$$Y_1 = g_1(Y_1, Y_2), \quad Y_2 = g_2(Y_1, Y_2), \quad \text{and} \quad Y_1 + Y_2 = g(Y_1, Y_2),$$

where $g_1(y_1, y_2) = y_1, g_2(y_1, y_2) = y_2$, and $g(y_1, y_2) = y_1 + y_2$. Consequently,

$$EY_1 = E[g_1(Y_1, Y_2)] = \sum_{y_1}\sum_{y_2} y_1 f_{Y_1, Y_2}(y_1, y_2),$$

$$EY_2 = E[g_2(Y_1, Y_2)] = \sum_{y_1}\sum_{y_2} y_2 f_{Y_1, Y_2}(y_1, y_2),$$

and

$$E(Y_1 + Y_2) = E[g(Y_1, Y_2)]$$

$$= \sum_{y_1}\sum_{y_2}(y_1 + y_2)f_{Y_1, Y_2}(y_1, y_2)$$

$$= \sum_{y_1}\sum_{y_2} y_1 f_{Y_1, Y_2}(y_1, y_2) + \sum_{y_1}\sum_{y_2} y_2 f_{Y_1, Y_2}(y_1, y_2).$$

Thus,

$$E(Y_1 + Y_2) = EY_1 + EY_2. \tag{4}$$

EXAMPLE 2.11   A roulette player bets the amount b on a subcollection A of $\{1, \ldots, 36, 0, 00\}$. Determine the expected amount W won by the player.

Solution   The probability function of W is given by

$$P\left(W = b\frac{36 - \#(A)}{\#(A)}\right) = \frac{\#(A)}{38} \quad \text{and} \quad P(W = -b) = 1 - \frac{\#(A)}{38}.$$

Thus,

$$EW = (-b)\left(1 - \frac{\#(A)}{38}\right) + b\frac{36 - \#(A)}{\#(A)}\frac{\#(A)}{38} = -\frac{1}{19}b,$$

which does not depend on the size of A.

Alternatively, let I denote the indicator random variable for winning the bet. Then $EI = P(I = 1) = \#(A)/38$. If $I = 1$, then

$$W = b\frac{36 - \#(A)}{\#(A)};$$

if $I = 0$ or, equivalently, if $1 - I = 1$, then $W = -b$. Thus,

$$W = b\frac{36 - \#(A)}{\#(A)}I - b(1 - I) = b\left(\frac{36}{\#(A)}I - 1\right)$$

and hence by Equations (3) and (4) and the solution to Example 2.1,

$$EW = b\left(\frac{36}{\#(A)}EI - 1\right) = b\left(\frac{36}{38} - 1\right) = -\frac{1}{19}b. \quad \blacksquare$$

EXAMPLE 2.12   A roulette player bets \$2 on red and \$1 on $\{1, \ldots, 12\}$ on the same trial. Let W_1 be the amount won by the player based on the \$2 bet on red and let W_2 be the amount won based on the \$1 bet on $\{1, \ldots, 12\}$. Then $W = W_1 + W_2$ is the overall amount won on the trial. Determine EW_1 and EW_2 and use these results to determine EW.

Solution   It follows from the solution to Example 2.11 that $EW_1 = -2/19$ and $EW_2 = -1/19$. Thus, by Equation (4),

$$EW = E(W_1 + W_2) = EW_1 + EW_2 = -\frac{2}{19} - \frac{1}{19} = -\frac{3}{19},$$

as we have already seen in the beginning of this section. $\blacksquare$

EXAMPLE 2.13   Determine the expected total number of spots that show when two fair dice are rolled in an honest manner.

Solution   Let Y_1 be the number of spots that show on the first die and let Y_2 be the number showing on the second die. Then $Y = Y_1 + Y_2$ is the total number of spots that show on the two dice. It follows from the solution to Example 2.4 that $EY_1 = EY_2 = 7/2$.

Thus, by Equation (4),

$$EY = E(Y_1 + Y_2) = EY_1 + EY_2 = \frac{7}{2} + \frac{7}{2} = 7.$$

Alternatively, the probability function of Y is given in the solution to Example 1.24 in Section 1.5. Thus,

$$EY = \sum_y y f_Y(y)$$

$$= 2 \cdot \frac{1}{36} + 3 \cdot \frac{2}{36} + 4 \cdot \frac{3}{36} + 5 \cdot \frac{4}{36} + 6 \cdot \frac{5}{36} + 7 \cdot \frac{6}{36}$$

$$+ 8 \cdot \frac{5}{36} + 9 \cdot \frac{4}{36} + 10 \cdot \frac{3}{36} + 11 \cdot \frac{2}{36} + 12 \cdot \frac{1}{36}$$

$$= 7. \quad \blacksquare$$

EXAMPLE 2.14 Determine the expected number of tosses up to and including the first occurrence of heads when a fair coin is repeatedly tossed in an honest manner.

Solution Let Y be the indicated number of tosses. Then $Y - 1$ has the geometric distribution with parameter $1/2$ (see the solution to Example 1.66 in Section 1.8). Thus, $E(Y - 1) = 1$ by the solution to Example 2.5 and hence

$$EY = E(Y - 1 + 1) = E(Y - 1) + E1 = 1 + 1 = 2. \quad \blacksquare$$

EXAMPLE 2.15 Let Y be a nonnegative, discrete random variable. Show that

(a) $EY \geq 0$; **(b)** if $P(Y > 0) > 0$, then $EY > 0$.

Solution Let f denote the probability function of Y. Then $f(y) = 0$ for $y < 0$. Consequently, $y f(y) \geq 0$ for all y.

(a) Since the sum of nonnegative quantities is nonnegative, we conclude that $EY = \sum_y y f(y) \geq 0$.

(b) Now $P(Y > 0) = \sum_{y > 0} f(y)$; so if $P(Y > 0) > 0$, then $f(y) > 0$ for some $y > 0$. Thus, at least one of the quantities $y f(y)$ is positive. Since all of these quantities are nonnegative, we conclude that their sum is positive; that is, $EY = \sum_y y f(y) > 0$. $\blacksquare$

Continuous Random Variables

Let Y now be a $\mathcal{Y}$-valued random variable having density function f_Y. By analogy with Equation (2), we get

$$E[g(Y)] = \int_{\mathcal{Y}} g(y) f_Y(y) \, dy. \tag{5}$$

In particular,

$$EY = \int_{-\infty}^{\infty} y f_Y(y) \, dy. \tag{6}$$

EXAMPLE 2.16   Let Y be uniformly distributed on $[\,c, d\,]$. Determine EY.

Solution   The density function of Y is given by $f(y) = 1/(d-c)$ for $c < y < d$ and $f(y) = 0$ otherwise. Thus, by Equation (6),

$$EY = \int_{c}^{d} y f(y) \, dy = \frac{1}{d-c} \int_{c}^{d} y \, dy = \frac{1}{d-c} \frac{y^2}{2} \Big|_{c}^{d}$$

$$= \frac{1}{d-c} \left(\frac{d^2 - c^2}{2} \right) = \frac{c+d}{2}. \quad \blacksquare$$

Recall the *integration by parts* formula $\int u \, dv = uv - \int v \, du$, which can also be written as $\int uv' = uv - \int vu'$.

EXAMPLE 2.17   Let Y be exponentially distributed with scale parameter β. Determine EY.

Solution   The density function of Y is given by $f(y) = 0$ for $y \le 0$ and

$$f(y) = \frac{1}{\beta} e^{-y/\beta}, \qquad y > 0.$$

Thus, by Equation (6),

$$EY = \int_{0}^{\infty} y \frac{1}{\beta} e^{-y/\beta} \, dy.$$

Using integration by parts, we get

$$EY = \int_{0}^{\infty} y \, d(-e^{-y/\beta}) = -y e^{-y/\beta} \Big|_{0}^{\infty} + \int_{0}^{\infty} e^{-y/\beta} \, dy$$

$$= \int_{0}^{\infty} e^{-y/\beta} \, dy = -\beta e^{-y/\beta} \Big|_{0}^{\infty} = \beta. \quad \blacksquare$$

In light of the solution to this example, we will refer to the exponential distribution with scale parameter β as the exponential distribution with mean β.

EXAMPLE 2.18   Let Y be exponentially distributed. Determine $E(e^Y)$

(a) if Y has mean $1/3$;   **(b)** if Y has mean 1.

Solution **(a)** It follows from Equation (5) that if Y has mean $1/3$, then

$$E(e^Y) = \int_0^\infty e^y 3 e^{-3y}\, dy = 3 \int_0^\infty e^{-2y}\, dy = -\frac{3}{2} e^{-2y}\Big|_0^\infty = \frac{3}{2}.$$

(b) Similarly, if Y has mean 1, then

$$E(e^Y) = \int_0^\infty e^y e^{-y}\, dy = \int_0^\infty dy = \infty. \quad \blacksquare$$

EXAMPLE 2.19 Let U be uniformly distributed on $(0, 1)$. Determine $E(1/U)$.

Solution By Equation (5),

$$E\left(\frac{1}{U}\right) = \int_0^1 \frac{1}{u}\, du = \log u \Big|_0^1 = \log 1 - \log 0 = 0 - (-\infty) = \infty.$$

Alternatively, the distribution function of $Y = 1/U$ is given by $F_Y(y) = 0$ for $y \le 1$ and

$$F_Y(y) = P(Y \le y) = P\left(\frac{1}{U} \le y\right) = P\left(U \ge \frac{1}{y}\right) = 1 - \frac{1}{y}, \qquad y > 1.$$

Thus, the density function of Y is given by $f_Y(y) = 0$ for $y \le 1$ and $f_Y(y) = 1/y^2$ for $y > 1$. Consequently, by Equation (6),

$$EY = \int_1^\infty y \frac{1}{y^2}\, dy = \int_1^\infty \frac{1}{y}\, dy = \log y \Big|_1^\infty = \log \infty - \log 1 = \infty - 0 = \infty. \quad \blacksquare$$

EXAMPLE 2.20 Let Y have the standard Cauchy distribution. Determine EY.

Solution The density function of Y is given by

$$f(y) = \frac{1}{\pi(1 + y^2)}, \qquad y \in \mathbb{R}.$$

Thus, by Equation (6),

$$EY = \int_{-\infty}^\infty yf(y)\, dy = \int_{-\infty}^\infty y \frac{1}{\pi(1 + y^2)}\, dy = \frac{1}{\pi} \int_{-\infty}^\infty \frac{y}{1 + y^2}\, dy.$$

Observe that

$$\int_{-\infty}^\infty \frac{y}{1 + y^2}\, dy = \int_{-\infty}^0 \frac{y}{1 + y^2}\, dy + \int_0^\infty \frac{y}{1 + y^2}\, dy.$$

Now

$$\int_0^\infty \frac{y}{1 + y^2}\, dy = \frac{\log(1 + y^2)}{2}\Big|_0^\infty = \infty,$$

and

$$\int_{-\infty}^0 \frac{y}{1 + y^2}\, dy = \frac{\log(1 + y^2)}{2}\Big|_{-\infty}^0 = -\infty.$$

Thus,

$$\int_{-\infty}^{\infty} \frac{y}{1+y^2} \, dy = \infty - \infty$$

(which is indeterminate) and hence $EY = \infty - \infty$.   ∎

Problems

2.1 A box has nine balls, which are labeled from 1 to 9. Determine the mean value of the label of a ball selected at random from the box.

2.2 A box contains ten tickets, one of which has the value 1, two of which have the value 2, three of which have the value 3, and four of which have the value 4. Determine the expected value of a ticket selected at random from the box.

2.3 Let Y_1 denote the number of spots that show on the first roll of a die and let Y_2 denote the number that show on the second roll. Determine $E|Y_1 - Y_2|$.

2.4 Two fair dice are rolled in an honest manner. Determine the mean of each of the following random variables: **(a)** the minimum number of spots showing on the two dice; **(b)** the maximum number of spots showing on the two dice; **(c)** the sum of the minimum number of spots showing on the two dice and the maximum number.

2.5 A roulette player bets \$1 on red. If red shows, he wins \$1. If not, he bets \$2 on red on the next spin of the wheel. If red shows, he wins $\$2 - \$1 = \$1$. If not, he bets \$4 on red on the following spin. If red shows, he wins $\$4 - \$1 - \$2 = \$1$. If not, he loses $\$1 + \$2 + \$4 = \$7$. Determine the expected amount won by the player.

2.6 A newspaper advertisement promotes a 1-year investment. According to the ad, for each \$1000 invested, the probability is .4 of getting back \$2000 at the end of the year (that is, of doubling the original investment), .5 of getting back \$1000 (that is, of breaking even), and .1 of getting back nothing (that is, of losing the original investment). Under the dubious assumption that the information in the ad is correct, determine the expected profit as a percentage of the original investment.

2.7 Let f have the probability function f given by

$$f(y) = \frac{1}{(-y)(1-y)}, \qquad y = -1, -2, \ldots,$$

and $f(y) = 0$ elsewhere. Determine EY.

2.8 Let X_1 and X_2 each be $\{-1, 1\}$-valued random variables and suppose X_1, X_2, and $X_1 X_2$ all have mean 0. Show that **(a)** X_1 and X_2 are each uniformly distributed on $\{-1, 1\}$; **(b)** X_1 and X_2 are independent random variables.

2.9 Determine the mean of a random variable having the density function f given by $f(y) = 6y(1 - y)$ for $0 < y < 1$ and $f(y) = 0$ elsewhere.

2.10 Let Y have the density function given by $f(y) = \alpha y^{\alpha-1}$ for $0 < y < 1$ and $f(y) = 0$ elsewhere, where $\alpha > 0$. Determine **(a)** EY; **(b)** $E(1/Y)$.

2.11 Let U be uniformly distributed on $(-2, 0)$. Determine $E(1/U)$.

2.12 Let Y have the exponential distribution with mean β. Determine $E(e^Y)$.

2.13 Let Y be uniformly distributed on the interval $[0, 3\pi]$. Determine the expected value of $\sin Y$.

2.14 Let W be uniformly distributed on a disk $\mathcal{W}$ having radius 1. Determine the expected distance from W to the center of the disk.

2.2
Properties of Expectation

In advanced probability theory, an expected value (or expectation, mean, or first moment) is assigned to every real-valued random variable Y. This expected value can be finite, ∞, $-\infty$, or indeterminate ($\infty - \infty$). The expected value of Y depends only on the distribution of Y; that is, if two random variables have the same distribution, then they have the same expected value. The mean of Y is also referred to as the mean of the distribution of Y or of its probability function or density function.

Let $\mathbf{Y}$ be a $\mathcal{Y}$-valued random variable and let g be a real-valued function on $\mathcal{Y}$. Then $E[g(\mathbf{Y})]$ is the mean of the random variable $g(\mathbf{Y})$. Since the distribution of $g(\mathbf{Y})$ depends only on g and the distribution of $\mathbf{Y}$, $E[g(\mathbf{Y})]$ also depends only on g and the distribution of $\mathbf{Y}$. The properties of expectation in the following theorem, motivated by the informal discussion in Section 2.1, are proven rigorously in advanced probability theory.

THEOREM 2.1   **(a)** If $P(Y = a) = 1$ for some $a \in \mathbb{R}$, then $E(Y) = a$.

(b) If Y has finite mean, then $E(bY) = bEY$ for $b \in \mathbb{R}$.

(c) If Y_1 and Y_2 each have finite mean, then $E(Y_1 + Y_2) = EY_1 + EY_2$.

(d) If $P(Y \geq 0) = 1$, then $EY \geq 0$.

(e) If $P(Y \geq 0) = 1$ and $EY = 0$, then $P(Y = 0) = 1$.

(f) If $\mathbf{Y}$ is a $\mathcal{Y}$-valued random variable having probability function $f_{\mathbf{Y}}$ and g is a real-valued function on $\mathcal{Y}$, then $E[g(\mathbf{Y})] = \sum_{\mathbf{y}} g(\mathbf{y}) f_{\mathbf{Y}}(\mathbf{y})$.

(g) If $\mathcal{Y} \subset \mathbb{R}$, Y is a $\mathcal{Y}$-valued random variable having density function f_Y and g is a real-valued function on $\mathcal{Y}$, then $E[g(Y)] = \int_{\mathcal{Y}} g(y) f_Y(y)\, dy$.

For a precise interpretation of Theorem 2.1(f), let S_- represent the summation of $g(\mathbf{y}) f_{\mathbf{Y}}(\mathbf{y})$ over all values of $\mathbf{y}$ for which the indicated product is negative and let S_+ denote the summation over all values of $\mathbf{y}$ for which it is positive. If S_- and S_+ are finite, then $E(g(\mathbf{Y})) = S_- + S_+$; if S_- is finite and $S_+ = \infty$, then $E[g(\mathbf{Y})] = \infty$; if $S_- = -\infty$ and S_+ is finite, then $E(g(\mathbf{Y})) = -\infty$; and if $S_- = -\infty$ and $S_+ = \infty$, then $E[g(\mathbf{Y})]$ is indeterminate. Theorem 2.1(g) is interpreted in a similar manner, with summation being replaced by integration.

The next result follows from parts (a) and (e) of Theorem 2.1.

COROLLARY 2.1   Let Y be a nonnegative random variable. Then $EY = 0$ if and only if $P(Y = 0) = 1$.

The next result follows from parts (f) and (g) of Theorem 2.1.

COROLLARY 2.2   If Y has probability function f_Y, then $EY = \sum_y y f_Y(y)$; if Y has density function f_Y, then $EY = \int_{-\infty}^{\infty} y f_Y(y)\,dy$.

COROLLARY 2.3   If $Y_1, \ldots, Y_n$ each have finite mean, then

$$E(b_1 Y_1 + \cdots + b_n Y_n) = b_1 EY_1 + \cdots + b_n EY_n, \qquad b_1, \ldots, b_n \in \mathbb{R}.$$

Proof   It follows from Theorem 2.1(b) that $E(b_1 Y_1) = b_1 EY_1$ and $E(b_2 Y_2) = b_2 EY_2$ and hence from Theorem 2.1(c) that

$$E(b_1 Y_1 + b_2 Y_2) = E(b_1 Y_1) + E(b_2 Y_2) = b_1 E(Y_1) + b_2 E(Y_2).$$

The general result now follows by induction.

The next result is a special case of Corollary 2.3.

COROLLARY 2.4   If Y_1 and Y_2 each have finite mean, then $E(Y_1 - Y_2) = EY_1 - EY_2$.

EXAMPLE 2.21   Let Y_1, Y_2, and Y_3 have means 1, 4, and -9, respectively. Determine

(a) $E(-Y_2)$;   (b) $E(Y_1 - Y_2)$;   (c) $E(Y_1 - Y_2 + 5Y_3)$.

Solution   (a) $E(-Y_2) = -EY_2 = -4$.
(b) $E(Y_1 - Y_2) = EY_1 - EY_2 = 1 - 4 = -3$.
(c) $E(Y_1 - Y_2 + 5Y_3) = EY_1 - EY_2 + 5EY_3 = 1 - 4 + (5)(-9) = -48$.   ∎

EXAMPLE 2.22   Consider a random temperature whose mean is 20 on the Celsius scale. Determine its mean on the Fahrenheit scale.

Solution   Let T denote the random temperature on the Celsius scale. The corresponding temperature on the Fahrenheit scale is $32 + \frac{9}{5}T$, whose mean is given by

$$E\left(32 + \frac{9}{5}T\right) = 32 + \frac{9}{5}(ET) = 32 + \frac{9}{5}(20) = 68.   \blacksquare$$

COROLLARY 2.5   If Y has finite mean μ, then $E(Y - \mu) = 0$.

Proof   Since $E\mu = \mu$, we conclude from Corollary 2.4 that

$$E(Y - \mu) = EY - E\mu = \mu - \mu = 0.   \blacksquare$$

Let Y be a $\mathcal{Y}$-valued random variable having probability function f and finite mean μ. Then $\sum_{y \in \mathcal{Y}} (y - \mu)f(y) = 0$ by Corollary 2.5. Thus, if weight $f(y)$ is placed

FIGURE 2.1

Weights .4, .2, and .4 at locations 1, 2, and 8, respectively, are balanced at location $\mu = 1(.4) + 2(.2) + 8(.4) = 4$.

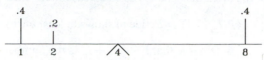

at location y on a weightless plank for $y \in \mathcal{Y}$, the plank will be balanced at location μ (Figure 2.1).

Recall that a $\{0, 1\}$-valued random variable is referred to as an indicator random variable.

COROLLARY 2.6   Let I be an indicator random variable. Then $EI = P(I = 1)$.

Proof   Let f denote the probability function of I. Then, by the first conclusion to Corollary 2.2,

$$EI = 0 \cdot f(0) + 1 \cdot f(1) = f(1) = P(I = 1). \quad \blacksquare$$

Let $\mathbf{Y}$ be a $\mathcal{Y}$-valued random variable, let B be a subset of $\mathcal{Y}$, and let $I = \text{ind}(\mathbf{Y} \in B)$ be the indicator random variable corresponding to the event that $\mathbf{Y} \in B$. Then $P(I = 1) = P(\mathbf{Y} \in B)$, so we conclude from Corollary 2.6 that

$$E[\text{ind}(\mathbf{Y} \in B)] = P(\mathbf{Y} \in B). \tag{7}$$

EXAMPLE 2.23   A fair die is rolled ten times in an honest manner. Determine

(a) the probability that side 1 shows at least once;

(b) the expected number of sides that show at least once.

Solution   The probability that side 1 never shows is $(5/6)^{10}$, so the probability that this side shows at least once is $1 - (5/6)^{10}$. Similarly, the probability that side m shows at least once is $1 - (5/6)^{10}$ for $m = 1, \ldots, 6$, so the expected number of sides that appear at least once is $6[1 - (5/6)^{10}]$.

Alternatively, we can express this solution in terms of random variables. Let Y_i denote the number of spots that show on the ith roll. Then $Y_1, \ldots, Y_{10}$ are independent, and each of these random variables is uniformly distributed on $\{1, \ldots, 6\}$.

(a) The probability that side 1 never shows is given by

$$P(Y_1 \neq 1, \ldots, Y_{10} \neq 1) = P(Y_1 \neq 1) \cdots P(Y_{10} \neq 1) = \left(\frac{5}{6}\right)^{10}.$$

Thus, the probability that side 1 shows at least once is given by

$$1 - P(Y_1 \neq 1, \ldots, Y_{10} \neq 1) = 1 - \left(\frac{5}{6}\right)^{10}.$$

(b) More generally, the probability that side m shows at least once is $1 - (\frac{5}{6})^{10}$ for $m = 1, \ldots, 6$. Let $I_m = \text{ind}(\text{side } m \text{ shows at least once})$ be the indicator random variable for the event that side m shows at least once. Then

$$E(I_m) = P(I_m = 1) = P(\text{side } m \text{ shows at least once}) = 1 - \left(\frac{5}{6}\right)^{10},$$

for $m = 1, \ldots, 6$. Observe that $I_1 + \cdots + I_6$ is the number of sides that show at least once. Thus, the expected number of sides that show at least once is given by

$$E(I_1 + \cdots + I_6) = E(I_1) + \cdots + E(I_6) = 6\left[1 - \left(\frac{5}{6}\right)^{10}\right]. \quad \blacksquare$$

COROLLARY 2.7   If Y_1 and Y_2 each have finite mean and $P(Y_1 \geq Y_2) = 1$, then $EY_1 \geq EY_2$.

Proof   Since $P(Y_1 - Y_2 \geq 0) = 1$, we conclude from Theorem 2.1(d) that $E(Y_1 - Y_2) \geq 0$. Thus, by Corollary 2.4, $EY_1 - EY_2 = E(Y_1 - Y_2) \geq 0$, and hence $EY_1 \geq EY_2$. $\quad \blacksquare$

EXAMPLE 2.24   Let Y be a nonnegative random variable having finite mean and let $c > 0$.

(a) Show that

$$P(Y \geq c) \leq \frac{EY}{c}. \tag{8}$$

(b) Show that

$$P(Y \geq c) = \frac{EY}{c} \tag{9}$$

if and only if $P(Y \in \{0, c\}) = 1$.

Solution   **(a)** Now

$$\text{ind}(Y \geq c) \leq \frac{Y}{c}. \tag{10}$$

[To verify (10), note that if $Y \geq c$, then the left side of (10) equals 1 and the right side is at least 1 and that if $0 \leq Y < c$, then the left side equals 0 and the right side is nonnegative.] It follows from Equation (7), Property (10), Theorem 2.1(b), and Corollary 2.7 that

$$P(Y \geq c) = E[\,\text{ind}(Y \geq c)\,] \leq E\left(\frac{Y}{c}\right) = \frac{EY}{c}.$$

(b) Consider the nonnegative random variable

$$W = \frac{Y}{c} - \text{ind}(Y \geq c).$$

Observe that (9) holds if and only if $EW = 0$ and hence, by Corollary 2.1, if and only if $P(W = 0) = 1$. Now $W = 0$ if and only if $Y = c\,\text{ind}(Y \geq c)$ and hence if and only if $Y \in \{0, c\}$. Thus, (9) holds if and only if $P(Y \in \{0, c\}) = 1$. $\quad \blacksquare$

The result in Property (8) is known as *Markov's inequality*.

EXAMPLE 2.25   Let Y be a random variable such that $E(Y^4) \leq 100$. Use this information to determine an upper bound to $P(Y \geq 5)$.

Solution   Set $W = Y^4$. Then W is a nonnegative random variable whose mean is at most 100. Observe that if $Y \geq 5$, then $W \geq 625$. Thus, by Markov's inequality,

$$P(Y \geq 5) \leq P(W \geq 625) \leq \frac{EW}{625} \leq \frac{100}{625}. \quad \blacksquare$$

COROLLARY 2.8   The mean of a random variable having a density function that is symmetric about 0 is either 0 or indeterminate.

Proof   Let f be a density function on $\mathbb{R}$ that is symmetric about zero. Then

$$\int_{-\infty}^{\infty} yf(y)\,dy = \int_{-\infty}^{0} yf(y)\,dy + \int_{0}^{\infty} yf(y)\,dy,$$

and

$$\int_{-\infty}^{0} yf(y)\,dy = \int_{\infty}^{0} yf(-y)\,dy = -\int_{0}^{\infty} yf(-y)\,dy = -\int_{0}^{\infty} yf(y)\,dy.$$

Thus,

$$\int_{-\infty}^{\infty} yf(y)\,dy$$

is 0 or indeterminate according as

$$\int_{0}^{\infty} yf(y)\,dy$$

is finite or ∞. The desired result now follows from Corollary 2.2. $\blacksquare$

EXAMPLE 2.26   Determine the mean of the standard logistic distribution.

Solution   The standard logistic distribution has density function

$$f(y) = \frac{e^y}{(1 + e^y)^2}, \qquad y \in \mathbb{R},$$

which is symmetric about 0 (see Example 1.42 in Section 1.6). Since

$$\int_{0}^{\infty} yf(y)\,dy = \int_{0}^{\infty} y\frac{e^y}{(1 + e^y)^2}\,dy \leq \int_{0}^{\infty} ye^{-y}\,dy = 1 < \infty,$$

the mean is not indeterminate. Thus, by Corollary 2.8, the mean is 0. $\blacksquare$

The next theorem contains two more properties of expectation, which are proven in advanced probability theory.

THEOREM 2.2   **(a)** A random variable Y has finite mean if and only if $|Y|$ has finite mean.

(b) If Y_1 and Y_2 are random variables such that Y_2 has finite mean and $P(|Y_1| \leq Y_2) = 1$, then Y_1 has finite mean.

A random variable Y is referred to as a *bounded random variable* if there is a (finite) nonnegative number M such that $P(|Y| \leq M) = 1$.

COROLLARY 2.9   Every bounded random variable has finite mean.

Proof   Let Y_1 be a bounded random variable and let M be a nonnegative number such that $P(|Y_1| \leq M) = 1$. Since the random variable $Y_2 = M$ has finite mean M, we conclude from Theorem 2.2(b) that Y_1 has finite mean. ∎

Let Y_1 and Y_2 be independent random variables having probability functions f_1 and f_2, respectively. Then their joint probability function f is given by

$$f(y_1, y_2) = f_1(y_1)f_2(y_2), \qquad y_1, y_2 \in \mathbb{R}.$$

Suppose Y_1 and Y_2 each have finite mean. Then

$$EY_1 = \sum_{y_1} y_1 f_1(y_1) \quad \text{and} \quad EY_2 = \sum_{y_2} y_2 f_2(y_2).$$

It follows from Theorem 2.1(f) and properties of summation that

$$E(Y_1 Y_2) = \sum_{y_1} \sum_{y_2} y_1 y_2 f(y_1, y_2)$$

$$= \sum_{y_1} \sum_{y_2} y_1 y_2 f_1(y_1) f_2(y_2)$$

$$= \left(\sum_{y_1} y_1 f_1(y_1) \right) \left(\sum_{y_2} y_2 f_2(y_2) \right)$$

and hence that

$$E(Y_1 Y_2) = (EY_1)(EY_2). \tag{11}$$

It is shown in advanced probability theory that (11) holds for independent random variables having finite means even if they do not have probability functions.

EXAMPLE 2.27   Let U_1 and U_2 be independent random variables, each uniformly distributed on $[0, 1]$. Determine the expected area of the random rectangle B having vertices at $(0, 0)$, $(U_1, 0)$, (U_1, U_2), and $(0, U_2)$.

Solution   The area of B is $U_1 U_2$ (Figure 2.2). According to the solution to Example 2.16 in Section 2.1, $EU_1 = EU_2 = 1/2$. Thus, the expected area of B is given by

$$E(U_1 U_2) = (EU_1)(EU_2) = \frac{1}{2} \cdot \frac{1}{2} = \frac{1}{4}. \qquad \blacksquare$$

FIGURE 2.2

Let Y_1 and Y_2 be independent random variables, each exponentially distributed with mean $1/3$. Determine $E(e^{Y_1+Y_2})$.

Solution   Now $E(e^{Y_1}) = E(e^{Y_2}) = \frac{3}{2}$ according to the solution to Example 2.18(a) in Section 2.1. Since e^{Y_1} and e^{Y_2} are transforms of independent random variables, they are independent. Applying Equation (5) with Y_1 and Y_2 replaced by e^{Y_1} and e^{Y_2}, respectively, we conclude that

$$E(e^{Y_1+Y_2}) = E(e^{Y_1}e^{Y_2}) = [\,E(e^{Y_1})\,][\,E(e^{Y_2})\,] = \frac{3}{2} \cdot \frac{3}{2} = \frac{9}{4}. \quad \blacksquare$$

Let Y_1, Y_2, and Y_3 be independent random variables, each having a finite mean. Then Y_1 and Y_2 are independent, and hence Equation (11) holds. Also, $Y_1 Y_2$ is independent of Y_3, so $E(Y_1 Y_2 Y_3) = [\,E(Y_1 Y_2)\,](EY_3) = (EY_1)(EY_2)(EY_3)$. More generally, we get the following result by using induction.

THEOREM 2.3   Let $Y_1, \ldots, Y_n$ be independent random variables each having finite mean. Then

$$E(Y_1 \cdots Y_n) = EY_1 \cdots EY_n.$$

EXAMPLE 2.29   Give an example to illustrate that the conclusion of Theorem 2.3 is *not* generally valid without the assumption of independence.

Solution   Let Y have the Bernoulli distribution with parameter $\pi \in (0, 1)$ and set $Y_1 = Y$ and $Y_2 = Y$. Then $EY_1 = EY_2 = EY = \pi$, so $(EY_1)(EY_2) = \pi^2$. Since Y is $\{0, 1\}$-valued, $Y^2 = Y$. Thus, $E(Y_1 Y_2) = E[\,Y^2\,] = EY = \pi \neq \pi^2 = (EY_1)(EY_2)$. $\quad \blacksquare$

Problems

2.15   Let Y_1, Y_2, and Y_3 have means 6, -5, and 3, respectively. Determine the mean of $2Y_1 + 7Y_2 - 4Y_3$.

2.16   Determine the mean of the total number of spots that show when ten fair dice are rolled in an honest manner.

2.17   A fair die is rolled ten times in an honest manner. Determine **(a)** the probability that side 1 appears at least once; **(b)** the expected number of sides that appear at least once.

2.18   Let $Y_1, \ldots, Y_n$ have common finite mean μ, let $b_1, \ldots, b_n \in \mathbb{R}$, and set $\hat{\mu} = b_1 Y_1 + \cdots + b_n Y_n$. **(a)** Show that if $b_1 + \cdots + b_n = 1$, then $E\hat{\mu} = \mu$. **(b)** Show that if $E\hat{\mu} = \mu$ for some nonzero value of μ, then $b_1 + \cdots + b_n = 1$. [It follows from parts (a) and (b) that $E\hat{\mu} = \mu$ regardless of μ if and only if $b_1 + \cdots + b_n = 1$.]

2.19   Let Y be uniformly distributed on $(0, 1)$. Show that $\log(Y/(1 - Y))$ has mean 0.

2.20   Consider the color that shows (that is, of the landing slot of the ball) on each of four consecutive spins of the roulette wheel. Determine **(a)** the probability that red shows at least once; **(b)** the probability that green shows at least once; **(c)** the expected number of colors that show at least once.

2.21   A roulette player bets \$2 on red and \$1 on $\{1, \ldots, 12\}$ on the same play. Let I_1 denote the indicator random variable for the event that the landing slot is red and let I_2 denote the indicator random variable for the event that the landing slot is a member of $\{1, \ldots, 12\}$. Also, let W_1 denote the amount of money won on the bet on red, W_2 the amount won on the bet on $\{1, \ldots, 12\}$, and $W = W_1 + W_2$ the combined amount won on the two bets. **(a)** Express W_1 in terms of I_1 and W_2 in terms of I_2. **(b)** Determine EI_1 and EI_2. **(c)** Use the answers to parts (a) and (b) to determine EW_1, EW_2, and EW. **(d)** Express W in terms of I_1 and I_2. **(e)** Use the answers to parts (b) and (d) to determine EW.

2.22   Let Y be a random variable such that $P(a \leq Y \leq b) = 1$, where a and b are finite real numbers. Show that Y has finite mean.

2.23   Give an example illustrating that the formula $E(|Y_1 - Y_2|) = |EY_1 - EY_2|$ is *not* generally valid for random variables Y_1 and Y_2 having finite means.

2.24   Let Y have the density function f given by $f(y) = 2y^{-3}$ for $y > 1$ and $f(y) = 0$ for $y \leq 1$. Determine **(a)** the distribution function of Y; **(b)** $P(Y \geq 10)$; **(c)** EY; **(d)** the upper bound to $P(Y \geq 10)$ given by Markov's inequality.

2.25   Let Y have the exponential distribution with mean β. **(a)** Determine $P(Y \geq 10\beta)$. **(b)** Use Markov's inequality to obtain an upper bound to $P(Y \geq 10\beta)$.

2.26   Let Y have the exponential distribution with mean β. Determine $E(e^{tY})$ for $t \in \mathbb{R}$.

2.27   (Continued.) Use Markov's inequality applied to the random variable $W = e^{tY}$ to verify that

$$P(Y \geq c\beta) \leq \frac{1}{e^{c\beta t}(1 - \beta t)}, \qquad c \in \mathbb{R} \text{ and } 0 \leq t < 1/\beta. \tag{12}$$

2.28   (Continued.) **(a)** Show that, as a function of t for fixed $c \geq 1$, the right side of (12) is minimized by choosing $t = (c - 1)/(c\beta)$. **(b)** Use (12) and the result of part (a) to verify that

$$P(Y \geq c\beta) \leq ce^{1-c}, \qquad c \geq 1.$$

(c) Use the result of part (b) to obtain an upper bound to $P(Y \geq 10\beta)$.

2.29   Let Y_1, Y_2, and Y_3 be independent random variables, each having finite mean μ. Determine the mean of $Y_1 Y_2 + Y_3$.

2.30   Consider an $\mathcal{X}$-valued random variable $\mathbf{X}$ and a $\mathcal{Y}$-valued random variable $\mathbf{Y}$. Show that $\mathbf{X}$ and $\mathbf{Y}$ are independent if and only if

$$E[\,g(\mathbf{X})h(\mathbf{Y})\,] = E[\,g(\mathbf{X})\,]E[\,h(\mathbf{Y})\,]$$

for all bounded, real-valued functions g on $\mathcal{X}$ and h on $\mathcal{Y}$.

2.3
Variance

Let Y be a real-valued random variable. The quantity $E(Y^2)$ is referred to as the *second moment* of Y (or of its distribution, probability function, or density function). Since $Y^2 \geq 0$, we see that $0 \leq E(Y^2) \leq \infty$. The second moment of a random variable can be regarded as an overall measure of its magnitude. In particular, the second moment equals 0 if and only if the random variable equals 0 with probability 1 (apply Corollary 2.1 in Section 2.2 to the square of the random variable).

If Y has probability function f_Y then, by Theorem 2.1(f) in Section 2.2 with $g(y) = y^2$,

$$E\left(Y^2\right) = \sum_y y^2 f_Y(y). \tag{13}$$

Similarly, if Y has density function f_Y, then, by part (g) of Theorem 2.1,

$$E\left(Y^2\right) = \int_{-\infty}^{\infty} y^2 f_Y(y)\, dy. \tag{14}$$

EXAMPLE 2.30   Detemine the second moment of the Bernoulli distribution with parameter π.

Solution   By (13), the desired second moment is given by $0^2(1 - \pi) + 1^2\pi = \pi$.   ∎

EXAMPLE 2.31   Determine the second moment of the exponential distribution with mean β.

Solution   The mean of the distribution is given by

$$\int_0^{\infty} y \frac{1}{\beta} e^{-y/\beta}\, dy = \beta.$$

Thus, using integration by parts, we see that the desired second moment is given by

$$\int_0^{\infty} y^2 \frac{1}{\beta} e^{-y/\beta}\, dy = -y^2 e^{-y/\beta}\Big|_0^{\infty} + \int_0^{\infty} 2y e^{-y/\beta}\, dy = 2\beta \int_0^{\infty} y \frac{1}{\beta} e^{-y/\beta}\, dy$$

$$= 2\beta^2. \quad ∎$$

The next elementary result is needed to prove the theorem that follows it.

LEMMA 2.1   Let a_1 and a_2 be real numbers. Then

$$|a_1 a_2| \leq \frac{1}{2}(a_1^2 + a_2^2) \quad \text{and} \quad (a_1 + a_2)^2 \leq 2(a_1^2 + a_2^2).$$

Proof   Observe that $0 \leq (a_1 - a_2)^2 = a_1^2 + a_2^2 - 2a_1 a_2$ and hence that

$$a_1 a_2 \leq \frac{1}{2}(a_1^2 + a_2^2). \tag{15}$$

Similarly, from the observation that $(a_1 + a_2)^2 \geq 0$, we conclude that

$$-a_1 a_2 \leq \frac{1}{2}(a_1^2 + a_2^2). \tag{16}$$

The first conclusion of the lemma follows from (15) and (16). It follows from (15) that

$$(a_1 + a_2)^2 = a_1^2 + a_2^2 + 2a_1 a_2 \leq a_1^2 + a_2^2 + a_1^2 + a_2^2$$

and hence that the second conclusion of the lemma is valid.   ∎

THEOREM 2.4   (a) If Y has finite second moment, then it has finite mean.

(b) If Y_1 and Y_2 each have finite second moment, then $Y_1 + Y_2$ has finite second moment, and $Y_1 Y_2$ has finite mean.

Proof   (a) Since Y has finite second moment, Y^2 has finite mean. Thus, $Y^2 + 1$ has finite mean and hence so does $(Y^2 + 1)/2$. According to Lemma 2.1, $|Y| = |Y \cdot 1| \leq (Y^2 + 1)/2$. Thus, we conclude from Theorem 2.2 in Section 2.2 that Y has finite mean.

(b) It follows from Theorem 2.1 in Section 2.2 that $b(Y_1^2 + Y_2^2)$ has finite mean for $b \in \mathbb{R}$. According to Lemma 2.1,

$$(Y_1 + Y_2)^2 \leq 2(Y_1^2 + Y_2^2) \quad \text{and} \quad |Y_1 Y_2| \leq \frac{1}{2}(Y_1^2 + Y_2^2);$$

thus, we conclude from Theorem 2.2 in Section 2.2 that $Y_1 + Y_2$ has finite second moment and that $Y_1 Y_2$ has finite mean.   ∎

Let Y have finite second moment and let μ denote the mean of Y, which is finite by Theorem 2.4(a). Then the second moment of $Y - EY = Y - \mu$ is finite and nonnegative. We refer to this quantity as the *variance* of Y (or of its distribution, and so forth) and to its nonnegative square root as the *standard deviation* of Y. The standard deviation of Y is denoted by σ or SD(Y), and the variance of Y is denoted by σ^2 or var(Y). Thus, $0 \leq \sigma < \infty$, $\sigma^2 = \text{var}(Y) = E[(Y - \mu)^2]$, and $\sigma = \text{SD}(Y) = \sqrt{\text{var}(Y)}$. If $E(Y^2) = \infty$, the standard deviation and variance of Y are defined to be ∞. Thus, Y has finite variance if and only if it has finite second moment. Consequently, by Theorem 2.4(a), if Y has finite variance, then it has finite mean.

Let Y have finite mean μ. If Y has probability function f_Y on $\mathcal{Y}$, then

$$\sigma^2 = \sum_y (y - \mu)^2 f_Y(y); \tag{17}$$

if Y has density function f_Y, then

$$\sigma^2 = \int_{-\infty}^{\infty} (y - \mu)^2 f_Y(y)\, dy. \tag{18}$$

These formulas are valid even if $\sigma^2 = \text{var}(Y) = \infty$.

THEOREM 2.5   The variance of a random variable Y is 0 if and only if Y is a constant random variable.

Proof   Suppose Y is a constant random variable; that is, $P(Y = a) = 1$ for some $a \in \mathbb{R}$. Then Y has mean $\mu = a$, and its probability function is given by $f(a) = 1$ and $f(y) = 0$ for $y \neq a$. Thus, by Equation (17), $\text{var}(Y) = \sum_y (y - a)^2 f(y) = (a - a)^2 = 0$.

Suppose, conversely, that the variance of Y is 0 and set $\mu = EY$. Then $(Y - \mu)^2$ is a nonnegative random variable whose mean (the variance of Y) is 0; thus,

$$P((Y - \mu)^2 = 0) = 1$$

by Theorem 2.1(e) in Section 2.2. Since $(Y - \mu)^2 = 0$ if and only if $Y = \mu$, we conclude that $P(Y = \mu) = 1$ and hence that Y is a constant random variable.   ∎

COROLLARY 2.10   The standard deviation of a random variable Y is 0 if and only if Y is a constant random variable.

EXAMPLE 2.32   Determine the variance of an indicator random variable I.

Solution   Set $\pi = P(I = 1)$. Then $P(I = 0) = 1 - \pi$. Also, $EI = \pi$ by Corollary 2.6 in Section 2.2. Thus,

$$\text{var}(I) = (0 - \pi)^2 (1 - \pi) + (1 - \pi)^2 \pi = \pi(1 - \pi) = P(I = 1)[\,1 - P(I = 1)\,].$$

In particular, the Bernoulli distribution with parameter π has variance $\pi(1 - \pi)$ and hence standard deviation $\sqrt{\pi(1 - \pi)}$.   ∎

EXAMPLE 2.33   A box contains six tickets labeled from 1 to 6. Tickets 1, 2, and 3 each have value 1, tickets 4 and 5 each have value 2, and ticket 6 has value 5. Determine the variance of the value Y of a ticket selected at random from the box.

Solution   Let L be the label of the ticket selected from the box, which is uniformly distributed on $\{1, \ldots, 6\}$. Write $Y = g(L)$ with $g(1) = g(2) = g(3) = 1$, $g(4) = g(5) = 2$, and $g(6) = 5$. Recall from the solution to Example 2.8 in Section 2.1 that $EY = 2$. Thus,

$$\text{var}(Y) = E[\,(Y - 2)^2\,] = E[\,(g(L) - 2)^2\,]$$
$$= 3(1 - 2)^2 \tfrac{1}{6} + 2(2 - 2)^2 \tfrac{1}{6} + (5 - 2)^2 \tfrac{1}{6}$$
$$= 2.   ∎$$

EXAMPLE **2.34**   A box contains N tickets, labeled from 1 to N. Suppose ticket l has value v_l for $1 \le l \le N$. Determine the variance of the value Y of a ticket selected at random from the box.

Solution   Let L be the label of the randomly selected ticket, which is uniformly distributed on $\{1, \ldots, N\}$. Then $Y = g(L)$, where $g(l) = v_l$ for $1 \le l \le N$. Recall from the solution to Example 2.9 in Section 2.1 that $EY = \mu$, where $\mu = N^{-1} \sum_l v_l$. Thus, Y has variance

$$\sigma^2 = E[(Y - \mu)^2] = E[(g(L) - \mu)^2] = \frac{1}{N} \sum_l (v_l - \mu)^2. \quad \blacksquare$$

EXAMPLE **2.35**   A roulette player bets \$2 on red and \$1 on $\{1, \ldots, 12\}$ on the same trial. Determine the variance of the net amount W won on the two bets.

Solution   Recall from the solution to Example 2.12 in Section 2.1 that $EW = -3/19$ and from the solution to Example 1.12 in Section 1.3 that the probability function f of W is given by $f(-3) = 7/19, f(0) = 3/19, f(1) = 6/19, f(4) = 3/19$, and $f(w) = 0$ for other values of w. Thus,

$$\begin{aligned}
\text{var}(W) &= \left(-3 + \frac{3}{19}\right)^2 \left(\frac{7}{19}\right) + \left(\frac{3}{19}\right)^2 \left(\frac{3}{19}\right) \\
&\quad + \left(1 + \frac{3}{19}\right)^2 \left(\frac{6}{19}\right) + \left(4 + \frac{3}{19}\right)^2 \left(\frac{3}{19}\right) \\
&= \frac{2214}{361}. \quad \blacksquare
\end{aligned}$$

THEOREM **2.6**   Let Y have finite variance. Then

$$\text{var}(Y) = E(Y^2) - (EY)^2, \tag{19}$$

$$\text{var}(a + Y) = \text{var}(Y), \qquad a \in \mathbb{R}, \tag{20}$$

$$\text{var}(bY) = b^2 \text{var}(Y), \qquad b \in \mathbb{R}, \tag{21}$$

and

$$\text{SD}(bY) = |b|\text{SD}(Y), \qquad b \in \mathbb{R}. \tag{22}$$

Proof   Set $\mu = EY$. Observe that

$$(Y - \mu)^2 = Y^2 - 2\mu Y + \mu^2.$$

Thus,

$$\begin{aligned}
\text{var}(Y) &= E[(Y - \mu)^2] \\
&= E(Y^2 - 2\mu Y + \mu^2) \\
&= E(Y^2) - 2\mu EY + \mu^2 \\
&= E(Y^2) - 2\mu^2 + \mu^2
\end{aligned}$$

$$= E(Y^2) - \mu^2$$
$$= E(Y^2) - (EY)^2,$$

so (19) is valid.

Let $a \in \mathbb{R}$. Then $E(a + Y) = a + EY$, so $a + Y - E(a + Y) = Y - EY$ and hence $\mathrm{var}(a + Y) = \mathrm{var}(Y)$. Thus, (20) is valid.

Let $b \in \mathbb{R}$. Then $E(bY) = b\mu$, and hence

$$bY - E(bY) = bY - b\mu = b(Y - \mu).$$

Consequently,

$$\mathrm{var}(bY) = E[\,(bY - E(bY))^2\,]$$
$$= E[\,(b(Y - \mu))^2\,]$$
$$= E[\,b^2(Y - \mu)^2\,]$$
$$= b^2 E[\,(Y - \mu)^2\,]$$
$$= b^2\mathrm{var}(Y),$$

so (21) is valid. Equation (22) follows from (21) by taking square roots. ∎

EXAMPLE 2.36 Determine the variance of the number Y of spots that show when a fair die is rolled in an honest manner.

Solution Now $EY = 7/2$ by Example 2.4 in Section 2.1. Since Y is uniformly distributed on $\{1, \ldots, 6\}$, we see that

$$E(Y^2) = \frac{1}{6}\sum_{y=1}^{6} y^2 = \frac{91}{6}.$$

Thus, by Equation (19),

$$\mathrm{var}(Y) = E(Y^2) - (EY)^2 = \frac{91}{6} - \left(\frac{7}{2}\right)^2 = \frac{35}{12}. \quad ∎$$

EXAMPLE 2.37 A roulette player bets the positive amount b on the subcollection A of $\{1, \ldots, 36, 0, 00\}$. Determine the variance of the amount W won by the player.

Solution Let I denote the indicator random variable for winning the bet. Then $EI = P(I = 1) = \#(A)/38$, and

$$W = b\left(\frac{36}{\#(A)}I - 1\right)$$

(see the solution to Example 2.11 in Section 2.1). Now

$$\mathrm{var}(I) = P(I = 1)[\,1 - P(I = 1)\,] = \frac{\#(A)}{38}\left(1 - \frac{\#(A)}{38}\right)$$

by the solution to Example 2.32, so

$$\text{var}(W) = \left(\frac{36b}{\#(A)}\right)^2 \text{var}(I) = \left(\frac{36b}{\#(A)}\right)^2 \frac{\#(A)}{38}\left(1 - \frac{\#(A)}{38}\right). \quad \blacksquare$$

EXAMPLE **2.38**   Let Y have the geometric distribution with parameter π. Determine

(a) $E[\, Y(Y-1)\,]$;   (b) $\text{var}(Y)$.

Solution   (a) Proceeding as in the solution to Example 2.5 in Section 2.1, we get

$$E[\, Y(Y-1)\,] = (1-\pi)\sum_{y=0}^{\infty} y(y-1)\pi^y$$

$$= \pi^2(1-\pi)\sum_{y=0}^{\infty} y(y-1)\pi^{y-2}$$

$$= \pi^2(1-\pi)\sum_{y=0}^{\infty} \frac{d^2}{d\pi^2}\pi^y$$

$$= \pi^2(1-\pi)\frac{d^2}{d\pi^2}\sum_{y=0}^{\infty} \pi^y$$

$$= \pi^2(1-\pi)\frac{d^2}{d\pi^2}\frac{1}{1-\pi}$$

$$= \frac{2\pi^2}{(1-\pi)^2}.$$

(b) Recall from the solution to Example 2.5 in Section 2.1 that $EY = \pi/(1-\pi)$. Thus, the second moment of Y is given by

$$E(Y^2) = E[\, Y(Y-1)\,] + EY = \frac{2\pi^2}{(1-\pi)^2} + \frac{\pi}{1-\pi} = \frac{\pi+\pi^2}{(1-\pi)^2}.$$

Consequently, the variance of Y is given by

$$\text{var}(Y) = E(Y^2) - (EY)^2 = \frac{\pi+\pi^2}{(1-\pi)^2} - \frac{\pi^2}{(1-\pi)^2} = \frac{\pi}{(1-\pi)^2}. \quad \blacksquare$$

EXAMPLE **2.39**   Johnny repeatedly tosses three coins until all three show heads. Determine the mean number of trials that are required.

Solution   Let N denote the number of trials that are required, where each toss of the three coins counts as one trial. Then $N-1$ has the geometric distribution with parameter $\pi = 7/8$ (see the solutions to Examples 1.66 and 1.67 in Section 1.8). Thus, $N-1$ has mean $\pi/(1-\pi) = 7$ (see the solution to Example 2.5 in Section 2.1), and it has variance $\pi/(1-\pi)^2 = 56$ [see the solution to Example 2.38(b)]. Thus, N has mean 8 and variance 56. $\blacksquare$

EXAMPLE **2.40**   Determine the variance of the uniform distribution on $[c, d]$.

Solution   Let Y be uniformly distributed on $[c, d]$. Then

$$U = \frac{Y - c}{d - c}$$

is uniformly distributed on $[0, 1]$ (see the solution to Example 1.52 in Section 1.7), and $Y = c + (d - c)U$. Now $EU = 1/2$ (see the solution to Example 2.16 in Section 2.1), so

$$\text{var}(U) = \int_0^1 \left(u - \frac{1}{2}\right)^2 du = \frac{(u - 1/2)^3}{3}\bigg|_0^1 = \frac{1}{24} + \frac{1}{24} = \frac{1}{12}.$$

Thus,

$$\text{var}(Y) = \text{var}(c + (d - c)U) = (d - c)^2\text{var}(U) = \frac{(d - c)^2}{12}. \quad \blacksquare$$

EXAMPLE **2.41**   Determine the variance and standard deviation of the exponential distribution with mean β.

Solution   Let Y have the exponential distribution with mean β. Then $E(Y^2) = 2\beta^2$ by the solution to Example 2.31. Thus, $\text{var}(Y) = E(Y^2) - (EY)^2 = 2\beta^2 - \beta^2 = \beta^2$, and hence $\text{SD}(Y) = \beta$. $\quad \blacksquare$

COROLLARY **2.11**   Let Y have finite variance. Then $\text{var}(-Y) = \text{var}(Y)$, and $\text{SD}(-Y) = \text{SD}(Y)$.

COROLLARY **2.12**   Let Y have mean μ and finite, positive standard deviation σ. Then $(Y - \mu)/\sigma$ has mean 0, variance 1, and standard deviation 1.

Proof   Since $E(Y - \mu) = 0$ by Corollary 2.5 in Section 2.2, we see that

$$E\left(\frac{Y - \mu}{\sigma}\right) = \frac{1}{\sigma}E(Y - \mu) = 0.$$

By Equations (20) and (21),

$$\text{var}\left(\frac{Y - \mu}{\sigma}\right) = \frac{1}{\sigma^2}\text{var}(Y - \mu) = \frac{1}{\sigma^2}\text{var}(Y) = \frac{\sigma^2}{\sigma^2} = 1,$$

so the standard deviation of $(Y - \mu)/\sigma$ equals 1. $\quad \blacksquare$

Let Y have mean μ and finite, positive standard deviation σ. Then the random variable $Z = (Y - \mu)/\sigma$, which has mean 0 and variance 1, is referred to as the *standardized random variable* corresponding to Y. Observe that $Y = \mu + \sigma Z$.

COROLLARY **2.13**   Let Y have mean μ and finite variance σ^2. Then

$$E[(Y - c)^2] = \sigma^2 + (\mu - c)^2, \qquad c \in \mathbb{R}. \tag{23}$$

Also, $E[(Y - c)^2]$ is uniquely minimized at $c = \mu$, and its minimum value is σ^2.

Proof   Observe that $E(Y - c) = \mu - c$ and $\text{var}(Y - c) = \text{var}(Y) = \sigma^2$. Thus, from Equation (19) with Y replaced by $Y - c$, we conclude that

$$\sigma^2 = \text{var}(Y - c) = E[(Y - c)^2] - [E(Y - c)]^2 = E[(Y - c)^2] - (\mu - c)^2$$

and hence that (23) holds. The second conclusion of the corollary follows from (23). ∎

Consider a distribution on $\mathbb{R}$ having positive mean μ and finite standard deviation σ. The quantity σ/μ is referred to as the *coefficient of variation* (CV) of the distribution. For an example of the use of this quantity, consider the mean μ and standard deviation σ of the distribution of annual family income in a country (that is, the distribution of the income of a family selected at random from the country). Then the corresponding coefficient of variation is a measure of the economic inequality in the country.

EXAMPLE 2.42   The data in Table 2.1 show the mean and standard deviation of annual family income in three countries. Which country has the highest measure of economic inequality, and which country has the lowest measure?

TABLE   2.1

Country	Mean Family Income	Standard Deviation
1	$20,000	$12,000
2	$10,000	$5,000
3	$5,000	$4,000

Solution   The coefficient of variation of annual family income in country 1 is .6, while in country 2 it is .5 and in country 3 it is .8. Thus, country 3 has the highest measure of economic inequality, and country 2 has the lowest measure. ∎

THEOREM 2.7   If $Y_1, \ldots, Y_n$ each have finite variance, then $Y_1 + \cdots + Y_n$ has finite variance.

Proof   Since $Y_1 + Y_2$ has finite second moment according to Theorem 2.4(b), it has finite variance. Similarly, $Y_1 + Y_2 + Y_3 = (Y_1 + Y_2) + Y_3$ has finite variance. The general result follows by induction. ∎

COROLLARY 2.14   Let $Y_1, \ldots, Y_n$ each have finite variance and let $b_1, \ldots, b_n \in \mathbb{R}$. Then $b_1 Y_1 + \cdots + b_n Y_n$ has finite variance.

Let Y_1 and Y_2 each have finite variance and set $\mu_1 = EY_1$ and $\mu_2 = EY_2$. Then $E(Y_1 + Y_2) = \mu_1 + \mu_2$ and hence

$$Y_1 + Y_2 - E(Y_1 + Y_2) = (Y_1 - \mu_1) + (Y_2 - \mu_2).$$

By squaring both sides and taking expectations, we conclude that

$$\text{var}(Y_1 + Y_2) = \text{var}(Y_1) + \text{var}(Y_2) + 2E[\,(Y_1 - \mu_1)(Y_2 - \mu_2)\,]. \tag{24}$$

[Note that $(Y_1 - \mu_1)(Y_2 - \mu_2)$ has finite mean by Theorem 2.4(b).]

Suppose Y_1 and Y_2 are independent. Since $Y_1 - \mu_1$ is a function of Y_1 and $Y_2 - \mu_2$ is a function of Y_2, we see that $Y_1 - \mu_1$ and $Y_2 - \mu_2$ are independent. Hence, it follows from Theorem 2.3 in Section 2.2 that

$$E[\,(Y_1 - \mu_1)(Y_2 - \mu_2)\,] = E(Y_1 - \mu_1)E(Y_2 - \mu_2) = 0.$$

Thus, we conclude from (24) that

$$\text{var}(Y_1 + Y_2) = \text{var}(Y_1) + \text{var}(Y_2). \tag{25}$$

It follows from (25) and Corollary 2.11 that

$$\text{var}(Y_1 - Y_2) = \text{var}(Y_1) + \text{var}(Y_2). \tag{26}$$

More generally, let $Y_1, \ldots, Y_n$ be independent random variables having finite variances. It follows from (25) and induction that

$$\text{var}(Y_1 + \cdots + Y_n) = \text{var}(Y_1) + \cdots + \text{var}(Y_n). \tag{27}$$

EXAMPLE **2.43**   Let $Y_1, \ldots, Y_n$ be independent random variables, each having mean μ and finite variance σ^2. Determine the mean, variance, and standard deviation of $Y_1 + \cdots + Y_n$.

Solution   The desired results are given by

$$E(Y_1 + \cdots + Y_n) = EY_1 + \cdots + EY_n = \mu + \cdots + \mu = n\mu,$$

$$\text{var}(Y_1 + \cdots + Y_n) = \text{var}(Y_1) + \cdots + \text{var}(Y_n) = \sigma^2 + \cdots + \sigma^2 = n\sigma^2,$$

and

$$\text{SD}(Y_1 + \cdots + Y_n) = \sqrt{\text{var}(Y_1 + \cdots + Y_n)} = \sigma\sqrt{n}. \quad \blacksquare$$

The following extension of Equation (27) follows from Equations (21) and (27).

THEOREM **2.8**   Let $Y_1, \ldots, Y_n$ be independent random variables each having finite variance. Then

$$\text{var}(b_1 Y_1 + \cdots + b_n Y_n) = b_1^2 \text{var}(Y_1) + \cdots + b_n^2 \text{var}(Y_n),$$

$$b_1, \ldots, b_n \in \mathbb{R}. \tag{28}$$

EXAMPLE **2.44**   Let Y_1, Y_2, and Y_3 be independent random variables having respective standard deviations 3, 5, and 7. Determine the variance of $Y_1 - 2Y_2 + 3Y_3$.

Solution   The desired variance is given by

$$\text{var}(Y_1 - 2Y_2 + 3Y_3) = \text{var}(Y_1) + 4\,\text{var}(Y_2) + 9\,\text{var}(Y_3)$$
$$= 3^2 + 4 \cdot 5^2 + 9 \cdot 7^2$$
$$= 550. \quad \blacksquare$$

EXAMPLE **2.45**   Give examples to illustrate that the following formulas for independent random variables Y_1 and Y_2 having finite variance are *not* generally valid:

(a) $\text{var}(Y_1 - Y_2) = \text{var}(Y_1) - \text{var}(Y_2)$;   **(b)** $\text{SD}(Y_1 - Y_2) = \text{SD}(Y_1) - \text{SD}(Y_2)$;
(c) $\text{SD}(Y_1 + Y_2) = \text{SD}(Y_1) + \text{SD}(Y_2)$.

Solution   Let Y_1 and Y_2 be independent random variables, each having an exponential distribution with mean 1. Then Y_1 and Y_2 each have variance 1 and standard deviation 1.

(a) By Equation (26), $\text{var}(Y_1 - Y_2) = \text{var}(Y_1) + \text{var}(Y_2) = 1 + 1 = 2$. Since $\text{var}(Y_1) - \text{var}(Y_2) = 1 - 1 = 0$, we conclude that $\text{var}(Y_1 - Y_2) \neq \text{var}(Y_1) - \text{var}(Y_2)$.

(b) It follows from the solution to part (a) that $\text{SD}(Y_1 - Y_2) = \sqrt{\text{var}(Y_1 - Y_2)} = \sqrt{2}$. Since $\text{SD}(Y_1) - \text{SD}(Y_2) = 1 - 1 = 0$, we conclude that $\text{SD}(Y_1 - Y_2) \neq \text{SD}(Y_1) - \text{SD}(Y_2)$.

(c) By Equation (25), $\text{var}(Y_1 + Y_2) = \text{var}(Y_1) + \text{var}(Y_2) = 1 + 1 = 2$, so $\text{SD}(Y_1 + Y_2) = \sqrt{2}$. Since $\text{SD}(Y_1) + \text{SD}(Y_2) = 1 + 1 = 2$, we conclude that $\text{SD}(Y_1 + Y_2) \neq \text{SD}(Y_1) + \text{SD}(Y_2)$. $\blacksquare$

The variance, standard deviation, and coefficient of variation of a real-valued random variable Y can be thought of as quantitative measures of its "randomness" or of the spread of its distribution. The variance and standard deviation are *location invariant*; that is, $\text{var}(a + Y) = \text{var}(Y)$, and $\text{SD}(a + Y) = \text{SD}(Y)$ for $a \in \mathbb{R}$. The coefficient of variation is *scale invariant*; that is, $\text{CV}(bY) = \text{CV}(Y)$ for $b > 0$. The standard deviation of Y has the same units of Y (such as feet, pounds, seconds, or dollars), whereas its coefficient of variation is unit-free.

Problems

2.31   A box has five balls, which are labeled from 1 to 5. Determine the variance of the label of a ball selected at random from the box.

2.32   Let Y be a $\mathcal{Y}$-valued random variable and let $B \subset \mathcal{Y}$. Determine the variance of $\text{ind}(Y \in B)$ in terms of $P(Y \in B)$.

2.33 Determine the variance of a random variable that is uniformly distributed on (a) $\{-1, 1\}$; (b) $\{-1, 0, 1\}$.

2.34 Let U be uniformly distributed on $[-1, 1]$. Determine (a) $E(U^2)$; (b) $E(U^4)$; (c) $\text{var}(U^2)$.

2.35 Let U be uniformly distributed on $\{-1, 0, 1\}$. Determine (a) $E(U^2)$; (b) $E(U^4)$; (c) $\text{var}(U^2)$.

2.36 Let U be uniformly distributed on $(0, 1)$. Determine the mean, second moment, and variance of $\log U$.

2.37 Determine the mean and variance of the number of rolls that are required to get one spot on repeated rolls of a fair die.

2.38 Determine the variance of the standard Cauchy distribution.

2.39 According to a suitable handbook,

$$\int_0^1 \left(\frac{\log w}{1+w}\right)^2 dw = \frac{\pi^2}{6}. \tag{29}$$

(a) Use (29) to verify that

$$\int_1^\infty \left(\frac{\log w}{1+w}\right)^2 dw = \frac{\pi^2}{6}. \tag{30}$$

It follows from (29) and (30) that

$$\int_0^\infty \left(\frac{\log w}{1+w}\right)^2 dw = \frac{\pi^2}{3}. \tag{31}$$

(b) Use (31) to show that the variance of the standard logistic distribution is $\pi^2/3$.

2.40 Give an example to illustrate that the formula $\text{var}(2Y) = 2\,\text{var}(Y)$ for random variables Y having finite variance is *not* generally valid.

2.41 Let Y_1 and Y_2 be independent random variables, each having finite variance. Show that $\text{var}(Y_1 Y_2) = (EY_1)^2\text{var}(Y_2) + (EY_2)^2\text{var}(Y_1) + \text{var}(Y_1)\text{var}(Y_2)$.

2.42 Suppose the length and width of a random plot of land are independently and exponentially distributed with mean 1 mile. Determine the mean and standard deviation of the area of a random plot of land.

2.43 Let Y_1, Y_2, and Y_3 be independent random variables having means 1, 2, and 3, respectively, and variances 4, 5, and 6, respectively, and set $W_1 = Y_1 + 2Y_2 + 3Y_3$ and $W_2 = 3Y_1 - 2Y_2 + Y_3$. Determine the mean and variance of $W_1 - W_2$.

2.44 Give examples to illustrate that the following formulas for independent random variables Y_1 and Y_2, each having finite variance, are *not* generally valid: (a) $\text{var}(2Y_1 + 3Y_2) = 2\,\text{var}(Y_1) + 3\,\text{var}(Y_2)$; (b) $\text{SD}(2Y_1 + 3Y_2) = 2\,\text{SD}(Y_1) + 3\,\text{SD}(Y_2)$.

2.45 Two fair dice are rolled in an honest manner. Determine the variance of each of the following random variables (see Problem 2.4 in Section 2.1 and Example 2.36): (a) the minimum number of spots showing on the two dice; (b) the maximum number of spots showing on the two dice; (c) the sum of the minimum number of spots showing on the two dice and the maximum number.

2.46 Consider a random variable Y having finite second moment. Determine a necessary and sufficient condition on Y in order that $E(Y^2) = (EY)^2$.

2.47 (Continued.) Consider a nonnegative random variable Y having finite mean. Determine a necessary and sufficient condition on Y in order that $E(\sqrt{Y}) = \sqrt{EY}$.

2.48 Let Y have infinite second moment. Show that $E[(Y-c)^2] = \infty$ for all $c \in \mathbb{R}$.

2.49 Determine the coefficient of variation of **(a)** the Bernoulli distribution with parameter π; **(b)** the geometric distribution with parameter π; **(c)** the exponential distribution with mean β.

2.50 A box has N tickets, labeled from 1 to N. Suppose ticket l has value v_l for $1 \le l \le N$, where $v_1, \ldots, v_N$. Let $\pi_1, \ldots, \pi_N$ be nonnegative numbers whose sum is 1, let L be the label of a ticket selected from the box by some random mechanism such that $P(L = l) = \pi_l$ for $1 \le l \le N$, and let Y be the value of the random ticket so selected. Then $Y = g(L)$, where $g(l) = v_l$ for $1 \le l \le N$. Show that **(a)** Y has mean $\mu = \sum_1^N \pi_l v_l$; **(b)** Y has variance $\sigma^2 = \sum_1^N \pi_l(v_l - \mu)^2$.

2.51 An investor can invest money for 1 year in either or both of two businesses. Let Y_1 and Y_2 denote the random returns per unit investment in the first and second business, respectively. Suppose Y_1 has positive mean μ_1 and finite, positive variance σ_1^2; Y_2 has positive mean μ_2 and finite, positive variance σ_2^2; and Y_1 and Y_2 are independent. Let b_1 denote the amount of money invested in the first business and let b_2 denote the amount invested in the second business. Then the return from the first business is $b_1 Y_1$ and that from the second business is $b_2 Y_2$, so the combined return is $b_1 Y_1 + b_2 Y_2$, which has mean $b_1 \mu_1 + b_2 \mu_2$ and variance $b_1^2 \sigma_1^2 + b_2^2 \sigma_2^2$. Let c be a finite positive constant and let b_1 and b_2 be chosen to maximize the mean of the combined return subject to the restriction that the variance of the combined return be at most c^2. Show that

$$\frac{b_2}{b_1} = \frac{\mu_2/\sigma_2^2}{\mu_1/\sigma_1^2}.$$

2.52 Let $Y_1, \ldots, Y_n$ be independent random variables, each uniformly distributed on $[0, \theta]$, where $0 < \theta < \infty$, and set $T = \max(Y_1, \ldots, Y_n)$. Determine **(a)** the distribution function and density function of T; **(b)** the mean, second moment, and variance of T.

2.4
Weak Law of Large Numbers

According to Corollary 2.10 in Section 2.3, if the standard deviation of a random variable Y is 0, then Y equals its mean with probability 1. From this result, it is plausible that if the standard deviation of Y is close to 0, then Y should be close to its mean with probability close to 1; that is, we should be able to find a precise form of the informal result that

$$P(Y \approx EY) \approx 1, \qquad \text{if } SD(Y) \approx 0. \tag{32}$$

Such a precise result will indeed be given in Property (34) and is based on the following result, known as *Chebyshev's inequality*.

THEOREM 2.9   Let Y have mean μ and finite standard deviation σ. Then

$$P(|Y - \mu| \geq c) \leq \frac{\sigma^2}{c^2}, \qquad c > 0. \tag{33}$$

Proof   The random variable $W = (Y - \mu)^2$ is nonnegative and has finite mean σ^2. Also, $W \geq c^2$ if and only if $|Y - \mu| \geq c$. Thus, by Markov's inequality,

$$P(|Y - \mu| \geq c) = P(W \geq c^2) \leq \frac{EW}{c^2} = \frac{\sigma^2}{c^2}. \quad \blacksquare$$

Since

$$P(|Y - \mu| < c) = 1 - P(|Y - \mu| \geq c),$$

it follows from Chebyshev's inequality that

$$P(|Y - \mu| < c) \geq 1 - \frac{\sigma^2}{c^2}, \qquad c > 0, \tag{34}$$

which is the precise form of Property (32) that was promised.

Let Y be as in Chebyshev's inequality. According to this inequality, σ^2/c^2 is an upper bound to the probability that $|Y - \mu| \geq c$. If the standard deviation of Y equals 0, then the upper bound and the indicated probability are both 0. The upper bound can be exactly equal to the indicated probability even if the standard deviation of Y is positive, as the next example shows.

EXAMPLE 2.46   Let Y have mean μ and finite standard deviation σ and let $c > 0$. Determine a necessary and sufficient condition on Y in order that

$$P(|Y - \mu| \geq c) = \frac{\sigma^2}{c^2}. \tag{35}$$

Solution   We are asked to determine when equality holds in Chebyshev's inequality. Examining the above proof of this inequality, we see that the desired equality holds if and only if it holds in the corresponding Markov inequality—that is, if and only if $P(W \geq c^2) = EW/c^2$, where $W = (Y - \mu)^2$. From Example 2.24(b) in Section 2.2, we see that the latter equality holds if and only if $P(W \in \{0, c^2\}) = 1$. Now $W \in \{0, c^2\}$ if and only if $Y - \mu$ equals $-c$, 0, or c and hence if and only if Y equals $\mu - c$, μ or $\mu + c$. Since Y has mean μ, it follows that (35) holds if and only if, for some $p \in [0, 1]$,

$$P(Y = \mu - c) = \frac{1-p}{2}, \quad P(Y = \mu) = p, \quad \text{and} \quad P(Y = \mu + c) = \frac{1-p}{2}. \tag{36}$$

[As a check, if (36) holds, then $\sigma^2 = c^2(1 - p)$ and $P(|Y - \mu| \geq c) = 1 - p = \sigma^2/c^2$.] $\blacksquare$

EXAMPLE **2.47**   Let Y have the standard logistic distribution.

(a) Use Chebyshev's inequality to obtain an upper bound to $P(|Y| \geq c)$ for $c > 0$.

(b) Compare the upper bound to $P(|Y| \geq 4)$ obtained from Chebyshev's inequality to the actual probability.

Solution   (a) Since $\text{var}(Y) = \pi^2/3$ according to Problem 2.39(b) in Section 2.3, we conclude from Chebyshev's inequality that

$$P(|Y| \geq c) \leq \frac{\pi^2}{3c^2}, \qquad c > 0. \tag{37}$$

(b) According to (37),

$$P(|Y| \geq 4) \leq \frac{\pi^2}{48} \approx .206.$$

The actual probability is given by

$$P(|Y| \geq 4) = 2\,P(Y \geq 4)$$
$$= 2[\,1 - P(Y \leq 4)\,]$$
$$= 2\left(1 - \frac{e^4}{1 + e^4}\right)$$
$$= \frac{2}{1 + e^4} \approx .036.$$

Thus, the upper bound given by Chebyshev's inequality is over five times as large as the actual probability.   ∎

Let $\mathbf{Y}_1, \ldots, \mathbf{Y}_n$ be independent random variables, each having a common distribution. Then these random variables are said to form a *random sample of size n* from this distribution.

Let $Y_1, \ldots, Y_n$ be a random sample from a distribution on $\mathbb{R}$ having mean μ and finite standard deviation σ. Then each of these random variables has mean μ and standard deviation σ. According to the solution to Example 2.43 in Section 2.3, the mean, variance, and standard deviation of the sum of these random variables are given by

$$E(Y_1 + \cdots + Y_n) = n\mu, \tag{38}$$

$$\text{var}(Y_1 + \cdots + Y_n) = n\sigma^2, \tag{39}$$

and

$$\text{SD}(Y_1 + \cdots + Y_n) = \sigma\sqrt{n}. \tag{40}$$

The random variable

$$\bar{Y} = \frac{Y_1 + \cdots + Y_n}{n}$$

is called the *sample mean*. It follows from (38)–(40), respectively, that

$$E\bar{Y} = \mu, \tag{41}$$

$$\text{var}(\bar{Y}) = \frac{\sigma^2}{n}, \tag{42}$$

and

$$\text{SD}(\bar{Y}) = \frac{\sigma}{\sqrt{n}}. \tag{43}$$

Thus, by Chebyshev's inequality with Y replaced by $\bar{Y}$,

$$P(|\bar{Y} - \mu| \geq c) \leq \frac{\sigma^2}{nc^2}, \qquad c > 0. \tag{44}$$

We can rewrite (44) as

$$P\left(\left|\frac{Y_1 + \cdots + Y_n}{n} - \mu\right| \geq c\right) \leq \frac{\sigma^2}{nc^2}, \qquad c > 0. \tag{45}$$

Since the left side of (45) is nonnegative and the right side tends to zero as $n \to \infty$, it follows from (45) that

$$\lim_{n \to \infty} P\left(\left|\frac{Y_1 + \cdots + Y_n}{n} - \mu\right| \geq c\right) = 0, \qquad c > 0. \tag{46}$$

An informal statement of (46) is that $P(\bar{Y} \approx \mu) \approx 1$ for $n \gg 1$; in words, when the sample size is large, the probability is nearly one that the sample mean is close to the true mean.

The above derivation of (46) depended on the assumption that the standard deviation of P is finite. In advanced probability theory it is shown that (46) holds even without this assumption; that is, that the following result, known as the *weak law of large numbers*, is valid.

THEOREM 2.10   Let P be a distribution on $\mathbb{R}$ having finite mean μ and let $Y_1, \ldots, Y_n$ be independent random variables, each having distribution P. Then (46) holds.

The weak law of large numbers is also known as the *law of averages*. There is a companion to the weak law that is referred to as the *strong law of large numbers*. The assumption of the strong law is essentially the same as that of the weak law, but its conclusion is that

$$P\left(\lim_{n \to \infty} \frac{Y_1 + \cdots + Y_n}{n} = \mu\right) = 1. \tag{47}$$

In words, (47) states that, with probability 1, the sample mean converges to the true mean as the sample size tends to infinity. The reason for the terminology *weak* and *strong* is that the conclusion of the strong law is stronger than that of the weak law; that is, (47) implies (46). The proof of the strong law and that its conclusion implies that of the weak law are given in advanced probability theory.

Let P be a distribution on $\mathcal{Y}$ and let $\mathbf{Y}_1, \ldots, \mathbf{Y}_n$ be a random sample from P. The sample distribution $\hat{P}$ of $\mathbf{Y}_1, \ldots, \mathbf{Y}_n$ is given by

$$\hat{P}(B) = \frac{1}{n}\#(\{\, i : 1 \leq i \leq n \text{ and } \mathbf{Y}_i \in B \,\}) = \frac{1}{n}\sum_{i=1}^{n} \text{ind}(\mathbf{Y}_i \in B), \qquad B \subset \mathcal{Y}.$$

Let $B \subset \mathcal{Y}$. Then the indicator random variables $\text{ind}(\mathbf{Y}_1 \in B), \ldots, \text{ind}(\mathbf{Y}_n \in B)$ are independent, and each has mean $\mu = P(B)$ and variance $\sigma^2 = P(B)\,[1 - P(B)]$. If $P(B)$ is 0 or 1, these indicator random variables are constant random variables; otherwise, they have the Bernoulli distribution with parameter $\pi = P(B)$. The mean, variance, and standard deviation of the relative frequency $\hat{P}(B)$ of B among $\mathbf{Y}_1, \ldots, \mathbf{Y}_n$ are given by

$$E(\hat{P}(B)) = P(B),$$

$$\text{var}(\hat{P}(B)) = \frac{P(B)\,[1 - P(B)]}{n},$$

and

$$\text{SD}(\hat{P}(B)) = \sqrt{\frac{P(B)\,[1 - P(B)]}{n}}.$$

It follows from Chebyshev's inequality [see Property (44)] that

$$P(|\hat{P}(B) - P(B)| \geq c) \leq \frac{P(B)\,[1 - P(B)]}{nc^2}, \qquad c > 0, \tag{48}$$

and hence that

$$\lim_{n \to \infty} P(|\hat{P}(B) - P(B)| \geq c) = 0, \qquad c > 0. \tag{49}$$

This result can be written informally as $P(\hat{P}(B) \approx P(B)) \approx 1$ for $n \gg 1$; that is, when the sample size is large, the probability is nearly 1 that the relative frequency of a set is close to its true probability. Thus, (49) is a mathematically precise form of the law of relative frequencies.

Estimation

Think of the distribution P as being less than completely known. Let τ be a real-valued parameter defined in terms of this distribution. For example, τ could be the probability $\pi = P(B)$ for a given subset B of $\mathcal{Y}$. Alternatively, it could be the mean μ, variance σ^2, standard deviation σ, or a quantile of a random variable $W = g(\mathbf{Y})$, where $\mathbf{Y}$ has distribution P and g is a known real-valued function on $\mathcal{Y}$.

Let $\hat{\tau}$ be a real-valued random variable defined in terms of $\mathbf{Y}_1, \ldots, \mathbf{Y}_n$, which is regarded as an estimate of τ. For example, τ could be the true probability $P(B)$ of B and $\hat{\tau}$ its relative frequency $\hat{P}(B)$ among $\mathbf{Y}_1, \ldots, \mathbf{Y}_n$. Alternatively, when $\mathcal{Y} \subset \mathbb{R}$, τ could be the true mean μ and $\hat{\tau}$ the sample mean $\bar{Y}$. We refer to $\hat{\tau} - \tau$ as the error of the estimate and to $\text{bias}(\hat{\tau}) = E(\hat{\tau} - \tau) = E\hat{\tau} - \tau$ as the *bias* of the estimate. The second moment of the error of the estimate, which is an overall measure of its magnitude, is referred to as the *mean squared error* (MSE) of the estimate. Thus, $\mathrm{MSE}(\hat{\tau}) = E[(\hat{\tau} - \tau)^2]$. It follows by applying the formula $E(Y^2) = \mathrm{var}(Y) + (EY)^2$ to the random variable $Y = \hat{\tau} - \tau$ that

$$\mathrm{MSE}(\hat{\tau}) = \mathrm{var}(\hat{\tau}) + \mathrm{bias}^2(\hat{\tau}). \tag{50}$$

EXAMPLE 2.48   Let $Y_1, \ldots, Y_n$ be a random sample of size n from the exponential distribution with mean β and set $\hat{\beta} = c\bar{Y}$, where $c \in \mathbb{R}$. Determine

(a) the bias, variance, and mean squared error of $\hat{\beta}$ as an estimate of β;

(b) the value of c that minimizes this mean squared error;

(c) the mean squared error of $\hat{\beta}$ corresponding to this choice of c.

Solution   (a) Now $E\hat{\beta} = E(c\bar{Y}) = cE\bar{Y} = c\beta$, so $\mathrm{bias}(\hat{\beta}) = (c-1)\beta$. The variance of $\hat{\beta}$ is given by

$$\mathrm{var}(\hat{\beta}) = \mathrm{var}(c\bar{Y}) = c^2\mathrm{var}(\bar{Y}) = \frac{c^2\sigma^2}{n} = \frac{c^2\beta^2}{n}$$

since the exponential distribution with mean β has variance $\sigma^2 = \beta^2$. Thus, the mean squared error of $\hat{\beta}$ is given by

$$\mathrm{MSE}(\hat{\beta}) = \mathrm{var}(\hat{\beta}) + \mathrm{bias}^2(\hat{\beta}) = \frac{c^2\beta^2}{n} + (c-1)^2\beta^2 = \left(\frac{c^2}{n} + (c-1)^2\right)\beta^2.$$

(b) Observe that $\mathrm{MSE}(\hat{\beta})$ is a quadratic function of c with positive second derivative. Thus, it is uniquely minimized at that value of c such that

$$0 = \frac{d}{dc}\left(\frac{c^2}{n} + (c-1)^2\right) = 2\left(\frac{c}{n} + c - 1\right);$$

that is, at $c = n/(n+1)$.

(c) When $c = n/(n+1)$, the mean squared error of $\hat{\beta}$ is given by

$$\mathrm{MSE}(\hat{\beta}) = \left(\frac{n}{(n+1)^2} + \frac{1}{(n+1)^2}\right)\beta^2 = \frac{\beta^2}{n+1}. \qquad \blacksquare$$

We say that $\hat{\tau}$ is an *unbiased estimate* of τ if we know (that is, if it follows from the assumptions under consideration) that $\mathrm{bias}(\hat{\tau}) = 0$ or, equivalently, that $E\hat{\tau} = \tau$; otherwise, $\hat{\tau}$ is said to be a *biased estimate* of τ. The estimate $\hat{\beta} = \frac{n}{n+1}\bar{Y}$ of β in Example 2.48 is a biased estimate; specifically, $\mathrm{bias}(\hat{\beta}) = -\beta/(n+1)$. The relative frequency $\hat{P}(B)$ is an unbiased estimate of $P(B)$. According to Equation (41), if P is a distribution on $\mathbb{R}$ having finite mean, then the sample mean is an unbiased estimate of the true mean.

Suppose we have a rule that, for every sufficiently large sample size n, defines the estimate $\hat{\tau}$ in terms of the random sample $Y_1, \ldots, Y_n$ of size n. The estimate is said to be *consistent* if $\lim_{n\to\infty} P(|\hat{\tau} - \tau| \geq c) = 0$ for $c > 0$ or, more informally, if $P(\hat{\tau} \approx \tau) \approx 1$ for $n \gg 1$. Consistency means that, when the sample size is large, the probability is nearly 1 that the estimate is close to the parameter being estimated. It follows from the weak law of large numbers that relative frequencies are consistent estimates of true probabilities and sample means are consistent estimates of (finite) true means.

Sample Variance

A *statistic* is a random variable whose definition does not involve unknown parameters. Suppose μ and σ^2 are both unknown. Then the random variable

$$\hat{\sigma}^2 = \frac{1}{n} \sum_{i=1}^{n} (Y_i - \mu)^2,$$

which has mean σ^2, is not a statistic since its definition involves the unknown parameter μ. The *sample variance* S^2 is defined (whenever the sample size is at least 2) by

$$S^2 = \frac{1}{n-1} \sum_{i=1}^{n} (Y_i - \bar{Y})^2;$$

here S, the *sample standard deviation*, is nonnegative. Observe that the sample variance and sample standard deviation are statistics.

To investigate the properties of the sample variance, we observe first that

$$\sum_{i=1}^{n} (Y_i - \bar{Y}) = 0. \tag{51}$$

To verify (51), note that

$$\sum_{i=1}^{n} (Y_i - \bar{Y}) = \sum_{i=1}^{n} Y_i - n\bar{Y} = n\left(\frac{1}{n} \sum_{i=1}^{n} Y_i - \bar{Y}\right) = n(\bar{Y} - \bar{Y}) = 0.$$

Let $c \in \mathbb{R}$. Then

$$\sum_{i=1}^{n} (Y_i - c)^2 = \sum_{i=1}^{n} (Y_i - \bar{Y})^2 + n(\bar{Y} - c)^2. \tag{52}$$

To verify (52), note that

$$(Y_i - c)^2 = (Y_i - \bar{Y} + \bar{Y} - c)^2 = (Y_i - \bar{Y})^2 + (\bar{Y} - c)^2 + 2(\bar{Y} - c)(Y_i - \bar{Y}).$$

Thus,

$$\sum_{i=1}^{n} (Y_i - c)^2 = \sum_{i=1}^{n} (Y_i - \bar{Y})^2 + n(\bar{Y} - c)^2 + 2(\bar{Y} - c) \sum_{i=1}^{n} (Y_i - \bar{Y}).$$

Equation (52) now follows from (51). It follows from (52) that

$$\sum_{i=1}^{n} (Y_i - c)^2$$

is uniquely minimized at $c = \bar{Y}$ and that the minimum value of this sum is

$$\sum_{i=1}^{n} (Y_i - \bar{Y})^2.$$

Suppose P has mean μ and finite variance σ^2. Then

$$E\left(\sum_{i=1}^{n}(Y_i - \bar{Y})^2 \right) = (n - 1)\sigma^2. \tag{53}$$

To verify (53), set $c = \mu$ in (52) to get that

$$\sum_{i=1}^{n}(Y_i - \mu)^2 = \sum_{i=1}^{n}(Y_i - \bar{Y})^2 + n(\bar{Y} - \mu)^2. \tag{54}$$

Now

$$E[\,(Y_i - \mu)^2\,] = \text{var}(Y_i) = \sigma^2, \qquad 1 \le i \le n,$$

and

$$E[\,(\bar{Y} - \mu)^2\,] = \text{var}(\bar{Y}) = \frac{\sigma^2}{n}.$$

Thus, we conclude from (54) by taking expectations that

$$n\sigma^2 = E\left(\sum_{i=1}^{n}(Y_i - \bar{Y})^2 \right) + \sigma^2$$

and hence that (53) holds.

Suppose $n \ge 2$. It follows from Equation (53) that

$$E(S^2) = E\left(\frac{1}{n-1} \sum_{i=1}^{n}(Y_i - \bar{Y})^2 \right) = \sigma^2;$$

that is, the sample variance is an unbiased estimate of σ^2. (This is the reason for having $n - 1$ instead of n in the denominator of the definition of the sample variance.) The sample standard deviation is not exactly unbiased (see Problem 2.62), but it is nearly so when the sample size is large; that is, $ES \approx \sigma$ for $n \gg 1$.

The sample variance is a consistent estimate of the true variance; that is,

$$P(S^2 \approx \sigma^2) \approx 1, \qquad n \gg 1. \tag{55}$$

Consequently, the sample standard deviation is a consistent estimate of the true standard deviation; that is, $P(S \approx \sigma) \approx 1$ for $n \gg 1$. To verify (55), consider the random variable

$$\hat{\sigma}^2 = \frac{1}{n} \sum_{i=1}^{n}(Y_i - \mu)^2,$$

which is a feasible estimate of the σ^2 when the μ is known. Since $(Y_i - \mu)^2$, $1 \le i \le n$, are independent random variables having a common distribution with mean σ^2, it follows from the weak law of large numbers that

$$P(\hat{\sigma}^2 \approx \sigma^2) \approx 1, \qquad n \gg 1. \tag{56}$$

It has already been pointed out that the sample mean is a consistent estimate of the true mean; that is,

$$P(\bar{Y} \approx \mu) \approx 1, \qquad n \gg 1. \tag{57}$$

It follows from Equation (54) and elementary algebra that

$$S^2 = \sigma^2 + (\hat{\sigma}^2 - \sigma^2) + \frac{\hat{\sigma}^2}{n-1} - \frac{n}{n-1}(\bar{Y} - \mu)^2. \tag{58}$$

Equation (55) follows from (56)–(58).

It is important to avoid confusing *variance* and *sample variance*. The variance of Y is a number that depends on the distribution of Y; the sample variance is a statistic that is defined in terms of a random sample from the distribution of Y and that is an unbiased estimate of the variance of Y.

Standard Error of the Sample Mean

Recall from Equation (43) that the standard deviation of the sample mean is given by $\mathrm{SD}(\bar{Y}) = \sigma/\sqrt{n}$. In practice, this quantity is typically unknown since σ is unknown. Estimation of σ by S leads to the corresponding estimate $S/\sqrt{n}$ of $\mathrm{SD}(\bar{Y})$, which is referred to as the *standard error* (SE) of $\bar{Y}$. Thus, $\mathrm{SE}(\bar{Y}) = S/\sqrt{N}$. We can regard $\mathrm{SE}(\bar{Y})$ as a rough measure of the inaccuracy of $\bar{Y}$ as an estimate of μ.

EXAMPLE **2.49**   The observed values of a random sample of size 10 are as follows: 88.7, 87.9, 77.8, 90.6, 82.1, 83.5, 82.6, 83.9, 86.0, 82.7. Determine

(a) the sample mean;   (b) the sample standard deviation;

(c) the standard error of the sample mean.

Solution   (a) The sample mean is given by

$$\bar{Y} = \frac{88.7 + \cdots + 82.7}{10} = 84.58.$$

(b) The sample variance is given by

$$S^2 = \frac{1}{9}[(88.7 - 84.58)^2 + \cdots + (82.7 - 84.58)^2] \doteq 14.162.$$

Thus, the sample standard deviation is given by $S \doteq \sqrt{14.162} \doteq 3.763$.

(c) The standard error of the sample mean is given by

$$\mathrm{SE}(\bar{Y}) \doteq \frac{3.763}{\sqrt{10}} \doteq 1.190. \quad \blacksquare$$

Let $Y_1, \ldots, Y_n$ be independent random variables, each having the Bernoulli distribution, with parameter π and hence mean $\mu = \pi$, variance $\sigma^2 = \pi(1 - \pi)$, and standard deviation $\sigma = \sqrt{\pi(1 - \pi)}$. Here $\hat{\pi} = \bar{Y}$ is an unbiased estimate of π, whose variance and standard deviation are given by $\mathrm{var}(\hat{\pi}) = \pi(1 - \pi)/n$ and

$$\mathrm{SD}(\hat{\pi}) = \sqrt{\frac{\pi(1 - \pi)}{n}}. \tag{59}$$

The sample variance of $Y_1, \ldots, Y_n$ is given by

$$S^2 = \frac{n}{n-1}\hat{\pi}(1-\hat{\pi})$$

[see Problem 2.61(c)]. Thus, the standard error $SE(\hat{\pi}) = S/\sqrt{n}$ of $\hat{\pi}$ is given by

$$SE(\hat{\pi}) = \sqrt{\frac{\hat{\pi}(1-\hat{\pi})}{n-1}}. \tag{60}$$

In this context, however, we use the slightly modified but more conventional definition of the standard error of $\hat{\pi}$ given by

$$SE(\hat{\pi}) = \sqrt{\frac{\hat{\pi}(1-\hat{\pi})}{n}}, \tag{61}$$

which is obtained by replacing π by $\hat{\pi}$ in Formula (59) for the standard deviation of $\hat{\pi}$. As the following example illustrates, the numerical difference between the right sides of (60) and (61) is negligible in practice.

EXAMPLE **2.50**   A bent coin shows heads 600 times when tossed 1000 times in an honest manner.

(a) Estimate the probability π that the coin will show heads on a single toss.

(b) Determine the standard error of this estimate.

Solution   (a) The probability π is estimated by $\hat{\pi} = 600/1000 = .6$.

(b) According to (30), the standard error of this estimate is given by

$$SE(\hat{\pi}) = \sqrt{\frac{(.6)(.4)}{1000}} \doteq .015492.$$

Alternatively, according to (29), the standard error of $\hat{\pi}$ is given by

$$SE(\hat{\pi}) = \sqrt{\frac{(.6)(.4)}{999}} \doteq .015500. \quad \blacksquare$$

Problems

2.53   A fair die is rolled ten times in an honest manner. Determine the variance of the sample mean of the numbers of spots showing on the ten rolls.

2.54   Let Y have the exponential distribution with mean β. Use Chebyshev's inequality to obtain an upper bound to $P(Y \geq 10\beta)$.

2.55   The failure times of two components are independent, and each has the exponential distribution with mean 100. Use Chebyshev's inequality to obtain an upper bound to the probability that the failure times differ from each other by more than 500.

2.56   Show that if Y has finite second moment, then $P(|Y| \geq c) \leq c^{-2}E(Y^2)$ for $c > 0$.

2.57   Let $\mathbf{Y}_1, \ldots, \mathbf{Y}_n$ be a random sample of size n from a distribution P on $\mathcal{Y}$ and let $B \subset \mathcal{Y}$. Use Property (48) to verify that

$$P(|\hat{P}(B) - P(B)| \geq c) \leq \frac{1}{4nc^2}, \qquad c > 0. \tag{62}$$

2.58   (Continued.) Determine the smallest positive integer n such that (62) implies that

$$P(|\hat{P}(B) - P(B)| \geq .03) \leq .05.$$

2.59   Let $Y_1, \ldots, Y_n$ be independent random variables having common mean μ and finite variance σ^2. Determine the second moment of $\bar{Y} = (Y_1 + \cdots + Y_n)/n$.

2.60   Let $Y_1, \ldots, Y_n$ be independent random variables, each having an exponential distribution, with mean β; let σ^2 denote the common variance of these random variables; set $\bar{Y} = (Y_1 + \cdots + Y_n)/n$. Show that $[n/(n+1)]\bar{Y}^2$ is an unbiased estimate of σ^2.

2.61   Let S^2 be the sample variance of $Y_1, \ldots, Y_n$. Verify that **(a)** $\sum_{i=1}^n (Y_i - \bar{Y})^2 = \sum_{i=1}^n Y_i^2 - n\bar{Y}^2$; **(b)** $S^2 = \frac{n}{n-1}\left(\frac{1}{n}\sum_{i=1}^n Y_i^2 - \bar{Y}^2\right)$; **(c)** if $Y_1, \ldots, Y_n$ are $\{0, 1\}$-valued, then $S^2 = \frac{n}{n-1}\bar{Y}(1 - \bar{Y})$; **(d)** if $n = 2$, then $S^2 = \sum_{i=1}^2 (Y_i - \bar{Y})^2 = (Y_1 - Y_2)^2/2$.

2.62   Let S denote the sample standard deviation based on a random sample of size $n \geq 2$ from a distribution having finite standard deviation σ. **(a)** Verify that $(ES)^2 + \text{var}(S) = \sigma^2$. **(b)** Use the result of part (a) to verify that $ES = \sigma$ if and only if $\text{var}(S) = 0$ or, equivalently, if and only if S is a constant random variable. [It can be shown that S is a constant random variable if and only if $\sigma = 0$, in which case $P(S = 0) = 1$.]

2.63   Let $Y_1, \ldots, Y_n$ be independent random variables having common finite variance σ^2 and set $\mu_i = EY_i$ for $1 \leq i \leq n$ and $\bar{\mu} = (\mu_1 + \cdots + \mu_n)/n$. Show that **(a)** $E\bar{Y} = \bar{\mu}$; **(b)** $\text{var}(Y_i - \bar{Y}) = \frac{n-1}{n}\sigma^2$ for $1 \leq i \leq n$; **(c)** $E[\sum_{i=1}^n (Y_i - \bar{Y})^2] = (n-1)\sigma^2 + \sum_{i=1}^n (\mu_i - \bar{\mu})^2$.

2.64   Let $Y_{11}, \ldots, Y_{1n_1}, Y_{21}, \ldots, Y_{2n_2}$ be independent random variables. Assume that $Y_{11}, \ldots, Y_{1n_1}$ have common mean μ_1 and finite variance σ_1^2 and $Y_{21}, \ldots, Y_{2n_2}$ have common mean μ_2 and finite variance σ_2^2. Set $\bar{Y}_1 = (Y_{11} + \cdots + Y_{1n_1})/n_1$ and $\bar{Y}_2 = (Y_{21} + \cdots + Y_{2n_2})/n_2$. Determine **(a)** $E(\bar{Y}_1 - \bar{Y}_2)$; **(b)** $\text{var}(\bar{Y}_1 - \bar{Y}_2)$.

In Problems 2.65 and 2.66, let $Y_1, \ldots, Y_n$ be independent random variables, each uniformly distributed on $[0, \theta]$, where θ is positive but otherwise unknown. Set $T = \max(Y_1, \ldots, Y_n)$. According to the answers to Problem 2.52(b) in Section 2.3,

$$ET = \frac{n}{n+1}\theta \quad \text{and} \quad \text{var}(T) = \frac{n}{(n+1)^2(n+2)}\theta^2.$$

2.65   Show that $\hat{\theta} = \frac{n+1}{n}T$ is an unbiased estimate of θ and determine its variance.

2.66   Set $\hat{\theta} = cT$, where $c \in \mathbb{R}$. Determine **(a)** the bias, variance, and mean squared error of $\hat{\theta}$ as an estimate of θ; **(b)** the value of c that minimizes this mean squared error; **(c)** the mean squared error of $\hat{\theta}$ corresponding to this choice of c.

2.5
Simulation and the Monte Carlo Method

Computers can be used to simulate a random sample from a distribution and thereby to estimate parameters of interest defined in terms of that distribution. This approach is referred to as the *Monte Carlo method*. An important advantage of this method over more classical analytical and numerical techniques is that it can readily be applied in complicated situations in which the classical techniques are infeasible.

We begin the discussion of the Monte Carlo method by illustrating the use of random samples to estimate various parameters.

Let a point (X, Y) be chosen at random from the square $[-1, 1] \times [-1, 1]$ and let $R = \sqrt{X^2 + Y^2}$ denote the distance from the chosen point to the center of the square. According to the solution to Example 1.21 in Section 1.4, the distribution function F of R is given by $F(r) = 0$ for $r \leq 0$, $F(r) = \pi r^2/4$ for $0 < r \leq 1$, $F(r) = \sqrt{r^2 - 1} + (\pi/4 - \text{arcsec } r)r^2$ for $1 < r < \sqrt{2}$, and $F(r) = 1$ for $r \geq \sqrt{2}$. Using this result in the solution to Example 1.44(b) in Section 1.6, we showed that the density function f of R is given by $f(r) = \pi r/2$ for $0 < r \leq 1$, $f(r) = (\pi/2 - 2 \text{ arcsec } r)r$ for $1 < r < \sqrt{2}$, and $f(r) = 0$ elsewhere. The mean and variance of R are given by

$$\mu = \int_0^{\sqrt{2}} rf(r)\, dr$$

and $\sigma^2 = E(R^2) - \mu^2$, where

$$E(R^2) = \int_0^{\sqrt{2}} r^2 f(r)\, dr.$$

Alternatively,

$$E(R^2) = E(X^2 + Y^2) = E(X^2) + E(Y^2).$$

Since X and Y are each uniformly distributed on $[-1, 1]$, they each have mean 0 and variance 1/3 (see the solution to Example 2.40 in Section 2.3). Thus, $E(X^2) = \text{var}(X) = 1/3$ and $E(Y^2) = 1/3$, and hence $E(R^2) = 2/3$. Consequently, $\sigma^2 = 2/3 - \mu^2$.

From the explicit formula for the density function of R, we get that

$$\mu = \int_0^1 \frac{\pi}{2} r^2 \, dr + \int_1^{\sqrt{2}} \left(\frac{\pi}{2} - 2 \text{ arcsec } r \right) r^2 \, dr$$

$$= \frac{\pi}{2} \int_0^{\sqrt{2}} r^2 \, dr - 2 \int_1^{\sqrt{2}} r^2 \text{ arcsec } r \, dr$$

$$= \frac{\pi \sqrt{2}}{3} - 2 \int_1^{\sqrt{2}} r^2 \text{ arcsec } r \, dr.$$

Using integration by parts, we see that

$$\int_1^{\sqrt{2}} r^2 \text{ arcsec } r \, dr = \frac{r^3}{3} \text{arcsec } r \Big|_1^{\sqrt{2}} - \frac{1}{3} \int_1^{\sqrt{2}} \frac{r^3}{r\sqrt{r^2 - 1}} \, dr$$

$$= \frac{2^{3/2}}{3} \frac{\pi}{4} - \frac{1}{3} \int_1^{\sqrt{2}} \frac{r^2}{\sqrt{r^2 - 1}} dr$$

$$= \frac{\pi \sqrt{2}}{6} - \frac{1}{3} \int_1^{\sqrt{2}} \frac{r^2}{\sqrt{r^2 - 1}} dr$$

and hence that

$$\mu = \frac{2}{3} \int_1^{\sqrt{2}} \frac{r^2}{\sqrt{r^2 - 1}} dr.$$

From a table of integrals, we find that

$$\int \frac{r^2}{\sqrt{r^2 - 1}} dr = \frac{r}{2}\sqrt{r^2 - 1} + \frac{1}{2} \log\left(r + \sqrt{r^2 - 1}\right)$$

and hence that

$$\int_1^{\sqrt{2}} \frac{r^2}{\sqrt{r^2 - 1}} dr = \frac{1}{2}[\sqrt{2} + \log(1 + \sqrt{2})].$$

Consequently,

$$\mu = \frac{1}{3}[\sqrt{2} + \log(1 + \sqrt{2})] \doteq 0.7652,$$

so

$$\sigma^2 = \frac{2}{3} - \mu^2 = \frac{2}{3} - \frac{1}{9}[\sqrt{2} + \log(1 + \sqrt{2})]^2 \doteq 0.08114$$

and hence $\sigma \doteq 0.2849$.

We can avoid all of this analysis and still estimate μ, σ^2, and σ by obtaining observed values (X_i, Y_i), $1 \le i \le n$, of a random sample of size n from the uniform distribution on the square $[-1, 1] \times [-1, 1]$ and setting $R_i = \sqrt{X_i^2 + Y_i^2}$ for $1 \le i \le n$. As an illustration, we obtained a random sample of size 10 from the indicated distribution and got the data shown in Table 2.2. The sample mean of the observed values of R is given by

$$\bar{R} = \frac{1}{10} \sum_{i=1}^{10} R_i \doteq \frac{7.160}{10} = 0.716,$$

their sample variance is given by

$$S^2 = \frac{1}{9} \sum_{i=1}^{10} (R_i - \bar{R})^2 \doteq \frac{0.714}{9} \doteq 0.079,$$

and their sample standard deviation is given by

$$S = \sqrt{S^2} \doteq \sqrt{0.079} \doteq 0.282.$$

Thus, the standard error of $\bar{R}$ as an estimate of the mean of R is given by

$$\text{SE}(\bar{R}) = \frac{S}{\sqrt{10}} \doteq \frac{0.282}{\sqrt{10}} \doteq 0.089.$$

TABLE   2.2

X	Y	R
−0.057	0.551	0.554
0.300	0.798	0.853
0.965	0.172	0.980
0.940	−0.197	0.960
−0.077	0.958	0.961
0.120	−0.289	0.313
0.741	−0.175	0.761
0.018	−0.233	0.234
−0.550	0.795	0.967
−0.374	−0.439	0.577

Let $p = F(.7)$ denote the probability that $R \leq 0.7$. We can estimate this probability by the sample proportion of the numbers $R_1, \ldots, R_{10}$ that are less than or equal to 0.7, which is given by $\hat{p} = 4/10 = .4$. The standard error of this estimate is given by

$$SE(\hat{p}) = \sqrt{\frac{\hat{p}(1 - \hat{p})}{10}} = \sqrt{\frac{(.4)(.6)}{10}} = \sqrt{.024} \doteq .155.$$

Let $R_{(1)}, \ldots, R_{(10)}$ denote the observed values $R_1, \ldots, R_{10}$ written in increasing order, so that $R_{(1)} \doteq 0.234$, $R_{(2)} \doteq 0.313$, $\ldots$, $R_{(10)} \doteq 0.980$. By plotting the points $(R_{(i)}, (i - 1/2)/10)$ for $1 \leq i \leq 10$, we get a rough estimate of the distribution function of R. Figure 2.3 shows this estimate along with the true distribution function of R. [The numbers $(1 - 1/2)/10 = .05, \ldots, (10 - 1/2)/10 = .95$, which range over $(0, 1)$ in a manner that is symmetric about .5, are more natural for plotting purposes than the numbers $.1, \ldots, 1$ (see Problem 2.67g)].

FIGURE   2.3

Distribution function of R and its estimate

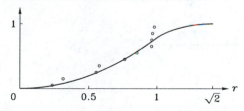

To obtain more accurate estimates of the various quantities defined in terms of the distribution of R, we should use a larger sample size. As an illustration, a (fresh) random sample of size 100 was obtained. Correspondingly, $\bar{R} \doteq 0.726$, $S^2 \doteq 0.084$, $S \doteq 0.290$, and $SE(\bar{R}) \doteq 0.029$. Also, $\hat{p} \doteq .47$, and $SE(\hat{p}) \doteq .045$. Figure 2.4 shows the points $(R_{(i)}, (i - 1/2)/100)$ for $1 \leq i \leq 100$ along with the distribution function.

F I G U R E   2.4

Distribution function of R and its estimate

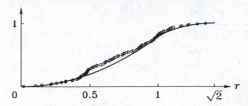

Using still larger sample sizes, we can get arbitrarily accurate estimates. Table 2.3 shows estimates of ER and the associated standard errors based on the two random samples referred to above and additional random samples of sizes 1000 and 10,000, with $n = \infty$ corresponding to the true value of ER. Similarly, Table 2.4 shows the estimates of $p = F(.7) = \pi(.7)^2/4$ and the standard errors of these estimates.

Next, let a point (X, Y, Z) be chosen at random from the cube $[-1, 1] \times [-1, 1] \times [-1, 1]$; let $R = \sqrt{X^2 + Y^2 + Z^2}$ denote the distance from the chosen point to the center of the cube; and let F, f, and F^{-1}, respectively, denote the distribution function, density function, and quantile function of R. Observe that $F(r) = 0$ for $r \le 0$, $F(r) = 1$ for $r \ge \sqrt{3}$, and $F(r) = \frac{1}{8} \cdot \frac{4}{3}\pi r^3 = \frac{1}{6}\pi r^3$ for $0 < r < 1$. Consequently $f(r) = 0$ for $r \in (-\infty, 0] \cup [\sqrt{3}, \infty)$, and $f(r) = \frac{1}{2}\pi r^2$ for $0 < r < 1$. Moreover, the pth quantile of R is given by $r_p = (6p/\pi)^{1/3}$ for $0 < p \le \pi/6$. Obtaining $F(r)$ and $f(r)$ for $1 < r < \sqrt{3}$, r_p for $\pi/6 < p < 1$, and the mean of R, however,

T A B L E   2.3

n	$\bar{R}$	SE
10	0.7160	0.0891
100	0.7256	0.0290
1,000	0.7670	0.0089
10,000	0.7659	0.0028
∞	0.7652	0.0000

T A B L E   2.4

n	$\hat{p}$	SE
10	.4000	.1549
100	.4700	.0499
1,000	.3820	.0154
10,000	.3801	.0049
∞	.3848	.0000

would appear to require considerably more effort here than for the corresponding two-dimensional problem.

On the other hand, using observed values (X_i, Y_i, Z_i), $1 \le i \le n$, of a random sample of size n from the uniform distribution on the indicated cube and setting $R_i = \sqrt{X_i^2 + Y_i^2 + Z_i^2}$ for $1 \le i \le n$, we can estimate various quantities defined in terms of the distribution of R just as easily for the three-dimensional problem as for its two-dimensional counterpart. In this manner, we get the estimates of ER and their standard errors as shown in Table 2.5. As a check on the reliability of the procedure, we get the estimates of $p = F(.7) = (\pi/6)(.7)^3$ and their standard errors as shown in Table 2.6.

T A B L E   2.5

n	$\bar{R}$	SE
10	1.0186	0.0819
100	0.9538	0.0271
1,000	0.9648	0.0090
10,000	0.9657	0.0028

T A B L E   2.6

n	$\hat{p}$	SE
10	.2000	.1265
100	.1800	.0384
1,000	.1930	.0125
10,000	.1722	.0038
∞	.1796	.0000

To give a rough estimate of the distribution function of R, Figure 2.5 shows the points $(R_{(i)}, (i - 1/2)/100)$, $1 \le i \le 100$, corresponding to the random sample of size 100. (There are several reasonable methods for estimating an unknown density function from sample data, but such methods will not be discussed in this book.)

F I G U R E   2.5
Estimated distribution function of R

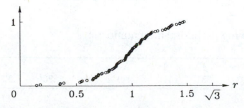

Suppose we have available a random sample (U_{1i}, U_{2i}), $1 \leq i \leq n$, from the uniform distribution on the square $[0, 1] \times [0, 1]$. Then we can obtain a random sample (X_i, Y_i), $1 \leq i \leq n$, from the uniform distribution on the square $[-1, 1] \times [-1, 1]$ by setting $X_i = 2U_{1i} - 1$ and $Y_i = 2U_{2i} - 1$.

Suppose we have available a random sample $U_1, \ldots, U_{2n}$ of size $2n$, from the uniform distribution on the interval $[0, 1]$. Then we can obtain a random sample (U_{1i}, U_{2i}), $1 \leq i \leq n$, from the uniform distribution on the square $[0, 1] \times [0, 1]$ by setting $U_{11} = U_1$, $U_{12} = U_2$, $U_{21} = U_3$, $U_{22} = U_4$, and so forth (see Example 5.17 in Section 5.5).

Thus, suppose we can generate a random sample of size n from the uniform distribution on the interval $[0, 1]$ for every positive integer n. Then we can generate a random sample of size n from the uniform distribution on the square $[-1, 1] \times [-1, 1]$ for every positive integer n. Similarly, we can generate a random sample of size n from the uniform distribution on the cube $[-1, 1] \times [-1, 1] \times [-1, 1]$ for every positive integer n.

We are led to the following fundamental question: How can we generate a random sample of size n from the uniform distribution on the interval $[0, 1]$? In practice, we have a more specific question: How can we use a digital computer to generate a random sample of size n from the uniform distribution on $[0, 1]$? Here we run into the apparent difficulty that, using such a computer in the usual manner, we get deterministic numbers, not random variables.

It turns out, however, that there are fast, computer-based methods for generating deterministic numbers $u_1, \ldots, u_n$ that *simulate* the behavior of observed values of a random sample of size n from the uniform distribution on $(0, 1)$ (so that, roughly speaking, we can safely act as if $u_1, \ldots, u_n$ actually are such the observed values from such a random sample). Such numbers $u_1, \ldots, u_n$ are referred to as *pseudorandom numbers*.

We will now briefly describe one method for generating pseudorandom numbers. Let m be a positive integer. Given a nonnegative integer x, the expression $x \,(\mathrm{mod}\, m)$ is defined as $x - km$, where $k = [x/m]$ is the greatest integer that is not greater than x/m; as examples, $0\,(\mathrm{mod}\,3) = 0$, $1\,(\mathrm{mod}\,3) = 1$, $2\,(\mathrm{mod}\,3) = 2$, $3\,(\mathrm{mod}\,3) = 0$, and $4\,(\mathrm{mod}\,3) = 1$. Let a be a positive integer and let c be a nonnegative integer. Let x_0 be a nonnegative integer, which is referred to as the *seed*, and let $x_1, x_2, \ldots$ be defined successively by $x_i = (ax_{i-1} + c) \,(\mathrm{mod}\, m)$ for $i = 1, 2, \ldots$. Set $u_i = x_i/m$ for $i = 1, 2, \ldots$. Under suitable conditions on m, a, c, and x_0, the numbers $u_1, \ldots, u_n$ form satisfactory pseudorandom numbers. This method of generating pseudorandom numbers is referred to as the *congruential method*. Further discussion of the congruential method, which would involve elementary number theory, is omitted. Computer programs that implement congruential and other fast and well-behaved pseudorandom number generators are readily available.

EXAMPLE **2.51**   Consider the following pseudorandom numbers:

i	1	2	3	4	5	6	7	8	9	10
u_i	.795	.303	.031	.720	.099	.480	.206	.337	.552	.192

Use these numbers to simulate the observed values of a random sample of size 5 from the uniform distribution on the square $[-1, 1] \times [-1, 1]$.

Solution   Setting $x_1 = 2u_1 - 1$, $y_1 = 2u_2 - 1$, $x_2 = 2u_3 - 1$, $y_2 = 2u_4 - 1$, and so forth, we get the following observed values, shown in Table 2.7, of the indicated simulated random sample.   ∎

TABLE   2.7

i	x_i	y_i
1	.590	−.394
2	−.938	.440
3	−.802	−.040
4	−.588	−.326
5	.104	−.616

Let U be a random variable that is uniformly distributed on $[0, 1]$. Then its distribution function is given by $F_U(u) = 0$ for $u \leq 0$, $F_U(u) = u$ for $0 < u < 1$, and $F_U(u) = 1$ for $u \geq 1$. Let P be any distribution function on $\mathbb{R}$ and let y_p be the corresponding pth quantile for $0 < p < 1$. Then the quantile function Q of the distribution is given by $Q(p) = y_p$ for $0 < p < 1$. Set $Y = Q(U)$. It can be shown, in complete generality, that Y has distribution P.

Suppose, in particular, that P has a density function that is continuous and positive on an interval (c, d) and equals zero outside that interval. Then $Q = F^{-1}$. Observe that

$$F^{-1}(u) \leq y \text{ if and only if } u \leq F(y), \qquad 0 < u < 1 \text{ and } y \in \mathbb{R}.$$

(See Figure 2.6.) Set $Y = Q(U) = F^{-1}(U)$. Then

$$F_Y(y) = P(Y \leq y) = P(F^{-1}(U) \leq y) = P(U \leq F(y)) = F_U(F(y)) = F(y)$$

for $y \in \mathbb{R}$, so Y has distribution function F and hence it has distribution P.

FIGURE   2.6
Distribution function

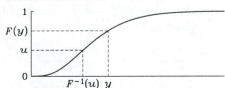

Let $U_1, \ldots, U_n$ be a random sample of size n from the uniform distribution on $(0, 1)$, let Q be the quantile function corresponding to a distribution P on $\mathbb{R}$, and set $Y_i = Q(U_i)$ for $1 \leq i \leq n$. Then $Y_1, \ldots, Y_n$ form a random sample of size n from P.

Similarly, if $u_1, \ldots, u_n$ are pseudorandom numbers and $y_i = Q(u_i)$ for $1 \leq i \leq n$, then $y_1, \ldots, y_n$ simulate observed values of a random sample of size n from P.

As an illustration, let Q be the quantile function of the standard logistic distribution, which is given by $Q(p) = \text{logit} \, p$ for $0 < p < 1$ (see Example 1.42 in Section 1.6), and let U be uniformly distributed on $(0, 1)$. Then $Y = Q(U) = \text{logit}(U)$ has the standard logistic distribution. More generally, let $U_1, \ldots, U_n$ be a random sample of size n from the uniform distribution on $(0, 1)$ and set $Y_i = \text{logit} \, U_i$ for $1 \leq i \leq n$. Then $Y_i, \ldots, Y_n$ form a random sample of size n from the standard logistic distribution. Similarly, if $u_1, \ldots, u_n$ are pseudorandom numbers and $y_i = \text{logit} \, u_i$ for $1 \leq i \leq n$, then $y_1, \ldots, y_n$ simulate observed values of a random sample of size n from the standard logistic distribution.

Let Y have the standard logistic distribution. Recall from Example 2.26 in Section 2.2 that Y has mean 0. In Problem 2.39 in Section 2.3, it is pointed out that its variance is given by $\sigma^2 = \text{var}(Y) = E(Y^2) = \pi^2/3 \doteq 3.290$. Suppose, however, that we did not realize that there was such an explicit formula for σ^2 and wanted to obtain a numerical approximation to this quantity. One approach would be to observe that

$$\sigma^2 = \int_{-\infty}^{\infty} y^2 \frac{e^y}{(1 + e^y)^2} \, dy$$

and approximate σ^2 by using Simpson's rule or some other numerical integration technique.

Another approach is to obtain observed values $y_1, \ldots, y_n$ of a simulated random sample from the standard logistic distribution and then estimate $\sigma^2 = E(Y^2)$ by the sample mean

$$\hat{\sigma}^2 = \frac{1}{n} \sum_{i=1}^{n} y_i^2$$

of $y_1^2, \ldots, y_n^2$. The corresponding standard error is given by $\text{SE}(\hat{\sigma}^2) = s/\sqrt{n}$, where

$$s^2 = \frac{1}{n-1} \sum_{i=1}^{n} (y_i^2 - \hat{\sigma}^2)^2$$

is the sample variance of $y_1^2, \ldots, y_n^2$.

To clarify this approach, consider the random variable $W = Y^2$, whose mean is given by $EW = E(Y^2) = \text{var}(Y) = \sigma^2$. Similarly, set $w_i = y_i^2$ for $1 \leq i \leq n$. Then $w_1, \ldots, w_n$ simulate a random sample of size n from the distribution of W. The sample mean of $w_1, \ldots, w_n$ is given by

$$\bar{w} = \frac{w_1 + \cdots + w_n}{n} = \frac{y_1^2 + \cdots + y_n^2}{n} = \hat{\sigma}^2.$$

The corresponding standard error is given by $\text{SE}(\hat{\sigma}^2) = \text{SE}(\bar{w}) = s/\sqrt{n}$, where

$$s^2 = \frac{1}{n-1} \sum_{i=1}^{n} (w_i - \bar{w})^2 = \frac{1}{n-1} \sum_{i=1}^{n} (y_i^2 - \hat{\sigma}^2)^2.$$

EXAMPLE 2.52   **(a)** Use the pseudorandom numbers from Example 2.51 to simulate a random sample of size 10 from the standard logistic distribution.

(b) Obtain the corresponding Monte Carlo estimate of the variance σ^2 of the standard logistic distribution as well as the standard error of this estimate.

Solution   **(a)** Setting $y_i = \text{logit}\, u_i = \log u_i/(1 - u_i)$, we get the following values:

i	1	2	3	4	5	6	7	8	9	10
y_i	1.355	−0.833	−3.442	0.944	−2.208	−0.080	−1.349	−0.677	0.209	−1.437

(b) Here

$$\hat\sigma^2 = \frac{1}{10}\sum_{i=1}^{10} y_i^2 \doteq 2.454,$$

and

$$s = \sqrt{\frac{1}{9}\sum_{i=1}^{10}(y_i^2 - \hat\sigma^2)^2} \doteq 3.596,$$

so $\text{SE}(\hat\sigma^2) = s/\sqrt{10} \doteq 1.137.$ ∎

Applying the same approach with $n = 10{,}000$, we get $\hat\sigma^2 \doteq 3.325$ and $S \doteq 5.913$, so $\text{SE}(\hat\sigma^2) \doteq 5.913/100 \doteq 0.059$. The actual error of this estimate is given by $\hat\sigma^2 - \sigma^2 \doteq 0.035$.

To illustrate how a discrete random variable can be obtained as a transform of a random variable U that is uniformly distributed on $[0, 1]$, consider the probability function on $\{0, 1, 2, 3\}$ and the corresponding distribution function given in Table 2.8. Figure 2.7 shows the graph of the distribution function F. The corresponding quantile function is given by $Q(p) = 0$ for $0 < p < .1$, $Q(.1) = .5$, $Q(p) = 1$ for $.1 < p < .3$, $Q(.3) = 1.5$, $Q(p) = 2$ for $.3 < p < .6$, $Q(.6) = 2.5$, and $Q(p) = 3$ for $.6 < p < 1$. Figure 2.8 shows the graph of Q. Set $Y = Q(U)$. Then

$$P(Y = 0) = P(0 \le U < .1) = .1,$$

$$P(Y = 1) = P(.1 \le U < .3) = .2,$$

$$P(Y = 2) = P(.3 \le U < .6) = .3,$$

and

$$P(Y = 3) = P(.6 \le U \le 1) = .4.$$

TABLE 2.8

y	0	1	2	3
$f(y)$	.1	.2	.3	.4
$F(y)$	.1	.3	.6	1.0

[Observe that $P(U \in \{0, .1, .3, .6, 1\}) = 0$.] Thus, Y has the indicated probability function.

FIGURE 2.7
Distribution function F

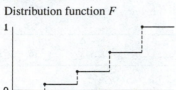

FIGURE 2.8
Quantile function Q

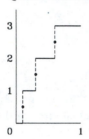

For a more general illustration, let a be a positive integer, let P be a distribution on $\{0, 1, \ldots, a\}$, let F be the corresponding distribution function, and let Y be given in terms of U as follows:

$$Y = \begin{cases} 0 & \text{if } 0 \le U < F(0), \\ 1 & \text{if } F(0) \le U < F(1), \\ \vdots & \qquad \vdots \\ a & \text{if } F(a-1) \le U \le 1. \end{cases}$$

Then Y has distribution P.

EXAMPLE 2.53   Use the pseudorandom numbers in Example 2.51 to simulate the numbers of spots that show when a fair die is repeatedly rolled in an honest manner.

Solution   We can simulate the number of spots y_i that show on the ith roll by setting $y_i = 1$ if $0 \le u_i < 1/6$, $y_i = 2$ if $1/6 \le u_i < 2/6, \ldots, y_i = 6$ if $5/6 \le u_i \le 1$. In this manner, we get results shown in Table 2.9.   ∎

EXAMPLE 2.54   Use the pseudorandom numbers in Example 2.51 to simulate the numbers of spots that show when two fair dice are repeatedly rolled in an honest manner.

TABLE 2.9

i	u_i	y_i	i	u_i	y_i
1	.795	5	6	.480	3
2	.303	2	7	.206	2
3	.031	1	8	.337	3
4	.720	5	9	.552	4
5	.099	1	10	.192	2

Solution   Let y_{1i} and y_{2i}, respectively, denote the number of spots that show on the first and second die on the ith role; then $y_i = y_{1i} + y_{2i}$ is the number of spots that show on the two dice on that roll. We can obtain y_{11} from u_1, y_{21} from u_2, y_{12} from u_3, y_{22} from u_4, and so forth, by following the prescription used in the solution to Example 2.53. In this manner, we get the results shown in Table 2.10.

Alternatively, the probability function of the number of spots that show when two fair dice are rolled in an honest manner is given in the solution to Example 1.24 in Section 1.5. The corresponding distribution function F is given by $F(1) = 0$, $F(2) = 1/36$, $F(3) = 3/36$, $F(4) = 6/36$, $F(5) = 10/36$, $F(6) = 15/36$, $F(7) = 21/36$, $F(8) = 26/36$, $F(9) = 30/36$, $F(10) = 33/36$, $F(11) = 35/36$, and $F(12) = 1$. We can simulate the number of spots y_i that show on the ith roll by setting $y_i = 2$ if $0 \leq u_i < 1/36$, $y_i = 3$ if $1/36 \leq u_i < 3/36$, $\ldots$, $y_i = 12$ if $35/36 \leq u_i \leq 1$. In this manner, we get the results shown in Table 2.11.   ∎

TABLE 2.10

i	u_{1i}	u_{2i}	y_{1i}	y_{2i}	y_i
1	.795	.303	5	2	7
2	.031	.720	1	5	6
3	.099	.480	1	3	4
4	.206	.337	2	3	5
5	.552	.192	4	2	6

TABLE 2.11

i	u_i	y_i	i	u_i	y_i
1	.795	9	6	.480	7
2	.303	6	7	.206	5
3	.031	3	8	.337	6
4	.720	8	9	.552	7
5	.099	4	10	.192	5

Consider a distribution P on $\{0, 1, 2, \ldots\}$, let F be the corresponding distribution function, and let Y be given in terms of U as follows:

$$Y = \begin{cases} 0 & \text{if } 0 \leq U < F(0), \\ 1 & \text{if } F(0) \leq U < F(1), \\ 2 & \text{if } F(1) \leq U < F(2), \\ \vdots & \quad\quad \vdots \end{cases}$$

Then Y has distribution P.

EXAMPLE **2.55**   Use the pseudorandom numbers in Example 2.51 to simulate a random sample of size 10 from the distribution of the number of trials that are required to get one spot when a fair die is repeatedly rolled in an honest manner.

Solution   Let N denote the number of rolls that are required to get one spot. According to the solution to Example 1.67 in Section 1.8, $N - 1$ has the geometric distribution with parameter $\pi = 5/6$. Let F denote the distribution function of N. It follows from the solution to Example 1.46 in Section 1.6 that $F(n) = 1 - (5/6)^n$ for $n = 1, 2, \ldots$. [Alternatively, observe that $N > n$ if and only if two, three, four, five, or six spots show on each of the first n trials, so $P(N > n) = (5/6)^n$ and hence $F(n) = P(N \leq n) = 1 - P(N > n) = 1 - (5/6)^n$.] We can simulate the number of rolls n_i that are required to get one spot on the ith run by setting $y_i = n$ if $1 - (5/6)^{n-1} \leq u_i < 1 - (5/6)^n$. In this manner, we get the results shown in Table 2.12.   ∎

TABLE   **2.12**

i	u_i	y_i	i	u_i	y_i
1	.795	9	6	.480	4
2	.303	2	7	.206	2
3	.031	1	8	.337	3
4	.720	7	9	.552	5
5	.099	1	10	.192	2

Problems

2.67   Given $a \in [0, 1/2]$ and the positive integer n, set

$$p_i = \frac{i - a}{n + 1 - 2a}, \qquad 1 \leq i \leq n.$$

Show that $p_{n+1-i} = 1 - p_i$ for $1 \leq i \leq n$.

2.68   Let μ and σ^2, respectively, denote the mean and variance of the distance from a point chosen at random from the cube $[-1, 1] \times [-1, 1] \times [-1, 1]$ to the center of the cube. Express σ^2 in terms of μ.

2.69 Let U be uniformly distributed on $[0, 1]$. Determine an increasing transform Y of U such that **(a)** Y is uniformly distributed on $[c, d]$, where $-\infty < c < d < \infty$; **(b)** Y has the exponential distribution with mean β.

2.70 Let Y have the exponential distribution with mean 1. **(a)** Use the pseudorandom numbers in Example 2.51 to simulate a random sample of size 10 from the distribution of Y. **(b)** Obtain the Monte Carlo estimate of the mean of Y and obtain the standard error of this estimate. **(c)** Obtain the Monte Carlo estimate of $\pi = P(Y \geq 1)$ and obtain the standard error of this estimate.

2.71 Use the pseudorandom numbers from Example 2.51 to simulate a random sample of size 10 from the distribution having the density function f given by $f(y) = 2y$ for $0 < y < 1$ and $f(y) = 0$ elsewhere.

2.72 Let U be uniformly distributed on $[0, 1]$. **(a)** Use U to construct a random variable having the density function f given by $f(y) = 2y$ for $0 < y < 1$ and $f(y) = 0$ otherwise. **(b)** Use U to construct a random variable having the standard Cauchy distribution.

2.73 Let Y have a density function that is continuous and positive on an interval (c, d) and is zero outside that interval, let F be the distribution function of Y, and set $U = F(Y)$. Show that U is uniformly distributed on $(0, 1)$.

2.74 Let W have the exponential distribution with mean 1. Explain how W can be transformed to yield **(a)** a random variable U having the uniform distribution on $(0, 1)$; **(b)** a random variable Y having the standard logistic distribution. [*Hints:* In part (a) use Problem 2.73, and in part (b) use part (a).]

2.75 (Continued.) Explain how a simulated random sample $w_1, \ldots, w_n$ of size n from the exponential distribution with mean 1 can be used to simulate a random sample of size n from the standard logistic distribution.

2.76 Explain how pseudorandom numbers $u_1, \ldots, u_n$ can be used to simulate a random sample of size n from **(a)** the Bernoulli distribution with parameter π; **(b)** the uniform distribution on $\{1, 2, 3\}$.

2.77 Let U be uniformly distributed on $(0, 1)$. Explain how U can be transformed to yield a random variable Y having the geometric distribution with parameter π.

2.78 (Continued.) Explain how pseudorandom numbers $u_1, \ldots, u_n$ can be used to simulate a random sample of size n from the geometric distribution with parameter π.

2.79 Let Y have the geometric distribution with parameter $\pi = .8$. **(a)** Use the pseudorandom numbers in Example 2.51 to simulate a random sample of size 10 from the distribution of Y. **(b)** Obtain the Monte Carlo estimate of the mean of Y and obtain the standard error of this estimate. **(c)** Obtain the Monte Carlo estimate of $\pi = P(Y \geq 5)$ and obtain the standard error of this estimate.

2.80 Suppose $w_1, \ldots, w_n$ simulate a random sample of size n from the exponential distribution with mean 1. Explain how these numbers can be used to simulate a random sample of size n from the geometric distribution with parameter π. [*Hint:* See Example 1.35 in Section 1.5.]

3

Special Continuous Models

3.1
Gamma and Beta Distributions

In this section, we will study some probability functions and density functions of particular importance in probability theory and its applications. These special models involve the factorials $0!, 1!, 2!, \ldots$, which are defined by $0! = 1, 1! = 1, 2! = 2 \cdot 1 = 2, 3! = 3 \cdot 2 \cdot 1$, and, more generally, by $n! = n(n-1) \cdots 1$ for n a positive integer. Observe that $n! = n \cdot (n-1)!$ for n a positive integer and that $n! = n \cdot (n-1) \cdot (n-2)!$ for $n \geq 2$. For convenience (especially in the context of Section 4.1), we make the convention that

$$\frac{1}{n!} = 0 \quad \text{if } n \text{ is a negative integer.} \tag{1}$$

Gamma Function

The factorials can be expressed in terms of the *gamma function*, which is defined by

$$\Gamma(\alpha) = \int_0^\infty w^{\alpha-1} e^{-w} \, dw$$

for $\alpha > 0$. An elementary but useful property of the gamma function is that

$$\Gamma(\alpha + 1) = \alpha \Gamma(\alpha). \tag{2}$$

To verify (2), note that, by integration by parts,

$$\Gamma(\alpha + 1) = \int_0^\infty w^\alpha e^{-w}\, dw$$

$$= \int_0^\infty w^\alpha\, d(-e^{-w})$$

$$= -w^\alpha e^{-w}\Big|_0^\infty + \alpha \int_0^\infty w^{\alpha-1} e^{-w}\, dw$$

$$= \alpha\Gamma(\alpha).$$

Now $\Gamma(1) = \int_0^\infty e^{-w}\, dw = 1$, so it follows from (2) that $\Gamma(2) = 1 \cdot \Gamma(1) = 1!$, $\Gamma(3) = 2 \cdot \Gamma(2) = 2 \cdot 1! = 2!$, $\Gamma(4) = 3 \cdot \Gamma(3) = 3 \cdot 2! = 3!$, and, more generally, that $\Gamma(n) = (n-1)!$ for n a positive integer. It also follows from (2), with α replaced by $\alpha + 1$, that $\Gamma(\alpha + 2) = (\alpha + 1)\Gamma(\alpha + 1)$ and hence that

$$\Gamma(\alpha + 2) = (\alpha + 1)\alpha\Gamma(\alpha). \tag{3}$$

A useful approximation to the gamma function is given by $\Gamma(\alpha + 1) \approx \sqrt{2\pi\alpha}\,(\alpha/e)^\alpha$ for $\alpha \gg 1$, which is known as *Stirling's formula*. The absolute error of this approximation tends to infinity as $\alpha \to \infty$, but its relative error tends to zero; that is,

$$\frac{\sqrt{2\pi\alpha}\,(\alpha/e)^\alpha}{\Gamma(\alpha + 1)} \approx 1, \qquad \alpha \gg 1. \tag{4}$$

It follows from (4) that

$$\frac{\sqrt{2\pi n}\,(n/e)^n}{n!} \approx 1, \qquad n \gg 1. \tag{5}$$

The approximation $\sqrt{2\pi n}\,(n/e)^n$ to $n!$ is also known as Stirling's formula. The corresponding relative errors are small even for moderate values of n. In particular, we have the following table:

n	1	2	3	4	5	6	7	8
Left Side of (5)	.922	.960	.973	.979	.983	.986	.988	.990

Gamma Distribution

Let α and β be positive numbers. Making the change of variable $y = \beta w$, we see that

$$\int_0^\infty y^{\alpha-1} e^{-y/\beta}\, dy = \int_0^\infty (\beta w)^{\alpha-1} e^{-w}\beta\, dw = \beta^\alpha \int_0^\infty w^{\alpha-1} e^{-w}\, dw$$

and hence that

$$\int_0^\infty y^{\alpha-1} e^{-y/\beta}\, dy = \beta^\alpha \Gamma(\alpha). \tag{6}$$

Consequently,

$$\int_0^\infty \frac{y^{\alpha-1}}{\beta^\alpha \Gamma(\alpha)} e^{-y/\beta} \, dy = 1. \tag{7}$$

It follows from (7) that the function f defined by $f(y) = 0$ for $y \le 0$ and

$$f(y) = \frac{y^{\alpha-1}}{\beta^\alpha \Gamma(\alpha)} e^{-y/\beta}, \qquad y > 0,$$

is a density function. The distribution having this density function is referred to as the *gamma distribution* with shape parameter α and scale parameter β. The corresponding distribution function is given by $F(y) = 0$ for $y < 0$ and

$$F(y) = \int_0^y \frac{z^{\alpha-1}}{\beta^\alpha \Gamma(\alpha)} e^{-z/\beta} \, dz, \qquad y \ge 0.$$

The gamma distribution with shape parameter 1 and scale parameter β coincides with the exponential distribution with mean β, so its distribution function is given by $F(y) = 1 - e^{-y/\beta}$ for $y \ge 0$ and its quantiles are given by

$$y_p = \beta \log\left(\frac{1}{1-p}\right), \qquad 0 < p < 1.$$

Suppose next that $\alpha = 2$ and let $y \ge 0$. Using integration by parts, we get that

$$F(y) = \int_0^y \frac{z}{\beta^2} e^{-z/\beta} \, dz$$

$$= \int_0^y \frac{z}{\beta} \, d(-e^{-z/\beta})$$

$$= -\frac{z}{\beta} e^{-z/\beta} \Big|_0^y + \int_0^y \frac{1}{\beta} e^{-z/\beta} \, dz$$

$$= 1 - e^{-y/\beta} - \frac{y}{\beta} e^{-y/\beta}.$$

Thus,

$$F(y) = 1 - e^{-y/\beta}\left(1 + \frac{y}{\beta}\right), \qquad y \ge 0. \tag{8}$$

There is no explicit formula (in terms of standard elementary functions) for the quantiles of this distribution.

By repeated use of integration by parts, the gamma distribution function can be determined explicitly whenever α is a positive integer. The general result is as follows.

THEOREM 3.1 Let n be a positive integer. The distribution function of the gamma distribution with shape parameter n and scale parameter β is given by $F(y) = 0$ for $y < 0$ and

$$F(y) = 1 - e^{-y/\beta} \sum_{m=0}^{n-1} \frac{y^m}{\beta^m m!}, \qquad y \ge 0. \tag{9}$$

Proof Equation (9) is true for $n = 1$. Let n be a positive integer such that (9) holds. We will show that (9) also holds for $n + 1$, the general result then following by induction. Let F denote the distribution function of the gamma distribution with shape parameter $n + 1$ and scale parameter β and let $y \geq 0$. Then

$$F(y) = \int_0^y \frac{z^n}{\beta^{n+1} n!} e^{-z/\beta} \, dz$$

$$= \int_0^y \frac{z^n}{\beta^n n!} \, d(-e^{-z/\beta})$$

$$= -\frac{z^n}{\beta^n n!} e^{-z/\beta} \Big|_0^y + \int_0^y \frac{z^{n-1}}{\beta^n (n-1)!} e^{-z/\beta} \, dz.$$

By the induction hypothesis,

$$F(y) = 1 - e^{-y/\beta} \sum_{m=0}^{n-1} \frac{y^m}{\beta^m m!} - e^{-y/\beta} \frac{y^n}{\beta^n n!} = 1 - e^{-y/\beta} \sum_{m=0}^{n} \frac{y^m}{\beta^m m!}.$$

Thus, Equation (9) holds for $n + 1$, as desired. ■

Let Y have the gamma distribution with shape parameter α and scale parameter β. Then

$$EY = \int_0^\infty y \frac{y^{\alpha-1}}{\beta^\alpha \Gamma(\alpha)} e^{-y/\beta} \, dy = \alpha\beta \int_0^\infty \frac{y^{\alpha+1-1}}{\beta^{\alpha+1} \Gamma(\alpha+1)} e^{-y/\beta} \, dy.$$

We conclude from Equation (7) with α replaced by $\alpha + 1$ (that is, by using the fact that the integral over $[0, \infty)$ of the density function of the gamma distribution with shape parameter $\alpha + 1$ and scale parameter β equals 1) that $EY = \alpha\beta$. Observe next that

$$E(Y^2) = \int_0^\infty y^2 \frac{y^{\alpha-1}}{\beta^\alpha \Gamma(\alpha)} e^{-y/\beta} dy = \alpha(\alpha+1)\beta^2 \int_0^\infty \frac{y^{\alpha+2-1}}{\beta^{\alpha+2} \Gamma(\alpha+2)} e^{-y/\beta} \, dy.$$

We conclude from (7) with α replaced by $\alpha + 2$ that $E(Y^2) = \alpha(\alpha + 1)\beta^2$. Consequently,

$$\mathrm{var}(Y) = E(Y^2) - (EY)^2 = \alpha(\alpha+1)\beta^2 - (\alpha\beta)^2 = \alpha\beta^2,$$

and hence $\mathrm{SD}(Y) = \beta\sqrt{\alpha}$.

In summary, the gamma distribution with shape parameter α and scale parameter β has mean $\alpha\beta$, variance $\alpha\beta^2$, and standard deviation $\beta\sqrt{\alpha}$. In particular, the exponential distribution with scale parameter β has mean β, variance β^2, and standard deviation β, as we already know.

Consider a density function that has a unique local maximum that is also its unique global maximum. The location of the maximum is referred to as the unique *mode* of the density function, and the density function is said to be *unimodal*.

Let f be the density function of the gamma distribution with shape parameter α and scale parameter β. Suppose $0 < \alpha \leq 1$. Then f is a decreasing function on $[0, \infty)$. If $0 < \alpha < 1$, then $f(y)$ approaches ∞ as y approaches 0 from the right; if

$\alpha = 1$, then $f(y)$ approaches $1/\beta$ as y approaches 0 from the right. Suppose instead that $\alpha > 1$. Then f is continuous at the origin and has value 0 there. As a function on $[0, \infty)$, it increases to its maximum at its unique mode, which equals $\beta(\alpha - 1)$, and then it decreases to 0. [The proof that $\beta(\alpha - 1)$ is the unique mode is left as a problem.] Figure 3.1 shows graphs of f for $\alpha = 1/2, 1, 2, 4$. (Here $\beta = 1/\alpha$, so that the density functions all have mean 1.)

FIGURE   3.1
Gamma density functions with mean 1.

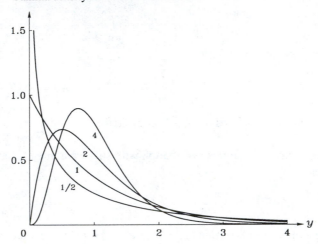

THEOREM 3.2   Let W have the gamma distribution with shape parameter α and scale parameter β and let $b > 0$. Then $Y = bW$ has the gamma distribution with shape parameter α and scale parameter $b\beta$.

Proof   According to Corollary 1.3 in Section 1.7 and the formula for the gamma density function, $f_Y(y) = 0$ for $y \leq 0$ and

$$f_Y(y) = \frac{1}{b}f_W(y/b) = \frac{(y/b)^{\alpha-1}}{b\beta^\alpha \Gamma(\alpha)}e^{-(y/b)/\beta} = \frac{y^{\alpha-1}}{(b\beta)^\alpha \Gamma(\alpha)}e^{-y/(b\beta)}, \qquad y > 0,$$

so Y has the indicated gamma distribution.   ∎

THEOREM 3.3   Let W have the gamma distribution with shape parameter α and scale parameter β. Then $Y = \sqrt{W}$ has the density function f_Y given by $f_Y(y) = 0$ for $y \leq 0$ and

$$f_Y(y) = \frac{2y^{2\alpha-1}}{\beta^\alpha \Gamma(\alpha)}e^{-y^2/\beta}, \qquad y > 0.$$

Proof   It follows from Theorem 1.5 in Section 1.7 and the formula for the density function of the gamma distribution that $f_Y(y) = 0$ for $y \leq 0$ and

$$f_Y(y) = 2yf_W(y^2) = \frac{2y^{2\alpha-1}}{\beta^\alpha \Gamma(\alpha)}e^{-y^2/\beta}, \qquad y > 0.   ∎$$

The next result will be obtained as part of Theorem 5.21 in Section 5.6.

THEOREM 3.4   Let W_1 and W_2 be independent, positive random variables having density functions f_{W_1} and f_{W_2}, respectively. Then the random variable $Y = W_1 + W_2$ has the density function given by $f_Y(y) = 0$ for $y \leq 0$ and

$$f_Y(y) = \int_0^y f_{W_1}(w) f_{W_2}(y - w) \, dw, \qquad y > 0.$$

Theorem 3.4 will now be used to determine the distribution of the sum of independent random variables having gamma distributions with a common scale parameter.

THEOREM 3.5   Let W_1 and W_2 be independent random variables having gamma distributions with shape parameters α_1 and α_2, respectively, and common scale parameter β. Then $Y = W_1 + W_2$ has the gamma distribution with shape parameter $\alpha_1 + \alpha_2$ and scale parameter β.

Proof   It follows from Theorem 3.4 that Y has a density function f_Y, which equals zero on $(-\infty, 0]$. Let $y > 0$. By Theorem 3.4 and the formula for the gamma density function,

$$f_Y(y) = \int_0^y \frac{w^{\alpha_1 - 1}}{\beta^{\alpha_1} \Gamma(\alpha_1)} e^{-w/\beta} \frac{(y - w)^{\alpha_2 - 1}}{\beta^{\alpha_2} \Gamma(\alpha_2)} e^{-(y-w)/\beta} \, dw$$

$$= \frac{e^{-y/\beta}}{\beta^{\alpha_1 + \alpha_2} \Gamma(\alpha_1) \Gamma(\alpha_2)} \int_0^y w^{\alpha_1 - 1} (y - w)^{\alpha_2 - 1} \, dw.$$

Making the change of variables $t = w/y$, we get that

$$\int_0^y w^{\alpha_1 - 1} (y - w)^{\alpha_2 - 1} \, dw = y^{\alpha_1 + \alpha_2 - 1} \int_0^1 t^{\alpha_1 - 1} (1 - t)^{\alpha_2 - 1} \, dt.$$

Thus,

$$f_Y(y) = \frac{y^{\alpha_1 + \alpha_2 - 1}}{c \beta^{\alpha_1 + \alpha_2}} e^{-y/\beta}, \qquad y > 0, \tag{10}$$

where

$$c = \frac{\Gamma(\alpha_1) \Gamma(\alpha_2)}{\int_0^1 t^{\alpha_1 - 1} (1 - t)^{\alpha_2 - 1} \, dt}. \tag{11}$$

Since f_Y is a density function, it integrates to one. We now conclude from (10) that

$$c = \frac{1}{\beta^{\alpha_1 + \alpha_2}} \int_0^\infty y^{\alpha_1 + \alpha_2 - 1} e^{-y/\beta} \, dy$$

and hence from (6) with $\alpha = \alpha_1 + \alpha_2$ that

$$c = \Gamma(\alpha_1 + \alpha_2). \tag{12}$$

It follows from (10) and (12) that f_Y is the density function of the gamma distribution with parameters $\alpha_1 + \alpha_2$ and β.   ∎

The next result follows from Theorem 3.5 and induction in the usual manner.

THEOREM **3.6**   Let $W_1, \ldots, W_n$ be independent random variables, each having gamma distributions with shape parameters $\alpha_1, \ldots, \alpha_n$, respectively, and common scale parameter β. Then $W_1 + \cdots + W_n$ has the gamma distribution with shape parameter $\alpha_1 + \cdots + \alpha_n$ and scale parameter β.

By letting $\alpha_1 = \cdots = \alpha_n = 1$ in Theorem 3.6, we get the following consequence.

COROLLARY **3.1**   Let $W_1, \ldots, W_n$ be independent random variables, each having the exponential distribution with mean β. Then $W_1 + \cdots + W_n$ has the gamma distribution with shape parameter n and scale parameter β.

We see from this result that gamma distributions with positive-integer shape parameters arise naturally as distributions of sums of independent exponential random variables with a common mean. In Section 7.2, we will see that, for ν a positive integer, gamma distributions with shape parameter $\nu/2$ arise naturally as distributions of sums of squares of normal random variables with mean 0 and common variance.

EXAMPLE **3.1**   Suppose the lifetime (in years) of a certain type of outdoor light bulb is exponentially distributed with mean 1. Frankie and Johnny receive two such light bulbs (along with a new house) as a wedding present. They install one of the light bulbs on their wedding day and plan to install the next one as soon as the first one burns out. Determine the probability that both light bulbs will have burned out by the couple's first wedding anniversary.

Solution   Let Y_1 and Y_2 denote the lifetimes of the first and second light bulbs, respectively. Then Y_1 and Y_2 each have the exponential distribution with mean 1. We make the reasonable assumption that these random variables are independent. Then, by Corollary 3.1, $Y_1 + Y_2$ has the gamma distribution with shape parameter 2 and scale parameter 1. Thus, we conclude from Equation (8) that $P(Y_1 + Y_2 \le 1) = 1 - 2/e \doteq .264$. ∎

EXAMPLE **3.2**   Gizmos, Inc. is the leading manufacturer of widgets. Its factory operates two 8-hour shifts, 5 days per week, 50 weeks per year. At the beginning of each such shift, a batch of ten widgets is made and turned over to the quality control department, where they are put under a stress test to determine how long they survive under extreme conditions. Let $Y_1, \ldots, Y_{10}$ denote the failure times (in minutes) of the widgets in the batch and let

$\bar{Y} = (Y_1 + \cdots + Y_{10})/10$ denote the corresponding sample mean. The procedure used by the quality control department is to accept the batch as satisfactory and get on with the manufacturing of more widgets if $\bar{Y} \geq c$ and to reject the batch as being unsatisfactory and to inspect the manufacturing setup for defects if $\bar{Y} < c$.

(a) Determine c such that if $Y_1, \ldots, Y_{10}$ are independent random variables, each having the exponential distribution with mean 1, then the probability that the batch is rejected equals .01.

(b) Let c be as in part (a). Determine the probability that the batch is rejected when $Y_1, \ldots, Y_{10}$ are independent random variables, each having the exponential distribution with mean 1/2.

Solution   Suppose $Y_1, \ldots, Y_{10}$ are independent random variables each having the exponential distribution with mean β. By Corollary 3.1, $Y_1 + \cdots + Y_{10}$ has the gamma distribution with shape parameter 10 and scale parameter β. Thus, by Theorem 3.2, $\bar{Y}$ has the gamma distribution with shape parameter 10 and scale parameter $\beta/10$.

(a) Here $\beta = 1$, so $\bar{Y}$ has the gamma distribution with shape parameters 10 and scale parameter 1/10. Observe that $P(\bar{Y} < c) = .01$ if and only if c is the lower 1st percentile of the distribution of $\bar{Y}$—that is, if and only if $F(c) = .01$, where F is given by Equation (9) with $n = 10$ and $\beta = 1/10$. We could determine c numerically by writing and carrying out a simple computer program. Alternatively, we can use an existing program for calculating quantiles of the gamma distribution. Taking the latter approach, we find that $c \doteq 0.4130$. [We can check this result, of course, by verifying that $F(0.4130) \doteq .01$.]

(b) Here $\beta = 1/2$, so $\bar{Y}$ has the gamma distribution with shape parameter 10 and scale parameter 1/20. By using Equation (9) with $n = 10$ and $\beta = 1/20$ or by using a suitable software package, we find that

$$P(\bar{Y} < c) \doteq P(\bar{Y} < 0.4130) = F(0.4130) \doteq .3162. \quad \blacksquare$$

Beta Distribution

It follows from Equations (11) and (12), with t replaced by y in the definite integral in (11), that

$$\int_0^1 \frac{\Gamma(\alpha_1 + \alpha_2)}{\Gamma(\alpha_1)\Gamma(\alpha_2)} y^{\alpha_1 - 1}(1 - y)^{\alpha_2 - 1} \, dy = 1. \tag{13}$$

Thus, the function f defined by

$$f(y) = \frac{\Gamma(\alpha_1 + \alpha_2)}{\Gamma(\alpha_1)\Gamma(\alpha_2)} y^{\alpha_1 - 1}(1 - y)^{\alpha_2 - 1}, \qquad 0 < y < 1,$$

and $f(y) = 0$ elsewhere is a density function. The corresponding distribution is referred to as the *beta distribution* with (shape) parameters α_1 and α_2, the motivation for the name being that the function

$$B(\alpha_1, \alpha_2) = \frac{\Gamma(\alpha_1)\Gamma(\alpha_2)}{\Gamma(\alpha_1 + \alpha_2)}, \qquad \alpha_1, \alpha_2 > 0,$$

is called the *beta function*. Observe that the beta distribution with parameters $\alpha_1 = 1$ and $\alpha_2 = 1$ coincides with the uniform distribution on $[0, 1]$. Beta distributions are commonly used in practice to model the distribution of a random probability (such as the probability that a randomly selected male student at some college or university would be successful in one free-throw attempt in basketball).

Let f be the density function of the beta distribution with parameters α_1 and α_2. Then

$$\lim_{y \downarrow 0} f(y) = \begin{cases} \infty & \text{if } \alpha_1 < 1 \\ \alpha_2 & \text{if } \alpha_1 = 1 \\ 0 & \text{if } \alpha_1 > 1 \end{cases}$$

and

$$\lim_{y \uparrow 1} f(y) = \begin{cases} \infty & \text{if } \alpha_2 < 1 \\ \alpha_1 & \text{if } \alpha_2 = 1 \\ 0 & \text{if } \alpha_2 > 1 \end{cases}$$

If $\alpha_1 > 1$ and $\alpha_2 > 1$, then f is unimodal, and its mode equals $(\alpha_1 - 1)/(\alpha_1 + \alpha_2 - 1)$ (the details are left as a problem). Figure 3.2 shows the graph of f for $\alpha_1 = \alpha_2 = \alpha$ with $\alpha = 1/2, 1, 2, 4$.

FIGURE 3.2
Beta density functions with $\alpha_1 = \alpha_2 = \alpha$

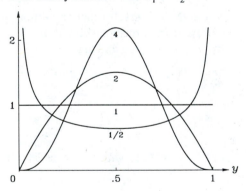

Let Y have the beta distribution with parameters α_1 and α_2. Then

$$EY = \int_0^1 y \frac{\Gamma(\alpha_1 + \alpha_2)}{\Gamma(\alpha_1)\Gamma(\alpha_2)} y^{\alpha_1 - 1}(1 - y)^{\alpha_2 - 1} \, dy$$

$$= \int_0^1 \frac{\Gamma(\alpha_1 + \alpha_2)}{\Gamma(\alpha_1)\Gamma(\alpha_2)} y^{\alpha_1 + 1 - 1}(1 - y)^{\alpha_2 - 1} \, dy.$$

It follows from Equation (2), with α successively replaced by α_1 and α_2, that

$\Gamma(\alpha_1 + 1) = \alpha_1\Gamma(\alpha_1)$ and $\Gamma(\alpha_1 + \alpha_2 + 1) = (\alpha_1 + \alpha_2)\Gamma(\alpha_1 + \alpha_2)$. Consequently,

$$EY = \frac{\alpha_1}{\alpha_1 + \alpha_2} \int_0^1 \frac{\Gamma(\alpha_1 + 1 + \alpha_2)}{\Gamma(\alpha_1 + 1)\Gamma(\alpha_2)} y^{\alpha_1 + 1 - 1}(1 - y)^{\alpha_2 - 1} \, dy = \frac{\alpha_1}{\alpha_1 + \alpha_2}.$$

[In the last step, we have applied Equation (13) with α_1 replaced by $\alpha_1 + 1$.] Similarly, using Equations (3) and (13), we get that

$$E(Y^2) = \int_0^1 \frac{\Gamma(\alpha_1 + \alpha_2)}{\Gamma(\alpha_1)\Gamma(\alpha_2)} y^{\alpha_1 + 1}(1 - y)^{\alpha_2 - 1} \, dy$$

$$= \frac{\alpha_1(\alpha_1 + 1)}{(\alpha_1 + \alpha_2)(\alpha_1 + \alpha_2 + 1)} \int_0^1 \frac{\Gamma(\alpha_1 + 2 + \alpha_2)}{\Gamma(\alpha_1 + 2)\Gamma(\alpha_2)} y^{\alpha_1 + 2 - 1}(1 - y)^{\alpha_2 - 1} \, dy$$

$$= \frac{\alpha_1(\alpha_1 + 1)}{(\alpha_1 + \alpha_2)(\alpha_1 + \alpha_2 + 1)}$$

and hence that

$$\mathrm{var}(Y) = E(Y^2) - (EY)^2$$

$$= \frac{\alpha_1(\alpha_1 + 1)}{(\alpha_1 + \alpha_2)(\alpha_1 + \alpha_2 + 1)} - \left(\frac{\alpha_1}{\alpha_1 + \alpha_2}\right)^2$$

$$= \frac{\alpha_1\alpha_2}{(\alpha_1 + \alpha_2)^2(\alpha_1 + \alpha_2 + 1)}.$$

In summary, if Y has the beta distribution with parameters α_1 and α_2, then its mean and variance are given by

$$EY = \frac{\alpha_1}{\alpha_1 + \alpha_2} \quad \text{and} \quad \mathrm{var}(Y) = \frac{\alpha_1\alpha_2}{(\alpha_1 + \alpha_2)^2(\alpha_1 + \alpha_2 + 1)}.$$

Values of the Gamma Function at the Half-Integers

The values $\Gamma(1/2)$, $\Gamma(3/2)$, ... of the gamma function can be determined explicitly. To this end, choose $\alpha_1 = \alpha_2 = 1/2$ in Equation (13) to get that

$$\int_0^1 \frac{1}{\sqrt{y(1 - y)}} \, dy = \left[\Gamma\left(\frac{1}{2}\right)\right]^2. \tag{14}$$

The definite integral in (14) can also be evaluated by making the substitution $y = \sin^2(t)$, where $0 < t < \pi/2$. Correspondingly,

$$\frac{dy}{dt} = 2\sin t \cos t = 2\sqrt{y(1 - y)},$$

and hence

$$\frac{1}{\sqrt{y(1 - y)}} \, dy = 2 \, dt.$$

Consequently,

$$\int_0^1 \frac{1}{\sqrt{y(1 - y)}} \, dy = 2 \int_0^{\pi/2} dt = \pi.$$

We now conclude from (14) that $[\,\Gamma(1/2)\,]^2 = \pi$ and hence that

$$\Gamma\left(\frac{1}{2}\right) = \sqrt{\pi}. \tag{15}$$

It follows from Equations (2) and (15) that

$$\Gamma\left(\frac{3}{2}\right) = \frac{1}{2}\sqrt{\pi} \quad \text{and} \quad \Gamma\left(\frac{5}{2}\right) = \frac{3 \cdot 1}{2 \cdot 2}\sqrt{\pi}.$$

More generally, if n is an odd positive integer with $n \geq 3$, then

$$\Gamma\left(\frac{n}{2}\right) = \left(\frac{1}{2}\right)\left(\frac{3}{2}\right)\cdots\left(\frac{n-2}{2}\right)\sqrt{\pi} = \frac{1 \cdot 3 \cdots (n-2)}{2^{(n-1)/2}}\sqrt{\pi}.$$

Since

$$1 \cdot 3 \cdots (n-2) = \frac{1 \cdot 2 \cdots (n-1)}{2 \cdot 4 \cdots (n-1)}$$

$$= \frac{(n-1)!}{2^{(n-1)/2}(1)(2)\cdots((n-1)/2)}$$

$$= \frac{(n-1)!}{2^{(n-1)/2}((n-1)/2)!},$$

we conclude that

$$\Gamma\left(\frac{n}{2}\right) = \frac{(n-1)!}{2^{n-1}((n-1)/2)!}\sqrt{\pi}, \qquad n = 1, 3, 5, \ldots. \tag{16}$$

Problems

3.1 Express the shape parameter $\alpha > 1$ and scale parameter β of the gamma distribution in terms of the mean μ and variance σ^2 of the distribution.

3.2 Consider three independent random variables each having the exponential distribution with mean 1. Determine the probability that the sum of these random variables is greater than 3.

3.3 Let Y have the gamma distribution with shape parameter $\alpha > 1$ and scale parameter β. **(a)** Show that $\beta(\alpha - 1)$ is the unique mode of the density function of Y. **(b)** Show that

$$E\left(\frac{1}{Y}\right) = \frac{1}{\beta(\alpha - 1)};$$

that is, the mean of the reciprocal of Y is the reciprocal of the mode of Y. [If $0 < \alpha \leq 1$, then $E(Y^{-1}) = \infty$.] **(c)** Show that if $\alpha > 2$, then

$$E\left(\frac{1}{Y^2}\right) = \frac{1}{\beta^2(\alpha - 1)(\alpha - 2)} \quad \text{and} \quad \text{var}\left(\frac{1}{Y}\right) = \frac{1}{\beta^2(\alpha - 1)^2(\alpha - 2)}.$$

[If $\alpha \leq 2$, then var$(1/Y) = \infty$.]

3.4 Let W have the exponential distribution with mean β and set $Y = \sqrt{W}$. Determine **(a)** $E(Y)$; **(b)** var(Y).

3.5 Let W have the gamma distribution with shape parameter α and scale parameter β and set $Y = \sqrt{W}$. Determine **(a)** $E(Y)$; **(b)** $\text{var}(Y)$.

3.6 Use Stirling's formula to approximate the value of the density function of the gamma distribution with shape parameter $\alpha \gg 1$ and scale parameter β at the mode of the density function.

3.7 Let Y have the beta distribution with parameters α_1 and α_2. Show that $1 - Y$ has the beta distribution with parameters α_2 and α_1.

3.8 Show that
$$\int_0^y w^{\alpha_1 - 1}(y - w)^{\alpha_2 - 1}\, dw = y^{\alpha_1 + \alpha_2 - 1}\frac{\Gamma(\alpha_1)\Gamma(\alpha_2)}{\Gamma(\alpha_1 + \alpha_2)}, \qquad \alpha_1, \alpha_2, y > 0.$$

3.9 Let W have the beta distribution with parameters α_1 and α_2. Determine the density function of $Y = cW$ for $c > 0$.

3.10 Show that if μ and σ^2 are the mean and variance, respectively, of a beta distribution, then $\sigma^2 < \mu(1 - \mu)$.

3.11 Express the parameters α_1 and α_2 of the beta distribution in terms of the mean μ and variance σ^2 of the distribution.

3.12 Show that the beta density function with parameters $\alpha_1 > 1$ and $\alpha_2 > 1$ has a unique mode at $(\alpha_1 - 1)/(\alpha_1 + \alpha_2 - 2)$.

3.13 Let W have the beta distribution with parameters α_1 and α_2, and set $Y = \frac{W}{1-W}$. Show that **(a)** $F_Y(y) = F_W(y/(1 + y))$ for $y \in \mathbb{R}$; **(b)** $f_Y(y) = (1 + y)^{-2}f_W(y/(1 + y))$ for $y \in \mathbb{R}$; **(c)** $f_Y(y) = 0$ for $y \le 0$ and
$$f_Y(y) = \frac{\Gamma(\alpha_1 + \alpha_2)y^{\alpha_1 - 1}}{\Gamma(\alpha_1)\Gamma(\alpha_2)(1 + y)^{\alpha_1 + \alpha_2}}, \qquad y > 0.$$

3.2
The Normal Distribution

The two-parameter family of normal distributions, which will be discussed in this section, is far and away the single most important family of distributions in probability and statistics. Several reasons account for this importance. The mathematical theory of this family is tractable and elegant, as is the theory of its multidimensional extension discussed in Section 5.7. It will be seen in Section 3.3 that the normal family provides an accurate approximation to a gamma distribution with a large shape parameter, and it will be seen in Chapter 4 that it also provides accurate approximations to certain discrete distributions that are important in theory and practice. The family also provides accurate approximations to the distribution of many chance phenomena in the real world, and it plays a key role in statistical theory and methodology as will be seen in Chapters 7–13.

The *standard normal* density function, denoted by ϕ, is defined as
$$\phi(z) = \frac{1}{c}e^{-z^2/2}, \qquad -\infty < z < \infty,$$

where

$$c = \int_{-\infty}^{\infty} e^{-z^2/2}\,dz = 2\int_0^{\infty} e^{-z^2/2}\,dz.$$

To evaluate the last integral, make the change of variables $w = z^2/2$; correspondingly, $z = (2w)^{1/2}$ and $dz = (2w)^{-1/2}dw$. Thus,

$$c = 2\int_0^{\infty} (2w)^{-1/2}e^{-w}\,dw = \sqrt{2}\int_0^{\infty} w^{-1/2}e^{-w}\,dw = \sqrt{2}\,\Gamma\!\left(\tfrac{1}{2}\right).$$

Now $\Gamma(1/2) = \sqrt{\pi}$ according to Equation (15) in Section 3.1, so $c = \sqrt{2\pi}$. Consequently,

$$\phi(z) = \frac{1}{\sqrt{2\pi}}e^{-z^2/2}, \qquad -\infty < z < \infty.$$

The standard normal density function is symmetric about zero, has a unique mode at zero, and decreases to zero very rapidly as $|z| \to \infty$ (see Figure 3.3).

FIGURE   3.3

The standard normal density function

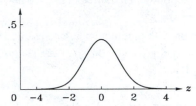

Some numerical values of this density function are as follows:

z	0	1	2	3	4
$\phi(z)$	0.3989	0.2420	0.0540	0.0044	0.0001

The distribution having density function ϕ is referred to as the *standard normal distribution*. The corresponding distribution function Φ (see Figure 3.4) is given by

$$\Phi(z) = \int_{-\infty}^{z} \phi(t)\,dt = \int_{-\infty}^{z} \frac{1}{\sqrt{2\pi}}e^{-t^2/2}\,dt.$$

FIGURE   3.4

The standard normal distribution function

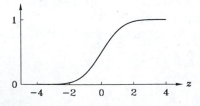

Since the standard normal density function is symmetric about zero, Equation (48) in Section 1.6 is applicable; thus,

$$\Phi(-z) = 1 - \Phi(z), \qquad -\infty < z < \infty. \tag{17}$$

Some numerical values of the standard normal distribution function are as follows:

z	0.0	0.5	1.0	1.5	2.0	2.5	3.0	3.5	4.0
$\Phi(z)$	.5000	.6915	.8413	.9332	.9773	.9938	.9987	.9998	1.0000

The pth quantile z_p of the standard normal distribution is given by $z_p = \Phi^{-1}(p)$. Since ϕ is symmetric about zero, we conclude from Equation (49) in Section 1.6 that

$$z_{1-p} = -z_p, \qquad 0 < p < 1. \tag{18}$$

Some numerical values for various standard normal quantiles are as follows:

p	.5000	.7500	.9000	.9500	.9750	.9900	.9950	.9990
z_p	0.0000	0.6745	1.2816	1.6449	1.9600	2.3263	2.5758	3.0902

More extensive tables of the standard normal distribution function and quantile function are found in Appendix E. These functions cannot be evaluated in terms of more familiar functions, but most statistical and many other software packages provide quick and accurate numerical approximations to them.

Since the density function of the standard normal distribution is symmetric about 0, its mean equals 0. (The mean is easily seen to be finite.) Observe that

$$\int_{-\infty}^{\infty} z^2 \frac{1}{\sqrt{2\pi}} e^{-z^2/2}\, dz = \int_{-\infty}^{\infty} \frac{1}{\sqrt{2\pi}} z\, d(-e^{-z^2/2})$$

$$= -\frac{1}{\sqrt{2\pi}} z e^{-z^2/2}\Big|_{-\infty}^{\infty} + \int_{-\infty}^{\infty} \frac{1}{\sqrt{2\pi}} e^{-z^2/2}\, dz$$

$$= 1.$$

Thus, the second moment of the standard normal distribution equals 1, so its variance equals 1 and hence its standard deviation equals 1.

EXAMPLE 3.3   Let Z be a standard normal random variable and let $0 < \alpha < 1$. Determine the value of the constant c such that

$$P(|Z| < c) = P(-c < Z < c) = 1 - \alpha.$$

Solution   By Equation (17),

$$P(-c < Z < c) = \Phi(c) - \Phi(-c) = \Phi(c) - [1 - \Phi(c)] = 2\Phi(c) - 1.$$

Thus, $P(-c < Z < c) = 1 - \alpha$ if and only if $2\Phi(c) - 1 = 1 - \alpha$ or, equivalently, $\Phi(c) = 1 - \alpha/2$. Therefore, the desired value of c is given by $c = \Phi^{-1}(1 - \alpha/2) = z_{1-\alpha/2}$.   ∎

Let Z be a standard normal random variable. According to the solution to Example 3.3,

$$P(|Z| < z_{1-\alpha/2}) = P(-z_{1-\alpha/2} < Z < z_{1-\alpha/2}) = 1 - \alpha. \tag{19}$$

In particular, choosing $\alpha = .05$, we conclude from (19) that

$$P(|Z| < z_{.975}) = P(-z_{.975} < Z < z_{.975}) = .95.$$

Since $z_{.975} \doteq 1.96$, we get that

$$P(|Z| < 1.96) = P(-1.96 < Z < 1.96) \doteq .95. \tag{20}$$

Consider a large number of independent trials of an experiment whose outcome is a standard normal random variable. According to (20) and the weak law of large numbers, the outcome will be between -1.96 and 1.96 in approximately 95% of the trials.

Let a and b be real numbers with $b > 0$ and set $Y = a + bZ$. The mean μ and standard deviation σ of Y are given by $\mu = a + bEZ = a$ and $\sigma = b \,\mathrm{SD}(Z) = b$, so $Y = \mu + \sigma Z$. According to Theorem 1.4 in Section 1.7, the density function of Y is given (see Figure 3.5) by

$$f(y) = \frac{1}{\sigma}\phi\left(\frac{y-\mu}{\sigma}\right) = \frac{1}{\sigma\sqrt{2\pi}}e^{-(y-\mu)^2/2\sigma^2}, \qquad y \in \mathbb{R},$$

its distribution function is given (see Figure 3.6) by

$$F(y) = \Phi\left(\frac{y-\mu}{\sigma}\right), \qquad y \in \mathbb{R},$$

and its pth quantile is given by $y_p = \mu + \sigma z_p$. The distribution of Y is referred to as the *normal distribution* with mean μ and variance σ^2.

FIGURE   3.5
The normal density function with mean μ and variance σ^2

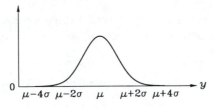

FIGURE   3.6
The normal distribution function with mean μ and variance σ^2

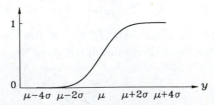

THEOREM 3.7   Let Y be normally distributed with mean μ and variance σ^2 and let $a, b \in \mathbb{R}$ with $b \neq 0$. Then $a + bY$ is normally distributed with mean $a + b\mu$ and variance $b^2\sigma^2$.

Proof   Write $Y = \mu + \sigma Z$, where Z has the standard normal distribution. Then

$$a + bY = a + b(\mu + \sigma Z) = a + b\mu + b\sigma Z.$$

If $b > 0$, then Y is clearly normally distributed with mean $a + b\mu$ and variance $b^2\sigma^2$. Suppose instead that $b < 0$. Using Equation (17), we get that

$$P(-Z \leq z) = P(Z \geq -z) = 1 - \Phi(-z) = \Phi(z) = P(Z \leq z), \qquad z \in \mathbb{R}.$$

Thus, $-Z$ has the same distribution function as Z and hence the same distribution. Consequently, the distribution of

$$a + bY = a + b\mu + b\sigma Z = a + b\mu + (-b)\sigma(-Z) = a + b\mu + |b|\sigma(-Z)$$

is the same as that of $a + b\mu + |b|\sigma Z$, which is normal with mean $a + b\mu$ and variance $(|b|\sigma)^2 = b^2\sigma^2$.   ∎

Let Y be a random variable having mean μ and finite, positive standard deviation σ. Recall that the corresponding standardized random variable $Z = (Y - \mu)/\sigma$ has mean 0 and variance 1 and that $Y = \mu + \sigma Z$.

COROLLARY 3.2   Let Y be normally distributed with mean μ and variance σ^2. Then $Z = (Y - \mu)/\sigma$ has the standard normal distribution.

Let Y be normally distributed with mean μ and variance σ^2 and let $Z = (Y - \mu)/\sigma$ be the corresponding standardized random variable, which has the standard normal distribution. Also, let $0 < \alpha < 1$. Then, by Equation (19),

$$P\left(\left|\frac{Y - \mu}{\sigma}\right| < z_{1-\alpha/2}\right) = P(|Z| < z_{1-\alpha/2}) = 1 - \alpha,$$

so

$$P(|Y - \mu| < z_{1-\alpha/2}\sigma) = 1 - \alpha. \tag{21}$$

In particular,

$$P(|Y - \mu| < 1.96\sigma) \doteq .95. \tag{22}$$

Consider a large number of independent trials of an experiment whose outcomes are normally distributed with mean μ and standard deviation σ. According to (22) and the weak law of large numbers, the outcome will differ from its mean by less than 1.96σ in approximately 95% of the trials.

Let Y have mean μ and finite, positive standard deviation σ. We refer to the normal distribution with mean μ and variance σ^2 as the usual form of the *normal approximation* to the distribution of Y. Similarly, we refer to the distribution function $\Phi((y - \mu)/\sigma)$, $y \in \mathbb{R}$, and quantiles $\mu + \sigma z_q$ of the indicated normal distribution as the normal approximation to the distribution function and quantiles of Y.

EXAMPLE **3.4**   The normal distribution typically provides a good approximation to the height of an individual selected at random from a large homogeneous population of adults. Suppose the height in inches of such a randomly selected individual has mean 68 and standard deviation 2. Use normal approximation to determine

(a) the probability that the height of such a randomly selected individual is between $5\frac{1}{2}$ feet and 6 feet;

(b) the quartiles of the distribution of height.

Solution   Let Y denote the height in inches of a randomly selected individual. Suppose Y is normally distributed with mean $\mu = 68$ and standard deviation $\sigma = 2$. Write $Y = \mu + \sigma Z = 68 + 2Z$, where $Z = (Y - 68)/2$ has the standard normal distribution.

(a) Now $5\frac{1}{2}$ feet equals 66 inches and 6 feet equals 72 inches, so the desired probability is given by

$$P(66 \le Y \le 72) = P(66 \le 68 + 2Z \le 72)$$

$$= P(-1 \le Z \le 2) = \Phi(2) - \Phi(-1).$$

Since $\Phi(2) \doteq .977$ and $\Phi(-1) = 1 - \Phi(1) \doteq .159$, we get that

$$P(66 \le Y \le 72) \doteq .819.$$

Alternatively, since $(Y - 68)/2$ has the standard normal distribution, we have that

$$P(66 \le Y \le 72) = P\left(\frac{66 - 68}{2} \le \frac{Y - 68}{2} \le \frac{72 - 68}{2}\right)$$

$$= P\left(-1 \le \frac{Y - 68}{2} \le 2\right)$$

$$= \Phi(2) - \Phi(-1),$$

and so forth. (Here, throughout the chain $66 \le Y \le 72$ of inequalities, we have subtracted the mean of Y and then divided by its standard deviation.)

(b) The quantiles of Y are given by $y_p = \mu + \sigma z_p = 68 + 2z_p$. Now $z_{.5} = 0$, $z_{.75} \doteq 0.674$, and $z_{.25} = -z_{.75} \doteq -0.674$. Thus, the quartiles of Y are given by $y_{.25} = 68 + 2z_{.25} \doteq 68 + 2(-0.674) \doteq 66.7$, $y_{.5} = 68$, and $y_{.75} = 68 + 2z_{.75} \doteq 68 + 2(0.674) \doteq 69.3$. ∎

Consider a positive random variable having mean μ (which must be positive) and finite, positive standard deviation σ. The probability assigned to $(-\infty, 0)$ by the normal approximation to the distribution of this random variable equals $\Phi(-\mu/\sigma)$. Since the probability that a positive random variable is less than zero equals zero, a necessary condition for the accuracy of this normal approximation is that $\Phi(-\mu/\sigma) \approx 0$ or, equivalently, that $\mu/\sigma \gg 1$. Thus, a necessary condition for the accuracy of normal approximation to the distribution of a positive random variable is that its coefficient of variation σ/μ be small. In Example 3.4, the coefficient of variation is given by $2/68 \doteq .029$, which is indeed small; thus, the necessary condition for the accuracy of normal approximation is satisfied. Although a small coefficient of variation suggests that the normal approximation will be accurate, it is no guarantee of such accuracy; that is, there are positive random variables having mean μ and standard deviation σ with $\sigma/\mu \approx 0$ whose distributions are not accurately approximated by the normal distribution with mean μ and variance σ^2. When the coefficient of variation is not small, it is likely that normal approximation to the distribution of $\log Y$ is more accurate than normal approximation to the distribution of Y itself. (If $\log Y$ has exactly a normal distribution, then Y is said to have a *lognormal* distribution.)

THEOREM 3.8 Let Y_1 and Y_2 be independent random variables such that Y_1 is normally distributed with mean μ_1 and variance σ_1^2 and Y_2 is normally distributed with mean μ_2 and variance σ_2^2; let b_1 and b_2 be real numbers at least one of which is nonzero. Then $b_1 Y_1 + b_2 Y_2$ is normally distributed with mean $b_1 \mu_1 + b_2 \mu_2$ and variance $b_1^2 \sigma_1^2 + b_2^2 \sigma_2^2$.

Proof Now $Z_1 = (Y_1 - \mu_1)/\sigma_1$ and $Z_2 = (Y_2 - \mu_2)/\sigma_2$ are independent, standard normal random variables. Also,

$$b_1 Y_1 + b_2 Y_2 = b_1(\mu_1 + \sigma_1 Z_1) + b_2(\mu_2 + \sigma Z_2)$$
$$= b_1 \mu_1 + b_2 \mu_2 + b_1 \sigma_1 Z_1 + b_2 \sigma_2 Z_2,$$

so

$$b_1 Y_1 + b_2 Y_2 = b_1 \mu_1 + b_2 \mu_2 + \sqrt{b_1^2 \sigma_1^2 + b_2^2 \sigma_2^2}\,(b_1' Z_1 + b_2' Z_2), \tag{23}$$

where

$$b_1' = \frac{b_1 \sigma_1}{\sqrt{b_1^2 \sigma_1^2 + b_2^2 \sigma_2^2}} \quad \text{and} \quad b_2' = \frac{b_2 \sigma_2}{\sqrt{b_1^2 \sigma_1^2 + b_2^2 \sigma_2^2}}. \tag{24}$$

Observe that $(b_1')^2 + (b_2')^2 = 1$. Thus, by Theorem 5.24 in Section 5.7, $b_1' Z_1 + b_2' Z_2$ has the standard normal distribution. The desired result now follows from (23) and Theorem 3.7. ∎

COROLLARY 3.3 Let Y_1 and Y_2 be independent random variables such that Y_1 is normally distributed with mean μ_1 and variance σ_1^2 and Y_2 is normally distributed with mean μ_2 and variance σ_2^2. Then

(a) $Y_1 + Y_2$ is normally distributed with mean $\mu_1 + \mu_2$ and variance $\sigma_1^2 + \sigma_2^2$;

(b) $Y_1 - Y_2$ is normally distributed with mean $\mu_1 - \mu_2$ and variance $\sigma_1^2 + \sigma_2^2$.

THEOREM 3.9 Let $Y_1, \ldots, Y_n$ be independent random variables such that Y_i is normally distributed with mean μ_i and variance σ_i^2 for $1 \leq i \leq n$, and let $b_1, \ldots, b_n$ be real numbers at least one of which is nonzero. Then $b_1 Y_1 + \cdots + b_n Y_n$ is normally distributed with mean $b_1 \mu_1 + \cdots + b_n \mu_n$ and variance $b_1^2 \sigma_1^2 + \cdots + b_n^2 \sigma_n^2$.

Proof After deleting all values of i, if any, such that $b_i = 0$, we can assume that $b_1, \ldots, b_n$ are all nonzero. When $n = 1$, the desired result now reduces to Theorem 3.7 and, when $n = 2$, it reduces to Theorem 3.8. Suppose $n = 3$. Observe that $b_1 Y_1 + b_2 Y_2$ and $b_3 Y_3$ are independent random variables. By Theorem 3.8, $b_1 Y_1 + b_2 Y_2$ is normally distributed with mean $b_1 \mu_1 + b_2 \mu_2$ and variance $b_1^2 \sigma_1^2 + b_2^2 \sigma_2^2$. By Theorem 3.7, $b_3 Y_3$ is normally distributed with mean $b_3 \mu_3$ and variance $b_3^2 \sigma_3^2$. Thus, by Corollary 3.3(a), $b_1 Y_1 + b_2 Y_2 + b_3 Y_3$ is normally distributed with mean $b_1 \mu_1 + b_2 \mu_2 + b_3 \mu_3$ and variance $b_1^2 \sigma_1^2 + b_2^2 \sigma_2^2 + b_3^2 \sigma_3^2$. Therefore, the desired result holds when $n = 3$. The result for general n follows by induction.   ∎

Scores on standard tests of ability are approximately normally distributed. This could be so because ability itself is approximately normally distributed. Ability, as usually defined, however, is a qualitative notion, and it only makes sense to talk about the distribution of a particular quantitative measure of ability. Whether the distribution of scores on a particular test of ability is approximately normally distributed depends on the spread of difficulty of the various tasks in the test and on how the test is scored. The most likely explanation for the approximate normality of test scores is that test makers construct tests with approximate normality as one of their goals. (A test that purports to measure a specific ability may or may not be a "valid" test of that ability. One reason that tests of human intelligence are controversial is that it is not clear how to determine the validity of such tests.)

EXAMPLE **3.5**   Suppose the score of a random woman on a certain test of physical strength is normally distributed with mean 140 and standard deviation 30 and that the score of a random man on the same test is normally distributed with mean 160 and standard deviation 40. Suppose a random woman and a random man take the test. Find the probabilities of each of the following events:

(a) the woman's score is above the man's score;

(b) their scores differ by less than 20;   **(c)** their average score is above 180.

Solution   Let Y_1 and Y_2, respectively, denote the woman's score and man's score on the test, let $\mu_1 = 140$ and $\sigma_1 = 30$ denote the mean and standard deviation of Y_1, and let $\mu_2 = 160$ and $\sigma_2 = 40$ denote the mean and standard deviation of Y_2. We also make the reasonable assumption that Y_1 and Y_2 are independent. By Corollary 3.3(b), $Y_1 - Y_2$ is normally distributed with mean $\mu_1 - \mu_2 = -20$ and variance $\sigma_1^2 + \sigma_2^2 = 2500$. Thus, $Y_1 - Y_2 = -20 + 50Z$, where Z has the standard normal distribution.

(a) Observe that

$$P(Y_1 > Y_2) = P(Y_1 - Y_2 > 0)$$

$$= P(-20 + 50Z > 0) = P(Z > .4) = 1 - \Phi(.4).$$

Now $\Phi(.4) \doteq .655$, so $P(Y_1 > Y_2) \doteq 1 - .655 = .345$.

(b) Since

$$P(|Y_1 - Y_2| < 20) = P(|-20 + 50Z| < 20)$$

$$= P(-20 < -20 + 50Z < 20)$$

$$= P(0 < Z < .8),$$

we see that

$$P(|Y_1 - Y_2| < 20) = \Phi(.8) - \Phi(0) \doteq .788 - .5 = .288.$$

(c) By Theorem 3.8, $(Y_1 + Y_2)/2 = .5Y_1 + .5Y_2$ is normally distributed with mean $(.5)(140) + (.5)(160) = 150$, variance $(.25)(30)^2 + (.25)(40)^2 = 625$, and standard deviation $\sqrt{625} = 25$. Thus, $(Y_1 + Y_2)/2 = 150 + 25Z$, where Z has the standard normal distribution. Consequently,

$$P\left(\frac{Y_1 + Y_2}{2} \geq 180\right) = P(150 + 25Z \geq 180)$$

$$= P(Z \geq 1.2)$$

$$= 1 - \Phi(1.2)$$

$$\doteq 1 - .885$$

$$= .115. \quad \blacksquare$$

COROLLARY 3.4 Let $Y_1, \ldots, Y_n$ be independent random variables, each normally distributed with mean μ and variance σ^2, and set $\bar{Y} = (Y_1 + \cdots + Y_n)/n$. Then $Y_1 + \cdots + Y_n$ is normally distributed with mean $n\mu$ and variance $n\sigma^2$, $\bar{Y}$ is normally distributed with mean μ and variance σ^2/n,

$$\frac{Y_1 + \cdots + Y_n - n\mu}{\sqrt{n}} = \frac{\bar{Y} - \mu}{1/\sqrt{n}}$$

is normally distributed with mean 0 and variance σ^2, and

$$\frac{Y_1 + \cdots + Y_n - n\mu}{\sigma\sqrt{n}} = \frac{\bar{Y} - \mu}{\sigma/\sqrt{n}}$$

has the standard normal distribution.

THEOREM 3.10 Let W be normally distributed with mean 0 and variance σ^2. Then $Y = W^2$ has the gamma distribution with shape parameter $1/2$ and scale parameter $2\sigma^2$.

Proof According to Theorem 1.7 in Section 1.7, Y has the density function f_Y given by $f_Y(y) = 0$ for $y \leq 0$ and

$$f_Y(y) = \frac{1}{\sqrt{y}} f_W(\sqrt{y}) = \frac{1}{\sigma\sqrt{2\pi y}} \exp(-y/(2\sigma^2)), \qquad y > 0.$$

Since $\Gamma(\frac{1}{2}) = \sqrt{\pi}$, we see that f_Y is the density function of the indicated gamma distribution. $\quad \blacksquare$

THEOREM 3.11 Let $W_1, \ldots, W_\nu$ be independent random variables having a common normal distribution with mean 0 and variance σ^2. Then $W_1^2 + \cdots + W_\nu^2$ has the gamma distribution with shape parameter $\nu/2$ and scale parameter $2\sigma^2$.

Proof   Now $W_1^2, \ldots, W_\nu^2$ are independent random variables and, by Theorem 3.10, W_i^2 has the gamma distribution with shape parameter $1/2$ and scale parameter $2\sigma^2$ for $1 \le i \le \nu$. The desired conclusion now follows from Theorem 3.6 in Section 3.1.   ∎

COROLLARY 3.5   Let W_1 and W_2 be independent random variables each normally distributed with mean 0 and variance σ^2. Then $W_1^2 + W_2^2$ has the exponential distribution with mean $2\sigma^2$.

EXAMPLE 3.6   Let W_1 and W_2 be independent random variables, each normally distributed with mean 0 and variance σ^2, and set $S = \sqrt{W_1^2 + W_2^2}$. Determine

(a) the density function of S;   (b) the mean and variance of S.

Solution   (a) According to Corollary 3.5 and the solution to Example 1.53 in Section 1.7, the density function of S is given by $f_S(s) = 0$ for $s \le 0$ and

$$f_S(s) = \frac{s}{\sigma^2} \exp(-s^2/2\sigma^2), \qquad s > 0.$$

(b) According to the answer to Problem 3.4 in Section 3.1, S has mean $\sigma\sqrt{\pi/2}$ and variance $\sigma^2(2 - \pi/2)$.   ∎

Problems

3.14   Determine the probability that a normally distributed random variable with mean 1 and variance 4 is positive.

3.15   Let Y be normally distributed with mean μ and variance σ^2. Determine c such that
(a) $P(Y < \mu + c\sigma) = .95$;   (b) $P(|Y - \mu| < c\sigma) = .9$;   (c) $P(|Y - \mu| < c\sigma) = .5$;
(d) $P(|Y - \mu| \ge c\sigma) = .01$.

3.16   Suppose measurements of lengths are unbiased and independently and normally distributed with standard deviation 1 inch. If two tables differ by exactly 1 inch in length, what is the probability that the measurements of their lengths will differ by more than 1 inch?

3.17   A certain elevator has a rated capacity of 1000 pounds. Suppose the distribution of the weight in pounds of a randomly selected individual wanting to use the elevator is normally distributed with mean 150 and standard deviation 20. (a) Determine the distribution of the total weight of six such (independently and) randomly selected individuals. (b) Determine the probability that this total weight exceeds the rated capacity of the elevator.

3.18   Let Y_1, Y_2, Y_3, Y_4, Y_5 be independent random variables, each having the normal distribution with mean 1 and variance 1. Determine $P(Y_1 + Y_2 + Y_3 \ge Y_4 + Y_5)$.

3.19   Let $Y_1, Y_2,$ and Y_3 be independent, normally distributed random variables having means 1, 2, and 3, respectively, and variances 4, 5, and 6, respectively. Determine $P(Y_1 + 2Y_2 + 3Y_3 > 3Y_1 - 2Y_2 + Y_3)$.

3.20 Let Z have the standard normal distribution and let $\mu \in \mathbb{R}$ and $\sigma > 0$. Then $X = \mu + \sigma Z$ is normally distributed with mean μ and variance σ^2. Determine **(a)** $E(Z^3)$; **(b)** $E(Z^4)$; **(c)** $\text{var}(Z^2)$; **(d)** $E(X^2)$; **(e)** $E(X^3)$ (the *third moment* of X); **(f)** $E(X^4)$ (the *fourth moment* of X); **(g)** $\text{var}(X^2)$.

3.21 Let (W_1, W_2, W_3) denote the velocity vector of a randomly selected molecule of a homogeneous gas. Maxwell showed that W_1, W_2, and W_3 are independent, normally distributed random variables, each having mean 0 and variance σ^2. The common variance σ^2 is proportional to T/m, where T is the temperature of the gas and m is its molecular weight. The speed of the randomly selected molecule is given by $S = (W_1^2 + W_2^2 + W_3^2)^{\frac{1}{2}}$. Determine **(a)** the distribution of $S^2 = W_1^2 + W_2^2 + W_3^2$; **(b)** the density function of S; **(c)** the mean of S; **(d)** the variance of S.

3.22 Let W be normally distributed with mean μ and variance σ^2. Then $Y = e^W$ has a lognormal distribution. Determine the corresponding **(a)** distribution function; **(b)** density function; **(c)** pth quantile.

3.23 Let Y have a lognormal distribution. Show that bY^c has a lognormal distribution for $b > 0$ and $c \neq 0$.

3.24 Let $Y_1, \ldots, Y_n$ be independent random variables, each having a lognormal distribution. Show that $Y_1 \cdots Y_n$ has a lognormal distribution.

3.25 Let W_1 and W_2 be independent random variables, each normally distributed with mean 0 and variance σ^2. Determine the median of $\sqrt{W_1^2 + W_2^2}$.

3.26 Show that

$$1 - \Phi(z) < \frac{1}{z\sqrt{2\pi}} e^{-z^2/2}, \qquad z > 0.$$

[*Hint:* Observe first that

$$1 - \Phi(z) = \frac{1}{\sqrt{2\pi}} \int_z^\infty e^{-t^2/2}\, dt < \frac{1}{\sqrt{2\pi}} \int_z^\infty \frac{t}{z} e^{-t^2/2}\, dt, \qquad z > 0.]$$

3.27 Show that

$$1 - \Phi(z) > \frac{1}{\sqrt{2\pi}} \left(\frac{1}{z} - \frac{1}{z^3} \right) e^{-z^2/2}, \qquad z > 0.$$

[*Hint:* Observe first that, for $z > 0$,

$$1 - \Phi(z) = \frac{1}{\sqrt{2\pi}} \int_z^\infty \frac{1}{t} \, d(-e^{-t^2/2})$$

$$= \frac{1}{z\sqrt{2\pi}} e^{-z^2/2} - \frac{1}{\sqrt{2\pi}} \int_z^\infty \frac{1}{t^2} e^{-t^2/2}\, dt$$

$$> \frac{1}{z\sqrt{2\pi}} e^{-z^2/2} - \frac{1}{\sqrt{2\pi}} \int_z^\infty \frac{t}{z^3} e^{-t^2/2}\, dt.]$$

3.28 Use the inequalities in Problems 3.26 and 3.27 to get lower and upper bounds to $1 - \Phi(4)$.

3.3

Normal Approximation and the Central Limit Theorem

Normal Approximation to the Gamma Distribution

Consider the gamma distribution with shape parameter α and scale parameter β, which has mean $\mu = \alpha\beta$, variance $\sigma^2 = \alpha\beta^2$, and standard deviation $\sigma = \beta\sqrt{\alpha}$. Let f denote the density function of the distribution, F its distribution function, and y_p its pth quantile. Then

$$f(y) = \frac{y^{\alpha-1}}{\beta^\alpha \Gamma(\alpha)} e^{-y/\beta}, \qquad y > 0,$$

and $f(y) = 0$ for $y < 0$. Using Stirling's formula to approximate the gamma function, we can show that if the shape parameter is large, then

$$f(y) \approx \frac{1}{\sigma\sqrt{2\pi}} e^{-(y-\mu)^2/2\sigma^2} = \frac{1}{\sigma}\phi\left(\frac{y-\mu}{\sigma}\right), \qquad y \in \mathbb{R},$$

and hence

$$F(y) = \int_{-\infty}^{y} f(t)\, dt \approx \int_{-\infty}^{y} \frac{1}{\sigma}\phi\left(\frac{t-\mu}{\sigma}\right) dt = \int_{-\infty}^{(y-\mu)/\sigma} \phi(z)\, dz = \Phi\left(\frac{y-\mu}{\sigma}\right)$$

for $y \in \mathbb{R}$. Let $0 < p < 1$. It follows from the latter result that if the shape parameter is large, then $y_p \approx \mu + \sigma z_p$. In summary, the normal approximations to the density function, distribution function, and quantiles of the gamma distribution are accurate when the shape parameter of the distribution is large.

Consider, in particular, the gamma distribution with shape parameter 10 and scale parameter 1. Figure 3.7 shows graphs of the density function and its normal approximation. Figure 3.8 shows graphs of the distribution function and its normal approximation.

Let $W_1, \ldots, W_n$ be independent random variables, each having the exponential distribution with mean β. Then $W_1 + \cdots + W_n$ has the gamma distribution with shape parameter n and scale parameter β and hence mean $n\beta$, variance $n\beta^2$, and standard deviation $\beta\sqrt{n}$. Thus, the normal approximations to the density function, distribution function, and quantiles of this random variable are accurate when n is large.

FIGURE   3.7

Gamma density function with shape parameter 10 and scale parameter 1 and its normal approximation (dashed curve)

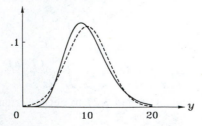

FIGURE 3.8

Gamma distribution function with shape parameter 10 and scale parameter 1 and its normal approximation (dashed curve)

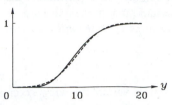

EXAMPLE 3.7 Let $Y_1, \ldots, Y_{10}$ denote the failure times of ten widgets, which are independent and exponentially distributed with mean 1, and let $\bar{Y} = (Y_1 + \cdots + Y_{10})/10$ denote their sample mean. Use the normal approximation to the gamma distribution to determine

(a) $P\left(\bar{Y} < \frac{1}{2}\right)$;

(b) the value of c such that $P(\bar{Y} < c) = .01$;

(c) the value of c such that $P(|\bar{Y} - 1| > c) = .01$.

Solution According to Corollary 3.1 in Section 3.1, $Y_1 + \cdots + Y_{10}$ has the gamma distribution with shape parameter 10 and scale parameter 1. Thus, by Theorem 3.2 in Section 3.1, $\bar{Y}$ has the gamma distribution with shape parameter 10 and scale parameter $1/10$, which has mean 1 and variance $1/10$.

(a) The normal approximation to $P(\bar{Y} < \frac{1}{2})$ is given by

$$P\left(\bar{Y} < \frac{1}{2}\right) \approx \Phi\left(\frac{\frac{1}{2} - 1}{\sqrt{1/10}}\right) \doteq \Phi(-1.581) \doteq .057.$$

(The actual value is given by $P(\bar{Y} < \frac{1}{2}) \doteq .032$.)

(b) The normal approximation to the value of c such that $P(\bar{Y} < c) = .01$ is given by

$$c \approx 1 + \sqrt{1/10}\, z_{.01} \doteq 1 + \sqrt{1/10}\,(-2.326) \doteq 0.264.$$

[The actual value, obtained in the solution to Example 3.2(a) in Section 3.1, is given by $c \doteq 0.413$.]

(c) The normal approximation to $P(|\bar{Y} - 1| > c)$ is given by

$$P(|\bar{Y} - 1| > c) = P(\bar{Y} - 1 < -c) + P(\bar{Y} - 1 > c)$$

$$= P\left(\frac{\bar{Y} - 1}{\sqrt{1/10}} < -\frac{c}{\sqrt{1/10}}\right) + P\left(\frac{\bar{Y} - 1}{\sqrt{1/10}} > \frac{c}{\sqrt{1/10}}\right)$$

$$\approx 2\Phi(-c\sqrt{10}),$$

which equals .01 if and only if $c\sqrt{10} = z_{.995}$ and hence if and only if

$$c = \frac{z_{.995}}{\sqrt{10}} \doteq \frac{2.576}{\sqrt{10}} \doteq 0.8145.$$

[Actually, $P(|\bar{Y} - 1| > 0.8145) \doteq .0142$, while $P(|\bar{Y} - 1| > 0.8783) \doteq .01$.] ∎

Central Limit Theorem

There are many important results in advanced probability theory to the effect that, under appropriate conditions, certain random variables are approximately normally distributed. Such results are known as *central limit theorems*, the most famous of which is Theorem 3.12.

THEOREM 3.12   Let $Y_1, Y_2, \ldots$ be independent random variables having a common distribution with mean μ and finite, positive standard deviation σ. Then

$$\lim_{n \to \infty} P\left(\frac{Y_1 + \cdots + Y_n - n\mu}{\sigma \sqrt{n}} \le z\right) = \Phi(z), \qquad z \in \mathbb{R}. \quad \blacksquare$$

Let $Y_1, Y_2, \ldots$ be as in the statement of the central limit theorem. Given the positive integer n, set $\bar{Y} = (Y_1 + \cdots + Y_n)/n$. The random variable $Y_1 + \cdots + Y_n$ has mean $n\mu$ and standard deviation $\sigma \sqrt{n}$, and the random variable $\bar{Y}$ has mean μ and standard deviation $\sigma/\sqrt{n}$. These random variables have the common standardized random variable Z given by

$$Z = \frac{Y_1 + \cdots + Y_n - n\mu}{\sigma \sqrt{n}} = \frac{\bar{Y} - \mu}{\sigma/\sqrt{n}}.$$

According to the central limit theorem, $P(Z \le z) \approx \Phi(z)$ for $n \gg 1$ and $z \in \mathbb{R}$. This implies that, for $-\infty \le a \le b \le \infty$,

$$P(a \le Z \le b) \approx \Phi(b) - \Phi(a), \qquad n \gg 1,$$

and that this approximation holds with one or both of the inequalities in $P(a \le Z \le b)$ replaced by the corresponding strict inequality. [Here, $\Phi(-\infty) = 0$ and $\Phi(\infty) = 1$.] Another consequence of the central limit theorem is that, for $0 < p < 1$, the pth quantile of Z is approximately equal to the pth quantile z_p of the standard normal distribution when $n \gg 1$. In summary, the normal approximations to the distribution function and quantiles of Z are accurate when n is large.

EXAMPLE 3.8   The weights of the employees at a certain company have mean 150 pounds and standard deviation 20 pounds. The elevator in the company's building has a rated capacity of 1500 pounds. If 9 randomly selected employees get on the elevator at the same time, what is the probability that their combined weight will exceed the capacity of the elevator?

Solution   We assume that the company has many employees (say, 100 or more), so that we can think of the employees as being selected by sampling with replacement. Then their weights are independent and identically distributed. Even though the normal approximation to the distribution of employee weight may not be very accurate, especially in the upper tail, it follows from a combination of the central limit theorem and practical experience, that the normal approximation to the distribution of the sum Y of the weights of the 9 randomly selected employees should be quite satisfactory for the purpose at hand. Now Y has mean $9 \cdot 150 = 1350$ and standard deviation $3 \cdot 20 = 60$. Thus, $Y = 1350 + 60Z$, where $Z = (Y - 1350)/60$ is the standardized random variable corresponding to Y. Consequently,

$$P(Y > 1500) = P(1350 + 60Z > 1500) = P(Z > 2.5) \approx 1 - \Phi(2.5) \doteq .0062. \quad \blacksquare$$

Let $W_1, \ldots, W_n$ be independent random variables such that W_i has mean μ_i and variance σ_i^2 for $1 \le i \le n$, let $b_0, \ldots, b_n \in \mathbb{R}$, and set $Y = b_0 + b_1 W_1 + \cdots + b_n W_n$. Then Y has mean $\mu = b_0 + b_1 \mu_1 + \cdots + b_n \mu_n$ and variance $\sigma^2 = b_1^2 \sigma_1^2 + \cdots + b_n^2 \sigma_n^2$. It can be shown that if W_i is approximately normally distributed with mean μ_i and variance σ_i^2 for $1 \le i \le n$, then Y is approximately normally distributed with mean μ and variance σ^2. In particular, if W_1 and W_2 are approximately normally distributed with the indicated means and variances, then $W_2 - W_1$ is approximately normally distributed with mean $\mu_2 - \mu_1$ and variance $\sigma_1^2 + \sigma_2^2$.

EXAMPLE 3.9   Let $X_1, \ldots, X_{50}, Y_1, \ldots, Y_{100}$ be independent random variables, each having an exponential distribution with mean 1, and set $\bar{X} = (X_1 + \cdots + X_{50})/50$ and $\bar{Y} = (Y_1 + \cdots + Y_{100})/100$. Use normal approximation to determine the value of c such that $P(|\bar{Y} - \bar{X}| > c) \approx .01$.

Solution   The random variables $\bar{X}$ and $\bar{Y}$ are independent. Moreover, $\bar{X}$ has mean 1 and variance $1/50$, $\bar{Y}$ has mean 1 and variance $1/100$, and $\bar{X}$ and $\bar{Y}$ are approximately normally distributed with the indicated means and variances. Thus, $\bar{Y} - \bar{X}$ is approximately normally distributed with mean 0 and variance

$$\sigma^2 = \frac{1}{50} + \frac{1}{100} = \frac{3}{100}.$$

Therefore, $P(|\bar{Y} - \bar{X}| > \sigma z_{.995}) \approx .01$, so the desired result holds with

$$c = \sigma z_{.995} \doteq \frac{\sqrt{3}}{10}(2.576) \doteq 0.446. \quad \blacksquare$$

EXAMPLE 3.10   Let X and Y be independent random variables having a common gamma distribution with shape parameter α. Use normal approximation to determine α such that $P(Y \ge 1.2X) \approx .1$.

Solution   Let β denote the scale parameter of the gamma distribution. Then $Y - 1.2X$ has mean $-0.2\alpha\beta$ and variance $\alpha\beta^2(1.2^2 + 1) = 2.44\alpha\beta^2$, so

$$P(Y \ge 1.2X) = P(Y - 1.2X \ge 0)$$

$$\approx 1 - \Phi\left(\frac{0.2\alpha\beta}{\beta\sqrt{2.44\alpha}}\right)$$

$$= 1 - \Phi\left(\frac{1}{5\sqrt{2.44}}\sqrt{\alpha}\right).$$

Thus, $P(Y \geq 1.2X) \approx .1$ if

$$\frac{1}{5\sqrt{2.44}}\sqrt{\alpha} = z_{.9}$$

or, equivalently, if $\alpha = 25(2.44)z_{.9}^2 \doteq 61(1.282)^2 \doteq 100.2$.   ∎

Problems

3.29 Consider ten independent random variables having a common exponential distribution. Use normal approximation to determine the probability that the sum of these random variables is less than one-half of its mean.

3.30 A firm buys 500 identical components whose failure times are independent and exponentially distributed with mean 100 hours. Use normal approximation to determine the probability that the sample mean of the 500 failure times will exceed 90 hours.

3.31 Let Y have a gamma distribution with shape parameter α. Use normal approximation to determine α so that $P(Y > 1.1EY) = .1$.

3.32 Let $Z_1, \ldots, Z_{50}$ be independent random variables, each having the standard normal distribution. Determine **(a)** the distribution of $Y = Z_1^2 + \cdots + Z_{50}^2$; **(b)** the mean and variance of this distribution; **(c)** the normal approximation to the probability that $45 \leq Y \leq 60$.

3.33 In the context of Example 3.8, use normal approximation to determine the probability that the average weight of the nine employees differs from the mean weight of all the company's employees by less than 10 pounds.

3.34 Four hundred numbers are each rounded off to the nearest multiple of 10^{-7} before being added together. Suppose the roundoff errors are independent and uniformly distributed on $[-0.5 \times 10^{-7}, 0.5 \times 10^{-7}]$. Use normal approximation to determine the probability that the sum of the rounded-off values is within 0.5×10^{-6} of the sum of the original numbers.

3.35 Let $Y_1, \ldots, Y_n$ be a random sample of size n from a distribution having mean μ and finite, positive variance σ^2 and let $\bar{Y}$ denote the corresponding sample mean. Determine the smallest value of n such that $P(|\bar{Y} - \mu| \geq \sigma/10) \leq .05$ by using **(a)** Chebyshev's inequality; **(b)** normal approximation.

3.36 Consider 100 independent random plots of land, each distributed as in Problem 2.42 in Section 2.3. Determine the probability that the sum of the areas of the 100 plots exceeds 120 square miles.

3.37 A runner, believing that each of his paces is 1 meter long, counts off 100 paces for a 100-meter race. Suppose his paces are independently distributed with mean 96 cm and standard deviation 5 cm. Determine the probability that the actual length of the race differs from 100 meters by less than 5 meters.

3.38 Let X and Y be independent random variables having gamma distributions with shape parameter $\alpha \gg 1$ and suppose that X has scale parameter β_1 and Y has scale parameter β_2. Determine the normal approximation to $P(Y < cX)$ for $c > 0$.

3.39 A manufacturer orders 100 identical components from each of two suppliers. Suppose the failure times of the 200 components are independently and exponentially distributed and that those from the first supplier have mean β_1 and those from the second supplier have mean β_2. Let $\bar{X}$ denote the sample mean of the failure times of the 100 components from the first supplier and let $\bar{Y}$ denote the sample mean corresponding to the components from the second supplier. Determine the normal approximation to $P(\bar{X}/1.3 \le \bar{Y} \le 1.3\bar{X})$ **(a)** if $\beta_1 = \beta_2$ (see Problem 3.38 above and Problem 1.115 in Section 1.8); **(b)** if $\beta_2 = 1.5\beta_1$.

3.40 Let $U_1, \ldots, U_n$ be independent random variables, each uniformly distributed on $(0, 1)$, and let $c > 0$. Use the central limit theorem to show that

$$P(\exp(-n - c\sqrt{n}) \le U_1 \cdots U_n \le \exp(-n + c\sqrt{n})) \approx 2\Phi(c) - 1, \qquad n \gg 1.$$

Special Discrete Models

4.1
Combinatorics

Recall that #(A) denotes the size (cardinality, number of members) of a finite set A. If A has small size (for example, if A is the collection of prime numbers less than 10), it is easy to determine the size of A by first listing its members and then counting them one by one. If A has large size, however (for example, if A is the collection of prime numbers less than 10^{100}), it might not be feasible to count the members of A in this manner. Also, the definition of A may involve various parameters, and what may be needed is a formula for the size of A in terms of these parameters, which may be difficult to obtain. Fortunately, the counting problems that arise in this book are easily handled by means of a few elementary rules. *Combinatorics*, the mathematical subject dealing with counting rules, contains many results that are simply and elegantly stated but that require considerable cleverness to prove.

Let A_1 and A_2 be finite sets. Then they have the same size if and only if there is a one-to-one correspondence between them; that is, if and only if there is a one-to-one function from A_1 onto A_2. Let $A_1, \ldots, A_y$ be finite sets. Then the size of their union equals the sum of their sizes if and only if these sets are disjoint. Recall that the Cartesian product $A_1 \times \cdots \times A_y$ of these sets is the collection of ordered y-tuples

$(\mathbf{a}_1, \ldots, \mathbf{a}_y)$, where $\mathbf{a}_1 \in A_1, \ldots, \mathbf{a}_y \in A_y$. The size of this Cartesian product is the product of the individual sizes; that is, $\#(A_1 \times \cdots \times A_y) = \#(A_1) \cdots \#(A_y)$.

Consider the set $\{1, 2, 3, 4\}$ of size 4. There are $4^2 = 16$ ordered pairs of members of this set (Table 4.1). There are $4 \cdot 3 = 12$ ordered pairs of distinct members of $\{1, 2, 3, 4\}$ (Table 4.2). More generally, we have the result stated in Theorem 4.1.

TABLE 4.1

(1,1)	(1,2)	(1,3)	(1,4)
(2,1)	(2,2)	(2,3)	(2,4)
(3,1)	(3,2)	(3,3)	(3,4)
(4,1)	(4,2)	(4,3)	(4,4)

TABLE 4.2

	(1,2)	(1,3)	(1,4)
(2,1)		(2,3)	(2,4)
(3,1)	(3,2)		(3,4)
(4,1)	(4,2)	(4,3)	

THEOREM 4.1 Let n and y be positive integers and consider a set of size n.

(a) There are n^y ordered y-tuples of members of the set.

(b) There are $n \cdots (n - y + 1) = n!/(n - y)!$ ordered y-tuples of distinct members of the set (which equals zero if $y > n$).

Proof Let A denote the set of size n.

(a) The desired result is obviously true when $y = 1$. Suppose next that $y = 2$. Then the collection of ordered pairs of members of A can be written as $A \times A$, which has size $\#(A)\#(A) = n^2$. Suppose next that $y = 3$. Then the collection of ordered triples of members of A can be written as $A \times A \times A$, which has size $\#(A)\#(A)\#(A) = n^3$. A similar argument works for larger values of y.

(b) The desired result is obviously true when $y = 1$. Suppose that $y = 2$. To determine the number of members of the set $\{(\mathbf{a}_1, \mathbf{a}_2) : \mathbf{a}_1, \mathbf{a}_2 \in A$ and $\mathbf{a}_2 \neq \mathbf{a}_1\}$, we observe that, for each of the n choices of $\mathbf{a}_1 \in A$, there are $n - 1$ choices of $\mathbf{a}_2 \in A$ such that $\mathbf{a}_2 \neq \mathbf{a}_1$. Thus, the indicated set has $n(n - 1)$ members as desired. Suppose next that $y = 3$. To determine the number of members of the set

$$\{(\mathbf{a}_1, \mathbf{a}_2, \mathbf{a}_3) : \mathbf{a}_1, \mathbf{a}_2, \mathbf{a}_3 \in A, \ \mathbf{a}_2 \neq \mathbf{a}_1 \text{ and } \mathbf{a}_3 \notin \{\mathbf{a}_1, \mathbf{a}_2\}\},$$

we observe that, for each of the $n(n - 1)$ choices of $\mathbf{a}_1, \mathbf{a}_2 \in A$ such that $\mathbf{a}_2 \neq \mathbf{a}_1$, there are $n - 2$ choices of $\mathbf{a}_3 \in A$ such that $\mathbf{a}_3 \notin \{\mathbf{a}_1, \mathbf{a}_2\}$. Thus, the indicated set has $n(n - 1)(n - 2)$ members as desired. A similar argument works for larger values of y. (Observe that if $y > n$, then there are no ordered y-tuples of distinct members of A.)   ∎

EXAMPLE **4.1**   A box contains n tickets, labeled from 1 to n. Let y tickets be selected from the box, one at a time, at random, and with replacement of each ticket before the next one is selected. Determine the probability that y different tickets are selected on the y trials.

Solution   Let L_i denote the label of the ticket selected on the ith trial for $1 \le i \le y$. We assume that $L_1, \ldots, L_y$ are independent random variables, each uniformly distributed on $\{1, \ldots, n\}$ or, equivalently, that $(L_1, \ldots, L_y)$ is uniformly distributed on the set of ordered y-tuples of members of $\{1, \ldots, n\}$, which has size n^y by Theorem 4.1(a). Observe that y different tickets are selected on the y trials if and only if $(L_1, \ldots, L_y)$ is a member of the set of ordered y-tuples of distinct members of $\{1, \ldots, n\}$, which has size $n \cdots (n - y + 1)$ by Theorem 4.1(b). Thus, the desired probability equals

$$\frac{n \cdots (n - y + 1)}{n^y} = \left(1 - \frac{1}{n}\right) \cdots \left(1 - \frac{y-1}{n}\right). \quad ∎$$

EXAMPLE **4.2**   Should a student in a class of 25 students be surprised upon learning that 2 students in the class have the same birthday?

Solution   More generally, let $y \ge 2$ denote the number of students in the class. We can think of these students as being labeled from 1 to y in some order. Suppose, for simplicity, that there are 365 days in a year, which we think of as being labeled from 1 to 365 in the natural order. Then the y birthdays form an ordered y-tuple of (not necessarily distinct) members of $\{1, \ldots, 365\}$. Suppose the y-tuple of birthdays for the class in question is uniformly distributed over the collection of all such ordered y-tuples. Then, according to the solution to Example 4.1, the probability p_y that no two students in the class have the same birthday is given by

$$p_y = \left(1 - \frac{1}{365}\right) \cdots \left(1 - \frac{y-1}{365}\right).$$

The (approximate) values of p_y for $y = 5, 10, \ldots, 50$ are as follows:

y	5	10	15	20	25	30	35	40	45	50
p_y	.973	.883	.747	.589	.431	.294	.186	.109	.059	.030

In particular, for $y = 25$, the probability is .431 that no two students in the class have the same birthday. Thus, the probability is $1 - .431 = .569$ that there are two students in the class having the same birthday; hence, a student in a class of size 25 should not be surprised upon learning this fact.   ∎

Let $(\mathbf{a}_1, \ldots, \mathbf{a}_y)$ be an ordered y-tuple whose entries are all the members of a set of size y. Then every member of the set appears exactly once as an entry of

$(\mathbf{a}_1, \ldots, \mathbf{a}_y)$. We refer to such an ordered y-tuple as a *permutation* of the set. According to Theorem 4.1(b), there are $y!$ permutations of the members of a set of size y.

EXAMPLE **4.3**   Determine the permutations of the members of $\{1, 2, 3\}$.

Solution   There are $3! = 6$ such permutations: $(1, 2, 3)$, $(1, 3, 2)$, $(2, 1, 3)$, $(2, 3, 1)$, $(3, 1, 2)$, $(3, 2, 1)$.   ∎

Consider the set $\{1, 2, 3, 4, 5\}$ of size 5. By Theorem 4.1(b), the collection of ordered triples of distinct members of this set has size $5!/(5-3)! = 5!/2! = 120/2 = 60$. Consider, in particular, the 6 ordered such triples $(1,2,3)$, $(1,3,2)$, $(2,1,3)$, $(2,3,1)$, $(3,1,2)$, $(3,2,1)$, which together make up the collection of all $3! = 6$ permutations of the members of the subset $B = \{1, 2, 3\}$ of $\{1, 2, 3, 4, 5\}$ of size 3. By letting B be any other subset of $\{1, 2, 3, 4, 5\}$ of size 3, we get 6 different ordered triples of distinct members of $\{1, 2, 3, 4, 5\}$. In this manner, by letting B range over all subsets of $\{1, 2, 3, 4, 5\}$ of size 3, we get all 60 ordered triples of distinct members of this set, as shown in Table 4.3. Observe that $60 = 10 \cdot 6$, where 60 is the number of ordered triples of distinct members of $\{1, 2, 3, 4, 5\}$, 10 is the number of subsets of $\{1, 2, 3, 4, 5\}$ of size 3, and 6 is the number of permutations of the members in each such subset.

Given a positive integer n and a nonnegative integer $y \le n$, let $\binom{n}{y}$ denote the number of subsets of $\{1, \ldots, n\}$ of size y. [We have just seen, for example, that $\binom{5}{3} = 10$.] Consider any population of n distinct objects. Then there are $\binom{n}{y}$ ways of choosing y of the objects [so we refer to $\binom{n}{y}$ as "n choose y"]. To verify this result, label the objects from 1 to n in some order. Then the various ways of choosing y of the objects in the population are in one-to-one correspondence with the various subsets of $\{1, \ldots, n\}$ of size y. A set consisting of certain objects is sometimes referred to as a *combination* of these objects. Correspondingly, we refer to $\binom{n}{y}$ as the number of combinations of n objects taken y at a time.

TABLE   **4.3**

B	Ordered Triples					
$\{1, 2, 3\}$	(1,2,3)	(1,3,2)	(2,1,3)	(2,3,1)	(3,1,2)	(3,2,1)
$\{1, 2, 4\}$	(1,2,4)	(1,4,2)	(2,1,4)	(2,4,1)	(4,1,2)	(4,2,1)
$\{1, 2, 5\}$	(1,2,5)	(1,5,2)	(2,1,5)	(2,5,1)	(5,1,2)	(5,2,1)
$\{1, 3, 4\}$	(1,3,4)	(1,4,3)	(3,1,4)	(3,4,1)	(4,1,3)	(4,3,1)
$\{1, 3, 5\}$	(1,3,5)	(1,5,3)	(3,1,5)	(3,5,1)	(5,1,3)	(5,3,1)
$\{1, 4, 5\}$	(1,4,5)	(1,5,4)	(4,1,5)	(4,5,1)	(5,1,4)	(5,4,1)
$\{2, 3, 4\}$	(2,3,4)	(2,4,3)	(3,2,4)	(3,4,2)	(4,2,3)	(4,3,2)
$\{2, 3, 5\}$	(2,3,5)	(2,5,3)	(3,2,5)	(3,5,2)	(5,2,3)	(5,3,2)
$\{2, 4, 5\}$	(2,4,5)	(2,5,4)	(4,2,5)	(4,5,2)	(5,2,4)	(5,4,2)
$\{3, 4, 5\}$	(3,4,5)	(3,5,4)	(4,3,5)	(4,5,3)	(5,3,4)	(5,4,3)

EXAMPLE 4.4   Consider a population of five balls. Determine the number of ways of painting three of the balls red and two of them white.

Solution   There are $\binom{5}{3} = 10$ ways of choosing which three of the balls to paint red, with the other two balls being painted white.   ∎

THEOREM 4.2   Let n be a positive integer. Then

$$\binom{n}{y} = \frac{n!}{y!\,(n-y)!}, \qquad y = 0, 1, \ldots, n.$$

Proof   Since the empty set is the only subset of $\{1, \ldots, n\}$ of size 0 and

$$\frac{n!}{0!\,n!} = 1,$$

the desired result is true when $y = 0$. Suppose now that $1 \le y \le n$. Then each of the subsets B of $\{1, \ldots, n\}$ of size y corresponds to $y!$ ordered y-tuples of distinct members of $\{1, \ldots, n\}$—namely, the $y!$ permutations of B. In this manner, as B ranges over the $\binom{n}{y}$ subsets of $\{1, \ldots, n\}$ of size y, we get all ordered y-tuples of distinct members of $\{1, \ldots, n\}$ with no such y-tuple occurring more than once. Consequently, by Theorem 4.1(b),

$$\binom{n}{y} y! = \frac{n!}{(n-y)!},$$

and hence

$$\binom{n}{y} = \frac{n!}{y!\,(n-y)!}.   ∎$$

In accordance with Theorem 3.2 and Equation (1) in Section 3.1, we make the convention that $\binom{n}{y}$ equals zero for every integer y that is negative or greater than n. This is compatible with the fact that there are no subsets of $\{1, \ldots, n\}$ of size y if y is negative or greater than n.

EXAMPLE 4.5   Simplify each of the following (where n is a positive integer):

(a) $\binom{n}{0}$;   **(b)** $\binom{n}{1}$;   **(c)** $\binom{n}{2}$;   **(d)** $\binom{n}{n-1}$;   **(e)** $\binom{n}{n-y}$;   **(f)** $\binom{n}{n}$.

Solution   **(a)** $\binom{n}{0} = \frac{n!}{0!\,n!} = 1$;   **(b)** $\binom{n}{1} = \frac{n!}{1!\,(n-1)!} = n$;

(c) $\binom{n}{2} = \frac{n!}{2!\,(n-2)!} = \frac{n(n-1)}{2}$;   **(d)** $\binom{n}{n-1} = \frac{n!}{(n-1)!\,1!} = n$;

(e) $\binom{n}{n-y} = \frac{n!}{(n-y)!\,y!} = \binom{n}{y}$;   **(f)** $\binom{n}{n} = \binom{n}{0} = 1$.   ∎

EXAMPLE 4.6   Recall that there are 12 ordered pairs of distinct members of $\{1, 2, 3, 4\}$, which were shown just prior to Theorem 4.1. Determine the subsets of $\{1, 2, 3, 4\}$ of size 2.

Solution   There are

$$\binom{4}{2} = \frac{4!}{2!\,2!} = \frac{24}{4} = 6$$

subsets of $\{1, 2, 3, 4\}$ of size 2: $\{1, 2\}, \{1, 3\}, \{1, 4\}, \{2, 3\}, \{2, 4\}, \{3, 4\}$. ■

THEOREM 4.3 Let n be a positive integer and let y be an integer with $0 \le y \le n$. Then there are $\binom{n}{y}$-ordered n-tuples of 0's and 1's whose sum equals y.

Proof There is a natural one-to-one correspondence between subsets B of $\{1, \ldots, n\}$ of size y and ordered n-tuples $(i_1, \ldots, i_n)$ of 0's and 1's whose sum equals y, according to which the entry in the lth position of $(i_1, \ldots, i_n)$ equals 1 if and only if l is a member of B. [An example of this correspondence with $n = 5$ and $y = 4$ is given by $\{1, 2, 4, 5\} \sim (1, 1, 0, 1, 1)$.] Since there are $\binom{n}{y}$ subsets of $\{1, \ldots, n\}$ of size y, there are $\binom{n}{y}$ ordered n-tuples of 0's and 1's whose sum is y. ■

THEOREM 4.4 Let n be a positive integer. Then

$$(a + b)^n = \sum_{y=0}^{n} \binom{n}{y} a^y b^{n-y}, \qquad a, b \in \mathbb{R}. \tag{1}$$

Proof Write $(a + b)^n$ as the product of n identical factors: $(a + b)^n = (a + b) \cdots (a + b)$. To obtain a term of the form $a^y b^{n-y}$, where $0 \le y \le n$, we must choose a from y of the factors and b from the remaining factors. There is a one-to-one correspondence between such choices and subsets B of $\{1, \ldots, n\}$ of size y: choose a from the lth factor if $l \in B$; otherwise, choose b from that factor. Since there are $\binom{n}{y}$ such subsets, $\binom{n}{y}$ is the coefficient of $a^y b^{n-y}$ in the expansion of $(a + b)^n$. ■

Equation (1) is known as the *binomial formula*. (In this formula, it is understood that a^0 and b^0 equal 1 even if a or b equals 0.)

EXAMPLE 4.7 Use the binomial formula to expand $(a + b)^4$.

Solution Observe that

$$\frac{4!}{0! \, 4!} = 1, \quad \frac{4!}{1! \, 3!} = 4 \quad \text{and} \quad \frac{4!}{2! \, 2!} = 6.$$

Thus, by the binomial formula, $(a + b)^4 = a^4 + 4a^3 b + 6a^2 b^2 + 4ab^3 + b^4$. ■

EXAMPLE 4.8 Use Stirling's formula to approximate $\binom{2n}{n}$ for $n \gg 1$.

Solution According to Stirling's formula,

$$\binom{2n}{n} = \frac{(2n)!}{(n!)^2} \approx \frac{\sqrt{4\pi n}\, \dfrac{(2n)^{2n}}{e^{2n}}}{2\pi n \dfrac{n^{2n}}{e^{2n}}} = \frac{2^{2n}}{\sqrt{\pi n}}. \qquad ■$$

THEOREM 4.5   Let y_1, y_2, and y_3 be nonnegative integers adding up to the positive integer n. Then there are

$$\frac{n!}{y_1!\, y_2!\, y_3!}$$

ordered triples (B_1, B_2, B_3) of subsets B_1, B_2, and B_3 of $\{1, \ldots, n\}$ such that B_1, B_2, and B_3 form a partition of $\{1, \ldots, n\}$ and these sets have sizes y_1, y_2, and y_3, respectively.

Proof   If $y_1 = n$ and hence $y_2 = y_3 = 0$, the desired result is trivially valid. Suppose, instead, that $y_1 < n$. Then there are

$$\binom{n}{y_1}$$

subsets B_1 of $\{1, \ldots, n\}$ of size y_1. For each such subset, there are

$$\binom{n - y_1}{y_2}$$

subsets B_2 of $\{1, \ldots, n\} \backslash B_1$ of size y_2. Correspondingly,

$$B_1, \ B_2, \ \text{and } B_3 = \{1, \ldots, n\} \backslash (B_1 \cup B_2)$$

form a partition of $\{1, \ldots, n\}$ into sets having sizes y_1, y_2, and y_3, respectively, and every such partition arises uniquely in this manner. Thus, the number of such partitions is

$$\binom{n}{y_1}\binom{n - y_1}{y_2} = \frac{n!}{y_1!\,(n - y_1)!}\,\frac{(n - y_1)!}{y_2!\,(n - y_1 - y_2)!}$$

$$= \frac{n!}{y_1!\, y_2!\,(n - y_1 - y_2)!}$$

$$= \frac{n!}{y_1!\, y_2!\, y_3!}. \qquad \blacksquare$$

EXAMPLE 4.9   Determine the ordered triples (B_1, B_2, B_3) such that B_1, B_2, and B_3 form a partition of $\{1, 2, 3, 4\}$ and these sets have sizes 1, 1, and 2, respectively.

Solution   According to Theorem 4.5 with $n = 4$, $y_1 = 1$, $y_2 = 1$, and $y_3 = 2$, there are

$$\frac{4!}{1!\,1!\,2!} = 12$$

such ordered triples, shown in Table 4.4.   ∎

Consider a population of n balls and let y_1, y_2, and y_3 be nonnegative integers adding up to n. According to Theorem 4.5, there are

$$\frac{n!}{y_1!\, y_2!\, y_3!}$$

TABLE 4.4

B_1	B_2	B_3
{1}	{2}	{3, 4}
{1}	{3}	{2, 4}
{1}	{4}	{2, 3}
{2}	{1}	{3, 4}
{2}	{3}	{1, 4}
{2}	{4}	{1, 3}
{3}	{1}	{2, 4}
{3}	{2}	{1, 4}
{3}	{4}	{1, 2}
{4}	{1}	{2, 3}
{4}	{2}	{1, 3}
{4}	{3}	{1, 2}

ways of painting y_1 of the balls red, y_2 of them white, and y_3 of them blue. (Label the balls from 1 to n in some order and let B_1, B_2, and B_3 denote the labels of the balls that are to be painted red, white, and blue, respectively.)

EXAMPLE **4.10**   Consider a population of ten balls. Determine the number of ways of painting five of the balls red, three of them white, and two of them blue.

Solution   According to Theorem 4.5, the number of ways is

$$\frac{10!}{5!\,3!\,2!} = 2520. \quad \blacksquare$$

THEOREM **4.6**   Let y_1, y_2, and y_3 be nonnegative integers adding up to the positive integer n. Then there are

$$\frac{n!}{y_1!\,y_2!\,y_3!}$$

ordered n-tuples of the integers 1, 2, and 3 such that the integer 1 appears y_1 times, the integer 2 appears y_2 times, and the integer 3 appears y_3 times in the ordered n-tuple.

Proof   Observe that there is a one-to-one correspondence between the indicated ordered n-tuples $(m_1, \dots, m_n)$ and partitions of $\{1, \dots, n\}$ into disjoint sets B_1, B_2, and B_3 having sizes y_1, y_2, and y_3, respectively:

$$B_1 = \{\, l : 1 \le l \le n \text{ and } m_l = 1 \,\};$$

$$B_2 = \{\, l : 1 \le l \le n \text{ and } m_l = 2 \,\};$$

$$B_3 = \{\, l : 1 \le l \le n \text{ and } m_l = 3 \,\}.$$

The desired result now follows from Theorem 4.5. $\blacksquare$

THEOREM 4.7   Let n be a positive integer. Then

$$(a + b + c)^n = \sum_{y_1} \sum_{y_2} \sum_{y_3} \frac{n!}{y_1! \, y_2! \, y_3!} a^{y_1} b^{y_2} c^{y_3}, \qquad a, b, c \in \mathbb{R}, \qquad \text{(2)}$$

where the sum is over the nonnegative integers y_1, y_2, and y_3 adding up to n.

Proof   Write $(a + b + c)^n$ as the product of n identical factors:

$$(a + b + c)^n = (a + b + c) \cdots (a + b + c).$$

Let y_1, y_2, and y_3 be nonnegative integers adding up to n. To obtain a term of the form $a^{y_1} b^{y_2} c^{y_3}$, we must choose a from y_1 of the factors, b from y_2 different factors, and c from the remaining y_3 factors. By Theorem 4.5 or Theorem 4.6, there are

$$\frac{n!}{y_1! \, y_2! \, y_3!}$$

such choices.   ∎

EXAMPLE 4.11   Use the trinomial formula (2) to expand $(a + b + c)^3$.

Solution   Observe that

$$\frac{3!}{3! \, 0! \, 0!} = 1, \qquad \frac{3!}{2! \, 1! \, 0!} = 3, \quad \text{and} \quad \frac{3!}{1! \, 1! \, 1!} = 6.$$

Thus, by the trinomial formula,

$$(a + b + c)^3 = a^3 + b^3 + c^3$$
$$+ \, 3a^2 b + 3ab^2 + 3a^2 c + 3ac^2 + 3b^2 c + 3bc^2 + 6abc. \quad ∎$$

Arguing as in the proof of Theorem 4.5 or using induction, we get the following generalization of this theorem.

THEOREM 4.8   Let $y_1, \ldots, y_M$ be nonnegative integers adding up to the positive integer n. Then there are

$$\frac{n!}{y_1! \cdots y_M!}$$

ordered M-tuples $(B_1, \ldots, B_M)$ such that $B_1, \ldots, B_M$ form a partition of $\{1, \ldots, n\}$ and these sets have sizes $y_1, \ldots, y_M$, respectively.

EXAMPLE 4.12   Consider a population of ten balls. Determine the number of ways of painting four of the balls red, three of them white, two of them blue, and one of them yellow.

Solution   According to Theorem 4.8, the number of ways is

$$\frac{10!}{4! \, 3! \, 2! \, 1!} = 12,600. \quad ∎$$

We will now give a generalization of Theorem 4.6.

THEOREM **4.9**   Let $y_1, \ldots, y_M$ be nonnegative integers adding up to the positive integer n. Then there are

$$\frac{n!}{y_1! \cdots y_M!}$$

ordered n-tuples $(m_1, \ldots, m_n)$ of integers in $\{1, \ldots, M\}$ such that the integer m appears y_m times among $m_1, \ldots, m_n$ for $1 \leq m \leq M$.

Proof   There is a one-to-one correspondence between the indicated ordered n-tuples and partitions of $\{1, \ldots, n\}$ into disjoint sets $B_1, \ldots, B_M$ having sizes $y_1, \ldots, y_M$ respectively: $B_m = \{l : 1 \leq l \leq n \text{ and } m_l = m\}$ for $1 \leq m \leq M$. The desired result now follows from Theorem 4.8.   ∎

THEOREM **4.10**   Let n and M be positive integers. Then

$$(a_1 + \cdots + a_M)^n = \sum \frac{n!}{y_1! \cdots y_M!} a_1^{y_1} \cdots a_M^{y_M}, \qquad a_1, \ldots, a_M \in \mathbb{R}, \qquad \textbf{(3)}$$

where the sum is over the collection of ordered M-tuples $(y_1, \ldots, y_M)$ of nonnegative integers adding up to n.

Proof   Write $(a_1 + \cdots + a_M)^n$ as the product of n identical factors:

$$(a_1 + \cdots + a_M)^n = (a_1 + \cdots + a_M) \cdots (a_1 + \cdots + a_M).$$

Let $y_1, \ldots, y_M$ be nonnegative integers whose sum is n. To obtain a term of the form $a_1^{y_1} \cdots a_M^{y_M}$, we must choose a_1 from y_1 of the factors, a_2 from y_2 of the remaining factors and so forth. By Theorem 4.8 or Theorem 4.9, there are

$$\frac{n!}{y_1! \cdots y_M!}$$

such choices.   ∎

We refer to (3) as the *multinomial formula*.

Problems

4.1   Determine the number of ways of choosing three tickets out of a box of ten tickets.

4.2   A random sample of size 2 is selected by sampling without replacement from a box having N tickets, labeled from 1 to N. Let L be the larger of the two labels in the sample. Determine **(a)** the distribution function of L; **(b)** the probability function of L.

4.3   A drawer contains 12 loose socks, of which 2 are black, 4 are blue, and 6 are brown. A student gets up early in the morning to study for an exam in probability theory. Not wanting to wake up his roommate, he randomly grabs 2 socks out of the drawer without turning on the lights to check their color. Determine the probability that the 2 socks are of the same color.

4.4 A fair die is rolled six times in an honest manner. Determine the probability that each side shows once.

4.5 Use the binomial formula to expand $(a + b)^5$.

4.6 Let n be a nonnegative integer and let y be an integer. Verify that **(a)** $\binom{n}{n-y} = \binom{n}{y}$; **(b)** $\binom{n}{y-1} + \binom{n}{y} = \binom{n+1}{y}$.

4.7 Let n be a nonnegative integer. Verify that **(a)** $\binom{n}{0} + \binom{n}{1} + \cdots + \binom{n}{n} = 2^n$; **(b)** $\binom{n}{0} - \binom{n}{1} + \cdots + (-1)^n \binom{n}{n} = 0$.

4.8 A fair die is rolled six times in an honest manner. **(a)** Determine the number of distinct outcomes in which sides 1, 2, and 3 each appear twice. **(b)** Determine the probability that sides 1, 2, and 3 each appear twice. **(c)** Determine the probability that three (unspecified) sides each appear twice.

4.9 Determine the number of ordered six-tuples of the letters a, b, and c, containing three a's, two b's, and one c.

4.10 Use Stirling's formula to approximate $(3n)!/(n!)^3$ for $n \gg 1$.

4.11 Determine the number of ways of putting ten balls into four boxes so that **(a)** there are four balls in the first box, three in the second box, two in the third box, and one in the fourth box; **(b)** some box has four balls, some box has three balls, some box has two balls, and some box has one ball.

4.12 Consider the expansion of $(a + b + c)^5$ given by the trinomial formula. Determine the coefficient of **(a)** a^5; **(b)** a^4b; **(c)** a^3b^2; **(d)** a^3bc; **(e)** a^2b^2c.

4.13 Let n be a nonnegative integer and let M be a positive integer. Verify that

$$\sum \frac{1}{y_1! \cdots y_M!} = \frac{M^n}{n!},$$

where the sum is over the collection of ordered M-tuples $(y_1, \ldots, y_M)$ of nonnegative integers adding up to n.

4.2
The Binomial Distribution

Consider an experiment whose random outcomes can be either of two types, which we refer to as "success" and "failure." Let π denote the probability of success, where $0 < \pi < 1$. Then $1 - \pi$ is the probability of failure. Let I denote the indicator random variable for success, which has the Bernoulli distribution with parameter π. Then I has the probability function f given by $f(0) = 1 - \pi$ and $f(1) = \pi$, and it has mean π, variance $\pi(1 - \pi)$ and standard deviation $\sqrt{\pi(1 - \pi)}$.

Consider, for example, the experiment of tossing a fair coin in an honest manner and think of heads as a success and of tails as a failure. Then the probability of a success is given by $\pi = 1/2$.

Consider, next, the experiment of rolling two fair dice in an honest manner, let W denote the total number of spots showing on the two dice, and think of 7 and

11 as successes and any other total number of spots as constituting a failure. Then the probability of success is given by $P(W = 7) + P(W = 11) = \frac{6}{36} + \frac{2}{36} = \frac{8}{36} = \frac{2}{9}$ (see the solution to Example 1.24 in Section 1.5).

Consider, instead, a newly installed component, whose failure time W has the exponential distribution with mean 1000 days. Let us regard the component as being successful if it lasts at least 500 days. Then the probability of success is given by $\pi = P(W \geq 500) = e^{-500/1000} = e^{-1/2} \doteq .607$.

Consider now n trials of the experiment and let I_i denote the indicator random variable for success on the ith trial. Then $I_1, \ldots, I_n$ are independent random variables, each having the Bernoulli distribution with parameter π and hence mean π and variance $\pi(1 - \pi)$. Let $Y = I_1 + \cdots + I_n$ denote the total number of successes on the n trials. Note that Y is a $\{0, \ldots, n\}$-valued random variable and that

$$EY = E\left(\sum_{i=1}^{n} I_i\right) = \sum_{i=1}^{n} EI_i = n\pi,$$

$$\text{var}(Y) = \text{var}\left(\sum_{i=1}^{n} I_i\right) = \sum_{i=1}^{n} \text{var}(I_i) = n\pi(1 - \pi),$$

and

$$SD(Y) = \sqrt{n\pi(1 - \pi)}.$$

The distribution of Y is referred to as the *binomial distribution* with parameters n and π. Observe the the binomial distribution with parameters $n = 1$ and π is the same as the Bernoulli distribution with parameter π.

THEOREM 4.11 The probability function f of the binomial distribution with parameters n and π is given by

$$f(y) = \binom{n}{y}\pi^y(1 - \pi)^{n-y}, \qquad y \in \{0, \ldots, n\},$$

and $f(y) = 0$ otherwise.

Proof Choose $y \in \{0, \ldots, n\}$ and let $i_1, \ldots, i_n$ be a sequence of y 1's and $n - y$ 0's. Then

$$P(I_1 = i_1, \ldots, I_n = i_n) = P(I_1 = i_1) \cdots P(I_n = i_n)$$
$$= \pi^{i_1}(1 - \pi)^{1-i_1} \cdots \pi^{i_n}(1 - \pi)^{1-i_n}$$
$$= \pi^{i_1 + \cdots + i_n}(1 - \pi)^{n-(i_1 + \cdots + i_n)}$$
$$= \pi^y(1 - \pi)^{n-y}.$$

Observe that $P(Y = y)$ is the sum of $P(I_1 = i_1, \ldots, I_n = i_n)$ over all such sequences of 0's and 1's. Since there are $\binom{n}{y}$ such sequences by Theorem 4.3 in Section 4.1, we conclude that

$$P(Y = y) = \binom{n}{y}\pi^y(1 - \pi)^{n-y}$$

as desired. ∎

EXAMPLE **4.13**   Determine the probability function f of the binomial distribution with parameters n and π when n equals

(a) 1   (b) 2   (c) 3   (d) 4   (e) 5

Solution   (a) The probability function is given by $f(0) = 1 - \pi$ and $f(1) = \pi$.

(b) The probability function is given by $f(0) = (1 - \pi)^2$, $f(1) = 2\pi(1 - \pi)$, and $f(2) = \pi^2$.

(c) The probability function is given by $f(0) = (1 - \pi)^3$, $f(1) = 3\pi(1 - \pi)^2$, $f(2) = 3\pi^2(1 - \pi)$, and $f(3) = \pi^3$.

(d) The probability function is given by $f(0) = (1 - \pi)^4$, $f(1) = 4\pi(1 - \pi)^3$, $f(2) = 6\pi^2(1 - \pi)^2$, $f(3) = 4\pi^3(1 - \pi)$, and $f(4) = \pi^4$.

(e) The probability function is given by $f(0) = (1 - \pi)^5$, $f(1) = 5\pi(1 - \pi)^4$, $f(2) = 10\pi^2(1 - \pi)^3$, $f(3) = 10\pi^3(1 - \pi)^2$, $f(4) = 5\pi^4(1 - \pi)$, and $f(5) = \pi^5$. ∎

EXAMPLE **4.14**   Determine the probability that at least two heads show when three fair coins are tossed in an honest manner.

Solution   Let Y denote the number of heads that show, which has the binomial distribution with parameters $n = 3$ and $\pi = 1/2$, and let f denote the probability function of Y. Then the desired probability is given by

$$P(Y \geq 2) = f(2) + f(3) = 3\left(\frac{1}{2}\right)^2 \left(\frac{1}{2}\right) + \left(\frac{1}{2}\right)^3 = \frac{1}{2}. \quad ∎$$

EXAMPLE **4.15**   Determine the probability that 7 or 11 shows at least twice when two fair dice are rolled together four times in an honest manner.

Solution   As we have already seen, the probability of getting a 7 or 11 on a single trial is 2/9. Let Y denote the number of times that 7 or 11 shows in the four rolls, which has the binomial distribution with parameters $n = 4$ and $\pi = 2/9$, and let f denote the probability function of Y. Then the desired probability is given by

$$P(Y \geq 2) = f(2) + f(3) + f(4) = 6\left(\frac{2}{9}\right)^2 \left(\frac{7}{9}\right)^2 + 4\left(\frac{2}{9}\right)^3 \left(\frac{7}{9}\right) + \left(\frac{2}{9}\right)^4 \doteq .216. \quad ∎$$

EXAMPLE **4.16**   A system is composed of five components, at least three of which must work for the system to work properly. Suppose the failure time of the components are independent random variables, each having the exponential distribution with mean 1000 days. Determine the probability that the system works properly for at least 500 days.

Solution   As we have already seen, the probability π that an individual component works properly for at least 500 days is given by $\pi = e^{-500/1000} = e^{-1/2}$. Let Y denote the number of components that last at least 500 days, which has the binomial distribution with parameters $n = 5$ and $\pi = e^{-1/2}$, and let f denote the probability function of

Y. Then the desired probability is given by $P(Y \geq 3) = f(3) + f(4) + f(5)$. Now

$$f(3) = \binom{5}{3}\pi^3(1-\pi)^2 = 10e^{-3/2}(1-e^{-1/2})^2 \doteq .3454,$$

$$f(4) = \binom{5}{4}\pi^4(1-\pi) = 5e^{-2}(1-e^{-1/2}) \doteq .2663,$$

and $f(5) = \pi^5 = e^{-5/2} \doteq .0821$. Thus, the probability that the system works properly for at least 500 days is given by $P(Y \geq 3) \doteq .3454 + .2663 + .0821 = .6938$. ∎

EXAMPLE 4.17 Frankie and Johnny go to Las Vegas so Johnny can play roulette. Suppose Johnny makes 100 consecutive \$1 bets each on a single slot. Determine the probability that Johnny will be ahead as a result of the 100 bets.

Solution Let Y denote the number of bets that Johnny wins, which has the binomial distribution with parameters $n = 100$ and $\pi = 1/38$. On each of the Y winning bets, Johnny wins \$35; on each of the $100 - Y$ losing bets, he loses \$1. Thus, his net winnings W on the 100 bets is given by $W = 35Y - (100 - Y) = 36Y - 100$. Consequently, Johnny will be ahead as a result of the 100 bets if and only if $Y \geq 3$. Now

$$P(Y = y) = \binom{100}{y}\left(\frac{1}{38}\right)^y\left(\frac{37}{38}\right)^{100-y}.$$

In particular,

$$P(Y = 0) = \left(\frac{37}{38}\right)^{100} \doteq .0695,$$

$$P(Y = 1) = 100\left(\frac{1}{38}\right)\left(\frac{37}{38}\right)^{99} \doteq .1878,$$

and

$$P(Y = 2) = \frac{100 \cdot 99}{2}\left(\frac{1}{38}\right)^2\left(\frac{37}{38}\right)^{98} \doteq .2512.$$

Therefore, the probability that Johnny will be ahead as a result of the 100 bets is given by

$$P(Y \geq 3) = 1 - P(Y = 0) - P(Y = 1) - P(Y = 2)$$

$$\doteq 1 - .0695 - .1878 - .2512$$

$$\doteq .4916. \quad ∎$$

Let f denote the probability function of the binomial distribution with parameters n and π. Then

$$\frac{f(y)}{f(y-1)} = \frac{\dfrac{n!}{y!\,(n-y)!}\pi^y(1-\pi)^{n-y}}{\dfrac{n!}{(y-1)!\,(n+1-y)!}\pi^{y-1}(1-\pi)^{n+1-y}}, \qquad y = 1, \ldots, n,$$

which simplifies to

$$\frac{f(y)}{f(y-1)} = \frac{n+1-y}{y} \frac{\pi}{1-\pi}, \qquad y = 1, \ldots, n. \tag{4}$$

It follows from (4) that

$$f(y) = \frac{n+1-y}{y} \frac{\pi}{1-\pi} f(y-1), \qquad y = 1, \ldots, n. \tag{5}$$

Equation (5) suggests one reasonable way to compute the binomial probability function: first, compute $f(0) = (1 - \pi)^n$; then successively compute $f(y)$ from $f(y-1)$ for $y = 1, \ldots, n$ by using (5).

Every probability function on a set $\mathcal{Y}$ has a global maximum, which is achieved at one or more points of $\mathcal{Y}$. Consider a probability function on a set $\mathcal{Y}$ of consecutive integers. A value of $y \in \mathcal{Y}$ for which the probability function achieves a relative maximum (in comparison with values in $\{y-1, y+1\} \cap \mathcal{Y}$) is referred to as a mode of the probability function. A value of y for which the probability function achieves a global maximum is necessarily a mode of the probability function.

Think of the probability function f of the binomial distribution with parameters n and π as a probability function on $\{0, \ldots, n\}$. It follows from Equation (4) or (5) that $f(y)$ is greater than, equal to, or less than $f(y-1)$ according as y is less than, equal to, or greater than $r = (n+1)\pi$. Suppose r is a positive integer. Then $r-1$ and r are the only modes and the only locations of global maxima of f. Also, $f(y)$ is strictly increasing in y for $y \in \{0, \ldots, r-1\}$, $f(r-1) = f(r)$, and $f(y)$ is strictly decreasing in y for $y \in \{r, \ldots, n\}$. Suppose, instead, that r is not a positive integer. Then $[r]$ (the greatest integer in r) is the unique mode and the unique location of the global maximum of f. Also, $f(y)$ is strictly increasing in y for $y \in \{0, \ldots, [r]\}$, and it is strictly decreasing in y for $y \in \{[r], \ldots, n\}$.

Consider, in particular, the probability function f of the binomial distribution with parameters n and $\pi = 1/3$. If $n = 5$, then $r = (n+1)\pi = 2$, so f has modes at 1 and 2. If $n = 6$, then $r = (n+1)\pi = 7/3$, so f has a unique mode at 2. Figure 4.1 illustrates these results.

EXAMPLE **4.18**   Let Y denote the number of times that one spot shows when a fair die is rolled 36 times in an honest manner, which has the binomial distribution with parameters $n = 36$ and $\pi = 1/6$. Suppose that the observed value of Y equals 2. Note that

$$P(Y = 2) = \binom{36}{2}\left(\frac{1}{6}\right)^2\left(\frac{5}{6}\right)^{34} \doteq .0356.$$

How unusual is it for the observed value of Y to have a probability this small or smaller?

Solution   Let f denote the probability function of Y. We want to find the sum of $f(y)$ over all values of y such that $f(y) \le 2$. To this end, set $r = (n+1)\pi = 37/6$. Then $[r] = 6$, so $f(y)$ is a strictly increasing function of y as y increases from 0 to 6 and it is a strictly decreasing function of y as y increases from 6 to 36. By computer,

$$f(10) = \binom{36}{10}\left(\frac{1}{6}\right)^{10}\left(\frac{5}{6}\right)^{26} \doteq .0367,$$

FIGURE 4.1

Binomial probability functions with $\pi = \frac{1}{3}$

$n = 5$

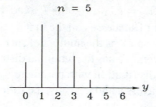

$n = 6$

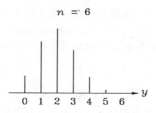

and

$$f(11) = \binom{36}{11}\left(\frac{1}{6}\right)^{11}\left(\frac{5}{6}\right)^{25} \doteq .0174.$$

Thus, $f(y) \leq f(2)$ if and only if $y \leq 2$ or $y \geq 11$. Consequently, the sum of $f(y)$ over all values of y such that $f(y) \leq f(2)$ is given by

$$\sum_{y=0}^{2}\binom{36}{y}\left(\frac{1}{6}\right)^{y}\left(\frac{5}{6}\right)^{36-y} + \sum_{y=11}^{36}\binom{36}{y}\left(\frac{1}{6}\right)^{y}\left(\frac{5}{6}\right)^{36-y} \doteq .0756 \doteq \frac{1}{13}.$$

Therefore, in a large number of trials of the experiment of rolling the die 36 times, the probability of the observed value of Y will be less than or equal to $f(2)$ in approximately 1/13 of the trials. ∎

EXAMPLE 4.19 A fair die is repeatedly rolled in an honest manner until one spot shows for the third time. Determine the probability that (exactly) 20 rolls are required.

Solution Observe that 20 rolls are required if and only if one spot shows twice on the first 19 rolls and one spot also shows on the 20th roll. The number of times that one spot shows on the first 19 rolls has the binomial distribution with parameters $n = 19$ and $\pi = 1/6$, so the probability that one spot shows twice during the first 19 rolls equals

$$\binom{19}{2}\left(\frac{1}{6}\right)^{2}\left(\frac{5}{6}\right)^{17}.$$

The probability that one spot shows on the 20th roll equals 1/6. Since the first 19 rolls and the 20th roll are independent, the probability that 20 rolls are required is given by

$$\binom{19}{2}\left(\frac{1}{6}\right)^{2}\left(\frac{5}{6}\right)^{17}\left(\frac{1}{6}\right) = \binom{19}{2}\left(\frac{1}{6}\right)^{3}\left(\frac{5}{6}\right)^{17} \doteq .0357. \quad ∎$$

Let n_1 and n_2 be positive integers and set $n = n_1 + n_2$. Also let $I_1, \ldots, I_n$ be independent random variables, each having a Bernoulli distribution with parameter π, and set $Y_1 = I_1 + \cdots + I_{n_1}$, $Y_2 = I_{n_1+1} + \cdots + I_n$, and $Y = I_1 + \cdots + I_n = Y_1 + Y_2$. (Think of Y_1 as the number of successes on the first n_1 trials, Y_2 as the number of successes on the next n_2 trials, and Y as the number of successes on all n trials.) Observe that Y_1 and Y_2 are independent, Y_1 has the binomial distribution with parameters n_1 and π, Y_2 has the binomial distribution with parameters n_2 and π, and Y has the binomial distribution with parameters n and π. Thus, the following result is valid.

THEOREM 4.12 Let Y_1 and Y_2 be independent random variables such that Y_1 has the binomial distribution with parameters n_1 and π and Y_2 has the binomial distribution with parameters n_2 and π. Then $Y_1 + Y_2$ has the binomial distribution with parameters $n_1 + n_2$ and π.

Let Y_1, Y_2, and Y_3 be independent random variables such that Y_i has the binomial distribution with parameters n_i and π for $i = 1, 2, 3$. It follows from Theorem 4.12 that $Y_1 + Y_2$ has the binomial distribution with parameters $n_1 + n_2$ and π. Since $Y_1 + Y_2$ and Y_3 are independent, it follows by another application of Theorem 4.12 that $Y_1 + Y_2 + Y_3$ has the binomial distribution with parameters $n_1 + n_2 + n_3$ and π. Similarly, or by induction, we have the following result.

THEOREM 4.13 Let $Y_1, \ldots, Y_N$ be independent random variables such that Y_i has the binomial distribution with parameters n_i and π for $1 \le i \le N$. Then $Y_1 + \cdots + Y_N$ has the binomial distribution with parameters $n = n_1 + \cdots + n_N$ and π.

Given random variables $Y_1, \ldots, Y_n$, the corresponding *order statistics* $Y_{(1)}, \ldots, Y_{(n)}$ are obtained by rewriting $Y_1, \ldots, Y_n$ in nondecreasing order. In particular, $Y_{(1)}$ is the minimum value among $Y_1, \ldots, Y_n$, and $Y_{(n)}$ is the maximum of these random variables. We refer to $Y_{(i)}$ as the ith order statistic for $1 \le i \le n$. As an example, if $n = 4$, $Y_1 = 110$, $Y_2 = -20$, $Y_3 = 120$, and $Y_4 = 110$, then $Y_{(1)} = -20$, $Y_{(2)} = 110$, $Y_{(3)} = 110$, and $Y_{(4)} = 120$.

THEOREM 4.14 Let $U_1, \ldots, U_n$ be independent random variables, each uniformly distributed on $[0, 1]$, and let $U_{(1)}, \ldots, U_{(n)}$ denote the corresponding order statistics. Then $U_{(i)}$ has the beta distribution with parameters i and $n + 1 - i$ for $1 \le i \le n$.

Proof Let $1 \le i \le n$ and $0 < u < 1$. Observe that the ith order statistic $U_{(i)}$ is less than or equal to u if and only if at least i of the random variables $U_1, \ldots, U_n$ are less than or equal to u. Thus, let Y denote the number of random variables $U_1, \ldots, U_n$ that are less than or equal to u. Then Y has the binomial distribution with parameters n and $\pi = u$, so

$$P(U_{(i)} \le u) = P(Y \ge i) = \sum_{y=i}^{n} \binom{n}{y} u^y (1 - u)^{n-y}.$$

Consequently [see Problem 4.25(c)], the distribution function of $U_{(i)}$ coincides with the distribution function of the beta distribution with parameters i and $n + 1 - i$. [It is easily seen directly that $P(U_{(i)} \leq u) = 0$ for $u \leq 0$ and $P(U_{(i)} \leq u) = 1$ for $u \geq 1$.] Therefore, $U_{(i)}$ has the indicated beta distribution. ∎

Normal Approximation to the Binomial Distribution

Let Y have the binomial distribution with parameters n and π and hence mean $\mu = n\pi$, variance $\sigma^2 = n\pi(1 - \pi)$, and standard deviation $\sigma = \sqrt{n\pi(1 - \pi)}$. The probability function f of Y is given by

$$f(y) = \binom{n}{y}\pi^y(1 - \pi)^{n-y} = \frac{n!}{y!\,(n - y)!}\pi^y(1 - \pi)^{n-y}, \qquad y \in \{0, \ldots, n\},$$

and $f(y) = 0$ for other values of y. Using Stirling's formula to approximate the factorials appearing in the formula for the probability function, we can show that if $\sigma^2 = n\pi(1 - \pi)$ is large, then

$$f(y) \approx \frac{1}{\sigma}\phi\left(\frac{y - \mu}{\sigma}\right), \qquad y \text{ an integer.} \tag{6}$$

Figure 4.2 shows a plot of the binomial probability function together with its normal approximation, given by (6), when $n = 40$ and $\pi = .25$.

FIGURE 4.2
Binomial probability function with parameters $n = 40$ and $\pi = .25$ and its normal approximation

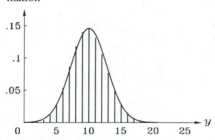

EXAMPLE 4.20   Determine the probability of getting 100 heads and 100 tails in 200 tosses of a fair coin by using

(a) the formula for the binomial probability function;

(b) Stirling's formula to approximate the factorials in the answer to part (a);

(c) normal approximation to the binomial probability function.

Solution   (a) The probability of getting 100 heads and 100 tails in the 200 tosses is

$$\binom{200}{100}\frac{1}{2^{200}} = \frac{200!}{(100!)^2 2^{200}} \doteq .05635.$$

(b) Using Stirling's formula (see the solution to Example 4.8 in Section 4.1), we get

$$\binom{200}{100} \approx \frac{2^{200}}{\sqrt{100\pi}} = \frac{2^{200}}{10\sqrt{\pi}}$$

(where $\pi \doteq 3.14159$) and hence that

$$\binom{200}{100}2^{-200} \approx \frac{1}{10\sqrt{\pi}} \doteq .05642.$$

(c) Applying Property (6) with $\mu = 100$, $\sigma = \sqrt{200(1/2)(1/2)} = 5\sqrt{2}$, and $y = 100$ and noting that $\phi(0) = 1/\sqrt{2\pi}$, we get

$$\binom{200}{100}2^{-200} \approx \frac{1}{5\sqrt{2}\sqrt{2\pi}} = \frac{1}{10\sqrt{\pi}} \doteq .05642.$$

(Here normal approximation coincides with the approximation given by Stirling's formula.) ∎

Let a and b be integers with $a \le b$. It follows from (6) that

$$P(a \le Y \le b) = \sum_{y=a}^{b} f(y) \approx \sum_{y=a}^{b} \frac{1}{\sigma}\phi\left(\frac{y-\mu}{\sigma}\right).$$

Using the definition of a definite integral as the limit of certain sums, we can show that

$$P(a \le Y \le b) \approx \int_{a}^{b} \frac{1}{\sigma}\phi\left(\frac{y-\mu}{\sigma}\right) dy = \Phi\left(\frac{b-\mu}{\sigma}\right) - \Phi\left(\frac{a-\mu}{\sigma}\right) \qquad (7)$$

(see Figure 4.3). Alternatively,

$$P(a \le Y \le b) = P\left(a - \tfrac{1}{2} \le Y \le b + \tfrac{1}{2}\right)$$

$$\approx \int_{a-\frac{1}{2}}^{b+\frac{1}{2}} \frac{1}{\sigma}\phi\left(\frac{y-\mu}{\sigma}\right) dy$$

$$= \Phi\left(\frac{b + \frac{1}{2} - \mu}{\sigma}\right) - \Phi\left(\frac{a - \frac{1}{2} - \mu}{\sigma}\right). \qquad (8)$$

(Here the integral is over all values of y that differ from some integer in $\{a, \ldots, b\}$ by one-half or less.) We refer to the right side of (7) as normal approximation to

FIGURE 4.3

Normal approximation

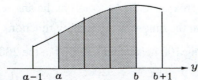

$P(a \le Y \le b)$ and to the right side of (8) as normal approximation with the *half-integer correction* (see Figure 4.4).

FIGURE 4.4
Normal approximation with the half-integer correction

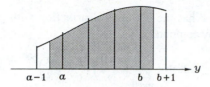

Consider, in particular, a positive probability of the form $P(Y = a)$. Observe that normal approximation, as given by Equation (7), yields the unreasonable approximation

$$P(Y = a) = P(a \le Y \le a) \approx \Phi\Big(\frac{a - \mu}{\sigma}\Big) - \Phi\Big(\frac{a - \mu}{\sigma}\Big) = 0,$$

while normal approximation with the half-integer correction, as given by Equation (8), yields the reasonable approximation

$$P(Y = a) = P\Big(a - \tfrac{1}{2} \le Y \le a + \tfrac{1}{2}\Big) \approx \Phi\Big(\frac{a + \tfrac{1}{2} - \mu}{\sigma}\Big) - \Phi\Big(\frac{a - \tfrac{1}{2} - \mu}{\sigma}\Big). \quad \text{(9)}$$

EXAMPLE **4.21**   Determine the probability of getting 100 heads and 100 tails in 200 tosses of a fair coin by using normal approximation with the half-integer correction.

Solution   Let Y be the number of heads in 200 tosses. Then Y has mean 100 and standard deviation $5\sqrt{2}$, as we saw in the solution to Example 4.20(c). Applying normal approximation with the half-integer correction, given by (9), we get

$$P(Y = 100) = P\Big(100 - \tfrac{1}{2} \le Y \le 100 + \tfrac{1}{2}\Big)$$

$$\approx \Phi\Big(\frac{100 + \tfrac{1}{2} - 100}{5\sqrt{2}}\Big) - \Phi\Big(\frac{100 - \tfrac{1}{2} - 100}{5\sqrt{2}}\Big)$$

$$= \Phi\Big(\frac{1}{10\sqrt{2}}\Big) - \Phi\Big(-\frac{1}{10\sqrt{2}}\Big)$$

$$= 2\Phi\Big(\frac{1}{10\sqrt{2}}\Big) - 1$$

$$\doteq .05637,$$

which is remarkably close to the actual probability .05635 that was obtained in the solution to Example 4.20(a). (Normal approximation with the half-integer correction is generally not this accurate.) ∎

EXAMPLE **4.22**   Let Y denote the number of times that one spot shows when a fair die is rolled 600 times in an honest manner. Determine $P(80 \leq Y \leq 120)$ by using

(a) the formula for the binomial probability function;

(b) normal approximation;

(c) normal approximation with the half-integer correction.

Solution   (a) The random variable Y has the binomial distribution with parameters $n = 600$ and $\pi = 1/6$. Thus (by computer),

$$P(80 \leq Y \leq 120) = \sum_{y=80}^{120} \binom{600}{y} \left(\frac{1}{6}\right)^y \left(\frac{5}{6}\right)^{600-y} \doteq .9754.$$

(b) Now Y has mean $\mu = 100$ and standard deviation

$$\sigma = \sqrt{600 \cdot \frac{1}{6} \cdot \frac{5}{6}} = \frac{5}{3}\sqrt{30} \doteq 9.129.$$

Thus, by normal approximation,

$$P(80 \leq Y \leq 120) \approx \Phi\left(\frac{120 - 100}{9.129}\right) - \Phi\left(-\frac{80 - 100}{9.129}\right)$$

$$\doteq \Phi(2.191) - \Phi(-2.191)$$

$$= 2\Phi(2.191) - 1$$

$$\doteq .9715.$$

(c) Using the normal approximation with the half-integer correction, we get

$$P(80 \leq Y \leq 120) = P\left(80 - \tfrac{1}{2} \leq Y \leq 120 + \tfrac{1}{2}\right)$$

$$\approx \Phi\left(\frac{120 + \tfrac{1}{2} - 100}{9.129}\right) - \Phi\left(-\frac{80 - \tfrac{1}{2} - 100}{9.129}\right)$$

$$\doteq \Phi(2.246) - \Phi(-2.246)$$

$$= 2\Phi(2.246) - 1$$

$$\doteq .9753,$$

which again is remarkably accurate.   ∎

EXAMPLE **4.23**   A company purchases 100 light bulbs, each of which is guaranteed to last for 5000 hours of continuous use. Under the assumption that the lifetimes of the light bulbs are independent and have the gamma distribution with shape parameter 5 and mean 10,000 hours, determine the probability that at least 85 of the light bulbs fulfill their guarantees by using

(a) the formula for the binomial probability function;

(b) normal approximation;

(c) normal approximation with the half-integer correction.

Solution   The gamma distribution with shape parameter 5 and scale parameter β has mean 5β. Since the mean equals 10,000, we conclude that $\beta = 2000$. The probability π that a random variable W having such a distribution exceeds 5000 is given by

$$\pi = P(W \geq 5000) = e^{-5/2} \sum_{m=0}^{4} \frac{(5/2)^m}{m!} \doteq .8912$$

(see Theorem 3.1 in Section 3.1).

(a) Let Y denote the number of light bulbs that fulfill the guarantee, which has the binomial distribution with parameters $n = 100$ and $\pi \doteq .8912$. Thus,

$$P(Y \geq 85) = \sum_{y=85}^{100} \binom{100}{y}(.8912)^y(1 - .8912)^{100-y} \doteq .9256.$$

(b) Now Y has mean $\mu = 100\pi \doteq 89.12$ and standard deviation

$$\sigma = \sqrt{100\pi(1 - \pi)} \doteq \sqrt{100(.8912)(.1088)} \doteq 3.114.$$

Thus, by normal approximation [with $a = 85$ and $b = \infty$ in Equation (7)],

$$P(Y \geq 85) \approx 1 - \Phi\left(\frac{85 - 89.12}{3.114}\right) \doteq 1 - \Phi(-1.322) = \Phi(1.322) \doteq .9070.$$

(c) Using normal approximation with the half-integer correction, we get

$$P(Y \geq 85) = P\left(Y \geq 85 - \tfrac{1}{2}\right)$$

$$\approx 1 - \Phi\left(\frac{85 - \tfrac{1}{2} - 89.12}{3.114}\right)$$

$$\doteq 1 - \Phi(-1.483)$$

$$= \Phi(1.483)$$

$$\doteq .9309. \quad \blacksquare$$

Let $\hat{\pi} = Y/n$ denote the sample proportion of successes on the n trials, which has mean π, variance $\pi(1 - \pi)/n$, and standard deviation $\sqrt{\pi(1 - \pi)/n}$. Applying normal approximation to the distribution of $\hat{\pi}$ is equivalent to applying normal approximation to the distribution of Y.

EXAMPLE 4.24   Use normal approximation to solve each of the following:

(a) Suppose $n = 400$ and $\pi = .5$. Determine $P(.45 \leq \hat{\pi} \leq .55)$.

(b) Suppose $\pi = .5$. Determine n such that $P(.47 \leq \hat{\pi} \leq .53) \approx .95$.

(c) Determine n such that $P(\pi - .03 \leq \hat{\pi} \leq \pi + .03) \geq .95$ for all π.

Solution   (a) Observe that

$$SD(\hat{\pi}) = \sqrt{\frac{\pi(1 - \pi)}{n}} = \sqrt{\frac{1}{1600}} = \frac{1}{40} = .025,$$

so that $\hat{\pi} = \pi + \text{SD}(\hat{\pi})Z = .5 + .025Z$, where the standardized random variable Z corresponding to $\hat{\pi}$ has approximately the standard normal distribution. Consequently,

$$P(.45 \leq \hat{\pi} \leq .55) = P(.45 \leq .5 + .025Z \leq .55)$$

$$= P(-2 \leq Z \leq 2)$$

$$\approx \Phi(2) - \Phi(-2).$$

Since

$$\Phi(2) - \Phi(-2) = 2\Phi(2) - 1 \doteq 2(.9773) - 1 = .9545,$$

we conclude that $P(.45 \leq \hat{\pi} \leq .55) \approx .9545$.

(b) Observe that

$$P(.47 \leq \hat{\pi} \leq .53) = P(.47 \leq .5 + \text{SD}(\hat{\pi})Z \leq .53)$$

$$= P\left(-\frac{.03}{\text{SD}(\hat{\pi})} \leq Z \leq \frac{.03}{\text{SD}(\hat{\pi})}\right)$$

$$\approx \Phi\left(\frac{.03}{\text{SD}(\hat{\pi})}\right) - \Phi\left(-\frac{.03}{\text{SD}(\hat{\pi})}\right)$$

$$= 2\Phi\left(\frac{.03}{\text{SD}(\hat{\pi})}\right) - 1.$$

Thus, $P(.47 \leq \hat{\pi} \leq .53) \approx .95$ if and only if $\Phi(.03/\text{SD}(\hat{\pi})) \approx .975$, which holds if and only if $.03/\text{SD}(\hat{\pi}) \approx z_{.975}$ or, equivalently, if and only if $\text{var}(\hat{\pi}) \approx (.03/z_{.975})^2$. This leads to the equation

$$\frac{\pi(1 - \pi)}{n} \approx \left(\frac{.03}{z_{.975}}\right)^2,$$

whose solution (when $\pi = .5$) is given by

$$n \approx \pi(1 - \pi)\left(\frac{z_{.975}}{.03}\right)^2 \doteq \frac{1}{4}\left(\frac{1.96}{.03}\right)^2 \doteq 1067.$$

(c) Observe that

$$P(\pi - .03 \leq \hat{\pi} \leq \pi + .03) = P(\pi - .03 \leq \pi + \text{SD}(\hat{\pi})Z \leq \pi + .03)$$

$$= P\left(-\frac{.03}{\text{SD}(\hat{\pi})} \leq Z \leq \frac{.03}{\text{SD}(\hat{\pi})}\right)$$

$$\approx \Phi\left(\frac{.03}{\text{SD}(\hat{\pi})}\right) - \Phi\left(-\frac{.03}{\text{SD}(\hat{\pi})}\right)$$

$$= 2\Phi\left(\frac{.03}{\text{SD}(\hat{\pi})}\right) - 1.$$

Thus (approximately),

$$P(\pi - .03 \leq \hat{\pi} \leq \pi + .03) \geq .95$$

if and only if

$$\Phi(.03/\text{SD}(\hat{\pi})) \geq .975$$

or, equivalently, if and only if $.03/SD(\hat{\pi}) \geq z_{.975}$. We are led to the inequality

$$\frac{\pi(1-\pi)}{n} \leq \left(\frac{.03}{z_{.975}}\right)^2,$$

whose solution is given by

$$n \geq \pi(1-\pi)\left(\frac{z_{.975}}{.03}\right)^2.$$

This inequality holds for all $\pi \in [0, 1]$ if and only if it holds for the value of π that maximizes $\pi(1-\pi)$, $0 \leq \pi \leq 1$. It is an easy problem in calculus to check that $\pi(1-\pi)$, $0 \leq \pi \leq 1$ is uniquely maximized at $\pi = .5$. Thus, the desired value of n is given by

$$n \approx \frac{1}{4}\left(\frac{z_{.975}}{.03}\right)^2 \doteq 1067,$$

which coincides with the solution to part (b).   ∎

Consider an opinion survey to determine the proportion π of people in a given country who, when asked in a specified manner, would state that they believed their leader to be doing a good job of leading the country. Let n denote the size of the survey and let I_i equal 1 or 0 according as the ith person in the survey does or does not state the indicated belief. If the survey is carried out in a suitable manner and if n is a small fraction of the total population of the country, it is reasonable to assume that $I_1, \ldots, I_n$ are independent random variables, each having the Bernoulli distribution with parameter π, and hence that $Y = I_1 + \cdots + I_n$ has the binomial distribution with parameters n and π. (In other words, the distinction between sampling with replacement and sampling without replacement can be ignored if $n/N \approx 0$, where n is the sample size and N is the population size.) According to the solution to Example 4.24(c), if the size of the survey is 1067 or more the probability is at least .95 that the sample proportion $\hat{\pi}$ will be within .03 of the true proportion π.

EXAMPLE **4.25**   At a party, Johnny boasted that he has extrasensory perception (ESP). To justify his claim, he had Frankie shuffle a deck of 52 playing cards, of which 26 are red and 26 are black. Frankie saw each card, Johnny tried to read her mind, and then he guessed the color of the card. After Frankie went through the deck, she reshuffled it, and the process was repeated. Altogether, Frankie went through the deck ten times. Jimmy kept track of the number of correct guesses and, at the end of the experiment, announced that Johnny had guessed correctly 290 times out of the 520 trials and that the proportion of correct guesses, $290/520 \doteq .5577$, was significantly greater than .5. Does this justify Johnny's claim of having ESP?

Solution   For $1 \leq i \leq n = 520$, let I_i equal 1 or 0 according as Johnny guesses correctly or incorrectly on the ith trial. Suppose Johnny does not have ESP and that $I_1, \ldots, I_n$ are independent random variables, each having a Bernoulli distribution with parameter

$\pi = .5$. Let $\hat{\pi}$ denote the sample proportion of correct guesses. Then

$$\mathrm{SD}(\hat{\pi}) = \sqrt{\frac{(.5)(.5)}{520}} \doteq .0219.$$

Using normal approximation, we get

$$P(\hat{\pi} \geq .5577) = P(.5 + \mathrm{SD}(\hat{\pi})Z \geq .5577)$$

$$\doteq P\left(Z \geq \frac{.0577}{.0219}\right)$$

$$\doteq P(Z \geq 2.63)$$

$$\approx 1 - \Phi(2.63),$$

so $P(\hat{\pi} \geq .5577) \approx .0043$. Therefore, if Johnny does not have ESP and if our other suppositions are valid, there is less than 1 chance in 200 that Johnny would guess correctly 290 or more times on 520 trials. Since Johnny did guess correctly 290 or more times, one of the following must be valid: (1) Johnny has ESP; (2) Johnny does not have ESP but not all our suppositions are valid: (3) Johnny was just lucky.

There are a number of possibilities besides ESP that would invalidate our suppositions. Jimmy could have made an incorrect announcement, accidently or on purpose. Frankie, Jimmy, or someone else or something else (a mirror or marks on the backs of the cards) could have tipped Johnny off about the color of the card before each guess. Johnny might have been informed about the color of the card after each guess and used this knowledge to determine the number of red and black cards remaining in the deck and thereby to increase his probability of making a correct guess. Unless all such possibilities can be ruled out, the cited results should not be regarded as providing solid evidence that Johnny has ESP.   ∎

ESP refers to perception without any physical basis. Many people have claimed to have demonstrated the existence of ESP, but no such claim has generally been accepted by scientists. Indeed, a number of such claims have been explicitly refuted by showing that the results were or easily could have been obtained by secret cues, sleight of hand, or other means of deception. (Sometimes the results were obtained by a magician and detected by another magician). Other claims have been dismissed as having been due to highly repetitive testing. For example, if each of 200 people makes 520 guesses in the same manner as Johnny, it is probable that one or more of them will guess right at least 290 times.

Problems

4.14   Determine the probability that, when a fair die is rolled six times in an honest manner, side 1 shows at least twice.

4.15   Let Y have the binomial distribution with parameters n and π. Show that **(a)** $\mathrm{var}(Y) \leq n/4$; **(b)** $\mathrm{var}(Y) = n/4$ if and only if $\pi = 1/2$.

4.16 Three pennies and three quarters are tossed. Determine the probability that the number of pennies that show heads coincides with the number of quarters that do so.

4.17 A box has one red ball and two white balls. On each of six trials, a ball is selected at random from the box, its color is noted, and then the ball is returned to the box. Determine the probability that the red ball is selected on two of the six trials.

4.18 On each of five trials, two fair coins are tossed in an honest manner. Determine the probability that both coins show heads on two of the five trials (and at least one coin shows tails on each of the other three trials).

4.19 Two fair dice are repeatedly rolled in an honest manner until 7 or 11 occurs for the fourth time. Determine the probability that 20 such rolls of the two dice are required.

4.20 Consider an experiment having probability π of success, where $0 < \pi < 1$. Let the experiment be performed repeatedly until r failures occur, where r is a positive integer, and let Y be the number of successes that have occurred by then. Determine the probability function of Y. (The distribution of Y is referred to as the *negative binomial distribution* with parameters r and π; when $r = 1$, it coincides with the geometric distribution with parameter π.)

4.21 Let Y have the binomial distribution with parameters n and $1/2$, where n is an even positive integer. Use Stirling's formula to show that $P(Y = n/2) \approx \sqrt{2/(n\pi)}$ for $n \gg 1$.

4.22 Let Y denote the number of heads that show when 12 fair coins are tossed in an honest manner. **(a)** Determine $P(Y = 10)$. **(b)** Suppose that the observed value of Y equals 10. How unusual is it for the observed value of Y to have such a small probability?

4.23 Suppose the height in inches of an individual selected at random from a large homogeneous population of adults is normally distributed with mean 68 and standard deviation 2. Let $Y_1, \ldots, Y_{10}$ be the heights of ten such randomly selected individuals. (It is assumed that these random variables form a random sample from the indicated normal distribution.) Let N denote the number of individuals in the sample that are between $5\frac{1}{2}$ feet and 6 feet in height. **(a)** Determine the distribution of N. **(b)** Determine the mean, variance, and standard deviation of N.

4.24 Let $f(\cdot; n, \pi)$ denote the probability function of the binomial distribution with parameters n and π and let r be a positive integer. Show that $r - 1$ and r are modes of $f(\cdot; n, \pi)$ if and only if $f(r; n, \pi) = f(r; n + 1, \pi)$.

4.25 Let $F(y; \alpha_1, \alpha_2), y \in \mathbb{R}$, denote the distribution function of the beta distribution with parameters α_1 and α_2. **(a)** Use integration by parts to show that if $\alpha_2 > 1$, then for $0 < y < 1$,

$$F(y; \alpha_1, \alpha_2) = \frac{\Gamma(\alpha_1 + \alpha_2)}{\Gamma(\alpha_1 + 1)\Gamma(\alpha_2)} y^{\alpha_1}(1 - y)^{\alpha_2 - 1} + F(y; \alpha_1 + 1, \alpha_2 - 1).$$

(b) Let l and m be positive integers. Show that

$$F(y; l, m) = \sum_{k=l}^{l+m-1} \binom{l + m - 1}{k} y^k (1 - y)^{l+m-1-k}, \qquad 0 < y < 1.$$

(c) Let W have the binomial distribution with parameters n and π. Show that $P(W \geq w) = F(\pi; w, n+1-w)$ for $w = 1, \ldots, n$.

4.26   Determine the mean and variance of the ith order statistic corresponding to a random sample of size n from the uniform distribution on $[0, 1]$.

4.27   A candidate for mayor commissions a survey to estimate the proportion of the city's eligible voters who favor her candidacy. To obtain adequate financial support for her campaign, she needs a favorable response from a majority of those being surveyed. Determine the minimal size of the survey such that if 55% of the eligible voters favor her candidacy, then, with probability .99, so will a majority of those in the survey.

4.28   A crapshooter claimed that the croupier had given him a loaded die and, in particular, that the probability that the die shows one spot when it lands is greater than $1/6$. To settle the dispute, they decide to perform an experiment in which the die is rolled in an honest manner n times. Assume that the number of times Y that one spot shows on the n trials has a binomial distribution with parameters n and π. The crapshooter and croupier agree to choose y such that if $\pi = 1/6$, then $P(Y \geq y) \approx .01$. If $Y \geq y$, the croupier will give the crapshooter 100 blue chips that can be cashed in; otherwise, the crapshooter will withdraw his claim that the die is loaded. **(a)** Determine y when $n = 1000$ (by normal approximation). **(b)** When $n = 1000$, y is as in part (a) and $\pi = 1/5$, determine $P(Y \geq y)$. **(c)** Determine n and y such that $P(Y \geq y) \approx .01$ when $\pi = 1/6$ and $P(Y \geq y) \approx .99$ when $\pi = 1/5$.

4.29   A fair die is rolled 600 times in an honest manner. **(a)** Determine the mean and variance of the total number of spots that show. **(b)** Use normal approximation to determine the probability that the total number of spots that show is at least 2200.

4.30   A firm buys 500 identical components, whose failure times are independent and exponentially distributed with mean 100 hours. Determine the probability that at least 125 of the components will survive 150 hours or more by using **(a)** normal approximation; **(b)** normal approximation with the half-integer correction.

4.31   Suppose 60% of the (large number of) citizens in a certain country would express a favorable opinion of their leader when surveyed. Each of two survey organizations chooses 480 people at random from the population of citizens and finds the sample proportion of those expressing a favorable opinion of their leader. Determine the approximate probability that the two sample proportions differ by .05 or more.

4.3
The Multinomial Distribution

In the previous section, we considered experiments whose outcomes are classified into two types, which we referred to as success and failure. Sometimes outcomes are classified into three or more types. For example, we could classify a randomly selected individual in a certain population as having AIDS, being HIV-positive but not having AIDS, and not being HIV-positive.

EXAMPLE **4.26**   A box contains ten balls, of which five are red, three are white, and two are blue. Six times, a ball is selected at random from the box and then returned to the box. Determine the probability that a red ball is selected three times, a white ball twice, and a blue ball once.

Solution   Observe that on each trial a red ball is drawn out with probability .5, a white ball is drawn out with probability .3, and a blue ball is drawn out with probability .2. Since the sampling is done with replacement, the results on the six trials are independent of each other. Now the specific sequence red, red, red, white, white, blue has probability $(.5)(.5)(.5)(.3)(.3)(.2) = (.5)^3(.3)^2(.2)$. More generally, this is the probability of any specific sequence involving three reds, two whites, and one blue. The indicated sequence corresponds in an obvious manner to the partition $\{1, 2, 3\}, \{4, 5\}, \{6\}$ of the collection of six trials into three sets having sizes 3, 2, and 1, respectively (red on trials 1, 2, and 3; white on trials 4 and 5; and blue on trial 6). According to Theorem 4.5 in Section 4.1, there are

$$\frac{6!}{3!\,2!\,1!}$$

partitions of $\{1, 2, 3, 4, 5, 6\}$ into three sets having sizes 3, 2, and 1, respectively. These partitions are in one-to-one correspondence with the sequences of reds, whites, and blues involving three reds, two whites, and one blue in some order, and each such sequence has probability $(.5)^3(.3)^2(.2)$ of occurring. Thus, the probability of getting three reds, two whites, and one blue in some order is

$$\frac{6!}{3!\,2!\,1!}(.5)^3(.3)^2(.2) = .135. \quad \blacksquare$$

More generally, consider a random experiment whose outcome is one of M (mutually exclusive) types, where $M \geq 2$. Let π_m denote the probability that the outcome is of type m for $1 \leq m \leq M$. Then $\pi_1 + \cdots + \pi_M = 1$. Consider now n independent trials of the experiment. For $1 \leq m \leq M$, let Y_m denote the number of these trials in which the outcome is of type m, which has the binomial distribution with parameters n and π_m. Observe also that $Y_1 + \cdots + Y_M = n$.

THEOREM **4.15**   The joint probability function of $Y_1, \ldots, Y_M$ is given by

$$P(Y_1 = y_1, \ldots, Y_M = y_M) = \frac{n!}{y_1!\cdots y_M!}\,\pi_1^{y_1}\cdots\pi_M^{y_M},$$

where $y_1, \ldots, y_M$ are nonnegative integers whose sum is n, and

$$P(Y_1 = y_1, \ldots, Y_M = y_M) = 0$$

otherwise.

Proof   Let $m_1, \ldots, m_n$ be a sequence of integers in $\{1, \ldots, M\}$ such that there are y_1 1's, $\ldots, y_M$ M's in the sequence. Then the probability that the first outcome is of type m_1, the second outcome is of type m_2, and so forth, equals

$$\pi_{m_1}\cdots\pi_{m_n} = \pi_1^{y_1}\cdots\pi_M^{y_M}.$$

According to Theorem 4.9 in Section 4.1, there are

$$\frac{n!}{y_1! \cdots y_M!}$$

such sequences. Therefore,

$$P(Y_1 = y_1, \ldots, Y_M = y_M) = \frac{n!}{y_1! \cdots y_M!} \pi_1^{y_1} \cdots \pi_M^{y_M}. \quad \blacksquare$$

The joint distribution of the random variables $Y_1, \ldots, Y_M$ in Theorem 4.15 is referred to as the *multinomial distribution* with parameters n and $\pi_1, \ldots, \pi_M$. Here n is a positive integer and $\pi_1, \ldots, \pi_M$ are nonnegative numbers whose sum is 1. For some purposes, it is convenient to require that the parameters $\pi_1, \ldots, \pi_M$ all be positive.

EXAMPLE 4.27   Use Theorem 4.15 to solve Example 4.26.

Solution   Theorem 4.15 is applicable with red balls being type 1, white balls being type 2, and blue balls being type 3. Here $M = 3$, $\pi_1 = .5$, $\pi_2 = .3$, $\pi_3 = .2$, $n = 6$, Y_1 is the number of times a red ball is selected, Y_2 is the number of times a white ball is selected, Y_3 is the number of times a blue ball is selected, $y_1 = 3$, $y_2 = 2$, and $y_3 = 1$. Thus, the probability that a red ball is selected three times, a white ball twice, and a blue ball once is given by

$$P(Y_1 = 3, Y_2 = 2 \text{ and } Y_3 = 1) = \frac{6!}{3! \, 2! \, 1!}(.5)^3(.3)^2(.2) \doteq .135. \quad \blacksquare$$

EXAMPLE 4.28   Consider three trials of sampling with replacement from the box containing five red balls, three white balls, and two blue balls. Determine the probability that

(a) the same color shows on all three trials;

(b) a different color shows on each trial.

Solution   Let Y_1 denote the number of times red shows on the three trials, Y_2 the number of times white shows, and Y_3 the number of times blue shows. Then Y_1, Y_2, and Y_3 have the trinomial joint distribution with parameters $n = 3$, $\pi_1 = .5$, $\pi_2 = .3$, and $\pi_3 = .2$.

(a) The probability that the same color shows on all three trials is given by

$$P(Y_1 = 3, Y_2 = 0, Y_3 = 0) + P(Y_1 = 0, Y_2 = 3, Y_3 = 0)$$
$$+ P(Y_1 = 0, Y_2 = 0, Y_3 = 3),$$

which equals

$$P(Y_1 = 3) + P(Y_2 = 3) + P(Y_3 = 3) = (.5)^3 + (.3)^3 + (.2)^3 = .16.$$

(b) The probability that a different color shows on each trial is given by

$$P(Y_1 = 1, Y_2 = 1, Y_3 = 1) = \frac{3!}{1! \, 1! \, 1!}(.5)(.3)(.2) = .18. \quad \blacksquare$$

EXAMPLE **4.29**   Consider repeated sampling with replacement from the box containing five red balls, three white balls, and two blue balls. Let the sampling continue until blue has shown five times. Determine the probability that, by then, red will have shown (exactly) eight times and white will have shown six times.

Solution   The stated conditions hold if and only if (1) on the first 18 trials, red shows eight times, white shows six times, and blue shows four times and (2) blue shows on the 19th trial. Thus, the desired probability is given by

$$\frac{18!}{8!\,6!\,4!}(.5)^8(.3)^6(.2)^4(.2) = \frac{18!}{8!\,6!\,4!}(.5)^8(.3)^6(.2)^5. \quad \blacksquare$$

EXAMPLE **4.30**   Suppose the lifetime of a certain type of light bulb has the gamma distribution with shape parameter $\alpha = 2$ and mean 1 year. Seven such bulbs are purchased and installed immediately. Determine the probability that two of them fail during the first 6 months, two fail during the next 6 months, and the other three last more than 1 year.

Solution   Let W have the gamma distribution with shape parameter $\alpha = 2$, scale parameter β, and mean $\mu = 1$. Then $1 = \mu = \alpha\beta = 2\beta$, so $\beta = 1/2$. Let F denote the distribution function of W. Then [see Equation (8) in Section 3.1] $F(w) = 1 - e^{-2w}(1 + 2w)$ for $w \geq 0$.

Let Y_1 denote the number of light bulbs that fail during the first 6 months (type 1 outcome), let Y_2 denote the number that fail during the next 6 months (type 2 outcome), and let Y_3 denote the number that last more than 1 year (type 3 outcome). Then Y_1, Y_2, and Y_3 have the trinomial joint distribution with parameters $n = 7$,

$$\pi_1 = P(W \leq .5) = F(.5) = 1 - 2e^{-1} \doteq .264,$$

$$\pi_2 = P(.5 < W \leq 1) = F(1) - F(.5) = 2e^{-1} - 3e^{-2} \doteq .330,$$

and

$$\pi_3 = P(W > 1) = 1 - F(1) = 3e^{-2} \doteq .406.$$

Thus, the probability that two light bulbs fail during the first 6 months, two fail during the next 6 months, and the other three last more than 1 year is given by

$$P(Y_1 = 2,\ Y_2 = 2,\ \text{and}\ Y_3 = 3) = \frac{7!}{2!\,2!\,3!}\pi_1^2\pi_2^2\pi_3^3$$

$$\doteq 210(.264)^2(.330)^2(.406)^3 \doteq .107. \quad \blacksquare$$

EXAMPLE **4.31**   A fair die is rolled 14 times in an honest manner. Determine the probability that

(a) side 1 shows (exactly) twice;   **(b)** sides 1 through 5 each show twice.

Solution   Let $Y_1, \ldots, Y_6$, respectively, denote the number of times that sides $1, \ldots, 6$ show on the 14 rolls. Then these random variables have the multinomial joint distribution with parameters $n = 14$, $\pi_1 = 1/6, \ldots, \pi_6 = 1/6$.

(a) The number of times Y_1 that side 1 shows has the binomial distribution with parameters $n = 14$ and $\pi = 1/6$. Thus, the probability that side 1 shows twice is given by

$$P(Y_1 = 2) = \binom{14}{2}\left(\frac{1}{6}\right)^2\left(\frac{5}{6}\right)^{12} \doteq .284.$$

(b) It follows from Theorem 4.15 that the probability that sides 1 through 5 each show twice and hence that side 6 shows four times is given by

$$P(Y_1 = 2, Y_2 = 2, Y_3 = 2, Y_4 = 2, Y_5 = 2, Y_6 = 4)$$

$$= \frac{14!}{(2!)^5 4!}\left(\frac{1}{6}\right)^{14} \doteq .00145. \quad \blacksquare$$

EXAMPLE **4.32**   A box contains six tickets, three of which have value 1, two of which have value 2, and one of which has value 5. Tickets are selected one at a time from the box by sampling with replacement. Determine the probability that the sum of the values on the first four tickets so selected equals 8.

Solution   There are two ways that the sum of the values of the four tickets can equal 8: (1) four 2's; (2) three 1's and one 5. The probability of getting four 2's is given by $(1/3)^4 = 1/81$, and the probability of getting three 1's and one 5 is given by

$$\frac{4!}{3!\,0!\,1!}\left(\frac{1}{2}\right)^3\left(\frac{1}{3}\right)^0\left(\frac{1}{6}\right)^1 = \frac{1}{12}.$$

Thus, the desired probability is given by $1/81 + 1/12 = 31/324.$ $\blacksquare$

EXAMPLE **4.33**   On each of ten trials, five fair dice are rolled in an honest manner, and the number y of dice showing one spot is noted. The observed results on the ten trials are summarized in the following table:

y	0	1	2	3 or more
Frequency	3	4	2	1

Determine the probability of getting this table of observed results on the ten trials.

Solution   Here we think of the categories 0, 1, 2, 3 or more as being fixed in advance. The probability that y of the five dice shows one spot on a single trial equals

$$\binom{5}{y}\left(\frac{1}{6}\right)^y\left(\frac{5}{6}\right)^{5-y}.$$

Thus, the probability that none (0) of the five dice shows one spot is given by

$$\pi_1 = \left(\frac{5}{6}\right)^5 \doteq .4019,$$

the probability that 1 of them shows one spot is given by

$$\pi_2 = 5\left(\frac{1}{6}\right)\left(\frac{5}{6}\right)^4 \doteq .4019,$$

and the probability that 2 of them show one spot is given by

$$\pi_3 = \binom{5}{2}\left(\frac{1}{6}\right)^2\left(\frac{5}{6}\right)^3 \doteq .1608,$$

so the probability that 3 or more of them show one spot is given by

$$\pi_4 = 1 - \pi_1 - \pi_2 - \pi_3 \doteq .0355.$$

Consequently, the probability that $y = 0$ occurs on three trials, $y = 1$ occurs on four trials, $y = 3$ occurs on two trials, and $y \geq 3$ occurs on one trial is given by

$$\frac{10!}{3!\,4!\,2!\,1!}\pi_1^3\pi_2^4\pi_3^2\pi_4 \doteq \frac{10!}{3!\,4!\,2!}(.4019)^7(.1608)^2(.0355) \doteq .0196. \quad \blacksquare$$

EXAMPLE 4.34 The manuscript for a book has n typos. Frankie and Johnny separately check the manuscript for typos. Frankie has probability p_1 of finding any given typo, and Johnny has probability p_2 of finding it. Let Y_1 be the number of typos found by both Frankie and Johnny, let Y_2 be the number found by Frankie but missed by Johnny, let Y_3 be the number found by Johnny but missed by Frankie, and let Y_4 be the number missed by both Frankie and Johnny. Under the relevant assumptions of independence, determine the joint distribution of Y_1, Y_2, Y_3, and Y_4.

Solution Think of the n typos as corresponding to the n trials of an experiment. The outcome of each trial can be one of four types: Type 1 means that the typo was found by both Frankie and Johnny; type 2 means that the typo was found by Frankie but missed by Johnny; type 3 means that the typo was found by Johnny but missed by Frankie; type 4 means that the typo was missed by both Frankie and Johnny. The probability of a type 1 outcome is given by $\pi_1 = p_1 p_2$, the probability of a type 2 outcome is given by $\pi_2 = p_1(1 - p_2)$, the probability of a type 3 outcome is given by $\pi_3 = p_2(1 - p_1)$, and the probability of a type 4 outcome is given by $\pi_4 = (1 - p_1)(1 - p_2)$. Observe that Y_m is the number of type m outcomes for $m = 1, \ldots, 4$. Thus, the joint distribution of Y_1, Y_2, Y_3, and Y_4 is multinomial with parameters $n, p_1 p_2, p_1(1 - p_2), p_2(1 - p_1)$, and $(1 - p_1)(1 - p_2)$. $\quad \blacksquare$

Problems

4.32 A box has ten balls, of which four are red, three are white, two are blue, and one is black. Four times, a ball is selected at random from the box and then returned to the box. Determine the probability that each color is selected once.

4.33 A deck of playing cards has 13 clubs, 13 diamonds, 13 hearts, and 13 spades. Thirteen times, a card is selected at random from the deck and then returned to the deck. Determine the probability that a club is selected five times, a diamond four times, a heart three times, and a spade once.

4.34 Let $W_1, \ldots, W_{10}$ be independent random variables, each having the exponential distribution with mean 1; let Y_1 be the number of these random variables that lie in the interval $[\,0, \log 2\,]$; and let Y_2 be the number lying in the interval $(\log 2, \log 3\,]$. Determine $P(Y_1 = 5, Y_2 = 2)$.

4.35   A fair die is rolled $6n$ times in an honest manner. **(a)** Determine the probability that each side shows exactly n times. **(b)** Use Stirling's formula to obtain an approximation to the answer to part (a).

4.36   Suppose an individual's scores on repetitions of a certain task are independently and normally distributed with mean 100 and standard deviation 20. A score is said to be low if it is below 70, high if it is above 120, and medium if it is between 70 and 120. Determine the probability that the individual scores two lows, five mediums, and three highs in ten repetitions of the task.

4.37   On each of nine trials, a fair coin is repeatedly tossed until heads shows for the first time and the required number of tosses is noted. Determine the probability that on three trials one toss is required; on four trials two tosses are required; and on the remaining two trials, three or more tosses are required.

4.38   A box contains ten tickets, four of which have value 1, three of which have value 2, two of which have value 3, and one of which has value 4. Tickets are selected one at a time from the box by sampling with replacement. Determine the probability that the sum of the values on the first four tickets so selected equals 8.

4.39   Consider a random experiment whose outcome can be one of three types. Let π_1, π_2, and π_3 denote the probabilities of type 1, type 2, and type 3 outcomes, respectively, which are all assumed to be positive. Consider independent trials of this experiment. Let Y_1 and Y_2 denote the numbers of type 1 and type 2 outcomes, respectively, before the rth type 3 outcome, where r is a positive integer. The joint distribution of Y_1 and Y_2 is a special case of the *negative multinomial distribution*. Determine the joint probability function of Y_1 and Y_2.

4.40   A fair die is repeatedly rolled in an honest manner until side 6 appears for the rth time, where r is a positive integer. Determine the probability that each of the other five sides will also have appeared exactly r times by then.

4.41   A fair die is rolled three times in an honest manner. Determine the probability that a different side shows on each roll.

4.42   A fair die is repeatedly rolled in an honest manner until some side appears twice. Determine the probability that four rolls are required.

4.43   Balls are selected one at a time by sampling with replacement from the box having five red balls, three white balls, and two blue balls. Determine the probability that, by the time a blue ball is selected for the first time, a red ball will have been selected (exactly) three times and a white ball twice and, by the time a blue ball is selected for the second time, a red ball will have been selected eight times (in total) and a white ball six times.

4.4

The Poisson Distribution

Recall that the power series expansion for the exponential function is given by

$$e^t = 1 + t + \frac{t^2}{2!} + \frac{t^3}{3!} + \cdots = \sum_{y=0}^{\infty} \frac{t^y}{y!}, \qquad -\infty < t < \infty.$$

EXAMPLE **4.35**   Let $\lambda > 0$. Determine a probability function f on $\{0, 1, 2, \ldots\}$ of the form

$$f(y) = \frac{1}{a} \frac{\lambda^y}{y!}, \qquad y = 0, 1, 2, \ldots .$$

Solution   According to the power series expansion of the exponential function,

$$\sum_{y=0}^{\infty} \frac{1}{a} \frac{\lambda^y}{y!} = \frac{1}{a} \sum_{y=0}^{\infty} \frac{\lambda^y}{y!} = \frac{e^\lambda}{a},$$

which equals 1 if and only if $a = e^\lambda$. Since

$$\frac{\lambda^y}{y!} e^{-\lambda} \geq 0, \qquad y = 0, 1, 2, \ldots,$$

the desired probability function is given by

$$f(y) = \frac{\lambda^y}{y!} e^{-\lambda}, \qquad y = 0, 1, 2, \ldots, \tag{10}$$

and $f(y) = 0$ elsewhere. In particular,

$$f(0) = e^{-\lambda}, \quad f(1) = \lambda e^{-\lambda}, \quad f(2) = \frac{\lambda^2}{2} e^{-\lambda} \quad \text{and} \quad f(3) = \frac{\lambda^3}{6} e^{-\lambda}. \quad \blacksquare$$

Let $\lambda > 0$. The distribution on $\{0, 1, 2, \ldots\}$ having the probability function given by (10) is referred to as the *Poisson distribution* with parameter λ.

THEOREM **4.16**   The Poisson distribution with parameter λ has mean λ, variance λ, and standard deviation $\sqrt{\lambda}$.

Proof   Let Y have the Poisson distribution with parameter λ. Then the mean of Y is given by

$$EY = \lambda e^{-\lambda} + 2 \frac{\lambda^2}{2!} e^{-\lambda} + 3 \frac{\lambda^3}{3!} e^{-\lambda} + \cdots = \lambda e^{-\lambda} \left(1 + \lambda + \frac{\lambda^2}{2!} + \cdots \right) = \lambda e^{-\lambda} e^\lambda,$$

so $EY = \lambda$. A simple way to determine the second moment of Y is to start out by obtaining $E[\,Y(Y-1)\,]$. Now

$$E[\,Y(Y-1)\,] = 2 \cdot 1 \frac{\lambda^2}{2!} e^{-\lambda} + 3 \cdot 2 \frac{\lambda^3}{3!} e^{-\lambda} + 4 \cdot 3 \frac{\lambda^4}{4!} e^{-\lambda} + \cdots$$

$$= \lambda^2 e^{-\lambda} \left(1 + \lambda + \frac{\lambda^2}{2!} + \cdots \right)$$

$$= \lambda^2 e^{-\lambda} e^{\lambda}$$

$$= \lambda^2,$$

so the second moment of Y is given by $E(Y^2) = E[\, Y(Y-1)\,] + EY = \lambda^2 + \lambda$. Consequently, the variance of Y is given by

$$\text{var}(Y) = E(Y^2) - (EY)^2 = \lambda^2 + \lambda - \lambda^2 = \lambda,$$

so $\text{SD}(Y) = \sqrt{\lambda}$. ∎

In light of the first conclusion of Theorem 4.16, we refer to the Poisson distribution with parameter λ as the Poisson distribution with mean λ. Observe that the variance of a Poisson distribution coincides with its mean.

EXAMPLE 4.36   Suppose the number N of patent applications submitted by a company during a 1-year period is a random variable having the Poisson distribution with mean λ and the various applications independently have probability π (where $0 < \pi < 1$) of eventually being approved. Determine the distribution of the number of patent applications during the 1-year period that are eventually approved.

Solution   We assume that the distribution of the number of applications during the 1-year period that are eventually approved coincides with the distribution of $S_N = I_1 + \cdots + I_N$, where $I_1, I_2, \ldots$ are random variables, each having the Bernoulli distribution with parameter π, and the random variables $N, I_1, I_2, \ldots$ are independent. (Here $S_0 = 0$.) Observe that $S_n = I_1 + \cdots + I_n$ has the binomial distribution with parameters n and π for $n \geq 1$, that N and S_n are independent for $n \geq 0$, and that $N \geq S_N$. Thus, for y a nonnegative integer,

$$P(S_N = y) = \sum_{n=y}^{\infty} P(N = n, S_N = y)$$

$$= \sum_{n=y}^{\infty} P(N = n, S_n = y)$$

$$= \sum_{n=y}^{\infty} P(N = n)P(S_n = y)$$

$$= \sum_{n=y}^{\infty} \frac{\lambda^n}{n!} e^{-\lambda} \frac{n!}{y!\,(n-y)!} \pi^y (1 - \pi)^{n-y}$$

$$= \frac{(\lambda\pi)^y}{y!} e^{-\lambda} \sum_{n=y}^{\infty} \frac{[\,\lambda(1-\pi)\,]^{n-y}}{(n-y)!}$$

$$= \frac{(\lambda\pi)^y}{y!} e^{-\lambda} \sum_{m=0}^{\infty} \frac{[\,\lambda(1-\pi)\,]^m}{m!}$$

$$= \frac{(\lambda\pi)^y}{y!}e^{-\lambda}e^{\lambda(1-\pi)}$$

$$= \frac{(\lambda\pi)^y}{y!}e^{-\lambda\pi}.$$

Consequently, S_N has the Poisson distribution with mean $\lambda\pi$. ∎

The next example is a modification of Example 4.34 in Section 4.3.

EXAMPLE **4.37** The manuscript for a book has N typos, where N has the Poisson distribution with mean λ. Frankie and Johnny separately check the manuscript for typos. Frankie has probability p_1 of finding any given typo, and Johnny has probability p_2 of finding it. Let Y_1 be the number of typos found by both Frankie and Johnny, let Y_2 be the number found by Frankie but missed by Johnny, let Y_3 be the number found by Johnny but missed by Frankie, and let Y_4 be the number missed by both Frankie and Johnny. Under the relevant assumptions of independence, determine the joint distribution of Y_1, Y_2, Y_3, and Y_4.

Solution Let $N, I_1, J_1, I_2, J_2, \ldots$ be independent random variables such that N has the Poisson distribution with mean λ, I_i has the Bernoulli distribution with parameter p_1, and J_i has the Bernoulli distribution with parameter p_2 for all i. Given the nonnegative integer n, set

$$S_{1n} = I_1 J_1 + \cdots + I_n J_n,$$

$$S_{2n} = I_1(1 - J_1) + \cdots + I_n(1 - J_n),$$

$$S_{3n} = (1 - I_1)J_1 + \cdots + (1 - I_n)J_n,$$

and

$$S_{4n} = (1 - I_1)(1 - J_1) + \cdots + (1 - I_n)(1 - J_n).$$

(Here $S_{10} = S_{20} = S_{30} = S_{40} = 0$.) It follows from the solution to Example 4.34 in Section 4.3 that, for $n \geq 1$, the joint distribution of S_{1n}, S_{2n}, S_{3n}, and S_{4n} is multinomial with parameters n, $p_1 p_2$, $p_1(1 - p_2)$, $(1 - p_1)p_2$, and $(1 - p_1)(1 - p_2)$. In determining the joint distribution of Y_1, Y_2, Y_3, and Y_4, we can assume that $Y_1 = S_{1N}$, $Y_2 = S_{2N}$, $Y_3 = S_{3N}$, and $Y_4 = S_{4N}$. In this context, we are thinking of I_i as denoting the indicator random variable for Frankie finding the ith typo and of J_i as the indicator random variable for Johnny finding that typo. Observe that $Y_1 + Y_2 + Y_3 + Y_4 = n$. Let y_1, y_2, y_3, and y_4 be nonnegative integers and set $n = y_1 + y_2 + y_3 + y_4$. Then

$$P(Y_1 = y_1, Y_2 = y_2, Y_3 = y_3, Y_4 = y_4)$$

$$= P(N = n, Y_1 = y_1, Y_2 = y_2, Y_3 = y_3, Y_4 = y_4)$$

$$= P(N = n, S_{1n} = y_1, S_{2n} = y_2, S_{3n} = y_3, S_{4n} = y_4)$$

$$= P(N = n)P(S_{1n} = y_1, S_{2n} = y_2, S_{3n} = y_3, S_{4n} = y_4)$$

$$= \frac{\lambda^n}{n!} e^{-\lambda} \frac{n!}{y_1! y_2! y_3! y_4!} (p_1 p_2)^{y_1} [p_1(1 - p_2)]^{y_2}$$

$$\times [(1 - p_1)p_2]^{y_3} [(1 - p_1)(1 - p_2)]^{y_4}$$

$$= \frac{(\lambda p_1 p_2)^{y_1}}{y_1!} e^{-\lambda p_1 p_2} \frac{[\lambda p_1(1 - p_2)]^{y_2}}{y_2!} e^{-\lambda p_1(1 - p_2)}$$

$$\times \frac{[\lambda(1 - p_1)p_2]^{y_3}}{y_3!} e^{-\lambda(1 - p_1)p_2} \frac{[\lambda(1 - p_1)(1 - p_2)]^{y_4}}{y_4!} e^{-\lambda(1 - p_1)(1 - p_2)}.$$

Thus, Y_1, Y_2, Y_3, and Y_4 are independent Poisson random variables having means $\lambda p_1 p_2$, $\lambda p_1(1 - p_2)$, $\lambda(1 - p_1)p_2$, and $\lambda(1 - p_1)(1 - p_2)$, respectively.   ■

Let f denote the probability function of the Poisson distribution with mean λ. Then

$$\frac{f(y)}{f(y - 1)} = \frac{\dfrac{\lambda^y}{y!} e^{-\lambda}}{\dfrac{\lambda^{y-1}}{(y - 1)!} e^{-\lambda}}, \qquad y = 1, 2, \ldots,$$

which simplifies to

$$\frac{f(y)}{f(y - 1)} = \frac{\lambda}{y}, \qquad y = 1, 2, \ldots. \tag{11}$$

It follows from (11) that

$$f(y) = \frac{\lambda}{y} f(y - 1), \qquad y = 1, 2, \ldots. \tag{12}$$

Equation (12) suggests one reasonable way to compute the Poisson probability function: First, compute $f(0) = e^{-\lambda}$; then, successively compute $f(y)$ from $f(y - 1)$ for $y = 1, 2, \ldots$ by using (12).

Think of f as a probability function on $\{0, 1, 2, \ldots\}$. It follows from (11) or (12) that $f(y)$ is greater than, equal to, or less than $f(y - 1)$ according as y is less than, equal to, or greater than λ. Suppose λ is a positive integer. Then $\lambda - 1$ and λ are the only modes and the only global maxima of f. Also $f(y)$ is strictly increasing in y for $y \in \{0, \ldots, \lambda - 1\}, f(\lambda - 1) = f(\lambda)$, and $f(y)$ is strictly decreasing in y for $y \in \{\lambda, \lambda + 1, \ldots\}$. Suppose, instead, that λ is not a positive integer. Then $[\lambda]$ (the greatest integer in λ) is the unique mode and the unique location of the global maximum of f. Also $f(y)$ is strictly increasing in y for $y \in \{0, \ldots, [\lambda]\}$ and it is strictly decreasing in y for $y \in \{[\lambda], [\lambda + 1], \ldots\}$. Figure 4.5 illustrates these results.

EXAMPLE **4.38**   Let $Y_1, \ldots, Y_n$ be independent random variables having Poisson distributions with means $\lambda_1, \ldots, \lambda_n$, respectively. Determine the joint probability function of these random variables.

FIGURE 4.5

Poisson probability functions

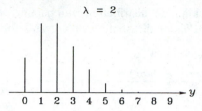

$\lambda = 2$

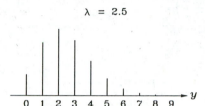

$\lambda = 2.5$

Solution   The joint probability function of $Y_1, \ldots, Y_n$ is given by

$$f_{Y_1, \ldots, Y_n}(y_1, \ldots, y_n) = \frac{\lambda_1^{y_1}}{y_1!} e^{-\lambda_1} \cdots \frac{\lambda_n^{y_n}}{y_n!} e^{-\lambda_n} = \frac{\lambda_1^{y_1} \cdots \lambda_n^{y_n}}{y_1! \cdots y_n!} e^{-(\lambda_1 + \cdots + \lambda_n)}$$

when $y_1, \ldots, y_n$ are nonnegative integers, and it equals zero otherwise.   ∎

THEOREM 4.17   Let $Y_1, \ldots, Y_n$ be independent random variables having Poisson distributions with means $\lambda_1, \ldots, \lambda_n$ respectively. Then $Y_1 + \cdots + Y_n$ has the Poisson distribution with mean $\lambda_1 + \cdots + \lambda_n$.

Proof   Clearly, $Y = Y_1 + \cdots + Y_n$ is a nonnegative, integer-valued random variable. Let y be a nonnegative integer. It follows from the solution to Example 4.38 that

$$f_Y(y) = \sum \frac{\lambda_1^{y_1} \cdots \lambda_n^{y_n}}{y_1! \cdots y_n!} e^{-(\lambda_1 + \cdots + \lambda_n)} = \frac{e^{-(\lambda_1 + \cdots + \lambda_n)}}{y!} \sum \frac{y!}{y_1! \cdots y_n!} \lambda_1^{y_1} \cdots \lambda_n^{y_n},$$

where the summations are over the collection of ordered n-tuples $(y_1, \ldots, y_n)$ of nonnegative integers adding up to y. By the multinomial formula,

$$\sum \frac{y!}{y_1! \cdots y_n!} \lambda_1^{y_1} \cdots \lambda_n^{y_n} = (\lambda_1 + \cdots + \lambda_n)^y.$$

Thus,

$$f_Y(y) = \frac{(\lambda_1 + \cdots + \lambda_n)^y}{y!} e^{-(\lambda_1 + \cdots + \lambda_n)}$$

for y a nonnegative integer and it equals zero otherwise, so Y has the Poisson distribution with mean $\lambda_1 + \cdots + \lambda_n$. (Alternatively, we could use the same argument for $n = 2$ and then get the result for general n by induction.)   ∎

Poisson Approximation

THEOREM 4.18   Let $0 < \lambda < \infty$. Then the binomial distribution with parameters n and $\pi = \lambda/n$ is approximately equal to the Poisson distribution with mean λ for $n \gg 1$.

Proof   Let y be a nonnegative integer. Then

$$\frac{n!}{n^y(n-y)!} = \frac{n}{n}\frac{n-1}{n}\cdots\frac{n-y+1}{n} = \left(1 - \frac{1}{n}\right)\cdots\left(1 - \frac{y-1}{n}\right),$$

so

$$\frac{\dfrac{n!}{y!\,(n-y)!}\left(\dfrac{\lambda}{n}\right)^y\left(1 - \dfrac{\lambda}{n}\right)^{n-y}}{\dfrac{\lambda^y}{y!}e^{-\lambda}}$$

$$= \left(1 - \frac{1}{n}\right)\cdots\left(1 - \frac{y-1}{n}\right)\left(1 - \frac{\lambda}{n}\right)^{-y}\left(1 - \frac{\lambda}{n}\right)^n e^\lambda. \quad \textbf{(13)}$$

Now

$$\lim_{n\to\infty}\left(1 - \frac{1}{n}\right)\cdots\left(1 - \frac{y-1}{n}\right) = 1 \quad \text{and} \quad \lim_{n\to\infty}\left(1 - \frac{\lambda}{n}\right)^{-y} = 1.$$

Also, as shown in calculus,

$$\lim_{n\to\infty}\left(1 - \frac{\lambda}{n}\right)^n = e^{-\lambda}.$$

Consequently, the right side of (13) approaches 1 as $n \to \infty$, which yields the desired result.   ∎

EXAMPLE 4.39   Recall Example 4.17 in Section 4.2, according to which Johnny makes 100 consecutive $1 bets, each on a single slot. Use the Poisson approximation to the binomial distribution to determine the probability that Johnny will be ahead as a result of the 100 bets.

Solution   Let Y denote the number of bets that Johnny wins, which has the binomial distribution with parameters $n = 100$ and $\pi = 1/38$. Then Y has approximately the Poisson distribution with mean $\lambda = 100/38 \doteq 2.632$. Recall from the solution to Example 4.17 in Section 4.2 that Johnny will be ahead as a result of the 100 bets if and only if $Y \geq 3$. Now

$$P(Y = 0) \approx e^{-\lambda} \doteq e^{-2.632} \doteq .0720,$$

$$P(Y = 1) \approx \lambda e^{-\lambda} \doteq (2.632)e^{-2.632} \doteq .1894,$$

and

$$P(Y = 2) \approx \frac{\lambda^2}{2}e^{-\lambda} \doteq \frac{(2.632)^2}{2}e^{-2.632} \doteq .2492,$$

so

$$P(Y \geq 3) = 1 - P(Y = 0) - P(Y = 1) - P(Y = 2)$$

$$\approx 1 - .0720 - .1894 - .2492$$

$$= .4895,$$

which is close to the value .4916 given in the solution to the cited example. ∎

EXAMPLE **4.40** Consider a given 10-gram mass of homogeneous radioactive material. Let Y denote the number of particle decays detected by a Geiger counter during a given time period. Suppose $EY = 10$. Explain why it is reasonable to assume a Poisson distribution for Y.

Solution Let n denote the number of particles that have the potential for decay and let π denote the probability that a given such particle decays during the time period and its decay is detected by the Geiger counter. From physics, it is known that n is extremely large and that the decay times are independent of each other. Thus, Y has a binomial distribution with parameters n and π. Set $\lambda = n\pi = EY = 10$. Since $n \gg 1$, the Poisson approximation to the distribution of Y is very accurate. ∎

Let $I_1, \ldots, I_n$ be independent random variables such that I_i has the Bernoulli distribution with parameter π_i for $1 \leq i \leq n$ and set $Y = I_1 + \cdots + I_n$. Then Y has mean $\lambda = \pi_1 + \cdots + \pi_n$. If $\pi_1 = \cdots = \pi_n$, then Y has a binomial distribution; hence, under a further condition, Y has approximately the Poisson distribution with mean λ, as we have already seen. More generally, even when the π_i's do not coincide, it can be shown that if $\pi_1, \ldots, \pi_n$ are all close to zero, then the Poisson distribution with mean λ is a good approximation to the distribution of Y.

EXAMPLE **4.41** A computer software company has a "telephone hotline" to provide assistance to its 100,000 customers. Suppose the mean number of customers who call one or more times during a given 1-hour period is 20. Explain why it is reasonable to assume a Poisson distribution for the number of customers that call during such a period.

Solution Let n denote the number of the company's customers, let π_i denote the probability that the ith customer will call during the given period, and let I_i denote the indicator random variable that is 1 if the ith customer calls during that period and is 0 otherwise. Then I_i has a Bernoulli distribution with parameter π_i. Let Y denote the total number of customers that call during the given period. Then $Y = I_1 + \cdots + I_n$. It is reasonable to assume that $I_1, \ldots, I_n$ are independent and that $\pi_1, \ldots, \pi_n$ are all close to 0. Thus, the Poisson distribution with parameter $\lambda = \pi_1 + \cdots + \pi_n$ should provide a good approximation to the distribution of Y. ∎

Normal Approximation to the Poisson Distribution

Let Y have the Poisson distribution with mean $\mu = \lambda$, which has variance $\sigma^2 = \lambda$ and standard deviation $\sigma = \sqrt{\lambda}$. The probability function f of Y is given by

$$f(y) = \frac{\lambda^y}{y!} e^{-\lambda}, \qquad y \text{ a nonnegative integer,}$$

and $f(y) = 0$ for other values of y. Using Stirling's formula to approximate $y!$, we can show that if $\sigma^2 = \lambda$ is large, then Properties (6)–(8) in Section 4.2 are valid in the present context. Figure 4.6 shows a plot of the Poisson probability function together with its normal approximation when $\lambda = 10$.

FIGURE   4.6

Poisson probability function with mean 10 and its normal approximation

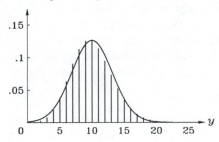

EXAMPLE 4.42   Use normal approximation to determine λ such that if Y has the Poisson distribution with mean λ, then $P(0.9\lambda \leq Y \leq 1.1\lambda) \approx .9$.

Solution   By normal approximation,

$$P(0.9\lambda \leq Y \leq 1.1\lambda) \approx \Phi\left(\frac{1.1\lambda - \lambda}{\sqrt{\lambda}}\right) - \Phi\left(\frac{0.9\lambda - \lambda}{\sqrt{\lambda}}\right) = 2\Phi\left(\frac{\sqrt{\lambda}}{10}\right) - 1 = .9$$

if $\sqrt{\lambda}/10 = z_{.95}$ and hence if $\lambda = 100 z_{.95}^2 \doteq 100(1.645)^2 \doteq 270.6$.   ∎

Problems

4.44   Let Y have the Poisson distribution with parameter $\lambda = 1$. **(a)** Determine the numerical values of the probability function of Y rounded off to the nearest multiple of .001. **(b)** Use these numerical values to determine $P(2 \leq Y \leq 4)$. **(c)** Use these values to determine $P(Y > 3)$.

4.45   Determine the maximum value of the probability function of the Poisson distribution with mean 5.

4.46 Let $Y_1, \ldots, Y_n$ be independent random variables, each having a Poisson distribution with mean λ. Show that $\bar{Y} = (Y_1 + \cdots + Y_n)/n$ is an unbiased estimate of the common variance of $Y_1, \ldots, Y_n$ and determine the variance of this estimate.

4.47 Let Y_1 and Y_2 be independent random variables, each having the Poisson distribution with mean 1, and set $W = \min(Y_1, Y_2)$. Determine $P(W = 1)$.

4.48 Let Y have the Poisson distribution with mean λ. Determine the probability that **(a)** Y is even (as a hint, add the power series for e^λ and $e^{-\lambda}$); **(b)** Y is odd.

4.49 According to a theoretical model for a certain random variable, it has the Poisson distribution with mean 10, and hence its probability function is given by

$$f(y) = \frac{10^y}{y!} e^{-10}, \qquad y = 0, 1, 2, \ldots .$$

The observed value of Y equals 17. According to the theoretical model, the probability of getting this observed value is given by $f(17) \doteq .013$. Under the theoretical model, determine the probability that $f(Y) \leq f(17)$. (Roughly speaking, to the extent that this answer is small, we are led to doubt the accuracy of the theoretical model.)

4.50 Suppose the probability of getting a rare, noncontagious disease in a given year is 10^{-4}. Consider a town of size 10,000. Determine the probability that **(a)** five or more of the townspeople come down with the disease in a given year; **(b)** over a 100-year period there are two or more years when five or more townspeople come down with the disease.

4.51 The number of work-related injuries at a certain small company in a given week has the Poisson distribution with mean 1/2, and the numbers of injuries for different weeks are independent. Determine the probability of getting the results in the following table for a 1-year (52-week) period:

y	0	1	2 or more
Frequency	30	16	6

4.52 Consider the setup of Example 4.36 and its solution. Let Y_1 denote the number of patent applications during the 1-year period that are eventually approved and let $Y_2 = N - Y_1$ denote the number that are not approved. According to the solution to Example 4.36, Y_1 has the Poisson distribution with mean $\lambda\pi$. Similarly, Y_2 has the Poisson distribution with mean $\lambda(1 - \pi)$. Show that Y_1 and Y_2 are independent. [*Hint:* Note that, for y_1 and y_2 nonnegative integers, $Y_1 = y_1$ and $Y_2 = y_2$ if and only if $N = y_1 + y_2$ and $S_{y_1 + y_2} = y_1$.]

4.53 Use Stirling's formula to show that if Y has a Poisson distribution with parameter λ, where λ is a large positive integer, then $P(Y = \lambda) \approx 1/\sqrt{2\pi\lambda}$.

4.54 Let Y denote the number of times the ball lands in slot 00 in 100 trials in roulette. Determine $P(2 \leq Y \leq 5)$ by using **(a)** the formula for the binomial probability function; **(b)** Poisson approximation.

4.55 Let Y have the Poisson distribution with mean 25. Determine the normal approximation to **(a)** $P(Y = 25)$; **(b)** $P(Y \leq 30)$; **(c)** the upper decile of Y.

4.5
The Poisson Process

Let Y have the gamma distribution with shape parameter α and scale parameter β. We refer to $\lambda = 1/\beta$ as the *inverse-scale parameter* of the distribution. The density function of Y can be written in terms of α and λ as $f(y) = 0$ for $y \le 0$ and

$$f(y) = \frac{\lambda^\alpha y^{\alpha-1}}{\Gamma(\alpha)} e^{-\lambda y}, \qquad y > 0.$$

Similarly, the mean and variance of Y are given by $EY = \alpha/\lambda$ and $\mathrm{var}(Y) = \alpha/\lambda^2$.

Let n be a positive integer. It follows from Theorem 3.1 in Section 3.1 that the distribution function of the gamma distribution with shape parameter n and inverse-scale parameter λ is given by $F(y) = 0$ for $y \le 0$ and

$$F(y) = 1 - e^{-\lambda y} \sum_{m=0}^{n-1} \frac{(\lambda y)^m}{m!}, \qquad y > 0. \tag{14}$$

As a special case, the density function of the exponential distribution with inverse-scale parameter λ is given by $f(y) = 0$ for $y \le 0$ and $f(y) = \lambda e^{-\lambda y}$ for $y > 0$, its distribution function is given by $F(y) = 0$ for $y \le 0$ and $F(y) = 1 - e^{-\lambda y}$ for $y > 0$, and its mean and variance are given by $EY = 1/\lambda$ and $\mathrm{var}(Y) = 1/\lambda^2$.

Let $W_1, W_2, \ldots$ be independent random variables, each having the exponential distribution with inverse-scale parameter λ, and set $T_n = W_1 + \cdots + W_n$ for $n \ge 1$. It follows from Corollary 3.1 in Section 3.1 that T_n has the gamma distribution with shape parameter n and inverse-scale parameter λ.

Think of W_i as the failure time of the ith component once it is installed. Suppose the first component is installed at time zero and a fresh component is installed as soon as the one currently installed fails. Then T_n is the failure time of the nth component that is installed. For $t \ge 0$, let $N(t)$ denote the number of components that have been installed and failed by time t. Then $N(0) = 0$ with probability 1. Let $t > 0$ and let n be a nonnegative integer. Then $N(t) \ge n+1$ if and only if $T_{n+1} \le t$ [that is, at least $n+1$ components have been installed and have failed by time t if and only if the $(n+1)$st component was installed and has failed by time t.] Thus by Equation (14) with n replaced by $n+1$ and y replaced by t,

$$P(N(t) \ge n+1) = P(T_{n+1} \le t) = 1 - \sum_{m=0}^{n} \frac{(\lambda t)^m}{m!} e^{-\lambda t},$$

so

$$P(N(t) \le n) = \sum_{m=0}^{n} \frac{(\lambda t)^m}{m!} e^{-\lambda t}.$$

Since $P(N(t) = n) = P(N(t) \le n) - P(N(t) \le n - 1)$, we see that

$$P(N(t) = n) = \frac{(\lambda t)^n}{n!} e^{-\lambda t}, \qquad n = 0, 1, 2, \ldots.$$

Therefore, $N(t)$ has the Poisson distribution with mean λt.

We have seen that the random variables $N(t)$, $t \geq 0$, satisfy the following two properties:

(i) $N(0) = 0$ with probability 1;

(ii) $N(t)$ has the Poisson distribution with mean λt for $t > 0$.

In advanced probability theory, it is shown that these random variables satisfy two additional properties, the first of which is a generalization of (ii):

(iii) $N(t) - N(s)$ has the Poisson distribution with mean $\lambda(t - s)$ for $0 \leq s < t$;

(iv) $N(t_1)$, $N(t_2) - N(t_1)$, ..., $N(t_n) - N(t_{n-1})$ are independent for $0 \leq t_1 < \cdots < t_n$.

Consider events occurring in time, starting at time zero. For $t \geq 0$, let $N(t)$ denote the number of events that have occurred by time t. Then, for $0 \leq s < t$, $N(t) - N(s)$ is the number of events that occur after time s but by time t. We say that the events occur according to the *Poisson process* with *rate* λ if $N(t)$, $t \geq 0$, satisfy properties (i), (iii), and (iv). (Sometimes we refer to λ as the mean rate, as in "accidents occur at the mean rate of ten per year.")

Let this be the case. Then, for fixed $t \geq 0$, the probability that an event occurs exactly at time t equals zero. Moreover, with probability 1, no two events occur simultaneously. For $n \geq 1$, let T_n denote the time until the nth event occurs. Also set $W_1 = T_1$ and $W_n = T_n - T_{n-1}$ for $n \geq 2$. Then W_n is the waiting time between the $(n - 1)$th event to occur and the nth event. Figure 4.7 illustrates the relationship between W_1, W_2, W_3 and T_1, T_2, T_3.

FIGURE 4.7

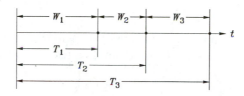

It can be shown that the random variables $W_1, W_2, \ldots$ are independent and that each has the exponential distribution with inverse-scale parameter λ. Since $T_n = W_1 + \cdots + W_n$, we conclude that T_n has the gamma distribution with shape parameter n and inverse-scale parameter λ.

EXAMPLE **4.43**   Consider events occurring in time, starting at time zero, according to a Poisson process with rate λ. Given $s, t \in \mathbb{R}$ with $0 \leq s < t$, determine the probability that no events occur during the time interval $(s, t]$.

Solution   The number $N(t) - N(s)$ of events that occur during the time interval $(s, t]$ has the Poisson distribution with mean $\lambda(t - s)$. Thus, the probability that this number equals zero is $e^{-\lambda(t-s)}$.   ∎

The Poisson process could provide a reasonable model for the occurrences in time of a noncommunicable disease (such as a particular form of cancer) in a given community. It would not be appropriate, however, for the occurrences of a communicable disease such as the flu, which tend to cluster in time. Similarly, it would not be completely appropriate to model the occurrences of earthquakes, which tend to trigger other earthquakes as aftershocks.

Consider as events the vehicular crossings of a particular point on a particular lane of a busy street. There are a number of sources of inaccuracy in using the Poisson process to model such events, including spacings between vehicles, traffic lights, variation in traffic with time of day and day of week, sporting and other special events, accidents, and flat tires.

EXAMPLE 4.44   In the R & D department of a certain firm, each of the ten engineers obtains patents at the mean rate of one per year. Determine the probability that some engineer in the department will get four or more patents during a given year.

Solution   We assume that the patents obtained by the different engineers occur independently of each other and each according to a Poisson process with rate $\lambda = 1$. Let $N_1, \ldots, N_{10}$ be the numbers of patents for the ten engineers during the year, which are independent random variables each having the Poisson distribution with mean one. Observe that

$$P(N_i \leq 3) = e^{-1}\left(1 + 1 + \frac{1}{2} + \frac{1}{3!}\right) \doteq .981, \qquad i = 1, \ldots, 10.$$

Thus, $P(N_i \leq 3 \text{ for } 1 \leq i \leq 10) \doteq .981^{10} \doteq .826$, so the probability that some engineer in the department will get four or more patents during the year is about $1 - .826 = .174$.   ∎

So far, we have been thinking of t as representing time and as ranging from zero to infinity. In some applications, however, t may represent another quantity or have a different range.

EXAMPLE 4.45   Flaws in a particular type of wire occur at the mean rate of one flaw per 100 feet. Determine the probability of two or more flaws in a 250-foot roll of wire.

Solution   We model the flaws as occurring according to a Poisson process with rate $\lambda = .01$, where t represents number of feet from the beginning of the roll, which ranges from 0 to 250. Then the number of flaws in the 250-foot roll has the Poisson distribution with mean $(.01)(250) = 2.5$. Consequently, the probability of less than two flaws in the roll equals $e^{-2.5}(1 + 2.5) \doteq .287$, so the probability of two or more flaws in the roll is about $1 - .287 = .713$.   ∎

EXAMPLE 4.46   Suppose accidental injuries at a new manufacturing plant occur according to a Poisson process with a mean rate of 10 events per month. Determine the probability that there are more than 25 accidental injuries during the first 2 months by using

(a) the formula for the Poisson probability function;

(b) normal approximation;

(c) normal approximation with the half-integer correction.

Solution   Let Y denote the number of accidental injuries during the first 2 months, which has the Poisson distribution with mean 20.

(a) The probability of getting more than 25 accidental injuries during the first 2 months is given by

$$P(Y \geq 26) = \sum_{y=26}^{\infty} \frac{20^y}{y!} e^{-20} \doteq .112.$$

(b) Using normal approximation, we get

$$P(Y \geq 26) \approx 1 - \Phi\left(\frac{26 - 20}{\sqrt{20}}\right) \doteq 1 - \Phi(1.342) \doteq .090.$$

(c) Using normal approximation with the half-integer correction, we get

$$P(Y \geq 26) \approx 1 - \Phi\left(\frac{26 - \frac{1}{2} - 20}{\sqrt{20}}\right) \doteq 1 - \Phi(1.230) \doteq .109. \quad \blacksquare$$

EXAMPLE **4.47**   (Continued.) Use normal approximation to the gamma distribution to determine the probability that there are more than 25 accidental injuries during the first 2 months.

Solution   Let T_n denote the time (in months) when the nth accidental injury occurs, which has the gamma distribution with shape parameter n and inverse-scale parameter 10. Thus, T_n has mean $n/10$, variance $n/100$, and standard deviation $\sqrt{n}/10$. Observe that more than 25 accidental injuries occur during the first 2 months if and only if $T_{26} \leq 2$. Therefore, the desired probability is given by

$$P(T_{26} \leq 2) \approx \Phi\left(\frac{2 - 26/10}{\sqrt{26}/10}\right) \doteq \Phi(-1.177) \doteq .120. \quad \blacksquare$$

Problems

4.56   Suppose calls coming into a telephone exchange form a Poisson process with rate $\lambda = 2$ (calls per minute). Determine the probability of more than three calls during a given 3-minute period.

4.57   A Geiger counter detects radioactive particles at the mean rate of one per minute. Determine the probability that it detects three particles during the first 5 minutes and one particle during the next 1/2 minute.

4.58   An industrial company carries out its operations 50 weeks per year. Accidental injuries occur at the company according to a Poisson process with the mean rate of three per week. Determine the probability that in a given year there will be at least one accidental injury during every week of operation.

4.59   Suppose typos made by a secretary on a typing test form a Poisson process with the mean rate of λ typos per minute and that the test lasts for 5 minutes. Determine the probability that the secretary makes two errors during the first 3 minutes and three errors during the last 3 minutes. (Note that the first 3 minutes and the last 3 minutes overlap.)

Problems 4.60 through 4.66 involve events occurring in time, starting at time zero, according to a Poisson process with rate λ. Let $N(t)$ denote the number of events that have occurred by time t and let T_n denote the time of the nth event.

4.60   Determine the probability that the first two events occur within time t of each other.

4.61   Determine the distribution of the waiting time from time $t \geq 0$ to the first event after time t.

4.62   Show that $\hat{\lambda} = N(t)/t$ is an unbiased estimate of λ for $t > 0$ and determine the variance of this estimate.

4.63   Show that $\hat{\lambda} = (n-1)/T_n$ is an unbiased estimate of λ for $n \geq 2$ and determine the variance of this estimate.

4.64   Use normal approximation to choose t (as a function of λ) such that $P(N(t) \geq 1000) \approx .9$.

4.65   Observe that $P(T_n \leq t) = P(N(t) \geq n)$. **(a)** Use normal approximation to the Poisson distribution to approximate the probability $P(N(t) \geq n)$ when $\lambda t \gg 1$. **(b)** Use normal approximation to the gamma distribution to approximate the probability $P(T_n \leq t)$ when $n \gg 1$. **(c)** Are the answers to parts (a) and (b) close to each other when $\lambda t \gg 1$ and $n \gg 1$?

4.66   Determine **(a)** $P(N(t_1) = 0$ and $N(t_2) \leq 1)$ for $0 < t_1 < t_2$; **(b)** $P(T_1 > t_1$ and $T_2 > t_2)$ for $0 < t_1 < t_2$.

4.67   Suppose the locations of certain objects in the plane form a Poisson process with rate λ. Specifically, suppose that (1) the number $N(B)$ of objects in the region B of the plane has the Poisson distribution with mean λ area(B); and (2) if $B_1, \ldots, B_n$ are disjoint regions of the plane, then $N(B_1), \ldots, N(B_n)$ are independent. Let R denote the distance from the origin to the object nearest to the origin. Observe that $R > r$ if and only if there are no objects in the disk having radius r and centered at the origin. Determine the distribution function, density function, quantiles, mean, second moment, and variance of R.

5

Dependence

5.1

Covariance, Linear Prediction, and Correlation

Covariance

Let X and Y have finite variance. Then $(X - EX)(Y - EY)$ has finite mean by Theorem 2.4(b) in Section 2.3. This mean is referred to as the *covariance* between X and Y and denoted by $\text{cov}(X, Y)$; thus,

$$\text{cov}(X, Y) = E[(X - EX)(Y - EY)]. \tag{1}$$

The covariance between two random variables is a measure of their "co-variability." Observe that $(X - EX)(Y - EY)$ is positive if X and Y differ from their means in the same direction and negative if they differ in opposite directions. Thus, roughly speaking, a positive covariance suggests that X and Y are more likely than not to differ from their means in the same direction, while a negative covariance suggests that they are more likely than not to differ in opposite directions (see Problem 5.91 in Section 5.7).

The covariance between a random variable and itself coincides with the variance of the random variable; that is,

$$\text{cov}(Y, Y) = \text{var}(Y). \tag{2}$$

This formula follows from (1) with $X = Y$.

Three additional properties of covariance are that

$$\text{cov}(Y, X) = \text{cov}(X, Y), \tag{3}$$

$$\text{cov}(a + X, b + Y) = \text{cov}(X, Y), \qquad a, b \in \mathbb{R}, \tag{4}$$

and

$$\text{cov}(aX, bY) = ab \, \text{cov}(X, Y), \qquad a, b \in \mathbb{R}. \tag{5}$$

Equation (3) follows from (1) and the commutativity of multiplication. Equation (4) follows from (1) and the formulas

$$(a + X) - E(a + X) = X - EX \quad \text{and} \quad (b + Y) - E(b + Y) = Y - EY.$$

Since

$$(aX) - E(aX) = a(X - EX) \quad \text{and} \quad (bY) - E(bY) = b(Y - EY),$$

we conclude from (1) that

$$\text{cov}(aX, bY) = E[\, ab(X - EX)(Y - EY)\,]$$
$$= ab \, E[\, (X - EX)(Y - EY)\,]$$
$$= ab \, \text{cov}(X, Y)$$

and hence that (5) is valid.

An alternative formula for the covariance between X and Y is given by

$$\text{cov}(X, Y) = E(XY) - (EX)(EY). \tag{6}$$

To verify (6), observe that

$$(X - EX)(Y - EY) = XY - (EX)Y - X(EY) + (EX)(EY)$$

and hence that

$$\text{cov}(X, Y) = E(XY) - (EX)(EY) - (EX)(EY) + (EX)(EY)$$
$$= E(XY) - (EX)(EY).$$

We can rewrite Equation (24) in Section 2.3 as

$$\text{var}(X + Y) = \text{var}(X) + \text{var}(Y) + 2 \, \text{cov}(X, Y). \tag{7}$$

It follows from (5) and (7) that

$$\text{var}(aX + bY) = a^2 \, \text{var}(X) + b^2 \, \text{var}(Y) + 2ab \, \text{cov}(X, Y), \qquad a, b \in \mathbb{R}. \tag{8}$$

EXAMPLE 5.1　Let $Y_1, \ldots, Y_M$ have the multinomial joint distribution with parameters n and π_1, $\ldots, \pi_M$. Determine

(a) $\text{cov}(Y_l, Y_m)$ for $l \neq m$;

(b) $\text{var}(Y_l - Y_m)$ for $l \neq m$.

Solution　(a) Since Y_m has the binomial distribution with parameters n and π_m, $\text{var}(Y_m) = n\pi_m(1 - \pi_m)$. Similarly, $\text{var}(Y_l) = n\pi_l(1 - \pi_l)$. Suppose that $l \neq m$. Then $Y_l + Y_m$ has the binomial distribution with parameters n and $\pi_l + \pi_m$ (combine types l and m into one new type), so $\text{var}(Y_l + Y_m) = n(\pi_l + \pi_m)(1 - \pi_l - \pi_m)$. Thus, by (7),

$$\text{cov}(Y_l, Y_m) = \frac{1}{2}[\,\text{var}(Y_l + Y_m) - \text{var}(Y_l) - \text{var}(Y_m)\,]$$

$$= \frac{n}{2}[\,(\pi_l + \pi_m)(1 - \pi_l - \pi_m) - \pi_l(1 - \pi_l) - \pi_m(1 - \pi_m)\,]$$

$$= -n\pi_l\pi_m.$$

(b) Let $l \neq m$. By (8) and the solution to part (a),

$$\text{var}(Y_l - Y_m) = \text{var}(Y_l) + \text{var}(Y_m) - 2\,\text{cov}(Y_l, Y_m)$$

$$= n[\,\pi_l(1 - \pi_l) + \pi_m(1 - \pi_m) + 2\pi_l\pi_m\,]$$

$$= n[\,\pi_l + \pi_m - (\pi_l - \pi_m)^2\,]. \quad \blacksquare$$

EXAMPLE 5.2　Let Y_m denote the number of times that side m shows when a fair die is rolled n times in an honest manner. Determine $\text{var}(Y_1 - Y_2)$.

Solution　The random variables $Y_1, \ldots, Y_6$ have the multinomial joint distribution with parameters n and $\pi_1, \ldots, \pi_6$, where $\pi_1 = \cdots = \pi_6 = 1/6$. Thus, $\text{var}(Y_1 - Y_2) = n/3$ by the answer to Example 5.1(b). $\quad \blacksquare$

EXAMPLE 5.3　Let $\mathbf{W}$ be a $\mathcal{W}$-valued random variable, and let A_1 and A_2 be subsets of $\mathcal{W}$.

(a) Determine the covariance between the indicator random variables

$$I_1 = \text{ind}(\mathbf{W} \in A_1) \quad \text{and} \quad I_2 = \text{ind}(\mathbf{W} \in A_2).$$

(b) Show that I_1 and I_2 are independent if and only if their covariance equals zero.

Solution　(a) Observe that $\text{cov}(I_1, I_2) = E(I_1 I_2) - (EI_1)(EI_2)$. Now $EI_1 = P(\mathbf{W} \in A_1)$ and $EI_2 = P(\mathbf{W} \in A_2)$. Also $I_1 I_2 = 1$ if and only if $I_1 = 1$ and $I_2 = 1$ and hence if and only if $\mathbf{W} \in A_1$ and $\mathbf{W} \in A_2$. Thus, $I_1 I_2 = 1$ if and only if $\mathbf{W} \in A_1 \cap A_2$, so $I_1 I_2$ is the indicator random variable corresponding to the event that $\mathbf{W} \in A_1 \cap A_2$ and hence $E(I_1 I_2) = P(\mathbf{W} \in A_1 \cap A_2)$. Consequently,

$$\text{cov}(I_1, I_2) = P(\mathbf{W} \in A_1 \cap A_2) - P(\mathbf{W} \in A_1)P(\mathbf{W} \in A_2).$$

(b) Recall from Example 1.56 in Section 1.8 that I_1 and I_2 are independent if and only if $P(\mathbf{W} \in A_1 \cap A_2) = P(\mathbf{W} \in A_1)P(\mathbf{W} \in A_2)$. The desired result now follows from the solution to part (a). $\quad \blacksquare$

EXAMPLE 5.4   A roulette player bets \$2 on red and \$1 on $\{1, \ldots, 12\}$ on the same trial. Let W_1 denote the amount won on the bet on red, W_2 the amount won on the bet on $\{1, \ldots, 12\}$, and $W = W_1 + W_2$ the combined amount won on the two bets. Determine

(a) $\mathrm{var}(W_1)$ and $\mathrm{var}(W_2)$;   **(b)** $\mathrm{cov}(W_1, W_2)$;   **(c)** $\mathrm{var}(W)$.

Solution   **(a)** It follows from the solution to Example 2.37 in Section 2.3 that

$$\mathrm{var}(W_1) = \left(\frac{36 \cdot 2}{18}\right)^2 \cdot \frac{18}{38} \cdot \frac{20}{38} = \frac{1440}{361}$$

and

$$\mathrm{var}(W_2) = \left(\frac{36}{12}\right)^2 \cdot \frac{12}{38} \cdot \frac{26}{38} = \frac{702}{361}.$$

(b) Let $A_1 = \{1, 3, 5, 7, 9, 12, \ldots\}$ denote the labels of the 18 red slots and set $A_2 = \{1, \ldots, 12\}$. Let $\mathbf{L}$ denote the label of the random landing slot of the ball, which is uniformly distributed on $\mathcal{L} = \{1, \ldots, 36, 0, 00\}$. Also, let $I_1 = \mathrm{ind}(\mathbf{L} \in A_1)$ and $I_2 = \mathrm{ind}(\mathbf{L} \in A_2)$ denote the indicator random variables corresponding to winning the bets on red and $\{1, \ldots, 12\}$, respectively. According to the solution to Example 5.3(a),

$$\mathrm{cov}(I_1, I_2) = P(\mathbf{L} \in A_1 \cap A_2) - P(\mathbf{L} \in A_1)P(\mathbf{L} \in A_2) = \frac{6}{38} - \frac{18}{38} \cdot \frac{12}{38},$$

so $\mathrm{cov}(I_1, I_2) = 3/361$. Moreover (see the solution to Example 2.11 in Section 2.1), $W_1 = 2(2I_1 - 1)$ and $W_2 = 3I_2 - 1$. Thus,

$$\mathrm{cov}(W_1, W_2) = \mathrm{cov}(2(2I_1 - 1), 3I_2 - 1) = 12\,\mathrm{cov}(I_1, I_2) = \frac{36}{361}.$$

(c) It follows from Equation (7) and the solutions to parts (a) and (b) that

$$\begin{aligned}
\mathrm{var}(W) &= \mathrm{var}(W_1 + W_2) \\
&= \mathrm{var}(W_1) + \mathrm{var}(W_2) + 2\,\mathrm{cov}(W_1, W_2) \\
&= \frac{1440}{361} + \frac{702}{361} + \frac{72}{361} \\
&= \frac{2214}{361},
\end{aligned}$$

which agrees with the solution to Example 2.35 in Section 2.3.   ■

THEOREM 5.1   Let X and Y have finite variance.

(a) If X or Y is a constant random variable, then $\mathrm{cov}(X, Y) = 0$.

(b) If X and Y are independent, then $\mathrm{cov}(X, Y) = 0$.

Proof   **(a)** Suppose X is a constant random variable. Then $X - EX = 0$ with probability 1, so $(X - EX)(Y - EY) = 0$ with probability 1 and hence the expected value $\mathrm{cov}(X, Y)$ of this random product is 0. Similarly, if Y is a constant random variable, then $\mathrm{cov}(X, Y) = 0$.

(b) If X and Y are independent, then

$$\text{cov}(X, Y) = E[\,(X - EX)(Y - EY)\,] = [\,E(X - EX)][E(Y - EY)\,] = 0. \quad \blacksquare$$

The converse to Theorem 5.1(b) is not valid; that is, there are dependent random variables X and Y such that $\text{cov}(X, Y) = 0$.

EXAMPLE 5.5 Construct an example of random variables X and Y such that $\text{cov}(X, Y) = 0$, but X and Y are dependent.

Solution Let (X, Y) be uniformly distributed on the set in the plane consisting of the following four points: $(1, 0)$, $(0, 1)$, $(-1, 0)$, and $(0, -1)$ (Figure 5.1). Then $P(X = -1) = 1/4$, $P(X = 0) = 1/2$, and $P(X = 1) = 1/4$, so $EX = 0$. Similarly, $EY = 0$. Since $P(XY = 0) = 1$ and hence $E(XY) = 0$, we conclude that

$$\text{cov}(X, Y) = E(XY) - (EX)(EY) = 0.$$

To see that X and Y are dependent, observe that $P(X = 0, Y = 0) = 0$ whereas

$$P(X = 0)P(Y = 0) = \left(\frac{1}{2}\right)\left(\frac{1}{2}\right) = \frac{1}{4}. \quad \blacksquare$$

FIGURE 5.1

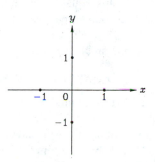

Linear Prediction

In many applications, we want to predict one random variable based on other random variables. For example, we may want to predict a certain interest rate 1 year from now based on the present value of various economic indicators, or we may want to predict a student's grade point average (GPA) in college from his GPA in high school and score on an aptitude test. Consider random variables $X_1, \ldots, X_m, Y$. A *predictor* $\hat{Y}$ of Y based on $X_1, \ldots, X_m$ is a random variable of the form $\hat{Y} = g(X_1, \ldots, X_m)$. We refer to $Y - \hat{Y}$ as the error of prediction and to $(Y - \hat{Y})^2$ as its squared error. The second moment of the error of prediction, which is an overall measure of its magnitude, is referred to as the mean squared error of prediction. Thus, $\text{MSE}(\hat{Y}) = E[\,(Y - \hat{Y})^2\,]$. Since the second moment of a random variable equals the sum of its variance and the square of its mean, we see that

$$\text{MSE}(\hat{Y}) = \text{var}(Y - \hat{Y}) + [E(Y - \hat{Y})]^2 \tag{9}$$

and hence that

$$\text{MSE}(\hat{Y}) = \text{var}(Y - \hat{Y}) \quad \text{if } E(Y - \hat{Y}) = 0. \tag{10}$$

The root mean squared error (RMSE) of $\hat{Y}$ is defined by

$$\text{RMSE}(\hat{Y}) = \sqrt{\text{MSE}(\hat{Y})} \, ,$$

which has the same units as Y. Given a collection of predictors, we refer to a predictor in the collection as being a *best* such predictor if there is no other predictor in the collection having smaller mean squared error.

Consider random variables X and Y, each having finite variance. A *linear predictor* $\hat{Y}$ of Y based on X is a random variable of the form $\hat{Y} = a + bX$. Suppose X and Y have finite variance. We refer to such a linear predictor as *a best linear predictor* of Y based on X if there is no linear predictor of Y based on X having a smaller mean squared error and as *the* best linear predictor of Y based on X if every other linear predictor of Y based on X has a larger mean squared error.

According to (9), the mean squared error of the linear predictor $\hat{Y} = a + bX$ is given by

$$\text{MSE}(\hat{Y}) = \text{var}(Y - bX) + (a - EY + bEX)^2.$$

Given any value of b, the right side of this equation is uniquely minimized by choosing $a = EY - bEX$ in order that its second term equal zero. Making this choice of a, we get that $\hat{Y} = EY + b(X - EX)$, $E\hat{Y} = EY$, $E(Y - \hat{Y}) = 0$, and $\text{MSE}(\hat{Y}) = E[(Y - \hat{Y})^2] = \text{var}(Y - \hat{Y}) = \text{var}(Y - bX)$ and hence that

$$\text{MSE}(\hat{Y}) = \text{var}(Y) - 2b \, \text{cov}(X, Y) + b^2 \, \text{var}(X). \tag{11}$$

As we have just seen, in finding the best linear predictor of Y based on X, we can confine our attention to predictors of the form $\hat{Y} = EY + b(X - EX)$. To get the best such predictor, we need to choose b to minimize the right side of (11). Suppose first that the variance of X equals zero and hence, by Theorem 5.1(a), that the covariance between X and Y equals zero. Then $\text{MSE}(\hat{Y}) = \text{var}(Y)$ regardless of the choice of b. Thus, $\hat{Y} = EY + \beta(X - EX)$ is a best linear predictor of Y based on X for any choice of $\beta \in \mathbb{R}$.

Suppose instead that the variance of X is positive. To minimize the right side of (11), we differentiate it with respect to b, getting

$$\frac{d}{db}\text{MSE}(\hat{Y}) = 2[b \, \text{var}(X) - \text{cov}(X, Y)]. \tag{12}$$

Since the second derivative of the right side of (11) equals $2 \, \text{var}(X)$, which is positive, the right side of (11) is uniquely minimized by setting the right side of (12) equal to zero. Consequently, the unique best linear predictor of Y based on X is given by

$$\hat{Y} = EY + \beta(X - EX),$$

where $\beta = \text{cov}(X, Y)/\text{var}(X)$. The mean squared error of this predictor is given by

$$\text{MSE}(\hat{Y}) = \text{var}(Y) - 2\beta \, \text{cov}(X, Y) + \beta^2 \, \text{var}(X)$$

and hence by

$$\text{MSE}(\hat{Y}) = \text{var}(Y) - \frac{[\,\text{cov}(X, Y)\,]^2}{\text{var}(X)} = \text{var}(Y) - \beta \, \text{cov}(X, Y). \tag{13}$$

THEOREM 5.2   Let $W = Y - \hat{Y}$ denote the error of a linear predictor $\hat{Y} = EY + \beta(X - EX)$ of Y based on X. Then $\hat{Y}$ is a best linear predictor of Y based on X if and only if $\text{cov}(X, W) = 0$.

Proof   Suppose first that the variance of X equals zero. Then, for all choices of β, $\hat{Y}$ is a best linear predictor of Y based on X and $\text{cov}(X, W) = 0$. Consequently, the desired conclusion is valid. Suppose, instead, that X has positive variance. Then

$$\text{cov}(X, W) = \text{cov}(X, \, Y - EY - \beta(X - EX))$$

$$= \text{cov}(X, \, Y - \beta X)$$

$$= \text{cov}(X, Y) - \beta \, \text{var}(X).$$

Thus, $\text{cov}(X, W) = 0$ if and only if $\beta = \text{cov}(X, Y)/\text{var}(X)$ and hence if and only if $\hat{Y}$ is the best linear predictor of Y based on X.   ∎

Correlation

Suppose the variance of X is positive. Then, according to Equation (13), the mean squared error of the best linear predictor of Y based on X equals

$$\text{var}(Y) - \frac{[\,\text{cov}(X, Y)\,]^2}{\text{var}(X)}.$$

Since this mean squared error, being the second moment of a random variable, is necessarily nonnegative, we conclude that $[\,\text{cov}(X, Y)\,]^2 \le \text{var}(X) \, \text{var}(Y)$ and hence that

$$|\,\text{cov}(X, Y)\,| \le \text{SD}(X) \, \text{SD}(Y). \tag{14}$$

If the variance of X equals zero, then both sides of (14) equal zero, so this inequality is again satisfied. The result in (14) is known as the *Schwarz inequality*.

Let X and Y each have positive variance. The *correlation coefficient* $\rho = \text{cor}(X, Y)$ between X and Y is defined by

$$\rho = \text{cor}(X, Y) = \frac{\text{cov}(X, Y)}{\text{SD}(X) \, \text{SD}(Y)}.$$

Observe that

$$\text{cov}(X, Y) = \rho \, \text{SD}(X) \, \text{SD}(Y) \tag{15}$$

It follows from the Schwarz inequality that $-1 \le \rho \le 1$. Observe also that the best linear predictor of Y based on X is given by $\hat{Y} = EY + \beta(X - EX)$, where

$$\beta = \frac{\text{cov}(X, Y)}{\text{var}(X)} = \frac{\text{cov}(X, Y)}{\text{SD}(X)\,\text{SD}(Y)} \frac{\text{SD}(Y)}{\text{SD}(X)} = \rho\,\frac{\text{SD}(Y)}{\text{SD}(X)}.$$

Thus, the sign of β coincides with the sign of ρ. According to Equation (13), the mean squared error of this predictor can be written as

$$\text{MSE}(\hat{Y}) = (1 - \rho^2)\,\text{var}(Y). \tag{16}$$

Observe that the mean squared error of the best linear predictor $\hat{Y}$ equals zero if and only if $\hat{Y} = Y = EY + \beta(X - EX)$ with probability 1. Alternatively, by (16), this mean squared error equals zero if and only if $\rho^2 = 1$ or, equivalently, $|\rho| = 1$ or $\rho = \pm 1$.

The square ρ^2 or absolute value $|\rho|$ of the correlation coefficient between X and Y can be regarded as a measure of the utility of using X in a linear manner to predict Y: values of either measure that are close to 1 indicate high utility (accurate prediction) and values close to 0 indicate low utility (inaccurate prediction).

The relative root mean squared error of a predictor $\hat{Y}$ is defined as the ratio

$$\frac{\text{RMSE}(\hat{Y})}{\text{SD}(Y)},$$

which does not depend on the scale in which Y is measured. It follows from (16) that the relative root mean squared error of the best linear predictor of Y based on X equals $\sqrt{1 - \rho^2}$. Values of $\sqrt{1 - \rho^2}$ that are close to 0 correspond to high utility of using X in a linear manner to predict Y, and values that are close to 1 correspond to low utility. It is clear from Table 5.1 that $\sqrt{1 - \rho^2}$ is not close to 0 unless $|\rho|$ is very close to 1.

Set $\mu_1 = EX$, $\sigma_1 = \text{SD}(X)$, $\mu_2 = EY$, and $\sigma_2 = \text{SD}(Y)$ and let $\rho = \text{cor}(X, Y)$ as above. Then the best linear predictor of Y based on X can be written as

$$\hat{Y} = \mu_2 + \rho\frac{\sigma_2}{\sigma_1}(X - \mu_1), \tag{17}$$

and its mean squared error can be written as

$$\text{MSE}(\hat{Y}) = (1 - \rho^2)\,\sigma_2^2. \tag{18}$$

TABLE 5.1

Relative root mean squared error of linear prediction

$\lvert\rho\rvert$	0.00	.10	.20	.30	.40	.50	.60	.70	.80	.90	.99	1.00
$\sqrt{1-\rho^2}$	1.00	.99	.98	.95	.92	.87	.80	.71	.60	.44	.14	0.00

EXAMPLE 5.6   In a certain population the height Y (in inches) of a random adult male has mean 70 and standard deviation 2, the height X of a random 2-year-old boy has mean 35 and

standard deviation 0.8, and the correlation coefficient among males between height at 2 years and adult height equals 0.8.

(a) Determine the best linear predictor of the height of a random adult male from his height as a 2-year-old.

(b) Determine the root mean squared error of prediction.

Solution   (a) According to (17), the best linear predictor of Y based on X is given by

$$\hat{Y} = 70 + .8\left(\frac{2}{0.8}\right)(X - 35) = 2X;$$

that is, the best linear way to predict adult height based on height at 2 years is to double the height at 2 years. (This rule is actually used in practice.)

(b) According to (18), the root mean squared error of prediction is given by $2\sqrt{1 - .8^2} = 1.2$ inches.   ∎

The random variables X and Y are said to be *uncorrelated* if $\operatorname{cor}(X, Y) = 0$ and *correlated* otherwise. We see from Theorem 5.1(b) that if X and Y are independent, then they are uncorrelated. The converse is not true, as Example 5.5 shows; that is, there are random variables that are uncorrelated but dependent.

Problems

5.1   Show that if Y_1 and Y_2 have finite variance, then **(a)** $\operatorname{cov}(Y_1, Y_2) = [\operatorname{var}(Y_1 + Y_2) - \operatorname{var}(Y_1 - Y_2)]/4$; **(b)** $\operatorname{var}(Y_1 + Y_2) + \operatorname{var}(Y_1 - Y_2) = 2[\operatorname{var}(Y_1) + \operatorname{var}(Y_2)]$.

5.2   Let X and Y, respectively, denote the minimum and maximum number of spots that show when two fair dice are rolled in an honest manner. Determine $\operatorname{cov}(X, Y)$ **(a)** directly; **(b)** by using the answers to Problem 2.45 in Section 2.3.

5.3   A roulette player bets \$2 on red and \$1 on $\{1, \ldots, 12\}$ during each of 100 consecutive spins of the wheel (see Example 2.12 in Section 2.1 and Example 5.4 in this section). Let W be the net amount won from both bets during the 100 spins. **(a)** Determine the mean, variance, and standard deviation of W. **(b)** Use normal approximation, as justified by the central limit theorem, to determine $P(W > 0)$. **(c)** Redo the solution to part (b) using the half-integer correction.

5.4   Consider the roulette player who bets \$2 on red and \$1 on $\{1, \ldots, 12\}$ on the same trial. Determine the correlation coefficient between the amount W_1 won on the bet on red and the amount W_2 won on the bet on $\{1, \ldots, 12\}$.

5.5   A box has ten balls, of which four are red, three are white, two are blue, and one is black. Four times, a ball is selected at random from the box and then returned to the box. Let X denote the number of times a red ball is selected during the four trials and let Y denote the number of times a white ball is selected during these trials. Determine the covariance and correlation between X and Y.

5.6 Let X and W be independent random variables, each having finite, positive variance; suppose that X is a positive random variable; and set $Y = XW$. Show that X and $(Y - X)/X$ are uncorrelated.

5.7 Show that if $\mathrm{var}(X) = \mathrm{var}(Y)$ and $\mathrm{cor}(X, Y) < 1$, then $\mathrm{cor}(X, Y - X) < 0$.

5.8 Let X and Y have finite, positive variance and let $a_1, a_2, b_1, b_2 \in \mathbb{R}$ with $b_1, b_2 \neq 0$. Show that

$$\mathrm{cor}(a_1 + b_1 X, a_2 + b_2 Y) = \mathrm{cor}(X, Y) \quad \text{if } b_1 b_2 > 0$$

and

$$\mathrm{cor}(a_1 + b_1 X, a_2 + b_2 Y) = -\mathrm{cor}(X, Y) \quad \text{if } b_1 b_2 < 0.$$

5.9 Let (X, Y) be uniformly distributed on the set in the plane consisting of the three points $(-1, 0)$, $(1, 0)$, and $(0, 1)$. Show that X and Y are dependent but uncorrelated.

5.10 Let Y_1 denote the weight (in pounds) of a randomly selected man upon arising in the morning and let Y_2 denote his weight just before retiring at night. Suppose that $\mathrm{SD}(Y_1) = \mathrm{SD}(Y_2) = 20$ and $\mathrm{SD}(Y_2 - Y_1) = 1$. Determine the correlation between Y_1 and Y_2.

5.11 Let I_1 denote the indicator random variable for getting red on a spin of the roulette wheel and let I_2 denote the indicator random variable for getting black on the same spin. Determine **(a)** the correlation coefficient between I_1 and I_2; **(b)** the best linear predictor of I_2 based on I_1; **(c)** the mean squared error of this predictor.

5.12 Let X and Y have finite, positive variance and set $\mu_1 = EX$, $\sigma_1 = \mathrm{SD}(X)$, $\mu_2 = EY$, $\sigma_2 = \mathrm{SD}(Y)$, and $\rho = \mathrm{cor}(X, Y)$. Also, let $\hat{Y} = \alpha + \beta X$ denote the best linear predictor of Y based on X and let σ^2 denote the mean squared error of this predictor (which is the variance of the error $Y - \hat{Y}$ of prediction). Determine μ_2, σ_2, and ρ in terms of $\mu_1, \sigma_1, \alpha, \beta$, and σ^2.

5.13 (Continued.) Determine the best linear predictor $\hat{X}$ of X based on Y and the mean squared error of this predictor in terms of $\mu_1, \sigma_1, \alpha, \beta$, and σ^2.

5.14 Let X and Y have finite, positive variance, let ρ denote the correlation coefficient between X and Y, and let $\hat{Y}$ be the best linear predictor of Y based on X. Verify each of the following formulas (earlier formulas can be used to verify later ones): **(a)** $\mathrm{cov}(\hat{Y}, Y - \hat{Y}) = 0$; **(b)** $\mathrm{var}(Y) = \mathrm{var}(\hat{Y}) + \mathrm{var}(Y - \hat{Y})$; **(c)** $\mathrm{var}(\hat{Y}) = \rho^2 \mathrm{var}(Y)$; **(d)** $\mathrm{var}(Y - \hat{Y}) = (1 - \rho^2) \mathrm{var}(Y)$; **(e)** $\mathrm{cov}(\hat{Y}, Y) = \rho^2 \mathrm{var}(Y)$; **(f)** $\mathrm{cor}(\hat{Y}, Y) = |\rho|$ if $\rho \neq 0$.

5.15 Let $W_1, \ldots, W_n$ be independent random variables, each having mean 0 and finite, positive variance; let $1 \leq m \leq n$; and set $X = W_1 + \cdots + W_m$ and $Y = W_1 + \cdots + W_n$. Show that the best linear predictor $\hat{Y}$ of Y based on X is given by $\hat{Y} = X$.

5.16 Let X and Y have means μ_1 and μ_2, respectively; finite, positive variances σ_1^2 and σ_2^2, respectively; and correlation coefficient ρ. Think of μ_1 as being known and of μ_2 as being unknown. Then $\hat{\mu}_2 = Y - b(X - \mu_1)$ is an unbiased estimate of μ_2 for $b \in \mathbb{R}$. Determine **(a)** $\mathrm{var}(\hat{\mu}_2)$ as a function of b; **(b)** the value of b that minimizes $\mathrm{var}(\hat{\mu}_2)$; **(c)** the minimum value of $\mathrm{var}(\hat{\mu}_2)$.

5.17 Let $Y_1, \ldots, Y_n$ be independent random variables having common mean μ and finite, positive variances $\sigma_1^2, \ldots, \sigma_n^2$, respectively. Set $\hat{\mu} = \beta_1 Y_1 + \cdots + \beta_n Y_n$, where

$$\beta_i = \frac{\sigma_i^{-2}}{\sigma_1^{-2} + \cdots + \sigma_n^{-2}}, \qquad 1 \le i \le n.$$

Observe that $\beta_1 + \cdots + \beta_n = 1$. Given real numbers $b_1, \ldots, b_n$ with $b_1 + \cdots + b_n = 1$, consider the linear estimate $T = b_1 Y_1 + \cdots + b_n Y_n$ of μ based on $Y_1, \ldots, Y_n$. Show that **(a)** T is an unbiased estimate of μ and, in particular, $\hat{\mu}$ is an unbiased estimate of μ; **(b)** $\mathrm{var}(\hat{\mu}) = \frac{1}{\sigma_1^{-2} + \cdots + \sigma_n^{-2}}$; **(c)** $\mathrm{cov}(\hat{\mu}, T - \hat{\mu}) = 0$; **(d)** $\mathrm{var}(T) = \mathrm{var}(\hat{\mu}) + \mathrm{var}(T - \hat{\mu})$; **(e)** $\mathrm{var}(T - \hat{\mu}) = 0$ if and only if $b_1 = \beta_1, \ldots, b_n = \beta_n$; **(f)** $\hat{\mu}$ is the unique best linear unbiased estimate (BLUE) of μ based on $Y_1, \ldots, Y_n$, where "best" means having the smallest possible variance.

5.2
Multivariate Expectation

The study of dependent random variables is simplified by using them to form random vectors or matrices. (The relevant properties of vectors and matrices are summarized in Appendix A.) Consider (real-valued) random variables $Y_1, \ldots, Y_n$. We can use them to form a random ordered n-tuple or random row vector $[Y_1, \ldots, Y_n]$. In this chapter, however, it is generally more convenient to use them to form the random column vector

$$\mathbf{Y} = [Y_1, \ldots, Y_n]^T = \begin{bmatrix} Y_1 \\ \vdots \\ Y_n \end{bmatrix}.$$

More generally, let n and m be positive integers, let Y_{ij}, $1 \le i \le n$ and $1 \le j \le m$, be random variables, and let

$$\mathbf{Y} = \begin{bmatrix} Y_{11} & \cdots & Y_{1m} \\ \vdots & & \vdots \\ Y_{n1} & \cdots & Y_{nm} \end{bmatrix}$$

denote the $n \times m$ random matrix having the entry Y_{ij} in row i and column j. If $m = 1$, then $\mathbf{Y}$ is an n-dimensional random column vector; if $n = 1$, then $\mathbf{Y}$ is an m-dimensional random row vector; and if $m = n = 1$, then $\mathbf{Y} = [Y_{11}]$ can be identified with the random variable Y_{11}.

The random matrix $\mathbf{Y}$ is said to have finite expectation (mean) if each of its entries has finite expectation, in which case the expectation $E\mathbf{Y}$ of $\mathbf{Y}$ is defined as the $n \times m$ matrix having entry EY_{ij} in row i and column j; that is,

$$E\mathbf{Y} = \begin{bmatrix} EY_{11} & \cdots & EY_{1m} \\ \vdots & & \vdots \\ EY_{n1} & \cdots & EY_{nm} \end{bmatrix}.$$

It follows easily from this definition that $E(\mathbf{Y} - E\mathbf{Y}) = 0$.

EXAMPLE **5.7**   Let W_1 and W_2 be independent random variables, each having mean 0 and variance σ^2, and set $\mathbf{W} = [\,W_1, W_2\,]^T$. Determine

(a) $E\left(\mathbf{W}^T\mathbf{W}\right)$;

(b) $E\left(\mathbf{W}\mathbf{W}^T\right)$.

Solution   (a) Observe that

$$\mathbf{W}^T\mathbf{W} = [\,W_1, W_2\,]\begin{bmatrix} W_1 \\ W_2 \end{bmatrix} = W_1^2 + W_2^2.$$

Thus,

$$
\begin{aligned}
E\left(\mathbf{W}^T\mathbf{W}\right) &= E(W_1^2 + W_2^2) \\
&= E(W_1^2) + E(W_2^2) \\
&= \mathrm{var}(W_1) + \mathrm{var}(W_2) \\
&= \sigma^2 + \sigma^2 \\
&= 2\sigma^2.
\end{aligned}
$$

(b) Observe that

$$\mathbf{W}\mathbf{W}^T = \begin{bmatrix} W_1 \\ W_2 \end{bmatrix}[\,W_1, W_2\,] = \begin{bmatrix} W_1^2 & W_1 W_2 \\ W_1 W_2 & W_2^2 \end{bmatrix}$$

and hence that

$$
\begin{aligned}
E\left(\mathbf{W}\mathbf{W}^T\right) &= E\begin{bmatrix} W_1^2 & W_1 W_2 \\ W_1 W_2 & W_2^2 \end{bmatrix} \\[2mm]
&= \begin{bmatrix} E(W_1^2) & E(W_1 W_2) \\ E(W_1 W_2) & E(W_2^2) \end{bmatrix} \\[2mm]
&= \begin{bmatrix} \mathrm{var}(W_1) & 0 \\ 0 & \mathrm{var}(W_2) \end{bmatrix}.
\end{aligned}
$$

Consequently,

$$E\left(\mathbf{W}\mathbf{W}^T\right) = \begin{bmatrix} \sigma^2 & 0 \\ 0 & \sigma^2 \end{bmatrix} = \sigma^2\begin{bmatrix} 1 & 0 \\ 0 & 1 \end{bmatrix} = \sigma^2\mathbf{I},$$

where $\mathbf{I}$ denotes the 2×2 identity matrix.   ∎

THEOREM **5.3**   Let $\mathbf{X}$ and $\mathbf{Y}$ be random $n \times m$ matrices having finite expectation. Then

$$E(\mathbf{X} + \mathbf{Y}) = E\mathbf{X} + E\mathbf{Y}.$$

Proof   The entry in row i and column j of $E(\mathbf{X} + \mathbf{Y})$ is $E(X_{ij} + Y_{ij}) = EX_{ij} + EY_{ij}$, which equals the entry in row i and column j of $E\mathbf{X} + E\mathbf{Y}$.   ∎

By a constant random matrix, we mean one whose entries are all constant random variables. Thus, $\mathbf{Y}$ is a constant random matrix if and only if, for some nonrandom matrix $\mathbf{A} = [\,a_{ij}\,]$, $P(\mathbf{Y} = \mathbf{A}) = 1$ or, equivalently, $P(Y_{ij} = a_{ij}) = 1$ for all i, j. Let this be the case. Then $EY_{ij} = a_{ij}$ for all i, j and hence $EY = A$. A number of additional properties of expectation of random matrices are given in the following result (in which A and B are nonrandom matrices).

THEOREM 5.4   Let Y be a random $n \times m$ matrix having finite expectation. Then

(a) $E\left(Y^T\right) = (EY)^T$;

(b) $E(bY) = bEY$ for $b \in \mathbb{R}$;

(c) $E(A + Y) = A + EY$, where A has n rows and m columns;

(d) $E(BY) = B(EY)$, where B has n columns;

(e) $E(YB) = (EY)B$, where B has m rows.

Proof   The verification of parts (a), (b), and (c) is left as a problem. In proving part (d), let b_{li} denote the entry in row l and column i of B. Then the entry in row l and column j of BY equals $\sum_i b_{li} Y_{ij}$, whose mean is $\sum_i b_{li} E(Y_{ij})$. Since this mean is the entry in row l and column j of $B(EY)$, part (d) is valid. The proof of part (e) is similar.   ∎

COROLLARY 5.1   Let $Y = [\, Y_1, \ldots, Y_n \,]^T$ have finite expectation. Then

(a) $E(a + Y) = a + EY$ for $a \in \mathbb{R}^n$;   (b) $E(b^T Y) = b^T EY$ for $b \in \mathbb{R}^n$.

The first result in this corollary is a special case of Theorem 5.4(c), and the second result is a special case of Theorem 5.4(d). Observe that the second result can be written as

$$E(b_1 Y_1 + \cdots + b_n Y_n) = b_1 EY_1 + \cdots + b_n EY_n,$$

which agrees with Corollary 2.3 in Section 2.2.

EXAMPLE 5.8   Let W_1, W_2, and W_3 be random variables having means 1, 2, and 3, respectively. Set $Y_1 = W_2$, $Y_2 = W_3$, $Y_3 = W_1$, $W = [\, W_1, W_2, W_3 \,]^T$, and $Y = [\, Y_1, Y_2, Y_3 \,]^T$.

(a) Determine EW and EY.

(b) Determine a matrix A such that $Y = AW$ and check that $EY = A(EW)$.

(c) Determine a matrix B such that $W = BY$ and check that $EW = B(EY)$.

(d) Check that $B = A^{-1}$.

Solution   (a) We have that

$$EW = E \begin{bmatrix} W_1 \\ W_2 \\ W_3 \end{bmatrix} = \begin{bmatrix} 1 \\ 2 \\ 3 \end{bmatrix}$$

and

$$EY = E \begin{bmatrix} Y_1 \\ Y_2 \\ Y_3 \end{bmatrix} = E \begin{bmatrix} W_2 \\ W_3 \\ W_1 \end{bmatrix} = \begin{bmatrix} 2 \\ 3 \\ 1 \end{bmatrix}.$$

(b) Now

$$\mathbf{Y} = \begin{bmatrix} Y_1 \\ Y_2 \\ Y_3 \end{bmatrix} = \begin{bmatrix} W_2 \\ W_3 \\ W_1 \end{bmatrix} = \begin{bmatrix} 0 & 1 & 0 \\ 0 & 0 & 1 \\ 1 & 0 & 0 \end{bmatrix} \begin{bmatrix} W_1 \\ W_2 \\ W_3 \end{bmatrix} = \mathbf{AW},$$

where

$$\mathbf{A} = \begin{bmatrix} 0 & 1 & 0 \\ 0 & 0 & 1 \\ 1 & 0 & 0 \end{bmatrix}.$$

Moreover,

$$\mathbf{A}E\mathbf{W} = \begin{bmatrix} 0 & 1 & 0 \\ 0 & 0 & 1 \\ 1 & 0 & 0 \end{bmatrix} \begin{bmatrix} 1 \\ 2 \\ 3 \end{bmatrix} = \begin{bmatrix} 2 \\ 3 \\ 1 \end{bmatrix} = E\mathbf{Y}.$$

(c) Now

$$\mathbf{W} = \begin{bmatrix} W_1 \\ W_2 \\ W_3 \end{bmatrix} = \begin{bmatrix} Y_3 \\ Y_1 \\ Y_2 \end{bmatrix} = \begin{bmatrix} 0 & 0 & 1 \\ 1 & 0 & 0 \\ 0 & 1 & 0 \end{bmatrix} \begin{bmatrix} Y_1 \\ Y_2 \\ Y_3 \end{bmatrix} = \mathbf{BY},$$

where

$$\mathbf{B} = \begin{bmatrix} 0 & 0 & 1 \\ 1 & 0 & 0 \\ 0 & 1 & 0 \end{bmatrix}.$$

Moreover,

$$\mathbf{B}E\mathbf{Y} = \begin{bmatrix} 0 & 0 & 1 \\ 1 & 0 & 0 \\ 0 & 1 & 0 \end{bmatrix} \begin{bmatrix} 2 \\ 3 \\ 1 \end{bmatrix} = \begin{bmatrix} 1 \\ 2 \\ 3 \end{bmatrix} = E\mathbf{W}.$$

(d) Since

$$\mathbf{AB} = \begin{bmatrix} 0 & 1 & 0 \\ 0 & 0 & 1 \\ 1 & 0 & 0 \end{bmatrix} \begin{bmatrix} 0 & 0 & 1 \\ 1 & 0 & 0 \\ 0 & 1 & 0 \end{bmatrix} = \begin{bmatrix} 1 & 0 & 0 \\ 0 & 1 & 0 \\ 0 & 0 & 1 \end{bmatrix},$$

we see that $\mathbf{B} = \mathbf{A}^{-1}$. ∎

COROLLARY **5.2** Let $\mathbf{Y}$ be a random $n \times m$ matrix having finite expectation, let $\mathbf{A}$ be a matrix having n columns, and let $\mathbf{B}$ be a matrix having m rows. Then

$$E(\mathbf{AYB}) = \mathbf{A}(E\mathbf{Y})\mathbf{B}.$$

Proof By parts (d) and (e) of Theorem 5.3,

$$E(\mathbf{AYB}) = E[\,(\mathbf{AY})\mathbf{B}\,] = [\,E(\mathbf{AY})\,]\mathbf{B} = [\,\mathbf{A}(E\mathbf{Y})\,]\mathbf{B} = \mathbf{A}(E\mathbf{Y})\mathbf{B}. ∎$$

EXAMPLE **5.9** Let W_1 and W_2 be independent random variables, each having mean 0 and variance σ^2. Set $Y_1 = W_1$, $Y_2 = W_1 + W_2$, and $\mathbf{Y} = [\,Y_1, Y_2\,]^T$. Determine $E\left(\mathbf{YY}^T\right)$.

Solution   It follows as in the solution to Example 5.7(b) that

$$E\left(\mathbf{Y}\mathbf{Y}^T\right) = \begin{bmatrix} E(Y_1^2) & E(Y_1Y_2) \\ E(Y_1Y_2) & E(Y_2^2) \end{bmatrix}.$$

Now

$$E(Y_1^2) = E(W_1^2) = \mathrm{var}(W_1) = \sigma^2,$$

$$E(Y_1Y_2) = E[\, W_1(W_1 + W_2)\,] = E(W_1^2) = \sigma^2,$$

and

$$E(Y_2^2) = E[\, (W_1 + W_2)^2\,] = \mathrm{var}(W_1 + W_2) = \mathrm{var}(W_1) + \mathrm{var}(W_2) = 2\sigma^2.$$

Thus,

$$E\left(\mathbf{Y}\mathbf{Y}^T\right) = \begin{bmatrix} \sigma^2 & \sigma^2 \\ \sigma^2 & 2\sigma^2 \end{bmatrix} = \sigma^2 \begin{bmatrix} 1 & 1 \\ 1 & 2 \end{bmatrix}.$$

Alternatively, set $\mathbf{W} = [\,W_1, W_2\,]^T$ and recall from the solution to Example 5.7(b) that $E\left(\mathbf{W}\mathbf{W}^T\right) = \sigma^2 \mathbf{I}$. Observe that

$$\mathbf{Y} = \begin{bmatrix} Y_1 \\ Y_2 \end{bmatrix} = \begin{bmatrix} W_1 \\ W_1 + W_2 \end{bmatrix} = \begin{bmatrix} 1 & 0 \\ 1 & 1 \end{bmatrix} \begin{bmatrix} W_1 \\ W_2 \end{bmatrix} = \mathbf{B}\mathbf{W},$$

where $\mathbf{B}$ is the lower triangular 2×2 matrix

$$\begin{bmatrix} 1 & 0 \\ 1 & 1 \end{bmatrix}.$$

Thus, by Corollary 5.2,

$$E\left(\mathbf{Y}\mathbf{Y}^T\right) = E\left(\mathbf{B}\mathbf{W}\mathbf{W}^T\mathbf{B}^T\right) = \mathbf{B}E\left(\mathbf{W}\mathbf{W}^T\right)\mathbf{B}^T = \mathbf{B}(\sigma^2\mathbf{I})\mathbf{B}^T = \sigma^2\mathbf{B}\mathbf{B}^T,$$

which is simply $\sigma^2 \begin{bmatrix} 1 & 1 \\ 1 & 2 \end{bmatrix}$.   ∎

Let $\mathbf{Y} = [\,Y_1, \ldots, Y_n\,]^T$, where Y_i has mean μ_i for $1 \le i \le n$. Then $E\mathbf{Y} = [\,EY_1, \ldots, EY_n\,]^T$ is commonly written as $\mu = [\,\mu_1, \ldots, \mu_n\,]^T$ and referred to as the *mean vector* of $\mathbf{Y}$ or of its distribution.

Partitioned vectors and matrices are commonly employed in the study of dependent random variables. As an illustration, let X_1, X_2, W_1, and W_2 be random variables with $EX_1 = \mu_1$, $EX_2 = \mu_2$, $EW_1 = 0$, and $EW_2 = 0$. Set $Y_1 = \alpha_1 + b_{11}X_1 + b_{12}X_2 + W_1$ and $Y_2 = \alpha_2 + b_{21}X_1 + b_{22}X_2 + W_2$. Then

$$EY_1 = \alpha_1 + b_{11}\mu_1 + b_{12}\mu_2 \quad \text{and} \quad EY_2 = \alpha_2 + b_{21}\mu_1 + b_{22}\mu_2.$$

Thus,

$$E\begin{bmatrix} X_1 \\ X_2 \\ Y_1 \\ Y_2 \end{bmatrix} = \begin{bmatrix} \mu_1 \\ \mu_2 \\ \alpha_1 + b_{11}\mu_1 + b_{12}\mu_2 \\ \alpha_2 + b_{21}\mu_1 + b_{22}\mu_2 \end{bmatrix}.$$

Alternatively, set

$$\mathbf{X} = \begin{bmatrix} X_1 \\ X_2 \end{bmatrix}, \quad \mathbf{W} = \begin{bmatrix} W_1 \\ W_2 \end{bmatrix} \quad \text{and} \quad \mathbf{Y} = \begin{bmatrix} Y_1 \\ Y_2 \end{bmatrix};$$

also set

$$\mu = \begin{bmatrix} \mu_1 \\ \mu_2 \end{bmatrix}, \quad \alpha = \begin{bmatrix} \alpha_1 \\ \alpha_2 \end{bmatrix} \quad \text{and} \quad \mathbf{B} = \begin{bmatrix} b_{11} & b_{12} \\ b_{21} & b_{22} \end{bmatrix}.$$

Then $E\mathbf{X} = \mu$, $E\mathbf{W} = \mathbf{0}$, and $\mathbf{Y} = \alpha + \mathbf{BX} + \mathbf{W}$, so $E\mathbf{Y} = \alpha + \mathbf{B}E\mathbf{X} + E\mathbf{W} = \alpha + \mathbf{B}\mu$ and hence

$$E \begin{bmatrix} X_1 \\ X_2 \\ Y_1 \\ Y_2 \end{bmatrix} = E \begin{bmatrix} \mathbf{X} \\ \mathbf{Y} \end{bmatrix} = \begin{bmatrix} \mu \\ \alpha + \mathbf{B}\mu \end{bmatrix}.$$

Problems

5.18 A box has ten balls, of which four are red, three are white, two are blue, and one is black. Four times, a ball is selected at random from the box and then returned to the box. Let X denote the number of times a red ball is selected during the four trials, let Y denote the number of times a white ball is selected during these trials, and let Z denote the number of times a blue ball is selected. Determine the mean vector of $[X, Y, Z]^T$.

5.19 Let X have mean μ_1 and set $Y = \alpha_1 + \beta_1 X$ and $Z = \alpha_2 + \beta_2 X + \beta_3 Y$. Determine the mean vector of $[X, Y, Z]^T$.

5.20 Let W_1, W_2, and W_3 be independent random variables, each having mean 0 and variance σ^2, and set $\mathbf{W} = [W_1, W_2, W_3]^T$. Determine **(a)** $E(\mathbf{W}^T\mathbf{W})$; **(b)** $E(\mathbf{W}\mathbf{W}^T)$.

5.21 Let W_1, W_2, and W_3 be independent random variables, each having mean 0 and variance σ^2. Set $Y_1 = W_1$, $Y_2 = W_1 + W_2$, $Y_3 = W_1 + W_2 + W_3$, and $\mathbf{Y} = [Y_1, Y_2, Y_3]^T$. Determine **(a)** $E(\mathbf{Y}^T\mathbf{Y})$; **(b)** $E(\mathbf{Y}\mathbf{Y}^T)$.

5.22 Let U be uniformly distributed on $[0, 1]$ and set $\mathbf{Y} = [1, U]^T$. Determine **(a)** $E(\mathbf{Y}^T\mathbf{Y})$; **(b)** $E(\mathbf{Y}\mathbf{Y}^T)$.

5.23 Let Z be a standard normal random variable and set $\mathbf{Y} = [1, Z, Z^2]^T$. Determine **(a)** $E(\mathbf{Y}^T\mathbf{Y})$; **(b)** $E(\mathbf{Y}\mathbf{Y}^T)$.

5.24 Verify parts (a), (b), and (c) of Theorem 5.4.

5.25 Let W_1 and W_2 be random variables having mean μ and set $Y_1 = W_1 + W_2$ and $Y_2 = W_1 - W_2$. Also set $\mathbf{W} = [W_1, W_2]^T$ and $\mathbf{Y} = [Y_1, Y_2]^T$. **(a)** Determine $E\mathbf{W}$. **(b)** Determine $E\mathbf{Y}$. **(c)** Determine a matrix $\mathbf{A}$ such that $\mathbf{Y} = \mathbf{A}\mathbf{W}$ and check that $E\mathbf{Y} = \mathbf{A}(E\mathbf{W})$. **(d)** Express W_1 and W_2 in terms of Y_1 and Y_2. **(e)** Determine a matrix $\mathbf{B}$ such that $\mathbf{W} = \mathbf{B}\mathbf{Y}$ and check that $E\mathbf{W} = \mathbf{B}(E\mathbf{Y})$. **(f)** Check that $\mathbf{B} = \mathbf{A}^{-1}$.

5.26 Let W_1, W_2, and W_3 be random variables having finite mean μ and set $Y_1 = W_1$, $Y_2 = W_1 + W_2$, and $Y_3 = W_1 + W_2 + W_3$. Also set

$$\mathbf{W} = [W_1, W_2, W_3]^T \quad \text{and} \quad \mathbf{Y} = [Y_1, Y_2, Y_3]^T.$$

(a) Determine EW. (b) Determine EY. (c) Determine a matrix $\mathbf{A}$ such that $\mathbf{Y} = \mathbf{AW}$ and check that $EY = \mathbf{A}(EW)$. (d) Express W_1, W_2, and W_3 in terms of Y_1, Y_2, and Y_3. (e) Determine a matrix B such that $\mathbf{W} = \mathbf{BY}$ and check that $EW = \mathbf{B}(EY)$. (f) Check that $\mathbf{B} = \mathbf{A}^{-1}$.

5.27 Let W_1, W_2, and W_3 be random variables having finite means μ_1, μ_2, and μ_3, respectively, and let $a_1, a_2, a_3, b_1, b_2, b_3 \in \mathbb{R}$. Set $Y_1 = a_1 + b_1 W_1$, $Y_2 = a_2 + b_2 W_2$, and $Y_3 = a_3 + b_3 W_3$. Also, set $\mathbf{W} = [\, W_1, W_2, W_3\,]^T$ and $Y = [\, Y_1, Y_2, Y_3\,]^T$. (a) Determine a vector $\mathbf{a}$ and a matrix $\mathbf{B}$ such that $\mathbf{Y} = \mathbf{a} + \mathbf{BW}$. (b) Check that $EY = \mathbf{a} + \mathbf{B}(EW)$.

5.28 Consider the random vector $\mathbf{W} = [\, W_1, W_2, W_3\,]^T$ and the matrix

$$\mathbf{B} = \begin{bmatrix} b_{11} & b_{12} & b_{13} \\ b_{21} & b_{22} & b_{23} \\ b_{31} & b_{32} & b_{33} \end{bmatrix}.$$

Let $\mathbf{Y} = [\, Y_1, Y_2, Y_3\,]^T$ be defined by $\mathbf{Y} = \mathbf{BW}$. Also, set $\mathbf{W}_1 = [\, W_1, W_2\,]^T$, $\mathbf{W}_2 = [\, W_3\,]$, $\mathbf{Y}_1 = [\, Y_1, Y_2\,]^T$, $\mathbf{Y}_2 = [\, Y_3\,]$,

$$\mathbf{B}_{11} = \begin{bmatrix} b_{11} & b_{12} \\ b_{21} & b_{22} \end{bmatrix}, \quad \mathbf{B}_{12} = \begin{bmatrix} b_{13} \\ b_{23} \end{bmatrix},$$

$$\mathbf{B}_{21} = [\, b_{31}, b_{32}\,], \quad \mathbf{B}_{22} = [\, b_{33}\,].$$

Then we can write $\mathbf{B}$, $\mathbf{W}$, and $\mathbf{Y}$ as

$$\mathbf{B} = \begin{bmatrix} \mathbf{B}_{11} & \mathbf{B}_{12} \\ \mathbf{B}_{21} & \mathbf{B}_{22} \end{bmatrix}, \quad \mathbf{W} = \begin{bmatrix} \mathbf{W}_1 \\ \mathbf{W}_2 \end{bmatrix}, \quad \text{and} \quad \mathbf{Y} = \begin{bmatrix} \mathbf{Y}_1 \\ \mathbf{Y}_2 \end{bmatrix}.$$

Check that $\mathbf{Y}_1 = \mathbf{B}_{11}\mathbf{W}_1 + \mathbf{B}_{12}\mathbf{W}_2$ and $\mathbf{Y}_2 = \mathbf{B}_{21}\mathbf{W}_1 + \mathbf{B}_{22}\mathbf{W}_2$.

5.3
Covariance and Variance–Covariance Matrices

In this section, we obtain generalizations of the variance of a random variable and the covariance between two random variables that apply to random vectors.

Covariance Between Random Vectors

Consider random vectors $\mathbf{X} = [\, X_1, \ldots, X_m\,]^T$ and $\mathbf{Y} = [\, Y_1, \ldots, Y_n\,]^T$, where the entries of both have finite variance. The *covariance matrix* $\mathrm{cov}(\mathbf{X}, \mathbf{Y})$ between $\mathbf{X}$ and $\mathbf{Y}$ is the $m \times n$ matrix the entry in row i and column j of which is given by

$$\mathrm{cov}(X_i, Y_j) = E[\, (X_i - EX_i)(Y_j - EY_j)\,].$$

As a consequence of this definition and the formula for $\mathbf{ab}^T$ when $\mathbf{a}$ and $\mathbf{b}$ are column vectors, we get

$$\mathrm{cov}(\mathbf{X}, \mathbf{Y}) = E[\, (\mathbf{X} - E\mathbf{X})(\mathbf{Y} - E\mathbf{Y})^T\,]. \tag{19}$$

It follows from (19) or directly from the definition of the covariance matrix that

$$\text{cov}(\mathbf{Y}, \mathbf{X}) = [\, \text{cov}(\mathbf{X}, \mathbf{Y}) \,]^T. \tag{20}$$

These and other properties of the covariance matrix between random vectors are very similar to the corresponding properties of the covariance between random variables.

THEOREM 5.5   Let the entries of $\mathbf{X}$ and $\mathbf{Y}$ have finite variance.

(a) If $\mathbf{X}$ or $\mathbf{Y}$ is a constant random vector, then $\text{cov}(\mathbf{X}, \mathbf{Y}) = \mathbf{0}$.

(b) If $\mathbf{X}$ and $\mathbf{Y}$ are independent, then $\text{cov}(\mathbf{X}, \mathbf{Y}) = \mathbf{0}$.

Proof   (a) Suppose $\mathbf{X}$ is a constant random vector; that is, X_i is a constant random variable for $1 \leq i \leq m$. Then, by Theorem 5.1(a) in Section 5.1, $\text{cov}(X_i, Y_j) = 0$ for $1 \leq i \leq m$ and $1 \leq j \leq n$ and hence $\text{cov}(\mathbf{X}, \mathbf{Y}) = \mathbf{0}$.

(b) Suppose $\mathbf{X}$ and $\mathbf{Y}$ are independent and let $1 \leq i \leq m$ and $1 \leq j \leq n$. Since X_i is a transform of $\mathbf{X}$ and Y_j is a transform of $\mathbf{Y}$, $\text{cov}(X_i, Y_j) = 0$ by Theorem 5.1(b) in Section 5.1. Therefore, $\text{cov}(\mathbf{X}, \mathbf{Y}) = \mathbf{0}$.   ■

THEOREM 5.6   Let the entries of $\mathbf{X}$ and $\mathbf{Y}$ have finite variance. Then

(a) $\text{cov}(\mathbf{a} + \mathbf{X}, \mathbf{b} + \mathbf{Y}) = \text{cov}(\mathbf{X}, \mathbf{Y})$ for $\mathbf{a} \in \mathbb{R}^m$ and $\mathbf{b} \in \mathbb{R}^n$;

(b) $\text{cov}(\mathbf{AX}, \mathbf{BY}) = \mathbf{A} \, \text{cov}(\mathbf{X}, \mathbf{Y}) \mathbf{B}^T$, where $\mathbf{A}$ has m columns and $\mathbf{B}$ has n columns;

(c) $\text{cov}(\mathbf{a}^T \mathbf{X}, \mathbf{b}^T \mathbf{Y}) = \mathbf{a}^T \text{cov}(\mathbf{X}, \mathbf{Y}) \mathbf{b}$ for $\mathbf{a} \in \mathbb{R}^m$ and $\mathbf{b} \in \mathbb{R}^n$;

(d) $\text{cov}(a\mathbf{X}, b\mathbf{Y}) = ab \, \text{cov}(\mathbf{X}, \mathbf{Y})$ for $a, b \in \mathbb{R}$.

Proof   (a) It follows from Corollary 5.1 in Section 5.2 that $E(\mathbf{a} + \mathbf{X}) = \mathbf{a} + E\mathbf{X}$ and hence that $\mathbf{a} + \mathbf{X} - E(\mathbf{a} + \mathbf{X}) = \mathbf{X} - E\mathbf{X}$. Similarly, $\mathbf{b} + \mathbf{Y} - E(\mathbf{b} + \mathbf{Y}) = \mathbf{Y} - E\mathbf{Y}$. Consequently, by Equation (19),

$$\text{cov}(\mathbf{a} + \mathbf{X}, \mathbf{b} + \mathbf{Y}) = E[\, (\mathbf{X} - E\mathbf{X})(\mathbf{Y} - E\mathbf{Y})^T \,] = \text{cov}(\mathbf{X}, \mathbf{Y}).$$

(b) It follows from Theorem 5.4(d) in Section 5.2 that $E(\mathbf{AX}) = \mathbf{A}E\mathbf{X}$ and hence that

$$\mathbf{AX} - E(\mathbf{AX}) = \mathbf{AX} - \mathbf{A}E\mathbf{X} = \mathbf{A}(\mathbf{X} - E\mathbf{X}).$$

Similarly,

$$\mathbf{BY} - E(\mathbf{BY}) = \mathbf{B}(\mathbf{Y} - E\mathbf{Y}).$$

Thus, by (19) together with Corollary 5.2 in Section 5.2,

$$\text{cov}(\mathbf{AX}, \mathbf{BY}) = E\{\mathbf{A}(\mathbf{X} - E\mathbf{X})[\, \mathbf{B}(\mathbf{Y} - E\mathbf{Y}) \,]^T\}$$
$$= E[\, \mathbf{A}(\mathbf{X} - E\mathbf{X})(\mathbf{Y} - E\mathbf{Y})^T \mathbf{B}^T \,]$$
$$= \mathbf{A}E[\, (\mathbf{X} - E\mathbf{X})(\mathbf{Y} - E\mathbf{Y})^T \,]\mathbf{B}^T$$
$$= \mathbf{A} \, \text{cov}(\mathbf{X}, \mathbf{Y})\mathbf{B}^T.$$

(c) This result follows by applying part (b) with $\mathbf{A} = \mathbf{a}^T$ and $\mathbf{B} = \mathbf{b}^T$.

(d) This result follows by applying part (b) with $\mathbf{A} = a\mathbf{I}_m$ and $\mathbf{B} = b\mathbf{I}_n$. ∎

Rewriting Theorem 5.6(c) in summation form, we get the following result.

COROLLARY 5.3 Let $X_1, \ldots, X_m$, $Y_1, \ldots, Y_n$ each have finite variance and let $a_1, \ldots, a_m$, $b_1, \ldots, b_n \in \mathbb{R}$. Then

$$\mathrm{cov}\left(\sum_i a_i X_i, \sum_j b_j Y_j \right) = \sum_{i,j} a_i b_j \, \mathrm{cov}(X_i, Y_j).$$

As a special case of Corollary 5.3, we see that if X_1, X_2, Y_1, and Y_2 each have finite variance, then

$$\mathrm{cov}(X_1 + X_2, Y_1 + Y_2) = \mathrm{cov}(X_1, Y_1) + \mathrm{cov}(X_1, Y_2)$$
$$+ \mathrm{cov}(X_2, Y_1) + \mathrm{cov}(X_2, Y_2).$$

More generally, we have the following result.

THEOREM 5.7 Let the entries of $\mathbf{X}_1$, $\mathbf{X}_2$, $\mathbf{Y}_1$, and $\mathbf{Y}_2$ each have finite variance, let $\mathbf{X}_1$ and $\mathbf{X}_2$ be m-dimensional, and let $\mathbf{Y}_1$ and $\mathbf{Y}_2$ be n-dimensional. Then

$$\mathrm{cov}(\mathbf{X}_1 + \mathbf{X}_2, \mathbf{Y}_1 + \mathbf{Y}_2) = \mathrm{cov}(\mathbf{X}_1, \mathbf{Y}_1) + \mathrm{cov}(\mathbf{X}_1, \mathbf{Y}_2)$$
$$+ \mathrm{cov}(\mathbf{X}_2, \mathbf{Y}_1) + \mathrm{cov}(\mathbf{X}_2, \mathbf{Y}_2).$$

Proof   Now $E(\mathbf{X}_1 + \mathbf{X}_2) = E\mathbf{X}_1 + E\mathbf{X}_2$, so

$$\mathbf{X}_1 + \mathbf{X}_2 - E(\mathbf{X}_1 + \mathbf{X}_2) = (\mathbf{X}_1 - E\mathbf{X}_1) + (\mathbf{X}_2 - E\mathbf{X}_2).$$

Similarly,

$$\mathbf{Y}_1 + \mathbf{Y}_2 - E(\mathbf{Y}_1 + \mathbf{Y}_2) = (\mathbf{Y}_1 - E\mathbf{Y}_1) + (\mathbf{Y}_2 - E\mathbf{Y}_2).$$

Thus,

$$\mathrm{cov}(\mathbf{X}_1 + \mathbf{X}_2, \mathbf{Y}_1 + \mathbf{Y}_2)$$
$$= E\{[(\mathbf{X}_1 - E\mathbf{X}_1) + (\mathbf{X}_2 - E\mathbf{X}_2)][(\mathbf{Y}_1 - E\mathbf{Y}_1)^T + (\mathbf{Y}_2 - E\mathbf{Y}_2)^T]\}$$
$$= E[(\mathbf{X}_1 - E\mathbf{X}_1)(\mathbf{Y}_1 - E\mathbf{Y}_1)^T] + E[(\mathbf{X}_1 - E\mathbf{X}_1)(\mathbf{Y}_2 - E\mathbf{Y}_2)^T]$$
$$+ E[(\mathbf{X}_2 - E\mathbf{X}_2)(\mathbf{Y}_1 - E\mathbf{Y}_1)^T] + E[(\mathbf{X}_2 - E\mathbf{X}_2)(\mathbf{Y}_2 - E\mathbf{Y}_2)^T]$$
$$= \mathrm{cov}(\mathbf{X}_1, \mathbf{Y}_1) + \mathrm{cov}(\mathbf{X}_1, \mathbf{Y}_2) + \mathrm{cov}(\mathbf{X}_2, \mathbf{Y}_1) + \mathrm{cov}(\mathbf{X}_2, \mathbf{Y}_2). \quad ∎$$

Variance–Covariance Matrix

Let $\mathbf{Y} = [Y_1, \ldots, Y_n]^T$, where $Y_1, \ldots, Y_n$ have finite variance. The *variance–covariance* matrix $\mathrm{VC}(\mathbf{Y})$ of $\mathbf{Y}$ is the $n \times n$ matrix having entry $\mathrm{cov}(Y_i, Y_j)$ in row i

and column j for $1 \le i, j \le n$. Thus,

$$\mathrm{VC}(\mathbf{Y}) = \begin{bmatrix} \mathrm{cov}(Y_1, Y_1) & \cdots & \mathrm{cov}(Y_1, Y_n) \\ \vdots & & \vdots \\ \mathrm{cov}(Y_n, Y_1) & \cdots & \mathrm{cov}(Y_n, Y_n) \end{bmatrix} = \begin{bmatrix} \mathrm{var}(Y_1) & \cdots & \mathrm{cov}(Y_1, Y_n) \\ \vdots & & \vdots \\ \mathrm{cov}(Y_n, Y_1) & \cdots & \mathrm{var}(Y_n) \end{bmatrix}.$$

We also refer to $\mathrm{VC}(\mathbf{Y})$ as the variance–covariance matrix of $Y_1, \ldots, Y_n$ or of their joint distribution. If $n = 1$, then $\mathrm{VC}(\mathbf{Y}) = [\,\mathrm{var}(Y_1)\,]$ can be identified with the number $\mathrm{var}(Y_1)$. If $n = 2$, then

$$\mathrm{VC}(\mathbf{Y}) = \begin{bmatrix} \mathrm{var}(Y_1) & \mathrm{cov}(Y_1, Y_2) \\ \mathrm{cov}(Y_1, Y_2) & \mathrm{var}(Y_2) \end{bmatrix}.$$

If $Y_1, \ldots, Y_n$ are independent, then their variance–covariance matrix is a diagonal matrix having diagonal entries $\mathrm{var}(Y_1), \ldots, \mathrm{var}(Y_n)$.

It follows from the definition of the variance–covariance matrix that

$$\mathrm{VC}(\mathbf{Y}) = \mathrm{cov}(\mathbf{Y}, \mathbf{Y}). \tag{21}$$

We conclude from (20) and (21) that the variance–covariance matrix is symmetric; that is,

$$[\,\mathrm{VC}(\mathbf{Y})\,]^T = \mathrm{VC}(\mathbf{Y}). \tag{22}$$

A number of additional properties involving the variance–covariance matrix of Y, which follow easily from (21) and the properties of the covariance matrix, are given in the next result.

THEOREM 5.8   (a) $\mathrm{VC}(\mathbf{Y}) = 0$ if and only if $\mathbf{Y}$ is a constant random vector;

(b) $\mathrm{VC}(\mathbf{a} + \mathbf{Y}) = \mathrm{VC}(\mathbf{Y})$ for $\mathbf{a} \in \mathbb{R}^n$;

(c) $\mathrm{VC}(b\mathbf{Y}) = b^2 \mathrm{VC}(\mathbf{Y})$ for $b \in \mathbb{R}$;

(d) $\mathrm{cov}(\mathbf{AY}, \mathbf{BY}) = \mathbf{A}\,\mathrm{VC}(\mathbf{Y})\mathbf{B}^T$, where $\mathbf{A}$ and $\mathbf{B}$ have n columns;

(e) $\mathrm{cov}(\mathbf{a}^T\mathbf{Y}, \mathbf{b}^T\mathbf{Y}) = \mathbf{a}^T \mathrm{VC}(\mathbf{Y})\mathbf{b}$ for $\mathbf{a}, \mathbf{b} \in \mathbb{R}^n$;

(f) $\mathrm{VC}(\mathbf{BY}) = \mathbf{B}[\,\mathrm{VC}(\mathbf{Y})\,]\mathbf{B}^T$, where $\mathbf{B}$ has n columns;

(g) $\mathrm{var}(\mathbf{b}^T\mathbf{Y}) = \mathbf{b}^T \mathrm{VC}(\mathbf{Y})\mathbf{b}$ for $\mathbf{b} \in \mathbb{R}^n$.

EXAMPLE 5.10   Let $X_1, V, W_1, W_2,$ and W_3 be independent random variables, each having variance 1, and set $X_2 = 2X_1 + V$, $Y_1 = 3X_1 + W_1$, $Y_2 = X_1 - 4X_2 + W_2$, and $Y_3 = X_1 + X_2 + W_3$. Determine

(a) the covariance matrix between $\mathbf{X} = [\,X_1, X_2\,]^T$ and $\mathbf{Y} = [\,Y_1, Y_2, Y_3\,]^T$;

(b) the variance–covariance matrices of $\mathbf{X}$ and $\mathbf{Y}$.

Solution   **(a)** Observe that

$$\mathbf{X} = \begin{bmatrix} X_1 \\ X_2 \end{bmatrix} = \begin{bmatrix} X_1 \\ 2X_1 + V \end{bmatrix} = \begin{bmatrix} 1 & 0 & 0 & 0 & 0 \\ 2 & 1 & 0 & 0 & 0 \end{bmatrix} \begin{bmatrix} X_1 \\ V \\ W_1 \\ W_2 \\ W_3 \end{bmatrix}$$

and

$$\mathbf{Y} = \begin{bmatrix} Y_1 \\ Y_2 \\ Y_3 \end{bmatrix}$$

$$= \begin{bmatrix} 3X_1 + W_1 \\ X_1 - 4X_2 + W_2 \\ X_1 + X_2 + W_3 \end{bmatrix}$$

$$= \begin{bmatrix} 3X_1 + W_1 \\ -7X_1 - 4V + W_2 \\ 3X_1 + V + W_3 \end{bmatrix}$$

$$= \begin{bmatrix} 3 & 0 & 1 & 0 & 0 \\ -7 & -4 & 0 & 1 & 0 \\ 3 & 1 & 0 & 0 & 1 \end{bmatrix} \begin{bmatrix} X_1 \\ V \\ W_1 \\ W_2 \\ W_3 \end{bmatrix}.$$

Since the variance–covariance matrix of X_1, V, W_1, W_2, and W_3 is the 5×5 identity matrix, we conclude from Theorem 5.8(d) that

$$\text{cov}(\mathbf{X}, \mathbf{Y}) = \begin{bmatrix} 1 & 0 & 0 & 0 & 0 \\ 2 & 1 & 0 & 0 & 0 \end{bmatrix} \begin{bmatrix} 3 & -7 & 3 \\ 0 & -4 & 1 \\ 1 & 0 & 0 \\ 0 & 1 & 0 \\ 0 & 0 & 1 \end{bmatrix} = \begin{bmatrix} 3 & -7 & 3 \\ 6 & -18 & 7 \end{bmatrix}.$$

(b) By Theorem 5.8(f), the variance–covariance matrices of $\mathbf{X}$ and $\mathbf{Y}$ are given by

$$\text{VC}(\mathbf{X}) = \begin{bmatrix} 1 & 0 & 0 & 0 & 0 \\ 2 & 1 & 0 & 0 & 0 \end{bmatrix} \begin{bmatrix} 1 & 2 \\ 0 & 1 \\ 0 & 0 \\ 0 & 0 \\ 0 & 0 \end{bmatrix} = \begin{bmatrix} 1 & 2 \\ 2 & 5 \end{bmatrix}$$

and

$$\text{VC}(\mathbf{Y}) = \begin{bmatrix} 3 & 0 & 1 & 0 & 0 \\ -7 & -4 & 0 & 1 & 0 \\ 3 & 1 & 0 & 0 & 1 \end{bmatrix} \begin{bmatrix} 3 & -7 & 3 \\ 0 & -4 & 1 \\ 1 & 0 & 0 \\ 0 & 1 & 0 \\ 0 & 0 & 1 \end{bmatrix} = \begin{bmatrix} 10 & -21 & 9 \\ -21 & 66 & -25 \\ 9 & -25 & 11 \end{bmatrix}. \quad \blacksquare$$

EXAMPLE **5.11**   Let X_1, W_1, and W_2 be independent random variables, each having variance 1. Determine the variance–covariance matrix of X_1, $X_2 = X_1 + W_1$, and $Y = X_1 + X_2 + W_2$.

Solution   Now $Y = X_1 + X_1 + W_1 + W_2 = 2X_1 + W_1 + W_2$, so

$$\begin{bmatrix} X_1 \\ X_2 \\ Y \end{bmatrix} = \begin{bmatrix} 1 & 0 & 0 \\ 1 & 1 & 0 \\ 2 & 1 & 1 \end{bmatrix} \begin{bmatrix} X_1 \\ W_1 \\ W_2 \end{bmatrix}.$$

Since the variance–covariance matrix of X_1, W_1, and W_2 is the 3×3 identity matrix, we conclude from Theorem 5.8(f) that the variance–covariance matrix of X_1, X_2, and Y is given by

$$\begin{bmatrix} 1 & 0 & 0 \\ 1 & 1 & 0 \\ 2 & 1 & 1 \end{bmatrix} \begin{bmatrix} 1 & 1 & 2 \\ 0 & 1 & 1 \\ 0 & 0 & 1 \end{bmatrix} = \begin{bmatrix} 1 & 1 & 2 \\ 1 & 2 & 3 \\ 2 & 3 & 6 \end{bmatrix}. \quad \blacksquare$$

The results in parts (e) and (g) of Theorem 5.8 can be expressed in summation form as follows.

COROLLARY **5.4**   Let $Y_1, \ldots, Y_n$ each have finite variance and let $a_1, \ldots, a_n \in \mathbb{R}$ and $b_1, \ldots, b_n \in \mathbb{R}$. Then

$$\mathrm{cov}\left(\sum_j a_j Y_j, \sum_j b_j Y_j \right) = \sum_{i,j} a_i b_j \, \mathrm{cov}(Y_i, Y_j)$$

$$= \sum_j a_j b_j \, \mathrm{var}(Y_j) + \sum_{i \neq j} a_i b_j \, \mathrm{cov}(Y_i, Y_j)$$

and

$$\mathrm{var}\left(\sum_j b_j Y_j \right) = \sum_{i,j} b_i b_j \, \mathrm{cov}(Y_i, Y_j) = \sum_j b_j^2 \, \mathrm{var}(Y_j) + \sum_{i \neq j} b_i b_j \, \mathrm{cov}(Y_i, Y_j).$$

The double sum over $i \neq j$ appearing in the conclusions of Corollary 5.4 can also be written as twice the double sum over $i < j$; that is,

$$\sum_{i \neq j} = 2 \sum_{i < j}.$$

Corollary 5.4, in turn has the following consequences, the second of which agrees with Theorem 2.8 in Section 2.3.

COROLLARY **5.5**   Let $Y_1, \ldots, Y_n$ be independent random variables having finite variance and let $a_1, \ldots, a_n, b_1, \ldots, b_n$ be real numbers. Then

$$\mathrm{cov}\left(\sum_j a_j Y_j, \sum_j b_j Y_j \right) = \sum_j a_j b_j \, \mathrm{var}(Y_j)$$

and

$$\operatorname{var}\left(\sum_j b_j Y_j \right) = \sum_j b_j^2 \operatorname{var}(Y_j).$$

The next result follows from Theorem 5.7.

THEOREM 5.9 Let $\mathbf{Y}_1$ and $\mathbf{Y}_2$ be n-dimensional random vectors whose entries have finite variance. Then

$$\operatorname{VC}(\mathbf{Y}_1 + \mathbf{Y}_2) = \operatorname{VC}(\mathbf{Y}_1) + \operatorname{cov}(\mathbf{Y}_1, \mathbf{Y}_2) + \operatorname{cov}(\mathbf{Y}_2, \mathbf{Y}_1) + \operatorname{VC}(\mathbf{Y}_2).$$

In particular, if $\mathbf{Y}_1$ and $\mathbf{Y}_2$ are independent, then

$$\operatorname{VC}(\mathbf{Y}_1 + \mathbf{Y}_2) = \operatorname{VC}(\mathbf{Y}_1) + \operatorname{VC}(\mathbf{Y}_2).$$

THEOREM 5.10 Every variance–covariance matrix is positive semidefinite.

Proof Let $\mathbf{Y}$ be an n-dimensional random column vector, the entries of which have finite variance. Then, by Theorem 5.8(g),

$$\mathbf{b}^T \operatorname{VC}(\mathbf{Y})\mathbf{b} = \operatorname{var}(\mathbf{b}^T \mathbf{Y}) \geq 0, \qquad \mathbf{b} \in \mathbb{R}^n,$$

so $\operatorname{VC}(\mathbf{Y})$ is positive semidefinite. ∎

THEOREM 5.11 Every positive, semidefinite, symmetric matrix $\boldsymbol{\Sigma}$ is a variance–covariance matrix.

Proof Let $\mathbf{LL}^T$ be a Cholesky decomposition of $\boldsymbol{\Sigma}$; let $W_1, \ldots, W_n$ be independent random variables, each having variance 1; and set $\mathbf{W} = [\, W_1, \ldots, W_n \,]^T$. Then $\operatorname{VC}(\mathbf{W}) = \mathbf{I}$, so

$$\operatorname{VC}(\mathbf{LW}) = \mathbf{L}[\, \operatorname{VC}(\mathbf{W}) \,]\mathbf{L}^T = \mathbf{LIL}^T = \mathbf{LL}^T = \boldsymbol{\Sigma}$$

and hence $\boldsymbol{\Sigma}$ is a variance–covariance matrix. ∎

THEOREM 5.12 Let $\mathbf{Y}$ be an n-dimensional random vector having a finite variance–covariance matrix and let $\mathbf{b} \in \mathbb{R}^n$. Then the following conditions are equivalent:

(a) $\mathbf{b}^T \mathbf{Y}$ is a constant random variable; **(b)** $\mathbf{b}^T \operatorname{VC}(\mathbf{Y})\mathbf{b} = 0$;
(c) $\operatorname{VC}(\mathbf{Y})\mathbf{b} = \mathbf{0}$.

Proof Now $\mathbf{b}^T \mathbf{Y}$ is a constant random variable if and only if its variance equals zero. Thus, since $\operatorname{var}(\mathbf{b}^T \mathbf{Y}) = \mathbf{b}^T \operatorname{var}(\mathbf{Y})\mathbf{b}$, we see that parts (a) and (b) are equivalent. Clearly, part (c) implies part (b). Suppose, conversely, that part (b) holds and let $\mathbf{LL}^T$ be a Cholesky decomposition of $\operatorname{VC}(\mathbf{Y})$. Then $0 = \mathbf{b}^T \mathbf{LL}^T \mathbf{b} = (\mathbf{L}^T \mathbf{b})^T \mathbf{L}^T \mathbf{b}$, so $\mathbf{L}^T \mathbf{b} = \mathbf{0}$ and hence $\operatorname{VC}(\mathbf{Y})\mathbf{b} = \mathbf{LL}^T \mathbf{b} = \mathbf{0}$; therefore, part (c) holds. ∎

   A variance–covariance matrix is positive definite if and only if it is invertible or, equivalently, if and only if its determinant is positive. Recall also that an $n \times n$

matrix $\mathbf{A}$ is noninvertible if and only if there is a nonzero vector $\mathbf{b} \in \mathbb{R}^n$ such that $\mathbf{Ab} = \mathbf{0}$.

COROLLARY **5.6**   Let $Y_1, \ldots, Y_n$ have finite variance. Then their variance–covariance matrix is non-invertible if and only if there are constants $b_1, \ldots, b_n \in \mathbb{R}$, not all zero, such that $b_1 Y_1 + \cdots + b_n Y_n$ is a constant random variable.

Proof   Set $\mathbf{Y} = [Y_1, \ldots, Y_n]^T$. Suppose there are constants $b_1, \ldots, b_n \in \mathbb{R}$, not all zero, such that $b_1 Y_1 + \cdots + b_n Y_n$ is a constant random variable, and set $\mathbf{b} = [b_1, \ldots, b_n]^T$. Then $\mathbf{b}$ is a nonzero vector and part (a) of Theorem 5.12 holds, so part (c) of this theorem also holds and hence VC($\mathbf{Y}$) is noninvertible. Suppose, conversely, that VC($\mathbf{Y}$) is noninvertible. Then there is a nonzero vector $\mathbf{b} = [b_1, \ldots, b_n]^T \in \mathbb{R}^n$ such that part (c) of Theorem 5.12 holds. Thus, $b_1, \ldots, b_n$ are not all zero. Also, part (a) of Theorem 5.12 holds, so that $b_1 Y_1 + \ldots + b_n Y_n$ is a constant random variable. ∎

EXAMPLE **5.12**   Let $n \geq 2$. Give a geometric interpretation to the statement that there are constants $b_1, \ldots, b_n \in \mathbb{R}$, not all zero, such that $b_1 Y_1 + \cdots + b_n Y_n$ is a constant random variable.

Solution   Consider the indicated statement with $n = 2$: there are constants $b_1, b_2 \in \mathbb{R}$, not both zero, such that $b_1 Y_1 + b_2 Y_2$ is a constant random variable. Suppose this statement holds and let a be such that $P(b_1 Y_1 + b_2 Y_2 = a) = 1$. Let $\mathcal{L}$ be the line in $\mathbb{R}^2$ defined by $\mathcal{L} = \{(y_1, y_2) \in \mathbb{R}^2 : b_1 y_1 + b_2 y_2 = a\}$. Then

$$P((Y_1, Y_2) \in \mathcal{L}) = 1. \tag{23}$$

Conversely, if there is a line $\mathcal{L}$ in $\mathbb{R}^2$ that satisfies (23), then the indicated statement holds. Therefore, the indicated statement holds if and only if there is a line $\mathcal{L}$ in $\mathbb{R}^2$ that satisfies (23).

Consider, next, the indicated statement with $n = 3$: there are constants b_1, $b_2, b_3 \in \mathbb{R}$, not all zero, such that $b_1 Y_1 + b_2 Y_2 + b_3 Y_3$ is a constant random variable. Suppose this statement holds and let a be such that $P(b_1 Y_1 + b_2 Y_2 + b_3 Y_3 = a) = 1$. Let $\mathcal{P}$ be the plane in $\mathbb{R}^3$ defined by $\mathcal{P} = \{(y_1, y_2, y_3) \in \mathbb{R}^3 : b_1 y_1 + b_2 y_2 + b_3 y_3 = a\}$. Then

$$P((Y_1, Y_2, Y_3) \in \mathcal{P}) = 1. \tag{24}$$

Conversely, if there is a plane $\mathcal{P}$ in $\mathbb{R}^3$ that satisfies (24), then the indicated statement holds. Therefore, the indicated statement holds if and only if there is a plane $\mathcal{P}$ in $\mathbb{R}^3$ that satisfies (24).

Consider now the indicated statement with $n \geq 2$. Suppose the statement holds and let a be such that $P(b_1 Y_1 + \cdots + b_n Y_n = a) = 1$. Let $\mathcal{H}$ be the hyperplane in $\mathbb{R}^n$ defined by

$$\mathcal{H} = \{(y_1, \ldots, y_n) \in \mathbb{R}^n : b_1 y_1 + \cdots + b_n y_n = a\}. \tag{25}$$

[A *hyperplane* in $\mathbb{R}^n$ is a set of the form given by (25) where $b_1, \ldots, b_n$ are not all zero.] Then

$$P((Y_1, \ldots, Y_n) \in \mathcal{H}) = 1. \tag{26}$$

Conversely, if there is a hyperplane $\mathcal{H}$ in $\mathbb{R}^n$ that satisfies (26), then the indicated statement holds. Therefore, the indicated statement holds if and only if there is a hyperplane $\mathcal{H}$ in $\mathbb{R}^n$ that satisfies (26). ∎

EXAMPLE 5.13 Let W_1 and W_2 be independent random variables, each having variance 1, and set $Y_1 = W_1 + W_2$, $Y_2 = W_1 - W_2$, and $Y_3 = 4W_1 + 2W_2$. [It is clear from the definition of Y_1, Y_2, and Y_3 as linear functions of the two random variables W_1 and W_2 that (Y_1, Y_2, Y_3) lies in a fixed plane.]

(a) Determine the variance–covariance matrix of Y_1, Y_2, and Y_3 and show directly that this matrix fails to be invertible.

(b) Choose $b_1, b_2 \in \mathbb{R}$ such that $Y_3 - b_1 Y_1 - b_2 Y_2$ is a constant random variable.

Solution **(a)** The variance–covariance matrix of W_1 and W_2 is the 2×2 identity matrix. Also,

$$\begin{bmatrix} Y_1 \\ Y_2 \\ Y_3 \end{bmatrix} = \begin{bmatrix} 1 & 1 \\ 1 & -1 \\ 4 & 2 \end{bmatrix} \begin{bmatrix} W_1 \\ W_2 \end{bmatrix},$$

so the variance–covariance matrix of Y_1, Y_2, and Y_3 is given by

$$\begin{bmatrix} 1 & 1 \\ 1 & -1 \\ 4 & 2 \end{bmatrix} \begin{bmatrix} 1 & 1 & 4 \\ 1 & -1 & 2 \end{bmatrix} = \begin{bmatrix} 2 & 0 & 6 \\ 0 & 2 & 2 \\ 6 & 2 & 20 \end{bmatrix}.$$

Since this matrix has determinant

$$2 \begin{vmatrix} 2 & 2 \\ 2 & 20 \end{vmatrix} + 6 \begin{vmatrix} 0 & 2 \\ 6 & 2 \end{vmatrix} = 2(36) + 6(-12) = 0,$$

it is noninvertible.

(b) Observe that

$$Y_3 - b_1 Y_1 - b_2 Y_2 = 4W_1 + 2W_2 - b_1(W_1 + W_2) - b_2(W_1 - W_2)$$
$$= (4 - b_1 - b_2)W_1 + (2 - b_1 + b_2)W_2,$$

which is a constant random variable if and only if $b_1 + b_2 = 4$ and $b_1 - b_2 = 2$ or, equivalently, if and only if $b_1 = 3$ and $b_2 = 1$. Alternatively, by the equivalence of parts (a) and (c) in Theorem 5.12 and the solution to part (a), $Y_3 - b_1 Y_1 - b_2 Y_2$ is a constant random variable if and only if

$$\begin{bmatrix} 0 \\ 0 \\ 0 \end{bmatrix} = \begin{bmatrix} 2 & 0 & 6 \\ 0 & 2 & 2 \\ 6 & 2 & 20 \end{bmatrix} \begin{bmatrix} -b_1 \\ -b_2 \\ 1 \end{bmatrix} = \begin{bmatrix} -2b_1 + 6 \\ -2b_2 + 2 \\ -6b_1 - 2b_2 + 20 \end{bmatrix}$$

or, equivalently, if and only if $b_1 = 3$ and $b_2 = 1$. ∎

THEOREM 5.13   Let $Y_1, \ldots, Y_n$ be independent random variables, each having finite, positive variance. Then their variance–covariance matrix is positive definite.

Proof   The variance–covariance matrix of $Y_1, \ldots, Y_n$ is a diagonal matrix whose diagonal entries $\mathrm{var}(Y_1), \ldots, \mathrm{var}(Y_n)$ are all positive, so it is a positive definite matrix.   ∎

THEOREM 5.14   Let $\mathbf{W}$ have a positive definite variance–covariance matrix and let $\mathbf{Y} = \mathbf{BW}$, where $\mathbf{B}$ is an invertible matrix. Then $\mathbf{Y}$ has a positive definite variance–covariance matrix.

Proof   Now $\mathrm{VC}(\mathbf{Y}) = \mathbf{B}\,\mathrm{VC}(\mathbf{W})\,\mathbf{B}^T$. Since the variance–covariance matrix of $\mathbf{W}$ is positive definite, it is invertible; since $\mathbf{B}$ is invertible, so is $\mathbf{B}^T$. Therefore, $\mathrm{VC}(\mathbf{Y})$ is invertible (the product of invertible matrices is invertible).   ∎

Correlation

EXAMPLE 5.14   Let Y_1 and Y_2 each have finite variance. Determine when their variance–covariance matrix is positive definite.

Solution   The variance–covariance matrix

$$\begin{bmatrix} \mathrm{var}(Y_1) & \mathrm{cov}(Y_1, Y_2) \\ \mathrm{cov}(Y_1, Y_2) & \mathrm{var}(Y_2) \end{bmatrix}$$

of Y_1 and Y_2 is positive definite if and only if its determinant

$$\mathrm{var}(Y_1)\,\mathrm{var}(Y_2) - [\,\mathrm{cov}(Y_1, Y_2)\,]^2$$

is positive or, equivalently, if and only if $\mathrm{var}(Y_1) > 0$, $\mathrm{var}(Y_2) > 0$ and

$$-1 < \mathrm{cor}(Y_1, Y_2) < 1.   ∎$$

Let $Y_1, \ldots, Y_n$ each have finite, positive variance. They are said to be uncorrelated if Y_i and Y_j are uncorrelated for $i \neq j$ or, equivalently, if $\mathrm{cov}(Y_i, Y_j) = 0$ for $i \neq j$. If $Y_1, \ldots, Y_n$ are independent, then they are uncorrelated; these random variables, however, can be uncorrelated without being independent.

Suppose $Y_1, \ldots, Y_n$ are uncorrelated. Then the off-diagonal entries of their variance–covariance matrix are all zero, so this matrix is the diagonal matrix whose diagonal entries are the variances of $Y_1, \ldots, Y_n$; that is,

$$\mathrm{VC}(\mathbf{Y}) = \begin{bmatrix} \mathrm{var}(Y_1) & \ldots & 0 \\ \vdots & & \vdots \\ 0 & \ldots & \mathrm{var}(Y_n) \end{bmatrix}.$$

Suppose, in addition, that $Y_1, \ldots, Y_n$ have common variance σ^2. Then $\mathrm{VC}(\mathbf{Y}) = \sigma^2 \mathbf{I}$. In particular, if $Y_1, \ldots, Y_n$ are uncorrelated random variables having variance 1, then $\mathrm{VC}(\mathbf{Y}) = \mathbf{I}$.

The converses of these results also hold. If $\mathrm{VC}(\mathbf{Y})$ is a diagonal matrix, then $Y_1, \ldots, Y_n$ are uncorrelated. If $\mathrm{VC}(\mathbf{Y}) = \sigma^2 \mathbf{I}$, then $Y_1, \ldots, Y_n$ are uncorrelated random variables having common variance σ^2. In particular, if $VC(\mathbf{Y}) = \mathbf{I}$, then $Y_1, \ldots, Y_n$ are uncorrelated random variables, each having variance 1.

Observe that $cov(Y_i, Y_j) = SD(Y_i) SD(Y_j) cor(Y_i, Y_j)$ for $1 \le i, j \le n$ and that $cor(Y_i, Y_i) = 1$ for $1 \le i \le n$. The $n \times n$ matrix having entry $cor(Y_i, Y_j)$ in row i and column j for $1 \le i, j \le n$ is referred to as the *correlation matrix* of $\mathbf{Y}$ or of $Y_1, \ldots, Y_n$ or their joint distribution; it coincides with the variance–covariance matrix of $Y_1/SD(Y_1), \ldots, Y_n/SD(Y_n)$.

Problems

5.29 Determine the variance–covariance matrix for the random variables X, Y, and Z in Problem 5.18 in Section 5.2.

5.30 Consider the setup of Example 5.10, and set $Z_1 = Y_1$, $Z_2 = Y_1 + Y_2$, $Z_3 = Y_1 + Y_2 + Y_3$, and $\mathbf{Z} = [Z_1, Z_2, Z_3]^T$. Determine **(a)** $cov(\mathbf{Y}, \mathbf{Z})$; **(b)** $VC(\mathbf{Z})$.

5.31 Let W_1, W_2, and W_3 be independent random variables, each having variance 1, and set $Y_1 = W_1 + W_2$, $Y_2 = W_1 + W_3$, $Y_3 = W_2 + W_3$, and $Y_4 = W_1 + 5W_2 + 2W_3$. **(a)** Determine the variance–covariance matrix of Y_1, Y_2, Y_3, and Y_4. **(b)** Is this matrix invertible (why or why not)? **(c)** Choose $b_1, b_2, b_3 \in \mathbb{R}$ such that $Y_4 - b_1 Y_1 - b_2 Y_2 - b_3 Y_3$ is a constant random variable.

5.32 Let X_2, W_1, and W_2 be independent random variables having variances 2, 1/2, and 1, respectively. Determine the variance–covariance matrix of $X_1 = (1/2)X_2 + W_1$, X_2, and $Y = X_1 + X_2 + W_2$.

5.33 Let $Y_1, \ldots, Y_n$ be random variables, each having variance 1. Show that the variance of $\bar{Y} = (Y_1 + \cdots + Y_n)/n$ is at most 1.

5.34 Let $Y_1, \ldots, Y_n$ have a positive definite variance–covariance matrix and let $1 \le m < n$. Show that the variance–covariance matrix of $Y_1, \ldots, Y_m$ is positive definite.

5.35 Let W_1 and W_2 be independent random variables, each having mean 0 and variance 1; let σ_1 and σ_2 be finite positive constants; let $-1 \le \rho \le 1$; and set $X = \sigma_1 W_1$ and $Y = \sigma_2(\rho W_1 + \sqrt{1 - \rho^2} W_2)$. Show that $var(X) = \sigma_1^2$, $var(Y) = \sigma_2^2$ and $cor(X, Y) = \rho$.

5.36 Determine the validity of the following conjecture: If the first of three random variables is positively correlated with the second and the second is positively correlated with the third, then the first is positively correlated with the third. [*Hint:* Let W_1 and W_2 be independent random variables, each having variance 1, and set $Y_1 = 2W_1 + W_2$, $Y_2 = W_1 + 2W_2$, and $Y_3 = W_2 - W_1$.]

5.37 Suppose $Y_1, Y_2, \ldots$ have variance σ^2 and that $cor(Y_i, Y_j) = \rho \ge 0$ for $i \ne j$. Determine **(a)** the variance of $\bar{Y} = (Y_1 + \cdots + Y_n)/n$; **(b)** the limit of this variance as $n \to \infty$.

5.38 Let X and Y have finite, positive variance and set $\mu_1 = EX$ and $\sigma_1 = SD(X)$. Also, let $\hat{Y} = \alpha + \beta X$ denote the best linear predictor of Y based on X and let σ^2 denote the mean squared error of this predictor (which is the variance of the error $Y - \hat{Y}$ of prediction). In terms of μ_1, σ_1, α, β, and σ^2, determine **(a)** the mean vector of X

and Y; **(b)** the variance–covariance matrix of X and Y. [*Hint:* See Problem 5.12 in Section 5.1.]

5.39 Let Y_1 and Y_2 be nonconstant random variables having finite, positive variances σ_1^2 and σ_2^2, respectively, and let ρ denote the correlation coefficient between Y_1 and Y_2.
(a) Show that the variance–covariance matrix of Y_1 and Y_2 can be written as

$$\begin{bmatrix} \sigma_1^2 & \rho\sigma_1\sigma_2 \\ \rho\sigma_1\sigma_2 & \sigma_2^2 \end{bmatrix} = \begin{bmatrix} \sigma_1 & 0 \\ 0 & \sigma_2 \end{bmatrix}\begin{bmatrix} 1 & \rho \\ \rho & 1 \end{bmatrix}\begin{bmatrix} \sigma_1 & 0 \\ 0 & \sigma_2 \end{bmatrix}.$$

(b) Determine the Cholesky decomposition $\mathbf{L}\mathbf{L}^T$ of this variance–covariance matrix such that $\mathbf{L}$ has nonnegative diagonal entries.

5.40 Let W_0, W_1, W_2, and W_3 be uncorrelated random variables, each having variance 1; let $0 \le \rho \le 1$; and set $Y_i = \sqrt{\rho}\,W_0 + \sqrt{1-\rho}\,W_i$ for $i = 1, 2, 3$. Determine the variance–covariance matrix of Y_1, Y_2, and Y_3.

5.41 Let $\mathbf{Y} = [\,Y_1, \ldots, Y_n\,]^T$ have a positive definite variance–covariance matrix. Consider the Cholesky decomposition $\mathbf{L}\mathbf{L}^T$ of this variance–covariance matrix, with $\mathbf{L}$ being an invertible lower triangular matrix, and let $\mathbf{W} = [\,W_1, \ldots, W_n\,]^T$ be defined by $\mathbf{W} = \mathbf{L}^{-1}\mathbf{Y}$. Show that $W_1, \ldots, W_n$ are uncorrelated random variables, each having variance 1.

5.42 Let X and Y have finite variance. Show that $|\mathrm{cov}(X, Y)| = \mathrm{SD}(X)\,\mathrm{SD}(Y)$ if and only if there are real numbers a and b, not both zero, such that $aX + bY$ is a constant random variable.

5.4
Multiple Linear Prediction

The theory of linear prediction of Y based on X that was developed in Section 5.1 can be extended in a straightforward manner to handle the linear prediction of Y based on any number of predictor variables. Let $X_1, \ldots, X_m, Y$ be random variables having finite variance and suppose the variance–covariance matrix of $\mathbf{X} = [\,X_1, \ldots, X_m\,]^T$ is positive definite or, equivalently, that it is invertible. Consider a linear predictor $\hat{Y} = a + b_1X_1 + \cdots + b_mX_m$ of Y based on $X_1, \ldots, X_m$. Arguing as in Section 5.1, we find that for fixed $b_1, \ldots, b_m$, the unique choice of a that minimizes the mean squared error of prediction is given by $a = EY - b_1EX_1 - \cdots - b_mEX_m$. Making this choice of a, we get that

$$\hat{Y} = EY + b_1(X_1 - EX_1) + \cdots + b_m(X_m - EX_m),$$

$E\hat{Y} = EY$, and $E(Y - \hat{Y}) = 0$. Set $\mathbf{b} = [\,b_1, \ldots, b_m\,]^T$. Then we can rewrite the predictor as $\hat{Y} = EY + \mathbf{b}^T(\mathbf{X} - E\mathbf{X})$. Let $W = Y - \hat{Y} = Y - EY - \mathbf{b}^T(\mathbf{X} - E\mathbf{X})$ denote the error of prediction. Then the natural generalization of Theorem 5.2 in Section 5.1 turns out to be valid in the present context; that is, $\hat{Y}$ is a best linear predictor of Y based on $\mathbf{X}$ if and only if $\mathrm{cov}(\mathbf{X}, W) = \mathbf{0}$. Here, however, it is more convenient to proceed in reverse, beginning with the following general result.

THEOREM 5.15   Set $W = Y - EY - \beta^T(\mathbf{X} - E\mathbf{X})$, where $\beta \in \mathbb{R}^m$. Then

(a) $EW = 0$;

(b) the unique choice of β such that $\text{cov}(\mathbf{X}, W) = \mathbf{0}$ [or, equivalently, such that $\text{cov}(X_i, W) = 0$ for $1 \le i \le m$] is given by

$$\beta = [\text{VC}(\mathbf{X})]^{-1}[\text{cov}(\mathbf{X}, Y)]; \tag{27}$$

(c) for the choice of β given by (27),

$$\text{var}(W) = \text{var}(Y) - [\text{cov}(\mathbf{X}, Y)]^T[\text{VC}(\mathbf{X})]^{-1}\text{cov}(\mathbf{X}, Y)$$
$$= \text{var}(Y) - \beta^T\text{cov}(\mathbf{X}, Y).$$

Proof   (a) We have that

$$EW = E[Y - EY - \beta^T(\mathbf{X} - E\mathbf{X})] = E(Y - EY) - \beta^TE(\mathbf{X} - E\mathbf{X}) = 0.$$

(b) Now

$$\text{cov}(\mathbf{X}, W) = \text{cov}(\mathbf{X}, Y - EY - \beta^T(\mathbf{X} - E\mathbf{X}))$$
$$= \text{cov}(\mathbf{X}, Y - \beta^T\mathbf{X})$$
$$= \text{cov}(\mathbf{X}, Y) - \text{VC}(\mathbf{X})\beta.$$

Thus, $\text{cov}(\mathbf{X}, W) = \mathbf{0}$ if and only if $\text{VC}(\mathbf{X})\beta = \text{cov}(\mathbf{X}, Y)$ and hence if and only if (27) holds.

(c) Observe that

$$\text{var}(W) = \text{var}(Y - EY - \beta^T(\mathbf{X} - E\mathbf{X}))$$
$$= \text{var}(Y - \beta^T\mathbf{X})$$
$$= \text{cov}(Y - \beta^T\mathbf{X}, Y - \beta^T\mathbf{X})$$
$$= \text{cov}(Y, Y) - \beta^T\text{cov}(\mathbf{X}, Y) - \text{cov}(Y, \mathbf{X})\beta + \beta^T\text{VC}(\mathbf{X})\beta$$
$$= \text{var}(Y) - 2\beta^T\text{cov}(\mathbf{X}, Y) + \beta^T\text{VC}(\mathbf{X})\beta.$$

It now follows from (27) that

$$\text{var}(W) = \text{var}(Y) - 2[\text{cov}(\mathbf{X}, Y)]^T[\text{VC}(\mathbf{X})]^{-1}\text{cov}(\mathbf{X}, Y)$$
$$+ [\text{cov}(\mathbf{X}, Y)]^T[\text{VC}(\mathbf{X})]^{-1}\text{VC}(\mathbf{X})[\text{VC}(\mathbf{X})]^{-1}\text{cov}(\mathbf{X}, Y)$$
$$= \text{var}(Y) - [\text{cov}(\mathbf{X}, Y)]^T[\text{VC}(\mathbf{X})]^{-1}\text{cov}(\mathbf{X}, Y)$$
$$= \text{var}(Y) - \beta^T\text{cov}(\mathbf{X}, Y). \quad \blacksquare$$

THEOREM 5.16   Consider the linear predictor $\tilde{Y} = EY + \mathbf{b}^T(\mathbf{X} - E\mathbf{X})$ of Y based on $\mathbf{X}$, where $\mathbf{b} \in \mathbb{R}^m$. Then

$$\text{MSE}(\tilde{Y}) = \text{var}(Y) - [\text{cov}(\mathbf{X}, Y)]^T[\text{VC}(\mathbf{X})]^{-1}\text{cov}(\mathbf{X}, Y)$$
$$+ (\beta - \mathbf{b})^T\text{VC}(\mathbf{X})(\beta - \mathbf{b}), \tag{28}$$

where β is given by (27).

Proof   Set $W = Y - EY - \beta^T(\mathbf{X} - E\mathbf{X})$. It follows from Theorem 5.15 that $EW = 0$, $\text{cov}(\mathbf{X}, W) = \mathbf{0}$, and $\text{var}(W) = \text{var}(Y) - [\,\text{cov}(\mathbf{X}, Y)\,]^T[\,\text{VC}(\mathbf{X})\,]^{-1}\text{cov}(\mathbf{X}, Y)$. Now

$$Y = EY + \beta^T(\mathbf{X} - E\mathbf{X}) + W,$$

so

$$Y - \tilde{Y} = (\beta - \mathbf{b})^T(\mathbf{X} - E\mathbf{X}) + W.$$

Consequently,

$$
\begin{aligned}
\text{MSE}(\tilde{Y}) &= \text{var}(Y - \tilde{Y}) \\
&= \text{var}\left((\beta - \mathbf{b})^T\mathbf{X} + W\right) \\
&= \text{var}(W) + 2\,\text{cov}((\beta - \mathbf{b})^T\mathbf{X}, W) + \text{var}((\beta - \mathbf{b})^T\mathbf{X}) \\
&= \text{var}(W) + 2(\beta - \mathbf{b})^T\text{cov}(\mathbf{X}, W) + (\beta - \mathbf{b})^T\text{VC}(\mathbf{X})(\beta - \mathbf{b}) \\
&= \text{var}(W) + (\beta - \mathbf{b})^T\text{VC}(\mathbf{X})(\beta - \mathbf{b}),
\end{aligned}
$$

which yields (28).   ∎

COROLLARY **5.7**   Set $\hat{Y} = EY + \beta^T(\mathbf{X} - E\mathbf{X})$, where β is given by Equation (27). Then $\hat{Y}$ is the unique best linear predictor of Y based on $\mathbf{X}$, $\text{cov}(\hat{Y}, Y - \hat{Y}) = 0$, and the mean squared error of this predictor is given by

$$
\begin{aligned}
\text{MSE}(\hat{Y}) &= \text{var}(Y) - [\,\text{cov}(\mathbf{X}, Y)\,]^T[\,\text{VC}(\mathbf{X})\,]^{-1}\text{cov}(\mathbf{X}, Y) \\
&= \text{var}(Y) - \beta^T\text{cov}(\mathbf{X}, Y). \tag{29}
\end{aligned}
$$

Proof   Set $W = Y - \hat{Y} = Y - EY - \beta^T(\mathbf{X} - E\mathbf{X})$. It follows from Theorem 5.15 that

$$\text{cov}(\hat{Y}, Y - \hat{Y}) = \text{cov}(\beta^T\mathbf{X}, W) = \beta^T\text{cov}(\mathbf{X}, W) = 0.$$

Since the variance–covariance matrix of $\mathbf{X}$ is positive definite, the term

$$(\beta - \mathbf{b})^T\text{VC}(\mathbf{X})(\beta - \mathbf{b})$$

in the right side of (28) is nonnegative and equals zero if and only if $\mathbf{b} = \beta$. The remaining conclusions now follow from Theorem 5.16.   ∎

EXAMPLE **5.15**   Let X_1, X_2, and Y have mean vector $[\,8, 5, 33\,]^T$ and variance–covariance matrix

$$
\begin{bmatrix}
9 & -6 & -9 \\
-6 & 16 & 42 \\
-9 & 42 & 121
\end{bmatrix},
$$

and set $\mathbf{X} = [\,X_1, X_2\,]^T$. Determine

(a) $E\mathbf{X}$, $\text{VC}(\mathbf{X})$, and $\text{cov}(\mathbf{X}, Y)$;

(b) the best linear predictor of Y based on $\mathbf{X}$;

(c) the mean squared error of this predictor.

Solution   **(a)** From the given information, we see that

$$EX = \begin{bmatrix} 8 \\ 5 \end{bmatrix}, \quad VC(\mathbf{X}) = \begin{bmatrix} 9 & -6 \\ -6 & 16 \end{bmatrix}, \quad \text{and} \quad \text{cov}(\mathbf{X}, Y) = \begin{bmatrix} -9 \\ 42 \end{bmatrix}.$$

(b) The best linear predictor of Y based on $\mathbf{X}$ is given by $\hat{Y} = EY + \boldsymbol{\beta}^T(\mathbf{X} - EX)$, where EX is given in the solution to part (a), $EY = 33$, and

$$\boldsymbol{\beta} = [\,VC(\mathbf{X})\,]^{-1}\text{cov}(\mathbf{X}, Y) = \begin{bmatrix} 4/27 & 1/18 \\ 1/18 & 1/12 \end{bmatrix} \begin{bmatrix} -9 \\ 42 \end{bmatrix} = \begin{bmatrix} 1 \\ 3 \end{bmatrix}.$$

Thus, $\hat{Y} = 33 + (X_1 - 8) + 3(X_2 - 5) = 10 + X_1 + 3X_2$.

(c) The mean squared error of this predictor is given by

$$\text{MSE}(\hat{Y}) = \text{var}(Y) - \boldsymbol{\beta}^T\text{cov}(\mathbf{X}, Y) = 121 - [\,1, 3\,]\begin{bmatrix} -9 \\ 42 \end{bmatrix} = 4. \quad \blacksquare$$

THEOREM 5.17   Let

$$\hat{Y} = EY + \beta_1(X_1 - EX_1) + \cdots + \beta_{m-1}(X_{m-1} - EX_{m-1}) + \beta_m(X_m - EX_m)$$

be the best linear predictor of Y based on $X_1, \ldots, X_m$ and let $\hat{X}_m$ be the best linear predictor of X_m based on $X_1, \ldots, X_{m-1}$. Then

$$\hat{Y}_0 = EY + \beta_1(X_1 - EX_1) + \cdots + \beta_{m-1}(X_{m-1} - EX_{m-1}) + \beta_m(\hat{X}_m - EX_m)$$

is the best linear predictor of Y based on $X_1, \ldots, X_{m-1}$.

Proof   It follows from Theorem 5.15 and Corollary 5.7 that $\text{cov}(X_i, Y - \hat{Y}) = 0$ for $1 \le i \le m$ and $\text{cov}(X_i, X_m - \hat{X}_m) = 0$ for $1 \le i \le m - 1$. Observe that $\hat{Y} - \hat{Y}_0 = \beta_m(X_m - \hat{X}_m)$ and hence that $Y - \hat{Y}_0 = Y - \hat{Y} + \hat{Y} - \hat{Y}_0 = Y - \hat{Y} + \beta_m(X_m - \hat{X}_m)$. Thus, $\text{cov}(X_i, Y - \hat{Y}_0) = 0$ for $1 \le i \le m - 1$. Consequently, by another application of Theorem 5.15 and Corollary 5.7, $\hat{Y}_0$ is the best linear predictor of Y based on $X_1, \ldots, X_{m-1}$. $\blacksquare$

The proof of the following consequence of Theorem 5.17 is left as a problem.

COROLLARY 5.8   Let $\hat{Y} = EY + \beta_1(X_1 - EX_1) + \beta_2(X_2 - EX_2)$ be the best linear predictor of Y based on X_1 and X_2 and let $\hat{Y}_0 = EY + \beta_{10}(X_1 - EX_1)$ be the best linear predictor of Y based on X_1. Then

$$\beta_{10} = \beta_1 + \frac{\text{cov}(X_1, X_2)}{\text{var}(X_1)}\beta_2.$$

EXAMPLE 5.16   Let X_1, X_2, and Y be as in Example 5.15. Determine

(a) the best linear predictor $\hat{X}_2$ of X_2 based on X_1;

(b) the best linear $\hat{Y}_0$ of Y based on X_1.

Solution   (a) The best linear predictor of X_2 based on X_1 is given by

$$\hat{X}_2 = EX_2 + \beta(X_1 - EX_1),$$

where $EX_1 = 8$, $EX_2 = 5$, and

$$\beta = \frac{\text{cov}(X_1, X_2)}{\text{var}(X_1)} = \frac{-6}{9} = -\frac{2}{3}.$$

Thus, $\hat{X}_2 = 5 - \frac{2}{3}(X_1 - 8) = \frac{31}{3} - \frac{2}{3}X_1$.

(b) According to Corollary 5.8 and the solution to Example 5.15(b), the best linear predictor of Y based on X_1 is given by $\hat{Y}_0 = EY + \beta_{10}(X_1 - EX_1)$, where $EX_1 = 8$, $EY = 33$, and

$$\beta_{10} = \beta_1 + \frac{\text{cov}(X_1, X_2)}{\text{var}(X_1)}\beta_2 = 1 + \frac{-6}{9} \cdot 3 = -1.$$

[Alternatively, we can obtain β_{10} more directly as

$$\beta_{10} = \frac{\text{cov}(X_1, Y)}{\text{var}(X_1)} = -\frac{9}{9} = -1.]$$

Thus, $\hat{Y}_0 = 33 - (X_1 - 8) = 41 - X_1$.

A related approach is to use Theorem 5.17 and the solutions to part (a) and Example 5.15(b) to conclude that

$$\hat{Y} = 33 + (X_1 - 8) + 3(\hat{X}_2 - 5) = 33 + (X_1 - 8) + 3\left(\frac{31}{3} - \frac{2}{3}X_1 - 5\right)$$

and hence that $\hat{Y} = 41 - X_1$.   ∎

Problems

5.43   Let X, Y, and Z be as in Problem 5.18 in Section 5.2 and Problem 5.29 in Section 5.3. Determine the best linear predictor of Z based on X and Y and its mean squared error.

5.44   Let X, Y, Z, and W have the multinomial distribution with parameters n and $\pi_1, \pi_2, \pi_3, \pi_4$. Determine the best linear predictor of Z based on X and Y and its mean squared error.

5.45   Let $\hat{Y} = \alpha + \beta^T \mathbf{X}$ denote the best linear predictor of Y based on $\mathbf{X}$ and let σ^2 denote the mean squared error of this predictor (which is the variance of the error $Y - \hat{Y}$ of prediction). In terms of $E\mathbf{X}$, $\text{VC}(\mathbf{X})$, α, β, and σ^2, determine (a) EY; (b) $\text{cov}(\mathbf{X}, Y)$; (c) $\text{var}(Y)$.

5.46   Let

$$\hat{Y} = EY + \beta_1(X_1 - EX_1) + \beta_2(X_2 - EX_2)$$
$$+ \beta_3(X_3 - EX_3) + \beta_4(X_4 - EX_4)$$

be the best linear predictor of Y based on $X_1, X_2, X_3,$ and X_4. Also, let $\hat{X}_3$ and $\hat{X}_4$ be the best linear predictors of X_3 and X_4, respectively, on X_1 and X_2. Show that

$$\hat{Y}_0 = EY + \beta_1(X_1 - EX_1) + \beta_2(X_2 - EX_2)$$
$$+ \beta_3(\hat{X}_3 - EX_3) + \beta_4(\hat{X}_4 - EX_4)$$

is the best linear predictor of Y based on X_1 and X_2.

5.47 In the context of Theorem 5.17, show that $\mathrm{MSE}(\hat{Y}_0) = \mathrm{MSE}(\hat{Y}) + \beta_m^2 \mathrm{MSE}(\hat{X}_m)$.

5.48 Verify Corollary 5.8.

5.49 Let $\hat{Y}$ be the best linear predictor of Y based on $\mathbf{X}$. Verify that **(a)** $\mathrm{cov}(\hat{Y}, Y) = \mathrm{var}(\hat{Y})$; **(b)** $\mathrm{var}(Y) = \mathrm{var}(\hat{Y}) + \mathrm{var}(Y - \hat{Y})$.

5.50 (Continued.) Suppose the variance of Y is positive. The quantity

$$\rho^2 = \frac{\mathrm{var}(\hat{Y})}{\mathrm{var}(Y)}$$

is referred to as the *squared multiple correlation coefficient*, and its nonnegative square root ρ is referred to as the *multiple correlation coefficient*. Verify that **(a)** $\mathrm{var}(Y - \hat{Y}) = (1 - \rho^2)\,\mathrm{var}(Y)$; **(b)** $\mathrm{cov}(\hat{Y}, Y) = \rho^2\,\mathrm{var}(Y)$; **(c)** $\mathrm{cor}(\hat{Y}, Y) = \rho$ if $\rho \neq 0$.

5.51 (Continued.) Determine the squared multiple correlation coefficient in the context of Example 5.15.

5.52 Let X_1, X_2, and Y have finite, positive variances σ_1^2, σ_2^2, and σ_3^2, respectively; set $\rho_{12} = \mathrm{cor}(X_1, X_2)$, $\rho_{13} = \mathrm{cor}(X_1, Y)$, and $\rho_{23} = \mathrm{cor}(X_2, Y)$; and suppose $\rho_{12}^2 < 1$. Let $\hat{Y} = EY + \beta_1(X_1 - EX_1) + \beta_2(X_2 - EX_2)$ be the best linear predictor of Y based on X_1 and X_2. Show that **(a)** $\beta_1 = \frac{\rho_{13} - \rho_{12}\rho_{23}}{1 - \rho_{12}^2}\frac{\sigma_3}{\sigma_1}$ and $\beta_2 = \frac{\rho_{23} - \rho_{12}\rho_{13}}{1 - \rho_{12}^2}\frac{\sigma_3}{\sigma_2}$; **(b)** $\hat{Y}$ coincides with the best linear predictor of Y based on X_1 if and only if $\rho_{23} = \rho_{12}\rho_{13}$.

5.53 Let $\hat{Y}$ be the best linear predictor of Y based on $\mathbf{X}$. Show that if $X_1, \ldots, X_m$ are uncorrelated, then **(a)** $\hat{Y} = EY + \sum_{i=1}^{m} \frac{\mathrm{cov}(X_i, Y)}{\mathrm{var}(X_i)}(X_i - EX_i)$; **(b)** $\mathrm{MSE}(\hat{Y}) = \mathrm{var}(Y) - \sum_{i=1}^{m} \frac{[\mathrm{cov}(X_i, Y)]^2}{\mathrm{var}(X_i)}$.

5.54 Let X_1, X_2, and Y, respectively, denote the undergraduate GPA, score on the Law School Aptitude Test (LSAT), and first-year GPA in law school for a random student in a certain law school. Suppose $EX_1 = 3.2$, $\mathrm{SD}(X_1) = 0.4$, $EX_2 = 500$, $\mathrm{SD}(X_2) = 100$, $\mathrm{cor}(X_1, X_2) = .8$, $EY = 3$, $\mathrm{SD}(Y) = 0.5$, $\mathrm{cor}(X_1, Y) = .55$, and $\mathrm{cor}(X_2, Y) = .6$. Determine **(a)** the variance–covariance matrix of X_1, X_2, and Y; **(b)** the best linear predictor of Y based on X_1 and X_2 and the mean squared error of this predictor; **(c)** the best linear predictor of Y based on X_1 and the mean squared error of this predictor; **(d)** the best linear predictor of Y based on X_2 and the mean squared error of this predictor.

5.55 Let X_1, X_2, and Y, respectively, denote the undergraduate GPA, score on the LSAT, and first-year GPA in law school for a random student in a certain law school. Suppose $EX_1 = 3.2$, $\mathrm{SD}(X_1) = 0.4$, $EX_2 = 500$, $\mathrm{SD}(X_2) = 100$, $\mathrm{cor}(X_1, X_2) = .8$, $EY = 3$, and $\mathrm{SD}(Y) = 0.5$. Suppose also that the best linear predictor of Y based on X_1 and X_2 is given by $\hat{Y} = 1.04 + 0.3X_1 + 0.002X_2$. Determine **(a)** the variance–covariance matrix of X_1, X_2, and Y; **(b)** the mean squared error of $\hat{Y}$; **(c)** the best linear predictor of Y based on X_1 and the mean squared error of this predictor; **(d)** the best linear predictor of Y based on X_2 and the mean squared error of this predictor.

5.5
Multivariate Density Functions

Although we have considered probability functions on an arbitrary set, so far we have considered density functions only on $\mathbb{R}$. In this section, we consider multivariate density functions—that is, density functions on $\mathbb{R}^n$ for some positive integer n.

The n-dimensional volume $\mathrm{vol}_n(B)$ of a subset B of $\mathbb{R}^n$ can be obtained as the multiple integral of $d\mathbf{y} = dy_1 \cdots dy_n$ over B. Observe that one-dimensional volume is length and two-dimensional volume is area. Let g be a function on B. It is known from integral calculus that

$$\int_B g(\mathbf{y})\,d\mathbf{y} = 0 \quad \text{if } \mathrm{vol}_n(B) = 0 \tag{30}$$

and that

$$\int_B g(\mathbf{y})\,d\mathbf{y} > 0 \quad \text{if } \mathrm{vol}_n(B) > 0 \text{ and } g > 0 \text{ on } B.$$

A function f on $\mathbb{R}^n$ is said to be a density function on $\mathbb{R}^n$ if $f \geq 0$ on $\mathbb{R}^n$ and

$$\int_{\mathbb{R}^n} f(\mathbf{y})\,d\mathbf{y} = 1.$$

Let f be a density function on $\mathbb{R}^n$. Then

$$P(B) = \int_B f(\mathbf{y})\,d\mathbf{y}, \qquad B \subset \mathbb{R}^n,$$

defines a distribution on $\mathbb{R}^n$ and f is referred to as the density function of P. Let $\mathbf{Y} = [Y_1, \ldots, Y_n]^T$ be a random vector having P as its distribution. Then

$$P(\mathbf{Y} \in B) = P(B) = \int_B f(\mathbf{y})\,d\mathbf{y}, \qquad B \subset \mathbb{R}^n, \tag{31}$$

and f is referred to as the density function of $\mathbf{Y}$ or as the *joint density function* of $Y_1, \ldots, Y_n$. We sometimes write this density function as $f_{\mathbf{Y}}$ or as $f_{Y_1, \ldots, Y_n}$. Set $\mathcal{Y} = \{\mathbf{y} \in \mathbb{R}^n : f(\mathbf{y}) > 0\}$. Then $\mathrm{vol}_n(\mathcal{Y}) > 0$ and $P(\mathbf{Y} \in \mathcal{Y}) = 1$. It follows from (30) and (31) that

$$P(\mathbf{Y} \in B) = 0 \quad \text{for } B \subset \mathbb{R}^n \text{ with } \mathrm{vol}_n(B) = 0. \tag{32}$$

In particular, $P(\mathbf{Y} = \mathbf{y}) = 0$ for $\mathbf{y} \in \mathbb{R}^n$.

The density function of a random vector is not uniquely defined. Let $\mathbf{Y}$ be a random vector having density function f on $\mathbb{R}^n$ and let f_1 be a nonnegative function on $\mathbb{R}^n$. Then f_1 is a density function of $\mathbf{Y}$ if and only if the set $\{\mathbf{y} \in \mathbb{R}^n : f_1(\mathbf{y}) \neq f(\mathbf{y})\}$ has zero n-dimensional volume.

Let $\mathbf{Y}$ be a random vector having a density function on $\mathbb{R}^n$ and let $\mathcal{Y}$ be a subset of $\mathbb{R}^n$ such that $P(\mathbf{Y} \in \mathcal{Y}) = 1$. Then $\mathbf{Y}$ has a density function f on $\mathbb{R}^n$ such that $f = 0$ on $\mathcal{Y}^c$. Suppose $\mathcal{Y}$ is an open subset of $\mathbb{R}^n$. Then $\mathbf{Y}$ has at most one density function f such that f is continuous on $\mathcal{Y}$ and $f = 0$ on $\mathcal{Y}^c$. In particular, $\mathbf{Y}$ has at most one density function that is continuous on $\mathbb{R}^n$.

Let $\mathcal{Y} \subset \mathbb{R}^n$ with $0 < \text{vol}_n(\mathcal{Y}) < \infty$. The uniform distribution P on $\mathcal{Y}$ is defined by

$$P(B) = \frac{\text{vol}_n(B)}{\text{vol}_n(\mathcal{Y})}, \qquad B \subset \mathcal{Y}.$$

Since

$$\frac{\text{vol}_n(B)}{\text{vol}_n(\mathcal{Y})} = \int_B \frac{1}{\text{vol}_n(\mathcal{Y})} \, d\mathbf{y}, \qquad B \subset \mathcal{Y},$$

the uniform distribution on $\mathcal{Y}$ has the density function f given by $f(\mathbf{y}) = 1/\text{vol}_n(\mathcal{Y})$ for $\mathbf{y} \in \mathcal{Y}$ and $f(\mathbf{y}) = 0$ for $\mathbf{y} \notin \mathcal{Y}$.

Let Y_1 and Y_2 have joint density function f_{Y_1, Y_2}. Then Y_1 has the density function f_{Y_1} given by

$$f_{Y_1}(y_1) = \int_{-\infty}^{\infty} f_{Y_1, Y_2}(y_1, y_2) \, dy_2, \qquad y_1 \in \mathbb{R}. \tag{33}$$

To verify this formula, observe that

$$P(Y_1 \in B) = P(Y_1 \in B, Y_2 \in \mathbb{R}) = \int_B \left(\int_{-\infty}^{\infty} f_{Y_1, Y_2}(y_1, y_2) \, dy_2 \right) dy_1, \quad B \subset \mathbb{R}.$$

Similarly, Y_2 has the density function f_{Y_2} given by

$$f_{Y_2}(y_2) = \int_{-\infty}^{\infty} f_{Y_1, Y_2}(y_1, y_2) \, dy_1, \qquad y_2 \in \mathbb{R}. \tag{34}$$

If

$$f_{Y_1, Y_2}(y_1, y_2) = f_{Y_1}(y_1) f_{Y_2}(y_2), \qquad y_1, y_2 \in \mathbb{R}, \tag{35}$$

then Y_1 and Y_2 are independent. To see this, observe that if (35) holds, then

$$P(Y_1 \in B_1, Y_2 \in B_2) = \int_{B_1} \left(\int_{B_2} f_{Y_1, Y_2}(y_1, y_2) \, dy_2 \right) dy_1$$

$$= \int_{B_1} \left(\int_{B_2} f_{Y_1}(y_1) f_{Y_2}(y_2) \, dy_2 \right) dy_1$$

$$= \int_{B_1} f_{Y_1}(y_1) \, dy_1 \int_{B_2} f_{Y_2}(y_2) \, dy_2$$

$$= P(Y_1 \in B_1) P(Y_2 \in B_2)$$

for $B_1, B_2 \subset \mathbb{R}$. Suppose, conversely, that Y_1 and Y_2 are independent. It can be shown that if y_1 is a continuity point of f_{Y_1}, y_2 is a continuity point of f_{Y_2}, and (y_1, y_2) is a continuity point of f_{Y_1, Y_2}, then (35) holds.

Let Y_1 and Y_2 be independent random variables having density functions f_{Y_1} and f_{Y_2}, respectively. Then the right side of (35) defines a joint density function of Y_1 and Y_2. Thus, in effect, (35) is necessary and sufficient for Y_1 and Y_2 to be independent.

The next result is a continuous analog of Theorem 1.8 in Section 1.8.

EXAMPLE **5.17**   Let $\mathcal{Y}_1$ and $\mathcal{Y}_2$ be intervals in $\mathbb{R}$ having finite, positive length, let Y_1 be a $\mathcal{Y}_1$-valued random variable, and let Y_2 be a $\mathcal{Y}_2$-valued random variable. Show that (Y_1, Y_2) is uniformly distributed on $\mathcal{Y}_1 \times \mathcal{Y}_2$ if and only if Y_1 is uniformly distributed on $\mathcal{Y}_1$, Y_2 is uniformly distributed on $\mathcal{Y}_2$, and Y_1 and Y_2 are independent.

Solution   Suppose (Y_1, Y_2) is uniformly distributed on $\mathcal{Y} = \mathcal{Y}_1 \times \mathcal{Y}_2$; that is, Y_1 and Y_2 have the joint density function f_{Y_1, Y_2} given by

$$f_{Y_1, Y_2}(y_1, y_2) = \frac{1}{\text{area}(\mathcal{Y})} = \frac{1}{\text{length}(\mathcal{Y}_1)\, \text{length}(\mathcal{Y}_2)}, \qquad y_1 \in \mathcal{Y}_1 \text{ and } y_2 \in \mathcal{Y}_2,$$

and $f_{Y_1, Y_2}(y_1, y_2) = 0$ otherwise. Then, by (33),

$$f_{Y_1}(y_1) = \int_{\mathcal{Y}_2} f_{Y_1, Y_2}(y_1, y_2)\, dy_2 = \frac{1}{\text{length}(\mathcal{Y}_1)}, \qquad y_1 \in \mathcal{Y}_1,$$

and $f_{Y_1}(y_1) = 0$ otherwise; thus, Y_1 is uniformly distributed on $\mathcal{Y}_1$. Similarly, Y_2 is uniformly distributed on $\mathcal{Y}_2$. Moreover, (35) holds, and hence Y_1 and Y_2 are independent.

Suppose, conversely, that Y_1 is uniformly distributed on $\mathcal{Y}_1$, Y_2 is uniformly distributed on $\mathcal{Y}_2$, and Y_1 and Y_2 are independent. Then Y_1 and Y_2 have density functions, which are given as follows:

$$f_{Y_1}(y_1) = \frac{1}{\text{length}(\mathcal{Y}_1)}, \qquad y_1 \in \mathcal{Y}_1,$$

and $f_{Y_1}(y_1) = 0$ otherwise;

$$f_{Y_2}(y_2) = \frac{1}{\text{length}(\mathcal{Y}_2)}, \qquad y_2 \in \mathcal{Y}_2,$$

and $f_{Y_2}(y_2) = 0$ otherwise. Since Y_1 and Y_2 are independent, we can choose their joint density function such that (35) holds; that is, Y_1 and Y_2 have the joint density function given by

$$f_{Y_1, Y_2}(y_1, y_2) = \frac{1}{\text{length}(\mathcal{Y}_1)\, \text{length}(\mathcal{Y}_2)} = \frac{1}{\text{area}(\mathcal{Y})}, \qquad y_1 \in \mathcal{Y}_1 \text{ and } y_2 \in \mathcal{Y}_2,$$

and $f_{Y_1, Y_2}(y_1, y_2) = 0$ otherwise. This is the density function of the uniform distribution on $\mathcal{Y}_1 \times \mathcal{Y}_2$.   ∎

EXAMPLE **5.18**   Let X and Y be independent random variables having exponential distributions with inverse-scale parameters λ_1 and λ_2, respectively. Determine $P(Y \geq X)$. (We can think of X and Y as the failure times of components made by two different processes and want to determine the probability that the component made by the second process lasts at least as long as the component made by the first process.)

Solution   The random variables X and Y have the joint density function f given by

$$f(x, y) = \lambda_1 e^{-\lambda_1 x} \lambda_2 e^{-\lambda_2 y}, \qquad x > 0 \text{ and } y > 0,$$

and $f(x, y) = 0$ elsewhere. Thus,

$$P(Y \geq X) = \int_0^\infty \left(\int_x^\infty f(x, y) \, dy \right) dx$$

$$= \int_0^\infty \lambda_1 e^{-\lambda_1 x} \left(\int_x^\infty \lambda_2 e^{-\lambda_2 y} \, dy \right) dx$$

$$= \int_0^\infty \lambda_1 e^{-(\lambda_1 + \lambda_2)x} \, dx$$

$$= \frac{\lambda_1}{\lambda_1 + \lambda_2}. \qquad \blacksquare$$

EXAMPLE 5.19   Let X and Y have the joint density function given by

$$f_{X,Y}(x, y) = \frac{x^3}{2} e^{-x(1+y)}, \qquad x, y > 0,$$

and $f_{X,Y} = 0$ elsewhere.

(a) Determine the density functions of X and Y.

(b) Show that X and Y are dependent by working directly with density functions.

Solution   **(a)** The density function of X is given by $f_X(x) = 0$ for $x \leq 0$ and

$$f_X(x) = \int_0^\infty \frac{x^3}{2} e^{-x(1+y)} \, dy = \frac{x^2}{2} e^{-x} \int_0^\infty x e^{-xy} \, dy = \frac{x^2}{2} e^{-x}, \qquad x > 0.$$

Thus, X has the gamma distribution with shape parameter 3 and scale parameter 1. The density function of Y is given by $f_Y(y) = 0$ for $y \leq 0$ and, by Equation (6) in Section 3.1,

$$f_Y(y) = \int_0^\infty \frac{x^3}{2} e^{-x(1+y)} \, dx = \frac{\Gamma(4)}{2(1+y)^4} = \frac{3}{(1+y)^4}, \qquad y > 0.$$

(b) Observe that f_X is continuous at $x = 1$, f_Y is continuous at $y = 1$, and $f_{X,Y}$ is continuous at $x = 1$ and $y = 1$. Since $f_{X,Y}(1, 1) = e^{-2}/2$, $f_X(1) = e^{-1}/2$, and $f_Y(1) = 3/16$, $f_{X,Y}(1, 1) \neq f_X(1) f_Y(1)$, and hence X and Y are dependent. $\blacksquare$

Let Y_1, Y_2, and Y_3 have a joint density function f_{Y_1, Y_2, Y_3}. The density functions of Y_1, Y_2, and Y_3 are given by

$$f_{Y_1}(y_1) = \int_{-\infty}^\infty \int_{-\infty}^\infty f_{Y_1, Y_2, Y_3}(y_1, y_2, y_3) \, dy_2 \, dy_3, \qquad y_1 \in \mathbb{R},$$

$$f_{Y_2}(y_2) = \int_{-\infty}^\infty \int_{-\infty}^\infty f_{Y_1, Y_2, Y_3}(y_1, y_2, y_3) \, dy_1 \, dy_3, \qquad y_2 \in \mathbb{R},$$

and

$$f_{Y_3}(y_3) = \int_{-\infty}^{\infty} \int_{-\infty}^{\infty} f_{Y_1,Y_2,Y_3}(y_1, y_2, y_3)\, dy_1\, dy_2, \qquad y_3 \in \mathbb{R}.$$

It is also straightforward to show that

$$f_{Y_1,Y_2}(y_1, y_2) = \int_{-\infty}^{\infty} f_{Y_1,Y_2,Y_3}(y_1, y_2, y_3)\, dy_3, \qquad y_1, y_2 \in \mathbb{R},$$

$$f_{Y_1,Y_3}(y_1, y_3) = \int_{-\infty}^{\infty} f_{Y_1,Y_2,Y_3}(y_1, y_2, y_3)\, dy_2, \qquad y_1, y_3 \in \mathbb{R},$$

and

$$f_{Y_2,Y_3}(y_2, y_3) = \int_{-\infty}^{\infty} f_{Y_1,Y_2,Y_3}(y_1, y_2, y_3)\, dy_1, \qquad y_2, y_3 \in \mathbb{R}.$$

In general, let $\mathbf{Y} = [Y_1, \ldots, Y_n]^T$ have density function $f_{\mathbf{Y}}$, let $1 \leq m < n$, and let $\mathbf{Y}_1$ be the m-dimensional random vector consisting of m specified entries of $\mathbf{Y}$. Then $\mathbf{Y}_1$ has the density function $f_{\mathbf{Y}_1}$ given by

$$f_{\mathbf{Y}_1}(\mathbf{y}_1) = \int_{\mathbb{R}^{(n-m)}} f_Y(\mathbf{y})\, d\mathbf{y}_2;$$

here $\mathbf{y} = [y_1, \ldots, y_n]^T$, $\mathbf{y}_1$ consists of the m specified entries of $\mathbf{y}$, and $\mathbf{y}_2$ consists of the remaining entries. For example, if $n = 4$, then

$$f_{Y_2,Y_3}(y_2, y_3) = \int_{-\infty}^{\infty} \int_{-\infty}^{\infty} f_{Y_1,Y_2,Y_3,Y_4}(y_1, y_2, y_3, y_4)\, dy_1\, dy_4.$$

Let $Y_1, \ldots, Y_n$ have density functions $f_{Y_1}, \ldots, f_{Y_n}$, respectively, and joint density function $f_{Y_1,\ldots,Y_n}$. If

$$f_{Y_1,\ldots,Y_n}(y_1, \ldots, y_n) = f_{Y_1}(y_1) \cdots f_{Y_n}(y_n) \tag{36}$$

for $y_1, \ldots, y_n \in \mathbb{R}$, then $Y_1, \ldots, Y_n$ are independent. Suppose, conversely, that $Y_1, \ldots, Y_n$ are independent and let $y_1, \ldots, y_n \in \mathbb{R}$ be such that y_i is a continuity point of f_{Y_i} for $1 \leq i \leq n$ and $(y_1, \ldots, y_n)$ is a continuity point of $f_{Y_1,\ldots,Y_n}$. Then (36) holds.

Let $Y_1, \ldots, Y_n$ be independent random variables having density functions $f_{Y_1}, \ldots, f_{Y_n}$, respectively. Then the right side of (36) defines a joint density function of $Y_1, \ldots, Y_n$. Thus, in effect, (36) is necessary and sufficient for $Y_1, \ldots, Y_n$ to be independent.

EXAMPLE **5.20**   Determine the joint density function of n independent, standard normal random variables.

Solution   Let $Z_1, \ldots, Z_n$ be independent, standard normal random variables and set $\mathbf{Z} = [Z_1, \ldots, Z_n]^T$. Choose $\mathbf{z} = [z_1, \ldots, z_n]^T \in \mathbb{R}^n$. Then

$$f_{\mathbf{Z}}(\mathbf{z}) = f_{Z_1,\ldots,Z_n}(z_1, \ldots, z_n)$$

$$= f_{Z_1}(z_1) \cdots f_{Z_n}(z_n)$$

$$= \prod_{i=1}^{n} \frac{1}{(2\pi)^{1/2}} \exp(-\tfrac{1}{2}z_i^2)$$

$$= \frac{1}{(2\pi)^{n/2}} \exp(-\tfrac{1}{2} \textstyle\sum_{i=1}^{n} z_i^2).$$

Since $\sum_{1}^{n} z_i^2 = \mathbf{z}^T \mathbf{z}$, the joint density function of $Z_1, \ldots, Z_n$ can be written as

$$f_{\mathbf{Z}}(\mathbf{z}) = \frac{1}{(2\pi)^{n/2}} \exp(-\tfrac{1}{2}\mathbf{z}^T\mathbf{z}). \quad \blacksquare \tag{37}$$

EXAMPLE 5.21   Let U_1 and U_2 be independent random variables, each uniformly distributed on $[0, 1]$. Consider the random rectangle B having vertices at $(0, 0)$, $(U_1, 0)$, (U_1, U_2), and $(0, U_2)$. Which is greater, the mean of the area of B or its median?

Solution   According to the solution to Example 2.27 in Section 2.2, $E(\text{area}(B)) = E(U_1 U_2) = 1/4$. To compute the median of $U_1 U_2$, we observe that, for $0 < y < 1$ (see Figure 5.2),

$$P(U_1 U_2 \geq y) = \int_y^1 \left(\int_{y/u_1}^1 du_2 \right) du_1$$

$$= \int_y^1 \left(1 - \frac{y}{u_1} \right) du_1$$

$$= \left. (u_1 - y \log u_1) \right|_y^1$$

$$= 1 - y(1 - \log y).$$

In particular,

$$P\left(U_1 U_2 \geq \frac{1}{4} \right) = 1 - \frac{1}{4}(1 + \log 4) \doteq .403,$$

so the median of $U_1 U_2$ is less than $1/4$ and hence less than its mean. $\blacksquare$

FIGURE   5.2

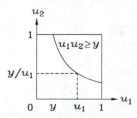

Theorem 2.1(g) in Section 2.2 has an obvious analog for random vectors.

THEOREM 5.18   Let $\mathbf{Y}$ be a $\mathcal{Y}$-valued random vector having density function $f_{\mathbf{Y}}$ and let g be a real-valued function on $\mathcal{Y}$. Then

$$E[g(\mathbf{Y})] = \int_{\mathcal{Y}} g(\mathbf{y}) f_{\mathbf{Y}}(\mathbf{y}) \, d\mathbf{y}.$$

EXAMPLE 5.22   Let X and Y be as in Example 5.19.

(a) Determine the mean and variance of X.

(b) Determine the mean and variance of Y.

(c) Determine the covariance and correlation coefficient between X and Y.

(d) Determine the best linear predictor of Y based on X and determine the mean squared error of this predictor.

Solution   (a) According to the solution to Example 5.19, X has the gamma distribution with shape parameter 3 and scale parameter 1. Thus, $EX = 3 \cdot 1 = 3$ and $\text{var}(X) = 3 \cdot 1^2 = 3$.

(b) The mean of Y is given by

$$\begin{aligned}
EY &= \int_0^\infty \frac{3y}{(1+y)^4} \, dy \\
&= \int_0^\infty \frac{3(1+y-1)}{(1+y)^4} \, dy \\
&= 3 \int_0^\infty \frac{1}{(1+y)^3} \, dy - 3 \int_0^\infty \frac{1}{(1+y)^4} \, dy \\
&= \frac{3}{2} - 1 = \frac{1}{2}.
\end{aligned}$$

The second moment of Y is given by

$$E(Y^2) = \int_0^\infty \frac{3y^2}{(1+y)^4} \, dy.$$

Since $y^2 = (1+y-1)^2 = (1+y)^2 - 2(1+y) + 1$, we conclude that

$$\begin{aligned}
E(Y^2) &= \int_0^\infty \frac{3[(1+y)^2 - 2(1+y) + 1]}{(1+y)^4} \, dy \\
&= \int_0^\infty \frac{3}{(1+y)^2} \, dy - \int_0^\infty \frac{6}{(1+y)^3} \, dy + \int_0^\infty \frac{3}{(1+y)^4} \, dy \\
&= 3 - 3 + 1 = 1.
\end{aligned}$$

Alternatively, by Theorem 5.18,

$$\begin{aligned}
EY &= \int_{-\infty}^\infty \int_{-\infty}^\infty y f_{X,Y}(x,y) \, dx \, dy \\
&= \int_0^\infty \left(\int_0^\infty y \frac{x^3}{2} e^{-x(1+y)} \, dy \right) dx
\end{aligned}$$

$$= \frac{1}{2} \int_0^\infty x^3 e^{-x} \left(\int_0^\infty y e^{-xy} \, dy \right) dx.$$

Since

$$\int_0^\infty y e^{-xy} \, dy = \frac{1}{x^2}$$

[by Equation (6) in Section 3.1], we conclude that

$$EY = \frac{1}{2} \int_0^\infty x e^{-x} \, dx = \frac{1}{2}.$$

Similarly,

$$E(Y^2) = \int_{-\infty}^\infty \int_{-\infty}^\infty y^2 f_{X,Y}(x, y) \, dx \, dy$$

$$= \int_0^\infty \left(\int_0^\infty y^2 \frac{x^3}{2} e^{-x(1+y)} \, dy \right) dx$$

$$= \frac{1}{2} \int_0^\infty x^3 e^{-x} \left(\int_0^\infty y^2 e^{-xy} \, dy \right) dx.$$

Since

$$\int_0^\infty y^2 e^{-xy} \, dy = \frac{2}{x^3},$$

we conclude that

$$E(Y^2) = \int_0^\infty e^{-x} \, dx = 1.$$

The variance of Y is given by

$$\mathrm{var}(Y) = E(Y^2) - (EY)^2 = 1 - \frac{1}{4} = \frac{3}{4}.$$

(c) By Theorem 5.18,

$$E(XY) = \int_{-\infty}^\infty \int_{-\infty}^\infty xy f_{X,Y}(x, y) \, dx \, dy$$

$$= \int_0^\infty \left(\int_0^\infty xy \frac{x^3}{2} e^{-x(1+y)} \, dy \right) dx$$

$$= \frac{1}{2} \int_0^\infty x^4 e^{-x} \left(\int_0^\infty y e^{-xy} \, dy \right) dx$$

$$= \frac{1}{2} \int_0^\infty x^2 e^{-x} \, dx$$

$$= 1.$$

Consequently, $\mathrm{cov}(X, Y) = E(XY) - (EX)(EY) = 1 - 3/2 = -1/2$. Therefore,

$$\rho = \mathrm{cor}(X, Y) = \frac{\mathrm{cov}(X, Y)}{\mathrm{SD}(X) \, \mathrm{SD}(Y)} = \frac{-1/2}{\sqrt{3 \cdot \frac{3}{4}}} = -\frac{1}{3}.$$

(d) The best linear predictor of Y based on X is given by

$$\hat{Y} = EY + \frac{\text{cov}(Y)}{\text{var}(X)}(X - EX) = \frac{1}{2} - \frac{1}{6}(X - 3) = 1 - \frac{X}{6}.$$

The mean squared error of this predictor is given by

$$\text{MSE}(\hat{Y}) = (1 - \rho^2)\,\text{var}(Y) = \left(1 - \frac{1}{9}\right)\left(\frac{3}{4}\right) = \frac{2}{3}.$$

(The linear predictor $\hat{Y} = 1 - X/6$ is somewhat defective in that it is negative when $X > 6$ even though Y is a positive random variable. The way to remove this defect is to consider nonlinear predictors, which will be done in Section 6.6.)   ∎

Let Y_1, Y_2, and Y_3 be random variables having finite variance and a joint density function and let $\mathcal{P}$ be a plane in $\mathbb{R}^3$. Since $\mathcal{P}$ has zero volume, we conclude from Equation (32) that

$$P((Y_1, Y_2, Y_3) \in \mathcal{P}) = 0. \tag{38}$$

The variance–covariance matrix of Y_1, Y_2, and Y_3 must be positive definite. To see this, suppose to the contrary that this variance–covariance matrix were noninvertible. Then, by Corollary 5.6 in Section 5.3 and the solution to Example 5.12 in that section, there would be a plane $\mathcal{P}$ in $\mathbb{R}^3$ such that $P((Y_1, Y_2, Y_3) \in \mathcal{P}) = 1$; but this is impossible by (38).

More generally, we have the following result.

THEOREM 5.19   Let $Y_1, \ldots, Y_n$ be random variables having finite variance and a joint density function. Then the variance–covariance matrix of $Y_1, \ldots, Y_n$ is positive definite.

The next result follows from Theorem 5.19 above and Example 5.14 in Section 5.3.

COROLLARY 5.9   Let X and Y be random variables having finite variance and a joint density function. Then $-1 < \text{cor}(X, Y) < 1$.

Problems

5.56   **(a)** Determine the number a such that the function f on $\mathbb{R}^2$ given by $f(x, y) = (x + y)/a$ for $0 < x, y < 1$ and $f(x, y) = 0$ elsewhere is a bivariate density function. **(b)** If X and Y have the bivariate density function in part (a), determine $P(X + Y \le 1)$.

5.57   Let X and Y have the joint density function given by

$$f_{X,Y}(x, y) = \frac{2 + x + y}{8}, \qquad -1 < x < 1 \text{ and } -1 < y < 1,$$

and $f_{X,Y} = 0$ elsewhere. **(a)** Determine the common density function of X and Y. **(b)** Show that X and Y are dependent by working directly with density functions.

5.58 (Continued.) Determine **(a)** $P(X \geq 0, Y \geq 0)$; **(b)** $P(X \geq 0, Y \geq 0, X + Y \leq 1)$.

5.59 (Continued.) Determine EY, $E(Y^2)$, and $\text{var}(Y)$ **(a)** by using the formula for f_Y in the solution to Problem 5.57(a); **(b)** by using Theorem 5.18.

5.60 (Continued.) Determine **(a)** $E(XY)$; **(b)** $\text{cov}(X, Y)$; **(c)** $\text{cor}(X, Y)$; **(d)** the best linear predictor of Y based on X; **(e)** the mean squared error of this predictor.

5.61 Let X and Y denote the Cartesian coordinates of a random point that is uniformly distributed on the triangle in $\mathbb{R}^2$ having vertices $(0, 0)$, $(8, 0)$, and $(0, 4)$. Determine **(a)** the density functions of X and Y; **(b)** the means and variances of X and Y; **(c)** $E(XY)$; **(d)** $\text{cov}(X, Y)$; **(e)** $\text{cor}(X, Y)$; **(f)** the best linear predictor of Y based on X; **(g)** the mean squared error of this predictor.

5.62 Show that if Y_1 and Y_2 have a joint density function on $\mathbb{R}^2$, then $P(Y_1 = Y_2) = 0$.

5.63 Let Y_1, Y_2, Y_3, and Y_4 have a joint density function on $\mathbb{R}^4$. Show that the random matrix

$$\begin{bmatrix} Y_1 & Y_2 \\ Y_3 & Y_4 \end{bmatrix}$$

is invertible with probability 1.

5.64 Let $\alpha_1, \alpha_2, \alpha_3 > 0$. **(a)** Show that

$$\int_0^{1-y_1} y_2^{\alpha_2-1} (1 - y_1 - y_2)^{\alpha_3-1} \, dy_2 = \frac{\Gamma(\alpha_2)\Gamma(\alpha_3)}{\Gamma(\alpha_2 + \alpha_3)} (1 - y_1)^{\alpha_2+\alpha_3-1}, \quad 0 < y_1 < 1.$$

[*Hint:* Make the change of variables $t = y_2/(1 - y_1)$.] **(b)** Set

$$f(y_1, y_2) = \frac{\Gamma(\alpha_1 + \alpha_2 + \alpha_3)}{\Gamma(\alpha_1)\Gamma(\alpha_2)\Gamma(\alpha_3)} y_1^{\alpha_1-1} y_2^{\alpha_2-1} (1 - y_1 - y_2)^{\alpha_3-1}$$

for $y_1 > 0$, $y_2 > 0$ and $y_1 + y_2 < 1$, and $f(y_1, y_2) = 0$ otherwise. Use the result of part (a) together with Equation (13) in Section 3.1 to show that f is a density function. [The corresponding distribution is referred to as the *Dirichlet distribution* with parameters $\alpha_1, \alpha_2, \alpha_3$. The Dirichlet distribution with parameters $\alpha_1 = \alpha_2 = \alpha_3 = 1$ coincides with the uniform distribution on the triangle having vertices at $(0, 0)$, $(1, 0)$, and $(0, 1)$. Consider a random distribution on $\{1, 2, 3\}$ and let Y_1, Y_2, and Y_3, respectively, be the probabilities assigned to $\{1\}$, $\{2\}$, and $\{3\}$. Then these random variables are nonnegative and their sum is 1. The Dirichlet distribution can be used to model the joint distribution of Y_1 and Y_2.]

5.65 (Continued.) Let Y_1 and Y_2 be jointly distributed according to the Dirichlet distribution with parameters $\alpha_1, \alpha_2, \alpha_3$. Show that the distribution of Y_1 is the beta distribution with parameters α_1 and $\alpha_2 + \alpha_3$. (Similarly the distribution of Y_2 is the beta distribution with parameters α_2 and $\alpha_1 + \alpha_3$.)

5.66 (Continued.) Show that

$$EY_1 = \frac{\alpha_1}{\alpha_1 + \alpha_2 + \alpha_3} \quad \text{and} \quad \text{var}(Y_1) = \frac{\alpha_1(\alpha_2 + \alpha_3)}{(\alpha_1 + \alpha_2 + \alpha_3)^2(\alpha_1 + \alpha_2 + \alpha_3 + 1)}.$$

Similarly,

$$EY_2 = \frac{\alpha_2}{\alpha_1 + \alpha_2 + \alpha_3} \quad \text{and} \quad \text{var}(Y_2) = \frac{\alpha_2(\alpha_1 + \alpha_3)}{(\alpha_1 + \alpha_2 + \alpha_3)^2(\alpha_1 + \alpha_2 + \alpha_3 + 1)}.$$

5.67   (Continued.) Show that

(a) $E(Y_1 Y_2) = \frac{\alpha_1 \alpha_2}{(\alpha_1 + \alpha_2 + \alpha_3)(\alpha_1 + \alpha_2 + \alpha_3 + 1)}$;   **(b)** $\text{cov}(Y_1, Y_2) = -\frac{\alpha_1 \alpha_2}{(\alpha_1 + \alpha_2 + \alpha_3)^2(\alpha_1 + \alpha_2 + \alpha_3 + 1)}$;

(c) $\text{cor}(Y_1, Y_2) = -\sqrt{\frac{\alpha_1 \alpha_2}{(\alpha_1 + \alpha_3)(\alpha_2 + \alpha_3)}}$.

5.68   (Continued.) Show that the best linear predictor of Y_2 based on Y_1 is given by

$$\hat{Y}_2 = \frac{\alpha_2}{\alpha_2 + \alpha_3}(1 - Y_1),$$

and determine the mean squared error of this predictor.

5.6
Invertible Transformations

In this section, we develop and apply a formula for the density function of a random vector that is obtained from another random vector by means of an invertible, not necessarily linear, transformation.

Let $\mathbf{W} = [W_1, \ldots, W_n]^T$ have density function $f_{\mathbf{W}}$ on $\mathbb{R}^n$ and let $\mathcal{W}$ be an open set in $\mathbb{R}^n$ such that $f_{\mathbf{W}} = 0$ on $\mathcal{W}^c$. Let $g_1, \ldots, g_n$ be real-valued functions on $\mathcal{W}$ and set $\mathbf{g} = [g_1, \ldots, g_n]^T$. Let $\mathbf{Y} = [Y_1, \ldots, Y_n]^T$ be the transform of $\mathbf{W}$ defined by $\mathbf{Y} = \mathbf{g}(\mathbf{W})$. Then $Y_i = g_i(\mathbf{W}) = g_i(W_1, \ldots, W_n)$ for $1 \leq i \leq n$. Under appropriate conditions on $\mathbf{g}$, $\mathbf{Y}$ has a density function, which can be given explicitly in terms of $\mathbf{g}$ and the density function of $\mathbf{W}$.

It is assumed, first of all, that $\mathbf{g}$ is a one-to-one mapping from $\mathcal{W}$ onto an open set $\mathcal{Y}$ in $\mathbb{R}^n$ and hence that $\mathbf{g}$ has an inverse mapping $\mathbf{h} = \mathbf{g}^{-1} = [h_1, \ldots, h_n]^T$ from $\mathcal{Y}$ to $\mathcal{W}$. Given $\mathbf{y} \in \mathcal{Y}$, the quantity $\mathbf{w} = \mathbf{h}(\mathbf{y})$ is the unique solution to the equation $\mathbf{g}(\mathbf{w}) = \mathbf{y}$. It is assumed that $\mathbf{g}$ is continuously differentiable on $\mathcal{W}$; that is, that for $1 \leq i, j \leq n$, the partial derivative of $y_i = g_i(\mathbf{w}) = g_i(w_1, \ldots, w_n)$ with respect to w_j exists for $\mathbf{w} \in \mathcal{W}$ and is a continuous function of $\mathbf{w}$. The *derivative* (matrix) of $\mathbf{g}$ at $\mathbf{w} = [w_1, \ldots, w_n]^T$ is the matrix

$$\begin{bmatrix} \frac{\partial y_1}{\partial w_1} & \cdots & \frac{\partial y_1}{\partial w_n} \\ \vdots & & \vdots \\ \frac{\partial y_n}{\partial w_1} & \cdots & \frac{\partial y_n}{\partial w_n} \end{bmatrix}$$

of these partial derivatives. The *Jacobian* of $\mathbf{g}$ at $\mathbf{w}$, denoted by

$$\frac{\partial(y_1, \ldots, y_n)}{\partial(w_1, \ldots, w_n)} = \begin{vmatrix} \frac{\partial y_1}{\partial w_1} & \cdots & \frac{\partial y_1}{\partial w_n} \\ \vdots & & \vdots \\ \frac{\partial y_n}{\partial w_1} & \cdots & \frac{\partial y_n}{\partial w_n} \end{vmatrix},$$

is the determinant of the derivative of **g** at **w**. It is assumed that the derivative of **g** is invertible everywhere on $\mathcal{W}$ or, equivalently, that its Jacobian is nonzero everywhere on $\mathcal{W}$. Then **h** is continuously differentiable on $\mathcal{Y}$. It follows from the multivariate chain rule that the derivative

$$
\begin{bmatrix}
\dfrac{\partial w_1}{\partial y_1} & \cdots & \dfrac{\partial w_1}{\partial y_n} \\[1em]
\vdots & & \vdots \\[1em]
\dfrac{\partial w_n}{\partial y_1} & \cdots & \dfrac{\partial w_n}{\partial y_n}
\end{bmatrix}
$$

of **h** at $\mathbf{y} = [\, y_1, \ldots, y_n \,]^T$ is the inverse of the derivative of **g** at $\mathbf{w} = \mathbf{h}(\mathbf{y})$. Consequently, the Jacobian

$$
\frac{\partial(w_1, \ldots, w_n)}{\partial(y_1, \ldots, y_n)}
$$

of **h** at $\mathbf{y} = [\, y_1, \ldots, y_n \,]^T$ is the reciprocal of the Jacobian of **g** at $\mathbf{w} = \mathbf{h}(\mathbf{y})$. Conversely, if **g** is a one-to-one mapping from $\mathcal{W}$ onto $\mathcal{Y}$ having a continuously differentiable inverse function **h** whose Jacobian is nonzero everywhere on $\mathcal{Y}$, then **g** is continuously differentiable, and its Jacobian is nonzero everywhere on $\mathcal{W}$.

Under the stated assumptions on **g** or the equivalent assumptions on the inverse mapping **h**, the following result is valid. [Here (39) involves the absolute value of the real-valued Jacobian.]

THEOREM 5.20   The transform $\mathbf{Y} = \mathbf{g}(\mathbf{W})$ has the density function on $\mathbb{R}^n$ given by

$$
f_{\mathbf{Y}}(\mathbf{y}) = \left| \frac{\partial(w_1, \ldots, w_n)}{\partial(y_1, \ldots, y_n)} \right| f_{\mathbf{W}}(\mathbf{h}(\mathbf{y})), \qquad \mathbf{y} \in \mathcal{Y}, \tag{39}
$$

and $f_{\mathbf{Y}}(\mathbf{y}) = 0$ for $\mathbf{y} \notin \mathcal{Y}$.

Proof   To verify the desired result, we need to show that

$$
P(\mathbf{Y} \in B) = \int_B f(\mathbf{y}) \, d\mathbf{y}, \qquad B \subset \mathcal{Y}, \tag{40}
$$

where f is given by the right side of (39). To this end, let $A = \{\, \mathbf{w} \in \mathcal{W} : \mathbf{g}(\mathbf{w}) \in B \,\}$ be the inverse image of B. Then

$$
P(\mathbf{Y} \in B) = P(\mathbf{W} \in A) = \int_A f_{\mathbf{W}}(\mathbf{w}) \, d\mathbf{w}.
$$

It follows from the change of variables formula for multiple integrals that

$$
\int_A f_{\mathbf{W}}(\mathbf{w}) \, d\mathbf{w} = \int_B \left| \frac{\partial(w_1, \ldots, w_n)}{\partial(y_1, \ldots, y_n)} \right| f_{\mathbf{W}}(\mathbf{h}(\mathbf{y})) \, d\mathbf{y} = \int_B f(\mathbf{y}) \, d\mathbf{y}
$$

and hence that (40) does indeed hold. ∎

Equation (39) can be rewritten as

$$
f_{\mathbf{Y}}(\mathbf{y}) = \left| \frac{\partial(w_1, \ldots, w_n)}{\partial(y_1, \ldots, y_n)} \right| f_{\mathbf{W}}(\mathbf{w}), \qquad \mathbf{y} \in \mathcal{Y}, \tag{41}
$$

where $\mathbf{w}$ is the unique solution to the equation $\mathbf{g}(\mathbf{w}) = \mathbf{y}$.

EXAMPLE **5.23** Let W_1 and W_2 be independent random variables, each having the exponential distribution with mean 1. Then $\mathbf{W} = [\,W_1, W_2\,]^T$ is a $\mathcal{W}$-valued random vector, where $\mathcal{W} = \{\,[\,w_1, w_2\,]^T : w_1 > 0 \text{ and } w_2 > 0\,\}$. The joint density function of W_1 and W_2 is given by

$$f_{\mathbf{W}}(\mathbf{w}) = f_{W_1, W_2}(w_1, w_2) = f_{W_1}(w_1) f_{W_2}(w_2) = e^{-w_1} e^{-w_2} = e^{-(w_1 + w_2)}$$

for $\mathbf{w} = [\,w_1, w_2\,]^T \in \mathcal{W}$ and $f_{\mathbf{W}}(\mathbf{w}) = f_{W_1, W_2}(w_1, w_2) = 0$ elsewhere. Set $Y_1 = W_1 + W_2$ and $Y_2 = W_1 - W_2$. Then $\mathbf{Y} = [\,Y_1, Y_2\,]^T$ is a $\mathcal{Y}$-valued random vector, where $\mathcal{Y} = \{\,[\,y_1, y_2\,]^T : y_1 > |y_2|\,\}$. Determine

(a) the joint density function of Y_1 and Y_2;

(b) the density function of Y_2.

Solution (a) For $y_1 > 0$ and $0 < |y_2| < y_1$, the unique solution to the equations

$$y_1 = w_1 + w_2$$
$$y_2 = w_1 - w_2$$

is given by

$$w_1 = \frac{y_1 + y_2}{2}$$

$$w_2 = \frac{y_1 - y_2}{2}.$$

The derivative of the inverse mapping is given by

$$\begin{bmatrix} \dfrac{\partial w_1}{\partial y_1} & \dfrac{\partial w_1}{\partial y_2} \\[2mm] \dfrac{\partial w_2}{\partial y_1} & \dfrac{\partial w_2}{\partial y_2} \end{bmatrix} = \begin{bmatrix} \frac{1}{2} & \frac{1}{2} \\[2mm] \frac{1}{2} & -\frac{1}{2} \end{bmatrix},$$

so its Jacobian is given by

$$\frac{\partial(w_1, w_2)}{\partial(y_1, y_2)} = \begin{vmatrix} \frac{1}{2} & \frac{1}{2} \\[2mm] \frac{1}{2} & -\frac{1}{2} \end{vmatrix} = -\frac{1}{2}.$$

We conclude from (41) that

$$f_{Y_1, Y_2}(y_1, y_2) = \frac{1}{2} f_{W_1, W_2}\!\left(\frac{y_1 + y_2}{2}, \frac{y_1 - y_2}{2}\right) = \frac{1}{2} e^{-y_1}, \qquad [\,y_1, y_2\,]^T \in \mathcal{Y},$$

and $f_{\mathbf{Y}}(\mathbf{y}) = f_{Y_1, Y_2}(y_1, y_2) = 0$ elsewhere.

(b) The density function of Y_2 is given by

$$f_{Y_2}(y_2) = \int_{-\infty}^{\infty} f_{Y_1, Y_2}(y_1, y_2)\, dy_1 = \int_{|y_2|}^{\infty} \frac{1}{2} e^{-y_1}\, dy_1 = \frac{1}{2} e^{-|y_2|}, \qquad y_2 \in \mathbb{R}.$$

[Recall from the description of Example 5.23 that $\mathcal{Y} = \{(y_1, y_2)^T : y_1 > |y_2|\}$.] ∎

Let W_1 and W_2 be random variables having joint density function f_{W_1, W_2} and suppose we want to determine the density function of a real-valued transform of W_1 and W_2. One approach is to introduce a second such real-valued transform, referred to as the auxiliary transform, denote the two transforms by Y_1 and Y_2, and determine their joint density function f_{Y_1, Y_2} from Theorem 5.20. If Y_1 is the transform of interest, its density function can then be obtained from the formula

$$f_{Y_1}(y_1) = \int_{\mathbb{R}} f_{Y_1, Y_2}(y_1, y_2)\, dy_2, \qquad y_1 \in \mathbb{R}.$$

Similarly, if Y_2 is the transform of interest, its density function can be obtained from the formula

$$f_{Y_2}(y_2) = \int_{\mathbb{R}} f_{Y_1, Y_2}(y_1, y_2)\, dy_1, \qquad y_2 \in \mathbb{R}.$$

This approach is illustrated in the next two examples. In the first, $W_1 + W_2$ is the transform of interest, and W_1 is the auxiliary transform; in the second, W_1/W_2 is the transform of interest, and W_2 is the auxiliary transform.

EXAMPLE 5.24 Let W_1 and W_2 have joint density function f_{W_1, W_2}.

(a) Determine the joint density function of W_1 and $W_1 + W_2$.

(b) Determine the density function of $W_1 + W_2$.

Solution **(a)** Set $Y_1 = W_1$ and $Y_2 = W_1 + W_2$. For $y_1, y_2 \in \mathbb{R}$, the unique solution to the equations

$$y_1 = w_1$$
$$y_2 = w_1 + w_2$$

is given by

$$w_1 = y_1$$
$$w_2 = y_2 - y_1.$$

The derivative of the inverse mapping is given by

$$\begin{bmatrix} \dfrac{\partial w_1}{\partial y_1} & \dfrac{\partial w_1}{\partial y_2} \\[2ex] \dfrac{\partial w_2}{\partial y_1} & \dfrac{\partial w_2}{\partial y_2} \end{bmatrix} = \begin{bmatrix} 1 & 0 \\ -1 & 1 \end{bmatrix},$$

so its Jacobian is given by

$$\frac{\partial(w_1, w_2)}{\partial(y_1, y_2)} = \begin{vmatrix} 1 & 0 \\ -1 & 1 \end{vmatrix} = 1.$$

We conclude from Equation (41) that $f_{Y_1, Y_2}(y_1, y_2) = f_{W_1, W_2}(y_1, y_2 - y_1)$ for $y_1, y_2 \in \mathbb{R}$. Replacing the symbols y_1 and y_2 by w and y, respectively, we get

$$f_{W_1, W_1 + W_2}(w, y) = f_{W_1, W_2}(w, y - w), \qquad w, y \in \mathbb{R}. \tag{42}$$

(b) The density function of $W_1 + W_2$ can be written as

$$f_{W_1 + W_2}(y) = \int_{-\infty}^{\infty} f_{W_1, W_1 + W_2}(w, y) \, dw, \qquad y \in \mathbb{R}.$$

Thus, by (42),

$$f_{W_1 + W_2}(y) = \int_{-\infty}^{\infty} f_{W_1, W_2}(w, y - w) \, dw, \qquad y \in \mathbb{R}. \quad \blacksquare \tag{43}$$

THEOREM 5.21   Let W_1 and W_2 be independent random variables having density functions f_{W_1} and f_{W_2}, respectively. Then $W_1 + W_2$ has the density function given by

$$f_{W_1 + W_2}(y) = \int_{-\infty}^{\infty} f_{W_1}(w) f_{W_2}(y - w) \, dw, \qquad y \in \mathbb{R}. \tag{44}$$

In particular, if W_1 and W_2 are positive random variables, then $W_1 + W_2$ has the density function given by $f_{W_1 + W_2}(y) = 0$ for $y \leq 0$ and

$$f_{W_1 + W_2}(y) = \int_{0}^{y} f_{W_1}(w) f_{W_2}(y - w) \, dw, \qquad y > 0. \tag{45}$$

Proof   By independence, $f_{W_1, W_2}(w, y - w) = f_{W_1}(w) f_{W_2}(y - w)$ for $w, y \in \mathbb{R}$; thus (44) follows from (43). Suppose W_1 and W_2 are positive random variables. Then $W_1 + W_2$ is a positive random variable, so it has a density function that is zero on $(-\infty, 0]$. Let $y > 0$. Since f_{W_1} and f_{W_2} are zero on $(-\infty, 0]$, the integrand in (44) is zero unless $0 < w < y$. Thus, (44) reduces to (45). $\blacksquare$

EXAMPLE 5.25   Let W_1 and W_2 have joint density function f_{W_1, W_2}.

(a) Determine the joint density function of W_1/W_2 and W_2.

(b) Determine the density function of W_1/W_2.

Solution   Since W_1 and W_2 have a joint density function, W_2 has a density function and hence $P(W_2 \neq 0) = 1$.

(a) Set $Y_1 = W_1/W_2$ and $Y_2 = W_2$. For $y_1 \in \mathbb{R}$ and $y_2 \neq 0$, the unique solution to the equations

$$y_1 = w_1/w_2$$
$$y_2 = w_2$$

is given by

$$w_1 = y_1 y_2$$

$$w_2 = y_2.$$

The derivative of the inverse mapping is given by

$$\begin{bmatrix} \frac{\partial w_1}{\partial y_1} & \frac{\partial w_1}{\partial y_2} \\[2mm] \frac{\partial w_2}{\partial y_1} & \frac{\partial w_2}{\partial y_2} \end{bmatrix} = \begin{bmatrix} y_2 & y_1 \\ 0 & 1 \end{bmatrix},$$

so its Jacobian is given by

$$\frac{\partial(w_1, w_2)}{\partial(y_1, y_2)} = \begin{vmatrix} y_2 & y_1 \\ 0 & 1 \end{vmatrix} = y_2.$$

We conclude from Equation (41) that

$$f_{Y_1, Y_2}(y_1, y_2) = |y_2| f_{W_1, W_2}(y_1 y_2, y_2), \qquad y_1, y_2 \in \mathbb{R}.$$

Replacing the symbols y_1 and y_2 by y and w, respectively, we get

$$f_{W_1/W_2, W_2}(y, w) = |w| f_{W_1, W_2}(yw, w), \qquad w, y \in \mathbb{R}. \tag{46}$$

(b) The density function of W_1/W_2 can be written as

$$f_{W_1/W_2}(y) = \int_{-\infty}^{\infty} f_{W_1/W_2, W_2}(y, w)\, dw, \qquad y \in \mathbb{R}.$$

Thus, by (46),

$$f_{W_1/W_2}(y) = \int_{-\infty}^{\infty} |w| f_{W_1, W_2}(yw, w)\, dw, \qquad y \in \mathbb{R}. \quad \blacksquare \tag{47}$$

THEOREM 5.22 Let W_1 and W_2 be independent random variables having density functions f_{W_1} and f_{W_2}, respectively. Then W_1/W_2 has the density function given by

$$f_{W_1/W_2}(y) = \int_{-\infty}^{\infty} |w| f_{W_1}(yw) f_{W_2}(w)\, dw, \qquad y \in \mathbb{R}. \tag{48}$$

In particular, if W_2 is a positive random variable, then W_1/W_2 has the density function given by

$$f_{W_1/W_2}(y) = \int_0^{\infty} w f_{W_1}(yw) f_{W_2}(w)\, dw, \qquad y \in \mathbb{R};$$

and if W_1 and W_2 are positive random variables, then W_1/W_2 has the density function given by $f_{W_1/W_2}(y) = 0$ for $y \leq 0$ and

$$f_{W_1/W_2}(y) = \int_0^{\infty} w f_{W_1}(yw) f_{W_2}(w)\, dw, \qquad y > 0.$$

Proof By independence, $f_{W_1, W_2}(yw, w) = f_{W_1}(yw) f_{W_2}(w)$ for $w, y \in \mathbb{R}$; thus (48) follows from (47). The remaining special cases are obvious consequences of (48). $\blacksquare$

THEOREM **5.23**   Let $\mathbf{W}$ have density function $f_{\mathbf{W}}$ on $\mathbb{R}^n$, let $\mathbf{a} \in \mathbb{R}^n$, let $\mathbf{B}$ be an invertible $n \times n$ matrix, and set $\mathbf{Y} = \mathbf{a} + \mathbf{BW}$. Then $\mathbf{Y}$ has the density function on $\mathbb{R}^n$ given by

$$f_{\mathbf{Y}}(\mathbf{y}) = \frac{1}{|\det(\mathbf{B})|} f_{\mathbf{W}}(\mathbf{B}^{-1}(\mathbf{y} - \mathbf{a})), \qquad \mathbf{y} \in \mathbb{R}^n.$$

Proof   Here $\mathbf{Y} = \mathbf{g}(\mathbf{W})$, where $\mathbf{g}(\mathbf{w}) = \mathbf{a} + \mathbf{Bw}$ for $\mathbf{w} \in \mathbb{R}^n$. Consequently,

$$y_i = a_i + \sum_{i=1}^{n} b_{ij} w_j, \qquad 1 \le i \le n,$$

where a_i is the ith entry of $\mathbf{a}$ and b_{ij} is the entry in row i and column j of $\mathbf{B}$. Thus,

$$\frac{\partial y_i}{\partial w_j} = b_{ij}, \qquad 1 \le i, j \le n,$$

and hence the derivative of $\mathbf{g}$ at $\mathbf{w}$ is the matrix $\mathbf{B}$; therefore, the corresponding Jacobian is $\det(\mathbf{B})$, which is nonzero since $\mathbf{B}$ is invertible. The unique solution to the equation $\mathbf{y} = \mathbf{a} + \mathbf{Bw}$ is given by $\mathbf{w} = \mathbf{B}^{-1}(\mathbf{y} - \mathbf{a}) = \mathbf{h}(\mathbf{y})$. The derivative of $\mathbf{h}$ at $\mathbf{y}$ is $\mathbf{B}^{-1}$, so the corresponding Jacobian is $\det(\mathbf{B}^{-1}) = 1/\det(\mathbf{B})$. The desired result now follows from Equation (41). ∎

The next result is an extension of Example 1.52 in Section 1.7.

EXAMPLE **5.26**   Let $\mathbf{W}$ be uniformly distributed on an open set $\mathcal{W}$ in $\mathbb{R}^n$, let $\mathbf{a} \in \mathbb{R}^n$, let $\mathbf{B}$ be an invertible $n \times n$ matrix, and set $\mathbf{Y} = \mathbf{a} + \mathbf{BW}$. Show that $\mathbf{Y}$ is uniformly distributed on $\mathcal{Y} = \{\mathbf{a} + \mathbf{Bw} : \mathbf{w} \in \mathcal{W}\}$.

Solution   The density function of $\mathbf{W}$ is given by $f_{\mathbf{W}}(\mathbf{w}) = 1/\mathrm{vol}_n(\mathcal{W})$ for $\mathbf{w} \in \mathcal{W}$ and $f_{\mathbf{W}}(\mathbf{w}) = 0$ for $\mathbf{w} \notin \mathcal{W}$. Thus, by Theorem 5.23, the density function of $\mathbf{Y}$ is given by

$$f_{\mathbf{Y}}(\mathbf{y}) = \frac{1}{|\det(\mathbf{B})|} f_{\mathbf{W}}(\mathbf{B}^{-1}(\mathbf{y} - \mathbf{a})) = \frac{1}{|\det(\mathbf{B})|\mathrm{vol}_n(\mathcal{W})}, \qquad \mathbf{y} \in \mathcal{Y},$$

and $f_{\mathbf{Y}}(\mathbf{y}) = 0$ for $\mathbf{y} \notin \mathcal{Y}$. Since $f_{\mathbf{Y}}$ is a density function, we see that

$$1 = \int_{\mathbb{R}^n} f_{\mathbf{Y}}(\mathbf{y}) \, d\mathbf{y} = \int_{\mathcal{Y}} \frac{1}{|\det(\mathbf{B})| \, \mathrm{vol}_n(\mathcal{W})} \, d\mathbf{y} = \frac{\mathrm{vol}_n(\mathcal{Y})}{|\det(\mathbf{B})| \, \mathrm{vol}_n(\mathcal{W})}$$

and hence that $|\det(\mathbf{B})|\mathrm{vol}_n(\mathcal{W}) = \mathrm{vol}_n(\mathcal{Y})$. Consequently, $f_{\mathbf{Y}}(\mathbf{y}) = 1/\mathrm{vol}_n(\mathcal{Y})$ for $\mathbf{y} \in \mathcal{Y}$. Therefore, $f_{\mathbf{Y}}$ is the density function of the uniform distribution on $\mathcal{Y}$. ∎

Suppose now that $n = 1$ and $\mathcal{W} = (c, d)$. Let g be a differentiable function from $\mathcal{W}$ onto an open interval $\mathcal{Y} = (c_1, d_1)$ whose derivative g' is continuous and nonzero on $\mathcal{W}$. Then either $g' > 0$ on $\mathcal{W}$ and hence g is an increasing function on $\mathcal{W}$ or $g' < 0$ on $\mathcal{W}$ and hence g is a decreasing function on $\mathcal{W}$. The derivative of g at $w \in \mathcal{W}$ is the 1×1 matrix whose single entry is $dy/dw = g'(w)$, and its determinant is dy/dw. [Here $y = g(w)$.] It follows from Equation (41) that $Y = g(W)$ has the density function f_Y given by

$$f_Y(y) = \left|\frac{dw}{dy}\right| f_W(w), \qquad c_1 < y < d_1, \tag{49}$$

and $f_Y(y) = 0$ elsewhere. In particular, if g is increasing on $\mathcal{W}$, then

$$f_Y(y) = \frac{dw}{dy} f_W(w), \qquad c_1 < y < d_1. \tag{50}$$

In the expression $f_W(w)$ in (49) and (50), w is the unique solution to the equation $g(w) = y$.

EXAMPLE **5.27** Let W have density function f_W on $\mathbb{R}$. Determine the density function of $Y = a + bW$, where $b \neq 0$.

Solution Here $g(w) = a + bw$, $w \in \mathbb{R}$, which is a continuously differentiable one-to-one function of $\mathbb{R}$ onto itself. The unique solution to the equation $a + bw = y$ is given by $w = (y - a)/b$. Observe that $dw/dy = 1/b$. If $b > 0$, we conclude from (50) that

$$f_Y(y) = \frac{1}{b} f_W\left(\frac{y - a}{b}\right), \qquad y \in \mathbb{R}.$$

This agrees with a result in Theorem 1.4 in Section 1.7. If $b < 0$ (and also if $b > 0$), we conclude from (49) that

$$f_Y(y) = \frac{1}{|b|} f_W\left(\frac{y - a}{b}\right), \qquad y \in \mathbb{R}. \qquad \blacksquare$$

EXAMPLE **5.28** Let W be a positive random variable having density function f_W. Determine the density function of $Y = W^a$, where $a \neq 0$.

Solution Here $g(w) = w^a$, $w > 0$, which is a continuously differentiable one-to-one function of $(0, \infty)$ onto itself. The unique positive solution to the equation $w^a = y$, where $y > 0$, is given by $w = y^{1/a}$. Observe that

$$\frac{dw}{dy} = \frac{d}{dy}(y^{1/a}) = \frac{1}{a} y^{(1/a)-1}.$$

Thus, by (49),

$$f_Y(y) = \frac{1}{|a|} y^{(1/a)-1} f_W(y^{1/a}), \qquad y > 0. \tag{51}$$

In particular, if $a > 0$, then

$$f_Y(y) = \frac{1}{a} y^{(1/a)-1} f_W(y^{1/a}), \qquad y > 0. \tag{52}$$

Equation (52) with $a = 1/2$ agrees with a result in Theorem 1.5 in Section 1.7. It follows from (51) with $a = -1$ that the random variable $Y = 1/W$ has the density function given by

$$f_Y(y) = \frac{1}{y^2} f_W\left(\frac{1}{y}\right), \qquad y > 0.$$

This agrees with a result in Theorem 1.6 in Section 1.7. ∎

When $n = 1$, the technique of Section 1.7 is generally preferable to the application of Theorem 5.20. The former technique yields formulas for the distribution function and quantiles of Y as well as its density function, and it can be used even when g is not one-to-one (see the proof of Theorem 1.7 in Section 1.7).

In trying to remember that (50) should be written as stated rather than with dw/dy mistakenly replaced by dy/dw, it is useful to rewrite this equation as

$$f_Y(y)dy = f_W(w)dw. \tag{53}$$

Alternatively, we can think of W as the "old" variable and Y as the "new" variable and write (53) as

$$f_{\text{NEW}}(\text{new})d(\text{new}) = f_{\text{OLD}}(\text{old})d(\text{old})$$

or, equivalently, as

$$f_{\text{NEW}}(\text{new}) = \frac{d(\text{old})}{d(\text{new})} f_{\text{OLD}}(\text{old}). \tag{54}$$

Equation (54), which is valid when the new variable is an increasing function of the old variable, is a helpful mnemonic for remembering (50). More generally, when the new variable is a monotonic function of the old variable,

$$f_{\text{NEW}}(\text{new}) = \left|\frac{d(\text{old})}{d(\text{new})}\right| f_{\text{OLD}}(\text{old}), \tag{55}$$

which is a helpful mnemonic for remembering (49). The multivariate extension of (55), given by

$$f_{\text{NEW}}(\mathbf{new}) = \left|\frac{\partial(\mathbf{old})}{\partial(\mathbf{new})}\right| f_{\text{OLD}}(\mathbf{old}), \tag{56}$$

is a helpful mnemonic for remembering (41).

EXAMPLE **5.29** Let X and Y be random variables having joint density function $f_{X,Y}$ and let R and Θ denote the polar coordinates of (X, Y), where $0 \le \Theta < 2\pi$. Then $X = R\cos\Theta$, $Y = R\sin\Theta$, and $X^2 + Y^2 = R^2$. Determine the joint density function of R and Θ.

Solution   Here X and Y are the old random variables and R and Θ are the new random variables. The old variables x and y can be expressed in terms of the new variables r and θ by $x = r\cos\theta$ and $y = r\sin\theta$, as seen in Figure 5.3. [Since Y has a density function, $P(Y = 0) = 0$. Thus, we can assume that (X, Y) ranges over the plane with the nonnegative portion of the x-axis removed, so that (r, θ) is a one-to-one function of (x, y).] The derivative of the old variables with respect to the new variables is

FIGURE **5.3**

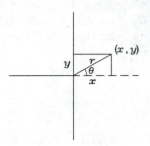

given by

$$
\begin{bmatrix} \frac{\partial x}{\partial r} & \frac{\partial x}{\partial \theta} \\[2mm] \frac{\partial y}{\partial r} & \frac{\partial y}{\partial \theta} \end{bmatrix} = \begin{bmatrix} \cos\theta & -r\sin\theta \\ \sin\theta & r\cos\theta \end{bmatrix},
$$

whose determinant is given by

$$
\frac{\partial(x, y)}{\partial(r, \theta)} = \begin{vmatrix} \cos\theta & -r\sin\theta \\ \sin\theta & r\cos\theta \end{vmatrix} = r(\cos^2\theta + \sin^2\theta) = r. \quad \blacksquare
$$

Thus, the joint density function of R and Θ is given by

$$
f_{R,\Theta}(r, \theta) = r f_{X,Y}(r\cos\theta, r\sin\theta), \qquad r > 0 \text{ and } 0 < \theta < 2\pi, \tag{57}
$$

and $f_{R,\Theta}(r, \theta) = 0$ elsewhere.

EXAMPLE 5.30   Let Z_1 and Z_2 be independent, standard normal random variables. Determine the joint distribution of the corresponding polar coordinates R and Θ.

Solution   According to (57) above and (37) in Section 5.5,

$$
f_{R,\Theta}(r, \theta) = r f_{Z_1, Z_2}(r\cos\theta, r\sin\theta) = \frac{r}{2\pi} e^{-r^2/2}, \qquad r > 0 \text{ and } 0 < \theta < 2\pi,
$$

and $f_{R,\Theta} = 0$ elsewhere. Thus, R and Θ are independent random variables, Θ is uniformly distributed on $(0, 2\pi)$, and R has the density function given by $f_R(r) = re^{-r^2/2}$ for $r > 0$ and $f_R(r) = 0$ for $r \le 0$. Alternatively, $R^2 = Z_1^2 + Z_2^2$ has the chi-square distribution with 2 degrees of freedom, which coincides with the exponential distribution with mean 2, so R is the square root of a random variable having such a distribution (see Example 1.53 in Section 1.7). $\blacksquare$

The solution to Example 5.30 suggests the *Box–Mueller method* for generating a pair of independent, standard normal random variables from a pair U_1, U_2 of independent, standard uniform random variables. Set $\Theta = 2\pi U_1$ and $V = 2\log(1/(1 - U_2))$. Then Θ and V are independent random variables, Θ is uniformly distributed on $(0, 2\pi)$, and V has the exponential distribution with mean

2. Set $R = \sqrt{V}$, $Z_1 = R \cos \Theta$, and $Z_2 = R \sin \Theta$. Then Z_1 and Z_2 are independent, standard normal random variables. More generally, given independent, standard uniform random variables $U_1, \ldots, U_{2n}$, we can generate independent, standard normal random variables $Z_1, \ldots, Z_{2n}$ by grouping the U_i's into n pairs and applying the Box–Mueller method to each pair.

Problems

5.69   Let Z_1 and Z_2 be independent, standard normal random variables. Determine the density function of Z_1/Z_2 and thereby determine its distribution.

5.70   Let X and Y have the bivariate density function given by $f_{X,Y}(x, y) = x + y$ for $0 < x, y < 1$ and $f_{X,Y}(x, y) = 0$ elsewhere. **(a)** Determine the joint density function of the random variables $W = X + Y$ and $Z = Y/X$. **(b)** Are W and Z independent?

5.71   Let W_1 and W_2 be positive random variables having joint density function f_{W_1,W_2}. Determine a formula for the joint density function of $Y_1 = W_1 W_2$ and $Y_2 = W_1/W_2$.

5.72   Let W_1 and W_2 be positive random variables having joint density function f_{W_1,W_2}. Determine a formula for the joint density function of $Y_1 = W_1 + W_2$ and $Y_2 = W_1^2/W_2$.

5.73   (Continued.) Let W_1 and W_2 be independent random variables having gamma distributions with shape parameters α_1 and α_2, respectively, and common scale parameter β. Show that $W_1 + W_2$ and W_1/W_2 are independent.

5.74   Let W_1 and W_2 be positive random variables having joint density function f_{W_1,W_2}. Determine a formula for the joint density function of $Y_1 = W_1/(W_1 + W_2)$ and $Y_2 = W_1 + W_2$.

5.75   (Continued.) Let W_1 and W_2 be independent random variables having gamma distributions with shape parameters α_1 and α_2, respectively, and common scale parameter β. Show that $W_1/(W_1 + W_2)$ and $W_1 + W_2$ are independent and that $W_1/(W_1 + W_2)$ has the beta distribution with parameters α_1 and α_2.

5.76   Let W_1 and W_2 be jointly distributed according to the Dirichlet distribution with parameters $\alpha_1, \alpha_2, \alpha_3$, which is defined in Problem 5.64 in Section 5.5. Show that $W_1 + W_2$ has the beta distribution with parameters $\alpha_1 + \alpha_2$ and α_3.

5.77   Let W_1, W_2, and W_3 be positive random variables having joint density function f_{W_1,W_2,W_3}. Determine a formula for the joint density function of $Y_1 = W_1/(W_1 + W_2 + W_3)$, $Y_2 = W_2/(W_1 + W_2 + W_3)$, and $Y_3 = W_1 + W_2 + W_3$.

5.78   (Continued.) Let W_1, W_2, W_3 be independent random variables having gamma distributions with shape parameters $\alpha_1, \alpha_2, \alpha_3$, respectively, and common scale parameter β. Show that the joint distribution of $W_1/(W_1 + W_2 + W_3)$ and $W_2/(W_1 + W_2 + W_3)$ is Dirichlet with parameters $\alpha_1, \alpha_2, \alpha_3$, which is defined in Problem 5.64 in Section 5.5.

5.79   Let W_1, W_2, W_3 be independent and exponentially distributed with mean β. Determine the joint density function of $Y_1 = W_1$, $Y_2 = W_1 + W_2$, and $Y_3 = W_1 + W_2 + W_3$.

5.80 Let U_1 and U_2 be independent random variables, each uniformly distributed on $(0, 1)$, and set $V_2 = U_2^{1/2}$ and $V_1 = U_1 V_2$. Determine the joint density function of V_1 and V_2.

5.81 Let U_1, U_2, U_3 be independent random variables, each uniformly distributed on $(0, 1)$, and set $V_3 = U_3^{1/3}$, $V_2 = U_2^{1/2} V_3$, and $V_1 = U_1 V_2$. Determine the joint density function of V_1, V_2, V_3.

5.82 Let W_1 and W_2 be independent random variables, each having the density function f given by $f(w) = 3w^{-4}$ for $w > 1$ and $f(w) = 0$ for $w \le 1$. Determine **(a)** the joint density function of W_1 and $W_1 W_2$; **(b)** the density function of $W_1 W_2$; **(c)** the mean and variance of $W_1 W_2$.

5.83 Let X and Y be random variables having a joint density function $f_{X,Y}$ such that $f_{X,Y}(x, y)$ depends only on the distance from (x, y) to the origin; that is, $f_{X,Y}(x, y) = g(\sqrt{x^2 + y^2})$ for $x, y \in \mathbb{R}$, where g is some function on $[0, \infty)$. Let R and Θ denote the polar coordinates of (X, Y) as introduced in Example 5.29. Show that **(a)** Θ is uniformly distributed on $[0, 2\pi)$; **(b)** R has the density function given by $f_R(r) = 2\pi r g(r)$ for $r > 0$ and $f_R(r) = 0$ for $r \le 0$; **(c)** R and Θ are independent.

5.84 Let $\mathbf{W}$ have a density function $f_{\mathbf{W}}$ on $\mathbb{R}^n$, let $\mathbf{a} \in \mathbb{R}^n$ and $b > 0$, and set $\mathbf{Y} = \mathbf{a} + b\mathbf{W}$. Show that $\mathbf{Y}$ has the density function $f_{\mathbf{Y}}$ on $\mathbb{R}^n$ given by $f_{\mathbf{Y}}(\mathbf{y}) = b^{-n} f_{\mathbf{W}}(b^{-1}(\mathbf{y} - \mathbf{a}))$ for $\mathbf{y} \in \mathbb{R}^n$.

5.7
The Multivariate Normal Distribution

In this section, we investigate the multivariate normal distribution, which is an extension of the univariate normal distribution that applies to random vectors. The multivariate normal distribution is sometimes an accurate model for the joint distribution of dependent measurements on a randomly selected individual, it has many elegant theoretical properties, and, especially in the context of large samples, it provides a good approximation to the joint distribution of estimates of several unknown parameters.

Let n be a positive integer, let $Z_1, \dots, Z_n$ be independent, standard normal random variables, and set $\mathbf{Z} = [Z_1, \dots, Z_n]^T$. Then $\mathbf{Z}$ has mean vector $\mathbf{0}$ and variance–covariance matrix $\mathbf{I}$. According to Equation (37) in Section 5.5, the density function of $\mathbf{Z}$ is given by

$$f_{\mathbf{Z}}(\mathbf{z}) = \frac{1}{(2\pi)^{n/2}} \exp(-\tfrac{1}{2}\mathbf{z}^T \mathbf{z}), \qquad \mathbf{z} \in \mathbb{R}^n. \tag{58}$$

Let $\mu \in \mathbb{R}^n$, let $\mathbf{B}$ be an invertible $n \times n$ matrix, and set $\mathbf{Y} = \mu + \mathbf{B}\mathbf{Z}$. Then $\mathbf{Y}$ has mean vector $E\mathbf{Y} = E(\mu + \mathbf{B}\mathbf{Z}) = \mu + \mathbf{B}(E\mathbf{Z}) = \mu$ and variance–covariance matrix

$$\Sigma = \text{VC}(\mu + \mathbf{B}\mathbf{Z}) = \mathbf{B}[\text{VC}(\mathbf{Z})]\mathbf{B}^T = \mathbf{B}\mathbf{I}\mathbf{B}^T = \mathbf{B}\mathbf{B}^T.$$

According to Theorem 5.23 in Section 5.6, the density function of $\mathbf{Y}$ is given by

$$f_{\mathbf{Y}}(\mathbf{y}) = \frac{1}{|\det(\mathbf{B})|} f_{\mathbf{Z}}(\mathbf{B}^{-1}(\mathbf{y} - \mu)), \qquad \mathbf{y} \in \mathbb{R}^n.$$

Now $\det(\Sigma) = \det(\mathbf{B}\mathbf{B}^T) = \det(\mathbf{B})\det(\mathbf{B}^T) = [\det(\mathbf{B})]^2 > 0$ and hence $|\det(\mathbf{B})| = \sqrt{\det(\Sigma)}$. By (58),

$$f_{\mathbf{Z}}(\mathbf{B}^{-1}(\mathbf{y} - \mu)) = \frac{1}{(2\pi)^{n/2}} \exp\{-\tfrac{1}{2}[\mathbf{B}^{-1}(\mathbf{y} - \mu)]^T \mathbf{B}^{-1}(\mathbf{y} - \mu)\}.$$

Observe that $[\mathbf{B}^{-1}(\mathbf{y} - \mu)]^T \mathbf{B}^{-1}(\mathbf{y} - \mu) = (\mathbf{y} - \mu)^T (\mathbf{B}^{-1})^T \mathbf{B}^{-1}(\mathbf{y} - \mu)$. Since

$$(\mathbf{B}^{-1})^T \mathbf{B}^{-1} = (\mathbf{B}^T)^{-1} \mathbf{B}^{-1} = (\mathbf{B}\mathbf{B}^T)^{-1} = \Sigma^{-1},$$

we conclude that

$$[\mathbf{B}^{-1}(\mathbf{y} - \mu)]^T \mathbf{B}^{-1}(\mathbf{y} - \mu) = (\mathbf{y} - \mu)^T \Sigma^{-1}(\mathbf{y} - \mu)$$

and hence that the density function of $\mathbf{Y}$ is given by

$$f_{\mathbf{Y}}(\mathbf{y}) = \frac{1}{(2\pi)^{n/2} \sqrt{\det(\Sigma)}} \exp\{-\tfrac{1}{2}(\mathbf{y} - \mu)^T \Sigma^{-1}(\mathbf{y} - \mu)\}, \qquad \mathbf{y} \in \mathbb{R}^n. \quad \textbf{(59)}$$

The distribution of $\mathbf{Y}$ is referred to as the *multivariate normal distribution* with mean vector μ and variance–covariance matrix Σ.

Given any positive definite, symmetric $n \times n$ matrix Σ, consider its Cholesky decomposition $\Sigma = \mathbf{L}\mathbf{L}^T$, with $\mathbf{L}$ being an invertible $n \times n$ lower triangular matrix. Then the construction leading to (59) is applicable with $\mathbf{B} = \mathbf{L}$. Therefore, for every n-dimensional vector μ and every positive definite, symmetric $n \times n$ matrix Σ, the multivariate normal distribution with mean vector μ and variance–covariance matrix Σ is well defined and has the density function given by (59).

Observe, in particular, that if $\mu = \mathbf{0}$ and $\Sigma = \mathbf{I}$, then $\det(\Sigma) = 1$ and hence the density function given by (59) coincides with the density function given by (58). Thus, if the joint distribution of $Y_1, \ldots, Y_n$ is multivariate normal with mean vector $\mathbf{0}$ and variance–covariance matrix $\mathbf{I}$, then $Y_1, \ldots, Y_n$ are independent, standard normal random variables.

THEOREM 5.24   Let Z_1 and Z_2 be independent, standard normal random variables and let $b_1, b_2 \in \mathbb{R}$ with $b_1^2 + b_2^2 = 1$. Then $b_1 Z_1 + b_2 Z_2$ and $-b_2 Z_1 + b_1 Z_2$ are independent, standard normal random variables.

Proof   Set $Y_1 = b_1 Z_1 + b_2 Z_2$, $Y_2 = -b_2 Z_1 + b_1 Z_2$, $Z = [Z_1, Z_2]^T$, and

$$Y = \begin{bmatrix} Y_1 \\ Y_2 \end{bmatrix} = \begin{bmatrix} b_1 Z_1 + b_2 Z_2 \\ -b_2 Z_1 + b_1 Z_2 \end{bmatrix} = \begin{bmatrix} b_1 & b_2 \\ -b_2 & b_1 \end{bmatrix} \begin{bmatrix} Z_1 \\ Z_2 \end{bmatrix} = \mathbf{B}\mathbf{Z},$$

where

$$\mathbf{B} = \begin{bmatrix} b_1 & b_2 \\ -b_2 & b_1 \end{bmatrix}.$$

Then **Y** has the bivariate normal distribution with mean vector **0** and variance–covariance matrix

$$\mathbf{BB}^T = \begin{bmatrix} b_1 & b_2 \\ -b_2 & b_1 \end{bmatrix} \begin{bmatrix} b_1 & -b_2 \\ b_2 & b_1 \end{bmatrix} = \begin{bmatrix} b_1^2 + b_2^2 & 0 \\ 0 & b_1^2 + b_2^2 \end{bmatrix} = \begin{bmatrix} 1 & 0 \\ 0 & 1 \end{bmatrix} = \mathbf{I},$$

so Y_1 and Y_2 are independent, standard normal random variables. ∎

EXAMPLE **5.31**   Let Z_1 and Z_2 be independent, standard normal random variables. Also, let $\mu_1, \mu_2 \in \mathbb{R}$, $\sigma_1, \sigma_2 > 0$, and $-1 < \rho < 1$ and set

$$X = \mu_1 + \sigma_1 Z_1 \quad \text{and} \quad Y = \mu_2 + \sigma_2 \left(\rho Z_1 + \sqrt{1 - \rho^2} Z_2 \right).$$

Show that X and Y have the bivariate normal joint distribution with $EX = \mu_1$, $\text{var}(X) = \sigma_1^2$, $EY = \mu_2$, $\text{var}(Y) = \sigma_2^2$, and $\text{cor}(X, Y) = \rho$.

Solution   It is easily seen that X and Y have the indicated means and variances. Moreover, $\text{cov}(X, Y) = \rho \sigma_1 \sigma_2$, so

$$\text{cor}(X, Y) = \frac{\text{cov}(X, Y)}{\text{SD}(X) \, \text{SD}(Y)} = \rho.$$

Set

$$\mu = \begin{bmatrix} \mu_1 \\ \mu_2 \end{bmatrix}, \quad \mathbf{B} = \begin{bmatrix} \sigma_1 & 0 \\ \rho \sigma_2 & \sqrt{1 - \rho^2} \sigma_2 \end{bmatrix}, \quad \text{and} \quad \mathbf{Z} = \begin{bmatrix} Z_1 \\ Z_2 \end{bmatrix}.$$

Then **B** is an invertible matrix. Since $[X, Y]^T = \mu + \mathbf{BZ}$, we conclude that X and Y have a bivariate normal joint distribution. ∎

THEOREM **5.25**   Let $Y_1, \ldots, Y_n$ be independent random variables such that Y_i is normally distributed with mean μ_i and variance σ_i^2 for $1 \leq i \leq n$. Then $\mathbf{Y} = [Y_1, \ldots, Y_n]^T$ has the multivariate normal distribution with mean vector $\mu = [\mu_1, \ldots, \mu_n]^T$ and variance–covariance matrix $\text{diag}(\sigma_1^2, \ldots, \sigma_n^2)$.

Proof   Set $Z_i = (Y_i - \mu_i)/\sigma_i$ for $1 \leq i \leq n$ and $\mathbf{Z} = [Z_1, \ldots, Z_n]^T$. Then $Z_1, \ldots, Z_n$ are independent, standard normal random variables. Observe that $Y_i = \mu_i + \sigma_i Z_i$ for $1 \leq i \leq n$ and hence that

$$\mathbf{Y} = \mu + \begin{bmatrix} \sigma_1 & & 0 \\ & \ddots & \\ 0 & & \sigma_n \end{bmatrix} \mathbf{Z}.$$

Consequently, **Y** has the multivariate normal distribution with mean vector μ and variance–covariance matrix

$$\begin{bmatrix} \sigma_1 & & 0 \\ & \ddots & \\ 0 & & \sigma_n \end{bmatrix} \begin{bmatrix} \sigma_1 & & 0 \\ & \ddots & \\ 0 & & \sigma_n \end{bmatrix} = \begin{bmatrix} \sigma_1^2 & & 0 \\ & \ddots & \\ 0 & & \sigma_n^2 \end{bmatrix}. \quad ∎$$

COROLLARY 5.10   Let $Y_1, \ldots, Y_n$ be independent random variables such that Y_i is normally distributed with mean μ_i and variance σ^2 for $1 \leq i \leq n$. Then $\mathbf{Y} = [Y_1, \ldots, Y_n]^T$ has the multivariate normal distribution with mean vector $\boldsymbol{\mu} = [\mu_1, \ldots, \mu_n]^T$ and variance–covariance matrix $\sigma^2 \mathbf{I}$.

THEOREM 5.26   Let $\mathbf{Y} = [Y_1, \ldots, Y_n]^T$ have a multivariate normal distribution, let $a \in \mathbb{R}$, and let $\mathbf{b} = [b_1, \ldots, b_n]^T \in \mathbb{R}^n$ with $\mathbf{b} \neq \mathbf{0}$. Then $a + \mathbf{b}^T\mathbf{Y} = a + b_1 Y_1 + \cdots + b_n Y_n$ is normally distributed.

Proof   Now $\mathbf{Y} = \boldsymbol{\mu} + \mathbf{BZ}$, where $\boldsymbol{\mu} = E\mathbf{Y} \in \mathbb{R}^n$, $\mathbf{B}$ is an invertible $n \times n$ matrix, and $\mathbf{Z} = [Z_1, \ldots, Z_n]^T$ with $Z_1, \ldots, Z_n$ being independent, standard normal random variables. Thus,

$$a + \mathbf{b}^T\mathbf{Y} = a + \mathbf{b}^T(\boldsymbol{\mu} + \mathbf{BZ}) = a + \mathbf{b}^T\boldsymbol{\mu} + \mathbf{b}^T\mathbf{BZ} = c + \mathbf{v}^T\mathbf{Z},$$

where $c = a + \mathbf{b}^T\boldsymbol{\mu} \in \mathbb{R}$ and $\mathbf{v} = [v_1, \ldots, v_n]^T = \mathbf{B}^T\mathbf{b} \in \mathbb{R}^n$. Since $\mathbf{b} \neq \mathbf{0}$ and $\mathbf{B}^T$ is invertible, we conclude that $\mathbf{v} \neq \mathbf{0}$; that is, not all of the numbers $v_1, \ldots, v_n$ are zero. Therefore, by Theorem 3.9 in Section 3.2,

$$a + \mathbf{b}^T\mathbf{Y} = c + \mathbf{v}^T\mathbf{Z} = c + v_1 Z_1 + \cdots + v_n Z_n$$

is normally distributed.   ∎

COROLLARY 5.11   Let $\mathbf{Y} = [Y_1, \ldots, Y_n]^T$ have a multivariate normal distribution. Then Y_i is normally distributed for $1 \leq i \leq n$.

Proof   Given i with $1 \leq i \leq n$, let $\mathbf{b}$ be the n-dimensional column vector whose ith entry equals one and whose other entries equal 0. Then $Y_i = \mathbf{b}^T\mathbf{Y}$ is normally distributed by Theorem 5.26.   ∎

THEOREM 5.27   Let $\mathbf{Y}$ have a multivariate normal distribution, let $\mathbf{a} \in \mathbb{R}^n$, and let $\mathbf{B}$ be an invertible $n \times n$ matrix. Then $\mathbf{a} + \mathbf{BY}$ has a multivariate normal distribution.

Proof   Write $\mathbf{Y} = \boldsymbol{\mu} + \mathbf{B}_1\mathbf{Z}$, where $\boldsymbol{\mu} = E\mathbf{Y} \in \mathbb{R}^n$, $\mathbf{B}_1$ is an invertible $n \times n$ matrix, and $\mathbf{Z} = [Z_1, \ldots, Z_n]^T$ with $Z_1, \ldots, Z_n$ being independent, standard normal random variables. Then $\mathbf{a} + \mathbf{BY} = \mathbf{a} + \mathbf{B}(\boldsymbol{\mu} + \mathbf{B}_1\mathbf{Z}) = \mathbf{a} + \mathbf{B}\boldsymbol{\mu} + \mathbf{BB}_1\mathbf{Z}$. Since $\mathbf{a} + \mathbf{B}\boldsymbol{\mu} \in \mathbb{R}^n$ and $\mathbf{BB}_1$ is an invertible $n \times n$ matrix, $\mathbf{a} + \mathbf{BY}$ has a multivariate normal distribution.   ∎

EXAMPLE 5.32   Let X and Y have a bivariate normal joint distribution. Obtain $P(X \geq EX, \, Y \geq EY)$ as a function of $\rho = \text{cor}(X, Y)$.

Solution   Set $X' = (X - EX)/\text{SD}(X)$ and $Y' = (Y - EY)/\text{SD}(Y)$. Then X' and Y' have the bivariate normal joint distribution with zero means, unit variances, and correlation coefficient ρ. Also, $P(X \geq EX, \, Y \geq EY) = P(X' \geq 0, \, Y' \geq 0)$.

Let Z_1 and Z_2 be independent, standard normal random variables and let R and Θ denote the corresponding polar coordinates, where $-\pi < \Theta < \pi$. Then, by a slight modification of the solution to Example 5.29 in Section 5.6, Θ is uniformly distributed on $(-\pi, \pi)$. According to the solution to Example 5.31, Z_1 and $\rho Z_1 + \sqrt{1 - \rho^2} Z_2$ have the same joint distribution as X' and Y'. Thus (see Figure 5.4),

$$P(X \geq EX, Y \geq EY) = P(X' \geq 0, Y' \geq 0)$$

$$= P\left(Z_1 \geq 0, \rho Z_1 + \sqrt{1 - \rho^2} Z_2 \geq 0 \right)$$

$$= P\left(Z_1 \geq 0, Z_2 \geq -\frac{\rho}{\sqrt{1 - \rho^2}} Z_1 \right)$$

$$= P\left(- \arcsin \rho < \Theta < \frac{\pi}{2} \right)$$

$$= \frac{1}{2\pi} \left(\frac{\pi}{2} + \arcsin \rho \right)$$

$$= \frac{1}{4} + \frac{1}{2\pi} \arcsin \rho. \quad \blacksquare$$

FIGURE 5.4
$\angle A = \frac{\pi}{2} + \arcsin \rho$

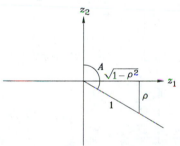

EXAMPLE 5.33   Let $\mathbf{Y} = [\, Y_1, Y_2, Y_3 \,]^T$ have a trivariate normal distribution. Show that $[\, Y_1, Y_3, Y_2 \,]^T$ has a trivariate normal distribution.

Solution   Since

$$\begin{bmatrix} Y_1 \\ Y_3 \\ Y_2 \end{bmatrix} = \begin{bmatrix} 1 & 0 & 0 \\ 0 & 0 & 1 \\ 0 & 1 & 0 \end{bmatrix} \begin{bmatrix} Y_1 \\ Y_2 \\ Y_3 \end{bmatrix},$$

the desired result follows from Theorem 5.27. $\blacksquare$

Example 5.33 has an obvious extension: If $\mathbf{Y} = [\, Y_1, \ldots, Y_n \,]^T$ has a multivariate normal distribution and $\mathbf{W}$ is obtained from $\mathbf{Y}$ by permuting its entries in any manner, then $\mathbf{W}$ has a multivariate normal distribution.

The next result is an extension of Corollary 5.11.

THEOREM 5.28   Let $\mathbf{Y}$ have a multivariate normal distribution. Then every subvector $\mathbf{Y}_1$ of $\mathbf{Y}$ has a multivariate normal distribution.

Proof   Let $\mathbf{Y}_1$ be a subvector of $\mathbf{Y}$. By permuting the entries of $\mathbf{Y}$ if necessary, we can assume that $\mathbf{Y} = [\, Y_1, \ldots, Y_n\,]^T$ and $\mathbf{Y}_1 = [\, Y_1, \ldots, Y_m\,]^T$, where $1 \leq m < n$. Set $\mu = [\,\mu_1, \ldots, \mu_n\,]^T = E\mathbf{Y}$ and consider the Cholesky decomposition $\mathrm{VC}(\mathbf{Y}) = \mathbf{L}\mathbf{L}^T$, where

$$\mathbf{L} = \begin{bmatrix} l_{11} & \cdots & 0 \\ \vdots & & \vdots \\ l_{n1} & \cdots & l_{nn} \end{bmatrix}$$

is an invertible lower-triangular $n \times n$ matrix. Observe that $l_{ij} = 0$ for $j > i$. Observe also that $0 \neq \det(L) = l_{11} \cdots l_{nn}$ and hence that $l_{ii} \neq 0$ for $1 \leq i \leq n$.

Now $\mathbf{Y} = \mu + \mathbf{L}\mathbf{Z}$, where $\mathbf{Z} = [\, Z_1, \ldots, Z_n\,]^T$ with $Z_1, \ldots, Z_n$ being independent, standard normal random variables. In expanded form,

$$\begin{bmatrix} Y_1 \\ \vdots \\ Y_n \end{bmatrix} = \begin{bmatrix} \mu_1 \\ \vdots \\ \mu_n \end{bmatrix} + \begin{bmatrix} l_{11} & \cdots & 0 \\ \vdots & & \vdots \\ l_{n1} & \cdots & l_{nn} \end{bmatrix} \begin{bmatrix} Z_1 \\ \vdots \\ Z_n \end{bmatrix},$$

and hence

$$\begin{bmatrix} Y_1 \\ \vdots \\ Y_m \end{bmatrix} = \begin{bmatrix} \mu_1 \\ \vdots \\ \mu_m \end{bmatrix} + \begin{bmatrix} l_{11} & \cdots & 0 \\ \vdots & & \vdots \\ l_{m1} & \cdots & l_{mm} \end{bmatrix} \begin{bmatrix} Z_1 \\ \vdots \\ Z_m \end{bmatrix}.$$

Set

$$\mu_1 = \begin{bmatrix} \mu_1 \\ \vdots \\ \mu_m \end{bmatrix}, \quad \mathbf{L}_1 = \begin{bmatrix} l_{11} & \cdots & 0 \\ \vdots & & \vdots \\ l_{m1} & \cdots & l_{mm} \end{bmatrix}, \quad \text{and} \quad \mathbf{Z}_1 = \begin{bmatrix} Z_1 \\ \vdots \\ Z_m \end{bmatrix}.$$

Since the lower-triangular matrix $\mathbf{L}_1$ has nonzero diagonal entries, it is an invertible matrix. Thus, $\mathbf{Y}_1 = \mu_1 + \mathbf{L}_1\mathbf{Z}_1$ has a multivariate normal distribution.   ∎

THEOREM 5.29   Let $\mathbf{Y} = [\, Y_1, \ldots, Y_n\,]^T$ have a multivariate normal distribution. Then $Y_1, \ldots, Y_n$ are independent if and only if they are uncorrelated.

Proof   Independent random variables having finite variance are uncorrelated, but uncorrelated random variables need not be independent. The point of the theorem is that uncorrelated random variables $Y_1, \ldots, Y_n$ having a multivariate normal joint distribution are independent. To prove this result, we observe first that Y_i is normally distributed with mean μ_i and variance σ_i^2 for $1 \leq i \leq n$ (see Corollary 5.11) and that the variance–covariance matrix $\mathbf{\Sigma}$ of these random variables is given by $\mathbf{\Sigma} = \mathrm{diag}(\sigma_1^2, \ldots, \sigma_n^2)$. Now $\mathbf{\Sigma}^{-1} = \mathrm{diag}(\sigma_1^{-2}, \ldots, \sigma_n^{-2})$ and hence

$$(\mathbf{y} - \boldsymbol{\mu})^T \boldsymbol{\Sigma}^{-1}(\mathbf{y} - \boldsymbol{\mu}) = [\, y_1 - \mu_1, \ldots, y_n - \mu_n\,] \begin{bmatrix} \sigma_1^{-2} & \cdots & 0 \\ \vdots & & \vdots \\ 0 & \cdots & \sigma_n^{-2} \end{bmatrix} \begin{bmatrix} y_1 - \mu_1 \\ \vdots \\ y_n - \mu_n \end{bmatrix}$$

$$= \sum_{i=1}^{n} \left(\frac{y_i - \mu_i}{\sigma_i} \right)^2$$

for $\mathbf{y} = [\, y_1, \ldots, y_n\,]^T \in \mathbb{R}^n$. Since $\sqrt{\det(\boldsymbol{\Sigma})} = \sigma_1 \cdots \sigma_n$, we conclude from Equation (59) that

$$f_{Y_1}, \ldots, Y_n(y_1, \ldots, y_n) = \sum_{i=1}^{n} \frac{1}{\sigma_i \sqrt{2\pi}} \exp\left(-\frac{1}{2} \left(\frac{y_i - \mu_i}{\sigma_i} \right)^2 \right)$$

$$= \prod_{i=1}^{n} f_{Y_i}(y_i),$$

for $y_1, \ldots, y_n \in \mathbb{R}$, and hence that $Y_1, \ldots, Y_n$ are independent. ∎

Given random variables $X_1, \ldots, X_m, Y_1, \ldots, Y_n$, set

$$\mathbf{X} = [\, X_1, \ldots, X_m\,]^T \quad \text{and} \quad \mathbf{Y} = [\, Y_1, \ldots, Y_n\,]^T.$$

Suppose $\mathbf{X}$ and $\mathbf{Y}$ have a joint density function. Then $\mathbf{X}$ has a density function $f_{\mathbf{X}}$, and $\mathbf{Y}$ has a density function $f_{\mathbf{Y}}$. The random vectors $\mathbf{X}$ and $\mathbf{Y}$ are independent if and only if

$$f_{\mathbf{X}, \mathbf{Y}}(\mathbf{x}, \mathbf{y}) = f_{\mathbf{X}}(\mathbf{x}) f_{\mathbf{Y}}(\mathbf{y}), \qquad \mathbf{x} \in \mathbb{R}^m \text{ and } \mathbf{y} \in \mathbb{R}^n, \tag{60}$$

determines a joint density function of $\mathbf{X}$ and $\mathbf{Y}$. Suppose $\mathbf{X}$ has a continuous density function $f_{\mathbf{X}}$, $\mathbf{Y}$ has a continuous density function $f_{\mathbf{Y}}$, and $\mathbf{X}$ and $\mathbf{Y}$ have a continuous joint density function $f_{\mathbf{X}, \mathbf{Y}}$. Then $\mathbf{X}$ and $\mathbf{Y}$ are independent if and only if (60) holds.

THEOREM 5.30   Let $\mathbf{X} = [\, X_1, \ldots, X_m\,]^T$ and $\mathbf{Y} = [\, Y_1, \ldots, Y_n\,]^T$ be independent random vectors, each having a multivariate normal distribution. Then they have a multivariate normal joint distribution.

Proof   We can write $\mathbf{X} = \boldsymbol{\mu}_1 + \mathbf{B}_1 \mathbf{Z}_1$ and $\mathbf{Y} = \boldsymbol{\mu}_2 + \mathbf{B}_2 \mathbf{Z}_2$, where $\mathbf{Z}_1 = [\, Z_1, \ldots, Z_m\,]^T$, $\mathbf{Z}_2 = [\, Z_{m+1}, \ldots, Z_{m+n}\,]^T$, $Z_1, \ldots, Z_{m+n}$, are independent, standard normal random variables, $\mathbf{B}_1$ is an invertible $m \times m$ matrix, and $\mathbf{B}_2$ is an invertible $n \times n$ matrix. Thus,

$$\begin{bmatrix} \mathbf{X} \\ \mathbf{Y} \end{bmatrix} = \begin{bmatrix} \boldsymbol{\mu}_1 \\ \boldsymbol{\mu}_2 \end{bmatrix} + \begin{bmatrix} \mathbf{B}_1 & \mathbf{0} \\ \mathbf{0} & \mathbf{B}_2 \end{bmatrix} \begin{bmatrix} \mathbf{Z}_1 \\ \mathbf{Z}_2 \end{bmatrix} = \boldsymbol{\mu} + \mathbf{B}\mathbf{Z},$$

where

$$\boldsymbol{\mu} = \begin{bmatrix} \boldsymbol{\mu}_1 \\ \boldsymbol{\mu}_2 \end{bmatrix} \quad \text{and} \quad \mathbf{B} = \begin{bmatrix} \mathbf{B}_1 & \mathbf{0} \\ \mathbf{0} & \mathbf{B}_2 \end{bmatrix}.$$

Since $\mathbf{B}$ is invertible, $\mathbf{X}$ and $\mathbf{Y}$ have a multivariate normal joint distribution. ∎

THEOREM 5.31   Let $\mathbf{X}$ and $\mathbf{Y}$ have a multivariate normal joint distribution. Then $\mathbf{X}$ and $\mathbf{Y}$ are independent if and only if $\mathrm{cov}(\mathbf{X}, \mathbf{Y}) = \mathbf{0}$.

Proof   It follows from Theorem 5.5(b) in Section 5.3 that if $\mathbf{X}$ and $\mathbf{Y}$ are independent, then $\mathrm{cov}(\mathbf{X}, \mathbf{Y}) = \mathbf{0}$. (This conclusion does not depend on the assumption of joint multivariate normality.)

Suppose, conversely, that $\mathbf{X}$ and $\mathbf{Y}$ have a multivariate normal joint distribution with $\mathrm{cov}(\mathbf{X}, \mathbf{Y}) = \mathbf{0}$. Set $\mu_1 = E\mathbf{X}$, $\Sigma_1 = \mathrm{VC}(\mathbf{X})$, $\mu_2 = E\mathbf{Y}$, and $\Sigma_2 = \mathrm{VC}(\mathbf{Y})$. Then the mean vector μ and variance–covariance matrix Σ of

$$\begin{bmatrix} \mathbf{X} \\ \mathbf{Y} \end{bmatrix}$$

can be written in partitioned form as

$$\mu = \begin{bmatrix} \mu_1 \\ \mu_2 \end{bmatrix} \quad \text{and} \quad \Sigma = \begin{bmatrix} \Sigma_1 & \mathbf{0} \\ \mathbf{0} & \Sigma_2 \end{bmatrix}.$$

Let $\mathbf{X}$ be m-dimensional and let $\mathbf{Y}$ be n-dimensional. Then, for $\mathbf{x} \in \mathbb{R}^m$ and $\mathbf{y} \in \mathbb{R}^n$,

$$[(\mathbf{x} - \mu_1)^T, (\mathbf{y} - \mu_2)^T]\Sigma^{-1}\begin{bmatrix} \mathbf{x} - \mu_1 \\ \mathbf{y} - \mu_2 \end{bmatrix}$$

equals

$$[(\mathbf{x} - \mu_1)^T, (\mathbf{y} - \mu_2)^T]\begin{bmatrix} \Sigma_1^{-1} & \mathbf{0} \\ \mathbf{0} & \Sigma_2^{-1} \end{bmatrix}\begin{bmatrix} \mathbf{x} - \mu_1 \\ \mathbf{y} - \mu_2 \end{bmatrix},$$

which is simply

$$(\mathbf{x} - \mu_1)^T\Sigma_1^{-1}(\mathbf{x} - \mu_1) + \left(\mathbf{y} - \mu_2\right)^T\Sigma_2^{-1}(\mathbf{y} - \mu_2).$$

Thus, the joint density function of $\mathbf{X}$ and $\mathbf{Y}$ is given by

$$\begin{aligned} f_{\mathbf{X},\mathbf{Y}}(\mathbf{x}, \mathbf{y}) &= \frac{1}{(2\pi)^{(m+n)/2}\sqrt{\det(\Sigma)}} \\ &\quad \times \exp\left(-\tfrac{1}{2}[(\mathbf{x} - \mu_1)^T, (\mathbf{y} - \mu_2)^T]\Sigma^{-1}\begin{bmatrix} \mathbf{x} - \mu_1 \\ \mathbf{y} - \mu_2 \end{bmatrix}\right) \\ &= \frac{1}{(2\pi)^{(m+n)/2}\sqrt{\det(\Sigma)}} \\ &\quad \times \exp\left(-\tfrac{1}{2}(\mathbf{x} - \mu_1)^T\Sigma_1^{-1}(\mathbf{x} - \mu_1) - \tfrac{1}{2}(\mathbf{y} - \mu_2)^T\Sigma_2^{-1}(\mathbf{y} - \mu_2)\right) \\ &= \frac{1}{(2\pi)^{m/2}\sqrt{\det(\Sigma_1)}}\exp\left(-\tfrac{1}{2}(\mathbf{x} - \mu_1)^T\Sigma_1^{-1}(\mathbf{x} - \mu_1)\right) \\ &\quad \times \frac{1}{(2\pi)^{n/2}\sqrt{\det(\Sigma_2)}}\exp\left(-\tfrac{1}{2}(\mathbf{y} - \mu_2)^T\Sigma_2^{-1}(\mathbf{y} - \mu_2)\right) \\ &= f_{\mathbf{X}}(\mathbf{x})f_{\mathbf{Y}}(\mathbf{y}). \end{aligned}$$

Therefore, $\mathbf{X}$ and $\mathbf{Y}$ are independent.   ∎

THEOREM 5.32   Let $\mathbf{X}$ and Y have a multivariate normal joint distribution and set

$$\beta = [\,VC(\mathbf{X})\,]^{-1}cov(\mathbf{X}, Y),$$

$\alpha = EY - \beta^T E\mathbf{X}$, and $W = Y - \alpha - \beta^T \mathbf{X}$. Then

(a) $\mathbf{X}$ and W are independent;

(b) W has the normal distribution with mean 0 and variance

$$var(Y) - [cov(\mathbf{X}, Y)]^T [\,VC(\mathbf{X})\,]^{-1}cov(\mathbf{X}, Y).$$

Proof   **(a)** It follows from Theorem 5.15 in Section 5.4 that $cov(\mathbf{X}, W) = \mathbf{0}$. Now

$$\begin{bmatrix} \mathbf{X} \\ W \end{bmatrix} = \begin{bmatrix} \mathbf{0} \\ -\alpha \end{bmatrix} + \begin{bmatrix} \mathbf{I} & 0 \\ -\beta^T & 1 \end{bmatrix} \begin{bmatrix} \mathbf{X} \\ Y \end{bmatrix}.$$

Since

$$\begin{bmatrix} \mathbf{I} & 0 \\ -\beta^T & 1 \end{bmatrix}$$

is a lower-triangular matrix, all of whose diagonal entries equal 1, it is an invertible matrix. Thus, by Theorem 5.27, $\mathbf{X}$ and W have a multivariate normal joint distribution. Consequently, by Theorem 5.31, $\mathbf{X}$ and W are independent.

(b) Since $\mathbf{X}$ and W have a multivariate normal joint distribution, it follows from Corollary 5.11 that W has a normal distribution. Moreover, by Theorem 5.15 in Section 5.4, W has mean 0 and the indicated variance.   ∎

The next result, which follows from Theorem 5.32 above and Corollary 5.7 in Section 5.4, will be used in Section 6.7.

COROLLARY 5.12   Let $\mathbf{X}$ and Y have a multivariate normal joint distribution and set

$$\beta = [\,VC(\mathbf{X})\,]^{-1}cov(\mathbf{X}, Y)$$

and $\alpha = EY - \beta^T E\mathbf{X}$, so that $\hat{Y} = \alpha + \beta^T \mathbf{X}$ is the best linear predictor of Y based on $\mathbf{X}$. Also, set $W = Y - \hat{Y} = Y - \alpha - \beta^T \mathbf{X}$. Then $Y = \hat{Y} + W = \alpha + \beta^T \mathbf{X} + W$, where $\mathbf{X}$ and W are independent and W is normally distributed with mean 0 and variance $var(Y) - [cov(\mathbf{X}, Y)]^T [\,VC(\mathbf{X})\,]^{-1}cov(\mathbf{X}, Y)$.

EXAMPLE 5.34   Let $\mathbf{X}$ and W be independent, where $\mathbf{X} = [X_1, \ldots, X_m]^T$ has a multivariate normal distribution and W is normally distributed with mean 0; let $\alpha \in \mathbb{R}$ and $\beta \in \mathbb{R}^m$; and set $Y = \alpha + \beta^T \mathbf{X} + W$. Determine the joint distribution of $\mathbf{X}$ and Y.

Solution   It follows from Theorem 5.30 that $\mathbf{X}$ and W have a multivariate normal joint distribution. Since

$$\begin{bmatrix} \mathbf{X} \\ Y \end{bmatrix} = \begin{bmatrix} 0 \\ \alpha \end{bmatrix} + \begin{bmatrix} \mathbf{I} & 0 \\ \beta^T & 1 \end{bmatrix} \begin{bmatrix} \mathbf{X} \\ W \end{bmatrix},$$

we conclude from Theorem 5.27 that $\mathbf{X}$ and Y have a multivariate normal joint distribution.

Observe that $EY = E(\alpha + \beta^T X + W) = \alpha + \beta^T EX$. Thus, the mean vector of the joint distribution of X and Y is given by

$$E\begin{bmatrix} X \\ Y \end{bmatrix} = \begin{bmatrix} EX \\ EY \end{bmatrix} = \begin{bmatrix} EX \\ \alpha + \beta^T EX \end{bmatrix}.$$

Observe next that

$$\text{cov}(X, Y) = \text{cov}(X, \alpha + \beta^T X + W) = \text{cov}(X, \beta^T X) = \text{VC}(X)\beta$$

and hence that $\text{cov}(Y, X) = \beta^T \text{VC}(X)$. Since X and W are independent, we conclude from Theorem 5.8(g) in Section 5.3 that

$$\text{var}(Y) = \text{var}(\alpha + \beta^T X + W) = \text{var}(\beta^T X) + \text{var}(W) = \beta^T \text{VC}(X)\beta + \text{var}(W).$$

The joint variance–covariance matrix of X and Y can be written in partitioned form as

$$\begin{bmatrix} \text{VC}(X) & \text{cov}(X, Y) \\ \text{cov}(Y, X) & \text{var}(Y) \end{bmatrix} = \begin{bmatrix} \text{VC}(X) & \text{VC}(X)\beta \\ \beta^T \text{VC}(X) & \beta^T \text{VC}(X)\beta + \text{var}(W) \end{bmatrix}.$$

Therefore, X and Y have the multivariate normal joint distribution with mean vector

$$\begin{bmatrix} EX \\ \alpha + \beta^T EX \end{bmatrix}$$

and variance–covariance matrix

$$\begin{bmatrix} \text{VC}(X) & \text{VC}(X)\beta \\ \beta^T \text{VC}(X) & \beta^T \text{VC}(X)\beta + \text{var}(W) \end{bmatrix}. \quad \blacksquare$$

Problems

5.85 Let X and Y have a bivariate normal distribution with $EX = 5$, $EY = -2$, $\text{var}(X) = 4$, $\text{var}(Y) = 9$, and $\text{cov}(X, Y) = -3$. Determine the joint distribution of $W = 3X + 4Y$ and $Z = 5X - 6Y$.

5.86 Let X and Y have a bivariate normal joint distribution, let μ_1 and μ_2 denote the respective means of X and Y, let σ_1 and σ_2 denote their respective standard deviations, and let ρ denote the correlation coefficient between these random variables. Show that the joint density function of X and Y is given explicitly by

$$f(x, y) = \frac{1}{2\pi \sigma_1 \sigma_2 \sqrt{1 - \rho^2}} \exp\left(-\frac{1}{2(1 - \rho^2)} Q(x, y) \right),$$

where

$$Q(x, y) = \left(\frac{x - \mu_1}{\sigma_1} \right)^2 - 2\rho \left(\frac{x - \mu_1}{\sigma_1} \right)\left(\frac{y - \mu_2}{\sigma_2} \right) + \left(\frac{y - \mu_2}{\sigma_2} \right)^2.$$

5.87 Let Z_1, Z_2, and Z_3 be independent, standard normal random variables and set

$$Y_1 = 1 + Z_1, \ Y_2 = 2 + Z_1 + Z_2, \ Y_3 = 3 + Z_1 + Z_2 + Z_3, \ \text{and} \ \mathbf{Y} = [\, Y_1, Y_2, Y_3 \,]^T.$$

(a) Determine the mean vector μ and variance–covariance matrix Σ of $\mathbf{Y}$. **(b)** Determine the joint distribution of $\mathbf{Y}$. **(c)** Determine a lower-triangular matrix $\mathbf{L}$ having positive diagonal entries such that $\mathbf{LL}^T$ is a Cholesky decomposition of $\mathbf{\Sigma}$. **(d)** Determine the distribution of $Y_1 + Y_2 + Y_3$ by using the answer to part (b). **(e)** Determine the distribution of $Y_1 + Y_2 + Y_3$ by writing this random variable directly in terms of $Z_1, Z_2,$ and Z_3. **(f)** Determine $P(Y_1 + Y_2 + Y_3 \geq 0)$.

5.88 Verify the "if" part of Theorem 5.31 in the special case $m = n = 1$; that is, show that uncorrelated random variables having a bivariate normal distribution are independent.

5.89 Let $Z_0, Z_1, Z_2,$ and Z_3 be independent, standard normal random variables and let $0 \leq \rho < 1$. Set $Y_i = \sqrt{\rho}Z_0 + \sqrt{1 - \rho}Z_i$ for $1 \leq i \leq 3$. Determine the joint distribution of $Y_1, Y_2,$ and Y_3.

5.90 Show that if $\mathbf{Y}$ has a multivariate normal distribution and $E\mathbf{Y} = 0$, then $-\mathbf{Y}$ has the same distribution as $\mathbf{Y}$.

5.91 Let X and Y have a bivariate normal joint distribution. Show that if $\mathrm{cor}(X, Y) > 0$, then

$$P((X - EX)(Y - EY) > 0) > \tfrac{1}{2} > P((X - EX)(Y - EY) < 0);$$

if $\mathrm{cor}(X, Y) = 0$, then

$$P((X - EX)(Y - EY) > 0) = \tfrac{1}{2} = P((X - EX)(Y - EY) < 0);$$

and if $\mathrm{cor}(X, Y) < 0$, then

$$P((X - EX)(Y - EY) > 0) < \tfrac{1}{2} < P((X - EX)(Y - EY) < 0).$$

These results justify (at least in the context of random variables having a bivariate normal joint distribution) the interpretation of "covariance" given at the beginning of this chapter. [*Hint:* Use the solution to Example 5.32 and the result in Problem 5.90.]

5.92 State and give a direct proof of a simplified version of Theorem 5.32 that is applicable to random variables X and Y having a bivariate normal joint distribution.

5.93 State a simplified version of Corollary 5.12 that is applicable to random variables X and Y having a bivariate normal joint distribution.

5.94 Let X and W be independent, normally distributed random variables with $EW = 0$. Set $Y = \alpha + \beta X + W$, where $\alpha, \beta \in \mathbb{R}$. Determine the joint distribution of X and Y.

5.95 Let Z_1 and Z_2 be independent, standard normal random variables and set $X_1 = Z_1$ and $Y_1 = X_1 + Z_2$. Also, let W_1 and W_2 be independent, normally distributed random variables having mean 0 and variances σ_1^2 and σ_2^2, respectively, and set $Y_2 = W_1$ and $X_2 = \beta Y_2 + W_2$. Determine β, σ_1^2 and σ_2^2 in order that the joint distribution of X_2 and Y_2 coincides with the joint distribution of X_1 and Y_1.

5.96 Let X_1, W_1, and W_2 be independent, normally distributed random variables, each having mean 0 and variance 1. Determine the joint distribution of $X_1, X_2 = X_1 + W_1$, and $Y = X_1 + X_2 + W_2$.

5.97 Let X_2, W_1, and W_2 be independent, normally distributed random variables, each having mean 0 and having variances 2, 1/2, and 1, respectively. Determine the joint distribution of $X_1 = (1/2)X_2 + W_1, X_2$, and $Y = X_1 + X_2 + W_2$.

6

Conditioning

6.1

Conditional Distributions

We start with an example that will be used to motivate the definition of conditional distributions.

EXAMPLE **6.1**   Let $\mathbf{X}$ be an $\mathcal{X}$-valued random variable, let A be a subset of $\mathcal{X}$ such that $P(\mathbf{X} \in A) > 0$, and let $\mathbf{Y}$ be a $\mathcal{Y}$-valued random variable. Also, let $(\mathbf{X}_1, \mathbf{Y}_1)$, $(\mathbf{X}_2, \mathbf{Y}_2)$, ... be independent random ordered pairs, with each ordered pair having the same distribution as $(\mathbf{X}, \mathbf{Y})$, and let N be the first trial such that $\mathbf{X}_i \in A$. Determine the distribution of $\mathbf{Y}_N$—that is, the distribution of $\mathbf{Y}_i$ on the first trial i such that $\mathbf{X}_i \in A$.

Solution   Let n be a positive integer. Observe that $N = n$ if and only if $\mathbf{X}_1 \notin A, \dots, \mathbf{X}_{n-1} \notin A$ and $\mathbf{X}_n \in A$. Let B be a subset of $\mathcal{Y}$. Then $\mathbf{Y}_N \in B$ if and only if, for some positive integer n, $\mathbf{X}_1 \notin A, \ldots, \mathbf{X}_{n-1} \notin A, \mathbf{X}_n \in A$, and $\mathbf{Y}_n \in B$. Thus, by countable additivity and independence,

$$P(\mathbf{Y}_N \in B) = \sum_{n=1}^{\infty} P(N = n, \mathbf{Y}_N \in B)$$

$$= \sum_{n=1}^{\infty} P(\mathbf{X}_1 \notin A, \ldots, \mathbf{X}_{n-1} \notin A, \mathbf{X}_n \in A, \mathbf{Y}_n \in B)$$

$$= \sum_{n=1}^{\infty} P(\mathbf{X}_1 \notin A) \cdots P(\mathbf{X}_{n-1} \notin A) P(\mathbf{X}_n \in A, \mathbf{Y}_n \in B)$$

$$= \sum_{n=1}^{\infty} [P(\mathbf{X} \notin A)]^{n-1} P(\mathbf{X} \in A, \mathbf{Y} \in B).$$

According to the formula for the sum of an infinite geometric series,

$$\sum_{n=1}^{\infty} [P(\mathbf{X} \notin A)]^{n-1} = \frac{1}{1 - P(\mathbf{X} \notin A)} = \frac{1}{P(\mathbf{X} \in A)}.$$

Consequently, the distribution of $\mathbf{Y}_N$ is given by

$$P(\mathbf{Y}_N \in B) = \frac{P(\mathbf{X} \in A, \mathbf{Y} \in B)}{P(\mathbf{X} \in A)}, \qquad B \subset \mathcal{Y}. \quad \blacksquare$$

The distribution of $\mathbf{Y}_N$ in the solution to Example 6.1 is referred to as the *conditional distribution* of $\mathbf{Y}$ given that $\mathbf{X}$ is in A. The probability that $\mathbf{Y}_N$ is in B is referred to as the *conditional probability* that $\mathbf{Y}$ is in B given that $\mathbf{X}$ is in A and denoted by $P(\mathbf{Y} \in B \mid \mathbf{X} \in A)$. Thus, for $A \subset \mathcal{X}$ with $P(\mathbf{X} \in A) > 0$,

$$P(\mathbf{Y} \in B \mid \mathbf{X} \in A) = \frac{P(\mathbf{X} \in A, \mathbf{Y} \in B)}{P(\mathbf{X} \in A)}, \qquad B \subset \mathcal{Y}. \tag{1}$$

It follows from (1) that

$$P(\mathbf{X} \in A, \mathbf{Y} \in B) = P(\mathbf{X} \in A)\, P(\mathbf{Y} \in B \mid \mathbf{X} \in A),$$

$$A \subset \mathcal{X} \text{ with } P(\mathbf{X} \in A) > 0 \text{ and } B \subset \mathcal{Y}. \tag{2}$$

Recall that $\mathbf{X}$ and $\mathbf{Y}$ are independent if and only if

$$P(\mathbf{X} \in A, \mathbf{Y} \in B) = P(\mathbf{X} \in A)P(\mathbf{Y} \in B), \qquad A \subset \mathcal{X} \text{ and } B \subset \mathcal{Y}. \tag{3}$$

This condition will now be rephrased in terms of conditional probabilities.

THEOREM **6.1**   The random variables $\mathbf{X}$ and $\mathbf{Y}$ are independent if and only if

$$P(\mathbf{Y} \in B \mid \mathbf{X} \in A) = P(\mathbf{Y} \in B)$$

$$\text{for } A \subset \mathcal{X} \text{ with } P(\mathbf{X} \in A) > 0 \text{ and } B \subset \mathcal{Y}. \tag{4}$$

Proof   Suppose first that $\mathbf{X}$ and $\mathbf{Y}$ are independent and let $A \subset \mathcal{X}$ with $P(A) > 0$ and $B \subset \mathcal{Y}$. Then, by (1) and (3),

$$P(\mathbf{Y} \in B \mid \mathbf{X} \in A) = \frac{P(\mathbf{X} \in A, \mathbf{Y} \in B)}{P(\mathbf{X} \in A)} = \frac{P(\mathbf{X} \in A)P(\mathbf{Y} \in B)}{P(\mathbf{X} \in A)} = P(\mathbf{Y} \in B).$$

Therefore (4) holds.

Suppose, conversely, that (4) holds and let $A \subset \mathcal{X}$ and $B \subset \mathcal{Y}$. If $P(\mathbf{X} \in A) > 0$, then, by (2) and (4),

$$P(\mathbf{X} \in A, \mathbf{Y} \in B) = P(\mathbf{X} \in A)P(\mathbf{Y} \in B \mid \mathbf{X} \in A) = P(\mathbf{X} \in A)P(\mathbf{Y} \in B);$$

if $P(\mathbf{X} \in A) = 0$, then $P(\mathbf{X} \in A, \mathbf{Y} \in B) = 0$ and hence the equation

$$P(\mathbf{X} \in A, \mathbf{Y} \in B) = P(\mathbf{X} \in A)P(\mathbf{Y} \in B)$$

again holds. Therefore, this equation holds both when $P(\mathbf{X} \in A) > 0$ and when $P(\mathbf{X} \in A) = 0$. Therefore, (3) holds.   ∎

The definitions of conditional probabilities and conditional distributions apply when $\mathbf{X}$ coincides with $\mathbf{Y}$. Since

$$P(\mathbf{Y} \in A, \mathbf{Y} \in B) = P(\mathbf{Y} \in A \cap B),$$

we see that

$$P(\mathbf{Y} \in B \mid \mathbf{Y} \in A) = \frac{P(\mathbf{Y} \in A \cap B)}{P(\mathbf{Y} \in A)}, \qquad B \subset \mathcal{Y}. \tag{5}$$

In particular,

$$P(\mathbf{Y} \in A \mid \mathbf{Y} \in A) = 1. \tag{6}$$

Thus the conditional distribution of $\mathbf{Y}$ given that $\mathbf{Y} \in A$ can be regarded as a distribution on A. (Alternatively, in the context of Example 6.1 with $\mathbf{X}_i = \mathbf{Y}_i$ for $i \geq 1$, since N is the first trial such that $\mathbf{Y}_i \in A$ it is obvious that $\mathbf{Y}_N \in A$). It follows from (5) that

$$P(\mathbf{Y} \in B \mid \mathbf{Y} \in A) = \frac{P(\mathbf{Y} \in B)}{P(\mathbf{Y} \in A)}, \qquad B \subset A. \tag{7}$$

EXAMPLE **6.2**   A fair coin is tossed three times in an honest manner. Given that heads shows at least once, what is the probability that it shows all three times?

Solution   Let Y denote the total number of heads on the three tosses, which has the binomial distribution with parameters $n = 3$ and $\pi = 1/2$. By (7),

$$P(Y = 3 \mid Y \geq 1) = \frac{P(Y = 3)}{P(Y \geq 1)} = \frac{P(Y = 3)}{1 - P(Y = 0)} = \frac{1/8}{1 - 1/8} = \frac{1}{7}.   ∎$$

EXAMPLE **6.3**   Let $\mathbf{Y}$ be uniformly distributed on $\mathcal{Y}$ and let A be a subset of $\mathcal{Y}$ having positive size. Determine the conditional distribution of $\mathbf{Y}$ given that $\mathbf{Y} \in A$.

Solution   The distribution of **Y** is given by

$$P(\mathbf{Y} \in B) = \frac{\text{size}(B)}{\text{size}(\mathcal{Y})}, \qquad B \subset \mathcal{Y}.$$

Choose $B \subset A$. Then, by (7),

$$P(\mathbf{Y} \in B \mid \mathbf{Y} \in A) = \frac{P(\mathbf{Y} \in B)}{P(\mathbf{Y} \in A)} = \frac{\text{size}(B)/\text{size}(\mathcal{Y})}{\text{size}(A)/\text{size}(\mathcal{Y})} = \frac{\text{size}(B)}{\text{size}(A)}.$$

Thus, the conditional distribution of **Y** given that $\mathbf{Y} \in A$ is the uniform distribution on A.   ∎

EXAMPLE 6.4   Let T have the geometric distribution with parameter π and let t be a nonnegative integer. Show that the conditional distribution of $T - t$ given that $T \geq t$ coincides with the distribution of T.

Solution   Recall from (32) and (33) in Section 1.5 that $P(T = t) = (1 - \pi)\pi^t$ and $P(T \geq t) = \pi^t$. Let s be a nonnegative integer. Observe that if $T - t = s$, then $T = t + s \geq t$. Thus, by (1),

$$
\begin{aligned}
P(T - t = s \mid T \geq t) &= \frac{P(T - t = s, T \geq t)}{P(T \geq t)} \\[2mm]
&= \frac{P(T - t = s)}{P(T \geq t)} \\[2mm]
&= \frac{P(T = t + s)}{P(T \geq t)} \\[2mm]
&= \frac{(1 - \pi)\pi^{t+s}}{\pi^t} \\[2mm]
&= (1 - \pi)\pi^s.
\end{aligned}
$$

Consequently, the conditional distribution of $T - t$ given that $T \geq t$ is geometric with parameter π, which coincides with the distribution of T.   ∎

The result in Example 6.4 is referred to as the *memoryless property* of the geometric distribution. (It can be shown by elementary mathematics that no other distribution on the set of nonnegative integers has this property.)

EXAMPLE 6.5   Let T have the exponential distribution with inverse-scale parameter λ and let $t \geq 0$. Show that the conditional distribution of $T - t$ given that $T \geq t$ coincides with the distribution of T.

Solution   Choose $s \geq 0$. Then, by Equation (1),

$$P(T - t \leq s \mid T \geq t) = \frac{P(T - t \leq s, T \geq t)}{P(T \geq t)}$$

$$= \frac{P(t \leq T \leq t+s)}{P(T \geq t)}$$

$$= \frac{e^{-\lambda t} - e^{-\lambda(t+s)}}{e^{-\lambda t}}$$

$$= 1 - e^{-\lambda s}.$$

Thus, the distribution function of the conditional distribution of $T - t$ given that $T \geq t$ coincides with the distribution function of the exponential distribution with inverse-scale parameter λ. Since a distribution on $\mathbb{R}$ is uniquely determined by its distribution function, we conclude that the conditional distribution of $T - t$ given that $T \geq t$ is exponential with inverse-scale parameter λ, which coincides with the distribution of T. ∎

The result in Example 6.5 is referred to as the memoryless property of the exponential distribution. [No other distribution on $(0, \infty)$ has this property.]

Discrete Random Variables

Suppose now that $\mathbf{X}$ and $\mathbf{Y}$ are discrete random variables and, for simplicity, that $P(\mathbf{X} = \mathbf{x}) > 0$ for $\mathbf{x} \in \mathcal{X}$. Then

$$P(\mathbf{Y} = \mathbf{y} \mid \mathbf{X} = \mathbf{x}) = \frac{P(\mathbf{X} = \mathbf{x}, \mathbf{Y} = \mathbf{y})}{P(\mathbf{X} = \mathbf{x})}, \qquad \mathbf{x} \in \mathcal{X} \text{ and } \mathbf{y} \in \mathcal{Y}, \tag{8}$$

from which we see that

$$P(\mathbf{X} = \mathbf{x}, \mathbf{Y} = \mathbf{y}) = P(\mathbf{X} = \mathbf{x})P(\mathbf{Y} = \mathbf{y} \mid \mathbf{X} = \mathbf{x}), \qquad \mathbf{x} \in \mathcal{X} \text{ and } \mathbf{y} \in \mathcal{Y}. \tag{9}$$

Since $P(\mathbf{Y} = \mathbf{y}) = \sum_{\mathbf{x}} P(\mathbf{X} = \mathbf{x}, \mathbf{Y} = \mathbf{y})$, we conclude from (9) that

$$P(\mathbf{Y} = \mathbf{y}) = \sum_{\mathbf{x}} P(\mathbf{X} = \mathbf{x})P(\mathbf{Y} = \mathbf{y} \mid \mathbf{X} = \mathbf{x}), \qquad \mathbf{y} \in \mathcal{Y}. \tag{10}$$

The random variables $\mathbf{X}$ and $\mathbf{Y}$ are independent if and only if

$$P(\mathbf{X} = \mathbf{x}, \mathbf{Y} = \mathbf{y}) = P(\mathbf{X} = \mathbf{x})P(\mathbf{Y} = \mathbf{y}), \qquad \mathbf{x} \in \mathcal{X} \text{ and } \mathbf{y} \in \mathcal{Y}. \tag{11}$$

It follows from (9) and (11) that $\mathbf{X}$ and $\mathbf{Y}$ are independent if and only if

$$P(\mathbf{Y} = \mathbf{y} \mid \mathbf{X} = \mathbf{x}) = P(\mathbf{Y} = \mathbf{y}), \qquad \mathbf{x} \in \mathcal{X} \text{ and } \mathbf{y} \in \mathcal{Y}. \tag{12}$$

EXAMPLE **6.6**   Let two fair dice, labeled die 1 and die 2, be rolled independently and in an honest manner. Determine the conditional distribution of the number of spots showing on die 1, given that the total number of spots showing on the two dice is five.

Solution   Let Y_1 and Y_2 denote the number of spots showing on die 1 and die 2, respectively. Then Y_1 and Y_2 are independent random variables, each of which is uniformly distributed on $\{1, \ldots, 6\}$. Thus,

$$P(Y_1 + Y_2 = 5) = \sum_{y_1=1}^{4} P(Y_1 = y_1, Y_2 = 5 - y_1)$$

$$= \sum_{y_1=1}^{4} P(Y_1 = y_1)P(Y_2 = 5 - y_1)$$

$$= \sum_{y_1=1}^{4} \left(\frac{1}{6}\right)\left(\frac{1}{6}\right)$$

$$= \frac{4}{36}$$

$$= \frac{1}{9},$$

as we already know from the solution to Example 1.24 in Section 1.5. Let $y_1 \in \{1, 2, 3, 4\}$. Then

$$P(Y_1 = y_1 \mid Y_1 + Y_2 = 5) = \frac{P(Y_1 = y_1, Y_1 + Y_2 = 5)}{P(Y_1 + Y_2 = 5)}$$

$$= \frac{P(Y_1 = y_1, Y_2 = 5 - y_1)}{P(Y_1 + Y_2 = 5)}$$

$$= \frac{P(Y_1 = y_1)P(Y_2 = 5 - y_1)}{P(Y_1 + Y_2 = 5)}$$

$$= \frac{\left(\frac{1}{6}\right)\left(\frac{1}{6}\right)}{\frac{1}{9}}$$

$$= \frac{1}{4}.$$

Thus, the conditional distribution of Y_1 given that $Y_1 + Y_2 = 5$ is uniform on $\{1, 2, 3, 4\}$. This is just what we would expect, as Figure 6.1 shows. ■

FIGURE 6.1

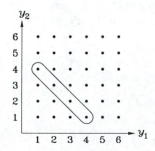

EXAMPLE 6.7   Let Y_1, Y_2, Y_3 have the trinomial joint distribution with parameters n and π_1, π_2, π_3 and let y_1 be a nonnegative integer with $y_1 < n$. Determine the conditional distribution of Y_2 given that $Y_1 = y_1$.

Solution   Now Y_1 has the binomial distribution with parameters n and π_1, so

$$P(Y_1 = y_1) = \binom{n}{y_1}\pi_1^{y_1}(1 - \pi_1)^{n-y_1} = \frac{n!}{y_1!\,(n - y_1)!}\pi_1^{y_1}(1 - \pi_1)^{n-y_1}.$$

Let y_2 be a nonnegative integer such that $y_1 + y_2 \le n$ and set $y_3 = n - y_1 - y_2$. Since $Y_1 + Y_2 + Y_3 = n$, we see that if $Y_1 = y_1$ and $Y_2 = y_2$, then $Y_3 = n - y_1 - y_2$. Thus,

$$P(Y_1 = y_1, Y_2 = y_2) = P(Y_1 = y_1, Y_2 = y_2, Y_3 = n - y_1 - y_2)$$

$$= \frac{n!}{y_1!\,y_2!\,(n - y_1 - y_2)!}\pi_1^{y_1}\pi_2^{y_2}(1 - \pi_1 - \pi_2)^{n-y_1-y_2}.$$

Consequently,

$$P(Y_2 = y_2 \mid Y_1 = y_1) = \frac{P(Y_1 = y_1, Y_2 = y_2)}{P(Y_1 = y_1)}$$

$$= \frac{\dfrac{n!}{y_1!\,y_2!\,(n - y_1 - y_2)!}\pi_1^{y_1}\pi_2^{y_2}(1 - \pi_1 - \pi_2)^{n-y_1-y_2}}{\dfrac{n!}{y_1!\,(n - y_1)!}\pi_1^{y_1}(1 - \pi_1)^{n-y_1}}$$

$$= \frac{(n - y_1)!}{y_2!\,(n - y_1 - y_2)!}\left(\frac{\pi_2}{1 - \pi_1}\right)^{y_2}\left(1 - \frac{\pi_2}{1 - \pi_1}\right)^{n-y_1-y_2}$$

$$= \binom{n - y_1}{y_2}\left(\frac{\pi_2}{1 - \pi_1}\right)^{y_2}\left(1 - \frac{\pi_2}{1 - \pi_1}\right)^{n-y_1-y_2}.$$

Thus, the conditional distribution of Y_2 given that $Y_1 = y_1$ is binomial with parameters $n - y_1$ and $\pi_2/(1 - \pi_1)$. [This result is intuitively plausible: Given that there are y_1 outcomes of type 1, the remaining $n - y_1$ outcomes must be of type 2 or type 3; the conditional probability that an outcome is of type 2 given that it is not of type 1 is $\pi_2/(1 - \pi_1)$.] ∎

EXAMPLE 6.8   Let $Y_1, \ldots, Y_5$ have the multinomial joint distribution with parameters n and $\pi_1, \ldots, \pi_5$ and let y_1 and y_2 be nonnegative integers whose sum is at most n. Determine the conditional joint distribution of Y_3, Y_4, and Y_5 given that $Y_1 = y_1$ and $Y_2 = y_2$.

Solution   The random variables Y_1, Y_2, and $Y_3 + Y_4 + Y_5$ have the trinomial joint distribution with parameters n and $\pi_1, \pi_2, \pi_3 + \pi_4 + \pi_5$. Thus,

$$P(Y_1 = y_1, Y_2 = y_2) = P(Y_1 = y_1, Y_2 = y_2, Y_3 + Y_4 + Y_5 = n - y_1 - y_2)$$

$$= \frac{n!}{y_1!\,y_2!\,(n - y_1 - y_2)!}\pi_1^{y_1}\pi_2^{y_2}(\pi_3 + \pi_4 + \pi_5)^{n-y_1-y_2}.$$

Let y_3, y_4, and y_5 be nonnegative integers whose sum equals $n - y_1 - y_2$. Then

$$P(Y_3 = y_3, Y_4 = y_4, Y_5 = y_5 \mid Y_1 = y_1, Y_2 = y_2)$$

$$= \frac{P(Y_1 = y_1, \ldots, Y_5 = y_5)}{P(Y_1 = y_1, Y_2 = y_2)}$$

$$= \frac{\dfrac{n!}{y_1! y_2! y_3! y_4! y_5!} \pi_1^{y_1} \pi_2^{y_2} \pi_3^{y_3} \pi_4^{y_4} \pi_5^{y_5}}{\dfrac{n!}{y_1! y_2! (n - y_1 - y_2)!} \pi_1^{y_1} \pi_2^{y_2} (\pi_3 + \pi_4 + \pi_5)^{y_3 + y_4 + y_5}}$$

$$= \frac{(n - y_1 - y_2)!}{y_3! y_4! y_5!} \left(\frac{\pi_3}{\pi_3 + \pi_4 + \pi_5} \right)^{y_3} \left(\frac{\pi_4}{\pi_3 + \pi_4 + \pi_5} \right)^{y_4} \left(\frac{\pi_5}{\pi_3 + \pi_4 + \pi_5} \right)^{y_5}.$$

Thus, the conditional joint distribution of Y_3, Y_4, and Y_5 given that $Y_1 = y_1$ and $Y_2 = y_2$ is trinomial with parameters $n - y_1 - y_2$ and

$$\frac{\pi_3}{\pi_3 + \pi_4 + \pi_5}, \qquad \frac{\pi_4}{\pi_3 + \pi_4 + \pi_5}, \qquad \frac{\pi_5}{\pi_3 + \pi_4 + \pi_5}. \qquad \blacksquare$$

We might have been able to guess the result in the solution to Example 6.8. Think of $\pi_1, \ldots, \pi_5$ as the probabilities that the random outcome of an experiment will be of types $1, \ldots, 5$, respectively, and of $Y_1, \ldots, Y_5$ as the numbers of outcomes of types $1, \ldots, 5$, respectively, in n trials of the experiment. Given that $Y_1 = y_1$ and $Y_2 = y_2$, the remaining $n - y_1 - y_2$ outcomes must be distributed over types 3, 4, and 5. The respective conditional probabilities of types 3, 4, and 5 on a single trial, given that the outcome is one of these three types, are

$$\frac{\pi_3}{\pi_3 + \pi_4 + \pi_5}, \qquad \frac{\pi_4}{\pi_3 + \pi_4 + \pi_5}, \qquad \frac{\pi_5}{\pi_3 + \pi_4 + \pi_5}.$$

EXAMPLE 6.9 Let Y_1 and Y_2 be independent random variables having Poisson distributions with means λ_1 and λ_2, respectively. Set $Y = Y_1 + Y_2$ and let n be a positive integer. Determine the conditional distribution of Y_1 given that $Y = n$.

Solution Recall from Theorem 4.17 in Section 4.4 that Y has the Poisson distribution with mean $\lambda = \lambda_1 + \lambda_2$. Let y_1 be a nonnegative integer. Then

$$P(Y_1 = y_1 \mid Y = n) = \frac{P(Y_1 = y_1, Y = n)}{P(Y = n)}$$

$$= \frac{P(Y_1 = y_1, Y_1 + Y_2 = n)}{P(Y = n)}$$

$$= \frac{P(Y_1 = y_1, Y_2 = n - y_1)}{P(Y = n)}$$

$$= \frac{P(Y_1 = y_1) P(Y_2 = n - y_1)}{P(Y = n)}$$

$$= \frac{\dfrac{\lambda_1^{y_1}}{y_1!}e^{-\lambda_1}\dfrac{\lambda_2^{n-y_1}}{(n-y_1)!}e^{-\lambda_2}}{\dfrac{\lambda^n}{n!}e^{-\lambda}}$$

$$= \binom{n}{y_1}\left(\frac{\lambda_1}{\lambda}\right)^{y_1}\left(\frac{\lambda_2}{\lambda}\right)^{n-y_1}.$$

Thus, the conditional distribution of Y_1 given that $Y = n$ is binomial with parameters n and λ_1/λ.   ■

Bayes' Theorem

It follows from Equation (9) with the roles of **X** and **Y** reversed that

$$P(\mathbf{X} = \mathbf{x}, \mathbf{Y} = \mathbf{y}) = P(\mathbf{Y} = \mathbf{y})P(\mathbf{X} = \mathbf{x} \,|\, \mathbf{Y} = \mathbf{y}).$$

Thus, by another application of (9),

$$P(\mathbf{X} = \mathbf{x} \,|\, \mathbf{Y} = \mathbf{y}) = \frac{P(\mathbf{X} = \mathbf{x})P(\mathbf{Y} = \mathbf{y} \,|\, \mathbf{X} = \mathbf{x})}{P(\mathbf{Y} = \mathbf{y})}, \qquad \mathbf{x} \in \mathcal{X} \text{ and } \mathbf{y} \in \mathcal{Y}.$$

This result is known as *Bayes' theorem*.

EXAMPLE 6.10   There are three boxes: Box 1 contains one red ball and three white balls; box 2 contains two red balls and two white balls; box 3 contains three red balls and one white ball. A box is selected at random, and then a ball is chosen at random from the selected box. Determine the conditional probability that box 1 was selected, given that a red ball is chosen.

Solution   Let X denote the label of the selected box and let Y denote the indicator random variable for choosing a red ball. Then $P(X = 1) = P(X = 2) = P(X = 3) = 1/3$. Also, $P(Y = 1 \,|\, X = 1) = 1/4$, $P(Y = 1 \,|\, X = 2) = 2/4$, and $P(Y = 1 \,|\, X = 3) = 3/4$. Thus, the probability of choosing a red ball is given by

$$P(Y = 1) = P(X = 1)P(Y = 1 \,|\, X = 1) + P(X = 2)P(Y = 1 \,|\, X = 2)$$

$$+ P(X = 3)P(Y = 1 \,|\, X = 3)$$

$$= \frac{1}{3} \cdot \frac{1}{4} + \frac{1}{3} \cdot \frac{2}{4} + \frac{1}{3} \cdot \frac{3}{4}$$

$$= \frac{1}{2}.$$

Therefore, by Bayes' theorem, the conditional probability that box 1 was selected, given that a red ball is chosen, is given by

$$P(X = 1 \,|\, Y = 1) = \frac{P(X = 1)P(Y = 1 \,|\, X = 1)}{P(Y = 1)} = \frac{(1/3)(1/4)}{(1/2)} = \frac{1}{6}.   \blacksquare$$

EXAMPLE **6.11**   There are two boxes, each containing one red ball and one white ball. At step 1, a ball is selected at random from the first box and transferred to the second box. At step 2, a ball is selected at random from the second box and transferred back to the first box. Determine the conditional probability that the ball selected at step 1 was red, given that the ball selected at step 2 was red.

Solution   Let X denote the indicator random variable for selecting a red ball at step 1, and let Y denote the indicator random variable for selecting a red ball at step 2. Suppose that at step 1 a white ball is transferred to the second box. Then that box will have one red ball and two white balls. Thus, the conditional probability of selecting a red ball from the second box at step 2, given that a white ball was transferred to this box at step 1, equals one-third. Similarly, the conditional probability of selecting a red ball from the second box at step 2, given that red ball was transferred to this box at step 1, equals two-thirds. In short,

$$P(Y = 1 \mid X = 0) = \frac{1}{3} \quad \text{and} \quad P(Y = 1 \mid X = 1) = \frac{2}{3}.$$

Hence, the probability that a red ball is selected at step 2 is given by

$$P(Y = 1) = P(X = 0)P(Y = 1 \mid X = 0) + P(X = 1)P(Y = 1 \mid X = 1)$$
$$= \frac{1}{2} \cdot \frac{1}{3} + \frac{1}{2} \cdot \frac{2}{3}$$
$$= \frac{1}{2}.$$

Consequently, by Bayes' theorem, the probability that a red ball was selected at step 1, given that a red ball is selected at step 2, is given by

$$P(X = 1 \mid Y = 1) = \frac{P(X = 1)P(Y = 1 \mid X = 1)}{P(Y = 1)} = \frac{(1/2)(2/3)}{1/2} = \frac{2}{3}. \quad \blacksquare$$

The joint probability function of $\mathbf{X}$ and $\mathbf{Y}$ is given by

$$f_{\mathbf{X},\mathbf{Y}}(\mathbf{x}, \mathbf{y}) = P(\mathbf{X} = \mathbf{x}, \mathbf{Y} = \mathbf{y}), \qquad \mathbf{x} \in \mathcal{X} \text{ and } \mathbf{y} \in \mathcal{Y}.$$

The probability functions of $\mathbf{X}$ and $\mathbf{Y}$ are given in terms of their joint probability function by

$$f_{\mathbf{X}}(\mathbf{x}) = \sum_{\mathbf{y}} f_{\mathbf{X},\mathbf{Y}}(\mathbf{x}, \mathbf{y}), \qquad \mathbf{x} \in \mathcal{X},$$

and

$$f_{\mathbf{Y}}(\mathbf{y}) = \sum_{\mathbf{x}} f_{\mathbf{X},\mathbf{Y}}(\mathbf{x}, \mathbf{y}), \qquad \mathbf{y} \in \mathcal{Y}.$$

The *conditional probability function* of $\mathbf{Y}$, given that $\mathbf{X} = \mathbf{x}$, is defined by

$$f_{\mathbf{Y} \mid \mathbf{X}}(\mathbf{y} \mid \mathbf{x}) = \frac{f_{\mathbf{X},\mathbf{Y}}(\mathbf{x}, \mathbf{y})}{f_{\mathbf{X}}(\mathbf{x})} = \frac{P(\mathbf{X} = \mathbf{x}, \mathbf{Y} = \mathbf{y})}{P(\mathbf{X} = \mathbf{x})} = P(\mathbf{Y} = \mathbf{y} \mid \mathbf{X} = \mathbf{x}) \qquad \textbf{(13)}$$

for $\mathbf{y} \in \mathcal{Y}$. The conditional probability function possesses the properties of a probability function. Moreover, the conditional distribution of $\mathbf{Y}$ given that $\mathbf{X} = \mathbf{x}$ is given in terms of its conditional probability function by

$$P(\mathbf{Y} \in B \mid \mathbf{X} = \mathbf{x}) = \sum_{\mathbf{y} \in B} f_{\mathbf{Y}|\mathbf{X}}(\mathbf{y} \mid \mathbf{x}), \qquad B \subset \mathcal{Y}. \tag{14}$$

[The proof of (14) is left as a problem.] It follows from (13) that

$$f_{\mathbf{X},\mathbf{Y}}(\mathbf{x}, \mathbf{y}) = f_{\mathbf{X}}(\mathbf{x}) f_{\mathbf{Y}|\mathbf{X}}(\mathbf{y} \mid \mathbf{x}), \qquad \mathbf{x} \in \mathcal{X} \text{ and } \mathbf{y} \in \mathcal{Y}. \tag{15}$$

Recall that $\mathbf{X}$ and $\mathbf{Y}$ are independent if and only if

$$f_{\mathbf{X},\mathbf{Y}}(\mathbf{x}, \mathbf{y}) = f_{\mathbf{X}}(\mathbf{x}) f_{\mathbf{Y}}(\mathbf{y}), \qquad \mathbf{x} \in \mathcal{X} \text{ and } \mathbf{y} \in \mathcal{Y}.$$

We conclude from (15) that $\mathbf{X}$ and $\mathbf{Y}$ are independent if and only if

$$f_{\mathbf{Y}|\mathbf{X}}(\mathbf{y} \mid \mathbf{x}) = f_{\mathbf{Y}}(\mathbf{y}), \qquad \mathbf{x} \in \mathcal{X} \text{ and } \mathbf{y} \in \mathcal{Y}.$$

Bayes' theorem can be written as

$$f_{\mathbf{X}|\mathbf{Y}}(\mathbf{x} \mid \mathbf{y}) = \frac{f_{\mathbf{X}}(\mathbf{x}) f_{\mathbf{Y}|\mathbf{X}}(\mathbf{y} \mid \mathbf{x})}{f_{\mathbf{Y}}(\mathbf{y})}, \qquad \mathbf{x} \in \mathcal{X} \text{ and } \mathbf{y} \in \mathcal{Y}.$$

Problems

6.1 A box has six balls, of which three are red, two are white, and one is blue. Four times, a ball is selected at random from the box and then returned to the box. Let X be the number of times a red or ball white ball is selected and let Y be the number of times a white or blue ball is selected. Determine $P(Y = 3 \mid X = 3)$.

6.2 Let two fair dice, referred to as die 1 and die 2, be rolled in an honest manner. Determine the conditional distribution of the number of spots showing on die 1, given that the total number of spots showing on the two dice is eight.

6.3 Let X and Y have a bivariate normal joint distribution with $\mathrm{cor}(X, Y) = \frac{1}{2}$. Determine $P(Y \geq EY \mid X \geq EX)$. [*Hint:* Use the solution to Example 5.32 in Section 5.7.]

6.4 Let X and Y have a bivariate normal joint distribution. Show that **(a)** $P(Y \geq EY \mid X \geq EX) > \frac{1}{2}$ if $\mathrm{cor}(X, Y) > 0$; **(b)** $P(Y \geq EY \mid X \geq EX) = \frac{1}{2}$ if $\mathrm{cor}(X, Y) = 0$; **(c)** $P(Y \geq EY \mid X \geq EX) < \frac{1}{2}$ if $\mathrm{cor}(X, Y) < 0$. [*Hint:* Use the solution to Example 5.32 in Section 5.7.]

6.5 Let Y_1, Y_2, and Y_3 be independent random variables having Poisson distributions with means λ_1, λ_2, and λ_3, respectively. Set $\lambda = \lambda_1 + \lambda_2 + \lambda_3$ and $Y = Y_1 + Y_2 + Y_3$ and let n be a positive integer. Show the conditional distribution of Y_1, Y_2, and Y_3, given that $Y = n$ is trinomial with parameters n and $\lambda_1/\lambda, \lambda_2/\lambda, \lambda_3/\lambda$.

6.6 Let N and Y be nonnegative random variables such that (i) N has the Poisson distribution with mean λ; (ii) if $N = 0$, then $Y = 0$; (iii) the conditional distribution of Y, given that $N = n \geq 1$, is binomial with parameters n and π. Show that **(a)** $P(N - Y = m, Y = y) = \frac{[\lambda(1-\pi)]^m}{m!} e^{-\lambda(1-\pi)} \frac{(\lambda\pi)^y}{y!} e^{-\lambda\pi}$ for y and m nonnegative integers; **(b)** Y has the Poisson distribution with mean $\lambda\pi$; **(c)** $N - Y$ has the Poisson distribution with mean $\lambda(1 - \pi)$; **(d)** $N - Y$ and Y are independent.

6.7 Consider events occurring in time, starting at time zero, according to a Poisson process with rate λ, and let $t > 0$. Determine the conditional probability that one event occurs during each of the time periods $(0, t/3]$, $(t/3, 2t/3]$, and $(2t/3, t]$, given that three events occur during the time period $(0, t]$.

6.8 A chest has three drawers. The first drawer contains two gold coins, the second drawer contains one gold coin and one silver coin, and the third drawer contains two silver coins. A drawer is chosen at random, and then a coin is selected at random from the chosen drawer. Determine **(a)** the probability that a gold coin is selected; **(b)** the conditional probability that the other coin in the chosen drawer is gold, given that the selected coin is gold.

6.9 There are three boxes: Box 1 contains one red ball and one white ball; box 2 contains two red balls and one white ball; box 3 contains three red balls and one white ball. A box is selected at random, and then a ball is chosen at random from the selected box. Determine **(a)** the probability a red ball is chosen; **(b)** the conditional probability that box 1 selected, given that a red ball is chosen.

6.10 There are two boxes: One box has three red balls and two white balls, and the other box has five red balls and four white balls. A ball is chosen at random from each box. Determine the conditional probability that both balls are red, given that they have the same color.

6.11 Two fair dice are rolled in an honest manner. Let X and Y denote, respectively, the maximum and minimum of the numbers of spots showing on the two dice. Determine $P(Y = y \mid X = x)$ for $1 \le y \le x \le 6$.

6.12 Verify Equation (14).

6.13 Consider an $\mathcal{X}$-valued random variable $\mathbf{X}$, a $\mathcal{Y}$-valued random variable $\mathbf{Y}$, and a $\mathcal{Z}$-valued random variable $\mathbf{Z}$. Let A, B, and C be subsets of $\mathcal{X}$, $\mathcal{Y}$, and $\mathcal{Z}$, respectively. Show that if $P(\mathbf{X} \in A, \mathbf{Y} \in B) > 0$, then

$$P(\mathbf{X} \in A, \mathbf{Y} \in B, \mathbf{Z} \in C) = P(\mathbf{X} \in A) P(\mathbf{Y} \in B \mid \mathbf{X} \in A) P(\mathbf{Z} \in C \mid \mathbf{X} \in A, \mathbf{Y} \in B).$$

6.14 (Continued.) Suppose $P(\mathbf{X} \in A, \mathbf{Y} \in B) > 0$ and $P(\mathbf{Y} \in B, \mathbf{Z} \in C) > 0$. Show that the following two statements are equivalent: **(a)** $P(\mathbf{Z} \in C \mid \mathbf{X} \in A, \mathbf{Y} \in B) = P(\mathbf{Z} \in C \mid \mathbf{Y} \in B)$; **(b)** $P(\mathbf{X} \in A \mid \mathbf{Y} \in B, \mathbf{Z} \in C) = P(\mathbf{X} \in A \mid \mathbf{Y} \in B)$.

6.15 Let T be a positive random variable having a density function that is continuous and positive on $(0, \infty)$ and equals zero elsewhere and let $\lambda(t)$, $t > 0$ be the hazard function of T as defined in Problem 1.85 in Section 1.6. Show that

$$\lim_{s \downarrow 0} \frac{1}{s} P(t \le T \le t + s \mid T \ge t) = \lambda(t), \qquad t > 0.$$

6.2
Sampling Without Replacement

In this section and the next, we will use conditioning to study sampling without replacement. To motivate the general approach, we first consider a box having six balls that are labeled from 1 to 6 and let a ball be selected at random from the box. Then its label is uniformly distributed on $\{1, \ldots, 6\}$. Suppose ball 4 is selected and removed from the box and let a second ball be chosen at random from the remaining five balls in the box. Then its label is uniformly distributed on $\{1, 2, 3, 5, 6\}$. Suppose ball 6 is chosen on the second trial and removed from the box and let a third ball be chosen at random from the four balls now remaining in the box. Then its label is uniformly distributed on $\{1, 2, 3, 5\}$. Similarly, on each of the later trials, the label of the ball selected on that trial is uniformly distributed on the set of labels of the balls that have not been removed from the box during the previous trials.

In general, consider a population of $N \geq 2$ objects, let these objects be labeled from 1 to N in some order, and let $\mathcal{L} = \{1, \ldots, N\}$ denote the set of labels. Then the label L_1 of an object selected at random from the population is uniformly distributed on $\mathcal{L}$.

Let a second object be selected at random from the remaining $N - 1$ objects—that is, without replacing the object selected on the first trial. Then, given the object selected on the first trial, the label of the object selected on the second trial is uniformly distributed over the set of labels of the remaining $N - 1$ objects. In other words, the conditional distribution of the label L_2 of the object selected on the second trial, given that $L_1 = l_1 \in \mathcal{L}$, is the uniform distribution on the set $\mathcal{L} \backslash \{l_1\} = \{l : l \in \mathcal{L} \text{ and } l \neq l_1\}$, which has size $N - 1$.

Let l_1 and l_2 be distinct members of $\mathcal{L}$. Then

$$P(L_2 = l_2 \mid L_1 = l_1) = \frac{1}{N - 1}.$$

Thus,

$$P(L_1 = l_1, L_2 = l_2) = P(L_1 = l_1)P(L_2 = l_2 \mid L_1 = l_1) = \frac{1}{N}\frac{1}{N-1} = \frac{1}{N(N-1)}.$$

Consequently, the joint distribution of L_1 and L_2 is the uniform distribution on the collection $\mathcal{L}^{(2)}$ of ordered pairs of distinct members of $\mathcal{L}$. To obtain the distribution of L_2, choose $l_2 \in \mathcal{L}$. Then

$$P(L_2 = l_2) = \sum_{l_1} P(L_1 = l_1, L_2 = l_2) = \sum_{l_1 \neq l_2} \frac{1}{N(N-1)} = \frac{1}{N}.$$

Therefore, L_2 has the uniform distribution on $\mathcal{L}$. Note the important distinction: The (unconditional) distribution of the label of the object selected on the second trial is the uniform distribution on the set of all N labels, but the conditional distribution of the label of the object selected on the second trial, given the object selected on the first trial, is the uniform distribution on the set of labels of the remaining $N - 1$ objects.

Suppose now that $N \geq 3$. Let a third object be selected at random from the remaining $N - 2$ objects—that is, without replacing the objects selected on the first two trials. Then, given the objects selected on the first two trials, the label L_3 of the object selected on the third trial is uniformly distributed over the set of labels of the remaining $N - 2$ objects. Thus, let l_1 and l_2 be distinct members of $\mathcal{L}$. Then the conditional distribution of L_3, given that $L_1 = l_1$ and $L_2 = l_2$, is the uniform distribution on the set $\mathcal{L} \backslash \{l_1, l_2\}$, which has size $N - 2$.

Let l_1, l_2, and l_3 be distinct members of $\mathcal{L}$. Then

$$P(L_3 = l_3 \mid L_1 = l_1 \text{ and } L_2 = l_2) = \frac{1}{N - 2},$$

so

$$P(L_1 = l_1, L_2 = l_2, L_3 = l_3)$$

$$= P(L_1 = l_1, L_2 = l_2) \, P(L_3 = l_3 \mid L_1 = l_1 \text{ and } L_2 = l_2)$$

$$= \frac{1}{N(N-1)} \frac{1}{N-2}$$

$$= \frac{1}{N(N-1)(N-2)}.$$

Consequently, the joint distribution of L_1, L_2, and L_3 is the uniform distribution on the collection $\mathcal{L}^{(3)}$ of ordered triples of distinct members of $\mathcal{L}$.

To obtain the joint distribution of L_2 and L_3, let l_2 and l_3 be distinct members of $\mathcal{L}$. Then

$$P(L_2 = l_2 \text{ and } L_3 = l_3) = \sum_{l_1} P(L_1 = l_1, L_2 = l_2, L_3 = l_3)$$

$$= \sum_{l_1 \notin \{l_2, l_3\}} \frac{1}{N(N-1)(N-2)}$$

$$= \frac{1}{N(N-1)}.$$

Thus, the joint distribution of L_2 and L_3 is the uniform distribution on $\mathcal{L}^{(2)}$. Similarly, the joint distribution of L_1 and L_3 is the uniform distribution on $\mathcal{L}^{(2)}$. It follows from either of these results that L_3 is uniformly distributed on $\mathcal{L}$; that is, the (unconditional) distribution of the label of the object selected on the third trial is the uniform distribution on the set of all N labels.

In general, on the nth trial for $2 \leq n \leq N$, let the object be selected at random from the $N - (n - 1)$ objects in the population that remain after the objects selected on the first $n - 1$ trials are removed, let L_n denote the label of the object selected on that trial, and let $l_1, \ldots, l_{n-1}$ be distinct members of $\mathcal{L}$. The conditional distribution of L_n, given that $L_1 = l_1, \ldots, L_{n-1} = l_{n-1}$, is the uniform distribution on $\mathcal{L} \setminus \{l_1, \ldots, l_{n-1}\}$. It follows by induction that the joint distribution of $L_1, \ldots, L_n$ is the uniform distribution on the collection $\mathcal{L}^{(n)}$ of ordered n-tuples of distinct members of $\mathcal{L}$.

Still more generally, let $i_1, \ldots, i_n$ be distinct members of $\{1, \ldots, N\}$. Then the joint distribution of $L_{i_1}, \ldots, L_{i_n}$ is the uniform distribution on $\mathcal{L}^{(n)}$. In particular, for $1 \leq i \leq N$, the distribution of L_i is the uniform distribution on $\mathcal{L}$; and, for $1 \leq i, j \leq N$ with $i \neq j$, the joint distribution of L_i and L_j is the uniform distribution on $\mathcal{L}^{(2)}$.

Means, Variances, and Covariances

Suppose each object in the population has a value associated with it (such as weight). Let $v(l) = v_l$ denote the value of the object having label l and let L denote the label of an object selected at random from the population. Then $Y = v(L)$ is the value of the randomly selected object. Since L is uniformly distributed on $\mathcal{L} = \{1, \ldots, N\}$, Y has mean

$$\mu = \frac{1}{N} \sum_l v_l$$

and variance

$$\sigma^2 = \frac{1}{N} \sum_l (v_l - \mu)^2.$$

For $1 \leq i \leq N$, let $Y_i = v(L_i)$ be the value of the object selected on the ith trial. Since L_i has the same distribution as L (namely, the uniform distribution on $\mathcal{L}$), Y_i has the same distribution as Y, and hence it has mean μ and variance σ^2.

When $\sigma^2 > 0$, the random variables $Y_1, \ldots, Y_N$ are dependent. In particular,

$$\text{cov}(Y_i, Y_j) = -\frac{\sigma^2}{N-1}, \qquad 1 \leq i, j \leq N \text{ with } i \neq j. \tag{16}$$

In order to verify (16), observe first that $\sum_l (v_l - \mu) = \sum_l v_l - N\mu$ and hence that

$$\sum_l (v_l - \mu) = 0. \tag{17}$$

Let $1 \leq i, j \leq N$ with $i \neq j$. Since the joint distribution of L_i and L_j is the uniform distribution on $\mathcal{L}^{(2)}$, it follows from (17) that

$$
\begin{aligned}
\text{cov}(Y_i, Y_j) &= E[\,(Y_i - \mu)(Y_j - \mu)\,] \\
&= E\{[\,v(L_i) - \mu\,][\,v(L_j) - \mu\,]\} \\
&= \frac{1}{N(N-1)} \sum_{l_1} \sum_{l_2 \neq l_1} (v_{l_1} - \mu)(v_{l_2} - \mu) \\
&= \frac{1}{N(N-1)} \left(\sum_{l_1} \sum_{l_2} (v_{l_1} - \mu)(v_{l_2} - \mu) - \sum_l (v_l - \mu)^2 \right) \\
&= \frac{1}{N(N-1)} \left(\left(\sum_l (v_l - \mu) \right)^2 - \sum_l (v_l - \mu)^2 \right) \\
&= -\frac{1}{N(N-1)} \sum_l (v_l - \mu)^2
\end{aligned}
$$

and hence that (16) holds as desired.

We refer to the objects selected on the first n trials as forming the random sample of size n that is selected by sampling without replacement. Observe that $Y_1 + \cdots + Y_n$ is the sum of the values of the objects in this random sample. Since $EY_1 = \cdots = EY_n = \mu$, we see that

$$E(Y_1 + \cdots + Y_n) = n\mu. \tag{18}$$

The variance of this sum is given by

$$\mathrm{var}(Y_1 + \cdots + Y_n) = n\sigma^2 \frac{N-n}{N-1}. \tag{19}$$

To verify this result, observe that

$$\mathrm{var}(Y_1 + \cdots + Y_n) = \sum_{i=1}^{n} \mathrm{var}(Y_i) + 2\sum_{i=1}^{n-1}\sum_{j=i+1}^{n} \mathrm{cov}(Y_i, Y_j)$$

$$= n\sigma^2 - \frac{n(n-1)}{N-1}\sigma^2$$

$$= n\sigma^2\left(1 - \frac{n-1}{N-1}\right),$$

which yields (19).

Consider, instead, sampling with replacement. Here $L_1, L_2, \ldots$ are independent and uniformly distributed on $\mathcal{L}$, so $Y_1, Y_2, \ldots$ are independent and identically distributed random variables, each having mean μ and variance σ^2. Equation (18) is valid in this context, but (19) is replaced by

$$\mathrm{var}(Y_1 + \cdots + Y_n) = n\sigma^2. \tag{20}$$

When $n/N \approx 0$, the right side of (19) is approximately equal to the right side of (20); that is, the result (20) for sampling with replacement is approximately correct for sampling without replacement. Informally, we say that the results for sampling with replacement are applicable to sampling without replacement from an "infinite" population.

EXAMPLE **6.12**   A box contains six tickets, three of which have the value 1, two of which have the value 2, and one of which has the value 5. The tickets are selected from the box one at a time by sampling without replacement.

(a) Determine the mean and variance of the value of the ticket selected on the ith trial for $1 \leq i \leq 6$.

(b) Determine the mean and variance of the sum of the values of the tickets selected on the first two trials.

(c) Determine the answers to parts (a) and (b) for sampling with replacement.

Solution   Let Y_i denote the value of the ticket selected on the ith trial for $1 \leq i \leq 6$. Then $Y_1, \ldots, Y_6$ have common mean μ and common variance σ^2.

(a) We see from the solution to Example 2.8 in Section 2.1 that $\mu = 2$ and from the solution to Example 2.33 in Section 2.3 that $\sigma^2 = 2$.

(b) Observe that $Y_1 + Y_2$ is the sum of the values of the tickets selected on the first two trials. According to Equation (18), with $n = 2$ and $\mu = 2$, the mean of this sum is given by $E(Y_1 + Y_2) = 2 \cdot 2 = 4$. According to Equation (19), with

$N = 6, n = 2$, and $\sigma^2 = 2$, the variance of this sum is given by

$$\text{var}(Y_1 + Y_2) = 2 \cdot 2 \cdot \frac{4}{5} = \frac{16}{5}.$$

(c) Consider sampling with replacement. Here part (a) is applicable, $E(Y_1 + Y_2) = 4$ as in the solution to part (b), and $\text{var}(Y_1 + Y_2) = 4$ by Equation (20), with $n = 2$ and $\sigma^2 = 2$. ∎

Consider a population of N objects. Suppose the object labeled l has value v_l for $1 \le l \le N$, where $v_1, \ldots, v_N$ are unknown and hence $\mu = (v_1 + \cdots + v_N)/N$ is unknown. Let Y_i be the value of the ith object selected under either sampling without replacement or sampling with replacement. Think of $\hat{\mu} = \bar{Y} = (Y_1 + \cdots + Y_n)/n$ as an estimate of μ based on the random sample of size n. This estimate is unbiased both under sampling without replacement and under sampling with replacement. Under sampling without replacement, its variance is given by

$$\text{var}(\hat{\mu}) = \frac{\sigma^2}{n} \frac{N - n}{N - 1}, \qquad 1 \le n \le N, \tag{21}$$

where

$$\sigma^2 = \frac{1}{N} \sum_l (v_l - \mu)^2;$$

under sampling with replacement, its variance is given by

$$\text{var}(\hat{\mu}) = \frac{\sigma^2}{n}. \tag{22}$$

If $n = 1$, then there is no distinction between sampling without replacement and sampling with replacement. Suppose now that $2 \le n \le N$. If $\sigma = 0$ (that is, if $v_1 = \cdots = v_N$), then $\text{var}(\hat{\mu}) = 0$ for either sampling scheme. Otherwise, we see from (21) and (22) that the variance of $\hat{\mu}$ is smaller under sampling without replacement than under sampling with replacement or, in other words, that sampling without replacement yields a more accurate estimate of μ than does sampling with replacement. This is just what we should expect: Under sampling with replacement, we might pick out the same object two or more times, but after the first such time we get no additional information about the values of the objects in the box.

A genuine sample without replacement from a finite population is commonly referred to as a simple random sample. In actual practice it may be too expensive to examine the objects in such a sample. (For example, it would be very expensive to travel to all the shops in a simple random sample of 200 automobile repair shops from the population of all such shops in a large country.) "Stratification," "clustering," and other sampling techniques have been devised to obtain samples and estimates that are both as accurate as and more cost effective than simple random sampling in various applications.

Problems

6.16 As a special case of Equation (19), $\text{var}(Y_1 + \cdots + Y_N) = 0$. Give a direct proof of this result; that is, explain directly why $Y_1 + \cdots + Y_N$ is a constant random variable.

6.17 Suppose the value of a randomly selected object in a population of size $N \geq 2$ has mean μ and positive variance σ^2. Let two objects be selected from the population by sampling without replacement, let Y_1 be the value of the first object selected, and let Y_2 be the value of the second object selected. Determine **(a)** $\text{cor}(Y_1, Y_2)$; **(b)** the best linear predictor of Y_2 based on Y_1; **(c)** the mean squared error of this predictor.

6.18 A deck of 52 playing cards has four aces, four kings, four queens, and four jacks. In the context of bridge, an ace is worth 4 honor points, a king is worth 3 such points, a queen is worth 2 points, a jack is worth 1 point, and the remaining 36 cards are each worth 0 points. Determine **(a)** the mean, variance, and standard deviation of the number of honor points of a card selected at random from the deck; **(b)** the mean, variance, and standard deviation of the total number of honor points in a bridge hand, which has 13 cards.

6.19 In the context of sampling without replacement, verify that

$$\text{var}(Y_i - Y_j) = 2\sigma^2 \frac{N}{N-1}, \qquad 1 \leq i, j \leq N \text{ with } i \neq j.$$

6.20 For given n, how large need N be to guarantee that the variance of $\hat{\mu}$ under sampling without replacement, as given by Equation (21), is at least 90% as large as the corresponding variance under sampling with replacement, as given by Equation (22)?

6.21 Determine the (positive integer) values n that maximize the right side of Equation (19), where $N \geq 2$ and $0 < \sigma < \infty$.

6.22 An urn starts out with r red balls and b black balls. According to *Polya's urn scheme*, on each trial a ball is selected at random from the urn, and then it and c new balls of the same color are returned to the urn. Let C_i denote the color of the ball selected on the ith trial. Show that **(a)** $P(C_2 = \text{red}) = P(C_1 = \text{red}) = \frac{r}{r+b}$; **(b)** $P(C_1 = \text{red} \mid C_2 = \text{red}) = P(C_2 = \text{red} \mid C_1 = \text{red}) = \frac{r+c}{r+b+c}$; **(c)** $P(C_2 = \text{red}, C_3 = \text{red}) = P(C_1 = \text{red}, C_2 = \text{red}) = \frac{r(r+c)}{(r+b)(r+b+c)}$; **(d)** $P(C_1 = \text{black} \mid C_2 = \text{red}, C_3 = \text{red}) = P(C_3 = \text{black} \mid C_1 = \text{red}, C_2 = \text{red}) = \frac{b}{r+b+2c}$.

6.23 An urn starts out with two red balls and two black balls. On each trial, a ball is selected at random from the urn and replaced by a ball of the opposite color. Determine the probability function f of the number of red balls in the urn at the end of the first two trials.

6.24 An urn starts out with three red balls, three white balls, and three blue balls. On each trial, a ball is selected at random from the urn and then replaced by two balls, one of each of the other two colors. Determine the probability that at the end of three such trials, the urn contains three red balls, three white balls, and three blue balls.

6.25 An urn starts out with three red balls, three white balls, and three blue balls. On each trial, a ball is selected at random from the urn and then replaced by a ball of a different color as follows: Red is replaced by white, white is replaced by blue, and blue is replaced by red. Determine the probability that at the end of three such trials, the urn contains three red balls, three white balls, and three blue balls.

6.26   An urn starts out with three red balls and three black balls. On each of six trials, Frankie selects a ball at random from the urn, notes its color, and removes it from the urn. Prior to the start of each trial, she guesses the color of the ball that will be selected on that trial using the following strategy: If there are more black balls than red balls currently in the urn, guess black; otherwise, guess red. Determine the probability that Frankie correctly guesses the color of the ball selected from the urn on all six trials.

6.27   In the context of Problem 6.26, Johnny guesses before any balls are drawn from the urn that red balls will be selected on the first three trials and black balls on the last three trials. Determine the probability that Johnny correctly guesses the color of the ball selected from the urn on all six trials.

6.28   Each of two urns contains three red balls and three black balls. On each of three trials, a ball is selected at random from each urn, and then each ball is put into the other urn. Determine the probability that, at the end of the three trials, all the red balls will be in one urn and all the black balls in the other urn.

6.29   Frankie and Johnny cannot determine on their own whether they are sufficiently compatible to get married, so they decide to let the Fates make the choice. Specifically, they obtain separate boxes, each containing three red balls and three black balls, from which they independently select three balls, one at a time, by sampling *without* replacement. If on each of the three trials the ball selected by Johnny has the same color as the one selected by Frankie, they will get married; otherwise, they will not get married. Determine the probability that Frankie and Johnny will get married. (As a check on the reasonableness of your answer, you may want to compare its numerical value with that for sampling *with* replacement.)

6.3
Hypergeometric Distribution

Consider a box having seven red balls and three black balls and let a sample of size 6 be selected from the box by sampling without replacement. What is the probability of getting four red balls and two black balls in the sample?

To answer this question, we observe that there are $\binom{10}{6}$ ways of choosing six balls out of ten. Moreover, there are $\binom{7}{4}$ ways of choosing four red balls out of the seven red balls in the box and $\binom{3}{2}$ ways of choosing two black balls out of the three black balls in the box. Thus, there are $\binom{7}{4}\binom{3}{2}$ ways of choosing four red balls and two black balls, so the desired probability equals

$$\frac{\binom{7}{4}\binom{3}{2}}{\binom{10}{6}}.$$

In general, consider a population of N objects, labeled from 1 to N. Let $1 \le n \le N$ and let $l_1, \ldots, l_n$ be distinct members of $\{1, \ldots, N\}$. Since the joint distribution of $L_1, \ldots, L_n$ is the uniform distribution on the set of ordered n-tuples of distinct members of $\{1, \ldots, N\}$, which has size

$$N \cdots (N - n + 1) = \frac{N!}{(N - n)!},$$

we see that

$$P(L_1 = l_1, \ldots, L_n = l_n) = \frac{(N - n)!}{N!}.$$

Note that $\{L_1, \ldots, L_n\} = \{l_1, \ldots, l_n\}$ if and only if $(L_1, \ldots, L_n)$ is a permutation of $(l_1, \ldots, l_n)$. Since there are $n!$ such permutations, we conclude that

$$P(\{L_1, \ldots, L_n\} = \{l_1, \ldots, l_n\}) = \frac{n! \, (N - n)!}{N!} = \frac{1}{\binom{N}{n}}.$$

Therefore, $\{L_1, \ldots, L_n\}$ is uniformly distributed over the collection of $\binom{N}{n}$ subsets of $\{1, \ldots, N\}$ of size n. Observe that $\{L_1, \ldots, L_n\}$ is the collection of labels of the objects in the random sample of size n that is obtained by sampling without replacement. Thus, the random sample of size n is uniformly distributed over the collection of all $\binom{N}{n}$ ways of choosing n of the objects out of the population of size N.

Let N_1 be an integer such that $0 \le N_1 \le N$. Suppose that N_1 of the N objects in the population are of type 1 and that the remaining $N_2 = N - N_1$ objects are of type 2. Let Y denote the number of type 1 objects in the random sample of size n that is obtained by sampling without replacement. Then

$$P(Y = y) = \frac{\binom{N_1}{y}\binom{N_2}{n-y}}{\binom{N}{n}}. \tag{23}$$

This probability is zero if $y < 0$, $y > N_1$, $y < n - N_2$, or $y > n$; otherwise, the probability is positive. In order to verify (23), suppose the probability is positive; that is, that

$$\max(0, n - N_2) \le y \le \min(n, N_1).$$

There are $\binom{N_1}{y}$ ways of choosing y objects out of the N_1 type 1 objects and $\binom{N_2}{n-y}$ ways of choosing $n - y$ type 2 objects out of the N_2 type 2 objects. Thus, there are

$$\binom{N_1}{y}\binom{N_2}{n-y}$$

ways of choosing n objects out of N so that y objects are of type 1 and $n - y$ are of type 2. Since each such way has probability $1/\binom{N}{n}$, (23) is valid. The distribution of Y is referred to as the *hypergeometric distribution* with parameters n, N_1, and N_2.

EXAMPLE 6.13   Determine the mean and variance of the hypergeometric distribution with parameters n, N_1, and N_2.

Solution   Suppose N_1 of the $N = N_1 + N_2$ objects in the population are of type 1 and that the remaining N_2 objects are not of type 1. Let the value $v(l) = v_l$ of the object having label l be 1 or 0 according as the object is or is not of type 1. Then

$$\mu = \frac{1}{N}\sum_l v_l = \frac{N_1}{N}$$

is the proportion of type 1 objects in the population, which we denote by π. The corresponding variance is given by

$$\sigma^2 = \frac{1}{N}\sum_l (v_l - \pi)^2 = \frac{1}{N}\sum_l v_l^2 - \pi^2 = \frac{1}{N}\sum_l v_l - \pi^2 = \pi - \pi^2 = \pi(1 - \pi).$$

Here $Y_i = v(L_i)$ is the indicator random variable for getting a type 1 object on the ith trial. Let $Y = Y_1 + \cdots + Y_n$ denote the number of type 1 objects in the random sample of size n. Then, under sampling without replacement, Y has the hypergeometric distribution with parameters n, N_1, and N_2. It follows from Equations (18) and (19) in Section 6.2 that Y has mean $n\pi$ and variance $n\pi(1 - \pi)\frac{N-n}{N-1}$.   ∎

An ordinary deck of *playing cards* (without jokers) consists of 52 distinct cards, which are divided into four suits—clubs, diamonds, hearts, and spades—each consisting of 13 cards; clubs and spades are black, while diamonds and hearts are red. The 13 cards in each suit are labeled as 2, 3, 4, 5, 6, 7, 8, 9, 10, jack, queen, king, ace. We refer to 2, . . . , ace as "kinds." Thus, there are 13 kinds in the deck and there are 4 cards of each kind, one in each of the four suits. In the context of the game of bridge, the 52 cards, after being thoroughly shuffled, are distributed one at a time to each of the four players until all the cards are distributed and each player has 13 cards in her hand. In effect, the cards in any given hand can be thought of as a sample of size 13 obtained by sampling without replacement from the entire deck. In the context of poker, the hands are of various sizes, but we will think of a poker hand as consisting of 5 cards, obtained by sampling without replacement from the deck.

EXAMPLE 6.14   Let a poker hand be drawn at random from a deck of playing cards. Determine the probability that all of the cards in the hand

(a) are clubs;

(b) form a flush (belong to the same suit).

Solution   (a) The number of five-card hands is

$$\binom{52}{5} = \frac{52!}{5!\,47!} = \frac{52 \cdot 51 \cdot 50 \cdot 49 \cdot 48}{5 \cdot 4 \cdot 3 \cdot 2 \cdot 1},$$

and the number of five-card hands consisting only of clubs is

$$\binom{13}{5} = \frac{13!}{5!\,8!} = \frac{13 \cdot 12 \cdot 11 \cdot 10 \cdot 9}{5 \cdot 4 \cdot 3 \cdot 2 \cdot 1}.$$

Thus, the probability that all of the cards in the hand are clubs equals

$$\frac{\binom{13}{5}}{\binom{52}{5}} = \frac{13 \cdot 12 \cdot 11 \cdot 10 \cdot 9}{52 \cdot 51 \cdot 50 \cdot 49 \cdot 48} = \frac{33}{66640} \doteq .0005.$$

(b) By the obvious extension of the solution to part (a), the probability that all of the cards in the hand form a flush is

$$4 \frac{\binom{13}{5}}{\binom{52}{5}} = \frac{33}{16660} \doteq .002. \quad \blacksquare$$

EXAMPLE 6.15 Determine the distribution, mean, variance, and standard deviation of the number of clubs in a bridge hand.

Solution Let Y denote the number of clubs in the hand (which we assume is obtained by sampling without replacement). Then Y has the hypergeometric distribution with parameters $n = 13$, $N_1 = 13$, and $N_2 = 39$. Thus, by the solution to Example 6.13,

$$EY = 13 \cdot \frac{1}{4} = \frac{13}{4} = 3.25,$$

$$\text{var}(Y) = 13 \cdot \frac{1}{4} \cdot \frac{3}{4} \cdot \frac{39}{51} = \frac{507}{272} \doteq 1.864,$$

and

$$\text{SD}(Y) = \sqrt{\frac{507}{272}} \doteq 1.365. \quad \blacksquare$$

EXAMPLE 6.16 Let Y_1 and Y_2 be independent random variables having binomial distributions with parameters n_1 and n_2, respectively, and π. Recall from Theorem 4.13 in Section 4.2 that the random variable $Y = Y_1 + Y_2$ has the binomial distribution with parameters $n = n_1 + n_2$ and π. Determine the conditional distribution of Y_1 given that $Y = y \in \{0, \ldots, n\}$.

Solution Let y_1 be a nonnegative integer. Then

$$P(Y_1 = y_1 \mid Y = y) = \frac{P(Y_1 = y_1, Y = y)}{P(Y = y)}.$$

Since $Y = Y_1 + Y_2$, we see that $Y_1 = y_1$ and $Y = y$ if and only if $Y_1 = y_1$ and $Y_2 = y - y_1$. Thus,

$$P(Y_1 = y_1 \mid Y = y) = \frac{P(Y_1 = y_1, Y_2 = y - y_1)}{P(Y = y)}$$

$$= \frac{P(Y_1 = y_1)P(Y_2 = y - y_1)}{P(Y = y)}$$

$$= \frac{\binom{n_1}{y_1}\pi^{y_1}(1-\pi)^{n_1-y_1} \binom{n_2}{y-y_1}\pi^{y-y_1}(1-\pi)^{n_2-y+y_1}}{\binom{n}{y}\pi^y(1-\pi)^{n-y}}$$

$$= \frac{\binom{n_1}{y_1}\binom{n_2}{y-y_1}}{\binom{n}{y}}.$$

Consequently, the conditional distribution of Y_1, given that $Y = y$, is the hypergeometric distribution with parameters y, n_1, and n_2. (The conditional distribution of Y_1, given that $Y = 0$, is the distribution that assigns probability 1 to the value 0.) ∎

Let f be the probability function of the hypergeometric distribution with parameters n, N_1, and N_2 and let y be a positive integer such that

$$\max(0, n - N_2) \leq y - 1 < y \leq \min(n, N_1).$$

It follows easily from Equation (23) that

$$\frac{f(y)}{f(y-1)} = \frac{N_1 - y + 1}{y} \frac{n - y + 1}{N_2 - n + y}. \tag{24}$$

We can use (24) to compute the hypergeometric probability function in an effective manner. It also follows from (24) that $f(y)$ is greater than, equal to, or less than $f(y-1)$ according as y is less than, equal to, or greater than the quantity

$$r = (n+1)\frac{N_1 + 1}{N_1 + N_2 + 2}.$$

Consequently, if r is a positive integer, then $r - 1$ and r are the only modes and the only locations of global maxima of f; also, $f(y)$ is strictly increasing in y for

$$y \in \{\max(0, n - N_2), \ldots, r - 1\},$$

$f(r-1) = f(r)$, and $f(y)$ is strictly decreasing in y for $y \in \{r, \ldots, \min(n, N_1)\}$. If r is not a positive integer, then $[r]$ (the greatest integer in r) is the unique location of the global maximum of f; also, $f(y)$ is strictly increasing in y for $y \in \{\max(0, n - N_2), \ldots, r]\}$, and it is strictly decreasing in y for $y \in \{[r], \ldots, \min(n, N_1)\}$.

EXAMPLE **6.17**   Determine the mode(s) of the probability function of the number of red balls in a random sample without replacement of size n from a box having three red balls and seven white balls.

Solution   Here $N_1 = 3$ and $N_2 = 7$. Set

$$r = (n+1)\frac{N_1 + 1}{N_1 + N_2 + 2} = \frac{n+1}{3}.$$

If n equals 2, 5, or 8, then r is a positive integer, so $r - 1$ and r are the two modes of the probability function of the number of red balls in the sample of size n. Otherwise, $[r]$ is the unique mode of the probability function. Specifically, we have the results in the following table:

n	1	2	3	4	5	6	7	8	9	10
Mode(s)	0	0,1	1	1	1,2	2	2	2,3	3	3

∎

Multivariate Hypergeometric Distribution

Let $N_1, \ldots, N_M$ be nonnegative integers having sum N, where $M \geq 2$. Suppose N_m of the objects in the population are of type m for $1 \leq m \leq M$. Consider a sample of size n from the population that is obtained by sampling without replacement. For $1 \leq m \leq M$, let Y_m denote the number of type m objects in the sample. It follows by an extension of the argument used to verify Equation (23) that

$$P(Y_1 = y_1, \ldots, Y_M = y_M) = \frac{\binom{N_1}{y_1} \cdots \binom{N_M}{y_M}}{\binom{N}{n}},$$

where $y_1, \ldots, y_n$ are nonnegative integers adding up to n. The joint distribution of $Y_1, \ldots, Y_M$ is referred to as the *multivariate hypergeometric distribution* with parameters $n, N_1, \ldots, N_M$.

EXAMPLE 6.18   (a) Determine the probability that a bridge hand contains five clubs, three diamonds, three hearts, and two spades.

(b) Determine the conditional probability that the hand contains five clubs, three diamonds, three hearts, and two spades, given that it contains exactly six red cards.

(c) Determine the probability of getting a 5–3–3–2 split; that is, the hand contains five cards of some suit, three cards of each of two other suits, and two cards of the remaining suit.

Solution   (a) Let Y_1, Y_2, Y_3, and Y_4, respectively, denote the numbers of clubs, diamonds, hearts, and spades in the hand. The joint distribution of these random variables is multivariate hypergeometric with parameters $n = 13$ and $N_1 = N_2 = N_3 = N_4 = 13$. Thus, the desired probability is given by

$$P(Y_1 = 5, Y_2 = 3, Y_3 = 3, Y_4 = 2) = \frac{\binom{13}{5}\binom{13}{3}\binom{13}{3}\binom{13}{2}}{\binom{52}{13}} \doteq .0129.$$

(b) The number $Y_2 + Y_3$ of red cards in the hand has the hypergeometric distribution with parameters $n = 13$ and $N_1 = N_2 = 26$, so

$$P(Y_2 + Y_3 = 6) = \frac{\binom{26}{6}\binom{26}{7}}{\binom{52}{13}}.$$

Observe that if $Y_1 = 5$, $Y_2 = 3$, $Y_3 = 3$, and $Y_4 = 2$, then $Y_2 + Y_3 = 6$. Thus, the desired conditional probability is given by

$$P(Y_1 = 5, Y_2 = 3, Y_3 = 3, Y_4 = 2 \mid Y_2 + Y_3 = 6)$$

$$= \frac{P(Y_1 = 5, Y_2 = 3, Y_3 = 3, Y_4 = 2)}{P(Y_2 + Y_3 = 6)}$$

$$= \frac{\binom{13}{5}\binom{13}{3}\binom{13}{3}\binom{13}{2}}{\binom{26}{6}\binom{26}{7}}$$

$$\doteq .0542.$$

(c) According to the solution to part (a), the probability of getting five clubs, three diamonds, three hearts, and two spades is given by

$$p = \frac{\binom{13}{5}\binom{13}{3}\binom{13}{3}\binom{13}{2}}{\binom{52}{13}} \doteq .0129.$$

Similarly, p is the probability of getting five clubs, three diamonds, three spades, two hearts, and so forth. Thus, the desired probability equals Np, where N is the number of ways of partitioning the four suits into three ordered sets, the first set having size 1, the second set having size 2, and the third set having size 1; the suit in the first set will be a five-card suit, the two suits in the second set will be three-card suits, and the suit in the third set will be a two-card suit. By Theorem 4.5 in Section 4.1,

$$N = \frac{4!}{1!\,2!\,1!} = 12.$$

Consequently, the probability of getting a 5-3-3-2 split is given by

$$Np = 12\frac{\binom{13}{5}\binom{13}{3}\binom{13}{3}\binom{13}{2}}{\binom{52}{13}} \doteq 12(.0129) \doteq .155. \quad\blacksquare$$

EXAMPLE **6.19**   A sample of size n is selected by sampling without replacement from a population of $N_1 + N_2 + N_3$ objects, of which N_1 are of type 1, N_2 are of type 2, and N_3 are of type 3. Suppose we know that exactly y_1 of the objects in the sample are of type 1, where $y_1 < n$. Then the remaining $n - y_1$ objects must be selected from the N_2 objects of type 2 and the N_3 objects of type 3. Thus, it is natural to conjecture that, conditioned on the known information, the distribution of the number of type 2 objects in the sample is hypergeometric with parameters $n - y_1$, N_2 and N_3. Verify this conjecture.

Solution   Let Y_1, Y_2, and Y_3 have the trivariate hypergeometric distribution with parameters n, N_1, N_2, and N_3 and let y_1 be an integer such that $P(Y_1 = y_1) > 0$. We want to show that the conditional distribution of Y_2, given that $Y_1 = y_1$, is hypergeometric with parameters $n - y_1$, N_2, and N_3. To this end, observe first that Y_1 has the hypergeometric distribution with parameters n, N_1, and $N_2 + N_3$, so

$$P(Y_1 = y_1) = \frac{\binom{N_1}{y_1}\binom{N_2+N_3}{n-y_1}}{\binom{N_1+N_2+N_3}{n}}.$$

Observe next that if $Y_1 = y_1$ and $Y_2 = y_2$, then $Y_3 = n - y_1 - y_2$. Thus,

$$P(Y_1 = y_1, Y_2 = y_2) = P(Y_1 = y_1, Y_2 = y_2, Y_3 = n - y_1 - y_2)$$

$$= \frac{\binom{N_1}{y_1}\binom{N_2}{y_2}\binom{N_3}{n-y_1-y_2}}{\binom{N_1+N_2+N_3}{n}}.$$

Consequently,

$$P(Y_2 = y_2 \mid Y_1 = y_1) = \frac{P(Y_1 = y_1, Y_2 = y_2)}{P(Y_1 = y_1)} = \frac{\binom{N_2}{y_2}\binom{N_3}{n-y_1-y_2}}{\binom{N_2+N_3}{n-y_1}},$$

which yields the desired result. ∎

Problems

6.30 Five balls are selected by sampling without replacement from a box having seven red balls and four white balls. Determine the probability that there are three red balls in the sample.

6.31 Determine the probability that a poker hand is a *flush*; that is, all five cards in the hand belong to the same suit.

6.32 A manufacturer receives a supply of 100 parts from a supplier. Five of these parts, selected at random by sampling without replacement, are inspected. Under the assumption that 20 of the entire supply of 100 parts are defective, determine the probability that none of the inspected parts are defective.

6.33 A sample of size N_1 is selected by sampling without replacement from a population of N_1 males and N_2 females. Determine the probability that there are at least $N_1 - 1$ males in the sample.

6.34 A box has seven red balls and three white balls. Balls are selected from the box one at a time by sampling without replacement. Determine the probability that white is selected for the second time on the eighth trial.

6.35 A box contains six red balls and eight white balls. A sample of size 8 is selected from the box by sampling without replacement, and then a sample of size 4 is selected from the remaining six balls in the box by sampling without replacement. Determine the probability that each of the two samples contains as many red balls as white balls.

6.36 A box contains N_1 red balls and N_2 white balls. Frankie selects the balls out of the box one-at-a-time by sampling without replacement. Johnny guesses which N_1 trials will yield a red ball and which N_2 trials will yield a white ball. (In making his guess on each trial, he does not have the benefit of any information about the results on previous trials.) Let W be the number of his red guesses that are correct and let Y be his total number of correct guesses. Explain why **(a)** W has the hypergeometric distribution with parameters N_1, N_1, and N_2; **(b)** $Y = N_2 - N_1 + 2W$.

6.37 A fair die is rolled 100 times in an honest manner. Determine the conditional probability that side 1 shows four times during the first 20 trials, given that this side shows 20 times during the 100 trials.

6.38 Let Y_1 and Y_2 be independent random variables such that Y_1 has the binomial distribution with parameters 9 and π and Y_2 has the binomial distribution with parameters 4 and π. For which value(s) of y_1 is $P(Y_1 = y_1 \mid Y_1 + Y_2 = 8)$ maximized?

6.39 **(a)** Determine the probability that a bridge hand contains four clubs, three diamonds, three hearts, and three spades. **(b)** Determine the conditional probability that the hand contains four clubs, three diamonds, three hearts, and three spades, given that it contains exactly six red cards.

6.40 **(a)** Determine the probability that a poker hand has three kings and two nines. **(b)** Determine the probability that the hand has the form of a full house; that is, it contains three cards of one kind and two of another kind.

6.41 **(a)** Determine the probability that a poker hand has two kings, two nines, and one card that is neither a king nor a nine. **(b)** Determine the probability that a poker hand has two pairs; that is, two cards are of one kind, two are of a second kind, and one is of a third kind.

6.42 **(a)** Determine the probability that a poker hand has one two, one three, one four, one five, and one six. **(b)** Determine the probability that the cards in the hand are of distinct kinds.

6.43 A box contains four red balls, three white balls, two blue balls, and one yellow ball. Each red ball has value 1, each white ball has value 2, each blue ball has value 3, and the yellow ball has value 4. Four balls are selected from the box by sampling without replacement. Determine the probability that **(a)** there are two red balls, one white ball, and one yellow ball in the sample; **(b)** the sum of the values of the balls in the sample equals 8.

6.44 A box has eight red balls, seven white balls, and six blue balls. Balls are selected from the box one-at-a-time by sampling without replacement. Determine the probability that, by the time a blue ball is selected for the third time, red will have been selected five times and white four times.

6.45 Let Y_1, Y_2, Y_3, and Y_4 have the multivariate hypergeometric joint distribution with parameters n, N_1, N_2, N_3, and N_4 and let y_4 be a nonnegative integer such that $y_4 < n$ and $P(Y_4 = y_4) > 0$. Show that the conditional joint distribution of Y_1, Y_2, and Y_3, given that $Y_4 = y_4$, is trivariate hypergeometric with parameters $n - y_4$, N_1, N_2, and N_3.

6.46 Let Y_1, Y_2, and Y_3 be independent random variables having binomial distributions with parameters n_1, n_2, and n_3, respectively, and π. Recall from Theorem 4.13 in Section 4.2 that the random variable $Y = Y_1 + Y_2 + Y_3$ has the binomial distribution with parameters $n = n_1 + n_2 + n_3$ and π. Show that, for $y \in \{1, \ldots, n\}$, the conditional joint distribution of Y_1, Y_2, and Y_3, given that $Y = y$, is trivariate hypergeometric with parameters y, n_1, n_2, and n_3.

6.4
Conditional Density Functions

Let $\mathbf{X} = [X_1, \ldots, X_m]^T$ and Y have joint density function $f_{\mathbf{X}, Y}$ and let $\mathcal{X}$ be a nonempty open set in $\mathbb{R}^m$ such that $P(\mathbf{X} \in \mathcal{X}) = 1$. We can assume that $f_{\mathbf{X}, Y}(\mathbf{x}, y) = 0$ for $\mathbf{x} \notin \mathcal{X}$ and $y \in \mathbb{R}$. The density function of $\mathbf{X}$ is given by

$$f_{\mathbf{X}}(\mathbf{x}) = \int_{\mathbb{R}} f_{\mathbf{X}, Y}(\mathbf{x}, y) \, dy, \qquad \mathbf{x} \in \mathcal{X},$$

and $f_{\mathbf{X}}(\mathbf{x}) = 0$ for $\mathbf{x} \notin \mathcal{X}$. We assume that $f_{\mathbf{X}}(\mathbf{x}) > 0$ for $\mathbf{x} \in \mathcal{X}$. The density function of Y is given by

$$f_Y(y) = \int_{\mathcal{X}} f_{\mathbf{X}, Y}(\mathbf{x}, y)\, d\mathbf{x}, \qquad y \in \mathbb{R}.$$

Let $B \subset \mathbb{R}$ and $\mathbf{x} \in \mathcal{X}$. We now give an approach to the definition of $P(Y \in B \mid \mathbf{X} = \mathbf{x})$ that is applicable in the present context. (The definition of conditional probability given in Section 6.1 is inapplicable since $P(\mathbf{X} = \mathbf{x}) = 0$.) Let $B_\epsilon(\mathbf{x})$ denote the open ball in $\mathbb{R}^m$ having center $\mathbf{x} = [x_1, \ldots, x_m]^T$ and radius ϵ, which consists of all $[x_1', \ldots, x_m']^T \in \mathbb{R}^m$ such that $(x_1' - x_1)^2 + \cdots + (x_m' - x_m)^2 < \epsilon^2$. It seems reasonable to define $P(Y \in B \mid \mathbf{X} = \mathbf{x})$ as the limit as $\epsilon \to 0$ of

$$P(Y \in B \mid \mathbf{X} \in B_\epsilon(\mathbf{x})) = \frac{P(\mathbf{X} \in B_\epsilon(\mathbf{x}),\, Y \in B)}{P(\mathbf{X} \in B_\epsilon(\mathbf{x}))}$$

(that is, as the limit as $\epsilon \to 0$ of the conditional probability that Y lies in B, given that $\mathbf{X}$ is within distances ϵ of $\mathbf{x}$), provided that the indicated limit exists. Now

$$P(\mathbf{X} \in B_\epsilon(\mathbf{x})) = \int_{B_\epsilon(\mathbf{x})} f_{\mathbf{X}}(\mathbf{x}')\, d\mathbf{x}'.$$

Suppose $f_{\mathbf{X}}$ is continuous at $\mathbf{x}$. Then

$$\lim_{\epsilon \to 0} \frac{P(\mathbf{X} \in B_\epsilon(\mathbf{x}))}{\mathrm{vol}_m(B_\epsilon(\mathbf{x}))} = f_{\mathbf{X}}(\mathbf{x}). \tag{25}$$

Observe that

$$P(\mathbf{X} \in B_\epsilon(\mathbf{x}),\, Y \in B) = \int_{B_\epsilon(\mathbf{x})} \left(\int_B f_{\mathbf{X}, Y}(\mathbf{x}', y)\, dy \right) d\mathbf{x}'.$$

Under the additional assumptions that B is a bounded set and that $f_{\mathbf{X}, Y}$ is a continuous function, it follows from standard properties of integration that

$$\lim_{\epsilon \to 0} \frac{P(\mathbf{X} \in B_\epsilon(\mathbf{x}),\, Y \in B)}{\mathrm{vol}_m(B_\epsilon(\mathbf{x}))} = \int_B f_{\mathbf{X}, Y}(\mathbf{x}, y)\, dy. \tag{26}$$

According to (25) and (26),

$$\lim_{\epsilon \to 0} \frac{P(\mathbf{X} \in B_\epsilon(\mathbf{x}),\, Y \in B)}{P(\mathbf{X} \in B_\epsilon(\mathbf{x}))} = \frac{\int_B f_{\mathbf{X}, Y}(\mathbf{x}, y)\, dy}{f_{\mathbf{X}}(\mathbf{x})}.$$

This suggests setting

$$P(Y \in B \mid \mathbf{X} = \mathbf{x}) = \frac{\int_B f_{\mathbf{X}, Y}(\mathbf{x}, y)\, dy}{f_{\mathbf{X}}(\mathbf{x})} = \int_B \frac{f_{\mathbf{X}, Y}(\mathbf{x}, y)}{f_{\mathbf{X}}(\mathbf{x})}\, dy. \tag{27}$$

The end result (27) of this approach is appropriate, but the various assumptions made along the way are unnatural.

Here is a better approach, which will be used in the remainder of this chapter. (A more general approach, which avoids the use of probability functions or density functions, is used in advanced probability theory.) Let $\mathbf{x} \in \mathcal{X}$. In light of (27) and by analogy with conditional probability functions, we define the *conditional density function* of Y, given that $\mathbf{X} = \mathbf{x}$, by

$$f_{Y|\mathbf{X}}(y \mid \mathbf{x}) = \frac{f_{\mathbf{X},Y}(\mathbf{x}, y)}{f_{\mathbf{X}}(\mathbf{x})}, \qquad y \in \mathbb{R};$$

it is a density function in the usual sense. Let $B \subset \mathbb{R}$. As suggested by (27), the conditional probability that $Y \in B$, given that $\mathbf{X} = \mathbf{x}$, is defined by

$$P(Y \in B \mid \mathbf{X} = \mathbf{x}) = \int_B f_{Y|\mathbf{X}}(y \mid \mathbf{x}) \, dy. \tag{28}$$

As a function of B, $P(Y \in B \mid \mathbf{X} = \mathbf{x})$ is a distribution on $\mathbb{R}$, which is referred to as the conditional distribution of Y, given that $\mathbf{X} = \mathbf{x}$. The joint density function of $\mathbf{X}$ and Y is determined by the density function of $\mathbf{X}$ and the conditional density function of Y, given that $\mathbf{X} = \mathbf{x}$:

$$f_{\mathbf{X},Y}(\mathbf{x}, y) = \begin{cases} f_{\mathbf{X}}(\mathbf{x}) f_{Y|\mathbf{X}}(y \mid \mathbf{x}), & \mathbf{x} \in \mathcal{X} \text{ and } y \in \mathbb{R}, \\ 0, & \mathbf{x} \notin \mathcal{X} \text{ and } y \in \mathbb{R}. \end{cases} \tag{29}$$

If

$$f_{Y|\mathbf{X}}(y \mid \mathbf{x}) = f_Y(y), \qquad \mathbf{x} \in \mathcal{X} \text{ and } y \in \mathbb{R}, \tag{30}$$

then $\mathbf{X}$ and Y are independent; for it follows from (29) and (30) that

$$f_{\mathbf{X},Y}(\mathbf{x}, y) = f_{\mathbf{X}}(\mathbf{x}) f_Y(y), \qquad \mathbf{x} \in \mathbb{R}^m \text{ and } y \in \mathbb{R}.$$

If $f_{\mathbf{X}}, f_Y$, and $f_{\mathbf{X},Y}$ are all continuous functions, then (30) is a necessary and sufficient condition for $\mathbf{X}$ and Y to be independent. Suppose $\mathbf{X}$ and Y are independent and, in fact, that (30) holds. Then by (28) and (30),

$$P(Y \in B \mid \mathbf{X} = \mathbf{x}) = P(Y \in B), \qquad \mathbf{x} \in \mathcal{X} \text{ and } B \subset \mathbb{R}.$$

The mean, variance, and standard deviation of the conditional distribution of Y, given that $\mathbf{X} = \mathbf{x}$, are denoted by $\mu(\mathbf{x}) = E(Y \mid \mathbf{X} = \mathbf{x})$, $\sigma^2(\mathbf{x}) = \mathrm{var}(Y \mid \mathbf{X} = \mathbf{x})$, and $\sigma(\mathbf{x}) = \sqrt{\sigma^2(\mathbf{x})} = \mathrm{SD}(Y \mid \mathbf{X} = \mathbf{x})$, respectively. [Think of $\mu(\mathbf{x})$ as an abbreviation for $\mu_{Y|\mathbf{X}}(\mathbf{x})$ and so forth.] In particular,

$$\mu(\mathbf{x}) = \int_{-\infty}^{\infty} y f_{Y|\mathbf{X}}(y \mid \mathbf{x}) \, dy.$$

If $\mu(\mathbf{x})$ is finite, then

$$\sigma^2(\mathbf{x}) = \int_{-\infty}^{\infty} [y - \mu(\mathbf{x})]^2 f_{Y|\mathbf{X}}(y \mid \mathbf{x}) \, dy.$$

We refer to $\mu(\mathbf{x})$, $\sigma^2(\mathbf{x})$, and $\sigma(\mathbf{x})$, respectively, as the conditional mean, conditional variance, and conditional standard deviation of Y, given that $\mathbf{X} = \mathbf{x}$. The function $\mu(\mathbf{x})$, $\mathbf{x} \in \mathcal{X}$, is referred to as the *regression function* of Y on $\mathbf{X}$. (The choice of the word *regression* will be discussed at the end of Section 6.5.)

EXAMPLE **6.20**   Let (X, Y) be uniformly distributed on the triangle having the vertices $(0, 0)$, $(8, 0)$, and $(0, 4)$ and let $0 < x < 8$. Determine

(a) the conditional distribution of Y, given that $X = x$;

(b) the mean and variance of this conditional distribution.

Solution   The joint density function of X and Y is given by

$$f_{X,Y}(x, y) = \frac{1}{16}, \qquad 0 < x < 8 \text{ and } 0 < y < 4 - \frac{x}{2},$$

and $f_{X,Y}(x, y) = 0$ elsewhere. Thus, the density function of X is given by

$$f_X(x) = \int_0^{4-x/2} \frac{1}{16}\, dy = \frac{4 - \frac{x}{2}}{16}, \qquad 0 < x < 8,$$

and $f_X(x) = 0$ elsewhere. Let $0 < x < 8$.

(a) The conditional density function of Y, given that $X = x$, is given by

$$f_{Y|X}(y \mid x) = \frac{1}{4 - \frac{x}{2}}, \qquad 0 < y < 4 - \frac{x}{2},$$

and $f_{Y|X}(y \mid x) = 0$ elsewhere. Thus, the conditional distribution of Y, given that $X = x$, is the uniform distribution on $[\,0, 4 - x/2\,]$.

(b) The mean and variance of this conditional distribution are given by

$$\mu(x) = 2 - \frac{x}{4} \quad \text{and} \quad \sigma^2(x) = \frac{1}{12}\left(4 - \frac{x}{2}\right)^2$$

(see Example 2.16 in Section 2.1 and Example 2.40 in Section 2.3).   ∎

EXAMPLE 6.21   Let X and Y have the joint density function given by

$$f_{X,Y}(x, y) = \frac{x^3}{2} e^{-x(1+y)}, \qquad x, y > 0,$$

and $f_{X,Y} = 0$ elsewhere and let $x > 0$. Determine

(a) the conditional distribution of Y, given that $X = x$;

(b) the mean and variance of this conditional distribution.

Solution   According to the solution to Example 5.19(a) in Section 5.5, the density function of X is given by $f_X(x) = 0$ for $x \le 0$ and $f_X(x) = x^2 e^{-x}/2$ for $x > 0$.

(a) The conditional density function of Y, given that $X = x$, is given by

$$f_{Y|X}(y \mid x) = \frac{f_{X,Y}(x, y)}{f_X(x)} = \frac{x^3 e^{-x(1+y)}/2}{x^2 e^{-x}/2} = xe^{-xy}, \qquad y > 0,$$

and $f_{Y|X}(y \mid x) = 0$ for $y \le 0$; thus, the conditional distribution of Y, given that $X = x$, is the exponential distribution with mean $1/x$.

(b) The mean and variance of this conditional distribution are given by

$$\mu(x) = \frac{1}{x} \quad \text{and} \quad \sigma^2(x) = \frac{1}{x^2}. \quad ∎$$

EXAMPLE 6.22   Consider the function f on $\mathbb{R}^2$ defined by

$$f(x, y) = \begin{cases} 4xy - 2x - 2y + 2, & 0 < x, y < 1, \\ 0 & \text{elsewhere.} \end{cases}$$

(a) Show that f is a bivariate density function.

Let X and Y have f as their joint density function. Determine

(b) the distribution of X;

(c) the conditional distribution of Y, given that $X = x$, for $0 < x < 1$;

(d) the mean and variance of this conditional distribution.

Solution   **(a)** Observe that

$$4xy - 2x - 2y + 2 = 2xy + 2xy - 2x - 2y + 2 = 2xy + 2(1 - x)(1 - y) > 0$$

for $0 < x, y < 1$ and hence that f is nonnegative. Moreover,

$$\int_0^1 \left(\int_0^1 f(x, y) \, dy \right) dx = 4 \int_0^1 x \, dx \int_0^1 y \, dy - 2 \int_0^1 x \, dx - 2 \int_0^1 y \, dy + 2$$

$$= 4 \cdot \frac{1}{2} \cdot \frac{1}{2} - 2 \cdot \frac{1}{2} - 2 \cdot \frac{1}{2} + 2$$

$$= 1,$$

so f is a density function.

(b) The density function of X is given by

$$f_X(x) = \int_0^1 (4xy - 2x - 2y + 2) \, dy$$

$$= 4x \int_0^1 y \, dy - 2x - 2 \int_0^1 y \, dy + 2$$

$$= 2x - 2x - 1 + 2$$

$$= 1$$

for $0 < x < 1$ and $f_X(x) = 0$ elsewhere, so X is uniformly distributed on $(0, 1)$.

(c) According to the answer to part (b), the conditional density function of Y, given that $X = x \in (0, 1)$, is given by

$$f_{Y|X}(y \mid x) = \begin{cases} 4xy - 2x - 2y + 2, & 0 < y < 1, \\ 0 & \text{elsewhere.} \end{cases}$$

(d) The mean of this conditional density function is given by

$$\mu(x) = \int_0^1 y(4xy - 2x - 2y + 2) \, dy$$

$$= 4x \int_0^1 y^2 \, dy - 2x \int_0^1 y \, dy - 2 \int_0^1 y^2 \, dy + 2 \int_0^1 y \, dy$$

$$= \frac{4x}{3} - x - \frac{2}{3} + 1$$

$$= \frac{1}{3}(1 + x).$$

Its second moment is given by

$$\int_0^1 y^2(4xy - 2x - 2y + 2)\, dy$$

$$= 4x \int_0^1 y^3\, dy - 2x \int_0^1 y^2\, dy - 2 \int_0^1 y^3\, dy + 2 \int_0^1 y^2\, dy$$

$$= x - \frac{2x}{3} - \frac{1}{2} + \frac{2}{3}$$

$$= \frac{1}{6}(1 + 2x).$$

Thus, its variance is given by

$$\sigma^2(x) = \frac{1}{6}(1 + 2x) - \frac{(1+x)^2}{9} = \frac{1}{18}(1 + 2x - 2x^2). \quad \blacksquare$$

In advanced probability theory, there is a general result to the effect that if $\mathbf{X}$ and W are independent, then the conditional distribution of $Y = g(\mathbf{X}, W)$, given that $\mathbf{X} = \mathbf{x}$ coincides with the distribution of $g(\mathbf{x}, W)$. As a special case, we have the following result.

THEOREM 6.2   Let $\mathbf{X}$ be $\mathcal{X}$-valued and have a density function $f_{\mathbf{X}}$ that is positive on $\mathcal{X}$, let W have a density function f_W on $\mathbb{R}$, and suppose $\mathbf{X}$ and W are independent. Let g be a continuously differentiable, real-valued function on $\mathcal{X}$ and set $Y = g(\mathbf{X}) + W$. Then, for $\mathbf{x} \in \mathcal{X}$, the conditional distribution of Y, given that $\mathbf{X} = \mathbf{x}$, coincides with the distribution of $g(\mathbf{x}) + W$.

Proof   Since $\mathbf{X}$ and W are independent, their joint density function is given by

$$f_{\mathbf{X}, W}(\mathbf{x}, w) = f_{\mathbf{X}}(\mathbf{x})f_W(w), \qquad \mathbf{x} \in \mathbb{R}^m \text{ and } w \in \mathbb{R}.$$

Think of $(\mathbf{X}, Y)$ as a transform of $(\mathbf{X}, W)$. The solution to the equations

$$\mathbf{x} = \mathbf{x}$$

$$y = g(\mathbf{x}) + w$$

is given by

$$\mathbf{x} = \mathbf{x}$$

$$w = y - g(\mathbf{x})$$

or, equivalently, by

$$x_1 = x_1$$

$$\vdots$$

$$x_m = x_m$$

$$w = y - g_1(x_1, \ldots, x_m).$$

The derivative of the variables $x_1, \ldots, x_m, w$ with respect to the variables $x_1, \ldots, x_m, y$ is a lower-triangular matrix all of whose diagonal entries equal one. Thus, the Jacobian equals 1. Consequently,

$$f_{\mathbf{X},Y}(\mathbf{x}, y) = f_{\mathbf{X},W}(\mathbf{x}, y - g(\mathbf{x})) = f_{\mathbf{X}}(\mathbf{x})f_W(y - g(\mathbf{x})), \qquad \mathbf{x} \in \mathcal{X} \text{ and } y \in \mathbb{R}.$$

Therefore,

$$f_{Y|\mathbf{X}}(y \mid \mathbf{x}) = \frac{f_{\mathbf{X},Y}(\mathbf{x}, y)}{f_{\mathbf{X}}(\mathbf{x})} = f_W(y - g(\mathbf{x})), \qquad \mathbf{x} \in \mathcal{X} \text{ and } y \in \mathbb{R}.$$

Since $f_W(y - g(\mathbf{x}))$, $y \in \mathbb{R}$, is the density function of $g(\mathbf{x}) + W$ (see Corollary 1.2 in Section 1.7), the desired result is valid. ∎

EXAMPLE 6.23   Consider independent random variables X and W, where X has a positive density function on $\mathbb{R}$ and W is normally distributed with mean 0 and variance σ^2. Determine the conditional distribution of $Y = \alpha + \beta X + W$, given that $X = x$.

Solution   According to Theorem 6.2, the conditional distribution of Y, given that $X = x$, coincides with the distribution of $\alpha + \beta x + W$, which is normal with mean $\alpha + \beta x$ and variance σ^2. ∎

Problems

6.47   Let X and Y have the bivariate density function f on $\mathbb{R}^2$ given by $f(x, y) = x + y$ for $0 < x, y < 1$ and $f(x, y) = 0$ elsewhere. Determine the conditional density function of Y, given that $X = x$, where $0 < x < 1$.

6.48   Let X and Y be random variables having a joint density function $f_{X,Y}$. **(a)** Verify that

$$f_Y(y) = \int_{-\infty}^{\infty} f_X(x)f_{Y|X}(y \mid x)\, dx, \qquad y \in \mathbb{R}.$$

(b) Let y be such that $f_Y(y) > 0$. Verify that

$$f_{X|Y}(x \mid y) = \frac{f_X(x)f_{Y|X}(y \mid x)}{\int_{-\infty}^{\infty} f_X(x)f_{Y|X}(y \mid x)\, dx}, \qquad x \in \mathbb{R}.$$

6.49   Let (X, Y) be distributed as in Example 6.20. Determine **(a)** the conditional distribution of X, given that $Y = y$, for $0 < y < 4$; **(b)** the mean and variance of this conditional distribution.

6.50   Let (X, Y) be distributed as in Example 6.21. Determine **(a)** the conditional distribution of X, given that $Y = y$, for $y > 0$; **(b)** the mean and variance of this conditional distribution.

6.51   Let X and Y be positive random variables having the joint density function f given by

$$f(x, y) = \frac{\tau^\gamma}{\Gamma(\alpha)\Gamma(\gamma)} x^{\alpha-1} y^{\alpha+\gamma-1} e^{-(\tau+x)y}, \qquad x, y > 0,$$

where α, γ, and τ are positive parameters. Determine **(a)** the density function of X; **(b)** the conditional distribution of Y, given that $X = x > 0$.

6.52   Let X and Y be as in Problem 6.51. Determine **(a)** the distribution of Y; **(b)** the conditional distribution of X, given that $Y = y$.

6.53   Consider the function f on $\mathbb{R}^2$ defined by

$$f(x, y) = \begin{cases} 4e^{-2x-2y} - 2e^{-2x-y} - 2e^{-x-2y} + 2e^{-x-y}, & 0 < x, y < \infty, \\ 0 & \text{elsewhere.} \end{cases}$$

(a) Show that f is a bivariate density function. Let X and Y have f as their joint density function. Determine **(b)** the distribution of X; **(c)** the conditional density function of Y, given that $X = x$, for $x > 0$; **(d)** the mean and variance of this conditional density function.

6.54   Let the joint distribution of Y_1 and Y_2 be Dirichlet with parameters α_1, α_2, and α_3 (see Problems 5.64 and 5.65 in Section 5.5). Show that the conditional distribution of Y_2, given that $Y_1 = y_1 \in (0, 1)$, coincides with the distribution of $(1 - y_1)W$, where W has the beta distribution with parameters α_2 and α_3.

6.55   Determine, explicitly, the derivative appearing in the proof of Theorem 6.2 when $X = \mathbb{R}^2$ and $g(x_1, x_2) = \beta_0 + \beta_1 x_1 + \beta_2 x_2$ for $x_1, x_2 \in \mathbb{R}$.

6.56   Consider independent random variables X and W, where X has a positive density function on $\mathbb{R}$ and W is normally distributed with mean 0 and variance σ^2. Determine the conditional distribution of $Y = \beta_0 + \beta_1 X + \beta_2 X^2 + W$, given that $X = x$.

6.57   Let $\mathbf{X}$ be X-valued and have a density function that is positive on X, let W have a density function f_W on $\mathbb{R}$, and suppose $\mathbf{X}$ and W are independent. Let g be a continuously differentiable, positive function on X and set $Y = g(\mathbf{X})W$. Show that, for $\mathbf{x} \in X$, the conditional distribution of Y, given that $\mathbf{X} = \mathbf{x}$, coincides with the distribution of $g(\mathbf{x})W$.

6.58   (Continued.) Determine the conditional distribution of Y given that $\mathbf{X} = \mathbf{x}$ **(a)** if W has the gamma distribution with parameters α and β; **(b)** if W is normally distributed with mean 0 and variance σ^2.

6.5
Conditional Expectation

Consider random variables $X_1, \ldots, X_m, Y$; set $\mathbf{X} = [X_1, \ldots, X_m]^T$; and let X be a subset of $\mathbb{R}^m$ such that $P(\mathbf{X} \in X) = 1$. We assume, for simplicity, either that $X_1, \ldots, X_m$ and Y are discrete random variables or that they are continuous random variables having a joint density function. Let $f_{\mathbf{X}}$ denote the probability function or density function of $\mathbf{X}$ and let $f_{\mathbf{X}, Y}$ denote the joint probability function or joint density function of $\mathbf{X}$ and Y. It is assumed that $f_{\mathbf{X}}(\mathbf{x}) > 0$ for $\mathbf{x} \in X$. In the context of continuous random variables, it is assumed that $f_{\mathbf{X}}(\mathbf{x}) = 0$ for $\mathbf{x} \notin X$ and that $f_{\mathbf{X}, Y}(\mathbf{x}, y) = 0$ for $\mathbf{x} \notin X$ and $y \in \mathbb{R}$. (These conditions are automatically satisfied in the context of discrete random variables.) Let $f_{Y|\mathbf{X}}(y \,|\, \mathbf{x}) = f_{\mathbf{X}, Y}(\mathbf{x}, y)/f_{\mathbf{X}}(\mathbf{x})$, $y \in \mathbb{R}$, denote the conditional probability function or conditional density function of Y, given that $\mathbf{X} = \mathbf{x} \in X$.

As in the previous section, the mean, variance, and standard deviation of the conditional distribution of Y, given that $\mathbf{X} = \mathbf{x}$, are denoted by

$$\mu(\mathbf{x}) = E(Y \mid \mathbf{X} = \mathbf{x}),$$

$$\sigma^2(x) = \text{var}(Y \mid \mathbf{X} = \mathbf{x}),$$

and

$$\sigma(x) = \sqrt{\sigma^2(x)} = \text{SD}(Y \mid \mathbf{X} = \mathbf{x}),$$

respectively. It is assumed that $\mu(\mathbf{x})$ and $\sigma^2(\mathbf{x})$ are finite for $\mathbf{x} \in \mathcal{X}$. If Y is a discrete random variable, then

$$\mu(\mathbf{x}) = \sum_y y f_{Y \mid \mathbf{X}}(y \mid \mathbf{x}), \qquad \mathbf{x} \in \mathcal{X},$$

and

$$\sigma^2(\mathbf{x}) = \sum_y [y - \mu(\mathbf{x})]^2 f_{Y \mid \mathbf{X}}(y \mid \mathbf{x}), \qquad \mathbf{x} \in \mathcal{X};$$

if Y is a continuous random variable, then

$$\mu(\mathbf{x}) = \int_{\mathbb{R}} y f_{Y \mid \mathbf{X}}(y \mid \mathbf{x}) \, dy, \qquad \mathbf{x} \in \mathcal{X},$$

and

$$\sigma^2(\mathbf{x}) = \int_{\mathbb{R}} [y - \mu(\mathbf{x})]^2 f_{Y \mid \mathbf{X}}(y \mid \mathbf{x}) \, dy, \qquad \mathbf{x} \in \mathcal{X}.$$

The random variable $\mu(\mathbf{X})$ is commonly denoted by $E(Y \mid \mathbf{X})$ and referred to as the *conditional expectation* of Y given $\mathbf{X}$, and the random variable $\sigma^2(\mathbf{X})$ is commonly denoted by $\text{var}(Y \mid \mathbf{X})$ and referred to as the *conditional variance* of Y given $\mathbf{X}$. Observe that $E(Y \mid \mathbf{X})$ and $\text{var}(Y \mid \mathbf{X})$ are both transforms of $\mathbf{X}$.

In the proofs of Theorems 6.3 and 6.4, it is assumed that the random variables under consideration are discrete. If these random variables are continuous, then the same proofs work with summation replaced by integration.

THEOREM 6.3   Let g be a real-valued function on $\mathcal{X}$. If $g(\mathbf{X})Y$ has finite mean, then

$$E[\, g(\mathbf{X})Y \,] = E[\, g(\mathbf{X})E(Y \mid \mathbf{X}) \,]. \tag{31}$$

In particular, if Y and $g(\mathbf{X})$ have finite variance, then (31) holds.

Proof   If $g(\mathbf{X})Y$ has finite mean, then

$$E[\, g(\mathbf{X})Y \,] = \sum_{\mathbf{x}} \sum_y g(\mathbf{x}) y f_{\mathbf{X}, Y}(\mathbf{x}, y)$$

$$= \sum_{\mathbf{x}} \sum_y g(\mathbf{x}) y f_{\mathbf{X}}(\mathbf{x}) f_{Y \mid \mathbf{X}}(y \mid \mathbf{x})$$

$$= \sum_{\mathbf{x}} g(\mathbf{x}) \left(\sum_y y f_{Y \mid \mathbf{X}}(y \mid \mathbf{x}) \right) f_{\mathbf{X}}(\mathbf{x})$$

$$= \sum_{\mathbf{x}} g(\mathbf{x})\mu(\mathbf{x})f_{\mathbf{X}}(\mathbf{x})$$

$$= E[\,g(\mathbf{X})\mu(\mathbf{X})\,]$$

$$= E[\,g(\mathbf{X})E(Y\,|\,\mathbf{X})\,],$$

so (31) holds. If Y and $g(\mathbf{X})$ have finite variance, then $g(\mathbf{X})Y$ has finite mean by Theorem 2.4(b) in Section 2.3 and hence (31) holds. $\blacksquare$

Equation (31) is often written as $E\{g(\mathbf{X})[\,Y - E(Y\,|\,\mathbf{X})\,]\} = 0$. By setting $g = 1$ in Theorem 6.3, we get the following result.

COROLLARY 6.1   Let Y have finite mean. Then $EY = E[\,E(Y\,|\,\mathbf{X})\,]$.

THEOREM 6.4   Let g be a real-valued function on $\mathfrak{X}$. Then

$$E\{[\,Y - g(\mathbf{X})\,]^2\} = E[\,\mathrm{var}(Y\,|\,\mathbf{X})\,] + E\{[\,E(Y\,|\,\mathbf{X}) - g(\mathbf{X})\,]^2\}.$$

Proof   Observe that

$$\sum_{y}[\,y - g(\mathbf{x})\,]^2 f_{Y|\mathbf{X}}(y\,|\,\mathbf{x}) = \sigma^2(\mathbf{x}) + [\,\mu(\mathbf{x}) - g(\mathbf{x})\,]^2, \qquad \mathbf{x} \in \mathfrak{X}.$$

[Apply Equation (23) in Section 2.3 with $c = g(\mathbf{x})$ to a random variable having $f_{Y|\mathbf{X}}(\cdot\,|\,\mathbf{x})$ as its probability function.] Thus,

$$E\{[\,Y - g(\mathbf{X})\,]^2\} = \sum_{\mathbf{x}}\sum_{y}[\,y - g(\mathbf{x})\,]^2 f_{\mathbf{X},Y}(\mathbf{x},y)$$

$$= \sum_{\mathbf{x}}\sum_{y}[\,y - g(\mathbf{x})\,]^2 f_{Y|\mathbf{X}}(y\,|\,\mathbf{x})f_{\mathbf{X}}(\mathbf{x})$$

$$= \sum_{\mathbf{x}}\{\sigma^2(\mathbf{x}) + [\,\mu(\mathbf{x}) - g(\mathbf{x})\,]^2\}f_{\mathbf{X}}(\mathbf{x})$$

$$= E[\,\sigma^2(\mathbf{X})\,] + E\{[\,\mu(\mathbf{X}) - g(\mathbf{X})\,]^2\}$$

$$= E[\,\mathrm{var}(Y\,|\,\mathbf{X})\,] + E\{[\,E(Y\,|\,\mathbf{X}) - g(\mathbf{X})\,]^2\}. \quad \blacksquare$$

Choosing $g = EY$ in Theorem 6.4 and using Corollary 6.1, we get the next result.

COROLLARY 6.2   Let Y have finite mean. Then $\mathrm{var}(Y) = E[\,\mathrm{var}(Y\,|\,\mathbf{X})\,] + \mathrm{var}(E(Y\,|\,\mathbf{X}))$.

Corollary 6.2, in turn, has the following consequence.

COROLLARY 6.3   Let Y have finite variance. Then $E(Y\,|\,\mathbf{X})$ has finite variance, and $\mathrm{var}(Y\,|\,\mathbf{X})$ has finite mean.

Letting the function g in Theorem 6.4 be the regression function of Y on $\mathbf{X}$, we get the following result.

COROLLARY 6.4 $E\{[\,Y - E(Y\,|\,\mathbf{X})\,]^2\} = E[\,\text{var}(Y\,|\,\mathbf{X})\,].$

In advanced probability theory, it is shown that Theorems 6.3 and 6.4 and Corollaries 6.1–6.4 hold without requiring that the random variables under consideration either be discrete or have a joint density function.

EXAMPLE 6.24   Let $W_1, W_2, \ldots$ be independent random variables, each having mean μ and finite variance σ^2. For n a nonnegative integer, set $S_n = W_1 + \cdots + W_n$ (with $S_0 = 0$). Then S_n has mean $n\mu$ and variance $n\sigma^2$. Let N be a nonnegative integer-valued random variable having finite mean and variance and which is independent of W_1, $W_2, \ldots$. Determine

(a) the conditional mean and variance of S_N, given that $N = n$;

(b) the mean and variance of S_N.

Solution   (a) Intuitively, we should have that

$$E(S_N\,|\,N = n) = E(S_n) = n\mu \tag{32}$$

and

$$\text{var}(S_N\,|\,N = n) = \text{var}(S_n) = n\sigma^2. \tag{33}$$

These results are clearly true when $n = 0$. Suppose, instead, that n is a positive integer. Suppose also (for simplicity) that $W_1, W_2, \ldots$ are discrete random variables. Let y be a possible value of S_N. Then

$$
\begin{aligned}
P(S_N = y\,|\,N = n) &= \frac{P(N = n, S_N = y)}{P(N = n)} \\[2mm]
&= \frac{P(N = n, S_n = y)}{P(N = n)} \\[2mm]
&= \frac{P(N = n)P(S_n = y)}{P(N = n)} \\[2mm]
&= P(S_n = y).
\end{aligned}
$$

Thus,

$$E(S_N\,|\,N = n) = \sum_y yP(S_N = y\,|\,N = n) = \sum_y yP(S_n = y) = E(S_n) = n\mu,$$

so

$$
\begin{aligned}
\text{var}(S_N\,|\,N = n) &= \sum_y (y - n\mu)^2 P(S_N = y\,|\,N = n) \\[2mm]
&= \sum_y (y - n\mu)^2 P(S_n = y)
\end{aligned}
$$

$$= \text{var}(S_n)$$

$$= n\sigma^2.$$

Consequently, (32) and (33) are indeed valid.

(b) Using (32) and Corollary 6.1, we get

$$E(S_N) = E[\, E(S_N \,|\, N)\,]$$

$$= \sum_{n=0}^{\infty} E(S_N \,|\, N = n) P(N = n)$$

$$= \sum_{n=0}^{\infty} (n\mu) P(N = n)$$

$$= \mu \sum_{n=0}^{\infty} n P(N = n)$$

$$= \mu E N.$$

Using (33) and Corollary 6.2, we get

$$\text{var}(S_N) = E[\, \text{var}(S_N \,|\, N)\,] + \text{var}[\, E(S_N \,|\, N)\,]$$

$$= \sum_{n=0}^{\infty} \text{var}(S_n \,|\, N = n) P(N = n) + \text{var}(N\mu)$$

$$= \sum_{n=0}^{\infty} n\sigma^2 P(N = n) + \mu^2 \text{var}(N)$$

$$= \sigma^2 \sum_{n=0}^{\infty} n P(N = n) + \mu^2 \text{var}(N)$$

$$= \sigma^2 E N + \mu^2 \text{var}(N). \quad \blacksquare$$

EXAMPLE 6.25 Accidental injuries at a certain factory occur according to a Poisson process with the mean rate of ten injuries per month, and the cost per injury has mean \$1000 and standard deviation \$500. Determine the mean and standard deviation of the total cost of all the accidental injuries occurring at the factory during a 1-month period.

Solution Let N denote the number of accidental injuries occurring at the factory during the 1-month period, which has the Poisson distribution with mean 10 and hence variance 10, and let Y denote the total cost of these accidents. We assume that $Y = S_N = W_1 + \cdots + W_N$, where $W_1, W_2, \ldots$ are independent random variables having common mean $\mu = 1000$ and standard deviation $\sigma = 500$ and N is independent of $W_1, W_2, \ldots$. Then, by the solution to Example 6.24(b),

$$EY = ES_N = \mu E N = 1000 \cdot 10 = \$10,000,$$

and

$$\text{var}(Y) = \text{var}(S_N) = \sigma^2 E N + \mu^2 \text{var}(N) = 10 \cdot 500^2 + 10 \cdot 1000^2 = 50 \cdot 500^2,$$

so
$$SD(Y) = 500\sqrt{50} = \$3536. \quad \blacksquare$$

Let $X_1, \ldots, X_m$ and Y have finite variance. Recall that
$$\mathrm{cov}(\mathbf{X}, Y) = E[(\mathbf{X} - E\mathbf{X})(Y - EY)].$$
Since $E[(\mathbf{X} - E\mathbf{X})EY] = (EY)[E(\mathbf{X} - E\mathbf{X})] = 0$, we see that
$$\mathrm{cov}(\mathbf{X}, Y) = E[(\mathbf{X} - E\mathbf{X})Y]. \tag{34}$$

COROLLARY 6.5 If $X_1, \ldots, X_m$ and Y have finite variance, then
$$\mathrm{cov}(\mathbf{X}, Y) = \mathrm{cov}(\mathbf{X}, E(Y \mid \mathbf{X})).$$

Proof By (34) and Theorem 6.3,
$$\mathrm{cov}(\mathbf{X}, Y) = E[(\mathbf{X} - E\mathbf{X})Y] = E[(\mathbf{X} - E\mathbf{X})E(Y \mid \mathbf{X})] = \mathrm{cov}(\mathbf{X}, E(Y \mid \mathbf{X})). \quad \blacksquare$$

COROLLARY 6.6 Suppose $X_1, \ldots, X_m$ and Y have finite variance and that the regression function of Y on $\mathbf{X}$ has the form $\mu(\mathbf{x}) = \alpha + \boldsymbol{\beta}^T \mathbf{x}$, $\mathbf{x} \in \mathcal{X}$, where $\alpha \in \mathbb{R}$ and $\boldsymbol{\beta} \in \mathbb{R}^m$ (that is, it is a linear function of $\mathbf{x}$). Then $\hat{Y} = E(Y \mid \mathbf{X}) = \alpha + \boldsymbol{\beta}^T \mathbf{X}$ is the best linear predictor of Y based on $\mathbf{X}$.

Proof By Corollary 6.1,
$$EY = E[E(Y \mid \mathbf{X})] = E[\mu(\mathbf{X})] = E(\alpha + \boldsymbol{\beta}^T \mathbf{X}) = \alpha + \boldsymbol{\beta}^T E\mathbf{X},$$
so $\alpha = EY - \boldsymbol{\beta}^T E\mathbf{X}$ and hence $\hat{Y} = EY + \boldsymbol{\beta}^T(\mathbf{X} - E\mathbf{X})$. Moreover, by Corollary 6.5,
$$\mathrm{cov}(\mathbf{X}, Y) = \mathrm{cov}(\mathbf{X}, E(Y \mid \mathbf{X}))$$
$$= \mathrm{cov}(\mathbf{X}, \mu(\mathbf{X}))$$
$$= \mathrm{cov}(\mathbf{X}, \alpha + \boldsymbol{\beta}^T \mathbf{X})$$
$$= \mathrm{cov}(\mathbf{X}, \mathbf{X})\boldsymbol{\beta}$$
$$= \mathrm{VC}(\mathbf{X})\boldsymbol{\beta}$$

and hence $\boldsymbol{\beta} = [\mathrm{VC}(\mathbf{X})]^{-1}\mathrm{cov}(\mathbf{X}, Y)$. The desired result now follows from Corollary 5.7 in Section 5.4. $\blacksquare$

EXAMPLE 6.26 Let X and Y have the joint density function
$$f(x, y) = \begin{cases} 4xy - 2x - 2y + 2, & 0 < x, y < 1, \\ 0 & \text{elsewhere.} \end{cases}$$
Determine

(a) the mean and variance of X and of Y;

(b) the covariance and correlation between X and Y;

(c) the best linear predictor of Y based on X and the mean squared error of this predictor.

Solution (a) According to the solution to Example 6.22(b) in Section 6.4, X is uniformly distributed on $(0,1)$, so it has mean $1/2$ and variance $1/12$ (see Example 2.16 in Section 2.1 and Example 2.40 in Section 2.3). Similarly, Y has the same distribution and hence the same mean and variance.

(b) Now

$$E(XY) = \int_0^1 \left(\int_0^1 xy(4xy - 2x - 2y + 2)\, dy \right) dx$$

$$= 4 \int_0^1 x^2\, dx \int_0^1 y^2\, dy - 2 \int_0^1 x^2\, dx \int_0^1 y\, dy$$

$$\qquad - 2 \int_0^1 x\, dx \int_0^1 y^2\, dy + 2 \int_0^1 x\, dx \int_0^1 y\, dy$$

$$= 4 \cdot \frac{1}{3} \cdot \frac{1}{3} - 2 \cdot \frac{1}{3} \cdot \frac{1}{2} - 2 \cdot \frac{1}{2} \cdot \frac{1}{3} + 2 \cdot \frac{1}{2} \cdot \frac{1}{2}$$

$$= \frac{4}{9} - \frac{1}{3} - \frac{1}{3} + \frac{1}{2}$$

$$= \frac{5}{18},$$

so $\mathrm{cov}(X, Y) = E(XY) - (EX)(EY) = \frac{5}{18} - \frac{1}{2} \cdot \frac{1}{2} = \frac{1}{36}$.

Alternatively, by the solution to Example 6.22(d) in Section 6.4, the regression function of Y on X is given by

$$\mu(x) = \frac{1}{3}(1 + x), \qquad x > 0.$$

Thus, by Corollary 6.5,

$$\mathrm{cov}(X, Y) = \mathrm{cov}(X, \mu(X)) = \mathrm{cov}\left(X, \frac{1}{3}(1 + X)\right) = \frac{1}{3}\mathrm{var}(X) = \frac{1}{36}.$$

The correlation between X and Y is given by

$$\mathrm{cor}(X, Y) = \frac{\mathrm{cov}(X, Y)}{\mathrm{SD}(X)\mathrm{SD}(Y)} = \frac{1/36}{1/12} = \frac{1}{3}.$$

(c) The best linear predictor of Y based on X is given [see Equation (17) in Section 5.1] by

$$\hat{Y} = EY + \rho\frac{\mathrm{SD}(Y)}{\mathrm{SD}(X)}(X - EX) = \frac{1}{2} + \frac{1}{3}\left(X - \frac{1}{2}\right) = \frac{1}{3}(1 + X).$$

The mean squared error of this predictor is given [see Equation (18) in Section 5.1] by

$$\mathrm{MSE}(\hat{Y}) = (1 - \rho^2)\mathrm{var}(Y) = \frac{8}{9} \cdot \frac{1}{12} = \frac{2}{27}.$$

Alternatively, since X is a $[0, 1]$-valued random variable and the regression function of Y on X is linear on $[0, 1]$, we conclude from Corollary 6.6 that the best linear predictor of Y based on X is given by $\hat{Y} = \mu(X) = \frac{1}{3}(1 + X)$. According to the solution to Example 6.22(d) in Section 6.4, $\text{var}(Y \mid X) = \frac{1}{18}(1 + 2X - 2X^2)$. Thus, by Corollary 6.4,

$$
\begin{aligned}
\text{MSE}(\hat{Y}) &= E\{[\, Y - E(Y \mid X)\,]^2\} \\
&= E[\,\text{var}(Y \mid X)\,] \\
&= \frac{1}{18} E(1 + 2X - 2X^2) \\
&= \frac{1}{18}[\,1 + 2EX - 2E(X^2)\,] \\
&= \frac{1}{18}\left[1 + 2\left(\frac{1}{2}\right) - 2\left(\left(\frac{1}{2}\right)^2 + \frac{1}{12}\right)\right] \\
&= \frac{2}{27}. \quad \blacksquare
\end{aligned}
$$

Now we will discuss the choice of the word *regression* in *regression function*. Suppose X and Y have finite variance and let ρ denote the correlation between these random variables. Since X and Y are assumed to have a joint density function, their variance–covariance matrix is positive definite (see Theorem 5.19 in Section 5.5), so $-1 < \rho < 1$ (see the solution to Example 5.14 in Section 5.3). Suppose the regression function of Y on X is linear; that is, it has the form $\mu(x) = \alpha + \beta x$, $x \in \mathcal{X}$. Then, by Corollary 6.6, $\alpha + \beta X$ is the best linear predictor of Y based on X. Thus,

$$
\alpha + \beta X = EY + \rho \frac{\text{SD}(Y)}{\text{SD}(X)}(X - EX),
$$

and hence

$$
\mu(x) = EY + \rho \frac{\text{SD}(Y)}{\text{SD}(X)}(x - EX), \qquad x \in \mathcal{X}. \tag{35}
$$

Suppose, in particular, that X and Y have common mean μ and common variance. Then (35) simplifies to

$$
\mu(x) = \mu + \rho(x - \mu), \qquad x \in \mathcal{X}. \tag{36}
$$

It follows from (36) (the details being left as a problem) that

$$
\mu(x) > x \text{ for } x < \mu, \quad \mu(x) = x \text{ for } x = \mu \quad \text{and} \quad \mu(x) < x \text{ for } x > \mu. \tag{37}
$$

Suppose, further, that $\rho > 0$. Then, by (36) and (37),

$$
x < \mu(x) < \mu \text{ for } x < \mu, \quad \mu(x) = \mu \text{ for } x = \mu \quad \text{and} \quad \mu < \mu(x) < x \text{ for } x > \mu. \tag{38}
$$

We summarize (38) by saying that the regression function *regresses to the mean* (see Figure 6.2).

FIGURE 6.2

When $EX = EY = \mu$, $\text{var}(X) = \text{var}(Y)$, $0 < \rho < 1$, and the regression function is linear, then it regresses to the mean; that is, it lies between the line $y = \mu$ and the line $y = x$.

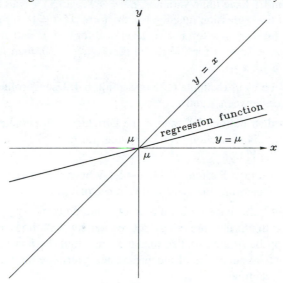

Consider, for example, a random father–son pair in some population. Let X represent the height (in inches) of the father and Y the adult height of the son. Suppose $EX = EY = 70$, $\text{SD}(X) = \text{SD}(Y)$, $\rho = \text{cor}(X, Y) = .8$, and that the regression function $\mu(\cdot)$ of Y on X is linear. Then $\mu(x) = 70 + .8(x - 70)$. In particular, the expected adult height of the son of a 6-foot father is given by $\mu(72) = 71.6$, which is greater than the overall mean but less than the height of the father.

Problems

6.59 Let X and Y have the bivariate density function f on $\mathbb{R}^2$ given by $f(x, y) = x + y$ for $0 < x, y < 1$ and $f(x, y) = 0$ elsewhere. Determine $E(Y \mid X = x)$ for $0 < x < 1$ (see Problem 6.47 in Section 6.4).

6.60 Two fair dice are rolled in an honest manner. Let X and Y denote, respectively, the maximum and minimum of the numbers of spots showing on the two dice. Determine $E(Y \mid X = x)$ for $1 \leq x \leq 6$.

6.61 Let g be a real-valued function on $\mathcal{X}$. Show that if $g(\mathbf{X})[Y - E(Y \mid \mathbf{X})]$ has finite mean, then its mean is zero. [*Hint:* Imitate the proof of Theorem 6.3.]

6.62 (Continued.) Show that if $Y - E(Y \mid \mathbf{X})$ has finite mean, then **(a)** $E[Y - E(Y \mid \mathbf{X})] = 0$; **(b)** $\text{var}(Y - E(Y \mid \mathbf{X})) = E[\text{var}(Y \mid \mathbf{X})]$.

6.63 Show that if $Y - E(Y \mid \mathbf{X})$ fails to have finite mean, then $\text{var}(Y - E(Y \mid \mathbf{X})) = E[\text{var}(Y \mid \mathbf{X})] = \infty$.

6.64 In the context of Example 6.24, determine **(a)** $\text{cov}(N, S_N)$; **(b)** the best linear predictor of S_N based on N; **(c)** the mean squared error of this predictor.

6.65 Suppose X and Y have finite variance, $EX = EY$, $\text{var}(X) = \text{var}(Y)$, $-1 < \text{cor}(X, Y) < 1$ and that the regression function has the form $\mu(x) = \alpha + \beta x$, $x \in \mathcal{X}$. Show that (for $x \in \mathcal{X}$) **(a)** $\mu(x) > x$ for $x < \mu$, $\mu(x) = x$ for $x = \mu$ and $\mu(x) < x$ for $x > \mu$; **(b)** $|\mu(x) - \mu| < |x - \mu|$ for $x \neq \mu$; **(c)** if $\text{cor}(X, Y) > 0$, then $\mu(x) < \mu$ for $x < \mu$ and $\mu(x) > \mu$ for $x > \mu$.

6.66 Let X and Y be as in Example 6.21 in Section 6.4. Use Corollary 6.5 to determine the covariance between X and Y.

6.67 Let the joint distribution of Y_1 and Y_2 be Dirichlet with parameters α_1, α_2, and α_3 (see Problem 6.54 in Section 6.4). Determine **(a)** $E(Y_2 \mid Y_1 = y_1)$; **(b)** $\text{var}(Y_2 \mid Y_1 = y_1)$; **(c)** $\text{cov}(Y_1, Y_2)$.

6.68 Consider the setup of Example 6.24, and let N have the geometric distribution with parameter π. Determine the mean and variance of S_N.

6.69 Suppose light bulbs have probability π of being defective (not working when first installed) and that if they are not defective their survival times are exponential with mean β. Light bulbs are installed one at a time until the first defect occurs. Under the appropriate assumptions of independence, determine the mean and variance of the total survival time.

6.70 Consider a Poisson process with rate λ and let $N(t)$ denote the number of events that have occurred by time t. Also, let T be a positive random variable that is independent of the Poisson process. Then $E(N(T) \mid T = t) = \lambda t$ and $\text{var}(N(t) \mid T = t) = \lambda t$ for $t \geq 0$. Use these results to determine **(a)** $E(N(T))$; **(b)** $\text{var}(N(T))$; **(c)** $\text{cov}(T, N(T))$; **(d)** the best linear predictor of $N(T)$ based on T; **(e)** the mean squared error of this predictor.

6.6
Prediction

In Section 5.4 we found the linear predictor of Y based on $X_1, \ldots, X_m$ that is best among all linear predictors. Here we find that predictor that is best among all predictors, linear or not. It might seem impossible to solve such a general problem, but properties of conditional expectation that were developed in the previous section readily yield the best predictor.

Thus, let $X_1, \ldots, X_m$ and Y be as in Section 6.5. Consider the task of predicting Y based on $\mathbf{X} = [X_1, \ldots, X_m]^T$—that is, by using a predictor of the form $g(\mathbf{X})$. Then $Y - g(\mathbf{X})$ is the error of prediction, and $\text{MSE}(g(\mathbf{X})) = E\{[Y - g(\mathbf{X})]^2\}$ is the mean squared error. By Theorem 6.4 in Section 6.5,

$$\text{MSE}(g(\mathbf{X})) = E[\sigma^2(\mathbf{X})] + E\{[\mu(\mathbf{X}) - g(\mathbf{X})]^2\}, \tag{39}$$

so

$$\text{MSE}(g(\mathbf{X})) \geq E[\sigma^2(\mathbf{X})] \tag{40}$$

and

$$\text{MSE}(g(\mathbf{X})) = E[\sigma^2(\mathbf{X})] \quad \text{if and only if} \quad P(g(\mathbf{X}) = \mu(\mathbf{X})) = 1. \tag{41}$$

It follows from (40) that if $E[\sigma^2(\mathbf{X})] = \infty$, then there is no predictor of Y based on $\mathbf{X}$ having finite mean squared error.

Suppose $E[\sigma^2(\mathbf{X})] < \infty$. It follows from (40) and (41) that $\hat{Y} = \mu(\mathbf{X})$ is the unique best predictor of Y based on $\mathbf{X}$, where *best* means having the smallest possible mean squared error and *unique* means that if $g(\mathbf{X})$ is also a best predictor of Y based on $\mathbf{X}$, then $P(g(\mathbf{X}) = \hat{Y}) = 1$. We have now established the following result.

THEOREM 6.5   Suppose $E[\sigma^2(\mathbf{X})] < \infty$. Then $\hat{Y} = \mu(\mathbf{X})$ is the unique best predictor of Y based on $\mathbf{X}$, and its mean squared error equals $E[\sigma^2(\mathbf{X})]$.

Let Y have finite variance. Then the best constant predictor of Y is the mean of its distribution. Using this result, we get the following interpretation of Theorem 6.5: To get the best predictor of Y based on $\mathbf{X}$, determine the conditional distribution of Y, given $\mathbf{X}$, and then use the best constant predictor of a random variable having this conditional distribution as its distribution.

EXAMPLE 6.27   Let (X, Y) be uniformly distributed on the triangle having the vertices $(0, 0)$, $(8, 0)$, and $(0, 4)$. Determine the best linear predictor of Y on X and the mean squared error of this predictor.

Solution   Here X is an $\mathcal{X}$-valued random variable, where $\mathcal{X} = (0, 8)$. According to the solution to Example 6.20 in Section 6.4, the density function of X is given by

$$f_X(x) = \frac{4 - \frac{x}{2}}{16}, \qquad 0 < x < 8,$$

and the mean and variance of the conditional distribution of Y, given that $X = x$, are given by

$$\mu(x) = 2 - \frac{x}{4} \quad \text{and} \quad \sigma^2(x) = \frac{1}{12}\left(4 - \frac{x}{2}\right)^2, \qquad 0 < x < 8.$$

Since the regression function is linear, it follows from Corollary 6.6 in Section 6.5 that the best linear predictor of Y based on X is given by $\hat{Y} = \mu(X) = 2 - X/4$. The mean squared error of this predictor is given by

$$\mathrm{MSE}(\hat{Y}) = E[\sigma^2(X)]$$

$$= \int_0^8 \sigma^2(x) f_X(x)\, dx$$

$$= \frac{1}{192} \int_0^8 \left(4 - \frac{x}{2}\right)^3 dx$$

$$= -\frac{1}{384}\left(4 - \frac{x}{2}\right)^4 \Big|_0^8$$

$$= \frac{2}{3}.$$

Alternatively, the best linear predictor of Y on X and its mean squared error can be obtained by using techniques in Sections 5.1 and 5.5 (see Problem 5.61 in Section 5.5). ▪

Let g be a real-valued function on $\mathcal{X}$. It follows from Theorem 6.4 in Section 6.5 and Theorem 6.5 in this section that the mean squared error of $g(\mathbf{X})$ as a predictor of Y based on $\mathbf{X}$ exceeds the mean squared error of the best such predictor $\mu(\mathbf{X})$ by the amount

$$\text{MSE}(g(\mathbf{X})) - \text{MSE}(\mu(\mathbf{X})) = E\{[\,\mu(\mathbf{X}) - g(\mathbf{X})\,]^2\} = \int_{\mathcal{X}} [\,\mu(\mathbf{x}) - g(\mathbf{x})\,]^2 f_{\mathbf{X}}(\mathbf{x})\,d\mathbf{x}.$$

Suppose, in particular, that $X_1, \ldots, X_m$ and Y have finite variance and let $a \in \mathbb{R}$ and $\mathbf{b} \in \mathbb{R}^m$. Then the mean squared error of the linear predictor $a + \mathbf{b}^T\mathbf{X}$ of Y based on $\mathbf{X}$ exceeds the mean squared error of the best predictor of Y based on $\mathbf{X}$ by the amount

$$\begin{aligned}\text{MSE}(a + \mathbf{b}^T\mathbf{X}) - \text{MSE}(\mu(\mathbf{X})) &= E\{[\,\mu(\mathbf{X}) - a - \mathbf{b}^T\mathbf{X}\,]^2\}\\ &= \int_{\mathcal{X}}[\,\mu(\mathbf{x}) - a - \mathbf{b}^T\mathbf{x}\,]^2 f_{\mathbf{X}}(\mathbf{x})\,d\mathbf{x}.\end{aligned} \tag{42}$$

Let $\alpha + \boldsymbol{\beta}^T\mathbf{X}$ be the best linear predictor of Y based on $\mathbf{X}$. We see from (42) that α and $\boldsymbol{\beta}$ are the unique values of a and $\mathbf{b}$, respectively, that minimize the quantity

$$E\{[\,\mu(\mathbf{X}) - a - \mathbf{b}^T\mathbf{X}\,]^2\} = \int_{\mathcal{X}}[\,\mu(\mathbf{x}) - a - \mathbf{b}^T\mathbf{x})\,]^2 f_{\mathbf{X}}(\mathbf{x})\,d\mathbf{x}.$$

In light of this result, we refer to $\alpha + \boldsymbol{\beta}^T\mathbf{x}$, $\mathbf{x} \in \mathcal{X}$, as the *best linear approximation* to the regression function of Y on $\mathbf{X}$. The excess of the mean squared error of the best linear predictor of Y based on $\mathbf{X}$ over that of the best predictor of Y based on $\mathbf{X}$ is given by

$$E\{[\,\mu(\mathbf{X}) - \alpha - \boldsymbol{\beta}^T\mathbf{X}\,]^2\} = \int_{\mathcal{X}}[\,\mu(\mathbf{x}) - \alpha - \boldsymbol{\beta}^T\mathbf{x}\,]^2 f_{\mathbf{X}}(\mathbf{x})\,d\mathbf{x}; \tag{43}$$

that is, as the integral with respect to the density function of $\mathbf{X}$ of the square of the difference between the regression function of Y on $\mathbf{X}$ and its best linear approximation.

EXAMPLE 6.28　Let X and Y have the joint density function given by

$$f_{X,Y}(x, y) = \frac{x^3}{2} e^{-x(1+y)}, \qquad x, y > 0,$$

and $f_{X,Y} = 0$ elsewhere.

(a) Determine the best predictor of Y based on X.

(b) Determine the mean squared error of this predictor.

(c) Determine the excess of the mean squared error of the best linear predictor of Y based on X over that of the best predictor of Y based on X.

(d) Show graphs of the density function of X, the regression function of Y on X, and its best linear approximation.

Solution
(a) According to the solution to Example 6.21(b) in Section 6.4, the regression function of Y on X is given by $\mu(x) = 1/x$ for $x > 0$. Thus, the best predictor of Y based on X is given by $\mu(X) = 1/X$. [Since $1/x > 0$ for $x > 0$, the best predictor does not suffer from the defect of the best linear predictor $Y = 1 - X/6$ that was pointed out at the end of the solution to Example 5.22(d) in Section 5.5.]

(b) Again by the solution to Example 6.21(b) in Section 6.4, the conditional variance of Y, given that $X = x$, is given by $\sigma^2(x) = 1/x^2$ for $x > 0$. According to the solution to Example 5.19(a) in Section 5.5, the density function of X is given by $f_X(x) = x^2 e^{-x}/2$ for $x > 0$ and $f_X = 0$ elsewhere. Thus, the mean squared error of the best predictor of Y based on X is given by

$$E[\sigma^2(X)] = \int_0^\infty \sigma^2(x) f_X(x)\, dx = \frac{1}{2}\int_0^\infty e^{-x}\, dx = \frac{1}{2}.$$

(c) According to the solution to Example 5.22(d) in Section 5.5, the best linear predictor of Y based on X is $1 - X/6$, and its mean squared error equals $2/3$. Thus, the excess of the mean squared error of $1 - X/6$ as a predictor of Y over that of $1/X$ is given by $2/3 - 1/2 = 1/6$. Alternatively, by Equation (43), this excess is given by

$$\int_0^\infty \left(\frac{1}{x} - 1 + \frac{x}{6}\right)^2 \frac{x^2}{2} e^{-x}\, dx$$

$$= \frac{1}{2}\int_0^\infty \left(1 - 2x + \frac{4x^2}{3} - \frac{x^3}{3} + \frac{x^4}{36}\right) e^{-x}\, dx$$

$$= \frac{1}{2}\left(1 - 2 + \frac{8}{3} - 2 + \frac{2}{3}\right)$$

$$= \frac{1}{6}.$$

(d) The density function of X is shown in Figure 6.3, and the regression function of Y on X and its best linear approximation are shown in Figure 6.4. ∎

EXAMPLE 6.29   Consider a box having four balls labeled from 1 to 4. Two balls are selected from the box by sampling without replacement. Let X and Y respectively denote the smaller and larger label of the balls in the sample. Determine

(a) the mean and variance of X;

(b) the mean and variance of Y;

(c) $\operatorname{cov}(X, Y)$ and $\operatorname{cor}(X, Y)$;

(d) the best linear predictor of Y based on X and the mean squared error of this predictor;

(e) the best predictor of Y based on X and the mean squared error of this predictor.

FIGURE   6.3

Density function of X

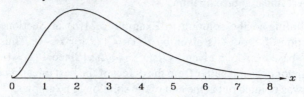

FIGURE   6.4

Regression function of Y on X and its best linear approximation

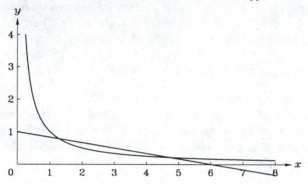

Solution   There are six possible ordered pairs of values for X and Y, which are as follows:

x	1	1	1	2	2	3
y	2	3	4	3	4	4

Each of these ordered pairs has probability $\frac{1}{6}$ of occurring.

(a) The mean and variance of X are given by

$$EX = \frac{1}{6}(3 \cdot 1 + 2 \cdot 2 + 1 \cdot 3) = \frac{5}{3}$$

and

$$\text{var}(X) = \frac{1}{6}\left(3\left(-\frac{2}{3}\right)^2 + 2\left(\frac{1}{3}\right)^2 + \left(\frac{4}{3}\right)^2\right) = \frac{5}{9}.$$

(b) The mean and variance of Y are given by

$$EY = \frac{1}{6}(1 \cdot 2 + 2 \cdot 3 + 3 \cdot 4) = \frac{10}{3}$$

and

$$\text{var}(Y) = \frac{1}{6}\left(1\left(-\frac{4}{3}\right)^2 + 2\left(-\frac{1}{3}\right)^2 + 3\left(\frac{2}{3}\right)^2\right) = \frac{5}{9}.$$

(c) Now

$$E(XY) = \frac{1}{6}(1 \cdot 2 + 1 \cdot 3 + 1 \cdot 4 + 2 \cdot 3 + 2 \cdot 4 + 3 \cdot 4) = \frac{35}{6}.$$

Thus, the covariance and correlation between X and Y are given by

$$\text{cov}(X, Y) = E(XY) - (EX)(EY) = \frac{35}{6} - \frac{5}{3} \cdot \frac{10}{3} = \frac{5}{18}$$

and

$$\text{cor}(X, Y) = \frac{\text{cov}(X, Y)}{\text{SD}(X)\text{SD}(Y)} = \frac{5/18}{5/9} = \frac{1}{2}.$$

(d) The best linear predictor of Y based on X is given by

$$\hat{Y} = \frac{10}{3} + \frac{1}{2}\left(X - \frac{5}{3}\right) = \frac{X+5}{2},$$

and the mean squared error of this predictor is given by

$$\text{MSE}(\hat{Y}) = \left(1 - \left(\frac{1}{2}\right)^2\right)\left(\frac{5}{9}\right) = \frac{5}{12}.$$

(e) The best predictor of Y based on X is given by $\hat{Y} = \mu(X)$, where

$$\mu(1) = \frac{2+3+4}{3} = 3, \quad \mu(2) = \frac{3+4}{2} = 3.5, \quad \text{and} \quad \mu(3) = 4.$$

Observe that $\mu(x)$ is a linear function of x for $x \in \{1, 2, 3\}$. Thus, the best predictor of Y based on X is a linear predictor, and hence it coincides with the best linear predictor of Y based on X, which was found in part (d). Consequently, the mean squared error of this predictor is given by $\text{MSE}(\hat{Y}) = \frac{5}{12}$. Alternatively,

$$\text{MSE}(\hat{Y}) = \frac{1}{6}\left\{(2-3)^2 + (3-3)^2 + (4-3)^2\right.$$

$$\left. + (3-3.5)^2 + (4-3.5)^2 + (4-4)^2\right\}$$

$$= \frac{5}{12}. \quad \blacksquare$$

Problems

6.71 A box has three balls, labeled from 1 to 3. Twice, a ball is selected at random from the box and then returned to the box. Let X denote the maximum label on the two trials and let Y denote the sum of the labels selected on these trials. Determine **(a)** the best predictor of Y based on X; **(b)** the mean squared error of this predictor.

6.72 Let X and Y be as in Examples 5.19 and 5.22 in Section 5.5, Example 6.21 and Problem 6.50 in Section 6.4, and Problem 6.66 in Section 6.5. Determine **(a)** the best linear predictor of X based on Y; **(b)** the mean squared error of this predictor.

6.73 (Continued.) Determine **(a)** the best predictor of X based on Y; **(b)** the mean squared error of this predictor.

6.74 (Continued.) Show the graphs of **(a)** the density function of Y; **(b)** the regression function of X on Y and its best linear approximation.

6.75   Let X and Y be as in Problem 6.51 in Section 6.4. Determine **(a)** the best predictor of Y based on X [see the answer to Problem 6.51(b) in Section 6.4]; **(b)** the mean squared error of this predictor.

6.76   Let X and Y be as in Problem 6.51 in Section 6.4. Determine **(a)** the best predictor of X based on Y [see the answer to Problem 6.52(b) in Section 6.4]; **(b)** the mean squared error of this predictor.

6.77   Let X and Y be as in Problem 6.53 in Section 6.4. Determine **(a)** $E(XY)$, $\text{cov}(X, Y)$, and $\text{cor}(X, Y)$; **(b)** the best linear predictor of Y based on X and the mean squared error of this predictor; **(c)** the best predictor of Y based on X and the mean squared error of this predictor.

6.78   Let X and Y, respectively, denote the minimum and maximum number of spots that show when two fair dice are rolled in an honest manner. Determine the best predictor of Y based on X.

6.79   Suppose the value of a randomly selected object in a population of size $N \geq 2$ has mean μ and positive variance σ^2. Let two objects be selected from the population by sampling without replacement, let Y_1 be the value of the first object selected, and let Y_2 be the value of the second object selected. Determine **(a)** the best predictor of Y_2 based on Y_1; **(b)** the mean squared error of this predictor.

6.80   Consider a box with four balls having values 1, 2, 3, and 7. Two balls are selected from the box by sampling without replacement. Let X and Y, respectively, denote the smaller and larger value of the balls in the sample. Determine **(a)** the best predictor of Y based on X; **(b)** the mean squared error of this predictor.

6.81   Suppose $Y = h(\mathbf{X})W$, where $\mathbf{X}$ is an $\mathcal{X}$-valued random variable, h is a real-valued function on $\mathcal{X}$ such that $h(\mathbf{X})$ has finite variance, W is a real-valued random variable having mean 1 and finite variance, and $\mathbf{X}$ and W are independent. Show that **(a)** if g is a real-valued function on $\mathcal{X}$ such that $g(X)$ has finite variance, then

$$E\{[\, Y - g(\mathbf{X})\,]^2\} = E\{[\, Y - h(\mathbf{X})\,]^2\} + E\{[\, h(\mathbf{X}) - g(\mathbf{X})\,]^2\};$$

(b) $\hat{Y} = h(\mathbf{X})$ is the best predictor of Y based on $\mathbf{X}$.

6.7
Conditioning and the Multivariate Normal Distribution

Let $\mathbf{X}$ and $\mathbf{Y}$ have a multivariate normal joint distribution. Then the conditional distribution of $\mathbf{Y}$, given that $\mathbf{X} = \mathbf{x}$, is also normal. In particular, we have the following result.

THEOREM 6.6   Let $\mathbf{X}$ and Y have a multivariate normal joint distribution and set

$$\beta = [\,\text{VC}(\mathbf{X})\,]^{-1}\text{cov}(\mathbf{X}, Y) \quad \text{and} \quad \alpha = EY - \beta^T E\mathbf{X}.$$

Then

(a) the conditional distribution of Y, given that $\mathbf{X} = \mathbf{x}$, is normal with mean $\mu(\mathbf{x}) = \alpha + \beta^T \mathbf{x}$ and variance

$$\sigma^2 = \sigma^2(\mathbf{x}) = \text{var}(Y) - [\text{cov}(\mathbf{X}, Y)]^T [\text{VC}(\mathbf{X})]^{-1} \text{cov}(\mathbf{X}, Y);$$

(b) $\alpha + \beta^T \mathbf{X}$ is the best linear predictor and the best predictor of Y based on $\mathbf{X}$, and σ^2 is the mean squared error of this predictor.

Proof   (a) Set $W = Y - \alpha - \beta^T \mathbf{X}$. Then, by Corollary 5.12 in Section 5.7, $\mathbf{X}$ and W are independent, and W is normally distributed with mean 0 and variance σ^2. Since $Y = \alpha + \beta^T \mathbf{X} + W$, we conclude from Theorem 6.2 in Section 6.4 that the conditional distribution of Y, given that $\mathbf{X} = \mathbf{x}$, coincides with the distribution of $\alpha + \beta^T \mathbf{x} + W$, which is normal with mean $\alpha + \beta^T \mathbf{x}$ and variance σ^2.

(b) The desired result follows from part (a), Corollary 6.6 in Section 6.5, and Theorem 6.5 in Section 6.6.   ∎

COROLLARY 6.7   Let X and Y have a bivariate normal joint distribution and set $\beta = \text{cov}(X, Y)/\text{var}(X)$ and $\alpha = EY - \beta EX$. Then

(a) the conditional distribution of Y, given that $X = x$, is normal with mean $\mu(x) = \alpha + \beta x$ and variance $\sigma^2 = \sigma^2(x) = \text{var}(Y) - [\text{cov}(X, Y)]^2/\text{var}(X)$;

(b) $\alpha + \beta X$ is the best linear predictor and the best predictor of Y based on X, and σ^2 is the mean squared error of this predictor.

The next result is a converse to Theorem 6.6.

THEOREM 6.7   Suppose $\mathbf{X}$ has a multivariate normal distribution and the conditional distribution of Y, given that $\mathbf{X} = \mathbf{x}$, is normal with mean $\alpha + \beta^T \mathbf{x}$ and variance σ^2. Then $\mathbf{X}$ and Y have the multivariate normal joint distribution with mean vector

$$\begin{bmatrix} E\mathbf{X} \\ \alpha + \beta^T E\mathbf{X} \end{bmatrix}$$

and variance–covariance matrix

$$\begin{bmatrix} \text{VC}(\mathbf{X}) & \text{VC}(\mathbf{X})\beta \\ \beta^T \text{VC}(\mathbf{X}) & \beta^T \text{VC}(\mathbf{X})\beta + \sigma^2 \end{bmatrix}.$$

Proof   Since $\mathbf{X}$ and Y have the joint density function given by $f_{\mathbf{X},Y}(\mathbf{x}, y) = f_{\mathbf{X}}(\mathbf{x}) f_{Y|\mathbf{X}}(y \mid \mathbf{x})$, the joint distribution of $\mathbf{X}$ and Y is uniquely determined by the distribution of $\mathbf{X}$ and the conditional distribution of Y given $\mathbf{X}$. Thus, in proving the theorem, we can use any random variable Y such that the conditional distribution of Y given $\mathbf{X}$ is as indicated.

Let W be a random vector such that X and W are independent and W is normally distributed with mean 0 and variance σ^2 and set $Y = \alpha + \beta^T \mathbf{X} + W$. It follows from Theorem 6.2 in Section 6.4 that the conditional distribution of Y, given that $\mathbf{X} = \mathbf{x}$, coincides with the distribution of $\alpha + \beta^T \mathbf{x} + W$, which is normal with mean $\alpha + \beta^T \mathbf{x}$ and variance σ^2. The desired result now follows from the solution to Example 5.34 in Section 5.7.   ∎

COROLLARY 6.8   Suppose X has a normal distribution and the conditional distribution of Y, given that $X = x$, is normal with mean $\alpha + \beta x$ and variance σ^2. Then X and Y have the bivariate normal joint distribution with mean vector

$$\begin{bmatrix} EX \\ \alpha + \beta EX \end{bmatrix}$$

and variance–covariance matrix

$$\begin{bmatrix} \text{var}(X) & \beta \, \text{var}(X) \\ \beta \, \text{var}(X) & \beta^2 \text{var}(X) + \sigma^2 \end{bmatrix}.$$

EXAMPLE 6.30   Suppose X is normally distributed with mean μ and variance γ^2 and the conditional distribution of Y, given that $X = x$, is normal with mean x and variance σ^2. Determine

(a) the joint distribution of X and Y;

(b) the conditional distribution of X, given that $Y = y$.

Solution   (a) According to Corollary 6.8, X and Y have the bivariate normal distribution with mean vector $[\, \mu, \mu \,]^T$ and variance–covariance matrix

$$\begin{bmatrix} \gamma^2 & \gamma^2 \\ \gamma^2 & \sigma^2 + \gamma^2 \end{bmatrix}.$$

(b) Applying Corollary 6.7 with the roles of X and Y reversed and the solution to part (a), we find that the conditional distribution of X, given that $Y = y$, is normal with mean $\alpha + \beta y$ and variance

$$\text{var}(X) - \frac{[\,\text{cov}(X, Y)\,]^2}{\text{var}(Y)} = \gamma^2 - \frac{\gamma^4}{\sigma^2 + \gamma^2} = \frac{\sigma^2 \gamma^2}{\sigma^2 + \gamma^2},$$

where

$$\beta = \frac{\text{cov}(X, Y)}{\text{var}(Y)} = \frac{\gamma^2}{\sigma^2 + \gamma^2}$$

and

$$\alpha = EX - \beta EY = \mu(1 - \beta) = \frac{\sigma^2 \mu}{\sigma^2 + \gamma^2}.$$

Thus, the conditional distribution of X, given that $Y = y$, is normal with mean $(\sigma^2 \mu + \gamma^2 y)/(\sigma^2 + \gamma^2)$ and variance $\sigma^2 \gamma^2/(\sigma^2 + \gamma^2)$.   ∎

Causation

Theorems 6.6 and 6.7 will now be used to illustrate the crucial distinction between correlation (and other measures of association) and causation. To this end, consider a random variable Y whose distribution depends on deterministic variables x_1 and x_2. We refer to x_1 and x_2 as the levels of factors 1 and 2, respectively, and to Y as the *response variable*. (Think of the response variable as the random yield of a certain chemical in an industrial manufacturing process, of x_1 as the level of temperature applied during the process, and of x_2 as the amount of pressure applied.) Suppose the response variable is normally distributed with mean $\alpha + \beta_1 x_1 + \beta_2 x_2$ and a variance σ^2 that does not depend on x_1 and x_2. If we increase (the level of) factor 1 by 1 unit, with factor 2 held fixed, then the mean response increases by β_1 units. In other words, increasing factor 1 by 1 unit "causes" the mean response to increase by β_1 units. Similarly, increasing factor 2 by 1 unit causes the mean response to increase by β_2 units. Thus, we refer to the model as a *causal model* for the dependence of the distribution of the response variable on the two factors.

Let the deterministic levels x_1 and x_2 now be replaced by random variables X_1 and X_2 having a positive joint density function f_{X_1,X_2} on $\mathbb{R}^2$. Correspondingly, we make the reasonable assumption that the conditional distribution of Y, given that $X_1 = x_1$ and $X_2 = x_2$, is normal with mean $\alpha + \beta_1 x_1 + \beta_2 x_2$ and variance σ^2; that is,

$$f_{Y|X_1,X_2}(y \,|\, x_1, x_2) = \frac{1}{\sigma\sqrt{2\pi}} \exp\left(-\frac{(y - \alpha - \beta_1 x_1 - \beta_2 x_2)^2}{2\sigma^2}\right).$$

The joint density function of X_1, X_2, and Y is then given by

$$f_{X_1,X_2,Y}(x_1, x_2, y) = f_{X_1,X_2}(x_1, x_2) f_{Y|X_1,X_2}(y \,|\, x_1, x_2), \qquad x_1, x_2, y \in \mathbb{R}.$$

Conversely, from the joint density function of X_1, X_2, and Y, we can determine the joint density function of X_1 and X_2 and hence determine the dependence on x_1 and x_2 of the conditional density function of Y, given that $X_1 = x_1$ and $X_2 = x_2$. Thus, from the joint density function of X_1, X_2, and Y, we can determine the effect on the mean response caused by a unit increase in either factor with the other factor held fixed.

Suppose now that we know only the joint density function of X_1 and Y. From this knowledge, can we determine the effect on the mean response caused by a unit increase in factor 1 with factor 2 held fixed?

In answering this question, we assume for simplicity that X_1 and X_2 have a bivariate normal joint distribution. It then follows from Theorem 6.7 that X_1, X_2, and Y have a trivariate normal joint distribution and hence that X_1 and Y have a bivariate normal joint distribution. Consequently, the conditional distribution of Y, given that $X_1 = x_1$, is normal with mean $\alpha_0 + \beta_{10} x_1$ and variance σ_0^2 for some $\alpha_0, \beta_{10} \in \mathbb{R}$ and $\sigma_0^2 > 0$. It might seem tempting to conjecture from this result that a unit increase in factor 1 with factor 2 held fixed would cause the mean response to increase by β_{10} units. We know, however, that it would actually cause the mean response to increase by β_1 units, so our conjecture is valid if and only if $\beta_{10} = \beta_1$. By Theorem 6.6, $\alpha + \beta_1 X_1 + \beta_2 X_2$ is the best linear predictor of Y based on X_1 and X_2; by Corollary 6.7, $\alpha_0 + \beta_{10} X_1$ is the best linear predictor of Y based on X_1. Thus, by Corollary 5.8 in Section 5.4,

$$\beta_{10} = \beta_1 + \frac{\text{cov}(X_1, X_2)}{\text{var}(X_1)} \beta_2.$$

Consequently, $\beta_{10} = \beta_1$ if and only if (a) $\text{cov}(X_1, X_2) = 0$ or (b) $\beta_2 = 0$. Since X_1 and X_2 are assumed to have a bivariate normal joint distribution, condition (a) holds if and only if these random variables are independent. Condition (b) holds if and only if a change in factor 2 with factor 1 held fixed has no effect on the mean response.

It is impossible to determine the validity of conditions (a) and (b) from knowledge only of the joint distribution of X_1 and Y. Thus, from this knowledge, we cannot determine conclusively the effect on the mean response of a given change in factor 1 with factor 2 held fixed. In fact, from knowledge only of the joint distribution of X_1 and Y, we cannot even determine whether an increase in factor 1 with factor 2 held fixed causes the mean response to increase or causes it to decrease; that is, we cannot determine the sign of β_1. The contribution

$$\frac{\text{cov}(X_1, X_2)}{\text{var}(X_1)} \beta_2$$

to β_{10} should be thought of as a *spurious effect* of factor 1, which is actually due to the *true effect* of factor 2 and the correlation between X_1 and X_2. (If $\beta_2 \neq 0$ and X_1 and X_2 have nonzero correlation, then X_2 is referred to as a *confounder* in the context of determining the effect on the mean response of changes in x_1.)

EXAMPLE **6.31**   Suppose that when the temperature is held at level x_1 (on a suitable scale) and the pressure is held at level x_2, the yield Y is normally distributed with mean $10 + x_1 + 3x_2$ and variance 4. Suppose now that the temperature and pressure form random variables X_1 and X_2, respectively, having a bivariate normal joint distribution with $EX_1 = 8$, $\text{SD}(X_1) = 3$, $EX_2 = 5$, $\text{SD}(X_2) = 4$, and $\text{cor}(X_1, X_2) = -1/2$. Correspondingly, we assume that the conditional distribution of Y, given that $X_1 = x_1$ and $X_2 = x_2$, is normal with mean $10 + x_1 + 3x_2$ and variance 4. Determine

(a) the joint distribution of X_1, X_2, and Y;

(b) the best linear predictor of Y based on X_1 and X_2 and the corresponding mean squared error of prediction;

(c) the joint distribution of X_1 and Y;

(d) the best linear predictor of Y based on X_1 and the corresponding mean squared error of prediction;

(e) the correlation between X_1 and Y.

Solution   **(a)** Theorem 6.7 is applicable with $\alpha = 10$, $\beta = [1, 3]^T$, $\sigma^2 = 4$, $EX = [8, 5]^T$, and

$$\text{VC}(\mathbf{X}) = \begin{bmatrix} 9 & -6 \\ -6 & 16 \end{bmatrix}.$$

Note that

$$\beta^T EX = [1, 3] \begin{bmatrix} 8 \\ 5 \end{bmatrix} = 23,$$

$$\text{VC}(\mathbf{X})\beta = \begin{bmatrix} 9 & -6 \\ -6 & 16 \end{bmatrix} \begin{bmatrix} 1 \\ 3 \end{bmatrix} = \begin{bmatrix} -9 \\ 42 \end{bmatrix},$$

and

$$\beta^T \text{VC}(\mathbf{X})\beta = [1, 3] \begin{bmatrix} -9 \\ 42 \end{bmatrix} = 117.$$

Thus X_1, X_2, and Y have the trivariate normal joint distribution with mean vector $[\, 8, 5, 33\,]^T$ and variance–covariance matrix

$$\begin{bmatrix} 9 & -6 & -9 \\ -6 & 16 & 42 \\ -9 & 42 & 121 \end{bmatrix}.$$

(b) By Theorem 6.6, the best linear predictor of Y based on X_1 and X_2 is given by $\hat{Y} = 10 + X_1 + 3X_2$, and the corresponding mean squared error of prediction equals 4.

(c) The random variables X_1 and Y have the bivariate normal joint distribution with mean vector $[\, 8, 33\,]^T$ and variance–covariance matrix

$$\begin{bmatrix} 9 & -9 \\ -9 & 121 \end{bmatrix}.$$

(d) The best linear predictor of Y based on X_1 is given by

$$\hat{Y} = EY + \frac{\mathrm{cov}(X_1, Y)}{\mathrm{var}(X_1)}(X_1 - EX_1) = 33 + \frac{-9}{9}(X_1 - 8) = 33 - (X_1 - 8)$$

and hence by $\hat{Y} = 41 - X_1$. The corresponding mean squared error of prediction is given by

$$\mathrm{var}(Y) - \frac{[\,\mathrm{cov}(X_1, Y)\,]^2}{\mathrm{var}(X_1)} = 121 - \frac{(-9)^2}{9} = 121 - 9 = 112.$$

(e) The correlation between X_1 and Y is given by

$$\mathrm{cor}(X_1, Y) = \frac{\mathrm{cov}(X_1, Y)}{\mathrm{SD}(X_1)\mathrm{SD}(Y)} = \frac{-9}{3 \cdot 11} = -\frac{3}{11}. \qquad \blacksquare$$

In the context of Example 6.31, a unit increase in temperature causes the mean yield to increase by $\beta_1 = 1$ unit. From the solution to Example 6.31(d), however, we might be led to believe that a unit increase in temperature causes the mean yield to increase by $\beta_{10} = -1$ units or, equivalently, to decrease by 1 unit. This discrepancy is caused by the spurious effect

$$\frac{\mathrm{cov}(X_1, X_2)}{\mathrm{var}(X_1)}\beta_2 = \frac{-6}{9} \cdot 3 = -2$$

due to the true effect of pressure and the correlation between temperature and pressure.

In the same context, the correlation between temperature and yield is negative, but increasing the temperature with the pressure held fixed causes the mean yield to increase. This illustrates the important distinction between causation and correlation.

In this context, we can determine the effect on mean yield of a given change in temperature with pressure held fixed from the joint distribution of X_1, X_2, and Y. In practice, however, we could not rule out the possibility of an additional unobserved factor (such as the amount of catalyst applied) that may be correlated with temperature and/or pressure and may also have an effect on yield. Without being able to rule out the existence of such an additional variable, we cannot determine conclusively the effect on mean yield of a change in temperature with pressure held fixed from knowledge of the joint distribution of temperature, pressure, and yield.

There are many contexts in which the distinction between causation and correlation is practically important. In particular, in the medical context, let X denote the level of cholesterol in the blood in an adult selected at random from some population at a particular time and let Y denote the number of years that the individual lives beyond that time. The random variables X and Y are negatively correlated. This is due partly to the harmful effect of very high levels of cholesterol on health and partly to the combination of a positive correlation between age and level of cholesterol and the negative effect of age on additional life expectancy.

In a typical practical situation involving random variables X_1, X_2, and Y, we would not know their joint distribution. Instead, we might have the observed values of a random sample $(X_{11}, X_{12}, Y_1), \ldots, (X_{n1}, X_{n2}, Y_n)$ from this joint distribution. Under the assumption that X_1, X_2, and Y have a trivariate normal joint distribution, we could use the observational data to estimate the unknown parameters of the distribution. Clearly, however, the extent of our knowledge would be less than if we knew the exact joint distribution of X_1, X_2, and Y. Thus, no matter how large n may be, we could not confidently determine from the observational data alone the effects on the mean response of changes in one factor with the other factor held fixed. Ideally (ignoring such issues as feasibility and ethics), in order to determine such effects, we should chuck the observational data and conduct an experiment in which we set the two factors at various deterministic levels and observe the corresponding responses. In Chapter 8, we begin our study of the design and analysis of such experiments.

Confounding is not the only obstacle to the drawing of valid causal implications from observational data. Suppose there are two factors, having levels x_1 and x_2, and a response variable Y. Let the levels of the two factors now be random variables X_1 and X_2 and suppose that X_1, X_2, and Y have a trivariate normal joint distribution and that the conditional distribution of Y, given that $X_1 = x_1$ and $X_2 = x_2$, is normal with mean $\beta_1 x_1 + \beta_2 x_2$. Even if there are no confounding factors, the change in the mean response caused by a unit increase in factor 1 need not equal β_1.

To justify this claim, suppose first that, when the factors have the deterministic levels x_1 and x_2, the response variable is distributed as $\beta_1 x_1 + \beta_2 x_2 + W$, where W is normally distributed with mean 0. Now let the level of factor 2 depend in a random manner on the level of factor 1. Specifically, suppose that the level of factor 2 is distributed as $\gamma x_1 + V$, where $\gamma \in \mathbb{R}$, V is normally distributed, and V and W are independent. Then the response variable is distributed as $\beta_1 x_1 + \beta_2 (\gamma x_1 + V) + W = (\beta_1 + \beta_2 \gamma) x_1 + \beta_2 V + W$. Thus, a unit increase in factor 1 causes the mean response to change by the amount $\beta_1 + \beta_2 \gamma$. Here β_1 is the direct effect of factor 1 on the mean response, and $\beta_2 \gamma$ is the indirect effect of this factor, which is due to the direct effect of factor 2 on the mean response and the direct effect of factor 1 on factor 2. Suppose now that factor 1 is distributed as a normal random variable U, where U, V, and W are independent. Then the two factors and the response variable are jointly distributed as $X_1 = U$, $X_2 = \gamma X_1 + V$, and $Y = \beta_1 X_1 + \beta_2 X_2 + W$. Observe that X_1, X_2, and Y have a trivariate normal joint distribution and, by Theorem 6.2 in Section 6.4, that the conditional distribution of Y, given that $X_1 = x_1$ and $X_2 = x_2$, is normal with mean $\beta_1 x_1 + \beta_2 x_2$.

On the other hand, observational data can suggest causal implications that are worthy of further investigation (for example, that secondhand cigarette smoking causes lung cancer). In the context of practical decision making, we sometimes have to choose between using causal conclusions based on observational data and those based on ideology; here the former conclusions may be no more suspect than the latter ones. Observational data can also be useful in prediction.

Problems

6.82 Suppose X and Y have a bivariate normal distribution with $EX = 5$, $EY = -2$, $\text{var}(X) = 4$, $\text{var}(Y) = 9$, and $\text{cov}(X, Y) = -3$. Determine the conditional distribution of X, given that $Y = 1$.

6.83 Let X and Y have a bivariate normal joint distribution with $\mu_1 = EX$, $\sigma_1 = \text{SD}(X)$, $\mu_2 = EY$, $\sigma_2 = \text{SD}(Y)$, and $\rho = \text{cor}(X, Y)$. Determine **(a)** the conditional distribution of Y given that $X = x$; **(b)** the conditional distribution of X, given that $Y = y$.

6.84 Let X_1, X_2, and Y have mean vector $[\,100, 150, 400\,]^T$, variance–covariance matrix

$$\begin{bmatrix} 16 & 16 & 48 \\ 16 & 25 & 57 \\ 48 & 57 & 169 \end{bmatrix},$$

and a trivariate normal joint distribution. Determine **(a)** the conditional distribution of Y, given that $X_1 = x_1$; **(b)** the best predictor of Y based on X_1 and the mean squared error of this predictor; **(c)** the conditional distribution of Y, given that $X_1 = x_1$ and $X_2 = x_2$; **(d)** the best linear predictor of Y based on X_1 and X_2 and the mean squared error of this predictor.

6.85 Let W_1, W_2, and W_3 be independent random variables that are normally distributed with 0 means and variances γ_1^2, γ_2^2, and γ_3^2, respectively. Set $X_1 = W_1, X_2 = \lambda W_1 + W_2$, and $Y = \beta_1 X_1 + \beta_2 X_2 + W_3$, where $\beta_1, \beta_2, \lambda \in \mathbb{R}$. **(a)** Show that X_1, X_2, and Y have a trivariate normal joint distribution. **(b)** Determine the conditional distribution of Y, given that $X_1 = x_1$ and $X_2 = x_2$. **(c)** Show that $Y = (\beta_1 + \lambda\beta_2)X_1 + W$, where $W = \beta_2 W_2 + W_3$. **(d)** Use the result of part (c) to determine the conditional distribution of Y, given that $X_1 = x_1$.

6.86 Let X_1 and X_2 have the bivariate normal joint distribution with $EX_1 = \mu_1$, $\text{var}(X_1) = \sigma_1^2$, $EX_2 = \mu_2$, $\text{var}(X_2) = \sigma_2^2$, and $\text{cov}(X_1, X_2) = \sigma_{12}$. Suppose the conditional distribution of Y, given that $X_1 = x_1$ and $X_2 = x_2$, is normal with mean $\alpha + \beta_1 x_1 + \beta_2 x_2$ and variance σ^2. Use Theorem 6.7 to determine **(a)** EY; **(b)** $\text{cov}(X_1, Y)$ and $\text{cov}(X_2, Y)$; **(c)** $\text{var}(Y)$.

6.87 Suppose X is normally distributed with mean 100 and standard deviation 20 and the conditional distribution of Y, given that $X = x$, is normally distributed with mean x and standard deviation 10. Determine **(a)** EY, $\text{var}(Y)$, $\text{cov}(X, Y)$, and $\text{cor}(X, Y)$; **(b)** the joint distribution of X and Y; **(c)** the conditional distribution of X, given that $Y = y$; **(d)** the conditional probability that $X > 100$, given that $Y = 110$.

6.88   Suppose X is normally distributed with mean μ and variance σ_1^2 and the conditional distribution of Y, given that $X = x$, is normal with mean βx and variance σ^2. Determine **(a)** the joint distribution of X and Y; **(b)** the conditional distribution of X, given that $Y = y$.

6.89   Let X and Y be as in Problem 6.88. Determine μ, β, σ_1^2, and σ^2 in order that Y have variance 25 and the conditional distribution of X, given that $Y = y$, be normal with mean $0.72 + 0.16y$ and variance 0.36.

6.90   Suppose X and Y have a bivariate normal joint distribution; the conditional distribution of Y, given that $X = x$, is normal with mean βx and variance σ^2; and the conditional distribution of X, given that $Y = y$, is normal with mean αy and variance τ^2. Show that $\beta/\sigma^2 = \alpha/\tau^2$.

6.91   Let X_1, W_1, and W_2 be independent and normally distributed random variables having mean 0 and variances 1, 2, and 3, respectively, and set $X_2 = X_1 + W_1$ and $Y = X_1 + X_2 + W_2$. Determine **(a)** the joint distribution of X_1, X_2, and Y; **(b)** the conditional distribution of Y, given that $X_1 = x_1$ and $X_2 = x_2$; **(c)** the conditional distribution of X_1, given that $X_2 = x_2$ and $Y = y$.

6.92   Let X be the amount of money spent on health care for a random adult individual in the United States in 1990 and let Y denote the number of years that the individual lives beyond 1990. **(a)** Is the correlation between X and Y positive or negative? **(b)** What is the causal effect of increased expenditures for health care on life expectancy? **(c)** Explain any apparent contradiction between your answers to parts (a) and (b).

6.93   Consider the population of all families in some large industrial country whose eldest child has just graduated from high school. Let X_1 be the annual family income of a random family in this population, X_2 a measure of the social health of the family, and Y a measure of the academic performance of the eldest child. Suppose X_1, X_2, and Y have a trivariate normal joint distribution and the conditional distribution of Y, given that $X_1 = x_1$ and $X_2 = x_2$, is normal with mean $\alpha + \beta_1 x_1 + \beta_2 x_2$ and variance σ^2. Suppose also there are no other relevant variables. Is it then reasonable to believe that if the government had been transferring an additional annual amount C to all such families, then the mean of Y would have been larger by the amount $\beta_1 C$?

6.8
Random Parameters

We start with three incomplete examples involving random parameters, which will be completed later in the section.

EXAMPLE **6.32**   The probability π that a basketball player makes a basket on a free throw varies from one player to another according to a beta distribution with parameters $\alpha_1 = 2$ and $\alpha_2 = 3$. For a player with probability π of making a basket on a given free throw, the distribution of the number of baskets in n free throws has the binomial distribution with parameters n and π.

EXAMPLE **6.33**   The number of automobile accidents in a year for a given driver in a certain population has a Poisson distribution with mean λ, which is referred to as the *accident proneness* of the driver and which varies from driver to driver according to the gamma distribution with shape parameter 2 and inverse-scale parameter 10.

EXAMPLE **6.34**   The IQ of a randomly selected individual in a certain population is normally distributed with mean 100 and standard deviation 15. The score of the individual on an IQ test is unbiased and normally distributed with standard deviation 5.

Consider, more generally, a $\mathcal{Y}$-valued random variable Y having probability function or density function $f(y \,|\, \theta)$, $y \in \mathcal{Y}$, which depends on a real-valued parameter θ that ranges over a set Θ. Suppose a unique such parameter θ is associated with each individual (or object) in a certain population and, as θ varies from individual to individual in the population, it has a distribution with a density function on Θ. (In practical applications, of course, the population is finite, so θ must necessarily have a discrete distribution. We assume, however, that the population size is so large that this discrete distribution can be accurately approximated by a continuous distribution.) We refer to the density function of θ as its *prior density function* and denote it by f_{prior}, and we refer to the corresponding distribution as the *prior distribution* of θ.

When θ is viewed as a random variable, let it still be denoted by θ. Then θ has density function f_{prior} on Θ. We now think of $f(y \,|\, \theta)$ as the conditional probability function or conditional density function of Y given θ. If Y is a continuous random variable, then $f_{\text{prior}}(\theta)f(y \,|\, \theta)$, $\theta \in \Theta$ and $y \in \mathcal{Y}$, is the joint density function of θ and Y; otherwise, it is referred to as the joint *density–probability function* of θ and Y. The probability function or density function f_Y of Y, which is referred to as its *marginal probability function* or *marginal density function*, is given by

$$f_Y(y) = \int_{\Theta} f_{\text{prior}}(\theta)f(y \,|\, \theta) \, d\theta, \qquad y \in \mathcal{Y};$$

the corresponding distribution is referred to as the *marginal distribution* of Y. The conditional density function of θ, given that $Y = y$, is referred to as the *posterior density function* of θ and denoted by f_{post}; thus,

$$f_{\text{post}}(\theta) = \frac{f_{\text{prior}}(\theta)f(y \,|\, \theta)}{f_Y(y)}, \qquad \theta \in \Theta.$$

The corresponding distribution is referred to as the *posterior distribution* of θ.

The best predictor of θ based on Y is given by $\hat{\theta} = E(\theta \,|\, Y)$. We also refer to this predictor as the *Bayes estimate* of θ (based on Y) and denote it by $\hat{\theta}$. Thus, $\hat{\theta}$ is the mean of the posterior distribution of θ; that is,

$$\hat{\theta} = \int_{\Theta} \theta f_{\text{post}}(\theta) \, d\theta = \frac{\int_{\Theta} \theta f_{\text{prior}}(\theta)f(Y \,|\, \theta) \, d\theta}{\int_{\Theta} f_{\text{prior}}(\theta)f(Y \,|\, \theta) \, d\theta}.$$

In this general discussion, we have used θ as the generic symbol for the parameter of interest. In specific situations, we use the conventional symbol for this parameter.

Suppose, in particular, that Y has the binomial distribution with parameters n and π. Here

$$f(y \mid \pi) = \binom{n}{y} \pi^y (1 - \pi)^{n-y}, \qquad y = 0, \ldots, n.$$

Let the prior distribution of π be beta with parameters $\alpha_1 > 0$ and $\alpha_2 > 0$. Then the prior density function of π is given by

$$f_{\text{prior}}(\pi) = \frac{\Gamma(\alpha_1 + \alpha_2)}{\Gamma(\alpha_1)\Gamma(\alpha_2)} \pi^{\alpha_1 - 1}(1 - \pi)^{\alpha_2 - 1}, \qquad 0 < \pi < 1.$$

The marginal probability function of Y is given by

$$f_Y(y) = \int_0^1 \frac{\Gamma(\alpha_1 + \alpha_2)}{\Gamma(\alpha_1)\Gamma(\alpha_2)} \pi^{\alpha_1 - 1}(1 - \pi)^{\alpha_2 - 1} \binom{n}{y} \pi^y (1 - \pi)^{n-y} \, d\pi$$

$$= \frac{\Gamma(\alpha_1 + \alpha_2)}{\Gamma(\alpha_1)\Gamma(\alpha_2)} \binom{n}{y} \int_0^1 \pi^{\alpha_1 + y - 1}(1 - \pi)^{\alpha_2 + n - y - 1} \, d\pi.$$

Thus, by Equation (13) in Section 3.1,

$$f_Y(y) = \frac{\Gamma(\alpha_1 + \alpha_2)}{\Gamma(\alpha_1)\Gamma(\alpha_2)} \binom{n}{y} \frac{\Gamma(\alpha_1 + y)\Gamma(\alpha_2 + n - y)}{\Gamma(\alpha_1 + \alpha_2 + n)}, \qquad y = 0, \ldots, n. \qquad \textbf{(44)}$$

Consequently, the posterior density function of π is given by

$$f_{\text{post}}(\pi) = \frac{\dfrac{\Gamma(\alpha_1 + \alpha_2)}{\Gamma(\alpha_1)\Gamma(\alpha_2)} \binom{n}{Y} \pi^{\alpha_1 + Y - 1}(1 - \pi)^{\alpha_2 + n - Y - 1}}{\dfrac{\Gamma(\alpha_1 + \alpha_2)}{\Gamma(\alpha_1)\Gamma(\alpha_2)} \binom{n}{Y} \dfrac{\Gamma(\alpha_1 + Y)\Gamma(\alpha_2 + n - Y)}{\Gamma(\alpha_1 + \alpha_2 + n)}}$$

$$= \frac{\Gamma(\alpha_1 + \alpha_2 + n)}{\Gamma(\alpha_1 + Y)\Gamma(\alpha_2 + n - Y)} \pi^{\alpha_1 + Y - 1}(1 - \pi)^{\alpha_2 + n - Y - 1}$$

for $0 < \pi < 1$. Thus, the posterior distribution of π is the beta distribution with parameters $\alpha_1 + Y$ and $\alpha_2 + n - Y$, which has mean $(\alpha_1 + Y)/(\alpha_1 + \alpha_2 + n)$. Therefore, the Bayes estimate of π is given by

$$\hat{\pi} = \frac{\alpha_1 + Y}{\alpha_1 + \alpha_2 + n}. \qquad \textbf{(45)}$$

EXAMPLE **6.35**   (Completed Example 6.32). The probability π that a basketball player makes a basket on a free throw varies from one player to another according to a beta distribution with parameters $\alpha_1 = 2$ and $\alpha_2 = 3$. For a player with probability π of making a basket on a given free throw, the distribution of the number of baskets in n free throws has the binomial distribution with parameters n and π. Suppose that a randomly selected basketball player makes five baskets on six free throws. Determine

(a) the posterior distribution of the player's probability π of making a basket on a single free throw;

(b) the Bayes estimate of π.

Solution (a) Here $n = 6$ and $Y = 5$. The posterior distribution of π is beta with parameters $2 + 5 = 7$ and $3 + 6 - 5 = 4$.

(b) The Bayes estimate of π is given by $\hat{\pi} = 7/11$. ∎

Suppose next that Y has the Poisson distribution with mean λ. Here

$$f(y \mid \lambda) = \frac{\lambda^y}{y!} e^{-\lambda}, \qquad y = 0, 1, 2, \ldots .$$

Let the prior distribution of λ be gamma with shape parameter $\alpha > 0$ and inverse-scale parameter $\tau > 0$. Then the prior density function of λ is given by

$$f_{\text{prior}}(\lambda) = \frac{\tau^\alpha \lambda^{\alpha-1}}{\Gamma(\alpha)} e^{-\tau\lambda}, \qquad \lambda > 0.$$

The marginal probability function of Y is given by

$$f_Y(y) = \int_0^\infty \frac{\tau^\alpha \lambda^{\alpha-1}}{\Gamma(\alpha)} e^{-\tau\lambda} \frac{\lambda^y}{y!} e^{-\lambda} \, d\lambda = \frac{\tau^\alpha}{\Gamma(\alpha)y!} \int_0^\infty \lambda^{\alpha+y-1} e^{-(\tau+1)\lambda} \, d\lambda.$$

Thus, by Equation (6) in Section 3.1,

$$f_Y(y) = \frac{\tau^\alpha \Gamma(\alpha + y)}{\Gamma(\alpha)y! \, (\tau + 1)^{\alpha+y}}, \qquad y = 0, 1, 2, \ldots . \tag{46}$$

Consequently, the posterior density function of λ is given by

$$f_{\text{post}}(\lambda) = \frac{\dfrac{\tau^\alpha}{\Gamma(\alpha)Y!} \lambda^{\alpha+Y-1} e^{-(\tau+1)\lambda}}{\dfrac{\tau^\alpha \Gamma(\alpha + Y)}{\Gamma(\alpha)Y!(\tau + 1)^{\alpha+Y}}} = \frac{(\tau + 1)^{\alpha+Y} \lambda^{\alpha+Y-1} e^{-(\tau+1)\lambda}}{\Gamma(\alpha + Y)}$$

for $\lambda > 0$. Thus, the posterior distribution of λ is gamma with shape parameter $\alpha + Y$ and inverse-scale parameter $\tau + 1$, which has mean $(\alpha + Y)/(\tau + 1)$. Therefore, the Bayes estimate of λ is given by

$$\hat{\lambda} = \frac{\alpha + Y}{\tau + 1}. \tag{47}$$

EXAMPLE 6.36 (Completed Example 6.33). The number of automobile accidents in a year for a given driver in a certain population has a Poisson distribution with mean λ, where the accident proneness λ varies from driver to driver according to the gamma distribution with shape parameter $\alpha = 2$ and inverse-scale parameter $\tau = 9$. Suppose a randomly selected driver has no accidents during the year. Determine

(a) the posterior distribution of the driver's accident proneness λ;

(b) the Bayes estimate of λ.

Solution (a) Here $Y = 0$. The posterior distribution of λ is gamma with shape parameter 2 and inverse-scale parameter $9 + 1 = 10$.

(b) The Bayes estimate of λ is given by $\hat{\lambda} = 2/10 = 1/5$. ∎

Suppose now that Y is normally distributed with mean θ and variance σ^2. Here

$$f(y \mid \theta) = \frac{1}{\sigma\sqrt{2\pi}} \exp(-(y - \theta)^2/2\sigma^2), \qquad y \in \mathbb{R}.$$

Let the prior distribution of θ be normal with mean μ and variance γ^2. Then, by the solution to Example 6.30 in Section 6.7 with X replaced by θ, the posterior distribution of θ is normal with mean $(\sigma^2\mu + \gamma^2 Y)/(\sigma^2 + \gamma^2)$ and variance $\sigma^2\gamma^2/(\sigma^2 + \gamma^2)$. Thus, the Bayes estimate of θ is given by

$$\hat{\theta} = \frac{\sigma^2\mu + \gamma^2 Y}{\sigma^2 + \gamma^2}. \tag{48}$$

EXAMPLE **6.37**   (Completed Example 6.34.) The true intelligence quotient θ of a randomly selected individual in a certain population is normally distributed with mean 100 and standard deviation 15. The score of the individual on an IQ test is unbiased and normally distributed with standard deviation 5. Suppose a randomly selected individual scores 120 on an IQ test. Determine

(a) the posterior distribution of the individual's true intelligence quotient θ;

(b) the Bayes estimate of θ.

Solution   **(a)** Here $\mu = 100$, $\gamma = 15$, $\sigma = 5$, and $Y = 120$. The posterior distribution of θ is normal with mean

$$\frac{25 \cdot 100 + 225 \cdot 120}{25 + 225} = 118$$

and variance

$$\frac{25 \cdot 225}{25 + 225} = \frac{45}{2} = 22.5.$$

(b) The Bayes estimate of θ is given by $\hat{\theta} \doteq 118$.   ∎

Subjective Priors

So far, we have been thinking of θ as an attribute of an object in a population and of the prior distribution of θ as the distribution of the attribute of a random member of the population. Alternatively, the prior distribution can be subjectively determined.

EXAMPLE **6.38**   Let π denote the probability that Johnny makes a basket on a free throw. Frankie's subjective prior distribution of π has mean .4 and standard deviation .1. Suppose Johnny makes five baskets on six free throws. Using a beta distribution to model Frankie's prior, determine her Bayes estimate of π.

Solution   Let $\mu = .4$ and $\sigma = .1$ denote, respectively, the mean and standard deviation of Frankie's prior and let α_1 and α_2 be the parameters of the beta distribution having this mean and standard deviation. It follows from the answer to Problem 3.11 in

Section 3.1 that

$$\alpha_1 = \mu\left(\frac{\mu(1-\mu)}{\sigma^2} - 1\right) = .4\left(\frac{(.4)(.6)}{.01} - 1\right) = 0.92$$

and

$$\alpha_2 = (1-\mu)\left(\frac{\mu(1-\mu)}{\sigma^2} - 1\right) = .6\left(\frac{(.4)(.6)}{.01} - 1\right) = 1.38.$$

Here Equation (45) is applicable with $n = 6$ and $Y = 5$, so Frankie's Bayes estimate of π is given by

$$\hat{\pi} = \frac{0.92 + 5}{2.3 + 6} \doteq .713. \quad \blacksquare$$

EXAMPLE **6.39** Let λ denote the accident proneness of Frankie as an automobile driver. Johnny's subjective prior distribution of λ has mean 0.5 and standard deviation 0.5. Suppose Frankie has no accidents during a 1-year period. Using a gamma distribution to model Johnny's prior, determine his Bayes estimate of λ.

Solution   Let $\mu = 0.5$ and $\sigma = 0.5$ denote, respectively, the mean and standard deviation of Johnny's prior and let α and τ denote, respectively, the shape parameter and inverse-scale parameter of the gamma distribution having this mean and standard deviation. It follows from the answer to Problem 3.1 in Section 3.1 that $\alpha = (\mu/\sigma)^2 = 1$ and $\tau = \mu/\sigma^2 = 2$, which corresponds to the exponential distribution with inverse-scale parameter 2. Here Equation (47) is applicable with $Y = 0$, so Johnny's Bayes estimate of λ is given by $\hat{\lambda} = (1 + 0)/(2 + 1) = 1/3$. $\quad \blacksquare$

EXAMPLE **6.40** Let θ denote Johnny's true IQ. Frankie's subjective prior distribution of θ has mean 130 and standard deviation 10. Suppose Johnny scores 120 on an IQ test in which the error is normally distributed with mean 0 and standard deviation 5. Using a normal distribution to model Frankie's prior, determine her Bayes estimate of θ.

Solution   Here Equation (48) is applicable with $\mu = 130$, $\gamma = 10$, $\sigma = 5$, and $Y = 120$. Thus, Frankie's Bayes estimate of θ is given by

$$\hat{\theta} = \frac{25 \cdot 130 + 100 \cdot 120}{100 + 25} = 122. \quad \blacksquare$$

Negative Binomial Distribution

The probability function given by Equation (46) can be written as

$$f(y) = \frac{\Gamma(\alpha + y)}{\Gamma(\alpha)y!}(1 - \pi)^\alpha \pi^y, \qquad y = 0, 1, 2, \ldots, \tag{49}$$

where $\pi = 1/(\tau + 1)$ and hence $\tau = (1 - \pi)/\pi$. The distribution having the probability function given by (49) is referred to as the *negative binomial distribution* with parameters α and π (see Problem 4.19 in Section 4.2); when $\alpha = 1$, it coincides with the geometric distribution with parameter π.

Let Y have the negative binomial distribution with parameters α and π. Then its mean is given by

$$EY = \sum_{y=1}^{\infty} y \frac{\Gamma(\alpha+y)}{\Gamma(\alpha)y!}(1-\pi)^{\alpha}\pi^{y}$$

$$= \sum_{y=1}^{\infty} \frac{\Gamma(\alpha+y)}{\Gamma(\alpha)(y-1)!}(1-\pi)^{\alpha}\pi^{y}$$

$$= \alpha\frac{\pi}{1-\pi} \sum_{y=1}^{\infty} \frac{\Gamma(\alpha+1+y-1)}{\Gamma(\alpha+1)(y-1)!}(1-\pi)^{\alpha+1}\pi^{y-1}$$

$$= \alpha\frac{\pi}{1-\pi} \sum_{y=0}^{\infty} \frac{\Gamma(\alpha+1+y)}{\Gamma(\alpha+1)y!}(1-\pi)^{\alpha+1}\pi^{y}$$

$$= \alpha\frac{\pi}{1-\pi},$$

since the probability function of the negative binomial distribution with parameters $\alpha+1$ and π has sum 1.

By a similar argument, the details being left as a problem, we get

$$E[\,Y(Y-1)\,] = \alpha(\alpha+1)\frac{\pi^{2}}{(1-\pi)^{2}}.$$

Consequently,

$$E(Y^{2}) = E[\,Y(Y-1)\,] + EY = \alpha(\alpha+1)\frac{\pi^{2}}{(1-\pi)^{2}} + \alpha\frac{\pi}{1-\pi},$$

and hence

$$\text{var}(Y) = E(Y^{2}) - (EY)^{2} = \alpha\frac{\pi^{2}}{(1-\pi)^{2}} + \alpha\frac{\pi}{1-\pi} = \alpha\frac{\pi}{(1-\pi)^{2}}.$$

Problems

6.94 Let the prior distribution of π be beta with parameters α_1 and α_2, which are to be determined. Suppose that if Y has the Bernoulli distribution with parameter π, then the Bayes estimate of π based on Y is given by $\hat{\pi} = .49$ when $Y = 0$ and by $\hat{\pi} = .51$ when $Y = 1$. Determine **(a)** α_1 and α_2; **(b)** the Bayes estimate of π based on a random variable Y having the binomial distribution with parameters $n = 100$ and π when $Y = 0$.

6.95 Let Y have the binomial distribution with parameters n and π, where π has the uniform prior. Determine the Bayes estimate of π when $Y = n$.

6.96 Let λ denote the accident proneness of Johnny as an automobile driver and let $\hat{\lambda}$ denote Frankie's Bayes estimate of λ based on the number Y of accidents during a 1-year period and her gamma prior for λ. Suppose that $\hat{\lambda} = 1$ if $Y = 1$ and $\hat{\lambda} = 1.5$

if $Y = 2$. Determine **(a)** Frankie's prior for λ; **(b)** her posterior for λ for general Y; **(c)** $\hat{\lambda}$ for general Y.

6.97 Redo the solution to Example 6.37 using $\sigma = 10$ instead of $\sigma = 5$.

6.98 Redo Example 6.39 with "1-year period" replaced by "2-year period."

6.99 Suppose the lactescence of a randomly selected cow is normally distributed with mean 100 and standard deviation 20 and that the amount of milk Y yielded by a cow with lactescence θ is normally distributed with mean θ and standard deviation 10. Determine **(a)** the posterior distribution of lactescence; **(b)** the Bayes estimate of the lactescence of a randomly selected cow whose milk yield is 110.

6.100 Let $Y_1, \ldots, Y_n$ be independent random variables, each normally distributed with mean θ and variance σ^2, where θ has the normal prior with mean μ and variance γ^2. Determine the posterior distribution of θ based on $\bar{Y} = (Y_1 + \cdots + Y_n)/n$.

6.101 Let Y have the geometric distribution with parameter π, where the prior distribution of π is beta with parameters α_1 and α_2. Determine **(a)** the posterior distribution of π; **(b)** the Bayes estimate of π.

6.102 Let Y have the gamma distribution with inverse-scale parameter λ and known shape parameter α, where the prior distribution of λ is gamma with shape parameter γ and inverse-scale parameter τ. Determine **(a)** the posterior distribution of λ; **(b)** the Bayes estimate of λ.

6.103 Let Y be normally distributed with mean 0 and variance σ^2, where the prior distribution of $\theta = 1/\sigma^2$ is gamma with shape parameter α and inverse-scale parameter τ. Determine **(a)** the posterior distribution of θ; **(b)** the Bayes estimate of θ.

6.104 The distribution having the probability function given by Equation (44) is referred to as the *beta-binomial* distribution with parameters α_1, α_2, and n. **(a)** Identify this distribution when $\alpha_1 = \alpha_2 = 1$. **(b)** Determine the mean of the distribution.

6.105 Show that if Y has the negative binomial distribution with parameters α and π, then $E[Y(Y - 1)] = \alpha(\alpha + 1)\pi^2/(1 - \pi)^2$.

6.106 Determine the parameters α and π and the negative binomial distribution in terms of its mean μ and variance σ^2.

6.107 Let Y have the negative binomial distribution with parameters α and π, where the prior distribution of π is beta with parameters α_1 and α_2. Determine **(a)** the posterior distribution of π; **(b)** the Bayes estimate of π.

7

Normal Models

7.1
Introduction

Estimation, confidence intervals, and tests of hypotheses play a major role in the theory and applications of statistics. Estimation has already been introduced in Section 2.4. In the text and problems in this section, we introduce the fundamental concepts involving confidence intervals and tests of hypotheses in a simplified setting.

Consider an agricultural experiment whose goal is to compare the yields of potatoes grown using two brands of fertilizer, which we refer to as brand A and brand B. Suppose the yield Y of potatoes when brand A is applied to a randomly selected plot is normally distributed with mean μ_1 and variance σ^2 and the yield when brand B is applied is normally distributed with mean μ_2 and variance σ^2. Then switching from brand A to brand B causes the mean yield to change by the amount $\tau = \mu_2 - \mu_1$. In practice, μ_1, μ_2, and σ^2 would all be unknown; but suppose here, for simplicity, that σ^2 is known.

To estimate μ_1 and μ_2 and thereby to estimate τ, we apply brand A to n_1 randomly selected plots and brand B to n_2 other such plots. Let Y_{1i}, $1 \le i \le n_1$, be the yields from the plots to which brand A is applied and let Y_{2i}, $1 \le i \le n_2$, be the yields from those to which brand B is applied. We assume that Y_{1i} is normally distributed with mean μ_1 and variance σ^2 for $1 \leq i \leq n_1$, that Y_{2i} is normally distributed with mean μ_2 and variance σ^2 for $1 \leq i \leq n_2$, and that the $n = n_1 + n_2$ random variables $Y_{11}, \ldots, Y_{1n_1}, Y_{21}, \ldots, Y_{2n_2}$ are independent.

Let

$$\bar{Y}_1 = \frac{Y_{11} + \cdots + Y_{1n_1}}{n_1} \quad \text{and} \quad \bar{Y}_2 = \frac{Y_{21} + \cdots + Y_{2n_2}}{n_1}$$

denote the sample means of the yields from the plots corresponding to brand A and brand B, respectively. Then $\bar{Y}_1$ is normally distributed with mean μ_1 and variance σ^2/n_1, $\bar{Y}_2$ is normally distributed with mean μ_2 and variance σ^2/n_2, and $\bar{Y}_1$ and $\bar{Y}_2$ are independent. Observe that $\hat{\mu}_1 = \bar{Y}_1$ is an unbiased estimate of μ_1 and $\hat{\mu}_2 = \bar{Y}_2$ is an unbiased estimate of μ_2 and hence that $\hat{\tau} = \hat{\mu}_2 - \hat{\mu}_1$ is an unbiased estimate of τ. The variance and standard deviation of this estimate are given by

$$\mathrm{var}(\hat{\tau}) = \sigma^2 \left(\frac{1}{n_1} + \frac{1}{n_2} \right)$$

and

$$\mathrm{SD}(\hat{\tau}) = \sigma \sqrt{\frac{1}{n_1} + \frac{1}{n_2}}.$$

Since the estimate $\hat{\tau}$ is normally distributed, the corresponding standardized random variable $(\hat{\tau} - \tau)/\mathrm{SD}(\hat{\tau})$ has the standard normal distribution. Thus,

$$P\left(-z_{.975} < \frac{\hat{\tau} - \tau}{\mathrm{SD}(\hat{\tau})} < z_{.975} \right) = .95,$$

so

$$P\left(-z_{.975}\,\mathrm{SD}(\hat{\tau}) < \hat{\tau} - \tau < z_{.975}\,\mathrm{SD}(\hat{\tau}) \right) = .95$$

and hence

$$P\left(\hat{\tau} - z_{.975}\,\mathrm{SD}(\hat{\tau}) < \tau < \hat{\tau} + z_{.975}\,\mathrm{SD}(\hat{\tau}) \right) = .95.$$

In other words, the probability that the unknown parameter τ lies in the observed interval

$$\hat{\tau} \pm z_{.975}\,\mathrm{SD}(\hat{\tau}) = \left(\hat{\tau} - z_{.975}\,\mathrm{SD}(\hat{\tau}),\ \hat{\tau} + z_{.975}\,\mathrm{SD}(\hat{\tau}) \right)$$

equals .95. For this reason, we refer to the indicated interval as the 95% confidence interval for τ.

EXAMPLE **7.1** Suppose brand A is applied to ten plots and brand B to five plots and that the observed yields are as follows:

A	517	407	390	244	473	488	435	433	617	396
B	565	495	457	600	568					

It is known from considerable previous experimentation that $\sigma = 100$.

(a) Estimate the change τ in mean yield caused by switching from brand A to brand B.

(b) Determine the standard deviation of this estimate.

(c) Determine the 95% confidence interval for τ.

Solution   (a) The sample mean of the yields from the ten plots to which brand A is applied is given by $\bar{Y}_1 = (517 + \cdots + 396)/10 = 440$, and sample mean of the yields from the five plots to which brand B is applied is given by $\bar{Y}_2 = (565 + \cdots + 568)/5 = 537$. Thus, τ is estimated by $\hat{\tau} = 537 - 440 = 97$.

(b) The standard deviation of the estimate $\hat{\tau}$ is given by

$$\text{SD}(\hat{\tau}) = 100\sqrt{\tfrac{1}{10} + \tfrac{1}{5}} \doteq 54.772.$$

(c) Since $z_{.975} \doteq 1.96$, the 95% confidence interval for τ is given by

$$97 \pm (1.96)(54.772) \doteq 97 \pm 107.352 = (-10.352, 204.352). \quad \blacksquare$$

It may be that brand A and brand B are equally effective in growing potatoes; that is, $\mu_1 = \mu_2$ or, equivalently, $\tau = 0$. To test this hypothesis, we consider the statistic

$$Z = \frac{\hat{\tau}}{\text{SD}(\hat{\tau})}.$$

If the hypothesis is valid, then Z has the standard normal distribution, and hence $P(|Z| \geq z_{.975}) = .05$. Suppose $|Z| \geq z_{.975}$. Then either (i) an unusual event occurred whose probability is only .05 or (ii) the hypothesis is invalid. On the other hand, if the hypothesis were invalid, then $|Z| \geq z_{.975}$ would not be such an unusual event. Let us agree to reject the hypothesis that $\tau = 0$ if $|Z| \geq z_{.975}$ and to accept it otherwise. We refer to $|Z| \geq z_{.975}$ as the critical region (or rejection region) for the test of size .05 of the hypothesis and to the complement $|Z| < z_{.975}$ of the critical region as the acceptance region. If the hypothesis is valid, then the probability that Z lies in the critical region equals the size of the test.

Observe that Z lies in the acceptance region for the test of size .05 if and only if $|\hat{\tau}| < z_{.975}\,\text{SD}(\hat{\tau})$ and hence if and only if $-z_{.975}\,\text{SD}(\hat{\tau}) < \hat{\tau} < z_{.975}\,\text{SD}(\hat{\tau})$ or, equivalently, if and only if $\hat{\tau} - z_{.975}\,\text{SD}(\hat{\tau}) < 0 < \hat{\tau} + z_{.975}\,\text{SD}(\hat{\tau})$. Thus, Z lies in the acceptance region for the test of size .05 of the hypothesis that τ equals zero if and only if zero lies in the 95% confidence interval for τ.

EXAMPLE **7.2**   In the context of Example 7.1, carry out the test of size .05 of the hypothesis that brand A and brand B are equally effective fertilizers.

Solution   Here the Z statistic is given by $Z \doteq 97/54.772 \doteq 1.771$, which lies in the acceptance region $|Z| < 1.96$ for the test of size .05. Thus we accept the hypothesis that brand A and brand B are equally effective fertilizers. This is compatible with the solution to Example 7.1(c), since the 95% confidence interval for $\hat{\tau}$ found there contains zero.

Saying that the hypothesis that τ equals zero is accepted does not mean that τ necessarily equals zero or even that we should believe that it equals zero, but merely that the observed data are compatible with this value of τ. Actually, the 95% confidence interval $(-10.352, 204.352)$ for τ that was obtained in the solution to Example 7.1(c) can be interpreted as meaning that the observed data are compatible with any hypothesized value of τ from -10.352 to 204.352. ∎

We refer to the acceptance or rejection of the hypothesis as a decision. If the hypothesis is either valid and accepted or invalid and rejected, the decision is said to be correct; otherwise, it is said to be erroneous. When the hypothesis is valid, the probability of erroneously rejecting it equals the size of the test. When the hypothesis is invalid, the probability of erroneously accepting it is denoted by β, and the probability $1 - \beta$ of correctly rejecting it is referred to as the power of the test.

Suppose the hypothesis is invalid; that is, $\tau \neq 0$. Then the power of the test is given by

$$1 - \beta = P(|Z| \geq z_{.975})$$

$$= P(Z \leq -z_{.975}) + P(Z \geq z_{.975})$$

$$= P\left(\frac{\hat{\tau}}{\mathrm{SD}(\hat{\tau})} \leq -z_{.975}\right) + P\left(\frac{\hat{\tau}}{\mathrm{SD}(\hat{\tau})} \geq z_{.975}\right)$$

$$= P\left(\frac{\hat{\tau} - \tau}{\mathrm{SD}(\hat{\tau})} \leq -z_{.975} - \frac{\tau}{\mathrm{SD}(\hat{\tau})}\right) + P\left(\frac{\hat{\tau} - \tau}{\mathrm{SD}(\hat{\tau})} \geq z_{.975} - \frac{\tau}{\mathrm{SD}(\hat{\tau})}\right)$$

$$= \Phi\left(-z_{.975} - \frac{\tau}{\mathrm{SD}(\hat{\tau})}\right) + 1 - \Phi\left(z_{.975} - \frac{\tau}{\mathrm{SD}(\hat{\tau})}\right)$$

$$= \Phi\left(-z_{.975} - \frac{\tau}{\mathrm{SD}(\hat{\tau})}\right) + \Phi\left(-z_{.975} + \frac{\tau}{\mathrm{SD}(\hat{\tau})}\right).$$

Thus,

$$1 - \beta = \Phi(-z_{.975} - \delta) + \Phi(-z_{.975} + \delta), \quad \text{where } \delta = \frac{|\tau|}{\mathrm{SD}(\hat{\tau})}. \tag{1}$$

Observe from (1) that the power of the test approaches its size as $\delta \to 0$. It also follows easily from (1) by differentiating with respect to δ that the power of the test is a strictly increasing function of δ for $\delta > 0$ (see Problem 7.13). In particular, the power of the test is greater than its size when the hypothesis is invalid.

EXAMPLE 7.3 Suppose brand A is applied to ten plots and brand B to five plots and $\sigma = 100$. Consider the test of size .05 of the hypothesis that $\tau = 0$. Determine the power of the test when $\tau = 50$.

Solution Now $\mathrm{SD}(\hat{\tau}) \doteq 54.772$ by the solution to Example 7.1(b), so (1) is applicable with $\delta \doteq 50/54.772 \doteq 0.913$. Thus, the power of the test against the indicated alternative is given by

$$1 - \beta \doteq \Phi(-1.96 - 0.913) + \Phi(-1.96 + 0.913)$$

$$\doteq \Phi(-2.873) + \Phi(-1.047)$$

$$\doteq .002 + .148 = .150.$$

Hence the test is not very powerful against the alternative that $\tau = 50$. This result supports the remarks in the second paragraph of the solution to Example 7.2. ∎

It follows from (1) that if β is small, then δ must be moderately large, and hence $\Phi(-z_{.975} - \delta)$ must be negligible. Thus, $1 - \beta \approx \Phi(-z_{.975} + \delta)$, so $-z_{.975} + \delta \approx z_{1-\beta}$ and hence $\delta \approx z_{1-\beta} + z_{.975}$. We now conclude from the formula for δ in (1) that

$$\text{SD}(\hat{\tau}) \approx \frac{|\tau|}{z_{1-\beta} + z_{.975}} \quad \text{if } \beta \text{ is small.} \tag{2}$$

EXAMPLE **7.4**   Suppose brand A and brand B are each to be applied to r plots and $\sigma = 100$. Determine r such that the test of size .05 of the hypothesis that $\tau = 0$ has power .9 when $\tau = 50$.

Solution   Now $\text{SD}(\hat{\tau}) = 100\sqrt{2/r}$, so we conclude from (2) that

$$100\sqrt{\frac{2}{r}} \approx \frac{50}{z_{.9} + z_{.975}}$$

and hence that

$$r \approx 8(z_{.9} + z_{.975})^2 \doteq 8(1.282 + 1.96)^2 \doteq 84.$$

As a check on the accuracy of (2) in the present application, note that if $\tau = 50$ and $\text{SD}(\hat{\tau}) = 50/(z_{.9} + z_{.975})$, then $\delta = 50/\text{SD}(\hat{\tau}) = z_{.9} + z_{.975}$, so $\Phi(-z_{.975} - \delta) = \Phi(-2z_{.975} - z_{.9}) \doteq \Phi(-5.201) \doteq 10^{-7}$, which is indeed negligible.

One plausible interpretation of the solution to Example 7.2 is that, because of the small number of plots involved in the experiment, the test of the hypothesis that $\tau = 0$ is not powerful enough to detect departures from the hypothesis that are large enough to be of practical importance. ∎

Consider the test of size α of the hypothesis that $\tau = 0$, where $0 < \alpha < 1$. The critical region for the test is given by $|Z| \geq z_{1-\alpha/2}$ or, equivalently, by $\Phi(|Z|) \geq 1 - \alpha/2$. Thus, the hypothesis is rejected if $\alpha \geq 2[1 - \Phi(|Z|)]$ and accepted if $\alpha < 2[1 - \Phi(|Z|)]$. We refer to $2[1 - \Phi(|Z|)]$ as the P-value for the test. The hypothesis is rejected if its size is greater than or equal to the P-value, but not if its size is less than the P-value. In other words, the P-value is the smallest size of the test for which the hypothesis is rejected.

EXAMPLE **7.5**   Determine the *P*-value for the test in Example 7.2.

Solution   Now $Z \doteq 1.771$ according to the solution to Example 7.2. Thus, the *P*-value is given by

$$P\text{-value} \doteq 2[\, 1 - \Phi(1.771)\,] \doteq .077.$$

In particular, in the context of Example 7.2, if we had carried out the test of size .1 of the hypothesis that $\tau = 0$, we would have rejected the hypothesis.   ∎

The size of tests that is commonly used in practice is .05. Sometimes a smaller size is used; rarely is a larger size used. Similarly, 95% confidence intervals are commonly used in practice rather than, say, 90% or 99% confidence intervals.

The material that has been developed in this section is directly applicable only when σ is known. In practice, however, σ is unknown and must be estimated from data. It turns out, somewhat surprisingly, that refinements of the results in this section are exactly valid, under the assumption of normality, when σ is unknown. These refinements involve three distributions closely related to the normal distribution, which are discussed in Section 7.2. The remaining sections of the chapter are devoted to the refinements of the material developed here to obtain confidence intervals and tests of hypotheses that are applicable when σ is unknown.

Problems

7.1   When the actual voltage equals V, the successive measurements of a voltmeter are independent and normally distributed with mean $V + \tau$ and standard deviation 0.1; here τ is the bias of the voltmeter. Five measurements are made when the actual voltage equals 50, and the observed results are as follows: 50.13, 50.15, 49.97, 50.02, 50.23. **(a)** Estimate the bias τ. **(b)** Determine the standard deviation of this estimate. **(c)** Determine the 95% confidence interval for τ. **(d)** Carry out the test of size .05 of the hypothesis that $\tau = 0$. **(e)** Determine the *P*-value for the test.

7.2   In the context of Example 7.1, **(a)** determine the standard deviation of $\hat{\mu}_1$ (here $\tau = \mu_1$ and $\hat{\tau} = \hat{\mu}_1$); **(b)** determine the 95% confidence interval for μ_1; **(c)** carry out the test of size .05 of the hypothesis that $\mu_1 = 380$ and check that the result is compatible with the answer to part (b); **(d)** determine the *P*-value for the test in part (c) and check that the result is compatible with the answer to part (c).

7.3   Suppose brand A is applied to ten plots and $\sigma = 100$. Consider the test of size .05 of the hypothesis that $\mu_1 = 380$. Determine the power of the test when $\mu_1 = 430$.

7.4   Suppose brand A is to be applied to r plots and $\sigma = 100$. Determine r such that the test of size .05 of the hypothesis that $\mu_1 = 380$ has power .9 when $\mu_1 = 430$.

7.5   Suppose $\hat{\tau}$ is normally distributed with mean τ. Show that $P(\tau \geq \hat{\tau} - z_{.95}\,\mathrm{SD}(\hat{\tau})) = .95$. [In light of this result, we refer to $\hat{\tau} - z_{.95}\,\mathrm{SD}(\hat{\tau})$ as the 95% lower confidence bound for τ.]

7.6   In the context of Example 7.1, determine the 95% lower confidence bound for μ_1 (see Problem 7.5).

7.7   **(a)** Consider the hypothesis that $\tau \le \tau_0$ and set $Z = (\hat{\tau} - \tau_0)/\mathrm{SD}(\hat{\tau})$. Show that $P(Z \ge z_{.95}) < .05$ if $\tau < \tau_0$, $P(Z \ge z_{.95}) = .05$ if $\tau = \tau_0$, and $P(Z \ge z_{.95}) > .05$ if $\tau > \tau_0$. (We refer to $Z \ge z_{.95}$ as the critical or rejection region for the test of size .05 of the hypothesis and to $Z < z_{.95}$ as the acceptance region. According to the results of this problem, the size of the test is the maximum probability that Z lies in the critical region, where the maximum is taken over all values of τ that are compatible with the hypothesis—that is, such that $\tau \le \tau_0$.) **(b)** Show that Z lies in the acceptance region for the test if and only if the 95% lower confidence bound for τ, as defined in Problem 7.5, is less than τ_0.

7.8   In the context of Example 7.1, carry out the test of size .05 of the hypothesis that $\mu_1 \le 380$, as described in Problem 7.7, and check that the result is compatible with the answer to Problem 7.6 and the result in Problem 7.7(b).

7.9   The P-value for the test of the hypothesis that $\tau \le \tau_0$ is given by $P\text{-value} = 1 - \Phi(Z)$. As usual, the hypothesis is rejected if its size is greater than or equal to the P-value, but not if its size is less than the P-value. Determine the P-value for the test in Problem 7.8 and check that it is compatible with the answer to this problem.

7.10   **(a)** Show that if $\tau > \tau_0$, then the power of the test in Problem 7.7 is given by $1 - \beta = \Phi(-z_{.95} + \delta)$, where $\delta = (\tau - \tau_0)/\mathrm{SD}(\hat{\tau})$. **(b)** Use the result of part (a) to show that

$$\mathrm{SD}(\hat{\tau}) = \frac{\tau - \tau_0}{z_{1-\beta} + z_{.95}} \quad \text{if} \quad \tau > \tau_0.$$

7.11   Suppose brand A is applied to ten plots and brand B to five plots and $\sigma = 100$. Consider the test of size .05 of the hypothesis that $\mu_1 \le \mu_2$. Determine the power of the test when $\mu_2 - \mu_1 = 50$ [see Problem 7.10(a)].

7.12   Suppose brand A and brand B are each to be applied to r plots and $\sigma = 100$. Determine r such that the test of size .05 of the hypothesis that $\mu_1 \le \mu_2$ has power .9 when $\mu_2 - \mu_1 = 50$. [Use the result of Problem 7.10(b).]

7.13   Show that the quantity $\Phi(-z_{.975} - \delta) + \Phi(-z_{.975} + \delta)$ appearing in Equation (1) is a strictly increasing function of δ for $\delta \ge 0$.

7.2
Chi-Square, t, and F Distributions

In this section, we develop the properties of three continuous univariate distributions of fundamental importance in the theory and practice of statistics.

Chi-Square Distribution

Let $Z_1, \ldots, Z_\nu$ be independent, standard normal random variables and set $\chi^2 = Z_1^2 + \cdots + Z_\nu^2$ with $\chi \ge 0$. The distribution of the positive random variable χ^2 is referred to as the *chi-square distribution* with ν "degrees of freedom." (The prefix "chi" and the Greek letter "χ" are pronounced like "ki" as in "kite.") We use $\chi_\nu^2(y)$ to denote the value of the corresponding distribution function at y and $\chi_{p,\nu}^2$ to denote the pth quantile of the distribution. The distribution function equals zero on $(-\infty, 0]$, and the quantiles are all positive. It follows from Theorem 3.11 in Section 3.2 that the chi-square distribution with ν degrees of freedom coincides with the gamma distribution with shape parameter $\nu/2$ and scale parameter 2. Consequently, the distribution has mean ν, variance 2ν, and standard deviation $\sqrt{2\nu}$, and its density function is given by $f(y) = 0$ for $y \leq 0$ and

$$f(y) = \frac{y^{\nu/2-1}e^{-y/2}}{2^{\nu/2}\Gamma(\nu/2)}, \qquad y > 0.$$

EXAMPLE 7.6   Determine $\chi_1^2(y)$ for $y > 0$ and $\chi_{p,1}^2$ for $0 < p < 1$ in terms of the distribution function and quantiles of the standard normal distribution.

Solution   Let Z have the standard normal distribution. Then Z^2 has the chi-square distribution with 1 degree of freedom. Thus, we conclude from Theorem 1.7 in Section 1.7 that $\chi_1^2(y) = 2\Phi(\sqrt{y}) - 1$ for $y > 0$ and $\chi_{p,1}^2 = z_{(1+p)/2}^2$ for $0 < p < 1$.   ∎

The normal approximation to the distribution function of the chi-square distribution with ν degrees of freedom is given by

$$\chi_\nu^2(y) \approx \Phi\left(\frac{y - \nu}{\sqrt{2\nu}}\right), \qquad y > 0,$$

and the normal approximation to its pth quantile is given by $\chi_{p,\nu}^2 \approx \nu + z_p\sqrt{2\nu}$. These approximations are accurate when ν is large.

EXAMPLE 7.7   Determine the normal approximations to $\chi_{30}^2(40)$ and $\chi_{.95,30}^2$.

Solution   Here $\nu = 30$, so $\sqrt{2\nu} = \sqrt{60} \doteq 7.446$. Thus,

$$\chi_{30}^2(40) \approx \Phi\left(\frac{40 - 30}{7.746}\right) \doteq \Phi(1.291) \doteq .902,$$

and

$$\chi_{.95,30}^2 \approx 30 + (7.746)(1.645) \doteq 42.7.$$

[The actual values are given by $\chi_{30}^2(40) \doteq .895$ and $\chi_{.95,30}^2 \doteq 43.8$.]   ∎

EXAMPLE 7.8   Let $Y_1, \ldots, Y_\nu$ be independent random variables having a common normal distribution with mean μ and variance σ^2. Show that

$$\frac{1}{\sigma^2}\sum_{i=1}^{\nu}(Y_i - \mu)^2$$

has the chi-square distribution with ν degrees of freedom.

Solution   Set $Z_i = (Y_i - \mu)/\sigma$ for $1 \le i \le \nu$. Then $Z_1, \ldots, Z_\nu$ are independent, standard normal random variables. Consequently,

$$\frac{1}{\sigma^2} \sum_{i=1}^{\nu} (Y_i - \mu)^2 = \sum_{i=1}^{\nu} \left(\frac{Y_i - \mu}{\sigma}\right)^2 = Z_1^2 + \cdots + Z_\nu^2$$

has the chi-square distribution with ν degrees of freedom.   ∎

THEOREM 7.1   Let $\chi_1^2, \ldots, \chi_n^2$ be independent random variables having chi-square distributions with $\nu_1, \ldots, \nu_n$ degrees of freedom, respectively. Then $\chi_1^2 + \cdots + \chi_n^2$ has the chi-square distribution with $\nu_1 + \cdots + \nu_n$ degrees of freedom.

Proof   Now χ_i^2 has the gamma distribution with shape parameter $\nu_i/2$ and scale parameter 2 for $1 \le i \le n$. Since $\chi_1^2, \ldots, \chi_n^2$ are independent, we conclude from Theorem 3.6 in Section 3.1 that $\chi_1^2 + \cdots + \chi_n^2$ has the gamma distribution with shape parameter $(\nu_1 + \cdots + \nu_n)/2$ and scale parameter 2, which equals the chi-square distribution with $\nu_1 + \cdots + \nu_n$ degrees of freedom.   ∎

The *t* Distribution

Let W_1 and W_2 be independent random variables having density functions f_{W_1} and f_{W_2}, respectively, and suppose W_2 is a positive random variable. According to Theorem 5.22 in Section 5.6, W_1/W_2 has the density function given by

$$f_{W_1/W_2}(y) = \int_0^\infty w f_{W_1}(yw) f_{W_2}(w)\, dw, \qquad y \in \mathbb{R}. \tag{3}$$

THEOREM 7.2   Let Z and V be independent random variables such that Z has the standard normal distribution and V has the gamma distribution with shape parameter α and scale parameter β. Then $Y = Z/\sqrt{V}$ has the density function given by

$$f_Y(y) = \frac{\Gamma(\alpha + \frac{1}{2})}{\beta^\alpha \Gamma(\alpha)\sqrt{2\pi}} \left(\frac{1}{\beta} + \frac{y^2}{2}\right)^{-\alpha - 1/2}, \qquad y \in \mathbb{R}.$$

Proof   According to Theorem 3.3 in Section 3.1, the density function of the positive random variable $\sqrt{V}$ is given by

$$f_{\sqrt{V}}(w) = \frac{2w^{2\alpha-1}}{\beta^\alpha \Gamma(\alpha)} e^{-w^2/\beta}, \qquad w > 0.$$

Let $y \in \mathbb{R}$. Then, by (3),

$$f_Y(y) = \int_0^\infty w \frac{1}{\sqrt{2\pi}} e^{-(yw)^2/2} \frac{2w^{2\alpha-1}}{\beta^\alpha \Gamma(\alpha)} e^{-w^2/\beta}\, dw$$

$$= \frac{1}{\beta^\alpha \Gamma(\alpha)\sqrt{2\pi}} \int_0^\infty w^{2\alpha-1} \exp\left[-\left(\frac{1}{\beta} + \frac{y^2}{2}\right)w^2\right] 2w\, dw.$$

Making the change of variables $t = w^2$ with $dt = 2w\,dw$, we get

$$f_Y(y) = \frac{1}{\beta^\alpha \Gamma(\alpha)\sqrt{2\pi}} \int_0^\infty t^{\alpha-1/2} \exp\left[-\left(\frac{1}{\beta} + \frac{y^2}{2}\right)t \right] dt.$$

The desired result now follows from Equation (6) in Section 3.1. ∎

Let Z and χ^2 be independent random variables such that Z has the standard normal distribution and χ^2 has the chi-square distribution with ν degrees of freedom. Consider the random variable

$$t = \frac{Z}{\sqrt{\chi^2/\nu}}. \tag{4}$$

The distribution of t is referred to as the t *distribution* with ν degrees of freedom. The value at y of the corresponding distribution function is denoted by $t_\nu(y)$, and its pth quantile is denoted by $t_{p,\nu}$ for $0 < p < 1$. (The t distribution is also referred to as *Student's t distribution.*)

THEOREM 7.3 The t distribution with ν degrees of freedom has the density function

$$f(y) = \frac{\Gamma((\nu+1)/2)}{\sqrt{\nu\pi}\,\Gamma(\nu/2)} \left(1 + \frac{y^2}{\nu}\right)^{-(\nu+1)/2}, \qquad y \in \mathbb{R}. \tag{5}$$

Proof Now χ^2 has the gamma distribution with shape parameter $\nu/2$ and scale parameter 2, so χ^2/ν has the gamma distribution with shape parameter $\nu/2$ and scale parameter $2/\nu$. The desired result now follows from Theorem 7.2. ∎

Observe from (5) that the density function of the t distribution with ν degrees of freedom is symmetric about zero. We now conclude from Equation (48) and (49) in Section 1.6 that

$$t_\nu(-y) = 1 - t_\nu(y), \qquad y \in \mathbb{R}, \tag{6}$$

and

$$t_{1-p,\nu} = -t_{p,\nu}, \qquad 0 < p < 1. \tag{7}$$

COROLLARY 7.1 The t distribution with 1 degree of freedom coincides with the standard Cauchy distribution.

Proof Since $\Gamma(1/2) = \sqrt{\pi}$ and $\Gamma(1) = 1$, (5) with $\nu = 1$ reduces to

$$f(y) = \frac{1}{\pi(1 + y^2)}, \qquad y \in \mathbb{R}. \qquad ∎$$

EXAMPLE 7.9 Determine the mean of the t distribution with ν degrees of freedom.

Solution   Since the t distribution with 1 degree of freedom coincides with the standard Cauchy distribution, we conclude from the solution to Example 2.20 in Section 2.1 that its mean is of the indeterminate form $\infty - \infty$. Consider, instead, the t distribution with ν degrees of freedom with $\nu \geq 2$. According to Theorem 7.3, its density function is given by $f(y) = c_\nu(1 + y^2/\nu)^{-(\nu+1)/2}$ for $y \in \mathbb{R}$, where

$$c_\nu = \frac{\Gamma((\nu + 1)/2)}{\sqrt{\nu\pi}\,\Gamma(\nu/2)}.$$

Observe that

$$\int_0^\infty yf(y)\,dy = c_\nu \int_0^\infty y\left(1 + \frac{y^2}{\nu}\right)^{-(\nu+1)/2} dy$$

$$= -\frac{\nu c_\nu}{\nu - 1}\left(1 + \frac{y^2}{\nu}\right)^{-(\nu-1)/2}\Bigg|_0^\infty$$

$$= \frac{\nu c_\nu}{\nu - 1}$$

$$< \infty.$$

Since f is symmetric about zero, we now conclude that

$$\int_{-\infty}^0 yf(y)\,dy = -\frac{\nu c_\nu}{\nu - 1}$$

and hence that the mean of the t distribution with $\nu \geq 2$ degrees of freedom is given by

$$\int_{-\infty}^\infty yf(y)\,dy = \int_{-\infty}^0 yf(y)\,dy + \int_0^\infty yf(y)\,dy = -\frac{\nu c_\nu}{\nu - 1} + \frac{\nu c_\nu}{\nu - 1} = 0. \quad \blacksquare$$

Let $Z_1, \ldots, Z_\nu$ be independent, standard normal random variables. Then $\chi^2 = Z_1^2 + \cdots + Z_\nu^2$ has the chi-square distribution with ν degrees of freedom and

$$\frac{\chi^2}{\nu} = \frac{Z_1^2 + \cdots + Z_\nu^2}{\nu}.$$

Since the common mean of $Z_1^2, \ldots, Z_\nu^2$ (that is, the common variance of $Z_1, \ldots, Z_\nu$) equals 1, it follows from the weak law of large numbers that

$$P\left(\frac{\chi^2}{\nu} \approx 1\right) \approx 1, \qquad \nu \gg 1. \tag{8}$$

Let t have the t distribution with ν degrees of freedom. It follows from (4) and (8) that $P(t \approx Z) \approx 1$ for $\nu \gg 1$. Therefore, for $y \in \mathbb{R}$,

$$t_\nu(y) \approx \Phi(y), \qquad \nu \gg 1;$$

also, for $0 < p < 1$,

$$t_{p,\nu} \approx z_p, \qquad \nu \gg 1. \tag{9}$$

Consider a t distribution with ν degrees of freedom, where ν is not large. When $|y| \gg 1$, the value of its density function at y is large relative to the value $\phi(y)$ of the standard normal density function at y. Consequently, when p is close to 0 or 1, $t_{p,\nu}$ is significantly larger in magnitude than z_p. Selected quantiles of the t distribution, as shown in Table 7.1, illustrate the accuracy of the approximation in (9). [The quantiles for $\nu = \infty$ are the standard normal quantiles z_p. According to (7), $t_{.5,\nu} = 0$, $t_{.25,\nu} = -t_{.75,\nu}$, $t_{.1,\nu} = -t_{.9,\nu}$, and so forth.]

TABLE 7.1

Selected values of $t_{p,\nu}$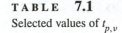

		p					
		.750	.900	.950	.975	.990	.995
	1	1.000	3.078	6.314	12.706	31.821	63.657
	2	0.816	1.886	2.920	4.303	6.965	9.925
	3	0.765	1.638	2.353	3.182	4.541	5.841
	4	0.741	1.533	2.132	2.776	3.747	4.604
	5	0.727	1.476	2.015	2.571	3.365	4.032
	10	0.700	1.372	1.812	2.228	2.764	3.169
	15	0.691	1.341	1.753	2.131	2.602	2.947
ν	20	0.687	1.325	1.725	2.086	2.528	2.845
	30	0.683	1.310	1.697	2.042	2.457	2.750
	40	0.681	1.303	1.684	2.021	2.423	2.704
	50	0.679	1.299	1.676	2.009	2.403	2.678
	100	0.677	1.290	1.660	1.984	2.364	2.626
	200	0.676	1.286	1.653	1.972	2.245	2.601
	∞	0.674	1.282	1.645	1.960	2.326	2.576

The F Distribution

THEOREM 7.4   Let W_1 and W_2 be independent random variables such that W_1 has the gamma distribution with shape parameter α_1 and scale parameter β_1 and W_2 has the gamma distribution with shape parameter α_2 and scale parameter β_2. Then W_1/W_2 has the density function given by $f_{W_1/W_2}(y) = 0$ for $y \leq 0$ and

$$f_{W_1/W_2}(y) = \frac{\beta_2 \Gamma(\alpha_1 + \alpha_2)}{\beta_1 \Gamma(\alpha_1)\Gamma(\alpha_2)} \left(\frac{\beta_2}{\beta_1}y\right)^{\alpha_1 - 1}\left(1 + \frac{\beta_2}{\beta_1}y\right)^{-(\alpha_1 + \alpha_2)}, \qquad y > 0.$$

Proof   Now W_1/W_2 is a positive random variable, so $f_{W_1/W_2}(y) = 0$ for $y \leq 0$. Let $y > 0$. Since W_2 is a positive random variable, it follows from Equation (3) that

$$f_{W_1/W_2}(y) = \int_0^\infty w \frac{(yw)^{\alpha_1 - 1}}{\beta_1^{\alpha_1}\Gamma(\alpha_1)} e^{-yw/\beta_1} \frac{w^{\alpha_2 - 1}}{\beta_2^{\alpha_2}\Gamma(\alpha_2)} e^{-w/\beta_2}\, dw$$

$$= \frac{1}{\beta_1^{\alpha_1}\beta_2^{\alpha_2}\Gamma(\alpha_1)\Gamma(\alpha_2)} y^{\alpha_1-1} \int_0^\infty w^{\alpha_1+\alpha_2-1} \exp\left[-\left(\frac{1}{\beta_2}+\frac{y}{\beta_1}\right)w\right]dw$$

$$= \frac{\Gamma(\alpha_1+\alpha_2)}{\beta_1^{\alpha_1}\beta_2^{\alpha_2}\Gamma(\alpha_1)\Gamma(\alpha_2)} y^{\alpha_1-1}\left(\frac{1}{\beta_2}+\frac{y}{\beta_1}\right)^{-(\alpha_1+\alpha_2)};$$

in the last step, we have used Equation (6) in Section 3.1 with $\alpha = \alpha_1 + \alpha_2$ and

$$\beta = \left(\frac{1}{\beta_2}+\frac{y}{\beta_1}\right)^{-1}.$$

The desired result now follows by elementary algebra.   ∎

Let χ_1^2 and χ_2^2 be independent random variables having chi-square distributions with ν_1 and ν_2 degrees of freedom, respectively. The distribution of the random variable

$$F = \frac{\chi_1^2/\nu_1}{\chi_2^2/\nu_2} \tag{10}$$

is referred to as the *F distribution* with ν_1 degrees of freedom in the numerator and ν_2 degrees of freedom in the denominator. The value at y of the corresponding distribution function is denoted by $F_{\nu_1,\nu_2}(y)$ and its pth quantile is denoted by F_{p,ν_1,ν_2}. Since F is a positive random variable, its distribution function equals zero on $(-\infty, 0]$, and its quantiles are all positive.

THEOREM 7.5   The F distribution with ν_1 degrees of freedom in the numerator and ν_2 degrees of freedom in the denominator has the density function f given by $f(y) = 0$ for $y \leq 0$ and

$$f(y) = \frac{\nu_1\Gamma((\nu_1+\nu_2)/2)}{\nu_2\Gamma(\nu_1/2)\Gamma(\nu_2/2)}\left(\frac{\nu_1 y}{\nu_2}\right)^{(\nu_1/2)-1}\left(1+\frac{\nu_1 y}{\nu_2}\right)^{-(\nu_1+\nu_2)/2}, \qquad y > 0.$$

Proof   Observe that χ_1^2/ν_1 has the gamma distribution with shape parameter $\nu_1/2$ and scale parameter $2/\nu_1$ and that χ_2^2/ν_2 has the gamma distribution with shape parameter $\nu_2/2$ and scale parameter $2/\nu_2$. Since $(2/\nu_2)/(2/\nu_1) = \nu_1/\nu_2$, the desired result follows from Theorem 7.4.   ∎

Let F have the F distribution with ν_1 degrees of freedom in the numerator and ν_2 degrees of freedom in the denominator. It follows from Equation (10) that $1/F$ has the F distribution with ν_2 degrees of freedom in the numerator and ν_1 degrees of freedom in the denominator. Thus, by Theorem 1.6 in Section 1.7,

$$F_{\nu_2,\nu_1}(y) = 1 - F_{\nu_1,\nu_2}(1/y), \qquad y > 0, \tag{11}$$

and

$$F_{p,\nu_2,\nu_1} = \frac{1}{F_{1-p,\nu_1,\nu_2}}, \qquad 0 < p < 1. \tag{12}$$

THEOREM 7.6   Let t have the t distribution with v degrees of freedom. Then $F = t^2$ has the F distribution with 1 degree of freedom in the numerator and v degrees of freedom in the denominator. Moreover,

$$1 - F_{1,v}(y^2) = 2[\, 1 - t_v(|y|)\,], \qquad y \in \mathbb{R}, \tag{13}$$

and

$$F_{p,1,v} = t^2_{(1+p)/2,v}, \qquad 0 < p < 1. \tag{14}$$

Proof   Write

$$t = \frac{Z}{\sqrt{\chi^2/v}},$$

where Z and χ^2 are independent random variables, Z has the standard normal distribution, and χ^2 has the chi-square distribution with v degrees of freedom. Then

$$F = t^2 = \frac{Z^2}{\chi^2/v}.$$

Now Z^2 has the chi-square distribution with 1 degree of freedom, and Z^2 and χ^2 are independent. Thus, F has the indicated F distribution, and the other conclusions follow from Theorem 1.7 in Section 1.7 and the symmetry of the density function of the t distribution.   ∎

     Let F [in the form of Equation (10)] have the F distribution with v_1 degrees of freedom in the numerator and v_2 degrees of freedom in the denominator. According to (8),

$$P\!\left(\frac{\chi^2_2}{v_2} \approx 1\right) \approx 1, \qquad v_2 \gg 1.$$

Thus, by (10),

$$P\!\left(F \approx \frac{\chi^2_1}{v_1}\right) \approx 1, \qquad v_2 \gg 1. \tag{15}$$

It follows from (15) that the distribution function of F is approximately equal to that of χ^2_1/v_1 for $v_2 \gg 1$; that is,

$$F_{v_1,v_2}(y) = P(F \leq y) \approx P\!\left(\frac{\chi^2_1}{v_1} \leq y\right) = P(\chi^2_1 \leq v_1 y), \qquad v_2 \gg 1,$$

and hence

$$F_{v_1,v_2}(y) \approx \chi^2_{v_1}(v_1 y), \qquad v_2 \gg 1 \quad \text{and} \quad y \in \mathbb{R}.$$

Let $0 < p < 1$. We conclude from (15) that the pth quantile of F is approximately equal to that of χ^2_1/v_1 for $v_2 \gg 1$; that is,

$$F_{p,v_1,v_2} \approx \frac{\chi^2_{p,v_1}}{v_1}, \qquad v_2 \gg 1.$$

This result is illustrated in Table 7.2 for $v_1 = 5$ degrees of freedom in the numerator.

TABLE   7.2

Selected values of $F_{p,5,\nu}$

		p					
	.010	.050	.100	.500	.900	.950	.990
1	0.062	0.151	0.246	1.894	57.240	230.162	5763.650
2	0.075	0.173	0.265	1.252	9.293	19.296	99.299
3	0.083	0.185	0.276	1.102	5.309	9.013	28.237
4	0.088	0.193	0.284	1.037	4.051	6.256	15.522
5	0.091	0.198	0.290	1.000	3.453	5.050	10.967
10	0.099	0.211	0.303	0.932	2.522	3.326	5.636
15	0.103	0.217	0.309	0.911	2.273	2.901	4.556
ν 20	0.105	0.219	0.312	0.900	2.158	2.711	4.103
30	0.107	0.222	0.315	0.890	2.049	2.534	3.699
40	0.108	0.224	0.317	0.885	1.997	2.449	3.514
50	0.108	0.225	0.318	0.882	1.966	2.400	3.408
100	0.110	0.227	0.320	0.876	1.906	2.305	3.206
200	0.110	0.228	0.321	0.873	1.876	2.259	3.110
∞	0.111	0.229	0.322	0.870	1.847	2.214	3.017

Problems

7.14 (a) Show that the chi-square distribution with 2 degrees of freedom coincides with the exponential distribution with mean 2. (b) Determine $\chi_2^2(y)$ for $y > 0$. (c) Determine $\chi_{p,2}^2$ for $0 < p < 1$.

7.15 Let Z_1, Z_2, Z_3, and Z_4 be independent, standard normal random variables. Determine $P(Z_1^2 + Z_2^2 + Z_3^2 + Z_4^2 \le 8)$.

7.16 Let Y have the gamma distribution with shape parameter $\nu/2$ and scale parameter β, where ν is a positive integer, and let F and y_p denote the distribution function and pth quantile of Y, respectively. Verify that (a) $2Y/\beta$ has the chi-square distribution with ν degrees of freedom; (b) $F(y) = \chi_\nu^2(2y/\beta)$ for $y \in \mathbb{R}$; (c) $y_p = (\beta/2)\chi_{p,\nu}^2$.

7.17 Let $Y_1, \ldots, Y_n$ be independent random variables, each having the exponential distribution with mean β. (a) Determine the distribution of $(Y_1 + \cdots + Y_n)/c$ for $c > 0$. (b) Determine c such that $(Y_1 + \cdots + Y_n)/c$ has a chi-square distribution.

7.18 Let $\chi = \sqrt{\chi^2}$, where χ^2 has the chi-square distribution with ν degrees of freedom. Determine (a) the density function of χ; (b) the mean of χ; (c) the variance of χ.

7.19 Use normal approximation to determine a value of ν such that $\chi_{.975,\nu}^2/\chi_{.025,\nu}^2 \approx 2$.

7.20 For which values of the positive integer ν does the t distribution with ν degrees of freedom have finite variance?

7.21 Let Y have the density function given by $f_Y(y) = 0$ for $y \le 0$ and

$$f_Y(y) = \frac{\Gamma(\alpha_1 + \alpha_2)}{\Gamma(\alpha_1)\Gamma(\alpha_2)} y^{\alpha_1 - 1}(1 + y)^{-(\alpha_1 + \alpha_2)}, \qquad y > 0.$$

Show that $W = Y/(1 + Y)$ has the beta distribution with parameters α_1 and α_2.

7.22   Let F have the F distribution with v_1 degrees of freedom in the numerator and v_2 degrees of freedom in the denominator. Use the result of Problem 7.21 to show that $Y = v_1 F/(v_2 + v_1 F)$ has the beta distribution with parameters $\alpha_1 = v_1/2$ and $\alpha_2 = v_2/2$.

7.23   (Continued.) Let y_p be the pth quantile of the beta distribution with parameters α_1 and α_2, where α_1 and α_2 are half-integers (so that $2\alpha_1$ and $2\alpha_2$ are positive integers). **(a)** Express y_p in terms of the pth quantile of a suitable F distribution. **(b)** Express the pth quantile of the F distribution in terms of y_p.

7.3
Confidence Intervals

In this section, quantiles of the t distribution are used to obtain confidence bounds and confidence intervals for an unknown mean in the context of the normal one-sample model or for a linear combination of unknown means in the context of the normal multisample model.

Normal One-Sample Model

Let $Y_1, \ldots, Y_n$ be independent random variables having a common normal distribution with mean μ and variance σ^2. (Here we think of μ and σ as being unknown.) We refer to this setup as the *normal one-sample model*. Under this model, the sample mean $\bar{Y} = (Y_1 + \cdots + Y_n)/n$ is normally distributed with mean μ, variance σ^2/n and standard deviation $\sigma/\sqrt{n}$.

In the present context, the *within sum of squares* (WSS) is defined by

$$\text{WSS} = \sum_{i=1}^{n}(Y_i - \bar{Y})^2.$$

According to Equation (53) in Section 2.4,

$$E(\text{WSS}) = E\left(\sum_{i=1}^{n}(Y_i - \bar{Y})^2 \right) = (n-1)\sigma^2. \tag{16}$$

If $n = 1$, then WSS $= 0$, and if $n \geq 2$, then WSS $= (n-1)S^2$, where

$$S^2 = \frac{1}{n-1}\sum_{i=1}^{n}(Y_i - \bar{Y})^2$$

is the sample variance. It follows from (16) that if $n \geq 2$, then $E(S^2) = \sigma^2$; that is, that S^2 is an unbiased estimate of σ^2. The next result will be verified in Section 10.3.

THEOREM 7.7   Under the assumptions of the normal one-sample model, $\bar{Y}$ and WSS are independent, and, if $n > 1$, then WSS $/\sigma^2$ has the chi-square distribution with $n - 1$ degrees of freedom.

COROLLARY **7.2**   Under the assumptions of the normal one-sample model with $n \geq 2$, $\bar{Y}$ and S^2 are independent, and $(n-1)S^2/\sigma^2$ has the chi-square distribution with $n-1$ degrees of freedom.

Normal Multisample Model

Let Y_{ki}, $1 \leq k \leq d$ and $1 \leq i \leq n_k$, be independent, normally distributed random variables having common variance σ^2 and such that the random variables Y_{ki}, $1 \leq i \leq n_k$, have common mean μ_k for $1 \leq k \leq d$; here d, n_1, ..., n_d are positive integers. This setup is referred to as the *normal d-sample model*.

We refer to Y_{ki}, $1 \leq i \leq n_k$, as forming the kth random sample. This random sample has size n_k, its sample mean is given by

$$\bar{Y}_k = \frac{1}{n_k} \sum_{i=1}^{n_k} Y_{ki},$$

its within sum of squares is given by

$$\text{WSS}_k = \sum_{i=1}^{n_k} (Y_{ki} - \bar{Y}_k)^2,$$

which equals zero when $n_k = 1$, and (for $n_k \geq 2$) its sample variance is given by

$$S_k^2 = \frac{\text{WSS}_k}{n_k - 1},$$

where $S_k \geq 0$.

The sample means $\bar{Y}_1, \ldots, \bar{Y}_d$ are independent random variables, and $\bar{Y}_k$ is normally distributed with mean μ_k and variance σ^2/n_k for $1 \leq k \leq d$. The random variables $\text{WSS}_1, \ldots, \text{WSS}_d$ are also independent. According to Equation (16),

$$E(\text{WSS}_k) = (n_k - 1)\sigma^2, \quad 1 \leq k \leq d. \tag{17}$$

If $n_k \geq 2$, then $\text{WSS}_k/\sigma^2 = (n_k - 1)S_k^2/\sigma^2$ has the chi-square distribution with $n_k - 1$ degrees of freedom.

Set $n = n_1 + \cdots + n_d$. Then $n \geq d$, and $n = d$ if and only if $n_1 = 1, \ldots, n_d = 1$. Observe that

$$\sum_{k=1}^{d} (n_k - 1) = n - d. \tag{18}$$

The within sum of squares for the combination of the d random samples is defined by

$$\text{WSS} = \sum_{k=1}^{d} \text{WSS}_k = \sum_{k=1}^{d} \sum_{i=1}^{n_k} (Y_{ki} - \bar{Y}_k)^2 = \sum_{k=1}^{d} (n_k - 1)S_k^2;$$

here $(n_k - 1)S_k^2$ is defined to be zero when $n_k = 1$. It follows from (17) and (18) that

$$E(\text{WSS}) = (n - d)\sigma^2.$$

If $n_1 = 1, \ldots, n_d = 1$, then $\text{WSS} = 0$.

THEOREM **7.8** Under the assumptions of the normal d-sample model, $\bar{Y}_1, \ldots, \bar{Y}_d$ and WSS are independent, and if $n > d$, then WSS $/\sigma^2$ has the chi-square distribution with $n - d$ degrees of freedom.

Proof According to Theorem 7.7, $\bar{Y}_k$ and WSS_k are independent for $1 \le k \le d$. Moreover, the d random samples are independent of each other. Since $(\bar{Y}_k, \mathrm{WSS}_k)$ is a transform of the kth random sample for $1 \le k \le d$, we conclude that $(\bar{Y}_k, \mathrm{WSS}_k)$, $1 \le k \le d$, are independent random pairs. Thus, $\bar{Y}_1, \mathrm{WSS}_1, \ldots, \bar{Y}_d, \mathrm{WSS}_d$ are independent, and hence $\bar{Y}_1, \ldots, \bar{Y}_d$ and $\mathrm{WSS} = \mathrm{WSS}_1 + \cdots + \mathrm{WSS}_d$ are independent.

In proving the second result, we can ignore any values of k such that $n_k = 1$ [see (18)] and thereby assume that $n_k \ge 2$ for $1 \le k \le d$. Since WSS_k /σ^2, $1 \le k \le d$, are independent random variables and, for $1 \le k \le d$, WSS_k /σ^2 has the chi-square distribution with $n_k - 1$ degrees of freedom, their sum WSS $/\sigma^2$ has the chi-square distribution with $n - d$ degrees of freedom. ∎

Consider the normal d-sample model with $n > d$ and set

$$S^2 = \frac{\mathrm{WSS}}{n - d} = \frac{1}{n - d} \sum_{k=1}^{d} (n_k - 1) S_k^2,$$

where $S \ge 0$. Then S^2 is an unbiased estimate of σ^2, which is said to be obtained by "pooling" the individual estimates $S_1^2, \ldots, S_d^2$ of σ^2.

COROLLARY **7.3** Under the assumptions of the normal d-sample model with $n > d$, $\bar{Y}_1, \ldots, \bar{Y}_d$, and S are independent and $(n - d)S^2/\sigma^2$ has the chi-square distribution with $n - d$ degrees of freedom.

For $1 \le k \le d$, $\hat{\mu}_k = \bar{Y}_k$ is an unbiased estimate of μ_k, and it is normally distributed with variance σ^2/n_k. The standard error $\mathrm{SE}(\hat{\mu}_k) = S/\sqrt{n_k}$ of $\hat{\mu}_k$ is a natural estimate of its standard deviation $\mathrm{SD}(\hat{\mu}_k) = \sigma/\sqrt{n_k}$. Observe that the random variable

$$\frac{\hat{\mu}_k - \mu_k}{\mathrm{SD}(\hat{\mu}_k)}$$

has the standard normal distribution.

More generally, given $c_1, \ldots, c_d \in \mathbb{R}$, consider the *linear parameter*

$$\tau = c_1 \mu_1 + \cdots + c_d \mu_d.$$

The linear estimate $\hat{\tau} = c_1 \hat{\mu}_1 + \cdots + c_d \hat{\mu}_d = c_1 \bar{Y}_1 + \cdots + c_d \bar{Y}_d$ of this parameter is normally distributed. Since

$$E\hat{\tau} = E(c_1 \hat{\mu}_1 + \cdots + c_d \hat{\mu}_d)$$

$$= c_1 E(\hat{\mu}_1) + \cdots + c_d E(\hat{\mu}_d)$$

$$= c_1 \mu_1 + \cdots + c_d \mu_d$$

$$= \tau,$$

we see that $\hat{\tau}$ is an unbiased estimate of τ. Moreover,

$$\text{var}(\hat{\tau}) = \text{var}(c_1\hat{\mu}_1 + \cdots + c_d\hat{\mu}_d) = c_1^2 \text{var}(\hat{\mu}_1) + \cdots + c_d^2 \text{var}(\hat{\mu}_d),$$

so

$$\text{var}(\hat{\tau}) = \sigma^2 \left(\frac{c_1^2}{n_1} + \cdots + \frac{c_d^2}{n_d} \right)$$

and hence

$$\text{SD}(\hat{\tau}) = \sigma \sqrt{\frac{c_1^2}{n_1} + \cdots + \frac{c_d^2}{n_d}}.$$

Suppose that $n > d$. Then the standard error

$$\text{SE}(\hat{\tau}) = S \sqrt{\frac{c_1^2}{n_1} + \cdots + \frac{c_d^2}{n_d}} = \frac{S}{\sigma} \text{SD}(\hat{\tau})$$

of $\hat{\tau}$ is a natural estimate of its standard deviation. It follows from Corollary 7.3 that $\hat{\tau}$ and S are independent and hence that $\hat{\tau}$ and $\text{SE}(\hat{\tau})$ are independent.

The linear parameter τ is said to be *trivial* if all of the coefficients $c_1, \ldots, c_d$ equal zero and *nontrivial* otherwise. Observe that τ is nontrivial if and only if $\hat{\tau}$ has positive variance. Moreover, if τ is nontrivial, then $\hat{\tau}$ is normally distributed, and the random variable $Z = (\hat{\tau} - \tau)/\text{SD}(\hat{\tau})$ has the standard normal distribution.

THEOREM 7.9   Let τ be a nontrivial linear parameter. Under the assumptions of the normal d-sample model with $n > d$, the random variable $(\hat{\tau} - \tau)/\text{SE}(\hat{\tau})$ has the t distribution with $n - d$ degrees of freedom.

Proof   According to Theorem 7.8, the random variable $\chi^2 = (n - d)S^2/\sigma^2 = \text{WSS}/\sigma^2$ has the chi-square distribution with $n - d$ degrees of freedom; also, Z and χ^2 are independent. Set $c = (c_1^2/n_1 + \ldots + c_d^2/n_d)^{1/2}$. Then $\text{SD}(\hat{\tau}) = c\sigma$ and $\text{SE}(\hat{\tau}) = cS$, so

$$\frac{Z}{\sqrt{\chi^2/(n-d)}} = \frac{(\hat{\tau} - \tau)/(c\sigma)}{S/\sigma} = \frac{\hat{\tau} - \tau}{cS} = \frac{\hat{\tau} - \tau}{\text{SE}(\hat{\tau})}$$

has the t distribution with $n - d$ degrees of freedom.   ∎

COROLLARY 7.4   Under the assumptions of the normal d-sample model with $n > d$, the random variable $(\hat{\mu}_k - \mu_k)/\text{SE}(\hat{\mu}_k)$ has the t distribution with $n - d$ degrees of freedom for $1 \leq k \leq d$.

Under the assumptions of the normal one-sample model, the sample mean $\hat{\mu} = \bar{Y}$ is an unbiased estimate of μ, $\text{SD}(\hat{\mu}) = \sigma/\sqrt{n}$, and $\text{SE}(\hat{\mu}) = S/\sqrt{n}$.

COROLLARY 7.5   Under the assumptions of the normal one-sample model with $n \geq 2$, the random variable $(\hat{\mu} - \mu)/\text{SE}(\hat{\tau})$ has the t distribution with $n - 1$ degrees of freedom.

Consider the normal two-sample model. The nontrivial linear parameter $\tau = \mu_2 - \mu_1$ can be written as $c_1\mu_1 + c_2\mu_2$, where $c_1 = -1$ and $c_2 = 1$. The estimate $\hat{\tau} = \hat{\mu}_2 - \hat{\mu}_1$ is unbiased, and its variance, standard deviation, and standard error are given by

$$\text{var}(\hat{\mu}_2 - \hat{\mu}_1) = \sigma^2 \left(\frac{1}{n_1} + \frac{1}{n_2} \right),$$

$$\text{SD}(\hat{\mu}_2 - \hat{\mu}_1) = \sigma \sqrt{\frac{1}{n_1} + \frac{1}{n_2}},$$

and

$$\text{SE}(\hat{\mu}_2 - \hat{\mu}_1) = S \sqrt{\frac{1}{n_1} + \frac{1}{n_2}}.$$

In this context, Theorem 7.9 yields the following result.

COROLLARY 7.6   Under the assumptions of the normal two-sample model with $n_1 + n_2 \geq 3$,

$$\frac{\hat{\mu}_2 - \hat{\mu}_1 - (\mu_2 - \mu_1)}{\text{SE}(\hat{\mu}_2 - \hat{\mu}_1)}$$

has the t distribution with $n - 2$ degrees of freedom.

More generally, consider the normal d-sample model with $n > d \geq 2$ and let $1 \leq j, k \leq d$ with $j \neq k$. Then $\hat{\mu}_k - \hat{\mu}_j = \bar{Y}_k - \bar{Y}_j$ is an unbiased estimate of $\mu_k - \mu_j$, whose variance, standard deviation, and standard error are given by

$$\text{var}(\hat{\mu}_k - \hat{\mu}_j) = \sigma^2 \left(\frac{1}{n_j} + \frac{1}{n_k} \right),$$

$$\text{SD}(\hat{\mu}_k - \hat{\mu}_j) = \sigma \sqrt{\frac{1}{n_j} + \frac{1}{n_k}},$$

and

$$\text{SE}(\hat{\mu}_k - \hat{\mu}_j) = S \sqrt{\frac{1}{n_j} + \frac{1}{n_k}}.$$

The extension of Corollary 7.6 to this context is as follows.

COROLLARY 7.7   Consider the normal d-sample model with $n > d \geq 2$ and let $1 \leq j, k \leq d$ with $j \neq k$. Then the random variable

$$\frac{\hat{\mu}_k - \hat{\mu}_j - (\mu_k - \mu_j)}{\text{SE}(\hat{\mu}_k - \hat{\mu}_j)}$$

has the t distribution with $n - d$ degrees of freedom.

Confidence Bounds and Confidence Intervals

Consider the normal d-sample model with $n > d$, let τ be a nontrivial linear parameter, and recall from Theorem 7.9 that the random variable $t = (\hat{\tau} - \tau)/\operatorname{SE}(\hat{\tau})$ has the t distribution with $n - d$ degrees of freedom.

Let $0 < \alpha < 1$. Then

$$P(t < t_{1-\alpha,n-d}) = 1 - \alpha,$$

as illustrated in Figure 7.1; that is,

$$P\left(\frac{\hat{\tau} - \tau}{\operatorname{SE}(\hat{\mu})} < t_{1-\alpha,n-d}\right) = 1 - \alpha.$$

Consequently,

$$P(\hat{\tau} - \tau < t_{1-\alpha,n-d}\operatorname{SE}(\hat{\tau})) = 1 - \alpha$$

and hence

$$P(\hat{\tau} - t_{1-\alpha,n-d}\operatorname{SE}(\hat{\tau}) < \tau) = 1 - \alpha. \tag{19}$$

In light of (19), we refer to $\hat{\tau} - t_{1-\alpha,n-d}\operatorname{SE}(\hat{\tau})$ as the $100(1 - \alpha)\%$ *lower confidence bound* for τ. The probability that τ is greater than its $100(1 - \alpha)\%$ lower confidence bound equals $1 - \alpha$.

Observe next that

$$P(t > t_{\alpha,n-d}) = 1 - \alpha.$$

Recall from Equation (7) in Section 7.2 that $t_{\alpha,n-d} = -t_{1-\alpha,n-d}$. Therefore, we conclude that

$$P(t > -t_{1-\alpha,n-d}) = 1 - \alpha,$$

as illustrated in Figure 7.2; that is,

$$P\left(\frac{\hat{\tau} - \tau}{\operatorname{SE}(\hat{\tau})} > -t_{1-\alpha,n-d}\right) = 1 - \alpha.$$

Consequently,

$$P\left(\hat{\tau} - \tau > -t_{1-\alpha,n-d}\operatorname{SE}(\hat{\tau})\right) = 1 - \alpha,$$

and hence

$$P\left(\tau < \hat{\tau} + t_{1-\alpha,n-d}\operatorname{SE}(\hat{\tau})\right) = 1 - \alpha. \tag{20}$$

FIGURE 7.1

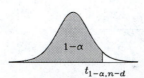

$1-\alpha$

$t_{1-\alpha,n-d}$

FIGURE 7.2

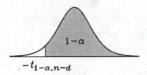

In light of (20), we refer to $\hat{\tau} + t_{1-\alpha, n-d}\, \mathrm{SE}(\hat{\tau})$ as the $100(1-\alpha)\%$ *upper confidence bound* for τ. The probability that τ is less than its $100(1-\alpha)\%$ upper confidence bound equals $1 - \alpha$.

Observe now that

$$P(t_{\alpha/2, n-d} < t < t_{1-\alpha/2, n-d}) = 1 - \alpha.$$

Since $t_{\alpha/2, n-d} = -t_{1-\alpha/2, n-d}$, we conclude that

$$P(-t_{1-\alpha/2, n-d} < t < t_{1-\alpha/2, n-d}) = 1 - \alpha,$$

as illustrated in Figure 7.3; that is,

$$P\!\left(-t_{1-\alpha/2, n-d} < \frac{\hat{\tau} - \tau}{\mathrm{SE}(\hat{\tau})} < t_{1-\alpha/2, n-d}\right) = 1 - \alpha,$$

so

$$P\!\left(-t_{1-\alpha/2, n-d}\,\mathrm{SE}(\hat{\tau}) < \hat{\tau} - \tau < t_{1-\alpha/2, n-d}\,\mathrm{SE}(\hat{\tau})\right) = 1 - \alpha$$

and hence

$$P\!\left(\hat{\tau} - t_{1-\alpha/2, n-d}\,\mathrm{SE}(\hat{\tau}) < \tau < \hat{\tau} + t_{1-\alpha/2, n-d}\,\mathrm{SE}(\hat{\tau})\right) = 1 - \alpha. \tag{21}$$

In light of (21), we refer to

$$\hat{\tau} \pm t_{1-\alpha/2, n-d}\,\mathrm{SE}(\hat{\tau}) = \left(\hat{\tau} - t_{1-\alpha/2, n-d}\,\mathrm{SE}(\hat{\tau}),\ \hat{\tau} + t_{1-\alpha/2, n-d}\,\mathrm{SE}(\hat{\tau})\right)$$

as the $100(1-\alpha)\%$ *confidence interval* for τ. The probability that τ lies in its $100(1-\alpha)\%$ confidence interval equals $1 - \alpha$. Note that the left end point of the $100(1-\alpha)\%$ confidence interval for τ equals its $100(1-\alpha/2)\%$ lower confidence bound, and the right end point equals its $100(1-\alpha/2)\%$ upper confidence bound. In particular, the left end point of the 95% confidence interval for τ equals its 97.5% lower confidence bound, and the right end point equals its 97.5% upper confidence bound. In practice, it is traditional to report 95% confidence bounds and confidence intervals—that is, those with $\alpha = .05$.

FIGURE 7.3

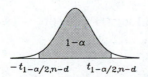

Sugar Beet Data

Consider the data in Table 7.3 on the percentage of sugar content of a standard variety (1) of sugar beet and four new varieties (2, 3, 4, 5). (Actually, the true experimental setup is more complicated than what is shown here; see Section 8.2 for a more complete description.)

TABLE   7.3
Sugar beet data

		Variety			
	1	2	3	4	5
	17.4	17.2	17.8	18.0	18.2
	18.6	18.0	18.8	18.7	18.9
	18.8	17.8	18.8	18.6	18.5
	18.6	18.1	18.7	18.1	18.2
	18.4	17.6	18.3	17.8	18.0
	17.6	17.4	17.8	17.4	17.6
	17.7	17.1	17.7	17.6	17.5
	17.8	17.2	18.2	17.8	18.0
	18.1	17.8	18.8	18.5	18.4
	18.4	17.6	18.3	18.1	18.0
	18.1				
Sample size	11	10	10	10	10
Sample mean	18.136	17.580	18.320	18.060	18.130
Sample SD	0.463	0.349	0.444	0.433	0.414

EXAMPLE 7.10   Consider the sugar beet data just on the standard variety.

(a) Verify the given values for the sample mean and standard deviation (Table 7.3).

(b) Determine the standard error of the sample mean.

(c) Under the assumptions of the normal one-sample model, determine the 97.5% lower and upper confidence bounds and the 95% confidence interval for the mean percentage of sugar content of the standard variety.

Solution   (a) The sample mean is given by

$$\bar{Y}_1 = \frac{17.4 + \cdots + 18.1}{11} = \frac{199.5}{11} \doteq 18.136.$$

The sample variance is given by

$$S_1^2 \doteq \frac{1}{10}[\,(17.4 - 18.136)^2 + \cdots + (18.1 - 18.136)^2\,]$$

$$\doteq \frac{2.145}{10} \doteq 0.215.$$

Thus, the sample standard deviation is given by $S_1 \doteq \sqrt{0.215} \doteq 0.463$.

(b) The standard error of the sample mean $\hat{\mu}_1 = \bar{Y}_1$ is given by

$$\text{SE}(\hat{\mu}_1) \doteq \frac{0.463}{\sqrt{11}} \doteq 0.140.$$

(c) Now $t_{.975,10} \doteq 2.228$ (see the table of quantiles of the t distribution in Appendix E). Thus, the 97.5% lower confidence bound for μ_1 is given by

$$18.136 - (2.228)(0.140) \doteq 18.136 - 0.311 = 17.825,$$

the 97.5% upper confidence bound for μ_1 is given by

$$18.136 + (2.228)(0.140) \doteq 18.136 + 0.311 \doteq 18.448,$$

and the 95% confidence interval for μ_1 is given by

$$18.136 \pm (2.228)(0.140) \doteq 18.136 \pm 0.311$$
$$\doteq (17.825, 18.448),$$

whose end points are the 97.5% lower and upper confidence bounds for μ_1. The proper interpretation of these results are as follows. Suppose the entire experiment (or, at least, that involving the standard variety) is repeated a very large number of times under independent but otherwise identical conditions and that the 97.5% lower and upper confidence bounds and the 95% confidence interval for μ_1 are obtained for each such repetition. Then approximately 97.5% of the lower confidence bounds for μ_1 will be below μ_1, approximately 97.5% of the upper confidence bounds will be above μ_1, and approximately 95% of the confidence intervals will contain μ_1. ∎

EXAMPLE **7.11** Consider the sugar beet data on varieties 1 and 2 (Table 7.3). Determine

(a) the within sum of squares and the pooled sample variance and standard deviation;

(b) the 95% confidence interval for μ_1 (under the assumptions of the normal two-sample model);

(c) the 95% confidence interval for $\mu_2 - \mu_1$.

Solution (a) The within sum of squares is given by

$$\text{WSS} \doteq [\,(17.4 - 18.136)^2 + \cdots + (18.1 - 18.136)^2\,]$$
$$+ [\,(17.2 - 17.58)^2 + \cdots + (17.6 - 17.58)^2\,]$$
$$\doteq 2.145 + 1.096$$
$$= 3.241.$$

The pooled sample variance is given by

$$S^2 = \frac{3.241}{19} \doteq 0.171,$$

so the pooled sample standard deviation is given by $S \doteq \sqrt{0.171} \doteq 0.413$.

(b) The standard error of $\hat{\mu}_1$ is given by

$$SE(\hat{\mu}_1) \doteq \frac{0.413}{\sqrt{11}} \doteq 0.125.$$

Now $t_{.975,19} \doteq 2.093$. Thus, the 95% confidence interval for μ_1 is given by

$$18.136 \pm (2.093)(0.125) \doteq 18.136 \pm 0.261 \doteq (17.876, 18.397),$$

which is somewhat shorter than that found in the solution to Example 7.10(c) because the pooled sample standard deviation is smaller than the sample standard deviation based on the first sample and there are now more degrees of freedom and hence the .975 quantile of the t distribution is smaller. (On the other hand, the assumptions of the normal two-sample model being used here are stronger than those of the normal one-sample model that were used in the solution to Example 7.10.)

(c) The standard error of $\hat{\mu}_2 - \hat{\mu}_1$ is given by

$$SE(\hat{\mu}_2 - \hat{\mu}_1) \doteq 0.413\sqrt{\tfrac{1}{11} + \tfrac{1}{10}} \doteq 0.180.$$

Thus, the 95% confidence interval for $\mu_2 - \mu_1$ is given by

$$17.580 - 18.136 \pm (2.093)(0.180) \doteq -0.556 \pm 0.378 \doteq (-0.934, -0.179). \quad \blacksquare$$

EXAMPLE 7.12   Consider the sugar beet data (Table 7.3) on all five varieties. Determine

(a) the within sum of squares and the pooled sample variance and standard deviation;

(b) the 95% confidence interval for $\mu_2 - \mu_1$ (under the assumptions of the normal five-sample model);

(c) the 95% confidence interval for

$$\frac{\mu_2 + \mu_3 + \mu_4 + \mu_5}{4} - \mu_1;$$

(d) the 95% confidence interval for

$$\frac{\mu_2 + \mu_3 + \mu_4 + \mu_5}{4} - \mu_2.$$

Solution   **(a)** The within sum of squares is given by

$$WSS \doteq [\,(17.4 - 18.136)^2 + \cdots + (18.1 - 18.136)^2\,] + \cdots$$

$$+ [\,(18.2 - 18.13)^2 + \cdots + (18.0 - 18.13)^2\,]$$

$$\doteq 2.145 + 1.096 + 1.776 + 1.684 + 1.541$$

$$= 8.242.$$

The pooled sample variance is given by

$$S^2 = \frac{8.242}{46} \doteq 0.179,$$

so the pooled sample standard deviation is given by $S \doteq \sqrt{0.179} \doteq 0.423$.

(b) The standard error of $\hat{\mu}_2 - \hat{\mu}_1$ is given by

$$\mathrm{SE}(\hat{\mu}_2 - \hat{\mu}_1) \doteq 0.423\sqrt{\tfrac{1}{11} + \tfrac{1}{10}} \doteq 0.185.$$

By computer (or by interpolating the table of quantiles of the t distribution), $t_{.975,46} \doteq 2.013$. Thus, the 95% confidence interval for $\mu_2 - \mu_1$ is given by

$$17.580 - 18.136 \pm (2.013)(0.185) \doteq -0.556 \pm 0.372 \doteq (-0.929, -0.184).$$

(c) The standard error of the estimate

$$\hat{\tau} = \frac{\hat{\mu}_2 + \hat{\mu}_3 + \hat{\mu}_4 + \hat{\mu}_5}{4} - \hat{\mu}_1$$

of the linear parameter

$$\tau = \frac{\mu_2 + \mu_3 + \mu_4 + \mu_5}{4} - \mu_1$$

is given by

$$\mathrm{SE}(\hat{\tau}) \doteq 0.423\sqrt{\tfrac{1}{11} + (\tfrac{1}{4})^2 \cdot \tfrac{1}{10} + (\tfrac{1}{4})^2 \cdot \tfrac{1}{10} + (\tfrac{1}{4})^2 \cdot \tfrac{1}{10} + (\tfrac{1}{4})^2 \cdot \tfrac{1}{10}}$$

$$\doteq 0.144.$$

Now

$$\hat{\tau} \doteq \frac{17.58 + 18.32 + 18.06 + 18.13}{4} - 18.136 \doteq -0.114.$$

Thus, the 95% confidence interval for τ is given by

$$-0.114 \pm (2.013)(0.144) \doteq -0.144 \pm 0.290 = (-0.404, 0.176).$$

(We can think of τ as the increase in the mean percentage of sugar content due to switching from the standard variety to a uniform mixture of varieties 2, 3, 4, and 5.)

(d) The standard error of the estimate

$$\hat{\tau} = \frac{\hat{\mu}_2 + \hat{\mu}_3 + \hat{\mu}_4 + \hat{\mu}_5}{4} - \hat{\mu}_2$$

of the linear parameter

$$\tau = \frac{\mu_2 + \mu_3 + \mu_4 + \mu_5}{4} - \mu_2$$

is given by

$$\mathrm{SE}(\hat{\tau}) \doteq 0.423\sqrt{(-\tfrac{3}{4})^2 \cdot \tfrac{1}{10} + (\tfrac{1}{4})^2 \cdot \tfrac{1}{10} + (\tfrac{1}{4})^2 \cdot \tfrac{1}{10} + (\tfrac{1}{4})^2 \cdot \tfrac{1}{10}}$$

$$\doteq 0.116.$$

Now

$$\hat{\tau} = \frac{17.58 + 18.32 + 18.06 + 18.13}{4} - 17.58 = 0.4425.$$

Thus, the 95% confidence interval for τ is given by

$$0.4425 \pm (2.013)(0.116) \doteq 0.4425 \pm 0.233 \doteq (0.209, 0.676). \quad \blacksquare$$

Problems

7.24 Consider the setup of Problem 7.1 in Section 7.1, but treat the standard deviation of the measurement as being unknown. Determine the 95% confidence interval for τ.

7.25 Let Y_1 and Y_2 be independent random variables, each having the normal distribution with mean μ and variance σ^2. Set $\bar{Y} = (Y_1 + Y_2)/2$, and let S be the nonnegative square root of $S^2 = (Y_1 - \bar{Y})^2 + (Y_2 - \bar{Y})^2$. **(a)** Use results from Section 5.7 to show that $\bar{Y}$ and $Y_1 - Y_2$ are independent. **(b)** Show that $S^2 = (Y_1 - Y_2)^2/2$. **(c)** Use the results of parts (a) and (b) to show that $\bar{Y}$ and S are independent. **(d)** Use the result of part (b) to show that S^2/σ^2 has the chi-square distribution with 1 degree of freedom.

7.26 Under the assumptions of the normal d-sample model with $n > d$, determine **(a)** the variance of S^2; **(b)** the mean of S [see Problem 7.18(b) in Section 7.2].

7.27 Consider the normal d-sample model. For each of the following definitions of the parameter τ, determine the variance, standard deviation, and standard error of $\hat{\tau}$. **(a)** $\tau = \mu_1 + \cdots + \mu_d$; **(b)** $\tau = n_1\mu_1 + \cdots + n_d\mu_d$.

7.28 In the context of Example 7.12, determine 95% confidence intervals for **(a)** μ_2; **(b)** $\mu_3 - \mu_2$; **(c)** $(\mu_2 + \mu_3 + \mu_4 + \mu_5)/4$; **(d)** $(\mu_4 + \mu_5)/2 - (\mu_2 + \mu_3)/2$.

7.29 Consider the sugar beet data on varieties 1, 2, and 3 (Table 7.3). Determine **(a)** the within sum of squares and the pooled sample variance and standard deviation; **(b)** the 95% confidence interval for $\tau = \frac{5}{10}\mu_1 + \frac{3}{10}\mu_2 + \frac{2}{10}\mu_3$ under the assumptions of the normal three-sample model. (Observe that τ is the mean percentage of sugar content for a mixture of 50% variety 1, 30% variety 2, and 20% variety 3.)

7.30 Consider the following data on yields of a new variety (1) of wheat and a standard variety (2):

1	22	26	
2	14	20	23

Determine **(a)** n_1, n_2, n, and d; **(b)** $\bar{Y}_1$ and $\bar{Y}_2$; **(c)** WSS_1, WSS_2, and WSS; **(d)** S_1^2, S_2^2, and S^2; **(e)** S_1, S_2, and S; **(f)** the 95% confidence interval for μ_1 (under the assumptions of the corresponding normal two-sample model); **(g)** the 95% confidence interval for $\mu_1 - \mu_2$.

7.31 Consider a manufacturing process designed to produce 25-ohm resistors. Suppose the actual resistances of the resistors made by the process are independent and normally distributed with mean μ and variance σ^2. In the course of carrying out a quality control program, ten resistors are made by the process, and their resistances determined. Table 7.4 shows the observed results. Determine **(a)** the sample mean and standard deviation; **(b)** the 95% lower confidence bound, 95% upper confidence bound, and 95% confidence interval for μ.

7.32 The company making 25-ohm resistors uses two machines: The resistances of resistors made by the first machine are normally distributed with mean μ_1 and variance σ^2, whereas those of resistors made by the second machine are normally distributed with mean μ_2 and variance σ^2. The observed resistances (see Table 7.5) in Problem 7.31 are based on output from the first machine and the observed resistances

TABLE   7.4

1	25.06	6	25.13
2	24.69	7	25.18
3	25.11	8	25.56
4	24.61	9	24.75
5	25.05	10	24.30

TABLE   7.5

1	25.23	6	25.10
2	24.70	7	25.37
3	25.47	8	26.01
4	25.08	9	25.42
5	24.42	10	25.82

of ten resistors made by the second machine. Determine **(a)** the sample mean and standard deviation for the resistances of the resistors made by the second machine; **(b)** the pooled sample variance and standard deviation; **(c)** 95% confidence intervals for μ_1, μ_2, and $\mu_2 - \mu_1$.

7.4
The *t* Test of an Inequality

Consider the normal d-sample model with $n > d$, a nontrivial linear parameter $\tau = c_1\mu_1 + \cdots + c_d\mu_d$, and its estimate $\hat{\tau} = c_1\hat{\mu}_1 + \cdots + c_d\hat{\mu}_d$. Recall that the standard deviation and standard error of this estimate are given by

$$\mathrm{SD}(\hat{\tau}) = \sigma\sqrt{\frac{c_1^2}{n_1} + \cdots + \frac{c_d^2}{n_d}}$$

and

$$\mathrm{SE}(\hat{\tau}) = S\sqrt{\frac{c_1^2}{n_1} + \cdots + \frac{c_d^2}{n_d}} = \frac{S}{\sigma}\,\mathrm{SD}(\hat{\tau}).$$

Recall also that $\hat{\tau}$ and S are independent, $(\hat{\tau} - \tau)/\mathrm{SD}(\hat{\tau})$ has the standard normal distribution, and $(n - d)S^2/\sigma^2$ has the chi-square distribution with $n - d$ degrees of freedom.

Given $\tau_0 \in \mathbb{R}$, consider the statistic

$$t = \frac{\hat{\tau} - \tau_0}{\mathrm{SE}(\hat{\tau})};\qquad\qquad (22)$$

we refer to this as a *t statistic* since if $\tau = \tau_0$, then it has the *t* distribution with $n - d$ degrees of freedom.

Testing the Inequality $\tau \leq \tau_0$

THEOREM 7.10   For fixed $c \in \mathbb{R}$, $P(t \geq c)$ is a continuous and strictly increasing function of $\delta = (\tau - \tau_0)/\operatorname{SD}(\hat{\tau})$.

Proof   Now $\operatorname{SE}(\hat{\tau}) = \frac{S}{\sigma}\operatorname{SD}(\hat{\tau})$, so

$$
\begin{aligned}
P(t \geq c) &= P\!\left(\frac{\hat{\tau} - \tau_0}{\operatorname{SE}(\hat{\tau})} \geq c\right) \\
&= P\!\left(\frac{\hat{\tau} - \tau_0}{\operatorname{SD}(\hat{\tau})} \geq c\,\frac{\operatorname{SE}(\hat{\tau})}{\operatorname{SD}(\hat{\tau})}\right) \\
&= P\!\left(\frac{\hat{\tau} - \tau}{\operatorname{SD}(\hat{\tau})} + \frac{\tau - \tau_0}{\operatorname{SD}(\hat{\tau})} \geq \frac{cS}{\sigma}\right) \\
&= P\!\left(\frac{\hat{\tau} - \tau}{\operatorname{SD}(\hat{\tau})} + \delta \geq \frac{cS}{\sigma}\right).
\end{aligned}
$$

Set

$$
V = \frac{cS}{\sigma} \quad \text{and} \quad W = \frac{\hat{\tau} - \tau}{\operatorname{SD}(\hat{\tau})} + \delta.
$$

Then V and W are independent random variables, W is normally distributed with mean δ and variance 1, and $P(t \geq c) = P(W \geq V)$. If $c = 0$, then $V = 0$ and hence

$$
P(t \geq c) = P(W \geq 0) = 1 - \Phi(-\delta) = \Phi(\delta),
$$

which is a continuous and strictly increasing function of δ. Suppose now that $c \neq 0$. Since $(n - d)S^2/\sigma^2$ has the chi-square distribution with $n - d$ degrees of freedom, we conclude that V has a density function f_V. Consequently (see Problem 7.33),

$$
P(t \geq c) = P(W \geq V) = \int_{-\infty}^{\infty} [\,1 - \Phi(v - \delta)\,]f_V(v)\,dv = \int_{-\infty}^{\infty} \Phi(\delta - v)f_V(v)\,dv,
$$

which is a continuous and strictly increasing function of δ.   ∎

COROLLARY 7.8   Let $0 < \alpha < 1$. Then

$$
\begin{aligned}
P(t \geq t_{1-\alpha,n-d}) &< \alpha \quad \text{if} \quad \tau < \tau_0, \\
P(t \geq t_{1-\alpha,n-d}) &= \alpha \quad \text{if} \quad \tau = \tau_0,
\end{aligned}
$$

and

$$
P(t \geq t_{1-\alpha,n-d}) > \alpha \quad \text{if} \quad \tau > \tau_0.
$$

Proof   If $\tau = \tau_0$, then t has the t distribution with $n - d$ degrees of freedom, so $P(t \geq t_{1-\alpha,n-d}) = \alpha$. The remaining conclusions now follow from Theorem 7.10.   ∎

Consider the possibility that $\tau \leq \tau_0$. We refer to this possibility as a *hypothesis*. (What we refer to as a hypothesis is commonly referred to as the *null hypothesis*, and its negation is commonly referred to as the *alternative hypothesis*.) Let $0 < \alpha < 1$. We refer to $t \geq t_{1-\alpha,n-d}$ as the *critical* (or *rejection*) *region* for the t test of *size* α of the hypothesis. According to Corollary 7.8, α is the maximum probability that the t statistic lies in the critical region, where the maximum is taken over all values of $\mu_1, \ldots, \mu_d$ that are compatible with the hypothesis and over all $\sigma > 0$.

Suppose $\tau = \tau_0$. Then t has the t distribution with $n - d$ degrees of freedom, and the probability that this statistic lies in the critical region equals α. Figure 7.4 illustrates this fact. (The actual density function shown in the figures in this section and the next is that of the t distribution with 10 degrees of freedom.)

The complement $t < t_{1-\alpha, n-d}$ of the critical region is referred to as the *acceptance region* for the test. It follows from Equation (22) that the t statistic lies in the acceptance region if and only if $(\hat{\tau} - \tau_0) / \text{SE}(\hat{\tau}) < t_{1-\alpha, n-d}$ or, equivalently, if and only if $\hat{\tau} - t_{1-\alpha, n-d} \text{SE}(\hat{\tau}) < \tau_0$; thus, the t statistic lies in the acceptance region if and only if the $100(1 - \alpha)\%$ lower confidence bound for τ is less than τ_0. In statistics, the phrase "the hypothesis is rejected" generally means that the data contradict the hypothesis. The phrase "the hypothesis is accepted" generally means that the data do not contradict the hypothesis, but not necessarily that they verify the hypothesis. On the other hand, to the extent that the data contradict the negation of a hypothesis, they can be thought of as verifying the hypothesis.

EXAMPLE 7.13   Consider the sugar beet data (Table 7.3 in Section 7.3). Carry out the t test of size .05 of the hypothesis that the mean percentage of sugar content for variety 3 is less than or equal to that of the standard variety.

Solution   First, we carry out the test using the data just on the two varieties of interest. We want to test the hypothesis that $\mu_3 \leq \mu_1$ or, equivalently, that $\tau \leq 0$, where $\tau = \mu_3 - \mu_1$. The pooled estimate of σ^2 is then given by

$$S^2 \doteq \frac{10(0.463)^2 + 9(0.444)^2}{19} \doteq 0.206,$$

so the corresponding estimate of σ is given by $S \doteq \sqrt{0.206} \doteq 0.454$. The estimate $\hat{\tau}$

FIGURE   7.4

The critical region for the t test of size α of the hypothesis that $\tau \leq \tau_0$ is given by $t \geq t_{1-\alpha, n-d}$. The shaded region under the graph of the density function of the t distribution with $n - d$ degrees of freedom has area α.

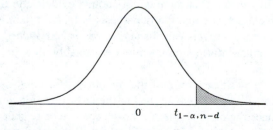

$$0 \qquad t_{1-\alpha, n-d}$$

of τ is given by $\hat{\tau} = \hat{\mu}_3 - \hat{\mu}_1 = \bar{Y}_3 - \bar{Y}_1 \doteq 18.320 - 18.136 = 0.184$, and its standard error is given by

$$\mathrm{SE}(\hat{\tau}) \doteq 0.454\sqrt{\tfrac{1}{11} + \tfrac{1}{10}} \doteq 0.198.$$

Thus, the relevant t statistic is given by

$$t = \frac{\hat{\tau}}{\mathrm{SE}(\hat{\tau})} \doteq \frac{0.184}{0.198} \doteq 0.925.$$

The acceptance region for the t test of size .05 of the hypothesis that $\tau \leq 0$ is given by $t < t_{.95,19} \doteq 1.729$. Since $t \doteq 0.925$ lies in the acceptance region, the hypothesis that $\tau \leq 0$ is accepted. Alternatively, the 95% lower confidence bound for τ is given by

$$\hat{\tau} - t_{.95,19}\,\mathrm{SE}(\hat{\tau}) \doteq 0.184 - (1.729)(0.198) \doteq -0.160,$$

which is less than zero, so the hypothesis that $\tau \leq 0$ is accepted.

Next, we carry out the test using the data on all five varieties to obtain the pooled estimate of variance. In this manner, we get the pooled estimate $S \doteq 0.423$ of σ [see the solution to Example 7.12(a) in Section 7.3]. Thus, the standard error of $\hat{\tau} = \bar{Y}_3 - \bar{Y}_1$ is given by

$$\mathrm{SE}(\hat{\tau}) \doteq 0.423\sqrt{\tfrac{1}{11} + \tfrac{1}{10}} \doteq 0.185,$$

so the relevant t statistic is given by

$$t = \frac{\hat{\tau}}{\mathrm{SE}(\hat{\tau})} \doteq \frac{0.184}{0.185} \doteq 0.993.$$

The acceptance region for the t test of size .05 for the hypothesis that $\tau \leq 0$ is given by $t < t_{.95,46} \doteq 1.679$. Since $t \doteq 0.993$ lies in the acceptance region, the hypothesis that $\tau \leq 0$ is accepted. Alternatively, the 95% lower confidence bound for τ is given by

$$\hat{\tau} - t_{.95,46}\,\mathrm{SE}(\hat{\tau}) \doteq 0.184 - (1.679)(0.185) \doteq -0.127,$$

which is less than zero, so the hypothesis that $\tau \leq 0$ is accepted. ∎

In the solution to Example 7.13, the statement that "the hypothesis that $\tau \leq 0$ is accepted" merely means that there is not enough evidence in the data to conclude with reasonable assurance that the hypothesis is false. It definitely does *not* mean that the hypothesis is proven, established, or confirmed. Indeed, $\hat{\tau}$ is positive, so in the absence of other information we might reasonably think it more likely than not that τ is positive.

Consider the t test of size α of the hypothesis that $\tau \leq \tau_0$. The hypothesis is said to be rejected if the t statistic lies in the critical region and accepted if it lies in the acceptance region. We refer to the acceptance or rejection of the hypothesis as a decision. If the hypothesis is either valid and accepted or invalid and rejected, the decision is said to be correct; otherwise, it is said to be erroneous. When the hypothesis is valid, the probability of erroneously rejecting it (sometimes referred to as the *type I error*) is at most α. When the hypothesis is invalid, the probability of erroneously accepting it (sometimes referred to as the *type II error*) is denoted by β, and the probability $1 - \beta$ of correctly rejecting it is referred to as the *power* of the test.

Suppose the hypothesis is invalid; that is, $\tau > \tau_0$. Then the power of the t test is given by

$$1 - \beta = P(t \geq t_{1-\alpha, n-d}) = P\left(\frac{\hat{\tau} - \tau_0}{\mathrm{SE}(\hat{\tau})} \geq t_{1-\alpha, n-d}\right).$$

Suppose $n - d \gg 1$. Then $t_{1-\alpha, n-d} \approx z_{1-\alpha} = -z_\alpha$ and $P(\mathrm{SE}(\hat{\tau})/\mathrm{SD}(\hat{\tau}) \approx 1) \approx 1$, so

$$1 - \beta \approx P\left(\frac{\hat{\tau} - \tau_0}{\mathrm{SD}(\hat{\tau})} \geq -z_\alpha\right)$$

$$= P\left(\frac{\hat{\tau} - \tau}{\mathrm{SD}(\hat{\tau})} \geq -z_\alpha - \frac{\tau - \tau_0}{\mathrm{SD}(\hat{\tau})}\right)$$

$$= 1 - \Phi(-z_\alpha - \delta)$$

$$= \Phi(z_\alpha + \delta),$$

where $\delta = (\tau - \tau_0)/\mathrm{SD}(\hat{\tau})$. Thus, $z_\alpha + \delta \approx z_{1-\beta}$, so $\delta \approx z_{1-\beta} - z_\alpha$, and hence

$$\mathrm{SD}(\hat{\tau}) \approx \frac{\tau - \tau_0}{z_{1-\beta} - z_\alpha}. \tag{23}$$

For an application of (23), let $n_1 = \cdots = n_d = r$, where the common sample size r is to be determined. The standard deviation of the estimate $\hat{\tau} = c_1\hat{\mu}_1 + \cdots + c_d\hat{\mu}_d$ of the linear parameter $\tau = c_1\mu_1 + \cdots + c_d\mu_d$ is given by

$$\mathrm{SD}(\hat{\tau}) = \frac{a\sigma}{\sqrt{r}}, \tag{24}$$

where $a = (c_1^2 + \cdots + c_d^2)^{1/2}$. It follows from (23) and (24) that if $\tau > \tau_0$ and $n - d \gg 1$, then

$$r \approx \left(\frac{a\sigma(z_{1-\beta} - z_\alpha)}{\tau - \tau_0}\right)^2. \tag{25}$$

Equation (25) can be used as a guide in determining the common sample size.

EXAMPLE **7.14**   Consider the normal two-sample model with $n_1 = n_2 = r$ and consider the hypothesis that $\mu_2 \leq \mu_1$. Determine r such that the power of the test of size .05 of the hypothesis is about .9 when $\mu_2 = \mu_1 + \sigma$.

Solution   The hypothesis that $\mu_2 \leq \mu_1$ is equivalent to the hypothesis that $\tau \leq 0$, where $\tau = \mu_2 - \mu_1$. Correspondingly, $c_1 = -1$ and $c_2 = 1$, so $a = \sqrt{2}$ in (24) and (25). Also, $\alpha = .05$, $\beta = .1$, $\tau_0 = 0$, and $\tau = \sigma$ in (23) and (25). Thus, by (25),

$$r = 2(z_{.9} - z_{.05})^2 \doteq 2(1.282 + 1.645)^2 \doteq 17.$$

[Here $n - d = 34 - 2 = 32$, which is large enough for the approximation in (25) to be reasonably accurate.]   ∎

THEOREM 7.11   Consider the t test of size α of the hypothesis that $\tau \le \tau_0$. The t statistic lies in the critical region if $\alpha \ge 1 - t_{n-d}(t)$ and in the acceptance region if $\alpha < 1 - t_{n-d}(t)$.

Proof   Suppose $\alpha \ge 1 - t_{n-d}(t)$. Then $t_{n-d}(t) \ge 1 - \alpha$ and hence $t \ge t_{1-\alpha, n-d}$, so the t statistic lies in the critical region. Suppose instead that $\alpha < 1 - t_{n-d}(t)$. Then $t_{n-d}(t) < 1 - \alpha$ and hence $t < t_{1-\alpha, n-d}$, so the t statistic lies in the acceptance region.   ■

We refer to $1 - t_{n-d}(t)$ as the *P-value* for the t test of the hypothesis that $\tau \le \tau_0$ (see Figure 7.5). Here by "t test" we mean the collection of t tests of size α as α ranges over $(0, 1)$. According to Theorem 7.11, the t statistic lies in the critical region for the t test of size α of the hypothesis if α is greater than or equal to the P-value and in the acceptance region if α is less than the P-value.

FIGURE 7.5

The P-value for the t test of the hypothesis that $\tau \le \tau_0$ equals the area of the shaded region under the graph of the density function of the t distribution with $n - d$ degrees of freedom.

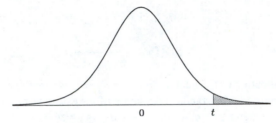

EXAMPLE 7.15   Consider the sugar beet data (Table 7.3 in Section 7.3). Determine and interpret the P-value for the t test (based on the data on all five varieties) of the hypothesis that the mean percentage of sugar content of variety 3 is less than or equal to that of the standard variety.

Solution   Here $n - d = 46$. Now $t \doteq 0.993$ according to the solution to Example 7.13, so the P-value is given by $1 - t_{46}(0.993) \doteq .163$. Therefore, t lies in the critical region for the test of size α if $\alpha \ge .163$ and in the acceptance region if $\alpha < .163$. In particular, it lies in the acceptance region for the test of size .05 (as we saw in the solution to Example 7.13) and even in the acceptance region for the test of size .15.   ■

Testing the Inequality $\tau \ge \tau_0$

Analogous results hold, of course, for testing a hypothesis of the form $\tau \ge \tau_0$. In particular,

$$P(t \le -t_{1-\alpha, n-d}) < \alpha \quad \text{if} \quad \tau > \tau_0,$$

$$P(t \le -t_{1-\alpha, n-d}) = \alpha \quad \text{if} \quad \tau = \tau_0,$$

and

$$P(t \le -t_{1-\alpha,n-d}) > \alpha \quad \text{if} \quad \tau < \tau_0.$$

(Note that $-t_{1-\alpha,n-d} = t_{\alpha,n-d}$.) Thus, the critical region $t \le -t_{1-\alpha,n-d}$ for the t test of the hypothesis that $\tau \ge \tau_0$ has size α (see Figure 7.6); that is, α is the maximum probability that the t statistic lies in the critical region, where the maximum is taken over $\tau \ge \tau_0$ and $\sigma > 0$.

FIGURE 7.6

The critical region for the t test of size α of the hypothesis that $\tau \ge \tau_0$ is given by $t \le -t_{1-\alpha,n-d}$. The shaded region under the graph of the density function of the t distribution with $n - d$ degrees of freedom has area α.

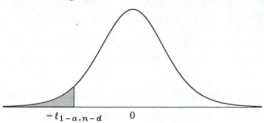

The t statistic lies in the acceptance region $t > -t_{1-\alpha,n-d}$ of the t test of the hypothesis that $\tau \ge \tau_0$ if and only if $(\hat{\tau} - \tau_0)/\operatorname{SE}(\hat{\tau}) > -t_{1-\alpha,n-d}$ or, equivalently, if and only if $\hat{\tau} + t_{1-\alpha,n-d}\operatorname{SE}(\hat{\tau}) > \tau_0$; thus, the t statistic lies in the acceptance region if and only if the $100(1-\alpha)\%$ upper confidence bound for τ is greater than τ_0.

Suppose the hypothesis is invalid; that is, $\tau < \tau_0$. Then the power of the t test (defined as before) is given by

$$1 - \beta \approx P\left(\frac{\hat{\tau} - \tau_0}{\operatorname{SD}(\hat{\tau})} \le -z_{1-\alpha}\right) = P\left(\frac{\hat{\tau} - \tau}{\operatorname{SD}(\hat{\tau})} \le -z_{1-\alpha} + \frac{\tau_0 - \tau}{\operatorname{SD}(\hat{\tau})}\right)$$
$$= \Phi(-z_{1-\alpha} + \delta),$$

where $\delta = |\tau - \tau_0|/\operatorname{SD}(\hat{\tau})$. Thus, $-z_{1-\alpha} + \delta \approx z_{1-\beta}$, so $\delta \approx z_{1-\beta} + z_{1-\alpha}$, and hence

$$\operatorname{SD}(\hat{\tau}) \approx \frac{|\tau - \tau_0|}{z_{1-\beta} + z_{1-\alpha}}, \tag{26}$$

which is a slight modification of (23) and has similar applications. In particular, (25) can again be used as a guide in determining the common sample size.

Consider the t test of size α of the hypothesis that $\tau \ge \tau_0$. The t statistic lies in the critical region if $\alpha \ge t_{n-d}(t)$ and in the acceptance region if $\alpha < t_{n-d}(t)$. We refer to $t_{n-d}(t)$ as the P-value for the t test of the hypothesis that $\tau \ge \tau_0$ (see Figure 7.7). Again, the t statistic lies in the critical region if α is greater than or equal to the P-value and in the acceptance region if α is less than the P-value.

FIGURE   7.7

The P-value for the t test of the hypothesis that $\tau \geq \tau_0$ is the area of the shaded region under the graph of the density function of the t distribution with $n - d$ degrees of freedom.

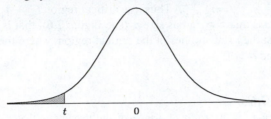

EXAMPLE **7.16**   Consider the sugar beet data (see Table 7.3 in Section 7.3), with that on all five varieties being used to obtain the pooled sample variance.

(a) Determine the critical region for the t test of size .05 of the hypothesis that the mean percentage of sugar content of variety 2 is greater than or equal to that of the standard variety.

(b) Does the t statistic lie in the critical region or the acceptance region?

(c) Determine the 95% upper confidence bound for $\mu_2 - \mu_1$.

(d) Check that the answers to parts (b) and (c) are compatible.

(e) Determine and interpret the P-value for the test.

Solution   We want to test the hypothesis that $\mu_2 \geq \mu_1$ or, equivalently, that $\tau \geq 0$, where $\tau = \mu_2 - \mu_1$. The relevant t statistic is given by $t = \hat{\tau} / \text{SE}(\hat{\tau})$, where $\hat{\tau} = \hat{\mu}_2 - \hat{\mu}_1$.

(a) The critical region for the t test of size .05 of the hypothesis is given by $t \leq -t_{.95,46} \doteq -1.679$.

(b) As we have already seen, the pooled sample standard deviation is given by $S \doteq 0.423$, so the standard error of $\hat{\tau}$ is given by

$$\text{SE}(\hat{\tau}) \doteq 0.423\sqrt{\tfrac{1}{11} + \tfrac{1}{10}} \doteq 0.185.$$

Since $\hat{\tau} \doteq 17.580 - 18.136 = -0.556$, the t statistic is given by

$$t \doteq -\frac{0.556}{0.185} \doteq -3.008 < -1.679.$$

Thus, the t statistic lies in the critical region.

(c) The 95% upper confidence bound for τ is given by

$$\hat{\tau} + t_{.95,46}\, \text{SE}(\hat{\tau}) \doteq -0.556 + (1.679)(0.185) \doteq -0.246.$$

(d) The upper confidence bound in part (c) is less than zero, which is compatible with rejecting the hypothesis that $\tau \geq 0$.

(e) The P-value for the test is given by $t_{46}(-3.008) \doteq .002$. Therefore, t lies in the critical region for the test of size α if $\alpha \geq .002$ and in the acceptance region if $\alpha < .002$. In particular, it lies in the critical region for the test of size .05, as we saw in the solution to part (b), and even in the critical region for the test of size .01, but it lies in the acceptance region for the test of size .001. ∎

In the solution to Example 7.16, we rejected the hypothesis that the mean percentage of sugar content of variety 2 is greater than or equal to that of the standard variety. Equivalently, we proved that the mean percentage of sugar content of variety 2 is less than that of the standard variety. More generally, if we want to use statistical methods to establish a hypothesis of the form $\tau < \tau_0$, we should test the hypothesis that $\tau \geq \tau_0$ and hopefully reject it.

Suppose government regulations prevent a company from marketing a new variety of sugar beet until the company "proves" that its mean sugar content is higher than that of the standard variety, in which case the government approves the variety for marketing. The solution to Example 7.13 with $\alpha = .05$ could then be interpreted as meaning that the analysis of the data does not prove that the mean sugar content of variety 3 is higher than that of the standard variety, so the company wishing to market variety 3 could not use this analysis to gain approval for their variety.

Suppose, instead, that a company could market a new variety of sugar beet until it was proven that its mean sugar content was less than that of the standard variety. The solution to Example 7.16 with $\alpha = .05$ could then be interpreted as meaning that the analysis of the data proves that the mean sugar content of variety 2 is less than that of the standard variety, so the government could use this analysis to order the company marketing this variety to withdraw it from the market. (Here, of course, *prove* is being used in a statistical, scientific, or legalistic rather than mathematical sense.)

Problems

7.33 The paper "Trifluoperazine ('Stelazine'): A controlled clinical trial in chronic schizophrenia" by W. J. Stanley and D. Walton, which was published in *Journal of Mental Science* **107** (1961), pages 250–257, described and analyzed an experiment to investigate the effect of the drug Stelazine on chronic schizophrenics. The experiment involved patients in two wards. Each patient was rated on a behavior scale that ranged from 1 to 5, with 5 representing normal behavior and lower scores representing increasingly severe deviations from normality. The patients in each ward were divided into pairs that were matched as closely as possible based on age, length of hospital stay, and behavior score. Within each such matched pair, one patient was given the drug, and the other was given a placebo, the choice (presumably) being made at random. The experiment was *double-blind* in that neither the patients nor the examining nurses knew who was given the drug and who the placebo. The dosage of the drug given to the treatment group in ward B was somewhat reduced from that given in ward A. Table 7.6 [taken from Problem 35 in Chapter 11 of *Mathematical Statistics and Data Analysis*, 2nd edition, by John A. Rice (Duxbury Press, Belmont, California, 1995)] gives the behavior rating scores at the beginning of the trial and after 3 months for nine matched pairs in ward A

TABLE   7.6

Ward A					Ward B			
Stelazine		Placebo			Stelazine		Placebo	
Before	After	Before	After		Before	After	Before	After
2.3	3.1	2.4	2.0		1.9	1.45	1.9	1.91
2.0	2.1	2.2	2.6		2.3	2.45	2.4	2.54
1.9	2.45	2.1	2.0		2.0	1.81	2.0	1.45
3.1	3.7	2.9	2.0		1.6	1.72	1.5	1.45
2.2	2.54	2.2	2.4		1.6	1.63	1.5	1.54
2.3	3.72	2.4	3.18		2.6	2.45	2.7	1.54
2.8	4.54	2.7	3.0		1.7	2.18	1.7	1.54
1.9	1.61	1.9	2.54					
1.1	1.63	1.3	1.72					

and seven matched pairs in ward B. Define the outcome variable Y for a given matched pair as the difference in the before–after change between placebo and treatment. [Thus, for the first matched pair in ward A, the outcome variable is given by $(3.1 - 2.3) - (2.0 - 2.4) = 1.2$.] Suppose Y is (approximately) normally distributed with unknown mean μ and variance σ^2. Then the treatment is effective if and only if $\mu > 0$. Determine the P-value for testing the hypothesis that $\mu \leq 0$ **(a)** for the patients in ward A; **(b)** for the patients in ward B; **(c)** for the patients in both wards (under the added assumption that μ and σ^2 do not differ between wards).

7.34   Let W and V be independent random variables having density functions f_W and f_V, respectively, and let F_W be the distribution function of W. Show that

$$P(W \geq V) = \int_{-\infty}^{\infty} \left(\int_{v}^{\infty} f_W(w)dw \right) f_V(v)dv = \int_{-\infty}^{\infty} [1 - F_W(v)] f_V(v)dv.$$

7.35   Consider the sugar beet data (see Table 7.3 in Section 7.3). **(a)** Carry out the test of size .05 of the hypothesis that $\mu_1 \leq (\mu_2 + \mu_3 + \mu_4 + \mu_5)/4 - 0.1$. **(b)** Determine the 95% lower confidence bound for $\mu_1 - (\mu_2 + \mu_3 + \mu_4 + \mu_5)/4$. **(c)** Check that the answers to parts (a) and (b) are compatible. **(d)** Determine and interpret the P-value for the test.

7.36   Use the sugar beet data (Table 7.3) on varieties 2 and 4 to establish that the mean percentage of sugar content of variety 2 is less than that of variety 4.

7.37   Consider the resistor data in Problem 7.31 in Section 7.3. **(a)** Determine the P-value for testing the hypothesis that the mean resistance is less than or equal to 24.7. **(b)** Can the data be used to prove that the mean resistance is less than 25.3?

7.38   Consider the resistor data in Problems 7.31 and 7.32 in Section 7.3. Can the data be used to prove that the mean resistance of the resistors made by the second machine is greater than that of the resistors made by the first machine?

7.39   Redo Example 7.14 using .95 instead of .9 as the power when $\mu_2 - \mu_1 = \sigma$.

7.40   Redo Example 7.14 using $\mu_2 = \mu_1 + \sigma/2$ instead of $\mu_2 = \mu_1 + \sigma$ as the alternative of interest.

7.41 Consider the normal three-sample model with $n_1 = n_2 = n_3 = r$ and consider the hypothesis that $(\mu_2 + \mu_3)/2 \leq \mu_1$. Determine r such that the power of the test of size .05 of the hypothesis is about .9 when $(\mu_2 + \mu_3)/2 = \mu_1 + \sigma$.

7.42 Consider the t test of size α of the hypothesis that $\tau \geq \tau_0$ and suppose this hypothesis invalid; that is, $\tau < \tau_0$. Show that if $n - d \gg 1$, then the power of the test is about $\Phi(-z_{1-\alpha} + \delta)$, where $\delta = |\tau - \tau_0|/\text{SD}(\hat{\tau})$.

7.5
The *t* Test of an Equality

Consider, still, the normal d-sample model with $n > d$, the estimate $\hat{\tau}$ of a nontrivial linear parameter τ, the real number τ_0, and the statistic $t = (\hat{\tau} - \tau_0)/\text{SE}(\hat{\tau})$, which has the t distribution with $n - d$ degrees of freedom when $\tau = \tau_0$. Consider now the hypothesis that $\tau = \tau_0$. If this hypothesis holds, then $P(|t| \geq t_{1-\alpha/2,n-d}) = \alpha$. Thus, the critical region $|t| \geq t_{1-\alpha/2,n-d}$ for the t test of this hypothesis has size α; that is, when the hypothesis holds, α is the probability of the critical region (see Figure 7.8).

The acceptance region for the t test of the hypothesis that $\tau = \tau_0$ is given by $|t| < t_{1-\alpha/2,n-d}$. The t statistic lies in the acceptance region if and only if

$$-t_{1-\alpha/2,n-d} < \frac{\hat{\tau} - \tau_0}{\text{SE}(\hat{\tau})} < t_{1-\alpha/2,n-d}$$

or, equivalently, if and only if

$$\hat{\tau} - t_{1-\alpha/2,n-d}\,\text{SE}(\hat{\tau}) < \tau_0 < \hat{\tau} + t_{1-\alpha/2,n-d}\,\text{SE}(\hat{\tau}).$$

Thus, the t statistic lies in the acceptance region if and only if τ_0 lies in the $100(1 - \alpha)\%$ confidence interval for τ.

THEOREM 7.12 Consider the t test of size α of the hypothesis that $\tau = \tau_0$. The t statistic lies in the critical region if $\alpha \geq 2[\,1 - t_{n-d}(|t|)\,]$ and in the acceptance region if $\alpha < 2[\,1 - t_{n-d}(|t|)\,]$.

FIGURE 7.8

The critical region for the t test of size α of the hypothesis that $\tau = \tau_0$ is given by $|t| \geq t_{1-\alpha/2,n-d}$. The shaded region under the graph of the density function of the t distribution with $n - d$ degrees of freedom has area α.

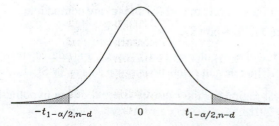

Proof   If $\alpha \geq 2[\, 1 - t_{n-d}(|t|)\,]$, then $t_{n-d}(|t|) \geq 1 - \alpha/2$ and hence $|t| \geq t_{1-\alpha/2,n-d}$, so t lies in the critical region. Alternatively, if $\alpha < 2[\, 1 - t_{n-d}(|t|)\,]$, then $t_{n-d}(|t|) < 1 - \alpha/2$ and hence $|t| < t_{1-\alpha/2,n-d}$, so t lies in the acceptance region.   ∎

We refer to $2[\, 1 - t_{n-d}(|t|)\,]$ as the *P*-value for the t test of the hypothesis that $\tau = \tau_0$ (see Figure 7.9). As before, the t statistic lies in the critical region if α is greater than or equal to the *P*-value and in the acceptance region if α is less than the *P*-value.

FIGURE   7.9

The *P*-value for the t test of the hypothesis that $\tau = \tau_0$ equals the area of the shaded region under the graph of the density function of the t distribution with $n - d$ degrees of freedom.

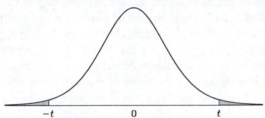

The *P*-value for the t test of size α of the hypothesis that $\tau = \tau_0$ is less than or equal to α if and only if the t statistic lies in the critical region for the test. Suppose the hypothesis is satisfied. Then the probability that the t statistic lies in the critical region equals α and hence $P(P\text{-value} \leq \alpha) = \alpha$. Since α is an arbitrary value in $(0, 1)$, we conclude that the *P*-value (viewed as a random variable) is uniformly distributed on $[\, 0, 1\,]$ when the hypothesis is satisfied.

EXAMPLE **7.17**   Consider the sugar beet data (see Table 7.3 in Section 7.3) (with those for all five varieties being used to obtain the pooled sample variance).

(a) Determine the critical region for the t test of size .05 of the hypothesis that the mean percentage of sugar content of variety 2 is equal to that of the standard variety.

(b) Does the t statistic lie in the critical region or the acceptance region?

(c) Determine the 95% confidence interval for $\mu_2 - \mu_1$.

(d) Check that the answers to part (b) and (c) are compatible.

(e) Determine and interpret the *P*-value for the test.

Solution   We want to test the hypothesis that $\mu_1 = \mu_2$ or, equivalently, that $\tau = 0$, where $\tau = \mu_2 - \mu_1$. The relevant t statistic is given by $t = \hat{\tau}/\,\mathrm{SE}(\hat{\tau})$, where $\hat{\tau} = \hat{\mu}_2 - \hat{\mu}_1$.

(a) The critical region for the t test of size .05 of the hypothesis is given by $|t| \geq t_{.975,46} \doteq 2.013$.

(b) As we have already seen, the pooled sample standard deviation is given by $S \doteq 0.423$, so the standard error of $\hat{\tau}$ is given by

$$\text{SE}(\hat{\tau}) \doteq 0.423\sqrt{\tfrac{1}{11} + \tfrac{1}{10}} \doteq 0.185.$$

Since $\hat{\tau} \doteq 17.580 - 18.136 = -0.556$, the t statistic is given by

$$t \doteq -\frac{0.556}{0.185} \doteq -3.008.$$

Thus, $|t| > 2.013$, and hence the t statistic lies in the critical region.

(c) The 95% confidence interval for τ is given by

$$\hat{\tau} \pm t_{.975,46}\,\text{SE}(\hat{\tau}) \doteq -0.556 \pm (2.013)(0.185)$$

$$\doteq -0.556 \pm 0.372$$

$$\doteq (-0.929, -0.184).$$

(d) The confidence interval in part (c) does not contain zero, which is compatible with rejecting the hypothesis that $\tau = 0$.

(e) The P-value for the test is given by $2[\,1 - t_{46}(3.008)\,] \doteq .004$. Therefore, t lies in the critical region for the test of size α if $\alpha \geq .004$ and in the acceptance region if $\alpha < .004$. In particular, it lies in the critical region for the test of size .05, as we saw in the solution to part (b), and even in the critical region for the test of size .01, but it lies in the acceptance region for the test of size .001.   ∎

In the solution to Example 7.17, the assumptions of the normal five-sample model corresponding to the sugar beet data are implicitly in effect. Moreover, it is implicitly assumed that the hypothesis that $\mu_1 = \mu_2$ was specified prior to seeing the data. Suppose, instead, that we had observed from the data that, among the four new varieties, the sample mean for variety 2 differed the most from the sample mean for the standard variety and that we had used this information to decide that $\mu_2 = \mu_1$ was the hypothesis to be tested. Then, strictly speaking, the analysis in the solution to Example 7.17 would not be valid.

Power

Consider the t test of size α of the hypothesis that $\tau = \tau_0$. When the hypothesis is valid, the probability of erroneously rejecting it equals α. Suppose the hypothesis is invalid; that is, $\tau \neq \tau_0$. Then the power of the test is given by

$$1 - \beta = P(|t| \geq t_{1-\alpha/2,n-d})$$

$$= P(t \leq -t_{1-\alpha/2,n-d}) + P(t \geq t_{1-\alpha/2,n-d})$$

$$= P\!\left(\frac{\hat{\tau} - \tau_0}{\text{SE}(\hat{\tau})} \leq -t_{1-\alpha/2,n-d}\right) + P\!\left(\frac{\hat{\tau} - \tau_0}{\text{SE}(\hat{\tau})} \geq t_{1-\alpha/2,n-d}\right).$$

Suppose $n - d \gg 1$. Then $t_{1-\alpha/2, n-d} \approx z_{1-\alpha/2} = -z_{\alpha/2}$ and $P(\mathrm{SE}(\hat{\tau})/\mathrm{SD}(\hat{\tau}) \approx 1) \approx 1$, so the power of the test is given approximately by

$$
1 - \beta \approx P\Big(\frac{\hat{\tau} - \tau_0}{\mathrm{SD}(\hat{\tau})} \leq z_{\alpha/2}\Big) + P\Big(\frac{\hat{\tau} - \tau_0}{\mathrm{SD}(\hat{\tau})} \geq -z_{\alpha/2}\Big)
$$

$$
= P\Big(\frac{\hat{\tau} - \tau}{\mathrm{SD}(\hat{\tau})} \leq z_{\alpha/2} - \frac{\tau - \tau_0}{\mathrm{SD}(\hat{\tau})}\Big) + P\Big(\frac{\hat{\tau} - \tau}{\mathrm{SD}(\hat{\tau})} \geq -z_{\alpha/2} - \frac{\tau - \tau_0}{\mathrm{SD}(\hat{\tau})}\Big)
$$

$$
= \Phi\Big(z_{\alpha/2} - \frac{\tau - \tau_0}{\mathrm{SD}(\hat{\tau})}\Big) + 1 - \Phi\Big(-z_{\alpha/2} - \frac{\tau - \tau_0}{\mathrm{SD}(\hat{\tau})}\Big)
$$

$$
= \Phi\Big(z_{\alpha/2} - \frac{\tau - \tau_0}{\mathrm{SD}(\hat{\tau})}\Big) + \Phi\Big(z_{\alpha/2} + \frac{\tau - \tau_0}{\mathrm{SD}(\hat{\tau})}\Big)
$$

$$
= \Phi(z_{\alpha/2} - \delta) + \Phi(z_{\alpha/2} + \delta),
$$

where $\delta = |\tau - \tau_0|/\mathrm{SD}(\hat{\tau})$. If α and β are both small, then $\Phi(z_{\alpha/2} - \delta)$ is negligible, and hence $1 - \beta \approx \Phi(z_{\alpha/2} + \delta)$. Thus, $z_{\alpha/2} + \delta \approx z_{1-\beta}$, so $\delta \approx z_{1-\beta} - z_{\alpha/2} = z_{1-\beta} + z_{1-\alpha/2}$, and hence

$$
\mathrm{SD}(\hat{\tau}) \approx \frac{|\tau - \tau_0|}{z_{1-\beta} + z_{1-\alpha/2}}. \tag{27}
$$

For an application of (27), let $n_1 = \cdots = n_d = r$, where the common number r again is to be determined. Let τ be a nontrivial linear parameter. Recall from Equation (24) in Section 7.4 that $\mathrm{SD}(\hat{\tau}) = a\sigma/\sqrt{r}$, where $a = (c_1^2 + \cdots + c_d^2)^{1/2}$. Thus, by (27), if $\tau \neq \tau_0$ and $n - d \gg 1$, then

$$
r \approx \Big(\frac{a\sigma(z_{1-\beta} + z_{1-\alpha/2})}{|\tau - \tau_0|}\Big)^2. \tag{28}
$$

EXAMPLE **7.18**   Consider the normal two-sample model with $n_1 = n_2 = r$ and the hypothesis that $\mu_1 = \mu_2$. Determine r such that the power of the test of size .05 of the hypothesis is about .9 when $|\mu_2 - \mu_1| = \sigma$.

Solution   It follows from (28) with $\alpha = .05$, $\beta = .1$, $\tau_0 = 0$, $|\tau| = \sigma$, and $a^2 = (-1)^2 + 1^2 = 2$ that $r = 2(z_{.9} + z_{.975})^2 \doteq 21$.   ■

Sleeping Drug Experiment

Next, we consider some data showing the different effects of the optical isomers of hyoscyamine hydrobromide in inducing sleep. As described by Student (William Sealy Gosset) in his famous article in *Biometrika* in 1908, the sleep of ten patients was measured without hypnotic and after treatment with dextrohyoscyamine hydrobromide (1) or levohyoscyamine hydromide (2). Table 7.7 lists the average number of hours of sleep gained over the duration of the experiment by the use of the two drugs.

TABLE   7.7

Patient	1 (Dextro-)	2 (Levo-)	Difference (2 − 1)
1	0.7	1.9	1.2
2	−1.6	0.8	2.4
3	−0.2	1.1	1.3
4	−1.2	0.1	1.3
5	−0.1	−0.1	0.0
6	3.4	4.4	1.0
7	3.7	5.5	1.8
8	0.8	1.6	0.8
9	0.0	4.6	4.6
10	2.0	3.4	1.4
Sample mean	0.75	2.33	1.58
Sample SD	1.789	2.002	1.230

Let μ_1 and μ_2, respectively, denote the mean amount of sleep gained by using drugs 1 and 2, respectively.

EXAMPLE 7.19   Under the assumptions of the corresponding normal one-sample model, determine the 95% confidence interval for

(a) μ_1;   (b) μ_2;   (c) $\mu_2 - \mu_1$.

Solution   (a) Suppose the ten responses (average numbers of hours of sleep gained) to drug 1 satisfy the assumptions of a normal one-sample model. The standard error of the estimate $\hat{\mu}_1 = 0.75$ of μ_1 is given by $SE(\hat{\mu}_1) \doteq 1.789/\sqrt{10} \doteq 0.566$. Now $t_{.975,9} \doteq 2.262$. Thus, the 95% confidence interval for μ_1 is given by

$$0.75 \pm (2.262)(0.566) \doteq 0.75 \pm 1.280 = (-0.530, 2.030).$$

(b) Suppose the ten responses to drug 2 satisfy the assumptions of the normal one-sample model. The standard error of the estimate $\hat{\mu}_2 = 2.33$ of μ_2 is given by $SE(\hat{\mu}_2) \doteq 2.002/\sqrt{10} \doteq 0.633$. Thus, the 95% confidence interval for μ_2 is given by

$$2.33 \pm (2.262)(0.633) \pm 2.33 \pm 1.432 = (0.898, 3.762).$$

(c) Suppose the ten differences themselves satisfy the assumptions of the normal one-sample model. The corresponding estimate of $\mu_2 - \mu_1$ is the average of the sample mean of the ten differences, which is given by $\hat{\mu}_2 - \hat{\mu}_1 = 1.58$, and the standard error of this estimate is given by $SE(\hat{\mu}_2 - \hat{\mu}_1) \doteq 1.230/\sqrt{10} \doteq 0.389$. Thus, the 95% confidence interval for $\mu_2 - \mu_1$ is given by

$$1.58 \pm (2.262)(0.389) \doteq 1.58 \pm 0.880 = (0.700, 2.460). \blacksquare$$

EXAMPLE **7.20**   Under the assumptions of the normal one-sample model for the ten differences, test the hypothesis that $\mu_1 = \mu_2$.

Solution   We want to test the hypothesis that $\tau = 0$, where $\tau = \mu_2 - \mu_1$. The corresponding t statistic is given by

$$t = \frac{\hat{\tau}}{\mathrm{SE}(\hat{\tau})} \doteq \frac{1.580}{0.389} \doteq 4.062.$$

Thus, the P-value for the t test of the hypothesis is given by $2[\,1 - t_9(4.062)\,] \doteq$ .003. [That this P-value is less than .05 is compatible with the 95% confidence interval in the solution to Example 7.19(c) not containing zero.] Therefore, we reject the hypothesis (in particular, for the t test of size .01) and conclude that $\mu_1 \neq \mu_2$. In practice it would be reasonable to be more specific and conclude that $\mu_2 > \mu_1$; that is, drug 2 is more effective than drug 1 in inducing sleep.   ■

An alternative design for the sleeping experiment is to obtain 20 patients, choose 10 of them at random to receive drug 1 and let the other 10 patients receive drug 2. The reason for using *randomization* to assign drugs to patients is to avoid the possibility of bias in the assignment (for example, the tendency to give drug 2 to the more responsive patients and drug 1 to the less responsive ones).

EXAMPLE **7.21**   Suppose the sleeping drug experiment had involved 20 patients, 10 of whom received drug 1 and 10 of received drug 2; the experimental data are given in Table 7.8 (to emphasize the distinction between the two experimental designs, the responses to the two treatments have been written in random order). Under the assumptions of the normal two-sample model,

TABLE   **7.8**

	1 (Dextro-)	2 (Levo-)
	−1.6	−0.1
	3.4	4.4
	3.7	1.1
	0.7	1.9
	0.8	1.6
	0.0	5.5
	2.0	3.4
	−0.2	0.8
	−1.2	4.6
	−0.1	0.1
Sample mean	0.75	2.33
Sample SD	1.789	2.002

(a) determine the 95% confidence interval for $\mu_2 - \mu_1$;

(b) test the hypothesis that $\mu_1 = \mu_2$.

Solution   (a) The pooled sample variance is given by

$$S^2 = \frac{9(1.789)^2 + 9(2.002)^2}{18} \doteq 3.605,$$

so the pooled sample standard deviation is given by $S \doteq \sqrt{3.605} \doteq 1.899$. The estimate $\hat{\tau} = \hat{\mu}_2 - \hat{\mu}_1$ of $\tau = \mu_2 - \mu_1$ is given by $\hat{\tau} = 2.33 - 0.75 = 1.58$, and the standard error of this estimate is given by

$$\mathrm{SE}(\hat{\tau}) \doteq 1.899\sqrt{\tfrac{1}{10} + \tfrac{1}{10}} \doteq 0.849.$$

Now $t_{.975,18} \doteq 2.101$. Thus, the 95% confidence interval for $\mu_2 - \mu_1$ is given by

$$1.58 \pm (2.101)(0.849) \doteq 1.58 \pm 1.784 = (-0.204, 3.364).$$

(b) The t statistic for testing the hypothesis that $\mu_1 = \mu_2$ or, equivalently, that $\tau = 0$ is given by

$$t = \frac{\hat{\tau}}{\mathrm{SE}(\hat{\tau})} \doteq \frac{1.580}{0.849} \doteq 1.861.$$

Thus, the P-value for the t test of the hypothesis is given by $2[\,1 - t_{18}(1.861)\,] \doteq .079$. Consequently, based on this analysis, the t test of size .05 accepts (fails to reject) the hypothesis that $\mu_1 = \mu_2$.   ∎

EXAMPLE **7.22**   Consider the actual sleeping drug experiment. Under the assumptions of the normal two-sample model, determine the 95% confidence interval for $\mu_2 - \mu_1$ and test the hypothesis that $\mu_1 = \mu_2$.

Solution   The solution to Example 7.21 is applicable without change. Thus, the estimate $\mu_2 - \mu_1$ is given by $\hat{\mu}_2 - \hat{\mu}_1 = 1.58$, the standard error of this estimate equals 0.849, the 95% confidence interval for $\mu_2 - \mu_1$ equals $(-0.204, 3.364)$, and the P-value for testing the hypothesis that $\mu_1 = \mu_2$ equals .079.   ∎

In the context of the actual sleeping drug experiment, then, we can apparently use either the normal two-sample model applied to the data on the two responses or the normal one-sample model applied to the response differences to analyze the experimental data. The corresponding estimates of $\mu_2 - \mu_1$ coincide, but the two analyses otherwise differ substantially. In particular, the two-sample approach yields a confidence interval for $\mu_2 - \mu_1$ of much greater length (3.568 versus 1.760) and a much larger P-value for testing the hypothesis that $\mu_1 = \mu_2$ (.079 versus .003). Which analysis is appropriate?

To answer this question, let us consider, more generally, an experiment involving the random responses Y_1 and Y_2 of a randomly selected subject to treatment 1 and treatment 2, respectively. Suppose Y_1 and Y_2 individually have normal distributions with means μ_1 and μ_2, respectively, and common variance σ^2. Suppose also that the joint distribution of Y_1 and Y_2 is bivariate normal, let ρ denote the correlation between Y_1 and Y_2, and let $\sigma_{12} = \rho\sigma^2$ denote the covariance between these two random variables. Then the difference $Y_2 - Y_1$ between the two responses is normally distributed with mean $\mu_2 - \mu_1$, and its variance is given by

$$\text{var}(Y_2 - Y_1) = \text{var}(Y_1) + \text{var}(Y_2) - 2\,\text{cov}(Y_1, Y_2) = 2\sigma^2(1 - \rho).$$

Consider an experiment in which the two treatments are applied to each of n randomly selected subjects (out of a population that is so large that the distinction between sampling with replacement and sampling without replacement can be ignored) and let Y_{1i} and Y_{2i} denote the responses of the ith subject to treatment 1 and treatment 2, respectively. We assume that $(Y_{11}, Y_{21}), \ldots, (Y_{1n}, Y_{2n})$ are independent pairs of random variables, each having the same bivariate distribution as (Y_1, Y_2). Then $Y_{21} - Y_{11}, \ldots, Y_{2n} - Y_{1n}$ are independent and normally distributed with mean $\mu_2 - \mu_1$ and variance $2\sigma^2(1 - \rho)$, so these n differences satisfy the assumptions of the normal one-sample model. Moreover, $Y_{11}, \ldots, Y_{1n}$ are independent and normally distributed with mean μ_1 and variance σ^2, and $Y_{21}, \ldots, Y_{2n}$ are independent and normally distributed with mean μ_2 and variance σ^2. However, the two random samples $(Y_{11}, \ldots, Y_{1n})$ and $(Y_{21}, \ldots, Y_{2n})$ are independent if and only if $\rho = 0$. Therefore, the assumptions of the normal two-sample model are satisfied if the correlation between Y_1 and Y_2 equals zero, but not otherwise.

Writing the difference $\bar{Y}_2 - \bar{Y}_1$ between the two sample means as

$$\bar{Y}_2 - \bar{Y}_1 = \frac{1}{n}\sum_{i=1}^{n}(Y_{2i} - Y_{1i}),$$

we see that

$$\text{var}(\hat{\mu}_2 - \hat{\mu}_1) = \text{var}(\bar{Y}_2 - \bar{Y}_1) = \frac{\text{var}(Y_2 - Y_1)}{n} = \frac{2\sigma^2(1 - \rho)}{n}$$

and hence that

$$\text{SD}(\hat{\mu}_2 - \hat{\mu}_1) = \sigma\sqrt{\frac{2(1 - \rho)}{n}}. \tag{29}$$

The estimate of this standard deviation that is used in the analysis based on the normal two-sample model is given by

$$\text{SE}(\hat{\mu}_2 - \hat{\mu}_1) = S\sqrt{\frac{2}{n}}. \tag{30}$$

Now the pooled sample variance S^2 is an unbiased and consistent estimate of σ^2 (even if $\rho \neq 0$). Since

$$\frac{\text{SE}(\hat{\mu}_2 - \hat{\mu}_1)}{\text{SD}(\hat{\mu}_2 - \hat{\mu}_1)} = \frac{S}{\sigma}\sqrt{\frac{1}{1 - \rho}}, \tag{31}$$

we see that, for large n, the standard error of $\hat{\mu}_2 - \hat{\mu}_1$ given by (30) is an accurate estimate of its standard deviation if and only if the correlation ρ between Y_1 and Y_2 is small in magnitude.

If ρ is small in magnitude, then the 95% confidence interval for $\mu_2 - \mu_1$ based on the normal one-sample model applied to the n differences should be in near agreement with that based on the normal two-sample model applied to the n pairs of responses. Similarly, the P-values for the corresponding tests of the hypothesis that $\mu_1 = \mu_2$ should be in near agreement.

If ρ is substantially positive, however, we see from (31) that the two-sample standard error of $\hat{\mu}_2 - \hat{\mu}_1$ given by (30) is too high and hence that the corresponding 95% confidence interval for $\mu_2 - \mu_1$ is too wide. Moreover, the t statistic for the two-sample test of the hypothesis that $\mu_1 = \mu_2$ is too small in magnitude, so the corresponding P-value is too large and hence the power of the test is too small.

Similarly, if ρ is substantially negative, the two-sample standard error of $\hat{\mu}_2 - \hat{\mu}_1$ is too low, so the corresponding 95% confidence interval for $\mu_2 - \mu_1$ is too narrow and hence the probability that this confidence interval actually contains $\mu_2 - \mu_1$ is much smaller than .95. Moreover, the t statistic for the two-sample test of the hypothesis that $\mu_1 = \mu_2$ is too large in magnitude, so the corresponding P-value is too small and hence, when the hypothesis is valid, the probability that the two-sample test of size α of the hypothesis rejects it is much larger than α.

The *sample covariance*

$$S_{12} = \frac{1}{n-1} \sum_{i=1}^{n} (Y_{1i} - \bar{Y}_1)(Y_{2i} - \bar{Y}_2)$$

is an unbiased estimate of the covariance σ_{12} between Y_1 and Y_2 (see Problem 7.49). The *sample correlation*

$$\hat{\rho} = \frac{S_{12}}{S_1 S_2} = \frac{\sum_1^n (Y_{1i} - \bar{Y}_1)(Y_{2i} - \bar{Y}_2)}{\sqrt{\sum_1^n (Y_{1i} - \bar{Y}_1)^2 \sum_1^n (Y_{2i} - \bar{Y}_2)^2}}$$

is a reasonable estimate of the correlation ρ between Y_1 and Y_2.

Figure 7.10 shows the pairs of responses for the ten patients in the sleeping drug experiment. It appears that the two responses are dependent and, specifically, that they are positively correlated. In fact, their sample covariance is given by $S_{12} \doteq 2.848$, so their sample correlation is given by

$$\hat{\rho} \doteq \frac{2.848}{(1.789)(2.002)} \doteq 0.795.$$

The point (0.0, 4.6) for patient 9 appears as an outlier, the other nine points lying near a straight line with slope close to 1. Thus, if a patient gains much sleep from one drug, he will most likely also gain much sleep from the other drug.

EXAMPLE **7.23**   In the context of the sleeping drug experiment, suppose Y_1 and Y_2 each have standard deviation 2 and that the correlation between these two random variables is .8. Determine

(a) the standard deviation $\hat{\mu}_2 - \hat{\mu}_1$;

(b) the value $\sigma \sqrt{2/10} = \sigma/\sqrt{5}$ for this standard deviation corresponding to the assumptions of the normal two-sample model (with $\rho = 0$).

FIGURE   7.10

Average number of hours of sleep gained by the use of the two drugs for each of the ten patients

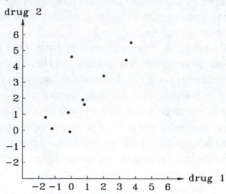

Solution   **(a)** According to Equation (29),

$$\text{SD}(\hat{\mu}_2 - \hat{\mu}_1) = 2\sqrt{\frac{2(1 - .8)}{10}} = 0.4,$$

which is close to the value of the standard error of $\hat{\mu}_2 - \hat{\mu}_1$ that was obtained in the solution to Example 7.19(c).

(b) Corresponding to the assumptions of the normal two-sample model, we calculate the standard deviation of $\hat{\mu}_2 - \hat{\mu}_1$ as

$$2\sqrt{\frac{2}{10}} = \frac{2}{\sqrt{5}} \doteq 0.894,$$

which is close to the value of the standard error of $\hat{\mu}_2 - \hat{\mu}_1$ that was noted in the solution to Example 7.22.   ∎

Suppose, now, that we apply treatment 1 to n randomly selected subjects and treatment 2 to n different random selected subjects. Then the n responses to treatment 1 and the n responses to treatment 2 form two independent random samples. Thus, under the additional assumptions of the normal two-sample model, the standard deviation of $\hat{\mu}_2 - \hat{\mu}_1$ is given by

$$\text{SD}(\hat{\mu}_2 - \hat{\mu}_1) = \sigma\sqrt{\frac{2}{n}}. \tag{32}$$

According to (29) and (32), if $\rho > 0$, we get a more accurate estimate of $\mu_2 - \mu_1$ (that is, one having smaller standard deviation) by applying both treatments to each of n individuals; if $\rho < 0$, we get a more accurate estimate by applying the two treatments to different randomly selected groups of n individuals; and if $\rho = 0$, the two designs are equally accurate. If the two treatments are similar, then ρ is probably positive since a subject is more likely to be responsive to both treatments or to neither treatment than to be responsive to one treatment and unresponsive to the other one. Thus, we should generally get a more accurate estimate of $\mu_2 - \mu_1$ by applying both treatments to each of n individuals than by applying them to different groups of n individuals. Also, it is generally less costly to involve one group of n individuals in an experiment than to involve two such groups.

Unfortunately, in many if not most applications, it is inappropriate to apply the two treatments under investigation to the same subject, the reason being that the first treatment may have a long-term or permanent effect on the subject and thereby on the response to the second treatment. (Think of teaching a student the same material by two different methods or trying out two potential cures for a disease on the same patient.) In such applications, there is an attractive alternative to pure randomization, which involves *matched pairs*. Specifically, we divide the $2n$ subjects into n groups of size 2, with the subjects in each group being as similar as possible with respect to features that may be relevant to their potential responses to the two treatments. Then we choose one subject at random within each of the n groups and assign treatment 1 to that subject and treatment 2 to the other member of the group. In this manner, we hope that the correlation between the responses of the two subjects in a randomly selected matched pair will be substantially positive and hence that the standard deviation of the estimate of $\mu_2 - \mu_1$ given by the sample mean of the differences between the paired responses will be substantially smaller than under pure randomization. (In the context of investigating two different teaching methods, we could group the students according to their GPAs or majors; in the context of trying out two potential cures for a disease, we could group the patients according to the severity of the disease.) *Blocking*, a generalization of pairing that involves matched groups of any given size, will be discussed in Section 11.4.

Problems

7.43 Consider the setup of Problem 7.1 in Section 7.1, but treat the standard deviation of the measurement as being unknown. **(a)** Carry out the t test of size .05 of the hypothesis that $\tau = 0$. **(b)** Determine the P-value for the test.

7.44 Consider the setup of Problem 7.33 in Section 7.4, assume that σ does not differ between wards, but allow the possibility that μ does differ between wards. **(a)** Carry out the t test of size .05 of the hypothesis that μ does not differ between wards. **(b)** Determine the P-value for the test.

7.45 Consider the sugar beet data (see Table 7.3 in Section 7.3) just on variety 3. **(a)** Determine the critical region for the t test of size .05 of the hypothesis that the mean percentage μ_3 of sugar content of variety 3 equals 18. **(b)** Does the t statistic lie in the critical region or the acceptance region? **(c)** Determine the 95% confidence interval for μ_3. **(d)** Check that the answers to parts (b) and (c) are compatible. **(e)** Determine the P-value for the test. **(f)** From the answer to part (e), what can be concluded about the 99% confidence interval for μ_3?

7.46 In the context of the sugar beet data, determine and interpret the P-value for testing the hypothesis that $(\mu_1 + \mu_2)/2 = (\mu_3 + \mu_4 + \mu_5)/3$.

7.47   In the context of the sleeping drug experiment, **(a)** determine the P-value for the t test of the hypothesis that $\mu_2 = 0$; **(b)** check that the answer to part (a) is compatible with the solution to Example 3.19(b).

7.48   Verify that **(a)** $\sum_{i=1}^{n}(Y_{1i} - c_1)(Y_{2i} - c_2) = \sum_{i=1}^{n}(Y_{1i} - \bar{Y}_1)(Y_{2i} - \bar{Y}_2) + n(\bar{Y}_1 - c_1)(\bar{Y}_2 - c_2)$ for $c_1, c_2 \in \mathbb{R}$ [see the proof of Equation (52) in Section 2.4]; **(b)** $\sum_{i=1}^{n}(Y_{1i} - \mu_1)(Y_{2i} - \mu_2) = \sum_{i=1}^{n}(Y_{1i} - \bar{Y}_1)(Y_{2i} - \bar{Y}_2) + n(\bar{Y}_1 - \mu_1)(\bar{Y}_2 - \mu_2)$.

7.49   In the context in which the sample covariance was introduced in the text, verify that **(a)** $\mathrm{cov}(\bar{Y}_1, \bar{Y}_2) = \sigma_{12}/n$; **(b)** $E\left(\sum_{i=1}^{n}(Y_{1i} - \bar{Y}_1)(Y_{2i} - \bar{Y}_2)\right) = (n-1)\sigma_{12}$ [use the result in Problem 7.48(b)]; **(c)** $E(S_{12}) = \sigma_{12}$.

7.50   Consider the sleeping drug data with that from patient 9 removed (because it appears as an outlier in Figure 7.10). Use the response differences for the remaining nine patients to determine **(a)** the estimate of $\mu_2 - \mu_1$ and its standard error; **(b)** the 95% confidence interval for $\mu_2 - \mu_1$; **(c)** the P-value for the t test of the hypothesis that $\mu_1 = \mu_2$.

7.51   Consider the sleeping drug data with that from patient 9 removed as in Problem 7.50. Use the data on the remaining nine patients to determine **(a)** $\bar{Y}_1, S_1, \bar{Y}_2, S_2$; **(b)** the sample covariance and sample correlation; **(c)** the pooled sample variance and sample standard deviation; **(d)** the estimate of $\mu_2 - \mu_1$ and its standard error based on the assumptions of the normal two-sample model (with $\rho = 0$); **(e)** the 95% confidence interval for $\mu_2 - \mu_1$ based on the assumptions of the normal two-sample model; **(f)** the P-value for the t test of the hypothesis that $\mu_1 = \mu_2$ based on the assumptions of the normal two-sample model; **(g)** the validity of the answers to parts (d) through (f).

7.52   The experimental data (Table 7.9) on the amount of wear on the soles of boys' shoes made with two synthetic materials, A and B, is taken from Table 4.3 of *Statistics for Experiments* by George E. P. Box, William G. Hunter, and J. Stuart Hunter (New York: Wiley, 1978). During the course of the experiment, each boy wore a special pair of shoes while carrying on his normal daily activity, which varied considerably from boy to boy. The sole of one shoe was made with material A and that of the other shoe with material B. In the table, (L) indicates that the material was used in

TABLE 7.9

Boy	A	B	Difference
1	13.2 (L)	14.0 (R)	0.8
2	8.2 (L)	8.8 (R)	0.6
3	10.9 (R)	11.2 (L)	0.3
4	14.3 (L)	14.2 (R)	−0.1
5	10.7 (R)	11.8 (L)	1.1
6	6.6 (L)	6.4 (R)	−0.2
7	9.5 (L)	9.8 (R)	0.3
8	10.8 (L)	11.3 (R)	0.5
9	8.8 (R)	9.3 (L)	0.5
10	13.3 (L)	13.6 (R)	0.3

the left sole and (R) that it was used in the right sole, the choice being determined by the flip of a coin. Material A was standard, and B was a less expensive substitute. The immediate purpose of the experiment was to test the hypothesis that switching from material A to material B would not increase the amount of wear. Let μ_1 and μ_2 be the mean wear on soles made with materials A and B, respectively.

Under the assumption of the normal one-sample model, determine (a) the estimate of $\mu_2 - \mu_1$ and the standard error of this estimate; (b) the 95% lower confidence bound for $\mu_2 - \mu_1$; (c) the P-value for the t test of the hypothesis that $\mu_2 \leq \mu_1$ (interpret the result).

7.53 Consider the boys' shoes experiment that was described in Problem 7.52, but suppose that five boys, chosen at random, wore shoes having soles made with material A and the other five boys wore shoes having soles made with material B. (a) Use the data from Problem 7.52 to assess the standard error of $\mu_2 - \mu_1$ that would ensue if the alternative design were employed. (b) Assess the power of the t test of size .05 of the hypothesis that $\mu_2 \leq \mu_1$ under the alternative design.

7.54 Consider an experiment such as the sleeping drug experiment that involves the response of randomly selected patients to either or both of two drugs. Let Y_1 denote the response of a randomly selected patient to drug 1 and let Y_2 denote the patient's response to drug 2. Suppose Y_1 and Y_2 have a bivariate normal joint distribution with means μ_1 and μ_2, respectively, common standard deviation 2, and correlation .8. Consider the t test of size .05 of the hypothesis that $\mu_1 = \mu_2$. Determine r so that the power of the test is about .9 if $|\mu_2 - \mu_1| = 1$ (a) if the experiment involves r patients each of whom gets both drugs (as was done with the actual sleeping drug experiment); (b) if r patients receive drug 1 and r other patients receive drug 2, as in Example 7.21.

7.55 Let Y_1 be the weight of a randomly selected man upon arising in the morning, let Y_2 denote his weight just prior to retiring at night, and let μ_1 and μ_2 denote the means of Y_1 and Y_2, respectively. Suppose Y_1 and Y_2 have a bivariate normal joint distribution, that Y_1 and Y_2 have common standard deviation 20, and that $Y_2 - Y_1$ has variance 1. (a) Determine the covariance and correlation between Y_1 and Y_2. (b) Suppose r randomly selected men are to be weighed both upon arising in the morning and just before retiring at night and that the sample mean of the r differences is to be used to estimate $\mu_2 - \mu_1$. Determine r such that this estimate has standard deviation 0.1. (c) Determine r such that the t test of size .05 of the hypothesis that $\mu_1 = \mu_2$ corresponding to the design in part (b) has power .9 when $\mu_2 - \mu_1 = 0.1$. (d) Suppose r randomly selected men are to be weighed upon arising in the morning, r other randomly selected men are to be weighed just before retiring at night, and the difference between the two sample means is used to estimate $\mu_2 - \mu_1$. Determine r such this estimate has standard deviation 0.1. (e) Determine r such that the t test of size .05 of the hypothesis that $\mu_1 = \mu_2$ corresponding to the design in part (d) has power .9 when $\mu_2 - \mu_1 = 0.1$.

7.56 Suppose the response of a patient to drug 1 is normally distributed with mean μ_1 and standard deviation 5 and the response to drug 2 is normally distributed with mean μ_2 and standard deviation 5. An experiment involving matched pairs is to be conducted, in which it is assumed that the paired responses will have a bivariate normal joint distribution with correlation ρ. It is desired to make the matching so close that the *t* test of size .05 of the hypothesis that $\mu_1 = \mu_2$ will have approximate power .9 when $|\mu_2 - \mu_1| = 1$ and 25 matched pairs are used. How large would ρ have to be to accomplish this task?

7.57   Consider an experiment involving matched pairs and let Y_1 and Y_2 be the responses to treatments 1 and 2, respectively, in a random matched pair. Suppose $Y_1 = \mu_1 + W + V_1$ and $Y_2 = \mu_2 + W + V_2$, where W, V_1 and V_2 are independent and normally distributed with mean 0, W has variance γ, and V_1, and V_2 each have variance δ. Determine **(a)** the joint distribution of Y_1 and Y_2; **(b)** the correlation between Y_1 and Y_2.

7.6
The *F* Test

Consider, as before, the normal *d*-sample model with $n > d$. The *hypothesis of homogeneity* is that the *d* means coincide; that is, $\mu_1 = \cdots = \mu_d$. In this section we develop a test of this hypothesis. Here we also develop a test of the hypothesis that the *d* means have a common specified value; that is, $\mu_1 = \cdots = \mu_d = \mu_0$, where μ_0 is a specified number. Moreover, it will be shown that the former test is equivalent to the *t* test when $d = 2$ and that the latter test is equivalent to the *t* test when $d = 1$.

THEOREM 7.13   The random variables

$$\chi^2 = \frac{1}{\sigma^2} \sum_{k=1}^{d} n_k (\bar{Y}_k - \mu_0)^2$$

and S^2 are independent, and if $\mu_1 = \cdots = \mu_d = \mu_0$, then χ^2 has the chi-square distribution with *d* degrees of freedom.

Proof   Since $\bar{Y}_1, \ldots, \bar{Y}_d$ and S^2 are independent by Corollary 7.3 in Section 7.3, χ^2 and S^2 are independent. Suppose $\mu_1 = \cdots = \mu_d = \mu_0$. Then

$$Z_k = \frac{\bar{Y}_k - \mu_0}{\sigma/\sqrt{n_k}} = \frac{\sqrt{n_k}}{\sigma}(\bar{Y}_k - \mu_0), \qquad 1 \le k \le d,$$

are independent, standard normal random variables, so

$$\chi^2 = \frac{1}{\sigma^2} \sum_{k=1}^{d} n_k (\bar{Y}_k - \mu_0)^2 = \sum_{k=1}^{d} Z_k^2$$

has the chi-square distribution with *d* degrees of freedom.   ∎

The overall sample mean $\bar{Y}$ is given by

$$\bar{Y} = \frac{1}{n} \sum_{k=1}^{d} \sum_{i=1}^{n_k} Y_{ki} = \frac{1}{n} \sum_{k=1}^{d} n_k \bar{Y}_k.$$

Observe that $\bar{Y}$ is normally distributed with mean

$$\frac{1}{n} \sum_{k=1}^{d} n_k \mu_k$$

and variance σ^2/n. In particular, if $\mu_1 = \cdots = \mu_d = \mu$, then $\bar{Y}$ is normally distributed with mean μ and variance σ^2/n.

THEOREM 7.14 Suppose $d \geq 2$. Then the random variables

$$\chi^2 = \frac{1}{\sigma^2} \sum_{k=1}^{d} n_k (\bar{Y}_k - \bar{Y})^2$$

and S^2 are independent, and if $\mu_1 = \cdots = \mu_d$, then χ^2 has the chi-square distribution with $d-1$ degrees of freedom.

Proof Since $\bar{Y}$ is a function of $\bar{Y}_1, \ldots, \bar{Y}_d$, so is χ^2. The independence of χ^2 and S^2 now follows from the independence of $\bar{Y}_1, \ldots, \bar{Y}_d$ and S^2.

The result that if $\mu_1 = \cdots = \mu_d$, then χ^2 has the chi-square distribution with $d-1$ degrees of freedom will be derived as a special case of a more general result in Section 10.8. In the special case $d = 2$, however, this result is easily established. To this end, we observe that

$$\sum_{k=1}^{2} n_k (\bar{Y}_k - \bar{Y})^2 = n_1 \left(\bar{Y}_1 - \frac{n_1 \bar{Y}_1 + n_2 \bar{Y}_2}{n_1 + n_2} \right)^2 + n_2 \left(\bar{Y}_2 - \frac{n_1 \bar{Y}_1 + n_2 \bar{Y}_2}{n_1 + n_2} \right)^2$$

$$= \frac{n_1 n_2}{n_1 + n_2} (\bar{Y}_2 - \bar{Y}_1)^2$$

$$= \frac{(\bar{Y}_2 - \bar{Y}_1)^2}{\frac{1}{n_1} + \frac{1}{n_2}}.$$

Suppose $\mu_1 = \mu_2$. Then $\bar{Y}_2 - \bar{Y}_1$ is normally distributed with mean 0 and variance

$$\sigma^2 \left(\frac{1}{n_1} + \frac{1}{n_2} \right),$$

so the random variable

$$Z = \frac{\bar{Y}_2 - \bar{Y}_1}{\sigma \sqrt{\frac{1}{n_1} + \frac{1}{n_2}}}$$

has the standard normal distribution. Consequently,

$$\chi^2 = \frac{1}{\sigma^2} \sum_{k=1}^{2} n_k (\bar{Y}_k - \bar{Y})^2 = \frac{(\bar{Y}_2 - \bar{Y}_1)^2}{\sigma^2 \left(\frac{1}{n_1} + \frac{1}{n_2} \right)} = Z^2$$

has the chi-square distribution with 1 degree of freedom. ∎

Consider the hypothesis that $\mu_1 = \cdots = \mu_d = \mu_0$ for some specified number μ_0 and the statistic

$$F = \frac{1}{S^2 d} \sum_{k=1}^{d} n_k (\bar{Y}_k - \mu_0)^2. \tag{33}$$

Alternatively, consider the hypothesis that $\mu_1 = \cdots = \mu_d$, where $d \geq 2$, and the statistic

$$F = \frac{1}{S^2(d-1)} \sum_{k=1}^{d} n_k (\bar{Y}_k - \bar{Y})^2. \tag{34}$$

In either case, the larger the value of F, the less plausible is the hypothesis. The next result follows from Theorems 7.13 and 7.14, Corollary 7.3 in Section 7.3, and the definition of the F distribution.

THEOREM 7.15   (a) If $\mu_1 = \cdots = \mu_d = \mu_0$, then the F statistic given by (33) has the F distribution with d degrees of freedom in the numerator and $n - d$ degrees of freedom in the denominator.

   (b) If $d \geq 2$ and $\mu_1 = \cdots = \mu_d$, then the F statistic given by (34) has the F distribution with $d - 1$ degrees of freedom in the numerator and $n - d$ degrees of freedom in the denominator.

In the context of the hypothesis that $\mu_1 = \cdots = \mu_d = \mu_0$ for some specified number μ_0, set $p_0 = 0$; in the context of the hypothesis that $\mu_1 = \cdots = \mu_d$, set $p_0 = 1$. Under either hypothesis, p_0 is the number of free parameters in the specification of $\mu_1, \ldots, \mu_d$. It follows from Theorem 7.15 that, under either hypothesis, the corresponding F statistic has the F distribution with $d - p_0$ degrees of freedom in the numerator and $n - d$ degrees of freedom in the denominator and hence (for $0 < \alpha < 1$) that $P(F \geq F_{1-\alpha, d-p_0, n-d}) = \alpha$. Thus, the critical region $F \geq F_{1-\alpha, d-p_0, n-d}$ for the F *test* of the hypothesis has size α; that is, when the hypothesis is satisfied, α is the probability of the critical region. This fact is illustrated in Figure 7.11. (The actual density function shown in the two figures in this section is that of the F distribution with 5 degrees of freedom in the numerator and 10 degrees of freedom in the denominator.) The acceptance region for the test is $F < F_{1-\alpha, d-p_0, n-d}$.

The proof of the next result is similar to that of Theorem 7.11 in Section 7.4, the details being left as a problem.

THEOREM 7.16   The F statistic for the test of size α lies in the critical region if $\alpha \geq 1 - F_{d-p_0, n-d}(F)$ and in the acceptance region if $\alpha < 1 - F_{d-p_0, n-d}(F)$.

We refer to $1 - F_{d-p_0, n-d}(F)$ as the *P-value* for the F test (see Figure 7.12). According to Theorem 7.16, the F statistic lies in the critical region for the F test of

FIGURE   7.11

The critical region for the F test of size α is $F \geq F_{1-\alpha,d-p_0,n-d}$. The shaded region under the graph of the density function of the F distribution with $d - p_0$ degrees of freedom in the numerator and $n - d$ degrees of freedom in the denominator has area α.

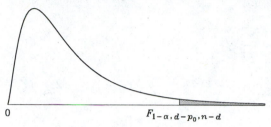

FIGURE   7.12

The P-value for the F test is the area of the shaded region under the graph of the density function of the F distribution with $d - p_0$ degrees of freedom in the numerator and $n - d$ degrees of freedom in the denominator.

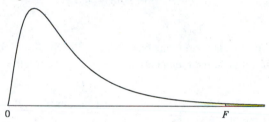

size α if α is greater than or equal to the P-value and in the acceptance region if α is less than the P-value. Moreover, if the hypothesis is valid, then the P-value (viewed as a random variable) is uniformly distributed on $[0,1]$.

The *between sum of squares* (BSS), defined by

$$\mathrm{BSS} = \sum_{k=1}^{d} n_k (\bar{Y}_k - \bar{Y})^2,$$

is a measure of the variation among the sample means $\bar{Y}_1, \ldots, \bar{Y}_d$. The F statistic in Equation (34) for testing the hypothesis of homogeneity can be rewritten as

$$F = \frac{\mathrm{BSS}/(d-1)}{\mathrm{WSS}/(n-d)}. \tag{35}$$

The *total sum of squares* (TSS), defined by

$$\mathrm{TSS} = \sum_{k=1}^{d} \sum_{i=1}^{n_k} (Y_{ki} - \bar{Y})^2,$$

is a measure of the variation among the n random variables Y_{ki}, $1 \leq k \leq d$ and

$1 \le i \le n_k$. Observe that

$$\text{TSS} = \sum_{k=1}^{d} \sum_{i=1}^{n_k} (Y_{ki} - \bar{Y})^2$$

$$= \sum_{k=1}^{d} \sum_{i=1}^{n_k} (\bar{Y}_k - \bar{Y} + Y_{ki} - \bar{Y}_k)^2$$

$$= \sum_{k=1}^{d} n_k (\bar{Y}_k - \bar{Y})^2 + \sum_{k=1}^{d} \sum_{i=1}^{n_k} (Y_{ki} - \bar{Y}_k)^2 + 2 \sum_{k=1}^{d} (\bar{Y}_k - \bar{Y}) \sum_{i=1}^{n_k} (Y_{ki} - \bar{Y}_k)$$

and that

$$\sum_{i=1}^{n_k} (Y_{ki} - \bar{Y}_k) = \sum_{i=1}^{n_k} Y_{ki} - n_k \bar{Y}_k = 0, \qquad 1 \le k \le d.$$

Consequently,

$$\text{TSS} = \text{BSS} + \text{WSS}. \tag{36}$$

EXAMPLE **7.24**   In the context of the sugar beet data (see Table 7.3 in Section 7.3), test the hypothesis of homogeneity; that is, the mean percentage of sugar content is the same for all five varieties.

Solution   Here WSS $\doteq$ 8.242 as we saw in the solution to Example 7.12(a) in Section 7.3. The overall sample mean is given by

$$\bar{Y} \doteq \frac{11(18.136) + \cdots + 10(18.130)}{51} \doteq 18.047.$$

Thus, the between sum of squares is given by

$$\text{BSS} \doteq 11(18.136 - 18.047)^2 + \cdots + 10(18.130 - 18.047)^2 \doteq 3.085.$$

Consequently, the F statistic for testing the hypothesis of homogeneity is given by

$$F \doteq \frac{3.085/4}{8.242/46} \doteq \frac{0.771}{0.179} \doteq 4.304,$$

and the corresponding P-value is given by $1 - F_{4,46}(4.304) \doteq .005$. Therefore, the data strongly suggest that the mean percentage of sugar content is not the same for all five varieties.

Table 7.10 shows these calculations of the P-value of the F test of homogeneity. In this table, SS is an abbreviation for *sum of squares*, DF is an abbreviation for degrees of freedom, and MS is an abbreviation for *mean square*. The row corresponding to the between sum of squares could have been given a more descriptive name such as "Varieties" instead of "Between." ∎

Tables such as the one shown in the solution to Example 7.24 are referred to as *analysis of variance* (ANOVA) tables.

TABLE   7.10

Source	SS	DF	MS	F	P-value
Between	3.085	4	0.771	4.304	.005
Within	8.242	46	0.179		
Total	11.327	50			

Equivalence of t Tests and F Tests

Consider the normal two-sample model with $n \geq 3$. The hypothesis of homogeneity is that $\mu_1 = \mu_2$. The F statistic for testing this hypothesis is given by

$$F = \frac{\mathrm{BSS}}{S^2} = \frac{(\bar{Y}_2 - \bar{Y}_1)^2}{S^2 \left(\frac{1}{n_1} + \frac{1}{n_2} \right)}$$

(see the proof of Theorem 7.14). Since the t statistic for testing this hypothesis is given by

$$t = \frac{\hat{\mu}_2 - \hat{\mu}_1}{\mathrm{SE}(\hat{\mu}_2 - \hat{\mu}_1)} = \frac{\bar{Y}_2 - \bar{Y}_1}{S\sqrt{\frac{1}{n_1} + \frac{1}{n_2}}},$$

we see that $F = t^2$. The P-value for the F test is given by $1 - F_{1,n-2}(F) = 1 - F_{1,n-2}(t^2)$, and the P-value for the t test equals $2[1 - t_{n-2}(|t|)]$. Since $1 - F_{1,n-2}(t^2) = 2[1 - t_{n-2}(|t|)]$ by Equation (13) in Section 7.2, the two P-values coincide. Similar results hold for the F and t tests of the hypothesis that $\mu = \mu_0$ in the context of the normal one-sample model, the details being left as a problem.

EXAMPLE 7.25   Consider the sugar beet data on varieties 1 and 2 (see Table 7.3 in Section 7.3). Test the hypothesis that the two varieties have the same mean percentage of sugar content using

(a) the t test;   (b) the F test.

Solution   (a) The hypothesis that $\mu_1 = \mu_2$ is equivalent to the hypothesis that $\tau = 0$, where $\tau = \mu_2 - \mu_1$. According to the solution to Example 7.11(c) in Section 7.3, $\hat{\tau} = -0.556$ and $\mathrm{SE}(\hat{\tau}) \doteq 0.180$. Thus, the t statistic for testing the hypothesis is given by

$$t \doteq -\frac{0.556}{0.180} \doteq -3.083.$$

The corresponding P-value is given by $2[1 - t_{19}(3.083)] \doteq .006$. Thus, the data on varieties 1 and 2 provide strong evidence that $\mu_1 \neq \mu_2$.

(b) Here $\mathrm{WSS} = 3.241$, as we saw in the solution to Example 7.11(a) in Section 7.3.

The overall sample mean for the data on the first two varieties is given by

$$\bar{Y} \doteq \frac{11(18.136) + 10(17.58)}{21} \doteq 17.871.$$

The corresponding between sum of squares is given by

$$\text{BSS} \doteq 11(18.136 - 17.871)^2 + 10(17.58 - 17.871)^2 \doteq 1.621.$$

Consequently, we get the *P*-value for the *F* test of the hypothesis $\mu_1 = \mu_2$ that is given in Table 7.11. Observe that the sum of the rounded values of BSS and WSS does not quite equal the rounded value of TSS and, more importantly, that the *P*-value in this table coincides with that given in the solution to part (a). ■

TABLE 7.11

Source	SS	DF	MS	F	P-value
Between	1.621	1	1.621	9.504	.006
Within	3.241	19	0.171		
Total	4.863	20			

Problems

7.58 Consider the setup of Problem 7.33 in Section 7.4, assume that σ does not differ between wards, but allow the possibility that μ does differ between wards. **(a)** Carry out the *F* test of size .05 of the hypothesis that μ does not differ between wards. **(b)** Determine the *P*-value for the test.

7.59 Consider the normal *d*-sample model with $d \geq 2$ and $n_1 = \cdots = n_d = r$. Use Theorem 7.7 in Section 7.3 to verify that the random variable

$$\frac{r}{\sigma^2} \sum_{k=1}^{d} (\bar{Y}_k - \bar{Y})^2$$

has the chi-square distribution with $d - 1$ degrees of freedom.

7.60 In the context of the sugar beet data (see Table 7.3 in Section 7.3), test the hypothesis that the mean percentage of sugar content does not differ among varieties 1, 3, 4, and 5. Show your calculations in the form of an ANOVA table.

7.61 In the context of the sugar beet data (see Table 7.3), test the hypothesis that the mean percentage of sugar content for each of the varieties 1, 3, 4, and 5 equals 18.

7.62 Verify Theorem 7.16.

7.63 Under the assumptions of the normal one-sample model for the ten differences in the context of the sleeping drug experiment, determine the *P*-value for the *F* test of the hypothesis that $\mu_1 = \mu_2$.

7.64 In the context of the sleeping drug experiment as described in Example 7.21 in Section 7.5, determine the P-value for the F test of the hypothesis that $\mu_1 = \mu_2$.

7.65 In the context of the boys' shoes experiment described in Problem 7.52 in Section 7.5, determine the P-value for the F test of the hypothesis that $\mu_1 = \mu_2$.

7.66 Consider the resistor data in Problems 7.31 and 7.32 in Section 7.3 and the hypothesis that the mean resistances of the resistors made by the two machines coincide. Determine the P-value for the corresponding **(a)** t test; **(b)** F test.

7.67 Consider the normal one-sample model with $n \geq 2$ and the hypothesis that $\mu = \mu_0$ for some specified number μ_0. **(a)** Determine the F statistic for testing the hypothesis. **(b)** Determine the t statistic for testing the hypothesis. **(c)** Verify that $F = t^2$. **(d)** Verify that the P-values for the F test and the t test of the hypothesis coincide.

7.68 Verify that

$$
\text{BSS} = \sum_{k=1}^{d} \frac{T_k^2}{n_k} - \frac{T^2}{n} \quad \text{and} \quad \text{TSS} = \sum_{k=1}^{d} \sum_{i=1}^{n_k} Y_{ki}^2 - \frac{T^2}{n},
$$

where

$$
T_k = n_k \bar{Y}_k = \sum_{i=1}^{n_k} Y_{ki} \quad \text{and} \quad T = n\bar{Y} = \sum_{k=1}^{d} \sum_{i=1}^{n_k} Y_{ki}.
$$

8

Introduction to Linear Regression

8.1
The Method of Least Squares

In this and the next three chapters, we will mainly be studying the normal linear regression model. This model includes the normal d-sample model as a special case, as we will see in Section 8.3.

We start out by considering some experimental data introduced on page 253 of *Statistical Methods in Engineering and Quality Assurance* by Peter W. M. John (New York: Wiley, 1990). The data, which come from a chemical engineering process for polymerization, are as follows:

T	220	230	240	250	260	270
Y	58.9	71.6	78.0	81.6	86.0	85.2

Here the one *factor* under the control of the investigator is the reactor temperature T, and the observed response Y is the yield of tetramer. The data can be viewed as consisting of six runs. On the ith run, the temperature is set at the level T_i and the observed yield is Y_i, where $T_1 = 220°C$, $Y_1 = 58.9$, $T_2 = 230°C$, $Y_2 = 71.6$, and so forth.

In analyzing the data, it is convenient to introduce the coded value $x = (T - 245)/5$ of temperature, so that $T = 245 + 5x$. The coded values of temperature on the $n = 6$ runs are then given by $x_1 = -5$, $x_2 = -3$, and so forth. The polymer data can now be shown as follows:

x	-5	-3	-1	1	3	5
Y	58.9	71.6	78.0	81.6	86.0	85.2

Consider a straight-line fit $\hat{Y} = b_0 + b_1 x$ to the data. We refer to $\hat{Y}_i = b_0 + b_1 x_i$ as the predicted value of Y_i corresponding to the fit, to $Y_i - \hat{Y}_i$ as the corresponding residual, and to RSS $= \sum_i (Y_i - \hat{Y}_i)^2$ as the residual sum of squares. The method of least squares is to choose b_0 and b_1 to minimize this residual sum of squares. To accomplish this task, we need to solve the "normal" equations $\partial\,\mathrm{RSS}\,/\partial b_0 = 0$ and $\partial\,\mathrm{RSS}\,/\partial b_1 = 0$ for b_0 and b_1.

Let $\hat{\beta}_0$ and $\hat{\beta}_1$ denote the minimizing values of b_0 and b_1. To motivate this notation, assume that the mean yield $\mu(x)$ at the coded level x of temperature is given by $\mu(x) = \beta_0 + \beta_1 x$, where β_0 and β_1 are unknown parameters. We refer to $\mu(\cdot)$ [that is, to $\mu(x)$ as x varies over its range] as the regression function, to $\hat{\beta}_0$ and $\hat{\beta}_1$ as the least-squares estimates of β_0 and β_1, respectively, and to $\hat{\mu}(x) = \hat{\beta}_0 + \hat{\beta}_1 x$ as the least-squares linear estimate of $\mu(x)$ or as the least-squares linear fit to the data.

Now RSS $= \sum_i (Y_i - b_0 - b_1 x_i)^2$, so the normal equations yield

$$\frac{\partial\,\mathrm{RSS}}{\partial b_0} = -2 \sum_i (Y_i - b_0 - b_1 x_i)$$

and

$$\frac{\partial\,\mathrm{RSS}}{\partial b_1} = -2 \sum_i x_i (Y_i - b_0 - b_1 x_i).$$

Thus,

$$n\hat{\beta}_0 + \hat{\beta}_1 \sum_i x_i = \sum_i Y_i$$

$$\hat{\beta}_0 \sum_i x_i + \hat{\beta}_1 \sum_i x_i^2 = \sum_i x_i Y_i.$$

In the present context, $\sum_i x_i = [\,(-5) + (-3) + (-1) + 1 + 3 + 5\,] = 0$, so the normal equations reduce to

$$n\hat{\beta}_0 = \sum_i Y_i$$

$$\hat{\beta}_1 \sum_i x_i^2 = \sum_i x_i Y_i,$$

whose unique solution is given by

$$\hat{\beta}_0 = \frac{1}{n} \sum_i Y_i = \bar{Y} \quad \text{and} \quad \hat{\beta}_1 = \frac{\sum_i x_i Y_i}{\sum_i x_i^2}.$$

Here $\sum_i Y_i = 461.3$, so $\hat{\beta}_0 = \bar{Y} = 461.3/6 \doteq 76.883$. Also,

$$\sum_i x_i^2 = [\,(-5)^2 + (-3)^2 + (-1)^2 + 1^2 + 3^2 + 5^2\,] = 70$$

and $\sum_i x_i Y_i = 178.3$, so $\hat{\beta}_1 = 178.3/70 \doteq 2.547$. Thus, the least-squares linear estimate of the regression function is given by $\hat{\mu}(x) \doteq 76.883 + 2.547x$. The predicted values $\hat{Y}_i = \hat{\mu}(x_i)$ from the linear fit are shown together with the polymer data in the following table:

x	-5	-3	-1	1	3	5
Y	58.9	71.6	78.0	81.6	86.0	85.2
$\hat{Y}$	64.1	69.2	74.3	79.4	84.5	89.6

Figure 8.1 shows the least-squares linear fit to the polymer data. The corresponding residual sum of squares is given by

$$\text{RSS} = \sum_i (Y_i - \hat{Y}_i)^2 \doteq (58.9 - 64.148)^2 + \cdots + (85.2 - 89.619)^2 \doteq 72.933.$$

Next, in light of the apparent curvature in the plot of the data as shown in Figure 8.1, we consider the least-squares quadratic fit to the data. Here the regression function is assumed to have the form $\mu(x) = \beta_0 + \beta_1 x + \beta_2 x^2$, and its least-squares estimate has the form $\hat{\mu}(x) = \hat{\beta}_0 + \hat{\beta}_1 x + \hat{\beta}_2 x^2$. The normal equations for $\hat{\beta}_0$, $\hat{\beta}_1$, and $\hat{\beta}_2$ (obtained by setting the partial derivatives of the residual sum of squares with respect to each of these quantities equal to zero) are given by

$$n\hat{\beta}_0 + \hat{\beta}_1 \sum_i x_i + \hat{\beta}_2 \sum_i x_i^2 = \sum_i Y_i,$$

$$\hat{\beta}_0 \sum_i x_i + \hat{\beta}_1 \sum_i x_i^2 + \hat{\beta}_2 \sum_i x_i^3 = \sum_i x_i Y_i,$$

$$\hat{\beta}_0 \sum_i x_i^2 + \hat{\beta}_1 \sum_i x_i^3 + \hat{\beta}_2 \sum_i x_i^4 = \sum_i x_i^2 Y_i.$$

FIGURE 8.1

Least-squares linear fit to the polymer data

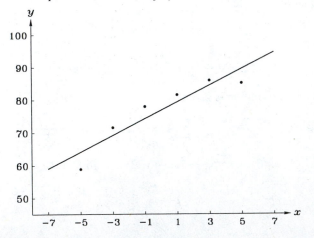

Since $\sum_i x_i = 0$ and $\sum_i x_i^3 = 0$, the normal equations reduce to

$$\hat{\beta}_0 + \hat{\beta}_2 \sum_i x_i^2 = \sum_i Y_i,$$

$$\hat{\beta}_1 \sum_i x_i^2 = \sum_i x_i Y_i,$$

$$\hat{\beta}_0 \sum_i x_i^2 + \hat{\beta}_2 \sum_i x_i^4 = \sum_i x_i^2 Y_i.$$

Thus, $\hat{\beta}_1 = \sum_i x_i Y_i / \sum_i x_i^2$, and $\hat{\beta}_0$ and $\hat{\beta}_2$ can be obtained by solving a system of two linear equations in these two unknowns. Using the polymer data (and a computer), we get that $\hat{\beta}_0 \doteq 80.816$, $\hat{\beta}_1 \doteq 2.547$, and $\hat{\beta}_2 \doteq -0.337$. Thus, the least-squares quadratic estimate of the regression function is given by $\hat{\mu}(x) \doteq 80.816 + 2.547x - 0.337x^2$. The polymer data and predicted values from the quadratic fit are as follows:

x	−5	−3	−1	1	3	5
Y	58.9	71.6	78.0	81.6	86.0	85.2
$\hat{Y}$	59.7	70.1	77.9	83.0	85.4	85.1

Figure 8.2 shows the least-squares quadratic fit to the polymer data. The corresponding residual sum of squares is given by RSS $\doteq 5.073$, which is substantially smaller than that for the least-squares linear fit.

FIGURE 8.2

Least-squares quadratic polynomial fit to the polymer data

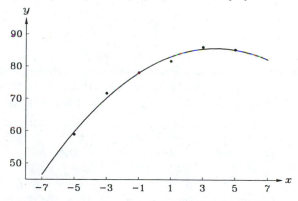

The least-squares quadratic estimate of the regression function appears to give a reasonably good fit to the polymer data. For illustrative purposes, however, we now consider the least-squares cubic polynomial fit to the data. Here the regression function is assumed to have the form $\mu(x) = \beta_0 + \beta_1 x + \beta_2 x^2 + \beta_3 x^3$, and its least-squares estimate has the form $\hat{\mu}(x) = \hat{\beta}_0 + \hat{\beta}_1 x + \hat{\beta}_2 x^2 + \hat{\beta}_3 x^3$. Solving the normal equations for $\hat{\beta}_0$, $\hat{\beta}_1$, $\hat{\beta}_2$, and $\hat{\beta}_3$, we get $\hat{\beta}_0 \doteq 80.816$, $\hat{\beta}_1 \doteq 2.166$, $\hat{\beta}_2 \doteq -0.337$, and $\hat{\beta}_3 \doteq 0.0189$. Thus, the least-squares cubic estimate of the regression function is given by $\hat{\mu}(x) \doteq 80.816 + 2.166x - 0.337x^2 + 0.0189x^3$. The polymer data and predicted values from the cubic fit are as follows:

x	−5	−3	−1	1	3	5
Y	58.9	71.6	78.0	81.6	86.0	85.2
$\hat{Y}$	59.2	70.8	78.3	82.7	84.8	85.6

Figure 8.3 shows the least-squares cubic fit to the polymer data. The corresponding residual sum of squares is given by RSS $\doteq 3.597$, which is somewhat smaller than that for the least-squares quadratic fit.

FIGURE 8.3

Least-squares cubic polynomial fit to the polymer data

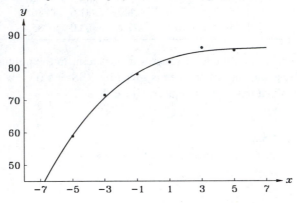

Similarly, the least-squares quartic polynomial estimate of the regression function is given by $\hat{\mu}(x) \doteq 79.814 + 2.166x - 0.00130x^2 + 0.0189x^3 - 0.0124x^4$. The polymer data and predicted values from the quartic fit are as follows:

x	−5	−3	−1	1	3	5
Y	58.9	71.6	78.0	81.6	86.0	85.2
$\hat{Y}$	58.9	71.8	77.6	82.0	85.8	85.2

Figure 8.4 shows the least-squares quartic fit to the polymer data. The corresponding residual sum of squares is given by RSS $\doteq 0.373$, which is substantially smaller than that for the least-squares cubic fit.

The least-squares quintic polynomial estimate of the regression function is given by $\hat{\mu}(x) \doteq 79.814 + 1.702x - 0.00130x^2 + 0.100x^3 - 0.0124x^4 - 0.00253x^5$ (see Figure 8.5). The residual sum of squares for this fit is exactly equal to zero; that is, $\hat{Y}_i = \hat{\mu}(x_i) = Y_i$ for all i. A function such as $\hat{\mu}(\cdot)$ that fits the experimental data exactly is referred to as an interpolating function. Since the residual sum of

FIGURE 8.4

Least-squares quartic polynomial fit to the polymer data

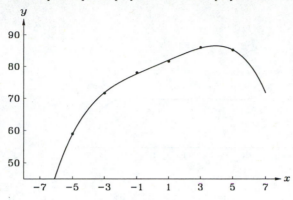

FIGURE 8.5

Interpolating quintic polynomial fit to the polymer data

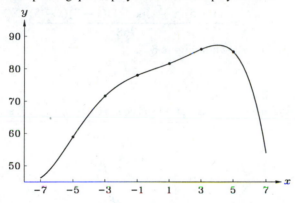

squares corresponding to an interpolating function equals zero, any such function is necessarily a least-squares fit.

In the context of the polymer data, the points $-5, -3, -1, 1, 3$, and 5 are referred to as the design points, and $\{-5, -3, -1, 1, 3, 5\}$ is referred to as the design set. For $1 \leq i \leq 6$, let g_i be the unique quintic polynomial that equals 1 at x_i and equals 0 at the other design points. Then $g_1(-5) = 1$ and $g_1(-3) = g_1(-1) = g_1(1) = g_1(3) = g_1(5) = 0$, so

$$g_1(x) = \frac{(x+3)(x+1)(x-1)(x-3)(x-5)}{(-5+3)(-5+1)(-5-1)(-5-3)(-5-5)}$$

$$= -\frac{(x+3)(x+1)(x-1)(x-3)(x-5)}{3840},$$

a plot of which is shown in Figure 8.6.

FIGURE 8.6

Quintic polynomial that equals 1 at -5 and equals 0 at $-3, -1, 1, 3,$ and 5

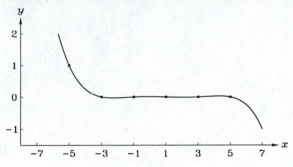

Similarly, the polynomial g_2 is given by

$$g_2(x) = \frac{(x+5)(x+1)(x-1)(x-3)(x-5)}{(-3+5)(-3+1)(-3-1)(-3-3)(-3-5)}$$
$$= \frac{(x+5)(x+1)(x-1)(x-3)(x-5)}{768},$$

and it has the plot shown in Figure 8.7.

FIGURE 8.7

Quintic polynomial that equals 1 at -3 and equals 0 at $-5, -1, 1, 3,$ and 5

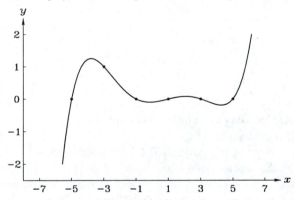

Moreover, g_3 is given by

$$g_3(x) = \frac{(x+5)(x+3)(x-1)(x-3)(x-5)}{(-1+5)(-1+3)(-1-1)(-1-3)(-1-5)}$$
$$= -\frac{(x+5)(x+3)(x-1)(x-3)(x-5)}{384},$$

and it has the plot shown in Figure 8.8. The polynomials g_4, g_5, and g_6 are given in terms of g_3, g_2, and g_1, respectively, by $g_4(x) = g_3(-x)$, $g_5(x) = g_2(-x)$, and

$g_6(x) = g_1(-x)$. The quintic interpolating polynomial $\hat{\mu}(\cdot)$ can be written explicitly in terms of $g_1, \dots, g_6$ as

$$\hat{\mu}(x) = Y_1 g_1(x) + \cdots + Y_6 g_6(x);$$

this is a special case of the Lagrange interpolating formula, which will be discussed in Section 9.3.

F I G U R E **8.8**

Quintic polynomial that equals 1 at -1 and 0 at -5, -3, 1, 3, and 5

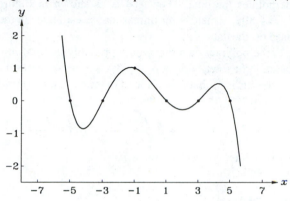

Let g be the sixth-degree polynomial that has roots at all the design points and, say, equals 1 at $x = 0$ (see Figure 8.9), which is given explicitly by

$$g(x) = -\frac{(x+5)(x+3)(x+1)(x-1)(x-3)(x-5)}{225}.$$

Also, let $\hat{\mu}(\cdot)$ be the unique quintic polynomial that interpolates the polymer data. Then, for $c \in \mathbb{R}$, $\hat{\mu}(\cdot) + cg$ is a sixth-degree polynomial that interpolates the polymer data; conversely, every such interpolating polynomial equals $\hat{\mu}(\cdot) + cg$ for some

F I G U R E **8.9**

The sixth-degree polynomial that has roots at -5, -3, -1, 1, 3, and 5 and equals 1 at 0

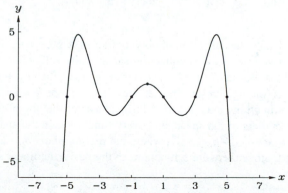

$c \in \mathbb{R}$. In particular, there are infinitely many sixth-degree interpolating polynomials and hence infinitely least-squares sixth-degree polynomial fits to the data.

Figure 8.9 illustrates an undesirable feature of interpolating polynomials. Consider the following as experimental data involving seven design points:

x	-5	-3	-1	0	1	3	5
Y	0	0	0	1	0	0	0

Let $\hat{\mu}(\cdot)$ be the least-squares sixth-degree polynomial fit to these data, which exactly interpolates the data. Then $\hat{\mu}(\cdot)$ is as shown in Figure 8.9. In particular, $\hat{\mu}(\pm 4.327) \doteq 4.807$, which is an unreasonable estimate of the mean response at ± 4.327 based on the data.

The fit to the polymer data that would probably be used in practice is the least-squares quadratic fit $\hat{\mu}(x) = \hat{\beta}_0 + \hat{\beta}_1 x + \hat{\beta}_2 x^2$. (The apparent improved accuracy of the higher-order fits is probably spurious.) In terms of the original temperature scale, this fit is given by

$$\hat{\mu}(T) = \hat{\beta}_0 + \hat{\beta}_1 \frac{T - 245}{5} + \hat{\beta}_2^2 \left(\frac{T - 245}{5}\right)^2,$$

the graph of which is shown in Figure 8.10.

F I G U R E   8.10

Least-squares quadratic polynomial fit to the polymer data

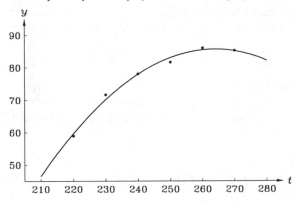

Let G denote the collection of polynomials of degree $p - 1$ (or less) on $\mathbb{R}$, where p is a positive integer. This collection satisfies the following two properties: (i) if $g \in G$ and $b \in \mathbb{R}$, then $bg \in G$; and (ii) if $g_1, g_2 \in G$, then $g_1 + g_2 \in G$. A collection of functions (on a given set) that satisfies these two properties is known as a linear space. The functions $1, x, \ldots, x^{p-1}$ form a basis of the polynomial space G; that is, they are members of the space, and every function g in the space can be written as $g = b_0 + b_1 x + \cdots + b_{p-1} x^{p-1}$, where the real numbers $b_0, \ldots, b_{p-1}$ are uniquely determined. Since there are p functions in the basis, the linear space G is said to be

p-dimensional. The relevant properties of finite-dimensional linear spaces and their bases will be discussed in Section 9.1.

Recall the design set $\{-5, -3, -1, 1, 3, 5\}$ that arose in the context of the polymer data. The linear space G is said to be nonidentifiable relative to the design set if there is a nonzero function $g \in G$ that equals zero on this set; otherwise, G is said to be identifiable. If the dimension of G is greater than the number of design points, then G is nonidentifiable; otherwise, in the context of polynomial spaces on $\mathbb{R}$, G is identifiable. The least-squares fit in G is unique if and only if G is identifiable. In particular, in the context of the polymer data, if $p \leq 6$, then G is identifiable and there is a unique least-squares fit; if $p \geq 7$, then G is nonidentifiable and there is not a unique least-squares fit. A detailed treatment of identifiability will be given in Section 9.2.

The linear space G is said to be saturated if, for every choice of the values of the response variable at the design points, there is an interpolating function in G; otherwise, G is said to be unsaturated. If G is saturated, then the residual sum of squares for the least-squares fit in this space to the polymer data equals zero. The space G of polynomials of degree $p - 1$ is saturated if and only if its dimension p is greater than or equal to the number of design points. In particular, in the context of the polymer data, G is saturated if and only if $p \geq 6$. The relevant properties of saturated and unsaturated spaces will be discussed in Section 9.3.

As we have just pointed out, the space G of polynomials of degree $p - 1$ is identifiable if and only if its dimension p is less than or equal to the number of design points, and it is saturated if and only if its dimension is greater than or equal to the number of design points. Thus, this polynomial space is identifiable and saturated if and only if its dimension equals the number of design points.

Problems

8.1 Determine the least-squares straight-line fit to the following data involving the first five prime numbers and determine the corresponding residual sum of squares:

x	1	2	3	4	5
y	2	3	5	7	11

8.2 Determine the least-squares quadratic fit to the data in Problem 8.1 and determine the corresponding residual sum of squares.

8.3 We say that x is a coded form of a variable t if $x = a + bt$ for some constants a and b with $b \neq 0$ (typically with $b > 0$). Suppose x is a coded form of t and $x = x_1$ when $t = t_1$ and $x = x_2$ when $t = t_2$, where $t_1 \neq t_2$ and hence $x_1 \neq x_2$. **(a)** Determine a and b in terms of the quantities t_1, t_2, x_1, and x_2. **(b)** Express x in terms of t and these quantities. **(c)** Express t in terms of x and these quantities.

8.4 Consider distinct design points $x_1, \ldots, x_n$, where $n \geq 2$, set $\bar{x} = (x_1 + \cdots + x_n)/n$, and observe that $\sum_i (x_i - \bar{x}) = 0$. Assume that the regression function is a linear function of x. Then it has the form $\mu(x) = \beta_0 + \beta_1 (x - \bar{x})$, where β_0 and β_1 are unknown parameters. Similarly, the least-squares estimate of the regression function can be written as $\hat{\mu}(x) = \hat{\beta}_0 + \hat{\beta}_1(x - \bar{x})$. **(a)** Determine the normal equations for $\hat{\beta}_0$ and $\hat{\beta}_1$. **(b)** Solve these equations.

8.5 Consider the polymer data. Assume that the regression function has the form of a piecewise linear function and, more specifically, that $\mu(x) = \beta_{10} + \beta_{11}(x + 3)$ for $x < c$ and $\mu(x) = \beta_{20} + \beta_{21}(x - 3)$ for $x \geq c$, where $-1 < c < 1$. Assume also that the regression function is continuous and hence that $\beta_{10} + \beta_{11}(c + 3) = \beta_{20} + \beta_{21}(c - 3)$. Similarly, we write $\hat{\mu}(x) = \hat{\beta}_{10} + \hat{\beta}_{11}(x + 3)$ for $x < \hat{c}$ and $\hat{\mu}(x) = \hat{\beta}_{20} + \hat{\beta}_{21}(x - 3)$ for $x \geq \hat{c}$, where $-1 < \hat{c} < 1$ and $\hat{c}$ is chosen so that $\hat{\mu}(\cdot)$ is continuous at $\hat{c}$. **(a)** Use the design points -5, -3, and -1 and the corresponding yields to determine $\hat{\beta}_{10}$ and $\hat{\beta}_{11}$ [use the answer to Problem 8.4(b)] and thereby to determine $\hat{\mu}(x)$ for $x < \hat{c}$. **(b)** Use the design points 1, 3, and 5 and the corresponding yields to determine $\hat{\beta}_{20}$ and $\hat{\beta}_{21}$ and thereby to determine $\hat{\mu}(x)$ for $x \geq \hat{c}$. **(c)** Choose $\hat{c}$ so that $\hat{\beta}_{10} + \hat{\beta}_{11}(\hat{c} + 3) = \hat{\beta}_{20} + \hat{\beta}_{21}(\hat{c} - 3)$. **(d)** Determine the residual sum of squares (for all six yields). **(e)** Show the graph of $\hat{\mu}(x)$, $-7 \leq x \leq 7$ together with the experimental data.

8.6 Let the design points be -5, -3, -1, 1, 3, and 5 as in the polymer experiment and assume that the mean response is a quadratic function of x. Then it can be written as $\mu(x) = \beta_0 + \beta_1 x + \beta_2(x^2 - 35/3)$, and its least-squares estimate can be written as $\hat{\mu}(x) = \hat{\beta}_0 + \hat{\beta}_1 x + \hat{\beta}_2(x^2 - 35/3)$. **(a)** Determine the normal equations for $\hat{\beta}_0$, $\hat{\beta}_1$, and $\hat{\beta}_2$. **(b)** Solve these equations. **(c)** Use the polymer data to determine $\hat{\beta}_0$, $\hat{\beta}_1$, $\hat{\beta}_2$, and $\hat{\mu}(x)$.

8.7 Let $\hat{\mu}(\cdot)$ denote the least-squares quadratic fit to the polymer data using the coded value of the temperature. Determine **(a)** the value $\hat{x}_{max}$ of x that maximizes $\hat{\mu}(x)$; **(b)** $\hat{\mu}(\hat{x}_{max})$.

8.8 (Continued.) Let $\hat{\mu}(\cdot)$ denote the least-squares quadratic fit to the polymer data using the original temperature scale. Determine **(a)** the value $\hat{T}_{max}$ of T that maximizes $\hat{\mu}(T)$; **(b)** $\hat{\mu}(\hat{T}_{max})$.

8.9 Suppose the regression function has the form $\mu(x) = \beta_0 + \beta_1 e^{\beta_2 x}$. Then its least-squares estimate has the form $\hat{\mu}(x) = \hat{\beta}_0 + \hat{\beta}_1 e^{\hat{\beta}_2 x}$. Determine the normal equations for $\hat{\beta}_0$, $\hat{\beta}_1$, and $\hat{\beta}_2$.

8.10 **(a)** Determine the quadratic polynomial $\hat{\mu}(\cdot)$ that interpolates the polymer data at $x_1 = -5$, $x_3 = -1$, and $x_6 = 5$. **(b)** Determine the value $\hat{x}_{max}$ that maximizes $\hat{\mu}(\cdot)$ and determine $\hat{\mu}(\hat{x}_{max})$. **(c)** Determine the temperature T_{max} corresponding to $\hat{x}_{max}$.

8.11 Let g be the quartic polynomial that has roots at -3, -1, 1, and 3 and equals 1 at 0, and let x_{min} be the value of x in $(0, 3)$ that minimizes g on $[-3, 3]$. Determine **(a)** an explicit formula for g; **(b)** x_{min}; **(c)** $g(\pm x_{min})$.

8.12 Let g be the quintic polynomial that has roots at -4, -2, 0, 2, and 4 and equals 1 at 1 and let x_{max} be the value of x in $[-4, 4]$ that maximizes g. Determine **(a)** an explicit formula for g; **(b)** x_{max}; **(c)** $g(x_{max})$.

8.13 Consider the design points -2, -1, 0, 1, and 2 and let G be the space of polynomials of degree $p - 1$ on $\mathbb{R}$. For which values of p is G **(a)** identifiable; **(b)** saturated; **(c)** identifiable and saturated?

8.2
Factorial Experiments

The polymer experiment, discussed in the previous section, involves temperature as a single factor. Before proceeding to the general notation and terminology that will be used to discuss statistical experiments, we first describe some experiments that involve two or more factors.

Sugar Beet Experiment

In 1929, Jerzy Neyman analyzed an experiment that compared four new varieties (2, 3, 4, 5) of sugar beet with a standard variety (1). The experiment involved 51 plots laid in a row, with the varieties assigned to plots in the following systematic manner: $1, 2, 3, 4, 5, \ldots, 1, 2, 3, 4, 5, 1$. Let the variety be denoted by x_1, which ranges from 1 to 5, and let the plot be denoted by x_2, which ranges from -25 to 25. The response variable was the percentage of sugar content of the beets grown on the given plot. The experimental results are given in Table 8.1. The sugar beet data introduced in Section 7.3 were obtained from this experiment by ignoring the plot variable x_2.

Pilot Plant Experiment

In Section 10.2 of *Statistics for Experimenters* by George E. P. Box, William G. Hunter, and J. Stuart Hunter (New York: Wiley, 1978), a simplified version of an actual pilot plant investigation of a process was introduced for illustrative purposes. According to the description of the experiment, there were two quantitative factors and one qualitative factor. The two quantitative factors were temperature, which took the two values 160°C and 180°C, and concentration, which took the two values 20% and 40%. The qualitative factor was a catalyst, and the two catalysts that were used were denoted by A and B. The response was a chemical yield in grams. Table 8.2 shows the data from the experiment (the individual responses being shown on page 320 of their book).

This experiment is complete in the sense that each of the $2^3 = 8$ combinations of one of the two levels of temperature, one of the two levels of concentration, and one of the two catalysts is employed. Also, there are two repetitions at each of these eight factor-level combinations, so the total number of runs is $8 \cdot 2 = 16$. The 16 runs were made in random temporal order as shown in parentheses following the individual responses. (We can think of putting 16 tickets, labeled from 1 to 16, in a box and then drawing them out of the box one at a time by sampling without replacement. In practice, pseudorandom numbers would most likely be used instead of actually drawing tickets out of a box.) In particular, on the first and last runs, the temperature was 160°C, the concentration was 40%, and catalyst A was used.

In describing and analyzing this experiment, it is convenient to use the coded values of -1 and 1 to denote the two levels of each of the factors. Thus, we let x_1 denote the temperature factor with $x_1 = -1$ corresponding to 160°C (low) and

TABLE 8.1

x_1	x_2	Y	x_1	x_2	Y
1	−25	17.4	1	0	17.6
2	−24	17.2	2	1	17.4
3	−23	17.8	3	2	17.8
4	−22	18.0	4	3	17.4
5	−21	18.2	5	4	17.6
1	−20	18.6	1	5	17.7
2	−19	18.0	2	6	17.1
3	−18	18.8	3	7	17.7
4	−17	18.7	4	8	17.6
5	−16	18.9	5	9	17.5
1	−15	18.8	1	10	17.8
2	−14	17.8	2	11	17.2
3	−13	18.8	3	12	18.2
4	−12	18.6	4	13	17.8
5	−11	18.5	5	14	18.0
1	−10	18.6	1	15	18.1
2	−9	18.1	2	16	17.8
3	−8	18.7	3	17	18.8
4	−7	18.1	4	18	18.5
5	−6	18.2	5	19	18.4
1	−5	18.4	1	20	18.4
2	−4	17.6	2	21	17.6
3	−3	18.3	3	22	18.3
4	−2	17.8	4	23	18.1
5	−1	18.0	5	24	18.0
			1	25	18.1

TABLE 8.2

Temperature (°C)	Concentration (%)	Catalyst	Individual Responses (Grams)	(Grams)	Response
160	20	A	59[6]	61[13]	60
180	20	A	74[2]	70[4]	72
160	40	A	50[1]	58[16]	54
180	40	A	69[5]	67[10]	68
160	20	B	50[8]	54[12]	52
180	20	B	81[9]	85[14]	83
160	40	B	46[3]	44[11]	45
180	40	B	79[7]	81[15]	80

$x_1 = 1$ corresponding to 180°C (high), we let x_2 denote the concentration factor with $x_2 = -1$ corresponding to 20% (low) and $x_2 = 1$ corresponding to 40% (high), and we let x_3 denote the catalyst factor with $x_3 = -1$ corresponding to catalyst A and $x_3 = 1$ corresponding to catalyst B. Table 8.3 lists the experimental data showing the coded values of the three factors.

TABLE 8.3

x_1	x_2	x_3	Individual Responses		Response
−1	−1	−1	59[6]	61[13]	60
1	−1	−1	74[2]	70[4]	72
−1	1	−1	50[1]	58[16]	54
1	1	−1	69[5]	67[10]	68
−1	−1	1	50[8]	54[12]	52
1	−1	1	81[9]	85[14]	83
−1	1	1	46[3]	44[11]	45
1	1	1	79[7]	81[15]	80

Lube Oil Experiment

On page 182 of *Statistical Design and Analysis of Experiments* by Peter W. M. John (New York: Macmillan, 1971), an interesting example of a regular fractional factorial experiment is described. The reference from which the example was taken is given as

> Vance, F. P. (1962). "Optimization Study of Lube Oil Treatment by Process 'X'." *Proc. of Symp. on Application of Statistics and Computers to Fuel and Lubricant Research Problems. Office of the Chief of Ordnance, U.S. Army,* March 13–15, 1962.

According to John's description, there were four major operating variables in the treatment of lube oil at a refinery, each of which was quantitative; these were taken as the factors in the experiment and referred to as A, B, C, and D. The objective of the experimental program was to find a set of operating conditions that would optimize a measure of quality in the lube oil, which was used as the response variable. Three levels of each factor were used, which we assign the coded values of −1 for "low", 0 for "medium," and 1 for "high." The coded levels of A, B, C, and D are denoted by x_1, x_2, x_3, and x_4, respectively.

A complete factorial experiment would involve $3^4 = 81$ factor-level combinations. Vance chose the one-third fraction of the complete factorial experiment containing the $3^{4-1} = 27$ factor-level combinations such that $x_1 + x_2 + x_3 + x_4 \equiv 2$ (mod 3) (that is, such that $x_1 + x_2 + x_3 + x_4 - 2$ is a multiple of 3 or, equivalently, such that $x_1 + x_2 + x_3 + x_4$ equals -4, -1, or 2). Accordingly, there was exactly one run at each combination of a level of A, a level of B, and a level of C. Specifically, if $x_1 + x_2 + x_3$ equals -3, 0, or 3, then $x_4 = -1$; if $x_1 + x_2 + x_3$ equals -1 or 2, then $x_4 = 0$; and if $x_1 + x_2 + x_3$ equals -2 or 1, then $x_4 = 1$. Thus, if we ignore factor D, then we get a complete factorial experiment involving the three factors A, B, and C, each having three levels, and one run at each factor-level combination. Similarly, there was exactly one run at each combination of a level of A, a level of B, and a level of D, and so forth. In other words, if we ignore any one of the four factors, we can view the experiment as a complete factorial experiment with one run at each combination of a level of the three remaining factors. Consequently, there are three runs at each of the nine combinations of levels of any two given factors, and there are nine runs at each of the three levels of each of the four factors.

Table 8.4 shows the factor-level combinations and reported responses.

TABLE   8.4

x_1	x_2	x_3	x_4	Y
-1	-1	-1	-1	4.2
-1	-1	0	1	5.9
-1	-1	1	0	8.2
-1	0	-1	1	13.1
-1	0	0	0	16.4
-1	0	1	-1	30.7
-1	1	-1	0	9.5
-1	1	0	-1	22.2
-1	1	1	1	31.0
0	-1	-1	1	7.7
0	-1	0	0	16.5
0	-1	1	-1	14.3
0	0	-1	0	11.0
0	0	0	-1	29.0
0	0	1	1	55.0
0	1	-1	-1	8.5
0	1	0	1	37.4
0	1	1	0	66.3
1	-1	-1	0	11.4
1	-1	0	-1	21.1
1	-1	1	1	57.9
1	0	-1	-1	13.5
1	0	0	1	51.6
1	0	1	0	76.5
1	1	-1	1	31.0
1	1	0	0	74.5
1	1	1	-1	85.1

Fertilizer Experiment

Another interesting example of a fractional factorial experiment is discussed on pages 90–91 and 199–207 of *The Design of Experiments* by Sir Ronald A. Fisher, 7th edition (New York: Hafner, 1960). In this experiment, six combinations of different quantities of nitrogeneous and phosphatic fertilizer were applied to $6^2 = 36$ plots of land forming a 6×6 square grid, in which potatoes were grown. We can think of the experiment as having involved four factors: a factor R having levels 1, 2, 3, 4, 5, 6, which indicates the row in which the plot lies; a factor C also having levels 1, 2, 3, 4, 5, 6, which indicates the column in which the plot lies; a factor N having levels -1 and 1, which indicates the amount of nitrogeneous fertilizer applied, with -1 and 1 corresponding, respectively, to 0 and 1 units; and a factor P having levels -1, 0, and 1, which are coded values for the amounts of phosphatic fertilizer applied with -1, 0, and 1 corresponding, respectively, to 0, 1, and 2 units of phosphatic fertilizer. The response variable was yield (in pounds) of potatoes. The six possible combinations of a level of N and a level of P, referred to as *treatments*, are conveniently labeled from A to F in Figure 8.11. We also think of these six treatments as being the possible levels of the treatment factor T, which is obtained by combining the factors N and P.

FIGURE 8.11

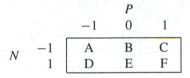

The experimental results are conveniently displayed in Figure 8.12. In row 1 and column 1, for example, treatment E (1 unit of nitrogeneous fertilizer and 1 unit of phosphatic fertilizer) was applied, and the yield was 633 pounds of potatoes. By inspection, the design has the form of a *Latin square*: Every treatment is applied exactly once in each row and once in each column.

Notation and Terminology

Consider, in general, a statistical experiment involving M factors and a random *response variable* Y, which is referred to as a *factorial experiment*. We use x_m to denote the *level* of the mth factor and $\mathbf{x} = (x_1, \ldots, x_M)$ to denote the *factor-level combination* corresponding to the levels $x_1, \ldots, x_M$ of factors $1, \ldots, M$, respectively. The distribution of Y is allowed to depend on $\mathbf{x}$. Let $\mathcal{X}_m$ denote the set of *allowable levels* of the mth factor. We assume that all factor-level combinations in $\mathcal{X} = \mathcal{X}_1 \times \cdots \times \mathcal{X}_M$ are allowable, and we refer to $\mathcal{X}$ as the *experimental region*. Observe that $\mathcal{X}$ is finite if and only if each of the sets $\mathcal{X}_1, \ldots, \mathcal{X}_M$ is finite, in which

FIGURE   8.12

	1	2	3	4	5	6
1	E 633	B 527	F 652	A 390	C 504	D 416
2	B 489	C 475	D 415	E 488	F 571	A 282
3	A 384	E 481	C 483	B 422	D 334	F 646
4	F 620	D 448	E 505	C 439	A 323	B 384
5	D 452	A 432	B 411	F 617	E 594	C 466
6	C 500	F 505	A 259	D 366	B 326	E 420

case $\#(\mathcal{X}) = \#(\mathcal{X}_1) \cdots \#(\mathcal{X}_M)$. The various factors are commonly denoted by capital letters $A, B, C, \ldots$ or by more suggestive letters such as T for temperature, P for pressure, and C for type of catalyst.

Suppose there are n *trials* (or *runs*) in the experiment. Let x_{im} denote the level of the mth factor on the ith trial and let Y_i denote the random response variable on that trial. Then $\mathbf{x}_i = (x_{i1}, \ldots, x_{iM})$ is the factor-level combination on the ith trial. Let d denote the number of distinct factor-level combinations in the experiment. Then $1 \le d \le n$, $d = n$ if and only if $\mathbf{x}_1, \ldots, \mathbf{x}_n$ are distinct, and $d = 1$ if and only if $\mathbf{x}_1, \ldots, \mathbf{x}_n$ coincide. We write the distinct factor-level combinations in some order as $\mathbf{x}'_1, \ldots, \mathbf{x}'_d$, and we refer to them as the *design points* and to $\mathcal{X}' = \{\mathbf{x}'_1, \ldots, \mathbf{x}'_d\}$ as the *design set*.

Let x'_{km} denote the level of the mth factor corresponding to the kth design point for $1 \le k \le d$ and $1 \le m \le M$. Then $\mathbf{x}'_k = (x'_{k1}, \ldots, x'_{kM})$ for $1 \le k \le d$. Let $\mathcal{X}'_m = \{x'_{km} : 1 \le k \le d\}$ denote the set of levels of the mth factor that are actually used in the experiment and let $d_m = \#(\mathcal{X}'_m)$ denote the number of such levels. Then $\mathcal{X}'_1 \times \cdots \times \mathcal{X}'_M$ has $d_1 \cdots d_M$ members, $\mathcal{X}' \subset \mathcal{X}'_1 \times \cdots \times \mathcal{X}'_M$, and $\mathcal{X}' = \mathcal{X}'_1 \times \cdots \times \mathcal{X}'_M$ if and only if $d = d_1 \cdots d_M$. The factorial experiment is said to be *complete* if $\mathcal{X}' = \mathcal{X}'_1 \times \cdots \times \mathcal{X}'_M$ and *fractional* otherwise; thus, the experiment is complete if and only if every factor-level combination of the form $(x_1, \ldots, x_M)$ with $x_m \in \mathcal{X}'_m$ for $1 \le m \le M$ is used at least once. If the experiment is complete and employs the same number of levels of each factor (that is, if $d_1 = \cdots = d_M$), then $d = d_1^M$.

For $1 \le k \le d$, let $\mathcal{I}_k = \{i : 1 \le i \le n \text{ and } \mathbf{x}_i = \mathbf{x}'_k\}$ denote the collection of trials on which the factor-level combination equals the kth design point (here $\mathcal{I}$ is the script form of I). We refer to $n_k = \#(\mathcal{I}_k)$ as the number of *repetitions* at the kth design point. Note that the total number of trials in the experiment is given by $n = n_1 + \cdots + n_d$. If $n_1 = \cdots = n_d = 1$ (that is, if $\mathbf{x}_1, \ldots, \mathbf{x}_n$ are distinct), then $n = d$, but if $n_k \ge 2$ for some k (that is, if there are two or more repetitions at some design point), then $n > d$. If there are r repetitions at each design point (that is, if $n_1 = \cdots = n_d = r$), then $n = rd$. In particular, if the experiment is complete and involves the same number of levels of each factor and if there are r repetitions at each design point, then $n = rd_1^M$; moreover, for any given factor, there are rd_1^{M-1} trials involving each level of that factor.

When there is a single factor ($M = 1$), the design is complete by definition. In this context, we denote the level of the factor by x, the levels on the n trials by $x_1, \ldots, x_n$ and the distinct design points by $x'_1, \ldots, x'_d$.

Consider, for example, the polymer experiment from Section 8.1. Here there is $M = 1$ factor, and there are $n = 6$ runs, which are at the levels $x_1 = -5$, $x_2 = -3$, $x_3 = -1, x_4 = 1, x_5 = 3$, and $x_6 = 5$. Thus, there are $d = 6$ design points, the design set is given by $X' = \{-5, -3, -1, 1, 3, 5\}$, and there is $r = 1$ run at each design point (that is, $n_1 = \cdots = n_6 = 1$). We write the design points in the natural order as $x'_1 = x_1 = -5$, $x'_2 = x_2 = -3$, $x'_3 = x_3 = -1$, $x'_4 = x_4 = 1$, $x'_5 = x_5 = 3$, and $x'_6 = x_6 = 5$. Then $\mathcal{I}_1 = \{1\}$, $\mathcal{I}_2 = \{2\}$, $\mathcal{I}_3 = \{3\}$, $\mathcal{I}_4 = \{4\}$, $\mathcal{I}_5 = \{5\}$, and $\mathcal{I}_6 = \{6\}$. The experimental region X contains the design set X'; otherwise, it is not precisely specified. We can think of X as some interval containing $[-5, 5]$ such that the temperature $T = 245 + 5x$ corresponding to each coded value $x \in X$ is physically realizable and otherwise reasonable in the context of the polymer experiment. If we are assuming, say, that the regression function is a quadratic polynomial on X, then X should be small enough that this assumption is reasonably accurate.

In the context of the pilot plant experiment, there are $M = 3$ factors. In the experiment, each factor takes the two levels -1 and 1, so $d_1 = d_2 = d_3 = 2$ and $X'_1 = X'_2 = X'_3 = \{-1, 1\}$. The design is complete, so there are $d = d_1 d_2 d_3 = 2^3 = 8$ design points; the design set is given by $X' = X'_1 \times X'_2 \times X'_3 = \{-1, 1\} \times \{-1, 1\} \times \{-1, 1\}$. There are $r = 2$ repetitions at each design point, so $n_1 = \cdots = n_8 = 2$ and there are $n = 2 \cdot 8 = 16$ runs. The factor-level combinations on these 16 runs are given by $\mathbf{x}_1 = (-1, 1, -1)$, $\mathbf{x}_2 = (1, -1, -1)$, $\mathbf{x}_3 = (-1, 1, 1)$, and so forth. Writing the eight design points in the order that the experimental data are displayed, we get $\mathbf{x}'_1 = (-1, -1, -1)$, $\ldots$, $\mathbf{x}'_8 = (1, 1, 1)$. Then the collection of two runs corresponding to the design point $\mathbf{x}'_1 = (-1, -1, -1)$ is given by $\mathcal{I}_1 = \{6, 13\}$. Similarly, $\mathcal{I}_2 = \{2, 4\}$, $\mathcal{I}_3 = \{1, 16\}$ and so forth.

In the context of the lube oil experiment, there are $M = 4$ factors. In the experiment, each factor takes the three levels -1, 0, and 1, so $d_1 = d_2 = d_3 = d_4 = 3$, and $X'_1 = X'_2 = X'_3 = X'_4 = \{-1, 0, 1\}$. The design is fractional and involves $d = 3^3 = 27$ factor-level combinations, which is one-third of the $3^4 = 81$ possible such combinations. There is one run at each design point, so $n_1 = \cdots = n_{27} = 1$ and there are $n = 27$ runs. Writing the 27 design points in the order that the experimental data are displayed, we get $\mathbf{x}'_1 = (-1, -1, -1, -1), \ldots, \mathbf{x}'_{27} = (1, 1, 1, -1)$.

We can think of the fertilizer experiment as involving $M = 4$ factors: row, column, nitrogen, and phosphorus. Each of the first two factors takes the six levels $1, \ldots, 6$, so $d_1 = d_2 = 6$ and $X'_1 = X'_2 = \{1, \ldots, 6\}$. The third factor takes the two levels -1 and 1, so $d_3 = 2$ and $X'_3 = \{-1, 1\}$. The fourth factor takes the three levels -1, 0, and 1, so $d_4 = 3$ and $X'_4 = \{-1, 0, 1\}$. Alternatively, we can think of the experiment as involving $M = 3$ factors: row, column, and treatment. It is convenient to think of the treatment factor as taking the six levels $1, \ldots, 6$ corresponding to treatments A, $\ldots$, F, respectively. Then $d_1 = d_2 = d_3 = 6$ and $X'_1 = X'_2 = X'_3 = \{1, \ldots, 6\}$. Either way, the design is fractional and involves $d = 6^2 = 36$ factor-level combinations, which is one-sixth of the $6^3 = 6^2 \cdot 2 \cdot 3 = 216$ possible such combinations. There is one trial at each design point, so $n_1 = \cdots = n_{36} = 1$ and there are $n = 36$ trials.

Problems

8.14 Consider a complete factorial experiment involving four factors, each at three levels, and two repetitions at each factor-level combination. Determine **(a)** the total number of runs; **(b)** the number of runs at each level of each factor.

8.15 Consider a complete factorial experiment involving five factors, each having four levels, and two repetitions at each factor-level combination. Determine **(a)** the total number of runs; **(b)** the number of runs at each level of each factor.

8.16 Consider a complete factorial experiment involving two repetitions at each factor-level combination and factors each at three levels: low, medium and high. Determine the maximum number of factors that can be accommodated in the experiment if the total number of runs cannot exceed 100.

8.17 Consider a complete factorial experiment involving two repetitions at each factor-level combination and four factors, with factors A and B each having two levels (low and high) and factors C and D each having three levels (low, medium, and high). Determine **(a)** the total number of runs; **(b)** the number of runs at each level of each factor.

8.18 Consider the setup of Problem 8.17, but suppose that factor D is replaced by a new factor E having two levels: E is at the low level if D is at the low or medium level, and E is at the high level if D is at the high level. Determine the number of runs at each level of E.

8.19 In John's description of the lube oil experiment, he assigned the coded values of 0 for low, 1 for medium, and 2 for high. Using $x_1, \ldots, x_4$ having these coded values, restate the criterion that determines the factor-level combinations included in the experiment.

8.20 Consider an experiment involving four factors each at three levels, which are assigned the coded values 0 (low), 1 (medium), and 2 (high), and let x_m be the coded value of the mth factor for $m = 1, 2, 3, 4$. Let the experiment consist of the one-ninth fraction of the complete factorial experiment containing the $3^{4-2} = 9$ factor-level combinations such that $x_1 + x_2 + x_3 \equiv 0 \pmod 3$ and $x_1 + 2x_2 + x_4 \equiv 0 \pmod 3$, with one run at each factor-level combination. Show the nine factor-level combinations used in the experiment in the form of a table and check that there is one run corresponding to each ordered pair of levels of each ordered pair of factors.

8.21 Consider the experiment in Problem 8.20, but let the three levels be assigned the coded values -1 (low), 0 (medium), and 1 (high). Using $x_1, \ldots, x_4$ having these coded values, restate the criterion that determines the factor-level combinations included in the experiment.

8.22 Consider the experiment in Problem 8.20 using the coded values as described in that problem, but let the experiment now consist of the one-ninth fraction of the complete factorial experiment containing the $3^{4-2} = 9$ factor-level combinations such that $x_1 + x_2 + x_3 \equiv 2 \pmod 3$ and $x_1 + 2x_2 + x_4 \equiv 1 \pmod 3$. Show the nine factor-level combinations used in the experiment in the form of a table and check that there is one run corresponding to each ordered pair of levels of each ordered pair of factors. (In a similar manner, we can construct seven additional one-ninth fractions of the complete factorial experiment each involving $3^{4-2} = 9$ runs. The combination of all nine one-ninth fractions contains all $3^4 = 81$ runs of the complete factorial experiment.)

8.23 Consider an experiment involving four factors each at two levels, which are assigned the coded values -1 (low) and 1 (high), and let x_m be the coded value of the mth factor for $m = 1,\ 2,\ 3,\ 4$. A complete factorial experiment would involve $2^4 = 16$ factor-level combinations. Let the experiment consist of the one-half fraction of the complete factorial experiment containing the $2^{4-1} = 8$ factor-level combinations such that $x_1 x_2 x_3 x_4 = 1$, with one run at each factor-level combination. Show the eight factor-level combinations used in the experiment in the form of a table and check that there is one run corresponding to each ordered triple of levels of each ordered triple of factors.

8.24 Consider the experiment in Problem 8.23, but let the two levels be assigned the coded values 0 (low) and 1 (high). Letting $x_1, \ldots, x_4$ have these coded values, show that the experiment consists of the one-half fraction of the complete factorial experiment containing the $2^{4-1} = 8$ factor-level combinations such that $x_1 + x_2 + x_3 + x_4 \equiv 0 \pmod 2$.

8.3
Input-Response and Experimental Models

Input-Response Model

Recall that in a factorial experiment the distribution of the response variable Y depends on the factor-level combination $\mathbf{x} = (x_1, \ldots, x_M)$, which varies over the experimental region $\mathcal{X} \in \mathbb{R}^M$. It is assumed that Y has finite mean and variance. Let $\mu(\mathbf{x}) = \mu(x_1, \ldots, x_M)$ denote the dependence of the mean of Y on the factor-level combination. The function $\mu(\mathbf{x})$, $\mathbf{x} \in \mathcal{X}$, is referred to as the *regression function*. We refer to this setup as the *input-response model* or as the input-response form of the *regression model*, and we sometimes refer to $\mathbf{x}$ as the *input vector*.

Let $\mathbf{x}_1, \mathbf{x}_2 \in \mathcal{X}$ be given. When the factor-level combination equals $\mathbf{x}_1$, the response variable has mean $\mu(\mathbf{x}_1)$, and when the factor-level combination equals $\mathbf{x}_2$, the response variable has mean $\mu(\mathbf{x}_2)$. Thus, a change in the factor-level combination from $\mathbf{x}_1$ to $\mathbf{x}_2$ "causes" the mean response to change from $\mu(\mathbf{x}_1)$ to $\mu(\mathbf{x}_2)$; hence, the change in factor-level combination causes the mean response to change by the amount $\mu(\mathbf{x}_2) - \mu(\mathbf{x}_1)$.

Let G be a p-dimensional linear space of functions on $\mathcal{X}$. The corresponding *linear regression model* consists of the regression model together with the assumption that the regression function is a member of G. Let $g_1, \ldots, g_p$ be a basis of G. If the regression function is in this space, it can be written as $\mu(\mathbf{x}) = \beta_1 g_1(\mathbf{x}) + \cdots +$

$\beta_p g_p(\mathbf{x})$. The uniquely determined coefficients $\beta_1, \ldots, \beta_p$ of the regression function relative to the given basis are commonly referred to as its *regression coefficients*. When G contains the constant functions, we commonly choose 1 as one of the basis functions of this space, write the basis as $1, g_1, \ldots, g_{p-1}$, and express the regression function in terms of this basis as $\mu(\mathbf{x}) = \beta_0 + \beta_1 g_1(\mathbf{x}) + \cdots + \beta_{p-1} g_{p-1}(\mathbf{x})$; here $\beta_0, \ldots, \beta_{p-1}$ are the regression coefficients.

EXAMPLE **8.1**   Suppose there is one factor and that the regression function is a linear function on $\mathbb{R}$; that is, $\mu(x) = \beta_0 + \beta_1 x$ for $x \in \mathbb{R}$. Interpret the regression coefficient β_1.

Solution   Let the factor change from level x to level $x + 1$. Then the mean response changes by the amount $\mu(x + 1) - \mu(x) = [\beta_0 + \beta_1(x + 1)] - [\beta_0 + \beta_1 x_1] = \beta_1$. Thus, β_1 is the change in the mean response caused by adding 1 unit to the level of the factor.   ∎

EXAMPLE **8.2**   Suppose there is one factor and that the regression function is a quadratic function on $\mathbb{R}$; that is, $\mu(x) = \beta_0 + \beta_1 x + \beta_2 x^2$ for $x \in \mathbb{R}$. Determine when there is a unique level $x_{\max}$ that maximizes the mean response and determine $x_{\max}$ when it is unique.

Solution   The derivative of the regression function is given by $d\mu/dx = \beta_1 + 2\beta_2 x$ for $x \in \mathbb{R}$. Suppose $\beta_2 \neq 0$. Then the equation $0 = d\mu/dx = \beta_1 + 2\beta_2 x$ has the unique solution $x = -\beta_1/(2\beta_2)$. If $\beta_2 < 0$, this solution maximizes the regression function; if $\beta_2 > 0$, the solution minimizes the mean regression function. Suppose, instead, that $\beta_2 = 0$. If $\beta_1 = 0$, then the regression function is a constant function, and hence every value of x maximizes this function; if $\beta_1 \neq 0$, then no value of x maximizes the regression function. In summary, the regression function has a unique maximum if and only if $\beta_2 < 0$, in which case the unique maximum is given by $x_{\max} = -\beta_1/(2\beta_2)$.   ∎

EXAMPLE **8.3**   Suppose there are two factors and that the regression function is a linear function on $\mathbb{R}^2$; that is, $\mu(x_1, x_2) = \beta_0 + \beta_1 x_1 + \beta_2 x_2$ for $x_1, x_2 \in \mathbb{R}$. Interpret the regression coefficients β_1 and β_2.

Solution   Let the first factor change from level x_1 to level $x_1 + 1$, with the second factor held fixed at level x_2. Then the mean response changes by the amount

$$\mu(x_1 + 1, x_2) - \mu(x_1, x_2) = [\beta_0 + \beta_1(x_1 + 1) + \beta_2 x_2] - [\beta_0 + \beta_1 x_1 + \beta_2 x_2] = \beta_1.$$

Thus, β_1 is the change in the mean response caused by adding 1 unit to the level of the first factor, with the second factor held fixed. Similarly, β_2 is the change in the mean response caused by adding 1 unit to the level of the second factor, with the first factor held fixed.   ∎

EXAMPLE **8.4**   Suppose there are two factors and the regression function is a quadratic polynomial on $\mathbb{R}^2$; that is, $\mu(x_1, x_2) = \beta_0 + \beta_1 x_1 + \beta_2 x_1^2 + \beta_3 x_2 + \beta_4 x_2^2 + \beta_5 x_1 x_2$ for $x_1, x_2 \in \mathbb{R}$. Interpret the regression coefficient β_5.

Solution   Let the first factor change from level x_1 to level $x_1 + 1$, the second factor held fixed at level x_2. Then the mean response changes by the amount

$$\mu(x_1 + 1, x_2) - \mu(x_1, x_2)$$

$$= [\beta_0 + \beta_1(x_1 + 1) + \beta_2(x_1 + 1)^2 + \beta_3 x_2 + \beta_4 x_2^2 + \beta_5(x_1 + 1)x_2]$$

$$- [\beta_0 + \beta_1 x_1 + \beta_2 x_1^2 + \beta_3 x_2 + \beta_4 x_2^2 + \beta_5 x_1 x_2]$$

$$= \beta_1 + \beta_2(2x_1 + 1) + \beta_5 x_2.$$

We refer to this amount as the effect of the change. Similarly, the effect of changing the first factor from level x_1 to level $x_1 + 1$, with the second factor held fixed at level $x_2 + 1$, is given by

$$\mu(x_1 + 1, x_2 + 1) - \mu(x_1, x_2 + 1) = \beta_1 + \beta_2(2x_1 + 1) + \beta_5(x_2 + 1).$$

Consequently,

$$[\mu(x_1 + 1, x_2 + 1) - \mu(x_1, x_2 + 1)] - [\mu(x_1 + 1, x_2) - \mu(x_1, x_2)] = \beta_5;$$

that is, β_5 is the change in the effect of a unit change in the level of the first factor caused by a unit change in the level of the second factor. Alternatively,

$$[\mu(x_1 + 1, x_2 + 1) - \mu(x_1 + 1, x_2)] - [\mu(x_1, x_2 + 1) - \mu(x_1, x_2)] = \beta_5;$$

that is, β_5 is the change in the effect of a unit change in the level of the second factor caused by a unit change in the level of the first factor. ■

The regression function is said to be *additive* if it can be written in the form $\mu(x_1, \ldots, x_M) = g_1(x_1) + \cdots + g_M(x_M)$ for some functions $g_1, \ldots, g_M$ on $\mathcal{X}_1, \ldots, \mathcal{X}_M$, respectively. Let this be the case. Then changing the mth factor from level x_{m1} to level x_{m2}, with the other factors held fixed, causes the mean response to change by the amount $g_m(x_{m2}) - g_m(x_{m1})$, which does not depend on the levels of the other factors. If the regression function is not additive, then, for some m and some choice of x_{m1} and x_{m2} in $\mathcal{X}_m$, the change in the mean response caused by changing the level of the mth factor from x_{m1} to x_{m2} with the other factors held fixed does depend on the levels of the other factors.

The regression function in Example 8.3 is additive. The regression function in Example 8.4 is nonadditive because of the presence of the term $\beta_5 x_1 x_2$, which is referred to as an *interaction* term.

Consider a regression function $\mu(x_1, x_2)$, $x_1 \in \mathcal{X}_1$ and $x_2 \in \mathcal{X}_2$, let x_{21} and x_{22} be distinct points in $\mathcal{X}_2$, and set $f_1(x_1) = \mu(x_1, x_{21})$, $x_1 \in \mathcal{X}_1$, and $f_2(x_1) = \mu(x_1, x_{22})$, $x_1 \in \mathcal{X}_1$. If the regression function is additive, then the difference between $f_1(x_1)$ and $f_2(x_1)$ is constant in x_1 [Figure 8.13(a)]. Suppose, instead, that the regression function is nonadditive. Then, for suitable choices of x_{21} and x_{22}, the difference between $f_1(x_1)$ and $f_2(x_1)$ is not constant in x_1 [Figure 8.13(b)].

The choice of the linear space G corresponds to the restrictive assumption we make about the dependence of the mean response on the factor-level combination. Making no such assumption is equivalent to choosing G to be the space of all functions on $\mathcal{X}$, which is finite-dimensional if and only if $\mathcal{X}$ is a finite set; if so, the

FIGURE   8.13

(a) The additive function $\mu(x_1, x_2) = x_1^2 + x_2^2$, $0 \le x_1 \le 1$, for $x_2 = 0$ and for $x_2 = 1$. Note that the difference between the two functions of x_1 is constant in x_1. (b) The nonadditive function $\mu(x_1, x_2) = x_1^2 + x_2^2 - x_1 x_2$, $0 \le x_1 \le 1$, for $x_2 = 0$ and for $x_2 = 1$. Note that the difference between the two functions of x_1 is not constant in x_1.

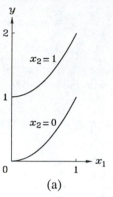 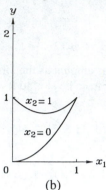

(a)   (b)

dimension p of G equals the number of points in $\mathcal{X}$. Suppose, then, that $\#(\mathcal{X}) = p$ and that G is the p-dimensional space of all functions on $\mathcal{X}$. Write the points in $\mathcal{X}$ in some order as $\mathbf{x}_1, \ldots, \mathbf{x}_p$. For $1 \le j \le p$, let $I_j(\mathbf{x}) = \mathrm{ind}(\mathbf{x} = \mathbf{x}_j)$ denote the indicator function corresponding to $\mathbf{x}_j$, which is given by $I_j(\mathbf{x}) = 1$ for $\mathbf{x} = \mathbf{x}_j$ and $I_j(\mathbf{x}) = 0$ for $\mathbf{x} \ne \mathbf{x}_j$. Then $I_1, \ldots, I_p$ form a basis of G, so the regression function can be written as $\mu(\mathbf{x}) = \beta_1 I_1(\mathbf{x}) + \cdots + \beta_p I_p(\mathbf{x})$. Alternatively, $1, I_1, \ldots, I_{p-1}$ form a basis of G, so the regression function can be written as $\mu(\mathbf{x}) = \beta_0 + \beta_1 I_1(\mathbf{x}) + \cdots + \beta_{p-1} I_{p-1}(\mathbf{x})$. (These results will be justified in Section 9.1.)

EXAMPLE **8.5**   Suppose there is one qualitative factor, which has the three levels 1, 2, and 3 (think of three brands of fertilizer), and let $I_j(x) = \mathrm{ind}(x = j)$ for $j = 1, 2, 3$.

(a) Let the regression function be written as $\mu(x) = \beta_1 I_1(x) + \beta_2 I_2(x) + \beta_3 I_3(x)$. Interpret β_1, β_2, β_3, and $\beta_2 - \beta_1$.

(b) Let the regression function be written as $\mu(x) = \beta_0 + \beta_1 I_1(x) + \beta_2 I_2(x)$. Interpret β_0, β_1, β_2, and $\beta_2 - \beta_1$.

Solution   (a) Observe that

$$\mu(1) = \beta_1 I_1(1) + \beta_2 I_2(1) + \beta_3 I_3(1) = \beta_1 \cdot 1 + \beta_2 \cdot 0 + \beta_3 \cdot 0 = \beta_1.$$

Thus, β_1 is the mean response when the factor is at level 1. Similarly, β_2 is the mean response when the factor is at level 2, and β_3 is the mean response when it is at level 3. Now $\mu(2) - \mu(1) = \beta_2 - \beta_1$. Thus, $\beta_2 - \beta_1$ is the change in the mean response caused by changing the level of the factor from 1 to 2.

(b) Here

$$\mu(3) = \beta_0 + \beta_1 I_1(3) + \beta_2 I_2(3) = \beta_0 + \beta_1 \cdot 0 + \beta_2 \cdot 0 = \beta_0,$$
$$\mu(1) = \beta_0 + \beta_1 I_1(1) + \beta_2 I_2(1) = \beta_0 + \beta_1 \cdot 1 + \beta_2 \cdot 0 = \beta_0 + \beta_1$$

and, similarly, $\mu(2) = \beta_0 + \beta_2$. Consequently, $\beta_1 = \mu(1) - \mu(3)$ and $\beta_2 = \mu(2) - \mu(3)$. Thus, β_0 is the mean response when the factor is at level 3, β_1 is the change in the mean response caused by changing the level of the factor from 3 to 1, and β_2 is the change in the mean response caused by changing its level from 3 to 2. Also, $\beta_2 - \beta_1 = \mu(2) - \mu(1)$; that is, $\beta_2 - \beta_1$ is the change in the mean response caused by changing the level of the factor from 1 to 2. ∎

Let $\sigma^2(\mathbf{x})$ denote the dependence of the variance of Y on $\mathbf{x}$. The *homoskedastic regression model* consists of the regression model together with the assumption that the variance of the response variable does not depend on the factor-level combination, so that $\sigma^2(\mathbf{x}) = \sigma^2$ for $\mathbf{x} \in \mathcal{X}$; a model that is not homoskedastic is said to be *heteroskedastic*. The *normal regression model* consists of the homoskedastic regression model together with the assumption that the response variable is normally distributed; accordingly, when the factor-level combination equals $\mathbf{x}$, the response variable Y is normally distributed with mean $\mu(\mathbf{x})$ and variance $\sigma^2 > 0$, and hence $Y = \mu(\mathbf{x}) + W$, where W is normally distributed with mean 0 and variance σ^2.

The *homoskedastic linear regression model* consists of the linear regression model together with the assumption that the variance of the response variable does not depend on the factor-level combination. The *normal linear regression model* consists of the homoskedastic linear regression model together with the assumption that the response variable is normally distributed; alternatively, it consists of the normal regression model together with the assumption that the regression function is a member of the specified linear space G.

Experimental Model

Recall that the experimental setup involves n trials (or runs), on the ith trial of which the factor-level combination equals x_i and the random response variable Y_i is observed. It is assumed that the distribution of Y_i coincides with the distribution of the random variable Y in the input-response model when the input vector equals $\mathbf{x}_i$. Accordingly, Y_i has mean $\mu(\mathbf{x}_i)$ and variance $\sigma^2(\mathbf{x}_i)$. It is also assumed that the random variables Y_i, $1 \le i \le n$, are independent. We refer to this setup as the *experimental model* or as the experimental version of the regression model.

According to the notation and terminology introduced in Section 8.2, there are d distinct factor-level combinations (design points) $\mathbf{x}'_1, \ldots, \mathbf{x}'_d$ among $\mathbf{x}_1, \ldots, \mathbf{x}_n$, where $1 \le d \le n$. The design set is given by $\mathcal{X}' = \{\mathbf{x}'_1, \ldots, \mathbf{x}'_d\}$. Let $1 \le k \le d$. Then $\mathcal{I}_k = \{i : \mathbf{x}_i = \mathbf{x}'_k\}$ is the collection of trials on which the factor-level combination equals $\mathbf{x}'_k$. There are $n_k = \#(\mathcal{I}_k)$ repetitions at the kth design point $\mathbf{x}'_k$; the corresponding n_k response variables Y_i, $i \in \mathcal{I}_k$, are independent, and each is distributed as the response variable Y in the input-response model when the factor-level combination equals $\mathbf{x}'_k$. In particular, they have common mean $\mu_k = \mu(\mathbf{x}'_k)$ and common variance $\sigma_k^2 = \sigma^2(\mathbf{x}'_k)$. Their sample mean

$$\bar{Y}_k = \frac{1}{n_k} \sum_{i \in \mathcal{I}_k} Y_i$$

has mean μ_k and variance σ_k^2/n_k. The within sum of squares

$$\text{WSS}_k = \sum_{i \in \mathcal{I}_k}(Y_i - \bar{Y}_k)^2$$

corresponding to the responses at the kth design point is a measure of the variation of these responses. It follows from Equation (53) in Section 2.4 that $E(\text{WSS}_k) = (n_k - 1)\sigma_k^2$. Observe that if $n_k = 1$, then $\text{WSS}_k = 0$.

The sample means $\bar{Y}_1, \ldots, \bar{Y}_d$ are independent random variables. Let $\bar{Y}$ denote the sample mean of all n response variables, which is given by

$$\bar{Y} = \frac{1}{n}\sum_i Y_i = \frac{1}{n}\sum_k n_k \bar{Y}_k.$$

Then $\bar{Y}$ has mean

$$\frac{1}{n}\sum_i \mu(x_i) = \frac{1}{n}\sum_k n_k \mu_k$$

and variance

$$\frac{1}{n^2}\sum_i \sigma^2(x_i) = \frac{1}{n^2}\sum_k n_k \sigma_k^2.$$

The random variables $\text{WSS}_1, \ldots, \text{WSS}_d$ are independent.

The within sum of squares

$$\text{WSS} = \sum_k \sum_{i \in \mathcal{I}_k}(Y_i - \bar{Y}_k)^2 = \sum_k \text{WSS}_k$$

is an overall measure of the variation of the response values on trials having a common factor-level combination. If $\mathbf{x}_1, \ldots, \mathbf{x}_n$ are distinct, then $\text{WSS} = 0$.

EXAMPLE 8.6   In the context of the pilot plant experiment described in Section 8.2, determine

(a) WSS_k, $1 \le k \le d$;   (b) WSS.

Solution   (a) Here $\bar{Y}_1 = (59 + 61)/2 = 60$, so $\text{WSS}_1 = (59 - 60)^2 + (61 - 60)^2 = 2$. Similarly, $\text{WSS}_2 = 8$, $\text{WSS}_3 = 32$, $\text{WSS}_4 = 2$, $\text{WSS}_5 = 8$, $\text{WSS} = 8$, $\text{WSS}_7 = 2$, and $\text{WSS}_8 = 2$.

(b) The within sum of squares is given by

$$\text{WSS} = 2 + 8 + 32 + 2 + 8 + 8 + 2 + 2 = 64. \quad \blacksquare$$

Under the experimental version of the homoskedastic regression model, Y_i has variance σ^2 for $1 \le i \le n$, $\bar{Y}_k$ has variance σ^2/n_k for $1 \le k \le d$, and $\bar{Y}$ has variance σ^2/n. Moreover, $E(\text{WSS}_k) = \sigma^2(n_k - 1)$ for $1 \le k \le d$, and $E(\text{WSS}) = \sigma^2(n - d)$. Suppose $n > d$ and set $S^2 = \text{WSS}/(n - d)$, where $S \ge 0$. Then S^2 is an unbiased estimate of σ^2. Under the experimental version of the normal regression model, Y_i is normally distributed with mean $\mu(\mathbf{x}_i)$ and variance σ^2 for $1 \le i \le n$, $\bar{Y}_k$ is normally distributed with mean μ_k and variance σ^2/n_k for $1 \le k \le d$, and $\bar{Y}$ is normally distributed with variance σ^2/n. It follows from Theorem 7.7 in Section 7.3 that $\bar{Y}_k$ and WSS_k are independent and that if $n_k > 1$, then WSS_k/σ^2 has the chi-square distribution with $n_k - 1$ degrees of freedom. Thus, we have the following results.

THEOREM 8.1   Under the assumptions of the normal regression model, $\bar{Y}_1, \ldots, \bar{Y}_d$ and WSS are independent random variables, and if $n > d$, then WSS/σ^2 has the chi-square distribution with $n - d$ degrees of freedom.

Proof   Think of Y_i, $i \in \mathcal{I}_k$, as forming the kth random sample for $1 \le k \le d$. Then these random variables satisfy the assumptions of the normal d-sample model, so the desired conclusion follows from Theorem 7.8 in Section 7.3. ∎

Consider the experimental version of the linear regression model corresponding to the linear space G, let $g_1, \ldots, g_p$ be a basis of G, and let $\beta_1, \ldots, \beta_p$ be the regression coefficients relative to this basis. Then $\mu(\mathbf{x}) = \beta_1 g_1(\mathbf{x}) + \cdots + \beta_p g_p(\mathbf{x})$, so the means of $Y_1, \ldots, Y_n$ are given by $EY_i = \mu(\mathbf{x}_i) = \beta_1 g_1(\mathbf{x}_i) + \cdots + \beta_p g_p(\mathbf{x}_i)$ for $1 \le i \le n$, and the means of $\bar{Y}_1, \ldots, \bar{Y}_d$ are given by $E\bar{Y}_k = \mu_k = \beta_1 g_1(\mathbf{x}'_k) + \cdots + \beta_p g_p(\mathbf{x}'_k)$ for $1 \le k \le d$.

According to the experimental version of the normal linear regression model, the random variables, $Y_1, \ldots, Y_n$ are independent and normally distributed with variance σ^2 and Y_i has mean $\mu(\mathbf{x}_i)$ for $1 \le i \le n$, where the regression function is a member of the linear space G. Alternatively, $Y_i = \mu(\mathbf{x}_i) + W_i$ for $1 \le i \le n$, where $W_1, \ldots, W_n$ are independent random variables, each of which is normally distributed with mean 0 and variance σ^2.

Suppose, for example, there is one factor and the regression function is a linear function on $\mathbb{R}$; that is, it has the form $\mu(x) = \beta_0 + \beta_1 x$, $x \in \mathbb{R}$. According to the corresponding normal linear regression model, the n responses are observed values of $Y_i = \beta_0 + \beta_1 x_i + W_i$, $1 \le i \le n$, where $W_1, \ldots, W_n$ are independent random variables, each having mean 0 and variance σ^2.

One-Sample and Multisample Models

The d-sample model can be viewed as a linear regression model involving a single factor with levels $1, \ldots, d$. To this end, write the n random variables Y_{ki}, $1 \le k \le d$ and $1 \le i \le n_k$, in some order as $Y_1, \ldots, Y_n$; set $x_i = k$ if Y_i is from the kth sample; and set $x'_k = k$ for $1 \le k \le d$. Then $\mathcal{I}_k = \{i : x_i = k\}$ is the set of trials on which the response variable is from the kth sample, and the values of n_k, $\bar{Y}_k$, and WSS_k in the experimental framework coincide with their values in the d-sample framework; in both frameworks, $n = n_1 + \cdots + n_d$. Here $\mathcal{X} = \mathcal{X}' = \{1, \ldots, d\}$, the factor is qualitative, and G is the space of all functions on $\{1, \ldots, d\}$. This space is d-dimensional, and it has the basis $I_1, \ldots, I_d$, where $I_k(x) = \text{ind}(x = k)$ for $1 \le k \le d$. The regression function can be written in terms of this basis as $\mu(x) = \mu_1 I_1(x) + \cdots + \mu_d I_d(x)$, where μ_k is the common mean of the random variables in the kth sample for $1 \le k \le d$. Thus, $\mu_1, \ldots, \mu_d$ are the coefficients of the regression function relative to the indicated basis.

Consider, in particular, the one-sample model. Here $X = X' = \{1\}$, G is the space of constant functions, 1 is the basis of G, and the regression function is given by $\mu(\cdot) = \mu$.

Consider, next, the two-sample model with $n_1 = 2$ and $n_2 = 3$. To view this as a linear regression model, we set $Y_1 = Y_{11}$, $Y_2 = Y_{12}$, $Y_3 = Y_{21}$, $Y_4 = Y_{22}$, and $Y_5 = Y_{23}$. Also, we set $x_1 = x_2 = 1$ and $x_3 = x_4 = x_5 = 2$ and we set $x_1' = 1$ and $x_2' = 2$. Then $\mathcal{I}_1 = \{1, 2\}$ and $\mathcal{I}_2 = \{3, 4, 5\}$. Here $X = X' = \{1, 2\}$, the factor is qualitative, and G is the space of all functions on $\{1, 2\}$. This space is two-dimensional, and it has the basis I_1, I_2, where $I_1(x) = \mathrm{ind}(x = 1)$ and $I_2(x) = \mathrm{ind}(x = 2)$. The regression function can be written in terms of this basis as $\mu(x) = \mu_1 I_1(x) + \mu_2 I_2(x)$, where μ_1 is the common mean of Y_{11} and Y_{12} and μ_2 is the common mean of Y_{21}, Y_{22}, and Y_{23}. The experimental data is conveniently shown in the form of a table:

x	1	1	2	2	2
Y	Y_{11}	Y_{12}	Y_{21}	Y_{22}	Y_{23}

Random Inputs

So far we have been thinking of the input as being nonrandom. Suppose, instead, that the input is an X-valued random vector $\mathbf{X}$. Here we think of $\mu(\mathbf{x})$ and $\sigma^2(\mathbf{x})$ as the mean and variance of the conditional distribution of the response variable Y given that $\mathbf{X} = \mathbf{x}$. Thus, $\mu(\mathbf{x})$, $\mathbf{x} \in X$, is the regression function in the sense of Section 6.4.

Suppose the conditional distribution of Y given that $\mathbf{X} = \mathbf{x}$ is normal with mean $\mu(\mathbf{x})$ and variance $\sigma^2(\mathbf{x}) = \sigma^2$ for $\mathbf{x} \in X$. Then $Y = \mu(\mathbf{X}) + W$, where $\mathbf{X}$ and W are independent and W is normally distributed with mean zero and variance σ^2 (see Theorem 6.2 in Section 6.4). Upon adding the assumption that the regression function is a member of the linear space G, we get the conditional form of the normal linear regression model.

Suppose now that the conditional distribution of Y given that $\mathbf{X} = \mathbf{x}$ is normal with mean $\mu(\mathbf{x})$ and variance σ^2. Then, for $\mathbf{x}_1, \ldots, \mathbf{x}_n \in X$, the conditional joint distribution of $Y_1, \ldots, Y_n$ given $\mathbf{X}_1 = \mathbf{x}_1, \ldots, \mathbf{X}_n = \mathbf{x}_n$ is that of n independent normal random variables having means $\mu(\mathbf{x}_1), \ldots, \mu(\mathbf{x}_n)$, respectively, and common variance σ^2. Consequently, $Y_i = \mu(\mathbf{X}_i) + W_i$ for $1 \leq i \leq n$, where $(\mathbf{X}_1, \ldots, \mathbf{X}_n)$, $W_1, \ldots, W_n$ are independent and W_i is normally distributed with mean 0 and variance σ^2 for $1 \leq i \leq n$. Upon adding the assumption that the regression function is a member of the linear space G, we get the normal linear regression model with random inputs. Except for causal implications, the results for the normal linear regression model with deterministic inputs that are obtained in this book have analogs for random inputs.

In particular, let $(\mathbf{X}_1, Y_1), \ldots, (\mathbf{X}_n, Y_n)$ be a random sample of size n from the joint distribution of $\mathbf{X}$ and Y. Then, for $\mathbf{x}_1, \ldots, \mathbf{x}_n \in X$, the conditional distribution of $Y_1, \ldots, Y_n$ given $\mathbf{X}_1 = \mathbf{x}_1, \ldots, \mathbf{X}_n = \mathbf{x}_n$ is that of n independent random variables having means $\mu(\mathbf{x}_1), \ldots, \mu(\mathbf{x}_n)$, respectively, and variances $\sigma^2(\mathbf{x}_1), \ldots, \sigma^2(\mathbf{x}_n)$, respectively. If $P(\mathbf{X} = \mathbf{x}) = 0$ for $\mathbf{x} \in X$ (which holds, in particular, whenever $\mathbf{X}$ has a density function), then $\mathbf{X}_1, \ldots, \mathbf{X}_n$ are distinct with probability 1.

Suppose, for example, X and Y have a bivariate normal distribution, X has mean μ_1 and variance σ_1^2, Y has mean μ_2 and variance σ_2^2, and ρ is the correlation between X and Y. Then the conditional distribution of Y given that $X = x$ is normal, with mean $\mu(x) = \mu_2 + \rho \frac{\sigma_2}{\sigma_1}(x - \mu_1)$ and variance $\sigma_2^2(1 - \rho^2)$. Thus,

$$Y = \mu(X) + W = \mu_2 + \rho \frac{\sigma_2}{\sigma_1}(X - \mu_1) + W,$$

where W is normally distributed with mean 0 and variance $\sigma_2^2(1 - \rho^2)$. Let

$$(X_1, Y_1), \ldots, (X_n, Y_n)$$

be a random sample of size n from the joint distribution of X and Y. Then

$$Y_i = \mu(X_i) + W_i = \mu_2 + \rho \frac{\sigma_2}{\sigma_1}(X_i - \mu_1) + W_i, \qquad 1 \le i \le n,$$

where $X_1, \ldots, X_n, W_1, \ldots, W_n$ are independent and normally distributed and X_i has mean μ_1 and variance σ_1^2 and W_i has mean 0 and variance $\sigma_2^2(1 - \rho^2)$ for $1 \le i \le n$.

Method of Least Squares

Consider the experimental version of the linear regression model associated with the linear space G. Given a function g, the quantity $Y_i - g(\mathbf{x}_i)$ is referred to as the *residual* of Y_i about g for $1 \le i \le n$, and the corresponding *residual sum of squares* (RSS) is defined by $\mathrm{RSS}(g) = \sum_i [\, Y_i - g(x_i)\,]^2$. The *method of least squares* is to estimate the regression function by the function $\hat{\mu}(\cdot)$ in G that has the minimum residual sum of squares—that is, such that $\mathrm{RSS}(\hat{\mu}(\cdot)) = \min_{g \in G} \mathrm{RSS}(g)$.

A convenient formula for the residual sum of squares, which will be used in Section 10.1 to obtain the least-squares estimate as a suitable orthogonal projection, will now be derived. To this end, observe first that $\sum_{i \in \mathcal{I}_k}(Y_i - \bar{Y}_k) = \sum_{i \in \mathcal{I}_k} Y_i - n_k \bar{Y}_k$ for $1 \le k \le d$ and hence that

$$\sum_{i \in \mathcal{I}_k}(Y_i - \bar{Y}_k) = 0, \qquad 1 \le k \le d. \tag{1}$$

Observe next that

$$\mathrm{RSS}(g) = \sum_i [\, Y_i - g(\mathbf{x}_i)\,]^2$$

$$= \sum_k \sum_{i \in \mathcal{I}_k} [\, Y_i - g(\mathbf{x}_i)\,]^2$$

$$= \sum_k \sum_{i \in \mathcal{I}_k} [\, Y_i - g(\mathbf{x}'_k)\,]^2$$

$$= \sum_k \sum_{i \in \mathcal{I}_k} [\, \bar{Y}_k - g(\mathbf{x}'_k) + Y_i - \bar{Y}_k\,]^2$$

and that

$$[\, \bar{Y}_k - g(\mathbf{x}'_k) + Y_i - \bar{Y}_k\,]^2 = [\, \bar{Y}_k - g(\mathbf{x}'_k)\,]^2 + (Y_i - \bar{Y}_k)^2 + 2[\, \bar{Y}_k - g(\mathbf{x}'_k)\,](Y_i - \bar{Y}_k).$$

Thus,

$$\text{RSS}(g) = \sum_k n_k [\, \bar{Y}_k - g(\mathbf{x}'_k)\,]^2 + \sum_k \sum_{i \in \mathcal{J}_k} (Y_i - \bar{Y}_k)^2$$

$$+ 2 \sum_{k=1}^{d} [\, \bar{Y}_k - g(\mathbf{x}'_k)\,] \sum_{i \in \mathcal{J}_k} (Y_i - \bar{Y}_k). \qquad (2)$$

The first term on the right side of (2) is referred to as the *lack-of-fit sum of squares* (LSS) corresponding to g and denoted by $\text{LSS}(g)$; thus,

$$\text{LSS}(g) = \sum_k n_k [\, \bar{Y}_k - g(\mathbf{x}'_k)\,]^2. \qquad (3)$$

This quantity is a measure of the discrepancy between the sample means of the responses at the various design points and the corresponding values of g. The second quantity on the right side of (2) is the within sum of squares WSS, which does not depend on g. It follows from (1) that the third term on the right side of (2) equals zero. Consequently, we conclude from (2) that

$$\text{RSS}(g) = \text{LSS}(g) + \text{WSS}. \qquad (4)$$

Therefore, the function $\hat{\mu}(\cdot)$ in G is the least-squares estimate in G of the regression function $\mu(\cdot)$ if and only if it has the minimum lack-of-fit sum of squares among all functions in G—that is, if and only if $\text{LSS}(\hat{\mu}(\cdot)) = \min_{g \in G} \text{LSS}(g)$.

Let $\bar{Y}(\cdot)$ be the (random) function on the design set defined by $\bar{Y}(\mathbf{x}'_k) = \bar{Y}_k$ for $1 \le k \le d$ (which we can refer to orally as the "Y-bar" function). Then $\bar{Y}(\mathbf{x}_i) = \bar{Y}_k$ for $1 \le k \le d$ and $i \in \mathcal{J}_k$; that is, $\bar{Y}(\mathbf{x}_i)$ is the sample mean of the responses on the trials having the same factor-level combination as that on the ith trial. If $\mathbf{x}_1, \dots, \mathbf{x}_n$ are distinct, then $\bar{Y}(\mathbf{x}_i) = Y_i$ for $1 \le i \le n$. Under the assumptions of the experimental model,

$$E[\,\bar{Y}(\mathbf{x}'_k)\,] = \mu_k = \mu(\mathbf{x}'_k), \qquad 1 \le k \le d,$$

and

$$E[\,\bar{Y}(\mathbf{x}_i)\,] = \mu(\mathbf{x}_i), \qquad 1 \le i \le n.$$

The within and lack-of-fit sums of squares can be expressed in terms of $\bar{Y}(\cdot)$. Indeed, since

$$\text{WSS} = \sum_k \sum_{i \in \mathcal{J}_k} (Y_i - \bar{Y}_k)^2 = \sum_k \sum_{i \in \mathcal{J}_k} [\, Y_i - \bar{Y}(\mathbf{x}_i)\,]^2,$$

we see that

$$\text{WSS} = \sum_i [\, Y_i - \bar{Y}(\mathbf{x}_i)\,]^2. \qquad (5)$$

Moreover, (3) can be written as

$$\text{LSS}(g) = \sum_k n_k [\, \bar{Y}(\mathbf{x}'_k) - g(\mathbf{x}'_k)\,]^2. \qquad (6)$$

In Sections 9.4 through 9.6, inner products and orthogonal projections will be studied and used to develop the method of least squares in the context of linear analysis. Statistical applications of the results obtained in these sections will be given in Chapters 10, 11, and 13.

Problems

8.25 Suppose the regression function is given by $\mu(x) = \beta_0 + \beta_1 x + \beta_2 x^2$ for $x \in \mathbb{R}$. Express β_0, β_1, and β_2 in terms of $\mu(-1)$, $\mu(0)$, and $\mu(1)$.

8.26 Let the regression function be given by

$$\mu(x) = \frac{x}{1+x^2}, \qquad x \in \mathbb{R}.$$

(a) Determine the levels $x_{\min}$ and $x_{\max}$ of the factor that minimize and maximize the mean of the response variable. **(b)** Determine the effect of changing the level of the factor from $x_{\min}$ to $x_{\max}$.

8.27 Let the regression function be a quadratic polynomial in x_1 and x_2 as in Example 8.4, and let $c_1, c_2 \in \mathbb{R}$. Determine **(a)** $[\mu(x_1 + c_1, x_2 + c_2) - \mu(x_1 + c_1, x_2)] - [\mu(x_1, x_2 + c_2) - \mu(x_1, x_2)]$; **(b)** $[\mu(x_1 + c_1, x_2 + c_2) - \mu(x_1, x_2 + c_2)] - [\mu(x_1 + c_1, x_2) - \mu(x_1, x_2)]$; **(c)** $\frac{\partial^2 \mu(x_1, x_2)}{\partial x_1 \partial x_2}$.

8.28 Let the regression function be a quadratic polynomial in x_1 and x_2 as in Example 8.4. **(a)** Determine a necessary and sufficient condition on the regression coefficients in order that there be a unique choice of x_1 and x_2 that minimizes the regression function. **(b)** Under the condition in part (a), determine the choice of x_1 and x_2 that minimizes the regression function.

8.29 Let the regression function be a quadratic polynomial in x, so that $\mu(x) = \beta_0 + \beta_1 x + \beta_2 x^2$. Express each of the following parameters in terms of the regression coefficients: **(a)** $\mu(2) - \mu(1)$; **(b)** $\frac{d\mu}{dx}(\frac{3}{2})$; **(c)** $\int_1^2 \mu(x)\,dx$; **(d)** $\int_1^2 \left(\frac{d\mu}{dx}\right)^2 dx$; **(e)** $\frac{\mu(x-1) - 2\mu(x) + \mu(x+1)}{2}$.

8.30 Suppose there are three factors and that the regression function is a linear function on $\mathbb{R}^3$; that is,

$$\mu(x_1, x_2, x_3) = \beta_0 + \beta_1 x_1 + \beta_2 x_2 + \beta_3 x_3, \qquad x_1, x_2, x_3 \in \mathbb{R}.$$

Interpret the regression coefficients β_1, β_2, and β_3.

8.31 Let the regression function be a cubic polynomial on $\mathbb{R}$, which we write as

$$\mu(x) = \beta_0 + \beta_1 x + \beta_2 x^2 + \beta_3 x^3.$$

(a) Determine the change in the mean response when the level of the factor is increased from 1 to 2. **(b)** Defining the new variables x_1, x_2, and x_3 by $x_1 = x$, $x_2 = x^2$, and $x_3 = x^3$, we rewrite the regression function as

$$\mu(x_1, x_2, x_3) = \beta_0 + \beta_1 x_1 + \beta_2 x_2 + \beta_3 x_3.$$

In light of the answer to Problem 8.30, is it now reasonable to interpret β_1 as the change in the mean response when we add 1 unit to x_1, with x_2 and x_3 held fixed?

8.32   Show that $Y_1, \ldots, Y_n$ satisfy the assumptions of the normal linear model corresponding to the space of constant functions if and only if they satisfy the assumptions of the normal one-sample model.

8.33   Consider the normal three-sample model viewed as a normal linear regression model with $\mathcal{X} = \{1, 2, 3\}$ and G the three-dimensional space of all functions on $\mathcal{X}$. Then the regression function can be written as $\mu(x) = \mu_1 I_1(x) + \mu_2 I_2(x) + \mu_3 I_3(x)$, where $I_k(x) = \mathrm{ind}(x = k)$ for $k = 1, 2, 3$. Alternatively, it can be written as $\mu(x) = \beta_0 + \beta_1 I_1(x) + \beta_2 I_2(x)$. Express **(a)** β_0, β_1, and β_2 in terms of μ_1, μ_2, and μ_3; **(b)** μ_1, μ_2, and μ_3 in terms of β_0, β_1, and β_2.

8.34   Suppose there are two factors: The level of the first factor ranges over $\{1, 2, 3\}$, and the level of the second factor ranges over $\mathbb{R}$. Let the regression function have the form $\mu(x_1, x_2) = \beta_0 + \beta_1 I_1(x_1) + \beta_2 I_2(x_1) + \beta_3 x_2$, where $I_1(x_1) = \mathrm{ind}(x_1 = 1)$ and $I_2(x_1) = \mathrm{ind}(x_1 = 2)$. Interpret β_1, β_2, $\beta_2 - \beta_1$, and β_3.

8.35   Describe explicitly how the normal three-sample model with $n_1 = 2$, $n_2 = 3$, and $n_3 = 2$ can be viewed as a normal linear regression model.

8.36   Suppose the assumptions of the homoskedastic regression model are satisfied. Determine the variance–covariance matrix of $\sqrt{n_1}\,\bar{Y}_1, \ldots, \sqrt{n_d}\,\bar{Y}_d$.

8.37   **(a)** Suppose there are two trials at each design point and let the responses at the kth design point be denoted by Y_{k1} and Y_{k2}. Verify that

$$\mathrm{WSS} = \frac{1}{2} \sum_k (Y_{k1} - Y_{k2})^2.$$

(b) Consider the pilot plant data. Determine WSS by using the formula in part (a).

8.38   Determine the within sum of squares for the following data:

x	−2	−2	−2	−1	−1	0	1	1	2	2	2
Y	14	10	15	11	13	15	13	17	20	19	18

9

Linear Analysis

9.1
Linear Spaces

In this section, we will define and discuss the properties of linear spaces and their bases. This material is mostly covered in standard elementary linear algebra texts, but the emphasis given here on spaces of functions rather than spaces of ordered n-tuples or abstract entities is needed for our treatment of statistics that resumes in Chapter 10.

Let $\mathcal{X}$ be a nonempty subset of $\mathbb{R}^M$ for some positive integer M. When we refer to a function g on $\mathcal{X}$, unless otherwise indicated, we mean that g is a real-valued function. Let g_1 and g_2 be functions on $\mathcal{X}$. By $g_1 = g_2$, we mean that $g_1(\mathbf{x}) = g_2(\mathbf{x})$ for all $\mathbf{x} \in \mathcal{X}$; by $g_1 \neq g_2$, we mean that there is at least one $\mathbf{x} \in \mathcal{X}$ such that $g_1(\mathbf{x}) \neq g_2(\mathbf{x})$.

The *zero function* on $\mathcal{X}$ is the function g given by $g(\mathbf{x}) = 0$ for $\mathbf{x} \in \mathcal{X}$. Every other function on $\mathcal{X}$ is said to be a *nonzero function*. Thus, a function g (on $\mathcal{X}$) is a nonzero function if and only if there is at least one $\mathbf{x} \in \mathcal{X}$ such that $g(\mathbf{x}) \neq 0$. Given $c \in \mathbb{R}$, we identify c with the *constant function* g defined by $g(\mathbf{x}) = c$ for $\mathbf{x} \in \mathcal{X}$. Thus, 0 is the zero function.

Given a function g and a real number b, the function bg is defined by

$$(bg)(\mathbf{x}) = bg(\mathbf{x}), \qquad \mathbf{x} \in \mathcal{X}.$$

Given functions g_1 and g_2, their sum $g_1 + g_2$ is defined by

$$(g_1 + g_2)(\mathbf{x}) = g_1(\mathbf{x}) + g_2(\mathbf{x}), \qquad \mathbf{x} \in \mathcal{X}.$$

Let q be a positive integer and let $g_1, \ldots, g_q$ be functions on $\mathcal{X}$. A function of the form $b_1 g_1 + \cdots + b_q g_q$ with $b_1, \ldots, b_q \in \mathbb{R}$ is said to be a *linear combination* of $g_1, \ldots, g_q$. The functions $g_1, \ldots, g_q$ are said to be *linearly dependent* if there exist $b_1, \ldots, b_q \in \mathbb{R}$, not all zero, such that $b_1 g_1 + \cdots + b_q g_q$ is the zero function; otherwise, $g_1, \ldots, g_q$ are said to be *linearly independent*. If $g_1, \ldots, g_q$ are linearly independent when viewed as functions on some subset of $\mathcal{X}$, then they are linearly independent when viewed as functions on $\mathcal{X}$. Suppose $g_1, \ldots, g_q$ are linearly independent. Then each of these functions is a nonzero function. Moreover, $g_1, \ldots, g_r$ are linearly independent for $1 \leq r \leq q$.

When g_1 is constant, as in Example 9.1, it is conventional to relabel the functions $g_1, \ldots, g_q$ and the coefficients $b_1, \ldots, b_q$ as $g_0, \ldots, g_{q-1}$ and $b_0, \ldots, b_{q-1}$, respectively.

EXAMPLE **9.1**   Show that $1, x, \ldots, x^{q-1}$ are linearly independent on $\mathcal{X} = (a, c)$ for q a positive integer.

Solution   Given $b_0, \ldots, b_{q-1} \in \mathbb{R}$, let g be the polynomial given by

$$g(x) = b_0 + b_1 x + \ldots + b_{q-1} x^{q-1}, \qquad x \in \mathcal{X}.$$

Suppose g is the zero function. Choose $x_0 \in \mathcal{X}$. Then

$$0 = \frac{d^{q-1}}{dx^{q-1}} g(x) \bigg|_{x=x_0} = (q-1)! \, b_{q-1}$$

and hence $b_{q-1} = 0$. If $q \geq 2$, we now conclude by evaluating the $(q-2)$th derivative of g at x_0 that $b_{q-2} = 0$. In a similar manner, we get that $b_0 = \cdots = b_{q-1} = 0$. Therefore $1, x \ldots, x^{q-1}$ are linearly independent on $\mathcal{X}$.   ∎

EXAMPLE **9.2**   Show that $1, x_1, \ldots, x_M$ are linearly independent on

$$\mathcal{X} = (a_1, c_1) \times \cdots \times (a_M, c_M).$$

Solution   Given $b_0, \ldots, b_M \in \mathbb{R}$, let g be defined by

$$g(x_1, \ldots, x_M) = b_0 + b_1 x_1 + \cdots + b_M x_M$$

on $\mathcal{X}$. Suppose g is the zero function. Then $0 = \partial g / \partial x_m = b_m$ on $\mathcal{X}$ for $1 \leq m \leq M$, so $b_1 = \cdots = b_M = 0$ and hence $b_0 = 0$. Therefore, the indicated functions are linearly independent.   ∎

Examples 9.1 and 9.2 have a common generalization. Consider *monomials* such as $1, x_1, x_1^2, x_1 x_2, x_1^3, x_1^2 x_2, x_1 x_2 x_3$ on $\mathbb{R}^3$. The *degree* of any such monomial is the sum of the powers of x_1, x_2, and x_3. Thus, 1 has degree 0, x_1 has degree 1, x_1^2 and $x_1 x_2$ have degree 2, and x_1^3, $x_1^2 x_2$ and $x_1 x_2 x_3$ have degree 3. Similarly, we can define monomials and their degrees on $\mathbb{R}^M$. A *polynomial* of degree s on $\mathbb{R}^M$ is a linear combination of monomials of degree at most s on $\mathbb{R}^M$.

THEOREM 9.1   Let $X = (a_1, c_1) \times \cdots \times (a_M, c_M)$. Then distinct monomials on X are linearly independent.

Proof   According to Example 9.1, the desired result is valid when $M = 1$. Suppose next that $M = 2$. Consider, for example, the monomials $1, x_1, x_2, x_1^2, x_1 x_2, x_1^2 x_2$ and consider the linear combination

$$b_0 + b_1 x_1 + b_2 x_2 + b_3 x_1^2 + b_4 x_1 x_2 + b_5 x_1^2 x_2$$
$$= (b_0 + b_2 x_2) + (b_1 + b_4 x_2)x_1 + (b_3 + b_5 x_2)x_1^2.$$

Suppose this function equals zero for $a_1 < x_1 < c_1$ and $a_2 < x_2 < c_2$. Fix $x_2 \in (a_2, c_2)$. Then

$$(b_0 + b_2 x_2) + (b_1 + b_4 x_2)x_1 + (b_3 + b_5 x_2)x_1^2 = 0, \qquad a_1 < x_1 < c_1.$$

Thus, by applying Example 9.1 to the monomials 1, x_1 and x_1^2, we see that $b_0 + b_2 x_2 = 0$, $b_1 + b_4 x_2 = 0$, and $b_3 + b_5 x_2 = 0$. Since x_2 is an arbitrary member of (a_2, c_2), we get that

$$b_0 + b_2 x_2 = 0, \quad b_1 + b_4 x_2 = 0, \quad \text{and} \quad b_3 + b_5 x_2 = 0, \qquad a_2 < x_2 < c_2.$$

Thus, by three applications of Example 9.2 to the monomials 1 and x_2, we conclude that $b_0 = \cdots = b_5 = 0$. Consequently, the monomials $1, x_1, x_2, x_1^2, x_1 x_2, x_1^2 x_2$ are linearly independent. The general proof for $M = 2$ follows in a similar manner, and the proof for general M follows by induction.   ∎

Given an initial collection of functions on X, their *span* is the collection G of all linear combinations of these functions—that is, the collection of all functions $b_1 g_1 + \cdots + b_q g_q$ as q ranges over the positive integers, $b_1, \ldots, b_q$ range over $\mathbb{R}$, and $g_1, \ldots, g_q$ range over the functions in initial collection. The functions in the initial collection are said to span G.

In particular, given functions $g_1, \ldots, g_q$ on X, their span G is the collection of linear combinations of these functions, and these functions are said to span G. If $g_1, \ldots, g_q$ are linearly independent and g_{q+1} is not in their span, then $g_1, \ldots, g_{q+1}$ are linearly independent.

Let G be the span of a collection of functions on X. Suppose that $g_1, \ldots, g_q$ are functions in the collection and that every function in the collection is a linear combination of $g_1, \ldots, g_q$. Then G is the span of $g_1, \ldots, g_q$.

A collection G of functions is said to be a *linear space* (or *vector space*) if G satisfies the following two properties: $bg \in G$ for $b \in \mathbb{R}$ and $g \in G$; and $g_1 + g_2 \in G$

for $g_1, g_2 \in G$. Let G be a linear space. Then G contains the zero function, and if $g_1, \ldots, g_q$ are in G, then G contains the span of these functions. The collection $\{0\}$, consisting only of the zero function, is a linear space. The span of a collection of functions is a linear space, which is referred to as the space spanned by these functions.

Let $g_1, \ldots, g_q \in G$ be linearly independent. If these functions span G, they are said to form a *basis* of G. If they do not span G, there is a function $g_{q+1} \in G$ such that $g_1, \ldots, g_{q+1}$ are linearly independent.

A linear space G is said to be *finite-dimensional* if $G = \{0\}$ or G is spanned by a finite collection of functions. Let G be a finite-dimensional linear space. If $G = \{0\}$, then G is said to have dimension zero. Otherwise, there is a positive integer p and there are functions $g_1, \ldots, g_p$ that form a basis of G. There are infinitely many such bases, but they all have the same number of functions. This common number is called the *dimension* of G and denoted by $\dim(G)$. If G is not finite-dimensional, it is said to be *infinite-dimensional* $(\dim(G) = \infty)$. The space G is infinite-dimensional if and only if, for any positive integer q, there are functions $g_1, \ldots, g_q \in G$ that are linearly independent. If $g_1, \ldots, g_q$ are linearly independent functions in G, then $\dim(G) \geq q$. If $g_1, \ldots, g_q$ are functions in G that span G, then $\dim(G) \leq q$.

The following are examples of infinite-dimensional linear spaces: the space of all functions on $\mathbb{R}$, the space of all continuous functions on $\mathbb{R}$, the space of all continuously differentiable functions on $\mathbb{R}$, and the space of all polynomials on $\mathbb{R}$.

Observe that the dimension of G is zero if and only if $G = \{0\}$. When we refer to G as being p-dimensional, it is to be understood that p is finite; when we refer to G as having a basis, it is to be understood that the dimension of G is finite and positive.

Let $g_1, \ldots, g_p$ form a basis of a linear space G. Then every function $g \in G$ can be written as a linear combination of these basis functions: $g = b_1 g_1 + \ldots + b_p g_p$, where $b_1, \ldots, b_p \in \mathbb{R}$ are uniquely determined. We refer to b_j as the *coefficient* of g_j and to $\mathbf{b} = [b_1, \ldots, b_p]^T$ as the *coefficient vector* of g relative to the given basis. (The use of "p" to denote the dimension of G is suggested by applications in which the coefficients $b_1, \ldots, b_p$ of the functions in G are referred to as "parameters.")

The collection G_{con} of constant functions on $\mathcal{X}$ is a one-dimensional linear space, and the constant function 1 is a basis of this space. If $\#(\mathcal{X}) = 1$, then $\{0\}$ and G_{con} are the only linear spaces of functions on $\mathcal{X}$. It follows from Example 9.1 that $1, x, \ldots, x^{p-1}$ form a basis of the space of polynomials $b_0 + b_1 x + \cdots + b_{p-1} x^{p-1}$ on $\mathbb{R}$ of degree $p - 1$ (or less) and hence that this space has dimension p. When $p = 1$, this space is the collection G_{con} of constant functions on $\mathbb{R}$; when $p = 2$, it is the collection of linear functions on $\mathbb{R}$; when $p = 3$, it is the collection of quadratic polynomials on $\mathbb{R}$; and when $p = 4$, it is the collection of cubic polynomials on $\mathbb{R}$. It follows from Example 9.2 that $1, x_1, \ldots, x_M$ form a basis of the space of linear functions $b_0 + b_1 x_1 + \cdots + b_M x_M$ on $\mathbb{R}^M$ and hence that this space has dimension $M + 1$.

It follows from Theorem 9.1 that $1, x_1, x_2, x_3, x_1^2, x_2^2, x_3^2, x_1 x_2, x_1 x_3, x_2 x_3$ are linearly independent monomials on $\mathbb{R}^3$, so they form a basis of the space of quadratic polynomials (polynomials of degree 2) on $\mathbb{R}^3$. Consequently, this space is ten-dimensional.

Let $g_1, \ldots, g_p$ be functions in a p-dimensional linear space G. Then these functions are linearly independent if and only if they span G. Thus, if they are either linearly independent or span G, then they form a basis of G.

Let $g_1, \ldots, g_q$ be functions in a p-dimensional linear space G. If these functions are linearly independent, then $q \leq p$. If these functions are linearly independent and $q < p$, then they can be extended to form a basis of G; that is, there are functions $g_{q+1}, \ldots, g_p$ such that $g_1, \ldots, g_p$ form a basis of G. If $g_1, \ldots, g_q$ span G, then $q \geq p$. If these functions span G and $q > p$, then there is a subcollection consisting of p of these functions that is a basis of G.

EXAMPLE 9.3   Let C be the collection of functions $(b_{10} + b_{11}x_1)(b_{20} + b_{21}x_2)$ on $\mathcal{X} = (a_1, c_1) \times (a_2, c_2)$ as $b_{10}, b_{11}, b_{20}, b_{21}$ range over $\mathbb{R}$. Show that

(a) $1, x_1, x_2, x_1 x_2$ form a basis of the span of C;

(b) C itself is not a linear space.

Solution   (a) By choosing $b_{10} = b_{20} = 1$ and $b_{11} = b_{21} = 0$, we see that $1 \in C$; and by choosing $b_{11} = b_{20} = 1$ and $b_{10} = b_{21} = 0$, we see that $x_1 \in C$. Similarly, $x_2 \in C$ and $x_1 x_2 \in C$. Now

$$(b_{10} + b_{11}x_1)(b_{20} + b_{21}x_2) = b_{10}b_{20} + b_{11}b_{20}x_1 + b_{10}b_{21}x_2 + b_{11}b_{21}x_1 x_2,$$

so every function in C is a linear combination of $1, x_1, x_2, x_1 x_2$. Consequently, every function in the span of C is a linear combination of these functions. According to Theorem 9.1, these functions are linearly independent. Therefore, they form a basis of the span of C.

(b) Suppose C is a linear space. Since 1 and $x_1 x_2$ are members of C, it follows from our assumption that $1 + x_1 x_2$ is a member of C. Thus, there are real numbers $b_{10}, b_{11}, b_{20}, b_{21}$ such that

$$1 + x_1 x_2 = (b_{10} + b_{11}x_1)(b_{20} + b_{21}x_2)$$
$$= b_{10}b_{20} + b_{11}b_{20}x_1 + b_{10}b_{21}x_2 + b_{11}b_{21}x_1 x_2$$

on $\mathcal{X}$ and hence

$$(b_{10}b_{20} - 1) + b_{11}b_{20}x_1 + b_{10}b_{21}x_2 + (b_{11}b_{21} - 1)x_1 x_2 = 0$$

on $\mathcal{X}$. Since $1, x_1, x_2, x_1 x_2$ are linearly independent on $\mathcal{X}$, we conclude that

$$b_{10}b_{20} = 1,$$
$$b_{11}b_{21} = 1,$$
$$b_{11}b_{20} = 0,$$
$$b_{10}b_{21} = 0.$$

It follows from the first two equations that $b_{11} \neq 0$ and $b_{20} \neq 0$ and hence that $b_{11}b_{20} \neq 0$, which contradicts the third equation. Therefore, C is not a linear space.   ∎

The indicator function $I(\mathbf{x}) = \text{ind}(\mathbf{x} \in A)$ corresponding to a subset A of $\mathcal{X}$ is defined by $I(\mathbf{x}) = 1$ for $\mathbf{x} \in A$ and $I(\mathbf{x}) = 0$ for $\mathbf{x} \notin A$. Similarly, the indicator function $I(\mathbf{x}) = \text{ind}(\mathbf{x} = \mathbf{x}_0)$ corresponding to a member $\mathbf{x}_0$ of $\mathcal{X}$ is defined by $I(\mathbf{x}_0) = 1$ and $I(\mathbf{x}) = 0$ for $\mathbf{x} \neq \mathbf{x}_0$.

Let $\mathcal{X}_1, \ldots, \mathcal{X}_p$ be a partition of $\mathcal{X}$ (that is, disjoint sets whose union is $\mathcal{X}$), and let $I_1, \ldots, I_p$ be the corresponding indicator functions; that is, set $I_j(\mathbf{x}) = \text{ind}(\mathbf{x} \in \mathcal{X}_j)$ for $1 \leq j \leq p$. Given $b_1, \ldots, b_p \in \mathbb{R}$, set $g = b_1 I_1 + \cdots + b_p I_p$. Then $g = b_j$ on $\mathcal{X}_j$ for $1 \leq j \leq p$. In particular, $I_1 + \cdots + I_p = 1$.

THEOREM 9.2   Let $\mathcal{X}_1, \ldots, \mathcal{X}_p$ be nonempty sets that form a partition of $\mathcal{X}$, let G be the collection of functions on $\mathcal{X}$ that are constant on each of the sets in the partition, and set $I_j(\mathbf{x}) = \text{ind}(\mathbf{x} \in \mathcal{X}_j)$ for $1 \leq j \leq p$. Then G is a p-dimensional linear space having basis $I_1, \ldots, I_p$.

Proof   Clearly, G is a linear space. Choose $g \in G$. Then g takes on a constant value b_j on $\mathcal{X}_j$ for $1 \leq j \leq p$ and hence $g = \sum_j b_j I_j$. Thus, $I_1, \ldots, I_p$ span G. It remains to prove that these functions are linearly independent. To this end, suppose $g = \sum_j b_j I_j$ is the zero function. Then, for $1 \leq j \leq p$, $0 = g(\mathbf{x}) = b_j$ for $\mathbf{x} \in \mathcal{X}_j$; thus, $b_1 = \cdots = b_p = 0$. Consequently, $I_1, \ldots, I_p$ are linearly independent and hence form a basis of G.   ∎

COROLLARY 9.1   Let G denote the collection of all functions on $\{\mathbf{x}_1, \ldots, \mathbf{x}_d\}$, where $\mathbf{x}_1, \ldots, \mathbf{x}_d$ are distinct and set $I_k(\mathbf{x}) = \text{ind}(\mathbf{x} = \mathbf{x}_k)$ for $1 \leq k \leq d$. Then G is a d-dimensional linear space having basis $I_1, \ldots, I_d$.

EXAMPLE 9.4   Let G denote the collection of all functions on $\{\mathbf{x}_1, \ldots, \mathbf{x}_d\}$, where $\mathbf{x}_1, \ldots, \mathbf{x}_d$ are distinct, and set $I_k(\mathbf{x}) = \text{ind}(\mathbf{x} = \mathbf{x}_k)$ for $1 \leq k \leq d$. Show that $1, I_1, \ldots, I_{d-1}$ form a basis of G.

Solution   By Corollary 9.1, G is a d-dimensional linear space having basis $I_1, \ldots, I_d$. Now $I_1 + \cdots + I_d = 1$, so $I_d = 1 - I_1 - \cdots - I_{d-1}$. Thus, each of the basis functions $I_1, \ldots, I_d$ is in the span of $1, I_1, \ldots, I_{d-1}$, so the latter set of d functions forms a basis of the d-dimensional space G.   ∎

THEOREM 9.3   Let G be the space of all functions on $\mathcal{X}$. Then G is finite-dimensional if and only if $\#(\mathcal{X}) < \infty$.

Proof   It follows from Corollary 9.1 that if $\#(\mathcal{X}) < \infty$, then G has finite dimension $\#(\mathcal{X})$. Suppose instead that $\#(\mathcal{X}) = \infty$. Given $q \geq 1$, let $\mathbf{x}_1, \ldots, \mathbf{x}_q$ be distinct points in $\mathcal{X}$ and set $I_j(\mathbf{x}) = \text{ind}(\mathbf{x} = \mathbf{x}_j)$ for $1 \leq j \leq q$. Then $I_1, \ldots, I_q$ are linearly independent functions in G (see the proof of Theorem 9.2). Since q can be made arbitrarily large, G is infinite-dimensional.   ∎

Let G_0 be a subset of a linear space G. If G_0 is also a linear space, then it is said to be a *subspace* of G. A subspace of G that is not equal to G is referred to as a *proper subspace*. Let G_0 be a subspace of G. Then $\dim(G_0) \leq \dim(G)$. Suppose $p = \dim(G) < \infty$ and set $p_0 = \dim(G_0) \leq p$. Then G_0 is a proper subspace of G if and only if $p_0 < p$. Suppose $1 \leq p_0 < p$. Then every basis $g_1, \ldots, g_{p_0}$ of G_0 can be extended to a basis $g_1, \ldots, g_p$ of G. Consequently, there is a basis $g_1, \ldots, g_p$ of G such that $g_1, \ldots, g_{p_0}$ is a basis of G_0.

For an illustration, let $1 \leq p_0 < p < \infty$, let G be the p-dimensional space of polynomials on $\mathbb{R}$ of degree $p - 1$, and let G_0 be the p_0-dimensional space of polynomials on $\mathbb{R}$ of degree $p_0 - 1$. Then G_0 is a subspace of G, and the basis $1, x, \ldots, x^{p_0 - 1}$ of G_0 can be extended to the basis $1, x, \ldots, x^{p-1}$ of G.

Change of Basis

Let $g_1, \ldots, g_p$ form a basis of G, and let $\tilde{g}_1, \ldots, \tilde{g}_q \in G$. If $g_1, \ldots, g_p$ are in the span of $\tilde{g}_1, \ldots, \tilde{g}_q$, then this span equals G.

Let $g_1, \ldots, g_p$ form a basis of G and let $\tilde{g}_1, \ldots, \tilde{g}_p \in G$. Then $\tilde{g}_1, \ldots, \tilde{g}_p$ form a basis of G if and only if $g_1, \ldots, g_p$ are in the span of $\tilde{g}_1, \ldots, \tilde{g}_p$.

EXAMPLE 9.5 Let $1, g_1, \ldots, g_{p-1}$ form a basis of G and let $c_1, \ldots, c_{p-1} \in \mathbb{R}$. Show that $1, g_1 - c_1, \ldots, g_{p-1} - c_{p-1}$ form a basis of G.

Solution Let $1 \leq j \leq p - 1$. Since $g_j = c_j + g_j - c_j = c_j \cdot 1 + 1 \cdot (g_j - c_j)$, we see that g_j is in the span of $1, g_j - c_j$ and hence in the span of $1, g_1 - c_1, \ldots, g_{p-1} - c_{p-1}$. Clearly, 1 is in the span of these functions. Thus, each of the functions $1, g_1, \ldots, g_{p-1}$ is in the span of $1, g_1 - c_1, \ldots, g_{p-1} - c_{p-1}$, so the functions in the latter sequence form a basis of G. ∎

Let $g_1, \ldots, g_p$ form a basis of G, which we refer to as the "old" basis, and let $\tilde{g}_1, \ldots, \tilde{g}_p$ form another basis, which we refer to as the "new" basis. Then each of the functions in the new basis is a unique linear combination of the functions in the old basis; that is, there are uniquely defined numbers a_{jl}, $1 \leq j, l \leq p$, such that

$$\tilde{g}_j = \sum_l a_{jl} g_l, \qquad 1 \leq j \leq p. \tag{1}$$

The $p \times p$ matrix

$$\mathbf{A} = \begin{bmatrix} a_{11} & \cdots & a_{1p} \\ \vdots & & \vdots \\ a_{p1} & \cdots & a_{pp} \end{bmatrix}$$

is referred to as the *coefficient matrix* of the new basis $\tilde{g}_1, \ldots, \tilde{g}_p$ relative to the old basis $g_1, \ldots, g_p$. Equation (1) can be written in matrix form as

$$
\begin{bmatrix} \tilde{g}_1 \\ \vdots \\ \tilde{g}_p \end{bmatrix} = \mathbf{A} \begin{bmatrix} g_1 \\ \vdots \\ g_p \end{bmatrix}.
\tag{2}
$$

Given $b_1, \ldots, b_p \in \mathbb{R}$, set $\mathbf{b} = [b_1, \ldots, b_p]^T$ and $g = b_1 g_1 + \cdots + b_p g_p$. Then

$$
g = \mathbf{b}^T \begin{bmatrix} g_1 \\ \vdots \\ g_p \end{bmatrix}.
\tag{3}
$$

Let $\tilde{b}_1, \ldots, \tilde{b}_p \in \mathbb{R}$ be the coefficients of g relative to the new basis and let $\tilde{\mathbf{b}} = [\tilde{b}_1, \ldots, \tilde{b}_p]^T$ denote the corresponding coefficient vector. Then

$$
g = \sum_j \tilde{b}_j \tilde{g}_j = \tilde{\mathbf{b}}^T \begin{bmatrix} \tilde{g}_1 \\ \vdots \\ \tilde{g}_p \end{bmatrix}.
$$

Thus, by (2),

$$
g = \tilde{\mathbf{b}}^T \mathbf{A} \begin{bmatrix} g_1 \\ \vdots \\ g_p \end{bmatrix}.
\tag{4}
$$

Since there is a unique coefficient vector of g relative to the old basis, we conclude from (3) and (4) that $\mathbf{b}^T = \tilde{\mathbf{b}}^T \mathbf{A}$ and hence that

$$
\mathbf{b} = \mathbf{A}^T \tilde{\mathbf{b}}.
\tag{5}
$$

If $\tilde{\mathbf{b}} \neq \mathbf{0}$, then g is a nonzero function and hence $\mathbf{b} \neq \mathbf{0}$. Thus, it follows from (5) that $\mathbf{A}^T$ is an invertible matrix and hence that $\mathbf{A}$ is invertible; moreover,

$$
\tilde{\mathbf{b}} = (\mathbf{A}^T)^{-1} \mathbf{b}.
\tag{6}
$$

In summary, the coefficient vectors $\mathbf{b}$ and $\tilde{\mathbf{b}}$ of g relative to the old basis and new basis, respectively, are related by (5) and (6), where $\mathbf{A}$ is the coefficient matrix of the new basis relative to the old basis.

It follows from (2) that

$$
\begin{bmatrix} g_1 \\ \vdots \\ g_1 \end{bmatrix} = \mathbf{A}^{-1} \begin{bmatrix} \tilde{g}_1 \\ \vdots \\ \tilde{g}_p \end{bmatrix};
\tag{7}
$$

that is, $\mathbf{A}^{-1}$ is the coefficient matrix of the old basis $g_1, \ldots, g_p$ relative to the new basis $\tilde{g}_1, \ldots, \tilde{g}_p$.

EXAMPLE **9.6**   (a) Determine the coefficient matrix $\mathbf{A}$ of the new basis $1, x - 1, (x - 1)^2$ of the space of quadratic polynomials on $\mathbb{R}$ relative to the old basis $1, x, x^2$.

(b) Determine the coefficient matrix of the old basis relative to the new basis and check that it equals $\mathbf{A}^{-1}$.

Solution   (a) Since $(x - 1)^2 = x^2 - 2x + 1$, the coefficient matrix of the new basis relative to the old basis is given by

$$\mathbf{A} = \begin{bmatrix} 1 & 0 & 0 \\ -1 & 1 & 0 \\ 1 & -2 & 1 \end{bmatrix}.$$

(b) Now $1 = 1$, $x = 1 + x - 1$, and $x^2 = (1 + x - 1)^2 = 1 + 2(x - 1) + (x - 1)^2$, so the coefficient matrix of the old basis relative to the new basis equals

$$\begin{bmatrix} 1 & 0 & 0 \\ 1 & 1 & 0 \\ 1 & 2 & 1 \end{bmatrix}.$$

To check that this matrix equals $\mathbf{A}^{-1}$, we observe that

$$\begin{bmatrix} 1 & 0 & 0 \\ -1 & 1 & 0 \\ 1 & -2 & 1 \end{bmatrix} \begin{bmatrix} 1 & 0 & 0 \\ 1 & 1 & 0 \\ 1 & 2 & 1 \end{bmatrix} = \begin{bmatrix} 1 & 0 & 0 \\ 0 & 1 & 0 \\ 0 & 0 & 1 \end{bmatrix}. \quad \blacksquare$$

EXAMPLE **9.7**   Let $\mathcal{X}_1$, $\mathcal{X}_2$, and $\mathcal{X}_3$ be nonempty sets that form a partition of $\mathcal{X}$. Consider the linear space G consisting of all functions on $\mathcal{X}$ that are constant on each of the sets $\mathcal{X}_1$, $\mathcal{X}_2$, and $\mathcal{X}_3$ and set $I_j(x) = \mathrm{ind}(x \in \mathcal{X}_j)$ for $1 \leq j \leq 3$. Then I_1, I_2, and I_3 form a basis of G by Theorem 9.2.

(a) Show that $1, I_1, I_2$ is a basis of G.

(b) Determine the coefficient matrix $\mathbf{A}$ of the new basis $1, I_1, I_2$ relative to the old basis I_1, I_2, I_3.

(c) Determine the coefficient matrix of the old basis relative to the new basis and check that it equals $\mathbf{A}^{-1}$.

Solution   (a) Observe that $I_1 + I_2 + I_3 = 1$ and hence that $I_3 = 1 - I_1 - I_2$. Thus, each of the functions I_1, I_2, I_3 is a linear combination of $1, I_1$, and I_2, so $1, I_1$, and I_2 form a basis of G.

(b) The coefficient matrix of the new basis relative to the old basis is given by

$$\mathbf{A} = \begin{bmatrix} 1 & 1 & 1 \\ 1 & 0 & 0 \\ 0 & 1 & 0 \end{bmatrix}.$$

(c) The coefficient matrix of the old basis relative to the new basis is

$$\begin{bmatrix} 0 & 1 & 0 \\ 0 & 0 & 1 \\ 1 & -1 & -1 \end{bmatrix}.$$

To check that this matrix equals $\mathbf{A}^{-1}$, we observe that

$$\begin{bmatrix} 1 & 1 & 1 \\ 1 & 0 & 0 \\ 0 & 1 & 0 \end{bmatrix}\begin{bmatrix} 0 & 1 & 0 \\ 0 & 0 & 1 \\ 1 & -1 & -1 \end{bmatrix} = \begin{bmatrix} 1 & 0 & 0 \\ 0 & 1 & 0 \\ 0 & 0 & 1 \end{bmatrix}. \quad \blacksquare$$

EXAMPLE **9.8**   Let G be the space of all functions on $X = \{-1, 0, 1\}$ and set $I_1(x) = \mathrm{ind}(x = -1)$, $I_2(x) = \mathrm{ind}(x = 0)$, and $I_3(x) = \mathrm{ind}(x = 1)$. Then I_1, I_2, and I_3 form a basis of G by Corollary 9.1.

(a) Show that 1, x, and x^2 form a basis of G.

(b) Determine the coefficient matrix $\mathbf{A}$ of 1, x, and x^2 relative to I_1, I_2, and I_3.

(c) Determine the coefficient matrix of I_1, I_2, and I_3 relative to 1, x, and x^2 and check that it equals $\mathbf{A}^{-1}$.

Solution   **(a)** Given $b_0, b_1, b_2 \in \mathbb{R}$, consider the quadratic polynomial $g = b_0 + b_1 x + b_2 x^2$. Suppose $g = 0$ on X. Then -1, 0, and 1 are roots of g. Since a nonzero quadratic polynomial has at most two distinct roots, we conclude that g is the zero polynomial and hence (see Example 9.1) that $b_0 = b_1 = b_2 = 0$. Therefore, 1, x, and x^2 are linearly independent as functions on X and hence form a basis of G.

(b) Now $1 = I_1 + I_2 + I_3$ on X, $x = (-1) \cdot I_1 + 0 \cdot I_2 + 1 \cdot I_3$ on X, and $x^2 = 1 \cdot I_1 + 0 \cdot I_2 + 1 \cdot I_3$ on X. Consequently,

$$\mathbf{A} = \begin{bmatrix} 1 & 1 & 1 \\ -1 & 0 & 1 \\ 1 & 0 & 1 \end{bmatrix}.$$

(c) According to the solution to part (b),

$$I_1 + I_2 + I_3 = 1,$$
$$-I_1 + I_3 = x,$$
$$I_1 + I_3 = x^2.$$

Thus, $I_1 = \frac{1}{2}(x^2 - x)$, $I_3 = \frac{1}{2}(x + x^2)$, and $I_2 = 1 - \frac{1}{2}(x^2 - x) - \frac{1}{2}(x^2 + x) = 1 - x^2$. Consequently, the coefficient matrix of I_1, I_2, and I_3 relative to 1, x, and x^2 equals

$$\begin{bmatrix} 0 & -\frac{1}{2} & \frac{1}{2} \\ 1 & 0 & -1 \\ 0 & \frac{1}{2} & \frac{1}{2} \end{bmatrix}.$$

To check that this matrix equals $\mathbf{A}^{-1}$, we observe that

$$\begin{bmatrix} 1 & 1 & 1 \\ -1 & 0 & 1 \\ 1 & 0 & 1 \end{bmatrix} \begin{bmatrix} 0 & -\frac{1}{2} & \frac{1}{2} \\ 1 & 0 & -1 \\ 0 & \frac{1}{2} & \frac{1}{2} \end{bmatrix} = \begin{bmatrix} 1 & 0 & 0 \\ 0 & 1 & 0 \\ 0 & 0 & 1 \end{bmatrix}. \quad \blacksquare$$

Problems

9.1 Are the functions 1, x, $\cos x$ on $\mathbb{R}$ linearly independent or linearly dependent?

9.2 Let $g_1, \ldots, g_q$ be functions on X and let X_0 be a subset of X. Show that if $g_1, \ldots, g_q$ are linearly independent on X_0, then they are linearly independent on X.

9.3 Let $g_1, \ldots, g_{q+1}$ be functions on X. Show that if $g_1, \ldots, g_q$ are linearly independent and g_{q+1} is *not* a linear combination of $g_1, \ldots, g_q$, then $g_1, \ldots, g_{q+1}$ are linearly independent.

9.4 Determine the coefficient vector of $(x_1 - 2x_2 + 3)^2$ relative to the basis 1, x_1, x_1^2, x_2, x_2^2, $x_1 x_2$ of the space of quadratic polynomials on $\mathbb{R}^2$.

9.5 Let G be a p-dimensional linear space and let g be a function that is *not* a member of G. Show that the span of $G \cup \{g\}$ is a $(p + 1)$-dimensional linear space.

9.6 Determine the dimension of the space of quadratic polynomials on **(a)** $\mathbb{R}^2$; **(b)** $\mathbb{R}^4$; **(c)** $\mathbb{R}^M$.

9.7 Determine the dimension of the space of cubic polynomials on **(a)** $\mathbb{R}^2$; **(b)** $\mathbb{R}^3$; **(c)** $\mathbb{R}^M$.

9.8 Let C be the collection of functions $(b_{10} + b_{11}x_1)(b_{20} + b_{21}x_2)(b_{30} + b_{31}x_3)$ on X as $b_{10}, b_{11}, b_{20}, b_{21}, b_{30}, b_{31}$ range over $\mathbb{R}$. Show that **(a)** 1, x_1, x_2, x_3, $x_1 x_2$, $x_1 x_3$, $x_2 x_3$, $x_1 x_2 x_3$ form a basis of the span of C; **(b)** C is not a linear space.

9.9 Let C be the collection of functions

$$(b_{10} + b_{11}x_1 + b_{12}x_1^2)(b_{20} + b_{21}x_2 + b_{22}x_2^2)$$

on X as $b_{10}, b_{11}, b_{12}, b_{20}, b_{21}, b_{22}$ range over $\mathbb{R}$. Show that **(a)** 1, x_1, x_1^2, x_2, x_2^2, $x_1 x_2$, $x_1^2 x_2$, $x_1 x_2^2$, $x_1^2 x_2^2$ form a basis of the span of C; **(b)** C is not a linear space.

9.10 **(a)** Determine the coefficient matrix $\mathbf{A}$ of the new basis 1, $x - 1$, $(x - 1)^2$, $(x - 1)^3$ of the space of cubic polynomials on $\mathbb{R}$ relative to the old basis 1, x, x^2, x^3. **(b)** Determine the coefficient matrix of the old basis relative to the new basis and check that it equals $\mathbf{A}^{-1}$.

9.11 Let $c \in \mathbb{R}$. **(a)** Determine the coefficient matrix $\mathbf{A}$ of the new basis $1, x, x^2 - c$ of the space of quadratic polynomials on $\mathbb{R}$ relative to the old basis $1, x, x^2$. **(b)** Determine the coefficient matrix of the old basis relative to the new basis and check that it equals $\mathbf{A}^{-1}$.

9.12 Consider the linear space consisting of all functions on $\{0, 1, 2\}$ and let I_1, I_2, and I_3 be the indicator functions corresponding to 0, 1, and 2, respectively. **(a)** Determine the coefficient matrix $\mathbf{A}$ of the new basis $1, x, x^2$ relative to the old basis I_1, I_2, I_3. **(b)** Determine the coefficient matrix of the old basis relative to the new basis and check that it equals $\mathbf{A}^{-1}$.

9.13 Let $\mathcal{X}_1, \mathcal{X}_2, \mathcal{X}_3$, and $\mathcal{X}_4$ be nonempty sets that form a partition of $\mathcal{X}$; consider the linear space G consisting of all functions on $\mathcal{X}$ that are constant on each of the sets $\mathcal{X}_1, \mathcal{X}_2, \mathcal{X}_3$, and $\mathcal{X}_4$; and set $I_j(\mathbf{x}) = \text{ind}(\mathbf{x} \in \mathcal{X}_j)$ for $1 \leq j \leq 4$. Then $I_1, \ldots, I_4$ form a basis of G. **(a)** Show that $1, I_1, I_2$, and I_3 form a basis of G. **(b)** Determine the coefficient matrix $\mathbf{A}$ of the new basis $1, I_1, I_2$, and I_3 relative to the old basis I_1, I_2, I_3, and I_4. **(c)** Determine the coefficient matrix of the old basis relative to the new basis and check that it equals $\mathbf{A}^{-1}$.

9.2
Identifiability

Recall the discussion in Section 8.1 of identifiability in the context of polynomial fits to the polymer data. More generally, suppose the response at the factor-level combination $\mathbf{x}$ has the nonrandom value $\mu(\mathbf{x})$, where $\mu(\cdot)$ is an unknown member of a linear space G of functions on $\mathcal{X}$. [Here "$\mu(\cdot)$" denotes the function $\mu(\mathbf{x})$, $\mathbf{x} \in \mathcal{X}$.] Suppose also that the values of this function are known at the design points $\mathbf{x}'_1, \ldots, \mathbf{x}'_d \in \mathcal{X}$. If this information is enough to determine $\mu(\cdot)$, the space G is said to be identifiable (relative to the design set); otherwise, G is said to be nonidentifiable.

Suppose G is nonidentifiable. Then there are distinct functions $\mu_1(\cdot)$ and $\mu_2(\cdot)$ in G having the same values at the design points. Thus, their difference $\mu_1(\cdot) - \mu_2(\cdot)$ is a nonzero function in G that equals zero on the design set $\mathcal{X}' = \{\mathbf{x}'_1, \ldots, \mathbf{x}'_d\}$. Suppose, conversely, that there is a nonzero function $g \in G$ that equals zero on the design set. Let $\mu(\cdot)$ be a function in G taking on the known values at the design points. Then $\mu(\cdot) + g$ is a different function in G that takes on the known values at the design points, so G is nonidentifiable. Therefore, G is nonidentifiable if and only if there is a nonzero function in G that equals zero on the design set.

Mathematically, it is preferable to use the above discussion as motivation only. Thus, starting from scratch, we say that the linear space G is *nonidentifiable* (relative to the design set) if there is a nonzero function in G that equals zero on the design set; otherwise, we say that G is *identifiable*. Observe that G is identifiable if and only if the only function in G that equals zero on the design set is the zero function. Alternatively, G is identifiable if and only if linearly independent functions in G remain linearly independent when we ignore all points in $\mathcal{X}$ that are not design points. Thus, G is identifiable if and only if, when we ignore points in $\mathcal{X}$ that are not design points, the dimension of G does not change. Every subspace of an identifiable linear space is identifiable. If X coincides with the design set, then every linear space of functions on X is identifiable. If G is identifiable relative to some subset of the design set, then G is identifiable relative to the entire design set.

THEOREM 9.4   The linear space G is identifiable if and only if the following condition holds: If g_1 and g_2 are in G and $g_1 = g_2$ on the design set, then $g_1 = g_2$ everywhere on X.

Proof   Suppose the indicated condition holds and let g be a function in G that equals zero on the design set. It follows from the condition by choosing $g_1 = g$ and $g_2 = 0$ that $g = 0$ on X. Consequently, G is identifiable.

Suppose, conversely, that G is identifiable. Let g_1 and g_2 be functions in G that are equal on the design set. Then $g_1 - g_2$ is a member of G that equals zero on the design set, so $g_1 - g_2$ equals zero on X and hence $g_1 = g_2$ on X. Consequently, the indicated condition holds.   ∎

COROLLARY 9.2   The linear space G is identifiable if and only if, for each choice of $c_1, \ldots, c_d \in \mathbb{R}$, there is at most one choice of $g \in G$ such that $g(x_1') = c_1, \ldots, g(x_d') = c_d$.

Proof   Suppose G is identifiable, let $c_1, \ldots, c_d \in \mathbb{R}$, and let g_1 and g_2 be functions in G such that $g_1(\mathbf{x}_1') = c_1, \ldots, g_1(\mathbf{x}_d') = c_d$ and also $g_2(\mathbf{x}_1') = c_1, \ldots, g_2(\mathbf{x}_d') = c_d$. Then $g_1(\mathbf{x}_1') = g_2(\mathbf{x}_1'), \ldots, g_1(\mathbf{x}_d') = g_2(\mathbf{x}_d')$. Thus, by Theorem 9.4, $g_1 = g_2$. Therefore, there is only one choice of $g \in G$ such that $g(\mathbf{x}_1') = c_1, \ldots, g(\mathbf{x}_d') = c_d$.

Suppose, instead, that G is nonidentifiable. Then, by Theorem 9.4, there are distinct functions g_1 and g_2 in G such that $g_1 = g_2$ on the design set. Let $c_1, \ldots, c_d \in \mathbb{R}$ be defined by $c_1 = g_1(\tilde{\mathbf{x}}_1'), \ldots, c_d = g_1(\mathbf{x}_d')$. Then $g_1(\mathbf{x}_1') = c_1, \ldots, g_1(\mathbf{x}_d') = c_d$ and $g_2(\mathbf{x}_1') = c_1, \ldots, g_2(\mathbf{x}_d') = c_d$. Therefore, there is more than one choice of $g \in G$ such that $g(\mathbf{x}_1') = c_1, \ldots, g(\mathbf{x}_d') = c_d$.   ∎

EXAMPLE 9.9   Let G denote the space of linear functions on $\mathbb{R}$.

(a) Show that G is nonidentifiable if and only if $d = 1$.

(b) Give an example with $d = 1$ and $x_1' = 1$ of a nonzero linear function that equals zero on the design set.

Solution   (a) Suppose G is nonidentifiable. Then there is a nonzero linear function $b_0 + b_1 x$ that equals zero on the design set—that is, such that $b_0 + b_1 x_k' = 0$ for $1 \leq k \leq d$. Consequently, $b_1 \neq 0$. (Otherwise, b_0 would equal zero, which would contradict the assumption that $b_0 + b_1 x$ is a nonzero function.) Thus, $x_k' = -b_0/b_1$ for $1 \leq k \leq d$. Since the design points are distinct, we conclude that $d = 1$. Suppose, conversely, that $d = 1$. Then $x - x_1'$ is a nonzero function on G that equals zero on the design set, so G is nonidentifiable.

(b) Let $d = 1$ and $x_1' = 1$. Then $g(x) = x - 1$ is an example of a nonzero linear function that equals zero on the design set (see Figure 9.1).   ∎

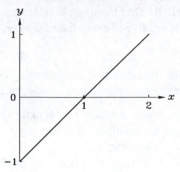

FIGURE   9.1

A nonzero linear function that equals zero at $x'_1 = 1$

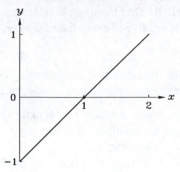

EXAMPLE **9.10**   Let G denote the space of quadratic polynomials on $\mathbb{R}$.

(a) Show that G is nonidentifiable if and only if $d \leq 2$.

(b) Give an example when $d = 2$, $x'_1 = 1$, and $x'_2 = 2$ of a nonzero quadratic polynomial that equals zero on the design set.

Solution   (a) Suppose G is nonidentifiable. Then there is a nonzero quadratic polynomial $b_0 + b_1 x + b_2 x^2$ that equals zero on the design set; that is, such that

$$b_0 + b_1 x'_k + b_2 (x'_k)^2 = 0, \qquad 1 \leq k \leq d.$$

If $b_2 = 0$, then $b_0 + b_1 x'_k = 0$ for $1 \leq k \leq d$, so it follows as in the solution to Example 9.9 that $d = 1$. Otherwise, we know from algebra that $b_0 + b_1 x + b_2 x^2$ has at most two distinct roots. (A derivation based on Rolle's theorem that a nonzero quadratic polynomial has at most two distinct real roots is left as a problem.) Since the design points are roots of this polynomial, we conclude that $d \leq 2$. Suppose, conversely, that $d \leq 2$. If $d = 1$, then $x - x'_1$ is a nonzero function in G that equals zero on the design set; if $d = 2$, then $(x - x'_1)(x - x'_2)$ is a nonzero function in G that equals zero on the design set. Thus, G is nonidentifiable.

(b) Suppose $d = 2$, $x'_1 = 1$, and $x'_2 = 2$. The nonzero quadratic polynomial g that equals zero on the design set and equals 1 at zero is given by $g(x) = (x - 1)(x - 2)/2$ (see Figure 9.2).   ∎

EXAMPLE **9.11**   Let G denote the space of linear functions on $\mathbb{R}^2$. Show that G is nonidentifiable if and only if the design points lie on a line in $\mathbb{R}^2$.

Solution   Suppose G is nonidentifiable. Then there is a nonzero linear function $b_0 + b_1 x_1 + b_2 x_2$ that equals zero on the design set—that is, such that $b_0 + b_1 x_{k1} + b_2 x_{k2} = 0$ for $1 \leq k \leq d$. The coefficients b_1 and b_2 cannot both be zero. (Otherwise, $b_0 = 0$, which would contradict the assumption that $b_0 + b_1 x_1 + b_2 x_2$ is a nonzero linear

F I G U R E 9.2

A nonzero quadratic polynomial that equals zero at $x'_1 = 1$ and $x'_2 = 2$

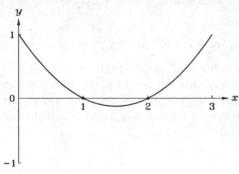

function.) Thus,

$$\{\,(x_1, x_2) \in \mathbb{R}^2 : b_0 + b_1 x_1 + b_2 x_2 = 0\,\} \qquad (8)$$

is a line in $\mathbb{R}^2$ that contains all of the design points.

Suppose, conversely, that the design points lie on a line in $\mathbb{R}^2$. This line can be written in the form of (8), where b_1 and b_2 are not both zero. Thus, $b_0 + b_1 x_1 + b_2 x_2$ is a nonzero function in G that equals zero on the design set, so G is nonidentifiable. ∎

When the design points lie on a line in $\mathbb{R}^2$, they are said to be *collinear*. According to Example 9.11, if G is the space of linear functions on $\mathbb{R}^2$, then it is nonidentifiable if and only if the design points are collinear.

EXAMPLE 9.12 Let G be the space of linear functions on $\mathbb{R}^2$ and let the design set consist of the three collinear points $(-1, -1)$, $(0, 0)$, and $(1, 1)$. Then G is nonidentifiable by Example 9.11. Determine an explicit nonzero function in G that equals zero on the design set.

Solution The function $x_1 - x_2$ is a nonzero function in G that equals zero on the design set. ∎

EXAMPLE 9.13 Let G denote the space of linear functions on $\mathbb{R}^3$. Show that G is nonidentifiable if and only if the design points lie on a plane in $\mathbb{R}^3$.

Solution Suppose G is nonidentifiable. Then there is a nonzero linear function $b_0 + b_1 x_1 + b_2 x_2 + b_3 x_3$ that equals zero on the design set—that is, such that

$$b_0 + b_1 x_{k1} + b_2 x_{k2} + b_3 x_{k3} = 0, \qquad 1 \le k \le d.$$

The coefficients b_1, b_2, and b_3 cannot all be zero. (Otherwise, $b_0 = 0$, which would contradict the assumption that $b_0 + b_1 x_1 + b_2 x_2 + b_3 x_3$ is a nonzero linear function.) Thus,

$$\{ (x_1, x_2, x_3) \in \mathbb{R}^3 : b_0 + b_1 x_1 + b_2 x_2 + b_3 x_3 = 0 \} \tag{9}$$

is a plane in $\mathbb{R}^3$ that contains all of the design points.

Suppose, conversely, that the design points lie on a plane in $\mathbb{R}^3$. This plane can be written in the form of (9), where b_1, b_2, and b_3 are not all zero. Thus, $b_0 + b_1 x_1 + b_2 x_2 + b_3 x_3$ is a nonzero function in G that equals zero on the design set, so G is nonidentifiable. ∎

When the design points lie on a plane in $\mathbb{R}^3$, they are said to be *coplanar*. According to Example 9.13, if G is the space of linear functions on $\mathbb{R}^3$, then it is nonidentifiable if and only if the design points are coplanar.

EXAMPLE 9.14　Let G denote the space of all functions on $\mathcal{X}$. Show that G is identifiable if and only if the design set coincides with $\mathcal{X}$—that is, if and only if every point in $\mathcal{X}$ is a design point.

Solution　Suppose the design set coincides with $\mathcal{X}$. Then, obviously, every function that equals zero on the design set equals zero on $\mathcal{X}$, so G is identifiable. Suppose, instead, that the design set is a proper subset of $\mathcal{X}$. Define the function g on $\mathcal{X}$ by $g(\mathbf{x}) = 0$ if $\mathbf{x}$ is a design point and $g(\mathbf{x}) = 1$ otherwise. Then g is a nonzero function in G that equals zero on the design set. Thus, G is nonidentifiable. Therefore, G is identifiable if and only if every point in $\mathcal{X}$ is a design point. ∎

Given a function g on $\mathcal{X}$, let g' denote the *restriction* of g to the design set, which is given by $g'(\mathbf{x}) = g(\mathbf{x})$ for $\mathbf{x} \in \mathcal{X}'$. Clearly, if $g = 0$, then $g' = 0$. Suppose that $g = b_1 g_1 + \cdots + b_q g_q$, where $g_1, \ldots, g_q$ are functions on $\mathcal{X}$. Then $g' = b_1 g_1' + \cdots + b_q g_q'$, where $g_1', \ldots, g_q'$ are the respective restrictions of $g_1, \ldots, g_q$ to the design set. Observe that the linear space G of functions on $\mathcal{X}$ is nonidentifiable if and only if there is a nonzero function $g \in G$ such that $g' = 0$. Let $G' = \{ g' : g \in G \}$ denote the linear space of the restrictions of the functions in G to the design set, which is necessarily identifiable. Since the dimension of G' is at most that of the space of all functions on the design set, which equals $d = \#(\mathcal{X}')$ (see Corollary 9.1 in Section 9.1), we see that $\dim(G') \leq d$.

THEOREM 9.5　Let $g_1, \ldots, g_q$ be functions on $\mathcal{X}$ and let $g_1', \ldots, g_q'$ be the respective restrictions of these functions to the design set. Then $g_1', \ldots, g_q'$ are linearly independent if and only if $g_1, \ldots, g_q$ are linearly independent and their span is identifiable.

Proof   Suppose $g'_1, \ldots, g'_q$ are linearly independent and let $g = b_1 g_1 + \cdots + b_q g_q$. Then $g' = b_1 g'_1 + \cdots + b_q g'_q$. If $g' = 0$, then $b_1 = \cdots = b_q = 0$ and hence $g = 0$. Consequently, the span of $g_1, \ldots, g_q$ is identifiable. Similarly, if $g = 0$, then $g' = 0$ and hence $b_1 = \cdots = b_q = 0$. Therefore, $g_1, \ldots, g_q$ are linearly independent.

Suppose, conversely, that $g_1, \ldots, g_q$ are linearly independent and that their span is identifiable. Suppose also that $b_1 g'_1 + \cdots + b_q g'_q = 0$ and hence that the restriction of $b_1 g_1 + \cdots + b_q g_q$ to the design set equals zero. Then $b_1 g_1 + \cdots + b_q g_q = 0$ (by identifiability) and hence $b_1 = \cdots = b_q = 0$ (by linear independence). Therefore, $g'_1, \ldots, g'_q$ are linearly independent. ∎

THEOREM 9.6   Let G' be the space of the restrictions of the functions in G to the design set. Then

(a) $\dim(G') \leq \dim(G)$;

(b) G is identifiable if and only if $\dim(G') = \dim(G)$.

Proof   (a) Set $p' = \dim(G')$, let $g'_1, \ldots, g'_{p'}$ be a basis of G', and let them be the respective restrictions of the functions $g_1, \ldots, g_{p'} \in G$ to the design set. Then $g_1, \ldots, g_{p'}$ are linearly independent by Theorem 9.5, so $\dim(G') = p' \leq \dim(G)$.

(b) Suppose $\dim(G') = \dim(G)$. Then the functions $g_1, \ldots, g_{p'}$ obtained in the proof of part (a) form a basis of G. In particular, their span is G. Since their span is identifiable by Theorem 9.5, we conclude that G is identifiable. Suppose, conversely, that G is identifiable. Let $g_1, \ldots, g_q$ be linearly independent functions in G and let $g'_1, \ldots, g'_q \in G'$ be their respective restrictions to the design set. Then $g'_1, \ldots, g'_q$ are linearly independent by Theorem 9.5, so $q \leq \dim(G')$. Therefore, $\dim(G) \leq \dim(G')$ and hence $\dim(G') = \dim(G)$ by (a). ∎

COROLLARY 9.3   If G is identifiable, then $\dim(G) \leq d$.

Proof   Suppose G is identifiable. Then, by Theorem 9.6(b), $\dim(G) = \dim(G') \leq d$. ∎

COROLLARY 9.4   Suppose G is identifiable, let $g_1, \ldots, g_p$ be a basis of G, and let $g'_1, \ldots, g'_p$ be the respective restrictions of $g_1, \ldots, g_p$ to the design set. Then $g'_1, \ldots, g'_p$ form a basis of G'.

Proof   It follows from Theorem 9.5 that $g'_1, \ldots, g'_p$ are linearly independent functions in G'. Since G' has dimension p by Theorem 9.6(b), these functions form a basis of G'. ∎

EXAMPLE 9.15   Let $X = \mathbb{R}^2$; let the design set consist of the $d = 3$ points $(-1, -1)$, $(1, -1)$, and $(-1, 1)$; and let G be the linear space spanned by $1, x_1, x_2$, and $x_1 x_2$. Since $1, x_1, x_2$, and $x_1 x_2$ are linearly independent by Theorem 9.1 in Section 9.1, they form a basis of G, so this space has dimension $p = 4$. Thus, by Corollary 9.3, G is nonidentifiable. Determine all the nonzero functions in G that equal zero on the design set.

Solution   Consider the function $g(x_1, x_2) = b_0 + b_1 x_1 + b_2 x_2 + b_3 x_1 x_2$. Observe that $g(-1, -1) = b_0 - b_1 - b_2 + b_3$, $g(1, -1) = b_0 + b_1 - b_2 - b_3$, and $g(-1, 1) = b_0 - b_1 + b_2 - b_3$. Thus, $g = 0$ on the design set if and only if

$$b_0 - b_1 - b_2 + b_3 = 0$$
$$b_0 + b_1 - b_2 - b_3 = 0$$
$$b_0 - b_1 + b_2 - b_3 = 0.$$

Using elementary algebra, we find that these three equations in the four unknowns b_0, b_1, b_2, and b_3 are satisfied if and only if $b_0 = b_1 = b_2 = b_3$, in which case $g(x_1, x_2) = b_0(1 + x_1 + x_2 + x_1 x_2) = b_0(1 + x_1)(1 + x_2)$. Thus, the nonzero functions in G that equal zero on the design set are the functions of the form $g(x_1, x_2) = b_0(1 + x_1)(1 + x_2)$ with $b_0 \neq 0$. ∎

Design Matrices

A matrix $\mathbf{X}$ having q columns is said to be *singular* if there is a nonzero q-dimensional column vector $\mathbf{b}$ such that $\mathbf{Xb} = \mathbf{0}$; otherwise, $\mathbf{X}$ is said to be *nonsingular*. A square matrix is nonsingular if and only if it is invertible.

Let $g_1, \ldots, g_q$ be functions on $\mathfrak{X}$. We refer to the $d \times q$ matrix

$$\mathbf{X} = \begin{bmatrix} g_1(\mathbf{x}'_1) & \cdots & g_q(\mathbf{x}'_1) \\ \vdots & & \vdots \\ g_1(\mathbf{x}'_d) & \cdots & g_q(\mathbf{x}'_d) \end{bmatrix}$$

as the $d \times q$ form of the *design matrix* corresponding to $g_1, \ldots, g_q$ (and relative to the design points). (An alternative form of the design matrix, having as many rows as trials of the experiment, will be introduced in Section 10.1.)

Let $g'_1, \ldots, g'_q$ be the respective restrictions of $g_1, \ldots, g_q$ to the design set. Given $b_1, \ldots, b_q \in \mathbb{R}$, set $\mathbf{b} = [b_1, \ldots, b_q]^T$ and $g = b_1 g_1 + \cdots + b_q g_q$. Then the restriction of g to the design set is given by $g' = b_1 g'_1 + \cdots + b_q g'_q$. Observe that $g'(\mathbf{x}'_k) = g(\mathbf{x}'_k) = b_1 g_1(\mathbf{x}'_k) + \cdots + b_q g_q(\mathbf{x}'_k)$ for $1 \leq k \leq d$. In matrix form,

$$\begin{bmatrix} g'(\mathbf{x}'_1) \\ \vdots \\ g'(\mathbf{x}'_d) \end{bmatrix} = \begin{bmatrix} g(\mathbf{x}'_1) \\ \vdots \\ g(\mathbf{x}'_d) \end{bmatrix} = \mathbf{Xb}. \tag{10}$$

THEOREM 9.7   Let $g_1, \ldots, g_q$ be functions on $\mathfrak{X}$, let $g'_1, \ldots, g'_q$ be the respective restrictions of these functions to the design set, and let $\mathbf{X}$ be the design matrix corresponding to $g_1, \ldots, g_q$. Then $\mathbf{X}$ is a nonsingular matrix if and only if $g'_1, \ldots, g'_q$ are linearly independent.

Proof Given $b_1, \ldots, b_q \in \mathbb{R}$, set $\mathbf{b} = [b_1, \ldots, b_q]^T$ and $g' = b_1 g'_1 + \cdots + b_q g'_q$. Suppose $\mathbf{X}$ is a nonsingular matrix. Suppose also that $\mathbf{b} \neq \mathbf{0}$ and hence that $\mathbf{Xb} \neq \mathbf{0}$. It then follows from (10) that g' is not the zero function on the design set. Consequently, $g'_1, \ldots, g'_q$ are linearly independent.

Suppose, conversely, that $g'_1, \ldots, g'_q$ are linearly independent. Suppose also that $\mathbf{b} \neq \mathbf{0}$ and hence that g' is not the zero function on the design set. It then follows from (10) that $\mathbf{Xb} \neq \mathbf{0}$. Therefore, $\mathbf{X}$ is a nonsingular matrix. ∎

The next two results follow from Theorems 9.5 and 9.7.

COROLLARY 9.5 Let $\mathbf{X}$ be the design matrix corresponding to $g_1, \ldots, g_q$. Then $\mathbf{X}$ is a nonsingular matrix if and only if the following two conditions are satisfied: $g_1, \ldots, g_q$ are linearly independent and the span of $g_1, \ldots, g_q$ is identifiable.

COROLLARY 9.6 Let $\mathbf{X}$ be the design matrix corresponding to a basis $g_1, \ldots, g_p$ of a linear space G of functions on $\mathcal{X}$. Then G is identifiable if and only if $\mathbf{X}$ is a nonsingular matrix.

EXAMPLE 9.16 (Continued.)

(a) Determine the design matrix $\mathbf{X}$ corresponding to the basis $1, x_1, x_2, x_1 x_2$ of G.

(b) Determine a nonzero vector $\mathbf{b} = [b_0, b_1, b_2, b_3]^T$ such that $\mathbf{Xb} = \mathbf{0}$.

Solution (a) The design matrix is given by

$$\mathbf{X} = \begin{bmatrix} 1 & -1 & -1 & 1 \\ 1 & 1 & -1 & -1 \\ 1 & -1 & 1 & -1 \end{bmatrix}.$$

(b) Motivated by the solution to Example 9.15, we choose $\mathbf{b} = [1, 1, 1, 1]^T$. Then $\mathbf{XB}$ is the sum of the four columns of $\mathbf{X}$, which equals $\mathbf{0}$. ∎

THEOREM 9.8 Let $\mathbf{X}$ be the design matrix corresponding to a basis $g_1, \ldots, g_p$ of G and let $\tilde{g}_1, \ldots, \tilde{g}_p$ be a basis of G having coefficient matrix $\mathbf{A}$ relative to the basis $g_1, \ldots, g_p$. Then the design matrix $\tilde{\mathbf{X}}$ relative to the basis $\tilde{g}_1, \ldots, \tilde{g}_p$ is given by $\tilde{\mathbf{X}} = \mathbf{XA}^T$.

Proof Write $\mathbf{A} = [a_{jl}]$. Then

$$\tilde{\mathbf{X}} = \begin{bmatrix} \tilde{g}_1(\mathbf{x}'_1) & \cdots & \tilde{g}_p(\mathbf{x}'_1) \\ \vdots & & \vdots \\ \tilde{g}_1(\mathbf{x}'_d) & \cdots & \tilde{g}_p(\mathbf{x}'_d) \end{bmatrix}$$

$$
= \begin{bmatrix} \sum a_{1l}g_l(\mathbf{x}_1') & \cdots & \sum a_{pl}g_l(\mathbf{x}_1') \\ \vdots & & \vdots \\ \sum a_{1l}g_l(\mathbf{x}_d') & \cdots & \sum a_{pl}g_l(\mathbf{x}_d') \end{bmatrix}
$$

$$
= \begin{bmatrix} g_1(\mathbf{x}_1') & \cdots & g_p(\mathbf{x}_1') \\ \vdots & & \vdots \\ g_1(\mathbf{x}_d') & \cdots & g_p(\mathbf{x}_d') \end{bmatrix} \begin{bmatrix} a_{11} & \cdots & a_{p1} \\ \vdots & & \vdots \\ a_{1p} & \cdots & a_{pp} \end{bmatrix}
$$

$$
= \mathbf{X}\mathbf{A}^T. \quad \blacksquare
$$

EXAMPLE **9.17** Let G be the space of quadratic polynomials on $\mathbb{R}$ and consider the design set $\{0, 1, 2\}$. Let $\mathbf{X}$ be the design matrix corresponding to the basis $1, x, x^2$ of G, let $\tilde{\mathbf{X}}$ be the design matrix corresponding to the alternative basis $1, x - 1, (x - 1)^2$ of this space, and let $\mathbf{A}$ be the coefficient matrix for the latter basis relative to the former one, which was determined in the solution to Example 9.6 in Section 9.1. Determine $\mathbf{X}$ and $\tilde{\mathbf{X}}$ and check that $\tilde{\mathbf{X}} = \mathbf{X}\mathbf{A}^T$.

Solution   Here

$$
\mathbf{X} = \begin{bmatrix} 1 & 0 & 0 \\ 1 & 1 & 1 \\ 1 & 2 & 4 \end{bmatrix}, \quad \tilde{\mathbf{X}} = \begin{bmatrix} 1 & -1 & 1 \\ 1 & 0 & 0 \\ 1 & 1 & 1 \end{bmatrix}, \quad \text{and} \quad \mathbf{A} = \begin{bmatrix} 1 & 0 & 0 \\ -1 & 1 & 0 \\ 1 & -2 & 1 \end{bmatrix}.
$$

Thus,

$$
\mathbf{X}\mathbf{A}^T = \begin{bmatrix} 1 & 0 & 0 \\ 1 & 1 & 1 \\ 1 & 2 & 4 \end{bmatrix} \begin{bmatrix} 1 & -1 & 1 \\ 0 & 1 & -2 \\ 0 & 0 & 1 \end{bmatrix} = \begin{bmatrix} 1 & -1 & 1 \\ 1 & 0 & 0 \\ 1 & 1 & 1 \end{bmatrix} = \tilde{\mathbf{X}}. \quad \blacksquare
$$

Problems

9.14   Is the space of linear functions on $\mathbb{R}^3$ identifiable or nonidentifiable relative to the design set consisting of the four points $(3, 0, 0), (0, 3, 0), (0, 0, 3), (1, 1, 1)$?

9.15   Use Rolle's theorem in calculus to show that a nonzero quadratic polynomial has at most two distinct roots.

9.16   Use Rolle's theorem to show that a nonzero cubic polynomial has at most three distinct real roots.

9.17   Let G denote the space of cubic polynomials on $\mathbb{R}$. **(a)** Show that G is nonidentifiable if and only if $d \leq 3$. **(b)** Suppose $d = 3$, $x_1' = 1$, $x_2' = 2$, and $x_3' = 3$. Give an example of a nonzero cubic polynomial that equals zero on the design set and sketch its graph over the interval $[0, 4]$.

9.18   Show that a nonzero polynomial of degree $p - 1$ has at most $p - 1$ distinct real roots.

9.19 Let G denote the space of polynomials on $\mathbb{R}$ of degree $p - 1$. Use the result of Problem 9.18 to show that G is identifiable if and only if $d \geq p$.

9.20 Let G be an identifiable linear space containing the constant functions and let $g \in G$. Show that G is constant on $\mathcal{X}$ if and only if it is constant on the design set.

9.21 Let $\mathcal{X}_1, \ldots, \mathcal{X}_p$ be nonempty sets that form a partition of $\mathcal{X}$ and let G denote the space of functions on $\mathcal{X}$ that are constant on each of the sets in the partition. Show that G is identifiable if and only if each of the sets in the partition contains one or more design points.

9.22 Consider an experiment having three factors and four design points, as shown in Table 9.1. Observe that $x_3 = 2x_1 + 3x_2$ on the design set. **(a)** Determine whether the space of linear functions

$$b_0 + b_1 x_1 + b_2 x_2 + b_3 x_3$$

of x_1, x_2, and x_3 is identifiable. **(b)** Determine whether the space of linear functions $b_0 + b_1 x_1 + b_2 x_2$ of x_1 and x_2 is identifiable.

 TABLE 9.1

x_1	x_2	x_3
-2	0	-4
0	1	3
1	1	5
1	-2	-4

9.23 Let $\mathbf{X}$ be a nonsingular matrix. If we permute the columns of $\mathbf{X}$, is the resulting matrix necessarily nonsingular (why, or why not)?

9.24 Let $\mathcal{X} = \mathbb{R}^2$; let the design set consist of the three points $(1, 0, 0)$, $(0, 1, 0)$, and $(0, 0, 1)$; and let G be the space of linear functions on $\mathbb{R}^3$. **(a)** Determine the design matrix X corresponding to the basis $1, x_1, x_2, x_3$ of G. **(b)** Determine a nonzero vector $\mathbf{b} = [\, b_0, b_1, b_2, b_3 \,]^T$ such that $\mathbf{Xb} = \mathbf{0}$.

9.25 Let G be the space of all functions on $\mathcal{X} = \{-1, 0, 1\}$ and set $I_1(x) = \mathrm{ind}(x = -1)$, $I_2(x) = \mathrm{ind}(x = 0)$, and $I_3(x) = \mathrm{ind}(x = 1)$. Let $\mathbf{X}$ be the design matrix corresponding to the basis I_1, I_2, I_3 of G; let $\tilde{\mathbf{X}}$ be the design matrix corresponding to the alternative basis $1, x, x^2$ of this space; and let $\mathbf{A}$ be the coefficient matrix for the latter basis relative to the former one, which was found in the solution to Example 9.8 in Section 9.1. **(a)** Determine $\mathbf{X}$. **(b)** Determine $\tilde{\mathbf{X}}$. **(c)** Check that $\tilde{\mathbf{X}} = \mathbf{XA}^T$.

9.3
Saturated Spaces

Recall the discussion in Section 8.1 of saturated spaces in the context of polynomial fits to the polymer data. More generally, let G be a linear space of functions on $\mathcal{X}$.

According to Corollary 9.2 in Section 9.2, G is identifiable if and only if, for each choice of $c_1, \ldots, c_d \in \mathbb{R}$, there is at most one choice of g in G such that

$$g(\mathbf{x}'_1) = c_1, \ldots, g(\mathbf{x}'_d) = c_d. \tag{11}$$

The space G is said to be *saturated* (relative to the design set) if, for each choice of $c_1, \ldots, c_d \in \mathbb{R}$, there is at least one choice of g in G such that (11) holds; otherwise, G is said to be *unsaturated*. Observe that G is saturated if and only if, when we ignore points in $\mathcal{X}$ that are not design points, G becomes the collection of all points on the design set. Moreover, G is identifiable and saturated if and only if, for each choice of $c_1, \ldots, c_d \in \mathbb{R}$, there is exactly one choice of $g \in G$ such that (11) holds.

Let G' be the space of the restrictions of the functions in G to the design set. Recall that $\dim(G') \le d$, where d is the number of design points, that $\dim(G') \le \dim(G)$, and that G is identifiable if and only if $\dim(G) = \dim(G')$. The space G is saturated if and only if G' is the space of all functions on the design set; hence, G is saturated if and only if $\dim(G') = d$. If G has a saturated subspace, then G itself is saturated. If G is saturated relative to the design set, then it is saturated relative to every nonempty subset of the design set.

EXAMPLE **9.18** Construct an example of a finite-dimensional linear space that is nonidentifiable and unsaturated.

Solution Let G be the space of linear functions on $\mathbb{R}^2$ and let the design set consist of the three points $(-1, -1)$, $(0, 0)$, $(1, 1)$. Then G is nonidentifiable according to Example 9.11 or Example 9.12 in Section 9.2. Moreover, G is unsaturated since there is no linear function $g(x_1, x_2) = b_0 + b_1 x_1 + b_2 x_2$ on $\mathbb{R}^2$ such that $g(-1, -1) = 0$, $g(0, 0) = 0$, and $g(1, 1) = 1$.

Alternatively, let G be the space of quadratic polynomials on $\mathbb{R}$ that have a root at zero (that is, the space of all functions on $\mathbb{R}$ of the form $b_1 x + b_2 x^2$) and let the design set consist of the two points 0 and 1. Then G is nonidentifiable since the nonzero function $x - x^2 = x(1 - x)$ is in G and equals zero at the design points. Moreover, G is unsaturated since there is no function $g \in G$ such that $g(0) = 1$ and $g(1) = 0$. ∎

THEOREM **9.9** (a) If the linear space G is saturated, then its dimension is at least d.

(b) If G is identifiable, then it is saturated if and only if its dimension equals d.

Proof (a) If G is saturated, then $\dim(G) \ge \dim(G') = d$ by Theorem 9.6(a) in Section 9.2.

(b) Suppose G is identifiable. Then $\dim(G) = \dim(G')$ by Theorem 9.6(b) in Section 9.2. Since G is saturated if and only if $\dim(G') = d$, we conclude that G is saturated if and only if $\dim(G) = d$. ∎

THEOREM **9.10** Suppose $\dim(G) = d$. Then G is identifiable if and only if it is saturated.

Proof   Since G is identifiable if and only if $\dim(G') = \dim(G)$, it is identifiable if and only if $\dim(G') = d$ and hence if and only if it is saturated. ■

THEOREM 9.11   The linear space G is saturated if and only if there are functions $g_1, \ldots, g_d \in G$ such that, for $1 \le k \le d$,

$$g_k(\mathbf{x}'_k) = 1 \quad \text{and} \quad g_k(\mathbf{x}'_l) = 0 \qquad \text{for } 1 \le l \le d \text{ with } l \ne k. \tag{12}$$

Proof   Suppose G is saturated and let $1 \le k \le d$. Set $c_k = 1$ and $c_l = 0$ for $1 \le l \le d$ with $l \ne k$. Then there is a function $g \in G$ satisfying (11). Thus, (12) holds with $g_k = g$.

Suppose, conversely, that (12) holds for $1 \le k \le d$. Choose $c_1, \ldots, c_d \in \mathbb{R}$ and set $g = c_1 g_1 + \cdots + c_d g_d$. Then

$$g(\mathbf{x}'_k) = \sum_l c_l g_l(\mathbf{x}'_k) = c_k g_k(\mathbf{x}'_k) + \sum_{l \ne k} c_l g_l(\mathbf{x}'_k) = c_k, \qquad 1 \le k \le d,$$

so g satisfies (11). Therefore, G is saturated. ■

EXAMPLE 9.19   Let G be the space of linear functions on $\mathbb{R}$. Show that G is saturated if and only if $d \le 2$.

Solution   Recall that G has dimension $p = 2$. Suppose G is saturated. Then $p \ge d$ by Theorem 9.9(a), so $d \le 2$.

Suppose, conversely, that $d \le 2$. In proving that G is saturated, we can assume that $d = 2$. (Recall that if G is saturated relative to the design set, then it is saturated relative to every nonempty subset of the design set.) Consider the linear functions g_1 and g_2 defined by

$$g_1(x) = \frac{x - x'_2}{x'_1 - x'_2} \quad \text{and} \quad g_2(x) = \frac{x - x'_1}{x'_2 - x'_1}, \qquad x \in \mathbb{R}.$$

Since these functions satisfy (12), G is saturated. ■

EXAMPLE 9.20   Let G be the space of quadratic polynomials on $\mathbb{R}$. Show that G is saturated if and only if $d \le 3$.

Solution   Recall that G has dimension $p = 3$. Thus, if G is saturated, then $d \le p = 3$.

Suppose, conversely, that $d \le 3$. In proving that G is saturated, we can assume that $d = 3$. Observe that the linear function

$$\frac{x - x'_2}{x'_1 - x'_2}$$

equals 1 at x'_1 and it equals 0 at x'_2. Similarly, the linear function

$$\frac{x - x'_3}{x'_1 - x'_3}$$

equals 1 at x_1' and it equals 0 at x_3'. Let g_1 denote the product of these two linear functions, which is the quadratic polynomial on $\mathbb{R}$ given by

$$g_1(x) = \frac{(x - x_2')(x - x_3')}{(x_1' - x_2')(x_1' - x_3')}.$$

Then g_1 equals 1 at x_1' and it equals 0 at x_2' and x_3'. With the same motivation, consider the quadratic polynomials g_2 and g_3 on $\mathbb{R}$ given by

$$g_2(x) = \frac{(x - x_1')(x - x_3')}{(x_2' - x_1')(x_2' - x_3')},$$

and

$$g_3(x) = \frac{(x - x_1')(x - x_2')}{(x_3' - x_1')(x_3' - x_2')}.$$

Since $g_1, g_2,$ and g_3 together satisfy (12), G is saturated.   ∎

Examples 9.19 and 9.20 have the following obvious generalization.

THEOREM 9.12   Let G be the space of polynomials on $\mathbb{R}$ of degree $p - 1$. Then G is saturated if and only if $d \leq p$.

Proof   Recall that G has dimension p. Thus, if G is saturated, then $d \leq p$.

Suppose, conversely, that $d \leq p$. In proving that G is saturated, we can assume that $d = p$. Consider the polynomials $g_1, \ldots, g_d$ of degree $p - 1$ defined by

$$g_k(x) = \prod_{l \neq k} \frac{x - x_l'}{x_k' - x_l'} = \frac{\prod_{l \neq k}(x - x_l')}{\prod_{l \neq k}(x_k' - x_l')}, \qquad x \in \mathbb{R}.$$

Since these polynomials satisfy (12), G is saturated.   ∎

The next result follows from Theorem 9.12 and the result of Problem 9.19 in Section 9.2.

COROLLARY 9.7   Let G be the space of polynomials on $\mathbb{R}$ of degree $p - 1$. Then G is identifiable and saturated if and only if $d = p$.

Let $x_1', \ldots, x_d' \in \mathbb{R}$ be distinct and let $c_1, \ldots, c_d \in \mathbb{R}$. Then there is a unique polynomial g of degree $d - 1$ such that $g(x_k') = c_k$ for $1 \leq k \leq d$, which is referred to as the *interpolation polynomial*. We see from the proof of Theorems 9.11 and 9.12 that

$$g(x) = \sum_k c_k \frac{\prod_{l \neq k}(x - x_l')}{\prod_{l \neq k}(x_k' - x_l')}, \qquad x \in \mathcal{X}. \tag{13}$$

The right side of (13) is referred to as the *Lagrange interpolation formula*.

EXAMPLE 9.21 (a) Determine the unique quadratic polynomial g such that $g(-1) = 1$, $g(0) = 2$, and $g(1) = 2$.

(b) Show the graph of $g(x)$, $-2 \le x \le 2$.

Solution (a) By (3) with $x'_1 = -1$, $x'_2 = 0$, and $x'_3 = 1$,

$$g(x) = \frac{x(x-1)}{(-1)(-2)} + 2\frac{(x+1)(x-1)}{(1)(-1)} + 2\frac{(x+1)x}{(2)(1)} = -\frac{x^2}{2} + \frac{x}{2} + 2.$$

(b) Figure 9.3 shows the graph of g. ∎

FIGURE 9.3

Quadratic interpolation polynomial

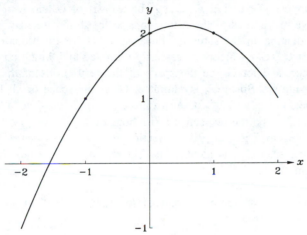

THEOREM 9.13 Let X be a subset of $\mathbb{R}^M$ containing the design set. Then there is an identifiable and saturated linear space of polynomials of degree $d - 1$ on X.

Proof We can assume that $X = \mathbb{R}^M$. If $d = 1$, then the space of constant functions is identifiable and saturated. Suppose, instead, that $d \ge 2$. We first construct a polynomial g_1 of degree $d - 1$ that equals 1 at $\mathbf{x}'_1$ and 0 at $\mathbf{x}'_l$ for $2 \le l \le d$. To this end, note first that since $\mathbf{x}'_2 \ne \mathbf{x}'_1$, there is a coordinate of $\mathbf{x}'_2$ that does not equal the corresponding coordinate of $\mathbf{x}'_1$; that is, there is an $m_2 \in \{1, \ldots, M\}$ such that $x'_{2m_2} \ne x'_{1m_2}$. Observe that the linear function

$$\frac{x_{m_2} - x'_{2m_2}}{x'_{1m_2} - x'_{2m_2}}$$

equals 1 at $\mathbf{x}'_1$ and 0 at $\mathbf{x}'_2$. Similarly, for $2 \le l \le M$, we can find a linear function

$$\frac{x_{m_l} - x'_{lm_l}}{x'_{1m_l} - x'_{lm_l}}$$

that equals 1 at $\mathbf{x}_1'$ and 0 at $\mathbf{x}_l'$. Multiplying these $d - 1$ linear functions together, we get a polynomial g_1 of degree $d - 1$ that equals 1 at $\mathbf{x}_1'$ and 0 at $\mathbf{x}_l'$ for $2 \leq l \leq d$. Similarly, for $2 \leq k \leq d$, we can construct a polynomial g_k of degree $d - 1$ that equals 1 at $\mathbf{x}_k'$ and 0 at $\mathbf{x}_l'$ for $l \neq k$. It follows from Theorems 9.10 and 9.11 that the span of $g_1, \ldots, g_d$ is an identifiable and saturated linear space of polynomials of degree $d - 1$. ∎

THEOREM 9.14   Suppose G is identifiable. Then there is an identifiable and saturated linear space G_{sat} of functions on $\mathfrak{X}$ containing G as a subspace.

Proof   If the dimension p of G equals d, then G is saturated by Theorem 9.9(b), so the desired conclusion holds with $G_{\mathrm{sat}} = G$.

Suppose $p < d$. Let $g_1, \ldots, g_p$ be a basis of G and let $g_1', \ldots, g_p'$ be the respective restrictions of these functions to the design set. By Theorem 9.13, there is an identifiable and saturated linear space G_1 of functions on $\mathfrak{X}$. Let G_2 be the span of $G \cup G_1$, which is necessarily saturated and which contains G as a proper subspace, and let G_2' be the space of the restrictions of the functions in G_2 to the design set. Since G_2 is saturated, G_2' is the space of all functions on the design set. Let $g_1', \ldots, g_d'$ be an extension of $g_1', \ldots, g_p'$ to a basis of G_2' and let $g_{p+1}', \ldots, g_d'$ be the respective restrictions of $g_{p+1}, \ldots, g_d \in G_2$ to the design set. According to Theorem 9.5 in Section 9.2, $g_1, \ldots, g_d$ are linearly independent, and their span G_{sat} is identifiable. The space G_{sat} satisfies the conclusion of the theorem. ∎

The next result is a consequence of Theorem 9.10 in this section and Corollary 9.6 in Section 9.2.

THEOREM 9.15   Let $\mathbf{X}$ be the $d \times d$ design matrix corresponding to a basis of a d-dimensional linear space G. Then the following conditions are equivalent:

(a) G is identifiable;   **(b)** G is saturated;   **(c)** $\mathbf{X}$ is invertible.

Consider now a d-dimensional linear space G and suppose that G is identifiable and hence saturated. Let $\mathbf{X}$ be the $d \times d$ design matrix corresponding to a basis $g_1, \ldots, g_d$ of G. Then $\mathbf{X}$ is invertible. Let $c_1, \ldots, c_d \in \mathbb{R}$ and set $\mathbf{c} = [c_1, \ldots, c_d]^T$. Consider the equation

$$g(\mathbf{x}_k') = c_k, \qquad g \in G \text{ and } 1 \leq k \leq d. \tag{14}$$

Write $g = b_1 g_1 + \cdots + b_d g_d$ and set $\mathbf{b} = [b_1, \ldots, b_d]^T$. Since

$$g(\mathbf{x}_k') = b_1 g_1(\mathbf{x}_k') + \cdots + b_d g_d(\mathbf{x}_k'), \qquad 1 \leq k \leq d,$$

(14) can be written in matrix form as $\mathbf{Xb} = \mathbf{c}$, the unique solution to which is given by

$$\mathbf{b} = \mathbf{X}^{-1}\mathbf{c}. \tag{15}$$

EXAMPLE **9.22** Consider the basis $1, x, x^2$ of the space of quadratic polynomials on $\mathbb{R}$ and the design set $\{-1, 0, 1\}$. Determine

(a) the design matrix $\mathbf{X}$;   (b) $\mathbf{X}^{-1}$;

(c) the unique quadratic polynomial such that $g(-1) = 1$, $g(0) = 2$, and $g(1) = 2$.

Solution (a) Label the design points in increasing order as $x_1' = -1$, $x_2' = 0$, and $x_3' = 1$. Correspondingly, the design matrix is given by

$$\mathbf{X} = \begin{bmatrix} 1 & -1 & 1 \\ 1 & 0 & 0 \\ 1 & 1 & 1 \end{bmatrix}.$$

(b) Now $\det(\mathbf{X}) = 2$, so

$$\mathbf{X}^{-1} = \frac{1}{2} \begin{bmatrix} 0 & 2 & 0 \\ -1 & 0 & 1 \\ 1 & -2 & 1 \end{bmatrix} = \begin{bmatrix} 0 & 1 & 0 \\ -\frac{1}{2} & 0 & \frac{1}{2} \\ \frac{1}{2} & -1 & \frac{1}{2} \end{bmatrix}.$$

(c) By (15), the coefficient vector of the desired quadratic polynomial

$$g(x) = b_0 + b_1 x + b_2 x^2$$

is given by

$$\begin{bmatrix} b_0 \\ b_1 \\ b_2 \end{bmatrix} = \begin{bmatrix} 0 & 1 & 0 \\ -\frac{1}{2} & 0 & \frac{1}{2} \\ \frac{1}{2} & -1 & \frac{1}{2} \end{bmatrix} \begin{bmatrix} 1 \\ 2 \\ 2 \end{bmatrix} = \begin{bmatrix} 2 \\ \frac{1}{2} \\ -\frac{1}{2} \end{bmatrix},$$

so the polynomial itself is given by $g(x) = -x^2/2 + x/2 + 2$. [This agrees with the solution to Example 9.21(a).] ∎

Problems

9.26 Is the space of quadratic polynomials on $\mathbb{R}^2$ saturated or unsaturated relative to the design set consisting of the four points $(-1, -1)$, $(0, 0)$, $(1, 1)$, $(2, 2)$?

9.27 Let $X_1, \ldots, X_p$ be nonempty sets that form a partition of X and let G denote the space of functions on X that are constant on each of the sets in the partition. Show that G is saturated if and only if each of the sets in the partition contains at most one design point.

9.28 Show that the space of functions on $\mathbb{R}$ spanned by $\sin x$ and $\cos x$ is saturated relative to the design set $\{x_1', x_2'\}$, where $0 \leq x_1' < x_2' < \pi$.

9.29 (a) Determine the unique cubic polynomial g on $\mathbb{R}$ such that $g(-3) = 1$, $g(-1) = 2$, $g(1) = 2$, and $g(3) = 0$. (b) Sketch the graph of $g(x)$, $-4 \leq x \leq 4$.

9.30 Let G be the four-dimensional space of linear functions on $\mathbb{R}^3$. Determine a subset X' of $\mathbb{R}^3$ of size 4 such that G is identifiable and hence saturated relative to X'.

9.31 Consider the basis 1, x, x^2, x^3 of the space of cubic polynomials on $\mathbb{R}$ and the design set $\{-3, -1, 1, 3\}$. **(a)** Determine the design matrix (label the design points in increasing order). **(b)** The inverse to the design matrix is given by

$$
\mathbf{X}^{-1} = \frac{1}{48}
\begin{bmatrix}
-3 & 27 & 27 & -3 \\
1 & -27 & 27 & -1 \\
3 & -3 & -3 & 3 \\
-1 & 3 & -3 & 1
\end{bmatrix}.
$$

Use Equation (15) to determine the unique cubic polynomial g such that $g(-3) = 1$, $g(-1) = 2$, $g(1) = 2$, and $g(3) = 0$. [The answer should agree with the answer to Problem 9.29(a).]

9.32 Let G be the space of linear functions on $\mathbb{R}^2$ and let the design set have three points. Determine a necessary and sufficient condition on these points for G to be saturated.

9.33 Let G be the space of linear functions on $\mathbb{R}^3$ and let the design set have four points. Determine a necessary and sufficient condition on these points for G to be saturated.

9.34 Let G be identifiable and saturated and let $\mathbf{X}$ be the design matrix corresponding to a basis $g_1, \ldots, g_d$ of G. For $1 \le k \le d$, let $\tilde{g}_k$ be the function in G such that $\tilde{g}_k(\mathbf{x}'_k) = 1$ and $\tilde{g}_k(\mathbf{x}'_l) = 0$ for $l \ne k$. Determine **(a)** the design matrix relative to $\tilde{g}_1, \ldots, \tilde{g}_d$; **(b)** the coefficient matrix of $\tilde{g}_1, \ldots, \tilde{g}_d$ relative to $g_1, \ldots, g_d$.

9.35 Let G be the space of linear functions on $\mathbb{R}^2$ and let the design set consist of the three points $\mathbf{x}'_1 = (1, 0)$, $\mathbf{x}'_2 = (0, 1)$, and $\mathbf{x}'_3 = (-1, 0)$. **(a)** Determine the design matrix relative to the basis $1, x_1, x_2$ of G. **(b)** For $1 \le k \le 3$, determine the function $g_k \in G$ such that $g_k(\mathbf{x}'_k) = 1$ and $g_k(\mathbf{x}'_l) = 0$ when $l \ne k$.

9.36 Let G be the four-dimensional space of functions on $\mathbb{R}^2$ spanned by $1, x_1, x_2, x_1 x_2$ and let the design set consist of the following four points: $\mathbf{x}'_1 = (0, 0)$, $\mathbf{x}'_2 = (1, 0)$, $\mathbf{x}'_3 = (0, 1)$, and $\mathbf{x}'_4 = (1, 1)$. **(a)** Determine the design matrix relative to the basis $1, x_1, x_2, x_1 x_2$ of G. **(b)** Show that G is identifiable and hence saturated.

9.37 In the context of the experimental design in Problem 9.22 in Section 9.2, determine whether the space of linear functions of x_1, x_2, and x_3 is saturated.

9.38 Consider the four design points $\mathbf{x}'_1 = (2, 1)$, $\mathbf{x}'_2 = (3, 2)$, $\mathbf{x}'_3 = (2, 3)$, and $\mathbf{x}'_4 = (1, 2)$. Determine quadratic polynomials g_1, g_2, g_3, and g_4 such that $g_k(\mathbf{x}'_k) = 1$ and $g_k(\mathbf{x}'_l) = 0$ for $1 \le k, l \le 4$ with $l \ne k$.

9.4
Inner Products

Conceptually, least-squares estimation is best thought of as orthogonal projection relative to a certain inner product. The appropriate inner product will be introduced in this section and the corresponding orthogonal projections in the next one.

In the context of least-squares estimation, we have measurements (generally with error) of the regression function only at the design points. To incorporate this setup within our general framework, it is convenient to let H denote the space of all functions whose domain is a subset of $\mathcal{X}$ that includes the design set. (Given functions $h_1, h_2 \in H$, the domain of their sum $h_1 + h_2$ is the intersection of the domain of h_1 and that of h_2.) If $\mathcal{X}$ coincides with the design set, then H is the space of all functions on this set. Observe that the function $\bar{Y}(\cdot)$ defined in Section 8.3 by $\bar{Y}(\mathbf{x}'_k) = \bar{Y}_k$ for $1 \leq k \leq d$ is a member of H.

Let $w_1, \ldots, w_d$ be positive numbers, which are referred to as *weights*; if $w_1 = \cdots = w_d$, we refer to them as *constant weights*, and if $w_1 = \cdots = w_d = 1$, we refer to them as *unit weights*. (In application to least-squares estimation, the weight w_k associated with the kth design point is the number n_k of repetitions at that design point. Here we get constant weights if and only if there is a common number of repetitions at the various design points, and we get unit weights if and only if there is one trial at each design point. Also the sum $w_1 + \cdots + w_d$ of the weights equals the total number $n = n_1 + \cdots + n_d$ of trials in the experiment.)

Given functions $h_1, h_2 \in H$, we define their *inner product* $\langle h_1, h_2 \rangle$ by

$$\langle h_1, h_2 \rangle = \sum_k w_k h_1(\mathbf{x}'_k) h_2(\mathbf{x}'_k).$$

[In applications to least-squares estimation, $\langle h_1, h_2 \rangle = \sum_i h_1(\mathbf{x}_i) h_2(\mathbf{x}_i)$.] Observe that $\langle h_1, h_2 \rangle$ depends only on the values of h_1 and h_2 on the design set. More generally, if $h_1, h_2, h_3, h_4 \in H$ and $h_1 h_2 = h_3 h_4$ on the design set, then $\langle h_1, h_2 \rangle = \langle h_3, h_4 \rangle$. Also, if $h_1 \in H$ and $h_1 = 0$ on the design set, then $\langle h_1, h_2 \rangle = 0$ for $h_2 \in H$.

The inner product, as defined above, satisfies the following four basic properties:

$$\langle h_1, h_2 \rangle = \langle h_2, h_1 \rangle; \tag{16}$$

$$\langle h_1 + h_2, h_3 \rangle = \langle h_1, h_3 \rangle + \langle h_2, h_3 \rangle; \tag{17}$$

$$\langle bh_1, h_2 \rangle = b\langle h_1, h_2 \rangle; \tag{18}$$

$$\langle h, h \rangle \geq 0. \tag{19}$$

It follows from (16) through (18) that

$$\left\langle \sum_j a_j g_j, \sum_l b_l h_l \right\rangle = \sum_j \sum_l a_j b_l \langle g_j, h_l \rangle. \tag{20}$$

In particular,

$$\left\langle \sum_l a_l h_l, \sum_l b_l h_l \right\rangle = \sum_j \sum_l a_j b_l \langle h_j, h_l \rangle \tag{21}$$

and

$$\left\langle \sum_l b_l h_l, \sum_l b_l h_l \right\rangle = \sum_j \sum_l b_j b_l \langle h_j, h_l \rangle. \tag{22}$$

The $q \times q$ matrix

$$\mathbf{M} = \begin{bmatrix} \langle h_1, h_1 \rangle & \cdots & \langle h_1, h_q \rangle \\ \vdots & & \vdots \\ \langle h_q, h_1 \rangle & \cdots & \langle h_q, h_q \rangle \end{bmatrix}$$

is referred to as the *Gram matrix* of $h_1, \ldots, h_q$. It follows from (16) that the Gram matrix is symmetric. Given $b_1, \ldots, b_q \in \mathbb{R}$, set $\mathbf{b} = [b_1, \ldots, b_q]^T$ and $h = b_1 h_1 + \cdots + b_q h_q$. According to (22),

$$\langle h, h \rangle = \mathbf{b}^T \mathbf{M} \mathbf{b}. \tag{23}$$

It follows from (19) and (23) that $\mathbf{M}$ is positive semidefinite and hence that its determinant is nonnegative. In particular, by choosing $q = 2$, we see that

$$\begin{bmatrix} \langle h_1, h_1 \rangle & \langle h_1, h_2 \rangle \\ \langle h_2, h_1 \rangle & \langle h_2, h_2 \rangle \end{bmatrix}$$

has a nonnegative determinant; Thus,

$$\langle h_1, h_1 \rangle \langle h_2, h_2 \rangle - (\langle h_1, h_2 \rangle)^2 \geq 0$$

and hence

$$(\langle h_1, h_2 \rangle)^2 \leq \langle h_1, h_1 \rangle \langle h_2, h_2 \rangle. \tag{24}$$

This inequality is known as the *Schwarz inequality*.

For $h \in H$, set $\|h\| = \sqrt{\langle h, h \rangle}$; we refer to $\|h\|$ as the *norm* of h and to

$$\|h\|^2 = \langle h, h \rangle = \sum_k w_k h^2(x'_k)$$

as its squared norm. Observe that $\|1\|^2 = w_1 1^2 + \cdots + w_d 1^2 = w_1 + \cdots + w_d$; in particular, if $w_1 = \cdots = w_d = r$, then $\|1\|^2 = rd$. [In applications to least-squares estimation, $\|h\|^2 = \sum_i h^2(\mathbf{x}_i)$; in particular, $\|1\|^2 = n$.] The Schwarz inequality can be written as

$$|\langle h_1, h_2 \rangle| \leq \|h_1\| \|h_2\|. \tag{25}$$

According to (22),

$$\left\| \sum_j b_j h_j \right\|^2 = \sum_j \sum_l b_j b_l \langle h_j, h_l \rangle$$

and hence

$$\left\| \sum_j b_j h_j \right\|^2 = \sum_j b_j^2 \|h_j\|^2 + \sum_j \sum_{l \neq j} b_j b_l \langle h_j, h_l \rangle. \tag{26}$$

In particular,

$$\|b_1 h_1 + b_2 h_2\|^2 = b_1^2 \|h_1\|^2 + b_2^2 \|h_2\|^2 + 2 b_1 b_2 \langle h_1, h_2 \rangle. \tag{27}$$

We refer to $\| \cdot \|$ as being a norm on H. Observe that

$$\|h\| \geq 0 \tag{28}$$

and

$$\|bh\| = |b| \|h\|. \tag{29}$$

Suppose $\|h\| > 0$. Then $h/\|h\|$ has norm 1 since, by (29),

$$\left\|\frac{1}{\|h\|}h\right\| = \frac{1}{\|h\|}\|h\| = 1.$$

Since

$$\|h\|^2 = \langle h, h \rangle = \sum_k w_k h^2(x'_k),$$

the norm of h is determined by the values of h on the design set; moreover, the norm of h equals zero if and only if h equals zero on the design set.

The norm satisfies the *triangle inequality*

$$\|h_1 + h_2\| \le \|h_1\| + \|h_2\|, \qquad h_1, h_2 \in H.$$

To verify this inequality, observe that, by (25) and (27),

$$\begin{aligned}
\|h_1 + h_2\|^2 &= \|h_1\|^2 + \|h_2\|^2 + 2\langle h_1, h_2 \rangle \\
&\le \|h_1\|^2 + \|h_2\|^2 + 2\|h_1\|\,\|h_2\| \\
&= (\|h_1\| + \|h_2\|)^2;
\end{aligned}$$

taking square roots, we get the desired result.

EXAMPLE **9.23**   Express the lack of fit sum of squares as a squared norm.

Solution   Set $w_k = n_k$ for $1 \le k \le d$. Then, by Equation (6) in Section 8.3,

$$\text{LSS}(g) = \sum_k w_k[\bar{Y}(x'_k) - g(x'_k)]^2 = \|\bar{Y}(\cdot) - g\|^2. \quad\blacksquare$$

Two functions, h_1 and h_2, are said to be *orthogonal* if $\langle h_1, h_2 \rangle = 0$. If h is orthogonal to $h_1, \ldots, h_q$, then h is orthogonal to every linear combination of $h_1, \ldots, h_q$. More generally, the functions $h_1, \ldots, h_q$ are said to be orthogonal if $\langle h_j, h_l \rangle = 0$ for $1 \le j, l \le q$ with $j \ne l$ or, equivalently, if the Gram matrix of $h_1, \ldots, h_q$ is a diagonal matrix. (By definition, if $q = 1$, then $h_1, \ldots, h_q$ are orthogonal.) Suppose $h_1, \ldots, h_q$ are orthogonal. Then, by (26), $\|\sum_j h_j\|^2 = \sum_j \|h_j\|^2$. This result is known as the *Pythagorean theorem* (see Figure 9.4). It follows from (26) or the Pythagorean theorem that

$$\left\|\sum_j b_j h_j\right\|^2 = \sum_j b_j^2 \|h_j\|^2. \tag{30}$$

It follows from (22) that

$$\left\langle \sum_j a_j h_j, \sum_j b_j h_j \right\rangle = \sum_j a_j b_j \|h_j\|^2. \tag{31}$$

FIGURE 9.4

The Pythagorean theorem. If h_1, h_2, and h_3 are orthogonal and $h = h_1 + h_2 + h_3$, then $\|h\|^2 = \|h_1\|^2 + \|h_2\|^2 + \|h_3\|^2$.

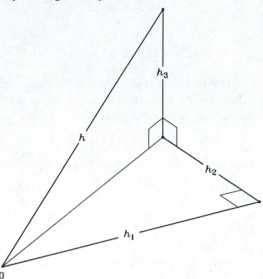

The functions $h_1, \ldots, h_q$ are said to be *orthonormal* if they are orthogonal functions each having norm 1 or, equivalently, if their Gram matrix is the $q \times q$ identity matrix. Suppose $h_1, \ldots, h_q$ are orthonormal. Then (30) and (31) reduce to

$$\left\| \sum_j b_j h_j \right\|^2 = \sum_j b_j^2 \tag{32}$$

and

$$\left\langle \sum_j a_j h_j, \sum_j b_j h_j \right\rangle = \sum_j a_j b_j. \tag{33}$$

THEOREM 9.16   Let $h_1, \ldots, h_q \in H$. If these functions are orthogonal and each of them has positive norm, then they are linearly independent. In particular, if these functions are orthonormal, then they are linearly independent.

Proof   Let $h_1, \ldots, h_q$ be orthogonal functions each having positive norm and let $b_1, \ldots, b_q \in \mathbb{R}$ be such that $b_1 h_1 + \cdots + b_q h_q = 0$. Then, by (30),

$$0 = \|b_1 h_1 + \cdots + b_q h_q\|^2 = b_1^2 \|h_1\|^2 + \cdots + b_q^2 \|h_q\|^2,$$

so $b_1 = \cdots = b_q = 0$. Therefore, $h_1, \ldots, h_q$ are linearly independent.   ∎

Let G be a linear space of functions on $\mathcal{X}$. Then $G \subset H$, so the inner product $\langle g_1, g_2 \rangle$ is well defined for $g_1, g_2 \in G$.

THEOREM 9.17   The linear space G is identifiable if and only if every nonzero function in G has positive norm.

Proof   Suppose G is identifiable and let g be a nonzero function in G. Then there is a k such that $g(x'_k) \neq 0$. Consequently,

$$\|g\|^2 = \langle g, g \rangle = \sum_k w_k g^2(\mathbf{x}'_k) > 0. \tag{34}$$

Suppose, conversely, that every nonzero function in G has positive norm and let g be a nonzero function in G. Then it follows from (34) that g does not equal zero on the design set. Therefore G is identifiable.   ∎

COROLLARY 9.8   Let $g_1, \ldots, g_p$ form a basis of an identifiable linear space. Then each of these functions has positive norm.

Proof   Since each function in a basis is a nonzero function, the desired result follows from Theorem 9.17.   ∎

THEOREM 9.18   Let $g_1, \ldots, g_p$ form a basis of G. Then their Gram matrix is invertible if and only if G is identifiable.

Proof   Suppose G is nonidentifiable. By Theorem 9.17, there is a nonzero function $g \in G$ such that $\langle g, g \rangle = 0$. Now $g = b_1 g_1 + \cdots + b_p g_p$, where $b_1, \ldots, b_p$ are not all zero. Set $\mathbf{b} = [b_1, \ldots, b_p]^T$ and let $\mathbf{M}$ denote the Gram matrix of $g_1, \ldots, g_p$. Then $\mathbf{b} \neq \mathbf{0}$ and $0 = \langle g, g \rangle = \mathbf{b}^T \mathbf{M} \mathbf{b}$ by (23), so $\mathbf{M}$ fails to be positive definite and hence fails to be invertible. Suppose, conversely, that $\mathbf{M}$ is noninvertible and hence fails to be positive definite. Then there is a nonzero vector $\mathbf{b} = [b_1, \ldots, b_p]^T$ such that $\mathbf{b}^T \mathbf{M} \mathbf{b} = 0$. Set $g = b_1 g_1 + \cdots + b_p g_p$. Then g is a nonzero function in G and $\langle g, g \rangle = \mathbf{b}^T \mathbf{M} \mathbf{b} = 0$ by (23). It now follows from Theorem 9.17 that G is nonidentifiable. Therefore, $\mathbf{M}$ is noninvertible if and only if G is nonidentifiable; equivalently, $\mathbf{M}$ is invertible if and only if G is identifiable.   ∎

EXAMPLE 9.24   Consider the basis $1, x_1, x_2$ of the space G of linear functions on $\mathbb{R}^2$; the design points $(-1, -1)$, $(0, 0)$, and $(1, 1)$; and unit weights. Recall from Example 9.12 in Section 9.2 that G is nonidentifiable. Determine the Gram matrix and show directly that it is noninvertible.

Solution   Observe that

$$\langle 1, 1 \rangle = 1 \cdot 1 + 1 \cdot 1 + 1 \cdot 1 = 3,$$

$$\langle 1, x_1 \rangle = 1 \cdot (-1) + 1 \cdot 0 + 1 \cdot 1 = 0,$$

$$\langle 1, x_2 \rangle = 1 \cdot (-1) + 1 \cdot 0 + 1 \cdot 1 = 0,$$

$$\langle x_1, x_1 \rangle = (-1)^2 + 0^2 + 1^2 = 2,$$

$$\langle x_1, x_2 \rangle = (-1) \cdot (-1) + 0 \cdot 0 + 1 \cdot 1 = 2,$$

and

$$\langle x_2, x_2 \rangle = (-1)^2 + 0^2 + 1^2 = 2.$$

Thus, the Gram matrix is given by

$$\mathbf{M} = \begin{bmatrix} 3 & 0 & 0 \\ 0 & 2 & 2 \\ 0 & 2 & 2 \end{bmatrix}.$$

Since the determinant of this matrix equals zero, it is noninvertible.   ∎

THEOREM 9.19   Let $\mathbf{M}$ denote the Gram matrix of a basis $g_1, \ldots, g_p$ of G, let $\tilde{\mathbf{M}}$ denote the Gram matrix of an alternative basis $\tilde{g}_1, \ldots, \tilde{g}_p$, and let $\mathbf{A}$ denote the coefficient matrix of $\tilde{g}_1, \ldots, \tilde{g}_p$ relative to $g_1, \ldots, g_p$. Then $\tilde{\mathbf{M}} = \mathbf{A}\mathbf{M}\mathbf{A}^T$.

Proof   To verify the desired result, observe that, for $1 \leq i, j \leq p$, the entry in row i and column j of $\tilde{\mathbf{M}}$ is given by

$$\langle \tilde{g}_i, \tilde{g}_j \rangle = \left\langle \sum_l a_{il} g_l, \sum_m a_{jm} g_m \right\rangle = \sum_l \sum_m a_{il} a_{jm} \langle g_l, g_m \rangle,$$

which is the entry in row i and column j of $\mathbf{A}\mathbf{M}\mathbf{A}^T$.   ∎

We refer to the $d \times d$ matrix $\mathbf{W} = \mathrm{diag}(w_1, \ldots, w_d)$ as the *weight matrix*. Observe that $w_1 = \cdots = w_d = r$ if and only if $\mathbf{W} = r\mathbf{I}_d$, where $\mathbf{I}_d$ is the $d \times d$ identity matrix. In particular, $w_1, \ldots, w_d$ are unit weights if and only if $\mathbf{W} = \mathbf{I}_d$.

THEOREM 9.20   Let $\mathbf{X}$ and $\mathbf{M}$ denote, respectively, the design matrix and Gram matrix corresponding to a basis $g_1, \ldots, g_p$ of G. Then $\mathbf{M} = \mathbf{X}^T \mathbf{W} \mathbf{X}$.

Proof   The desired conclusion follows from the formula

$$\langle g_j, g_l \rangle = \sum_k w_k g_j(x_k') g_l(x_k'), \qquad 1 \leq j, l \leq p.   ∎$$

Suppose G is identifiable and saturated. Then $p = d$ and $\mathbf{X}$ is invertible (see Theorems 9.9 and 9.15 in Section 9.3.) Consequently, by Theorem 9.20, $\mathbf{X}^{-1} = \mathbf{M}^{-1}\mathbf{X}^T\mathbf{W}$. If $\mathbf{M}$ is a block diagonal matrix with blocks of small size (in particular, if each basis function is orthogonal to all or all but one of the other basis functions), then this formula can be useful in hand computation of the inverse of the design matrix.

COROLLARY 9.9   Suppose $w_1 = \cdots = w_d = r$. Then $\mathbf{M} = r\mathbf{X}^T\mathbf{X}$. In particular, if $w_1, \ldots, w_d$ are unit weights, then $\mathbf{M} = \mathbf{X}^T\mathbf{X}$.

Orthogonal and Orthonormal Bases

Let G be a p-dimensional linear space, where $p \geq 1$. A basis of G consisting of orthogonal functions is said to be an *orthogonal basis*, and a basis consisting of orthonormal functions is said to be an *orthonormal basis*. Let $g_1, \ldots, g_p \in G$ be orthogonal functions having positive norms. Then these functions are linearly independent by Theorem 9.16. Thus, they form an orthogonal basis of G, and hence $g_1/\|g_1\|, \ldots, g_p/\|g_p\|$ form an orthonormal basis of this space.

THEOREM 9.21   Let $g_1, \ldots, g_p$ be orthogonal functions in a linear space G that span this space, with each of these functions having positive norm. Then G is identifiable and has dimension p, and $g_1, \ldots, g_p$ form an orthogonal basis of G.

Proof   Since $g_1, \ldots, g_p$ are linearly independent by Theorem 9.16, they form an orthogonal basis of G, and hence G has dimension p. Let g be a nonzero function in G. Then $g = b_1 g_1 + \cdots + b_p g_p$, where $b_1, \ldots, b_p \in \mathbb{R}$ are not all equal to zero. Thus, $\|g\|^2 = b_1^2 \|g_1\|^2 + \cdots + b_p^2 \|g_p\|^2 > 0$; that is, g has positive norm. We conclude from Theorem 9.17 that G is identifiable. ∎

COROLLARY 9.10   Let $\mathcal{X}_1, \ldots, \mathcal{X}_p$ form a partition of $\mathcal{X}$ such that each of the sets in the partition contains one or more design points, let G denote the space of functions on $\mathcal{X}$ that are constant on each of the sets in the partition, and set $I_j(\mathbf{x}) = \mathrm{ind}(\mathbf{x} \in \mathcal{X}_j)$ for $1 \leq j \leq p$. Then G is identifiable and $I_1, \ldots, I_p$ form an orthogonal basis of G.

Proof   By Theorem 9.2 in Section 9.1, $I_1, \ldots, I_p$ form a basis of G. Let $1 \leq j \leq p$. Then $\|I_j\|^2 = \sum_k w_k I_j^2(\mathbf{x}_k') = \sum_k w_k I_j(\mathbf{x}_k')$, which equals the summation of w_k over all k such that the kth design point is in $\mathcal{X}_k$. Since there is at least one such design point, I_j has positive norm. Let $1 \leq j, l \leq p$ with $j \neq l$. Then $I_j I_l = 0$, and hence

$$\langle I_j, I_l \rangle = \sum_k w_k I_j(x_k') I_l(x_k') = 0.$$

Thus, $I_1, \ldots, I_p$ are orthogonal, so they form an orthogonal basis of G. Moreover, it follows from Theorem 9.21 that G is identifiable. ∎

COROLLARY 9.11   Set $I_k(\mathbf{x}) = \mathrm{ind}(\mathbf{x} = \mathbf{x}_k')$ for $1 \leq k \leq d$. Then $I_1, \ldots, I_d$ form an orthogonal basis of the space of all functions on the design set, and the squared norms of these basis functions are given by $\|I_k\|^2 = w_k$ for $1 \leq k \leq d$.

Proof   The orthogonality of $I_1, \ldots, I_d$ is a special case of Corollary 9.10. The squared norms of the basis functions are given by $\|I_j\|^2 = \sum_k w_k I_j(\mathbf{x}_k') = w_j$ for $1 \leq j \leq d$ or, equivalently, by $\|I_k\|^2 = w_k$ for $1 \leq k \leq d$. ∎

EXAMPLE 9.25   Consider the design set $\{-1, 1\}$ and suppose $w_1 = w_2 = r$.

(a) Show that 1 and x form an orthogonal basis of the space G of linear functions on $\mathbb{R}$.

(b) Determine the squared norms of these functions.

Solution (a) We already know that 1 and x form a basis of G. Now

$$\langle 1, x \rangle = r(1)(-1) + r(1)(1) = 0,$$

so 1 and x are orthogonal.

(b) The squared norms of these functions are given by $\|1\|^2 = 2r$ and

$$\|x\|^2 = r(-1)^2 + r(1)^2 = 2r. \quad \blacksquare$$

EXAMPLE **9.26** Consider the design set $\{-1, 0, 1\}$ and suppose $w_1 = w_2 = w_3 = r$.

(a) Determine an orthogonal basis of the space G of quadratic polynomials on $\mathbb{R}$ having the form $1, x, x^2 - c$ for some $c \in \mathbb{R}$.

(b) Determine the squared norms of the functions in this basis.

(c) Show the graphs of the corresponding orthonormal functions when $r = 1$ (unit weights).

Solution (a) Since 1, x, and x^2 form a basis of G, so do 1, x, and $x^2 - c$ for $c \in \mathbb{R}$ (see Example 9.5 in Section 9.1). Now

$$\langle 1, x \rangle = r(1)(-1) + r(1)(0) + r(1)(1) = 0.$$

Also,

$$\langle x, x^2 \rangle = r(-1)(-1)^2 + r(0)(0)^2 + r(1)(1)^2 = 0,$$

so

$$\langle x, x^2 - c \rangle = \langle x, x^2 \rangle - c\langle 1, x \rangle = 0.$$

Moreover, $\langle 1, 1 \rangle = 3r$ and

$$\langle 1, x^2 \rangle = r(1)(-1)^2 + r(1)(0)^2 + r(1)(1)^2 = 2r,$$

so

$$\langle 1, x^2 - c \rangle = \langle 1, x^2 \rangle - c\langle 1, 1 \rangle = (2 - 3c)r,$$

which equals zero if and only if $c = 2/3$. Thus, 1, x, and $x^2 - 2/3$ form an orthogonal basis of G.

(b) The squared norms of these functions are given by $\|1\|^2 = 3r$,

$$\|x\|^2 = \langle x, x \rangle = \langle 1, x^2 \rangle = 2r,$$

and

$$\|x^2 - \tfrac{2}{3}\|^2 = r(\tfrac{1}{3})^2 + r(-\tfrac{2}{3})^2 + r(\tfrac{1}{3})^2 = \tfrac{2}{3}r.$$

(c) Figure 9.5 shows the graphs of the orthonormal functions $1/\sqrt{3}$, $x/\sqrt{2}$, and $(x^2 - 2/3)/\sqrt{2/3}$ corresponding to unit weights. $\quad \blacksquare$

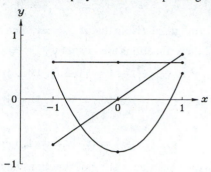

FIGURE 9.5

Orthonormal polynomials corresponding to unit weights

EXAMPLE 9.27 Consider the complete factorial design shown in Table 9.2, let $w_1 = w_2 = w_3 = w_4 = r$, and let G be the space of polynomials on $\mathbb{R}^2$ spanned by $1, x_1, x_2, x_1x_2$. It follows from Theorem 9.1 in Section 9.1 that $1, x_1, x_2, x_1x_2$ are linearly independent and hence form a basis of G.

(a) Show that $1, x_1, x_2, x_1x_2$ are orthogonal.

(b) Determine the squared norms of these functions.

(c) Show that G is identifiable and saturated.

(d) Determine the design matrix corresponding to the indicated basis.

(e) Determine the Gram matrix.

TABLE 9.2

x_1	x_2
-1	-1
1	-1
-1	1
1	1

Solution **(a)** Now

$$\langle 1, x_1 \rangle = r(1)(-1) + r(1)(1) + r(1)(-1) + r(1)(1) = 0,$$

$$\langle 1, x_2 \rangle = r(1)(-1) + r(1)(-1) + r(1)(1) + r(1)(1) = 0,$$

and

$$\langle 1, x_1x_2 \rangle = \langle x_1, x_2 \rangle = r(-1)(-1) + r(1)(-1) + r(-1)(1) + r(1)(1) = 0.$$

Moreover,

$$\langle x_1, x_1x_2 \rangle = \langle x_1^2, x_2 \rangle = \langle 1, x_2 \rangle = 0,$$

and

$$\langle x_2, x_1x_2 \rangle = \langle x_2^2, x_1 \rangle = \langle 1, x_1 \rangle = 0,$$

so $1, x_1, x_2, x_1x_2$ are orthogonal. (Note that if $x = \pm 1$, then $x^2 = 1$.)

(b) The squared norms of these functions are given by $\|1\|^2 = \langle 1, 1 \rangle = 4r$,

$$\|x_1\|^2 = \langle x_1, x_1 \rangle = \langle x_1^2, 1 \rangle = \langle 1, 1 \rangle = 4r,$$

$$\|x_2\|^2 = \langle x_2, x_2 \rangle = \langle x_2^2, 1 \rangle = \langle 1, 1 \rangle = 4r,$$

and

$$\|x_1x_2\|^2 = \langle x_1x_2, x_1x_2 \rangle = \langle x_1^2, x_2^2 \rangle = \langle 1, 1 \rangle = 4r.$$

(c) Since G has an orthogonal basis consisting of functions having positive norm, it is identifiable. Therefore, since the dimension of G coincides with the number of design points, this space is saturated.

(d) The design matrix is given by

$$\mathbf{X} = \begin{bmatrix} 1 & -1 & -1 & 1 \\ 1 & 1 & -1 & -1 \\ 1 & -1 & 1 & -1 \\ 1 & 1 & 1 & 1 \end{bmatrix}.$$

(e) Since the basis functions are orthogonal, their Gram matrix $\mathbf{M}$ is the diagonal matrix whose diagonal entries equal their respective squared norms. Thus, by the solution to part (b), $\mathbf{M} = 4r\mathbf{I}_4$. ∎

EXAMPLE 9.28 Consider the fractional factorial design shown in Table 9.3. (Observe that the design involves all ordered triples of $x_1, x_2, x_3 = \pm 1$ such that $x_1x_2x_3 = 1$.) Let $w_1 = w_2 = w_3 = w_4 = r$ and let G be the space of linear functions on $\mathbb{R}^3$, which has the basis $1, x_1, x_2, x_3$.

(a) Determine the design matrix.   **(b)** Determine the Gram matrix.

(c) Show that $1, x_1, x_2, x_3$ are orthogonal.

(d) Determine the squared norms of these functions.

(e) Show that G is identifiable and saturated.

TABLE 9.3

x_1	x_2	x_3
−1	−1	1
1	−1	−1
−1	1	−1
1	1	1

Solution **(a)** The design matrix is given by the solution to Example 9.27(d).

(b) Since the design matrix and the weight matrix in this example coincide with those in Example 9.27, we conclude from Theorem 9.20 that the Gram matrix **M** in this example coincides with that given in the solution to Example 9.27(e); namely, $\mathbf{M} = 4r\mathbf{I}_4$.

(c) Since the Gram matrix is a diagonal matrix, the basis functions are orthogonal.

(d) Since the squared norms of the basis functions equal the respective diagonal entries of the Gram matrix, we conclude that $\|1\|^2 = \|x_1\|^2 = \|x_2\|^2 = \|x_3\|^2 = 4r$.

(e) The proof that G is identifiable and saturated is identical to that given in the solution to Example 9.27(c). ∎

It is easy to give a direct solution to parts (a) and (b) of Example 9.28 that does not involve Example 9.27. The point of using Example 9.27 as we did is to illustrate (with $M = 2$) a method for constructing a practically useful fractional factorial design involving $2^M - 1$ factors each at two levels and 2^M design points starting with a complete factorial design involving M factors each at two levels.

THEOREM 9.22 Let $g_1, \ldots, g_p$ form an orthogonal basis of an identifiable linear space G and let

$$g = b_1 g_1 + \cdots + b_p g_p \in G.$$

Then

$$b_j = \frac{\langle g_j, g \rangle}{\|g_j\|^2}, \qquad 1 \le j \le p, \tag{35}$$

and

$$\|g\|^2 = \sum_j \frac{(\langle g_j, g \rangle)^2}{\|g_j\|^2}. \tag{36}$$

In particular, if $g_1, \ldots, g_p$ form an orthonormal basis of G, then

$$b_j = \langle g_j, g \rangle, \qquad 1 \le j \le p, \tag{37}$$

and

$$\|g\|^2 = \sum_j (\langle g_j, g \rangle)^2. \tag{38}$$

Proof Now

$$\langle g_j, g \rangle = \langle g_j, b_1 g_1 + \cdots + b_p g_p \rangle$$
$$= b_1 \langle g_j, g_1 \rangle + \cdots + b_p \langle g_j, g_p \rangle$$
$$= b_j \langle g_j, g_j \rangle$$
$$= b_j \|g_j\|^2$$

for $1 \le j \le p$, so (35) holds. It follows from the Pythagorean theorem [see (30)] and (35) that (36) holds. If $g_1, \ldots, g_p$ are orthonormal, then $\|g_j\| = 1$ for $1 \le j \le p$, so (35) and (36) reduce to (37) and (38), respectively. ∎

THEOREM **9.23**   Let **M** be the Gram matrix of a basis $g_1, \ldots, g_p$ of an identifiable linear space G, let $\mathbf{LL}^T$ be the Cholesky decomposition of **M** such that **L** has positive diagonal entries, and let $\tilde{g}_1, \ldots, \tilde{g}_p$ be the basis of G having the coefficient matrix $\mathbf{L}^{-1}$ relative to $g_1, \ldots, g_p$. Then

(a) $\tilde{g}_1, \ldots, \tilde{g}_p$ is an orthonormal basis of G;

(b) $\tilde{g}_j$ is a linear combination of $g_1, \ldots, g_j$ for $1 \le j \le p$;

(c) if $g_1, \ldots, g_{p_0}$ are orthonormal, where $1 \le p_0 \le p$, then $\tilde{g}_j = g_j$ for $1 \le j \le p_0$.

Proof   (a) According to Theorem 9.19, the Gram matrix of $\tilde{g}_1, \ldots, \tilde{g}_p$ is given by

$$\mathbf{L}^{-1}\mathbf{M}(\mathbf{L}^{-1})^T = \mathbf{L}^{-1}\mathbf{L}\mathbf{L}^T(\mathbf{L}^T)^{-1} = \mathbf{I}$$

(where **I** is the $p \times p$ identity matrix). Therefore, $\tilde{g}_1, \ldots, \tilde{g}_p$ are orthonormal.

(b) Since **L** is a lower triangular matrix, so is $\mathbf{L}^{-1}$. Thus, $\mathbf{L}^{-1} = [\, a_{jl} \,]$, where $a_{jl} = 0$ for $l > j$. Consequently, for $1 \le j \le p$,

$$\tilde{g}_j = \sum_l a_{jl} g_l = a_{j1} g_1 + \cdots + a_{jj} g_j,$$

which is a linear combination of $g_1, \ldots, g_j$.

(c) The diagonal entries of $\mathbf{L}^{-1}$, being the reciprocals of the corresponding diagonal entries of **L**, are positive. Suppose $g_1, \ldots, g_{p_0}$ are orthonormal. Now $a_{11} > 0$ and $\tilde{g}_1 = a_{11} g_1$, so $1 = \|\tilde{g}_1\|^2 = a_{11}^2 \|g_1\|^2 = a_{11}^2$; hence $a_{11} = 1$ and $\tilde{g}_1 = g_1$. Suppose $p_0 \ge 2$. Then $\tilde{g}_2 = a_{21} g_1 + a_{22} g_2$. Since g_1 and g_2 are orthogonal and $\tilde{g}_2$ is orthogonal to $\tilde{g}_1 = g_1$, we see that $0 = \langle \tilde{g}_2, g_1 \rangle = a_{21} \|g_1\|^2$; thus, $a_{21} = 0$ and hence $\tilde{g}_2 = a_{22} g_2$. Arguing as before, we find that $a_{22} = 1$ and $\tilde{g}_2 = g_2$. In a similar manner, we conclude that, for $1 \le j \le p_0$, $a_{jl} = 0$ for $l < j$ and $a_{jj} = 1$ and hence $\tilde{g}_j = g_j$. ∎

COROLLARY **9.12**   (a) G has an orthonormal basis.

(b) Let $\tilde{g}_1, \ldots, \tilde{g}_{p_0}$ be orthonormal functions in G, where $p_0 < p$. Then there are functions $\tilde{g}_{p_0+1}, \ldots, \tilde{g}_p$ in G such that $\tilde{g}_1, \ldots, \tilde{g}_p$ is an orthonormal basis of G.

Proof   (a) Let $g_1, \ldots, g_p$ be a basis of G. Applying Theorem 9.23(a) to this basis, we get an orthonormal basis of G.

(b) Let $\tilde{g}_1, \ldots, \tilde{g}_{p_0}$ be orthonormal functions in G, where $p_0 < p$. Then there are functions $g_{p_0+1}, \ldots, g_p$ in G such that $\tilde{g}_1, \ldots, \tilde{g}_{p_0}, g_{p_0+1}, \ldots, g_p$ form a basis of G. Applying parts (a) and (c) of Theorem 9.23 to this basis, we get an orthonormal basis $\tilde{g}_1, \ldots, \tilde{g}_{p_0}, \tilde{g}_{p_0+1}, \ldots, \tilde{g}_p$ of G. ∎

EXAMPLE **9.29**   Consider the design set $\{-1, 0, 1\}$, unit weights, and the basis $1, x, x^2$ of the space G of quadratic polynomials on $\mathbb{R}$.

(a) Determine the corresponding Gram matrix $\mathbf{M}$.

(b) Determine the Cholesky decomposition $\mathbf{LL}^T$ of $\mathbf{M}$ such that $\mathbf{L}$ has positive diagonal entries and determine $\mathbf{L}^{-1}$.

(c) Apply Theorem 9.23(a) to the indicated basis to obtain an orthonormal basis of G.

Solution   (a) According to the solution to parts (a) and (b) of Example 9.26, $\langle 1, 1 \rangle = 3$, $\langle 1, x \rangle = 0$, $\langle 1, x^2 \rangle = 2$, $\langle x, x \rangle = 2$, and $\langle x, x^2 \rangle = 0$. Also,

$$\langle x^2, x^2 \rangle = (-1)^2(-1)^2 + (0)^2(0)^2 + (1)^2(1)^2 = 2.$$

Thus, the Gram matrix is given by

$$\mathbf{M} = \begin{bmatrix} 3 & 0 & 2 \\ 0 & 2 & 0 \\ 2 & 0 & 2 \end{bmatrix}.$$

(b) The desired Cholesky decomposition of $\mathbf{M}$ is given by $\mathbf{M} = \mathbf{LL}^T$, where

$$\mathbf{L} = \begin{bmatrix} \sqrt{3} & 0 & 0 \\ 0 & \sqrt{2} & 0 \\ \frac{2}{3}\sqrt{3} & 0 & \sqrt{2/3} \end{bmatrix}.$$

Now $\det(\mathbf{L}) = 2$, so

$$\mathbf{L}^{-1} = \frac{1}{2}\begin{bmatrix} 2/\sqrt{3} & 0 & 0 \\ 0 & \sqrt{2} & 0 \\ -\frac{2}{3}\sqrt{6} & 0 & \sqrt{6} \end{bmatrix} = \begin{bmatrix} 1/\sqrt{3} & 0 & 0 \\ 0 & 1/\sqrt{2} & 0 \\ -\frac{1}{3}\sqrt{6} & 0 & \frac{1}{2}\sqrt{6} \end{bmatrix}.$$

(c) Using Theorem 9.23(a), we get the orthonormal basis of G consisting of

$$\frac{1}{\sqrt{3}}, \quad \frac{x}{\sqrt{2}} \quad \text{and} \quad -\frac{1}{3}\sqrt{6} + \frac{x^2}{2}\sqrt{6} = \frac{x^2 - 2/3}{\sqrt{2/3}}.$$

[It is easier to obtain this orthonormal basis from the solutions to parts (a) and (b) of Example 9.26.]   ∎

Problems

9.39   Consider the design set consisting of the points $(-1, -1)$, $(-1, 1)$, $(1, -1)$, $(1, 1)$ in $\mathbb{R}^2$ and set $w_1 = w_2 = 1$ and $w_3 = w_4 = 2$. Are the functions x_1 and x_2 orthogonal?

9.40   Verify each of the following: (a) $\langle h_1, h_2 \rangle = \frac{1}{4}(\|h_1 + h_2\|^2 - \|h_1 - h_2\|^2)$; (b) $\|h_1 + h_2\|^2 + \|h_1 - h_2\|^2 = 2(\|h_1\|^2 + \|h_2\|^2)$ (*parallelogram law*); (c) $\|h_1 + h_2\|^2 \leq 2(\|h_1\|^2 + \|h_2\|^2)$.

9.41　Suppose G is identifiable and saturated and hence that $p = d$. Let $g_1, \ldots, g_d$ be functions in G such that $g_j(x'_k) = 1$ for $j = k$ and $g_j(x'_k) = 0$ for $j \neq k$. **(a)** Show that $g_1, \ldots, g_d$ form an orthogonal basis of G. **(b)** Determine the squared norms of these functions.

9.42　Consider the design set $\{-3, -1, 1, 3\}$ and suppose $w_1 = w_2 = w_3 = w_4 = r$. **(a)** Determine an orthogonal basis of the space of cubic polynomials on $\mathbb{R}$ having the form 1, x, $x^2 - c_1$, $x^3 - c_2 x$ for some $c_1, c_2 \in \mathbb{R}$. **(b)** Determine the squared norms of the functions in this basis. **(c)** Sketch the graphs of the corresponding orthonormal functions on $[-3, 3]$ when $r = 1$.

9.43　Consider the design set $\{-2, -1, 0, 1, 2\}$ and suppose that $w_1 = \cdots = w_5 = r$. **(a)** Determine an orthogonal basis of the space of cubic polynomials on $\mathbb{R}$ having the form 1, x, $x^2 - c_1$, $x^3 - c_2 x$ for some $c_1, c_2 \in \mathbb{R}$. **(b)** Determine the squared norms of these functions. **(c)** Sketch the graphs of the corresponding orthonormal functions on $[-2, 2]$ when $r = 1$.

9.44　Consider the complete factorial design shown in Table 9.4, let $w_1 = \cdots = w_8 = r$, and let G be the space of polynomials on $\mathbb{R}^3$ spanned by $1, x_1, x_2, x_3, x_1 x_2, x_1 x_3, x_2 x_3, x_1 x_2 x_3$. **(a)** Show that the functions in the indicated basis are orthogonal. **(b)** Determine the squared norms of these functions. **(c)** Show that G is identifiable and saturated. **(d)** Determine the design matrix corresponding to the indicated basis. **(e)** Determine the Gram matrix.

TABLE　**9.4**

x_1	x_2	x_3
-1	-1	-1
1	-1	-1
-1	1	-1
1	1	-1
-1	-1	1
1	-1	1
-1	1	1
1	1	1

9.45　Consider the fractional factorial design shown in Table 9.5, let $w_1 = \cdots = w_8 = r$, and let G be the space of linear functions on $\mathbb{R}^7$, which has the basis $1, x_1, \ldots, x_7$. **(a)** Determine the design matrix corresponding to the indicated basis. **(b)** Determine the Gram matrix. [*Hint:* Use the result of Problem 9.44(e).] **(c)** Show that the functions in the indicated basis are orthogonal. **(d)** Determine the squared norms of these functions. **(e)** Show that G is identifiable and saturated.

9.46　Consider the fractional factorial design shown in Table 9.6. (Observe that the design involves all ordered 4-tuples of $x_1, x_2, x_3, x_4 = \pm 1$ such that $x_1 x_2 = x_4$.) Let $w_1 = \cdots = w_8 = r$ and let G be the space of polynomials on $\mathbb{R}^4$ spanned by $1, x_1, x_2, x_3, x_4, x_1 x_3, x_2 x_3, x_3 x_4$. **(a)** Determine the design matrix corresponding to

TABLE   9.5

x_1	x_2	x_3	x_4	x_5	x_6	x_7
-1	-1	-1	1	1	1	-1
1	-1	-1	-1	-1	1	1
-1	1	-1	-1	1	-1	1
1	1	-1	1	-1	-1	-1
-1	-1	1	1	-1	-1	1
1	-1	1	-1	1	-1	-1
-1	1	1	-1	-1	1	-1
1	1	1	1	1	1	1

TABLE   9.6

x_1	x_2	x_3	x_4
-1	-1	-1	1
1	-1	-1	-1
-1	1	-1	-1
1	1	-1	1
-1	-1	1	1
1	-1	1	-1
-1	1	1	-1
1	1	1	1

the indicated basis of G. **(b)** Determine the Gram matrix. [*Hint:* Use the result of Problem 9.44(e).] **(c)** Show that the functions in the indicated basis are orthogonal. **(d)** Determine the squared norms of these functions. **(e)** Show that G is identifiable and saturated.

9.47 Consider the design set $\{-3, -1, 1, 3\}$, unit weights, and the basis $1, x, x^2$ of the space of quadratic polynomials on $\mathbb{R}$. **(a)** Determine the corresponding Gram matrix **M**. **(b)** Determine the Cholesky decomposition $\mathbf{LL}^T$ of **M** such that **L** has positive diagonal entries and determine $\mathbf{L}^{-1}$. **(c)** Apply Theorem 9.23(a) to the indicated basis to obtain an orthonormal basis of G.

9.48 Let $g_1, \ldots, g_{p_0}$ be nonzero orthogonal functions in an identifiable p-dimensional linear space G, where $p_0 < p$. Show that there are functions $g_{p_0+1}, \ldots, g_p$ such that $g_1, \ldots, g_p$ is an orthogonal basis of G.

9.49 Let G be a linear space spanned by $g_1, \ldots, g_p$ and suppose the Gram matrix of $g_1, \ldots, g_p$ is positive definite (equivalently, invertible). Show that $g_1, \ldots, g_p$ is a basis of G.

9.50 Let g be a given function on X and let $a_1, \ldots, a_d \in \mathbb{R}$ be given. Show that

$$\langle g, h \rangle = \sum_k a_k h(x'_k), \qquad h \in H,$$

if and only if $g(x'_k) = a_k w_k^{-1}$ for $1 \leq k \leq d$.

9.51   Let g be a given function on X having positive norm and let $a_1, \ldots, a_d \in \mathbb{R}$ be given. **(a)** Show that

$$\frac{\langle g, h \rangle}{\|g\|^2} = \sum_k a_k h(x'_k), \qquad h \in H, \tag{39}$$

if and only if

$$g(x'_k) = a_k w_k^{-1} \|g\|^2, \qquad 1 \le k \le d. \tag{40}$$

(b) Use the result of part (a) to show that (39) holds if and only if

$$g(x'_k) = \frac{a_k w_k^{-1}}{\sum_k a_k^2 w_k^{-1}}, \qquad 1 \le k \le d. \tag{41}$$

[*Hint:* Show that if (40) or (41) holds, then $\|g\|^2 = 1 / \sum_k a_k^2 w_k^{-1}$.] **(c)** Show that if the weights are uniform, then (39) holds if and only if

$$g(x'_k) = \frac{a_k}{\sum_k a_k^2}, \qquad 1 \le k \le d.$$

9.52   Suppose G is identifiable, let $g_1, \ldots, g_p$ be a basis of G, and let $c_1, \ldots, c_p \in \mathbb{R}$. Determine the unique function $g \in G$ such that $\langle g, g_j \rangle = c_j$ for $1 \le j \le p$. (*Hint:* Write $g = b_1 g_1 + \cdots + b_p g_p$ and determine $\mathbf{b} = [b_1, \ldots, b_p]^T$ in terms of $\mathbf{c} = [c_1, \ldots, c_p]^T$ and the Gram matrix $\mathbf{M}$ of $g_1, \ldots, g_p$.)

9.53   Consider an experiment having three factors and four design points as shown in Table 9.7. Let the design points have unit weights. In each of the following contexts, determine whether the basis functions are orthogonal, whether the linear space G is identifiable, and whether it is saturated: **(a)** G is the space of linear functions of x_1, x_2, and x_3, and $1, x_1, x_2, x_3$ is the basis of this space; **(b)** G is the space of linear functions of x_1 and x_2, and $1, x_1, x_2$ is a basis of this space.

TABLE   9.7

x_1	x_2	x_3
-2	1	-3
-1	-1	-1
1	-1	1
2	1	3

9.5
Orthogonal Projections

Let G be a p-dimensional linear space. In this section, we will develop the properties of orthogonal projection onto G. These results will be applied to least-squares estimation starting in Section 10.1.

Identifiable Spaces

Until the end of this section, it is assumed that G is identifiable. Recall that H is the space of functions that are defined on the design set and possibly elsewhere as well.

THEOREM 9.24　Let $h \in H$. Then there is a unique function $h^* \in G$ such that $h - h^*$ is orthogonal to every function in G.

Proof　If $p = 0$, then 0 is the unique function in G and h is orthogonal to 0, so $h^* = 0$ is the unique function in G such that $h - h^*$ is orthogonal to every function in G.

Suppose, instead, that $p \geq 1$, let $g_1, \ldots, g_p$ be an orthonormal basis of G, and set $h^* = \sum_j \langle g_j, h \rangle g_j$. Choose $g = \sum b_j g_j \in G$. Then

$$\langle g, h^* \rangle = \left\langle \sum_j b_j g_j, \sum_j \langle g_j, h \rangle g_j \right\rangle = \sum_j b_j \langle g_j, h \rangle = \left\langle \sum_j b_j g_j, h \right\rangle = \langle g, h \rangle$$

and hence $\langle g, h - h^* \rangle = \langle g, h \rangle - \langle g, h^* \rangle = 0$. Thus, $h - h^*$ is orthogonal to every function in G. Suppose $\tilde{h} \in G$ and that $h - \tilde{h}$ is also orthogonal to every function in G. Then $\tilde{h} - h^* \in G$, so $h - h^*$ and $h - \tilde{h}$ are both orthogonal to $\tilde{h} - h^*$; that is,

$$\langle \tilde{h} - h^*, h - h^* \rangle = 0 \qquad \text{and} \qquad \langle \tilde{h} - h^*, h - \tilde{h} \rangle = 0.$$

Consequently,

$$0 = \langle \tilde{h} - h^*, h - h^* \rangle - \langle \tilde{h} - h^*, h - \tilde{h} \rangle = \langle \tilde{h} - h^*, \tilde{h} - h^* \rangle = \| \tilde{h} - h^* \|^2.$$

Since G is identifiable, we conclude from Theorem 9.17 in Section 9.4 that $\tilde{h} - h^* = 0$ and hence that $\tilde{h} = h^*$. Therefore, h^* is the unique function in G such that $h - h^*$ is orthogonal to every function in G.　∎

Let $h \in H$. The function h^* in Theorem 9.24 is referred to as the *orthogonal projection* of h onto G and denoted by $P_G h$ (see Figure 9.6). Thus,

$$\langle g, h - P_G h \rangle = 0, \qquad g \in G. \tag{42}$$

Since $P_G h \in G$, it follows from (42) that

$$\langle P_G h, h - P_G h \rangle = 0. \tag{43}$$

Now $h = P_G h + h - P_G h$. Thus, by (43) and the Pythagorean theorem,

$$\|h\|^2 = \|P_G h\|^2 + \|h - P_G h\|^2,$$

so

$$\|h - P_G h\|^2 = \|h\|^2 - \|P_G h\|^2. \tag{44}$$

Let $g_1, \ldots, g_p$ be an orthonormal basis of G. Then, by the proof of Theorem 9.24,

$$P_G h = \sum_j \langle g_j, h \rangle g_j; \tag{45}$$

FIGURE   9.6

The orthogonal projection $h^* = P_G h$ of h onto G. Note that $h - h^*$ is orthogonal to every function $g \in G$, including h^*.

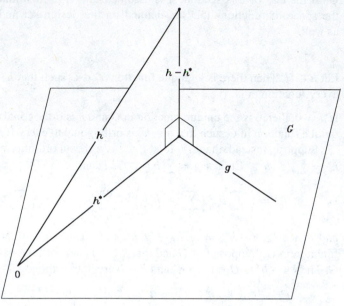

hence, by the Pythagorean theorem,

$$\|P_G h\|^2 = \sum_j (\langle g_j, h \rangle)^2. \tag{46}$$

By (44) and (46),

$$\|h - P_G h\|^2 = \|h\|^2 - \sum_j (\langle g_j, h \rangle)^2. \tag{47}$$

It follows easily from (45) that P_G is a *linear transformation* from H to G; that is,

$$P_G(bh) = b P_G h, \qquad b \in \mathbb{R} \text{ and } h \in H,$$

and

$$P_G(h_1 + h_2) = P_G h_1 + P_G h_2, \qquad h_1, h_2 \in H.$$

Moreover,

$$P_G g = g, \qquad g \in G,$$

and hence

$$P_G(P_G h) = P_G h, \qquad h \in H.$$

THEOREM 9.25   Let $g_1, \ldots, g_p$ form an orthogonal basis of G. Then

$$P_G h = \sum_j \frac{\langle g_j, h \rangle}{\|g_j\|^2} g_j, \qquad h \in H,$$

and

$$\|P_G h\|^2 = \sum_j \frac{(\langle g_j, h \rangle)^2}{\|g_j\|^2}, \qquad h \in H.$$

Proof   Set $\tilde{g}_1 = g_j / \|g_j\|$ for $1 \le j \le p$. Then $\tilde{g}_1, \ldots, \tilde{g}_p$ form an orthonormal basis of G. Let $h \in H$. Then, by (45),

$$P_G h = \sum_j \langle \tilde{g}_j, h \rangle \tilde{g}_j = \sum_j \frac{\langle g_j, h \rangle}{\|g_j\|^2} g_j$$

and, by (46),

$$\|P_G h\|^2 = \sum_j (\langle \tilde{g}_j, h \rangle)^2 = \sum_j \frac{(\langle g_j, h \rangle)^2}{\|g_j\|^2}. \qquad \blacksquare$$

EXAMPLE 9.30   Let $\mathcal{X}' = \mathcal{X} = \{-1, 0, 1\}$, let $w_1 = w_2 = w_3 = r$, let G be the space of all functions on $\mathcal{X}$, and let I_1 and I_2 be the indicator functions given by $I_1(x) = \mathrm{ind}(x = -1)$ and $I_2(x) = \mathrm{ind}(x = 0)$. Extend $1, I_1 - \frac{1}{3}$ to an orthogonal basis of G.

Solution   Observe that $I_1 - \frac{1}{3}$ and $I_2 - \frac{1}{3}$ are orthogonal to 1 and that $1, I_1 - \frac{1}{3}, I_2 - \frac{1}{3}$ is a basis of G (see Examples 9.4 and 9.5 in Section 9.1). To make this into an orthogonal basis, we need to replace $I_2 - \frac{1}{3}$ by $I_2 - \frac{1}{3} - P_{G_1}(I_2 - \frac{1}{3})$, where G_1 is the space spanned by 1 and $I_1 - \frac{1}{3}$. Since $I_2 - \frac{1}{3}$ is orthogonal to 1, we conclude from Theorem 9.25 that

$$P_{G_1}(I_2 - \tfrac{1}{3}) = \frac{\langle I_1 - \frac{1}{3}, I_2 - \frac{1}{3} \rangle}{\|I_1 - \frac{1}{3}\|^2}(I_1 - \tfrac{1}{3}).$$

Now

$$\langle I_1 - \tfrac{1}{3}, I_2 - \tfrac{1}{3} \rangle = r[\,(\tfrac{2}{3})(-\tfrac{1}{3}) + (-\tfrac{1}{3})(\tfrac{2}{3}) + (-\tfrac{1}{3})(-\tfrac{1}{3})\,] = -\tfrac{1}{3}r$$

and

$$\|I_1 - \tfrac{1}{3}\|^2 = r[\,(\tfrac{2}{3})^2 + (-\tfrac{1}{3})^2 + (-\tfrac{1}{3})^2\,] = \tfrac{2}{3}r,$$

so

$$P_{G_1}(I_2 - \tfrac{1}{3}) = -\tfrac{1}{2}(I_1 - \tfrac{1}{3})$$

and hence

$$I_2 - \tfrac{1}{3} - P_{G_1}(I_2 - \tfrac{1}{3}) = I_2 - \tfrac{1}{3} + \tfrac{1}{2}(I_1 - \tfrac{1}{3}) = \tfrac{1}{2}I_1 + I_2 - \tfrac{1}{2}.$$

Consequently, $1, I_1 - \frac{1}{3}, \frac{1}{2}I_1 + I_2 - \frac{1}{2}$ is an orthogonal basis of G. $\blacksquare$

A function $h^* \in G$ is said to be a (weighted) *least-squares approximation* in G to a given function $h \in H$ if

$$\|h - h^*\|^2 = \min_{g \in G} \|h - g\|^2 = \min_{g \in G} \sum_k w_k [\, h(\mathbf{x}'_k) - g(\mathbf{x}'_k)\,]^2.$$

If $h^* \in G$ and $\|h - h^*\|^2 < \|h - g\|^2$ for all $g \in G$ with $g \neq h^*$, then h^* is the unique least-squares approximation in G to h, $h - h^*$ is referred to as the error of approximation, and $\|h - h^*\|^2$ is referred to as the squared norm of the error of approximation.

THEOREM 9.26   Let $h \in H$. Then $P_G h$ is the unique least-squares approximation in G to h.

Proof   Set $h^* = P_G h$. Choose $g \in G$ such that $g \neq h^*$ and hence $\|h^* - g\|^2 > 0$. Since $h^* - g \in G$, we conclude from the definition of orthogonal projection that

$$\langle h^* - g, h - h^* \rangle = 0;$$

hence, by the Pythagorean theorem (see Figure 9.7),

$$\|h - g\|^2 = \|h - h^* + h^* - g\|^2 = \|h - h^*\|^2 + \|h^* - g\|^2 > \|h - h^*\|^2.$$

Consequently, h^* is the unique least-squares approximation in G to h.   ∎

[Starting in Section 10.1, we will apply least-squares approximation to statistics in two closely related ways: (i) $h = \bar{Y}(\cdot)$ and $\hat{\mu}(\cdot) = P_G h$ is the least-squares estimate in G of the regression function $\mu(\cdot)$; and (ii) $h = \mu(\cdot)$ and $\mu^*(\cdot) = P_G \mu(\cdot)$ is the least-squares approximation in G to the regression function, which is also the mean of the least-squares estimate $\hat{\mu}(\cdot)$.]

Set $w = \sum_k w_k = \|1\|^2 = \langle 1, 1 \rangle$. Given $h \in H$, consider the *weighted average*

$$\bar{h} = \frac{1}{w} \langle 1, h \rangle = \frac{1}{w} \sum_k w_k h(x'_k)$$

of the values of h at the design points. Observe that

$$\langle 1, h - \bar{h} \rangle = \langle 1, h \rangle - \langle 1, \bar{h} \rangle = \langle 1, h \rangle - \bar{h} \langle 1, 1 \rangle = \langle 1, h \rangle - w\bar{h}$$

and hence that

$$\langle 1, h - \bar{h} \rangle = 0, \qquad h \in H. \tag{48}$$

Also, $\|\bar{h}\|^2 = \bar{h}^2 \|1\|^2$ and hence

$$\|\bar{h}\|^2 = w\bar{h}^2, \qquad h \in H. \tag{49}$$

Since $h = \bar{h} + h - \bar{h}$, we conclude from (48), (49), and the Pythagorean theorem that

$$\|h\|^2 = w\bar{h}^2 + \|h - \bar{h}\|^2. \tag{50}$$

Recall that G_{con} denotes the space of constant functions.

FIGURE 9.7

The orthogonal projection $h^* = P_G h$ is a better approximation to h than any $g \in G$ with $g \neq h^*$.

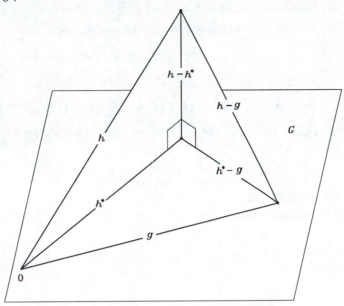

EXAMPLE 9.31   Given $h \in H$, determine

(a) $P_{G_{con}} h$;

(b) $\|P_{G_{con}} h\|^2$;

(c) $\|h - P_{G_{con}} h\|^2$.

Solution   **(a)** The constant function 1 constitutes an orthogonal basis of G_{con}. By Theorem 9.25,

$$P_{G_{con}} h = \frac{\langle 1, h \rangle}{\|1\|^2} 1 = \frac{1}{w} \langle 1, h \rangle = \bar{h}.$$

(b) By (49) and the solution to part (a), $\|P_{G_{con}} h\|^2 = \|\bar{h}\|^2 = w\bar{h}^2$.

(c) By (49) and the solution to part (b),

$$\|h - P_{G_{con}} h\|^2 = \|h - \bar{h}\|^2 = \|h\|^2 - w\bar{h}^2.$$

Alternatively,

$$\|h - P_{G_{con}} h\|^2 = \|h - \bar{h}\|^2 = \sum_k w_k [h(x'_k) - \bar{h}]^2.$$

EXAMPLE 9.32   Consider the design set $\{-3, -1, 1, 3\}$ and unit weights. Let h be a function such that $h(-3) = 1$, $h(-1) = 2$, $h(1) = 2$, and $h(3) = 0$.

(a) Determine $P_{G_{con}} h$, $\|P_{G_{con}} h\|^2$ and $\|h - P_{G_{con}} h\|^2$.

(b) Show the indicated values of $(x, h(x))$ together with the graph of $P_{G_{con}} h$ on $[-3, 3]$.

Solution   **(a)** Here, $w = 4$ and $\bar{h} = 5/4$. Thus, by the solution to Example 9.31, $P_{G_{con}} h = \bar{h} = 5/4$ and $\|P_{G_{con}} h\|^2 = w\bar{h}^2 = 25/4$. Also, $\|h\|^2 = 1^2 + 2^2 + 2^2 + 0^2 = 9$, so

$$\|h - P_{G_{con}} h\|^2 = \|h\|^2 - w\bar{h}^2 = 9 - \tfrac{25}{4} = \tfrac{11}{4}.$$

Alternatively,

$$\|h - P_{G_{con}} h\|^2 = (1 - \tfrac{5}{4})^2 + (2 - \tfrac{5}{4})^2 + (2 - \tfrac{5}{4})^2 + (0 - \tfrac{5}{4})^2 = \tfrac{11}{4}.$$

(b) Figure 9.8 shows the indicated values of $(x, h(x))$ together with the graph of $P_{G_{con}} h$ on $[-3, 3]$.   ∎

FIGURE   9.8

Least-squares constant approximation

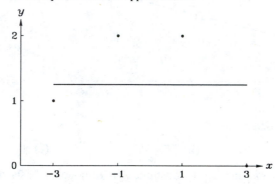

EXAMPLE 9.33   Consider the design set $\{-3, -1, 1, 3\}$, unit weights, and the orthonormal basis

$$g_0(x) = \frac{1}{2}, \quad g_1(x) = \frac{\sqrt{5}}{10}x, \quad g_2(x) = \frac{1}{8}(x^2 - 5), \quad \text{and} \quad g_3(x) = \frac{\sqrt{5}}{24}\left(x^3 - \frac{41}{5}x\right)$$

of the space of cubic polynomials on $\mathbb{R}$ (see the answer to Problem 9.42 in Section 9.4). Let h be a function such that $h(-3) = 1$, $h(-1) = 2$, $h(1) = 2$, and $h(3) = 0$. Determine the least-squares approximation to h in each of the following spaces of functions on $\mathbb{R}$ and determine the squared norm of the error of approximation:

(a) the space G_{con} of constant functions;

(b) the space G_1 of linear functions;

(c) the space G_2 of quadratic polynomials;

(d) the space G_3 of cubic polynomials.

Solution   Now $\|h\|^2 = 1^2 + 2^2 + 2^2 + 0^2 = 9$. Observe that

$$\langle g_0, h \rangle = \frac{1}{2}[1 \cdot 1 + 1 \cdot 2 + 1 \cdot 2 + 1 \cdot 0] = \frac{5}{2},$$

$$\langle g_1, h \rangle = \frac{\sqrt{5}}{10}[(-3) \cdot 1 + (-1) \cdot 2 + 1 \cdot 2 + 3 \cdot 0] = -\frac{3\sqrt{5}}{10},$$

$$\langle g_2, h \rangle = \frac{1}{8}[4 \cdot 1 + (-4) \cdot 2 + (-4) \cdot 2 + 4 \cdot 0] = -\frac{3}{2},$$

and

$$\langle g_3, h \rangle = \frac{\sqrt{5}}{24}\left[\left(-\frac{12}{5}\right) \cdot 1 + \frac{36}{5} \cdot 2 + \left(-\frac{36}{5}\right) \cdot 2 + \frac{12}{5} \cdot 0\right] = -\frac{\sqrt{5}}{10}.$$

Thus,

$$\langle g_0, h \rangle g_0 = \left(\frac{5}{2}\right)\left(\frac{1}{2}\right) = \frac{5}{4},$$

$$\langle g_1, h \rangle g_1 = \left(-\frac{3\sqrt{5}}{10}\right)\left(\frac{\sqrt{5}}{10}x\right) = -\frac{3}{20}x,$$

$$\langle g_2, h \rangle g_2 = \left(-\frac{3}{2}\right)\left(\frac{1}{8}\right)(x^2 - 5) = -\frac{3}{16}(x^2 - 5),$$

and

$$\langle g_3, h \rangle g_3 = \left(-\frac{\sqrt{5}}{10}\right)\left(\frac{\sqrt{5}}{24}\right)\left(x^3 - \frac{41}{5}x\right) = -\frac{1}{48}\left(x^3 - \frac{41}{5}x\right).$$

(a) Observe that g_0 is an orthonormal basis of G_{con}. Thus, the least-squares constant approximation to h is given by

$$P_{G_{\text{con}}}h = \langle g_0, h \rangle g_0 = \frac{5}{4}.$$

According to (47), the squared norm of the error of approximation is given by

$$\|h - P_{G_{\text{con}}}h\|^2 = \|h\|^2 - (\langle g_0, h \rangle)^2 = 9 - \left(\frac{5}{2}\right)^2 = \frac{11}{4} = 2.75.$$

[These results agree with the solution to Example 9.32(a).]

(b) Observe that g_0 and g_1 form an orthonormal basis of G_1. Thus, the least-squares linear approximation to h is given by

$$P_{G_1}h = \langle g_0, h \rangle g_0 + \langle g_1, h \rangle g_1 = \frac{5}{4} + \left(-\frac{3}{20}x\right) = -\frac{3}{20}x + \frac{5}{4}.$$

The squared norm of the error of approximation is given by

$$\|h - P_{G_1}h\|^2 = \|h\|^2 - (\langle g_0, h \rangle)^2 - (\langle g_1, h \rangle)^2$$

$$= \frac{11}{4} - \left(-\frac{3\sqrt{5}}{10}\right)^2 = \frac{23}{10} = 2.3.$$

(c) Observe that g_0, g_1, and g_2 form an orthonormal basis of G_2. Thus, the least-squares quadratic approximation to h is given by

$$P_{G_2}h = \langle g_0, h \rangle g_0 + \langle g_1, h \rangle g_1 + \langle g_2, h \rangle g_2$$

$$= -\frac{3}{20}x + \frac{5}{4} - \frac{3}{16}(x^2 - 5)$$

$$= -\frac{3}{16}x^2 - \frac{3}{20}x + \frac{35}{16}.$$

The squared norm of the error of approximation is given by

$$\|h - P_{G_2}h\|^2 = \|h\|^2 - (\langle g_0, h \rangle)^2 - (\langle g_1, h \rangle)^2 - (\langle g_2, h \rangle)^2$$

$$= \frac{23}{10} - \left(\frac{3}{2}\right)^2 = \frac{1}{20} = 0.05.$$

(d) Since $g_0, g_1, g_2,$ and g_3 form an orthogonal basis of G_3, the least-squares cubic approximation to h is given by

$$P_{G_3}h = \langle g_0, h \rangle g_0 + \langle g_1, h \rangle g_1 + \langle g_3, h \rangle g_2 + \langle g_3, h \rangle g_3$$

$$= -\frac{3}{16}x^2 - \frac{3}{20}x + \frac{35}{16} - \frac{1}{48}\left(x^3 - \frac{41}{5}x\right)$$

$$= -\frac{1}{48}(x^3 + 9x^2 - x - 105),$$

which agrees with the answer to Problem 9.29(a) in Section 9.3. The squared norm of the error of approximation is given by

$$\|h - P_{G_3}h\|^2 = \|h\|^2 - (\langle g_0, h \rangle)^2 - (\langle g_1, h \rangle)^2 - (\langle g_2, h \rangle)^2 - (\langle g_3, h \rangle)^2$$

$$= \tfrac{1}{20} - \tfrac{1}{20}$$

$$= 0. \quad \blacksquare$$

Since G is assumed to be identifiable, it follows from Theorem 9.9(b) in Section 9.3 that G is saturated if and only if $p = d$.

THEOREM 9.27   If G is saturated, then $P_G h = h$ on the design set and $\|h - P_G h\|^2 = 0$.

Proof   Suppose G is saturated. Then there is a function $g \in G$ such that $g(\mathbf{x}'_k) = h(\mathbf{x}'_k)$ for $1 \le k \le d$ and hence such that $\|h - g\|^2 = 0$. Thus, g is the least-squares approximation in G to h, so $g = P_G h$; hence, $(P_G h)(\mathbf{x}'_k) = h(\mathbf{x}'_k)$ for $1 \le k \le d$ and $\|h - P_G h\|^2 = 0$. $\blacksquare$

In the context of Example 9.33, the space G_3 of cubic polynomials is saturated relative to $\{-3, 1, 1, 3\}$. Thus, Theorem 9.27 explains why, in the solution to Example 9.33(d), $P_{G_3}h$ agrees with the answer to Problem 9.29(a) in Section 9.3 and $\|h - P_{G_3}h\| = 0$.

EXAMPLE 9.34   Let G be the space of linear functions on $\mathbb{R}$ and let $d = 2$.

(a) Determine the form of the least-squares approximation in G to h.

(b) Specialize the answer to part (a) to the design set $\{-1, 1\}$.

Solution **(a)** Since G is identifiable and saturated, $P_G h$ is the linear function that interpolates the values of h at the design points. Thus, by the Lagrange interpolation formula,

$$P_G h(x) = h(x_1') \frac{x - x_2'}{x_1' - x_2'} + h(x_2') \frac{x - x_1'}{x_2' - x_1'}.$$

(b) Let $x_1' = -1$ and $x_2' = 1$. Then

$$P_G h(x) = h(-1) \frac{x - 1}{-2} + h(1) \frac{x + 1}{2} = \frac{h(-1) + h(1)}{2} + \frac{h(1) - h(-1)}{2} x. \quad \blacksquare$$

THEOREM 9.28 Let G_1 and G_2 be identifiable linear spaces of functions on $\mathcal{X}$ such that $G_1' = G_2'$, where G_1' and G_2' are the spaces of the restrictions of the functions in G_1 and G_2, respectively, to the design set, and let $h \in H$. Then $P_{G_1} h = P_{G_2} h$ on the design set, and $\|h - P_{G_1} h\|^2 = \|h - P_{G_2} h\|^2$.

Proof We can ignore points in $\mathcal{X}$ that are not design points, whereupon the conclusions of the theorem are obviously valid. $\quad \blacksquare$

Subspaces

THEOREM 9.29 Let G_0 be a subspace of G and let $h \in H$. Then

(a) $P_{G_0} h$, $P_G h - P_{G_0} h$ and $h - P_G h$ are orthogonal,

(b) $P_{G_0} h = P_{G_0} (P_G h)$, and

(c) $P_{G_0} h = P_G h$ if and only if $P_G h \in G_0$.

Proof **(a)** By the definition of orthogonal projection onto G,

$$\langle g, h - P_G h \rangle = 0, \qquad g \in G. \tag{51}$$

Since $P_{G_0} h$ and $P_G h - P_{G_0} h$ are in G, it follows from (51) that $P_{G_0} h$ and $P_G h - P_{G_0} h$ are orthogonal to $h - P_G h$. It remains to show that $P_{G_0} h$ is orthogonal to $P_G h - P_{G_0} h$. To this end, we conclude from (51) that

$$\langle g, h - P_G h \rangle = 0, \qquad g \in G_0. \tag{52}$$

(Note that $G_0 \subset G$.) By the definition of orthogonal projection onto G_0,

$$\langle g, h - P_{G_0} h \rangle = 0, \qquad g \in G_0. \tag{53}$$

It follows from (52) and (53) that

$$\langle g, P_G h - P_{G_0} h \rangle = 0, \qquad g \in G_0. \tag{54}$$

Since $P_{G_0} h \in G_0$, we conclude from (54) that $P_{G_0} h$ is orthogonal to $P_G h - P_{G_0} h$ as desired.

(b) According to (54), $P_{G_0}h$ is the orthogonal projection of P_Gh onto G_0.

(c) Suppose $P_{G_0}h = P_Gh$. Since $P_{G_0}h \in G_0$, we conclude that $P_Gh \in G_0$. Suppose, conversely, $P_Gh \in G_0$. Then it follows from (52) that P_Gh is the orthogonal projection of h onto G_0; that is, $P_{G_0}h = P_Gh$. ∎

Let G_0 be a subspace of G. Recall from the definition of orthogonal projection and the Pythagorean theorem that

$$\|h\|^2 = \|P_Gh\|^2 + \|h - P_Gh\|^2, \qquad h \in H,$$

and

$$\|h\|^2 = \|P_{G_0}h\|^2 + \|h - P_{G_0}h\|^2, \qquad h \in H.$$

Similarly, by using Theorem 9.29(a) and the Pythagorean theorem (or by inspection of Figure 9.9), we obtain the following results.

COROLLARY 9.13 Let G_0 be a subspace of G. Then

$$\|P_Gh\|^2 = \|P_{G_0}h\|^2 + \|P_Gh - P_{G_0}h\|^2, \qquad h \in H,$$

and

$$\|h - P_{G_0}h\|^2 = \|P_Gh - P_{G_0}h\|^2 + \|h - P_Gh\|^2, \qquad h \in H.$$

The proof of the next result is left as a problem.

FIGURE 9.9

Orthogonal projections onto G and G_0. Here $h^* = P_Gh$ and $h_0^* = P_{G_0}h$.

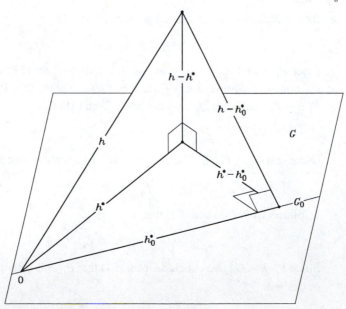

THEOREM 9.30   Let G_0 be a p_0-dimensional subspace of G, let $g_1, \ldots, g_p$ be an orthonormal basis of G such that $g_1, \ldots, g_{p_0}$ is an orthonormal basis of G_0, let $h \in H$, let

$$h^* = \sum_{j=1}^{p} b_j^* g_j$$

be the orthogonal projection of h onto G, and let h_0^* be the orthogonal projection of h onto G_0. Then

$$h_0^* = \sum_{j=1}^{p_0} b_j^* g_j,$$

$$h^* - h_0^* = \sum_{j=p_0+1}^{p} b_j^* g_j,$$

$$\|h^*\|^2 = \sum_{j=1}^{p} (b_j^*)^2,$$

$$\|h_0^*\|^2 = \sum_{j=1}^{p_0} (b_j^*)^2,$$

and

$$\|h^* - h_0^*\|^2 = \sum_{j=p_0+1}^{p} (b_j^*)^2.$$

Let $h^* = P_G h$ denote the least-squares approximation in G to h and let $h_0^* = P_{G_0} h$ denote the least-squares approximation in the subspace G_0 of G to h. By Theorem 9.29(b), $h_0^* = P_{G_0} h^*$; that is, h_0^* is the least-squares approximation in G_0 to h^*.

EXAMPLE 9.35   Consider the setup of Example 9.33, and let h^* denote the least-squares quadratic polynomial approximation to h.

(a) Determine the squared norm of h^*.

(b) Determine the least-squares linear approximation h_0^* to h^*.

(c) Determine the squared norm of h_0^*.

(d) Determine the squared norm of the error $h - h_0^*$ of approximation.

(e) Show the graphs of h^* and h_0^* on $[-3, 3]$.

Solution   (a) According to the solution to Example 9.33(c), the least-squares quadratic approximation to h is given by

$$h^*(x) = P_{G_2} h(x) = -\frac{3}{16}x^2 - \frac{3}{20}x + \frac{35}{16}.$$

The values of h^* at the design points are given by

$$h^*(-3) = \frac{19}{20}, \quad h^*(-1) = \frac{43}{20}, \quad h^*(1) = \frac{37}{20}, \quad \text{and} \quad h^*(3) = \frac{1}{20}.$$

Thus, the squared norm of this function is given by

$$\|h^*\|^2 = \left(\frac{19}{20}\right)^2 + \left(\frac{43}{20}\right)^2 + \left(\frac{37}{20}\right)^2 + \left(\frac{1}{20}\right)^2 = \frac{895}{100}.$$

(b) Consider the orthonormal basis

$$g_0(x) = \frac{1}{2} \quad \text{and} \quad g_1(x) = \frac{\sqrt{5}}{10}x$$

of the space of linear functions on $\mathbb{R}$. Observe that

$$\langle g_0, h^* \rangle = \frac{1}{2}\left(1 \cdot \frac{19}{20} + 1 \cdot \frac{43}{20} + 1 \cdot \frac{37}{20} + 1 \cdot \frac{1}{20}\right) = \frac{5}{2}$$

and

$$\langle g_1, h^* \rangle = \frac{\sqrt{5}}{10}\left[(-3) \cdot \frac{19}{20} + (-1) \cdot \frac{43}{20} + 1 \cdot \frac{37}{20} + 3 \cdot \frac{1}{20}\right] = -\frac{3\sqrt{5}}{10}.$$

Thus, the least-squares linear approximation to h^* is given by

$$h_0^* = P_{G_1} h^* = \langle g_0, h^* \rangle g_0 + \langle g_1, h^* \rangle g_1$$

$$= \left(\frac{5}{2}\right)\left(\frac{1}{2}\right) + \left(-\frac{3\sqrt{5}}{10}\right)\frac{\sqrt{5}}{10}x = -\frac{3}{20}x + \frac{5}{4},$$

which coincides with the least-squares linear approximation to h, as determined in the solution to Example 9.33(b).

(c) The values of h_0^* at the design points are given by

$$h_0^*(-3) = \frac{17}{10}, \quad h_0^*(-1) = \frac{14}{10}, \quad h_0^*(1) = \frac{11}{10}, \quad \text{and} \quad h_0^*(3) = \frac{8}{10}.$$

Thus, the squared norm of h_0^* is given by

$$\|h_0^*\|^2 = \left(\frac{17}{10}\right)^2 + \left(\frac{14}{10}\right)^2 + \left(\frac{11}{10}\right)^2 + \left(\frac{8}{10}\right)^2 = \frac{67}{10}.$$

Alternatively,

$$\|h_0^*\|^2 = (\langle g_0, h^* \rangle)^2 + (\langle g_1, h^* \rangle)^2 = \left(\frac{5}{2}\right)^2 + \left(-\frac{3\sqrt{5}}{10}\right)^2 = \frac{67}{10}.$$

(d) The squared norm of the error of approximation is given by

$$\|h^* - h_0^*\|^2 = \left(\frac{19}{20} - \frac{17}{10}\right)^2 + \left(\frac{43}{20} - \frac{14}{10}\right)^2 + \left(\frac{37}{20} - \frac{11}{10}\right)^2 + \left(\frac{1}{20} - \frac{8}{10}\right)^2$$

$$= \frac{9}{4}.$$

Alternatively,

$$\|h^* - h_0^*\|^2 = \|h^*\|^2 - \|h_0^*\|^2 = \frac{895}{100} - \frac{67}{10} = \frac{9}{4}.$$

(e) Figure 9.10 shows the graphs of h^* and h_0^* on $[-3, 3]$. ∎

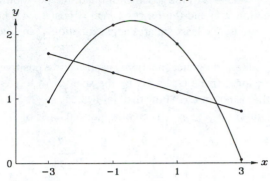

FIGURE 9.10

Least-squares linear and quadratic approximations

Nonidentifiable Spaces

For completeness, we temporarily drop the assumption that G is identifiable. Recall that the space G' of the restrictions g' of the functions $g \in G$ to the design set is an identifiable p_0-dimensional subspace of H, where $p_0 \le p$, and that $p = p_0$ if and only if G is identifiable. Given $h \in H$, $P_{G'}h$ is the unique least-squares approximation in G' to h. Let $g \in G$. Then g is a least-squares approximation in G to h if and only if $g' = P_{G'}h$. (This follows from the formula $\|h - g\| = \|h - g'\|$.) In particular, there is at least one least-squares approximation h^* in G to h. If G is identifiable, then h^* is the unique least-squares approximation in G to h, as we already know.

Suppose, instead, that G is nonidentifiable, let g be a nonzero function in G that equals zero on the design set, and let $b \in \mathbb{R}$. Then the restriction of $h^* + bg$ to the design set is given by $(h^* + bg)' = (h^*)' = P_{G'}h$. Thus, $h^* + bg$ is a least-squares approximation in G to h for $b \in \mathbb{R}$. As b ranges over $\mathbb{R}$, we get infinitely many least-squares approximations in G to h in this manner.

Problems

9.54 Consider the design set $\{-2, -1, 0, 1, 2\}$, unit weights, and the orthogonal basis $1, x$ of the space of linear functions on $\mathbb{R}$. Determine the orthogonal projection of x^3 onto this space.

9.55 Show that if G is identifiable, then **(a)** $\|h - P_G h\|^2 = \|h\|^2 - \langle P_G h, h \rangle$ for $h \in H$; **(b)** $\langle h_1, P_G h_2 \rangle = \langle P_G h_1, h_2 \rangle$ for $h_1, h_2 \in H$.

9.56 Show that if G is identifiable and $h \in H$ is orthogonal to every function in G, then $P_G h = 0$.

9.57 Consider the setup of Example 9.33. Determine **(a)** the value $x_{\max}$ of x in $[-3, 3]$ that maximizes $P_{G_2} h$ on that interval; **(b)** the value of $P_{G_2} h$ at $x_{\max}$.

9.58 Consider the setup of Example 9.33. Determine **(a)** the value $x_{\max}$ of x in $[-3, 3]$ that maximizes $P_{G_3} h$ on that interval; **(b)** the value of $P_{G_3} h$ at $x_{\max}$.

9.59 Consider the design set $\{-2, -1, 0, 1, 2\}$, unit weights, the orthogonal basis 1, $g_1(x) = x, g_2(x) = x^2 - 2$ of the space G of quadratic polynomials on $\mathbb{R}$ (see Problem 9.43 in Section 9.4) and the function $h(x) = \cos(\pi x/4)$. Determine **(a)** $\langle 1, h \rangle$, $\langle g_1, h \rangle$ and $\langle g_2, h \rangle$; **(b)** the least-squares approximation h^* in G to h; **(c)** $\|h\|^2$, $\|h^*\|^2$ and $\|h - h^*\|^2$.

9.60 Let $g_1, \dots, g_{p-1}$ be orthogonal functions, each having positive norm, in the identifiable p-dimensional space G; let G_0 be the span of $g_1, \dots, g_{p-1}$; and let g be a function in G that is not in G_0. Show that **(a)** $g_1, \dots, g_{p-1}, g - P_{G_0}g$ is an orthogonal basis of G; **(b)** if $g_1, \dots, g_{p-1}$ is an orthonormal basis of G_0, then

$$g_1, \dots, g_{p-1}, \frac{g - P_{G_0}g}{\|g - P_{G_0}g\|}$$

is an orthonormal basis of G.

9.61 Verify Theorem 9.30.

9.62 Show that $\langle h_1 - \bar{h}_1, h_2 - \bar{h}_2 \rangle = \langle h_1 - \bar{h}_1, h_2 \rangle = \langle h_1, h_2 \rangle - w\bar{h}_1\bar{h}_2$ for $h_1, h_2 \in H$.

9.63 Let G_1 and G_2 be identifiable linear spaces of functions on $\mathcal{X}$. Show that the following two statements are equivalent: **(a)** Every function in G_1 is orthogonal to every function in G_2; **(b)** $P_{G_1}(P_{G_2}h) = 0$ for $h \in H$.

9.64 Let G be the space of linear functions on $\mathbb{R}$ and suppose $d \geq 2$ and hence that G is identifiable. Show that **(a)** $1, x - \bar{x}$ form an orthogonal basis of G; **(b)** $P_G h(x) = \bar{h} + \frac{\langle x - \bar{x}, h \rangle}{\|x - \bar{x}\|^2}(x - \bar{x}) = \bar{h} + \frac{\langle x - \bar{x}, h - \bar{h} \rangle}{\|x - \bar{x}\|^2}(x - \bar{x})$ for $h \in H$.

9.65 Suppose G is identifiable, let G' be the space of the restrictions g' of the functions $g \in G$ to the design set, and let $h \in H$. Show that **(a)** $\langle (h - (P_G h))', g' \rangle = \langle h - P_G h, g \rangle = 0$ for $g \in G$; **(b)** $(P_G h)' = P_{G'} h$ (that is, the restriction to the design set of the orthogonal projection of h onto G equals the orthogonal projection of h onto G'.)

9.66 Suppose G is identifiable and saturated and hence that $p = d$. Let $g_1, \dots, g_d$ be functions in G such that $g_j(x_k') = 1$ for $j = k$ and $g_j(x_k') = 0$ for $j \neq k$. Then $g_1, \dots, g_d$ form an orthogonal basis of G and $\|g_k\|^2 = w_k$ for $1 \leq k \leq d$ [see the answer to Problem 9.41(b) in Section 9.4]. Let $h \in H$. **(a)** Determine $P_G h$ as a linear combination of $g_1, \dots, g_d$. **(b)** Use the answer to part (a) to verify that $\|P_G h\|^2 = \|h\|^2$ and hence that $\|h - P_G h\|^2 = 0$.

9.67 Let G be a one-dimensional linear space having basis function g with $\|g\| > 0$. Then g is an orthogonal basis of G. Show that **(a)** $P_G h = \frac{\langle g, h \rangle}{\|g\|^2} g$ for $h \in H$; **(b)** $\|P_G h\|^2 = \frac{(\langle G, h \rangle)^2}{\|g\|^2}$ for $h \in H$; **(c)** $\|h - P_G h\|^2 = \|h\|^2 - \frac{(\langle g, h \rangle)^2}{\|g\|^2}$ for $h \in H$.

9.68 Suppose G is identifiable, let $g_1, \dots, g_p$ be a basis of G, and let $c_1, \dots, c_p \in \mathbb{R}$. According to Problem 9.52 in Section 9.4, there is a unique function $g \in G$ such that $\langle g, g_j \rangle = c_j$ for $1 \leq j \leq p$. Let $1 \leq p_0 \leq p$ and let G_0 be in the space spanned by $g_1, \dots, g_{p_0}$. Determine, in terms of g and G_0, the unique function g_0 in G_0 such that $\langle g_0, g_j \rangle = c_j$ for $1 \leq j \leq p_0$.

9.69 In the context of Problem 9.53 in Section 9.4, **(a)** determine the orthogonal projection of x_3 onto the space of linear functions of x_1 and x_2; **(b)** use the result of part (a) to determine an orthogonal basis of the space of linear functions of x_1, x_2, and x_3.

9.6
Normal Equations

Let G be an identifiable p-dimensional linear space and let $h \in H$. As we saw in Section 9.5, the least-squares approximation h^* in G to h can be found by first finding an orthogonal basis of G. In this section, we will determine h^* by using an arbitrary basis of this space. To this end, we have the following result.

THEOREM 9.31 Let $g_1, \ldots, g_p$ form a basis of G and let $h \in H$. Then h is orthogonal to every function in G if and only if it is orthogonal to each of the functions $g_1, \ldots, g_p$.

Proof Suppose h is orthogonal to every function in G. Then, in particular, it is orthogonal to each of the functions $g_1, \ldots, g_p$. Suppose, conversely, h is orthogonal to each of the functions $g_1, \ldots, g_p$ and let $g \in G$. Then $g = b_1 g_1 + \cdots + b_p g_p$, where $b_1, \ldots, b_p \in \mathbb{R}$, so

$$\langle g, h \rangle = \left\langle \sum_j b_j g_j, \, h \right\rangle = \sum_j b_j \langle g_j, h \rangle = 0.$$

Therefore, h is orthogonal to every function in G. ∎

COROLLARY 9.14 Let $g_1, \ldots, g_p$ form a basis of an identifiable space G and let $h \in H$ and $h^* \in G$. Then $h^* = P_G h$ if and only if

$$\langle g_j, h^* \rangle = \langle g_j, h \rangle, \qquad 1 \le j \le p. \tag{55}$$

Proof Observe that (55) holds if and only if

$$\langle g_j, h - h^* \rangle = 0, \qquad 1 \le j \le p. \tag{56}$$

It follows from Theorem 9.31 that (56) holds if and only if $h - h^*$ is orthogonal to every function in G and hence if and only if $h^* = P_G h$. ∎

Let $g_1, \ldots, g_p$ be a basis of G and let $h^* \in G$. Then $h^* = b_1^* g_1 + \cdots + b_p^* g_p$, where $b_1^*, \ldots, b_p^* \in \mathbb{R}$ are uniquely determined. Now

$$\langle g_j, h^* \rangle = \langle g_j, b_1^* g_1 + \cdots + b_p^* g_p \rangle = \langle g_j, g_1 \rangle b_1^* + \cdots + \langle g_j, g_p \rangle b_p^*.$$

Thus, (55) can be written as

$$\langle g_1, g_1 \rangle b_1^* + \cdots + \langle g_1, g_p \rangle b_p^* = \langle g_1, h \rangle$$

$$\vdots \qquad \qquad \vdots \quad \vdots \tag{57}$$

$$\langle g_p, g_1 \rangle b_1^* + \cdots + \langle g_p, g_p \rangle b_p^* = \langle g_p, h \rangle$$

We refer to (57) as the system of *normal equations* for $b_1^*, \ldots, b_p^*$. The following result is now an immediate consequence of Theorem 9.31.

COROLLARY 9.15   The numbers $b_1^*, \ldots, b_p^*$ are the coefficients relative to the basis $g_1, \ldots, g_p$ of the least-squares approximation in G to h if and only if they satisfy the system (57) of normal equations.

EXAMPLE 9.36   Consider the design set $\{-3, -1, 1, 3\}$ and unit weights and let h be a function such that $h(-3) = 1$, $h(-1) = 2$, $h(1) = 2$, and $h(3) = 0$. Use the basis $1, x, x^2$ to determine the least-squares quadratic polynomial approximation h^* to h.

Solution   Write $h^*(x) = b_0^* + b_1^* x + b_2^* x^2$. The normal equations for b_0^*, b_1^*, and b_2^* can be written as

$$\langle 1, 1 \rangle b_0^* + \langle 1, x \rangle b_1^* + \langle 1, x^2 \rangle b_2^* = \langle 1, h \rangle,$$

$$\langle x, 1 \rangle b_0^* + \langle x, x \rangle b_1^* + \langle x, x^2 \rangle b_2^* = \langle x, h \rangle,$$

$$\langle x^2, 1 \rangle b_0^* + \langle x^2, x \rangle b_1^* + \langle x^2, x^2 \rangle b_2^* = \langle x^2, h \rangle.$$

Now $\langle 1, 1 \rangle = 4$, $\langle 1, x \rangle = 0$, $\langle 1, x^2 \rangle = \langle x, x \rangle = 20$, $\langle x, x^2 \rangle = 0$, and $\langle x^2, x^2 \rangle = 164$. Also, $\langle 1, h \rangle = 5$, $\langle x, h \rangle = -3$, and $\langle x^2, h \rangle = 13$. Thus, the normal equations are given by

$$4b_0^* + 20b_2^* = 5,$$

$$20b_1^* = -3,$$

$$20b_0^* + 164b_2^* = 13.$$

According to the second of these equations, $b_1^* = -3/20$. Solving the first and third equations for b_0^* and b_2^*, we get $b_0^* = 35/16$ and $b_2^* = -3/16$. Thus,

$$h^*(x) = -\frac{3}{16}x^2 - \frac{3}{20}x + \frac{35}{16},$$

which agrees with the solution to Example 9.33(c) in Section 9.5.   ∎

Let $\mathbf{M}$ denote the Gram matrix corresponding to $g_1, \ldots, g_p$ and set

$$\mathbf{U} = [\, \langle g_1, h \rangle, \ldots, \langle g_p, h \rangle \,]^T \tag{58}$$

and $\mathbf{b}^* = [\, b_1^*, \ldots, b_p^* \,]^T$. The system of normal equations can be written in matrix form as

$$\mathbf{M}\mathbf{b}^* = \mathbf{U}. \tag{59}$$

Since $\mathbf{M}$ is positive definite and hence invertible, (59) has the unique solution

$$\mathbf{b}^* = \mathbf{M}^{-1}\mathbf{U}. \tag{60}$$

The least-squares approximation h^* in G to h is given by

$$h^* = P_G h = \sum_j b_j^* g_j. \tag{61}$$

It follows from (59) together with (23) in Section 9.4 that

$$\|h^*\|^2 = \|P_G h\|^2 = (\mathbf{b}^*)^T \mathbf{M} \mathbf{b}^* = (\mathbf{b}^*)^T \mathbf{U}, \tag{62}$$

so the squared norm of the error of the least-squares approximation to h is given by

$$\|h - h^*\|^2 = \|h\|^2 - \|h^*\|^2 = \|h\|^2 - (\mathbf{b}^*)^T \mathbf{U}. \tag{63}$$

EXAMPLE 9.37   Consider the setup of Example 9.36.

(a) Determine the Gram matrix and its inverse.

(b) Determine **U**.

(c) Use (60) to determine $\mathbf{b}^*$.

(d) Use (63) to determine the squared norm of the error of the least-squares approximation to h.

Solution   (a) The Gram matrix is given by

$$\mathbf{M} = \begin{bmatrix} \langle 1,1 \rangle & \langle 1,x \rangle & \langle 1,x^2 \rangle \\ \langle x,1 \rangle & \langle x,x \rangle & \langle x,x^2 \rangle \\ \langle x^2,1 \rangle & \langle x^2,x \rangle & \langle x^2,x^2 \rangle \end{bmatrix} = \begin{bmatrix} 4 & 0 & 20 \\ 0 & 20 & 0 \\ 20 & 0 & 164 \end{bmatrix},$$

and its inverse is given by

$$\mathbf{M}^{-1} = \frac{1}{320} \begin{bmatrix} 205 & 0 & -25 \\ 0 & 16 & 0 \\ -25 & 0 & 5 \end{bmatrix}.$$

(b) The vector **U** is given by

$$\mathbf{U} = \begin{bmatrix} \langle 1,h \rangle \\ \langle x,h \rangle \\ \langle x^2,h \rangle \end{bmatrix} = \begin{bmatrix} 5 \\ -3 \\ 13 \end{bmatrix}.$$

(c) By (60),

$$\mathbf{b}^* = \mathbf{M}^{-1} \mathbf{U} = \frac{1}{320} \begin{bmatrix} 205 & 0 & -25 \\ 0 & 16 & 0 \\ -25 & 0 & 5 \end{bmatrix} \begin{bmatrix} 5 \\ -3 \\ 13 \end{bmatrix} = \begin{bmatrix} 35/16 \\ -3/20 \\ -3/16 \end{bmatrix},$$

which agrees with the solution to Example 9.36.

(d) Now $\|h\|^2 = 9$. Thus, by (63),

$$\|h - h^*\|^2 = \|h\|^2 - (\mathbf{b}^*)^T \mathbf{U} = 9 - \begin{bmatrix} \dfrac{35}{16}, & -\dfrac{3}{20}, & -\dfrac{3}{16} \end{bmatrix} \begin{bmatrix} 5 \\ -3 \\ 13 \end{bmatrix} = \frac{1}{20},$$

which agrees with the solution to Example 9.33(c) in Section 9.5.   ∎

Let $\mathbf{X}$ denote the $d \times p$ form of the design matrix corresponding to the basis $g_1, \ldots, g_p$ and let $\mathbf{W}$ denote the weight matrix. Recall from Theorem 9.20 in Section 9.4 that $\mathbf{M} = \mathbf{X}^T\mathbf{W}\mathbf{X}$. Set $\mathbf{h} = [\, h(\mathbf{x}_1'), \ldots, h(\mathbf{x}_d') \,]^T$ and let $\mathbf{U} = [\, U_1, \ldots, U_p \,]^T$ be given by (58). Then

$$U_j = \langle g_j, h \rangle = \sum_k w_k g_j(x_k') h(x_k'), \qquad 1 \le j \le p,$$

from which it follows that

$$\mathbf{U} = \mathbf{X}^T\mathbf{W}\mathbf{h}. \tag{64}$$

Thus, the matrix form (59) of the system of normal equations can be written as

$$\mathbf{X}^T\mathbf{W}\mathbf{X}\mathbf{b}^* = \mathbf{X}^T\mathbf{W}\mathbf{h}, \tag{65}$$

whose unique solution is given by

$$\mathbf{b}^* = (\mathbf{X}^T\mathbf{W}\mathbf{X})^{-1}\mathbf{X}^T\mathbf{W}\mathbf{h}. \tag{66}$$

If $w_1, \ldots, w_d$ are unit weights (equivalently, if $\mathbf{W} = \mathbf{I}$), then

$$\mathbf{M} = \mathbf{X}^T\mathbf{X}, \tag{67}$$

$$\mathbf{U} = \mathbf{X}^T\mathbf{h}, \tag{68}$$

and the matrix form of the system of normal equations can be written as

$$\mathbf{X}^T\mathbf{X}\mathbf{b}^* = \mathbf{X}^T\mathbf{h}, \tag{69}$$

whose solution is given by

$$\mathbf{b}^* = (\mathbf{X}^T\mathbf{X})^{-1}\mathbf{X}^T\mathbf{h}. \tag{70}$$

EXAMPLE 9.38   Consider the setup of Example 9.36.

(a) Determine the design matrix.

(b) Use (67) to determine the Gram matrix.

(c) Use (68) to determine $\mathbf{U}$.

Solution   (a) The design matrix is given by

$$\mathbf{X} = \begin{bmatrix} 1 & x_1' & (x_1')^2 \\ 1 & x_2' & (x_2')^2 \\ 1 & x_3' & (x_3')^2 \\ 1 & x_4' & (x_4')^2 \end{bmatrix} = \begin{bmatrix} 1 & -3 & 9 \\ 1 & -1 & 1 \\ 1 & 1 & 1 \\ 1 & 3 & 9 \end{bmatrix}.$$

(b) The Gram matrix is given by

$$\mathbf{M} = \mathbf{X}^T\mathbf{X} = \begin{bmatrix} 1 & 1 & 1 & 1 \\ -3 & -1 & 1 & 3 \\ 9 & 1 & 1 & 9 \end{bmatrix} \begin{bmatrix} 1 & -3 & 9 \\ 1 & -1 & 1 \\ 1 & 1 & 1 \\ 1 & 3 & 9 \end{bmatrix} = \begin{bmatrix} 4 & 0 & 20 \\ 0 & 20 & 0 \\ 20 & 0 & 164 \end{bmatrix}.$$

(c) Since $\mathbf{h} = [\,1, 2, 2, 0\,]^T$,

$$\mathbf{U} = \mathbf{X}^T\mathbf{h} = \begin{bmatrix} 1 & 1 & 1 & 1 \\ -3 & -1 & 1 & 3 \\ 9 & 1 & 1 & 9 \end{bmatrix} \begin{bmatrix} 1 \\ 2 \\ 2 \\ 0 \end{bmatrix} = \begin{bmatrix} 5 \\ -3 \\ 13 \end{bmatrix}. \qquad \blacksquare$$

The normal equations can be obtained by using calculus instead of linear algebra, as was done in Section 8.1. Suppose, say, that $p = 3$; let g_1, g_2, and g_3 form a basis of G; write the least-squares approximation to h as $h^* = b_1^*g_1 + b_2^*g_2 + b_3^*g_3$; and write an arbitrary function in G as $g = b_1g_1 + b_2g_2 + b_3g_3$. The squared norm of $h - g$ is given by

$$\|h - g\|^2 = \sum_k w_k[\,h(\mathbf{x}_k') - b_1g_1(\mathbf{x}_k') - b_2g_2(\mathbf{x}_k') - b_3g_3(\mathbf{x}_k')\,]^2.$$

To minimize $\|h - g\|^2$, we set its partial derivatives with respect to b_j equal to zero for $j = 1, 2, 3$ to get the normal equations. Since

$$\frac{\partial}{\partial b_j}\|h - g\|^2 = -2\sum_k w_kg_j(\mathbf{x}_k')[\,h(\mathbf{x}_k') - b_1g_1(\mathbf{x}_k') - b_2g_2(\mathbf{x}_k') - b_3g_3(\mathbf{x}_k')\,],$$

the normal equations for b_1^*, b_1^*, and b_1^* can be written as

$$\sum_k w_kg_j(\mathbf{x}_j')[\,h(\mathbf{x}_k') - b_1^*g_1(\mathbf{x}_k') - b_2^*g_2(\mathbf{x}_k') - b_3^*g_3(\mathbf{x}_k')\,] = 0, \qquad j = 1, 2, 3,$$

or, equivalently, as

$$b_1^*\sum_k w_kg_1(\mathbf{x}_k')g_1(\mathbf{x}_k') + b_2^*\sum_k w_kg_1(\mathbf{x}_k')g_2(\mathbf{x}_k') + b_3^*\sum_k w_kg_1(\mathbf{x}_k')g_3(\mathbf{x}_k')$$
$$= \sum_k w_kg_1(\mathbf{x}_k')h(\mathbf{x}_k'),$$

$$b_1^*\sum_k w_kg_2(\mathbf{x}_k')g_1(\mathbf{x}_k') + b_2^*\sum_k w_kg_2(\mathbf{x}_k')g_2(\mathbf{x}_k') + b_3^*\sum_k w_kg_2(\mathbf{x}_k')g_3(\mathbf{x}_k')$$
$$= \sum_k w_kg_2(\mathbf{x}_k')h(\mathbf{x}_k'),$$

and

$$b_1^*\sum_k w_kg_3(\mathbf{x}_k')g_1(\mathbf{x}_k') + b_2^*\sum_k w_kg_3(x_k')g_2(x_k') + b_3^*\sum_k w_kg_3(\mathbf{x}_k')g_3(\mathbf{x}_k')$$
$$= \sum_k w_kg_3(\mathbf{x}_k')h(\mathbf{x}_k').$$

Suppose $\mathcal{X} \subset \mathbb{R}$ and let $h \in H$. The function h is said to be an even function if $h(-x) = h(x)$ whenever x and $-x$ are both in $\mathcal{X}$, and it is said to be an odd function if $h(-x) = -h(x)$ whenever x and $-x$ are both in $\mathcal{X}$. A linear combination of even functions is even, and a linear combination of odd functions is odd. Observe that x^j is an even function if j is a nonnegative even integer and an odd function if j is an odd positive integer.

The design set $\{x'_1, \ldots, x'_d\}$ is said to be *symmetric* if

$$\{-x'_1, \ldots, -x'_d\} = \{x'_1, \ldots, x'_d\}.$$

Some examples of symmetric design sets are

$$\{0\}, \quad \{-1, 1\}, \quad \{-1, 0, 1\}, \quad \{-3, -1, 1, 3\}, \quad \{-2, -1, 0, 1, 2\}.$$

Suppose the design set is symmetric. Then $\sum_k h(x'_k) = 0$ for h an odd function. Under the additional assumption of constant weights, $\langle h_1, h_2 \rangle = 0$ for functions $h_1, h_2 \in H$ such that $h_1 h_2$ is an odd function; in particular, $\langle h_1, h_2 \rangle = 0$ if one of the functions $h_1, h_2 \in H$ is odd and the other function is even.

EXAMPLE 9.39 (a) Determine the normal equations relative to unit weights and the basis $1, x, x^2$ for least-squares quadratic polynomial approximation.

(b) Simplify these equations when the design set is symmetric and explain how the equations can easily be solved.

Solution (a) Let $h^* = b_0^* + b_1^* x + b_2^* x^2$ be the least-squares quadratic approximation to a function h. The normal equations for b_0^*, b_1^*, and b_2^* can be written as

$$b_0^* d + b_1^* \sum_k x'_k + b_2^* \sum_k (x'_k)^2 = \sum_k h(x'_k)$$

$$b_0^* \sum_k x'_k + b_1^* \sum_k (x'_k)^2 + b_2^* \sum_k (x'_k)^3 = \sum_k x'_k h(x'_k)$$

$$b_0^* \sum_k (x'_k)^2 + b_1^* \sum_k (x'_k)^3 + b_2^* \sum_k (x'_k)^4 = \sum_k (x'_k)^2 h(x'_k).$$

(b) Suppose the design set is symmetric. Then the normal equations simplify to

$$b_0^* d + b_2^* \sum_k (x'_k)^2 = \sum_k h(x'_k)$$

$$b_1^* \sum_k (x'_k)^2 = \sum_k x'_k h(x'_k)$$

$$b_0^* \sum_k (x'_k)^2 + b_2^* \sum_k (x'_k)^4 = \sum_k (x'_k)^2 h(x'_k).$$

The coefficient b_1^* is obtained immediately from the second of these equations, and b_0^* and b_2^* are easily obtained from the first and third equations. ■

In practical applications, least-squares approximation is commonly done on a computer.

EXAMPLE 9.40   Consider the design set $\{-1, -.9, -.8, \ldots, 1\}$ and unit weights and set

$$h(x) = \sin\left(\frac{3\pi x}{2}\right) + \cos\left(\frac{3\pi x}{2}\right).$$

(a) Use the basis $1, x, \ldots, x^5$ to determine the least-squares quintic polynomial approximation h^* to h.

(b) Determine the squared norm of the error of approximation.

(c) Determine the maximum value of the absolute error of approximation on the design set.

(d) Determine the maximum value of the absolute error of approximation on $[-1, 1]$.

(e) Determine the maximum value of the absolute error of approximation on $[-1.1, 1.1]$.

(f) Show the plots of h and h^* on $[-1.1, 1.1]$.

Solution   **(a)** Using a computer to obtain and solve the normal equations, we get

$$h^*(x) \doteq 0.7985 + 4.1604x - 6.7772x^2 - 12.3919x^3 + 6.1681x^4 + 7.2866x^5.$$

(b) The squared norm of the error of approximation is given by $\|h - h^*\|^2 \doteq 0.5807$.

(c) The maximum value of the absolute error of approximation on the design set is 0.2858, which occurs at $x = 0.9$.

(d) The maximum value of the absolute error of approximation on $[-1, 1]$ is 0.3135, which occurs at $x \doteq 0.8652$.

(e) The maximum value of the absolute error of approximation on $[-1.1, 1.1]$ is 1.8839, which occurs at $x = 1.1$.

(f) The plots of h and h^* on $[-1.1, 1.1]$, shown in Figure 9.11, illustrate the danger of *extrapolation*—that is, of using $h^*(x)$ for values of x that lie beyond the range of the design points.   ∎

Problems

9.70   Consider the design set $\{1, 2, 3\}$ and unit weights. Determine the least-squares linear approximation to x^2 by starting with the basis $1, x$ and solving the corresponding normal equations.

9.71   Consider the design set $\{-3, -1, 1, 3\}$ and unit weights. Let $h^*(x) = b_0^* + b_1^* x + b_2^* x^2$ be the least-squares quadratic polynomial approximation to $h(x) = \cos(\pi x/4)$. Determine **(a)** $h(-3), h(-1), h(1), h(3)$; **(b)** $\langle 1, h \rangle, \langle x, h \rangle, \langle x^2, h \rangle$; **(c)** the normal equations for b_0^*, b_1^*, b_2^*; **(d)** b_0^*, b_1^*, b_2^*; **(e)** $\|h\|^2$; **(f)** $\|h - h^*\|^2$.

9.72   Consider the design set $\{-2, -1, 0, 1, 2\}$ and unit weights. Let

$$h^*(x) = b_0^* + b_1^* x + b_2^* x^2 + b_3^* x^3$$

FIGURE   9.11

$\sin(3\pi x/2) + \cos(3\pi x/2)$ and its least-squares quintic polynomial approximation (dashed curve)

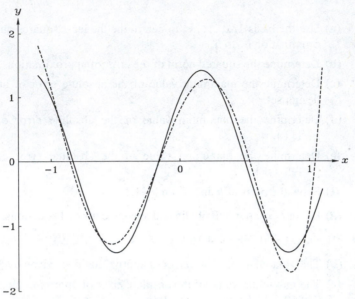

be the least-squares cubic polynomial approximation to $h(x) = \sin(\pi x/4)$. Determine **(a)** $h(-2)$, $h(-1)$, $h(0)$, $h(1)$, $h(2)$; **(b)** $\langle 1, h \rangle$, $\langle x, h \rangle$, $\langle x^2, h \rangle$, $\langle x^3, h \rangle$; **(c)** the normal equations for b_0^*, b_1^*, b_2^*, b_3^*; **(d)** b_0^*, b_1^*, b_2^*, b_3^*; **(e)** $\|h\|^2$; **(f)** $\|h - h^*\|^2$.

9.73 **(a)** Determine the normal equations relative to unit weights and the basis 1, x, x^2, x^3 for least-squares cubic polynomial approximation. **(b)** Simplify these equations when the design set is symmetric and explain how the equations can easily be solved.

9.74 Let G be an identifiable three-dimensional linear space containing the constant functions and let $1, g_1, g_2$ be a basis of G. Then $1, g_1 - \bar{g}_1, g_2 - \bar{g}_2$ is also a basis of G. Let $b_0^* + b_1^*(g_1 - \bar{g}_1) + b_2^*(g_2 - \bar{g}_2)$ be the least-squares approximation in G to $h \in H$. **(a)** Show that $b_0^* = \bar{h}$. **(b)** Determine the normal equations for b_1^* and b_2^*.

9.75 Let G be an identifiable p-dimensional linear space of functions on $\mathcal{X}$ containing the constant functions and let $1, g_1, \ldots, g_{p-1}$ be a basis of G such that $g_1, \ldots, g_{p-1}$ are each orthogonal to 1. Then $P_G h = b_0^* + b_1^* g_1 + \cdots + b_{p-1}^* g_{p-1}$, where $b_0^*, \ldots, b_{p-1}^* \in \mathbb{R}$ are uniquely determined. Show that $b_0^* = \bar{h}$.

9.76 Let h be a fixed real-valued function on $\mathbb{R}$. Consider the design points $-c, 0, c$ with $c > 0$ and unit weights and let G denote the space of linear functions on $\mathbb{R}$. Determine a simple analytic condition on h such that $\|h - P_G h\|^2 = 0$ for all $c > 0$ if and only if this condition is satisfied.

9.77 Let $g_{11}, \ldots, g_{1p_1}, g_{21}, \ldots, g_{2p_2}$ be a basis of the identifiable $(p_1 + p_2)$-dimensional linear space G and suppose that each of the functions $g_{11}, \ldots, g_{1p_1}$ is orthogonal to each of the functions $g_{21}, \ldots, g_{2p_2}$. Let G_1 be the span of $g_{11}, \ldots, g_{1p_1}$ and let G_2

be the span of $g_{21}, \ldots, g_{2p_2}$. Then G_1 and G_2 are subspaces of G, and each function in G_1 is orthogonal to every function in G_2. Moreover, each function $g \in G$ can be written in a unique manner as $g = g_1 + g_2$, where $g_1 \in G_1$ and $g_2 \in G_2$; specifically, g_1 and g_2 are the orthogonal projections of g onto G_1 and G_2, respectively. Let $h \in H$. Show that if h is orthogonal to every function in G_2, then $P_G h = P_{G_1} h$. (Similarly, if h is orthogonal to every function in G_1, then $P_G h = P_{G_2} h$.)

9.78   Suppose $\mathcal{X} \subset \mathbb{R}$ and that the design set is symmetric. Let

$$g_{11}, \ldots, g_{1p_1}, g_{21}, \ldots, g_{2p_2}$$

be a basis of the identifiable $(p_1 + p_2)$-dimensional linear space G and suppose the functions $g_{11}, \ldots, g_{1p_1}$ are odd and the functions $g_{21}, \ldots, g_{2p_2}$ are even. Let $h \in H$ and let $h^* = b_{11}^* g_{11} + \cdots + b_{1p_1}^* g_{1p_1} + b_{21}^* g_{21} + \cdots + b_{2p_2}^* g_{2p_2}$ be the orthogonal projection of h onto G. Use the result of Problem 9.77 to show that if h is an odd function, then $b_{21}^* = \cdots = b_{2p_2}^* = 0$. (Similarly, if h is an even function, then $b_{11}^* = \cdots = b_{1p_1}^* = 0$.)

9.79   Suppose $\mathcal{X} \subset \mathbb{R}$, $d \geq 2$, and the design set is symmetric. Given $h \in H$, let h^* be the orthogonal projection of h onto the space G of polynomials of degree $d - 2$. Use the result of Problem 9.78 to show that if either d is an odd integer and h is an odd function or d is even integer and h is an even function, then $\|h - P_G h\| = 0$.

9.80   Let $\mathbf{M}$ be the Gram matrix of the basis $g_1, \ldots, g_p$ of G and let $\mathbf{U}$ and $\mathbf{b}^*$ be given by (58) and (60) respectively. Let $\tilde{\mathbf{M}}$ be the Gram matrix of an alternative basis $\tilde{g}_1, \ldots, \tilde{g}_p$, set $\tilde{\mathbf{U}} = [\, \langle \tilde{g}_1, h \rangle, \ldots, \langle \tilde{g}_p, h \rangle \,]^T$ and $\tilde{\mathbf{b}}^* = \tilde{\mathbf{M}}^{-1} \tilde{\mathbf{U}}$, and let $\mathbf{A}$ be the coefficient matrix for $\tilde{g}_1, \ldots, \tilde{g}_p$ relative to $g_1, \ldots, g_p$. Recall from Theorem 9.18 in Section 9.4 that $\tilde{\mathbf{M}} = \mathbf{A}\mathbf{M}\mathbf{A}^T$. Verify that **(a)** $\tilde{\mathbf{U}} = \mathbf{A}\mathbf{U}$; **(b)** $\tilde{\mathbf{M}}^{-1} = (\mathbf{A}^T)^{-1}\mathbf{M}^{-1}\mathbf{A}^{-1}$; **(c)** $\tilde{\mathbf{b}}^* = (\mathbf{A}^T)^{-1}\mathbf{b}^*$. [In light of (6) in Section 9.1, we should expect part (c) to hold since $h^* = P_G h$ does not depend on the choice of the basis of G.]

9.81   In the context of Problem 9.53 in Section 9.4, determine the least-squares approximation to x_3^2 in the space G of linear functions of x_1, x_2, and x_3 **(a)** by solving the normal equations corresponding to the basis $1, x_1, x_2, x_3$ of G; **(b)** by using the orthogonal basis of G given by the answer to Problem 9.69(b) in Section 9.5.

10

Linear Regression

10.1
Least-Squares Estimation

In this section, we begin the development of the properties of the least-squares estimate of the regression function in an identifiable linear space G of functions on X by viewing this estimate as the orthogonal projection of $\bar{Y}(\cdot)$ onto G.

To this end, let h_1 and h_2 be members of H—that is, functions that are defined on the design set and possibly elsewhere on X as well. Then

$$\sum_i h_1(\mathbf{x}_i)h_2(\mathbf{x}_i) = \sum_k \sum_{i \in \mathcal{I}_k} h_1(\mathbf{x}_i)h_2(\mathbf{x}_i) = \sum_k n_k h_1(\mathbf{x}'_k)h_2(\mathbf{x}'_k),$$

where $\mathcal{I}_k = \{ i : \mathbf{x}_i = \mathbf{x}'_k \}$ for $1 \leq k \leq d$. Think of $n_1, \ldots, n_d$ as weights. Then the corresponding inner product and squared norm are given by

$$\langle h_1, h_2 \rangle = \sum_k n_k h_1(\mathbf{x}'_k)h_2(\mathbf{x}'_k) = \sum_i h_1(\mathbf{x}_i)h_2(\mathbf{x}_i) \tag{1}$$

and

$$\|h\|^2 = \langle h, h \rangle = \sum_k n_k h^2(\mathbf{x}'_k) = \sum_i h^2(\mathbf{x}_i). \tag{2}$$

Recall from Section 8.3 that $\bar{Y}(\cdot)$ is the member of H that is defined on the design set by

$$\bar{Y}(\mathbf{x}'_k) = \bar{Y}_k = \frac{1}{n_k} \sum_{i \in \mathcal{I}_k} Y_i, \qquad 1 \le k \le d.$$

Let $g \in G$. Then

$$\sum_i g(\mathbf{x}_i) Y_i = \sum_k \sum_{i \in \mathcal{I}_k} g(\mathbf{x}_i) Y_i$$

$$= \sum_k g(\mathbf{x}'_k) \sum_{i \in \mathcal{I}_k} Y_i$$

$$= \sum_k n_k g(\mathbf{x}'_k) \bar{Y}_k$$

$$= \sum_k n_k g(\mathbf{x}'_k) \bar{Y}(\mathbf{x}'_k).$$

Applying (1) to the functions $h_1 = g$ and $h_2 = \bar{Y}(\cdot)$, we see that

$$\sum_k n_k g(\mathbf{x}'_k) \bar{Y}(\mathbf{x}'_k) = \sum_i g(\mathbf{x}_i) \bar{Y}(\mathbf{x}_i).$$

Consequently,

$$\sum_i g(\mathbf{x}_i) Y_i = \sum_i g(\mathbf{x}_i) \bar{Y}(\mathbf{x}_i), \qquad g \in G, \tag{3}$$

so

$$\sum_i g(\mathbf{x}_i)[Y_i - \bar{Y}(\mathbf{x}_i)] = 0, \qquad g \in G. \tag{4}$$

Recall from (4) in Section 8.3 that

$$\text{RSS}(g) = \text{LSS}(g) + \text{WSS}, \qquad g \in G. \tag{5}$$

Here WSS is the within sum of squares, which is given by

$$\text{WSS} = \sum_k \sum_{i \in \mathcal{I}_k} (Y_i - \bar{Y}_k)^2 = \sum_i [Y_i - \bar{Y}(\mathbf{x}_i)]^2,$$

and $\text{LSS}(g)$ is the lack-of-fit sum of squares, which is given by

$$\text{LSS}(g) = \sum_k n_k [\bar{Y}_k - g(\mathbf{x}'_k)]^2 = \sum_i [\bar{Y}(\mathbf{x}_i) - g(\mathbf{x}_i)]^2$$

and hence by

$$\text{LSS}(g) = \|\bar{Y}(\cdot) - g\|^2, \qquad g \in G. \tag{6}$$

The *least-squares estimate* in G of the regression function, or the *least-squares fit* in G to the data, is the function in G having the smallest residual sum of squares. By (5) it is also the function in G having the smallest lack-of-fit sum of squares.

Thus, by (6), the least-squares estimate in G equals the least-squares approximation in G to $\bar{Y}(\cdot)$. Therefore, by Theorem 9.26 in Section 9.5, we have the following fundamental result.

THEOREM 10.1   There is a unique least-squares estimate $\hat{\mu}(\cdot)$ in G of the regression function, which is the orthogonal projection of $\bar{Y}(\cdot)$ onto G.

The least-squares estimate $\hat{\mu}(\cdot)$ in G of $\mu(\cdot)$ can be visualized as in Figure 10.1, which is a modification of Figure 9.6 in Section 9.5.

FIGURE   10.1
The least-squares estimate $\hat{\mu}(\cdot)$ in G of the regression function is the orthogonal projection of $\bar{Y}(\cdot)$ onto G.

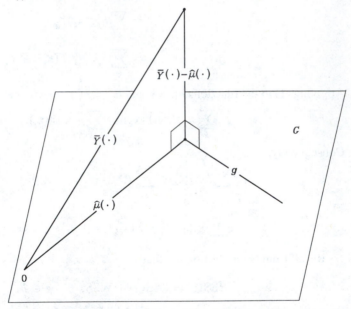

The uniqueness of the least-squares estimate depends on G being identifiable. If G were nonidentifiable, there would be infinitely many least-squares estimates in G (see the discussion at the end of Section 9.5). In particular, let $\hat{\mu}(\cdot)$ be a least-squares estimate in G and let g be a nonzero function in G that equals zero on the design set. Then $g(\mathbf{x}_i) = 0$ for $1 \leq i \leq n$. Given $b \in \mathbb{R}$, set $\hat{g} = \hat{\mu}(\cdot) + bg$. Then $\hat{g}(\mathbf{x}_i) = \hat{\mu}(\mathbf{x}_i) + bg(\mathbf{x}_i) = \hat{\mu}(\mathbf{x}_i)$ for $1 \leq i \leq n$, so $\mathrm{RSS}(\hat{g}) = \mathrm{RSS}(\hat{\mu}(\cdot))$ and hence $\hat{g}$ would also be a least-squares estimate. As b ranges over $\mathbb{R}$, we would get infinitely many least-squares estimates in this manner.

It is generally not sensible to try to obtain a least-squares estimate in G of the regression function when G is nonidentifiable. In particular, as Example 9.9 in Section 9.2 suggests, it is not sensible to try to obtain a linear estimate of the regression function when there is one factor and the design set consists of a single point, for in that case $x_1 = \cdots = x_n$ and hence the experimental data would contain no information to help us in determining the slope of the regression function. Similarly, as Example 9.10 in that section suggests, it is not sensible to try to obtain a quadratic estimate of the regression function when there is one factor and the design set contains only one or two points. Finally, as Example 9.11 in Section 9.2 suggests, it is not sensible to try to obtain a linear estimate of the regression function when there are two factors and the design points are collinear.

According to Theorem 10.1, the unique least-squares estimate in G of the regression function is given by $\hat{\mu}(\cdot) = P_G[\bar{Y}(\cdot)]$. Let $\text{LSS} = \text{LSS}(G)$ and $\text{RSS} = \text{RSS}(G)$, respectively, denote the lack-of-fit and residual sums of squares for this least-squares estimate. Then

$$\text{LSS} = \|\bar{Y}(\cdot) - \hat{\mu}(\cdot)\|^2 = \sum_k n_k[\bar{Y}(\mathbf{x}'_k) - \hat{\mu}(\mathbf{x}'_k)]^2 = \sum_i [\bar{Y}(\mathbf{x}_i) - \hat{\mu}(\mathbf{x}_i)]^2$$

and $\text{RSS} = \sum_i [Y_i - \hat{\mu}(\mathbf{x}_i)]^2 = \text{LSS} + \text{WSS}$. According to the next result, which is a consequence of Theorem 9.27 in Section 9.5, if G is saturated, then $\hat{\mu}(\mathbf{x}'_k) = \bar{Y}_k$ for $1 \le k \le d$.

THEOREM 10.2 Suppose G is saturated. Then $\hat{\mu}(\cdot) = \bar{Y}(\cdot)$ on the design set, $\text{LSS} = 0$ and $\text{RSS} = \text{WSS}$.

It is sometimes convenient to ignore points in $\mathcal{X}$ that are not design points. Then H is the d-dimensional space of all functions on the design set, which is identifiable and saturated. Moreover, $\bar{Y}(\cdot) \in H$, and hence $\bar{Y}(\cdot) = P_H[\bar{Y}(\cdot)]$. Thus, $\bar{Y}(\cdot)$ is the least-squares estimate in H of the regression function. The lack-of-fit sum of squares for $\bar{Y}(\cdot)$ equals zero, and its residual sum of squares equals WSS.

Normal Equations

Since $\hat{\mu}(\cdot)$ is the orthogonal projection of $\bar{Y}(\cdot)$ onto G, $\bar{Y}(\cdot) - \hat{\mu}(\cdot)$ is orthogonal to every function in G; that is, $\langle g, \bar{Y}(\cdot) - \hat{\mu}(\cdot) \rangle = 0$ for $g \in G$ or, equivalently,

$$\sum_i g(\mathbf{x}_i)[\bar{Y}(\mathbf{x}_i) - \hat{\mu}(\mathbf{x}_i)] = 0, \qquad g \in G. \tag{7}$$

Conversely, if $\hat{\mu}(\cdot)$ is a member of G and (7) holds, then $\hat{\mu}(\cdot)$ is the orthogonal projection of $\bar{Y}(\cdot)$ onto G. Thus, (7) is a necessary and sufficient condition for a function $\hat{\mu}(\cdot) \in G$ to be the least-squares estimate in G of the regression function. It follows from (4) that (7) is equivalent to

$$\sum_i g(\mathbf{x}_i)[Y_i - \hat{\mu}(\mathbf{x}_i)] = 0, \qquad g \in G, \tag{8}$$

and hence that (8) is necessary and sufficient for a function $\hat{\mu}(\cdot) \in G$ to be the least-squares estimate in G of the regression function. Suppose G contains the constant functions. We then conclude from (8) with $g = 1$ that the sum $\sum_i [Y_i - \hat{\mu}(\mathbf{x}_i)]$ of the residuals about the least-squares estimate of the regression function equals zero.

Equation (8) can be rewritten as

$$\sum_i g(\mathbf{x}_i)\hat{\mu}(\mathbf{x}_i) = \sum_i g(\mathbf{x}_i)Y_i, \qquad g \in G. \tag{9}$$

Let $g_1, \ldots, g_p$ be a basis of G. Then (9) holds if and only if

$$\sum_i g_1(\mathbf{x}_i)\hat{\mu}(\mathbf{x}_i) = \sum_i g_1(\mathbf{x}_i)Y_i$$

$$\vdots \qquad \vdots \qquad \vdots \tag{10}$$

$$\sum_i g_p(\mathbf{x}_i)\hat{\mu}(\mathbf{x}_i) = \sum_i g_p(\mathbf{x}_i)Y_i.$$

Let $\hat{\beta}_1, \ldots, \hat{\beta}_p$ be the coefficients of $\hat{\mu}(\cdot)$ relative to the basis $g_1, \ldots, g_p$, which we refer to as the least-squares estimates of the regression coefficients $\beta_1, \ldots, \beta_p$. Then $\hat{\mu}(\cdot) = \hat{\beta}_1 g_1 + \cdots + \hat{\beta}_p g_p$, so (10) can be written as

$$\hat{\beta}_1 \sum_i g_1^2(\mathbf{x}_i) + \cdots + \hat{\beta}_p \sum_i g_1(\mathbf{x}_i)g_p(\mathbf{x}_i) = \sum_i g_1(\mathbf{x}_i)Y_i$$

$$\vdots \qquad \vdots \qquad \vdots \tag{11}$$

$$\hat{\beta}_1 \sum_i g_p(\mathbf{x}_i)g_1(\mathbf{x}_i) + \cdots + \hat{\beta}_p \sum_i g_p^2(\mathbf{x}_i) = \sum_i g_p(\mathbf{x}_i)Y_i,$$

which we refer to as the *system of normal equations* for $\hat{\beta}_1, \ldots, \hat{\beta}_p$.

EXAMPLE 10.1   Determine

(a) the least-squares constant estimate of the regression function;

(b) its lack-of-fit sum of squares;   (c) its residual sum of squares.

Solution   (a) Now 1 is a basis of the space of constant functions. Thus, the least-squares constant estimate of the regression function is given by $\hat{\mu}(\cdot) = \hat{\beta}_0$, where $\hat{\beta}_0$ satisfies the normal equation $n\hat{\beta}_0 = \sum_i Y_i$, whose unique solution is given by $\hat{\beta}_0 = \bar{Y}$. Alternatively, by (52) in Section 2.4,

$$\text{RSS}(c) = \sum_i (Y_i - c)^2 = \sum_i (Y_i - \bar{Y})^2 + n(\bar{Y} - c)^2,$$

which is clearly minimized at $c = \bar{Y}$, so $\hat{\mu}(\cdot) = \bar{Y}$.

(b) The lack-of-fit sum of squares for $\bar{Y}$ is given by $\text{LSS} = \sum_k n_k (\bar{Y}_k - \bar{Y})^2$.

(c) The residual sum of squares is given by $\text{RSS} = \sum_i (Y_i - \bar{Y})^2$.   ∎

EXAMPLE 10.2   Use calculus to derive the system of normal equations for $\hat{\beta}_1$ and $\hat{\beta}_2$ directly from the definition of the least-squares estimate in G of the regression function when $p = 2$.

Solution The residual sum of squares about $g = b_1 g_1 + b_2 g_2 \in G$ is given by

$$\text{RSS}(g) = \sum_i [\, Y_i - b_1 g_1(\mathbf{x}_i) - b_2 g_2(\mathbf{x}_i) \,]^2.$$

At the values $\hat{\beta}_1$ and $\hat{\beta}_2$ of b_1 and b_2, respectively, that minimize $\text{RSS}(g)$, the partial derivatives of this quantity with respect to b_1 and b_2 equal zero. Since

$$\frac{\partial \text{RSS}(g)}{\partial b_1} = -2 \sum_i g_1(\mathbf{x}_i)[\, Y_i - b_1 g_1(\mathbf{x}_i) - b_2 g_2(\mathbf{x}_i) \,]$$

and

$$\frac{\partial \text{RSS}(g)}{\partial b_2} = -2 \sum_i g_2(\mathbf{x}_i)[\, Y_i - b_1 g_1(\mathbf{x}_i) - b_2 g_2(\mathbf{x}_i) \,],$$

we conclude that

$$\sum_i g_1(\mathbf{x}_i)[\, Y_i - \hat{\beta}_1 g_1(\mathbf{x}_i) - \hat{\beta}_2 g_2(\mathbf{x}_i) \,] = 0$$

and

$$\sum_i g_2(\mathbf{x}_i)[\, Y_i - \hat{\beta}_1 g_1(\mathbf{x}_i) - \hat{\beta}_2 g_2(\mathbf{x}_i) \,] = 0$$

and hence that $\hat{\beta}_1$ and $\hat{\beta}_2$ satisfy the system

$$\hat{\beta}_1 \sum_i g_1^2(\mathbf{x}_i) + \hat{\beta}_2 \sum_i g_1(\mathbf{x}_i) g_2(\mathbf{x}_i) = \sum_i g_1(\mathbf{x}_i) Y_i$$

$$\hat{\beta}_1 \sum_i g_2(\mathbf{x}_i) g_1(\mathbf{x}_i) + \hat{\beta}_2 \sum_i g_2^2(\mathbf{x}_i) = \sum_i g_2(\mathbf{x}_i) Y_i$$

of normal equations. ∎

Suppose $p = 2$ and that G contains the constant functions, let $1, g_1$ form a basis of G, and write the least-squares estimate in G of the regression function as $\hat{\mu}(\cdot) = \hat{\beta}_0 + \hat{\beta}_1 g_1$. Then the system of normal equations for $\hat{\beta}_0$ and $\hat{\beta}_1$ is given by

$$\hat{\beta}_0 n + \hat{\beta}_1 \sum_i g_1(\mathbf{x}_i) = \sum_i Y_i$$

$$\hat{\beta}_0 \sum_i g_1(\mathbf{x}_i) + \hat{\beta}_1 \sum_i g_1^2(\mathbf{x}_i) = \sum_i g_1(\mathbf{x}_i) Y_i.$$

EXAMPLE 10.3 Obtain the system of normal equations for the coefficients $\hat{\beta}_0$, $\hat{\beta}_1$ of $\hat{\mu}(\cdot)$ relative to the basis $1, x$ of the space of linear functions of x.

Solution Here $g_1(x_i) = x_i$, so the system of normal equations is given by

$$\hat{\beta}_0 n + \hat{\beta}_1 \sum_i x_i = \sum_i Y_i$$

$$\hat{\beta}_0 \sum_i x_i + \hat{\beta}_1 \sum_i x_i^2 = \sum_i x_i Y_i. \quad \blacksquare$$

EXAMPLE 10.4   (a) Obtain the normal equations for the coefficients $\hat{\beta}_0$ and $\hat{\beta}_1$ of $\hat{\mu}(\cdot)$ relative to the basis $1, x - \bar{x}$ of the space of linear functions of x, where $\bar{x} = (x_1 + \cdots + x_n)/n$.

(b) Solve these equations.

Solution   (a) Here $g_1(x_i) = x_i - \bar{x}$, so the system of normal equations is given by

$$\hat{\beta}_0 n + \hat{\beta}_1 \sum_i (x_i - \bar{x}) = \sum_i Y_i$$

$$\hat{\beta}_0 \sum_i (x_i - \bar{x}) + \hat{\beta}_1 \sum_i (x_i - \bar{x})^2 = \sum_i (x_i - \bar{x}) Y_i.$$

Since $\sum_i (x_i - \bar{x}) = \sum_i x_i - n\bar{x} = 0$, these equations simplify to

$$\hat{\beta}_0 n = \sum_i Y_i$$

$$\hat{\beta}_1 \sum_i (x_i - \bar{x})^2 = \sum_i (x_i - \bar{x}) Y_i.$$

(b) The unique solution to these equations is given by $\hat{\beta}_0 = \bar{Y}$ and

$$\hat{\beta}_1 = \frac{\sum_i (x_i - \bar{x}) Y_i}{\sum_i (x_i - \bar{x})^2}. \qquad \blacksquare$$

Set

$$\mathbf{X} = \begin{bmatrix} g_1(\mathbf{x}_1) & \cdots & g_p(\mathbf{x}_1) \\ \vdots & & \vdots \\ g_1(\mathbf{x}_n) & \cdots & g_p(\mathbf{x}_n) \end{bmatrix},$$

which we refer to as the $n \times p$ form of the design matrix relative to the basis $g_1, \ldots, g_p$ of G. In the present context, this form of the design matrix is generally more convenient (especially in computer applications) than the $d \times p$ form that was introduced in Section 9.3. Theorems 9.7 and 9.8 and Corollaries 9.5 and 9.6 in Section 9.2 hold for the $n \times p$ form of the design matrix. It follows as in the proof of Theorem 9.20 in Section 9.4 that

$$\mathbf{X}^T \mathbf{X} = \begin{bmatrix} \sum_i g_1^2(\mathbf{x}_i) & \cdots & \sum_i g_1(\mathbf{x}_i) g_p(\mathbf{x}_i) \\ \vdots & & \vdots \\ \sum_i g_p(\mathbf{x}_i) g_1(\mathbf{x}_i) & \cdots & \sum_i g_p^2(\mathbf{x}_i) \end{bmatrix}$$

is the Gram matrix of the basis $g_1, \ldots, g_p$ of G, so it is positive definite and symmetric and hence invertible.

EXAMPLE 10.5   Determine the design matrix $\mathbf{X}$ and Gram matrix $\mathbf{X}^T \mathbf{X}$ relative to

(a) the basis $1, x$ of the space of linear functions of x;

(b) the basis $1, x, x^2$ of the space of quadratic polynomials in x;

(c) the basis $1, x_1, x_2$ of the space of linear functions of x_1 and x_2.

Solution   (a) The design matrix is given by

$$\mathbf{X} = \begin{bmatrix} 1 & x_1 \\ \vdots & \vdots \\ 1 & x_n \end{bmatrix},$$

and the Gram matrix is given by

$$\mathbf{X}^T\mathbf{X} = \begin{bmatrix} n & \sum_i x_i \\ \sum_i x_i & \sum_i x_i^2 \end{bmatrix}.$$

(b) The design matrix is given by

$$\mathbf{X} = \begin{bmatrix} 1 & x_1 & x_1^2 \\ \vdots & \vdots & \vdots \\ 1 & x_n & x_n^2 \end{bmatrix},$$

and the Gram matrix is given by

$$\mathbf{X}^T\mathbf{X} = \begin{bmatrix} n & \sum_i x_i & \sum_i x_i^2 \\ \sum_i x_i & \sum_i x_i^2 & \sum_i x_i^3 \\ \sum_i x_i^2 & \sum_i x_i^3 & \sum_i x_i^4 \end{bmatrix}.$$

(c) The design matrix is given by

$$\mathbf{X} = \begin{bmatrix} 1 & x_{11} & x_{12} \\ \vdots & \vdots & \vdots \\ 1 & x_{n1} & x_{n2} \end{bmatrix},$$

and the Gram matrix is given by

$$\mathbf{X}^T\mathbf{X} = \begin{bmatrix} n & \sum_i x_{i1} & \sum_i x_{i2} \\ \sum_i x_{i1} & \sum_i x_{i1}^2 & \sum_i x_{i1}x_{i2} \\ \sum_i x_{i2} & \sum_i x_{i2}x_{i1} & \sum_i x_{i2}^2 \end{bmatrix}. \qquad \blacksquare$$

Let $\hat{\beta} = [\hat{\beta}_1, \ldots, \hat{\beta}_p]^T$ denote the coefficient vector of $\hat{\mu}(\cdot)$ relative to the basis $g_1, \ldots, g_p$ and let $\mathbf{Y} = [Y_1, \ldots, Y_n]^T$ be the random vector consisting of the n response variables. Then the system (11) of normal equations for $\hat{\beta}_1, \ldots, \hat{\beta}_p$ can be written in matrix form as

$$\mathbf{X}^T\mathbf{X}\hat{\beta} = \mathbf{X}^T\mathbf{Y}, \tag{12}$$

whose unique solution is given by

$$\hat{\beta} = (\mathbf{X}^T\mathbf{X})^{-1}\mathbf{X}^T\mathbf{Y}. \tag{13}$$

Let $\mathbf{U} = [U_1, \ldots, U_p]^T = \mathbf{X}^T\mathbf{Y}$ denote the right side of (12). Then

$$U_j = \langle g_j, \bar{Y}(\cdot) \rangle = \sum_k n_k g_j(\mathbf{x}'_k)\bar{Y}_k = \sum_i g_j(\mathbf{x}_i)Y_i, \qquad 1 \le j \le p. \qquad (14)$$

Thus, for $1 \le j \le p$, the statistic U_j can be obtained by multiplying the ith entry $g_j(\mathbf{x}_i)$ of the jth column of the $n \times p$ design matrix $\mathbf{X}$ by the ith entry Y_i of Y and adding up the resulting products. These computations are conveniently shown in tabular form:

g_1	$\cdots$	g_p	Y
$g_1(\mathbf{x}_1)$	$\cdots$	$g_p(\mathbf{x}_1)$	Y_1
$\cdot$		$\cdot$	$\cdot$
$\cdot$		$\cdot$	$\cdot$
$\cdot$		$\cdot$	$\cdot$
$g_1(\mathbf{x}_n)$	$\cdots$	$g_p(\mathbf{x}_n)$	Y_n
U_1	$\cdots$	U_p	

EXAMPLE 10.6   Consider the polymer experiment, which was introduced in Section 8.1, and let x be the coded value of temperature.

(a) Determine the design matrix $\mathbf{X}$ relative to the basis $1, x, x^2$ of the space of quadratic polynomials in x.

(b) Determine the corresponding Gram matrix $\mathbf{X}^T\mathbf{X}$.

(c) Determine the inverse of $\mathbf{X}^T\mathbf{X}$.   (d) Determine $\mathbf{X}^T\mathbf{Y}$.

(e) Use the answers to parts (c) and (d) to determine the least-squares quadratic polynomial estimate $\hat{\mu}(x) = \hat{\beta}_0 + \hat{\beta}_1 x + \hat{\beta}_2 x^2$ of the regression function.

Solution   (a) The design matrix is given by

$$\mathbf{X} = \begin{bmatrix} 1 & -5 & 25 \\ 1 & -3 & 9 \\ 1 & -1 & 1 \\ 1 & 1 & 1 \\ 1 & 3 & 9 \\ 1 & 5 & 25 \end{bmatrix}.$$

(b) Observe that

$$\sum_{i=1}^{6} x_i = 0 \quad \text{and} \quad \sum_{i=1}^{6} x_i^3 = 0.$$

Observe also that

$$\sum_{i=1}^{6} x_i^2 = (-5)^2 + (-3)^2 + (-1)^2 + 1^2 + 3^2 + 5^2 = 2(1 + 9 + 25) = 70$$

and

$$\sum_{i=1}^{6} x_i^4 = (-5)^4 + (-3)^4 + (-1)^4 + 1^4 + 3^4 + 5^4 = 2(1 + 81 + 625) = 1414.$$

Thus [see the solution to Example 9.5(b)], the Gram matrix is given by

$$\mathbf{X}^T\mathbf{X} = \begin{bmatrix} 6 & 0 & 70 \\ 0 & 70 & 0 \\ 70 & 0 & 1414 \end{bmatrix}.$$

(c) Observe that if we permute the first two rows of $\mathbf{X}^T\mathbf{X}$ and then its first two columns, then we get a block diagonal matrix. Applying the reverse permutations to the inverse of this block diagonal matrix, we get the inverse of $\mathbf{X}^T\mathbf{X}$. The determinant of the 2×2 matrix

$$\begin{bmatrix} 6 & 70 \\ 70 & 1414 \end{bmatrix}$$

equals $6 \cdot 1414 - 70^2 = 3584$, so the inverse of this matrix is given by

$$\frac{1}{3584} \begin{bmatrix} 1414 & -70 \\ -70 & 6 \end{bmatrix} \doteq \begin{bmatrix} 0.395 & -0.0195 \\ -0.0195 & 0.00167 \end{bmatrix}.$$

Since $1/70 \doteq 0.0143$, the inverse of $\mathbf{X}^T\mathbf{X}$ is given by

$$(\mathbf{X}^T\mathbf{X})^{-1} \doteq \begin{bmatrix} 0.395 & 0 & -0.0195 \\ 0 & 0.0143 & 0 \\ -0.0195 & 0 & 0.00167 \end{bmatrix}.$$

(d) Now

$$\sum_{i=1}^{6} Y_i = 58.9 + 71.6 + 78.0 + 81.6 + 86.0 + 85.2$$

$$= 461.3,$$

$$\sum_{i=1}^{6} x_i Y_i = (-5)(58.9) + (-3)(71.6) + (-1)(78.0)$$

$$+ 1(81.6) + 3(86.0) + 5(85.2)$$

$$= 178.3,$$

and

$$\sum_{i=1}^{6} x_i^2 Y_i = (-5)^2(58.9) + (-3)^2(71.6) + (-1)^2(78.0)$$

$$+ 1^2(81.6) + 3^2(86.0) + 5^2(85.2)$$

$$= 5180.5,$$

so $\mathbf{X}^T\mathbf{Y} = [461.3, 178.3, 5180.5]^T$. Table 10.1 shows this computation.

TABLE　10.1

1	x	x^2	Y
1	−5	25	58.9
1	−3	9	71.6
1	−1	1	78.0
1	1	1	81.6
1	3	9	86.0
1	5	25	85.2
461.3	178.3	5180.5	

(e) The coefficient vector relative to the indicated basis of the least-squares estimate of the regression function function is given by

$$
\begin{bmatrix} \hat{\beta}_0 \\ \hat{\beta}_1 \\ \hat{\beta}_2 \end{bmatrix} \doteq \begin{bmatrix} 0.395 & 0 & -0.0195 \\ 0 & 0.0143 & 0 \\ -0.0195 & 0 & 0.00167 \end{bmatrix} \begin{bmatrix} 461.3 \\ 178.3 \\ 5180.5 \end{bmatrix} \doteq \begin{bmatrix} 80.816 \\ 2.547 \\ -0.337 \end{bmatrix}.
$$

Thus, the least-squares quadratic polynomial estimate of the regression function is given by $\hat{\mu}(x) \doteq 80.816 + 2.547x - 0.337x^2$ as was seen in Section 8.1.　∎

The effect of a change in the basis of G on the design matrix is given next. The proof is essentially the same as the proof of Theorem 9.8 in Section 9.2.

THEOREM 10.3　Let $\tilde{g}_1, \ldots, \tilde{g}_p$ be a basis of G having coefficient matrix $\mathbf{A}$ relative to the basis $g_1, \ldots, g_p$. Then the design matrix $\tilde{\mathbf{X}}$ relative to the basis $\tilde{g}_1, \ldots, \tilde{g}_p$ is given by $\tilde{\mathbf{X}} = \mathbf{X}\mathbf{A}^T$.

EXAMPLE 10.7　Use Theorem 10.3 to rederive Theorem 9.19 in Section 9.4 in the present context.

Solution　Let $\tilde{g}_1, \ldots, \tilde{g}_p$, $\mathbf{A}$ and $\tilde{\mathbf{X}}$ be as in Theorem 10.3, and let $\mathbf{M}$ and $\tilde{\mathbf{M}}$ be the Gram matrices corresponding to $g_1, \ldots, g_p$ and $\tilde{g}_1, \ldots, \tilde{g}_p$, respectively. Then

$$
\tilde{\mathbf{M}} = \tilde{\mathbf{X}}^T\tilde{\mathbf{X}} = (\mathbf{X}\mathbf{A}^T)^T\mathbf{X}\mathbf{A}^T = \mathbf{A}\mathbf{X}^T\mathbf{X}\mathbf{A}^T = \mathbf{A}\mathbf{M}\mathbf{A}^T.　∎
$$

Problems

10.1　Obtain the normal equations for the coefficients $\hat{\beta}_0, \hat{\beta}_1, \hat{\beta}_2$ of $\hat{\mu}(\cdot)$ relative to the basis $1, x_1 - \bar{x}_1, x_2 - \bar{x}_2$ of the space of linear functions on $\mathbb{R}^2$, where $\bar{x}_1 = (x_{11} + \cdots + x_{n1})/n$ and $\bar{x}_2 = (x_{12} + \cdots + x_{n2})/n$.

10.2 Verify each of the following formulas: **(a)** $\sum_i [Y_i - \bar{Y}] = 0$; **(b)** $\sum_i [Y_i - \bar{Y}(\mathbf{x}_i)] = 0$; **(c)** $\sum_i [\bar{Y}(\mathbf{x}_i) - \bar{Y}] = 0$; **(d)** $\sum_k n_k (\bar{Y}_k - \bar{Y}) = 0$.

10.3 Verify that $\mathrm{LSS} = \|\bar{Y}(\cdot)\|^2 - \|\hat{\mu}(\cdot)\|^2$.

10.4 Verify that $\mathrm{RSS} = \sum_i Y_i^2 - \sum_i [\hat{\mu}(\mathbf{x}_i)]^2$.

10.5 Let G be a three-dimensional space of functions on $\mathcal{X}$; let 1, g_1, and g_2 form a basis of this space; and write the least-squares estimate in G of the regression function as $\hat{\mu}(\mathbf{x}) = \hat{\beta}_0 + \hat{\beta}_1 g_1(\mathbf{x}) + \hat{\beta}_2 g_2(\mathbf{x})$. Use calculus to derive the normal equations

$$n\hat{\beta}_0 + \hat{\beta}_1 \sum_i g_1(\mathbf{x}_i) + \hat{\beta}_2 \sum_i g_2(\mathbf{x}_i) = \sum_i Y_i$$

$$\hat{\beta}_0 \sum_i g_1(\mathbf{x}_i) + \hat{\beta}_1 \sum_i g_1^2(\mathbf{x}_i) + \hat{\beta}_2 \sum_i g_1(\mathbf{x}_i)g_2(\mathbf{x}_i) = \sum_i g_1(\mathbf{x}_i)Y_i$$

$$\hat{\beta}_0 \sum_i g_2(\mathbf{x}_i) + \hat{\beta}_1 \sum_i g_2(\mathbf{x}_i)g_1(\mathbf{x}_i) + \hat{\beta}_2 \sum_i g_2^2(\mathbf{x}_i) = \sum_i g_2(\mathbf{x}_i)Y_i$$

for $\hat{\beta}_0$, $\hat{\beta}_1$ and $\hat{\beta}_2$.

10.6 Show that if $g_1, \ldots, g_p$ are orthogonal or, equivalently, if $\mathbf{X}^T\mathbf{X}$ is a diagonal matrix, then

$$\hat{\beta}_j = \frac{\langle g_j, \bar{Y}(\cdot) \rangle}{\|g_j\|^2} = \frac{\sum_i g_j(\mathbf{x}_i)Y_i}{\sum_i g_j^2(\mathbf{x}_i)} = \frac{\sum_k n_k g_j(\mathbf{x}_k')\bar{Y}_k}{\sum_k n_k g_j^2(\mathbf{x}_k')}, \qquad 1 \le j \le p.$$

10.7 Let the d-sample model be viewed as a linear regression model as in Section 8.3. Then G is the identifiable and saturated space of all functions on $\mathcal{X} = \mathcal{X}' = \{1, \ldots, d\}$; $I_1, \ldots, I_d$ is an orthogonal basis of G, where $I_k(x) = \mathrm{ind}(x = k)$ for $1 \le k \le d$; and the regression function can be written in terms of this basis as $\mu(x) = \mu_1 I_1(x) + \cdots + \mu_d I_d(x)$. Determine **(a)** the squared norms of the basis functions; **(b)** the least-squares estimates $\hat{\mu}_1, \ldots, \hat{\mu}_d$ of the regression coefficients $\mu_1, \ldots, \mu_d$ respectively. [*Hint:* Use the result of Problem 10.6.]

10.8 Consider the polymer experiment. **(a)** Determine the design matrix $\mathbf{X}$ relative to the basis $1, x$ of the space of linear functions of x. **(b)** Determine the corresponding Gram matrix $\mathbf{X}^T\mathbf{X}$. **(c)** Determine the inverse of $\mathbf{X}^T\mathbf{X}$. **(d)** Determine $\mathbf{X}^T\mathbf{Y}$. **(e)** Use the answers to parts (c) and (d) to determine the least-squares linear estimate $\hat{\mu}(x) = \hat{\beta}_0 + \hat{\beta}_1 x$ of the regression function.

10.9 Consider the polymer experiment. **(a)** Determine the design matrix $\mathbf{X}$ relative to the basis $1, x, x^2 - \frac{35}{3}$ of the space of quadratic polynomials in x. **(b)** Determine the corresponding Gram matrix $\mathbf{X}^T\mathbf{X}$. **(c)** Determine the inverse of $\mathbf{X}^T\mathbf{X}$. **(d)** Determine $\mathbf{X}^T\mathbf{Y}$. **(e)** Use the answers to parts (c) and (d) to determine the least-squares quadratic polynomial estimate $\hat{\mu}(x) = \hat{\beta}_0 + \hat{\beta}_1 x + \hat{\beta}_2 (x^2 - \frac{35}{3})$ of the regression function.

10.10 Show that $\hat{\mu}(\cdot) \in G$ is the least-squares estimate in G of the regression function if and only if $\sum_k n_k g(\mathbf{x}_k')[\bar{Y}_k - \hat{\mu}(\mathbf{x}_k')] = 0$ for $g \in G$.

10.11 Verify that **(a)** $[\hat{\mu}(\mathbf{x}_1), \ldots, \hat{\mu}(\mathbf{x}_n)]^T = \mathbf{X}\hat{\beta}$; **(b)** $\mathrm{RSS} = (\mathbf{Y} - \mathbf{X}\hat{\beta})^T(\mathbf{Y} - \mathbf{X}\hat{\beta}) = \mathbf{Y}^T\mathbf{Y} - \mathbf{Y}^T\mathbf{X}(\mathbf{X}^T\mathbf{X})^{-1}\mathbf{X}^T\mathbf{Y}$.

10.12   Show that the system (11) of normal equations can be written as

$$\hat{\beta}_1 \sum_k n_k g_1^2(\mathbf{x}_k') + \cdots + \hat{\beta}_p \sum_k n_k g_1(\mathbf{x}_k') g_p(\mathbf{x}_k') = \sum_k n_k g_1(\mathbf{x}_k') \bar{Y}_k$$

$$\vdots \qquad \qquad \vdots \qquad \vdots$$

$$\hat{\beta}_1 \sum_k n_k g_p(\mathbf{x}_k') g_1(\mathbf{x}_k') + \cdots + \hat{\beta}_p \sum_k n_k g_p^2(\mathbf{x}_k') = \sum_k n_k g_p(\mathbf{x}_k') \bar{Y}_k$$

10.13   (Continued.) Consider again the $d \times p$ form

$$\mathbf{X} = \begin{bmatrix} g_1(\mathbf{x}_1') & \cdots & g_p(\mathbf{x}_1') \\ \vdots & & \vdots \\ g_1(\mathbf{x}_d') & \cdots & g_p(\mathbf{x}_d') \end{bmatrix}$$

of the design matrix and set $\mathbf{N} = \mathrm{diag}(n_1, \ldots, n_d)$ and $\bar{\mathbf{Y}} = [\bar{Y}_1, \ldots, \bar{Y}_d]^T$. Show that **(a)** the normal equations can be written as $\mathbf{X}^T \mathbf{N} \mathbf{X} \hat{\beta} = \mathbf{X}^T \mathbf{N} \bar{\mathbf{Y}}$ whose unique solution is given by $\hat{\beta} = (\mathbf{X}^T \mathbf{N} \mathbf{X})^{-1} \mathbf{X}^T \mathbf{N} \bar{\mathbf{Y}}$; **(b)** if there are r repetitions at each design point, then the normal equations can be written as $\mathbf{X}^T \mathbf{X} \hat{\beta} = \mathbf{X}^T \bar{\mathbf{Y}}$, whose unique solution is given by $\hat{\beta} = (\mathbf{X}^T \mathbf{X})^{-1} \mathbf{X}^T \bar{\mathbf{Y}}$; **(c)** if $p = d$ (and hence G is saturated), then $\hat{\beta} = \mathbf{X}^{-1} \bar{\mathbf{Y}}$; **(d)** if $p = d$, $g_k(\mathbf{x}_k') = 1$ for $1 \leq k \leq d$, and $g_l(\mathbf{x}_k') = 0$ for $1 \leq k, l \leq d$ with $l \neq k$, then $\hat{\beta} = \bar{\mathbf{Y}}$.

10.14   Let G_1 and G_2 be identifiable linear spaces of functions on $\mathcal{X}$ such that $G_1' = G_2'$, where G_1' and G_2' are the spaces of the restrictions of the functions in G_1 and G_2, respectively, to the design set, and let $\hat{\mu}_1(\cdot)$ and $\hat{\mu}_2(\cdot)$ be the least-squares estimates of the regression function in G_1 and G_2, respectively. Show that **(a)** $\hat{\mu}_1(\cdot) = \hat{\mu}_2(\cdot)$ on the design set; **(b)** $\mathrm{LSS}(\hat{\mu}_1(\cdot) = \mathrm{LSS}(\hat{\mu}_2(\cdot))$; **(c)** $\mathrm{RSS}(\hat{\mu}_1(\cdot)) = \mathrm{RSS}(\hat{\mu}_2(\cdot))$.

10.2
Sums of Squares

Let G be an identifiable p-dimensional linear space and let $\hat{\mu}(\cdot)$ be the least-squares estimate in G of the regression function. Then the residual sum of squares for this estimate is given by $\mathrm{RSS} = \sum_i [\, Y_i - \hat{\mu}(\mathbf{x}_i)\,]^2$, and its lack-of-fit sum of squares is given by

$$\begin{aligned} \mathrm{LSS} &= \|\bar{Y}(\cdot) - \hat{\mu}(\cdot)\|^2 \\ &= \sum_k n_k [\, \bar{Y}(\mathbf{x}_k') - \hat{\mu}(\mathbf{x}_k')\,]^2 \\ &= \sum_k n_k [\, \bar{Y}_k - \hat{\mu}(\mathbf{x}_k')\,]^2 \\ &= \sum_i [\, \bar{Y}(\mathbf{x}_i) - \hat{\mu}(\mathbf{x}_i)\,]^2. \end{aligned}$$

The residual sum of squares can be decomposed as

$$\text{RSS} = \text{LSS} + \text{WSS} . \tag{15}$$

If G is saturated, then $\text{LSS} = 0$ and $\text{RSS} = \text{WSS}$.

The total and between sums of squares are given by $\text{TSS} = \sum_i (Y_i - \bar{Y})^2$ and $\text{BSS} = \sum_k n_k (\bar{Y}_k - \bar{Y})^2$. Thus, the total sum of squares is the residual sum of squares for the least-squares constant estimate $\bar{Y}$ of the regression function, and the between sum of squares is the lack-of-fit sum of squares $\| \bar{Y}(\cdot) - \bar{Y} \|^2$ for this estimate. The formula

$$\text{TSS} = \text{BSS} + \text{WSS}, \tag{16}$$

given as (36) in Section 7.6, is now seen as the special case of (15) corresponding to the least-squares constant estimate of the regression function.

Suppose G contains the constant functions. Then $\hat{\mu}(\cdot) - \bar{Y}$ is a member of G. Now $\hat{\mu}(\cdot)$ is the orthogonal projection of $\bar{Y}(\cdot)$ onto G, so $\bar{Y}(\cdot) - \hat{\mu}(\cdot)$ is orthogonal to every function in G; in particular, it is orthogonal to $\hat{\mu}(\cdot) - \bar{Y}$. Since

$$\text{BSS} = \| \bar{Y}(\cdot) - \bar{Y} \|^2 = \| \hat{\mu}(\cdot) - \bar{Y} + \bar{Y}(\cdot) - \hat{\mu}(\cdot) \|^2,$$

we conclude from the Pythagorean theorem that

$$\text{BSS} = \| \hat{\mu}(\cdot) - \bar{Y} \|^2 + \| \bar{Y}(\cdot) - \hat{\mu}(\cdot) \|^2. \tag{17}$$

The second term on the right side of (17) is the lack-of-fit sum of squares for G. The first term on the right side of this equation is referred to as the *fitted sum of squares* (FSS) for G. Thus,

$$\text{FSS} = \text{FSS}(G) = \| \hat{\mu}(\cdot) - \bar{Y} \|^2 = \sum_k n_k [\, \hat{\mu}(\mathbf{x}_k') - \bar{Y} \,]^2 = \sum_i [\, \hat{\mu}(\mathbf{x}_i) - \bar{Y} \,]^2,$$

and (17) can be rewritten as

$$\text{BSS} = \text{FSS} + \text{LSS} . \tag{18}$$

It follows from (16) and (18) that

$$\text{TSS} = \text{FSS} + \text{LSS} + \text{WSS} . \tag{19}$$

We conclude from (15) and (19) that

$$\text{TSS} = \text{FSS} + \text{RSS} . \tag{20}$$

If G is saturated, then its lack-of-fit sum of squares equals zero, so it follows from (18) that its fitted sum of squares equals the between sum of squares.

Figure 10.2 is a diagram that serves as a mnemonic for the relationships involving the various sums of squares and their components.

Suppose also that $\text{TSS} > 0$; that is, $Y_1, \ldots, Y_n$ do not coincide. Then the quantity

$$R^2 = 1 - \frac{\text{RSS}}{\text{TSS}} = \frac{\text{FSS}}{\text{TSS}}$$

F I G U R E   10.2

Relationships involving sums of squares and their components

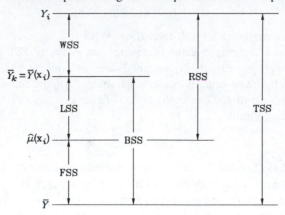

is referred to as the *squared multiple correlation coefficient*. Observe that $0 \leq R^2 \leq 1$. Also, $R^2 = 0$ if and only if FSS $= 0$ and hence if and only if $\hat{\mu}(\cdot) = \bar{Y}$. Moreover, $R^2 = 1$ if and only if RSS $= 0$. In particular, if WSS > 0, then RSS > 0 and hence $R^2 < 1$. The nonnegative square root R of R^2 is referred to as the *multiple correlation coefficient*.

Recall from Section 8.3 the description of the d-sample model as a normal linear regression model involving a single factor with levels $1, \ldots, d$. Correspondingly, G is the identifiable and saturated space of all functions on $\{1, \ldots, d\}$, and the regression function is given by $\mu(\cdot) = \mu_1 I_1 + \cdots + \mu_d I_d$, where $I_k(x) = \mathrm{ind}(x = k)$ for $1 \leq k \leq d$ and $x \in \mathcal{X} = \{1, \ldots, d\}$. Since $I_k(x)I_l(x) = 0$ for $k \neq l$ and $x \in \mathcal{X}$, the functions $I_1, \ldots, I_d$ form an orthogonal basis of the space G of all functions on $\mathcal{X}$. The squared norms of these functions are given by $\|I_k\|^2 = n_k$ for $1 \leq k \leq d$. The least-squares estimates of the regression coefficients $\mu_1, \ldots, \mu_d$ are given (see Problem 10.7 in Section 10.1) by $\hat{\mu}_k = \bar{Y}_k$ for $1 \leq k \leq d$. Thus, $\hat{\mu}(\cdot) = \bar{Y}_1 I_1 + \cdots + \bar{Y}_d I_d$, and hence

$$\hat{\mu}(x_i) = \bar{Y}_k, \qquad 1 \leq k \leq d \text{ and } i \in \mathcal{I}_k.$$

Since G is saturated, its residual sum of squares equals the within sum of squares, its fitted sum of squares equals the between sum of squares, and its the squared multiple correlation coefficient is given by $R^2 = $ BSS / TSS.

EXAMPLE 10.8   Consider the sugar beet data as treated in Example 7.24 in Section 7.6, in which the possible effect of the plot variable is ignored. Determine the squared multiple correlation coefficient.

Solution   According to the solution to Example 7.24 in Section 7.6, BSS $\doteq 3.085$ and TSS $\doteq 11.327$, so the squared multiple correlation coefficient is given by

$R^2 \doteq 3.085/11.327 \doteq .272$. In other words, 27.2% of the variation in sugar content is due to the apparent differences between the five varieties of sugar beet. ∎

EXAMPLE **10.9**   Consider the quadratic fit to the polymer data from Section 8.1. Determine

(a) TSS, RSS, and FSS;

(b) the squared multiple correlation coefficient.

Solution   (a) The sample mean of the responses is given by $\bar{Y} \doteq 76.883$. Thus, the total sum of squares is given by

$$\text{TSS} \doteq (58.9 - 76.883)^2 + \cdots + (85.2 - 76.883)^2 \doteq 527.088.$$

The residual sum of squares from the quadratic fit is given by

$$\text{RSS} = \sum_{i=1}^{6} [\, Y_i - \hat{\mu}(x_i)\,]^2 = \sum_{i=1}^{6}(Y_i - 80.816 - 2.547x_i + 0.337x_i^2)^2 \doteq 5.073.$$

Thus, the fitted sum of squares is given by

$$\text{FSS} = \text{TSS} - \text{RSS} \doteq 527.088 - 5.073 \doteq 522.016.$$

The numerical values of FSS, RSS, and TSS are conveniently shown together in an ANOVA table (Table 10.2).

(b) The squared multiple correlation is given by

$$R^2 = \frac{\text{FSS}}{\text{TSS}} \doteq \frac{522.016}{527.088} \doteq .990.$$

In other words, 99% of the variation in the yield of tetramer is explained by the quadratic dependence of yield on reactor temperature. ∎

TABLE **10.2**

Source	SS
Fit	522.016
Residuals	5.073
Total	527.088

Expected Values

Throughout the remainder of the section, the assumptions of the experimental form of the regression model are in effect; that is, $Y_1, \ldots, Y_n$ are independent random variables, each having finite mean and variance, and $EY_i = \mu(\mathbf{x}_i)$ and $\text{var}(Y_i) = \sigma^2(\mathbf{x}_i)$

for $1 \leq i \leq n$. Then $\bar{Y}_1, \ldots, \bar{Y}_d$ are independent random variables, and $E(\bar{Y}_k) = \mu(\mathbf{x}_k)$ for $1 \leq k \leq d$. Under the assumptions of the linear regression model, the regression function is a member of the linear space G; under the assumptions of the homoskedastic regression model, $\text{var}(Y_i) = \sigma^2$ for $1 \leq i \leq n$ and $\text{var}(\bar{Y}_k) = \sigma^2/n_k$ for $1 \leq k \leq d$.

THEOREM 10.4 $E[\langle g, \hat{\mu}(\cdot) \rangle] = \langle g, \mu(\cdot) \rangle$ for $g \in G$.

Proof   Let $g \in G$. Since $\hat{\mu}(\cdot) = P_G[\bar{Y}(\cdot)]$ is the orthogonal projection of $\bar{Y}(\cdot)$ onto G, we have that $\langle g, \bar{Y}(\cdot) - \hat{\mu}(\cdot) \rangle = 0$ and hence that

$$\langle g, \hat{\mu}(\cdot) \rangle = \langle g, \bar{Y}(\cdot) \rangle = \sum_k n_k g(\mathbf{x}_k')\bar{Y}_k. \tag{21}$$

Consequently,

$$E[\langle g, \hat{\mu}(\cdot) \rangle] = \sum_k n_k g(\mathbf{x}_k')\mu(\mathbf{x}_k') = \langle g, \mu(\cdot) \rangle. \quad \blacksquare$$

Let $\mu^*(\cdot) = P_G[\mu(\cdot)]$ denote the orthogonal projection of the regression function onto G or, equivalently, the least-squares approximation in G to this function.

THEOREM 10.5 $E[\hat{\mu}(\cdot)] = P_G[\mu(\cdot)] = \mu^*(\cdot).$

Proof   Let $g_1, \ldots, g_p$ be an orthonormal basis of G. Then $\hat{\mu}(\cdot) = \sum_j \langle g_j, \hat{\mu}(\cdot) \rangle g_j$. Thus, by Theorem 10.4,

$$E[\hat{\mu}(\cdot)] = \sum_j \langle g_j, \mu(\cdot) \rangle g_j = P_G[\mu(\cdot)] = \mu^*(\cdot). \quad \blacksquare$$

COROLLARY 10.1   If $\mu(\cdot) \in G$, then $E[\hat{\mu}(\cdot)] = \mu(\cdot)$.

THEOREM 10.6   Under the assumptions of the homoskedastic regression model,

$$\text{var}(\langle g, \hat{\mu}(\cdot) \rangle) = \sigma^2 \|g\|^2, \qquad g \in G, \tag{22}$$

$$\sum_i \text{var}(\hat{\mu}(\mathbf{x}_i)) = \sum_k n_k \text{var}(\hat{\mu}(\mathbf{x}_k')) = E[\|\hat{\mu}(\cdot) - \mu^*(\cdot)\|^2] = p\sigma^2, \tag{23}$$

$$E[\|\hat{\mu}(\cdot) - \mu(\cdot)\|^2] = \|\mu(\cdot) - \mu^*(\cdot)\|^2 + p\sigma^2, \tag{24}$$

and

$$E[\|\bar{Y}(\cdot) - \mu^*(\cdot)\|^2] = \|\mu(\cdot) - \mu^*(\cdot)\|^2 + d\sigma^2. \tag{25}$$

Proof   Let $g \in G$. Then, by (22),

$$\text{var}(\langle g, \hat{\mu}(\cdot) \rangle) = \sum_k n_k^2 g^2(\mathbf{x}_k')\text{var}(\bar{Y}_k) = \sigma^2 \sum_k n_k g^2(\mathbf{x}_k') = \sigma^2 \|g\|^2,$$

so (22) holds.

The first equality in (23) is obviously valid since n_k is the number of repetitions of $\mathbf{x}'_k$ among $\mathbf{x}_1, \ldots, \mathbf{x}_n$. By Theorem 10.5,

$$\sum_k n_k \operatorname{var}(\hat{\mu}(\mathbf{x}'_k)) = \sum_k n_k E[\,(\hat{\mu}(\mathbf{x}'_k) - \mu^*(\mathbf{x}'_k))^2\,]$$

$$= E\left(\sum_k n_k [\,\hat{\mu}(\mathbf{x}'_k) - \mu^*(\mathbf{x}'_k)\,]^2\right)$$

$$= E[\,\|\hat{\mu}(\cdot) - \mu^*(\cdot)\|^2\,],$$

so the second equality in (23) is valid. If $p = 0$, then the third equality in (23) is valid since each side of this equality equals zero. Suppose $p \geq 1$ and let $g_1, \ldots, g_p$ be an orthonormal basis of G. Then

$$\hat{\mu}(\cdot) = \sum_j \langle g_j, \hat{\mu}(\cdot)\rangle g_j \quad \text{and} \quad \mu^*(\cdot) = \sum_j \langle g_j, \mu(\cdot)\rangle g_j.$$

Thus,

$$\hat{\mu}(\cdot) - \mu^*(\cdot) = \sum_j [\,\langle g_j, \hat{\mu}(\cdot)\rangle - \langle g_j, \mu(\cdot)\rangle\,]g_j,$$

so

$$\|\hat{\mu}(\cdot) - \mu^*(\cdot)\|^2 = \sum_j [\,\langle g_j, \hat{\mu}(\cdot)\rangle - \langle g_j, \mu(\cdot)\rangle\,]^2$$

and hence, by (22) and Theorem 10.4,

$$E[\,\|\hat{\mu}(\cdot) - \mu^*(\cdot)\|^2\,] = \sum_j E\{[\,\langle g_j, \hat{\mu}(\cdot)\rangle - \langle g_j, \mu(\cdot)\rangle\,]^2\} = \sum_j \operatorname{var}(\langle g_j, \hat{\mu}(\cdot)\rangle) = p\sigma^2.$$

Therefore, the third equality in (23) is valid, and hence (23) is valid.

Since $\mu(\cdot) - \mu^*(\cdot)$ is orthogonal to every function in G, it is orthogonal to $\hat{\mu}(\cdot) - \mu^*(\cdot)$. Thus, by the Pythagorean theorem,

$$\|\hat{\mu}(\cdot) - \mu(\cdot)\|^2 = \|\mu^*(\cdot) - \mu(\cdot)\|^2 + \|\hat{\mu}(\cdot) - \mu^*(\cdot)\|^2.$$

We now conclude from (23) that

$$E[\,\|\hat{\mu}(\cdot) - \mu(\cdot)\|^2\,] = \|\mu^*(\cdot) - \mu(\cdot)\|^2 + E[\,\|\hat{\mu}(\cdot) - \mu^*(\cdot)\|^2\,]$$

$$= \|\mu^*(\cdot) - \mu(\cdot)\|^2 + p\sigma^2$$

and hence that (24) holds.

Since $\bar{Y}_k$ has mean $\mu(\mathbf{x}'_k)$ and variance σ^2/n_k for $1 \leq k \leq d$, we see that

$$E[\,\|\bar{Y}(\cdot) - \mu^*(\cdot)\|^2\,] = E\left(\sum_k n_k [\,\bar{Y}_k - \mu^*(\mathbf{x}'_k)\,]^2\right)$$

$$= \sum_k n_k [\,\mu(\mathbf{x}'_k) - \mu^*(\mathbf{x}'_k)\,]^2 + \sum_k n_k \operatorname{var}(\bar{Y}_k)$$

$$= \|\mu(\cdot) - \mu^*(\cdot)\|^2 + d\sigma^2.$$

Thus, (25) holds. ∎

Since $\mu^*(\cdot) = \mu(\cdot)$ under the assumptions of the homoskedastic linear regression model, the next result follows from (24).

COROLLARY 10.2 Under the assumptions of the homoskedastic linear regression model,

$$E[\,\|\hat{\mu}(\cdot) - \mu(\cdot)\|^2\,] = p\sigma^2.$$

Expected Sums of Squares

THEOREM 10.7 Under the assumptions of the homoskedastic regression model,

$$E(\text{LSS}) = \|\mu(\cdot) - \mu^*(\cdot)\|^2 + (d - p)\sigma^2, \tag{26}$$

$$E(\text{WSS}) = (n - d)\sigma^2, \tag{27}$$

and

$$E(\text{RSS}) = \|\mu(\cdot) - \mu^*(\cdot)\|^2 + (n - p)\sigma^2. \tag{28}$$

Proof Now $\hat{\mu}(\cdot)$ is the orthogonal projection of $\bar{Y}(\cdot)$ onto G, so $\bar{Y}(\cdot) - \hat{\mu}(\cdot)$ is orthogonal to every function in G. Since $\hat{\mu}(\cdot)$ and $\mu^*(\cdot)$ are members of G and hence so is their difference, $\bar{Y}(\cdot) - \hat{\mu}(\cdot)$ is orthogonal to $\hat{\mu}(\cdot) - \mu^*(\cdot)$. Thus, by the Pythagorean theorem,

$$\|\bar{Y}(\cdot) - \mu^*(\cdot)\|^2 = \|\bar{Y}(\cdot) - \hat{\mu}(\cdot)\|^2 + \|\hat{\mu}(\cdot) - \mu^*(\cdot)\|^2,$$

and hence

$$\|\bar{Y}(\cdot) - \hat{\mu}(\cdot)\|^2 = \|\bar{Y}(\cdot) - \mu^*(\cdot)\|^2 - \|\hat{\mu}(\cdot) - \mu^*(\cdot)\|^2.$$

Taking expectations and using (23) and (25), we get (26).

The within sum of squares is given by

$$\text{WSS} = \sum_k \sum_{i \in \mathcal{I}_k} (Y_i - \bar{Y}_k)^2.$$

It follows from (53) in Section 2.4 (the proof of this result requires that the random response variables be independent and have common mean and variance, but not that they be identically distributed) that

$$E\left(\sum_{i \in \mathcal{I}_k} (Y_i - \bar{Y}_k)^2\right) = (n_k - 1)\sigma^2, \qquad 1 \le k \le d,$$

and hence that

$$E(\text{WSS}) = \sum_k E\left(\sum_{i \in \mathcal{I}_k} (Y_i - \bar{Y}_k)^2\right) = \sum_k (n_k - 1)\sigma^2 = (n - d)\sigma^2.$$

Thus (27) is valid.

Since RSS = LSS + WSS, (28) follows from (26) and (27). ∎

Under the assumptions of the experimental form of the regression model, set

$$\bar{\mu} = E\bar{Y} = \frac{1}{n}\sum_k n_k\mu_k = \frac{1}{n}\sum_i \mu(\mathbf{x}_i).$$

It follows from the solution to Example 9.31(a) in Section 9.5 that $\bar{\mu}$ is the orthogonal projection of $\mu(\cdot)$ onto the space of constant functions.

COROLLARY 10.3   Under the assumptions of the homoskedastic regression model,

$$E(\text{BSS}) = \|\mu(\cdot) - \bar{\mu}\|^2 + (d-1)\sigma^2.$$

Proof   Since BSS is the lack-of-fit sum of squares for the least-squares constant fit, the desired result follows from (26) with $p = 1$ and $\mu^*(\cdot) = \bar{\mu}$.   ∎

COROLLARY 10.4   Suppose G contains the constant functions and that the assumptions of the homoskedastic regression model are satisfied. Then $E(\text{FSS}) = \|\mu^*(\cdot) - \bar{\mu}\|^2 + (p-1)\sigma^2$.

Proof   Recall that BSS = FSS + LSS and hence that FSS = BSS − LSS. Thus, by (26) and Corollary 10.3,

$$E(\text{FSS}) = \|\mu(\cdot) - \bar{\mu}\|^2 - \|\mu(\cdot) - \mu^*(\cdot)\|^2 + (p-1)\sigma^2.$$

Now $\mu(\cdot) - \mu^*(\cdot)$ is orthogonal to every function in G; in particular, it is orthogonal to $\mu^*(\cdot) - \bar{\mu}$. Thus,

$$\|\mu(\cdot) - \bar{\mu}\|^2 = \|\mu(\cdot) - \mu^*(\cdot)\|^2 + \|\mu^*(\cdot) - \bar{\mu}\|^2$$

by the Pythagorean theorem, so the desired result is valid.   ∎

COROLLARY 10.5   Under the assumptions of the homoskedastic linear regression model,

$$E(\text{LSS}) = (d-p)\sigma^2 \quad and \quad E(\text{RSS}) = (n-p)\sigma^2.$$

When $n > p$, set $S^2 = \text{RSS}/(n-p)$, where $S \geq 0$.

COROLLARY 10.6   Suppose the assumptions of the homoskedastic linear regression model are satisfied and that $n > p$. Then S^2 is an unbiased estimate of σ^2.

In light of Corollary 10.6, we estimate σ^2 by S^2 and σ by S.

COROLLARY 10.7   Suppose G contains the constant functions and that the assumptions of the homoskedastic linear regression model are satisfied. Then

$$E(\text{FSS}) = \|\mu(\cdot) - \bar{\mu}\|^2 + (p-1)\sigma^2.$$

Problems

10.15 Consider the linear fit to the polymer data from Section 8.1. Determine **(a)** TSS, RSS, and FSS; **(b)** the squared multiple correlation coefficient.

10.16 Verify that **(a)** $\sum_k n_k \hat{\mu}(\mathbf{x}'_k)[\,\bar{Y}_k - \hat{\mu}(\mathbf{x}'_k)\,] = 0$;
(b) LSS $= \sum_k n_k \bar{Y}_k^2 - \sum_k n_k [\,\hat{\mu}(\mathbf{x}'_k)\,]^2 = \sum_k n_k \bar{Y}_k^2 - \sum_i [\,\hat{\mu}(\mathbf{x}_i)\,]^2$.

10.17 Suppose G contains the constant functions. Verify that **(a)** $\sum_i [\,Y_i - \hat{\mu}(\mathbf{x}_i)\,] = 0$;
(b) $\sum_k n_k [\,\bar{Y}_k - \hat{\mu}(\mathbf{x}'_k)\,] = 0$; **(c)** $\sum_k n_k \hat{\mu}(\mathbf{x}'_k) = \sum_i \hat{\mu}(\mathbf{x}_i) = n\bar{Y}$.

10.18 (Continued.) Verify that

$$\text{FSS} = \|\hat{\mu}(\cdot)\|^2 - n\bar{Y}^2 = \sum_k n_k [\,\hat{\mu}(\mathbf{x}'_k)\,]^2 - n\bar{Y}^2 = \sum_i [\,\hat{\mu}(\mathbf{x}_i)\,]^2 - n\bar{Y}^2.$$

10.19 (Continued.) Verify that **(a)** $\sum_i [\,\hat{\mu}(\mathbf{x}_i) - \bar{Y}\,][\,Y_i - \hat{\mu}(\mathbf{x}_i)\,] = 0$;
(b) $\sum_i [\,\hat{\mu}(\mathbf{x}_i) - \bar{Y}\,](Y_i - \bar{Y}) = \text{FSS}$;

(c) $R = \sum_i [\,\hat{\mu}(\mathbf{x}_i) - \bar{Y}\,](Y_i - \bar{Y}) \big/ \sqrt{\sum_i [\,\hat{\mu}(\mathbf{x}_i) - \bar{Y}\,]^2 \sum_i (Y_i - \bar{Y})^2}$.

10.20 Let G be the space of linear functions on $\mathbb{R}$ and set $\bar{x} = (x_1 + \cdots + x_n)/n$. Write the least-squares estimate of the regression function as $\hat{\mu}(x) = \bar{Y} + \hat{\beta}(x - \bar{x})$. Show that
(a) $\hat{\mu}(x_i) - \bar{Y} = \hat{\beta}(x_i - \bar{x})$ for $1 \leq i \leq n$;

(b) $R^2 = \dfrac{[\sum_i (x_i - \bar{x})(Y_i - \bar{Y})]^2}{\sum_i (x_i - \bar{x})^2 \sum_i (Y_i - \bar{Y})^2} = \hat{\beta}^2 \dfrac{\sum_i (x_i - \bar{x})^2}{\sum_i (Y_i - \bar{Y})^2}$.

[*Hint:* For part (b) use the result in Problem 10.19(c) and the solution to Example 10.4(b) in Section 10.1.]

10.21 There are alternative definitions of the fitted and total sums of squares that are used when G does not contain the constant functions and sometimes even when it does contain the constant functions. Specifically, the total sum of squares is defined by TSS $= \sum_i Y_i^2$, and the fitted sum of squares for $\hat{\mu}(\cdot)$ is defined by FSS $= \|\hat{\mu}(\cdot)\|^2$. Using these alternative definitions of TSS and FSS, verify that **(a)** $\sum_i \hat{\mu}(\mathbf{x}_i)[\,Y_i - \hat{\mu}(\mathbf{x}_i)\,] = 0$; **(b)** FSS $= \sum_k n_k [\,\hat{\mu}(\mathbf{x}'_k)\,]^2 = \sum_i [\,\hat{\mu}(\mathbf{x}_i)\,]^2 = \sum_i \hat{\mu}(\mathbf{x}_i) Y_i$; **(c)** if G is saturated, then FSS $= \sum_k n_k \bar{Y}_k^2$; **(d)** TSS $=$ FSS $+$ RSS.

10.22 Suppose G contains the constant functions; let $1, g_1, \ldots, g_{p-1}$ be an orthogonal basis of G; and let $\hat{\beta}_0, \ldots, \hat{\beta}_{p-1}$ be the coefficients of $\hat{\mu}(\cdot)$ relative to this basis. Show that **(a)** $\hat{\beta}_0 = \bar{Y}$; **(b)** FSS $= \sum_{j=1}^{p-1} \hat{\beta}_j^2 \|g_j\|^2$.

10.23 Verify that **(a)** $\dfrac{\text{RSS}(g)}{S^2} = \sum_i \left(\dfrac{Y_i - \hat{\mu}(x_i)}{S}\right)^2$ [to interpret this equation, think of S on its right side as an estimate of $\text{SD}(Y_i) = \sigma$]; **(b)** $\dfrac{\text{LSS}(g)}{S^2} = \sum_k \left(\dfrac{\bar{Y}_k - \hat{\mu}(\mathbf{x}'_k)}{S/\sqrt{n_k}}\right)^2$ [to interpret this equation, think of $S/\sqrt{n_k}$ on its right side as an estimate of $\text{SD}(\bar{Y}_k) = \sigma/\sqrt{n_k}$].

10.24 Determine the squared multiple correlations for the least-squares constant, linear, quadratic, cubic, quartic and quintic fits to the polymer data from Section 8.1.

10.25 Suppose G is saturated. Show that $E[\,\hat{\mu}(\cdot)\,] = \mu(\cdot)$ on the design set.

10.26 Suppose the assumptions of the homoskedastic regression model are satisfied and that $\text{var}(\hat{\mu}(x'_k))$ does not depend on k. Show that $\text{var}(\hat{\mu}(x'_k)) = p\sigma^2/n$ for $1 \leq k \leq d$ or, equivalently, that $\text{var}(\hat{\mu}(x_i)) = p\sigma^2/n$ for $1 \leq i \leq n$.

10.27 Let $\hat{\tau}_1 = \langle h_1, \bar{Y}(\cdot) \rangle$ and $\hat{\tau}_2 = \langle h_2, \bar{Y}(\cdot) \rangle$, where $h_1, h_2 \in H$. Under the assumptions of the homoskedastic regression model, show that **(a)** $\text{cov}(\hat{\tau}_1, \hat{\tau}_2) = \sigma^2 \langle h_1, h_2 \rangle$; **(b)** if h_1 and h_2 are orthogonal, then $\text{cov}(\hat{\tau}_1, \hat{\tau}_2) = 0$.

10.3
Distribution Theory

In this section, we show that, under the assumptions of the normal linear regression model corresponding to an identifiable p-dimensional linear space, the least-squares estimate $\hat{\beta}$ of the coefficient vector β relative to a given basis is normally distributed with mean vector β and variance–covariance matrix $\sigma^2(\mathbf{X}^T\mathbf{X})^{-1}$, that RSS $/\sigma^2$ has the chi-square distribution with $n - p$ degrees of freedom, and that $\hat{\beta}$ and RSS are independent. These results are then used to get confidence intervals for nontrivial linear parameters. We start with a special case.

Normal One-Sample Model

THEOREM 10.8 Let $Z_1, \ldots, Z_n$ $(n \geq 2)$ be independent, standard normal random variables and set $\bar{Z} = (Z_1 + \cdots + Z_n)/n$. Then $\bar{Z}$ and $\sum_i (Z_i - \bar{Z})^2$ are independent, and $\sum_i (Z_i - \bar{Z})^2$ has the chi-square distribution with $n - 1$ degrees of freedom.

Proof Let G denote the n-dimensional space consisting of all functions on $\{1, \ldots, n\}$ (which is equivalent to the space of ordered n-tuples). The inner product on G corresponding to the design set $\{1, \ldots, n\}$ and unit weights is given by $\langle g_1, g_2 \rangle = \sum_i g_1(i)g_2(i)$, and the corresponding squared norm is given by $\|g\|^2 = \sum_i g^2(i)$. In particular, the squared norm of the constant function 1 is given by $\|1\|^2 = n$, so the function $g_1 = 1/\sqrt{n}$ has norm 1. By Corollary 9.12(b) in Section 9.4, there are functions $g_2, \ldots, g_n$ such that $g_1, \ldots, g_n$ is an orthonormal basis of G. The design matrix relative to this basis is the $n \times n$ matrix

$$\mathbf{X} = \begin{bmatrix} g_1(1) & \cdots & g_n(1) \\ \vdots & & \vdots \\ g_1(n) & \cdots & g_n(n) \end{bmatrix},$$

and the Gram matrix of $g_1, \ldots, g_n$ is given by $\mathbf{X}^T\mathbf{X} = \mathbf{I}_n$. Thus, $\mathbf{X}$ is invertible and $\mathbf{X}^{-1} = \mathbf{X}^T$.

Consider the (random) function $Z(\cdot)$ on $\{1, \ldots, n\}$ defined by $Z(i) = Z_i$ for $1 \leq i \leq n$. Let $W_1, \ldots, W_n$ be the (random) coefficients of this function relative to the orthonormal basis $g_1, \ldots, g_n$. Then $Z(\cdot) = \sum_j W_j g_j$, where

$$W_j = \langle g_j, Z(\cdot) \rangle = \sum_i g_j(i)Z_i, \qquad 1 \leq j \leq n. \tag{29}$$

In particular,

$$W_1 = \sum_i g_1(i)Z_i = \frac{1}{\sqrt{n}} \sum_i Z_i = \sqrt{n}\bar{Z},$$

so

$$\bar{Z} = \frac{W_1}{\sqrt{n}}. \tag{30}$$

Moreover,

$$Z(\cdot) - \bar{Z} = Z(\cdot) - W_1 g_1 = \sum_{j=2}^{n} W_j g_j,$$

and hence

$$\|Z(\cdot) - \bar{Z}\|^2 = \sum_{j=2}^{n} W_j^2.$$

Since $\|Z(\cdot) - \bar{Z}\|^2 = \sum_i (Z_i - \bar{Z})^2$, we conclude that

$$\sum_i (Z_i - \bar{Z})^2 = \sum_{j=2}^{n} W_j^2. \tag{31}$$

Set $\mathbf{Z} = [Z_1, \ldots, Z_n]^T$ and $\mathbf{W} = [W_1, \ldots, W_n]^T$. From (29) we find that $\mathbf{W} = \mathbf{X}^T \mathbf{Z}$. Since $Z_1, \ldots, Z_n$ are independent, standard normal random variables, $\mathbf{Z}$ has the multivariate normal distribution with mean vector 0 and variance–covariance matrix $\mathbf{I}_n$. Consequently, $\mathbf{W}$ has the multivariate normal distribution with mean vector 0 and variance–covariance matrix $\mathbf{X}^T \mathbf{I}_n \mathbf{X} = \mathbf{X}^T \mathbf{X} = \mathbf{I}_n$, so $W_1, \ldots, W_n$ are independent, standard normal random variables. The conclusions of the theorem now follow from (30) and (31). ■

Let $Y_1, \ldots, Y_n$ satisfy the assumptions of the normal one-sample model; that is, let them be independent random variables having a common normal distribution with mean μ and variance σ^2. Set $\bar{Y} = (Y_1 + \cdots + Y_n)/n$ and $\text{WSS} = \sum_i (Y_i - \bar{Y})^2$. Also, set $Z_i = (Y_i - \mu)/\sigma$ for $1 \leq i \leq n$. Then $Z_1, \ldots, Z_n$ are independent, standard normal random variables. Now $Y_i = \mu + \sigma Z_i$ for $1 \leq i \leq n$, so $\sum_i Y_i = \sum_i (\mu + \sigma Z_i) = n\mu + \sigma \sum_i Z_i$ and hence

$$\bar{Y} = \mu + \sigma \bar{Z}. \tag{32}$$

Also, $Y_i - \bar{Y} = \sigma(Z_i - \bar{Z})$ for $1 \leq i \leq n$, so $\text{WSS} = \sum_i (Y_i - \bar{Y})^2 = \sigma^2 \sum_i (Z_i - \bar{Z})^2$ and hence

$$\frac{\text{WSS}}{\sigma^2} = \sum_i (Z_i - \bar{Z})^2. \tag{33}$$

It follows from (32), (33) and Theorem 10.8 that $\bar{Y}$ and WSS are independent and that WSS/σ^2 has the chi-square distribution with $n - 1$ degrees of freedom. This verifies Theorem 7.7 in Section 7.3.

Normal Linear Regression Model

Let G be an identifiable p-dimensional linear space, let $g_1, \ldots, g_p$ be a basis of G, and let $\mathbf{X}$ be the $n \times p$ design matrix corresponding to this basis. Also, let $\mathbf{Y} = [Y_1, \ldots, Y_n]^T$ be the random vector composed of the n response variables, let $\hat{\mu}(\cdot) = \hat{\beta}_1 g_1 + \cdots + \hat{\beta}_p g_p$ be the least-squares fit in G, and let $\hat{\beta} = [\hat{\beta}_1, \ldots, \hat{\beta}_p]^T$ be the coefficient vector for the fit. Under the assumptions of the regression model, $\mathbf{Y}$ has mean vector $\mu = [\mu(\mathbf{x}_1), \ldots, \mu(\mathbf{x}_n)]^T$. Since $\hat{\beta} = (\mathbf{X}^T \mathbf{X})^{-1} \mathbf{X}^T \mathbf{Y}$ by (13) in Section 10.1, we conclude that $E\hat{\beta} = (\mathbf{X}^T \mathbf{X})^{-1} \mathbf{X}^T \mu$. In particular, $\hat{\beta}_1, \ldots, \hat{\beta}_p$ have finite mean.

Under the assumptions of the linear regression model, $\mu^*(\cdot) = P_G[\mu(\cdot)] = \mu(\cdot) = \beta_1 g_1 + \cdots + \beta_p g_p$. Under the assumptions of the normal regression model, $\mathbf{Y}$ has the multivariate normal distribution with mean vector μ and variance–covariance matrix $\sigma^2 \mathbf{I}_n$.

THEOREM 10.9   (a) Under the assumptions of the linear regression model, $\hat{\beta}$ has mean vector β.

(b) Under the assumptions of the homoskedastic regression model, $\hat{\beta}$ has variance–covariance matrix $\sigma^2 (\mathbf{X}^T \mathbf{X})^{-1}$.

Proof   (a) Since $\mu^*(\cdot) = \mu(\cdot)$ under the assumptions of the linear regression model, it follows from Theorem 10.5 in Section 10.2 that $E[\hat{\mu}(\mathbf{x})] = \mu(\mathbf{x})$ for $\mathbf{x} \in \mathcal{X}$. Thus,

$$\sum_j (E\hat{\beta}_j) g_j(\mathbf{x}) = E\left(\sum_j \hat{\beta}_j g_j(\mathbf{x}) \right) = E[\hat{\mu}(\mathbf{x})] = \mu(\mathbf{x}) = \sum_j \beta_j g_j(\mathbf{x}), \qquad \mathbf{x} \in \mathcal{X},$$

so $\sum_j (E\hat{\beta}_j) g_j = \sum_j \beta_j g_j$. Since $g_1, \ldots, g_p$ are linearly independent, we conclude that $E\hat{\beta}_j = \beta_j$ for $1 \leq j \leq p$ and hence that $E\hat{\beta} = \beta$.

(b) According to (13) in Section 10.1,

$$\begin{aligned} \mathrm{VC}(\hat{\beta}) &= \mathrm{VC}((\mathbf{X}^T \mathbf{X})^{-1} \mathbf{X}^T \mathbf{Y}) \\ &= (\mathbf{X}^T \mathbf{X})^{-1} \mathbf{X}^T \, \mathrm{VC}(\mathbf{Y}) \mathbf{X} (\mathbf{X}^T \mathbf{X})^{-1} \\ &= \sigma^2 (\mathbf{X}^T \mathbf{X})^{-1} \mathbf{X}^T \mathbf{X} (\mathbf{X}^T \mathbf{X})^{-1} \\ &= \sigma^2 (\mathbf{X}^T \mathbf{X})^{-1}. \quad \blacksquare \end{aligned}$$

Let $\mathbf{V} = [v_{jl}] = \sigma^2 (\mathbf{X}^T \mathbf{X})^{-1}$ denote the variance–covariance matrix of $\hat{\beta}$ under the assumptions of the homoskedastic regression model.

THEOREM 10.10   Under the assumptions of the normal linear regression model,

(a) $\hat{\beta}$ has a multivariate normal distribution;

(b) $\hat{\beta}$, LSS, and WSS are independent;

(c) if $d > p$, then LSS $/\sigma^2$ has the chi-square distribution with $d - p$ degrees of freedom.

Proof   **(a)** By ignoring points in $\mathfrak{X}$ that are not design points, we can assume that $\mathfrak{X}$ coincides with the design set. Then G is a subspace of the d-dimensional linear space H of all functions on this set. Let $\tilde{g}_1, \ldots, \tilde{g}_d$ be an orthonormal basis of H such that $\tilde{g}_1, \ldots, \tilde{g}_p$ is an orthonormal basis of G. Since $\bar{Y}(\cdot)$ is a member of H, we see that $\bar{Y}(\cdot) = P_H[\bar{Y}(\cdot)]$ and hence that $\bar{Y}(\cdot)$ is the least-squares estimate in H of the regression function. Observe that

$$\bar{Y}(\cdot) = \sum_{j=1}^{d} \hat{\gamma}_j \tilde{g}_j,$$

where

$$\hat{\gamma}_j = \langle \tilde{g}_j, \bar{Y}(\cdot) \rangle = \sum_{k} n_k \tilde{g}_j(\mathbf{x}'_k) \bar{Y}_k, \qquad 1 \le j \le d,$$

and hence that

$$\hat{\gamma}_j = \sum_{k} \sqrt{n_k}\, \tilde{g}_j(\mathbf{x}'_k) \sqrt{n_k} \bar{Y}_k, \qquad 1 \le j \le d. \tag{34}$$

Set $W_k = \sqrt{n_k} \bar{Y}_k$ for $1 \le k \le d$ and $\mathbf{W} = [W_1, \ldots, W_d]^T$. Then $W_1, \ldots, W_d$ are independent and normally distributed with common variance σ^2, so $\mathbf{W}$ has a multivariate normal distribution with variance–covariance $\sigma^2 \mathbf{I}_d$.

Set $c_{kj} = \sqrt{n_k} \tilde{g}_j(\mathbf{x}'_k)$ for $1 \le j, k \le d$ and

$$\mathbf{C} = \begin{bmatrix} c_{11} & \cdots & c_{1d} \\ \vdots & & \vdots \\ c_{d1} & \cdots & c_{dd} \end{bmatrix}.$$

We can rewrite (34) in matrix form as

$$\begin{bmatrix} \hat{\gamma}_1 \\ \vdots \\ \hat{\gamma}_d \end{bmatrix} = \mathbf{C}^T \mathbf{W},$$

from which we see that the variance–covariance matrix of $\hat{\gamma}_1, \ldots, \hat{\gamma}_d$ equals $\sigma^2 \mathbf{C}^T \mathbf{C}$. For $1 \le j, l \le d$, the entry in row j and column l of $\mathbf{C}^T \mathbf{C}$ is given by

$$\sum_{k} c_{kj} c_{kl} = \sum_{k} n_k \tilde{g}_j(\mathbf{x}'_k) \tilde{g}_l(\mathbf{x}'_k) = \langle \tilde{g}_j, \tilde{g}_l \rangle.$$

Thus, $\mathbf{C}^T \mathbf{C} = \mathbf{I}_d$ by the orthonormality of $\tilde{g}_1, \ldots, \tilde{g}_d$. In particular, $\mathbf{C}$ is invertible. Moreover, $\hat{\gamma}_1, \ldots, \hat{\gamma}_d$ have a multivariate normal joint distribution with variance–covariance matrix $\sigma^2 \mathbf{I}_d$, so they are independent, normal random variables, each having variance σ^2.

The least-squares estimate in G of the regression function is given by

$$\hat{\mu}(\cdot) = P_G[\bar{Y}(\cdot)] = \sum_{j=1}^{p} \hat{\gamma}_j \tilde{g}_j.$$

Let $\hat{\gamma} = [\hat{\gamma}_1, \ldots, \hat{\gamma}_p]^T$ be the coefficient vector of $\hat{\mu}(\cdot)$ relative to the basis $\tilde{g}_1, \ldots, \tilde{g}_p$ of G and let $\mathbf{A}$ be the coefficient matrix of the basis $\tilde{g}_1, \ldots, \tilde{g}_p$ relative to the basis $g_1, \ldots, g_p$. Then $\mathbf{A}$ is an invertible $p \times p$ matrix, and $\hat{\beta} = \mathbf{A}^T \hat{\gamma}$ by (5) in Section 9.1. Since $\hat{\gamma}$ has a multivariate normal distribution, so does $\hat{\beta}$.

(b) Now $\bar{Y}_1, \ldots, \bar{Y}_d$, WSS are independent by Theorem 8.1 in Section 8.3, and $[\hat{\gamma}_1, \ldots, \hat{\gamma}_d]^T$ is a transform of $\bar{Y}_1, \ldots, \bar{Y}_d$, so $[\hat{\gamma}_1, \ldots, \hat{\gamma}_d]^T$ and WSS are independent. Since $\hat{\gamma}_1, \ldots, \hat{\gamma}_d$ themselves are independent, we conclude that $\hat{\gamma}_1, \ldots, \hat{\gamma}_d$, WSS are independent. Observe that

$$\bar{Y}(\cdot) - \hat{\mu}(\cdot) = \sum_{j=p+1}^{d} \hat{\gamma}_j \tilde{g}_j.$$

Thus, the lack-of-fit sum of squares for G is given by

$$\text{LSS} = \|\bar{Y}(\cdot) - \hat{\mu}(\cdot)\|^2 = \sum_{j=p+1}^{d} \hat{\gamma}_j^2.$$

Moreover, $\hat{\gamma}_1, \ldots, \hat{\gamma}_p$, $\hat{\gamma}_{p+1}^2 + \cdots + \hat{\gamma}_d^2$, and WSS are independent, so $\hat{\gamma}$, LSS, and WSS are independent. Since $\hat{\beta}$ is a transform of $\hat{\gamma}$, we conclude that $\hat{\beta}$, LSS, and WSS are independent.

(c) Suppose $d > p$. From the proof of part (b), we see that

$$\frac{\text{LSS}}{\sigma^2} = \sum_{j=p+1}^{d} \left(\frac{\hat{\gamma}_j}{\sigma}\right)^2,$$

where $\hat{\gamma}_{p+1}/\sigma, \ldots, \hat{\gamma}_d/\sigma$ are independent, normal random variables, each having variance 1. Since $\mu(\cdot)$ is a member of G, $\langle \tilde{g}_j, \mu(\cdot) \rangle = 0$ for $p + 1 \leq j \leq d$; thus, by Theorem 10.4 in Section 10.2 applied to $\bar{Y}(\cdot)$ instead of $\hat{\mu}(\cdot)$,

$$E\hat{\gamma}_j = E[\langle \tilde{g}_j, \bar{Y}(\cdot) \rangle] = \langle \tilde{g}_j, \mu(\cdot) \rangle = 0, \qquad p + 1 \leq j \leq d.$$

Consequently, $\hat{\gamma}_{p+1}/\sigma, \ldots, \hat{\gamma}_d/\sigma$ are independent, standard normal random variables, so LSS/σ^2 has the chi-square distribution with $d - p$ degrees of freedom. ∎

COROLLARY 10.8   Under the assumptions of the normal linear regression model,

(a) $\hat{\beta}$ has the multivariate normal distribution with mean vector β and variance–covariance matrix $\sigma^2 (\mathbf{X}^T \mathbf{X})^{-1}$;

(b) $\hat{\beta}$ and RSS are independent;

(c) if $n > p$, then RSS/σ^2 has the chi-square distribution with $n - p$ degrees of freedom.

Proof   (a) The desired result follows from Theorem 10.9 and Theorem 10.10(a).

(b) Since $\text{RSS} = \text{LSS} + \text{WSS}$, the desired result follows from Theorem 10.10(b).

(c) Observe that $p \le d \le n$ (see Corollary 9.3 in Section 9.2). If $n > p$, then either $n > d$ or $d > p$. If $n = d$, then WSS $= 0$; if $n > d$, then WSS $/\sigma^2$ has the chi-square distribution with $n - d$ degrees of freedom (see Theorem 8.1 in Section 8.3). Also, if $d = p$, then LSS $= 0$ [G is saturated by Theorem 9.9(b) in Section 9.3]; and, according to Theorem 10.10(c), if $d > p$, then LSS $/\sigma^2$ has the chi-square distribution with $d - p$ degrees of freedom. Since LSS and WSS are independent and RSS $=$ LSS $+$ WSS, the desired result is valid. ∎

Recall from Corollary 10.6 in Section 10.2 that, under the assumptions of the homoskedastic linear regression model with $n > p$, $S^2 =$ RSS $/(n - p)$ is an unbiased estimate of σ^2.

COROLLARY **10.9**   Under the assumptions of the normal linear regression model with $n > p$, $\hat{\beta}$ and S are independent, and $(n - p)S^2/\sigma^2$ has the chi-square distribution with $n - p$ degrees of freedom.

Linear Parameters

Let $c_1, \ldots, c_p \in \mathbb{R}$, set $\mathbf{c} = [c_1, \ldots, c_p]^T$, and consider the *linear parameter* $\tau = \mathbf{c}^T\beta = c_1\beta_1 + \cdots + c_p\beta_p$. We refer to $\hat{\tau} = \mathbf{c}^T\hat{\beta} = c_1\hat{\beta}_1 + \cdots + c_p\hat{\beta}_p$ as the least-squares estimate of τ.

Let $\mathbf{x}_1, \mathbf{x}_2 \in X$ be given. Recall that changing the factor-level combination from $\mathbf{x}_1$ to $\mathbf{x}_2$ causes the mean response to change by the amount $\tau = \mu(\mathbf{x}_2) - \mu(\mathbf{x}_1)$. The quantity τ is a linear parameter, and its least-squares estimate is given by $\hat{\tau} = \hat{\mu}(\mathbf{x}_2) - \hat{\mu}(\mathbf{x}_1)$. More generally, we have the following result.

THEOREM **10.11**   Let $\mathbf{x}_1, \ldots, \mathbf{x}_L \in X$, let $a_1, \ldots, a_L \in \mathbb{R}$, and set $\tau = a_1\mu(\mathbf{x}_1) + \cdots + a_L\mu(\mathbf{x}_L)$. Then τ is a linear parameter, and its least-squares estimate is given by

$$\hat{\tau} = a_1\hat{\mu}(\mathbf{x}_1) + \cdots + a_L\hat{\mu}(\mathbf{x}_L).$$

Proof   Observe that

$$\tau = \sum_l a_l\mu(\mathbf{x}_l) = \sum_l a_l \sum_j \beta_j g_j(\mathbf{x}_l) = \sum_j \beta_j \sum_l a_l g_j(\mathbf{x}_l) = \sum_j c_j\beta_j,$$

where $c_j = \sum_l a_l g_j(x_l)$. Thus, τ is a linear parameter, and its least-squares estimate is given by

$$\hat{\tau} = \sum_j c_j\hat{\beta}_j = \sum_j \hat{\beta}_j \sum_l a_l g_j(\mathbf{x}_l) = \sum_l a_l \sum_j \hat{\beta}_j g_j(\mathbf{x}_l) = \sum_l a_l\hat{\mu}(\mathbf{x}_l). \quad ∎$$

EXAMPLE **10.10**   Let the d-sample model be viewed as a linear regression model as in Section 8.3. Then G is the identifiable and saturated space of all functions on $X = X' = \{1, \ldots, d\}$. Determine the least-squares estimates of

(a) $\mu(k)$ for $1 \leq k \leq d$;

(b) $\mu(k) - \mu(j)$ for $1 \leq j, k \leq d$ with $j \neq k$.

Solution   (a) Consider the orthogonal basis $I_1, \ldots, I_d$ of G, where $I_k(x) = \text{ind}(x = k)$ for $1 \leq k \leq d$. The regression function can be written in terms of this basis as $\mu(k) = \mu_1 I_1(k) + \cdots + \mu_d I_d(k)$, where μ_k is the common mean for the random variables in the kth sample. The least-squares estimates $\hat{\mu}_1, \ldots, \hat{\mu}_d$ of the regression coefficients $\mu_1, \ldots, \mu_d$, respectively, are given by $\hat{\mu}_k = \bar{Y}_k$ for $1 \leq k \leq d$ (see Problem 10.7 in Section 10.1). Since $\mu(k) = \mu_k$ for $1 \leq k \leq d$, we conclude that $\hat{\mu}(k) = \hat{\mu}_k = \bar{Y}_k$ for $1 \leq k \leq d$.

(b) Let $1 \leq j, k \leq d$ with $j \neq k$. According to the solution to part (a), the least-squares estimate of $\mu(k) - \mu(j)$ is given by $\hat{\mu}(k) - \hat{\mu}(j) = \hat{\mu}_k - \hat{\mu}_j = \bar{Y}_k - \bar{Y}_j$. ∎

Under the assumptions of the linear regression model,

$$E\hat{\tau} = E(\mathbf{c}^T \hat{\beta}) = \mathbf{c}^T E\hat{\beta} = \mathbf{c}^T \beta = \tau,$$

and hence $\hat{\tau}$ is an unbiased estimate of τ. Under the assumptions of the homoskedastic regression model, the variance of this estimate is given by

$$\text{var}(\hat{\tau}) = \text{var}(\mathbf{c}^T \hat{\beta}) = \mathbf{c}^T \, \text{VC}(\hat{\beta})\mathbf{c} = \sigma^2 \mathbf{c}^T (\mathbf{X}^T \mathbf{X})^{-1} \mathbf{c} = \mathbf{c}^T \mathbf{V} \mathbf{c} = \sum_j \sum_l c_j c_l v_{jl},$$

so its standard deviation is given by $\text{SD}(\hat{\tau}) = \sqrt{\mathbf{c}^T \mathbf{V} \mathbf{c}}$. The linear parameter τ is said to be *nontrivial* if at least one of the numbers $c_1, \ldots, c_p$ is nonzero or, equivalently, if $\mathbf{c} \neq \mathbf{0}$, and it is said to be *trivial* otherwise. Under the assumptions of the normal linear regression model, if $\tau = \mathbf{c}^T \beta$ is a nontrivial linear parameter, then its least-squares estimate $\hat{\tau} = \mathbf{c}^T \hat{\beta}$ is normally distributed with mean τ and positive variance $\mathbf{c}^T \mathbf{V} \mathbf{c}$. In particular, we have the following result.

COROLLARY 10.10   Let $\hat{\tau}$ be the least-squares estimate of a nontrivial linear parameter τ. Under the assumptions of the normal linear regression model, $(\hat{\tau} - \tau)/\text{SD}(\hat{\tau})$ has the standard normal distribution.

Confidence Intervals

Suppose that $n > p$. Then the variance–covariance matrix $\mathbf{V} = \sigma^2 (\mathbf{X}^T \mathbf{X})^{-1}$ of $\hat{\beta}$ can be estimated by $\hat{\mathbf{V}} = [\hat{v}_{jl}] = S^2 (\mathbf{X}^T \mathbf{X})^{-1}$. Observe that $\hat{\mathbf{V}} = (S/\sigma)^2 \mathbf{V}$. Under the assumptions of the homoskedastic linear regression model, S^2 is an unbiased estimate of σ^2, so $\hat{v}_{jl}$ is an unbiased estimate of v_{jl} for $1 \leq j, l \leq p$, and $\hat{\mathbf{V}}$ is an unbiased estimate of $\mathbf{V}$ (that is, $E\hat{\mathbf{V}} = \mathbf{V}$). The standard error of the least-squares estimate $\hat{\tau}$ of τ, defined by

$$\text{SE}(\hat{\tau}) = \sqrt{\mathbf{c}^T \hat{\mathbf{V}} \mathbf{c}} = \frac{S}{\sigma} \sqrt{\mathbf{c}^T \mathbf{V} \mathbf{c}} = \frac{S}{\sigma} \text{SD}(\hat{\tau}),$$

is an estimate of the standard deviation of τ. Note, in particular, that the standard errors of the least-squares estimates of the regression coefficients are given by $SE(\hat{\beta}_j) = \sqrt{\hat{v}_{jj}}$ for $1 \le j \le p$.

THEOREM 10.12   Let $\hat{\tau}$ be the least-squares estimate of a nontrivial linear parameter τ. Then, under the assumptions of the normal linear regression model with $n > p$, $\hat{\tau}$ and S are independent and $(\hat{\tau} - \tau)/SE(\hat{\tau})$ has the t distribution with $n - p$ degrees of freedom.

Proof   Since $\hat{\tau}$ is normally distributed with mean τ, $Z = (\hat{\tau} - \tau)/SD(\hat{\tau})$ has the standard normal distribution. According to Corollary 10.9, $\hat{\beta}$ and S are independent, and $\chi^2 = (n - p)S^2/\sigma^2$ has the chi-square distribution with $n - p$ degrees of freedom. Thus, $\hat{\tau} = \mathbf{c}^T\hat{\beta}$ and S are independent, so Z and χ^2 are independent and hence

$$\frac{Z}{\sqrt{\chi^2/(n-p)}} = \frac{(\hat{\tau} - \tau)/SD(\hat{\tau})}{S/\sigma} = \frac{\hat{\tau} - \tau}{SE(\hat{\tau})}$$

has the t distribution with $n - p$ degrees of freedom.   ∎

Let $0 < \alpha < 1$. It follows from Theorem 10.12 (see Section 7.3) that $\hat{\tau} - t_{1-\alpha,n-p} SE(\hat{\tau})$ is the $100(1 - \alpha)\%$ lower confidence bound for the parameter τ, $\hat{\tau} + t_{1-\alpha,n-p} SE(\hat{\tau})$ is the $100(1 - \alpha)\%$ upper confidence bound for this parameter, and

$$\hat{\tau} \pm t_{1-\alpha/2,n-p} SE(\hat{\tau}) = \left(\hat{\tau} - t_{1-\alpha/2,n-p} SE(\hat{\tau}),\ \hat{\tau} + t_{1-\alpha/2,n-p} SE(\hat{\tau})\right)$$

is its $100(1 - \alpha)\%$ confidence interval.

EXAMPLE 10.11   Consider the polymer experiment and the normal linear regression model corresponding to the assumption that $\mu(x) = \beta_0 + \beta_1 x + \beta_2 x^2$. Determine 95% confidence intervals for

(a) β_0, β_1, and β_2;   **(b)** $\mu(x), x \in \mathbb{R}$;   **(c)** $\mu(3) - \mu(-3)$;
(d) $\mu(5) - \mu(1)$;   **(e)** $\frac{d\mu}{dx}(5)$.

Solution   Here

$$\hat{\mu}(x) = \hat{\beta}_0 + \hat{\beta}_1 x + \hat{\beta}_2 x^2 \doteq 80.816 + 2.547x - 0.337x^2,$$

and RSS $\doteq 5.073$. Also, $n = 6$ and $p = 3$, so $S^2 \doteq 5.073/3 \doteq 1.691$ and hence $S \doteq \sqrt{1.691} \doteq 1.300$. Moreover,

$$(\mathbf{X}^T\mathbf{X})^{-1} \doteq \begin{bmatrix} 0.395 & 0 & -0.0195 \\ 0 & 0.0143 & 0 \\ -0.0195 & 0 & 0.00167 \end{bmatrix},$$

so

$$\hat{\mathbf{V}} \doteq 1.691 \begin{bmatrix} 0.395 & 0 & -0.0195 \\ 0 & 0.0143 & 0 \\ -0.0195 & 0 & 0.0067 \end{bmatrix} \doteq \begin{bmatrix} 0.667 & 0 & -0.0330 \\ 0 & 0.0242 & 0 \\ -0.0330 & 0 & 0.00283 \end{bmatrix}.$$

(a) The standard errors of $\hat{\beta}_0$, $\hat{\beta}_1$ and $\hat{\beta}_2$ are given by

$$SE(\hat{\beta}_0) \doteq \sqrt{0.667} \doteq 0.817,$$

$$SE(\hat{\beta}_1) \doteq \sqrt{0.0242} \doteq 0.155,$$

and

$$SE(\hat{\beta}_2) \doteq \sqrt{0.00283} \doteq 0.0532.$$

Since $t_{.975,3} \doteq 3.182$, the 95% confidence interval for β_0 is given by

$$80.816 \pm (3.182)(0.817) \doteq 80.816 \pm 2.599 \doteq (78.216, 83.415),$$

the 95% confidence interval for β_1 is given by

$$2.547 \pm (3.182)(0.155) \doteq 3.547 \pm 0.495 \doteq (2.053, 3.042),$$

and the 95% confidence interval for β_2 is given by

$$-0.337 \pm (3.182)(0.0532) \doteq -0.337 \pm 0.169 = (-0.506, -0.168).$$

(b) Choose $x \in \mathbb{R}$. The standard error of $\hat{\mu}(x)$ is the nonnegative square root of

$$[1, x, x^2]\hat{\mathbf{V}} \begin{bmatrix} 1 \\ x \\ x^2 \end{bmatrix} \doteq 0.667 + [0.0242 - 2(0.0330)]x^2 + 0.00283x^4$$

$$\doteq 0.667 - 0.0419x^2 + 0.00283x^4.$$

Thus, the 95% confidence interval for $\mu(x)$ is given by

$$80.816 + 2.547x - 0.337x^2 \pm 3.182\sqrt{0.667 - 0.0419x^2 + 0.00283x^4}.$$

Figure 10.3 shows the graphs of the lower and upper end points of this confidence interval, as a function of x. [The graph of $\hat{\mu}(\cdot)$ is shown in Figure 8.2 in Section 8.1.]

(c) Now

$$\hat{\mu}(3) - \hat{\mu}(-3) = \hat{\beta}_0 + 3\hat{\beta}_1 + 9\hat{\beta}_2 - (\hat{\beta}_0 - 3\hat{\beta}_1 + 9\hat{\beta}_2)$$

$$= 6\hat{\beta}_1$$

$$\doteq 6(2.547)$$

$$\doteq 15.263,$$

so

$$SE(\hat{\mu}(3) - \hat{\mu}(-3)) = 6\,SE(\hat{\beta}_1) \doteq 6(0.155) \doteq 0.933.$$

FIGURE   10.3

Pointwise confidence intervals

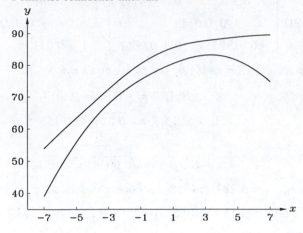

Thus, the 95% confidence interval for $\mu(3) - \mu(-3)$ is given by

$$15.263 \pm (3.182)(0.933) \doteq 15.263 \pm 2.968 = (12.315, 18.251).$$

(d) Observe that

$$\hat{\mu}(5) - \hat{\mu}(1) = \hat{\beta}_0 + 5\hat{\beta}_1 + 25\hat{\beta}_2 - (\hat{\beta}_0 + \hat{\beta}_1 + \hat{\beta}_2)$$

$$= 4\hat{\beta}_1 + 24\hat{\beta}_2$$

$$\doteq 4(2.547) - 24(0.337)$$

$$\doteq 2.099.$$

Since the entry in row 2 and column 3 of $\hat{\mathbf{V}}$ equals zero, we see that

$$\text{SE}(\hat{\mu}(5) - \hat{\mu}(1)) \doteq \sqrt{16(0.0242) + 576(0.00283)} \doteq 1.420.$$

Thus, the 95% confidence interval for $\mu(5) - \mu(1)$ is given by

$$2.099 \pm (3.182)(1.420) \doteq 2.099 \pm 4.520 \doteq (-2.420, 6.619).$$

(e) Now

$$\frac{d\hat{\mu}}{dx}(x) = \hat{\beta}_1 + 2x\hat{\beta}_2,$$

so

$$\frac{d\hat{\mu}}{dx}(5) = \hat{\beta}_1 + 10\hat{\beta}_2 \doteq 2.547 - 10(0.337) = -0.823$$

and hence

$$\text{SE}\left(\frac{d\hat{\mu}}{dx}(5)\right) \doteq \sqrt{0.0242 + 100(0.00283)} \doteq 0.5543.$$

Consequently, the 95% confidence interval for $\frac{d\mu}{dx}(5)$ is given by

$$-0.823 \pm (3.182)(0.543) \doteq -0.823 \pm 1.764 = (-2.587, 0.941). \quad \blacksquare$$

Problems

10.28 Let $\hat{\beta}_0 + \hat{\beta}_1 x$ be the least-squares fit in the space of linear functions on $\mathbb{R}$. Under the assumptions of the corresponding normal linear regression model, determine the distribution of $\hat{\beta}_1$.

10.29 Under the assumptions of the regression model, verify that $E\hat{\beta}_j = \beta_j^*$ for $1 \leq j \leq p$ and $E\hat{\beta} = \beta^*$, where $\beta_1^*, \ldots, \beta_p^*$ are the coefficients of $\mu^*(\cdot) = P_G[\mu(\cdot)]$ relative to the basis $g_1, \ldots, g_p$ of G and $\beta^* = [\beta_1^*, \ldots, \beta_p^*]^T$.

10.30 Suppose the assumptions of the homoskedastic regression model are satisfied, where G is the space of linear functions on $\mathbb{R}$. Let $\hat{\beta}_0, \hat{\beta}_1$ be the coefficients of $\hat{\mu}(\cdot)$ relative to the basis $1, x - \bar{x}$ of G (see Example 10.4 of Section 10.1). Verify that   **(a)** $\mathrm{var}(\hat{\beta}_0) = \sigma^2/n$;   **(b)** $\mathrm{var}(\hat{\beta}_1) = \sigma^2/\sum_i (x_i - \bar{x})^2$;   **(c)** $\mathrm{cov}(\hat{\beta}_0, \hat{\beta}_1) = 0$;   **(d)** $\mathrm{var}(\hat{\mu}(x)) = \sigma^2 \left(\frac{1}{n} + \frac{(x - \bar{x})^2}{\sum_i (x_i - \bar{x})^2} \right)$ for $x \in \mathbb{R}$.

10.31 (Continued.) According to Problem 10.30(b), to minimize $\mathrm{var}(\hat{\beta}_1)$, we need to maximize $\sum_i (x_i - \bar{x})^2$. Suppose $n \geq 2$, $x_1, \ldots, x_{n-1}$ are fixed, and x_n is to be chosen to maximize $\sum_i (x_i - \bar{x})^2$ subject to the constraint that $a \leq x_n \leq b$, where $-\infty < a < b < \infty$. Show that x_n equals a or b. [*Hint:* Show that the second derivative of $\sum_i (x_i - \bar{x})^2$ with respect to x_n is positive.]

10.32 (Continued.) Suppose we want to choose $x_1, \ldots, x_n$ to maximize $\sum_i (x_i - \bar{x})^2$ subject to the constraint that $a \leq x_i \leq b$ for $1 \leq i \leq n$. It follows by repeated application of the result of Problem 10.31 that at the maximum all of the x_i's equal a or b. Let m be the number of x_i's that equal a. Show that **(a)** $\sum_i (x_i - \bar{x})^2 = \frac{m(n-m)}{n} (b - a)^2$; **(b)** if n is even, then $\sum_i (x_i - \bar{x})^2$ is maximized when $m = n/2$; **(c)** if n is odd, then $\sum_i (x_i - \bar{x})^2$ is maximized when m equals $(n \pm 1)/2$.

10.33 Suppose the design set is $\{-1, 0, 1\}$ and that there are r repetitions at each design point. Under the setup of Problem 10.30, determine **(a)** $\mathrm{var}(\hat{\beta}_1)$ as a function of r; **(b)** the smallest value of r such that $\mathrm{var}(\hat{\beta}_1) \leq \sigma^2/16$.

10.34 (Continued.) Rework Problem 10.33 using $\{-2, 0, 2\}$ as the design set and compare the two answers to part (b).

10.35 Under the assumptions of the homoskedastic regression model, show that **(a)** $\hat{\beta}_1, \ldots, \hat{\beta}_p$ are uncorrelated if and only if $g_1, \ldots, g_p$ are orthogonal or, equivalently, if and only if $\mathbf{X}^T \mathbf{X}$ is a diagonal matrix; **(b)** if $g_1, \ldots, g_p$ are orthogonal, then

$$\mathrm{var}(\hat{\beta}_j) = \frac{\sigma^2}{\|g_j\|^2} = \frac{\sigma^2}{\sum_i g_j^2(\mathbf{x}_i)}, \qquad 1 \leq j \leq p,$$

and

$$\mathrm{var}(\hat{\mu}(\mathbf{x})) = \sigma^2 \sum_j \frac{g_j^2(\mathbf{x})}{\|g_j\|^2}, \qquad \mathbf{x} \in \mathcal{X}.$$

10.36 Let $\mathbf{X}$, $\mathbf{N}$, and $\bar{\mathbf{Y}}$ be as in Problem 10.13 in Section 10.1. Show that **(a)** $\mathrm{VC}(\hat{\beta}) = \sigma^2 (\mathbf{X}^T \mathbf{N} \mathbf{X})^{-1}$; **(b)** $\mathrm{var}(\mathbf{c}^T \hat{\beta}) = \sigma^2 \mathbf{c}^T (\mathbf{X}^T \mathbf{N} \mathbf{X})^{-1} \mathbf{c}$ for $\mathbf{c} \in \mathbb{R}^p$.

10.37   (Continued.) Suppose there are r repetitions at each design point. Show that
(a) $VC(\hat{\beta}) = \sigma^2 r^{-1}(\mathbf{X}^T\mathbf{X})^{-1}$;   (b) $var(\mathbf{c}^T\hat{\beta}) = \sigma^2 r^{-1}\mathbf{c}^T(\mathbf{X}^T\mathbf{X})^{-1}\mathbf{c}$ for $\mathbf{c} \in \mathbb{R}^p$.

10.38   Let $\mathbf{V}$ be the variance–covariance matrix of the least-squares estimate of the coefficient vector of the regression function relative to a basis $g_1, \ldots, g_p$ of G, let $\tilde{\mathbf{V}}$ be the variance–covariance matrix of the least-squares estimate of the coefficient vector relative to an alternative basis $\tilde{g}_1, \ldots, \tilde{g}_p$, and let $\mathbf{A}$ be the coefficient matrix of $\tilde{g}_1, \ldots, \tilde{g}_p$ relative to $g_1, \ldots, g_p$. Verify that $\tilde{\mathbf{V}} = (\mathbf{A}^T)^{-1}\mathbf{V}\mathbf{A}^{-1}$.

10.39   Consider the d-sample model, in which G is the space of all functions on $X = \{1, \ldots, d\}$ and consider the basis $1, I_1, \ldots, I_{d-1}$ of G, where $I_k(x) = \text{ind}(x = k)$ for $x \in X$. The regression function can be written in terms of this basis as $\mu(\cdot) = \beta_0 + \beta_1 I_1 + \cdots + \beta_{d-1}I_{d-1}$, and its least-squares estimate can be written as $\hat{\mu}(\cdot) = \hat{\beta}_0 + \hat{\beta}_1 I_1 + \cdots + \hat{\beta}_{d-1}I_{d-1}$. (a) Show that $\mu(d) = \beta_0$ and $\mu(k) - \mu(d) = \beta_k$ for $1 \leq k \leq d - 1$; similarly, show that $\hat{\mu}(d) = \hat{\beta}_0$ and $\hat{\mu}(k) - \hat{\mu}(d) = \hat{\beta}_k$ for $1 \leq k \leq d - 1$. (b) Determine the Gram matrix $\mathbf{X}^T\mathbf{X}$, where $\mathbf{X}$ is the design matrix. (c) Show that the normal equations for $\hat{\beta}_0, \ldots, \hat{\beta}_{d-1}$ can be written as

$$\hat{\beta}_0 + \frac{n_1}{n}\hat{\beta}_1 + \cdots + \frac{n_{d-1}}{n}\hat{\beta}_{d-1} = \bar{Y}$$

$$\hat{\beta}_0 + \hat{\beta}_1 = \bar{Y}_1$$

$$\vdots \quad \vdots \quad \vdots$$

$$\hat{\beta}_0 + \hat{\beta}_{d-1} = \bar{Y}_{d-1};$$

(d) Show that the least-squares estimates of $\beta_0, \ldots, \beta_{d-1}$ are given by $\hat{\beta}_0 = \bar{Y}_d$ and $\hat{\beta}_k = \bar{Y}_k - \bar{Y}_d$ for $1 \leq k \leq d - 1$. (e) Show that $\hat{\mu}(k) = \bar{Y}_k$ for $1 \leq k \leq d$. (f) Show that LSS $= 0$, FSS $=$ BSS, and RSS $=$ WSS.

10.40   Consider the polymer experiment as treated in Problem 10.8 in Section 10.1. The residual sum of squares for the least-squares linear fit $\hat{\mu}(x) = \hat{\beta}_0 + \hat{\beta}_1 x$ is given by RSS $\doteq 72.933$. Determine 95% confidence intervals for (a) β_0 and β_1; (b) $\mu(x)$, $x \in \mathbb{R}$; (c) $\mu(3) - \mu(-3)$; (d) $\mu(5) - \mu(1)$; (e) $\frac{d\mu}{dx}(5)$.

10.41   Consider the polymer experiment as treated in Problem 10.9 in Section 10.1. The residual sum of squares for the least-squares quadratic polynomial fit $\hat{\mu}(x) = \hat{\beta}_0 + \hat{\beta}_1 x + \hat{\beta}_2(x^2 - \frac{35}{3})$ is given by RSS $\doteq 5.073$. Determine 95% confidence intervals for (a) β_0, β_1, and β_2; (b) $\mu(x)$, $x \in \mathbb{R}$; (c) $\mu(3) - \mu(-3)$; (d) $\mu(5) - \mu(1)$; (e) $\frac{d\mu}{dx}(5)$.

10.4
Sugar Beet Experiment

Consider the sugar beet experiment as described in Section 8.2, where the variety is denoted by x_1 and the plot by x_2. The sugar beet data used in Chapter 7 were obtained from this experiment by ignoring the plot variable x_2. In the corresponding linear regression model, G is the space of all functions on $X_1 = \{1, \ldots, 5\}$. Consider the basis $I_1(x_1)$, $I_2(x_1)$, $I_3(x_1)$, $I_4(x_1)$, $I_5(x_1)$ of G, where $I_j(x_1) = \text{ind}(x_1 = j)$. The regression function can be written in terms of this basis as

$$\mu(x_1) = \beta_1 I_1(x_1) + \beta_2 I_2(x_1) + \beta_3 I_3(x_1) + \beta_4 I_4(x_1) + \beta_5 I_5(x_1).$$

Observe that $\beta_j = \mu(j)$ for $1 \le k \le 5$.

The least-squares estimate of the regression function is given by $\hat{\mu}(x_1) = \hat{\beta}_1 I_1(x_1) + \hat{\beta}_2 I_2(x_1) + \hat{\beta}_3 I_3(x_1) + \hat{\beta}_4 I_4(x_1) + \hat{\beta}_5 I_5(x_1)$, where

$$\hat{\beta}_1 = \hat{\mu}(1) = \bar{Y}_1 \doteq 18.136,$$

$$\hat{\beta}_2 = \hat{\mu}(2) = \bar{Y}_2 = 17.58,$$

$$\hat{\beta}_3 = \hat{\mu}(3) = \bar{Y}_3 = 18.32,$$

$$\hat{\beta}_4 = \hat{\mu}(4) = \bar{Y}_4 = 18.06,$$

$$\hat{\beta}_0 = \hat{\mu}(5) = \bar{Y}_5 = 18.13;$$

here $\bar{Y}_1, \dots, \bar{Y}_5$ are the respective average values of the response variable for varieties $1, \dots, 5$.

Let $R_1 = Y_1 - \hat{\mu}(x_{1,1}), \dots, R_{51} = Y_{51} - \hat{\mu}(x_{51,1})$ denote the respective residuals corresponding to the yields on plots $-25, \dots, 25$. Now $Y_1 = 17.4$ and $x_{1,1} = 1$, so $R_1 = Y_1 - \bar{Y}_1 \doteq 17.4 - 18.136 = -0.736$. Similarly,

$$R_2 = Y_2 - \hat{\mu}(x_{2,1}) = Y_2 - \bar{Y}_2 = 17.2 - 17.58 = -0.38,$$

and

$$R_{51} = Y_{51} - \hat{\mu}(x_{51,1}) = Y_{51} - \bar{Y}_1 \doteq 18.1 - 18.136 = -0.036.$$

The residual sum of squares is given by RSS $= R_1^2 + \cdots + R_{51}^2 \doteq 8.242$.

Figure 10.4 illustrates the unrevealing plot of residuals versus variety (here 2 and 3 respectively, denote two and three coinciding residuals), and Figure 10.5 shows the plot of the residuals versus the plot variable. An inspection of this plot suggests that there is a systematic plot effect: Roughly speaking, the mean percentage of sugar content increases from plot -25 to plot -16, then decreases to plot 3, then stays level to plot 9, then increases to plot 17, and then decreases to the end of the field. This dependence of mean response on plot could be modeled as a polynomial of degree 4 or higher.

Use of Quartic Polynomials

Let us first model the dependence of mean percentage of sugar content on variety and plot as the sum of an arbitary function of variety and a quartic polynomial function of plot; that is, let us assume that the regression function has the form

$$\mu(x_1, x_2) = g_1(x_1) + g_2(x_2)$$

for $x_1 \in \mathcal{X}_1 = \{1, \dots, 5\}$ and $x_2 \in \mathcal{X}_2 = \{-25, \dots, 25\}$, where g_1 is an arbitrary function on $\mathcal{X}_1$ and g_2 is a polynomial of degree 4. The corresponding linear space G of functions on $\mathcal{X} = \mathcal{X}_1 \times \mathcal{X}_2$ is nine-dimensional. (The space of all functions on

FIGURE   10.4
Residuals versus variety

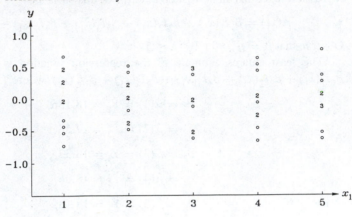

FIGURE   10.5
Residuals versus plot

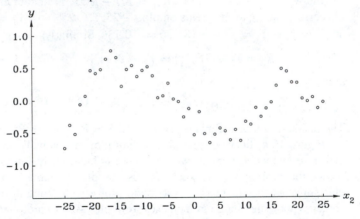

$\mathcal{X}_1$ is five-dimensional, and the space of polynomials of degree 4 is five-dimensional; each space, however, contains the constant functions, so G has dimension $5 + 5 - 1 = 9$. We will discuss such matters more systematically in Section 11.1.)

The functions $I_1(x_1), I_2(x_1), I_3(x_1), I_4(x_1), I_5(x_1), x_2, x_2^2, x_2^3, x_2^4$ form a basis of G. To avoid an excess of leading zeros in the least-squares estimates of the coefficients of x_2 and its various powers, however, we will use the basis

$$I_1(x_1), \ I_2(x_1), \ I_3(x_1), \ I_4(x_1), \ I_5(x_1), \ \frac{x_2}{25}, \ \left(\frac{x_2}{25}\right)^2, \ \left(\frac{x_2}{25}\right)^3, \ \left(\frac{x_2}{25}\right)^4$$

of G. Correspondingly, we write the regression function as

$$\mu(x_1, x_2) = \beta_1 I_1(x_1) + \beta_2 I_2(x_1) + \beta_3 I_3(x_1) + \beta_4 I_4(x_1) + \beta_5 I_5(x_1)$$

$$+ \beta_6 \frac{x_2}{25} + \beta_7 \left(\frac{x_2}{25}\right)^2 + \beta_8 \left(\frac{x_2}{25}\right)^3 + \beta_9 \left(\frac{x_2}{25}\right)^4$$

and its least-squares estimate as

$$\hat{\mu}(x_1, x_2) = \hat{\beta}_1 I_1(x_1) + \hat{\beta}_2 I_2(x_1) + \hat{\beta}_3 I_3(x_1) + \hat{\beta}_4 I_4(x_1) + \hat{\beta}_5 I_5(x_1)$$
$$+ \hat{\beta}_6 \frac{x_2}{25} + \hat{\beta}_7 \left(\frac{x_2}{25}\right)^2 + \hat{\beta}_8 \left(\frac{x_2}{25}\right)^3 + \hat{\beta}_9 \left(\frac{x_2}{25}\right)^4.$$

The design matrix is given by

$$\mathbf{X} = \begin{bmatrix} 1 & 0 & 0 & 0 & 0 & (-25/25) & (-25/25)^2 & (-25/25)^3 & (-25/25)^4 \\ 0 & 1 & 0 & 0 & 0 & (-24/25) & (-24/25)^2 & (-24/25)^3 & (-24/25)^4 \\ \vdots & \vdots & \vdots & \vdots & \vdots & \vdots & \vdots & \vdots & \vdots \\ 0 & 0 & 0 & 0 & 1 & (24/25) & (24/25)^2 & (24/25)^3 & (24/25)^4 \\ 1 & 0 & 0 & 0 & 0 & (25/25) & (25/25)^2 & (25/25)^3 & (25/25)^4 \end{bmatrix}.$$

Using a computer to solve the normal equations or otherwise to obtain the least-squares estimates of the regression coefficients, we get

$$\hat{\mu}(x_1, x_2) \doteq 17.839 I_1(x_1) + 17.241 I_2(x_1) + 17.953 I_3(x_1)$$
$$+ 17.676 I_4(x_1) + 17.740 I_5(x_1)$$
$$- 1.101 \frac{x_2}{25} + 2.994 \left(\frac{x_2}{25}\right)^2 + 1.532 \left(\frac{x_2}{25}\right)^3 - 3.162 \left(\frac{x_2}{25}\right)^4.$$

Table 10.3 shows the corresponding ANOVA table. The squared multiple correlation coefficient is given by

$$R^2 = \frac{\text{FSS}}{\text{TSS}} \doteq \frac{10.049}{11.327} \doteq .887.$$

Thus, 88.7% of the variation in sugar content can be explained by treating this quantity as the sum of an arbitrary function of variety and a quartic polynomial.

TABLE 10.3

Source	SS
Fit	10.049
Residuals	1.278
Total	11.327

The plot of the residuals against variety again is unrevealing. Figure 10.6 shows the plot of the residuals versus the plot variable. There is no longer a striking pattern in the dependence of the residuals on the plot variable. This suggests that quartic polynomials are sufficiently flexible to model the dependence of yield on the plot variable.

F I G U R E   10.6

Residuals from the quartic polynomial fit versus plot

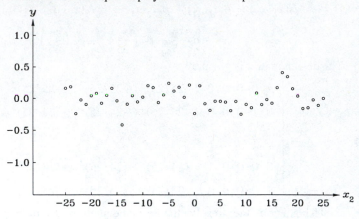

In Figure 10.7, we show the graph of

$$\hat{\beta}_6 \frac{x_2}{25} + \hat{\beta}_7 \left(\frac{x_2}{25}\right)^2 + \hat{\beta}_8 \left(\frac{x_2}{25}\right)^3 + \hat{\beta}_9 \left(\frac{x_2}{25}\right)^4 - \hat{c},$$

where the "centering constant" $\hat{c}$ is defined by

$$\hat{c} = \frac{1}{51} \sum_{x_2 = -25}^{25} \left[\hat{\beta}_6 \frac{x_2}{25} + \hat{\beta}_7 \left(\frac{x_2}{25}\right)^2 + \hat{\beta}_8 \left(\frac{x_2}{25}\right)^3 + \hat{\beta}_9 \left(\frac{x_2}{25}\right)^4 \right] \doteq 0.354.$$

(In this plot, x_2 is shown as ranging over $[-25, 25]$ even though it actually ranges only over $\{-25, \ldots, 25\}$.) This graph shows the up-down/up-down pattern that we should expect in light of the plot of the residuals from the initial least-squares fit against the plot variable.

F I G U R E   10.7

Centered least-square quartic polynomial fit

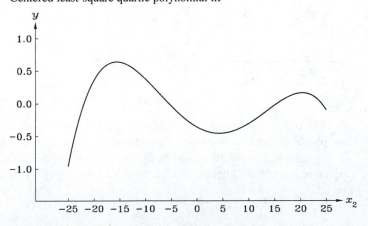

Under the corresponding homoskedastic regression model, the variance σ^2 is estimated by

$$S^2 = \frac{\text{RSS}}{n-p} \doteq \frac{1.278}{42} \doteq 0.0304,$$

so σ is estimated by $S \doteq \sqrt{0.0304} \doteq 0.174$.

The estimated variance–covariance $\hat{\mathbf{V}} = S^2(\mathbf{X}^T\mathbf{X})^{-1}$ of $\hat{\beta}_1, \ldots, \hat{\beta}_9$ is a 9×9 matrix. The estimated variance–covariance matrix $\hat{\mathbf{V}}_{11}$ of $\hat{\beta}_1, \ldots, \hat{\beta}_5$, which is the submatrix consisting of rows 1 through 5 and columns 1 through 5 of $\hat{\mathbf{V}}$, is given by

$$\hat{\mathbf{V}}_{11} \doteq \begin{bmatrix} 0.00421 & 0.00144 & 0.00145 & 0.00145 & 0.00144 \\ 0.00144 & 0.00458 & 0.00155 & 0.00154 & 0.00150 \\ 0.00145 & 0.00155 & 0.00461 & 0.00157 & 0.00154 \\ 0.00145 & 0.00154 & 0.00157 & 0.00461 & 0.00155 \\ 0.00144 & 0.00150 & 0.00154 & 0.00155 & 0.00458 \end{bmatrix}.$$

(Why is the first diagonal entry of $\hat{\mathbf{V}}_{11}$ somewhat smaller than its other diagonal entries?) Under the linear regression model, the change in the mean percentage of sugar content caused by switching from the standard variety to variety 2 on a given plot is given by $\mu(2, x_2) - \mu(1, x_2) = \beta_2 - \beta_1$, and the least-squares estimate of this change is given by

$$\hat{\mu}(2, x_2) - \hat{\mu}(1, x_2) = \hat{\beta}_2 - \hat{\beta}_1 \doteq 17.241 - 17.839 \doteq -0.599.$$

The standard error of this estimate is the nonnegative square root of

$$[-1, 1] \begin{bmatrix} 0.00421 & 0.00144 \\ 0.00144 & 0.00458 \end{bmatrix} \begin{bmatrix} -1 \\ 1 \end{bmatrix} \doteq 0.00421 + 0.00458 - 2(0.00144)$$

$$\doteq 0.00591,$$

so $\text{SE}(\hat{\beta}_2 - \hat{\beta}_1) \doteq \sqrt{0.00591} \doteq 0.0769$. Now $t_{.975,42} \doteq 2.018$. Thus, under the corresponding normal linear regression model, the 95% confidence interval for $\beta_2 - \beta_1$ is given by

$$-0.599 \pm (2.018)(0.0769) \doteq -0.599 \pm 0.155 = (-0.754, -0.444).$$

Use of Quintic Polynomials

Next, we model the dependence of mean percentage of sugar content on variety and plot as the sum of an arbitary function of variety and a quintic polynomial function of plot; that is, we assume that the regression function has the form

$$\mu(x_1, x_2) = g_1(x_1) + g_2(x_2)$$

for $x_1 \in \mathcal{X}_1 = \{1, \ldots, 5\}$ and $x_2 \in \mathcal{X}_2 = \{-25, \ldots, 25\}$, where g_1 is an arbitrary function on $\mathcal{X}_1$ and g_2 is a quintic polynomial. The corresponding linear space G of functions on $\mathcal{X} = \mathcal{X}_1 \times \mathcal{X}_2$ is ten-dimensional, and the functions

$$I_1(x_1),\ I_2(x_1),\ I_3(x_1),\ I_4(x_1),\ I_5(x_5),\ \frac{x_2}{25},\ \left(\frac{x_2}{25}\right)^2,\ \left(\frac{x_2}{25}\right)^3,\ \left(\frac{x_2}{25}\right)^4,\ \left(\frac{x_2}{25}\right)^5$$

form a basis of this space. We write the regression function as

$$\mu(x_1, x_2) = \beta_1 I_1(x_1) + \beta_2 I_2(x_1) + \beta_3 I_3(x_1) + \beta_4 I_4(x_1) + \beta_5 I_5(x_5)$$

$$+ \beta_6 \frac{x_2}{25} + \beta_7 \left(\frac{x_2}{25}\right)^2 + \beta_8 \left(\frac{x_2}{25}\right)^3 + \beta_9 \left(\frac{x_2}{25}\right)^4 + \beta_{10} \left(\frac{x_2}{25}\right)^5$$

and its least-squares estimate as

$$\hat{\mu}(x_1, x_2) = \hat{\beta}_1 I_1(x_1) + \hat{\beta}_2 I_2(x_1) + \hat{\beta}_3 I_3(x_1) + \hat{\beta}_4 I_4(x_1) + \hat{\beta}_5 I_5(x_1)$$

$$+ \hat{\beta}_6 \frac{x_2}{25} + \hat{\beta}_7 \left(\frac{x_2}{25}\right)^2 + \hat{\beta}_8 \left(\frac{x_2}{25}\right)^3 + \hat{\beta}_9 \left(\frac{x_2}{25}\right)^4 + \hat{\beta}_{10} \left(\frac{x_2}{25}\right)^5 .$$

Using a computer to obtain the least-squares estimates of the regression coefficients, we get that

$$\hat{\mu}(x_1, x_2) \doteq 17.839 I_1(x_1) + 17.231 I_2(x_1)$$

$$+ 17.950 I_3(x_1) + 17.669 I_4(x_1) + 17.749 I_5(x_1)$$

$$- 1.540 \frac{x_2}{25} + 2.994 \left(\frac{x_2}{25}\right)^2 + 3.509 \left(\frac{x_2}{25}\right)^3$$

$$- 3.162 \left(\frac{x_2}{25}\right)^4 - 1.718 \left(\frac{x_2}{25}\right)^5 .$$

Table 10.4 shows the corresponding ANOVA table. The squared multiple correlation coefficient is given by

$$R^2 = \frac{\text{FSS}}{\text{TSS}} \doteq \frac{10.310}{11.327} \doteq .910.$$

Thus, 91.0% of the variation in sugar content can be explained by treating this quantity as the sum of an arbitrary function of variety and a quintic polynomial in plot.

TABLE   10.4

Source	SS
Fit	10.310
Residuals	1.017
Total	11.327

The plot of the residuals versus variety and versus the plot variable again are unrevealing. In particular, Figure 10.8 shows the latter plot.

FIGURE   **10.8**

Residuals from the quintic polynomial fit versus plot

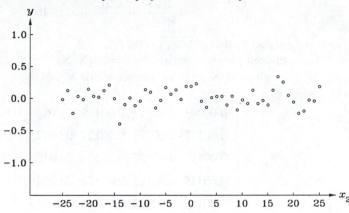

Figure 10.9 shows the graph of

$$\hat{\beta}_6 \frac{x_2}{25} + \hat{\beta}_7 \left(\frac{x_2}{25}\right)^2 + \hat{\beta}_8 \left(\frac{x_2}{25}\right)^3 + \hat{\beta}_9 \left(\frac{x_2}{25}\right)^4 + \hat{\beta}_{10} \left(\frac{x_2}{25}\right)^5 - \hat{c},$$

where

$$\hat{c} = \frac{1}{51} \sum_{x_2=-25}^{25} \left[\hat{\beta}_6 \frac{x_2}{25} + \hat{\beta}_7 \left(\frac{x_2}{25}\right)^2 + \hat{\beta}_8 \left(\frac{x_2}{25}\right)^3 + \hat{\beta}_9 \left(\frac{x_2}{25}\right)^4 + \hat{\beta}_{10} \left(\frac{x_2}{25}\right)^5 \right]$$

$$\doteq 0.354.$$

This graph again shows the up-down/up-down pattern.

FIGURE   **10.9**

Centered least-squares quintic polynomial fit

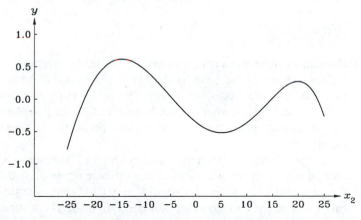

Under the homoskedastic regression model, the variance σ^2 is estimated by

$$S^2 = \frac{\text{RSS}}{n-p} \doteq \frac{1.017}{41} \doteq 0.0248,$$

so σ is estimated by $S \doteq \sqrt{0.0248} \doteq 0.157$.

The estimated variance–covariance $\hat{\mathbf{V}} = S^2(\mathbf{X}^T\mathbf{X})^{-1}$ of $\hat{\beta}_1, \ldots, \hat{\beta}_{10}$ is a 10×10 matrix. The estimated variance–covariance matrix $\hat{\mathbf{V}}_{11}$ of $\hat{\beta}_1, \ldots, \hat{\beta}_5$ is the submatrix of $\hat{\mathbf{V}}$ given by

$$\hat{\mathbf{V}}_{11} \doteq \begin{bmatrix} 0.00343 & 0.00117 & 0.00119 & 0.00119 & 0.00117 \\ 0.00117 & 0.00374 & 0.00127 & 0.00125 & 0.00122 \\ 0.00119 & 0.00127 & 0.00376 & 0.00128 & 0.00125 \\ 0.00119 & 0.00125 & 0.00128 & 0.00376 & 0.00127 \\ 0.00117 & 0.00122 & 0.00125 & 0.00127 & 0.00374 \end{bmatrix}.$$

Under the linear regression model, the change in the mean percentage of sugar content caused by switching from the standard variety to variety 2 on a given plot is given by $\mu(2, x_2) - \mu(1, x_2) = \beta_2 - \beta_1$, and its least-squares estimate is given by

$$\hat{\mu}(2, x_2) - \hat{\mu}(1, x_2) = \hat{\beta}_2 - \hat{\beta}_1 \doteq 17.231 - 17.839 = -0.608.$$

The standard error of this estimate is the nonnegative square root of

$$[-1, 1] \begin{bmatrix} 0.00343 & 0.00117 \\ 0.00117 & 0.00374 \end{bmatrix} \begin{bmatrix} -1 \\ 1 \end{bmatrix} \doteq 0.00343 + 0.00374 - 2(0.00117)$$

$$= 0.00483,$$

so $\text{SE}(\hat{\beta}_2 - \hat{\beta}_1) \doteq \sqrt{0.00483} \doteq 0.0695$. Now $t_{.975,41} \doteq 2.020$. Thus, under the corresponding normal linear regression model, the 95% confidence interval for $\beta_2 - \beta_1$ is given by

$$-0.608 \pm (2.020)(0.0695) \doteq -0.608 \pm 0.140 = (-0.748, -0.468).$$

Comparison of Fits

Overall, when we use a quintic polynomial instead of a quartic polynomial to model the dependence of the mean percentage of sugar content on plot, we get a slightly more complicated model, but apparently slightly more accurate least-squares estimates of and slightly shorter confidence intervals for parameters such as $\beta_2 - \beta_1$ that depend on the varieties of sugar beet but not on the 51 plots used to perform the experiment.

In the solution to Example 7.12(a) in Section 7.3, when we ignored the effect of plot on sugar content, we estimated the variance σ^2 by $S^2 \doteq 0.179$. In this section, we estimated σ^2 by $S^2 \doteq 0.0304$ and $S^2 \doteq 0.0248$ using quartic and quintic polynomials, respectively, to model the plot effect. In hindsight, it appears that by ignoring the plot effect in Section 7.3, we got an inflated estimate of σ^2.

Consider, more generally, the least-squares fit in a linear space G. Suppose this space does not contain the true regression function and, specifically, $\|\mu(\cdot) - \mu^*(\cdot)\| > 0$, where $\mu^*(\cdot)$ is the orthogonal projection of the regression function $\mu(\cdot)$ onto G. Then, by (28) in Section 10.2,

$$E(S^2) = E\left(\frac{\text{RSS}}{n-p}\right) = \sigma^2 + \frac{\|\mu(\cdot) - \mu^*(\cdot)\|^2}{n-p} > \sigma^2,$$

so S^2 tends to be an inflated estimate of σ^2.

An Alternative Design

In the sugar beet experiment, as described in Section 8.2, the varieties were assigned to the 51 plots in the systematic manner 1, 2, 3, 4, 5, ..., 1, 2, 3, 4, 5, 1. Suppose, instead, that variety 1 had been assigned to plots labeled from -25 to -15, variety 2 to plots labeled from -14 to -5, variety 3 to plots labeled from -4 to 5, variety 4 to plots labeled from 6 to 15, and variety 5 to plots labeled from 16 to 25. If we were to analyze the resulting experimental data without taking the plot effect into account, as was done in Chapter 7, the results would be substantially biased since some varieties would have systematically been assigned to plots that were relatively conducive to high sugar content and other varieties would have been assigned to plots that were relatively unconducive to high sugar content (see Section 11.4). If we would take the plot effect into account, however, as was done in this section, we could eliminate most of this bias. Nevertheless, the results would have been inferior to what we obtained from the actual data. To see this, suppose we again model the dependence of the mean percentage of sugar content on variety and plot as the sum of an arbitrary function of variety and a quintic function of plot, using the same ten basis functions as before. Also suppose, as before, that RSS $\doteq 1.017$ and hence that $S \doteq 0.157$. The estimated variance–covariance matrix $\hat{\mathbf{V}}_{11}$ of $\hat{\beta}_1, \ldots, \hat{\beta}_5$ would then be given by

$$\hat{\mathbf{V}}_{11} \doteq \begin{bmatrix} 0.05581 & 0.02403 & 0.00029 & -0.00404 & -0.00291 \\ 0.02403 & 0.01885 & -0.00036 & -0.00789 & -0.00395 \\ 0.00029 & -0.00036 & 0.00258 & 0.00131 & 0.00164 \\ -0.00404 & -0.00789 & 0.00131 & 0.02185 & 0.02716 \\ -0.00291 & -0.00395 & 0.00164 & 0.02716 & 0.05895 \end{bmatrix},$$

so the standard error for the least-squares estimate of the change in the mean percentage of sugar content caused by switching from the standard variety to variety 2 on a given plot would be given by

$$\text{SE}(\hat{\beta}_2 - \hat{\beta}_1) \doteq \sqrt{0.05581 + 0.01885 - 2(0.02403)} \doteq \sqrt{0.0266} \doteq 0.1631$$

and hence the 95% confidence interval for the change would be more than twice as wide as it was when based on the actual design. The explanation for the poor performance of the alternative design is that under the actual design the basis functions $I_j(x_1)$, $1 \le j \le 5$, are nearly orthogonal to the quintic polynomials in x_2, whereas under the alternative design they can be reasonably well approximated by such polynomials. (If one of these basis functions were exactly equal to a quintic polynomial in x_2, then the Gram matrix would have been noninvertible.)

Problems

10.42 Consider the least-squares fit to the sugar beet data having the form of the sum of an arbitrary function of variety and a quartic function of plot. Express the least-squares estimates of the regression coefficients relative to the basis $1, I_1(x_1), I_2(x_1), I_3(x_1), I_4(x_1), x_2, x_2^2, x_2^3, x_2^4$ of the corresponding linear space G in terms of the least-squares estimates $\hat{\beta}_1, \ldots, \hat{\beta}_9$ of the regression coefficients relative to the basis $I_1(x_1), I_2(x_1), I_3(x_1), I_4(x_1), I_5(x_1), x_2/25, (x_2/25)^2, (x_2/25)^3, (x_2/25)^4$ of this space.

10.43 Consider the least-squares fit to the sugar beet data having the form of the sum of an arbitrary function of variety and a quartic polynomial in plot. Let τ be the change in the mean percentage of sugar content caused by switching from variety 3 to variety 4 on a given plot and let $\hat{\tau}$ be the least-squares estimate of τ. Determine **(a)** $\hat{\tau}$; **(b)** $SE(\hat{\tau})$; **(c)** the 95% confidence interval for τ.

10.44 Consider the least-squares fit to the sugar beet data having the form of the sum of an arbitrary function of variety and a quintic polynomial function of plot. Let τ be the change in the mean percentage of sugar content caused by switching from variety 3 to variety 4 on a given plot and let $\hat{\tau}$ be the least-squares estimate of τ. Determine **(a)** $\hat{\tau}$; **(b)** $SE(\hat{\tau})$; **(c)** the 95% confidence interval for τ.

Problems 10.45–10.48 involve an experiment that compares a new variety (1) of wheat to a standard variety (2) on five plots laid in a row, with the varieties assigned to plots in the systematic manner 2, 1, 2, 1, 2. Let the variety be labeled by x_1, which takes on the values 1 and 2, and let the plot be labeled by x_2, which ranges from -2 to 2. The response variable is yield of wheat and the experimental data are shown in Table 10.5. The data used in Problem 7.30 in Section 7.3 was obtained from the above experimental data by ignoring the plot variable x_2.

TABLE   10.5

x_1	x_2	Y
2	-2	14
1	-1	22
2	0	20
1	1	26
2	2	23

10.45 Consider the linear regression model in which the plot variable is ignored and the mean yield is assumed to be an arbitrary function of variety. Here G is the space of all functions on $\mathcal{X}_1 = \{1, 2\}$. Consider the basis $1, I_1(x_1)$ of G, where $I_k(x_k) = \text{ind}(x_1 = k)$. The regression function can be written in terms of this basis as $\mu(x_1) = \beta_0 + \beta_1 I_1(x_1)$. Write the least-squares estimate of the regression function as $\hat{\mu}(x_1) = \hat{\beta}_0 + \hat{\beta}_1 I_1(x_1)$. **(a)** Interpret β_1. **(b)** Determine the design matrix $\mathbf{X}$, the Gram matrix $\mathbf{X}^T\mathbf{X}$, and its inverse. **(c)** Determine $(\mathbf{X}^T\mathbf{X})^{-1}\mathbf{X}^T$. **(d)** Determine $\hat{\beta}_0$ and $\hat{\beta}_1$. **(e)** Determine the residual sum of squares. **(f)** Plot the residuals against variety. What can you conclude from this plot? **(g)** Plot the residuals against the plot variable. What can you conclude from this plot?

Consider the normal linear regression model in which the mean yield is assumed to be the sum of an arbitrary function of variety and a linear function of plot; that is, it has the form $\mu(x_1, x_2) = g_1(x_1) + g_2(x)$ for $x_1 \in \mathcal{X}_1 = \{1, 2\}$ and $x_2 \in \mathcal{X}_2 = \{-2, -1, 0, 1, 2\}$, where g_1 an arbitrary function on $\mathcal{X}_1$ and g_2 is a linear function. The corresponding linear space G of functions on $\mathcal{X} = \mathcal{X}_1 \times \mathcal{X}_2$ is three-dimensional.

10.46 Consider the basis $1, I_1(x_1), x_2$ of G. Write the regression function in terms of this basis as $\mu(x_1, x_2) = \beta_0 + \beta_1 I_1(x_1) + \beta_2 x_2$ and its least-squares estimate as $\hat{\mu}(x_1, x_2) = \hat{\beta}_0 + \hat{\beta}_1 I_1(x_1) + \hat{\beta}_2 x_2$. **(a)** Interpret β_1. **(b)** Determine the design matrix $\mathbf{X}$, the Gram matrix $\mathbf{X}^T\mathbf{X}$, and its inverse. **(c)** Determine $(\mathbf{X}^T\mathbf{X})^{-1}\mathbf{X}^T$. **(d)** Determine $\hat{\beta}_0$, $\hat{\beta}_1$, and $\hat{\beta}_2$. **(e)** Show the corresponding ANOVA table. **(f)** Determine the squared multiple correlation coefficient. **(g)** Determine S^2 and S. **(h)** Determine $\text{SE}(\hat{\beta}_1)$ and the 95% confidence interval for β_1. **(i)** Plot the residuals against the plot variable. What can you conclude from this plot?

10.47 Consider the basis $1, x_1 - \frac{8}{5}, x_2$ of G. Write the regression function in terms of this basis as $\mu(x_1, x_2) = \beta_0 + \beta_1(x_1 - \frac{8}{5}) + \beta_2 x_2$ and its least-squares estimate as $\hat{\mu}(x_1, x_2) = \hat{\beta}_0 + \hat{\beta}_1(x_1 - \frac{8}{5}) + \hat{\beta}_2 x_2$. **(a)** Interpret β_1. **(b)** Show that the indicated basis is an orthogonal basis and determine the squared norms of the basis functions. **(c)** Determine $\hat{\beta}_0$, $\hat{\beta}_1$, and $\hat{\beta}_2$.

10.48 Consider the normal linear regression model in which the mean yield is assumed to be the sum of an arbitrary function of variety and a quadratic function of plot; that is, it has the form $\mu(x_1, x_2) = g_1(x_1) + g_2(x_2)$, for $x_1 \in \mathcal{X}_1 = \{1, 2\}$ and $x_2 \in \mathcal{X}_2 = \{-2, -1, 0, 1, 2\}$, where g_1 is an arbitrary function on $\mathcal{X}_1$ and g_2 is a quadratic polynomial. The corresponding linear space G of functions on $\mathcal{X} = \mathcal{X}_1 \times \mathcal{X}_2$ is four-dimensional. Consider the basis $1, x_1 - \frac{8}{5}, x_2, x_2^2 - 2$ of G. Write the regression function in terms of this basis as $\mu(x_1, x_2) = \beta_0 + \beta_1(x_1 - \frac{8}{5}) + \beta_2 x_2 + \beta_3(x_2^2 - 2)$ and its least-squares estimate as $\hat{\mu}(x_1, x_2) = \hat{\beta}_0 + \hat{\beta}_1(x_1 - \frac{8}{5}) + \hat{\beta}_2 x_2 + \hat{\beta}_3(x_2^2 - 2)$. **(a)** Interpret β_1. **(b)** Determine the design matrix $\mathbf{X}$, the Gram matrix $\mathbf{X}^T\mathbf{X}$, and its inverse. **(c)** Determine $(\mathbf{X}^T\mathbf{X})^{-1}\mathbf{X}^T$. **(d)** Determine $\hat{\beta}_0$, $\hat{\beta}_1$, $\hat{\beta}_2$, and $\hat{\beta}_3$. **(e)** Show the corresponding ANOVA table. **(f)** Determine the squared multiple correlation coefficient. **(g)** Determine S^2 and S. **(h)** Determine $\text{SE}(\hat{\beta}_1)$ and the 95% confidence interval for β_1. **(i)** Plot the residuals against the plot variable. What can you conclude from this plot?

10.5
Lube Oil Experiment

Consider the lube oil experiment as described in Section 8.2. Recall that there are four quantitative factors, referred to as A, B, C, and D, and that each factor takes on three levels, which we have assigned the coded values of -1 for low, 0 for medium, and 1 for high. We have also denoted the levels of A, B, C, and D by x_1, x_2, x_3, and x_4, respectively. The experiment consisted of the $3^{4-1} = 27$ factor-level combinations such that $x_1 + x_2 + x_3 + x_4 \equiv 2 \pmod 3$. If we ignore any of the four factors, then we can view the experiment as a complete factorial experiment with one run at each factor-level combination.

Quadratic Model Based on the First Three Factors

We first analyze the experiment by ignoring factor D and treating it as a complete factorial experiment with one run at each combination of a level of A, a level of B, and a level of C. Consider the linear regression model in which the model space is the ten-dimensional space G_0 of quadratic polynomials in the levels of the three factors. The first choice of a basis of G_0 that comes to mind is 1, x_1, x_1^2, x_2, x_2^2, x_3, x_3^2, $x_1 x_2$, $x_1 x_3$, $x_2 x_3$, but this is not an orthogonal basis. Indeed,

$$\langle 1, x_1^2 \rangle = \langle 1, x_2^2 \rangle = \langle 1, x_3^2 \rangle = 9[\,(-1)^2 + 0^2 + 1^2\,] = 18.$$

We can remedy the indicated lack of orthogonality by subtracting the common average value $\frac{2}{3}$ from each of the basis functions x_1^2, x_2^2, and x_3^2 [see the solution to Example 9.26(a) in Section 9.4]. This leads to the orthogonal basis 1, x_1, $x_1^2 - \frac{2}{3}$, x_2, $x_2^2 - \frac{2}{3}$, x_3, $x_3^2 - \frac{2}{3}$, $x_1 x_2$, $x_1 x_3$, $x_2 x_3$ of G_0. (Since there are nine runs at each level of A,

$$\langle 1, x_1 \rangle = 9[\,1 \cdot (-1) + 1 \cdot 0 + 1 \cdot 1\,] = 0$$

and

$$\langle 1, x_1^2 - \tfrac{2}{3} \rangle = 9[\,1 \cdot \tfrac{1}{3} + 1 \cdot (-\tfrac{2}{3}) + 1 \cdot \tfrac{1}{3}\,] = 0.$$

Since there are three runs at each combination of a level of A and a level of B,

$$\langle x_1, x_2 \rangle = 3\{(-1) \cdot (-1) + (-1) \cdot 0 + (-1) \cdot 1 + 0 \cdot (-1) + 0 \cdot 0 + 0 \cdot 1$$
$$+ 1 \cdot (-1) + 1 \cdot 0 + 1 \cdot 1\}$$
$$= 3(-1 + 0 + 1)(-1 + 0 + 1)$$
$$= 0$$

and

$$\langle x_1, x_2^2 - \tfrac{2}{3} \rangle = 3\{(-1) \cdot \tfrac{1}{3} + (-1) \cdot (-\tfrac{2}{3}) + (-1) \cdot \tfrac{1}{3} + 0 \cdot \tfrac{1}{3} + 0 \cdot (-\tfrac{2}{3}) + 0 \cdot \tfrac{1}{3}$$
$$+ 1 \cdot \tfrac{1}{3} + 1 \cdot (-\tfrac{2}{3}) + 1 \cdot \tfrac{1}{3}\}$$

$$= 3(-1 + 0 + 1)(\tfrac{1}{3} - \tfrac{2}{3} + \tfrac{1}{3})$$

$$= 0.$$

Replacing x_2 by x_3 and then x_1 by x_3, we see that the indicated basis is indeed an orthogonal basis. Such orthogonal relationships will be treated more systematically in Sections 11.2 and 11.3.) The squared norms of the basis functions are given by

$$\|1\|^2 = 27 \cdot 1 = 27,$$

$$\|x_1\|^2 = \|x_2\|^2 = \|x_3\|^3 = 9[(-1)^2 + 0^2 + 1^2] = 18,$$

$$\|x_1^2 - \tfrac{2}{3}\|^2 = \|x_2^2 - \tfrac{2}{3}\|^2 = \|x_3^2 - \tfrac{2}{3}\|^2 = 9[(\tfrac{1}{3})^2 + (-\tfrac{2}{3})^2 + (\tfrac{1}{3})^2] = 6,$$

and

$$\|x_1 x_2\|^2 = \|x_1 x_3\|^2$$

$$= \|x_2 x_3\|^2$$

$$= 3\{((-1)\cdot(-1))^2 + ((-1)\cdot 0)^2 + ((-1)\cdot 1)^2 + (0\cdot(-1))^2 + (0\cdot 0)^2$$

$$\quad + (0\cdot 1)^2 + (1\cdot(-1))^2 + (1\cdot 0)^2 + (1\cdot 1)^2\}$$

$$= 3[(-1)^2 + 0^2 + 1^2]^2$$

$$= 12.$$

Let $\beta_0, \ldots, \beta_9$ be the regression coefficients relative to the indicated orthogonal basis, so that (under the assumption that the regression function is in G_0)

$$\mu(x_1, x_2, x_3) = \beta_0 + \beta_1 x_1 + \beta_2(x_1^2 - \tfrac{2}{3}) + \beta_3 x_2 + \beta_4(x_2^2 - \tfrac{2}{3}) + \beta_5 x_3$$

$$\quad + \beta_6(x_3^2 - \tfrac{2}{3}) + \beta_7 x_1 x_2 + \beta_8 x_1 x_3 + \beta_9 x_2 x_3.$$

Let $\hat{\beta}_0, \ldots, \hat{\beta}_9$ be the least-squares estimates of the regression coefficients. Then

$$\hat{\mu}(x_1, x_2, x_3) = \hat{\beta}_0 + \hat{\beta}_1 x_1 + \hat{\beta}_2(x_1^2 - \tfrac{2}{3}) + \hat{\beta}_3 x_2 + \hat{\beta}_4(x_2^2 - \tfrac{2}{3}) + \hat{\beta}_5 x_3$$

$$\quad + \hat{\beta}_6(x_3^2 - \tfrac{2}{3}) + \hat{\beta}_7 x_1 x_2 + \hat{\beta}_8 x_1 x_3 + \hat{\beta}_9 x_2 x_3.$$

According to Problem 10.6 in Section 10.1,

$$\hat{\beta}_j = \frac{\langle g_j, \bar{Y}(\cdot)\rangle}{\|g_j\|^2} = \frac{\sum_i g_j(x_i) Y_i}{\|g_j\|^2}, \qquad 0 \le j \le 9,$$

where $g_0 = 1, g_1 = x_1, \ldots, g_9 = x_2 x_3$ are the basis functions in the indicated order. Using the experimental data in Section 8.2, we get the values shown in Table 10.6 for the statistics $\sum_i g_j(x_i) Y_i$, $0 \le j \le 9$. Thus,

$$\hat{\beta}_0 = \bar{Y} = \frac{809.5}{27} \doteq 29.981,$$

$$\hat{\beta}_1 = \frac{1}{\|x_1\|^2} \sum_i x_{i1} Y_i = \frac{281.4}{18} \doteq 15.633,$$

TABLE   10.6

Statistic	Value
$\sum_i Y_i$	809.500
$\sum_i x_{i1} Y_i$	281.400
$\sum_i (x_{i1}^2 - \frac{2}{3}) Y_i$	24.133
$\sum_i x_{i2} Y_i$	218.300
$\sum_i (x_{i2}^2 - \frac{2}{3}) Y_i$	−26.967
$\sum_i x_{i3} Y_i$	315.100
$\sum_i (x_{i3}^2 - \frac{2}{3}) Y_i$	−4.767
$\sum_i x_{i1} x_{i2} Y_i$	55.800
$\sum_i x_{i1} x_{i3} Y_i$	120.500
$\sum_i x_{i2} x_{i3} Y_i$	76.300

$$\hat{\beta}_2 = \frac{1}{\|x_1^2 - \frac{2}{3}\|^2} \sum_i (x_{i1}^2 - \tfrac{2}{3}) Y_i \doteq \frac{24.133}{6} \doteq 4.022,$$

$$\hat{\beta}_3 = \frac{1}{\|x_2\|^2} \sum_i x_{i2} Y_i = \frac{218.3}{18} \doteq 12.128,$$

$$\hat{\beta}_4 = \frac{1}{\|x_2^2 - \frac{2}{3}\|^2} \sum_i (x_{i2}^2 - \tfrac{2}{3}) Y_i \doteq -\frac{26.967}{6} \doteq -4.494,$$

$$\hat{\beta}_5 = \frac{1}{\|x_3\|^2} \sum_i x_{i3} Y_i = \frac{315.1}{18} \doteq 17.506,$$

$$\hat{\beta}_6 = \frac{1}{\|x_3^2 - \frac{2}{3}\|^2} \sum_i (x_{i3}^2 - \tfrac{2}{3}) Y_i \doteq -\frac{4.767}{6} \doteq -0.794,$$

$$\hat{\beta}_7 = \frac{1}{\|x_1 x_2\|^2} \sum_i x_{i1} x_{i2} Y_i = \frac{55.8}{12} = 4.65,$$

$$\hat{\beta}_8 = \frac{1}{\|x_1 x_3\|^2} \sum_i x_{i1} x_{i3} Y_i = \frac{120.5}{12} \doteq 10.042,$$

and

$$\hat{\beta}_9 = \frac{1}{\|x_2 x_3\|^2} \sum_i x_{i2} x_{i3} Y_i = \frac{76.3}{18} \doteq 6.358.$$

The total sum of squares is given by

$$\text{TSS} = \sum_i (Y_i - \bar{Y})^2 \doteq 15544.46,$$

and the fitted sum of squares is given by

$$\text{FSS} = \|\hat{\mu}(\cdot) - \bar{Y}\|^2$$
$$= \|\hat{\beta}_1 x_1 + \cdots + \hat{\beta}_9 x_2 x_3\|^2$$

$$= \hat{\beta}_1^2 \|x_1\|^2 + \cdots + \hat{\beta}_9^2 \|x_2 x_3\|^2$$

$$= 18(\hat{\beta}_1^2 + \hat{\beta}_3^2 + \hat{\beta}_5^2) + 6(\hat{\beta}_2^2 + \hat{\beta}_4^2 + \hat{\beta}_6^2) + 12(\hat{\beta}_7^2 + \hat{\beta}_8^2 + \hat{\beta}_9^2)$$

$$\doteq 14739.40.$$

Thus, by subtraction, we get the ANOVA table shown in Table 10.7. Under the homoskedastic linear regression model, the variance σ^2 is estimated by

$$S^2 = \frac{\text{RSS}}{n-p} \doteq \frac{805.06}{17} \doteq 47.356,$$

and σ is estimated by $S \doteq \sqrt{47.356} \doteq 6.882$.

Under the homoskedastic regression model, the variances of the least-squares estimates of $\beta_0, \ldots, \beta_9$ are given [see Problem 10.35(b) in Section 10.3] by

$$\text{var}(\hat{\beta}_0) = \text{var}(\bar{Y}) = \frac{\sigma^2}{27},$$

$$\text{var}(\hat{\beta}_5) = \text{var}(\hat{\beta}_3) = \text{var}(\hat{\beta}_1) = \frac{\sigma^2}{\|x_1\|^2} = \frac{\sigma^2}{18},$$

$$\text{var}(\hat{\beta}_6) = \text{var}(\hat{\beta}_4) = \text{var}(\hat{\beta}_2) = \frac{\sigma^2}{\|x_1^2 - \frac{2}{3}\|^2} = \frac{\sigma^2}{6},$$

and

$$\text{var}(\hat{\beta}_9) = \text{var}(\hat{\beta}_8) = \text{var}(\hat{\beta}_7) = \frac{\sigma^2}{\|x_1 x_2\|^2} = \frac{\sigma^2}{12}.$$

Under the homoskedastic linear regression model, the standard errors of $\hat{\beta}_0, \ldots, \hat{\beta}_9$ are given by $\text{SE}(\hat{\beta}_0) = S/\sqrt{27} \doteq 1.324$, $\text{SE}(\hat{\beta}_1) = S/\sqrt{18} \doteq 1.622$, and so forth.

Now $t_{.975,17} \doteq 2.110$. Under the normal linear regression model, the 95% confidence interval for β_0 is given by

$$\hat{\beta}_0 \pm t_{.975,17} \, \text{SE}(\hat{\beta}_0) \doteq 29.981 \pm (2.110)(1.324)$$

$$\doteq 29.981 \pm 2.794 \doteq (27.187, 32.776),$$

the 95% confidence interval for β_1 is given by

$$\hat{\beta}_1 \pm t_{.975,17} \, \text{SE}(\hat{\beta}_1) \doteq 15.633 \pm (2.110)(1.622)$$

$$\doteq 15.633 \pm 3.422 = (12.211, 19.055)$$

and so forth. In this manner, we get the results shown in Table 10.8.

TABLE 10.7

Source	SS
Fit	14739.40
Residuals	805.06
Total	15544.46

TABLE   10.8

Term	Coefficient	Estimate	SE	95% CI
1	β_0	29.981	1.324	(27.187, 32.776)
x_1	β_1	15.633	1.622	(12.211, 19.055)
$x_1^2 - 2/3$	β_2	4.022	2.809	(−1.905, 9.950)
x_2	β_3	12.128	1.622	(8.706, 15.550)
$x_2^2 - 2/3$	β_4	−4.494	2.809	(−10.422, 1.433)
x_3	β_5	17.506	1.622	(14.083, 20.928)
$x_3^2 - 2/3$	β_6	−0.794	2.809	(−6.722, 5.133)
$x_1 x_2$	β_7	4.650	1.987	(0.459, 8.841)
$x_1 x_3$	β_8	10.042	1.987	(5.850, 14.233)
$x_2 x_3$	β_9	6.358	1.987	(2.167, 10.550)

EXAMPLE **10.12**   Consider the lube oil experiment and the normal linear regression model corresponding to G_0.

(a) Determine the least-squares estimate of and 95% confidence interval for the change in the mean response when the first factor is switched from its low level to its high level with the other factors being kept at their medium levels.

(b) Determine the least-squares estimate of and 95% confidence interval for the change in the mean response when the first factor is switched from its low level to its high level with the other factors being kept at their high levels.

(c) Discuss the answers to parts (a) and (b).

Solution   (a) The change τ_0 in the mean response when x_1 is switched from −1 to 1 with $x_2 = x_3 = 0$ is given by $\tau_0 = 2\beta_1$. The least-squares estimate of this change is given by $\hat{\tau}_0 = 2\hat{\beta}_1 \doteq 31.267$, and the standard error of this estimate is given by $\text{SE}(\hat{\tau}_0) = 2\,\text{SE}(\hat{\beta}_1) \doteq 3.244$. Thus, the 95% confidence interval for τ_0 is given by

$$\hat{\tau}_0 \pm t_{.975,17}\,\text{SE}(\hat{\tau}_0) \doteq 31.267 \pm (2.110)(3.244)$$

$$\doteq 31.267 \pm 6.844 \doteq (24.422, 38.111).$$

(b) The change τ_1 in the mean response when x_1 is switched from −1 to 1 with $x_2 = x_3 = 1$ is given by $\tau_1 = 2\beta_1 + 2\beta_7 + 2\beta_8$. The least-squares estimate of this change is given by $\hat{\tau}_1 = 2\hat{\beta}_1 + 2\hat{\beta}_7 + 2\hat{\beta}_8 \doteq 60.650$. Since $\hat{\beta}_1$, $\hat{\beta}_7$, and $\hat{\beta}_8$ are uncorrelated [see Problem 10.35(a) in Section 10.3], $\hat{\beta}_1$ has variance $\sigma^2/18$, and $\hat{\beta}_7$ and $\hat{\beta}_8$ have common variance $\sigma^2/12$, we conclude that

$$\text{var}(\hat{\tau}_1) = 4[\,\text{var}(\hat{\beta}_1) + \text{var}(\hat{\beta}_7) + \text{var}(\hat{\beta}_8)\,] = 4\sigma^2\left(\frac{1}{18} + \frac{1}{12} + \frac{1}{12}\right) = \frac{8}{9}\sigma^2$$

and hence that $\mathrm{SE}(\hat{\tau}_1) = S\sqrt{8/9} \doteq 6.488$. Thus, the 95% confidence interval for τ_1 is given by

$$\hat{\tau}_1 \pm t_{.975,17}\,\mathrm{SE}(\hat{\tau}_1) \doteq 60.650 \pm (2.110)(6.488)$$

$$\doteq 60.650 \pm 13.689$$

$$\doteq (46.961, 74.339).$$

(c) The answers to parts (a) and (b) indicate that switching the first factor from its low level to its high level has a much greater effect on the mean response when the other factors are kept at their high levels than when they are kept at their medium levels. To pursue this issue further, consider the linear parameter $\tau_1 - \tau_0 = 2\beta_7 + 2\beta_8$. The least-squares estimate of this parameter is given by $\hat{\tau}_1 - \hat{\tau}_0 = 2\hat{\beta}_7 + 2\hat{\beta}_8 \doteq 29.838$, and its standard error is given by $\mathrm{SE}(\hat{\tau}_1 - \hat{\tau}_0) = S\sqrt{2/3} \doteq 5.619$. Consequently, the 95% confidence interval for $\tau_1 - \tau_0$ is given by

$$\hat{\tau}_1 - \hat{\tau}_0 \pm t_{.975,17}\,\mathrm{SE}(\hat{\tau}_1 - \hat{\tau}_0) \doteq 29.383 \pm (2.110)(5.619)$$

$$\doteq 29.383 \pm 11.855$$

$$\doteq (17.529, 41.238). \quad\blacksquare$$

Quadratic Model Based on All Four Factors

We now include all four factors in the analysis of the lube oil experiment and consider the linear regression model in which the model space is the 15-dimensional space G of quadratic polynomials in the levels of the four factors. Also, we extend the basis of the G_0 that was used in the previous analysis to the basis $1, x_1, x_1^2 - \frac{2}{3}$, $x_2, x_2^2 - \frac{2}{3}, x_3, x_3^2 - \frac{2}{3}, x_4, x_4^2 - \frac{2}{3}, x_1x_2, x_3x_4, x_1x_3, x_2x_4, x_1x_4, x_2x_3$ of G. Here x_1x_2 is *not* orthogonal to x_3x_4, x_1x_3 is *not* orthogonal to x_2x_4, and x_1x_4 is *not* orthogonal to x_2x_3. Indeed,

$$\langle x_1x_2, x_3x_4\rangle = \langle x_1x_3, x_2x_4\rangle = \langle x_1x_4, x_2x_3\rangle = \sum_i x_{i1}x_{i2}x_{i3}x_{i4} = -3. \tag{35}$$

These are the only exceptions to orthogonality. The squared norms of the functions in this basis are given by

$$\|1\|^2 = 27,$$

$$\|x_1\|^2 = \|x_2\|^2 = \|x_3\|^2 = \|x_4\|^2 = 18,$$

$$\left\|x_1^2 - \tfrac{2}{3}\right\|^2 = \left\|x_2^2 - \tfrac{2}{3}\right\|^2 = \left\|x_3^2 - \tfrac{2}{3}\right\|^2 = \left\|x_4^2 - \tfrac{2}{3}\right\|^2 = 6,$$

$$\|x_1x_2\|^2 = \|x_3x_4\|^2 = \|x_1x_3\|^2 = \|x_2x_4\|^2 = \|x_1x_4\|^2 = \|x_2x_3\|^2 = 12.$$

Let $\beta_0, \ldots, \beta_{14}$ be the regression coefficients relative to the indicated basis and let $\hat{\beta}_0, \ldots, \hat{\beta}_{14}$ respectively denote the least-squares estimates of these coefficients. Then

$$\mu(x_1, x_2, x_3, x_4) = \beta_0 + \beta_1 x_1 + \cdots + \beta_{14} x_2 x_3,$$

and

$$\hat{\mu}(x_1, x_2, x_3, x_4) = \hat{\beta}_0 + \hat{\beta}_1 x_1 + \cdots + \hat{\beta}_{14} x_2 x_3.$$

For simplicity in notation, let the 15 functions $1, \ldots, x_2 x_3$ in the indicated basis be denoted by $g_0, \ldots, g_{14}$, respectively. Since $\hat{\mu}(\cdot)$ is the orthogonal projection of $\bar{Y}(\cdot)$ onto G, we have $\langle g, \hat{\mu}(\cdot) \rangle = \langle g, \bar{Y}(\cdot) \rangle$ for $g \in G$ or, equivalently,

$$\langle g_l, \hat{\mu}(\cdot) \rangle = \langle g_l, \bar{Y}(\cdot) \rangle, \qquad 0 \le l \le 14.$$

Now

$$\langle g_l, \hat{\mu}(\cdot) \rangle = \left\langle g_l, \sum_j \hat{\beta}_j g_j \right\rangle = \sum_j \hat{\beta}_j \langle g_l, g_j \rangle, \qquad 0 \le l \le 14,$$

so the normal equations for $\hat{\beta}_0, \ldots, \hat{\beta}_{14}$ can be written as

$$\sum_j \hat{\beta}_j \langle g_l, g_j \rangle = \langle g_l, \bar{Y}(\cdot) \rangle, \qquad 0 \le l \le 14. \tag{36}$$

Let $0 \le j \le 8$. Then g_j is orthogonal to g_l for $0 \le l \le 14$ with $l \ne j$. We conclude from (36) that $\hat{\beta}_j \|g_j\|^2 = \langle g_j, \bar{Y}(\cdot) \rangle$ for $0 \le j \le 8$ and hence that

$$\hat{\beta}_j = \frac{\langle g_j, \bar{Y}(\cdot) \rangle}{\|g_j\|^2} = \frac{\sum_i g_j(x_i) Y_i}{\|g_j\|^2}, \qquad 0 \le j \le 8.$$

Consequently, for $0 \le j \le 8$, the coefficient of g_j for the least-squares fit in G coincides with the coefficient of g_j for the least-squares fit in G_0. (Note that some of these coefficients have been relabeled for convenience in the present analysis.)

The functions $g_9 = x_1 x_2$ and $g_{10} = x_3 x_4$ are not orthogonal to each other, but they are orthogonal to each of the remaining basis functions $g_0, \ldots, g_8, g_{11}, \ldots, g_{14}$. Thus, we conclude from (36) that

$$\|g_9\|^2 \hat{\beta}_9 + \langle g_9, g_{10} \rangle \hat{\beta}_{10} = \langle g_9, \bar{Y}(\cdot) \rangle$$

and

$$\langle g_9, g_{10} \rangle \hat{\beta}_9 + \|g_{10}\|^2 \hat{\beta}_{10} = \langle g_{10}, \bar{Y}(\cdot) \rangle;$$

that is,

$$12 \hat{\beta}_9 - 3 \hat{\beta}_{10} = \langle x_1 x_2, \bar{Y}(\cdot) \rangle,$$

and

$$-3 \hat{\beta}_9 + 12 \hat{\beta}_{10} = \langle x_3 x_4, \bar{Y}(\cdot) \rangle.$$

The unique solution to these two equations in the two unknowns $\hat{\beta}_9$ and $\hat{\beta}_{10}$ is given by

$$\hat{\beta}_9 = \frac{4}{45} \langle x_1 x_2, \bar{Y}(\cdot) \rangle + \frac{1}{45} \langle x_3 x_4, \bar{Y}(\cdot) \rangle$$

and

$$\hat{\beta}_{10} = \frac{1}{45} \langle x_1 x_2, \bar{Y}(\cdot) \rangle + \frac{4}{45} \langle x_3 x_4, \bar{Y}(\cdot) \rangle.$$

In the same manner, we conclude that the coefficients of x_1x_3 and x_2x_4 are given by

$$\hat{\beta}_{11} = \frac{4}{45}\langle x_1x_3, \bar{Y}(\cdot)\rangle + \frac{1}{45}\langle x_2x_4, \bar{Y}(\cdot)\rangle$$

and

$$\hat{\beta}_{12} = \frac{1}{45}\langle x_1x_3, \bar{Y}(\cdot)\rangle + \frac{4}{45}\langle x_2x_4, \bar{Y}(\cdot)\rangle$$

and that the coefficients of x_1x_4 and x_2x_3 are given by

$$\hat{\beta}_{13} = \frac{4}{45}\langle x_1x_4, \bar{Y}(\cdot)\rangle + \frac{1}{45}\langle x_2x_3, \bar{Y}(\cdot)\rangle$$

and

$$\hat{\beta}_{14} = \frac{1}{45}\langle x_1x_4, \bar{Y}(\cdot)\rangle + \frac{4}{45}\langle x_2x_3, \bar{Y}(\cdot)\rangle.$$

(Incidently, the derivation of the formulas for $\hat{\beta}_0, \ldots, \hat{\beta}_{14}$ implicitly shows that G is identifiable.)

Using the experimental data in Section 8.2, we get the values shown in Table 10.9 for those statistics $\sum_i g_j(x_i)Y_i$, $0 \le j \le 14$, that were not previously reported.

TABLE 10.9

Statistic	Value
$\sum_i x_{i4}Y_i$	62.000
$\sum_i (x_{i4}^2 - \frac{2}{3})Y_i$	-20.467
$\sum_i x_{i3}x_{i4}Y_i$	-11.800
$\sum_i x_{i2}x_{i4}Y_i$	-48.300
$\sum_i x_{i1}x_{i4}Y_i$	27.900

The coefficients $\hat{\beta}_0, \ldots, \hat{\beta}_6$ of 1, x_1, $x_1^2 - \frac{2}{3}$, x_2, $x_2^2 - \frac{2}{3}$, x_3, and $x_3^2 - \frac{2}{3}$, respectively, are as reported previously. The coefficients of x_4 and $x_4^2 - \frac{2}{3}$ are given, respectively, by

$$\hat{\beta}_7 = \frac{1}{\|x_4\|^2}\sum_i x_{i4}Y_i = \frac{62}{18} \doteq 3.444$$

and

$$\hat{\beta}_8 = \frac{1}{\|x_4 - \frac{2}{3}\|^2}\sum_i (x_{i4} - \frac{2}{3})Y_i \doteq -\frac{20.467}{6} \doteq -3.411.$$

The coefficients of x_1x_2, x_3x_4, x_1x_3, x_2x_4, x_1x_4, and x_2x_3 are given, respectively, by

$$\hat{\beta}_9 = \frac{4}{45}\sum_i x_{i1}x_{i2}Y_i + \frac{1}{45}\sum_i x_{i3}x_{i4}Y_i = \frac{4}{45}(55.8) - \frac{1}{45}(11.8) \doteq 4.698,$$

$$\hat{\beta}_{10} = \frac{1}{45}\sum_i x_{i1}x_{i2}Y_i + \frac{4}{45}\sum_i x_{i3}x_{i4}Y_i = \frac{1}{45}(55.8) - \frac{4}{45}(11.8) \doteq 0.191,$$

$$\hat{\beta}_{11} = \frac{4}{45} \sum_i x_{i1} x_{i3} Y_i + \frac{1}{45} \sum_i x_{i2} x_{i4} Y_i = \frac{4}{45}(120.5) - \frac{1}{45}(48.3) \doteq 9.638,$$

$$\hat{\beta}_{12} = \frac{1}{45} \sum_i x_{i1} x_{i3} Y_i + \frac{4}{45} \sum_i x_{i2} x_{i4} Y_i = \frac{1}{45}(120.5) - \frac{4}{45}(48.3) \doteq -1.616,$$

$$\hat{\beta}_{13} = \frac{4}{45} \sum_i x_{i1} x_{i4} Y_i + \frac{1}{45} \sum_i x_{i2} x_{i3} Y_i = \frac{4}{45}(27.9) + \frac{1}{45}(76.3) \doteq 4.176,$$

and

$$\hat{\beta}_{14} = \frac{1}{45} \sum_i x_{i1} x_{i4} Y_i + \frac{4}{45} \sum_i x_{i2} x_{i3} Y_i = \frac{1}{45} \cdot 27.9 + \frac{4}{45} \cdot 76.3 \doteq 7.402.$$

Note that the value 4.698 of the coefficient of $x_1 x_2$ for the least-squares fit in G differs slightly from its value 4.650 for the least-squares fit in G_0, the reason being that the new basis function $x_3 x_4$ is not orthogonal to $x_1 x_2$. Similarly, the values 9.638 and 7.402 of the coefficients of $x_1 x_3$ and $x_2 x_3$, respectively, for the least-squares fit in G differ somewhat from their values value 10.042 and 6.358, respectively, for the least-squares fit in G_0.

The fitted sum of squares is given by

$$\begin{aligned}
\text{FSS} &= \|\hat{\mu}(\cdot) - \bar{Y}\|^2 \\
&= \|\hat{\beta}_1 x_1 + \cdots + \hat{\beta}_{14} x_2 x_3\|^2 \\
&= \hat{\beta}_1^2 \|x_1\|^2 + \cdots + \hat{\beta}_8^2 \|x_4^2 - \tfrac{2}{3}\|^2 + \|\hat{\beta}_9 x_1 x_2 + \hat{\beta}_{10} x_3 x_4\|^2 \\
&\quad + \|\hat{\beta}_{11} x_1 x_3 + \hat{\beta}_{12} x_2 x_4\|^2 + \|\hat{\beta}_{13} x_1 x_4 + \hat{\beta}_{14} x_2 x_3\|^2.
\end{aligned}$$

Now

$$\begin{aligned}
\hat{\beta}_1^2 \|x_1\|^2 &+ \cdots + \hat{\beta}_8^2 \left\| x_4^2 - \tfrac{2}{3} \right\|^2 \\
&= 18(\hat{\beta}_1^2 + \hat{\beta}_3^2 + \hat{\beta}_5^2 + \hat{\beta}_7^2) + 6(\hat{\beta}_2^2 + \hat{\beta}_4^2 + \hat{\beta}_6^2 + \hat{\beta}_8^2).
\end{aligned}$$

Moreover,

$$\begin{aligned}
\|\hat{\beta}_9 x_1 x_2 + \hat{\beta}_{10} x_3 x_4\|^2 &= \hat{\beta}_9^2 \|x_1 x_2\|^2 + \hat{\beta}_{10}^2 \|x_3 x_4\|^2 + 2\hat{\beta}_9 \hat{\beta}_{10} \langle x_1 x_2, x_3 x_4 \rangle \\
&= 12(\hat{\beta}_9^2 + \hat{\beta}_{10}^2) - 6\hat{\beta}_9 \hat{\beta}_{10}.
\end{aligned}$$

Similarly,

$$\|\hat{\beta}_{11} x_1 x_3 + \hat{\beta}_{12} x_2 x_4\|^2 = 12(\hat{\beta}_{11}^2 + \hat{\beta}_{12}^2) - 6\hat{\beta}_{11} \hat{\beta}_{12},$$

and

$$\|\hat{\beta}_{12} x_1 x_4 + \hat{\beta}_{14} x_2 x_3\|^2 = 12(\hat{\beta}_{13}^2 + \hat{\beta}_{14}^2) - 6\hat{\beta}_{13} \hat{\beta}_{14}.$$

Thus,

$$\begin{aligned}
\text{FSS} &= 18(\hat{\beta}_1^2 + \hat{\beta}_3^2 + \hat{\beta}_5^2 + \hat{\beta}_7^2) + 6(\hat{\beta}_2^2 + \hat{\beta}_4^2 + \hat{\beta}_6^2 + \hat{\beta}_8^2) \\
&\quad + 12(\hat{\beta}_9^2 + \hat{\beta}_{10}^2 + \hat{\beta}_{11}^2 + \hat{\beta}_{12}^2 + \hat{\beta}_{13}^2 + \hat{\beta}_{14}^2) \\
&\quad - 6(\hat{\beta}_9 \hat{\beta}_{10} + \hat{\beta}_{11} \hat{\beta}_{12} + \hat{\beta}_{13} \hat{\beta}_{14}).
\end{aligned}$$

Using the numerical values of $\hat{\beta}_1, \ldots, \hat{\beta}_{14}$ to compute the fitted sum of squares and using the value of TSS already reported, we get the ANOVA table shown in Table 10.10. Under the homoskedastic linear regression model, the variance σ^2 is estimated by

$$S^2 = \frac{\text{RSS}}{n-p} \doteq \frac{295.77}{12} \doteq 24.647,$$

and σ is estimated by $S \doteq \sqrt{24.647} \doteq 4.965$.

TABLE 10.10

Source	SS
Fit	15248.69
Residuals	295.77
Total	15544.46

Under the homoskedastic regression model,

$$\text{var}(\hat{\beta}_j) = \frac{\sigma^2}{\|g_j\|^2}, \qquad 0 \le j \le 8,$$

and hence

$$\text{SE}(\hat{\beta}_j) = \frac{S}{\|g_j\|}, \qquad 0 \le j \le 8.$$

In particular, $\text{SE}(\hat{\beta}_0) = S/\sqrt{27} \doteq 0.955$,

$$\text{SE}(\hat{\beta}_1) = \text{SE}(\hat{\beta}_3) = \text{SE}(\hat{\beta}_5) = \text{SE}(\hat{\beta}_7) = S/\sqrt{18} \doteq 1.170,$$

and

$$\text{SE}(\hat{\beta}_2) = \text{SE}(\hat{\beta}_4) = \text{SE}(\hat{\beta}_6) = \text{SE}(\hat{\beta}_8) = S/\sqrt{6} \doteq 2.027.$$

Moreover,

$$\hat{\beta}_9 = \tfrac{4}{45}\langle x_1 x_2, \bar{Y}(\cdot)\rangle + \tfrac{1}{45}\langle x_3 x_4, \bar{Y}(\cdot)\rangle = \left\langle \tfrac{4}{45}x_1 x_2 + \tfrac{1}{45}x_3 x_4, \ \bar{Y}(\cdot)\right\rangle,$$

so [recall Equation (22) in Section 10.2]

$$\text{var}(\hat{\beta}_9) = \sigma^2 \|\tfrac{4}{45}x_1 x_2 + \tfrac{1}{45}x_3 x_4\|^2$$

$$= \frac{\sigma^2}{2025}\left(16\|x_1 x_2\|^2 + \|x_3 x_4\|^2 + 8\langle x_1 x_2, x_3 x_4\rangle\right)$$

$$= \frac{\sigma^2}{2025}(16 \cdot 12 + 12 - 8 \cdot 3)$$

$$= \frac{4}{45}\sigma^2$$

and hence

$$SE(\hat{\beta}_9) = S\sqrt{4/45} \doteq 1.480.$$

In the same manner, we get that

$$\text{var}(\hat{\beta}_j) = \frac{4}{45}\sigma^2, \qquad 9 \le j \le 14,$$

and hence that

$$SE(\hat{\beta}_j) = S\sqrt{4/45} \doteq 1.480, \qquad 9 \le j \le 14.$$

Now $t_{.975,12} \doteq 2.179$. Thus, for $9 \le j \le 14$, the 95% confidence interval for β_j is given by

$$\hat{\beta}_j \pm t_{.975,12}\, SE(\hat{\beta}_j) \doteq \hat{\beta}_j \pm (2.179)(1.480) \doteq \hat{\beta}_j \pm 3.225.$$

Table 10.11 shows the estimates, standard errors, and 95% confidence intervals corresponding to the various regression coefficients and their least-squares estimates.

TABLE 10.11

Term	Coefficient	Estimate	SE	95% CI
1	β_0	29.981	0.955	(27.900, 32.063)
x_1	β_1	15.633	1.170	(13.084, 18.183)
$x_1^2 - 2/3$	β_2	4.022	2.027	(−0.394, 8.438)
x_2	β_3	12.128	1.170	(9.578, 14.677)
$x_2^2 - 2/3$	β_4	−4.494	2.027	(−8.910, −0.078)
x_3	β_5	17.506	1.170	(14.956, 20.055)
$x_3^2 - 2/3$	β_6	−0.794	2.027	(−5.210, 3.622)
x_4	β_7	3.444	1.170	(0.895, 5.994)
$x_4^2 - 2/3$	β_8	−3.411	2.027	(−7.827, 1.005)
$x_1 x_2$	β_9	4.698	1.480	(1.473, 7.923)
$x_3 x_4$	β_{10}	0.191	1.480	(−3.034, 3.416)
$x_1 x_3$	β_{11}	9.638	1.480	(6.413, 12.863)
$x_2 x_4$	β_{12}	−1.616	1.480	(−4.841, 1.609)
$x_1 x_4$	β_{13}	4.176	1.480	(0.951, 7.401)
$x_2 x_3$	β_{14}	7.402	1.480	(4.177, 10.627)

EXAMPLE 10.13 Rework Example 10.12 with G being the space of quadratic polynomials in the levels of the four factors.

Solution **(a)** The change τ_0 in the mean response when x_1 is switched from -1 to 1 with $x_2 = x_3 = x_4 = 0$ is given by $\tau_0 = 2\beta_1$. The least-squares estimate of this change is given by $\hat{\tau}_0 = 2\hat{\beta}_1 \doteq 31.267$, and the standard error of this estimate is given by $\mathrm{SE}(\hat{\tau}_0) = 2\,\mathrm{SE}(\hat{\beta}_1) \doteq 2.340$. Thus, the 95% confidence interval for τ_0 is given by

$$\hat{\tau}_0 \pm t_{.975,12}\,\mathrm{SE}(\hat{\tau}_0) \doteq 31.267 \pm (2.179)(2.340)$$

$$\doteq 31.267 \pm 5.099$$

$$= (26.168, 36.366).$$

(b) The change τ_1 in the mean response when x_1 is switched from -1 to 1 with $x_2 = x_3 = x_4 = 1$ is given by $\tau_1 = 2\beta_1 + 2\beta_9 + 2\beta_{11} + 2\beta_{13}$. The least-squares estimate of this change is given by $\hat{\tau}_1 = 2\hat{\beta}_1 + 2\hat{\beta}_9 + 2\hat{\beta}_{11} + 2\hat{\beta}_{13} \doteq 68.289$. Now

$$\hat{\beta}_1 = \left\langle \tfrac{1}{18}x_1, \bar{Y}(\cdot) \right\rangle \quad \text{and} \quad \hat{\beta}_9 = \left\langle \tfrac{4}{45}x_1x_2 + \tfrac{1}{45}x_3x_4, \bar{Y}(\cdot) \right\rangle.$$

Since x_1 is orthogonal to x_1x_2 and to x_3x_4, we conclude under the assumptions of the homoskedastic regression model [see Problem 10.27(b) in Section 10.2] that $\mathrm{cov}(\hat{\beta}_1, \hat{\beta}_9) = 0$. In the same manner, we get that $\hat{\beta}_1$, $\hat{\beta}_9$, $\hat{\beta}_{11}$, and $\hat{\beta}_{13}$ are uncorrelated. Thus,

$$\mathrm{var}(\hat{\tau}_1) = 4[\,\mathrm{var}(\hat{\beta}_1) + \mathrm{var}(\hat{\beta}_9) + \mathrm{var}(\hat{\beta}_{11}) + \mathrm{var}(\hat{\beta}_{13})\,]$$

$$= 4\sigma^2(\tfrac{1}{18} + 3 \cdot \tfrac{4}{45})$$

$$= \tfrac{116}{90}\sigma^2$$

and hence

$$\mathrm{SE}(\hat{\tau}_1) = S\sqrt{116/90} \doteq 5.636.$$

Consequently, the 95% confidence interval for τ_1 is given by

$$\hat{\tau}_1 \pm t_{.975,12}\,\mathrm{SE}(\hat{\tau}_1) \doteq 68.289 \pm (2.179)(5.636)$$

$$\doteq 68.289 \pm 12.280$$

$$\doteq (56.008, 80.569).$$

(c) The answers to parts (a) and (b) indicate that switching the first factor from its low level to its high level has a much greater effect on the mean response when the other factors are kept at their high levels than when they are kept at their medium levels. To pursue this issue further, consider the linear parameter $\tau_1 - \tau_0 = 2\beta_9 + 2\beta_{11} + 2\beta_{13}$. The least-squares estimate of this parameter is given by $\hat{\tau}_1 - \hat{\tau}_0 = 2\hat{\beta}_9 + 2\hat{\beta}_{11} + 2\hat{\beta}_{13} \doteq 37.022$, and its standard error is given by $\mathrm{SE}(\hat{\tau}_1 - \hat{\tau}_0) = 4S/\sqrt{15} \doteq 5.127$. Consequently, the 95% confidence interval for $\tau_1 - \tau_0$ is given by

$$\hat{\tau}_1 - \hat{\tau}_0 \pm t_{.975,12}\,\mathrm{SE}(\hat{\tau}_1 - \hat{\tau}_0) \doteq 37.022 \pm (2.179)(5.127)$$

$$\doteq 37.022 \pm 11.172$$

$$\doteq (25.851, 48.194). \quad \blacksquare$$

Problems

10.49   Consider the least-squares fit to the lube oil data in the space of quadratic polynomials in the levels of the four factors. How do the least-squares estimates of the regression coefficients change when we switch from the basis $1, x_1, x_1^2 - \frac{2}{3}, x_2, x_2^2 - \frac{2}{3}, x_3, x_3^2 - \frac{2}{3}, x_4, x_4^2 - \frac{2}{3}, x_1 x_2, \ldots, x_2 x_3$ in the text to the basis $1, x_1, x_1^2, x_2, x_2^2, x_3, x_3^2, x_4, x_4^2, x_1 x_2, \ldots, x_2 x_3$?

10.50   Consider the lube oil experiment and the normal linear regression model corresponding to G. Determine the least-squares estimate of and 95% confidence interval for the change in the mean response when the first factor is switched from its low level to its high level with the other factors being kept at their low levels.

10.51   Let $\hat{\beta}_9, \ldots, \hat{\beta}_{14}$ be the respective coefficients of $x_1 x_2, \ldots, x_2 x_3$ for the least-squares fit in G to the lube oil data. Under the homoskedastic regression model, verify that $\text{cov}(\hat{\beta}_9, \hat{\beta}_{10}) = \text{cov}(\hat{\beta}_{11}, \hat{\beta}_{12}) = \text{cov}(\hat{\beta}_{13}, \hat{\beta}_{14}) = \sigma^2/45$. [*Hint:* Use the result of Problem 10.27(a) in Section 10.2.]

10.52   Consider the lube oil experiment, the normal linear regression model corresponding to the space G of linear functions in the levels of the four factors, and the basis $1, x_1, x_2, x_3, x_4$ of G. Let $\beta_0, \ldots, \beta_4$ be the regression coefficients relative to the indicated basis and let $\hat{\beta}_0, \ldots, \hat{\beta}_4$ be the least-squares estimates of these coefficients. **(a)** Determine $\hat{\beta}_0, \ldots, \hat{\beta}_4$. **(b)** Determine the fitted sum of squares. **(c)** Show the corresponding ANOVA table. **(d)** Determine the standard errors of $\hat{\beta}_0, \ldots, \hat{\beta}_4$. **(e)** Determine the 95% confidence intervals for $\beta_0, \ldots, \beta_4$. **(f)** Determine the least-squares estimate of and 95% confidence interval for the change in the mean response when the first factor is switched from its low level to its high level with the other three factors held fixed.

10.53   Consider an experiment involving two factors, let G be the space of quadratic polynomials on $\mathbb{R}^2$, consider the basis $1, x_1, x_2, x_1^2, x_2^2, x_1 x_2$ of this space, assume that the regression function is in this space, and let $\beta_0, \ldots, \beta_5$ denote the regression coefficients relative to the indicated basis. Determine the change in the mean response when x_1 is changed from level a to level b with x_2 kept at level c.

Problems 10.54–10.58 involve an experiment with three factors each at three levels, which are assigned the coded values -1 (low), 0 (medium), and 1 (high). Let x_m be the coded value of the mth factor for $m = 1, 2, 3$ and let the experiment consist of the one-third fraction of the complete factorial experiment containing the $3^{3-1} = 9$ factor-level combinations such that $x_1 + x_2 + x_3 \equiv 0 \pmod{3}$, with one run at each such factor-level combination. Table 10.12 shows the experimental data. Observe that there is one run corresponding to each ordered pair of levels of each ordered pair of factors. The total sum of squares is given by $\text{TSS} = \sum_i (Y_i - \bar{Y})^2 = 2496$. Table 10.13 shows some other useful statistics.

10.54   Let G be the space of quadratic polynomials in the levels of the first two factors and consider the basis $1, x_1, x_1^2 - \frac{2}{3}, x_2, x_2^2 - \frac{2}{3}, x_1 x_2$ of G. Let $\beta_0, \ldots, \beta_5$ denote the corresponding regression coefficients and let $\hat{\beta}_0, \ldots, \hat{\beta}_5$, respectively, denote the least-squares estimates of these coefficients. **(a)** Show that the indicated basis is an orthogonal basis and determine the squared norms of the functions in the basis.

TABLE 10.12

x_1	x_2	x_3	Y
−1	−1	−1	48
0	−1	1	74
1	−1	0	88
−1	0	1	48
0	0	0	50
1	0	−1	52
−1	1	0	42
0	1	−1	32
1	1	1	70

TABLE 10.13

Statistic	Value
$\sum_i Y_i$	504
$\sum_i x_{i1} Y_i$	72
$\sum_i (x_{i1}^2 - \frac{2}{3}) Y_i$	12
$\sum_i x_{i2} Y_i$	−66
$\sum_i (x_{i2}^2 - \frac{2}{3}) Y_i$	18
$\sum_i x_{i3} Y_i$	60
$\sum_i (x_{i3}^2 - \frac{2}{3}) Y_i$	−12
$\sum_i x_{i1} x_{i2} Y_i$	−12
$\sum_i x_{i1} x_{i3} Y_i$	18
$\sum_i x_{i2} x_{i3} Y_i$	12

(b) Determine the least-squares estimates of the regression coefficients. **(c)** Determine the fitted and residual sums of squares. **(d)** Determine the estimates S^2 and S of σ^2 and σ, respectively. **(e)** Determine the variances of $\hat{\beta}_0, \ldots, \hat{\beta}_5$. **(f)** Determine the 95% confidence intervals for $\beta_0, \ldots, \beta_5$. **(g)** Let τ denote the change in the mean response when the first factor is switched from its low level to its high level with the second factor being kept at its high level. Determine τ explicitly in terms of the regression coefficients. **(h)** Determine the variance of the least-squares estimate of τ, the numerical values of this estimate and its standard error, and the 95% confidence interval for τ.

10.55 Let G be the space of additive quadratic polynomials in the levels of the three factors and consider the orthogonal basis $1, x_1, x_1^2 - \frac{2}{3}, x_2, x_2^2 - \frac{2}{3}, x_3, x_3^2 - \frac{2}{3}$ of G. Determine **(a)** the least-squares estimates of the regression coefficients; **(b)** the fitted and residual sums of squares.

10.56 Let G be the space of functions in the levels of the three factors spanned by $1, x_1, x_2, x_3, x_1 x_2$. **(a)** Show that the indicated functions form an orthogonal basis of G

and determine the squared norms of these functions. **(b)** Determine the least-squares estimates $\hat{\beta}_0, \ldots, \hat{\beta}_4$ of the regression coefficients $\beta_0, \ldots, \beta_4$ relative to the indicated basis. **(c)** Determine the fitted and residual sums of squares. **(d)** Determine the estimates S^2 and S of σ^2 and σ, respectively. **(e)** Determine the variances of $\hat{\beta}_0, \ldots, \hat{\beta}_4$. **(f)** Determine the 95% confidence intervals for $\beta_0, \ldots, \beta_4$. **(g)** Let τ denote the change in the mean response when the first factor is switched from its low level to its high level with the other factors being kept at their high level. Determine τ explicitly in terms of the regression coefficients. **(h)** Determine the variance of the least-squares estimate of τ, the numerical values of this estimate and its standard error, and the 95% confidence interval for τ.

10.57 Let G be the space of functions in the levels of the three factors spanned by the functions $1, x_1, x_2, x_3, x_1x_2, x_1x_3$. **(a)** Determine the Gram matrix of these functions. **(b)** Show that these functions form a basis of G. [*Hint:* Use Problem 9.49 in Section 9.4.] **(c)** Is this basis an orthogonal basis? **(d)** Determine the least-squares estimates $\hat{\beta}_0, \ldots, \hat{\beta}_5$ of the regression coefficients $\beta_0, \ldots, \beta_5$ relative to the indicated basis. **(e)** Determine the fitted and residual sums of squares. **(f)** Determine the estimates S^2 and S of σ^2 and σ, respectively. **(g)** Determine the variances of $\hat{\beta}_0, \ldots, \hat{\beta}_5$. **(h)** Determine the 95% confidence intervals for $\beta_0, \ldots, \beta_5$. **(i)** Let τ denote the change in the mean response when the first factor is switched from its low level to its high level with the other factors being kept at their high level. Determine τ explicitly in terms of the regression coefficients. **(j)** Determine the variance of the least-squares estimate of τ, the numerical values of this estimate and its standard error, and the 95% confidence interval for τ.

10.58 Let G be the space of functions in the levels of the three factors spanned by the functions $1, x_1, x_2, x_3, x_1x_2, x_1x_3, x_2x_3$. **(a)** Determine the Gram matrix of these functions. **(b)** Show that these functions form a basis of G. [*Hint:* Use Problem 9.49 in Section 9.4.] **(c)** Is this basis an orthogonal basis? **(d)** Determine the least-squares estimates $\hat{\beta}_0, \ldots, \hat{\beta}_6$ of the regression coefficients $\beta_0, \ldots, \beta_6$ relative to the indicated basis. **(e)** Determine the fitted and residual sums of squares. **(f)** Determine the estimates S^2 and S of σ^2 and σ, respectively. **(g)** Determine the variances of $\hat{\beta}_0, \ldots, \hat{\beta}_6$. **(h)** Determine the 95% confidence intervals for $\beta_0, \ldots, \beta_6$. **(i)** Let τ denote the change in the mean response when the first factor is switched from its low level to its high level with the other factors being kept at their high level. Determine τ explicitly in terms of the regression coefficients. **(j)** Determine the variance of the least-squares estimate of τ, the numerical values of this estimate and its standard error, and the 95% confidence interval for τ.

10.6
The t Test

Consider a normal linear regression model in which G is an identifiable p-dimensional linear space with $p < n$. Let RSS denote the residual sum of squares for the least-squares estimate $\hat{\mu}(\cdot)$ in G of the regression function $\mu(\cdot)$ and let $S^2 = \text{RSS}/(n-p)$ be the corresponding unbiased estimate of σ^2. Consider a linear parameter $\tau = \mathbf{c}^T\beta$, where β is the coefficient vector of $\mu(\cdot)$ relative to a given basis of G and $\mathbf{c}$ is a nonzero p-dimensional column vector. Let $\hat{\tau} = \mathbf{c}^T \hat{\beta}$ denote the least-squares estimate of τ, and let $\tau_0 \in \mathbb{R}$. Observe that if $\tau = \tau_0$, then $t = (\hat{\tau} - \tau_0)/\operatorname{SE}(\hat{\tau})$ has the t distribution with $n - p$ degrees of freedom. The material on t tests in Sections 7.4 and 7.5 in the context of the normal d-sample model extends to the normal linear regression model in an obvious manner.

Consider first the hypothesis that $\tau \leq \tau_0$. Let $0 < \alpha < 1$. The critical (or rejection) region for the t test of size α of the hypothesis is given by $t \geq t_{1-\alpha, n-p}$. Here α is the maximum probability that the t statistic lies in the critical region, where the maximum is taken over all values in G of the regression function that are compatible with the hypothesis (that is, such that $\mathbf{c}^T \beta \leq \tau_0$) and over all $\sigma > 0$. In particular, if $\tau = \tau_0$, then the probability that the t statistic lies in the critical region equals α. The acceptance region for the test is the complement $t < t_{1-\alpha, n-p}$ of the critical region. The t statistic lies in the acceptance region if and only if the $100(1 - \alpha)\%$ lower confidence bound $\hat{\tau} - t_{1-\alpha, n-p} \operatorname{SE}(\hat{\tau})$ for τ is less than τ_0. The P-value for the test equals $1 - t_{n-p}(t)$. As usual, the t statistic lies in the critical region for the t test of size α if α is greater than or equal to the P-value and in the acceptance region if α is less than the P-value.

Consider next the hypothesis that $\tau \geq \tau_0$. The critical (or rejection) region for the t test of size α of the hypothesis is given by $t \leq -t_{1-\alpha, n-p}$. The t statistic lies in the acceptance region $t > -t_{1-\alpha, n-p}$ for the test if and only if the $100(1 - \alpha)\%$ upper confidence bound $\hat{\tau} + t_{1-\alpha, n-p} \operatorname{SE}(\hat{\tau})$ for τ is greater than τ_0. The P-value for the test equals $t_{n-p}(t)$.

Consider now the hypothesis that $\tau = \tau_0$. The critical region for the t test of size α of the hypothesis is given by $|t| \geq t_{1-\alpha/2, n-p}$. If the hypothesis is valid, then the probability that the t statistic lies in the critical region equals α. The t statistic lies in the acceptance region $|t| < t_{1-\alpha/2, n-p}$ for the test if and only if τ_0 lies in the $100(1 - \alpha)\%$ confidence interval $\tau \pm t_{1-\alpha/2, n-p} \operatorname{SE}(\hat{\tau})$ for τ. The P-value for the test equals $2[\,1 - t_{n-p}(|t|)\,]$.

Let G_0 be a subspace of G having dimension $p - 1$. The t test can be used to test the hypothesis that the regression function is in G_0. Specifically, let $g_1, \ldots, g_p$ be a basis of G such that $g_1, \ldots, g_{p-1}$ is a basis of G_0. (If $p = 1$ and hence $G_0 = \{0\}$, let g_1 be any nonzero function in G, which necessarily forms a basis of this space.) The regression function is given in terms of this basis by $\mu(\cdot) = \beta_1 g_1 + \cdots + \beta_p g_p$. The hypothesis that the regression function is in G_0 coincides with the hypothesis that $\beta_p = 0$. The t statistic for testing this hypothesis is given by $t = \hat{\beta}_p / \operatorname{SE}(\hat{\beta}_p)$, where $\hat{\beta}_p$ is the least-squares estimate of β_p.

Sugar Beet Experiment

EXAMPLE 10.14   Consider the sugar beet experiment with the dependence of the mean percentage of sugar content on plot being ignored. Test the hypothesis that the mean percentage of sugar content for variety 3 is less than or equal to that for the standard variety.

Solution   In the normal linear regression model corresponding to the test based on all five varieties, G is the space of all functions on $\mathcal{X} = \{1, 2, 3, 4, 5\}$, the residual sum of squares is given by $\operatorname{RSS} = \operatorname{WSS} \doteq 8.242$, and $S = \sqrt{\operatorname{RSS}/46} \doteq \sqrt{8.242/46} \doteq$

0.423. Consider the basis $I_1, \ldots, I_5$ of G. The regression function can be written in terms of this basis as $\mu(\cdot) = \beta_1 I_1 + \cdots + \beta_5 I_5$, where $\beta_1 = \mu_1, \ldots, \beta_5 = \mu_5$, and its least-squares estimate can be written as $\hat{\mu}(\cdot) = \hat{\beta}_1 I_1 + \cdots + \hat{\beta}_5 I_5$, where $\hat{\beta}_1 = \hat{\mu}_1 = \bar{Y}_1, \ldots, \hat{\beta}_5 = \hat{\mu}_5 = \bar{Y}_5$. The estimates $\hat{\beta}_1, \ldots, \hat{\beta}_5$ are independent, and their variances are given by $\text{var}(\hat{\beta}_1) = \sigma^2/11$ and $\text{var}(\hat{\beta}_2) = \cdots = \text{var}(\hat{\beta}_5) = \sigma^2/10$. The least-squares estimate of the linear parameter $\tau = \beta_3 - \beta_1 = \mu_3 - \mu_1$ is given by $\hat{\tau} = \hat{\beta}_3 - \hat{\beta}_1 = \bar{Y}_3 - \bar{Y}_1 \doteq 18.320 - 18.136 \doteq 0.184$, and its variance is given by

$$\text{var}(\hat{\tau}) = \text{var}(\hat{\beta}_3 - \hat{\beta}_1) = \text{var}(\hat{\beta}_1) + \text{var}(\hat{\beta}_3) = \sigma^2(\tfrac{1}{11} + \tfrac{1}{10}),$$

so its standard error is given by

$$\text{SE}(\hat{\tau}) = S\sqrt{\tfrac{1}{11} + \tfrac{1}{10}} \doteq 0.185.$$

The t statistic for testing the hypothesis that $\beta_3 \leq \beta_1$ or, equivalently, that $\tau \leq 0$ is given by

$$t = \frac{\hat{\tau}}{\text{SE}(\hat{\tau})} \doteq \frac{0.184}{0.185} \doteq 0.993.$$

The corresponding P-value is given by $1 - t_{46}(0.993) \doteq .163$, so the hypothesis is accepted for the test of any conventional size. This analysis is almost identical to the analysis in terms of the normal five-sample model, which was given in the solution to Examples 7.13 and 7.15 in Section 7.4.

Consider, instead, the basis $1, I_1, I_2, I_3, I_4$ of G. The regression function can be written in terms of this basis as $\mu(\cdot) = \beta_0 + \beta_1 I_1 + \beta_2 I_2 + \beta_3 I_3 + \beta_4 I_4$, where $\beta_0 = \mu_5$, $\beta_1 = \mu_1 - \mu_5$, $\beta_2 = \mu_2 - \mu_5$, $\beta_3 = \mu_3 - \mu_5$ and $\beta_4 = \mu_4 - \mu_5$ [see the solution to Example 8.5(b) in Section 8.3]. Similarly, the least-squares estimate of the regression function can be written as $\hat{\mu}(\cdot) = \hat{\beta}_0 + \hat{\beta}_1 I_1 + \hat{\beta}_2 I_2 + \hat{\beta}_3 I_3 + \hat{\beta}_4 I_4$, where $\hat{\beta}_0 = \bar{Y}_5 = 18.13$, $\hat{\beta}_1 = \bar{Y}_1 - \bar{Y}_5 \doteq 0.006$, $\hat{\beta}_2 = \bar{Y}_2 - \bar{Y}_5 = -0.55$, $\hat{\beta}_3 = \bar{Y}_3 - \bar{Y}_5 = 0.19$, and $\hat{\beta}_4 = \bar{Y}_4 - \bar{Y}_5 = -0.07$. The least-squares estimate of the linear parameter $\tau = \mu_3 - \mu_1 = \beta_3 - \beta_1$ is given by $\hat{\tau} = \hat{\beta}_3 - \hat{\beta}_1 \doteq 0.19 - 0.006 = 0.184$. Moreover, the Gram matrix is given by

$$\mathbf{X}^T\mathbf{X} = \begin{bmatrix} 51 & 11 & 10 & 10 & 10 \\ 11 & 11 & 0 & 0 & 0 \\ 10 & 0 & 10 & 0 & 0 \\ 10 & 0 & 0 & 10 & 0 \\ 10 & 0 & 0 & 0 & 10 \end{bmatrix},$$

whose inverse is given by

$$(\mathbf{X}^T\mathbf{X})^{-1} = \frac{1}{110} \begin{bmatrix} 11 & -11 & -11 & -11 & -11 \\ -11 & 21 & 11 & 11 & 11 \\ -11 & 11 & 22 & 11 & 11 \\ -11 & 11 & 11 & 22 & 11 \\ -11 & 11 & 11 & 11 & 22 \end{bmatrix}.$$

The variance–covariance matrix of $\hat{\beta}_0, \ldots, \hat{\beta}_4$ is given by

$$
\mathbf{V} = \frac{\sigma^2}{110}
\begin{bmatrix}
11 & -11 & -11 & -11 & -11 \\
-11 & 21 & 11 & 11 & 11 \\
-11 & 11 & 22 & 11 & 11 \\
-11 & 11 & 11 & 22 & 11 \\
-11 & 11 & 11 & 11 & 22
\end{bmatrix},
$$

which is estimated by

$$
\hat{\mathbf{V}} = \frac{S^2}{110}
\begin{bmatrix}
11 & -11 & -11 & -11 & -11 \\
-11 & 21 & 11 & 11 & 11 \\
-11 & 11 & 22 & 11 & 11 \\
-11 & 11 & 11 & 22 & 11 \\
-11 & 11 & 11 & 11 & 22
\end{bmatrix}.
$$

Thus, the variance–covariance matrix of $\hat{\beta}_1$ and $\hat{\beta}_3$ equals

$$
\frac{\sigma^2}{110}
\begin{bmatrix}
21 & 11 \\
11 & 22
\end{bmatrix},
$$

which is estimated by

$$
\frac{S^2}{110}
\begin{bmatrix}
21 & 11 \\
11 & 22
\end{bmatrix}.
$$

The variance of $\tau = \beta_3 - \beta_1$ equals

$$
\frac{\sigma^2}{110}[-1\ 1]
\begin{bmatrix}
21 & 11 \\
11 & 22
\end{bmatrix}
\begin{bmatrix}
-1 \\
1
\end{bmatrix}
= \frac{21}{110}\sigma^2.
$$

Thus,

$$
\text{SE}(\hat{\tau}) = S\sqrt{\frac{21}{110}} \doteq 0.423\sqrt{\frac{21}{110}} \doteq 0.185,
$$

so the t statistic for testing the hypothesis that $\tau \leq 0$ is given by

$$
t = \frac{\hat{\tau}}{\text{SE}(\hat{\tau})} \doteq \frac{0.184}{0.185} \doteq 0.993,
$$

as before. ■

Next, we modify Example 10.14 to take the systematic effect of the plot variable into account.

EXAMPLE 10.15   Consider again the sugar beet experiment, but suppose that the regression function is the sum of an arbitrary function of variety and a quintic polynomial function of plot.

Test the hypothesis that switching from the standard variety to variety 3 on a given plot does not cause the mean percentage of sugar content to increase.

Solution   As in Section 10.4, we write the regression function as

$$\mu(x_1, x_2) = \beta_0 + \beta_1 I_1(x_1) + \beta_2 I_2(x_1) + \beta_3 I_3(x_1) + \beta_4 I_4(x_1)$$
$$+ \beta_5 \frac{x_2}{25} + \beta_6 \left(\frac{x_2}{25}\right)^2 + \beta_7 \left(\frac{x_2}{25}\right)^3 + \beta_8 \left(\frac{x_2}{25}\right)^4 + \beta_9 \left(\frac{x_2}{25}\right)^5$$

and its least-squares estimate as

$$\hat{\mu}(x_1, x_2) = \hat{\beta}_0 + \hat{\beta}_1 I_1(x_1) + \hat{\beta}_2 I_2(x_1) + \hat{\beta}_3 I_3(x_1) + \hat{\beta}_4 I_4(x_1)$$
$$+ \hat{\beta}_5 \frac{x_2}{25} + \hat{\beta}_6 \left(\frac{x_2}{25}\right)^2 + \hat{\beta}_7 \left(\frac{x_2}{25}\right)^3 + \hat{\beta}_8 \left(\frac{x_2}{25}\right)^4 + \hat{\beta}_9 \left(\frac{x_2}{25}\right)^5$$
$$\doteq 17.749 + 0.090 I_1(x_1) - 0.518 I_2(x_1) + 0.201 I_3(x_1) - 0.070 I_4(x_1)$$
$$- 1.540 \frac{x_2}{25} + 2.994 \left(\frac{x_2}{25}\right)^2 + 3.509 \left(\frac{x_2}{25}\right)^3$$
$$- 3.162 \left(\frac{x_2}{25}\right)^4 - 1.718 \left(\frac{x_2}{25}\right)^5.$$

The change in the mean percentage of sugar content caused by switching from the standard variety to variety 3 on a given plot is given by $\tau = \mu(3, x_2) - \mu(1, x_2) = \beta_3 - \beta_1$, and its least-squares estimate is given by $\hat{\tau} = \hat{\beta}_3 - \hat{\beta}_1 \doteq 0.201 - 0.090 = 0.111$.

The estimated variance–covariance matrix of $\hat{\beta}_1, \ldots, \hat{\beta}_4$ was shown in Section 10.4. Accordingly, the standard error of $\hat{\tau} = \hat{\beta}_3 - \hat{\beta}_1$ is the nonnegative square root of

$$[-1, 1] \begin{bmatrix} 0.00483 & 0.00250 \\ 0.00250 & 0.00500 \end{bmatrix} \begin{bmatrix} -1 \\ 1 \end{bmatrix} \doteq 0.00482,$$

so $\mathrm{SE}(\hat{\tau}) \doteq \sqrt{0.00482} \doteq 0.0695$. The t statistic corresponding to the hypothesis $\tau \le 0$ of interest is given by

$$t = \frac{\hat{\tau}}{\mathrm{SE}(\hat{\tau})} \doteq \frac{0.111}{0.0695} \doteq 1.596.$$

The corresponding P-value is given by $1 - t_{41}(1.596) \doteq .059$. Therefore, t lies in the critical region for the test of size α if $\alpha \ge .059$ and in the acceptance region if $\alpha < .059$. In particular, it lies in the acceptance region for the test of size .05, but in the critical region for the test of size .06. ∎

EXAMPLE **10.16**   Consider the sugar beet experiment as treated in Example 10.15. Justify thinking of the component of the regression function involving plot as a quintic polynomial rather than as a quartic polynomial.

Solution   Recall that β_9 is the coefficient of the quintic term $(x_1/25)^5$. The t statistic for testing the hypothesis that $\beta_9 = 0$ is given by $t = \hat{\beta}_9 / \mathrm{SE}(\hat{\beta}_9)$, and the P-value for the test is given by $P\text{-value} = 2[1 - t_{41}(|t|)]$. Now $\hat{\beta}_9 \doteq -1.718$. Moreover, $\mathrm{SE}(\hat{\beta}_9)$ is the positive square root of the last diagonal entry of the estimated variance–covariance matrix $\hat{V}$ of $\hat{\beta}$. By computer, this entry equals 0.280, so $\text{SE}(\hat{\beta}_9) \doteq \sqrt{0.280} \doteq 0.529$. Consequently, $t \doteq -1.718/0.529 \doteq -3.246$, and hence the P-value is given by $2[\,1 - t_{41}(3.246)\,] \doteq .002$. Since this P-value is so close to zero, we are justified in keeping the quintic term in the model.   ∎

Lube Oil Experiment

Consider the lube oil experiment, the normal linear regression model in which G is the space of quadratic polynomials in the levels of the four factors, and the basis of G that was used in Section 10.5. According to that discussion, the least-squares estimate $\hat{\beta}_1$ of the coefficient β_1 of x_1 is given by $\hat{\beta}_1 \doteq 15.633$, and its standard error is given by $\text{SE}(\hat{\beta}_1) \doteq 1.170$. Thus, the t statistic for testing the hypothesis that $\beta_1 = 0$ is given by $t = \hat{\beta}_1 / \text{SE}(\hat{\beta}_1) \doteq 13.360$, and the P-value is given by $2[\,1 - t_{12}(13.360)\,] \doteq .000$. Similarly, we get the other new results in Table 10.14.

An inspection of this table suggests dropping the term $x_3 x_4$ from the model and obtaining the least-squares fit in the 14-dimensional space spanned by $1, x_1, x_1^2 - \frac{2}{3}, x_2, x_2^2 - \frac{2}{3}, x_3, x_3^2 - \frac{2}{3}, x_4, x_4^2 - \frac{2}{3}, x_1 x_2, x_1 x_3, x_2 x_4, x_1 x_4, x_2 x_3$. In so doing, we get the results shown in Table 10.15.

Note that the least-squares estimate of the coefficient β_9 of $x_1 x_2$ has changed, but that none of the other coefficients has changed (other than that of $x_3 x_4$, of course).

TABLE   10.14

Term	Coefficient	Estimate	SE	t	P-value
1	β_0	29.981	0.955	31.380	.000
x_1	β_1	15.633	1.170	13.360	.000
$x_1^2 - \frac{2}{3}$	β_2	4.022	2.027	1.985	.071
x_2	β_3	12.128	1.170	10.364	.000
$x_2^2 - \frac{2}{3}$	β_4	−4.494	2.027	−2.218	.047
x_3	β_5	17.506	1.170	14.960	.000
$x_3^2 - \frac{2}{3}$	β_6	−0.794	2.027	−0.392	.702
x_4	β_7	3.444	1.170	2.944	.012
$x_4^2 - \frac{2}{3}$	β_8	−3.411	2.027	−1.683	.118
$x_1 x_2$	β_9	4.698	1.480	3.174	.008
$x_3 x_4$	β_{10}	0.191	1.480	0.129	.899
$x_1 x_3$	β_{11}	9.638	1.480	6.511	.000
$x_2 x_4$	β_{12}	−1.616	1.480	−1.091	.296
$x_1 x_4$	β_{13}	4.176	1.480	2.821	.015
$x_2 x_3$	β_{14}	7.402	1.480	5.001	.000

TABLE   10.15

Term	Est. Coef.	SE	t	P-value
1	29.981	0.919	32.638	.000
x_1	15.633	1.125	13.896	.000
$x_1^2 - \frac{2}{3}$	4.022	1.949	2.064	.060
x_2	12.128	1.125	10.780	.000
$x_2^2 - \frac{2}{3}$	−4.494	1.949	−2.306	.038
x_3	17.506	1.125	15.560	.000
$x_3^2 - \frac{2}{3}$	−0.794	1.949	−0.408	.690
x_4	3.444	1.125	3.062	.009
$x_4^2 - \frac{2}{3}$	−3.411	1.949	−1.751	.104
$x_1 x_2$	4.650	1.378	3.375	.005
$x_1 x_3$	9.638	1.423	6.772	.000
$x_2 x_4$	−1.616	1.423	−1.135	.277
$x_1 x_4$	4.176	1.423	2.934	.012
$x_2 x_3$	7.402	1.423	5.202	.000

The explanation is that the functions $x_1 x_2$ and $x_3 x_4$ are not orthogonal to each other, but each of these functions is orthogonal to all the other functions in the original basis.

An inspection of the last table suggests removing the term $x_3^2 - \frac{2}{3}$ from the model. When we do so, none of the coefficients of the other nonconstant terms change. At the next step, we are led to removing the term $x_2 x_4$ (whereupon the coefficient of $x_1 x_3$ changes). Upon removing both of these terms, we get Table 10.16.

Among the remaining terms, $x_1^2 - \frac{2}{3}$ and $x_4^2 - \frac{2}{3}$ appear to be only marginally worthwhile, but we will stay with the current fit. The corresponding ANOVA table is shown in Table 10.17.

The method of sequentially removing basis functions whose coefficients are statistically insignificant is referred to as *stepwise deletion*. There are similar methods for the *stepwise addition* of basis functions. Also, there are computational shortcuts for implementing such stepwise model selection procedures on a computer, but their discussion would take us too far afield. It should be emphasized that P-values obtained in the course of such procedures are not to be taken literally. (A similar reservation follows the the solution to Example 7.17 in Section 7.5.)

Problems

10.59   Let $\hat{\beta}_0 + \hat{\beta}_1 x$ be the least-squares fit in the space of linear functions on $\mathbb{R}$. Under the assumptions of the corresponding normal linear regression model, determine **(a)** the t statistic for testing the hypothesis that $\beta_1 = 0$; **(b)** the corresponding P-value.

TABLE   10.16

Term	Est. Coef.	SE	t	P-value
1	29.981	0.902	33.248	.000
x_1	15.633	1.104	14.155	.000
$x_1^2 - \frac{2}{3}$	4.022	1.913	2.103	.053
x_2	12.128	1.104	10.981	.000
$x_2^2 - \frac{2}{3}$	−4.494	1.913	−2.350	.033
x_3	17.506	1.104	15.851	.000
x_4	3.444	1.104	3.119	.007
$x_4^2 - \frac{2}{3}$	−3.411	1.913	−1.783	.095
$x_1 x_2$	4.650	1.353	3.438	.004
$x_1 x_3$	10.042	1.353	7.424	.000
$x_1 x_4$	4.176	1.397	2.989	.009
$x_2 x_3$	7.402	1.397	5.299	.000

TABLE   10.17

Source	SS
Fit	15215.13
Residuals	329.33
Total	15544.46

10.60 Consider an experiment involving a single factor and r repetitions at each of the three levels -1, 0, and 1. Suppose the assumptions of the normal linear regression model are satisfied with the regression function being a quadratic polynomial, which we write as $\mu(x) = \beta_0 + \beta_1 x + \beta_2(x^2 - \frac{2}{3})$. (Observe that 1, x, and $x^2 - \frac{2}{3}$ form an orthogonal basis of the space of quadratic polynomials.) Let $\hat{\beta}_2$ be the least-squares estimate of β_2. **(a)** Determine $\|x^2 - \frac{2}{3}\|^2$, $\hat{\beta}_2$ and $\mathrm{var}(\hat{\beta}_2)$. **(b)** Consider the hypothesis that $\beta_2 = 0$. Determine r such that the power of the test of size .05 of the hypothesis is about .9 when $\beta_2 = \sigma$. [*Hint:* Recall Equation (28) in Section 7.5.]

10.61 Consider the sugar beet experiment as treated in Example 10.15. Let τ be the change in the mean percentage of sugar content caused by switching from the standard variety to variety 4 on a given plot. Determine **(a)** $\hat{\tau}$; **(b)** $\mathrm{SE}(\hat{\tau})$; **(c)** the 95% upper confidence bound for τ; **(d)** the P-value for the t test of the hypothesis that $\tau \geq 0$; **(e)** the 95% confidence interval for τ; **(f)** the P-value for the t test of the hypothesis that $\tau = 0$.

10.62 Consider the lube oil experiment as treated in Problem 10.52 in Section 10.5. Carry out a t test of the hypothesis that $\beta_4 = 0$; that is, factor D has no effect on the quality of lube oil.

10.63   Consider a normal linear regression model in which G is the space of linear functions on $\mathbb{R}^2$ and the regression function is given by

$$\mu(x_1, x_2) = \beta_0 + \beta_1 x_1 + \beta_2 x_2.$$

Suppose $n = 20$, $\hat{\beta}_1 = 10$, $\hat{\beta}_2 = 11$, and $\mathrm{SE}(\hat{\beta}_1) = \mathrm{SE}(\hat{\beta}_2) = 5$. **(a)** Determine the observed 95% confidence interval for β_1. **(b)** Carry out the t test of size .05 of the hypothesis that changes in x_1 with x_2 held fixed have no effect on the mean response. **(c)** Determine the observed 95% confidence interval for β_2. **(d)** Carry out the t test of size .05 of the hypothesis that changes in x_2 with x_1 held fixed have no effect on the mean response. **(e)** Is it reasonable to conclude from the solutions to parts (b) and (d) that the mean response depends on x_2 but not on x_1?

10.64   In the context of Problem 10.46 in Section 10.4, determine the P-value for the t test of the hypothesis that $\beta_1 = 0$.

10.65   In the context of Problem 10.47 in Section 10.4, determine the P-value for the t test of the hypothesis that $\beta_1 = 0$.

10.66   Consider the polymer experiment and the normal linear regression model corresponding to the assumption that the regression function has the form $\mu(x) = \beta_0 + \beta_1 x + \beta_2 x^2$. Determine the P-value for the t test of the hypothesis that **(a)** $\beta_2 = 0$; **(b)** $\frac{d\mu}{dx}(5) = 0$.

10.67   Consider the polymer experiment and the normal linear regression model corresponding to the assumption that the regression function has the form $\mu(x) = \beta_0 + \beta_1 x + \beta_2(x^2 - \frac{35}{3})$. Determine the P-value for the t test of the hypothesis that **(a)** $\beta_2 = 0$; **(b)** $\frac{d\mu}{dx}(5) = 0$.

10.68   Consider the polymer experiment and the normal linear regression model corresponding to the linear space G of cubic polynomials on $\mathbb{R}$. **(a)** Show that $1, x, x^2 - \frac{35}{3}, x^3 - \frac{101}{5}x$ is an orthogonal basis of G and determine the squared norms of the functions in this basis. **(b)** Determine the least-squares estimates $\hat{\beta}_0, \dots, \hat{\beta}_3$ of the corresponding regression coefficients $\beta_0, \dots, \beta_3$. **(c)** Determine the fitted sum of squares. **(d)** Show the corresponding ANOVA table. **(e)** Determine the standard error of $\hat{\beta}_3$ and the 95% confidence interval for β_3. **(f)** Carry out a t test of the hypothesis that $\beta_3 = 0$; that is, the regression function is a quadratic polynomial.

10.69   Consider the experimental data that was used in Problems 10.54–10.58 in Section 10.5. Let G be the six-dimensional space that was used in Problem 10.57 in Section 10.5 and let G_0 be the five-dimensional subspace of G that was used in Problem 10.56 in that section. Determine the P-value for the t test of the hypothesis that the regression function lies in G_0.

10.7
Submodels

Recall from (36) in Section 7.6 that the between, within, and total sums of squares are related by the formula $\mathrm{TSS} = \mathrm{BSS} + \mathrm{WSS}$. Let G be an identifiable linear space that contains the constant functions. The least-squares fit in this space is given by $\hat{\mu}(\cdot) = P_G[\bar{Y}(\cdot)]$. It was shown in Section 10.2 that

$$BSS = FSS + LSS, \quad RSS = LSS + WSS, \quad \text{and} \quad TSS = FSS + RSS,$$

where RSS, LSS, and FSS, respectively, denote the corresponding residual, lack-of-fit, and fitted sums of squares.

Let G_0 be a subspace of G that also contains the constant functions. The least-squares fit in G_0 is given by $\hat{\mu}_0(\cdot) = P_{G_0}[\bar{Y}(\cdot)]$, and it follows from Theorem 9.29(b) in Section 9.5 that $\hat{\mu}_0(\cdot) = P_{G_0}[\hat{\mu}(\cdot)]$. We denote the residual, lack-of-fit, and fitted sums of squares for the least-squares fit in G_0 by RSS_0, LSS_0, and FSS_0, respectively. Thus,

$$BSS = FSS_0 + LSS_0, \quad RSS_0 = LSS_0 + WSS, \quad \text{and} \quad TSS = FSS_0 + RSS_0.$$

Observe that

$$FSS - FSS_0 = (BSS - LSS) - (BSS - LSS_0) = LSS_0 - LSS$$

and

$$RSS_0 - RSS = (LSS_0 + WSS) - (LSS + WSS) = LSS_0 - LSS.$$

Therefore,

$$FSS - FSS_0 = LSS_0 - LSS = RSS_0 - RSS. \tag{37}$$

To obtain a formula for the common value of the three differences in (37), we first note that $\hat{\mu}(\cdot) - \bar{Y} = \hat{\mu}_0(\cdot) - \bar{Y} + \hat{\mu}(\cdot) - \hat{\mu}_0(\cdot)$. Since $\hat{\mu}(\cdot) - \hat{\mu}_0(\cdot) = \hat{\mu}(\cdot) - P_{G_0}[\hat{\mu}(\cdot)]$ is orthogonal to every function in G_0, it is orthogonal to $\hat{\mu}_0(\cdot) - \bar{Y}$. Thus, by the Pythagorean theorem,

$$FSS = \|\hat{\mu}(\cdot) - \bar{Y}\|^2 = \|\hat{\mu}_0(\cdot) - \bar{Y}\|^2 + \|\hat{\mu}(\cdot) - \hat{\mu}_0(\cdot)\|^2 = FSS_0 + \|\hat{\mu}(\cdot) - \hat{\mu}_0(\cdot)\|^2,$$

so

$$FSS - FSS_0 = \|\hat{\mu}(\cdot) - \hat{\mu}_0(\cdot)\|^2. \tag{38}$$

In summation form,

$$FSS - FSS_0 = \sum_k n_k [\hat{\mu}(\mathbf{x}'_k) - \hat{\mu}_0(\mathbf{x}'_k)]^2 = \sum_i [\hat{\mu}(\mathbf{x}_i) - \hat{\mu}_0(\mathbf{x}_i)]^2. \tag{39}$$

Since $TSS = FSS + LSS + WSS$, we see that

$$TSS = FSS_0 + (FSS - FSS_0) + LSS + WSS. \tag{40}$$

The quantity $FSS - FSS_0$ is referred to as the *sum of squares due to G after G_0* and denoted by $SS_{G|G_0}$. Thus, by (37),

$$SS_{G|G_0} = FSS - FSS_0 = LSS_0 - LSS = RSS_0 - RSS. \tag{41}$$

Also, (40) can be rewritten as

$$TSS = FSS_0 + SS_{G|G_0} + LSS + WSS. \tag{42}$$

Since $RSS = LSS + WSS$, it follows from (42) that

$$TSS = FSS_0 + SS_{G|G_0} + RSS. \tag{43}$$

EXAMPLE **10.17**   Consider the analysis of the sugar beet experiment in which G is the linear space consisting of all functions that can be written as the sum of an arbitrary function of variety and a quintic polynomial function of plot. The corresponding ANOVA table (Table 10.18) is as shown in Section 10.4. Similarly, by computer we get the ANOVA table (Table 10.19) for the least-squares fit in the subspace G_0 of G consisting of the quintic polynomial functions of plot.

(a) Determine the sum of squares due to varieties after plots—that is, due to G after G_0.

(b) Show the various quantities in (43) in the form of an ANOVA table and interpret the result.

 T A B L E   10.18

Source	SS
Fit	10.310
Residuals	1.017
Total	11.327

 T A B L E   10.19

Source	SS
Fit	7.268
Residuals	4.059
Total	11.327

Solution   (a) The sum of squares due to varieties after plots is given by

$$SS_{G|G_0} = FSS - FSS_0 \doteq 10.310 - 7.268 = 3.042.$$

Alternatively,

$$SS_{G|G_0} = RSS_0 - RSS = 4.059 - 1.017 = 3.042.$$

(b) The ANOVA table corresponding to (43) is as shown in Table 10.20. According to this table, 64% of the variation in sugar content can be explained in terms of a quintic polynomial function of plot, 27% can be explained as being due to the variation in varieties after the plot effect has been taken into account, and 9% is unexplained by the additive combination of plots and varieties. ∎

EXAMPLE **10.18**   (Continued.) Let G be as in Example 10.17, but let G_0 be the subspace of G consisting of all functions of variety.

TABLE 10.20

Source	SS
Plots (quintic)	7.268
Varieties after plots	3.042
Residuals	1.017
Total	11.327

(a) Determine the sum of squares due to plots after varieties.

(b) Show the various quantities in (43) in the form of an ANOVA table and interpret the result.

Solution (a) The ANOVA table corresponding to the least-squares fit in G_0 is shown in Table 10.21 (see the solution to Example 7.24 in Section 7.6). Thus,

$$SS_{G|G_0} = FSS - FSS_0 \doteq 10.310 - 3.085 \doteq 7.226.$$

Alternatively,

$$SS_{G|G_0} \doteq RSS_0 - RSS \doteq 8.242 - 1.017 \doteq 7.226.$$

TABLE 10.21

Source	SS
Fit	3.085
Residuals	8.242
Total	11.327

(b) The ANOVA table corresponding to (43) is shown in Table 10.22. According to this table 27% of the variation in sugar content can be explained by the variation in varieties, 64% can be explained as being due to the variation in plots after the effect of varieties is taken into account, and 9% is unexplained by the additive combination of varieties and plots. ∎

TABLE 10.22

Source	SS
Varieties	3.085
Plots (quintic) after varieties	7.226
Residuals	1.017
Total	11.327

The ANOVA tables in the solutions to Examples 10.17(b) and 10.18(b) are in agreement that about 64% of the variation in sugar content is due to plots, 27% is due to varieties, and 9% is unexplained by the additive combination of plots and varieties.

EXAMPLE **10.19**   Consider the lube oil experiment and let G be the space of quadratic polynomials in the levels of the four factors. The corresponding ANOVA table (Table 10.23) is as shown in Section 10.5. Let G_0 be the subspace of G consisting of the linear functions of the four factors.

(a) Determine the least-squares fit in G_0 to the lube oil data.

(b) Show the corresponding ANOVA table.

(c) Determine $SS_{G|G_0}$.

(d) Show the various quantities in (43) in the form of an ANOVA table and interpret the result.

TABLE   **10.23**

Source	SS
Fit	15248.69
Residuals	295.77
Total	15544.46

Solution   (a) Recall the basis $1, x_1, x_1^2 - \frac{2}{3}, x_2, x_2^2 - \frac{2}{3}, x_3, x_3^2 - \frac{2}{3}, x_4, x_4^2 - \frac{2}{3}, x_1x_2, x_3x_4, x_1x_3,$ x_2x_4, x_1x_4, x_2x_3 of G that was used in Section 10.5. Observe that $1, x_1, x_2, x_3, x_4$ is an orthogonal basis of G_0 and that the functions in this basis are orthogonal to each of the other functions in the basis of G. Thus, the coefficients of the functions in the indicated basis of G_0 for the least-squares fit in this subspace equal the respective coefficients of these functions for the least-squares fit in G. Consequently (see Section 10.5), the least-squares fit in G_0 is given by

$$\hat{\mu}_0(x_1, x_2, x_3, x_4) \doteq 29.891 + 15.633x_1 + 12.128x_2 + 17.506x_3 + 3.444x_4.$$

(b) Since $x_1, x_2, x_3,$ and x_4 are orthogonal functions, each having squared norm 18 and orthogonal to 1,

$$FSS_0 \doteq 18(15.633^2 + 12.128^2 + 17.506^2 + 3.444^2) \doteq 12776.27.$$

The corresponding ANOVA table is shown in Table 10.24.

(c) The sum of squares for the quadratic terms after the linear terms is given by

$$SS_{G|G_0} = FSS - FSS_0 \doteq 15248.69 - 12776.27 \doteq 2472.42.$$

(d) The ANOVA table corresponding to (43) is shown in Table 10.25. Accordingly, 82% of the variation in the quality of lube oil is due to the linear effects of the

TABLE 10.24

Source	SS
Fit	12776.27
Residuals	2768.19
Total	15544.46

TABLE 10.25

Source	SS
Linear	12776.27
Quadratic after linear	2472.42
Residuals	295.77
Total	15544.46

four factors, 16% is due to the quadratic effects, and only 2% of the variation is unexplained by the quadratic fit in the levels of the four factors. ∎

EXAMPLE **10.20** (Continued.) Let G_0 now be the space of quadratic polynomials in the levels of factors A, B, and C. The corresponding ANOVA table is shown in Table 10.26 (see Section 10.5):

(a) Determine the sum of squares due to factor D after taking into account the other three factors; that is, determine $SS_{G|G_0}$.

(b) Show the various quantities in (43) in the form of an ANOVA table and interpret the result.

TABLE 10.26

Source	SS
Fit	14739.40
Residuals	805.06
Total	15544.46

Solution **(a)** We have that

$$SS_{G|G_0} = FSS - FSS_0 \doteq 15248.69 - 14739.40 = 509.29.$$

(b) The ANOVA table corresponding to (43) is shown in Table 10.27. Accordingly, 95% of the variation in the quality of lube oil can be explained by the quadratic polynomial fit in the levels of *A*, *B*, and *C*; 3% of the variation can be explained as being due to the effect of *D* after taking into account the combined effect of *A*, *B*, and *C*; and 2% of the variation is unexplained by the combined effect of all four factors. This interpretation suggests either that factor *D* has a minor effect on the quality of lube oil, that the levels of *D* were too close together in the experiment for changes in the level of this factor to have much effect on quality, or that the medium level of the factor is nearly optimal. ∎

TABLE 10.27

Source	SS
A, B, C (quadratic)	14739.40
D after *A, B, C*	509.29
Residuals	295.77
Total	15544.46

Expected Sums of Squares

Let the linear spaces G and G_0 have dimensions p and p_0, respectively, and let $\mu^*(\cdot) = P_G[\mu(\cdot)] = E[\hat{\mu}(\cdot)]$ and $\mu_0^*(\cdot) = P_{G_0}[\mu(\cdot)] = E[\hat{\mu}_0(\cdot)]$ be the orthogonal projections of the regression function onto G and G_0, respectively.

THEOREM 10.13   Under the assumptions of the homoskedastic regression model,

$$E(\mathrm{RSS}_0 - \mathrm{RSS}) = \|\mu^*(\cdot) - \mu_0^*(\cdot)\|^2 + (p - p_0)\sigma^2.$$

Proof   According to (28) in Section 10.2,

$$E(\mathrm{RSS}) = \|\mu(\cdot) - \mu^*(\cdot)\|^2 + (n - p)\sigma^2.$$

Applying this formula with G replaced by G_0, we get

$$E(\mathrm{RSS}_0) = \|\mu(\cdot) - \mu_0^*(\cdot)\|^2 + (n - p_0)\sigma^2$$

and hence

$$E(\mathrm{RSS}_0 - \mathrm{RSS}) = \|\mu(\cdot) - \mu_0^*(\cdot)\|^2 - \|\mu(\cdot) - \mu^*(\cdot)\|^2 + (p - p_0)\sigma^2. \tag{44}$$

Since $\mu(\cdot) - \mu^*(\cdot)$ is orthogonal to every function in G, it is orthogonal to $\mu^*(\cdot) - \mu_0^*(\cdot)$. Thus, by the Pythagorean theorem,

$$\|\mu(\cdot) - \mu_0^*(\cdot)\|^2 = \|\mu(\cdot) - \mu^*(\cdot)\|^2 + \|\mu^*(\cdot) - \mu_0^*(\cdot))\|^2.$$

The desired result now follows from (44). ∎

COROLLARY 10.11 Under the assumptions of the homoskedastic linear regression model corresponding to G,

$$E(\text{RSS}_0 - \text{RSS}) = \|\mu(\cdot) - \mu_0^*(\cdot)\|^2 + (p - p_0)\sigma^2.$$

Problems

10.70 Consider an experiment involving a single factor that takes on the levels $-2, -1, 0, 1, 2$ with r runs at each of these levels and write the least-squares fit to the experimental data as $\hat{\mu}(x) = \hat{\beta}_0 + \hat{\beta}_1 x + \hat{\beta}_2 (x^2 - 2)$. Express the fitted sum of squares in terms of $\hat{\beta}_0, \hat{\beta}_1$, and $\hat{\beta}_2$.

10.71 Suppose G contains the constant functions, let $1, g_1, \ldots, g_{p-1}$ be an orthogonal basis of G, let $\hat{\beta}_0, \ldots, \hat{\beta}_{p-1}$ be the coefficients of $\hat{\mu}(\cdot)$ relative to this basis, and let G_0 be the p_0-dimensional subspace of G spanned by $1, g_1, \ldots, g_{p_0-1}$, where $1 \leq p_0 < p$. Show that

$$\text{FSS} - \text{FSS}_0 = \sum_{j=p_0}^{p-1} \hat{\beta}_j^2 \|g_j\|^2.$$

[*Hint:* Use Problem 10.22(b) in Section 10.2.]

10.72 Consider the lube oil experiment. Let G be the space of linear functions of the levels of the four factors and let G_0 be the subspace of G consisting of the linear functions of the levels of factors A, B, and C. **(a)** Determine the least-squares estimate in G_0 of the regression function. **(b)** Determine FSS_0. **(c)** Show the corresponding ANOVA table. **(d)** Determine $\text{SS}_{G|G_0}$. **(e)** Show and interpret the ANOVA table corresponding to (43).

10.73 Consider the sugar beet experiment. Let G be the linear space consisting of sums of arbitrary functions of variety and quintic polynomial functions of plot and let G_0 be the subspace of G consisting of sums of arbitrary functions of variety and quartic polynomial functions of plot. Recall that $\text{TSS} \doteq 11.327$ and $\text{RSS} \doteq 1.017$. By computer, $\text{RSS}_0 \doteq 1.278$. **(a)** Determine FSS_0. **(b)** Show the ANOVA table corresponding to the least-squares fit in G_0. **(c)** Determine $\text{SS}_{G|G_0}$. **(d)** Show and interpret the ANOVA table corresponding to (43).

10.74 Consider the polymer experiment. Let G_1 be the space of linear functions on $\mathbb{R}$, G_2 the space of quadratic polynomials, G_3 the space of cubic polynomials, G_4 the space of quartic polynomials, G_5 the space of quintic polynomials. Also, let $\text{FSS}_1, \ldots, \text{FSS}_5$ denote the corresponding fitted sums of squares and set $\text{SS}_{G_2|G_1} = \text{FSS}_2 - \text{FSS}_1$ and so forth. **(a)** Determine FSS_1, $\text{SS}_{G_2|G_1}$, $\text{SS}_{G_3|G_2}$, $\text{SS}_{G_4|G_3}$, and $\text{SS}_{G_5|G_4}$. **(b)** Incorporate the various sums of squares in part (a) into a suitable ANOVA table and interpret the result.

10.75 Consider the experimental data that was used in Problems 10.54–10.58 in Section 10.5. Let G be the six-dimensional space that was used in Problem 10.57 in Section 10.5, write the least-squares fit in this space as

$$\hat{\mu}(\cdot) = \hat{\beta}_0 + \hat{\beta}_1 x_1 + \hat{\beta}_2 x_2 + \hat{\beta}_3 x_3 + \hat{\beta}_4 x_1 x_2 + \hat{\beta}_5 x_1 x_3,$$

and let RSS be the corresponding residual sum of squares. Let G_0 be the five-dimensional subspace of G that was used in Problem 10.56 in Section 10.5, write the least-squares fit in this space as

$$\hat{\mu}_0(\cdot) = \hat{\beta}_{00} + \hat{\beta}_{10}x_1 + \hat{\beta}_{20}x_2 + \hat{\beta}_{30}x_3 + \hat{\beta}_{40}x_1x_2,$$

and let RSS_0 be the corresponding residual sum of squares. **(a)** Express $SS_{G|G_0} = RSS_0 - RSS$ in terms of the coefficients $\hat{\beta}_1, \ldots, \hat{\beta}_5$ and $\hat{\beta}_{10}, \ldots, \hat{\beta}_{40}$ of $\hat{\mu}(\cdot)$ and $\hat{\mu}_0(\cdot)$, respectively. **(b)** Show the ANOVA table corresponding to (43).

10.76 Suppose the assumptions of the homoskedastic regression model are satisfied, let $\hat{\sigma}^2$ be an unbiased estimate of σ^2, let G_0 be an identifiable p_0-dimensional linear space (which is not necessarily a subspace of G), let $\hat{\mu}_0(\cdot)$ be the least-squares estimate in G_0 of $\mu(\cdot)$, and let RSS_0 be the corresponding residual sum of squares. **(a)** Show that $RSS + 2p\hat{\sigma}^2$ is an unbiased estimate of $E[\, \|\hat{\mu}(\cdot) - \mu(\cdot)\|^2\,] + n\sigma^2$. [*Hint:* Use (24) and (28) in Section 10.2.] **(b)** Show that $(RSS + 2p\hat{\sigma}^2) - (RSS_0 + 2p_0\hat{\sigma}^2)$ is an unbiased estimate of $E[\, \|\hat{\mu}(\cdot) - \mu(\cdot)\|^2 - \|\hat{\mu}_0(\cdot) - \mu(\cdot)\|^2\,]$.

10.77 Let $\hat{\mu}(\cdot)$ be the least-squares fit in an identifiable linear space G that does not necessarily contain the constant functions, let $\hat{\mu}_0(\cdot)$ be the least-squares fit in a subspace G_0 of G, and let RSS and RSS_0 be the residual sums of squares for the least-squares fits in G and G_0, respectively. Show that $RSS_0 - RSS = LSS_0 - LSS = \|\hat{\mu}(\cdot) - \hat{\mu}_0(\cdot)\|^2$.

10.8
The F Test

Let G be an identifiable p-dimensional linear space of functions on $\mathcal{X}$ with $p < n$, let G_0 be a proper subspace of G, and let p_0 denote the dimension of G_0. Then $p_0 < p$. Consider the normal linear regression model corresponding to G and the hypothesis that the regression function is in G_0. In Section 10.6, we developed a t test of this hypothesis that is applicable whenever $p_0 = p - 1$. In this section, we will develop an F test of the hypothesis that is applicable for arbitrary $p_0 < p$. It will be shown that if $p_0 = p - 1$, then the two tests are equivalent; that is, they yield identical P-values. We will also see that if $n > d > p$, then the F test can be used to check the assumption that the regression function is in G.

As usual, let RSS denote the residual sum of squares for the least-squares estimate $\hat{\mu}(\cdot)$ in G of $\mu(\cdot)$, let RSS_0 denote the residual sum of squares for the least-squares estimate $\hat{\mu}_0(\cdot)$ in G_0, and set $\mu_0^*(\cdot) = P_{G_0}[\mu(\cdot)] = E[\hat{\mu}_0(\cdot)]$.

THEOREM 10.14 Under the assumptions of the normal linear regression model corresponding to G,

(a) $RSS_0 - RSS$ and RSS are independent;

(b) if $\mu(\cdot) \in G_0$, then $(RSS_0 - RSS)/\sigma^2$ has the chi-square distribution with $p - p_0$ degrees of freedom.

Proof Let $g_1, \ldots, g_p$ be an orthonormal basis of G such that $g_1, \ldots, g_{p_0}$ is a basis of G_0 and let $\beta_1, \ldots, \beta_p$ be the coefficients of $\mu(\cdot)$ relative to the indicated basis of G. Then

$$\mu(\cdot) = \sum_{j=1}^{p} \beta_j g_j$$

and, by Theorem 9.30 in Section 9.5,

$$\mu_0^*(\cdot) = \sum_{j=1}^{p_0} \beta_j g_j.$$

Observe that $\mu(\cdot) \in G_0$ if and only if $\beta_{p_0+1} = \cdots = \beta_p = 0$.

Let $\hat{\beta}_1, \ldots, \hat{\beta}_p$ be the coefficients of $\hat{\mu}(\cdot)$ relative to $g_1, \ldots, g_p$. Then $\hat{\beta}_1, \ldots, \hat{\beta}_p$, RSS are independent random variables, and, for $1 \le j \le p$, $\hat{\beta}_j$ is normally distributed with mean β_j and variance σ^2 (see Corollary 10.8 in Section 10.3). Since

$$\hat{\mu}(\cdot) = \sum_{j=1}^{p} \hat{\beta}_j g_j,$$

it follows from another application of Theorem 9.30 in Section 9.5 that

$$\|\hat{\mu}(\cdot) - \hat{\mu}_0(\cdot)\|^2 = \sum_{j=p_0+1}^{p} \hat{\beta}_j^2.$$

Now $\mathrm{RSS}_0 - \mathrm{RSS} = \|\hat{\mu}(\cdot) - \hat{\mu}_0(\cdot)\|^2$ [see (37) and (38) in Section 10.7], so

$$\mathrm{RSS}_0 - \mathrm{RSS} = \sum_{j=p_0+1}^{p} \hat{\beta}_j^2. \tag{45}$$

The first conclusion follows from (45) and the independence of $\hat{\beta}_1, \ldots, \hat{\beta}_p$, RSS.

Suppose $\mu(\cdot) \in G_0$ and hence that $\beta_{p_0+1} = \cdots = \beta_p = 0$. Then $\hat{\beta}_{p_0+1}/\sigma, \ldots, \hat{\beta}_p/\sigma$ are independent, standard normal random variables and, by (45),

$$\frac{\mathrm{RSS}_0 - \mathrm{RSS}}{\sigma^2} = \sum_{j=p_0+1}^{p} (\hat{\beta}_j/\sigma)^2.$$

Thus, the second conclusion of the theorem is valid. ∎

Under the assumptions of the homoskedastic linear regression model corresponding to G, $E(\mathrm{RSS}) = (n-p)\sigma^2$ (see Corollary 10.5 in Section 10.2) and

$$E(\mathrm{RSS}_0 - \mathrm{RSS}) = \|\mu(\cdot) - \mu_0^*(\cdot)\|^2 + (p - p_0)\sigma^2$$

(see Corollary 10.11 in Section 10.7). Consequently,

$$\frac{E[\,(\mathrm{RSS}_0 - \mathrm{RSS})/(p - p_0)\,]}{E[\,\mathrm{RSS}/(n-p)\,]} = 1 + \frac{1}{\sigma^2(p - p_0)} \|\mu(\cdot) - \mu_0^*(\cdot)\|^2.$$

Since $\|\mu(\cdot) - \mu_0^*(\cdot)\|^2 = 0$ if and only if $\mu(\cdot) \in G_0$, we conclude that the ratio

$$\frac{E[\,(\mathrm{RSS}_0 - \mathrm{RSS})/(p - p_0)\,]}{E[\,\mathrm{RSS}/(n - p)\,]}$$

equals 1 if $\mu(\cdot) \in G_0$ and that it is greater than 1 otherwise. In practice, of course, we cannot determine the indicated ratio of expected values. We can, however, determine the statistic

$$F = \frac{(\mathrm{RSS}_0 - \mathrm{RSS})/(p - p_0)}{\mathrm{RSS}/(n - p)}. \tag{46}$$

It seems reasonable to reject the hypothesis that $\mu(\cdot) \in G_0$ if this statistic is sufficiently large and to accept it otherwise.

The sum of squares due to G after G_0 is given by

$$\mathrm{SS}_{G|G_0} = \mathrm{RSS}_0 - \mathrm{RSS} = \mathrm{LSS}_0 - \mathrm{LSS};$$

if G_0 contains the constant functions, then $\mathrm{SS}_{G|G_0} = \mathrm{FSS} - \mathrm{FSS}_0$. We refer to $\mathrm{MS} = \mathrm{RSS}/(n - p)$ and $\mathrm{MS}_0 = \mathrm{SS}_{G|G_0}/(p - p_0)$ as the *mean squares* corresponding to RSS and $\mathrm{SS}_{G|G_0}$, respectively. The F statistic given by (46) can be written in terms of these quantities as $F = \mathrm{MS}_0/\mathrm{MS}$. The next result follows from Theorem 10.14 in this section and Corollary 10.8(c) in Section 10.3.

THEOREM **10.15**   Suppose the assumptions of the normal linear regression model are satisfied and that $\mu(\cdot) \in G_0$. Then the F statistic given by (46) has the F distribution with $p - p_0$ degrees of freedom in the numerator and $n - p$ degrees of freedom in the denominator.

Suppose the assumptions of the normal linear regression model corresponding to G are satisfied. It follows from Theorem 10.15 that $F \geq F_{1-\alpha, p-p_0, n-p}$ is the critical region for the F test of size α of the hypothesis that the regression function is a member of G_0. The acceptance region for the test is $F < F_{1-\alpha, p-p_0, n-p}$, and its P-value equals $1 - F_{p-p_0, n-p}(F)$.

EXAMPLE **10.21**   Consider the lube oil experiment and let G be the space of quadratic polynomials in the levels of the four factors. Test the hypothesis that the mean response is a linear function of the four factors.

Solution   The computation of the P-value is conveniently summarized (Table 10.28) by extending the ANOVA table in the solution to Example 10.19(d) in Section 10.7. Here $n = 27$, $p = 15$ and $p_0 = 5$, so $\mathrm{MS}_0 \doteq 2472.42/10 \doteq 247.24$, $\mathrm{MS} \doteq 295.77/12 \doteq 24.65$, $F \doteq 247.24/24.65 \doteq 10.031$, and hence the P-value is given by $1 - F_{10,12}(10.031) \doteq .000$. The hypothesis that the mean yield is a linear function of the four factors is clearly unacceptable. ∎

EXAMPLE **10.22**   (Continued.) Under the assumption that the mean response is a quadratic polynomial

TABLE   **10.28**

Source	SS	DF	MS	F	P-value
Linear	12776.27	4	3194.07		
Quadratic after linear	2472.42	10	247.24	10.031	.000
Residuals	295.77	12	24.65		
Total	15544.46	26			

in the levels of the four factors, test the hypothesis that the mean response does *not* depend on the level of factor D.

Solution   Extending the ANOVA table in the solution to Example 10.20(b) in Section 10.7, we get Table 10.29. Thus, the hypothesis that the mean response does not depend on the level of factor D is rejected by the F test of size .05 but accepted by the test of size .01 [see the interpretation in the solution to Example 10.20(b) in Section 10.7]. ∎

TABLE   **10.29**

Source	SS	DF	MS	F	P-value
A, B, C (quadratic)	14739.40	9	1637.71		
D after A, B, C	509.29	5	101.86	4.133	.020
Residuals	295.77	12	24.65		
Total	15544.46	26			

EXAMPLE **10.23**   Consider the sugar beet experiment. Test the assumption that the mean response is an additive function of variety and plot.

Solution   Let G_0 be the space of functions having the form of the sum of an arbitrary function of variety and a quintic polynomial function of plot. As noted in in Section 10.4, the residual sum of squares for the least-squares fit in G_0 is given by $\text{RSS}_0 \doteq 1.017$.

To test the assumption of additivity, we need to embed G_0 in a reasonable nonadditive space G of functions. To this end, we choose to let G be the space of functions having the form of a separate quintic polynomial in plot for each of the five varieties. The space G has dimension $p = 6 \cdot 5 = 30$. To obtain the corresponding residual sum of squares, we fit a quintic polynomial to the 11 plots on which variety 1 is grown and obtain the corresponding residual sum of squares $\text{RSS}^{(1)}$. Then we fit a quintic polynomial to the 10 plots on which variety 2 is grown and obtain the residual sum of squares $\text{RSS}^{(2)}$. Similarly, we obtain $\text{RSS}^{(3)}$, $\text{RSS}^{(4)}$, $\text{RSS}^{(5)}$, and $\text{RSS} = \text{RSS}^{(1)} + \cdots + \text{RSS}^{(5)}$. In this manner, we get that $\text{RSS}^{(1)} \doteq 0.094$, $\text{RSS}^{(2)} \doteq 0.187$, $\text{RSS}^{(3)} \doteq 0.067$, $\text{RSS}^{(4)} \doteq 0.065$, $\text{RSS}^{(5)} \doteq 0.119$ and $\text{RSS} \doteq 0.531$. Thus, we are led to the ANOVA table shown in Table 10.30, in which we think of $\text{RSS}_0 - \text{RSS} \doteq 0.485$ as a measure of the extent of interaction (departure from additivity) between plots and varieties, which we write as "Plots $\times$ Varieties." Thus, no evidence exists of an interaction between plots and varieties; that is, the analysis is compatible with the mean response being the sum of a function of variety and a function of plot. (Additivity and interactions will be treated systematically in Chapter 11.)   ■

T A B L E   10.30

Source	SS	DF	MS	F	P-value
Plots + Varieties	10.310	9	1.146		
Plots × Varieties	0.485	20	0.024	0.959	.536
Residuals	0.531	21	0.025		
Total	11.327	50			

EXAMPLE 10.24   Consider the normal linear regression model for the sugar beet experiment in which G is the space of all functions that can be written as the sum of an arbitrary function of variety and a quintic polynomial function of plot.

(a) Test the hypothesis that the mean response does not depend on the variety.

(b) Test the hypothesis that the mean response does not depend on plot.

(c) Show the *P*-values in the solution to parts (a) and (b) in the form of a single ANOVA table.

Solution   (a) Extending the ANOVA table in the solution to Example 10.17(b) in Section 10.7, we get Table 10.31. The hypothesis that the mean response does not depend on the variety is clearly untenable.

(b) Extending the ANOVA table in the solution to Example 10.18(b) in Section 10.7, we get Table 10.32 The hypothesis that the mean response does not depend on the plot is also clearly untenable.

(c) The *P*-values in the solution to parts (a) and (b) are shown in Table 10.33. The sums of squares for plots after varieties, varieties after plots, and residuals add up to 11.284, which does not quite equal the total sum of squares 11.327.   ■

T A B L E   10.31

Source	SS	DF	MS	F	P-value
Plots (quintic)	7.268	5	1.454		
Varieties after plots	3.042	4	0.760	30.658	.000
Residuals	1.017	41	0.025		
Total	11.327	50			

TABLE 10.32

Source	SS	DF	MS	F	P-value
Varieties	3.085	4	0.771		
Plots (quintic) after varieties	7.226	5	1.445	58.261	.000
Residuals	1.017	41	0.025		
Total	11.327	50			

TABLE 10.33

Source	SS	DF	MS	F	P-value
Plots (quintic) after varieties	7.226	5	1.445	58.261	.000
Varieties after plots	3.042	4	0.760	30.658	.000
Residuals	1.017	41	0.025		
Total		50			

Test of Constancy

Suppose the space of constant functions is a proper subspace of G and hence that $p \geq 2$. Since G is identifiable, it follows from the assumption that the regression function is a member of G that it is constant if and only if it is constant on the design set—that is, if and only if $\mu_1 = \cdots = \mu_d$.

EXAMPLE 10.25   Suppose the space of constant functions is a proper subspace of G. Under the assumption that the regression function is a member of G, determine the F statistic for testing the hypothesis that it is constant

(a) in general;   (b) if G is saturated.

Solution   (a) The residual sum of squares for the least-squares constant estimate of the regression function is given by $\text{RSS}_0 = \text{TSS}$. Thus, $\text{RSS}_0 - \text{RSS} = \text{TSS} - \text{RSS} = \text{FSS}$. Since the space of constant functions has dimension 1, the desired statistic is given by

$$F = \frac{\text{FSS}/(p-1)}{\text{RSS}/(n-p)}.$$

(b) Suppose G is saturated. Then $p = d$, $\text{FSS} = \text{BSS}$, and $\text{RSS} = \text{WSS}$, so

$$F = \frac{\text{BSS}/(d-1)}{\text{WSS}/(n-d)}.$$

Thus, in the present context, the F test of the hypothesis that the regression function is constant is equivalent to the F test of the hypothesis of homogeneity, which was described in Section 7.6. ∎

EXAMPLE **10.26**   Consider the sugar beet experiment and the assumption that the regression function is the sum of an arbitrary function of variety and a quintic polynomial function of plot. Test the hypothesis that the regression function is constant.

Solution   The *P*-value for the desired test is shown in the ANOVA table shown in Table 10.34. The regression function is clearly not constant.   ■

TABLE   10.34

Source	SS	DF	MS	F	P-value
Fit	10.310	9	1.146	46.185	.000
Residuals	1.017	41	0.025		
Total	11.327	50			

Goodness-of-Fit Test

Let LSS and RSS denote the lack-of-fit and residual sums of squares, respectively, for the least-squares fit in an identifiable *p*-dimensional linear space *G* of functions on $\mathcal{X}$, where $p < d < n$. Then the lack-of-fit sum of squares can be used to test the assumption that the regression function is a member of *G*.

To this end, suppose the assumptions of the normal regression model are satisfied but consider the possibility that the regression function may not be in *G*. By ignoring points of $\mathcal{X}$ that are not design points, we can think of *G* as a proper subspace of the *d*-dimensional space *H* of all functions on the design set. Then the regression function is a member of *H*, its least-squares estimate in *H* equals $\bar{Y}(\cdot)$, and the corresponding residual sum of squares equals the within sum of squares WSS. We can think of the assumption that the regression function is in *G* as a hypothesis. The statistic for testing this hypothesis is given by

$$F = \frac{(\text{RSS} - \text{WSS})/(d - p)}{\text{WSS}/(n - d)}.$$

Since $\text{RSS} = \text{LSS} + \text{WSS}$, we can rewrite this statistic as

$$F = \frac{\text{LSS}/(d - p)}{\text{WSS}/(n - d)}. \tag{47}$$

If the regression function is in *G*, then *F* has the *F* distribution with $d - p$ degrees of freedom in the numerator and $n - d$ degrees of freedom in the denominator. The *F* test of the assumption that the regression function is in *G* based on the *F* statistic defined by (47) is referred to as a *goodness-of-fit test*.

Alternatively, let G_{sat} be any identifiable and saturated linear space containing *G* as a proper subspace. Then (47) gives the *F* statistic for testing the hypothesis that the regression function is in *G* under the assumption that it is in G_{sat}.

EXAMPLE **10.27**   Consider the lube oil experiment but ignore factors C and D. Then the experimental data can be shown as in Table 10.35. Test the assumption that the mean response is the sum of a quadratic polynomial in x_1 and a quadratic polynomial in x_2.

TABLE   **10.35**

x_1	x_2	Y		
−1	−1	4.2	5.9	8.2
−1	0	13.0	16.4	30.7
−1	1	9.5	22.2	31.0
0	−1	7.7	16.5	14.3
0	0	11.0	29.0	55.0
0	1	8.5	37.4	66.3
1	−1	11.4	21.1	57.9
1	0	13.5	51.6	76.5
1	1	31.0	74.5	85.1

Solution   Here the within sum of squares is given by WSS $\doteq$ 7968.67, and the residual sum of squares for the indicated additive fit is given by RSS $\doteq$ 8279.48. Thus, the lack-of-fit sum of squares for this fit is given by LSS = RSS − WSS $\doteq$ 310.80. In the present context, $n = 27$, $d = 9$, and the linear space corresponding to the additive fit has dimension $p = 5$, so the within sum of squares has $27 − 9 = 18$ degrees of freedom, the lack-of-fit sum of squares has $9 − 5 = 4$ degrees of freedom, and the fitted sum of squares has $5 − 1 = 4$ degrees of freedom. Consequently, we get the ANOVA table shown in Table 10.36, in which the departure from the additive model is indicated as the interaction AB between A and B. Therefore, at least when we ignore factors C and D, we do not see any evidence of an AB interaction.   ■

TABLE   **10.36**

Source	SS	DF	MS	F	P-value
$A + B$	7264.98	4	1816.25		
AB	310.80	4	77.70	0.176	.948
Within	7968.67	18	442.70		
Total	15544.46	26			

Equivalence of t Tests and F tests

The regression function is given in terms of a basis $g_1, \ldots, g_p$ of G by $\mu(\cdot) = \sum_j \beta_j g_j$. Observe that this function is in the subspace G_0 of G spanned by $g_1, \ldots, g_{p-1}$ if and only if $\beta_p = 0$. The least-squares estimate in G of the regression function is given by $\hat{\mu}(\cdot) = \sum_j \hat{\beta}_j g_j$.

THEOREM 10.16   Let G_0 be the subspace of G consisting of all functions in G whose coefficient β_p of g_p equals zero, let

$$F = \frac{\text{RSS}_0 - \text{RSS}}{\text{RSS} / (n - p)}$$

be the F statistic for testing the hypothesis that $\mu(\cdot) \in G_0$ under the assumption that $\mu(\cdot) \in G$, and let $t = \hat{\beta}_p / \text{SE}(\hat{\beta}_p)$ be the t statistic for testing the equivalent hypothesis that $\beta_p = 0$. Then $F = t^2$.

Proof   Replacing g_p by $g_p - P_{G_0} g_p$ if necessary [which does not alter β_p, $\hat{\beta}_p$ or $\text{SE}(\hat{\beta}_p)$], we can assume that g_p is orthogonal to G_0 and hence to each of the functions $g_1, \ldots, g_{p-1}$. If g_p is replaced by g_p/a, where $a > 0$, then G_0 and hence F are unaltered; also, $\hat{\beta}_p$ is replaced by $a\hat{\beta}_p$ and $\text{SE}(\hat{\beta}_p)$ is replaced by $a\,\text{SE}(\hat{\beta}_p)$, so the t statistic is unaltered. Thus, without loss of generality, we can assume that $\|g_p\| = 1$. Replacing $g_1, \ldots, g_{p-1}$ by an orthonormal basis of G_0 if necessary [which again does not alter β_p, $\hat{\beta}_p$, or $\text{SE}(\hat{\beta}_p)$], we can assume that $g_1, \ldots, g_p$ is an orthonormal basis of G. Then $\text{VC}(\hat{\beta}) = \sigma^2 \mathbf{I}$, so $\text{var}(\hat{\beta}_p) = \sigma^2$ and hence $\text{SD}(\hat{\beta}_p) = \sigma$ and $\text{SE}(\hat{\beta}_p) = S$, where $S^2 = \text{RSS}/(n-p)$. Since $\text{RSS}_0 - \text{RSS} = \hat{\beta}_p^2$ by (45), we conclude that

$$F = \frac{\text{RSS}_0 - \text{RSS}}{\text{RSS}/(n-p)} = \frac{\hat{\beta}_p^2}{S^2} = \left(\frac{\hat{\beta}_p}{\text{SE}(\hat{\beta}_p)} \right)^2 = t^2. \qquad \blacksquare$$

According to the next result, when the F test and the t test are both applicable, they have equal P-values.

COROLLARY 10.12   Let G_0 and β_p be as in Theorem 10.16. Then the P-value for the F test of the hypothesis that $\mu(\cdot) \in G_0$ coincides with the P-value for the t test of the equivalent hypothesis that $\beta_p = 0$.

Proof   The P-value for the F test of the hypothesis that $\mu(\cdot) \in G_0$ equals $1 - F_{1,n-p}(F)$. Since $F = t^2$ by Theorem 10.16, we can write this P-value as $1 - F_{n-p}(t^2)$. The P-value for the t test of the hypothesis that $\beta_p = 0$ equals $2[1 - t_{n-p}(|t|)]$. Since $1 - F_{1,n-p}(t^2) = 2[1 - t_{n-p}(|t|)]$ by (13) in Section 7.2, the two P-values coincide. $\blacksquare$

EXAMPLE 10.28   Consider the sugar beet experiment and suppose that the regression function is the sum of an arbitrary function of variety and a function of plot. Use an F test to justify thinking of the component of the regression function involving plot as a quintic polynomial rather than as a quartic polynomial.

Solution Let G be the space of functions having the form of the sum of an arbitrary function of variety and a quintic polynomial function of plot and let G_0 be the subspace of G consisting of the functions having the form of the sum of an arbitrary function of variety and a quartic polynomial function of plot. Then RSS $\doteq 1.018$, and $\mathrm{RSS}_0 \doteq 1.278$, as we have already seen, so $\mathrm{RSS}_0 - \mathrm{RSS} \doteq 0.261$. Thus, we get the ANOVA table shown in Table 10.37, which justifies keeping the term $(x_1/25)^5$ in the model. In the solution to Example 10.16 in Section 10.6, we used the t test for the same purpose, getting $t \doteq -3.246$ and P-value $\doteq .002$. Since $(-3.246)^2 \doteq 10.538$, these results are compatible with Theorem 10.14 and Corollary 10.12. ∎

TABLE 10.37

Source	SS	DF	MS	F	P-value
Plots (quartic) + varieties	10.049	8	1.256		
$(x_1/25)^5$	0.261	1	0.261	10.538	.002
Residuals	1.017	41	0.025		
Total	11.327	50			

Problems

10.78 Let $\hat{\beta}_0 + \hat{\beta}_1 x$ be the least-squares fit in the space of linear functions on $\mathbb{R}$. Under the assumptions of the corresponding normal linear regression model, determine **(a)** the F statistic for testing the hypothesis that $\beta_1 = 0$; **(b)** the corresponding P-value.

10.79 Consider the lube oil experiment as treated in Problem 10.52 in Section 10.5 and Problem 10.62 in Section 10.6, in which G is the space of of linear functions of the levels of the four factors. Carry out an F test of the hypothesis that the regression function is a linear function of the levels of factors A, B, and C.

10.80 Consider the polymer experiment and the normal linear regression model corresponding to the space of cubic polynomials on $\mathbb{R}$. Carry out the F test of the hypothesis that the regression function is a quadratic polynomial.

10.81 Consider the polymer experiment and the normal linear regression model corresponding to the space of quartic polynomials on $\mathbb{R}$. Carry out the F test of the hypothesis that the regression function is a quadratic polynomial.

10.82 Consider the lube oil experiment but ignore factors C and D as in Example 10.27. Let $\hat{\beta}_0 + \hat{\beta}_1 x_1 + \hat{\beta}_2 (x_1^2 - \frac{2}{3}) + \hat{\beta}_3 x_2 + \hat{\beta}_4 (x_2^2 - \frac{2}{3})$ denote the least-squares fit to the data having the form of the sum of a quadratic polynomial in x_1 and a quadratic polynomial in x_2. **(a)** Explain why the numerical values of the coefficients $\hat{\beta}_0, \ldots, \hat{\beta}_4$ coincide with their values as shown in Section 10.5. **(b)** Express the fitted sum of squares in terms of these coefficients.

10.83 Consider the lube oil experiment but ignore factors C and D as in Example 10.27. Test the assumption that the mean response is a linear function of x_1 and x_2. Show your results in the form of an ANOVA table.

10.84   Consider the 41 trials of the sugar beet experiment involving varieties 1, 3, 4, and 5. The residual sum of squares for the least-squares fit to this data in the linear space consisting of sums of arbitrary functions of variety and quintic polynomial functions of plot is given by $\text{RSS} \doteq 0.655$, and the residual sum of squares for the least-squares fit in the subspace consisting of quintic polynomial functions of plot is given by $\text{RSS}_0 \doteq 1.078$. Use these results to test the hypothesis that the mean percentage of sugar content does not differ among the four varieties. Show your calculations in the form of an ANOVA table.

10.85   Consider the experimental data, the linear space G, and the subspace G_0 of G in Problem 10.75 in Section 10.7. Extend the ANOVA table in the answer to the problem to show the P-value for testing the hypothesis that the regression function is in G_0 under the assumption that it is in G.

10.86   Suppose G contains the constant functions as a proper subspace. Let F denote the F statistic for testing the hypothesis that the regression function is constant or, equivalently, that the mean response does not depend on the factor-level combination. Let R^2 denote the squared multiple correlation coefficient. Verify that **(a)** $F = \frac{n-p}{p-1} \frac{R^2}{1-R^2}$; **(b)** $R^2 = \frac{(p-1)F}{(n-p)+(p-1)F}$; **(c)** under the assumptions of the normal one-sample model (that is, under those of the normal linear regression model corresponding to the space of constant functions), R^2 has the beta distribution with parameters $\alpha_1 = (p-1)/2$ and $\alpha_2 = (n-p)/2$ (see Problem 7.22 in Section 7.2).

10.87   Suppose $G_0 \subset G$ and $1 < p_0 < p$, where p_0 and p are the dimensions of G_0 and G, respectively, and that G_0 contains the constant functions. Let R^2 and R_0^2 denote the coefficients of determination for the least-squares fits in G and G_0, respectively, and let F denote the F statistic for testing the hypothesis that the regression function is in G_0 under the assumption that it is in G. Express F in terms of n, p, p_0, R^2, and R_0^2.

10.88   Suppose the total sum of squares is positive. The squared multiple correlation coefficient can be written as $R^2 = 1 - \text{RSS}/\text{TSS}$. Under the assumptions of the homoskedastic linear regression model, $\text{RSS}/(n-p)$ has mean σ^2; under the homoskedastic regression model with constant regression function, $\text{TSS}/(n-1)$ has mean σ^2. The *adjusted squared multiple correlation coefficient* R_a^2 is defined as

$$R_a^2 = 1 - \frac{\text{RSS}/(n-p)}{\text{TSS}/(n-1)} = 1 - \frac{n-1}{n-p}\frac{\text{RSS}}{\text{TSS}} = 1 - \frac{n-1}{n-p}(1-R^2).$$

The purpose of the adjustment is to avoid spurious inflation of R^2 due to overfitting the data with a high-dimensional linear space G. (Note, in particular, that $R^2 = 1$ if $p = n$; that is, if there is one run per design point and G is saturated.) Let

$$R_{a0}^2 = 1 - \frac{\text{RSS}_0/(n-p_0)}{\text{TSS}/(n-1)}$$

be the adjusted squared multiple correlation coefficient corresponding to a linear space G_0 having dimension $p_0 < p$, suppose $\text{RSS} > 0$, and set

$$F = \frac{(\text{RSS}_0 - \text{RSS})/(p-p_0)}{\text{RSS}/(n-p)}.$$

Show that $R_a^2 > R_{a0}^2$ if and only if $F > 1$.

Orthogonal Arrays

11.1
Main Effects

Orthogonality

Two linear spaces G_1 and G_2 are said to be orthogonal if each function in G_1 is orthogonal to every function in G_2. Let a basis of G_1 and a basis of G_2 be specified. Then G_1 and G_2 are orthogonal if and only each basis function of G_1 is orthogonal to every basis function of G_2. More generally, linear spaces $G_1, \ldots, G_L$ are said to be orthogonal if G_j and G_l are orthogonal for $j \neq l$.

Let $G_1, \ldots, G_L$ be orthogonal linear spaces. Their *direct sum* $G_1 \oplus \cdots \oplus G_L$ is the linear space defined by

$$G_1 \oplus \cdots \oplus G_L = \{ g_1 + \cdots + g_L : g_1 \in G_1, \ldots, g_L \in G_L \}.$$

This operation is commutative; in particular, $G_1 \oplus G_2 = G_2 \oplus G_1$. The operation is also associative; in particular, $(G_1 \oplus G_2) \oplus G_3 = G_1 \oplus (G_2 \oplus G_3)$.

(The above definition of the direct sum is also used when $G_1, \ldots, G_L$ are not orthogonal, provided that they satisfy the condition of *independence*; that is, the only choice of $g_1 \in G_1, \ldots, g_L \in G_L$ such that $g_1 + \cdots + g_L = 0$ is given by $g_1 = 0, \ldots, g_L = 0$.) If this condition is not satisfied, we refer to $\{ g_1 + \cdots + g_L : g_1 \in G_1, \ldots, g_L \in G_L \}$ as the *sum* of the spaces $G_1, \ldots, G_L$, which is denoted by $G_1 + \cdots + G_L$. Properties of direct sums of spaces do not generally hold for sums.

For example, the dimension of the direct sum of spaces equals the sum of the individual dimensions, as we will see for orthogonal spaces in the next theorem, but the dimension of the sum of nonindependent, finite-dimensional spaces is less than the sum of the individual dimensions.)

THEOREM 11.1   Let $G_1, \ldots, G_L$ be orthogonal, identifiable linear spaces. For $1 \leq l \leq L$, let G_l have dimension p_l and let g_{lj}, $1 \leq j \leq p_l$, be an orthogonal basis of G_l. Then the linear space $G = G_1 \oplus \cdots \oplus G_L$ is identifiable, its dimension p is given by $p = p_1 + \cdots + p_n$, and the functions g_{lj}, $1 \leq l \leq L$ and $1 \leq j \leq p_l$, form a basis of G.

Proof   The functions g_{lj}, $1 \leq l \leq L$ and $1 \leq j \leq p_l$, have positive norms, are orthogonal to each other, and span G. The desired results now follow from Theorem 9.21 in Section 9.4.   ∎

In the context of Theorem 1.11, the orthogonal subspaces $G_1, \ldots, G_L$ of their direct sum are referred to as its *orthogonal components*.

THEOREM 11.2   Let G be an identifiable p-dimensional linear space of functions, let G_0 be a p_0-dimensional subspace of G, and let $G \ominus G_0$ be the collection of all functions in G that are orthogonal to every function in G_0. Then $G \ominus G_0$ is a $(p - p_0)$-dimensional subspace of G, G_0 and $G \ominus G_0$ are orthogonal spaces, and $G = G_0 \oplus (G \ominus G_0)$. Let $g_1, \ldots, g_p$ be an orthogonal basis of G such that $g_1, \ldots, g_{p_0}$ is a basis of G_0. Then $g_{p_0+1}, \ldots, g_p$ is a basis of $G \ominus G_0$.

Proof   Let $g_1, \ldots, g_p$ be an orthogonal basis of G such that $g_1, \ldots, g_{p_0}$ is a basis of G_0. Let $g = b_1 g_1 + \cdots + b_p g_p \in G$. Then g is orthogonal to every function in G_0 if and only if g is orthogonal to each of the functions $g_1, \ldots, g_{p_0}$ and hence if and only if $b_1 = \cdots = b_{p_0} = 0$. Thus, g is orthogonal to every function in G_0 if and only if g is a linear combination of $g_{p_0+1}, \ldots, g_p$. Consequently, $G \ominus G_0$ is the span of these functions, so it is a $(p - p_0)$-dimensional linear space having basis $g_{p_0+1}, \ldots, g_p$. It follows from the definition of $G \ominus G_0$ that G_0 and $G \ominus G_0$ are orthogonal. Moreover, $G = G_0 \oplus (G \ominus G_0)$.   ∎

The linear space $G \ominus G_0$ in Theorem 11.2 is referred to as the *orthogonal complement* of G_0 relative to G. According to the next result, the orthogonal projection onto a sum of orthogonal components equals the sum of the orthogonal projections onto the individual components.

THEOREM 11.3   Suppose $G = G_1 \oplus \cdots \oplus G_L$, where $G_1, \ldots, G_L$ are orthogonal, identifiable linear spaces. Then $P_G = P_{G_1} + \cdots + P_{G_L}$; that is,

$$P_G h = P_{G_1} h + \cdots + P_{G_L} h, \qquad h \in H. \tag{1}$$

Moreover,

$$\|P_G h\|^2 = \|P_{G_1} h\|^2 + \cdots + \|P_{G_L} h\|^2, \qquad h \in H. \tag{2}$$

Proof To verify (1), choose $g \in G$ and write $g = g_1 + \cdots + g_L$, where $g_1 \in G_1, \ldots, g_L \in G_L$. Let $h \in H$ and $1 \le j \le L$. Then $\langle g_j, h - P_{G_j} h \rangle = 0$ and hence

$$\langle g_j, P_{G_j} h \rangle = \langle g_j, h \rangle. \tag{3}$$

Observe next that

$$\langle g_j, P_{G_l} h \rangle = 0, \qquad l \ne j, \tag{4}$$

since G_j and G_l are orthogonal for $l \ne j$. It follows from (3) and (4) that

$$\langle g_j, P_{G_1} h + \cdots + P_{G_L} h \rangle = \langle g_j, h \rangle.$$

By summing on j, we get

$$\langle g, P_{G_1} h + \cdots + P_{G_L} h \rangle = \langle g, h \rangle.$$

Consequently,

$$\langle g, h - P_{G_1} h - \cdots - P_{G_L} h \rangle = 0, \qquad g \in G. \tag{5}$$

Since $P_{G_1} h + \cdots + P_{G_L} h \in G_1 \oplus \cdots \oplus G_L = G$, we conclude from (5) that (1) holds. Equation (2) follows from (1) and the Pythagorean theorem. ∎

Let G be an identifiable p-dimensional linear space containing the constant functions and let $G^0 = G \ominus G_{\mathrm{con}}$ be the $(p-1)$-dimensional subspace of G consisting of the functions in G that are orthogonal to 1 (and hence to all constant functions). Then $G = G_{\mathrm{con}} \oplus G^0$. It follows from Theorem 11.3 that

$$P_G = P_{G_{\mathrm{con}}} + P_{G^0} \tag{6}$$

and hence that

$$P_{G^0} = P_G - P_{G_{\mathrm{con}}}. \tag{7}$$

EXAMPLE 11.1 (a) Consider the design set $\{-1, 1\}$, suppose $n_1 = n_2$, and let G be the space of linear functions on $\mathbb{R}$. Determine a basis of $G^0 = G \ominus G_{\mathrm{con}}$.

(b) Consider the design set $\{-1, 0, 1\}$, suppose $n_1 = n_2 = n_3$, and let G be the space of quadratic polynomials on $\mathbb{R}$. Determine an orthogonal basis of $G^0 = G \ominus G_{\mathrm{con}}$.

Solution (a) Now 1 and x form an orthogonal basis of G, so x is a basis of G^0.

(b) According to the solution to Example 9.26(a) in Section 9.4, 1, x, and $x^2 - \frac{2}{3}$ form an orthogonal basis of G. Thus, by Theorem 11.2, x and $x^2 - \frac{2}{3}$ form an orthogonal basis of G^0. ∎

EXAMPLE **11.2**   **(a)** Consider the design set $\{-1, 1\}$, suppose $n_1 = n_2$, and let G be the space of all functions on $\{-1, 1\}$. Determine a basis of $G^0 = G \ominus G_{\text{con}}$.

(b) Consider the design set $\{-1, 0, 1\}$, suppose $n_1 = n_2 = n_3$, and let G be the space of all functions on $\{-1, 0, 1\}$. Determine an orthogonal basis of $G^0 = G \ominus G_{\text{con}}$.

Solution   **(a)** According to the solution to Example 11.1(a), x is a basis of G^0. Alternatively, 1, $I_1(x) - \frac{1}{2}$ is a an orthogonal basis of G, where $I_1(x) = \text{ind}(x = -1)$ (see Examples 9.4 and 9.5 in Section 9.1). Thus, $I_1(x) - \frac{1}{2}$ is a basis of G^0. [Note that $I_1(x) - \frac{1}{2} = -\frac{x}{2}$ on $\{-1, 1\}$.]

(b) According to the solution to Example 11.1(b), x and $x^2 - \frac{2}{3}$ form an orthogonal basis of G^0. Alternatively, according to the solution to Example 9.30 in Section 9.5, 1, $I_1(x) - \frac{1}{3}$, $\frac{1}{2}I_1(x) + I_2(x) - \frac{1}{2}$ form an orthogonal basis of G, where $I_1(x) = \text{ind}(x = -1)$ and $I_2(x) = \text{ind}(x = 0)$. Thus, by Theorem 11.2, $I_1(x) - \frac{1}{3}$ and $\frac{1}{2}I_1(x) + I_2(x) - \frac{1}{2}$ form an orthogonal basis of G^0.   ∎

Design Distributions

The probability function f on the design set given by $f(\mathbf{x}'_k) = n_k/n$ for $1 \le k \le d$ is referred to as the *design probability function*. Observe that $f(\mathbf{x})$ is the relative frequency of runs at the factor-level combination $\mathbf{x}$. The corresponding distribution on the design set is referred to as the *design distribution*. Let $\mathbf{X} = (X_1, \ldots, X_M)$ be distributed as the design distribution. Then X_m ranges over $\mathcal{X}'_m$ for $1 \le m \le M$, and $\mathbf{X}$ ranges over the design set $\mathcal{X}'$. Observe that $X_1, \ldots, X_M$ and $\mathbf{X}$ are "artificial" random variables, which we refer to as *design random variables*. In the context of the experimental form of the regression model, we continue to regard the inputs $\mathbf{x}_1, \ldots, \mathbf{x}_n$ as fixed, nonrandom quantities.

Let $h_1, h_2 \in H$. Then

$$\langle h_1, h_2 \rangle = \sum_k n_k h_1(\mathbf{x}'_k) h_2(\mathbf{x}'_k) = n \sum_k f(\mathbf{x}'_k) h_1(\mathbf{x}'_k) h_2(\mathbf{x}'_k)$$

and hence

$$\langle h_1, h_2 \rangle = nE[\, h_1(\mathbf{X}) h_2(\mathbf{X})\,]. \tag{8}$$

In particular, for $h \in H$,

$$\|h\|^2 = nE[\, h^2(\mathbf{X})\,]. \tag{9}$$

For $1 \le m \le M$, let G_m be an identifiable p_m-dimensional linear space of functions on $\mathcal{X}$ that depend only on x_m and suppose G_m contains the constant functions (for example, G_m could be the space of quadratic polynomials in x_m, which has the basis 1, x_m, x_m^2). Set $G_m^0 = G_m \ominus G_{\text{con}}$. Then, by (6) and (7),

$$P_{G_m} = P_{G_{\text{con}}} + P_{G_m^0} \tag{10}$$

and

$$P_{G_m^0} = P_{G_m} - P_{G_{\text{con}}}. \tag{11}$$

THEOREM 11.4 Suppose the design random variables X_l and X_m are independent, where $l \neq m$. Then G_l^0 and G_m^0 are orthogonal linear spaces.

Proof Choose members $g_l(x_l)$ and $g_m(x_m)$ of G_l^0 and G_m^0, respectively. Then, by (8),

$$E[\,g_l(X_l)\,] = \frac{1}{n}\langle 1, g_l(x_l)\rangle = 0 \quad \text{and} \quad E[\,g_m(X_m)\,] = \frac{1}{n}\langle 1, g_m(x_m)\rangle = 0.$$

Thus, by (8) and the independence of X_l and X_m,

$$\langle g_l(x_l), g_m(x_m)\rangle = nE[\,g_l(X_l)g_m(X_m)\,] = nE[\,g_l(X_l)\,]E[\,g_m(X_m)\,] = 0.$$

Consequently G_l^0 and G_m^0 are orthogonal. ∎

The design random variables $X_1, \ldots, X_M$ are said to be *pairwise-independent* if X_l and X_m are independent for $l \neq m$. The next result follows immediately from Theorem 11.4.

COROLLARY 11.1 Suppose the design random variables $X_1, \ldots, X_M$ are pairwise-independent. Then $G_{\mathrm{con}}, G_1^0, \ldots, G_M^0$ are orthogonal linear spaces.

Suppose the design random variables $X_1, \ldots, X_M$ are pairwise-independent and hence, by Corollary 11.1, that the linear spaces $G_{\mathrm{con}}, G_1^0, \ldots, G_M^0$ are orthogonal. Set $G = G_{\mathrm{con}} \oplus G_1^0 \oplus \cdots \oplus G_M^0$, which is the space of all functions g on $\mathcal{X}$ having the additive form $g(x_1, \ldots, x_M) = g_1(x_1) + \cdots + g_M(x_M)$, where $g_m \in G_m$ for $1 \leq m \leq M$. By Theorem 11.3,

$$P_G = P_{G_{\mathrm{con}}} + P_{G_1^0} + \cdots + P_{G_M^0}, \tag{12}$$

so

$$P_G - P_{G_{\mathrm{con}}} = P_{G_1^0} + \cdots + P_{G_M^0}$$

and hence

$$\|P_G h - P_{G_{\mathrm{con}}} h\|^2 = \|P_{G_1^0} h\|^2 + \cdots + \|P_{G_M^0} h\|^2, \qquad h \in H. \tag{13}$$

We will now give some statistical interpretations of the above results. First, since

$$P_{G_{\mathrm{con}}}[\bar{Y}(\cdot)] = \bar{Y}, \tag{14}$$

we conclude from (7) that

$$P_{G_m^0}[\bar{Y}(\cdot)] = P_{G_m}[\bar{Y}(\cdot)] - \bar{Y}, \qquad 1 \leq m \leq M, \tag{15}$$

and hence that

$$\|P_{G_m^0}[\bar{Y}(\cdot)]\|^2 = \|P_{G_m}[\bar{Y}(\cdot)] - \bar{Y}\|^2, \qquad 1 \leq m \leq M.$$

We refer to $P_{G_m^0}[\bar{Y}(\cdot)]$ as the estimated *main effect* of the mth factor and to the corresponding fitted sum of squares $\|P_{G_m^0}[\bar{Y}(\cdot)]\|^2$ as the sum of squares due to this main effect. Let $g_{mj}(x_m)$, $1 \le j \le p_m - 1$, be an orthogonal basis of G_m^0 and write

$$P_{G_m^0}[\bar{Y}(\cdot)] = \sum_{j=1}^{p_m-1} \hat{\beta}_j g_{mj}(x_m),$$

where

$$\hat{\beta}_j = \frac{\langle g_{mj}(x_m), \bar{Y}(\cdot) \rangle}{\|g_{mj}(x_m)\|^2} = \frac{\sum_{i=1}^n g_{mj}(x_{mi}) Y_i}{\sum_{i=1}^n g_{mj}^2(x_{mi})}.$$

Then

$$\|P_{G_m^0}[\bar{Y}(\cdot)]\|^2 = \sum_{j=1}^{p_m-1} \hat{\beta}_j^2 \|g_{mj}(x_m)\|^2. \tag{16}$$

Suppose the design random variables $X_1, \ldots, X_M$ are pairwise-independent and that $G = G_{\mathrm{con}} \oplus G_1^0 \oplus \cdots \oplus G_M^0$. Then, by (12), (14), and (15),

$$P_G[\bar{Y}(\cdot)] = \bar{Y} + P_{G_1^0}[\bar{Y}(\cdot)] + \cdots + P_{G_M^0}[\bar{Y}(\cdot)]$$

$$= \bar{Y} + \{P_{G_1}[\bar{Y}(\cdot)] - \bar{Y}\} + \cdots + \{P_{G_M}[\bar{Y}(\cdot)] - \bar{Y}\},$$

so

$$P_G[\bar{Y}(\cdot)] = P_{G_1}[\bar{Y}(\cdot)] + \cdots + P_{G_M}[\bar{Y}(\cdot)] - (M-1)\bar{Y}. \tag{17}$$

Let the factors be labeled $A, B, C, \ldots$. We denote the sum of squares due to the main effects of these factors as $\mathrm{SS}_A, \mathrm{SS}_B, \mathrm{SS}_C, \ldots$. Thus, $\mathrm{SS}_A = \|P_{G_1^0}[\bar{Y}(\cdot)]\|^2$, $\mathrm{SS}_B = \|P_{G_2^0}[\bar{Y}(\cdot)]\|^2$, and so forth. Let $\mathrm{FSS} = \|P_G[\bar{Y}(\cdot)] - \bar{Y}\|^2$ denote the fitted sum of squares for the least-squares fit in G. We conclude from (13) applied to $h = \bar{Y}(\cdot)$ that $\mathrm{FSS} = \mathrm{SS}_A + \mathrm{SS}_B + \cdots$. In particular, if $M = 2$, then $\mathrm{FSS} = \mathrm{SS}_A + \mathrm{SS}_B$; and if $M = 3$, then $\mathrm{FSS} = \mathrm{SS}_A + \mathrm{SS}_B + \mathrm{SS}_C$.

Under the additional assumption that the regression function is a member of G, it can be written in the additive form $\mu(x_1, \ldots, x_M) = \bar{\mu} + g_1(x_1) + \cdots + g_M(x_M)$, where $\bar{\mu} = P_{G_{\mathrm{con}}}[\mu(\cdot)]$ and $g_m(x_m)$ is the orthogonal projection of the regression function onto G_m^0 for $1 \le m \le M$. Let $x_{m1}, x_{m2} \in \mathcal{X}_m$ be given. Then changing the level of the mth factor from x_{m1} to x_{m2} with the levels of the other factors held fixed causes the mean response to change by the amount $\tau = g_m(x_{m2}) - g_m(x_{m1})$. In this formula for the linear parameter τ, the orthogonal projection $g_m(x_m)$ of the regression function onto G_m^0 can be replaced by the orthogonal projection of the regression function onto G_m since the two orthogonal projections differ by a constant.

Similarly, the least-squares estimate in G of the regression function can be written as $\hat{\mu}(x_1, \ldots, x_M) = \bar{Y} + \hat{g}_1(x_1) + \cdots + \hat{g}_M(x_M)$, where $\hat{g}_m(x_m)$ is the least-squares estimate in G_m^0 of the regression function—that is, the orthogonal projection of $\bar{Y}(\cdot)$ onto G_m^0. The least-squares estimate of the linear parameter $\tau = g_m(x_{m2}) - g_m(x_{m1})$ is given by $\hat{\tau} = \hat{g}_m(x_{m2}) - \hat{g}_m(x_{m1})$. In this formula for $\hat{\tau}$, the function $\hat{g}_m(x_m)$ can be replaced by the least-squares estimate in G_m of the regression function. Suppose, in particular, that G_m is saturated and that x_{m1} and x_{m2} are levels of the mth factor that are actually taken on in the experiment. Then $\hat{\tau}$ is the difference between the sample mean on all the runs in which the mth factor is at level x_{m1} and the sample mean on all the runs in which it is at level x_{m2}.

We say that G_m is a saturated space of functions of the level x_m of the mth factor if it is saturated in the usual sense when we ignore the other factors. Since G_m is assumed to be identifiable, it is saturated (in the sense just defined) if and only if the dimension p_m of G_m coincides with the number d_m of points in $\mathcal{X}'_m$ (that is, with the number of levels of the mth factor that are actually used in the experiment). Suppose G_m is saturated. If $\mathbf{x} = (x_1, \ldots, x_M) \in \mathcal{X}$ with $x_m \in \mathcal{X}'_m$, then the value of $P_{G_m}[\bar{Y}(\cdot)]$ at $\mathbf{x}$ equals the sample mean of the responses on all trials in which the level of the mth factor equals x_m. Moreover, the corresponding fitted sum of squares $\|P_{G^0_m}[\bar{Y}(\cdot)]\|^2$ equals the between sum of squares when we ignore all but the mth factor.

Suppose, in particular, that $M = 1$ and let $G = G_1$ be saturated. Then $\mathrm{SS}_A = \mathrm{BSS} = \sum_k n_k (\bar{Y}_k - \bar{Y})^2$.

When $M \geq 2$, it is convenient to label the d_m points in the set $\mathcal{X}'_m$ of levels of the mth factor in the design by $1, \ldots, d_m$ in some order. Suppose $M = 2$. It is convenient to let n_{ij} denote the number of trials in which the level of the first factor has label i and the level of the second factor has label j and to let $\bar{Y}_{ij}$ denote the sample mean of the responses on these n_{ij} trials. Also, we let $n_{i\cdot} = \sum_j n_{ij}$ denote the number of trials in which the level of the first factor has label i, and we let

$$\bar{Y}_{i\cdot} = \frac{1}{n_{i\cdot}} \sum_j n_{ij} \bar{Y}_{ij}$$

denote the sample mean of the responses on these $n_{i\cdot}$ trials. Similarly, we let $n_{\cdot j} = \sum_i n_{ij}$ denote the number of trials in which the level of the second factor has label j, and we let

$$\bar{Y}_{\cdot j} = \frac{1}{n_{\cdot j}} \sum_i n_{ij} \bar{Y}_{ij}$$

denote the sample mean of the responses on these $n_{\cdot j}$ trials. The sample mean of the responses on all n trials is given by

$$\bar{Y} = \frac{1}{n} \sum_i \sum_j n_{ij} \bar{Y}_{ij} = \frac{1}{n} \sum_i n_{i\cdot} \bar{Y}_{i\cdot} = \frac{1}{n} \sum_j n_{\cdot j} \bar{Y}_{\cdot j}.$$

(In this context, it is conventional to denote the overall sample mean $\bar{Y}$ by $\bar{Y}_{\cdot\cdot}$.) Suppose G_1 is saturated. If $\mathbf{x} = (x_1, x_2) \in \mathcal{X}$, where x_1 is the level of the first factor in $\mathcal{X}'_1$ having label i, then the value of $P_{G^0_1}[\bar{Y}(\cdot)]$ at $\mathbf{x}$ equals $\bar{Y}_{i\cdot}$. Moreover,

$$\mathrm{SS}_A = \sum_i n_{i\cdot} (\bar{Y}_{i\cdot} - \bar{Y})^2,$$

which is the between sum of squares when we ignore factor B. Similarly, suppose G_2 is saturated. If $\mathbf{x} = (x_1, x_2) \in \mathcal{X}$, where x_2 is the level of the second factor in $\mathcal{X}'_2$ having label j, then the value of $P_{G^0_2}[\bar{Y}(\cdot)]$ at $\mathbf{x}$ equals $\bar{Y}_{\cdot j}$. Moreover,

$$\mathrm{SS}_B = \sum_j n_{\cdot j} (\bar{Y}_{\cdot j} - \bar{Y})^2,$$

which is the between sum of squares when we ignore factor A.

Suppose the design random variables X_1 and X_2 are independent and that $G = G_{\mathrm{con}} \oplus G_1^0 \oplus G_2^0$, where G_1 and G_2 are saturated. It follows from (17) that if $\mathbf{x} = (x_1, x_2) \in \mathcal{X}$, where x_1 is the level of the first factor in $\mathcal{X}_1'$ having label i and x_2 is the level of the second factor in $\mathcal{X}_2'$ having label j, then the value of $P_G[\bar{Y}(\cdot)]$ at $\mathbf{x}$ equals $\bar{Y}_{i\cdot} + \bar{Y}_{\cdot j} - \bar{Y}$.

Similar notation is used for $M = 3$ (and higher values of M as well). Let $\mathbf{x} = (x_1, x_2, x_3) \in \mathcal{X}$. Suppose G_1 is saturated and x_1 is the level of the first factor in $\mathcal{X}_1'$ having label i. Then the value of $P_{G_1^0}[\bar{Y}(\cdot)]$ at $\mathbf{x}$ equals $\bar{Y}_{i\cdot\cdot}$. Moreover,

$$\mathrm{SS}_A = \sum_i n_{i\cdot\cdot}(\bar{Y}_{i\cdot\cdot} - \bar{Y})^2,$$

which is the between sum of squares when we ignore factors B and C. (In this context, $\bar{Y}$ is commonly denoted by $\bar{Y}_{\cdots}$.) Similarly, suppose G_2 is saturated and x_2 is the level of the second factor in $\mathcal{X}_2'$ having label j. Then the value of $P_{G_2^0}[\bar{Y}(\cdot)]$ at $\mathbf{x}$ equals $\bar{Y}_{\cdot j\cdot}$. Moreover,

$$\mathrm{SS}_B = \sum_j n_{\cdot j\cdot}(\bar{Y}_{\cdot j\cdot} - \bar{Y})^2,$$

which is the between sum of squares when we ignore factors A and C. Suppose G_3 is saturated and x_3 is the level of the third factor in $\mathcal{X}_3'$ having label k. Then the value of $P_{G_3^0}[\bar{Y}(\cdot)]$ at x equals $\bar{Y}_{\cdot\cdot k}$. Moreover,

$$\mathrm{SS}_C = \sum_k n_{\cdot\cdot k}(\bar{Y}_{\cdot\cdot k} - \bar{Y})^2,$$

which is the between sum of squares when we ignore factors A and B.

Suppose the design random variables X_1, X_2, and X_3 are pairwise-independent and that $G = G_{\mathrm{con}} \oplus G_1^0 \oplus G_2^0 \oplus G_3^0$, where G_1, G_2 and G_3 are saturated. It follows from (17) that if $\mathbf{x} = (x_1, x_2, x_3) \in \mathcal{X}$, where x_1 is the level of the first factor in $\mathcal{X}_1'$ having label i, x_2 is the level of the second factor in $\mathcal{X}_2'$ having label j, and x_3 is the level of the third factor in $\mathcal{X}_3'$ having label k, then the value of $P_G[\bar{Y}(\cdot)]$ at $\mathbf{x}$ equals $\bar{Y}_{i\cdot\cdot} + \bar{Y}_{\cdot j\cdot} + \bar{Y}_{\cdot\cdot k} - 2\bar{Y}$.

Orthogonal Arrays

The experimental design is said to have the form of an *orthogonal array* if the design random variables $X_1, \ldots, X_M$ are pairwise-independent and X_m is uniformly distributed on $\mathcal{X}_m'$ for $1 \le m \le M$. Recall that $d_m = \#(\mathcal{X}_m')$ is the number of distinct levels of the mth factor that are taken on in the experiment. Thus, X_m is uniformly distributed on $\mathcal{X}_m'$ if and only if each level in $\mathcal{X}_m'$ of the mth factor is taken on during n/d_m runs. Let $1 \le l, m \le M$ with $l \ne m$. suppose X_l is uniformly distributed on $\mathcal{X}_l'$ and X_m is uniformly distributed on $\mathcal{X}_m'$. Then X_l and X_m are independent if and only if each combination of a level in $\mathcal{X}_l'$ of the lth factor and a level in $\mathcal{X}_m'$ of the mth factor is taken on during $n/(d_l d_m)$ runs—that is, if and only if (X_l, X_m) is uniformly distributed on $\mathcal{X}_l' \times \mathcal{X}_m'$. Suppose, conversely, that (X_l, X_m) is uniformly distributed on $\mathcal{X}_l' \times \mathcal{X}_m'$. Then X_l and X_m are independent, X_l is uniformly distributed on $\mathcal{X}_l'$, and X_m is uniformly distributed on $\mathcal{X}_m'$ (see Theorem 1.8 in Section 1.8).

The lube oil experiment has the form of an orthogonal array involving 27 runs and four factors, each having the three levels -1, 0, and 1. Consider the additive fit to the experimental data in which G_m is the (identifiable and saturated) space of quadratic polynomials in the level of the mth factor for $m = 1, 2, 3, 4$. Table 11.1 is the corresponding ANOVA table.

TABLE 11.1

Source	SS	DF	MS	F	P-value
A	4,496.29	2	2,248.14	16.341	.000
B	2,768.69	2	1,384.35	10.063	.001
C	5,519.79	2	2,759.89	20.061	.000
D	283.37	2	141.68	1.030	.377
Residuals	2,476.32	18	137.57		
Total	15,544.46	26			

EXAMPLE 11.3 Verify the value given for SS_A in Table 11.1.

Solution Let the levels -1, 0, and 1 of each factor be labeled as 1, 2, and 3, respectively. Then

$$SS_A = \sum_{i=1}^{3} n_{i\cdots}(\bar{Y}_{i\cdots} - \bar{Y})^2 = 9\sum_{i=1}^{3}(\bar{Y}_{i\cdots} - \bar{Y})^2.$$

From the lube oil data in Section 8.2, we get

$$\bar{Y}_{1\cdots} = \frac{4.2 + 5.9 + 8.2 + 13.1 + 16.4 + 30.7 + 9.5 + 22.2 + 31.0}{9} \doteq 15.689.$$

Similarly, $\bar{Y}_{2\cdots} = 27.3$ and $\bar{Y}_{3\cdots} \doteq 46.956$, so $\bar{Y} \doteq (15.689 + 27.3 + 46.956)/3 \doteq 29.981$. Thus,

$$SS_A = 9[(15.689 - 29.981)^2 + (27.3 - 29.981)^2 + (46.956 - 29.981)^2]$$

$$\doteq 4496.29.$$

Alternatively, by the solution to Example 11.1(b), x_1 and $x_1^2 - \frac{2}{3}$ form an orthogonal basis of G_1^0. The squared norms of these basis functions are given by

$$\|x_1\|^2 = 9[(-1)^2 + 0^2 + 1^2] = 18$$

and

$$\|x_1^2 - \tfrac{2}{3}\|^2 = 9[(\tfrac{1}{3})^2 + (-\tfrac{2}{3})^2 + (\tfrac{1}{3})^2] = 6.$$

Thus,

$$P_{G_1^0}[\bar{Y}(\cdot)] = \hat{\beta}_1 x_1 + \hat{\beta}_2(x_1^2 - \tfrac{2}{3}),$$

where

$$\hat{\beta}_1 = \frac{\langle x_1, \bar{Y}(\cdot)\rangle}{\|x_1\|^2} = \frac{\bar{Y}_{3\cdots} - \bar{Y}_{1\cdots}}{2} \doteq 15.633$$

and

$$\hat{\beta}_2 = \frac{\langle x_1^2 - \frac{2}{3}, \bar{Y}(\cdot)\rangle}{\|x_1^2 - \frac{2}{3}\|^2} = \frac{\bar{Y}_{1\cdots} - 2\bar{Y}_{2\cdots} + \bar{Y}_{3\cdots}}{2} \doteq 4.022.$$

Consequently,

$$SS_A = \|P_{G_1^0}[\bar{Y}(\cdot)]\|^2$$
$$= \hat{\beta}_1^2\|x_1\|^2 + \hat{\beta}_2^2\|x_1^2 - \tfrac{2}{3}\|^2$$
$$\doteq 18(15.633)^2 + 6(4.022)^2$$
$$\doteq 4496.29. \quad \blacksquare$$

Latin square designs, such as the 6×6 design used in the fertilizer experiment described in Section 8.2, form orthogonal arrays. In the analysis of the experimental data obtained from a $c \times c$ Latin square design, it is customary to let G_1, G_2, and G_3 be the spaces of all functions of the levels of the row, column, and treatment factors, respectively, and to consider the corresponding additive space $G = G_{\text{con}} \oplus G_1^0 \oplus G_2^0 \oplus G_3^0$, whose indicated components are orthogonal. The row, column, and treatment sums of squares are given explicitly by

$$SS_R = c\sum_{i=1}^{c}(\bar{Y}_{i\cdot\cdot} - \bar{Y})^2, \quad SS_C = c\sum_{j=1}^{c}(\bar{Y}_{\cdot j\cdot} - \bar{Y})^2, \quad \text{and} \quad SS_T = c\sum_{k=1}^{c}(\bar{Y}_{\cdot\cdot k} - \bar{Y})^2.$$

Here, for a given value of i, $\bar{Y}_{i\cdot\cdot}$ is the average of the c responses in row i; for a given value of j, $\bar{Y}_{\cdot j\cdot}$ is the average of the c responses in column j; and for a given value of k, $\bar{Y}_{\cdot\cdot k}$ is the average of the c responses to treatment k. The least-squares estimate of the mean response when treatment k is applied to row i and column j equals $\bar{Y}_{i\cdot\cdot} + \bar{Y}_{\cdot j\cdot} + \bar{Y}_{\cdot\cdot k} - 2\bar{Y}$. The fitted sum of squares is given by FSS $= SS_R + SS_C + SS_T$, and the ANOVA table has the form shown in Table 11.2.

TABLE 11.2

Source	SS	DF
Rows	SS_R	$c - 1$
Columns	SS_C	$c - 1$
Treatments	SS_T	$c - 1$
Residuals	RSS	$(c-1)(c-2)$
Total	TSS	$c^2 - 1$

EXAMPLE 11.4   (a) Determine and interpret the ANOVA table for the fertilizer experiment.

(b) Determine the 95% confidence interval for the change in mean yield caused by switching from treatment 1 (low levels of nitrogen and phosphorous) to treatment 6 (high levels of nitrogen and phosphorous).

Solution   **(a)** Here we label the treatments from 1 to 6 instead of from A to F. Then treatment 5 was applied to the plot in row 1 and column 1, and the corresponding yield is given by $Y_{115} = 633$. Also,

$$\bar{Y}_{1..} = \frac{633 + 527 + 652 + 390 + 504 + 416}{6} \doteq 520.333,$$

$$\bar{Y}_{.1.} = \frac{633 + 489 + 384 + 620 + 452 + 500}{6} = 513,$$

and

$$\bar{Y}_{..1} = \frac{390 + 282 + 384 + 323 + 432 + 259}{6} = 345.$$

Table 11.3 is the ANOVA table. Accordingly, 15% of the total variation in the yield of potatoes is explained by the row effect, 7% is explained by the column effect, 69% is explained by the treatment effect, and 9% is unexplained by the three main effects. There is clear evidence of row and treatment effects, but only weak evidence of a column effect. (The P-value for columns corresponds to the F test of the hypothesis that the regression function lies in the subspace $G_{con} \oplus G_1^0 \oplus G_3^0$ of G; that is, the mean yield is an additive function of row and treatment. The P-values for rows and treatments have similar interpretations.)

TABLE   11.3

Source	SS	DF	MS	F	P-value
Rows	54,199	5	10,840	7.098	.001
Columns	24,467	5	4,893	3.205	.028
Treatments	248,180	5	49,636	32.504	.000
Residuals	30,541	20	1,527		
Total	357,387	35			

(b) Let τ be the change in the mean yield caused by switching from treatment 1 to treatment 6. The least-squares estimate of this parameter is given by $\hat{\tau} = \bar{Y}_{..6} - \bar{Y}_{..1}$. Now $\bar{Y}_{..1} = 345$ as we have already seen, and

$$\bar{Y}_{..6} = \frac{652 + 571 + 646 + 620 + 617 + 505}{6} \doteq 601.833,$$

so $\hat{\tau} \doteq 601.833 - 345 = 256.833$. Also $S \doteq \sqrt{1527} \doteq 39.077$, so the standard error of $\hat{\tau}$ is given by

$$\text{SE}(\hat{\tau}) = S\sqrt{\frac{1}{6} + \frac{1}{6}} \doteq \frac{39.077}{\sqrt{3}} \doteq 22.561$$

and hence $t_{.975,20}\, \text{SE}(\hat{\tau}) \doteq (2.086)(22.561) \doteq 47.062$. Consequently, the 95% confidence interval for τ is given by $256.833 \pm 47.062 \doteq (209.771, 303.896)$. ∎

The orthogonal array shown in Table 11.4 is for an experiment with eight runs and seven factors, each at the two levels -1 and 1. One method for arriving at this array will be given at the end of Section 11.2. The factors in such an experiment could be qualitative (such as type of catalyst) or quantitative (such as -1 for low and 1 for high). The experiment itself would typically be a *screening experiment*—that is, one designed for the preliminary purpose of determining the important factors from a larger collection of potentially important factors without having to use an excessive number of runs.

T A B L E   11.4

x_1	x_2	x_3	x_4	x_5	x_6	x_7
-1	-1	-1	1	1	1	-1
1	-1	-1	-1	-1	1	1
-1	1	-1	-1	1	-1	1
1	1	-1	1	-1	-1	-1
-1	-1	1	1	-1	-1	1
1	-1	1	-1	1	-1	-1
-1	1	1	-1	-1	1	-1
1	1	1	1	1	1	1

Using the array (Table 11.4) and the response data from the pilot plant experiment described in Section 8.2, we get the artificial experimental data shown in Table 11.5. We refer to $x_1, \ldots, x_7$ as the respective levels of factors $A, \ldots, G$. (We can think of x_1, x_2, and x_3 as the levels of temperature, pressure, and concentration, respectively, and of x_4 through x_7 as four additional factors, each taking on two levels in the pilot plant experiment.) In the experiment, there are $d = 8$ design points, $r = 2$ runs per design point, and $n = rd = 2 \cdot 8 = 16$ runs in total. To analyze the data, we set $\mathcal{X}_m = \{-1, 1\}$ for $m = 1, \ldots, 7$ and $\mathcal{X} = \mathcal{X}_1 \times \cdots \times \mathcal{X}_7$. For $1 \leq m \leq 7$, let G_m be the space of functions on $\mathcal{X}$ that depend only on x_m and hence that are linear functions of x_m (since x_m ranges over $\mathcal{X}_m$, which has only two points). We will assume that the regression function belongs to the space $G = G_{\text{con}} \oplus G_1^0 \oplus \cdots \oplus G_7^0$ of linear functions of $x_1, \ldots, x_7$. Each function in the basis $1, x_1, \ldots, x_7$ of G has absolute value 1 on every run and hence squared norm $n = 16$. It is also easily seen that these basis functions are orthogonal. (Such orthogonality will be discussed in a more general context in Section 11.2.) Since G is eight-dimensional and identifiable and there are eight design points, G is saturated.

The regression function is given by

$$\mu(x_1, \ldots, x_7) = \beta_0 + \beta_1 x_1 + \cdots \beta_7 x_7,$$

and its least-squares estimate is given by

$$\hat{\mu}(x_1, \ldots, x_7) = \hat{\beta}_0 + \hat{\beta}_1 x_1 + \cdots + \hat{\beta}_7 x_7.$$

TABLE 11.5

x_1	x_2	x_3	x_4	x_5	x_6	x_7	Individual Responses	Average Response
−1	−1	−1	1	1	1	−1	59, 61	60
1	−1	−1	−1	−1	1	1	74, 70	72
−1	1	−1	−1	1	−1	1	50, 58	54
1	1	−1	1	−1	−1	−1	69, 67	68
−1	−1	1	1	−1	−1	1	50, 54	52
1	−1	1	−1	1	−1	−1	81, 85	83
−1	1	1	−1	−1	1	−1	46, 44	45
1	1	1	1	1	1	1	79, 81	80

Here

$$\hat{\beta}_0 = \bar{Y} = \frac{1}{8} \sum_{k=1}^{8} \bar{Y}_k,$$

and

$$\hat{\beta}_m = \frac{\langle x_m, \bar{Y}(\cdot) \rangle}{\|x_m\|^2} = \frac{1}{16} \sum_{i=1}^{16} x_{im} Y_i = \frac{1}{8} \sum_{k=1}^{8} x'_{km} \bar{Y}_k, \qquad 1 \le m \le 7.$$

Also, by Problem 10.35(b) in Section 10.3, $\mathrm{var}(\hat{\beta}_m) = \sigma^2/16$ for $0 \le m \le 7$.
From the experimental data, we get

$$\hat{\beta}_0 = \frac{60 + 72 + 54 + 68 + 52 + 83 + 45 + 80}{8} = 64.25,$$

$$\hat{\beta}_1 = \frac{-60 + 72 - 54 + 68 - 52 + 83 - 45 + 80}{8} = 11.5,$$

and, similarly, that $\hat{\beta}_2 = -2.5$, $\hat{\beta}_3 = 0.75$, $\hat{\beta}_4 = 0.75$, $\hat{\beta}_5 = 5$, $\hat{\beta}_6 = 0$ and $\hat{\beta}_7 = 0.25$. Applying the formula for the within sum of squares in Problem 8.37(a) in Section 8.3, we get

$$\mathrm{WSS} = \frac{1}{2}\big\{(61 - 59)^2 + (70 - 74)^2 + (58 - 50)^2 + (67 - 69)^2$$

$$+ (54 - 50)^2 + (85 - 81)^2 + (44 - 46)^2 + (81 - 79)^2\big\}$$

$$= 64.$$

Since G is saturated, the residual sum of squares for the least-squares fit in this space is given by RSS = WSS, so σ^2 is estimated by $S^2 = \mathrm{WSS}/8 = 64/8 = 8$ and hence σ is estimated by $S = \sqrt{8} \doteq 2.828$. Since each estimated coefficient has variance $\sigma^2/16$, it has standard error $S/4 = \sqrt{2}/2 \doteq 0.707$. Thus, we get the results shown in Table 11.6. Since

$$\|P_{G_m^0}[\bar{Y}(\cdot)]\|^2 = \|\hat{\beta}_m x_m\|^2 = \hat{\beta}_m^2 \|x_m\|^2 = 16\hat{\beta}_m^2, \qquad 1 \le m \le 7,$$

we see that $SS_A = \|P_{G_1^0}[\bar{Y}(\cdot)]\|^2 = 16\hat{\beta}_1^2 = 16(11.5)^2 = 2116$ and, similarly, that $SS_B = 16\hat{\beta}_2^2 = 16(-2.5)^2 = 100$, and so forth. In this manner, we arrive at the ANOVA table shown in Table 11.7.

TABLE 11.6

Term	Factor	Coef.	Estimate	SE	t	P-value
x_1	A	β_1	11.50	0.707	16.263	.000
x_2	B	β_2	−2.50	0.707	−3.536	.008
x_3	C	β_3	0.75	0.707	1.061	.320
x_4	D	β_4	0.75	0.707	1.061	.320
x_5	E	β_5	5.00	0.707	7.071	.000
x_6	F	β_6	0.00	0.707	0.000	1.000
x_7	G	β_7	0.25	0.707	0.354	.733

TABLE 11.7

Source	SS	DF	MS	F	P-value
A	2116	1	2116	264.500	.000
B	100	1	100	12.500	.008
C	9	1	9	1.125	.320
D	9	1	9	1.125	.320
E	400	1	400	50.000	.000
F	0	1	0	0.000	1.000
G	1	1	1	0.125	0.733
Within	64	8	8		
Total	2699	15			

Observe that the P-values in Tables 11.6 and 11.7 coincide. According to either table, the statistically significant effects in decreasing magnitude are the main effects of A (temperature), E, and B (concentration). To maximize the estimated mean response, we should set the temperature factor A at 180°C ($x_1 = 1$) since $\hat{\beta}_1$ has positive sign, the concentration factor B at 20% ($x_2 = -1$) since $\hat{\beta}_2$ has negative sign, and factor E at the level having the coded value 1 since $\hat{\beta}_5$ has positive sign.

If we want to perform a follow-up experiment, we could restrict attention to factors A, B, and E, let the temperature factor A take on three levels near 180°C, let the concentration factor B take on three levels near 20%, and let factor E take on three levels near the level having coded value 1 in the present experiment (assuming that this factor is also quantitative). Alternatively, we could perform another screening experiment involving A, B, E, and a number of new factors of potential importance, again letting each factor have two levels to avoid having to make an excessive number of runs. (When screening experiments are carried out in practice, they commonly involve only one run per design point.)

The various factors in an orthogonal array need not all have the same number of levels. For example, the next orthogonal array involves 18 runs and eight factors with one factor having two levels and each of the other factors having three levels (Table 11.8).

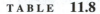

TABLE 11.8

x_1	x_2	x_3	x_4	x_5	x_6	x_7	x_8
−1	−1	−1	−1	−1	−1	−1	−1
−1	−1	0	0	0	0	0	0
−1	−1	1	1	1	1	1	1
−1	0	−1	−1	0	0	1	1
−1	0	0	0	1	1	−1	−1
−1	0	1	1	−1	−1	0	0
−1	1	−1	0	−1	1	0	1
−1	1	0	1	0	−1	1	−1
−1	1	1	−1	1	0	−1	0
1	−1	−1	1	1	0	0	−1
1	−1	0	−1	−1	1	1	0
1	−1	1	0	0	−1	−1	1
1	0	−1	0	1	−1	1	0
1	0	0	1	−1	0	−1	1
1	0	1	−1	0	1	0	−1
1	1	−1	1	0	1	−1	0
1	1	0	−1	1	−1	0	1
1	1	1	0	−1	0	1	−1

Problems

11.1 Consider an experiment involving two factors, each of which takes on the two levels −1 and 1. Let n_1, n_2, n_3, and n_4 denote the numbers of repetitions at the design points $(−1, −1)$, $(−1, 1)$, $(1, −1)$, and $(1, 1)$, respectively. Determine a necessary and sufficient condition on $n_1, \dots, n_4$ in order that the design random variables X_1 and X_2 be independent. [*Hint:* X_1 and X_2 are independent if and only if $P(X_1 = 1, X_2 = 1) = P(X_1 = 1)P(X_2 = 1)$.]

11.2 Let G_1, G_2, and G_3 be orthogonal linear spaces. Show that G_1 and $G_2 \oplus G_3$ are orthogonal.

11.3 Let $G = G_1 \oplus G_2$, where G_1 and G_2 are orthogonal, identifiable linear spaces. Show that $G_2 = G \ominus G_1$ and $G_1 = G \ominus G_2$.

11.4 Let G_0 be a subspace of an identifiable linear space G. Show that

$$G \ominus G_0 = \{g - P_{G_0}g : g \in G\}.$$

11.5 In the context of the sugar beet experiment, are the design random variables X_1 and X_2 independent?

11.6   Does the design of the experiment described in Problem 8.18 in Section 8.2 have the form of an orthogonal array?

11.7   Consider the design set $\{-1, 1\}$ and let the design points -1 and 1 be labeled as 1 and 2, respectively. suppose $n_1 = n_2 = r$, let G be the space of linear functions on $\mathbb{R}$, consider the basis x of $G^0 = G \ominus G_{con}$, and write $P_{G^0}[\bar{Y}(\cdot)] = \hat{\beta}x$. **(a)** Determine $\|x\|^2$. **(b)** Express $\hat{\beta}$ in terms of $\bar{Y}_1$ and $\bar{Y}_2$. **(c)** Express $\|P_{G^0}[\bar{Y}(\cdot)]\|^2$ in terms of $\hat{\beta}$.

11.8   Consider the design set $\{-1, 1\}$ and let the design points -1 and 1 be labeled as 1 and 2, respectively. Suppose $n_1 = n_2 = r$, let G be the space of all functions on the design set, consider the basis $g(x) = I_1(x) - \frac{1}{2}$ of $G^0 = G \ominus G_{con}$ with $I_1(x) = \mathrm{ind}(x = -1)$, and write $P_{G^0}[\bar{Y}(\cdot)] = \hat{\beta}g(x)$. **(a)** Determine the squared norm of the function $g(x)$. **(b)** Express $\hat{\beta}$ in terms of $\bar{Y}_1$ and $\bar{Y}_2$. **(c)** Express $\|P_{G^0}[\bar{Y}(\cdot)]\|^2$ in terms of $\hat{\beta}$.

11.9   Consider the design set $\{-1, 0, 1\}$ and let the design points -1, 0, and 1 be labeled as 1, 2, and 3, respectively. Suppose $n_1 = n_2 = n_3 = r$, let G be the space of all functions on the design set, consider the orthogonal basis $g_1(x) = I_1(x) - \frac{1}{3}$ and $g_2(x) = \frac{1}{2}I_1(x) + I_2(x) - \frac{1}{2}$ of $G^0 = G \ominus G_{con}$ with $I_1(x) = \mathrm{ind}(x = -1)$ and $I_2(x) = \mathrm{ind}(x = 0)$, and write $P_{G^0}[\bar{Y}(\cdot)] = \hat{\beta}_1 g_1 + \hat{\beta}_2 g_2$. **(a)** Determine the squared norms of the functions g_1 and g_2. **(b)** Express $\hat{\beta}_1$ and $\hat{\beta}_2$ in terms of $\bar{Y}_1$, $\bar{Y}_2$, and $\bar{Y}_3$. **(c)** Express $\|P_{G^0}[\bar{Y}(\cdot)]\|^2$ in terms of $\hat{\beta}_1$ and $\hat{\beta}_2$.

11.10   Consider the lube oil experiment and the additive model

$$G = G_{con} \oplus G_1^0 \oplus G_2^0 \oplus G_3^0 \oplus G_4^0,$$

where $\mathcal{X}_m = \{-1, 0, 1\}$ and G_m is the space of all functions of x_m. Consider the basis $g_1(x_1), g_2(x_1)$ of G_1^0, where g_1 and g_2 are defined in Problem 11.9, and write

$$P_{G_1^0}[\bar{Y}(\cdot)] = \hat{\beta}_1 g_1(x_1) + \hat{\beta}_2 g_2(x_1).$$

(a) Determine the squared norms of the functions $g_1(x_1)$ and $g_2(x_1)$. **(b)** Determine $\hat{\beta}_1$ and $\hat{\beta}_2$. **(c)** Express $\|P_{G_1^0}[\bar{Y}(\cdot)]\|^2$ in terms of $\hat{\beta}_1$ and $\hat{\beta}_2$. **(d)** Use the answers to parts (b) and (c) to verify that $SS_A \doteq 4496.29$.

11.11   Consider the fertilizer experiment as analyzed in this section. **(a)** Determine $\bar{Y}_{..k}$ for $k = 1, \ldots, 6$. **(b)** Determine $\bar{Y}$. **(c)** Use the answers to parts (a) and (b) to verify that $SS_T \doteq 248180$.

11.12   Consider the fertilizer experiment, but ignore the column effect. **(a)** Determine and interpret the corresponding ANOVA table. **(b)** Determine the least-squares estimate of the change τ in the mean yield caused by switching from treatment 1 to treatment 6 and determine the standard error of this estimate. **(c)** Determine the 95% confidence interval for τ.

11.13   Consider the fertilizer experiment, but ignore the row and column effects. **(a)** Determine and interpret the corresponding ANOVA table. **(b)** Determine the least-squares estimate of the change τ in the mean yield caused by switching from treatment 1 to treatment 6 and determine the standard error of this estimate. **(c)** Determine the 95% confidence interval for τ.

11.14   Consider the artificial pilot plant experiment involving 16 runs and seven factors each at the two levels -1 and 1 and consider all $2^7 = 128$ factor-level combinations in which the factors are restricted to these two levels. Determine **(a)** the factor-level combination (among the 128 combinations just described) that maximizes the least-squares estimate of the regression function that was obtained in the text; **(b)** the least-squares estimate of the value of the regression function at the factor-level combination obtained in part (a).

11.2
Interactions

Recall Example 8.4 in Section 8.3 and the ensuing discussion of the interaction term $\beta_5 x_1 x_2$ in the model for the regression function. In this section, we discuss interactions in a more systematic manner.

Let G_l be a (finite-dimensional) linear space of functions on X that depend only x_l and let G_m be a linear space of functions on X that depend only on x_m, where $l \neq m$. Consider the collection C of all functions on X of the form $g_l(x_l)g_m(x_m)$ as $g_l(x_l)$ and $g_m(x_m)$ range over G_l and G_m, respectively. As Example 9.3 in Section 9.1 illustrates, this collection is generally not a linear space. The span of the collection is a linear space, however, which is referred to as the *tensor product* of G_l and G_m and denoted by $G_l \otimes G_m$ or, more simply, by G_{lm}. We can think of the functions in $G_l \otimes G_m$ as being functions on $X_l \times X_m$.

THEOREM 11.5   Let $g(x_l, x_m) \in G_l \otimes G_m$, where $l \neq m$. Then

(a) for each $x_l \in X_l$, the function $g_m(x_m)$ defined by $g_m(x_m) = g(x_l, x_m)$, $x_m \in X_m$, is a member of G_m;

(b) for each $x_m \in X_m$, the function $g_l(x_l)$ defined by $g_l(x_l) = g(x_l, x_m)$, $x_l \in X_l$, is a member of G_l.

Proof   According to the definition of the tensor product of G_l and G_m, there is a positive number J, and there are functions $g_{l1}(x_l), \ldots, g_{lJ}(x_l) \in G_l$ and $g_{m1}(x_m), \ldots, g_{mJ}(x_m) \in G_m$ such that $g(x_l, x_m) = g_{l1}(x_l)g_{m1}(x_m) + \cdots + g_{lJ}(x_l)g_{mJ}(x_m)$. Given $x_l \in X_l$, let $g_m(x_m)$ be defined by

$$g_m(x_m) = g(x_l, x_m) = g_{l1}(x_l)g_{m1}(x_m) + \cdots + g_{lJ}(x_l)g_{mJ}(x_m), \qquad x_m \in X_m.$$

Since $g_m(x_m)$ is a linear combination of the functions $g_{m1}(x_m), \ldots, g_{mJ}(x_m) \in G_m$, it is a member of G_m. This completes the proof of part (a), and part (b) is proven in a similar manner.   ∎

THEOREM 11.6   Let $l \neq m$, let G_l have dimension p_l, let $g_{l1}(x_l), \ldots, g_{lp_l}(x_l)$ be a basis of G_l, let G_m have dimension p_m, and let $g_{m1}(x_m), \ldots, g_{mp_m}(x_m)$ be a basis of G_m. Then $G_l \otimes G_m$

has dimension $p_l p_m$, and the functions $g_{lj}(x_l)g_{mk}(x_m)$, $1 \le j \le p_l$ and $1 \le k \le p_m$, form a basis of $G_l \otimes G_m$.

Proof   Let G be the linear space spanned by the functions

$$g_{lj}(x_l)g_{mk}(x_m), \qquad 1 \le j \le p_l \text{ and } 1 \le k \le p_m.$$

Since each of these functions is in $G_l \otimes G_m$, we conclude that $G \subset G_l \otimes G_m$. Let

$$g_l(x_l) = \sum_j b_{lj}g_{lj}(x_l) \in G_l \quad \text{and} \quad g_m(x_m) = \sum_k b_{mk}g_{mk}(x_m) \in G_m.$$

Then

$$g_l(x_l)g_m(x_m) = \left(\sum_j b_{lj}g_{lj}(x_l) \right)\left(\sum_k b_{mk}g_{mk}(x_m) \right)$$

$$= \sum_j \sum_k b_{lj}b_{mk}g_{lj}(x_l)g_{mk}(x_m)$$

is a member of G. Since the functions of the form $g_l(x_l)g_m(x_m)$ with $g_l(x_l) \in G_l$ and $g_m(x_m) \in G_m$ span $G_l \otimes G_m$, we conclude that $G_l \otimes G_m \subset G$ and hence that $G_l \otimes G_m = G$.

It remains to verify that the functions that form the asserted basis are linearly independent. To this end, suppose

$$\sum_j \sum_k b_{jk}g_{lj}(x_l)g_{mk}(x_m) = 0, \qquad x_l \in \mathcal{X}_l \text{ and } x_m \in \mathcal{X}_m.$$

Choose $x_m \in \mathcal{X}_m$. Then

$$\sum_j \left(\sum_k b_{jk}g_{mk}(x_m) \right)g_{lj}(x_l) = 0, \qquad x_l \in \mathcal{X}_l.$$

Since the functions $g_{lj}(x_l)$, $1 \le j \le p_l$, are linearly independent, we see that

$$\sum_k b_{jk}g_{mk}(x_m) = 0, \qquad 1 \le j \le p_l.$$

Thus, for $1 \le j \le p_l$,

$$\sum_k b_{jk}g_{mk}(x_m) = 0, \qquad x_m \in \mathcal{X}_m,$$

so we conclude from the linear independence of $g_{mk}(x_m)$, $1 \le k \le p_m$, that $b_{jk} = 0$ for $1 \le k \le p_m$. Consequently, $b_{jk} = 0$ for $1 \le j \le p_l$ and $1 \le k \le p_m$. Therefore, the functions that form the asserted basis are indeed linearly independent. Since there are $p_l p_m$ functions in this basis, $G_l \otimes G_m$ has dimension $p_l p_m$.   ∎

EXAMPLE **11.5**   Let $M = 2$, let $\mathcal{X}_1 = \{1, \dots, p_1\}$, let G_1 be the space of all functions on $\mathcal{X} = \mathcal{X}_1 \times \mathcal{X}_2$ that depend only on x_1, and let G_2 be a linear space of functions on $\mathcal{X}$ that depend only on x_2. Show that $G_1 \otimes G_2$ consists of all functions g on $\mathcal{X}_1 \times \mathcal{X}_2$ having the form $g(x_1, x_2) = g_j(x_2)$ when $x_1 = j$, where $g_1(x_2), \dots, g_{p_1}(x_2)$ are arbitrary functions in G_2.

Solution Given the functions $g_1(x_2), \ldots, g_{p_1}(x_2) \in G_2$, define the function g on $X_1 \times X_2$ by $g(x_1, x_2) = g_j(x_2)$ when $x = j$. Then

$$g(x_1, x_2) = \text{ind}(x_1 = 1)g_1(x_2) + \cdots + \text{ind}(x_1 = p_1)g_{p_1}(x_2) \in G_1 \otimes G_2.$$

Conversely, it follows from Theorem 11.5 that every function in $G_1 \otimes G_2$ has the indicated form. ∎

EXAMPLE **11.6** Consider the sugar beet experiment and recall the space G used in the solution to Example 10.23 in Section 10.8 to test for the additivity of the regression function. Here $M = 2$ and $X = X_1 \times X_2$, where $X_1 = \{1, \ldots, 5\}$ and $X_2 = \{-25, \ldots, 25\}$. In light of Example 11.5, we can write $G = G_1 \otimes G_2$, where G_1 is the space of all functions on X that depend only on x_1 and G_2 is the space of all functions on X that are quintic polynomials in x_2. Determine

  (a) the dimension of $G_1 \otimes G_2$;   **(b)** a basis of $G_1 \otimes G_2$.

Solution **(a)** Since G_1 has dimension 5 and G_2 has dimension 6, we conclude from Theorem 11.6 that $G_1 \otimes G_2$ has dimension $5 \cdot 6 = 30$.

  (b) Now $1, I_1(x_1), I_2(x_1), I_3(x_1), I_4(x_1)$ form a basis of G_1, where $I_j(x_1) = \text{ind}(x_1 = j)$, and $1, x_2, x_2^2, x_2^3, x_2^4, x_2^5$ form a basis of G_2. Thus, by Theorem 11.6, $1, I_1(x_1), \ldots, I_4(x_1), x_2, I_1(x_1)x_2, \ldots, I_4(x_1)x_2, \ldots, x_2^5, I_1(x_1)x_2^5, \ldots, I_4(x_1)x_2^5$ form a basis of $G_1 \otimes G_2$. ∎

For another example of the use of tensor product spaces in statistics, suppose the mean response is a linear function of x_1, where x_1 ranges over $X_1 = \mathbb{R}$. Then the regression function can be written as $\beta_0 + \beta_1 x_1$. Suppose now that β_0 and β_1 are each quadratic polynomials in x_2, where x_2 ranges over $X_2 = \mathbb{R}$; that is,

$$\beta_0 = \beta_{00} + \beta_{01}x_2 + \beta_{02}x_2^2 \quad \text{and} \quad \beta_1 = \beta_{10} + \beta_{11}x_2 + \beta_{12}x_2^2.$$

Then, as a function of x_1 and x_2, the regression function can be written as

$$\mu(x_1, x_2) = (\beta_{00} + \beta_{01}x_2 + \beta_{02}x_2^2) + (\beta_{10} + \beta_{11}x_2 + \beta_{12}x_2^2)x_1$$
$$= \beta_{00} + \beta_{10}x_1 + \beta_{01}x_2 + \beta_{02}x_2^2 + \beta_{11}x_1x_2 + \beta_{12}x_1x_2^2.$$

Think of $\beta_{00}, \beta_{10}, \beta_{01}, \beta_{02}, \beta_{11}$, and β_{12} as being arbitrary unknown parameters. Then the regression function is an arbitrary unknown member of the six-dimensional space G of functions on $X_1 \times X_2 = \mathbb{R}^2$ spanned by $1, x_1, x_2, x_2^2, x_1x_2$, and $x_1x_2^2$. Thus, by Theorem 11.6, $G = G_1 \otimes G_2$, where G_1 is the space of functions on $\mathbb{R}^2$ that are linear functions of x_1 (which has basis $1, x_1$) and G_2 is the space of functions on $\mathbb{R}^2$ that are quadratic polynomials in x_2 (which has basis $1, x_2, x_2^2$).

THEOREM **11.7** Suppose X_l and X_m are finite sets, where $l \neq m$, and set $p_l = \#(X_l)$ and $p_m = \#(X_m)$. Let G_l be the space of all functions on X that depend only on x_l, which has dimension p_l, and let G_m be the space of all functions on X that depend only on x_m, which has dimension p_m. Then $G_l \otimes G_m$ is the space of all functions on $\mathfrak{X}$ that depend only on x_l and x_m, which has dimension $p_l p_m = \#(\mathfrak{X}_l \times \mathfrak{X}_m)$.

Proof   According to Theorem 11.6, $G_l \otimes G_m$ has dimension $p_l p_m$. Since this space is contained in the space G of all functions on $\mathfrak{X}$ that depend only on x_l and x_m, which also has dimension $p_l p_m$, we conclude that $G_l \otimes G_m = G$.   ∎

THEOREM 11.8   Let the design random variables X_l and X_m be independent, where $l \neq m$, let the functions $g_{l1}(x_l), \ldots, g_{lp_l}(x_l)$ be an orthogonal basis of G_l, and let $g_{m1}(x_m), \ldots, g_{mp_m}(x_m)$ be an orthogonal basis of G_m. Then

(a) the functions $g_{lj}(x_l)g_{mk}(x_m)$, $1 \leq j \leq p_l$ and $1 \leq k \leq p_m$, form an orthogonal basis of $G_l \otimes G_m$;

(b) the squared norms of these functions are given by

$$\|g_{lj}(x_l)g_{mk}(x_m)\|^2 = \frac{1}{n}\|g_{lj}(x_l)\|^2 \|g_{mk}(x_m)\|^2, \qquad 1 \leq j \leq p_l \text{ and } 1 \leq k \leq p_m.$$

Proof   (a) Let $1 \leq j_1, j_2 \leq p_l$ and $1 \leq k_1, k_2 \leq p_m$. Suppose $(j_1, k_1) \neq (j_2, k_2)$; that is, $j_1 \neq j_2$ or $k_1 \neq k_2$. We need to show that $g_{lj_1}(x_l)g_{mk_1}(x_m)$ and $g_{lj_2}(x_l)g_{mk_2}(x_m)$ are orthogonal; that is,

$$\langle g_{lj_1}(x_l)g_{mk_1}(x_m), g_{lj_2}(x_l)g_{mk_2}(x_m)\rangle = 0. \tag{18}$$

To this end, we conclude from (8) in Section 11.1 that

$$\langle g_{lj_1}(x_l)g_{mk_1}(x_m), g_{lj_2}(x_l)g_{mk_2}(x_m)\rangle$$
$$= nE[\, g_{lj_1}(X_l)g_{lj_2}(X_l)g_{mk_1}(X_m)g_{mk_2}(X_m)\,]$$
$$= nE[\, g_{lj_1}(X_l)g_{lj_2}(X_l)\,]E[\, g_{mk_1}(X_m)g_{mk_2}(X_m)\,].$$

Suppose $j_1 \neq j_2$. Then $g_{lj_1}(x_l)$ and $g_{lj_2}(x_l)$ are orthogonal. Thus, by another application of (8) in Section 11.1, we see that

$$0 = \langle g_{lj_1}(x_l), g_{lj_2}(x_l)\rangle = nE[\, g_{lj_1}(X_l)g_{lj_2}(X_l)\,]$$

and hence that (18) holds. Similarly, if $k_1 \neq k_2$, then (18) holds.

(b) Let $1 \leq j \leq p_l$ and $1 \leq k \leq p_m$. Then by the independence of X_l and X_m and three applications of (9) in Section 11.1,

$$\|g_{lj}(x_l)g_{mk}(x_m)\|^2 = nE[\, g_{lj}^2(X_l)g_{mk}^2(X_m)\,]$$
$$= nE[\, g_{lj}^2(X_l)\,]E[\, g_{mk}^2(X_m)\,]$$
$$= \frac{1}{n}\|g_{lj}(x_l)\|^2 \|g_{mk}(x_m)\|^2.   ∎$$

THEOREM 11.9   Suppose the design random variables X_l and X_m are independent, where $l \neq m$, and G_l and G_m are identifiable. Then $G_l \otimes G_m$ is identifiable.

Proof   Consider an orthogonal basis of G_l and an orthogonal basis of G_m. Since G_l and G_m are identifiable, the functions in these bases have positive norms. It now follows from Theorem 11.8 that $G_l \otimes G_m$ has an orthogonal basis consisting of functions with positive norm and hence from Theorem 9.21 in Section 9.4 that $G_l \otimes G_m$ is identifiable. ∎

For $1 \leq m \leq M$, let G_m be an identifiable, p_m-dimensional linear space of functions that depend only on x_m and suppose G_m contains the constant functions. Let G_m^0 be the $(p_m - 1)$-dimensional identifiable subspace of G_m given by $G_m^0 = G_m \ominus G_{con}$. Observe that G_{con} and G_m^0 are orthogonal.

Suppose the design random variables X_l and X_m are independent, where $l \neq m$. Recall that G_{lm} is the $p_l p_m$-dimensional linear space given by $G_{lm} = G_l \otimes G_m$. Let G_{lm}^0 be the identifiable $(p_l - 1)(p_m - 1)$-dimensional subspace of G_{lm} given by $G_{lm}^0 = G_l^0 \otimes G_m^0$. Let $1, g_{l1}(x_l), \ldots, g_{l,p_l-1}(x_l)$ be an orthogonal basis of G_l, and let $1, g_{m1}(x_m), \ldots, g_{m,p_m-1}(x_m)$ be an orthogonal basis of G_m. Then 1 is an (orthogonal) basis of G_{con}; $g_{l1}(x_l), \ldots, g_{l,p_l-1}(x_l)$ is an orthogonal basis of G_l^0; $g_{m1}(x_m), \ldots,$ $g_{m,p_m-1}(x_m)$ is an orthogonal basis of G_m^0; and, by Theorem 11.8,

$$g_{lj}(x_l)g_{mk}(x_m), \qquad 1 \leq j \leq p_l - 1 \text{ and } 1 \leq k \leq p_m - 1,$$

is an orthogonal basis of G_{lm}^0. By combining the indicated bases of G_{con}, G_l^0, G_m^0, and G_{lm}^0, we get an orthogonal basis of G_{lm}. As a consequence,

$$G_{lm} = G_{con} \oplus G_l^0 \oplus G_m^0 \oplus G_{lm}^0, \tag{19}$$

where G_{con}, G_l^0, G_m^0, and G_{lm}^0 are orthogonal components of G_{lm}. We refer to G_{lm}^0 as the *interaction subspace* of G_{lm}. Observe that G_{lm}^0 is the orthogonal complement of $G_{con} \oplus G_l^0 \oplus G_m^0$ relative to G_{lm}. Observe also that we have verified part (a) of Theorem 11.10.

THEOREM 11.10   (a) Suppose the design random variables X_j and X_k are independent, where j and k are distinct. Then G_j^0, G_k^0, and G_{jk}^0 are orthogonal.

(b) Suppose the design random variables X_j, X_k, and X_l are independent, where j, k, and l are distinct. Then G_j^0 and G_{kl}^0 are orthogonal, and G_{jk}^0 and G_{jl}^0 are orthogonal.

(c) Suppose the design random variables X_j, X_k, X_l, and X_m are independent, where j, k, l, and m are distinct. Then G_{jk}^0 and G_{lm}^0 are orthogonal.

Proof   For simplicity in notation, we take $j = 1$, $k = 2$, $l = 3$, and $m = 4$. Let $g_{11}(x_1)$, $\ldots, g_{1,p_1-1}(x_1)$ be an orthogonal basis of G_1^0, let $g_{21}(x_2), \ldots, g_{2,p_2-1}(x_2)$ be an orthogonal basis of G_2^0 and so forth.

(a) (Alternative proof.) Suppose X_1 and X_2 are independent. To show that G_1^0 and G_2^0 are orthogonal, we write a typical basis function of G_1^0 as $g_{1j}(x_1)$ and a typical basis function of G_2^0 as $g_{2k}(x_2)$. Then, according to (8) in Section 11.1,

$$\langle g_{1j}(x_1), g_{2k}(x_2) \rangle = nE[g_{1j}(X_1)g_{2k}(X_2)] = nE[g_{1j}(X_1)]E[g_{2k}(X_2)] = 0$$

since $E[g_{1j}(X_1)] = n^{-1}\langle 1, g_{1j}(x_1)\rangle = 0$ by the orthogonality of G_{con} and G_1^0. Thus, each basis function of G_1^0 is orthogonal to every basis function of G_2^0, so G_1^0 and G_2^0 are orthogonal as desired.

Similarly, to show that G_1^0 and G_{12}^0 are orthogonal, we write a typical basis function of G_1^0 as $g_{1j}(x_j)$ and a typical basis function of G_{12}^0 as $g_{1k}(x_1)g_{2l}(x_2)$. Then, again according to (8) in Section 11.1,

$$\langle g_{1j}(x_1), g_{1k}(x_1)g_{2l}(x_2)\rangle = nE[g_{1j}(X_1)g_{1k}(X_1)g_{2l}(X_2)]$$
$$= nE[g_{1j}(X_1)g_{1k}(X_1)]E[g_{2l}(X_2)]$$
$$= 0$$

since $E[g_{2l}(X_2)] = n^{-1}\langle 1, g_{2l}(x_2)\rangle = 0$ by the orthogonality of G_{con} and G_2^0.

(b) Suppose X_1, X_2, and X_3 are independent. To show that G_1^0 and G_{23}^0 are orthogonal, we write a typical basis function of G_1^0 as $g_{1j}(x_1)$ and a typical basis function of G_{23}^0 as $g_{2k}(x_2)g_{3l}(x_3)$. Then

$$\langle g_{1j}(x_1), g_{2k}(x_2)g_{3l}(x_3)\rangle = nE[g_{1j}(X_1)g_{2k}(X_2)g_{3l}(X_3)]$$
$$= nE[g_{1j}(X_1)]E[g_{2k}(X_2)]E[g_{3l}(X_3)]$$
$$= 0$$

since $E[g_{1j}(X_1)] = n^{-1}\langle 1, g_{1j}(x_1)\rangle = 0$ by the orthogonality of G_{con} and G_1^0. Thus, each basis function of G_1^0 is orthogonal to every basis function of G_{23}^0, so G_1^0 and G_{23}^0 are orthogonal as desired.

To show that G_{12}^0 and G_{13}^0 are orthogonal, we write a typical basis function of G_{12}^0 as $g_{1j}(x_1)g_{2k}(x_2)$ and a typical basis function of G_{13}^0 as $g_{1l}(x_1)g_{3m}(x_3)$. Then

$$\langle g_{1j}(x_1)g_{2k}(x_2), g_{1l}(x_1)g_{3m}(x_3)\rangle$$
$$= nE[g_{1j}(X_1)g_{2k}(X_2)g_{1l}(X_1)g_{3m}(X_3)]$$
$$= nE[g_{1j}(X_1)g_{1l}(X_1)]E[g_{2k}(X_2)]E[g_{3m}(X_3)]$$
$$= 0$$

since $E[g_{2k}(X_2)] = n^{-1}\langle 1, g_{2k}(X_2)\rangle = 0$ by the orthogonality of G_{con} and G_2^0. Thus, each basis function of G_{12}^0 is orthogonal to every basis function of G_{13}^0, so G_{12}^0 and G_{13}^0 are orthogonal

(c) We want to prove that if X_1, X_2, X_3, and X_4 are independent, then G_{12}^0 and G_{34}^0 are orthogonal. The details, which are similar to those given in the proofs of parts (a) and (b), are left as a problem. ∎

EXAMPLE **11.7**   Consider Table 11.9, an orthogonal array involving 18 runs and four factors that was obtained by using columns 1, 2, 3, and 4 of Table 11.8, an orthogonal array involving 18 runs and eight factors, in Section 11.1. Here factor A has two levels, and factors

TABLE   11.9

x_1	x_2	x_3	x_4
−1	−1	−1	−1
−1	−1	0	0
−1	−1	1	1
−1	0	−1	0
−1	0	0	1
−1	0	1	−1
−1	1	−1	1
−1	1	0	−1
−1	1	1	0
1	−1	−1	0
1	−1	0	1
1	−1	1	−1
1	0	−1	−1
1	0	0	0
1	0	1	1
1	1	−1	1
1	1	0	−1
1	1	1	0

B, C, and D each have three levels. Let G_1 be the space of linear functions of x_1 and let G_2, G_3, and G_4 the spaces of quadratic polynomials in x_2, x_3, and x_4, respectively. Consider the linear space

$$G = G_{\mathrm{con}} \oplus G_1^0 \oplus G_2^0 \oplus G_3^0 \oplus G_4^0 \oplus G_{12}^0 \oplus G_{13}^0 \oplus G_{14}^0.$$

(a) Explain why the indicated components $G_{\mathrm{con}}, \ldots, G_{14}^0$ are orthogonal.

(b) Show the corresponding ANOVA table.

Solution   **(a)** By definition G_{con} is orthogonal to each of the other components. Now the design random variables X_1, X_2, X_3, and X_4 are pairwise-independent. Thus, G_1^0, G_2^0, G_3^0, and G_4^0 are orthogonal by Theorem 11.10(a). Also, G_1^0 is orthogonal to each of the components G_{12}^0, G_{13}^0, and G_{14}^0; G_2^0 and G_{12}^0 are orthogonal; G_3^0 and G_{13}^0 are orthogonal; and G_4^0 and G_{14}^0 are orthogonal. Observe next that X_1, X_2, and X_3 are independent. Thus, it follows from Theorem 11.10(b) that G_2^0 and G_{13}^0 are orthogonal, G_3^0 and G_{12}^0 are orthogonal, and G_{12}^0 and G_{13}^0 are orthogonal. Similarly, X_1, X_2, and X_4 are independent, so G_2^0 and G_{14}^0 are orthogonal, G_4^0 and G_{12}^0 are orthogonal, and G_{12}^0 and G_{14}^0 are orthogonal. Moreover, X_1, X_3, and X_4 are independent, so G_3^0 and G_{14}^0 are orthogonal, G_4^0 and G_{13}^0 are orthogonal, and G_{13}^0 and G_{14}^0 are orthogonal.

(b) The ANOVA table is shown in Table 11.10. Here SS_A is the squared norm of the orthogonal projection of $\bar{Y}(\cdot)$ onto G_1^0, SS_{AB} is the squared norm of the orthogonal projection of $\bar{Y}(\cdot)$ onto G_{12}^0, and so forth. ∎

TABLE 11.10

Source	SS	DF
A	SS_A	1
B	SS_B	2
C	SS_C	2
D	SS_D	2
AB	SS_{AB}	2
AC	SS_{AC}	2
AD	SS_{AD}	2
Residuals	RSS	4
Total	TSS	17

Experiments with Factors Having Two Levels

Consider now a complete three-factor experiment in which each factor takes on two coded levels, -1 (low) and 1 (high), and there are r replications at each factor-level combination. Here $d_1 = d_2 = d_3 = 2$, $d = 2^3 = 8$, and $n = 8r$. The design random variables X_1, X_2, and X_3 are independent and each is uniformly distributed on $\{-1, 1\}$. Let G_1, G_2, and G_3 be the spaces of linear functions of x_1, x_2, and x_3, respectively. Then G_1^0, G_2^0, and G_3^0 are the one-dimensional spaces spanned by x_1, x_2, and x_3, respectively. Observe that G_{con}, G_1^0, G_2^0, G_3^0, G_{12}^0, G_{13}^0, and G_{23}^0 are orthogonal one-dimensional spaces having bases 1, x_1, x_2, x_3, x_1x_2, x_1x_3, and x_2x_3, respectively. Consider the seven-dimensional linear space

$$G = G_{\mathrm{con}} \oplus G_1^0 \oplus G_2^0 \oplus G_3^0 \oplus G_{12}^0 \oplus G_{13}^0 \oplus G_{23}^0.$$

The functions 1, x_1, x_2, x_3, x_1x_2, x_1x_3, and x_2x_3 form an orthogonal basis of G; since each function takes only the values ± 1, it has squared norm $n = 8r$.

Suppose the regression function is a member of G and let $\beta_0, \dots, \beta_6$ denote the regression coefficients relative to the indicated basis. The regression function is additive if and only if the three interaction coefficients β_4, β_5, and β_6 equal zero, and it is constant if and only if the three main effect coefficients β_1, β_2, and β_3 equal zero and the three interaction coefficients equal zero.

Let $\hat{\beta}_0, \dots, \hat{\beta}_6$ denote the least-squares estimates of the regression coefficients, which are given by

$$\hat{\beta}_0 = \frac{\langle 1, \bar{Y}(\cdot) \rangle}{n} = \frac{1}{8} \sum_{k=1}^{8} \bar{Y}_k = \bar{Y},$$

$$\hat{\beta}_1 = \frac{\langle x_1, \bar{Y}(\cdot) \rangle}{n} = \frac{1}{8} \sum_{k=1}^{8} x_{k1} \bar{Y}_k,$$

$$\hat{\beta}_2 = \frac{\langle x_2, \bar{Y}(\cdot) \rangle}{n} = \frac{1}{8} \sum_{k=1}^{8} x_{k2} \bar{Y}_k,$$

$$\hat{\beta}_3 = \frac{\langle x_3, \bar{Y}(\cdot)\rangle}{n} = \frac{1}{8}\sum_{k=1}^{8}x_{k3}\bar{Y}_k,$$

$$\hat{\beta}_4 = \frac{\langle x_1 x_2, \bar{Y}(\cdot)\rangle}{n} = \frac{1}{8}\sum_{k=1}^{n}x_{k1}x_{k2}\bar{Y}_k,$$

$$\hat{\beta}_5 = \frac{\langle x_1 x_3, \bar{Y}(\cdot)\rangle}{n} = \frac{1}{8}\sum_{k=1}^{8}x_{k1}x_{k3}\bar{Y}_k,$$

and

$$\hat{\beta}_6 = \frac{\langle x_2 x_3, \bar{Y}(\cdot)\rangle}{n} = \frac{1}{8}\sum_{k=1}^{8}x_{k2}x_{k3}\bar{Y}_k.$$

Under the assumption of homoskedasticity, each of the estimates $\hat{\beta}_0, \ldots, \hat{\beta}_6$ has variance σ^2/n [see Problem 10.35(b) in Section 10.3]. The corresponding ANOVA table is shown in Table 11.11. Here $P_{G_1^0}[\bar{Y}(\cdot)] = \hat{\beta}_1 x_1$, so $SS_A = \|P_{G_1^0}[\bar{Y}(\cdot)]\|^2 = \hat{\beta}_1^2\|x_1\|^2 = n\hat{\beta}_1^2$. Similarly, $SS_B = n\hat{\beta}_2^2$, $SS_C = n\hat{\beta}_3^2$, $SS_{AB} = n\hat{\beta}_4^2$, $SS_{AC} = n\hat{\beta}_5^2$, and $SS_{BC} = n\hat{\beta}_6^2$.

TABLE 11.11

Source	SS	DF
A	SS_A	1
B	SS_B	1
C	SS_C	1
AB	SS_{AB}	1
AC	SS_{AC}	1
BC	SS_{BC}	1
Residuals	RSS	$n-7$
Total	TSS	$n-1$

Pilot Plant Experiment

Recall the pilot plant experiment, which was described in Section 8.2. In this complete three-factor experiment, each factor has two levels, -1 and 1, and there are $r = 2$ replications at each of the $d = 2^3 = 8$ factor-level combinations, so the total number of runs is given by $n = 8 \cdot 2 = 16$.

Hand computation of the least-squares estimates of the regression coefficients is shown in Table 11.12. Here, to obtain these least-squares estimates, we compute the inner product of the corresponding column vector of (± 1)'s and the column vector of average responses and then divide by $n/r = d = 8$. The least-squares estimate of the regression function is given by

$$\hat{\mu}(x_1, x_2, x_3) = 64.25 + 11.5x_1 - 2.5x_2 + 0.75x_3 + 0.75x_1x_2 + 5x_1x_3.$$

TABLE   11.12

1	x_1	x_2	x_3	$x_1 x_2$	$x_1 x_3$	$x_2 x_3$	Average Response
1	−1	−1	−1	1	1	1	60
1	1	−1	−1	−1	−1	1	72
1	−1	1	−1	−1	1	−1	54
1	1	1	−1	1	−1	−1	68
1	−1	−1	1	1	−1	−1	52
1	1	−1	1	−1	1	−1	83
1	−1	1	1	−1	−1	1	45
1	1	1	1	1	1	1	80
64.25	11.50	−2.50	0.75	0.75	5.00	0.00	
$\hat{\beta}_0$	$\hat{\beta}_1$	$\hat{\beta}_2$	$\hat{\beta}_3$	$\hat{\beta}_4$	$\hat{\beta}_5$	$\hat{\beta}_6$	

The sums of squares corresponding to the various main effects and two-factor interactions are given by

$$SS_A = 16\hat{\beta}_1^2 = 16(11.5)^2 = 2116,$$

$$SS_B = 16\hat{\beta}_2^2 = 16(-2.5)^2 = 100,$$

$$SS_C = 16\hat{\beta}_3^2 = 16(0.75)^2 = 9,$$

$$SS_{AB} = 16\hat{\beta}_4^2 = 16(0.75)^2 = 9,$$

$$SS_{AC} = 16\hat{\beta}_5^2 = 16(5)^2 = 400,$$

and

$$SS_{BC} = 16\hat{\beta}_6^2 = 16(0)^2 = 0.$$

Thus, the fitted sum of squares is given by FSS $= 2116 + 100 + 9 + 9 + 400 = 2634$. Since the total sum of squares is given by TSS $= 2996$, we get the ANOVA table shown in Table 11.13. The standard deviation σ is estimated by $S \doteq \sqrt{7.222} \doteq 2.687$. Since each estimated coefficient has variance $\sigma^2/n = \sigma^2/16$, it has standard error $S/4 \doteq 0.672$. Thus, we get the results shown in Table 11.14. Observe that the *P*-values in Tables 11.13 and 11.14 coincide. According to either table, the statistically significant effects in decreasing magnitude are the main effect of temperature (*A*), the temperature × catalyst (*AC*) interaction, and the main effect of concentration (*B*).

Recall that $x_3 = -1$ corresponds to catalyst A and $x_3 = 1$ corresponds to catalyst B. Thus, if the catalyst is switched from A to B with the temperature and concentration held at the levels x_1 and x_2, respectively, then the mean yield increases by about

$$\hat{\mu}(x_1, x_2, 1) - \hat{\mu}(x_1, x_2, -1) = 1.5 + 10x_1.$$

In particular, if the temperature is 160°C ($x_1 = -1$), switching from catalyst A to catalyst B causes the mean yield to decrease by about 8.5; if the temperature is 180°C

TABLE   **11.13**

Source	SS	DF	MS	F	P-value
A	2116	1	2116.000	292.985	.000
B	100	1	100.000	13.846	.005
C	9	1	9.000	1.246	.293
AB	9	1	9.000	1.246	.293
AC	400	1	400.000	55.385	.000
BC	0	1	0.000	0.000	1.000
Residuals	65	9	7.222		
Total	2699	15			

TABLE   **11.14**

Term	Effect	Coef.	Estimate	SE	t	P-value
x_1	A	β_1	11.50	0.672	17.117	.000
x_2	B	β_2	−2.50	0.672	−3.721	.005
x_3	C	β_3	0.75	0.672	1.116	.293
x_1x_2	AB	β_4	0.75	0.672	1.116	.293
x_1x_3	AC	β_5	5.00	0.672	7.442	.000
x_2x_3	BC	β_6	0.00	0.672	0.000	1.000

($x_1 = 1$), however, making the indicated switch causes the mean yield to increase by about 11.5. The discussion at the end of Section 10.5 of *Statistics for Experimenters* (George E.P Box, William G. Hunter, and J. Stuart Hunter. New York: Wiley, 1978) is now quite illuminating:

> The result of most practical interest was the very different behaviors of the two "catalyst" types in response to temperature. The effect was unexpected, for although obtained from two different suppliers, the catalysts were supposedly identical. Also, the yield from catalyst B at 180°C was the highest that, up to that time, had been seen. The finding led to a very careful study of the catalyst in further iterations of the investigation.

Three-Factor Interactions

Three-factor and higher-order interactions are defined by the obvious extensions of the definitions involving two-factor interactions (but are not used much in practice). Thus, if k, l, and m are distinct, the tensor product $G_{klm} = G_k \otimes G_l \otimes G_m$ of G_k, G_l and G_m is defined as the span of all functions on $\mathcal{X}$ of the form $g_k(x_k)g_l(x_l)g_m(x_m)$ as $g_k(x_k)$, $g_l(x_l)$, and $g_m(x_m)$ range over G_k, G_l, and G_m, respectively. The obvious extension of Theorem 11.6 is valid. In particular, the dimension of the tensor product of G_k, G_l, and G_m is the product $p_k p_l p_m$ of the individual dimensions. The obvious extension of Theorems 11.7–11.9 are also valid.

Suppose the design random variables X_k, X_l, and X_m are independent. The identifiable $(p_k - 1)(p_l - 1)(p_m - 1)$ dimensional linear space $G_{klm}^0 = G_k^0 \otimes G_l^0 \otimes G_m^0$ is referred to as the three-factor interaction subspace of G_{klm}. Moreover,

$$G_{klm} = G_{con} \oplus G_k^0 \oplus G_l^0 \oplus G_m^0 \oplus G_{kl}^0 \oplus G_{km}^0 \oplus G_{lm}^0 \oplus G_{klm}^0,$$

and the indicated components $G_{con}, \ldots, G_{klm}^0$ of G_{klm} are orthogonal.

Consider again the complete three-factor experiment in which each factor takes on the two levels -1 and 1 and there are r replications at each of the eight factor-level combinations. Consider also the eight-dimensional linear space

$$G = G_{123} = G_{con} \oplus G_1^0 \oplus G_2^0 \oplus G_3^0 \oplus G_{12}^0 \oplus G_{13}^0 \oplus G_{23}^0 \oplus G_{123}^0,$$

which is identifiable and saturated. The one-dimensional orthogonal components $G_{con}, G_1^0, G_2^0, G_3^0, G_{12}^0, G_{13}^0, G_{23}^0, G_{123}^0$ of this space have the bases $1, x_1, x_2, x_3, x_1 x_2, x_1 x_3, x_1 x_3, x_2 x_3, x_1 x_2 x_3$, respectively, and these eight functions form an orthogonal basis of G. Each function has squared norm n.

Suppose the regression function is in G, let $\beta_0, \ldots, \beta_7$ denote the regression coefficients relative to the indicated basis, and let $\hat{\beta}_0, \ldots, \hat{\beta}_7$ denote the least-squares estimates of these coefficients. Then $\hat{\beta}_0, \ldots, \hat{\beta}_6$ are as given above, whereas

$$\hat{\beta}_7 = \frac{1}{8} \sum_{k=1}^{8} x_{k1} x_{k2} x_{k3} \bar{Y}_k.$$

Under the assumption of homoskedasticity, each of the estimates $\hat{\beta}_0, \ldots, \hat{\beta}_7$ has variance σ^2/n. Table 11.15 shows the corresponding ANOVA table. Here $SS_A, \ldots, SS_{BC}$ are as given above, whereas $SS_{ABC} = n\hat{\beta}_7^2$.

TABLE 11.15

Source	SS	DF
A	SS_A	1
B	SS_B	1
C	SS_C	1
AB	SS_{AB}	1
AC	SS_{AC}	1
BC	SS_{BC}	1
ABC	SS_{ABC}	1
Within	WSS	$n - 8$
Total	TSS	$n - 1$

In the context of the pilot plant experiment, the numerical values of the least-squares estimates of the regression functions are shown in Table 11.16. Observe that

TABLE   11.16

1	x_1	x_2	x_3	x_1x_2	x_1x_3	x_2x_3	$x_1x_2x_3$	Average Response
1	−1	−1	−1	1	1	1	−1	60
1	1	−1	−1	−1	−1	1	1	72
1	−1	1	−1	−1	1	−1	1	54
1	1	1	−1	1	−1	−1	−1	68
1	−1	−1	1	1	−1	−1	1	52
1	1	−1	1	−1	1	−1	−1	83
1	−1	1	1	−1	−1	1	−1	45
1	1	1	1	1	1	1	1	80
64.25	11.50	−2.50	0.75	0.75	5.00	0.00	0.25	
$\hat{\beta}_0$	$\hat{\beta}_1$	$\hat{\beta}_2$	$\hat{\beta}_3$	$\hat{\beta}_4$	$\hat{\beta}_5$	$\hat{\beta}_6$	$\hat{\beta}_7$	

the 8×8 matrix of ± 1's is the design matrix $\mathbf{X}$ in the form with d rows, which was introduced in Section 9.3. The least-squares estimate of the regression function is now given by

$$\hat{\mu}(x_1, x_2, x_3) = 64.25 + 11.5x_1 - 2.5x_2 + 0.75x_3$$
$$+ 0.75x_1x_2 + 5x_1x_3 + 0.25x_1x_2x_3.$$

The within sum of squares is given by WSS = 64 [see the solution to Example 8.6(b) in Section 8.3], so σ^2 is estimated by $S^2 = \text{WSS}/8 = 64/8 = 8$ and hence σ is estimated by $S = \sqrt{8} \doteq 2.828$. Since each estimated regression coefficient has variance $\sigma^2/n = \sigma^2/16$, it has standard error $S/4 = \sqrt{2}/2 \doteq 0.707$. Thus, we get the results in Table 11.17. Table 11.18 shows the corresponding ANOVA table. Since the least-squares estimate $\hat{\beta}_7$ of the three-factor interaction coefficient β_7 is insignificant, these results do not differ much from those given earlier for the model without the three-factor interaction.

TABLE   11.17

Term	Effect	Coef.	Estimate	SE	t	P-value
x_1	A	β_1	11.50	0.707	16.263	.000
x_2	B	β_2	−2.50	0.707	−3.536	.008
x_3	C	β_3	0.75	0.707	1.061	.320
x_1x_2	AB	β_4	0.75	0.707	1.061	.320
x_1x_3	AC	β_5	5.00	0.707	7.071	.000
x_2x_3	BC	β_6	0.00	0.707	0.000	1.000
$x_1x_2x_3$	ABC	β_7	0.25	0.707	0.354	.733

TABLE   11.18

Source	SS	DF	MS	F	P-value
A	2116	1	2116	264.500	.000
B	100	1	100	12.500	.008
C	9	1	9	1.125	.320
AB	9	1	9	1.125	.320
AC	400	1	400	50.000	.000
BC	0	1	0	0.000	1.000
ABC	1	1	1	0.125	0.733
Within	64	8	8		
Total	2699	15			

An Orthogonal Array Involving Seven Factors, Each at Two Levels

Table 11.19 shows the last seven columns of the 8×8 design matrix used in the analysis of the pilot plant experiment. We conclude from the orthogonality of the basis $1, x_1, x_2, x_3, x_1x_2, x_1x_3, x_2x_3, x_1x_2x_3$ used in the analysis of the experimental data and the fact that all three of the factors have the two levels -1 and 1 that the array shown in Table 11.19 is orthogonal [see Problem 2.8 in Section 2.1].

TABLE   11.19

−1	−1	−1	1	1	1	−1
1	−1	−1	−1	−1	1	1
−1	1	−1	−1	1	−1	1
1	1	−1	1	−1	−1	−1
−1	−1	1	1	−1	−1	1
1	−1	1	−1	1	−1	−1
−1	1	1	−1	−1	1	−1
1	1	1	1	1	1	1

At this point, you should note the similarity between the last analysis of the pilot plant experiment and the analysis of the artificial pilot plant experiment involving seven factors that was given in Section 11.1.

Problems

11.15   Let G_1 and G_2 be the spaces of functions on $\mathbb{R}^2$ that are linear functions of x_1 and x_2, respectively. Determine a basis of $G_1 \otimes G_2$.

11.16 Let G_1 be the space of functions on $\mathbb{R}^2$ that are quadratic polynomials in x_1 and let G_2 be the space of functions on $\mathbb{R}^2$ that are quadratic polynomials in x_2. Consider the basis $1, x_1, x_1^2$ of G_1 and the basis $1, x_2, x_2^2$ of G_2. Determine the corresponding basis of $G_1 \otimes G_2$.

11.17 Consider the orthogonal array involving 18 runs and eight factors as shown in Table 11.8 in Section 11.1. Show that the design random variables X_k, X_l, and X_m are dependent for k, l, and m distinct members of $\{3, \ldots, 8\}$.

11.18 (Continued.) List the ordered pairs (l, m) of distinct members of $\{3, \ldots, 8\}$ such that X_1, X_l, and X_m are independent.

11.19 Consider the orthogonal array involving 18 runs and eight factors as shown in Table 11.8 in Section 11.1. Let the first two factors (A and B) be replaced by a combined factor having the six levels $1, \ldots, 6$ and defined as $\frac{3}{2}x_1 + x_2 + \frac{7}{2}$ when the first factor has level x_1 and the second factor has level x_2:

		B		
		-1	0	1
	-1	1	2	3
A	1	4	5	6

Then we get the following array for a design involving 18 runs and seven factors with one factor having six levels and each of the other factors having three levels (Table 11.20). Explain why this is also an orthogonal array.

T A B L E   11.20

x_1	x_2	x_3	x_4	x_5	x_6	x_7
1	-1	-1	-1	-1	-1	-1
1	0	0	0	0	0	0
1	1	1	1	1	1	1
2	-1	-1	0	0	1	1
2	0	0	1	1	-1	-1
2	1	1	-1	-1	0	0
3	-1	0	-1	1	0	1
3	0	1	0	-1	1	-1
3	1	-1	1	0	-1	0
4	-1	1	1	0	0	-1
4	0	-1	-1	1	1	0
4	1	0	0	-1	-1	1
5	-1	0	1	-1	1	0
5	0	1	-1	0	-1	1
5	1	-1	0	1	0	-1
6	-1	1	0	1	-1	0
6	0	-1	1	-1	0	1
6	1	0	-1	0	1	-1

11.20   In the context of Example 11.7, show that G_{12}^0 and G_{34}^0 are *not* orthogonal, G_{13}^0 and G_{24}^0 are *not* orthogonal, and G_{14}^0 and G_{23}^0 are *not* orthogonal.

11.21   Consider the orthogonal array involving 18 runs and eight factors that was shown in Table 11.8 in Section 11.1. Recall that the first factor has two levels and that each of the other seven factors has three levels. Let G_1 be the space of linear functions of x_1; let $G_2, \ldots, G_8$ be the spaces of quadratic polynomials in $x_2, \ldots, x_8$, respectively; and set

$$G = G_{\text{con}} \oplus G_1^0 \oplus G_2^0 \oplus G_{12}^0 \oplus G_3^0 \oplus \cdots \oplus G_8^0.$$

(a) Explain why the indicated components $G_{\text{con}}, \ldots, G_8^0$ of G are orthogonal.
(b) Show the corresponding ANOVA table.

11.22   (Continued.) Show that G_{13}^0 and G_4^0 are not orthogonal.

11.23   Complete the proof of Theorem 11.10(c) by showing that if the design random variables X_1, X_2, X_3, and X_4 are independent, then G_{12}^0 and G_{34}^0 are orthogonal.

11.24   Consider a complete factorial experiment involving two factors, A and B, with each factor having two levels, -1 and 1, and r replications at each factor-level combination. Let G_1 and G_2 be the spaces of linear functions of x_1 and x_2, respectively. (a) Verify that $1, x_1, x_2$, and $x_1 x_2$ form an orthogonal basis of $G = G_1 \otimes G_2$ and that each function has squared norm $n = 4r$. (b) Show the corresponding ANOVA table, which includes the sums of squares for main effect of A, the main effect of B, and the interaction between A and B.

11.25   Consider the pilot plant experiment. (a) Analyze the data from the eight runs in which catalyst A is used. (b) Analyze the data from the eight runs in which catalyst B is used. (c) Discuss and compare the results of parts (a) and (b).

11.26   Explain how to construct an orthogonal array for an experiment involving 15 factors, each at two levels and 16 design points.

11.27   Consider an experiment involving five factors, each at two levels that are assigned the coded values -1 (low) and 1 (high) and let x_m be the coded value of the mth factor for $1 \le m \le 5$. A complete factorial experiment would involve $2^5 = 32$ factor-level combinations. Let the experiment consist of the one-quarter fraction of the complete factorial experiment containing the $2^{5-2} = 8$ factor-level combinations such that $x_3 = -x_1 x_2$ and $x_5 = -x_1 x_4$. (a) Show that the design is in the form of an orthogonal array. (b) Observe that the design random variables X_1, X_2, and X_3 are dependent and that X_1, X_4, and X_5 are dependent. Show that these are the only departures from independence three-at-a-time. (c) Show that no four design random variables are independent. (d) Show that the experiment can also be described as consisting of those factor-level combinations such that $x_2 = -x_1 x_3$ and $x_4 = -x_1 x_5$.

11.28   Suppose the design random variables X_l and X_m are independent, where $l \ne m$; let G_l and G_m be spaces of functions of x_l and x_m, respectively, each of which contains the constant functions; and let $G_{lm} = G_l \otimes G_m$, $G_l^0 = G_l \ominus G_{\text{con}}$, $G_m^0 = G_m \ominus G_{\text{con}}$, and $G_{lm}^0 = G_l^0 \otimes G_m^0$. Show that (a) $P_{G_l}[\bar{Y}(\cdot)] = \bar{Y} + P_{G_l^0}[\bar{Y}(\cdot)]$ and $P_{G_m}[\bar{Y}(\cdot)] = \bar{Y} + P_{G_m^0}[\bar{Y}(\cdot)]$; (b) $P_{G_{lm}}[\bar{Y}(\cdot)] = \bar{Y} + P_{G_l^0}[\bar{Y}(\cdot)] + P_{G_m^0}[\bar{Y}(\cdot)] + P_{G_{lm}^0}[\bar{Y}(\cdot)]$; (c) $P_{G_{lm}^0}[\bar{Y}(\cdot)] = P_{G_{lm}}[\bar{Y}(\cdot)] - P_{G_l}[\bar{Y}(\cdot)] - P_{G_m}[\bar{Y}(\cdot)] + \bar{Y}$.

11.3
Experiments with Factors Having Three Levels

In this section we apply the material in Sections 11.1 and 11.2 to experiments with factors having three levels and, in particular, to the lube oil experiment.

Orthogonal Bases

Consider an experiment in which the mth factor takes on the three levels -1, 0, and 1 and the design random variable X_m is uniformly distributed on $\mathcal{X}'_m = \{-1, 0, 1\}$. Let G_m be the space of quadratic polynomials in x_m and consider the orthogonal basis 1, x_m, $x_m^2 - \frac{2}{3}$ of this space. The squared norms of the functions in this basis are given by $\|1\|^2 = n$, $\|x_m\|^2 = nE(X_m^2) = \frac{2}{3}n$, and $\|x_m^2 - \frac{2}{3}\|^2 = nE[(X_m^2 - \frac{2}{3})^2] = \frac{2}{9}n$. Observe that x_m and $x_m^2 - \frac{2}{3}$ form an orthogonal basis of $G_m^0 = G_m \ominus G_{\text{con}}$.

Let the orthogonal projection of $\bar{Y}(\cdot)$ onto G_m^0 be written as

$$P_{G_m^0}[\bar{Y}(\cdot)] = \hat{\beta}_1 x_m + \hat{\beta}_2 (x_m^2 - \tfrac{2}{3}).$$

Then

$$\hat{\beta}_1 = \frac{\langle x_m, \bar{Y}(\cdot)\rangle}{\frac{2}{3}n}, \qquad \hat{\beta}_2 = \frac{\langle x_m^2 - \frac{2}{3}, \bar{Y}(\cdot)\rangle}{\frac{2}{9}n},$$

and

$$\|P_{G_m^0}[\bar{Y}(\cdot)]\|^2 = n(\tfrac{2}{3}\hat{\beta}_1^2 + \tfrac{2}{9}\hat{\beta}_2^2).$$

Under the assumptions of the homoskedastic regression model, $\hat{\beta}_1$ and $\hat{\beta}_2$ are uncorrelated, and

$$\text{var}(\hat{\beta}_1) = \frac{\sigma^2}{\frac{2}{3}n} \quad \text{and} \quad \text{var}(\hat{\beta}_2) = \frac{\sigma^2}{\frac{2}{9}n}$$

(see Problem 10.35 in Section 10.3).

Suppose now that the lth and mth factors each take on the three levels -1, 0, and 1, where $l \neq m$, that the design random variables X_l and X_m are independent, and that each of them is uniformly distributed on $\{-1, 0, 1\}$. Then $x_l x_m$, $x_l(x_m^2 - \frac{2}{3})$, $(x_l^2 - \frac{2}{3})x_m$, $(x_l^2 - \frac{2}{3})(x_m^2 - \frac{2}{3})$ is an orthogonal basis of the four-dimensional space $G_{lm}^0 = G_l^0 \otimes G_m^0$ (see Theorem 11.8 in Section 11.2). The squared norms of the functions in this basis are given by

$$\|x_l x_m\|^2 = nE(X_l^2)E(X_m^2) = \tfrac{2}{3} \cdot \tfrac{2}{3}n = \tfrac{4}{9}n,$$

$$\|x_l(x_m^2 - \tfrac{2}{3})\|^2 = \|(x_l^2 - \tfrac{2}{3})x_m\|^2 = nE[(X_l^2 - \tfrac{2}{3})^2]E(X_m^2) = \tfrac{2}{9} \cdot \tfrac{2}{3}n = \tfrac{4}{27}n,$$

and

$$\|(x_l^2 - \tfrac{2}{3})(x_m^2 - \tfrac{2}{3})\|^2 = nE[(X_l^2 - \tfrac{2}{3})^2]E[(X_m^2 - \tfrac{2}{3})^2] = \tfrac{2}{9} \cdot \tfrac{2}{9}n = \tfrac{4}{81}n.$$

Let the orthogonal projection of $\bar{Y}(\cdot)$ onto G_{lm}^0 be written as

$$P_{G_{lm}^0}[\bar{Y}(\cdot)] = \hat{\beta}_1 x_l x_m + \hat{\beta}_2 x_l(x_m^2 - \tfrac{2}{3}) + \hat{\beta}_3(x_l^2 - \tfrac{2}{3})x_m + \hat{\beta}_4(x_l^2 - \tfrac{2}{3})(x_m^2 - \tfrac{2}{3}).$$

Then

$$\hat{\beta}_1 = \frac{\langle x_l x_m, \bar{Y}(\cdot)\rangle}{\tfrac{4}{9}n},$$

$$\hat{\beta}_2 = \frac{\langle x_l(x_m^2 - \tfrac{2}{3}), \bar{Y}(\cdot)\rangle}{\tfrac{4}{27}n},$$

$$\hat{\beta}_3 = \frac{\langle (x_l^2 - \tfrac{2}{3})x_m, \bar{Y}(\cdot)\rangle}{\tfrac{4}{27}n},$$

$$\hat{\beta}_4 = \frac{\langle (x_l^2 - \tfrac{2}{3})(x_m^2 - \tfrac{2}{3}), \bar{Y}(\cdot)\rangle}{\tfrac{4}{81}n},$$

and

$$\|P_{G_{lm}^0}[\bar{Y}(\cdot)]\|^2 = n(\tfrac{4}{9}\hat{\beta}_1^2 + \tfrac{4}{27}\hat{\beta}_2^2 + \tfrac{4}{27}\hat{\beta}_3^2 + \tfrac{4}{81}\hat{\beta}_4^2).$$

Under the assumptions of the homoskedastic regression model, $\hat{\beta}_1$, $\hat{\beta}_2$, $\hat{\beta}_3$, and $\hat{\beta}_4$ are uncorrelated, and

$$\text{var}(\hat{\beta}_1) = \frac{\sigma^2}{\tfrac{4}{9}n}, \quad \text{var}(\hat{\beta}_2) = \text{var}(\hat{\beta}_3) = \frac{\sigma^2}{\tfrac{4}{27}n}, \quad \text{and} \quad \text{var}(\hat{\beta}_4) = \frac{\sigma^2}{\tfrac{4}{81}n}.$$

Lube Oil Experiment

Recall that the lube oil experiment has the form of an orthogonal array involving four factors, each of which takes on the three levels -1, 0, and 1. Moreover, there are 27 runs and the design random variables X_1, X_2, X_3, and X_4 are independent three-at-a-time; that is, there is one run at each combination of possible levels of any three of the factors. For $1 \le m \le 4$, let G_m be the space of quadratic polynomials in x_m and consider the orthogonal basis $1, x_m, x_m^2 - \tfrac{2}{3}$ of this space. The squared norms of the functions in this basis are given by $\|1\|^2 = 27$, $\|x_m\|^2 = 18$, and $\|x_m^2 - \tfrac{2}{3}\|^2 = 6$.

Let the orthogonal projection of $\bar{Y}(\cdot)$ onto G_1^0 be written as

$$P_{G_1^0}[\bar{Y}(\cdot)] = \hat{\beta}_1 x_1 + \hat{\beta}_2(x_1^2 - \tfrac{2}{3}).$$

Then

$$\hat{\beta}_1 = \frac{1}{18}\sum_{i=1}^{27} x_{i1} Y_i \doteq 15.633,$$

and

$$\hat{\beta}_2 = \frac{1}{6}\sum_{i=1}^{27}(x_1^2 - \tfrac{2}{3})Y_i \doteq 4.022,$$

so

$$SS_A = \|P_{G_1^0}[\bar{Y}(\cdot)]\|^2 = 18\hat{\beta}_1^2 + 6\hat{\beta}_2^2 \doteq 4496.29.$$

Here SS_A is referred to as the sum of squares due to the main effect of A.

Let the orthogonal projection of $\bar{Y}(\cdot)$ onto $G_{12}^0 = G_1^0 \otimes G_2^0$ be written as

$$P_{G_{12}^0}[\bar{Y}(\cdot)] = \hat{\beta}_1 x_1 x_2 + \hat{\beta}_2 x_1(x_2^2 - \tfrac{2}{3}) + \hat{\beta}_3(x_1^2 - \tfrac{2}{3})x_2 + \hat{\beta}_4(x_1^2 - \tfrac{2}{3})(x_2^2 - \tfrac{2}{3}).$$

The squared norms of the functions in the indicated orthogonal basis of G_{12}^0 are given by $\|x_1 x_2\|^2 = 12$, $\|x_1(x_2^2 - \tfrac{2}{3})\|^2 = \|(x_1^2 - \tfrac{2}{3})x_2\|^2 = 4$, and $\|(x_1^2 - \tfrac{2}{3})(x_2^2 - \tfrac{2}{3})\|^2 = \tfrac{4}{3}$. Thus,

$$\hat{\beta}_1 = \frac{1}{12}\sum_{i=1}^{27} x_{i1} x_{i2} Y_i \doteq 4.650,$$

$$\hat{\beta}_2 = \frac{1}{4}\sum_{i=1}^{27} x_{i1}(x_{i2}^2 - \tfrac{2}{3})Y_i \doteq 3.100,$$

$$\hat{\beta}_3 = \frac{1}{4}\sum_{i=1}^{27}(x_{i1}^2 - \tfrac{2}{3})x_{i2}Y_i \doteq -0.233,$$

and

$$\hat{\beta}_4 = \frac{3}{4}\sum_{i=1}^{27}(x_{i1}^2 - \tfrac{2}{3})(x_{i2}^2 - \tfrac{2}{3})Y_i \doteq 3.083,$$

so

$$SS_{AB} = \|P_{G_{12}^0}[\bar{Y}(\cdot)]\|^2 = 12\hat{\beta}_1^2 + 4\hat{\beta}_2^2 + 4\hat{\beta}_3^2 + \tfrac{4}{3}\hat{\beta}_4^2 \doteq 310.80.$$

Here SS_{AB} is referred to as the sum of squares due to the interaction between A and B.

In the same manner, we get the other orthogonal projections shown in Table 11.21. Similarly, we get the other sums of squares in Table 11.22.

The design random variables X_1, X_2, X_3, and X_4 are independent three-at-a-time, but they are not independent in the usual sense. This can be seen by noting, for example, that $P(X_1 = X_2 = X_3 = X_4 = 1) = 0$ since the factor-level combination $(1, 1, 1, 1)$ is not a member of the design set. Thus, we cannot use Theorem 11.10(c) in Section 11.2 to conclude that G_{jk}^0 and G_{lm}^0 are orthogonal when (j, k, l, m) is a permutation of $1, 2, 3, 4$. In fact, these spaces are not orthogonal; that is, G_{12}^0 and G_{34}^0 are not orthogonal, G_{13}^0 and G_{24}^0 are not orthogonal, and G_{23}^0 and G_{14}^0 are not orthogonal [see Equation (35) in Section 10.5]. These are the only departures from orthogonality, however, among the 11 spaces $G_{con}, G_1^0, \ldots, G_4^0, G_{12}^0, \ldots, G_{34}^0$ [see parts (a) and (b) of Theorem 11.10 in Section 11.2]. To obtain an orthogonal subcollection of these spaces, we can choose $G_{con}, G_1^0, \ldots, G_4^0$, one of the spaces G_{12}^0 and G_{34}^0, one of the spaces G_{13}^0 and G_{24}^0, and one of the spaces G_{23}^0 and G_{14}^0. According to Table 11.22, $SS_{AB} > SS_{CD}$, $SS_{AC} > SS_{BD}$, and $SS_{BC} > SS_{AD}$. This suggests choosing G_{12}^0 rather than G_{34}^0, G_{13}^0 rather than G_{24}^0, and G_{23}^0 rather than G_{14}^0. In this manner, we are led to considering the least-squares fit to the lube oil data

TABLE   11.21

Space	Orthogonal Projection of $\bar{Y}(\cdot)$
G_1^0	$15.633x_1 + 4.022(x_1^2 - \frac{2}{3})$
G_2^0	$12.128x_2 - 4.494(x_2^2 - \frac{2}{3})$
G_3^0	$17.506x_3 - 0.794(x_3^2 - \frac{2}{3})$
G_4^0	$3.444x_4 - 3.411(x_4^2 - \frac{2}{3})$
G_{12}^0	$4.650x_1x_2 + 3.100x_1(x_2^2 - \frac{2}{3}) - 0.233(x_1^2 - \frac{2}{3})x_2 + 3.083(x_1^2 - \frac{2}{3})(x_2^2 - \frac{2}{3})$
G_{13}^0	$10.042x_1x_3 - 2.225x_1(x_3^2 - \frac{2}{3}) - 0.842(x_1^2 - \frac{2}{3})x_3 - 0.442(x_1^2 - \frac{2}{3})(x_3^2 - \frac{2}{3})$
G_{14}^0	$2.325x_1x_4 - 8.625x_1(x_4^2 - \frac{2}{3}) - 6.908(x_1^2 - \frac{2}{3})x_4 + 3.808(x_1^2 - \frac{2}{3})(x_4^2 - \frac{2}{3})$
G_{23}^0	$6.358x_2x_3 - 4.458x_2(x_3^2 - \frac{2}{3}) - 4.892(x_2^2 - \frac{2}{3})x_3 - 2.642(x_2^2 - \frac{2}{3})(x_3^2 - \frac{2}{3})$
G_{24}^0	$-4.025x_2x_4 - 10.358x_2(x_4^2 - \frac{2}{3}) - 6.458(x_2^2 - \frac{2}{3})x_4 - 1.392(x_2^2 - \frac{2}{3})(x_4^2 - \frac{2}{3})$
G_{34}^0	$-0.983x_3x_4 - 3.517x_3(x_4^2 - \frac{2}{3}) - 0.483(x_3^2 - \frac{2}{3})x_4 + 6.783(x_3^2 - \frac{2}{3})(x_4^2 - \frac{2}{3})$

TABLE   11.22

Source	SS	DF	MS
A	4496.29	2	2248.14
B	2768.69	2	1384.35
C	5519.79	2	2759.89
D	283.37	2	141.68
AB	310.80	4	77.70
AC	1232.92	4	308.23
AD	572.67	4	143.17
BC	669.67	4	167.42
BD	793.01	4	198.25
CD	123.36	4	30.84

in the 21-dimensional linear space

$$G = G_{\text{con}} \oplus G_1^0 \oplus G_2^0 \oplus G_3^0 \oplus G_4^0 \oplus G_{12}^0 \oplus G_{13}^0 \oplus G_{23}^0, \tag{20}$$

the indicated components of which are orthogonal. The corresponding ANOVA table is shown in Table 11.23. Accordingly, the statistically significant contributions to the fitted sum of squares in decreasing order of importance are the main effect of C, the main effect of A, the main effect of B, and the AC interaction. The main effect of D is not statistically significant, and its sum of squares is relatively small in magnitude [see the interpretation in the solution to Example 10.20(b) in Section 10.7].

Combining the basis 1 of G_{con} and the bases of the other components of G, as described above, we get an orthogonal basis of G. Table 11.24 shows the least-squares estimate of the coefficients of these functions, along with their standard errors, t statistics, and P-values.

TABLE 11.23

Source	SS	DF	MS	F	P-value
A	4,496.29	2	2,248.14	51.301	.000
B	2,768.69	2	1,384.35	31.590	.001
C	5,519.79	2	2,759.89	62.979	.000
D	283.37	2	141.68	3.233	.111
AB	310.80	4	77.70	1.773	.253
AC	1,232.92	4	308.23	7.034	.019
BC	669.67	4	167.42	3.820	.071
Residuals	262.93	6	43.82		
Total	15,544.46	26			

TABLE 11.24

Term	Coef.	Estimate	SE	t	P-value
1	β_0	29.981	1.274	23.534	.000
x_1	β_1	15.633	1.560	10.019	.000
$x_1^2 - \frac{2}{3}$	β_2	4.022	2.703	1.488	.187
x_2	β_3	12.128	1.560	7.773	.000
$x_2^2 - \frac{2}{3}$	β_4	−4.494	2.703	−1.663	.147
x_3	β_5	17.506	1.560	11.219	.000
$x_3^2 - \frac{2}{3}$	β_6	−0.794	2.703	−0.294	.779
x_4	β_7	3.444	1.560	2.208	.069
$x_4^2 - \frac{2}{3}$	β_8	−3.411	2.703	−1.262	.254
$x_1 x_2$	β_9	4.650	1.911	2.433	.051
$x_1(x_2^2 - \frac{2}{3})$	β_{10}	3.100	3.310	0.937	.385
$(x_1^2 - \frac{2}{3})x_2$	β_{11}	−0.233	3.310	−0.070	.946
$(x_1^2 - \frac{2}{3})(x_2^2 - \frac{2}{3})$	β_{12}	3.083	5.733	0.538	.610
$x_1 x_3$	β_{13}	10.042	1.911	5.255	.002
$x_1(x_3^2 - \frac{2}{3})$	β_{14}	−2.225	3.310	−0.672	.526
$(x_1^2 - \frac{2}{3})x_3$	β_{15}	−0.842	3.310	−0.254	.808
$(x_1^2 - \frac{2}{3})(x_3^2 - \frac{2}{3})$	β_{16}	−0.442	5.733	−0.077	.941
$x_2 x_3$	β_{17}	6.358	1.911	3.327	.016
$x_2(x_3^2 - \frac{2}{3})$	β_{18}	−4.458	3.310	−1.347	.227
$(x_2^2 - \frac{2}{3})x_3$	β_{19}	−4.892	3.310	−1.478	.190
$(x_2^2 - \frac{2}{3})(x_3^2 - \frac{2}{3})$	β_{20}	−2.642	5.733	−0.461	.661

Under the assumptions of the normal linear regression model corresponding to G, $\hat{\beta}_0, \ldots, \hat{\beta}_{20}$ and RSS are independent, $\hat{\beta}_0, \ldots, \hat{\beta}_{20}$ are normally distributed with the respective means $\beta_0, \ldots, \beta_{20}$ and variances $\sigma^2/\|1\|^2, \ldots, \sigma^2/\|(x_2^2 - \frac{2}{3})(x_3^2 - \frac{2}{3})\|^2$, and RSS $/\sigma^2$ has the chi-square distribution with 6 degrees of freedom. Suppose $P_{G_1^0}[\mu(\cdot)] = 0$ or, equivalently, $\beta_1 = \beta_2 = 0$. Then

$$Z_1 = \frac{\hat{\beta}_1}{\text{SD}(\hat{\beta}_1)} = \frac{\|x_1\|}{\sigma}\hat{\beta}_1 = \frac{\sqrt{18}}{\sigma}\hat{\beta}_1$$

and

$$Z_2 = \frac{\hat{\beta}_2}{\text{SD}(\hat{\beta}_2)} = \frac{1}{\sigma}\|x_1^2 - \tfrac{2}{3}\|\hat{\beta}_2 = \frac{\sqrt{6}}{\sigma}\hat{\beta}_2$$

are independent, standard normal random variables, so $\text{SS}_A/\sigma^2 = Z_1^2 + Z_2^2$ has the chi-square distribution with 2 degrees of freedom and hence

$$F_A = \frac{\text{SS}_A/2}{\text{RSS}/6} = \frac{(Z_1^2 + Z_2^2)/2}{(\text{RSS}/\sigma^2)/6}$$

has the F distribution with 2 degrees of freedom in the numerator and 6 degrees of freedom in the denominator. If F_A is unusually large, we reject the hypothesis that $P_{G_1^0}[\mu(\cdot)] = 0$. The corresponding P-value equals $= 1 - F_{2,6}(F_A)$. Thus, we have justified the first P-value in Table 11.24. The remaining P-values in the table can be justified in a similar manner.

According to the P-values for the t statistics in Table 11.24, the coefficients of the six cubic terms $x_1(x_2^2 - \frac{2}{3})$, $(x_1^2 - \frac{2}{3})x_2$, $x_1(x_3^2 - \frac{2}{3})$, $(x_1^2 - \frac{2}{3})x_3$, $x_2(x_3^2 - \frac{2}{3})$, $(x_2^2 - \frac{2}{3})x_3$ and the three quartic terms $(x_1^2 - \frac{2}{3})(x_2^2 - \frac{2}{3})$, $(x_1^2 - \frac{2}{3})(x_3^2 - \frac{2}{3})$, $(x_2^2 - \frac{2}{3})(x_3^2 - \frac{2}{3})$ are statistically insignificant, and hence these terms can be deleted from the model, which would leave us with a proper subspace of the the quadratic model used in Section 10.5 to analyze the lube oil data (the $x_1 x_4$ term would be omitted). The full quadratic model uses just one term to model each two-factor interaction; that is, $x_1 x_2$ is used to model the AB interaction, $x_1 x_3$ is used to model the AC interaction, and so forth. Using the space of quadratic functions in the levels of the four factors to model the regression function has several advantages over using the space G defined in Equation (20): It is simpler and more familiar, the two-factor interactions are more easily interpreted and explained, and it includes all six of the two-factor interactions.

As a final remark, if we would try to augment the space G given by (20) by adding the four basis functions, say, from the space G_{34}^0 corresponding to the CD interaction, we would get a nonidentifiable model. Indeed, the four orthogonal basis functions of G_{12}^0 and the four orthogonal basis functions of G_{34}^0 are together linearly dependent on the design set. In particular, by using a linear regression program, we get $(x_3^2 - \frac{2}{3})(x_4^2 - \frac{2}{3})$ equals

$$\tfrac{1}{6}x_1 x_2 + \tfrac{1}{2}x_1(x_2^2 - \tfrac{2}{3}) + \tfrac{1}{2}(x_1^2 - \tfrac{2}{3})x_2 - \tfrac{1}{2}(x_1^2 - \tfrac{2}{3})(x_2^2 - \tfrac{2}{3}) + \tfrac{1}{3}x_3 x_4$$

on the design set.

Problems

11.29   Consider an experiment involving two factors, each of which takes on the three levels 0, 1, and 2, and suppose there are r repetitions at each factor-level combination. Let G_1 and G_2 be the spaces of quadratic polynomials in x_1 and x_2, respectively. Determine an orthogonal basis of the interaction space $G_{12}^0 = G_1^0 \otimes G_2^0$.

11.30   Consider the analysis in the text of the lube oil experiment corresponding to the 21-dimensional linear space G. Use the numerical values of $\hat\beta_{13}, \ldots, \hat\beta_{16}$ to determine SS_{AC}.

11.31   Consider the lube oil experiment and let $\mathcal{X}_m = \{-1, 0, 1\}$ for $1 \le m \le 4$. If we let G_m be the space of all functions of x_m for $1 \le m \le 4$, would we get the same numerical values for SS_A, SS_{AB}, and so forth, as we did by letting G_m be the space of quadratic polynomials in x_m for $1 \le m \le 4$?

11.32   Show the ANOVA table for the lube oil experiment that includes the four main effects and the AC interaction but ignores the other interactions.

11.33   Consider the lube oil experiment with factors C and D being ignored and let G_1 and G_2 be the spaces of quadratic polynomials in x_1 and x_2, respectively. Determine an orthogonal basis of $G_1 \otimes G_2$.

11.34   Consider a complete two-factor experiment with factor A having the two levels -1 and 1; factor B having the three levels $-1, 0$, and 1; and r replications per factor-level combination. Let G_1 be the space of linear functions of x_1 and let G_2 be the space of quadratic polynomials in x_2. Determine **(a)** the joint distribution of X_1 and X_2; **(b)** an orthogonal basis of G_1^0 and the squared norms of the functions in this basis; **(c)** an orthogonal basis of G_2^0 and the squared norms of the functions in this basis; **(d)** an orthogonal basis of $G_{12}^0 = G_1^0 \otimes G_2^0$ and the squared norms of the functions in this basis; **(e)** an orthogonal basis of $G_1 \otimes G_2$.

11.35   Consider a complete two-factor experiment with the following design set:

x_1	-1	1	1	-1	1	1	-1	1	1
x_2	-1	-1	-1	0	0	0	1	1	1

Suppose there are r runs per design point. Let G_1 be the space of linear functions of x_1 and let G_2 be the space of quadratic polynomials in x_2. Determine **(a)** the joint distribution of X_1 and X_2; **(b)** an orthogonal basis of G_1^0 and the squared norms of the functions in this basis; **(c)** an orthonormal basis of G_2^0 and the squared norms of the functions in this basis; **(d)** an orthogonal basis of $G_{12}^0 = G_1^0 \otimes G_2^0$ and the squared norms of the functions in this basis. **(e)** an orthogonal basis of $G_1 \otimes G_2$.

11.36   Consider the design of Problem 8.20 in Section 8.2. **(a)** Does the design have the form of an orthogonal array? **(b)** Let G_1, G_2, G_3, and G_4 be the spaces of quadratic polynomials in x_1, x_2, x_3, and x_4, respectively. Show that the space $G = G_{\mathrm{con}} \oplus G_1^0 \oplus G_2^0 \oplus G_3^0 \oplus G_4^0$ of additive quadratic polynomials in the levels of the four factors is identifiable and saturated.

11.37   Consider the experimental data shown in Table 11.25. The total sum of squares is given by $\mathrm{TSS} = 37284$. Let G_1, G_2, G_3, G_4, and G be as in Example 11.7 in

Section 11.2. The sums of squares corresponding to the four main effects and the three two-factor interactions involving factor A are shown in Table 11.26. **(a)** Complete and interpret the ANOVA table shown in Table 11.10 in Section 11.2. **(b)** Show and interpret the corresponding ANOVA table for the least-squares fit in the additive subspace $G_{con} \oplus G_1^0 \oplus G_2^0 \oplus G_3^0 \oplus G_4^0$ of G.

TABLE   11.25

x_1	x_2	x_3	x_4	Y
-1	-1	-1	-1	115
-1	-1	0	0	111
-1	-1	1	1	133
-1	0	-1	0	118
-1	0	0	1	147
-1	0	1	-1	145
-1	1	-1	1	159
-1	1	0	-1	161
-1	1	1	0	219
1	-1	-1	0	107
1	-1	0	1	133
1	-1	1	-1	131
1	0	-1	-1	125
1	0	0	0	164
1	0	1	1	229
1	1	-1	1	199
1	1	0	-1	197
1	1	1	0	275

TABLE   11.26

Source	SS
A	3528
B	19396
C	8419
D	1684
AB	1344
AC	837
AD	724

11.38   In the context of Problem 11.37, we get the orthogonal projections shown in Table 11.27. **(a)** Verify the formulas for the orthogonal projections of $\bar{Y}(\cdot)$ onto G_1^0, G_2^0 and G_{12}^0. **(b)** Use the formulas that were verified in part (a) to verify the values of SS_A, SS_B, and SS_{AB} that were given in Problem 11.37.

TABLE 11.27

Space	Orthogonal Projection of $\bar{Y}(\cdot)$
G_1^0	$14x_1$
G_2^0	$40x_2 + 7(x_2^2 - \frac{2}{3})$
G_3^0	$25.75x_3 + 10.75(x_3^2 - \frac{2}{3})$
G_4^0	$10.5x_4 - 9.5(x_4^2 - \frac{2}{3})$
G_{12}^0	$10x_1x_2 - 6x_1(x_2^2 - \frac{2}{3})$
G_{13}^0	$8.25x_1x_3 + 2.25x_1(x_3^2 - \frac{2}{3})$
G_{14}^0	$7.5x_1x_4 - 3.5x_1(x_4^2 - \frac{2}{3})$

11.39 In the context of the experimental data in Problem 11.37, obtain the least-squares estimate of the regression function in a reasonable, not unnecessarily complicated linear space.

11.40 After recoding the levels of the four factors in the experiment in Problem 11.37, we get the data shown in Table 11.28. Think of x_1 as ranging over $\{1, 2\}$ and of x_2, x_3 and x_4 as ranging over $\{1, 2, 3\}$. Correspondingly, let G_1, G_2, G_3, and G_4 be the spaces of all functions of x_1, x_2, x_3, and x_4, respectively. Then we get the orthogonal projections in Table 11.29. Here $I_1(x) = \text{ind}(x = 1)$ and $I_2(x) = \text{ind}(x = 2)$. **(a)** Show that 1 and $I_1(x_1) - \frac{1}{2}$ are orthogonal and that 1, $I_1(x_2) - \frac{1}{3}$ and $I_2(x_2) + \frac{1}{2}I_1(x_2) - \frac{1}{2}$

TABLE 11.28

x_1	x_2	x_3	x_4	Y
1	1	1	1	115
1	1	2	2	111
1	1	3	3	133
1	2	1	2	118
1	2	2	3	147
1	2	3	1	145
1	3	1	3	159
1	3	2	1	161
1	3	3	2	219
2	1	1	2	107
2	1	2	3	133
2	1	3	1	131
2	2	1	1	125
2	2	2	2	164
2	2	3	3	229
2	3	1	3	199
2	3	2	1	197
2	3	3	2	275

are orthogonal. **(b)** Verify the formulas in Table 11.29 for the orthogonal projections of $\bar{Y}(\cdot)$ onto G_1^0, G_2^0, and G_{12}^0. **(c)** Use the formulas that were verified in part (a) to verify the values of SS_A, SS_B, and SS_{AB} that were given in Problem 11.37.

TABLE   11.29

Space	Orthogonal Projection of $\bar{Y}(\cdot)$
G_1^0	$-28[\,I_1(x_1) - \frac{1}{2}\,]$
G_2^0	$-\frac{113}{2}[\,I_1(x_2) - \frac{1}{3}\,] - 47[\,I_2(x_2) + \frac{1}{2}I_1(x_2) - \frac{1}{2}\,]$
G_3^0	$-\frac{133}{4}[\,I_1(x_3) - \frac{1}{3}\,] - \frac{73}{2}[\,I_2(x_3) + \frac{1}{2}I_1(x_3) - \frac{1}{2}\,]$
G_4^0	$-\frac{41}{2}[\,I_1(x_4) - \frac{1}{3}\,] - [\,I_2(x_4) + \frac{1}{2}I_1(x_4) - \frac{1}{2}\,]$
G_{12}^0	$36[\,I_1(x_1) - \frac{1}{2}\,][\,I_1(x_2) - \frac{1}{3}\,] + 8[\,I_1(x_1) - \frac{1}{2}\,][\,I_2(x_2) + \frac{1}{2}I_1(x_2) - \frac{1}{2}\,]$
G_{13}^0	$\frac{45}{2}[\,I_1(x_1) - \frac{1}{2}\,][\,I_1(x_3) - \frac{1}{3}\,] + 21[\,I_1(x_1) - \frac{1}{2}\,][\,I_2(x_3) + \frac{1}{2}I_1(x_3) - \frac{1}{2}\,]$
G_{14}^0	$26[\,I_1(x_1) - \frac{1}{2}\,][\,I_1(x_4) - \frac{1}{3}\,] + 8[\,I_1(x_1) - \frac{1}{2}\,][\,I_2(x_4) + \frac{1}{2}I_1(x_4) - \frac{1}{2}\,]$

11.4
Randomization, Blocking, and Covariates

Sugar Beet Experiment

To illustrate the use of randomization and blocking, let us consider an idealized sugar beet experiment in which the regression function is truly the sum of an arbitrary function of variety and a quintic polynomial of plot. Suppose further that the regression function for the idealized experiment coincides with the least-squares estimate of this function based on the data from the actual sugar beet experiment, which was determined in Section 10.4. Then (see Problem 11.42) the regression function can be written uniquely in the form

$$\mu(x_1, x_2) = \beta_1 I_1(x_1) + \beta_2 I_2(x_1) + \beta_3 I_3(x_1) + \beta_4 I_4(x_1) + \beta_5 I_5(x_1) + g(x_2); \quad \textbf{(21)}$$

here g is a quintic polynomial such that

$$\sum_{x_2=-25}^{24} g(x_2) = 0$$

and hence $\bar{g} = 0$, where

$$\bar{g} = \frac{1}{50} \sum_{x_2=-25}^{24} g(x_2)$$

(The reason for letting x_2 range over $\{-25, \ldots, 24\}$ instead of $\{-25, \ldots, 25\}$ will become clear in the next paragraph.) Specifically, $\beta_1 \doteq 18.199$, $\beta_2 \doteq 17.591$, $\beta_3 \doteq 18.310$, $\beta_4 \doteq 18.039$, $\beta_5 \doteq 18.109$, and

$$g(x_2) \doteq -0.360 - 1.540\frac{x_2}{25} + 2.994\left(\frac{x_2}{25}\right)^2 + 3.509\left(\frac{x_2}{25}\right)^3$$

$$- 3.162\left(\frac{x_2}{25}\right)^4 - 1.718\left(\frac{x_2}{25}\right)^5.$$

Suppose also that the remaining assumptions of the normal linear regression model hold with σ coinciding with its estimate 0.157 based on the actual sugar beet data, as determined in Section 10.4.

In the actual sugar beet experiment, varieties were assigned to plots in the systematic manner $1, \ldots, 5, \ldots, 1, \ldots, 5, 1$. Suppose in the idealized experiment we assign varieties to plots in a different systematic manner—namely, that we assign the standard variety to plots $-25, \ldots, -16$; variety 2 to plots $-15, \ldots, -6$; variety 3 to the plots $-5, \ldots, 4$; variety 4 to the plots $5, \ldots, 14$; variety 5 to the plots $15, \ldots, 24$; and that we let plot 25 lie fallow. Under the given assumptions, an analysis of the experimental data as in Section 10.4 would certainly be appropriate. In this analyis, the plot variable would be referred to as a *covariate*.

Suppose, however, that we analyze the data as in Section 7.3, in which the effect of plot is ignored. Then β_1 is estimated by the sample mean $\hat{\beta}_1 = \bar{Y}_1$ of the responses on the first ten plots, β_2 is estimated by the sample mean $\hat{\beta}_2 = \bar{Y}_2$ of the responses on the next ten plots, and so forth. Under the given assumptions, the estimates $\hat{\beta}_1, \ldots, \hat{\beta}_5$ are independent, and each is normally distributed with variance $\sigma^2/10$ and hence standard deviation $\sigma/\sqrt{10} \doteq 0.0498$. These estimates are not unbiased, however. Indeed,

$$E(\hat{\beta}_1) = \beta_1 + \frac{1}{10}\sum_{x_2=-25}^{-16} g(x_2) \doteq \beta_1 + 0.072,$$

$$E(\hat{\beta}_2) = \beta_2 + \frac{1}{10}\sum_{x_2=-15}^{-6} g(x_2) \doteq \beta_2 + 0.433,$$

$$E(\hat{\beta}_3) = \beta_3 + \frac{1}{10}\sum_{x_2=-5}^{4} g(x_2) \doteq \beta_3 - 0.292,$$

$$E(\hat{\beta}_4) = \beta_4 + \frac{1}{10}\sum_{x_2=5}^{14} g(x_2) \doteq \beta_4 - 0.358,$$

and

$$E(\hat{\beta}_5) = \beta_5 + \frac{1}{10}\sum_{x_2=15}^{24} g(x_2) \doteq \beta_5 + 0.144.$$

Consider $\hat{\beta}_3 - \hat{\beta}_2$ as an estimate of the change $\beta_3 - \beta_2$ in the mean response when we switch from variety 2 to variety 3. This estimate is normally distributed with variance $\sigma^2/5$ and hence standard deviation $\sigma/\sqrt{5} \doteq 0.0704$. It is far from unbiased, however, since $E(\hat{\beta}_3 - \hat{\beta}_2) \doteq \beta_3 - \beta_2 - 0.725$. Thus, $\hat{\beta}_3 - \hat{\beta}_2$ substantially underestimates $\beta_3 - \beta_2 \doteq 0.719$, the reason being that, on average, the plots on which variety 2 is grown are significantly more conducive to high sugar content than those on which variety 3 is grown.

One way to avoid such bias is to use a *completely randomized design*—that is, to randomly assign varieties to plots. For example, we could select plots one at a time by sampling without replacement from the population of plots $-25, \ldots, 24$, assign the standard variety to the first ten plots so selected, variety 2 to the next ten plots selected, and so forth. When we employ such randomization and properly take it into account, we see that the estimates $\hat{\beta}_1 = \bar{Y}_1, \ldots, \hat{\beta}_5 = \bar{Y}_5$ of $\beta_1, \ldots, \beta_5$, respectively, are unbiased.

The disadvantage of a completely randomized design over a systematic design in the context of our idealized sugar beet experiment is that the variances of $\hat{\beta}_1, \ldots, \hat{\beta}_5$ and their differences are substantially inflated. Specifically, for $1 \le j \le 5$,

$$\text{var}(\hat{\beta}_j) = \frac{\sigma^2}{10} + \frac{40}{49} \frac{\sigma_1^2}{10} \doteq 0.0142,$$

where

$$\sigma_1^2 = \frac{1}{50} \sum_{x_2=-25}^{24} [\, g(x_2) - \bar{g}\,]^2 = \frac{1}{50} \sum_{x_2=-25}^{24} [\, g(x_2)\,]^2 \doteq 0.144$$

is the variance of $g(X_2)$ with X_2 being a randomly selected plot (see the answer to Problem 4a); thus $\text{SD}(\hat{\beta}_j) \doteq \sqrt{0.0142} \doteq 0.1193$. Let $1 \le i, l \le 5$ with $i \ne l$, and set $\tau = \beta_l - \beta_i$ and $\hat{\tau} = \hat{\beta}_l - \hat{\beta}_i = \bar{Y}_l - \bar{Y}_i$. Then $\hat{\tau}$ is an unbiased estimate of the variety difference τ, and its variance is given by

$$\text{var}(\hat{\tau}) = \text{var}(\hat{\beta}_l - \hat{\beta}_i) = \frac{\sigma^2}{5} + \frac{40}{49} \frac{\sigma_1^2}{5} + \frac{2}{49}\sigma_1^2 \doteq 0.0344$$

[see the answer to Problem 11.44(c)]; thus, $\text{SD}(\hat{\tau}) \doteq \sqrt{0.0344} \doteq 0.185$.

To illustrate the analyis of experiments having the form of a completely randomized design, we will now use data from a simulated sugar beet experiment having the form of such a design. (The details of the simulation will be discussed at the end of the section.) In the experiment, varieties are assigned to plots in the following manner (the first row indicates that varieties 2, 5, ..., 3 are assigned to plots -25, $-24, \ldots, -16$, respectively; the second row indicates that varieties 2, 2, ..., 3 are assigned to plots $-15, -14, \ldots, -6$, respectively; and so forth):

2	5	4	3	5	4	2	1	4	3
2	2	3	5	1	4	5	5	3	3
4	4	3	5	1	2	2	4	2	2
1	5	1	1	2	2	3	1	3	1
1	5	4	3	4	5	5	3	4	1

The simulated values of the response variable (percentage of sugar content) on the various plots are as follows:

17.0	17.4	17.9	18.3	18.1	18.2	17.9	18.8	18.7	19.0
18.3	18.0	18.8	18.7	18.8	18.6	18.5	18.3	18.4	18.5
17.7	18.0	18.3	18.1	17.6	17.2	17.1	17.7	16.9	16.9
17.9	17.3	17.5	17.7	17.3	17.3	18.1	18.1	18.4	18.1
17.9	18.4	18.3	18.5	18.3	18.5	18.6	18.7	18.3	18.3

Table 11.30 shows the response data grouped according to variety and some relevant summary statistics as in Section 7.3, and Table 11.31 shows the corresponding ANOVA table. The usual estimate of the standard deviation of the response variable is given by $S = \sqrt{\text{WSS}/45} \doteq \sqrt{7.900/45} \doteq 0.419$. Let $1 \le i, l \le 5$ with $i \ne l$. The least-squares estimate of the variety difference $\tau = \beta_l - \beta_i$ is given by $\hat{\tau} = \bar{Y}_l - \bar{Y}_i$, and the standard error of this estimate is given by $\text{SE}(\hat{\tau}) = S/\sqrt{5} \doteq 0.187$ (which is surprisingly close to the theoretical value of 0.185 that was obtained above based on the theoretical model for the idealized sugar beet experiment and the actual randomization employed in the completely randomized design).

TABLE   11.30

k	1	2	3	4	5
	18.8	17.0	18.3	17.9	17.4
	18.8	17.9	19.0	18.2	18.1
	17.6	18.3	18.8	18.7	18.7
	17.9	18.0	18.4	18.6	18.5
	17.5	17.2	18.5	17.7	18.3
	17.7	17.1	18.3	18.0	18.1
	18.1	16.9	18.1	17.7	17.3
	18.1	16.9	18.4	18.3	18.4
	17.9	17.3	18.5	18.3	18.5
	18.3	17.3	18.7	18.3	18.6
$\bar{Y}_k$	18.07	17.39	18.50	18.17	18.19
S_k	0.455	0.498	0.267	0.343	0.484

TABLE   11.31

Source	SS	DF	MS	F	P-value
Varieties	6.715	4	1.679	9.563	.000
Within	7.900	45	0.176		
Total	14.615	49			

To reduce the variance of the estimates of the various linear parameters without making these estimates biased, we can employ a *randomized complete block design* in which we divide up the 50 plots into 10 blocks each containing 5 consecutive plots and, within each block, assign the five varieties to the 5 plots at random. (If, instead, we would divide up the 50 plots into 25 blocks each containing 2 consecutive plots, the blocks would be incomplete in that we could not assign all five varieties to the same block. This *incomplete block design* would be somewhat more complicated to analyze than the complete block design under consideration.)

The usual model for analyzing the data from such a design is to assume that the mean response is the sum of an arbitrary function of variety and an arbitrary function of block. Think of the experiment as involving two qualitative factors: the varieties factor, which takes on the values $1, \ldots, 5$; and the blocking factor, which takes on the levels $1, \ldots, 10$. Viewed in this manner, the experiment has the form of a complete factorial design with 1 trial at each of the $5 \cdot 10 = 50$ factor-level combinations, so the corresponding design random variables are independent and each is uniformly distributed. Thus, the design has the form of an orthogonal array.

Let Y_{ij} denote the response (percentage of sugar content) when the ith variety is grown in the jth block. The least-squares estimate of the corresponding mean response equals $\bar{Y}_{i.} + \bar{Y}_{.j} - \bar{Y}$. Let $1 \leq i, l \leq 5$ with $i \neq l$. Then switching from the ith variety to the lth variety in a given block causes the mean response to change by the amount $\tau = \beta_l - \beta_i$. The least-squares estimate of this variety difference is given by $\hat{\tau} = \bar{Y}_{l.} - \bar{Y}_{i.}$. This estimate is unbiased, and its variance is given by

$$\text{var}(\hat{\tau}) = \frac{\sigma^2}{5} + \frac{1}{100} \sum_{j=1}^{10} \text{var}(g(X_{lj}) - g(X_{ij})), \qquad (22)$$

where X_{ij} is the plot in the block j on which variety i is grown and X_{lj} is the plot in this block on which variety l is grown (see Problem 11.45).

For $1 \leq j \leq 10$, let $\mathcal{B}_j$ denote the set of five plots in the jth block and set

$$\bar{g}_j = \frac{1}{5} \sum_{x_2 \in \mathcal{B}_j} g(x_2)$$

and

$$\sigma_{2j}^2 = \frac{1}{5} \sum_{x_2 \in \mathcal{B}_j} [\, g(x_2) - \bar{g}_j \,]^2,$$

which is a measure of the *heterogeneity* of the plots in the jth block with respect to the given response variable. It follows from Equation (16) in Section 6.2 (the details being left as a problem) that

$$\text{var}(g(X_{lj}) - g(X_{ij})) = 2\sigma_{2j}^2\left(1 + \frac{1}{4}\right) = \frac{5}{2}\sigma_{2j}^2, \qquad i \neq l. \qquad (23)$$

Consequently,

$$\text{var}(\hat{\tau}) = \frac{\sigma^2}{5} + \frac{\sigma_2^2}{4},$$

where

$$\sigma_2^2 = \frac{1}{10} \sum_{j=1}^{10} \sigma_{2j}^2.$$

We can assume that the blocks are ordered from left to right, so that

$$\mathcal{B}_1 = \{-25, \ldots, -21\}, \ldots, \mathcal{B}_{10} = \{20, \ldots, 24\}.$$

Then

$$\bar{g}_1 = \frac{1}{5} \sum_{x_2=-25}^{-21} g(x_2) \doteq -0.304,$$

and hence

$$\sigma_{21}^2 = \frac{1}{5} \sum_{x_2=-25}^{-21} [\, g(x_2) - \bar{g}_1 \,]^2 \doteq 0.0996.$$

Similarly, we get the other results shown in Table 11.32. Consequently, $\sigma_2^2 \doteq 0.0179$, so $\text{var}(\hat{\tau}) \doteq 0.00944$ and hence $\text{SD}(\hat{\tau}) \doteq 0.0971$, which is just over half as large as the corresponding standard deviation 0.185 in the context of the completely randomized design. The reason that the blocking leads to such a substantial improvement in accuracy over the completely randomized design is that the blocks are quite homogeneous; more precisely, σ_2^2 is only about one-eighth as large as σ_1^2.

TABLE 11.32

j	$\bar{g}_j$	σ_{2j}^2
1	−0.304	0.0996
2	0.448	0.0138
3	0.575	0.0015
4	0.292	0.0131
5	−0.134	0.0136
6	−0.450	0.0032
7	−0.488	0.0014
8	−0.227	0.0112
9	0.142	0.0075
10	0.147	0.0143

It is left as a problem to show that

$$\sigma_1^2 = \sigma_2^2 + \frac{1}{10} \sum_{j=1}^{10} (\bar{g}_j - \bar{g})^2 = \sigma_2^2 + \frac{1}{10} \sum_{j=1}^{10} \bar{g}_j^2. \tag{24}$$

According to (24), the fact that σ_2^2 is only about one-eighth as large as σ_1^2 is equivalent to the fact that about seven-eighths of the variation in $g(\cdot)$ from plot to plot is accounted for by the variation in the block averages $\bar{g}_j$ from block to block.

Consider the randomized complete block design shown in Table 11.33. Here block 1 consists of plots −25, −24, −23, −22, and −21, which are referred to as

TABLE   **11.33**

Block	Plot 1	2	3	4	5
1	5	4	3	2	1
2	4	5	2	1	3
3	5	3	2	1	4
4	2	1	5	3	4
5	2	1	4	5	3
6	3	1	2	5	4
7	4	2	3	5	1
8	3	4	2	1	5
9	3	2	5	4	1
10	2	5	3	4	1

plots 1, 2, 3, 4, and 5 in block 1 and which are assigned varieties 5, 4, 3, 2, and 1, respectively. Similarly, block 2 consists of plots -20, -19, -18, -17, and -16, which are assigned varieties 4, 5, 2, 1, and 3, respectively, and so forth.

Consider also the simulated response data shown in Table 11.34.

TABLE   **11.34**

Block	Plot 1	2	3	4	5
1	17.5	17.3	18.2	17.6	18.2
2	18.2	18.4	18.2	18.8	19.0
3	18.8	18.7	18.1	18.8	18.7
4	18.2	18.6	18.3	18.4	18.2
5	17.3	18.2	18.1	18.1	17.8
6	17.9	17.7	17.3	17.5	17.4
7	17.7	16.8	17.6	17.6	17.9
8	18.0	17.9	17.5	18.3	18.0
9	18.0	17.9	18.3	18.2	18.5
10	18.0	18.6	18.7	18.3	18.3

Ordering the response data within each block according to variety and calculating the various row and column averages and the sample mean of all 50 responses, we get the results shown in Table 11.35.

Consider the normal linear regression model corresponding to the assumption that the regression function is the sum of an arbitrary function of block and an arbitrary function of variety. The corresponding ANOVA table is shown in Table 11.36.

T A B L E   **11.35**

Block	1	2	3	4	5	Average
			Variety			
1	18.2	17.6	18.2	17.3	17.5	17.76
2	18.8	18.2	19.0	18.2	18.4	18.52
3	18.8	18.1	18.7	18.7	18.8	18.62
4	18.6	18.2	18.4	18.2	18.3	18.34
5	18.2	17.3	17.8	18.1	18.1	17.90
6	17.7	17.3	17.9	17.4	17.5	17.56
7	17.9	16.8	17.6	17.7	17.6	17.52
8	18.3	17.5	18.0	17.9	18.0	17.94
9	18.5	17.9	18.0	18.2	18.3	18.18
10	18.3	18.0	18.7	18.3	18.6	18.38
Average	18.33	17.69	18.23	18.00	18.11	18.072

T A B L E   **11.36**

Source	SS	DF	MS	F	P-value
Varieties	2.441	4	0.610	13.840	.000
Blocks	6.953	9	0.773	17.522	.000
Residuals	1.587	36	0.044		
Total	10.981	49			

Here

$$SS_{\text{varieties}} = 10 \sum_{i=1}^{5} (\bar{Y}_{i\cdot} - \bar{Y})^2 \quad \text{and} \quad SS_{\text{blocks}} = 5 \sum_{j=1}^{10} (\bar{Y}_{\cdot j} - \bar{Y})^2,$$

where $\bar{Y}_{i\cdot}$ is the sample mean of the responses for the ith variety and $\bar{Y}_{\cdot j}$ is the sample mean of the responses in the jth block. As usual, the standard deviation of the response variable is estimated by $S = \sqrt{RSS/36} \doteq \sqrt{0.044} \doteq 0.210$. Let $1 \le i, l \le 5$ with $i \ne l$ and let τ be the change in the mean response when the variety is switched from i to l in the same block. Then the least-squares estimate of the variety difference τ is given by $\hat{\tau} = \bar{Y}_{l\cdot} - \bar{Y}_{i\cdot}$, and the standard error of this estimate is given by $SE(\hat{\tau}) = S/\sqrt{5} \doteq 0.0939$ (which is rather close to the theoretical value of 0.0971 that was obtained above based on the theoretical model for the idealized sugar beet experiment and the actual randomization employed in the randomized complete block design).

General Discussion

Consider, more generally, an experiment in which each trial involves a distinct *experimental unit*. The experimental units could be plots of land, periods of time, pieces of cloth, animals, or people. In a completely randomized design, we assign the factor-level combinations to the experimental units completely at random. The purpose of using a random rather than systematic assignment is to avoid the possibility of a biased assignment—that is, one in which certain factor-level combinations are systematically assigned better experimental units than others vis-à-vis their effect on the mean response. Such a biased assignment can lead to substantial bias in the estimates of parameters of interest, as we have seen in the context of the idealized sugar beet experiment. It is generally difficult to assess the potential for bias in a systematic assignment of the factor-level combinations to the experimental units. Even if the experimenter is confident that such bias is negligible, others may be skeptical, and, for them, conclusions based on analysis of the experimental data would lack credibility.

The main adverse consequence of a completely randomized design is that the random effect of the experimental unit on the response variable can lead to a substantial increase in the variance of the estimates of parameters of interest. One way to reduce this variance to a more acceptable level is to increase the total number of runs by using more design points or more replications. Such an increase in the total number of runs, however, can be expensive and time-consuming; also, it may be infeasible because of the limited availability of experimental units.

Another way to increase the precision of a completely randomized design is to use experimental units that are very similar to each other in their effect on the mean response variable. For example, in an educational experiment involving students as subjects (experimental units), the experiment could be confined, say, to engineering students. The disadvantage of this approach, especially if there is an interaction between the experimental unit and the various factors, is that conclusions based on such an experiment may not generalize to the complete target population of experimental units. (A certain educational method may work well for engineering students but poorly for social science students.)

Alternatively, we can construct a randomized complete block design by combining the experimental units into blocks that, hopefully, are relatively homogeneous with respect to their effect on the response variable and then introduce block as an additional factor. When the blocks are of size 2, they are commonly referred to as *matched pairs*. In assigning the planned factor-level combinations to the individual units within each block, randomization should be employed; also, if there are more units in a block than will actually be used for the experiment, the units to be used preferably should be selected by sampling without replacement or some other suitable sampling scheme. (This applies to completely randomized designs as well.)

To create relatively homogeneous blocks, we should use whatever understanding we have of the effects of the experimental units on the response variable, but this understanding need not be very precise. In agricultural experiments, it is reasonable to group the plots into blocks based on their physical location. In industrial experiments, it is reasonable to think of time period as the experimental unit and to form blocks of consecutive time periods. In educational experiments involving students as subjects, the various majors could be used as the blocks. In animal experiments, the available litters could be thought of as distinct blocks. In experiments involving human subjects, it is occasionally feasible to use identical twins as matched pairs. In principle, several blocking factors can be introduced; the classical example here is to use longitude and latitude of plots (rows and columns) as blocking factors in agricultural experiments, as in the fertilizer experiment introduced in Section 8.2.

In most experiments, each experimental unit is used at most once. Sometimes, however, the experimental units are each used on several trials at different factor-level combinations. Here the various trials involving the same experimental unit may be treated as a block. Repeated use of the same experimental unit, however, typically has serious limitations. For example, in educational experiments involving several trials with each human subject, what a student learns on one trial might affect the student's performance on later trials. Also, performance on later trials can deteriorate because of fatigue or boredom in a manner that depends in an unknown manner on the factor-level combinations being used and their order of application.

In analyzing experiments based on randomized complete block designs, it is conventional to ignore interactions between the blocking factor and the primary factors. This is generally reasonable if the experimental units in the experiment are typical of those in the target population to which the conclusions of the analysis are to be applied. (The solution to Example 10.23 in Section 10.8 suggests that, in the actual sugar beet experiment, the interaction between plot and variety is negligible.)

Even when randomization is employed in the experiment, the analysis is customarily carried out under the assumptions of the normal linear regression model, which ignores the randomization. In the context of the idealized sugar beet experiment based either on a completely randomized design or a randomized complete block design, we have seen that conclusions based on the normal linear regression model (at least those involving the sizes of the standard deviations of the estimates of variety differences) are in reasonably close agreement with conclusions that take the randomization explicitly into account. There are also (specialized and rather complicated) theoretical results to the effect that conclusions based on the assumptions of the normal linear regression model are approximately applicable to experiments that involve randomization. In fact, the randomization tends to improve the accuracy of inference based on the normal linear regression model by ameliorating the adverse consequences of nonnormality, heteroskedasticity (departure from homoskedasticity), and dependence.

In the context of certain industrial experiments, randomization of the run order of the various factor-level combinations may be infeasible because certain changes in level are time-consuming or otherwise expensive to implement. (Think of making complicated adjustments to an intricate piece of machinery.) In such situations, the use of covariates may be a viable alternative to randomization and blocking. In particular, in the actual sugar beet experiment, the assignment of varieties to plots was done in a systematic manner. From the analysis in Section 10.4, it would appear that using this systematic assignment and quintic polynomials to model the additive contribution of plot as a covariate together leads to more accurate estimates of variety differences than those based on the randomized complete block design. Covariates can also be employed in experiments involving randomization and blocking to take into account environmental or other fluctuations that are uncontrollable but measurable and thereby to reduce the adverse effect of such fluctuations on the precision of the experiment.

Simulated Experiments

We now use the idealized sugar beet experiment to simulate data from completely randomized and randomized complete block designs. To this end, we first use pseudorandom numbers to simulate the observed values of 50 independent random variables $W_1, \ldots, W_{50}$, each having the normal distribution with mean 0 and standard deviation $\sigma \doteq 0.157$. The actual numerical values of these random variables that will be used in our illustrations are as follows:

0.149	−0.178	0.150	0.035	−0.118	−0.104	−0.056	0.111	0.095	0.083
0.121	−0.173	−0.128	−0.005	0.121	0.163	0.008	−0.101	−0.084	0.020
−0.333	0.057	0.155	0.252	−0.263	−0.057	−0.055	0.171	−0.147	−0.134
0.177	−0.244	−0.218	−0.022	0.115	0.051	0.137	0.172	0.240	−0.035
−0.277	0.179	0.064	−0.026	0.052	0.111	0.257	0.220	0.202	0.202

Here $W_1 = 0.149$, $W_2 = -0.178$, ..., $W_{20} = 0.020$, ..., and $W_{50} = 0.202$. Note that

$$\bar{W} \doteq 0.022 \quad \text{and} \quad \sqrt{\frac{1}{49} \sum_i (W_i - \bar{W})^2} \doteq 0.154,$$

which are not at all unusual for a random sample of size 50 from the indicated normal distribution. Also, none of the 50 individual observations is unusually large in magnitude relative to σ.

To obtain a completely randomized design, we first obtain the following 50 pseudorandom numbers $u_1, \ldots, u_{50}$:

.681	.599	.217	.193	.067	.585	.537	.661	.068	.431
.710	.607	.271	.809	.013	.159	.004	.928	.946	.353
.286	.235	.879	.292	.783	.436	.493	.365	.884	.565
.978	.451	.696	.019	.917	.598	.517	.354	.403	.905
.752	.265	.036	.401	.279	.563	.070	.411	.515	.318

Rewriting these numbers in increasing order and obtaining the corresponding indices, we get the following permutation of the integers $1, \ldots, 50$:

17	15	34	43	5	9	47	16	4	3
22	42	13	45	21	24	50	20	38	28
44	39	48	10	26	32	27	49	37	7
46	30	6	36	2	12	8	1	33	11
41	25	14	23	29	40	35	18	19	31

Here the first integer in the permutation is 17 because the smallest number among $u_1, \ldots, u_{50}$ is u_{17}, the next integer in the permutation is 15 because the second smallest number among $u_1, \ldots, u_{50}$ is u_{15}, and the last integer in the permutation is 31 because the largest number among $u_1, \ldots, u_{50}$ is u_{31}.

Next we apply this permutation to the initial sequence 1, 2, 3, 4, 5, 1, 2, 3, 4, 5, ..., 1, 2, 3, 4, 5 of variety labels of length 50 getting the final sequence of variety labels:

2	5	4	3	5	4	2	1	4	3
2	2	3	5	1	4	5	5	3	3
4	4	3	5	1	2	2	4	2	2
1	5	1	1	2	2	3	1	3	1
1	5	4	3	4	5	5	3	4	1

(The first integer in the permutation is 17; thus, the first label in the final sequence of variety labels is 2 since this is the label in position 17 of the initial sequence of labels. Similarly, the second label in the final sequence is 5 since this is the label in position 15 of the initial sequence, and the last label in the final sequence is 1 since this is the label in position 31 of the initial sequence.) Correspondingly, variety 2 is assigned to plot -25, variety 5 is assigned to plot -24, and variety 1 is assigned to plot 24.

The value of the response on plot -25 is defined as $\beta_2 + g(-25) + W_1$, the response value for plot -24 is defined as $\beta_5 + g(-24) + W_2$, and the response for plot 24 is defined as $\beta_1 + g(24) + W_{50}$. We then round off these values to the nearest multiple of .1, since this was apparently done in the actual sugar beet experiment. In this manner, we get the simulated response data corresponding to the completely randomized design that was analyzed earlier in this section.

We can use the same pseudorandom numbers $u_1, \ldots, u_{50}$ (in consecutive blocks of size 5) to construct the randomized complete block design, shown in Table 11.37, that was used earlier in this section. Here block 1 consists of plots -25, -24, -23, -22, and -21, which are referred to as plots 1, 2, 3, 4, and 5 in block 1 and which are assigned varieties 5, 4, 3, 2, and 1, respectively (since the numbers u_1, u_2, u_3, u_4,

TABLE 11.37

	Plot				
Block	1	2	3	4	5
1	5	4	3	2	1
2	4	5	2	1	3
3	5	3	2	1	4
4	2	1	5	3	4
5	2	1	4	5	3
6	3	1	2	5	4
7	4	2	3	5	1
8	3	4	2	1	5
9	3	2	5	4	1
10	2	5	3	4	1

and u_5 happen to be in decreasing order). Similarly, block 2 consists of plots -20, $-19, -18, -17$, and -16, which are assigned varieties 4, 5, 2, 1, and 3, respectively, based on the ordering of $u_6, \ldots, u_{10}$ (u_9, the smallest of these numbers, appears in position 4 on the list; u_{10}, the next smallest number, appears in position 5; and so forth).

The value of the response on plot -25 is defined as $\beta_5 + g(-25) + W_1$, the response value for plot -24 is defined as $\beta_4 + g(-24) + W_2$, and so forth. Again we round off these values to the nearest multiple of .1. In this manner we get the simulated response data corresponding to the randomized complete block design that was analyzed above.

Problems

11.41 An experiment to compare three treatments is carried out in the form of a randomized complete block design involving four blocks each of size three. In each block, the three treatments are randomly assigned to the three plots. The observed results are shown in Table 11.38. **(a)** Give formulas for the block and treatment sums of squares. **(b)** Test the hypothesis of no treatment differences and show the results in the form of an ANOVA table.

TABLE 11.38

	Treatment		
Block	1	2	3
1	225	280	215
2	119	176	125
3	217	278	255
4	119	186	205

11.42 In the context of the idealized sugar beet experiment, explain why the regression function can be written in the form of Equation (21), where g is a quintic polynomial such that $\bar{g} = 0$.

11.43 Consider the setup for sampling without replacement as described in Section 6.2. Specifically, consider a population of N objects that are labeled from 1 to N in some order, let the object labeled l have value $v(l) = v_l$, let L denote the label of an object selected at random from the population, and let $V = v(L)$ denote its value. Then V has mean $\mu = N^{-1} \sum_l v_l$ and variance $\sigma_1^2 = N^{-1} \sum_l (v_l - \mu)^2$. For $1 \le i \le N$, let L_i denote the label of the object selected on the ith trial and let $V_i = v(L_i)$ denote the value of this object. Then V_i has mean μ and variance σ_1^2 for $1 \le i \le N$. Also, according to Equation (16) in Section 6.2, $\mathrm{cov}(V_i, V_j) = -\sigma_1^2/(N-1)$ for $1 \le i, j \le N$ with $i \ne j$. Let n_1 and n_2 be positive integers with $n_1 + n_2 \le N$ and set

$$\bar{V}_1 = \frac{V_1 + \cdots + V_{n_1}}{n_1} \quad \text{and} \quad \bar{V}_2 = \frac{V_{n_1+1} + \cdots + V_{n_1+n_2}}{n_2}.$$

We can think of $\bar{V}_1$ as the sample mean of the values in a random sample of size n_1 without replacement from the population and of $\bar{V}_2$ as the sample mean of the values in a random sample of size n_2 without replacement from the $N - n_1$ objects that remain in the population after the first random sample is removed. Now $\bar{V}_1$ and $\bar{V}_2$ each have mean μ. Moreover,

$$\text{var}(\bar{V}_1) = \frac{\sigma_1^2}{n_1} \frac{N - n_1}{N - 1} \quad \text{and} \quad \text{var}(\bar{V}_2) = \frac{\sigma_1^2}{n_2} \frac{N - n_2}{N - 1}$$

by (21) in Section 6.2. Determine **(a)** $\text{cov}(\bar{V}_1, \bar{V}_2)$; **(b)** $\text{var}(\bar{V}_2 - \bar{V}_1)$.

11.44 Consider the setup of Problem 11.43, but suppose now that the values of the objects in the population are measured with error. Specifically, suppose the object selected on the ith trial is measured as $Y_i = V_i + W_i = v(L_i) + W_i$, where $W_1, \ldots, W_N$ are independent random variables, each having mean 0 and variance σ^2. Also suppose the measurement errors are independent of the order of selection; that is, $(L_1, \ldots, L_N), W_1, \ldots, W_N$ are independent. Set

$$\bar{Y}_1 = \frac{Y_1 + \cdots + Y_{n_1}}{n_1} = \bar{V}_1 + \bar{W}_1$$

and

$$\bar{Y}_2 = \frac{Y_{n_1+1} + \cdots + Y_{n_1+n_2}}{n_2} = \bar{V}_2 + \bar{W}_2,$$

where

$$\bar{W}_1 = \frac{W_1 + \cdots + W_{n_1}}{n_1} \quad \text{and} \quad \bar{W}_2 = \frac{W_{n_1+1} + \cdots + W_{n_1+n_2}}{n_2}.$$

Determine **(a)** $\text{var}(\bar{Y}_1)$ and $\text{var}(\bar{Y}_2)$; **(b)** $\text{cov}(\bar{Y}_1, \bar{Y}_2)$; **(c)** $\text{var}(\bar{Y}_2 - \bar{Y}_1)$.

11.45 In order to verify Equation (22), write the response variables as $Y_{ij} = \beta_i + g(X_{ij}) + W_{ij}$, where the random variables W_{ij}, $1 \le i \le 5$ and $1 \le j \le 10$, are independent of the random variables X_{ij}, $1 \le i \le 5$ and $1 \le j \le 10$, and each W_{ij} has mean 0 and variance σ^2. Show that **(a)** $\bar{Y}_{i\cdot} = \beta_i + \frac{1}{10} \sum_{j=1}^{10} g(X_{ij}) + \bar{W}_{i\cdot}$ for $1 \le i \le 5$; **(b)** $\hat{\tau} = \beta_l - \beta_i + \frac{1}{10} \sum_{j=1}^{10} [g(X_{lj}) - g(X_{ij})] + \bar{W}_{l\cdot} - \bar{W}_{i\cdot}$ for $1 \le i, l \le 5$; **(c)** (22) holds for $1 \le i, l \le 5$ with $l \ne i$.

11.46 Verify Equation (23).

11.47 Show that **(a)** $\sum_{x_2 \in B_j} [g(x_2) - \bar{g}]^2 = \sum_{x_2 \in B_j} [g(x_2) - \bar{g}_j]^2 + 5(\bar{g}_j - \bar{g})^2$ for $1 \le j \le 10$; **(b)** $\sum_{j=1}^{10} \sum_{x_2 \in B_j} [g(x_2) - \bar{g}]^2 = \sum_{i=1}^{10} \sum_{x_2 \in B_j} [g(x_2) - \bar{g}_j]^2 + 5 \sum_{j=1}^{10} (\bar{g}_j - \bar{g})^2$; **(c)** Equation (24) is valid.

11.48 Consider the simulated sugar beet experiment having the form of a completely randomized design. **(a)** Determine the 95% confidence interval for the change in the mean percentage of sugar content caused by switching from the standard variety to variety 2. **(b)** Does the interval in part (a) contain the "true" value of $\beta_2 - \beta_1$?

11.49   Rework Problem 11.48 for the simulated sugar beet experiment having the form of a randomized complete block design.

11.50   The data in Table 11.39, taken from page 68 of *The Design of Experiments* by R. A. Fisher (New York: Hafner, 1971), show a comparison of the yields of five varieties of barley in an experiment in the form of a randomized complete block design carried out in Minnesota during 1930–1931 and reported in the article "Statistical Determination of Barley Varietal Adaption" by F. R. Immer, H. K. Hayes, and Le Roy Powers (*Journal of the American Society of Agronomy* **26** (1934): 403–419).

T A B L E   11.39

	Variety					
Block	1	2	3	4	5	Average
1	81.0	105.4	119.7	109.7	98.3	102.82
2	80.7	82.3	80.4	87.2	84.2	82.96
3	146.6	142.0	150.7	191.5	145.7	155.30
4	100.4	115.5	112.2	147.7	108.1	116.78
5	82.3	77.3	78.4	131.3	89.6	91.78
6	103.1	105.1	116.5	139.9	129.6	118.84
7	119.8	121.4	124.0	140.8	124.8	126.16
8	98.9	61.9	96.2	125.5	75.7	91.64
9	98.9	89.0	69.1	89.3	104.1	90.08
10	66.4	49.9	96.7	61.9	80.3	71.04
11	86.9	77.1	78.9	101.8	96.0	88.14
12	67.7	66.7	67.4	91.8	94.1	77.54
Average	94.392	91.133	99.183	118.2	102.542	101.09

The actual experiment involved ten varieties, of which Fisher used five in his example. The blocks are 12 separate experiments carried out at six locations in the state during the 2 years (blocks 1 and 2 correspond to 1930 and 1931, respectively, at location 1, and so forth). Determine **(a)** the corresponding ANOVA table; **(b)** the estimate S of σ; **(c)** the least-squares estimate of the change in the mean yield caused by switching from variety 2 to variety 5 in a given block; **(d)** the standard error of the estimate in part (c); **(e)** the 95% confidence interval for the true change; **(f)** the P-value for the t test of the hypothesis that the true change equals zero.

11.51   Rework Problem 11.50 but ignore the block effect and compare the two sets of results.

Binomial and Poisson Models

In this chapter, we obtain estimates, confidence intervals, and P-values for tests in the context of binomial and Poisson d-sample models. Whereas we used the t distribution to obtain exact confidence intervals and P-values in the context of the normal d-sample model and the more general normal linear model, here we mainly use normal approximation. It turns out that there are several apparently reasonable ways of using normal approximation to obtain confidence intervals and P-values in the present context, and the P-values so obtained can differ considerably from each other. In Section 12.2, we discuss certain "exact" P-values and use them to gain insight regarding the merits of the various approaches to normal approximation. The general issue of evaluating different methods of normal approximation and of coming up with better methods has not been (and perhaps never will be) thoroughly resolved, and what is known is too specialized and complicated for an introductory text.

In Section 12.3, the binomial and Poisson distributions are written in a common form, referred to as a one-parameter exponential family. The unknown parameter θ of such a family is referred to as its canonical parameter. In the context of the binomial distribution with known positive integer n and unknown $\pi \in (0, 1)$, $\theta = \operatorname{logit} \pi = \log(\pi/(1 - \pi))$; in the context of the Poisson distribution with mean $n\lambda$, where n is a known positive number and $\lambda \in (0, \infty)$ is unknown, $\theta = \log \lambda$. At least in the few numerical illustrations that are examined in the text and problems, P-values based on normal approximation to the distribution of the estimates of canonical parameters and their differences are reasonably close to the corresponding exact P-values.

The methods treated in this chapter yield tests of specified values of π in the context of the binomial one-sample model and of λ in the context of the Poisson one-sample model. They also yield tests of the equality of π_1 and π_2 in the context of the binomial two-sample model and of λ_1 and λ_2 in the context of the Poisson two-sample model. In Section 12.2, we introduce the likelihood-ratio test, which can be used to test the equality of $\pi_1, \ldots, \pi_d$ in the context of the binomial d-sample model and of $\lambda_1, \ldots, \lambda_d$ in the context of the Poisson d-sample model when $d \geq 2$. The likelihood-ratio test, which is analogous to the F test in the context of normal models, is motivated by the Neyman–Pearson lemma as presented in Section 12.2.

Typically, results and numerical illustrations for binomial models are presented in detail in the text, while the corresponding results and illustrations for Poisson models are left as problems.

12.1
Nominal Confidence Intervals and Tests

Consider an independent sequence of successes and failures having common probability π of success on each trial, where $0 < \pi < 1$. Given a positive integer n, let Y denote the number of successes by "time" n—that is, up to and including the nth trial. Then Y has the binomial distribution with parameters n and π, which has mean $n\pi$ and variance $n\pi(1 - \pi)$.

Consider next a Poisson process with rate λ on $[0, \infty)$, let n be a positive number (not necessarily an integer), and let Y denote the number of events that occur by time n. Then Y has the Poisson distribution with parameter $n\lambda$, which has mean $n\lambda$ and variance $n\lambda$.

In the binomial context, set $\mu = \pi$ and $\sigma^2 = \pi(1 - \pi)$; in the Poisson context, set $\mu = \lambda$ and $\sigma^2 = \lambda$. In both contexts, n is referred to as the *time parameter*, Y has mean $n\mu$ and variance $n\sigma^2$, and $\bar{Y} = Y/n$ has mean μ and variance σ^2/n.

More generally, let $Y_1, \ldots, Y_d$ be independent random variables and such that either one of the following holds: (i) For $1 \leq k \leq d$, Y_k has the binomial distribution with known time parameter $n_k \geq 1$ and unknown parameter $\pi_k = \mu_k \in (0, 1)$; (ii) for $1 \leq k \leq d$, Y_k has the Poisson distribution with mean $n_k\lambda_k$, where the time parameter $n_k > 0$ is known and $\lambda_k = \mu_k > 0$ is unknown. In the context of (i), we refer to this setup as the binomial d-sample model; in the context of (ii), we refer to it as the Poisson d-sample model.

Let $1 \leq k \leq d$. Then the random variable Y_k has mean $n_k\mu_k$ and variance $n_k\sigma_k^2$. In the context of the binomial d-sample model, $\sigma_k^2 = \pi_k(1 - \pi_k)$; in the context of the Poisson d-sample model, $\sigma_k^2 = \lambda_k$. Observe that the random variable $\bar{Y}_k = Y_k/n_k$ has mean μ_k and variance σ_k^2/n_k. Thus, the estimate $\hat{\mu}_k = \bar{Y}_k$ of μ_k is unbiased. In the context of the binomial d-sample model, $\hat{\pi}_k = \bar{Y}_k$ is an unbiased estimate of π_k; in the context of the Poisson d-sample model, $\hat{\lambda}_k = \bar{Y}_k$ is an unbiased estimate of λ_k.

The random variables $\hat{\mu}_1, \ldots, \hat{\mu}_d$ are independent, and the random vector $\hat{\mu} = [\hat{\mu}_1, \ldots, \hat{\mu}_d]^T$ has mean vector $\mu = [\mu_1, \ldots, \mu_d]^T$ and variance–covariance matrix $\mathrm{diag}(\sigma_1^2/n_1, \ldots, \sigma_d^2/n_d)$. According to the central limit theorem, for $1 \leq k \leq d$, if $n_k \gg 1$ (more generally, if $n_k\sigma_k^2 \gg 1$), then $\hat{\mu}_k - \mu_k$ has approximately the normal distribution with mean 0 and variance σ_k^2/n_k. Thus, we have the following result.

THEOREM 12.1 If $n_1, \ldots, n_d \gg 1$, then $\hat{\mu} - \mu$ has approximately the multivariate normal distribution with mean vector $\mathbf{0}$ and variance–covariance matrix $\mathrm{diag}(\sigma_1^2/n_1, \ldots, \sigma_d^2/n_d)$.

Given $1 \leq j, k \leq d$ with $j \neq k$, consider the parameter $\tau = \mu_k - \mu_j$. The estimate $\hat{\tau} = \hat{\mu}_k - \hat{\mu}_j$ of this parameter is unbiased, and its variance is given by

$$\mathrm{var}(\hat{\tau}) = \frac{\sigma_j^2}{n_j} + \frac{\sigma_k^2}{n_k}.$$

More generally, given $c_1, \ldots, c_d \in \mathbb{R}$, consider the linear parameter $\tau = c_1\mu_1 + \cdots + c_d\mu_d$. The estimate $\hat{\tau} = c_1\hat{\mu}_1 + \cdots + c_d\hat{\mu}_d$ of τ is unbiased, and its variance is given by

$$\mathrm{var}(\hat{\tau}) = \sum_k \frac{c_k^2 \sigma_k^2}{n_k}.$$

It follows from Theorem 12.1 that if $n_1, \ldots, n_d \gg 1$ and $c_1, \ldots, c_d$ are not all zero, then $(\hat{\tau} - \tau)/\mathrm{SD}(\hat{\tau})$ has approximately the standard normal distribution.

In the context of binomial and Poisson models, we are also interested in estimating nonlinear functions of $\mu_1, \ldots, \mu_d$. Consider first the binomial one-sample model, in which Y has the binomial distribution with known n and unknown $\pi \in (0, 1)$. The estimate $\hat{\pi} = Y/n$ of π has mean π and variance $\pi(1 - \pi)/n$. Let h be a possibly nonlinear function on $(0, 1)$ that has a continuous, nonzero derivative on this interval and consider the parameter $\tau = h(\pi)$. The obvious estimate of this parameter is given by $\hat{\tau} = h(\hat{\pi})$. Observe that $\hat{\tau} - \tau = h(\hat{\pi}) - h(\pi)$. Applying the linear Taylor approximation to h about π, we get

$$\hat{\tau} - \tau \approx h'(\pi)(\hat{\pi} - \pi) \qquad \text{if } \hat{\pi} \approx \pi. \tag{1}$$

Suppose n is large. Then the distribution of $\hat{\pi} - \pi$ is approximately normal with mean 0 and variance $\pi(1 - \pi)/n$. Thus, by (1), the distribution of $\hat{\tau} - \tau$ should be approximately normal with mean 0 and variance $[h'(\pi)]^2\pi(1 - \pi)/n$. We refer to the variance of the indicated normal approximation to the distribution of $\hat{\tau}$ as the *asymptotic variance* (AV) of this estimate and to the nonnegative square root of the asymptotic variance as the *asymptotic standard deviation* (ASD) of the estimate. Thus,

$$\mathrm{AV}(\hat{\tau}) = [h'(\pi)]^2 \frac{\pi(1 - \pi)}{n}. \tag{2}$$

EXAMPLE 12.1 In the context of the binomial one-sample model, determine the asymptotic variance of the estimate $\hat{\tau} = \hat{\pi}/(1 - \hat{\pi})$ of the odds $\tau = \pi/(1 - \pi)$.

Solution Now $\tau = h(\pi)$ with $h(\pi) = \pi/(1 - \pi)$ for $0 < \pi < 1$. Observe that

$$h'(\pi) = \frac{1}{1 - \pi} + \frac{\pi}{(1 - \pi)^2} = \frac{1}{(1 - \pi)^2} = \frac{\pi}{(1 - \pi)^2} \frac{1}{1 - \pi} \frac{1}{\pi(1 - \pi)} = \tau \frac{1}{\pi(1 - \pi)}.$$

Thus by (2), the asymptotic variance of the estimated odds is given by

$$AV(\hat{\tau}) = \tau^2 \frac{1}{n\pi(1-\pi)}. \quad \blacksquare$$

Consider next the binomial two-sample model, in which Y_1 and Y_2 are independent random variables, Y_1 has the binomial distribution with known n_1 and unknown $\pi_1 \in (0, 1)$, and Y_2 has the binomial distribution with known n_2 and unknown $\pi_2 \in (0, 1)$. The estimate $\hat{\pi}_1 = Y_1/n_1$ of π_1 has mean π_1 and variance $\pi_1(1 - \pi_1)/n_1$, the estimate $\hat{\pi}_2 = Y_2/n_2$ of π_2 has mean π_2 and variance $\pi_2(1 - \pi_2)/n_2$, and these two estimates are independent. Consider the parameter $\tau = \pi_2 - \pi_1$. The estimate $\hat{\tau} = \hat{\pi}_2 - \hat{\pi}_1$ of this parameter is unbiased, and its variance is given by

$$\text{var}(\hat{\tau}) = \frac{\pi_1(1 - \pi_1)}{n_1} + \frac{\pi_2(1 - \pi_2)}{n_2}.$$

Thinking of π_1 and π_2 as *risks* of failure (rather than probabilities of success), we refer to the parameter $\tau = \pi_2/\pi_1$ as the relative risk. The obvious estimate of this parameter is given by $\hat{\tau} = \hat{\pi}_2/\hat{\pi}_1$. Similarly, the obvious estimate of the *odds ratio*

$$\tau = \frac{\pi_2/(1 - \pi_2)}{\pi_1/(1 - \pi_1)}$$

is given by

$$\hat{\tau} = \frac{\hat{\pi}_2/(1 - \hat{\pi}_2)}{\hat{\pi}_1/(1 - \hat{\pi}_1)}.$$

Observe that

$$\frac{\pi_2/(1 - \pi_2)}{\pi_1/(1 - \pi_1)} = \frac{\pi_2}{\pi_1} \frac{(1 - \pi_1)}{(1 - \pi_2)}.$$

Thus, if the risks π_1 and π_2 are small, then the odds ratio is close to the relative risk.

Consider, more generally, a parameter $\tau = h(\pi_1, \pi_2)$ and its estimate $\hat{\tau} = h(\hat{\pi}_1, \hat{\pi}_2)$, where h is a possibly nonlinear function on $(0, 1) \times (0, 1)$ having a nonzero, continuous derivative on this set—that is, such that the two partial derivatives

$$\frac{\partial h}{\partial \pi_1}(\pi_1, \pi_2) \qquad \text{and} \qquad \frac{\partial h}{\partial \pi_2}(\pi_1, \pi_2)$$

are continuous functions of π_1 and π_2 and do not both equal zero at any point (π_1, π_2). Observe that $\hat{\tau} - \tau = h(\hat{\pi}_1, \hat{\pi}_2) - h(\pi_1, \pi_2)$. Applying the linear Taylor approximation to the nonlinear function h about the point (π_1, π_2), we get

$$\hat{\tau} - \tau \approx \frac{\partial h}{\partial \pi_1}(\pi_1, \pi_2)(\hat{\pi}_1 - \pi_1)$$

$$+ \frac{\partial h}{\partial \pi_2}(\pi_1, \pi_2)(\hat{\pi}_2 - \pi_2) \qquad \text{if } \hat{\pi}_1 \approx \pi_1 \text{ and } \hat{\pi}_2 \approx \pi_2. \qquad (3)$$

Suppose n_1 and n_2 are large. Then the distribution of $\hat{\pi}_1 - \pi_1$ is approximately normal with mean 0 and variance $\pi_1(1 - \pi_1)/n_1$, the distribution of $\hat{\pi}_2 - \pi_2$ is approximately normal with mean 0 and variance $\pi_2(1 - \pi_2)/n_2$, and these two random variables are independent. Thus, by (3), the distribution of $\hat{\tau} - \tau$ should be approximately normal with mean 0 and the variance given by

$$\mathrm{AV}(\hat{\tau}) = \left(\frac{\partial h}{\partial \pi_1}(\pi_1, \pi_2)\right)^2 \frac{\pi_1(1 - \pi_1)}{n_1} + \left(\frac{\partial h}{\partial \pi_2}(\pi_1, \pi_2)\right)^2 \frac{\pi_2(1 - \pi_2)}{n_2}. \qquad (4)$$

EXAMPLE **12.2** In the context of the binomial two-sample model, determine the asymptotic variance of

(a) the estimate $\hat{\tau} = \hat{\pi}_2/\hat{\pi}_1$ of the relative risk $\tau = \pi_2/\pi_1$;

(b) the estimate

$$\hat{\tau} = \frac{\hat{\pi}_2/(1 - \hat{\pi}_2)}{\hat{\pi}_1/(1 - \hat{\pi}_1)}$$

of the odds ratio

$$\tau = \frac{\pi_2/(1 - \pi_2)}{\pi_1/(1 - \pi_1)}.$$

Solution (a) Here $h(\pi_1, \pi_2) = \pi_2/\pi_1$, so

$$\frac{\partial h}{\partial \pi_1}(\pi_1, \pi_2) = -\frac{\pi_2}{\pi_1^2} = -\frac{\pi_2}{\pi_1}\frac{1}{\pi_1} = -\tau\frac{1}{\pi_1}$$

and

$$\frac{\partial h}{\partial \pi_2}(\pi_1, \pi_2) = \frac{1}{\pi_1} = \frac{\pi_2}{\pi_1}\frac{1}{\pi_2} = \tau\frac{1}{\pi_2}.$$

Thus, by (4), the asymptotic variance of the estimated relative risk is given by

$$\mathrm{AV}(\hat{\tau}) = \tau^2\left(\frac{1 - \pi_1}{n_1\pi_1} + \frac{1 - \pi_2}{n_2\pi_2}\right).$$

(b) Here

$$h(\pi_1, \pi_2) = \frac{\pi_2/(1 - \pi_2)}{\pi_1/(1 - \pi_1)}.$$

Now

$$\frac{d}{d\pi}\frac{\pi}{1 - \pi} = \frac{1}{(1 - \pi)^2}$$

as we saw in the solution to Example 12.1. Thus, by the chain rule,

$$\frac{\partial h}{\partial \pi_1}(\pi_1, \pi_2) = -\frac{\pi_2/(1 - \pi_2)}{[\pi_1/(1 - \pi_1)]^2}\frac{1}{(1 - \pi_1)^2}$$

$$= -\frac{\pi_2/(1 - \pi_2)}{\pi_1/(1 - \pi_1)}\frac{1}{\pi_1(1 - \pi_1)}$$

$$= -\frac{\tau}{\pi_1(1 - \pi_1)},$$

and

$$\frac{\partial h}{\partial \pi_2}(\pi_1, \pi_2) = \frac{1}{\pi_1/(1 - \pi_1)} \frac{1}{(1 - \pi_2)^2}$$

$$= \frac{\pi_2/(1 - \pi_2)}{\pi_1/(1 - \pi_1)} \frac{1}{\pi_2(1 - \pi_2)}$$

$$= \frac{\tau}{\pi_2(1 - \pi_2)}.$$

Consequently, by (4), the asymptotic variance of the estimated odds ratio is given by

$$\mathrm{AV}(\hat{\tau}) = \tau^2 \Big(\frac{1}{n_1 \pi_1(1 - \pi_1)} + \frac{1}{n_2 \pi_2(1 - \pi_2)} \Big). \quad \blacksquare$$

In the still more general context of a binomial or Poisson d-sample model, consider a parameter $\tau = h(\mu) = h(\mu_1, \ldots, \mu_d)$, where h is a continuously differentiable, possibly nonlinear function of μ that is defined on an open set in $\mathbb{R}^d$ containing the possible values of μ. A reasonable estimate of τ is given by $\hat{\tau} = h(\hat{\mu}) = h(\hat{\mu}_1, \ldots, \hat{\mu}_d)$. [This estimate is undefined if $\hat{\mu}$ is not in the domain of h, and it is biased if $h(\mu_1, \ldots, \mu_d)$ is not a linear function of $\mu_1, \ldots, \mu_d$.] Observe that $\hat{\tau} - \tau = h(\hat{\mu}) - h(\mu)$. Applying the linear Taylor approximation to h about μ, we get

$$\hat{\tau} - \tau \approx \sum_k \frac{\partial h}{\partial \mu_k}(\mu)(\hat{\mu}_k - \mu_k) = [\nabla h(\mu)]^T(\hat{\mu} - \mu) \qquad \text{if } \hat{\mu} \approx \mu, \qquad (5)$$

where

$$\nabla h(\mu) = \Big[\frac{\partial h}{\partial \mu_1}(\mu), \ldots, \frac{\partial h}{\partial \mu_d}(\mu) \Big]^T$$

is the gradient of h at μ. Suppose $\nabla h(\mu) \neq 0$ and that $n_1, \ldots, n_k \gg 1$. Then by (5) and Theorem 12.1, $\hat{\tau} - \tau$ has approximately the normal distribution with mean 0 and the variance given by

$$\mathrm{AV}(\hat{\tau}) = \sum_k \Big(\frac{\partial h}{\partial \mu_k}(\mu) \Big)^2 \frac{\sigma_k^2}{n_k}. \qquad (6)$$

As above, we refer to $\mathrm{AV}(\hat{\tau})$ and its nonnegative square root $\mathrm{ASD}(\hat{\tau})$ as the asymptotic variance and asymptotic standard deviation, respectively, of $\hat{\tau}$. (If h is a linear function, then the asymptotic variance and asymptotic standard deviation of $\hat{\tau}$ coincide with its ordinary variance and standard deviation, respectively.) If $n_1, \ldots, n_k \gg 1$, then $\mathrm{AV}(\hat{\tau}) \approx 0$ and $\mathrm{ASD}(\hat{\tau}) \approx 0$. We have now derived the following result.

THEOREM 12.2   If $\nabla h(\mu) \neq 0$ and $n_1, \ldots, n_d \gg 1$, then $P(\hat{\tau} \approx \tau) \approx 1$, $\mathrm{ASD}(\hat{\tau}) \approx 0$, and $(\hat{\tau} - \tau)/\mathrm{ASD}(\hat{\tau})$ has approximately the standard normal distribution.

EXAMPLE 12.3 Given $1 \leq j, k \leq d$ with $j \neq k$, determine the asymptotic variance and standard error of the estimate $\hat{\tau} = \hat{\mu}_k / \hat{\mu}_j$ of the parameter $\tau = \mu_k / \mu_j$.

Solution Here $h(\mu_j, \mu_k) = \mu_k / \mu_j$, so

$$\frac{\partial h}{\partial \mu_j}(\mu_j, \mu_k) = -\frac{\mu_k}{\mu_j^2} = -\frac{\tau}{\mu_j} \quad \text{and} \quad \frac{\partial h}{\partial \mu_k}(\mu_j, \mu_k) = \frac{1}{\mu_j} = \frac{\tau}{\mu_k}.$$

Thus, by (6),

$$\mathrm{AV}(\hat{\tau}) = \tau^2 \left(\frac{\sigma_j^2}{n_j \mu_j^2} + \frac{\sigma_k^2}{n_k \mu_k^2} \right). \quad \blacksquare$$

Let $1 \leq k \leq d$. In the context of the binomial d-sample model, we estimate $\sigma_k^2 = \pi_k(1 - \pi_k)$ by $\hat{\sigma}_k^2 = \hat{\pi}_k(1 - \hat{\pi}_k)$ and $\sigma_k = \sqrt{\pi_k(1 - \pi_k)}$ by $\hat{\sigma}_k = \sqrt{\hat{\pi}(1 - \hat{\pi}_k)}$. In the context of the Poisson d-sample model, we estimate $\sigma_k^2 = \lambda_k$ by $\hat{\sigma}_k^2 = \hat{\lambda}_k$ and $\sigma_k = \sqrt{\lambda_k}$ by $\hat{\sigma}_k = \sqrt{\hat{\lambda}_k}$. Observe that $\hat{\mu}_k$ is a consistent estimate of μ_k, $\hat{\sigma}_k$ is a consistent estimate of σ_k, and

$$\frac{\partial h}{\partial \mu_k}(\hat{\mu})$$

is a consistent estimate of

$$\frac{\partial h}{\partial \mu_k}(\mu);$$

that is, if $n_1, \ldots, n_d \gg 1$, then for $1 \leq k \leq d$

$$P(\hat{\mu}_k \approx \mu_k) \approx 1, \quad P(\hat{\sigma}_k \approx \sigma_k) \approx 1 \quad \text{and} \quad P\left(\frac{\partial h}{\partial \mu_k}(\hat{\mu}) \approx \frac{\partial h}{\partial \mu_k}(\mu) \right) \approx 1.$$

We refer to the asymptotic standard deviation of $\hat{\tau}$ with the various quantities in the definition of $\mathrm{AV}(\hat{\tau})$ being replaced by their estimates, as defined above, as the standard error of $\hat{\tau}$. Thus,

$$\mathrm{SE}(\hat{\tau}) = \sqrt{ \sum_k \left(\frac{\partial h}{\partial \mu_k}(\hat{\mu}) \right)^2 \frac{\hat{\sigma}_k^2}{n_k} }.$$

Using Theorem 12.2, we get the following result.

THEOREM 12.3 If $\nabla h(\mu) \neq 0$ and $n_1, \ldots, n_d \gg 1$, then

$$P\left(\frac{\mathrm{SE}(\hat{\tau})}{\mathrm{ASD}(\hat{\tau})} \approx 1 \right) \approx 1,$$

and $(\hat{\tau} - \tau) / \mathrm{SE}(\hat{\tau})$ has approximately the standard normal distribution.

THEOREM 12.4 Suppose $\nabla h(\mu) \neq 0$ and $n_1, \ldots, n_d \gg 1$ and let $0 < \alpha < 1$. Then

$$P\left(\hat{\tau} - z_{1-\alpha} \, \mathrm{SE}(\hat{\tau}) < \tau\right) \approx 1 - \alpha,$$

$$P\left(\tau < \hat{\tau} + z_{1-\alpha} \, \mathrm{SE}(\hat{\tau})\right) \approx 1 - \alpha,$$

and

$$P\left(\hat{\tau} - z_{1-\alpha/2} \, \mathrm{SE}(\hat{\tau}) < \tau < \hat{\tau} + z_{1-\alpha/2} \, \mathrm{SE}(\hat{\tau})\right) \approx 1 - \alpha.$$

Proof   To verify the first conclusion, observe that, by Theorem 12.3,

$$P\left(\hat{\tau} - z_{1-\alpha} \, \mathrm{SE}(\hat{\tau}) < \tau\right) = P\left(\frac{\hat{\tau} - \tau}{\mathrm{SE}(\hat{\tau})} < z_{1-\alpha}\right) \approx \Phi(z_{1-\alpha}) = 1 - \alpha.$$

To verify the second conclusion, observe that

$$P\left(\tau < \hat{\tau} + z_{1-\alpha} \, \mathrm{SE}(\hat{\tau})\right) = P\left(\frac{\hat{\tau} - \tau}{\mathrm{SE}(\hat{\tau})} > -z_{1-\alpha}\right)$$

$$\approx 1 - \Phi(-z_{1-\alpha}) = \Phi(z_{1-\alpha}) = 1 - \alpha.$$

To verify the third conclusion, observe that

$$P\left(\hat{\tau} - z_{1-\alpha/2} \, \mathrm{SE}(\hat{\tau}) < \tau < \hat{\tau} + z_{1-\alpha/2} \, \mathrm{SE}(\hat{\tau}))\right)$$

$$= P\left(-z_{1-\alpha/2} < \frac{\hat{\tau} - \tau}{\mathrm{SE}(\hat{\tau})} < z_{1-\alpha/2}\right)$$

$$\approx \Phi(z_{1-\alpha/2}) - \Phi(-z_{1-\alpha/2})$$

$$\approx 1 - \alpha. \quad \blacksquare$$

In light of Theorem 12.4, we refer to $\hat{\tau} - z_{1-\alpha} \, \mathrm{SE}(\hat{\tau})$ as the *nominal* $100(1 - \alpha)\%$ lower confidence bound for the parameter τ, to $\hat{\tau} + z_{1-\alpha} \, \mathrm{SE}(\hat{\tau})$ as the nominal $100(1 - \alpha)\%$ upper confidence bound for this parameter, and to

$$\hat{\tau} \pm z_{1-\alpha/2} \, \mathrm{SE}(\hat{\tau}) = \left(\hat{\tau} - z_{1-\alpha/2} \, \mathrm{SE}(\hat{\tau}), \, \hat{\tau} + z_{1-\alpha/2} \, \mathrm{SE}(\hat{\tau})\right)$$

as its nominal $100(1 - \alpha)\%$ confidence interval.

Given $\tau_0 \in \mathbb{R}$, set $W = (\hat{\tau} - \tau_0)/\mathrm{SE}(\hat{\tau})$. We refer to W as the *Wald statistic* corresponding to the possible value τ_0 of τ.

THEOREM 12.5   If $\nabla h(\mu) \neq \mathbf{0}$ and $n_1, \dots, n_d \gg 1$, then

$$P(W \leq w) \approx \Phi\left(w - \frac{\tau - \tau_0}{\mathrm{ASD}(\hat{\tau})}\right), \qquad w \in \mathbb{R}.$$

Proof   Let $w \in \mathbb{R}$. By Theorems 12.2 and 12.3,

$$P(W \leq w) = P\left(\frac{\hat{\tau} - \tau_0}{\mathrm{SE}(\hat{\tau})} \leq w\right)$$

$$= P\left(\frac{\hat{\tau} - \tau_0}{\mathrm{ASD}(\hat{\tau})} \leq \frac{\mathrm{SE}(\hat{\tau})}{\mathrm{ASD}(\hat{\tau})} w\right)$$

$$= P\left(\frac{\hat{\tau} - \tau}{\mathrm{ASD}(\hat{\tau})} \leq \frac{\mathrm{SE}(\hat{\tau})}{\mathrm{ASD}(\hat{\tau})} w - \frac{\tau - \tau_0}{\mathrm{ASD}(\hat{\tau})}\right)$$

$$\approx P\left(\frac{\hat{\tau} - \tau}{\text{ASD}(\hat{\tau})} \le w - \frac{\tau - \tau_0}{\text{ASD}(\hat{\tau})}\right)$$

$$\approx \Phi\left(w - \frac{\tau - \tau_0}{\text{ASD}(\hat{\tau})}\right). \quad \blacksquare$$

Suppose the conditions of Theorem 12.5 are satisfied and let $0 < \alpha < 1$. The inequality $W \ge z_{1-\alpha}$ determines the critical region having nominal size α for the *Wald test* of the hypothesis that $\tau \le \tau_0$; that is, if $\tau = \tau_0$, then $P(W \ge z_{1-\alpha}) \approx \alpha$ and if $\tau < \tau_0$, then

$$P(W \ge z_{1-\alpha}) \approx 1 - \Phi\left(z_{1-\alpha} - \frac{\tau - \tau_0}{\text{ASD}(\hat{\tau})}\right) = \Phi\left(z_\alpha + \frac{\tau - \tau_0}{\text{ASD}(\hat{\tau})}\right) < \Phi(z_\alpha) = \alpha.$$

Observe that the Wald statistic lies in the acceptance region $W < z_{1-\alpha}$ if and only if $\hat{\tau} - z_{1-\alpha} \text{SE}(\hat{\tau}) < \tau_0$ or, equivalently, if and only if τ_0 is greater than the nominal $100(1 - \alpha)\%$ lower confidence bound $\hat{\tau} - z_{1-\alpha} \text{SE}(\hat{\tau})$ for τ. The (nominal) *P*-value for the test is defined as $1 - \Phi(W)$. The Wald statistic lies in the critical region $W \ge z_{1-\alpha}$ if α is greater than or equal to the *P*-value and in the acceptance region $W < z_{1-\alpha}$ if α is less than the *P*-value.

The inequality $W \le -z_{1-\alpha}$ determines the critical region having nominal size α for the Wald test of the hypothesis that $\tau \ge \tau_0$; that is, if $\tau = \tau_0$, then $P(W \le -z_{1-\alpha}) \approx \alpha$ and if $\tau > \tau_0$, then

$$P(W \le -z_{1-\alpha}) \approx \Phi\left(-z_{1-\alpha} - \frac{\tau - \tau_0}{\text{ASD}(\hat{\tau})}\right) < \Phi(-z_{1-\alpha}) = \alpha.$$

The Wald statistic lies in the acceptance region $W > -z_{1-\alpha}$ if and only if $\hat{\tau} + z_{1-\alpha}\text{SE}(\hat{\tau}) > \tau_0$ or, equivalently, if and only if τ_0 is less than the nominal $100(1 - \alpha)\%$ upper confidence bound $\hat{\tau} + z_{1-\alpha}\text{SE}(\hat{\tau})$ for τ. The *P*-value for the test is defined as $\Phi(W)$. The Wald statistic lies in the critical region $W \le -z_{1-\alpha}$ if $\alpha \ge$ *P*-value and in the acceptance region $W > -z_{1-\alpha}$ if $\alpha <$ *P*-value.

The inequality $|W| \ge z_{1-\alpha/2}$ determines the critical region having nominal size α for the Wald test of the hypothesis that $\tau = \tau_0$; that is, if $\tau = \tau_0$, then $P(|W| \ge z_{1-\alpha/2}) \approx \alpha$. The Wald statistic lies in the acceptance region $|W| < z_{1-\alpha/2}$ if and only if $\hat{\tau} - z_{1-\alpha/2} \text{SE}(\hat{\tau}) < \tau_0 < \hat{\tau} + z_{1-\alpha/2} \text{SE}(\hat{\tau})$ or, equivalently, if and only if τ_0 lies in the nominal $100(1 - \alpha)\%$ confidence interval for τ. The *P*-value for the test is defined as $2[\, 1 - \Phi(|W|)\,]$. The Wald statistic lies in the critical region $|W| \ge z_{1-\alpha/2}$ if $\alpha \ge$ *P*-value and in the acceptance region $|W| < z_{1-\alpha/2}$ if $\alpha <$ *P*-value.

EXAMPLE **12.4** In the context of the binomial one-sample model, determine the standard error of $\hat{\tau}$, where

(a) $\hat{\tau} = \hat{\pi}$;   **(b)** $\hat{\tau} = \hat{\pi}/(1 - \hat{\pi})$.

Solution **(a)** Here $\text{AV}(\hat{\tau}) = \text{var}(\hat{\tau}) = \text{var}(\hat{\pi}) = \pi(1 - \pi)/n$. Thus,

$$\text{SE}(\hat{\pi}) = \sqrt{\frac{\hat{\pi}(1 - \hat{\pi})}{n}}.$$

(b) According to the solution to Example 12.1,

$$AV(\hat{\tau}) = \tau^2 \frac{1}{n\pi(1-\pi)}.$$

Thus,

$$SE(\hat{\tau}) = \hat{\tau}\sqrt{\frac{1}{n\hat{\pi}(1-\hat{\pi})}}. \quad \blacksquare$$

EXAMPLE **12.5**   In the context of the binomial two-sample model, determine the standard error of $\hat{\tau}$, where

(a) $\hat{\tau} = \hat{\pi}_2 - \hat{\pi}_1;$ **(b)** $\hat{\tau} = \hat{\pi}_2/\hat{\pi}_1;$ **(c)** $\hat{\tau} = \dfrac{\hat{\pi}_2/(1-\hat{\pi}_2)}{\hat{\pi}_1/(1-\hat{\pi}_1)}.$

Solution   **(a)** Here

$$AV(\hat{\tau}) = var(\hat{\tau}) = var(\hat{\pi}_2 - \hat{\pi}_1) = \frac{\pi_1(1-\pi_1)}{n_1} + \frac{\pi_2(1-\pi_2)}{n_2}.$$

Thus,

$$SE(\hat{\tau}) = \sqrt{\frac{\hat{\pi}_1(1-\hat{\pi}_1)}{n_1} + \frac{\hat{\pi}_2(1-\hat{\pi}_2)}{n_2}}.$$

(b) According to the solution to Example 12.2(a),

$$AV(\hat{\tau}) = \tau^2 \left(\frac{1-\pi_1}{n_1\pi_1} + \frac{1-\pi_2}{n_2\pi_2} \right).$$

Thus,

$$SE(\hat{\tau}) = \hat{\tau}\sqrt{\frac{1-\hat{\pi}_1}{n_1\hat{\pi}_1} + \frac{1-\hat{\pi}_2}{n_2\hat{\pi}_2}}.$$

(c) According to the solution to Example 12.2(b),

$$AV(\hat{\tau}) = \tau^2 \left(\frac{1}{n_1\pi_1(1-\pi_1)} + \frac{1}{n_2\pi_2(1-\pi_2)} \right).$$

Thus,

$$SE(\hat{\tau}) = \hat{\tau}\sqrt{\frac{1}{n_1\hat{\pi}_1(1-\hat{\pi}_1)} + \frac{1}{n_2\hat{\pi}_2(1-\hat{\pi}_2)}}. \quad \blacksquare$$

EXAMPLE **12.6**   In the context of a certain agricultural experiment involving 100 pea plants, genetic theory predicts that the probability π of getting a plant that produces green (rather than yellow) seeds is $1/4$. It is observed that 35 of the plants produce green seeds. Assuming that the binomial model is applicable, determine

(a) the estimate of and nominal 95% confidence interval for π and the *P*-value for the Wald test of the hypothesis that $\pi = 1/4$;

(b) the estimate of and nominal 95% confidence interval for the odds $\tau = \pi/(1-\pi)$ of getting a plant that produces green peas and the *P*-value for the Wald test of the hypothesis that $\tau = 1/3$.

Solution **(a)** The probability π of getting a plant that produces green seeds is estimated by $\hat{\pi} = 35/100 = .35$. According to the solution to Example 12.4(a), the standard error of this estimate is given by

$$\text{SE}(\hat{\pi}) = \sqrt{\frac{(.35)(.65)}{100}} \doteq .0477.$$

Thus, the nominal 95% confidence interval for π is given by

$$.35 \pm (1.96)(.0477) \doteq .35 \pm .093 = (.257, .443).$$

The Wald statistic for testing the hypothesis that $\pi = 1/4$ is given by $W \doteq (.35 - .25)/.0477 \doteq 2.097$. The corresponding *P*-value is given by $2[1 - \Phi(2.097)] \doteq .0360$. Note that this *P*-value is less than .05, which is equivalent to the result that 1/4 lies in the exterior of the 95% confidence interval $(.257, .443)$ for π.

(b) The quantity $\tau = \pi/(1-\pi)$ is estimated by $\hat{\tau} = .35/.65 \doteq 0.538$. According to the solution to Example 12.4(b), the standard error of this estimate is given by

$$\text{SE}(\hat{\tau}) \doteq 0.538\sqrt{\frac{1}{100(.35)(.65)}} \doteq 0.113.$$

Thus, the nominal 95% confidence interval for τ is given by

$$0.538 \pm (1.96)(0.113) \doteq 0.538 \pm 0.221 \doteq (0.317, 0.760).$$

The Wald statistic for testing the hypothesis that $\tau = 1/3$ is given by $W \doteq (0.538 - 0.333)/0.113 \doteq 1.817$. The corresponding *P*-value is given by $2[1 - \Phi(1.817)] \doteq .0692$. Note that this *P*-value is greater than .05, which is equivalent to the result that 1/3 lies in the 95% confidence interval $(0.317, 0.760)$ for τ. Note also that the hypothesis that $\pi = 1/4$ is equivalent to the hypothesis that $\tau = 1/3$, but the *P*-value for the Wald test of the latter hypothesis differs somewhat from the *P*-value given in the solution to part (a) for the Wald test of the former hypothesis; the explanation for this discrepancy involves the nonlinearity of τ as a function of π, which leads to a somewhat different normal approximation. ■

EXAMPLE **12.7**   A Swiss study on the incidence of colds was discussed by Linus Pauling in "The Significance of the Evidence About Ascorbic Acid and the Common Cold," *Proceedings of the National Academy of Sciences* **68** (1971). In this study, 139 skiers received a supplement of 1 gram of vitamin C per day in their diets and 140 received no supplement; 17 skiers in the supplement group and 31 skiers in the control group caught colds during the study. (The data were also used in Example 5.2.2 of *The Statistical Analysis of Discrete Data* by Thomas J. Santner and Diane E. Duffy, Springer-Verlag, New York, 1989.) Let π_1 or π_2 denote the probability that a randomly selected skier given no supplement or the vitamin C supplement, respectively, would have caught a cold during the study.

(a) Determine the estimate of and nominal 95% confidence interval for $\pi_2 - \pi_1$ and the P-value for the Wald test of the hypothesis that $\pi_1 = \pi_2$.

(b) Determine the estimate of and nominal 95% confidence interval for the relative risk $\tau = \pi_2/\pi_1$ and the P-value for the Wald test of the hypothesis that $\tau = 1$.

(c) Determine the estimate of and nominal 95% confidence interval for the odds ratio

$$\tau = \frac{\pi_2/(1 - \pi_2)}{\pi_1/(1 - \pi_1)}$$

and the P-value for the Wald test of the hypothesis that $\tau = 1$.

(d) Discuss the results in the solutions to parts (a)–(c).

Solution   **(a)** Here $\hat{\pi}_1 = 31/140 \doteq .221$ and $\hat{\pi}_2 = 17/139 \doteq .122$, so $\pi_2 - \pi_1$ is estimated by $\hat{\pi}_2 - \hat{\pi}_1 \doteq -.099$. According to the solution to Example 12.5(a), the standard error of this estimate is given by

$$\text{SE}(\hat{\pi}_2 - \hat{\pi}_1) \doteq \sqrt{\frac{(.221)(.779)}{140} + \frac{(.122)(.878)}{139}} \doteq .0448.$$

Thus, the nominal 95% confidence interval for $\pi_2 - \pi_1$ is given by

$$-.099 \pm (1.96)(.0448) \doteq -.099 \pm .087 \doteq (-.187, -.011).$$

The Wald statistic for testing the hypothesis that $\pi_1 = \pi_2$ is given by $W \doteq -.099/.0448 \doteq -2.215$. The corresponding P-value is given by $2\Phi(-2.215) \doteq .0268$.

(b) The relative risk $\tau = \pi_2/\pi_1$ is estimated by $\hat{\tau} \doteq .122/.221 \doteq 0.552$. According to the solution to Example 12.5(b), the standard error of this estimate is given by

$$\text{SE}(\hat{\tau}) = 0.552\sqrt{\frac{.779}{140(.221)} + \frac{.878}{139(.122)}} \doteq 0.153.$$

Thus, the nominal 95% confidence interval for τ is given by

$$0.552 \pm (1.96)(0.153) \doteq 0.552 \pm 0.300 = (0.252, 0.852).$$

The Wald statistic for testing the hypothesis that $\tau = 1$ is given by $W \doteq (0.552 - 1)/0.153 \doteq -2.926$. The corresponding P-value is given by $2\Phi(-2.926) \doteq .0034$.

(c) The odds ratio $\tau = \frac{\pi_2/(1-\pi_2)}{\pi_1/(1-\pi_1)}$ is estimated by $\hat{\tau} \doteq \frac{.122/.878}{.221/.779} \doteq 0.490$. According to the solution to Example 12.5(c), the standard error of this estimate is given by

$$\text{SE}(\hat{\tau}) \doteq 0.490\sqrt{\frac{1}{140(.221)(.779)} + \frac{1}{139(.122)(.878)}} \doteq 0.161.$$

Thus, the nominal 95% confidence interval for τ is given by

$$0.490 \pm (1.96)(0.161) \doteq 0.490 \pm 0.316 = (0.174, 0.806).$$

The Wald statistic for testing the hypothesis that $\tau = 1$ is given by $W \doteq (0.490 - 1)/0.161 \doteq -3.161$. The corresponding P-value is given by $2\Phi(-3.161) \doteq .0016$.

(d) The P-value in the solution to part (a) is less than .05, which is equivalent to the result that zero lies in the exterior of the nominal 95% confidence interval for $\pi_2 - \pi_1$. According to this analysis, the experimental data provide some but not overwhelming evidence that taking the vitamin C supplement affects the probability of catching a cold under the indicated conditions. The hypotheses in parts (a), (b), and (c) are equivalent, but the P-value in the solutions to parts (b) and (c) are suspiciously smaller than that found in the solution to part (a). (We will obtain other P-values in Example 12.11 in Section 12.2 and Example 12.18 in Section 12.3.) ∎

EXAMPLE **12.8** In the context of the vitamin C experiment in Example 12.7,

(a) determine the nominal 95% upper confidence bound for $\pi_2 - \pi_1$ and the P-value for the Wald test of the hypothesis that $\pi_2 \geq \pi_1$;

(b) determine the nominal 95% upper confidence bound for the relative risk $\tau = \pi_2/\pi_1$ and the P-value for the Wald test of the hypothesis that $\tau \geq 1$;

(c) determine the nominal 95% upper confidence bound for the odds ratio

$$\tau = \frac{\pi_2/(1 - \pi_2)}{\pi_1/(1 - \pi_1)}$$

and the P-value for the Wald test of the hypothesis that $\tau \geq 1$.

Solution **(a)** The nominal 95% upper confidence bound for $\pi_2 - \pi_1$ is given by

$$-.099 + (1.645)(.0448) \doteq -.099 + .074 = -.025.$$

The P-value for the Wald test of the hypothesis that $\pi_2 \geq \pi_1$ or, equivalently, that $\pi_2 - \pi_1 \geq 0$ is given by $\Phi(-2.215) \doteq .0134$. Thus, the experimental data provides fairly strong evidence that $\pi_2 < \pi_1$; that is, taking the vitamin C supplement reduces the risk of catching a cold under the stated conditions. [An alternative P-value will be obtained in the solution to Example 12.11(b) in Section 12.2.]

(b) The nominal 95% upper confidence bound for the relative risk $\tau = \pi_2/\pi_1$ is given by $.552 + (1.645)(0.153) \doteq 0.552 + 0.252 = 0.804$. The P-value for the Wald test of the hypothesis that $\tau \geq 1$ is given by $\Phi(-2.926) \doteq .0017$.

(c) The nominal 95% upper confidence bound for the odds ratio

$$\tau = \frac{\pi_2/(1 - \pi_2)}{\pi_1/(1 - \pi_1)}$$

is given by $0.490 + (1.645)(0.161) \doteq 0.490 + 0.265 = .755$. The P-value for the Wald test of the hypothesis that $\tau \geq 1$ is given by $\Phi(-3.161) \doteq .0008$. ∎

EXAMPLE **12.9**   Let Y_1, Y_2, and Y_3 be independent random variables such that Y_1 has the binomial distribution with parameters 50 and π_1, Y_2 has the binomial distribution with parameters 50 and π_2, and Y_3 has the binomial distribution with parameters 50 and π_3. The observed values of these random variables are given by $Y_1 = 30$, $Y_2 = 25$, and $Y_3 = 15$. Determine the P-value for a reasonable test of the hypothesis that

(a) $\pi_1 = .5$;   (b) $\pi_2 = \pi_3$;   (c) $\pi_1 = .5$ and $\pi_2 = \pi_3$.

Solution   The usual estimates of π_1, π_2 and π_3 are given by $\hat{\pi}_1 = 30/50 = .6$, $\hat{\pi}_2 = 25/50 = .5$, and $\hat{\pi}_3 = 15/50 = .3$.

(a) The standard error of $\hat{\pi}_1$ is given by

$$SE(\hat{\pi}_1) = \sqrt{\frac{(.6)(.4)}{50}} \doteq .0693.$$

Thus, the Wald statistic for testing the hypothesis that $\pi_1 = .5$ is given by $W_1 \doteq .1/.0693 \doteq 1.443$. The corresponding P-value is given by $P\text{-value}_1 \doteq 2[\,1 - \Phi(1.443)\,] \doteq .149$.

(b) The standard error of the estimate $\hat{\pi}_2 - \hat{\pi}_3$ of $\pi_2 - \pi_3$ is given by

$$SE(\pi_2 - \pi_3) = \sqrt{\frac{(.5)(.5)}{50} + \frac{(.3)(.7)}{50}} \doteq .0959.$$

Thus, the Wald statistic for testing the hypothesis that $\pi_2 = \pi_3$ or, equivalently, that $\pi_2 - \pi_3 = 0$ is given by $W_2 \doteq .2/.0959 \doteq 2.085$. The corresponding P-value is given by $P\text{-value}_2 \doteq 2[\,1 - \Phi(2.085)\,] \doteq .0371$.

(c) Since Y_1, Y_2, and Y_3 are independent, W_1 is a transform of Y_1, and W_2 is a transform of Y_2 and Y_3, we see that W_1 and W_2 are independent random variables and hence that $P\text{-value}_1$ and $P\text{-value}_2$ are independent random variables. It may seem tempting to calculate the P-value for the combined hypothesis that $\pi_1 = .5$ and $\pi_2 = \pi_3$ as

$$P\text{-value} \doteq P(|W_1| \geq 1.443, |W_2| \geq 2.085)$$

$$= P(|W_1| \geq 1.443)P(|W_2| \geq 2.085)$$

$$\doteq P\text{-value}_1 P\text{-value}_2$$

$$\doteq (.149)(.0371)$$

$$\doteq .0055,$$

but such a calculation would be invalid.

A reasonable way to calculate the P-value for the combined hypothesis would be to note that if both individual hypotheses are valid, then each of the independent random variables W_1 and W_2 has approximately the standard normal distribution, so $W_1^2 + W_2^2$ has approximately the chi-square distribution with 2 degrees of freedom, which coincides with the exponential distribution with mean 2. In this manner, we get the P-value given by

$$P\text{-value} = \exp(-(W_1^2 + W_2^2)/2) \doteq \exp(-(1.443^2 + 2.085^2)/2) \doteq .0401.$$

The first approach amounted to multiplying the individual *P*-values. Why is this approach invalid? Consider the independent random variables $U_1 = P\text{-value}_1$ and $U_2 = P\text{-value}_2$. If the individual hypotheses are both satisfied, then each of these random variables has approximately the uniform distribution on $(0, 1)$. In particular, $P(U_1 \leq .05) \doteq .05$, and $P(U_2 \leq .05) \doteq .05$. It is clearly not true, however, that $U_1 U_2$ has approximately the uniform distribution on $(0, 1)$. In particular, $P(U_1 U_2 \leq .05) \doteq .05(1 + \log 20) \doteq .2$ and $P(U_1 U_2 \leq .0055) \doteq .0055(1 - \log .0055) \doteq .0341$ (see the solution to Example 5.21 in Section 5.5). ∎

Problems

12.1 An automatic, individual word recognizer has probability π of correctly recognizing a randomly selected word from the dictionary when uttered by a given speaker. Of the first 100 individual words uttered, 80 are correctly recognized. Determine **(a)** the nominal 95% confidence interval for π; **(b)** the nominal 95% lower confidence bound for π; **(c)** the *P*-value for the Wald test of the hypothesis that $\pi \leq .75$.

12.2 In the context of an investigation to improve the treatment of a certain type of cancer, 100 patients are given the standard treatment, and 50 patients are given a new treatment. It is observed that 75 patients given the standard treatment and 30 patients given the new treatment fail to survive at least 5 years. Let π_1 and π_2 denote the probability of failing to survive at least 5 years under the standard treatment and the new treatment, respectively. Assume that the binomial two-sample model is applicable. **(a)** Determine the estimate of and nominal 95% confidence interval for $\pi_2 - \pi_1$ and the *P*-value for the Wald test of the hypothesis that $\pi_1 = \pi_2$. **(b)** Determine the estimate of and nominal 95% confidence interval for the relative risk $\tau = \pi_2/\pi_1$ and the *P*-value for the Wald test of the hypothesis that $\tau = 1$. **(c)** Determine the estimate of and nominal 95% confidence interval for the odds ratio

$$\tau = \frac{\pi_2/(1 - \pi_2)}{\pi_1/(1 - \pi_1)}$$

and the *P*-value for the Wald test of the hypothesis that $\tau = 1$. **(d)** Discuss the results in the solutions to parts (a)–(c).

12.3 In the context of Problem 12.2, **(a)** determine the nominal 95% upper confidence bound for $\pi_2 - \pi_1$ and the *P*-value for the Wald test of the hypothesis that $\pi_2 \geq \pi_1$; **(b)** determine the nominal 95% upper confidence bound for the relative risk $\tau = \pi_2/\pi_1$ and the *P*-value for the Wald test of the hypothesis that $\tau \geq 1$; **(c)** determine the nominal 95% upper confidence bound for the odds ratio

$$\tau = \frac{\pi_2/(1 - \pi_2)}{\pi_1/(1 - \pi_1)}$$

and the *P*-value for the Wald test of the hypothesis that $\tau \geq 1$; **(d)** discuss the results in the solutions to parts (a)–(c).

12.4 In the context of the Poisson d-sample model, determine the asymptotic variance and standard error of (a) $\hat{\lambda}_k$; (b) $\hat{\lambda}_k - \hat{\lambda}_j$ for $j \neq k$; (c) $\hat{\lambda}_k / \hat{\lambda}_j$ for $j \neq k$.

12.5 The number of people getting a certain rare and noncontagious disease in a given country in a given year has a Poisson distribution with mean $n\lambda$, where λ is the rate of getting the disease in the given country (new cases per person-year) and n is the population of the country. In a certain country having 10 million people, there are 100 new cases in a given year. Determine the estimate of and nominal 95% confidence interval for the corresponding (a) rate λ; (b) mean $n\lambda$.

12.6 In country 1, which has 10 million people, there are 100 new cases of the disease (see Problem 12.5) in a given year. In country 2, which has 20 million people, there are 160 new cases of the disease during the year. Let λ_1 and λ_2 denote the rates of getting the disease in countries 1 and 2, respectively. (a) Determine the estimate of and nominal 95% confidence interval for $\lambda_2 - \lambda_1$ and the P-value for the Wald test of the hypothesis that $\lambda_1 = \lambda_2$. (b) Determine the estimate of and nominal 95% confidence interval for the relative rate $\tau = \lambda_2/\lambda_1$ and the P-value for the Wald test of the hypothesis that $\tau = 1$. (c) Discuss the results in the solutions to parts (a) and (b).

12.7 In the context of Problem 12.6, (a) determine the nominal 95% upper confidence bound for $\lambda_2 - \lambda_1$ and the P-value for the Wald test of the hypothesis that $\lambda_2 \geq \lambda_1$; (b) determine the nominal 95% upper confidence bound for the relative rate $\tau = \lambda_2/\lambda_1$ and the P-value for the Wald test of the hypothesis that $\tau = 1$; (c) discuss the results in the solutions to parts (a) and (b).

12.8 Let Y_{ki}, $1 \leq k \leq d$ and $1 \leq i \leq n_k$, be independent random variables such that Y_{ki} has the exponential distribution with unknown mean β_k for $1 \leq k \leq d$ and $1 \leq i \leq n_k$. We refer to this setup as the exponential d-sample model. The sample mean for the kth sample is given by $\bar{Y}_k = n_k^{-1} \sum_i Y_{ki}$. Observe that $\bar{Y}_1, \ldots, \bar{Y}_d$ are independent and that $\hat{\beta}_k = \bar{Y}_k$ has mean β_k and variance β_k^2/n_k for $1 \leq k \leq d$. Determine (a) the variance and standard error of $\hat{\beta}_k$; (b) the variance and standard error of $\hat{\beta}_k - \hat{\beta}_j$ for $j \neq k$; (c) the asymptotic variance and standard error of $\hat{\beta}_k/\hat{\beta}_j$ for $j \neq k$.

12.9 A manufacturing company uses a certain component in its product that it obtains from suppliers. It obtains a sample of size 100 of the component from supplier A and a sample of size 50 from supplier B, and then it determines the lifetimes of all 150 components so obtained, getting sample means of 120 hours for the sample from supplier A and 170 hours for the sample from supplier B. Under the assumptions of the exponential two-sample model (see Problem 12.8), can it be concluded from the observed data that the components obtained from the two suppliers differ in mean lifetimes? Specifically, let β_1 and β_2 be the mean lifetimes of the components obtained from suppliers A and B, respectively. (a) Determine the estimate of and nominal 95% confidence interval for $\beta_2 - \beta_1$ and the P-value for the Wald test of the hypothesis that $\beta_1 = \beta_2$. (b) Determine the estimate of and nominal 95% confidence interval for $\tau = \beta_2/\beta_1$ and the P-value for the Wald test of the hypothesis that $\tau = 1$. (c) Discuss the results in the solutions to parts (a) and (b).

12.10 The company (see Problem 12.9) has been using supplier A. Can it be concluded from the observed data that the components obtained from supplier B have a higher mean lifetime than those obtained from supplier A? Specifically, (a) determine the nominal 95% lower confidence bound for $\beta_2 - \beta_1$ and the P-value for the Wald test of the hypothesis that $\beta_2 \leq \beta_1$; **(b)** determine the nominal 95% lower confidence bound for $\tau = \beta_2/\beta_1$ and the *P*-value for the test of the hypothesis that $\tau \leq 1$; **(c)** discuss the results in the solutions to parts (a) and (b).

12.11 In the context of the binomial one-sample model, determine the asymptotic variance of the estimate $\arcsin\sqrt{\hat{\pi}}$ of $\arcsin\sqrt{\pi}$. [*Hint:* Recall that

$$\frac{d}{dt}\arcsin t = \frac{1}{\sqrt{1-t^2}}, \qquad -1 < t < 1.]$$

12.12 In the context of the Poisson one-sample model, determine the asymptotic variance of the estimate $\sqrt{\hat{\lambda}}$ of $\sqrt{\lambda}$.

12.13 Determine the asymptotic variance of the estimate $\log(\hat{\mu}_k/\hat{\mu}_j)$ of $\log(\mu_k/\mu_j)$ for $j \neq k$ **(a)** in the general context of this section; **(b)** in the context of the binomial d-sample model; **(c)** in the context of the Poisson d-sample model.

12.2
Exact *P*-Values

The formulas for the *P*-values corresponding to the Wald test do not take into account the inaccuracy of the standard normal approximation to the distribution of $(\hat{\tau} - \tau)/\mathrm{ASD}(\hat{\tau})$ or the inaccuracy of the standard error of $\hat{\tau}$ as an approximation to its asymptotic standard deviation. (These inaccuracies approach zero as the time parameters $n_1, \ldots, n_d$ tend to infinity, but they can be nonnegligible for small to moderately larger values of the time parameters.) In the context of binomial and Poisson models, however, it is possible to obtain "exact *P*-values."

Suppose first that Y has the binomial distribution with parameters n and π. Then Y has mean $n\pi$ and variance $n\pi(1-\pi)$, its probability function is given by

$$f(y; n, \pi) = \binom{n}{y}\pi^y(1-\pi)^{n-y}, \qquad y \in \{0, \ldots, n\},$$

and its distribution function is given by

$$F(y; n, \pi) = \sum_{z=0}^{y} f(z; n, \pi), \qquad y \in \{0, \ldots, n\}.$$

Let y be the observed value of Y and let π_0 be a given number in $(0, 1)$.

Consider first the hypothesis that $\pi \leq \pi_0$. The *exact P*-value for the test of this hypothesis is defined by

$$P\text{-value} = \sum_{z=y}^{n} f(z; n, \pi_0) = 1 - F(y-1; n, \pi_0),$$

the motivation being that sufficiently large values of Y make us doubt that π is less than or equal to π_0. Similarly, the exact *P*-value for the test of the hypothesis that $\pi \geq \pi_0$ is defined by

$$P\text{-value} = \sum_{z=0}^{y} f(z; n, \pi_0) = F(y; n, \pi_0),$$

the motivation being that sufficiently small values of Y make us doubt that π is greater than or equal to π_0.

Consider now the hypothesis that $\pi = \pi_0$. The exact P-value for the test of this hypothesis is defined as the sum of $f(z; n, \pi_0)$ over all values of z such that $f(z; n, \pi_0) \leq f(y; n, \pi_0)$. According to the discussion of the unimodality properties of the binomial probability function that followed from Equation (5) in Section 4.2, the exact P-value is given by

$$P\text{-value} = \sum_{z=0}^{c_1} f(z; n, \pi_0) + \sum_{z=c_2}^{n} f(z; n, \pi_0) = F(c_1; n, \pi_0) + 1 - F(c_2 - 1; n, \pi_0),$$

where the integers c_1 and c_2 depend on n, π_0, and y and one of these integers equals y.

Of course, we can use normal approximation (preferably with the half-integer correction) to evaluate any of exact P-values defined above, but then the word *exact* becomes slightly misleading.

EXAMPLE **12.10**   Determine the exact P-value and its normal approximation in the context of Example 12.6 in Section 12.1.

Solution   Here Y has the binomial distribution with parameters $n = 100$ and π, the hypothesis is that $\pi = .25$, and the observed value of Y equals 35. Under the hypothesis, Y has mean 25, variance 18.75, and standard deviation $\sqrt{18.75} \doteq 4.33$. Also, under the hypothesis, $f(z; 100, .25)$ is a strictly increasing function of z for $z \in \{0, \ldots, 25\}$ and a strictly decreasing function of z for $z \in \{25, \ldots, 100\}$. Now

$$f(35; 100, .25) \doteq .00702,$$

$$f(15; 100, .25) \doteq .00566,$$

and

$$f(16; 100, .25) \doteq .01003.$$

Thus, $f(z; 100, .25) \leq f(35; 100, .25)$ if and only if either $z \leq 15$ or $z \geq 35$. Consequently, the exact P-value is given by

$$P\text{-value} = \sum_{z=0}^{15} f(z; 100, .25) + \sum_{z=35}^{100} f(z; 100, .25)$$

$$= F(15; 100, .25) + 1 - F(34; 100, .25)$$

$$\approx \Phi\left(\frac{15 + \frac{1}{2} - 25}{4.33}\right) + 1 - \Phi\left(\frac{35 - \frac{1}{2} - 25}{4.33}\right).$$

Using a computer to evaluate the indicated quantities, we get the exact P-value and its normal approximation are given by P-value $\doteq .0275$ and P-value $\approx .0282$, respectively. Observe that the exact P-value and its normal approximation are in close agreement and that they are somewhat smaller than the value given in the solution to part (a) of Example 12.6 in Section 12.1 and considerably smaller than the value given in the solution to part (b) of that example.   ∎

Consider next the binomial two-sample model and set $Y = Y_1 + Y_2$. Under the hypothesis that $\pi_1 = \pi_2$, the conditional distribution of Y_1 given that $Y = y$ is hypergeometric with parameters y, n_1, and n_2 (see Example 6.16 in Section 6.3), which has mean $y\frac{n_1}{n}$ and variance $y\frac{n_1}{n}\frac{n_2}{n}\frac{n-y}{n-1}$, where $n = n_1 + n_2$. Let f denote the probability function of this hypergeometric distribution, which is given by

$$f(z) = \frac{\binom{n_1}{z}\binom{n_2}{y-z}}{\binom{n}{y}};$$

let F be the corresponding distribution function; and let y_1 and y denote the observed values of Y_1 and Y, respectively.

The exact P-value for the test of the hypothesis that $\pi_1 \leq \pi_2$ is defined by

$$P\text{-value} = \sum_{z \geq y_1} f(z) = 1 - F(y_1 - 1),$$

the motivation being that an observed value of Y_1 that is too large relative to the observed value of Y raises doubt about the possibility that $\pi_1 \leq \pi_2$. Similarly, the exact P-value for the test of the hypothesis that $\pi_1 \geq \pi_2$ is defined by

$$P\text{-value} = \sum_{z \leq y_1} f(z) = F(y_1),$$

the motivation being that an observed value of Y_1 that is too small relative to the observed value of Y raises doubt about the possibility that $\pi_1 \geq \pi_2$.

Consider now the hypothesis that $\pi_1 = \pi_2$. The exact P-value for the test of this hypothesis is defined as the sum of $f(z)$ over all values of z such that $f(z) \leq f(y_1)$. According to the discussion of the unimodality properties of the hypergeometric probability function that followed from Equation (24) in Section 6.3, the exact P-value is given by

$$P\text{-value} = \sum_{z \leq c_1} f(z) + \sum_{z \geq c_2} f(z) = F(c_1) + 1 - F(c_2 - 1),$$

where the integers c_1 and c_2 depend on n_1, n_2, y, and y_1 and one of these integers equals y_1.

EXAMPLE **12.11** Recall the vitamin C experiment that was discussed in Examples 12.7 and 12.8 in Section 12.1. Determine the exact P-value and its normal approximation for the test of the hypothesis that

(a) $\pi_1 = \pi_2$; (b) $\pi_2 \geq \pi_1$.

Solution Here $n_1 = 140$, $n_2 = 139$, the observed value of Y_1 is 31, and the observed value of Y_2 is 17. Thus, $n = n_1 + n_2 = 279$, and the observed value of $Y = Y_1 + Y_2$ is 48. Under the assumption that $\pi_1 = \pi_2$, the conditional distribution of Y_1 given that $Y = 48$ is hypergeometric with parameters 48, 140, and 139, which has mean $48 \cdot \frac{140}{279} = 24.086$, variance $48 \cdot \frac{140}{279} \cdot \frac{139}{279} \cdot \frac{231}{278} \doteq 9.971$, and standard deviation $\sqrt{9.971} \doteq 3.158$. Table 12.1 shows a portion of the corresponding probability function f and distribution function F.

TABLE   12.1

y	f(y)	F(y)	y	f(y)	F(y)
15	.0020	.0030	28	.0589	.9196
16	.0047	.0078	29	.0379	.9575
17	.0102	.0180	30	.0220	.9795
18	.0199	.0378	31	.0115	.9910
19	.0348	.0726	32	.0054	.9964
20	.0550	.1276	33	.0023	.9987

(a) The exact P-value for testing the hypothesis that $\pi_1 = \pi_2$ is given by $F(17) + 1 - F(30) \doteq .0385$. The normal approximation to this P-value is given by

$$\Phi\left(\frac{17 + \frac{1}{2} - 24.086}{3.158}\right) + 1 - \Phi\left(\frac{31 - \frac{1}{2} - 24.086}{3.158}\right) \doteq .0396.$$

Observe that the exact P-value and its normal approximation are larger than any of the P-values obtained in the solution to Example 12.7 in Section 12.1.

(b) The exact P-value for testing the hypothesis that $\pi_2 \geq \pi_1$ is given by $1 - F(30) \doteq .0205$. The normal approximation to this P-value is given by

$$1 - \Phi\left(\frac{31 - \frac{1}{2} - 24.086}{3.158}\right) \doteq .0211.$$

Observe that the exact P-value and its normal approximation are larger than any of the P-values obtained in the solution to Example 12.8 in Section 12.1.   ∎

Let Y have the binomial distribution with parameters n and π. Consider the hypothesis that $\pi \leq \pi_0$, where $0 < \pi_0 < 1$. The reasonable critical regions are of the form $Y \geq c$ for some $c \in \{1, \ldots, n\}$. It is left as a problem to verify the intuitively obvious result that $P(Y \geq c)$ is a strictly increasing function of π. Thus, the size of the test is given by $\alpha = P(Y \geq c)$ when $\pi = \pi_0$; that is, by

$$\alpha = \max_{\pi \leq \pi_0} \sum_{y=c}^{n} \binom{n}{y} \pi^y (1 - \pi)^{n-y} = \sum_{y=c}^{n} \binom{n}{y} \pi_0^y (1 - \pi_0)^{n-y}.$$

The exact P-value of the test is given by

$$P\text{-value} = \sum_{z=y}^{n} \binom{n}{z} \pi_0^z (1 - \pi_0)^{n-z},$$

where y is the observed value of Y. As usual, Y lies in the critical region for the test of size α if α is greater than or equal to the P-value and in the acceptance region $Y < c$ if α is less than the P-value. The power of the test when $\pi > \pi_0$ is given by

$$1 - \beta = \sum_{y=c}^{n} \binom{n}{y} \pi^y (1 - \pi)^{n-y}.$$

EXAMPLE **12.12**   Johnny claims that he has ESP (extrasensory perception) and, in particular, that he can correctly guess the outcome of coin tossing with probability $\pi > .5$. To confirm or falsify this claim, Frankie tosses a fair coin n times in an honest manner and Johnny guesses the outcome just as she tosses it. We assume that the number Y of correct guesses has the binomial distribution with parameters n and π. The claim is valid if and only if the hypothesis that $\pi \leq .5$ is false. Determine

(a) the exact P-value if Johnny guesses correctly in 18 out of 25 tosses;

(b) the power of the test of size $\alpha \approx .05$ when $n = 25$ and $\pi = .6$;

(c) n so that the test of approximate size .05 has approximate power .9 when $\pi = .6$.

Solution   (a) The exact P-value is given by

$$P\text{-value} = \sum_{y=18}^{25} \binom{25}{y}(.5)^{25} \doteq .0216.$$

This may make Johnny more convinced than ever that he has ESP, but Frankie could legitimately remain skeptical (if her subjective prior probability that Johnny had ESP were low).

(b) The critical region $Y \geq 17$ has size

$$\alpha = \sum_{y=17}^{25} \binom{25}{y}(.5)^{25} \doteq .0539$$

when $n = 25$. The power of the corresponding test when $\pi = .6$ is given by

$$1 - \beta = \sum_{y=17}^{25} \binom{25}{y}(.6)^{y}(.4)^{25-y} \doteq .2735.$$

(c) Consider, more generally the test of size α of the hypothesis that $\pi \leq \pi_0$. Using normal approximation to the binomial distribution, we see that the critical region is given by

$$\alpha \approx 1 - \Phi\left(\frac{c - n\pi_0}{\sqrt{n\pi_0(1 - \pi_0)}}\right),$$

so that

$$\frac{c - n\pi_0}{\sqrt{n\pi_0(1 - \pi_0)}} \approx z_{1-\alpha}$$

and hence

$$c \approx n\pi_0 + z_{1-\alpha}\sqrt{n\pi_0(1 - \pi_0)}.$$

Let $\pi > \pi_0$ be fixed. Then the power of the test is given by

$$1 - \beta \approx 1 - \Phi\left(\frac{c - n\pi}{\sqrt{n\pi(1 - \pi)}}\right),$$

so that

$$\frac{c - n\pi}{\sqrt{n\pi(1 - \pi)}} \approx z_{\beta} = -z_{1-\beta}$$

and hence

$$c \approx n\pi - z_{1-\beta}\sqrt{n\pi(1-\pi)}.$$

Consequently,

$$n\pi - z_{1-\beta}\sqrt{n\pi(1-\pi)} \approx n\pi_0 + z_{1-\alpha}\sqrt{n\pi_0(1-\pi_0)},$$

so

$$n(\pi - \pi_0) \approx \sqrt{n}[z_{1-\alpha}\sqrt{\pi_0(1-\pi_0)} + z_{1-\beta}\sqrt{\pi(1-\pi)}]$$

and hence

$$n \approx \left(\frac{z_{1-\alpha}\sqrt{\pi_0(1-\pi_0)} + z_{1-\beta}\sqrt{\pi(1-\pi)}}{\pi - \pi_0}\right)^2.$$

In the present context, $\pi_0 = .5, \alpha = .05, \pi = .6$, and $\beta = .1$, so

$$n \approx \left(\frac{(1.645)(.5) + 1.282\sqrt{.24}}{.1}\right)^2 \doteq 210.$$

Correspondingly, $c \approx (210)(.5) + 1.645\sqrt{210(.5)(.5)} \doteq 117$. As a check,

$$\sum_{y=117}^{210} \binom{210}{y}(.5)^{210} \doteq .0561 \quad \text{and} \quad \sum_{y=117}^{210} \binom{210}{y}(.6)^y(.4)^{210-y} \doteq .9091. \quad \blacksquare$$

Likelihood-Ratio Tests

Consider a discrete $\mathcal{Y}$-valued random variable $\mathbf{Y}$ that is known to have one of two positive probability functions f_0 or f_1 on $\mathcal{Y}$ and consider the hypothesis that the probability function equals f_0 and the alternative that it equals f_1. Let C be a critical region for a test of the null hypothesis based on $\mathbf{Y}$ and let $A = \mathcal{Y}\backslash C$ be the corresponding acceptance region. Then the size α and power $1 - \beta$ of the test are given by $\alpha = \sum_{\mathbf{y} \in C} f_0(\mathbf{y})$ and $1 - \beta = \sum_{\mathbf{y} \in C} f_1(\mathbf{y})$. The function f_1/f_0 is referred to as the likelihood ratio. The test is referred to as a *likelihood-ratio test* if A and C are both nonempty and the likelihood ratio at each point in the critical region is at least as large as it is in every point in the acceptance region. According to the next result, known as the *Neyman–Pearson lemma*, every likelihood-ratio test is a *most powerful test* of its size.

THEOREM 12.6   If the size of a test is no larger than the size of some likelihood-ratio test, then the power of the test is no larger than that of the likelihood-ratio test.

Proof   Let C be the critical region for the likelihood-ratio test and let α and $1 - \beta$ denote, respectively, the size and power of that test. Similarly, let C' and A', respectively, be the critical region and acceptance region for the competing test and let α' and $1 - \beta'$, respectively, denote the corresponding size and power.

Now

$$\alpha = \sum_{C \cap C'} f_0(\mathbf{y}) + \sum_{C \cap A'} f_0(\mathbf{y}) \quad \text{and} \quad \alpha' = \sum_{C \cap C'} f_0(\mathbf{y}) + \sum_{A \cap C'} f_0(\mathbf{y}),$$

so

$$\alpha' - \alpha = \sum_{A \cap C'} f_0(\mathbf{y}) - \sum_{C \cap A'} f_0(\mathbf{y}). \tag{7}$$

Let r be a positive number such that $f_1/f_0 \geq r$ on C and $f_1/f_0 \leq r$ on A. Then

$$1 - \beta = \sum_{C \cap C'} f_1(\mathbf{y}) + \sum_{C \cap A'} f_1(\mathbf{y})$$

$$= \sum_{C \cap C'} f_1(\mathbf{y}) + \sum_{C \cap A'} \frac{f_1(\mathbf{y})}{f_0(\mathbf{y})} f_0(\mathbf{y})$$

$$\geq \sum_{C \cap C'} f_1(\mathbf{y}) + r \sum_{C \cap A'} f_0(\mathbf{y}).$$

Similarly,

$$1 - \beta' \leq \sum_{C \cap C'} f_1(\mathbf{y}) + r \sum_{A \cap C'} f_0(\mathbf{y}), \tag{8}$$

the proof being left as a problem. Consequently,

$$(1 - \beta') - (1 - \beta) \leq r\left(\sum_{A \cap C'} f_0(\mathbf{y}) - \sum_{C \cap A'} f_0(\mathbf{y}) \right),$$

so we conclude from (7) that

$$(1 - \beta') - (1 - \beta) \leq r(\alpha' - \alpha). \tag{9}$$

It follows from (9) that if $\alpha' \leq \alpha$, then $1 - \beta' \leq 1 - \beta$. ∎

THEOREM 12.7   Let Y have the binomial distribution with parameters n and π, where n is known and π is unknown. Consider the hypothesis that $\pi = \pi_0$ and the alternative that $\pi = \pi_1$, where π_0 and π_1 are distinct specified numbers in the interval $(0, 1)$. Let C and A, respectively, denote the critical region and acceptance region for some test based on Y, where C and A are both nonempty.

(a) Suppose $\pi_1 > \pi_0$. Then the test is a likelihood-ratio test if and only if $C = \{c, \ldots, n\}$ for some $c \in \{1, \ldots, n\}$.

(b) Suppose $\pi_1 < \pi_0$. Then the test is a likelihood-ratio test if and only if $C = \{0, \ldots, c\}$ for some $c \in \{0, \ldots, n - 1\}$.

Proof   The probability function of Y is given by

$$f(y; n, \pi) = \binom{n}{y} \pi^y (1 - \pi)^{n-y}, \qquad y \in \{0, \ldots, n\}.$$

Thus, the likelihood ratio is given by

$$\frac{f(y; n, \pi_1)}{f(y; n, \pi_0)} = \left(\frac{\pi_1/(1 - \pi_1)}{\pi_0/(1 - \pi_0)} \right)^y \left(\frac{1 - \pi_1}{1 - \pi_0} \right)^n, \qquad y \in \{0, \ldots, n\}.$$

(a) Suppose $\pi_1 > \pi_0$. Since the indicated odds ratio is greater than 1, the likelihood ratio is a strictly increasing function of y, and hence the desired result is valid.

(b) Suppose $\pi_1 < \pi_0$. Since the odds ratio is less than 1, the likelihood ratio is a strictly decreasing function of y, and hence the desired result is valid. ∎

For an illustration of the Neyman–Pearson lemma and Theorem 12.7, let Y have the binomial distribution with parameters $n = 5$ and π. Consider the hypothesis that $\pi = \pi_0 = 1/4$ and the alternative that $\pi = \pi_1 = 3/4$ and let f_0 and f_1 denote the probability function of Y under the hypothesis and its alternative, respectively. Then

$$f_0(y) = \binom{5}{y}\left(\frac{1}{4}\right)^y\left(\frac{3}{4}\right)^{5-y}, \qquad y \in \{0, \ldots, 5\},$$

and

$$f_1(y) = \binom{5}{y}\left(\frac{3}{4}\right)^y\left(\frac{1}{4}\right)^{5-y}, \qquad y \in \{0, \ldots, 5\}.$$

Thus, the likelihood ratio is given by

$$\frac{f_1(y)}{f_0(y)} = \frac{9^y}{243},$$

which is a strictly increasing function of y. Consider the critical region $C = \{4, 5\}$ and the corresponding acceptance region $A = \{0, 1, 2, 3\}$. The size of the the test is given by

$$\alpha = \sum_{y=4}^{5} f_0(y) = 5\left(\frac{1}{4}\right)^4\left(\frac{3}{4}\right) + \left(\frac{1}{4}\right)^5 = \frac{1}{64},$$

and the power of the test is given by

$$1 - \beta = \sum_{y=4}^{5} f_1(y) = 5\left(\frac{3}{4}\right)^4\left(\frac{1}{4}\right) + \left(\frac{3}{4}\right)^5 = \frac{81}{128}.$$

According to the Neyman–Pearson lemma, there is no test of the indicated hypothesis against the indicated alternative whose size is at most $1/64$ and whose power is greater than $81/128$.

There is an obvious analog of likelihood-ratio tests for models involving density functions, and the Neyman–Pearson lemma applies to such tests.

EXAMPLE **12.13** Let Y be normally distributed with known variance σ^2 and unknown mean μ. Determine the form of the critical regions (having size less than 1 and positive power) for likelihood-ratio tests of the hypothesis that $\mu = \mu_0$ against the alternative that $\mu = \mu_1$

(a) when $\mu_0 < \mu_1$;   **(b)** when $\mu_0 > \mu_1$.

Solution   If $\mu = \mu_0$, then Y has the density function f_0 given by

$$f_0(y) = \frac{1}{\sigma\sqrt{2\pi}}\exp\left(-\frac{(y - \mu_0)^2}{2\sigma^2}\right)$$

$$= \frac{1}{\sigma\sqrt{2\pi}} \exp\left(-\frac{\mu_0^2}{2\sigma^2}\right) \exp\left(\frac{\mu_0 y}{\sigma^2}\right) \exp\left(-\frac{y^2}{2\sigma^2}\right),$$

for $y \in \mathbb{R}$. Similarly, if $\mu = \mu_1$, then Y has the density function f_1 given by

$$f_1(y) = \frac{1}{\sigma\sqrt{2\pi}} \exp\left(-\frac{(y - \mu_1)^2}{2\sigma^2}\right)$$

$$= \frac{1}{\sigma\sqrt{2\pi}} \exp\left(-\frac{\mu_1^2}{2\sigma^2}\right) \exp\left(\frac{\mu_1 y}{\sigma^2}\right) \exp\left(-\frac{y^2}{2\sigma^2}\right),$$

for $y \in \mathbb{R}$. Thus, the likelihood ratio is given by

$$\frac{f_1(y)}{f_0(y)} = \exp\left(\frac{\mu_0^2 - \mu_1^2}{2\sigma^2}\right) \exp\left(\frac{(\mu_1 - \mu_0)y}{\sigma^2}\right),$$

which is a strictly increasing function of y if $\mu_0 < \mu_1$ and a strictly decreasing function if $\mu_0 > \mu_1$.

(a) If $\mu_0 < \mu_1$, the critical regions for likelihood-ratio tests of f_0 against f_1 are of the form $[c, \infty)$ or (c, ∞) for $c \in \mathbb{R}$.

(b) If $\mu_0 > \mu_1$, the critical regions are of the form $(-\infty, c]$ or $(-\infty, c)$ for $c \in \mathbb{R}$. ∎

Problems

12.14 In the context of Problem 12.1 in Section 12.1, determine the exact P-value for the test of the hypothesis that $\pi \leq .75$ using the following abbreviated table of the distribution function of the binomial distribution with parameters 100 and 75:

y	78	79	80	81	82
$F(y)$	.7886	.8512	.9005	.9370	.9624

12.15 Consider the setup of Problem 7.33 in Section 7.4, but drop the assumption that Y is normally distributed. Set $I = \text{ind}(Y > 0)$ and $\pi = P(I = 1) = P(Y > 0)$. Think of the treatment as being effective if and only if $\pi > 1/2$. A test of the hypothesis that $\pi \leq 1/2$ is known as a *sign test*. Determine the exact P-value for the sign test of the hypothesis that $\pi \leq 1/2$ **(a)** for the patients in ward A; **(b)** for the patients in ward B; **(c)** for the patients in both wards.

12.16 Let π denote the probability that a possibly loaded die shows side 1 when rolled in an honest manner and consider the hypothesis that $\pi = 1/6$. The die is rolled 24 times in an honest manner, and it is observed that side 1 shows eight times. The values of the probability function f and distribution function F of the binomial distribution with parameters $n = 24$ and $\pi = 1/6$ are shown in Table 12.2. Determine the exact P-value for the test of the indicated hypothesis by using the values of **(a)** f; **(b)** F.

12.17 Consider the setup of Problems 12.2 and 12.3 in Section 12.1. Table 12.3 shows a portion of the probability function f and distribution function F of the hypergeometric distribution with parameters 105, 100, and 50. Determine the exact P-value

TABLE 12.2

y	$f(y)$	$F(y)$	y	$f(y)$	$F(y)$
0	.0126	.0126	7	.0557	.9646
1	.0604	.0730	8	.0237	.9882
2	.1389	.2118	9	.0084	.9967
3	.2037	.4155	10	.0025	.9992
4	.2139	.6294	11	.0006	.9998
5	.1711	.8005	12	.0001	1.0000
6	.1084	.9088	13	.0000	1.0000

TABLE 12.3

y	$f(y)$	$F(y)$	y	$f(y)$	$F(y)$
62	.0013	.0017	72	.1115	.8279
63	.0042	.0058	73	.0784	.9062
64	.0112	.0171	74	.0482	.9544
65	.0255	.0425	75	.0259	.9803
66	.0491	.0916	76	.0122	.9924
67	.0810	.1726	77	.0050	.9974

and its normal approximation (with the half-integer correction) for the test of the hypothesis that **(a)** $\pi_1 = \pi_2$; **(b)** $\pi_2 \geq \pi_1$.

12.18 Let Y have the Poisson distribution with mean $n\lambda$, where n is a known positive number and λ is an unknown positive number, and let $f(\cdot; n, \lambda)$ and $F(\cdot; n, \lambda)$ denote the probability function and distribution function, respectively, of Y. Recall the discussion of the properties of the Poisson probability function that followed Equation (12) in Section 4.4. Define a reasonable exact P-value for testing the hypothesis that $\lambda = \lambda_0$ for a given positive number λ_0.

12.19 Consider the Poisson two-sample model and set $Y = Y_1 + Y_2$. Then the conditional distribution of Y_1 given that $Y = y$ is binomial with parameters y and $n_1\lambda_1/(n_1\lambda_1 + n_2\lambda_2)$ (see Example 6.9 in Section 6.1). Recall the discussion of the properties of the binomial probability function that followed from Equation (5) in Section 4.2. Define a reasonable exact P-value for testing the hypothesis that **(a)** $\lambda_1 = \lambda_2$; **(b)** $\lambda_2 \geq \lambda_1$.

12.20 Use the results of Problem 12.19 to determine the exact P-value in the context of Problems 12.6 and 12.7 in Section 12.1 for the test of the hypothesis that **(a)** $\lambda_1 = \lambda_2$; **(b)** $\lambda_2 \geq \lambda_1$. For convenience, Table 12.4 shows a portion of the probability function f and distribution function F of the binomial distribution with parameters $n = 260$ and $\pi = 1/3$.

12.21 Consider the Poisson process with rate λ on $[0, \infty)$, let λ_0 be a given positive number, and consider the hypothesis $\lambda \leq \lambda_0$ and the critical region $Y \geq c$ having approximate size α, where Y is the number of events that have occurred by time n.

TABLE   **12.4**

y	f(y)	F(y)	y	f(y)	F(y)
71	.0062	.0217	97	.0206	.9220
72	.0081	.0298	98	.0172	.9391
73	.0104	.0402	99	.0140	.9532
74	.0132	.0533	100	.0113	.9645
75	.0163	.0696	101	.0089	.9734
76	.0199	.0895	102	.0070	.9804

Use normal approximation to determine n such that the power of the test is about $1 - \beta$ for some fixed $\lambda > \lambda_0$.

12.22 Let $U_1, \ldots, U_n$ be independent random variables, each uniformly distributed on $(0, 1)$; let $0 < \pi_1 < \pi_2 < 1$; and set $Y_1 = \#(\{i : 1 \le i \le n \text{ and } U_i \le \pi_1\})$ and $Y_2 = \#(\{i : 1 \le i \le n \text{ and } U_i \le \pi_2\})$. Also, let $c \in \{1, \ldots, n\}$. Explain why **(a)** Y_1 has the binomial distribution with parameters n and π_1 and Y_2 has the binomial distribution with parameters n and π_2; **(b)** $P(Y_2 \ge c) > P(Y_1 \ge c)$; **(c)** $\sum_{y=c}^{n} \binom{n}{y} \pi_2^y (1 - \pi_2)^{n-y} > \sum_{y=c}^{n} \binom{n}{y} \pi_1^y (1 - \pi_1)^{n-y}$.

12.23 Let $0 < \lambda_1 < \lambda_2$, let Y_1 and W be independent random variables such that Y_1 has the Poisson distribution with mean λ_1 and W has the Poisson distribution with mean $\lambda_2 - \lambda_1$, and set $Y_2 = Y_1 + W$. Let c be a positive integer. Explain why **(a)** Y_2 has the Poisson distribution with mean λ_2; **(b)** $P(Y_2 \ge c) > P(Y_1 \ge c)$; **(c)** $\sum_{y=c}^{\infty} \frac{\lambda_2^y}{y!} \exp(-\lambda_2) > \sum_{y=c}^{\infty} \frac{\lambda_1^y}{y!} \exp(-\lambda_1)$.

12.24 Verify Inequality (8).

12.25 Let Y have the Poisson distribution with mean $n\lambda$, where n is a known positive number and λ is unknown. Consider the hypothesis $\lambda = \lambda_0$ and the alternative $\lambda = \lambda_1$, where λ_0 and λ_1 are distinct specified positive numbers. Let C and A, respectively, denote the critical region and acceptance region for some test based on Y, where C and A are both nonempty. **(a)** Suppose $\lambda_1 > \lambda_0$. Show that the test is a likelihood-ratio test if and only if $C = \{c, c+1, c+2, \ldots\}$ for some positive integer c. **(b)** Suppose $\lambda_1 < \lambda_0$. Show that the test is a likelihood-ratio test if and only if $C = \{0, \ldots, c\}$ for some nonnegative integer c.

12.26 Let Y have the gamma distribution with known shape parameter α and unknown inverse-scale parameter λ. Determine the form of the critical regions (having size less than 1 and positive power) for likelihood-ratio tests of the hypothesis that $\lambda = \lambda_0$ against the alternative that $\lambda = \lambda_1$, where λ_0 and λ_1 are specified positive numbers with **(a)** $\lambda_0 > \lambda_1$; **(b)** $\lambda_0 < \lambda_1$.

12.27 Let $\mathbf{Y}$ have the multivariate normal distribution with known $n \times n$ variance–covariance matrix Σ and unknown mean vector $\boldsymbol{\mu}$. Determine the form of the critical regions (having size less than 1 and positive power) for likelihood-ratio tests of the hypothesis that $\boldsymbol{\mu} = \boldsymbol{\mu}_0$ against the alternative that $\boldsymbol{\mu} = \boldsymbol{\mu}_1$, where $\boldsymbol{\mu}_0$ and $\boldsymbol{\mu}_1$ are distinct specified vectors.

12.3

One-Parameter Exponential Families

A number of special one-parameter families of univariate distributions, including binomial, Poisson, negative binomial, normal, and gamma distributions, can be written in a common form, that of the one-parameter exponential family. Here we use binomial and Poisson distributions to illustrate the general properties of such families. The exponential family form for binomial and Poisson distributions will play a crucial role in our treatment of logistic regression and Poisson regression in Chapter 13. Also, one-parameter and multiparameter exponential families are fundamental to advanced statistical theory.

Let Y have the binomial distribution with parameters n and π, whose probability function is given by

$$f(y) = \binom{n}{y}\pi^y(1-\pi)^{n-y}, \qquad y \in \{0, \ldots, n\},$$

and $f(y) = 0$ elsewhere. We can rewrite this probability function as

$$f(y) = \left(\frac{\pi}{1-\pi}\right)^y (1-\pi)^n r(y; n), \qquad y \in \mathcal{Y}_n,$$

where $\mathcal{Y}_n = \{0, \ldots, n\}$ and $r(y; n) = \binom{n}{y}$ for $y \in \mathcal{Y}_n$. Consider the parameter

$$\theta = \operatorname{logit}\pi = \log\frac{\pi}{1-\pi},$$

which ranges over all of $\mathbb{R}$ as π ranges over $(0, 1)$. Now

$$\frac{\pi}{1-\pi} = e^\theta,$$

so

$$\left(\frac{\pi}{1-\pi}\right)^y = e^{\theta y}, \qquad y \in \mathcal{Y}_n,$$

$$\pi = \frac{e^\theta}{1+e^\theta},$$

and

$$1 - \pi = \frac{1}{1+e^\theta}.$$

Consequently,

$$(1-\pi)^n = e^{-n\log(1+e^\theta)} = e^{-nC(\theta)},$$

where

$$C(\theta) = \log(1 + e^\theta), \qquad \theta \in \mathbb{R}. \tag{10}$$

The probability function of the binomial distribution can now be written as

$$f(y; n, \theta) = e^{\theta y - nC(\theta)} r(y; n), \qquad y \in \mathcal{Y}_n. \tag{11}$$

Let Y now have the Poisson distribution with mean λn, whose probability function is given by

$$f(y) = \frac{(\lambda n)^y}{y!} e^{-\lambda n}, \qquad y \in \{0, 1, 2, \ldots\},$$

and $f(y) = 0$ elsewhere. We can rewrite this probability function as

$$f(y) = \lambda^y e^{-\lambda n} r(y; n), \qquad y \in \mathcal{Y}_n,$$

where $\mathcal{Y}_n = \{0, 1, 2, \ldots\}$ (which actually does not depend on n) and $r(y; n) = n^y / y!$ for $y \in \mathcal{Y}_n$. Consider the parameter $\theta = \log \lambda$, which ranges over all of $\mathbb{R}$ as λ ranges over $(0, \infty)$. Now $\lambda = e^\theta$, so $\lambda^y = e^{\theta y}$ and

$$e^{-\lambda n} = e^{-n e^\theta} = e^{-n C(\theta)},$$

where

$$C(\theta) = e^\theta, \qquad \theta \in \mathbb{R}. \tag{12}$$

The probability function of the Poisson distribution can now be written as in (11).

Distributions having probability functions or density functions that can be written in the form given by (11) are referred to as one-parameter *exponential families*. We refer to n as the (known) time parameter of the distribution, to θ as the (unknown) *canonical parameter*, and to $C(\theta)$, $\theta \in \mathbb{R}$, as the *normalizing function*. [We multiply the function $e^{\theta y} r(y; n)$, $y \in \mathcal{Y}_n$, by $e^{-n C(\theta)}$ in order to normalize it—that is, to make it have sum 1.] As we have already seen, binomial and Poisson distributions are examples of such families. Some other examples are the gamma distribution with time parameter $n = \alpha$ and canonical parameter $\theta = -\lambda \in (-\infty, 0)$, where λ is the inverse-scale parameter of the distribution, and the normal distribution with mean $n\mu$ and variance $n\sigma^2$ (σ known), which has time parameter n and canonical parameter $\theta = \mu / \sigma^2 \in \mathbb{R}$.

Consider now either the binomial distribution or the Poisson distribution, let its probability function be written in the exponential family form (11), and let Y be a random variable having this probability function. Since $\sum_y f(y; n, \theta) = 1$ for $\theta \in \mathbb{R}$, we conclude from (11) that $\sum_y e^{\theta y} r(y; n) < \infty$ for $\theta \in \mathbb{R}$ and, explicitly, that

$$\sum_y e^{\theta y} r(y; n) = e^{n C(\theta)}, \qquad \theta \in \mathbb{R}. \tag{13}$$

Thus,

$$n C(\theta) = \log \left(\sum_y e^{\theta y} r(y; n) \right), \qquad \theta \in \mathbb{R}. \tag{14}$$

It follows from (14) or directly from the formulas for the normalizing functions for the binomial and Poisson families that the normalizing function is infinitely differentiable; that is, all its derivatives exist and are continuous. Moreover, it follows from (13) and (14) by differentiation that

$$n C'(\theta) = \frac{\sum_y y e^{\theta y} r(y; n)}{\sum_y e^{\theta y} r(y; n)} = \sum_y y e^{\theta y - n C(\theta)} r(y; n), \qquad \theta \in \mathbb{R}, \tag{15}$$

and hence that Y has mean $n\mu$, where

$$\mu = C'(\theta) \tag{16}$$

is the mean of Y when $n = 1$. Differentiating (15), we get

$$nC''(\theta) = \frac{\sum_y y^2 e^{\theta y} r(y; n)}{\sum_y e^{\theta y} r(y; n)} - \left(\frac{\sum_y y e^{\theta y} r(y; n)}{\sum_y e^{\theta y} r(y; n)} \right)^2$$

$$= \sum_y y^2 e^{\theta y - nC(\theta)} r(y; n) - \left(\sum_y y e^{\theta y - nC(\theta)} r(y; n) \right)^2$$

$$= E(Y^2) - (EY)^2$$

and hence that Y has variance $n\sigma^2$, where

$$\sigma^2 = C''(\theta) \tag{17}$$

is the variance of Y when $n = 1$.

Since binomial and Poisson distributions have positive variance (we require that $n \geq 1$ and $0 < \pi < 1$ in the definition of the binomial distribution and that $n > 0$ and $\lambda > 0$ in the definition of the Poisson distribution), we conclude from (17) that

$$C''(\theta) > 0, \qquad \theta \in \mathbb{R}.$$

Thus, $C'(\cdot)$ is a continuous and strictly increasing function on $\mathbb{R}$. Consequently, it has an inverse function $(C')^{-1}(\cdot)$, whose domain is an open interval in $\mathbb{R}$ (the interval of all possible values of the mean μ of the distribution when $n = 1$) and whose range is all of $\mathbb{R}$. This function is referred to as the *link function* because it links the canonical parameter θ of the distribution to its mean μ when $n = 1$; that is, by (16),

$$\theta = (C')^{-1}(\mu). \tag{18}$$

EXAMPLE 12.14   Consider the binomial distribution.

(a) Determine $C'(\cdot)$, show directly that this function is strictly increasing, and determine its range.

(b) Determine $C''(\cdot)$ and show directly that this function is positive.

(c) Determine the link function.

(d) Use (16) and (17) to show that the binomial distribution with parameters n and π has mean $n\pi$ and variance $n\pi(1 - \pi)$.

Solution   (a) It follows from (10) that

$$C'(\theta) = \frac{e^\theta}{1 + e^\theta}, \qquad \theta \in \mathbb{R}, \tag{19}$$

which can be rewritten as

$$C'(\theta) = 1 - \frac{1}{1 + e^\theta}, \qquad \theta \in \mathbb{R}. \tag{20}$$

As θ increases, e^θ increases, so $1/(1 + e^\theta)$ decreases and hence $C'(\theta)$ increases. Now $C'(\theta) \to 0$ as $\theta \to -\infty$, and $C'(\theta) \to 1$ as $\theta \to \infty$. Since $C'(\cdot)$ is continuous and increasing, it ranges over $(0, 1)$ as θ ranges over $\mathbb{R}$.

(b) It follows from (19) or (20) that

$$C''(\theta) = \frac{e^\theta}{(1 + e^\theta)^2} = \left(\frac{e^\theta}{1 + e^\theta}\right)\left(1 - \frac{e^\theta}{1 + e^\theta}\right), \qquad \theta \in \mathbb{R},$$

and hence that $C''(\theta) > 0$ for $\theta \in \mathbb{R}$.

(c) Since $C'(\theta)$ ranges over $(0, 1)$ as θ ranges over $\mathbb{R}$, the domain of the link function $(C')^{-1}(\cdot)$ is $(0, 1)$. Given $t \in (0, 1)$, the equation $C'(\theta) = t$ specializes to $e^\theta/(1 + e^\theta) = t$, whose unique solution is given by $\theta = \text{logit}\, t$. Thus, the link function is given by $(C')^{-1}(t) = \text{logit}\, t$ for $0 < t < 1$.

(d) The distribution has mean

$$n\mu = nC'(\theta) = n\frac{e^\theta}{1 + e^\theta} = n\pi$$

and variance

$$n\sigma^2 = nC''(\theta) = n\frac{e^\theta}{1 + e^\theta}\left(1 - \frac{e^\theta}{1 + e^\theta}\right) = n\pi(1 - \pi). \qquad \blacksquare$$

EXAMPLE **12.15** Consider the Poisson distribution.

(a) Determine $C'(\cdot)$, show directly that this function is strictly increasing, and determine its range.

(b) Determine $C''(\cdot)$ and show directly that this function is positive.

(c) Determine the link function.

(d) Use (16) and (17) to show that the Poisson distribution with parameters n (time parameter) and λ has mean $n\lambda$ and variance $n\lambda$.

Solution (a) It follows from (12) that

$$C'(\theta) = e^\theta, \qquad \theta \in \mathbb{R}, \tag{21}$$

which is a strictly increasing function of θ. Now $C'(\theta) \to 0$ as $\theta \to -\infty$, and $C'(\theta) \to \infty$ as $\theta \to \infty$. Since $C'(\cdot)$ is continuous and increasing, $C'(\theta)$ ranges over $(0, \infty)$ as θ ranges over $\mathbb{R}$.

(b) It follows from (21) that $C''(\theta) = e^\theta$ for $\theta \in \mathbb{R}$ and hence that $C''(\theta) > 0$ for $\theta \in \mathbb{R}$.

(c) Since $C'(\theta)$ ranges over $(0, \infty)$ as θ ranges over $\mathbb{R}$, the domain of the link function $(C')^{-1}(\cdot)$ is $(0, \infty)$. Given $t \in (0, \infty)$, the equation $C'(\theta) = t$ specializes to $e^\theta = t$, the unique solution to which is given by $\theta = \log t$. Thus, the link function is given by $(C')^{-1}(t) = \log t$ for $0 < t < \infty$.

(d) The distribution has mean $n\mu = nC'(\theta) = ne^\theta = n\lambda$ and variance $n\sigma^2 = nC''(\theta) = ne^\theta = n\lambda$. $\blacksquare$

Inference Involving Canonical Parameters

Consider now the binomial or Poisson d-sample model and let $1 \leq k \leq d$. Then the mean μ_k and variance σ_k^2 can be expressed in terms of the canonical parameter $\theta_k = (C')^{-1}(\mu_k)$ by $\mu_k = C'(\theta_k)$ and $\sigma_k^2 = C''(\theta_k)$.

The canonical parameter θ_k is estimated by $\hat{\theta}_k = (C')^{-1}(\hat{\mu}_k) = (C')^{-1}(\bar{Y}_k)$. To obtain the asymptotic variance and standard error of this estimate, we need to determine the derivative of the link function $(C')^{-1}(\cdot)$. To this end, we observe that $C'((C')^{-1}(\mu)) = \mu$ [where μ ranges over $(0, 1)$ in the context of binomial models and over $(0, \infty)$ in the context of Poisson models]. Thus, by the chain rule,

$$C''((C')^{-1}(\mu))\frac{d}{d\mu}(C')^{-1}(\mu) = 1,$$

and hence

$$\frac{d}{d\mu}(C')^{-1}(\mu) = \frac{1}{C''((C')^{-1}(\mu))}.$$

According to (6) in Section 12.1, the asymptotic variance of $\hat{\theta}_k$ is now given by

$$\mathrm{AV}(\hat{\theta}_k) = \frac{1}{\left[C''((C')^{-1}(\mu_k)) \right]^2} \frac{\sigma_k^2}{n_k}$$

and hence by

$$\mathrm{AV}(\hat{\theta}_k) = \frac{1}{n_k \sigma_k^2}.$$

Thus, the standard error of $\hat{\theta}_k$ is given by

$$\mathrm{SE}(\hat{\theta}_k) = \frac{1}{\hat{\sigma}_k \sqrt{n_k}}, \tag{22}$$

where $\hat{\sigma}_k^2 = C''(\hat{\theta}_k)$.

Suppose $d \geq 2$ and let $1 \leq j, k \leq d$ with $j \neq k$. Then the asymptotic variance of the estimate $\hat{\theta}_k - \hat{\theta}_j$ of $\theta_k - \theta_j$ is given by

$$\mathrm{AV}(\hat{\theta}_k - \hat{\theta}_j) = \frac{1}{n_j \sigma_j^2} + \frac{1}{n_k \sigma_k^2},$$

so its standard error is given by

$$\mathrm{SE}(\hat{\theta}_k - \hat{\theta}_j) = \sqrt{\frac{1}{n_j \hat{\sigma}_j^2} + \frac{1}{n_k \hat{\sigma}_k^2}}.$$

EXAMPLE **12.16** In the context of the binomial d-sample model, determine the asymptotic variance and standard error of

(a) $\hat{\theta}_k$;   **(b)** $\hat{\theta}_k - \hat{\theta}_j$ for $j \neq k$.

Solution **(a)** Here

$$\text{AV}(\hat{\theta}_k) = \frac{1}{n_k \sigma_k^2} = \frac{1}{n_k \pi_k (1 - \pi_k)},$$

so

$$\text{SE}(\hat{\theta}_k) = \sqrt{\frac{1}{n_k \hat{\pi}_k (1 - \hat{\pi}_k)}}.$$

(b) Here

$$\text{AV}(\hat{\theta}_k - \hat{\theta}_j) = \frac{1}{n_j \sigma_j^2} + \frac{1}{n_k \sigma_k^2} = \frac{1}{n_j \pi_j (1 - \pi_j)} + \frac{1}{n_k \pi_k (1 - \pi_k)},$$

so

$$\text{SE}(\hat{\theta}_k - \hat{\theta}_j) = \sqrt{\frac{1}{n_j \hat{\pi}_j (1 - \hat{\pi}_j)} + \frac{1}{n_k \hat{\pi}_k (1 - \hat{\pi}_k)}}. \qquad \blacksquare$$

The usual nominal $100(1 - \alpha)\%$ confidence interval for θ_k equals $\hat{\theta}_k \pm z_{1-\alpha/2} \, \text{SE}(\hat{\theta}_k)$, where $\text{SE}(\hat{\theta}_k)$ is given by (22); that is,

$$P\Big(\hat{\theta}_k - z_{1-\alpha/2} \, \text{SE}(\hat{\theta}_k) < \theta_k < \hat{\theta}_k + z_{1-\alpha/2} \, \text{SE}(\hat{\theta}_k)\Big) \approx 1 - \alpha \qquad \text{for } n_k \gg 1. \quad \textbf{(23)}$$

Recall that $C'(\cdot)$ is a continuous and strictly increasing function. Thus,

$$C'(\hat{\theta}_k - z_{1-\alpha/2} \, \text{SE}(\hat{\theta}_k)) < C'(\theta_k) < C'(\hat{\theta}_k + z_{1-\alpha/2} \, \text{SE}(\hat{\theta}_k))$$

if and only if

$$\hat{\theta}_k - z_{1-\alpha/2} \, \text{SE}(\hat{\theta}_k) < \theta_k < \hat{\theta}_k + z_{1-\alpha/2} \, \text{SE}(\hat{\theta}_k). \qquad \textbf{(24)}$$

Since $\mu_k = C'(\theta_k)$, we conclude that

$$C'(\hat{\theta}_k - z_{1-\alpha/2} \, \text{SE}(\hat{\theta}_k)) < \mu_k < C'(\hat{\theta}_k + z_{1-\alpha/2} \, \text{SE}(\hat{\theta}_k))$$

if and only if (24) holds. Consequently, by (23),

$$P\Big(C'(\hat{\theta}_k - z_{1-\alpha/2} \, \text{SE}(\hat{\theta}_k)) < \mu_k < C'(\hat{\theta}_k + z_{1-\alpha/2} \, \text{SE}(\hat{\theta}_k))\Big)$$
$$\approx 1 - \alpha \qquad \text{for } n_k \gg 1. \quad \textbf{(25)}$$

According to (25), we can regard

$$\Big(C'(\hat{\theta}_k - z_{1-\alpha/2} \, \text{SE}(\hat{\theta}_k)), \; C'(\hat{\theta}_k + z_{1-\alpha/2} \, \text{SE}(\hat{\theta}_k))\Big)$$

as an *indirect* nominal $100(1 - \alpha)\%$ confidence interval for μ_k.

The *direct* nominal $100(1 - \alpha)\%$ confidence interval for μ_k equals

$$\Big(\hat{\mu}_k - z_{1-\alpha/2} \, \text{SE}(\hat{\mu}_k), \; \hat{\mu}_k + z_{1-\alpha/2} \, \text{SE}(\hat{\mu}_k)\Big).$$

Now $\hat{\mu}_k = C'(\hat{\theta}_k)$, and $\text{SE}(\hat{\mu}_k) = C''(\hat{\theta}_k) \, \text{SE}(\hat{\theta}_k)$ (the justification being left as a problem), so the direct nominal $100(1 - \alpha)\%$ confidence interval for μ_k can be written as

$$\left(C'(\hat{\theta}_k) - z_{1-\alpha/2} C''(\hat{\theta}_k) \, \mathrm{SE}(\hat{\theta}_k), \; C'(\hat{\theta}_k) + z_{1-\alpha/2} C''(\hat{\theta}_k) \, \mathrm{SE}(\hat{\theta}_k) \right).$$

Thus, the direct confidence interval for μ_k can be obtained from the indirect confidence interval by replacing $C'(\cdot)$ by its linear Taylor approximation about $\hat{\theta}_k$:

$$C'(\hat{\theta}_k + t) \approx C'(\hat{\theta}_k) + C''(\hat{\theta}_k)t.$$

When $n_k \gg 1$, the two confidence intervals for μ_k will nearly coincide, but if n_k is small or only moderately large, the discrepancy between the two confidence intervals can be of practical importance. (Similar remarks apply to confidence bounds.)

EXAMPLE **12.17**   Consider the pea plant experiment that was discussed in Example 12.6 in Section 12.1 and set $\theta = \mathrm{logit}\,\pi$.

(a) Determine the nominal 95% confidence interval for θ.

(b) Use the solution to part (a) to determine nominal 95% confidence intervals for π and $\pi/(1 - \pi)$.

(c) Determine the P-value for the Wald test of the hypothesis that $\theta = \mathrm{logit}(1/4) = \log(1/3)$, which is equivalent to the hypothesis that $\pi = 1/4$.

(d) Discuss the solutions to parts (b) and (c).

Solution   (a) Now $n = 100$ and $\hat{\pi} = .35$ according to Example 12.6 in Section 12.1 and its solution. Thus, $\hat{\theta} = \mathrm{logit}\,\hat{\pi} = \log(.35/.65) \doteq -0.619$, and, by the solution to Example 12.16(a),

$$\mathrm{SE}(\hat{\theta}) = \frac{1}{\sqrt{100(.35)(.65)}} \doteq 0.210.$$

Consequently, the nominal 95% confidence interval for θ is given by

$$-0.619 \pm (1.96)(0.210) \doteq -0.619 \pm 0.411 = (-1.030, -0.208).$$

(b) Using the nominal 95% confidence interval for θ obtained in part (a), we get the nominal 95% confidence interval

$$\left(\frac{\exp(-1.030)}{1 + \exp(-1.030)}, \; \frac{\exp(-0.208)}{1 + \exp(-0.208)} \right) \doteq (.263, .448)$$

for π. Since $\pi/(1 - \pi)$ is a continuous and strictly increasing function of π, we get the corresponding nominal 95% confidence interval

$$\left(\frac{.263}{1 - .263}, \; \frac{.448}{1 - .448} \right) \doteq (0.357, 0.812)$$

for $\pi/(1 - \pi)$.

(c) The Wald statistic for testing the hypothesis that $\theta = \mathrm{logit}(1/4) = -\log 3$ is given by

$$W \doteq \frac{-0.619 + \log 3}{0.210} \doteq 2.287,$$

so the P-value for the test is given by $2[\,1 - \Phi(2.287)\,] \doteq .0222$.

(d) The confidence interval $(.263, .448)$ for π obtained in the solution to part (b) is close to the interval $(.257, .443)$ obtained in the solution to Example 12.6(a) in Section 12.1. The confidence interval $(0.357, 0.812)$ for the odds $\pi/(1 - \pi)$ obtained in the solution to part (c) differs moderately from the interval $(0.317, 0.760)$ obtained in the solution to Example 12.6(b) in Section 12.1.

We have by now obtained four P-values for testing the hypothesis that $\pi = 1/4$ or its equivalent. Specifically, in the solution to Example 12.6 in Section 12.1, we obtained the P-value of .0360 for the Wald test of the hypothesis that $\pi = 1/4$ and the P-value of .0692 for the Wald test of the hypothesis that $\pi/(1 - \pi) = 1/3$. In the solution to Example 12.17(c), we obtained the P-value of .0222 for the Wald test of the hypothesis that $\theta = \text{logit}(1/4)$. Finally, in the solution to Example 12.10 in Section 12.2, we obtained the exact P-value of .0275. Finding the exact P-value initially more credible than those corresponding to the three Wald tests, we are led to discarding the P-value based on the Wald test of the hypothesis that $\pi/(1 - \pi) = 1/3$ and viewing the P-values for the other two Wald tests as being reasonable alternatives to the exact P-value. This suggests that we also discard the confidence interval $(0.317, 0.760)$ for $\pi/(1 - \pi)$ obtained in the solution to Example 12.6(b) in Section 12.1 and use either the confidence interval $(0.357, 0.812)$ obtained in the solution to Example 12.17(b) or the confidence interval

$$\left(\frac{.257}{1 - .257}, \frac{.443}{1 - .443} \right) \doteq (0.345, 0.797)$$

for $\pi/(1 - \pi)$ that is obtained by transforming the confidence interval $(.257, .443)$ for π that was found in the solution to Example 12.6(a) in Section 12.1. ∎

Let $1 \le j, k \le d$ with $j \ne k$. In the context of the binomial d-sample model,

$$\theta_j = \log \frac{\pi_j}{1 - \pi_j} \quad \text{and} \quad \theta_k = \log \frac{\pi_k}{1 - \pi_k},$$

so

$$\theta_k - \theta_j = \log \frac{\pi_k/(1 - \pi_k)}{\pi_j/(1 - \pi_j)}$$

and hence the odds ratio is given by

$$\frac{\pi_k/(1 - \pi_k)}{\pi_j/(1 - \pi_j)} = \exp(\theta_k - \theta_j).$$

In the context of the Poisson d-sample model, $\theta_j = \log \lambda_j$ and $\theta_k = \log \lambda_j$, so $\theta_k - \theta_j = \log(\lambda_k/\lambda_j)$, and hence the *relative rate* $\lambda_k/\lambda_j = \mu_k/\mu_j$ is given by $\lambda_k/\lambda_j = \exp(\theta_k - \theta_j)$. Since the exponential function is continuous and strictly increasing, confidence bounds and confidence intervals for $\theta_k - \theta_j$ can be used to obtain confidence bounds and confidence intervals for $\exp(\theta_k - \theta_j)$. Thus, they can be used to obtain confidence bounds and confidence intervals for odds ratios in the context of multisample binomial models and for relative rates in the context of multisample Poisson models.

EXAMPLE **12.18**   Consider the vitamin C experiment that was introduced in Example 12.7 in Section 12.1 and set $\theta_1 = \operatorname{logit} \pi_1$ and $\theta_2 = \operatorname{logit} \pi_2$.

(a) Determine the nominal 95% confidence interval for $\theta_2 - \theta_1$.

(b) Use the solution to part (a) to determine a nominal 95% confidence interval for the odds ratio

$$\frac{\pi_2/(1 - \pi_2)}{\pi_1/(1 - \pi_1)}.$$

(c) Determine the P-value for the Wald test of the hypothesis that $\theta_1 = \theta_2$, which is equivalent to the hypothesis that $\pi_1 = \pi_2$.

(d) Discuss the solutions to parts (b) and (c).

Solution   (a) Here $n_1 = 140$, $n_2 = 139$, $\hat{\pi}_1 \doteq .221$, and $\hat{\pi}_2 \doteq .122$. Thus, $\hat{\theta}_1 \doteq \log \frac{.221}{.779} \doteq -1.257$ and $\hat{\theta}_2 \doteq \log \frac{.122}{.878} \doteq -1.971$, so $\theta_2 - \theta_1$ is estimated by $\hat{\theta}_2 - \hat{\theta}_1 \doteq -0.713$. According to the solution to Example 12.16(b), the standard error of this estimate is given by

$$\operatorname{SE}(\hat{\theta}_2 - \hat{\theta}_1) \doteq \sqrt{\frac{1}{140(.221)(.779)} + \frac{1}{139(.122)(.878)}} \doteq 0.329.$$

Thus, the nominal 95% confidence interval for $\theta_2 - \theta_1$ is given by

$$-0.713 \pm (1.96)(0.329) \doteq -0.713 \pm 0.645 \doteq (-1.359, -0.068).$$

(b) The nominal 95% confidence interval for the odds ratio corresponding to the confidence interval for $\theta_2 - \theta_1$ obtained in the solution to part (a) is given by

$$\Big(\exp(-1.359), \ \exp(-0.068) \Big) \doteq (0.257, 0.934).$$

(c) The Wald statistic for testing the hypothesis that $\theta_1 = \theta_2$ is given by

$$W \doteq -\frac{0.713}{0.329} = -2.166,$$

so the P-value for the test is given by P-value $\doteq 2\Phi(-2.166) \doteq .030$.

(d) The P-value of .030 in the solution to part (c) is less than .05, which is equivalent to the result that zero lies in the exterior of the nominal 95% confidence interval for $\theta_2 - \theta_1$. According to this analysis, the experimental data provide weak evidence that taking the vitamin C supplement affects the probability of catching a cold under the stated conditions. The P-value .030 found here, based on the Wald test corresponding to the estimate $\hat{\theta}_2 - \hat{\theta}_1$ of the difference between the canonical parameters θ_1 and θ_2, is smaller than the exact P-value of .0385 that was obtained in the solution to Example 12.11(a) in Section 12.2, but larger than any of the three P-values (.0268, .0034, and .0016) corresponding to the various Wald tests that were obtained in the solution to Example 12.7 in Section 12.1. Thus, at least in this example, we are led to thinking of the P-value for the Wald test based directly on the canonical parameters θ_1 and θ_2 as more credible than the P-values for Wald tests based directly on π_1 and π_2. This suggests that we also use the nominal confidence interval for the odds ratio found in the solution to part (b) rather than the interval $(0.174, 0.806)$ found in the solution to Example 12.7(c) in Section 12.1. ∎

Problems

12.28 In the context of Problem 12.1 in Section 12.1, set $\theta = \text{logit}\, \pi$. **(a)** Determine the nominal 95% confidence interval for θ and use the result to get a nominal 95% confidence interval for π. **(b)** Determine the nominal 95% lower confidence bound for θ and use the result to get a nominal 95% lower confidence bound for π. **(c)** Determine the P-value for the Wald test of the hypothesis that $\theta \le \log 3$ or, equivalently, that $\pi \le .75$.

12.29 Given $\pi \in (0, 1)$, set $\theta = \text{logit}\, \pi$. **(a)** Verify that $e^\theta = \frac{\pi}{1-\pi}$. **(b)** Verify that $\pi = (1 - \pi)e^\theta$. **(c)** Verify that the following are equivalent: $\pi \approx 0$; $1 - \pi \approx 1$; $\pi/e^\theta \approx 1$.

12.30 Given $\pi_1, \pi_2 \in (0, 1)$, set $\theta_1 = \text{logit}\, \pi_1$ and $\theta_2 = \text{logit}\, \pi_2$. Verify that **(a)**

$$\frac{\pi_2}{\pi_1} = \frac{1 - \pi_2}{1 - \pi_1} \exp(\theta_2 - \theta_1);$$

(b) if $\pi_1 \approx 0$ and $\pi_2 \approx 0$, then $\pi_2/\pi_1 \approx \exp(\theta_2 - \theta_1)$.

12.31 (Continued.) Suppose $\pi_1 = 1/N$, where $N \gg 1$, and $\theta_2 - \theta_1 = \log 2 \doteq 0.693$. According to Problem 12.30(b), $\pi_2/\pi_1 \approx 2$. Determine the actual value of π_2/π_1.

12.32 In the context of the Poisson d-sample model, determine the asymptotic variance and standard error of **(a)** $\hat{\theta}_k$; **(b)** $\hat{\theta}_k - \hat{\theta}_j$ for $j \neq k$.

12.33 Justify the formulas **(a)** $\hat{\mu}_k = C'(\hat{\theta}_k)$; **(b)** $\hat{\sigma}_k^2 = C''(\hat{\theta}_k)$; **(c)** $\text{AV}(\hat{\mu}_k) = [C''(\theta_k)]^2 \, \text{AV}(\hat{\theta}_k) = \sigma_k^4 \, \text{AV}(\hat{\theta}_k)$; **(d)** $\text{SE}(\hat{\mu}_k) = C''(\hat{\theta}_k)\, \text{SE}(\hat{\theta}_k) = \hat{\sigma}_k^2 \, \text{SE}(\hat{\theta}_k)$.

12.34 In the context of the vitamin C experiment that was introduced in Example 12.7 in Section 12.1, **(a)** determine the nominal 95% upper confidence bound for $\theta_2 - \theta_1$; **(b)** use the solution to part (a) to obtain a nominal 95% upper confidence bound for the odds ratio

$$\frac{\pi_2/(1 - \pi_2)}{\pi_1/(1 - \pi_1)};$$

(c) determine the P-value for the Wald test of the hypothesis that $\theta_2 \ge \theta_1$, which is equivalent to the hypothesis that $\pi_2 \ge \pi_1$; **(d)** discuss the solutions to parts (b) and (c), especially in light of the solutions to Example 12.8 in Section 12.1 and Example 12.11(b) in Section 12.2.

12.35 Consider the setup of Problem 12.6 in Section 12.1, which involves the Poisson two-sample model, and set $\theta_1 = \log \lambda_1$ and $\theta_2 = \log \lambda_2$. **(a)** Determine the nominal 95% confidence interval for $\theta_2 - \theta_1$. [Use the answer to Problem 12.32(b).] **(b)** Use the solution to part (a) to determine a nominal 95% confidence interval for the relative rate λ_2/λ_1. **(c)** Determine the P-value for the Wald test of the hypothesis that $\theta_1 = \theta_2$, which is equivalent to the hypothesis that $\lambda_1 = \lambda_2$ or that $\lambda_2/\lambda_1 = 1$.

(d) Discuss the solutions to parts (b) and (c), especially in light of the solutions to Problem 12.6 in Section 12.1 and Problem 12.20(a) in Section 12.2.

12.36   Consider the gamma distribution with shape parameter $n = \alpha > 0$ and inverse-scale parameter $\lambda > 0$, whose density function is given by

$$f(y) = \frac{\lambda^n y^{n-1} e^{-\lambda y}}{\Gamma(n)}, \qquad y > 0.$$

This density function can be written in the exponential family form (11) with $r(y; n) = y^{n-1}/\Gamma(n)$ for $y \in \mathcal{Y}_n = (0, \infty)$. **(a)** Determine the canonical parameter θ in terms of λ and determine the range of θ. **(b)** Determine λ in terms of θ. **(c)** Determine the normalizing function $C(\cdot)$. **(d)** Determine $C'(\cdot)$, show directly that this function is strictly increasing, and determine its range. **(e)** Determine $C''(\cdot)$ and show directly that this function is positive. **(f)** Determine the link function. **(g)** Use (16) and (17) to determine the mean and variance of the gamma distribution with shape parameter n and inverse-scale parameter λ.

12.37   Consider the normal distribution with mean $n\mu$ and variance $n\sigma^2$, where $n > 0$ and $\sigma^2 > 0$ are known. The corresponding density function is given by

$$f(y) = \frac{1}{\sigma\sqrt{2\pi n}} \exp\left(-\frac{(y - n\mu)^2}{2n\sigma^2}\right), \qquad -\infty < y < \infty.$$

This density function can be written in the exponential family form (11) with

$$r(y; n) = \frac{1}{\sigma\sqrt{2\pi n}} \exp\left(-\frac{y^2}{2n\sigma^2}\right), \qquad y \in \mathcal{Y}_n = (-\infty, \infty).$$

(a) Determine the canonical parameter θ in terms of μ (and σ^2) and determine the range of θ. **(b)** Determine μ in terms of θ. **(c)** Determine the normalizing function $C(\cdot)$. **(d)** Determine $C'(\cdot)$, show directly that this function is strictly increasing, and determine its range. **(e)** Determine $C''(\cdot)$ and show directly that this function is positive. **(f)** Determine the link function. **(g)** Use (16) and (17) to confirm that the indicated normal distribution does indeed have mean $n\mu$ and variance $n\sigma^2$.

12.38   Let $f(\cdot; \theta)$, $\theta \in \Theta$, be probability functions having the exponential family form (11) with fixed $n > 0$, where Θ is an open interval in $\mathbb{R}$ and $\#(\mathcal{Y}_n) \geq 2$. **(a)** Show that

$$\frac{f(y_2; \theta_2)/f(y_2; \theta_1)}{f(y_1; \theta_2)/f(y_1; \theta_1)} = \exp((\theta_2 - \theta_1)(y_2 - y_1)), \quad y_1, y_2 \in \mathcal{Y}_n \text{ and } \theta_1, \theta_2 \in \Theta.$$

(b) Use the result of part (a) to show that the family satisfies the *monotone likelihood-ratio property*: $f(y; \theta_2)/f(y; \theta_1)$ is a strictly increasing function of y for $\theta_1, \theta_2 \in \Theta$ with $\theta_1 < \theta_2$. **(c)** Use the result of part (a) to show that if $\theta_1 \neq \theta_2$, then $f(\cdot; \theta_2)$ is *not* the same probability function as $f(\cdot; \theta_1)$. (We refer to this property by saying that the parameter θ is identifiable.)

13

Logistic Regression and Poisson Regression

We turn to the study of response variables having a binomial or Poisson distribution whose unknown parameter π or λ depends on the levels of various factors. Consider first a response variable having a binomial distribution with π depending on the level x of a single factor. (Think of x as the number of cigarettes smoked per day by an individual and of π as the probability that the individual gets lung cancer during a given year.) Let $\pi(x)$ denote the dependence of π on x and let $\theta(x)$ similarly denote the dependence of the canonical parameter $\theta = \text{logit}\,\pi$ on x. We refer to $\pi(\cdot)$ as the risk function and to $\theta(\cdot)$ as the canonical regression function. We could attempt to model π as a linear function of x: $\pi(x) = \beta_0 + \beta_1 x$ for $x \in \mathbb{R}$. A serious defect of this approach, however, is that if $\beta_1 \neq 0$, then $\beta_0 + \beta_1 x$ is greater than 1 for some values of x and less than 0 for other values of x, whereas $\pi(x)$ must lie between 0 and 1 for all values of x.

A reasonable way to avoid this defect is to model the canonical parameter as a linear function of x: $\theta(x) = \beta_0 + \beta_1 x$ for $x \in \mathbb{R}$. The motivation here is that the resulting model for the risk function will at least have the property of ranging over $(0, 1)$. Specifically, since the solution to the equation $\theta = \text{logit}\,\pi$ is given by $\pi = (\exp\theta)/(1 + \exp\theta)$, we get

$$\pi(x) = \frac{\exp(\beta_0 + \beta_1 x)}{1 + \exp(\beta_0 + \beta_1 x)},$$

which does indeed range over $(0, 1)$ as x ranges over $\mathbb{R}$.

Similarly, consider a response variable having a Poisson distribution with parameter λ depending on x. (Think of x as the amount of alcohol consumed per day by a motorist, which is assumed not to vary over time, and of λ as the mean number of automobile accidents that the motorist has per year.) Let $\lambda(x)$ denote the dependence of λ on x and let $\theta(x)$ denote the dependence of the canonical parameter $\theta = \log \lambda$ on x. We refer to $\lambda(\cdot)$ as the rate function and, as before, to $\theta(\cdot)$ as the canonical regression function. We could model λ as a linear function of x: $\lambda(x) = \beta_0 + \beta_1 x$ for $x \in \mathbb{R}$. This approach is defective, however, in that if $\beta_1 \neq 0$, then $\beta_0 + \beta_1 x$ is negative for some values of x, whereas $\lambda(x)$ must be positive for all values of x. Again, one remedy is to model the canonical parameter as a linear function of x. Since the solution to the equation $\theta = \log \lambda$ is given by $\lambda = e^{\theta}$, we get $\lambda(x) = \exp(\beta_0 + \beta_1 x)$, which ranges over $(0, \infty)$ as x ranges over $\mathbb{R}$.

As was illustrated in the examples and problems in Section 12.3, confidence intervals and tests based on normal approximation in the context of binomial and Poisson models are generally more accurate when applied directly to estimates of canonical parameters than when applied to estimates of probabilities or rates or, especially, ratios of such quantities. This gives another reason for defining the response model directly in terms of the canonical regression function.

In Section 13.1, we describe the input-response and experimental models corresponding to binomial and Poisson response variables and the assumption that the canonical regression function belongs to a given identifiable linear space G. Letting $g_1, \ldots, g_p$ be a basis of G, we write the canonical regression function as $\theta(\cdot) = \beta_1 g_1 + \cdots + \beta_p g_p$, where the unknown parameters $\beta_1, \ldots, \beta_p$ are referred to as the regression coefficients.

In Section 13.2, we describe the maximum-likelihood method, which yields an estimate $\hat{\theta}(\cdot) = \hat{\beta}_1 g_1 + \cdots + \hat{\beta}_p g_p$ of the canonical regression function. The maximum-likelihood estimates $\hat{\beta}_1, \ldots, \hat{\beta}_p$ of the regression coefficients satisfy a system of p nonlinear equations in p unknowns, which can be solved explicitly only when G is saturated. Otherwise, they must be solved numerically. Existence and uniqueness of the maximum-likelihood estimates is discussed in Section 13.3. In Section 13.4, an iterative method for solving the maximum-likelihood equations is presented; in each iteration, a linear system of normal equations must be solved, the system changing from one iteration to the next. [The various desirable properties of maximum-likelihood estimation that are established in Sections 13.3 and 13.4 do not apply if the regression function, $\pi(\cdot)$ or $\lambda(\cdot)$, rather than the canonical regression function $\theta(\cdot)$ is modeled as an unknown member of a known linear space.] Under suitable conditions, the maximum-likelihood estimates of the regression coefficients have an approximately multivariate normal joint distribution. This result is derived in Section 13.5 and then used to obtain nominal confidence intervals for parameters of interest and nominal P-values for the correponding Wald tests. In Section 13.6, we describe the likelihood-ratio test, which plays a role here similar to that of the F test in the context of the normal linear model.

Throughout this chapter, following the pattern of Chapter 12, we emphasize binomial response variables in the text and Poisson response variables in the problems.

13.1

Input-Response and Experimental Models

Input-Response Model

Consider a binomial or Poisson response variable Y. According to the corresponding input-response model, there is a function $\theta(\cdot)$ on $\mathcal{X}$ such that, when the input vector equals $\mathbf{x} \in \mathcal{X}$ and the time parameter equals n, Y has probability function $f(\cdot; n, \theta(\mathbf{x}))$ on $\mathcal{Y}_n$. Here

$$f(y; n, \theta) = e^{\theta y - nC(\theta)} r(y; n), \qquad y \in \mathcal{Y}_n;$$

in the context of a binomial response variable, $\mathcal{Y}_n = \{0, \ldots, n\}$, $r(y; n) = \binom{n}{y}$ for $y \in \mathcal{Y}_n$, and $C(\theta) = \log(1 + e^{\theta})$ for $\theta \in \mathbb{R}$; in the context of a Poisson response variable, $\mathcal{Y}_n = \{0, 1, 2, \ldots\}$, $r(y; n) = n^y/y!$ for $y \in \mathcal{Y}_n$, and $C(\theta) = e^{\theta}$ for $\theta \in \mathbb{R}$. The response variable Y has mean $n\mu(\mathbf{x})$ and variance $n\sigma^2(\mathbf{x})$, where $\mu(\mathbf{x}) = C'(\theta(\mathbf{x}))$ and $\sigma^2(\mathbf{x}) = C''(\theta(\mathbf{x}))$. We refer to $\mu(\cdot)$ as the regression function, to $\sigma^2(\cdot)$ as the *variance function*, to $\theta(\cdot)$ as the *canonical regression function*, and to the overall setup as the input-response version of the *canonical regression model*.

Consider, in particular, a binomial response variable. When the input vector equals $\mathbf{x}$ and the time parameter equals n, the response variable has the binomial distribution with parameters n and

$$\pi(\mathbf{x}) = \mu(\mathbf{x}) = C'(\theta(\mathbf{x})) = \frac{\exp \theta(\mathbf{x})}{1 + \exp \theta(\mathbf{x})}, \qquad \mathbf{x} \in \mathcal{X}.$$

Here

$$\sigma^2(\mathbf{x}) = \pi(\mathbf{x})[1 - \pi(\mathbf{x})], \qquad \mathbf{x} \in \mathcal{X},$$

and

$$\theta(\mathbf{x}) = \operatorname{logit} \pi(\mathbf{x}), \qquad \mathbf{x} \in \mathcal{X}.$$

In this context, we refer to the canonical regression function $\theta(\cdot) = \operatorname{logit} \pi(\cdot)$ as the *logistic regression function* and to the canonical regression model as the *logistic regression model*. Sometimes, we refer to the regression function $\mu(\cdot) = \pi(\cdot)$ as the *risk function*. [Let the input vector equal $\mathbf{x} \in \mathcal{X}$ and think of the response variable as the number of failures in n trials. Then $\pi(\mathbf{x})$ is the risk (probability) of failure on a single trial.]

Let $\mathbf{x}_1, \mathbf{x}_2 \in \mathcal{X}$ be fixed. Then

$$\frac{\pi(\mathbf{x}_1)}{1 - \pi(\mathbf{x}_1)} = \exp \theta(\mathbf{x}_1) \quad \text{and} \quad \frac{\pi(\mathbf{x}_2)}{1 - \pi(\mathbf{x}_2)} = \exp \theta(\mathbf{x}_2)$$

are the odds when the input vectors equal $\mathbf{x}_1$ and $\mathbf{x}_2$, respectively, and

$$\frac{\pi(\mathbf{x}_2)/(1 - \pi(\mathbf{x}_2))}{\pi(\mathbf{x}_1)/(1 - \pi(\mathbf{x}_1))} = \exp(\theta(\mathbf{x}_2) - \theta(\mathbf{x}_1))$$

is the corresponding odds ratio. Suppose, for example that $\theta(\mathbf{x}_2) - \theta(\mathbf{x}_1) = \log 2 \doteq 0.693$. Then the odds ratio equals 2, so changing the input vector from $\mathbf{x}_1$ to $\mathbf{x}_2$ causes the odds to double.

Also, $\pi(\mathbf{x}_1)$ and $\pi(\mathbf{x}_2)$ are the risks when the input vector equals $\mathbf{x}_1$ and $\mathbf{x}_2$, respectively, and $\pi(\mathbf{x}_2)/\pi(\mathbf{x}_1)$ is the corresponding relative risk. This relative risk can be written as

$$\frac{\pi(\mathbf{x}_2)}{\pi(\mathbf{x}_1)} = \frac{1 - \pi(\mathbf{x}_2)}{1 - \pi(\mathbf{x}_1)} \exp(\theta(\mathbf{x}_2) - \theta(\mathbf{x}_1)).$$

If $\pi(\mathbf{x}_1) \approx 0$ and $\pi(\mathbf{x}_2) \approx 0$, then $(1 - \pi(\mathbf{x}_2))/(1 - \pi(\mathbf{x}_1)) \approx 1$ and hence the relative risk is given approximately by

$$\frac{\pi(\mathbf{x}_2)}{\pi(\mathbf{x}_1)} \approx \exp(\theta(\mathbf{x}_2) - \theta(\mathbf{x}_1))$$

(see Problem 12.30 in Section 12.3). Suppose, for example, that $\pi(\mathbf{x}_1) \approx 0$ and $\theta(\mathbf{x}_2) - \theta(\mathbf{x}_1) = \log 2 \doteq 0.693$. Then changing the input vector from $\mathbf{x}_1$ to $\mathbf{x}_2$ causes the risk approximately to double (see Problem 12.31 in Section 12.3).

Accurate estimation of small probabilities and their dependence on various factors is important in many fields, including epidemiology, environmental and industrial safety, manufacturing, marketing, insurance, and banking.

EXAMPLE **13.1**   *Dose-response model.* Consider an experiment in which insects of a certain type receive various dosages of an insecticide. Let the input variable x be the log-dose (natural logarithm of weight in a specified unit) of the insecticide and let the response on a single trial be defined by $I = 1$ if the insect dies within a specified time after receiving the dosage and $I = 0$ otherwise. Let T be the random variable denoting the "tolerance" of a randomly selected insect to the insecticide, defined so that the insect is killed by a log-dose of T or more but survives a log-dose of less than T (that is, $I = 1$ if $x \geq T$, but $I = 0$ if $x < T$).

(a) Express the distribution function F of T in terms of the canonical regression function.

(b) Express the density function of F in terms of the canonical regression function when the latter function is continuously differentiable.

Solution   (a) Now $F(x) = P(T \leq x) = P(I = 1) = \pi(x)$, so

$$F(x) = C'(\theta(x)) = \frac{\exp \theta(x)}{1 + \exp \theta(x)}, \qquad x \in \mathbb{R}.$$

(b) Suppose the canonical regression function is continuously differentiable. Then, by the solution to part (a), F has the density function f given by

$$f(x) = \theta'(x)C''(\theta(x)) = \frac{\theta'(x) \exp \theta(x)}{[1 + \exp \theta(x)]^2}, \qquad x \in \mathbb{R}. \quad \blacksquare$$

Consider next a Poisson response variable. When the input vector equals $\mathbf{x}$ and the time parameter equals n, the response variable has the Poisson distribution with mean $n\lambda(\mathbf{x})$, where

$$\lambda(\mathbf{x}) = \mu(\mathbf{x}) = C'(\theta(\mathbf{x})) = e^{\theta(\mathbf{x})}, \qquad \mathbf{x} \in \mathcal{X}.$$

Here

$$\sigma^2(\mathbf{x}) = \lambda(\mathbf{x}) = C''(\theta(\mathbf{x})) = e^{\theta(\mathbf{x})}, \qquad \mathbf{x} \in \mathcal{X},$$

and

$$\theta(\mathbf{x}) = \log \lambda(\mathbf{x}), \qquad \mathbf{x} \in \mathcal{X}.$$

In this context, we refer to the regression function $\mu(\cdot) = \lambda(\cdot)$ as the *rate function*, to the canonical regression function $\theta(\cdot) = \log \lambda(\cdot)$ as the *Poisson regression function*, and to the canonical regression model as the *Poisson regression model*.

Let G be a p-dimensional linear space of functions on $\mathcal{X}$. The corresponding *linear canonical regression model* consists of the canonical regression model together with the assumption that $\theta(\cdot) \in G$. Let $g_1, \ldots, g_p$ be a basis of G. Under the assumption that $\theta(\cdot) \in G$,

$$\theta(\mathbf{x}) = \beta_1 g_1(\mathbf{x}) + \cdots + \beta_p g_p(\mathbf{x}), \qquad \mathbf{x} \in \mathcal{X},$$

where the regression coefficients $\beta_1, \ldots, \beta_p$ are uniquely determined. Correspondingly,

$$\mu(\mathbf{x}) = C'(\beta_1 g_1(\mathbf{x}) + \cdots + \beta_p g_p(\mathbf{x})), \qquad \mathbf{x} \in \mathcal{X}.$$

(Observe that $\mu(\mathbf{x})$ is a nonlinear function of $\beta_1, \ldots, \beta_p$.) In the context of the *linear logistic regression model*,

$$\pi(\mathbf{x}) = \mu(\mathbf{x}) = \frac{\exp(\beta_1 g_1(\mathbf{x}) + \cdots + \beta_p g_p(\mathbf{x}))}{1 + \exp(\beta_1 g_1(\mathbf{x}) + \cdots + \beta_p g_p(\mathbf{x}))}, \qquad \mathbf{x} \in \mathcal{X};$$

in the context of the *linear Poisson regression model*,

$$\lambda(\mathbf{x}) = \mu(\mathbf{x}) = \exp(\beta_1 g_1(\mathbf{x}) + \cdots + \beta_p g_p(\mathbf{x})), \qquad \mathbf{x} \in \mathcal{X}.$$

EXAMPLE 13.2 Consider the linear logistic regression model in which G is the space of linear functions on $\mathbb{R}$. Let the logistic regression function be written as $\theta(x) = \beta_0 + \beta_1 x$, $x \in \mathbb{R}$.

(a) Express $\pi(\cdot)$ in terms of β_0 and β_1.

(b) Suppose $\beta_1 > 0$. Given $p \in (0, 1)$, determine x_p such that $\pi(x_p) = p$.

(c) Let T be a random variable such that $\pi(x) = P(T \le x)$ for $x \in \mathbb{R}$. Determine the distribution function and density function of T.

(d) Interpret the solution to part (b) in terms of the distribution of T.

Solution (a) Here

$$\pi(x) = \frac{\exp(\beta_0 + \beta_1 x)}{1 + \exp(\beta_0 + \beta_1 x)}, \qquad x \in \mathbb{R}.$$

(b) The equation $\pi(x) = p$ can be written as

$$\frac{\exp(\beta_0 + \beta_1 x)}{1 + \exp(\beta_0 + \beta_1 x)} = p$$

or, equivalently, as $\exp(\beta_0 + \beta_1 x) = p/(1-p)$ or as $\beta_0 + \beta_1 x = \text{logit } p$. Thus, the unique solution to the equation $\pi(x) = p$ is given by

$$x_p = \frac{\text{logit } p - \beta_0}{\beta_1}.$$

(c) According to the solution to Example 13.1, the distribution function of T is given by

$$F(x) = \frac{\exp(\beta_0 + \beta_1 x)}{1 + \exp(\beta_0 + \beta_1 x)}, \qquad x \in \mathbb{R},$$

and its density function is given by

$$f(x) = \frac{\beta_1 \exp(\beta_0 + \beta_1 x)}{[1 + \exp(\beta_0 + \beta_1 x)]^2}, \qquad x \in \mathbb{R}.$$

Thus, T has the logistic distribution with location parameter $-\beta_0/\beta_1$ and inverse-scale parameter β_1.

(d) The quantity x_p obtained in the solution to part (b) is the pth quantile of the distribution of T. [Think of the random variable T as the minimum dose required to get a specified response from an organism selected at random from a certain population. In such a context, the median $x_{.5}$ is sometimes denoted by ED50; here ED stands for *effective dose*. A similar notation is used for other quantiles. Thus, ED90 $= x_{.9}$ is the dose required to get a 90% response rate—that is, a response from 90% of the population. If the organism is an insect, the drug is an insecticide, and a response occurs when the insect is killed by the insecticide, then LD (*lethal dose*) is used instead of ED.] ▪

EXAMPLE 13.3   Consider an experiment designed to compare the probabilities of curing a certain disease using the standard treatment and two new treatments. There are two factors in the experiment. Factor 1 is the treatment factor, which has three levels: levels 0, 1, and 2 correspond to the standard treatment, treatment 1, and treatment 2, respectively. Factor 2 is a quantitative factor (such as the amount of time between the onset of the disease and the beginning of the treatment), which is referred to as the covariate. The response Y at levels x_1 and x_2 of factors 1 and 2, respectively, has the binomial distribution with parameters n and $\pi(x_1, x_2)$. Here n is the number of patients with the disease to whom the indicated factor-level combination is applied, Y is the number of these patients that are cured of the disease, and $\pi(x_1, x_2)$ is the corresponding probability of cure. Suppose the logistic regression function is an additive function of x_1 and x_2 that is linear in x_2. Then it can be written as

$$\theta(x_1, x_2) = \beta_0 + \beta_1 \, \text{ind}(x_1 = 1) + \beta_2 \, \text{ind}(x_1 = 2) + \beta_3 x_2.$$

(a) Interpret the indicated regression coefficients.

(b) Express $\theta(2, x_2) - \theta(1, x_2)$ in terms of the regression coefficients.

(c) Express the hypothesis that the two new treatments are equally effective in curing the disease in terms of the regression coefficients.

Solution **(a)** Here the intercept β_0 is the logit of the probability that the patient is cured of the disease by the standard treatment when the covariate equals zero. The regression coefficient β_1 is the log of the odds ratio for cure (change in the logit of the probability of cure) when we switch from the standard treatment to treatment 1 without changing the covariate, and β_2 is the log of the odds ratio for cure when we switch from the standard treatment to treatment 2 without changing the covariate. The coefficient β_3 is the change in the logit of the probability of cure when we add one unit to the covariate without changing the treatment.

(b) It follows from the formula for the logistic regression function that $\theta(2, x_2) - \theta(1, x_2) = \beta_2 - \beta_1$.

(c) According to the answer to part (b), the two new treatments are equally effective in curing the disease if and only if $\beta_1 = \beta_2$. ∎

Experimental Model

According to the experimental version of the canonical regression model, there are d input vectors $\mathbf{x}_1, \ldots, \mathbf{x}_d$ and corresponding time parameters $n_1, \ldots, n_d$ and response variables $Y_1, \ldots, Y_d$. Set $n = n_1 + \cdots + n_d$. The response variables are independent, and, for $1 \leq k \leq d$, Y_k has probability function $f(\cdot; n_k, \theta(\mathbf{x}_k))$ on $\mathcal{Y}_{n_k}$ (where $\mathcal{Y}_{n_k}$ equals $\{0, \ldots, n_k\}$ in the context of a binomial response variable and it equals $\{0, 1, 2, \ldots\}$ in the context of a Poisson response variable). We assume that the input vectors are distinct (although this assumption is not required for most of our results), and we refer to $\{\mathbf{x}_1, \ldots, \mathbf{x}_d\}$ as the design set.

Let $1 \leq k \leq d$ and set $\theta_k = \theta(\mathbf{x}_k)$, $\mu_k = \mu(\mathbf{x}_k) = C'(\theta(\mathbf{x}_k)) = C'(\theta_k)$, and $\sigma_k^2 = \sigma^2(\mathbf{x}_k) = C''(\theta(\mathbf{x}_k)) = C''(\theta_k)$. Then $\theta_k = (C')^{-1}(\mu_k)$. The random variable Y_k has probability function $f(\cdot; n_k, \theta_k)$, mean $n_k \mu_k$, and variance $n_k \sigma_k^2$, and the random variable $\bar{Y}_k = Y_k / n_k$ has mean μ_k and variance σ_k^2 / n_k. The random variables $\bar{Y}_1, \ldots, \bar{Y}_d$ are independent, and the random variable

$$\bar{Y} = \frac{1}{n} \sum_k Y_k = \frac{1}{n} \sum_k n_k \bar{Y}_k$$

has mean

$$\frac{1}{n} \sum_k n_k \mu_k$$

and variance

$$\frac{1}{n^2} \sum_k n_k \sigma_k^2.$$

Consider, in particular, the experimental version of the logistic regression model. Let $1 \leq k \leq d$ and set

$$\pi_k = \mu_k = C'(\theta_k) = \frac{\exp \theta_k}{1 + \exp \theta_k}.$$

Then $\theta_k = \text{logit}\,\pi_k$ and $\sigma_k^2 = \pi_k(1 - \pi_k)$. The random variable Y_k has the binomial distribution with parameters n_k and π_k and hence mean $n_k\pi_k$ and variance $n_k\pi_k(1 - \pi_k)$, the random variable $\bar{Y}_k$ has mean π_k and variance $\pi_k(1 - \pi_k)/n_k$, and the random variable $\bar{Y}$ has mean

$$\frac{1}{n}\sum_k n_k\pi_k$$

and variance

$$\frac{1}{n^2}\sum_k n_k\pi_k(1 - \pi_k).$$

Consider, instead, the experimental version of the Poisson regression model. Let $1 \le k \le d$ and set $\lambda_k = \mu_k = C'(\theta_k) = \exp\theta_k$. Then $\theta_k = \log\lambda_k$ and $\sigma_k^2 = \lambda_k$. The random variable Y_k has the Poisson distribution with mean $n_k\lambda_k$ and hence variance $n_k\lambda_k$, the random variable $\bar{Y}_k$ has mean λ_k and variance λ_k/n_k, and the random variable $\bar{Y}$ has mean

$$\frac{1}{n}\sum_k n_k\lambda_k$$

and variance

$$\frac{1}{n^2}\sum_k n_k\lambda_k.$$

In either case, set

$$\mathcal{Y} = \{\,[y_1,\ldots,y_d]^T : y_k \in \mathcal{Y}_{n_k} \text{ for } 1 \le k \le d\,\} = \mathcal{Y}_{n_1} \times \cdots \times \mathcal{Y}_{n_d}.$$

Given $\mathbf{y} = [y_1,\ldots,y_d]^T \in \mathcal{Y}$, set $\bar{y}_k = y_k/n_k$ for $1 \le k \le d$ and

$$\bar{y} = \frac{1}{n}\sum_k y_k = \frac{1}{n}\sum_k n_k\bar{y}_k.$$

The probability function of the $\mathcal{Y}$-valued random vector $\mathbf{Y} = [Y_1,\ldots,Y_d]^T$ is given by

$$f(\mathbf{y};\mathbf{n},\theta(\cdot)) = \prod_k f(y_k;n_k,\theta(\mathbf{x}_k)), \qquad \mathbf{y} \in \mathcal{Y},$$

where $\mathbf{n} = [n_1,\ldots,n_d]^T$. Thus,

$$f(\mathbf{y};\mathbf{n},\theta(\cdot)) = \prod_k \exp(\theta(\mathbf{x}_k)y_k - n_k C(\theta(\mathbf{x}_k))r(y_k;n_k))$$

$$= \left(\prod_k \exp(\theta(\mathbf{x}_k)y_k - n_k C(\theta(\mathbf{x}_k)))\right)r(\mathbf{y};\mathbf{n})$$

for $\mathbf{y} \in \mathcal{Y}$, where

$$r(\mathbf{y};\mathbf{n}) = \prod_k r(y_k;n_k), \qquad \mathbf{y} \in \mathcal{Y}.$$

Alternatively,

$$f(\mathbf{y};\mathbf{n},\theta(\cdot)) = \exp\left(\sum_k n_k[\theta(\mathbf{x}_k)\bar{y}_k - C(\theta(\mathbf{x}_k))]\right)r(\mathbf{y};\mathbf{n}), \qquad \mathbf{y} \in \mathcal{Y}. \tag{1}$$

Since $\#(\mathcal{Y}) \geq 2$ and $f(\mathbf{y}; \mathbf{n}, \theta(\cdot))$, $\mathbf{y} \in \mathcal{Y}$, is a positive probability function on $\mathcal{Y}$, we see that $0 < f(\mathbf{y}; \mathbf{n}, \theta(\cdot)) < 1$ for $\mathbf{y} \in \mathcal{Y}$.

Given a vector $\mathbf{v} = [v_1, \ldots, v_d]^T$ and a function f of a single real variable whose domain includes the entries of $\mathbf{v}$, we define $f(\mathbf{v})$ to be the vector that is obtained by applying f to each entry of $\mathbf{v}$; that is, $f(\mathbf{v}) = [f(v_1), \ldots, f(v_d)]^T$.

Set $\boldsymbol{\theta} = [\theta_1, \ldots, \theta_d]^T$, $\boldsymbol{\mu} = [\mu_1, \ldots, \mu_d]^T$, and $\boldsymbol{\sigma}^2 = [\sigma_1^2, \ldots, \sigma_d^2]^T$. Then $\boldsymbol{\mu} = C'(\boldsymbol{\theta})$, $\boldsymbol{\sigma}^2 = C''(\boldsymbol{\theta})$, and $\boldsymbol{\theta} = (C')^{-1}(\boldsymbol{\mu})$. In the context of logistic regression, set $\boldsymbol{\pi} = [\pi_1, \ldots, \pi_d]^T$. Then $\boldsymbol{\mu} = \boldsymbol{\pi} = (\exp \boldsymbol{\theta})/(1 + \exp \boldsymbol{\theta})$, $\boldsymbol{\sigma}^2 = [\pi_1(1 - \pi_1), \ldots, \pi_d(1 - \pi_d)]^T$, and $\boldsymbol{\theta} = \text{logit}\,\boldsymbol{\pi}$. In the context of Poisson regression, set $\boldsymbol{\lambda} = [\lambda_1, \ldots, \lambda_d]^T$. Then $\boldsymbol{\sigma}^2 = \boldsymbol{\mu} = \boldsymbol{\lambda} = \exp \boldsymbol{\theta}$ and $\boldsymbol{\theta} = \log \boldsymbol{\mu}$.

Let $g_1, \ldots, g_p$ be a basis of G. The corresponding $d \times p$ design matrix is given by

$$\mathbf{X} = \begin{bmatrix} g_1(\mathbf{x}_1) & \cdots & g_p(\mathbf{x}_1) \\ \vdots & & \vdots \\ g_1(\mathbf{x}_d) & \cdots & g_p(\mathbf{x}_d) \end{bmatrix}.$$

Suppose the canonical regression function is in G and let $\boldsymbol{\beta}$ be its coefficient vector relative to the indicated basis. Then $\boldsymbol{\theta} = \mathbf{X}\boldsymbol{\beta}$. [Observe that $\sum_j \beta_j g_j(\mathbf{x}_k) = (\mathbf{X}\boldsymbol{\beta})_k$ for $1 \leq k \leq d$.]

Suppose G is identifiable and saturated or, equivalently, that $\mathbf{X}$ is an invertible $d \times d$ matrix. Then

$$\boldsymbol{\beta} = \mathbf{X}^{-1}\boldsymbol{\theta} = \mathbf{X}^{-1}[(C')^{-1}(\boldsymbol{\mu})]. \tag{2}$$

In particular, in the context of logistic regression, $\boldsymbol{\beta} = \mathbf{X}^{-1}\text{logit}(\boldsymbol{\pi})$; in the context of Poisson regression, $\boldsymbol{\beta} = \mathbf{X}^{-1}\log(\boldsymbol{\lambda})$.

The binomial and Poisson d-sample models can be viewed as identifiable and saturated linear canonial regression models. To this end, we set $\mathcal{X} = \{1, \ldots, d\}$ and let G be the space of all functions on $\mathcal{X}$. Choosing the basis $I_1, \ldots, I_d$ of G, where $I_k(x) = \text{ind}(x = k)$, we can write the canonical regression function as $\theta(\cdot) = \beta_1 I_1 + \cdots + \beta_d I_d$. Letting the design points be given by $x_k = k$ for $1 \leq k \leq d$, we get $\mathbf{X} = \mathbf{I}$ (the $d \times d$ identity matrix) and hence from (2) that $\boldsymbol{\beta} = \boldsymbol{\theta} = (C')^{-1}(\boldsymbol{\mu})$ [that is, $\beta_k = \theta_k = (C')^{-1}(\mu_k)$ for $1 \leq k \leq d$].

Normal linear models, linear logistic regression models, and linear Poisson regression models are included in the class of models treated in the book *Generalized Linear Models* by P. McCullagh and J. A. Nelder, 2nd edition (London: Chapman and Hall, 1989).

Random Inputs

So far we have been thinking of the input as being nonrandom. Suppose, instead, that the input is an $\mathcal{X}$-valued random vector $\mathbf{X}$ and that, for $\mathbf{x} \in \mathcal{X}$, the conditional distribution of Y given that $\mathbf{X} = \mathbf{x}$ is, say, Bernoulli with parameter

$$\pi(\mathbf{x}) = P(Y = 1 | \mathbf{X} = \mathbf{x}) = \frac{\exp \theta(\mathbf{x})}{1 + \exp \theta(\mathbf{x})}.$$

We refer to such setup as a logistic regression model with a random input vector. Upon adding the assumption that $\theta(\cdot) \in G$, we get a linear logistic regression model with a random input vector. Given a basis $g_1, \ldots, g_p$ of G, we write the logistic regression function as $\theta(\cdot) = \beta_1 g_1 + \cdots + \beta_p g_p$. Then

$$\pi(\mathbf{x}) = \frac{\exp(\beta_1 g_1(\mathbf{x}) + \cdots + \beta_p g_p(\mathbf{x}))}{1 + \exp(\beta_1 g_1(\mathbf{x}) + \cdots + \beta_p g_p(\mathbf{x}))}, \qquad \mathbf{x} \in \mathcal{X}.$$

Let $(\mathbf{X}_1, Y_1), \ldots, (\mathbf{X}_n, Y_n)$ be a random sample of size n from the joint distribution of $\mathbf{X}$ and Y. Then, for $\mathbf{x}_1, \ldots, \mathbf{x}_n \in \mathcal{X}$ (not necessarily distinct), the conditional distribution of $Y_1, \ldots, Y_n$, given that $\mathbf{X}_1 = \mathbf{x}_1, \ldots, \mathbf{X}_n = \mathbf{x}_n$, is that of n independent random variables, the ith of which is Bernoulli with parameter $\pi(\mathbf{x}_i)$. Except for causal implications, the results for the logistic regression model with deterministic inputs that will be obtained in this chapter have analogs for random inputs. In particular, we could use the sample data to obtain maximum-likelihood estimates $\hat{\beta}_1, \ldots, \hat{\beta}_p$ of $\beta_1, \ldots, \beta_p$ respectively (see Section 13.2) and thereby to obtain the maximum-likelihood estimate

$$\hat{\pi}(\mathbf{x}) = \frac{\exp(\hat{\beta}_1 g_1(\mathbf{x}) + \cdots + \hat{\beta}_p g_p(\mathbf{x}))}{1 + \exp(\hat{\beta}_1 g_1(\mathbf{x}) + \cdots + \hat{\beta}_p g_p(\mathbf{x}))}, \qquad \mathbf{x} \in \mathcal{X},$$

of $\pi(\cdot)$.

As an application, let $Y = 1$ if a borrower defaults on his loan and $Y = 0$ otherwise and let $\mathbf{X}$ involve various characteristics of the borrower: age, sex, marital status, education, annual income, total amount of debt, and so forth. Then $\pi(\mathbf{x})$ is the risk that a borrower having characteristic vector $\mathbf{x}$ would default on his loan. An accurate estimate $\hat{\pi}(\mathbf{x})$ of the risk of default for a loan applicant having characteristic vector $\mathbf{x}$ could help the loan officer in deciding whether to make the loan.

In a similar manner, logistic regression models with random inputs can be applied to problems in medical diagnosis and in optical and speech recognition. For example, we might let $Y = 1$ if a patient entering a hospital complaining of chest pains is going to have a myocardial infarction (heart attack) within a specified time and $Y = 0$ otherwise. An accurate estimate $\hat{\pi}(\mathbf{x})$ of the risk $\pi(\mathbf{x}) = P(Y = 1 | \mathbf{X} = \mathbf{x})$ of myocardial infarction for a patient with input vector $\mathbf{x}$ could help the hospital staff in deciding whether to admit the patient into its coronary care unit.

Beetle Experiment

Next we turn to the first of three studies that will be used later on to illustrate logistic and Poisson regression. Table 13.1 reports the kills of adult flour beetles following 5-hour exposures to known concentrations (milligrams/liter) of carbon disulphide (CS_2). The data, taken from more extensive experiments, were used for illustrative purposes in "The Calculation of the Dosage-Mortality Curve," by C. L. Bliss, *Annals of Applied Biology*, 22 (1935), 134–167. More recently, they have been used for such

TABLE 13.1
Kills of Adult Flour Beetles

CS_2	No. Insects	No. Killed	% Killed
49.06	59	6	10.2
52.99	60	13	21.7
56.91	62	18	29.0
60.84	56	28	50.0
64.76	63	52	82.5
68.69	59	53	89.8
72.61	62	61	98.4
76.54	60	60	100.0

purposes in *An Introduction to Statistical Modelling* by Annette J. Dobson (London: Chapman and Hall, 1990); in "Generalized Logistic Models," by Therese A. Stukel, *Journal of the American Statistical Association*, 83 (1988), 426–432; and in "A General Model for Estimating ED_{100p} for Binary Response Dose-Response Data," by Therese A. Stukel, *American Statistician*, 44 (1990), 19–22.

Coronary Heart Disease Study

In 1948 a cohort study was begun in Framingham, Massachusetts, to determine which of a number of potential risk factors are related to the occurrence of coronary heart disease (CHD). At the start of the study, a large proportion of the town's inhabitants were examined for the presence of CHD. Those individuals found initially to be free of the disease were followed for 12 years, and those who developed CHD during that period were identified. Table 13.2 lists summary data, taken from page 224 of *Modelling Binary Data* by D. Collett (London: Chapman Hall, 1991), which were adapted from J. Truett, J. Cornfield, and W. Kannel, "A Multivariate Analysis of the Risk of Coronary Heart Disease in Framingham," *Journal of Chronic Diseases*, 20 (1967) 511–524. In Table 13.2, age in years and serum cholesterol level (mg / 100 ml) are those at the beginning of the study.

TABLE 13.2
Proportions of Cases of Coronary Heart Disease

Sex	Age Group	Serum Cholesterol Level			
		< 190	190–219	220–249	≥ 250
Male	30–49	13/340	18/408	40/421	57/362
	50–62	13/123	33/176	35/174	49/183
Female	30–49	6/542	5/552	10/412	18/357
	50–62	9/58	12/135	21/218	48/395

Lung Cancer Study

Table 13.3 lists data taken from E. L. Frome, "The Analysis of Rates Using Poisson Regression Models," *Biometrics*, 39 (1983), 665–674. The original source is R. Doll and A. B. Hill, "Mortality of British Doctors in Relation to Smoking: Observations on Coronary Thrombosis, in *Epidemiological Study of Cancer and Other Chronic Diseases*, ed. W. Haenszel, National Cancer Institute Monograph 19, 205–268 (Washington, D.C.: Government Printing Office, 1966).

TABLE 13.3

Man-Years at Risk, Numbers of Cases of Lung Cancer (in parentheses)

					C			
T	Non-smoker 0	1–9 5.2	10–14 11.2	15–19 15.9	20–24 20.4	25–34 27.4	35+ 40.8
15–19	10366 (1)	3121	3577	4317	5683	3042	670
20–24	8162	2937	3286 (1)	4214	6385 (1)	4050 (1)	1166
25–29	5969	2288	2546 (1)	3185	5483 (1)	4290 (4)	1482
30–34	4496	2015	2219 (2)	2560 (4)	4687 (6)	4268 (9)	1580 (4)
35–39	3512	1648 (1)	1826	1893	3646 (5)	3529 (9)	1336 (6)
40–44	2201	1310 (2)	1386 (1)	1334 (2)	2411 (12)	2424 (11)	924 (10)
45–49	1421	927	988 (2)	849 (2)	1567 (9)	1409 (10)	556 (7)
50–54	1121	710 (3)	684 (4)	470 (2)	857 (7)	663 (5)	255 (4)
55–59	826 (2)	606	449 (3)	280 (5)	416 (7)	284 (3)	104 (1)

The study was restricted to men, and the smokers were further restricted to those who smoked constant amounts and started smoking at ages 15 to 24 years. Here T denotes the number of years of smoking, which is estimated as age minus 20 years, C denotes the number of cigarettes consumed per day, and the mean number of cigarettes smoked per day within the indicated ranges of C is also shown. [Consider, for example, two doctors in the study, each of whom smoked between 25 and 34 cigarettes a day: one doctor, who smoked 25 cigarettes a day spent 5 years in the study in the 70–74 age group ($50 \leq T \leq 54$) and did not get lung cancer while in this age group; the other doctor, who smoked 30 cigarettes a day, spent 3 years in the study in the 70–74 age group and did get lung cancer while in this age group. Considering only these two doctors and the cell in the table determined by $50 \leq T \leq 54$ and $25 \leq C \leq 30$, the number of man-years at risk is $5 + 3 = 8$, the number of cases of lung cancer is 1, and the mean number of cigarettes smoked per day is $(5 \cdot 25 + 3 \cdot 30)/8 \doteq 26.9$.]

The beetle experiment and coronary heart disease study will be used to illustrate logistic regression, and the lung cancer study will be used to illustrate Poisson regression.

Problems

13.1 Consider the setup of Example 13.1, but suppose that T is normally distributed with mean τ and variance σ^2. Determine LD50 and LD90 in terms of μ and σ.

13.2 Consider the setup of Example 13.2, and suppose that ED $50 = 10$ and ED $90 = 20$. **(a)** Determine β_0 and β_1. **(b)** Determine ED99. **(c)** Determine the probability of a response when $x = 25$. **(d)** Sketch the graph of $\pi(x), 0 \le x \le 25$.

13.3 Suppose $\pi(x) = P(T \le x)$ for $x \in \mathbb{R}$, where T is normally distributed with mean β_0 and standard deviation β_1. Determine ED_{100p}; that is, for $0 < p < 1$, determine the value x_p such that $\pi(x_p) = p$.

13.4 Consider a logistic regression model in which $\pi(\mathbf{x})$ is the risk of getting the disease for a person with input vector $\mathbf{x}$ and the logistic regression function has the form $\theta(\mathbf{x}) = \beta_0 + \beta_1 x_1 + \cdots + \beta_M x_M$, where the factors $x_1, \dots, x_M$ each vary over $\mathbb{R}$. Under what additional conditions does a change in the mth factor from x_m to $x_m + c$ with the other factors held fixed cause the risk of getting the disease approximately to be multiplied by the factor $\exp(\beta_m c)$?

13.5 Let $\pi(x)$ be the probability of getting a certain noncontagious disease, where x is a behavioral factor. Suppose that the corresponding logistic regression function is given by $\theta(x) = x - 5$. Consider two societies. In the first society the behavior of 60% of the population corresponds to $x = 0$, 25% corresponds to $x = 1$, 10% corresponds to $x = 2$, and 5% corresponds to $x = 3$. In the second society, the behavior of 25% of the population corresponds to $x = 0$, 25% corresponds to $x = 1$, 25% to $x = 2$, and 25% to $x = 3$. Let $\pi^{(1)}$ be the probability that a randomly selected individual in the first society gets the disease and let $\pi^{(2)}$ denote the probability that a randomly selected individual in the second society gets the disease. Determine the relative risk $\pi^{(2)}/\pi^{(1)}$.

13.6 Determine $r(\mathbf{y}; \mathbf{n})$, $\mathbf{y} \in \mathcal{Y}$, in the context of **(a)** logistic regression; **(b)** Poisson regression.

13.7 Consider the Poisson regression model with $\theta(\mathbf{x}) = \beta_0 + \beta_1 x_1 + \cdots + \beta_M x_M$ for $\mathbf{x} \in \mathbb{R}^M$. Determine the effect on the rate of a change in the mth factor from level x_m to level $x_m + c$, with the other factors held fixed.

13.8 Suppose the Poisson regression function is an additive function of the various factors. Determine the form of the corresponding rate function.

13.9 Consider the setup of Example 13.3 and suppose, specifically, that

$$\theta(x_1, x_2) = \tfrac{1}{5} + \tfrac{2}{5} \mathrm{ind}(x_1 = 1) + \tfrac{4}{5} \mathrm{ind}(x_1 = 2) - \tfrac{3}{5} x_2.$$

Determine **(a)** $\theta(0, 0)$ and $\pi(0, 0)$; **(b)** $\theta(1, x_2) - \theta(0, x_2)$; **(c)** $\dfrac{\pi(1,x_2)/[1-\pi(1,x_2)]}{\pi(0,x_2)/[1-\pi(0,x_2)]}$; **(d)** $\dfrac{\pi(1,x_2)}{\pi(0,x_2)}$; **(e)** $\theta(x_1, x_2 + 1) - \theta(x_1, x_2)$; **(f)** $\dfrac{\pi(x_1,x_2+1)/[1-\pi(x_1,x_2+1)]}{\pi(x_1,x_2)/[1-\pi(x_1,x_2)]}$; **(g)** $\dfrac{\pi(x_1,x_2+1)}{\pi(x_1,x_2)}$; **(h)** $\theta(2, x_2) - \theta(1, x_2)$.

13.10 Consider an identifiable and saturated linear canonical regression model and let $g_1, \dots, g_d$ be an orthogonal basis of G relative to the inner product with unit weights; that is, such that $\sum_k g_j(\mathbf{x}_k) g_l(\mathbf{x}_k) = 0$ for $j \ne l$. Let $\|g_j\|^2 = \sum_k g_j^2(\mathbf{x}_k)$,

$1 \leq j \leq d$, be the squared norms of the basis functions relative to the indicated inner product. Then $\mathbf{X}^T\mathbf{X} = \mathbf{D}$, where $\mathbf{D} = \mathrm{diag}(\|g_1\|^2, \ldots, \|g_d\|^2)$. Show that $\beta = \mathbf{D}^{-1}\mathbf{X}^T\theta = \mathbf{D}^{-1}\mathbf{X}^T[(C')^{-1}(\mu)]$.

13.11 Consider the linear canonical regression model in which there is one factor and G is the space of linear functions in the level of that factor. Suppose the experiment consists of two runs: one at the level -1 and one at the level 1. Observe that $1, x$ is an orthogonal basis of G relative to the inner product with unit weights (see Problem 13.10). Use the result of Problem 13.10 to express the regression coefficients β_0 and β_1 in terms of θ_1 and θ_2.

13.12 Consider the linear canonical regression model in which there are three factors and G is the space of linear functions in the levels of these factors and consider the fractional factorial design shown in Table 13.4. Observe that $1, x_1, x_2, x_3$ is an orthogonal basis of G relative to the inner product with unit weights (see Problem 13.10) and that each of these functions has squared norm 4. Use the result of Problem 13.10 to express the regression coefficients $\beta_0, \beta_1, \beta_2, \beta_3$ in terms of $\theta_1, \theta_2, \theta_3, \theta_4$.

TABLE   13.4

x_1	x_2	x_3
-1	-1	1
1	-1	-1
-1	1	-1
1	1	1

13.2
Maximum-Likelihood Estimation

To motivate the maximum-likelihood method, we start out by considering a random variable Y having the binomial distribution with parameters $n = 5$ and unknown π. Let $f(y; \pi)$ denote the probability function of Y, which is given by $f(y; \pi) = \binom{5}{y}\pi^y(1-\pi)^{5-y}$ for $y \in \{0, \ldots, 5\}$. It is observed, say, that $Y = 4$. Let $L(\pi)$, $0 < \pi < 1$, denote the probability that Y takes on this observed value as a function of the unknown probability π. Then $L(\pi) = 5\pi^4(1 - \pi)$ for $0 < \pi < 1$. We refer to $L(\cdot)$ as the "likelihood function."

Observe, for example, that $L(.01) = 5(.01)^4(.99) \doteq 4.95 \times 10^{-8}$; that is, if the true value of π were $.01$, there would be less than 1 chance in 10 million of getting the observed value of Y. Thus, $.01$ is a highly implausible value for the unknown parameter π when the observed value of Y equals 4. Observe next that $L(.99) = 5(.99)^4(.01) \doteq .048$; that is, if the true value of π were $.99$, there would be about 1 chance in 21 of getting the observed value of Y. Thus, $.99$ is a somewhat implausible value of π, but it is much more plausible than $.01$. Similarly, $L(.3) = 5(.3)^4(.7) \doteq .028$. Thus, $.3$ is less plausible than $.99$ as the true value of π, but much more plausible than $.01$.

Figure 13.1 shows the graph of the likelihood function $L(\pi)$, $0 < \pi < 1$. According to this graph, .8 is the most plausible value of π; that is, it maximizes the likelihood function. This result is easily seen using calculus. Indeed,

$$\frac{d}{d\pi}L(\pi) = 20\pi^3 - 25\pi^4,$$

which equals 0 if and only if $\pi = .8$. Since the likelihood function is continuous and it approaches 0 as π approaches 0 or 1, we conclude that this function has a unique maximum at .8. In light of this result, we refer to $\hat{\pi} = .8$ as the "maximum-likelihood estimate" of π. Observe that this estimate coincides with the estimate 4/5 that would have been used in Chapter 12.

FIGURE 13.1

Likelihood function when $Y = 4$

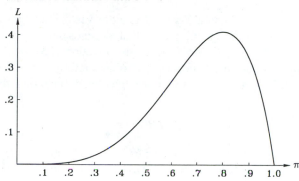

The function

$$l(\pi) = \log L(\pi) = 4\log \pi + \log(1 - \pi) + \log 5, \qquad 0 < \pi < 1,$$

is referred to as the "log-likelihood function." Figure 13.2 shows the graph of this function. Since the "log" function is continuous and strictly increasing on the interval $(0, \infty)$, it is clear that the location of the maximum of the likelihood function and that of the log-likelihood function coincide and hence that the log-likelihood function has a unique maximum at .8. Alternatively, by calculus

$$\frac{d}{d\pi}l(\pi) = \frac{4}{\pi} - \frac{1}{1 - \pi},$$

which equals zero if and only if $\pi = .8$. Also

$$\frac{d^2}{d\pi^2}l(\pi) = -\frac{4}{\pi^2} - \frac{1}{(1 - \pi)^2} < 0, \qquad 0 < \pi < 1,$$

so the log-likelihood function is uniquely maximized at .8. Moreover, since the second derivative of the log-likelihood function is negative everywhere on $(0, 1)$, this function is "strictly concave" on $(0, 1)$; that is, a line segment connecting any two points on the graph of the function lies below the graph between the two points.

FIGURE   **13.2**

FIGURE   **13.2**

The log-likelihood function

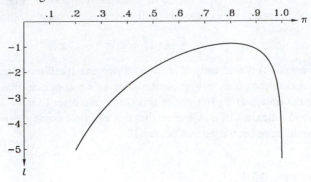

In general, let **Y** be a $\mathcal{Y}$-valued random vector having a probability function or density function $f(\mathbf{y}; \theta)$, $\mathbf{y} \in \mathcal{Y}$, that depends on an unknown vector-valued parameter $\theta \in \Theta \subset \mathbb{R}^p$. We refer to the function $L(\theta) = f(\mathbf{Y}; \theta)$, $\theta \in \Theta$, as the *likelihood function*. An estimate $\hat{\theta} \in \Theta$ of θ is referred to as a *maximum-likelihood estimate* (MLE) if it maximizes the likelihood function—that is, if it satisfies the equation

$$L(\hat{\theta}) = \max_{\theta \in \Theta} L(\theta).$$

We refer to the function $l(\theta) = \log L(\theta) = \log f(\mathbf{Y}; \theta)$, $\theta \in \Theta$, as the *log-likelihood function*. Observe that $\hat{\theta} \in \Theta$ is a maximum-likelihood estimate of θ if and only if it maximizes the log-likelihood function—that is, if and only if it satisfies the equation

$$l(\hat{\theta}) = \max_{\theta \in \Theta} l(\theta).$$

In the context of the normal linear regression model, the maximum-likelihood estimates of the regression coefficients coincide with their least-squares estimates and the maximum-likelihood estimate of σ^2 is given by $\hat{\sigma}^2 = \text{RSS}/n$, which differs somewhat from the unbiased estimate $S^2 = \text{RSS}/(n - p)$ (see Problem 13.21).

The Binomial Distribution

Let Y have the binomial distribution with parameters n and $\pi \in (0, 1)$, whose probability function is given by

$$f(y; n, \pi) = \binom{n}{y}\pi^y(1 - \pi)^{n-y}, \qquad y \in \mathcal{Y}_n.$$

The corresponding likelihood and log-likelihood functions are given by

$$L(\pi) = \binom{n}{Y}\pi^Y(1 - \pi)^{n-Y}, \qquad 0 < \pi < 1,$$

and

$$l(\pi) = Y \log \pi + (n - Y) \log(1 - \pi) + \log \binom{n}{Y}, \qquad 0 < \pi < 1.$$

Observe that

$$l'(\pi) = \frac{Y}{\pi} - \frac{n - Y}{1 - \pi}, \qquad 0 < \pi < 1,$$

and

$$l''(\pi) = -\left(\frac{Y}{\pi^2} + \frac{n - Y}{(1 - \pi)^2}\right), \qquad 0 < \pi < 1.$$

Set $\bar{Y} = Y/n$. Then

$$l(\pi) = n[\,\bar{Y} \log \pi + (1 - \bar{Y}) \log(1 - \pi)\,] + \log \binom{n}{Y}, \qquad 0 < \pi < 1,$$

$$l'(\pi) = n\left(\frac{\bar{Y}}{\pi} - \frac{1 - \bar{Y}}{1 - \pi}\right), \qquad 0 < \pi < 1,$$

and

$$l''(\pi) = -n\left(\frac{\bar{Y}}{\pi^2} + \frac{1 - \bar{Y}}{(1 - \pi)^2}\right) < 0, \qquad 0 < \pi < 1.$$

Suppose $0 < \bar{Y} < 1$ and let $\hat{\pi} \in (0, 1)$. If $\hat{\pi}$ maximizes the log-likelihood function, then

$$0 = l'(\hat{\pi}) = n\left(\frac{\bar{Y}}{\hat{\pi}} - \frac{1 - \bar{Y}}{1 - \hat{\pi}}\right)$$

and hence $\hat{\pi} = \bar{Y}$. Since $l''(\cdot) < 0$ on $(0, 1)$, we conclude that $\hat{\pi} = \bar{Y}$ is the unique maximum-likelihood estimate of π.

Suppose next that $\bar{Y} = 0$. Then the log-likelihood function is given by $l(\pi) = n \log(1 - \pi)$, which is a strictly decreasing function of π. This function does not have a maximum on $(0, 1)$. It does have a unique maximum on $[0, 1)$, however, which is given by $\hat{\pi} = 0$. Thus, we can think of the maximum-likelihood estimate of π either as being undefined or as being given by $\hat{\pi} = \bar{Y} = 0$.

Suppose instead that $\bar{Y} = 1$. Then the log-likelihood function is given by $l(\pi) = n \log \pi$, which is a strictly increasing function of π. This function does not have a maximum on $(0, 1)$. It does have a unique maximum on $(0, 1]$, however, which is given by $\hat{\pi} = 1$. Thus, we can think of the maximum-likelihood estimate of π either as being undefined or as being given by $\hat{\pi} = \bar{Y} = 1$.

We can think of the maximum-likelihood estimate of π as always being defined by $\hat{\pi} = \bar{Y}$, which is unbiased. If $Y = I_1 + \cdots + I_n$, where $I_1, \ldots, I_n$ are independent random variables each having the Bernoulli distribution with parameter π, then $\hat{\pi}$ is the relative frequency of ones among $I_1, \ldots, I_n$.

The maximum-likelihood estimate of the rate parameter λ of a Poisson distribution can be obtained in a similar manner, the details being left as a problem. As preparation for the study of maximum-likelihood estimation in the context of the canonical regression model, however, we need to consider maximum-likelihood estimation of the canonical parameter of an exponential family.

Exponential Families

Consider now either the binomial or the Poisson distribution and let the probability function be written in the form of an exponential family as

$$f(y; n, \theta) = \exp[\,\theta y - nC(\theta)\,]r(y; n) = \exp\{n[\,\theta\bar{y} - C(\theta)\,]\}r(y; n), \qquad y \in \mathcal{Y}_n,$$

where $\bar{y} = y/n$. The corresponding likelihood and log-likelihood functions are given by

$$L(\theta) = \exp\{n[\,\theta\bar{Y} - C(\theta)\,]\}r(Y; n), \qquad \theta \in \mathbb{R},$$

and

$$l(\theta) = n[\,\theta\bar{Y} - C(\theta)\,] + \log r(Y; n), \qquad \theta \in \mathbb{R}.$$

Observe that

$$l'(\theta) = n[\,\bar{Y} - C'(\theta)\,], \qquad \theta \in \mathbb{R},$$

and

$$l''(\theta) = -nC''(\theta) < 0, \qquad \theta \in \mathbb{R}.$$

THEOREM 13.1   Let θ be the canonical parameter of the exponential family. Then there is at most one $\hat{\theta} \in \mathbb{R}$ such that

$$C'(\hat{\theta}) = \bar{Y}. \tag{3}$$

If there is a solution $\hat{\theta}$ to (3), then $\hat{\theta} = (C')^{-1}(\bar{Y})$, and it is the unique maximum-likelihood estimate of θ. Otherwise, there is no (finite) maximum-likelihood estimate of θ.

Proof   Since $C'(\theta)$ is a strictly increasing function of θ, there is at most one solution to (3). Suppose there is a solution $\hat{\theta} \in \mathbb{R}$ to (1). Then $\hat{\theta} = (C')^{-1}(\bar{Y})$. Moreover, $l'(\hat{\theta}) = 0$. Since $l''(\theta) < 0$ for $\theta \in \mathbb{R}$, $\hat{\theta}$ is the unique maximum-likelihood estimate of θ. Suppose, instead, that there is no solution to (3). Then there is no $\hat{\theta} \in \mathbb{R}$ such that $l'(\hat{\theta}) = 0$, so there is no maximum-likelihood estimate in $\mathbb{R}$ of θ.   ∎

Suppose there is a maximum-likelihood estimate $\hat{\theta}$ of θ. Recalling that $\mu = C'(\theta)$, we refer to $\hat{\mu} = C'(\hat{\theta}) = C'((C')^{-1}(\bar{Y})) = \bar{Y}$ as the maximum-likelihood estimate of μ. We can think of $\hat{\mu} = \bar{Y}$ as being the maximum-likelihood estimate of μ even when there is no maximum-likelihood estimate of θ. Then $\hat{\mu}$ is an unbiased estimate of μ.

COROLLARY 13.1   Let θ be the canonical parameter of the binomial distribution. If $\bar{Y}$ equals 0 or 1, there is no maximum-likelihood estimate of θ. Otherwise, there is a unique maximum-likelihood estimate of θ, which is the unique solution to (3).

Proof According to the solution to Example 12.14(a) in Section 12.3, $C'(\theta)$ ranges over $(0, 1)$ as θ ranges over $\mathbb{R}$. Consequently, there is a $\hat{\theta} \in \mathbb{R}$ that satisfies (3) if and only if $0 < \bar{Y} < 1$. The conclusions of Corollary 13.1 now follow from Theorem 13.1.

[If $\bar{Y} = 0$, then $l(\theta) = -nC(\theta) = -n\log(1 + e^\theta)$, $\theta \in \mathbb{R}$, is a strictly decreasing function of θ. We can think of this function as being uniquely maximized at $\hat{\theta} = -\infty$, which corresponds to $\hat{\pi} = \exp(-\infty)/[1 + \exp(-\infty)] = 0$. Similarly, if $\bar{Y} = 1$, then

$$l(\theta) = n[\theta - C(\theta)] = n[\theta - \log(1 + e^\theta)], \qquad \theta \in \mathbb{R},$$

is a strictly increasing function. We can think of this function as being uniquely maximized at $\hat{\theta} = \infty$, which corresponds to $\hat{\pi} = (\exp\infty)/(1 + \exp\infty) = 1$. Having the extension to the linear canonical regression model in mind, however, we prefer to restrict attention to finite-valued maximum-likelihood estimates of the canonical parameter.] ∎

The proof of the next result is left as a problem.

COROLLARY **13.2** Let θ be the canonical parameter of the Poisson distribution. If $\bar{Y} = 0$, there is no maximum-likelihood estimate of θ. Otherwise, there is a unique maximum-likelihood estimate of θ, which is the unique solution to (3).

Linear Canonical Regression

Consider the canonical regression model. Recall from (1) in Section 13.1 that the probability function of $\mathbf{Y}$ can be written as

$$f(\mathbf{y}; \mathbf{n}, \theta(\cdot)) = \exp\left(\sum_k n_k[\theta(\mathbf{x}_k)\bar{y}_k - C(\theta(\mathbf{x}_k))]\right) r(\mathbf{y}; \mathbf{n}), \qquad \mathbf{y} \in \mathcal{Y}.$$

Consider now the linear canonical regression model, which involves the added assumption that the canonical regression function lies in a specified p-dimensional linear space G. The corresponding likelihood and log-likelihood functions can be written as

$$L(\theta(\cdot)) = \exp\left(\sum_k n_k[\theta(\mathbf{x}_k)\bar{Y}_k - C(\theta(\mathbf{x}_k))]\right) r(\mathbf{Y}; \mathbf{n}), \qquad \theta(\cdot) \in G,$$

and

$$l(\theta(\cdot)) = \sum_k n_k[\theta(\mathbf{x}_k)\bar{Y}_k - C(\theta(\mathbf{x}_k))] + \log r(\mathbf{Y}; \mathbf{n}), \qquad \theta(\cdot) \in G.$$

If there is a function $\hat{\theta}(\cdot) \in G$ such that $l(\hat{\theta}(\cdot)) = \max_{\theta(\cdot) \in G} l(\theta(\cdot))$, then this function is referred to as a maximum-likelihood estimate in G of the canonical regression function. Otherwise, we say that there is no maximum-likelihood estimate in G of this function or that the maximum-likelihood estimate is undefined or does not exist. In Section 13.3, we will investigate the existence and uniqueness of such maximum-likelihood estimates.

Recall that the regression function and variance function are given in terms of the canonical regression function by $\mu(\mathbf{x}) = C'(\theta(\mathbf{x}))$ and $\sigma^2(\mathbf{x}) = C''(\theta(\mathbf{x}))$ for $\mathbf{x} \in \mathcal{X}$. When the maximum-likelihood estimate $\hat{\theta}(\cdot)$ in G of the canonical regression function exists and is unique, we refer to $\hat{\mu}(\mathbf{x}) = C'(\hat{\theta}(\mathbf{x}))$, $\mathbf{x} \in \mathcal{X}$, as the maximum-likelihood estimate of the regression function and to $\hat{\sigma}^2(\mathbf{x}) = C''(\hat{\theta}(\mathbf{x}))$, $\mathbf{x} \in \mathcal{X}$, as the maximum-likelihood estimate of the variance function. In the context of logistic regression,

$$\hat{\pi}(\mathbf{x}) = \hat{\mu}(\mathbf{x}) = \frac{\exp \hat{\theta}(\mathbf{x})}{1 + \exp \hat{\theta}(\mathbf{x})}, \qquad \mathbf{x} \in \mathcal{X},$$

and

$$\hat{\sigma}^2(\mathbf{x}) = \hat{\pi}(\mathbf{x})[\, 1 - \hat{\pi}(\mathbf{x})\,], \qquad \mathbf{x} \in \mathcal{X};$$

in the context of Poisson regression,

$$\hat{\sigma}^2(\mathbf{x}) = \hat{\mu}(\mathbf{x}) = \hat{\lambda}(\mathbf{x}) = \exp \hat{\theta}(\mathbf{x}), \qquad \mathbf{x} \in \mathcal{X}.$$

Let $g_1, \ldots, g_p$ be a basis of G. Then the canonical regression function can be written as $\theta(\cdot) = \beta_1 g_1 + \cdots + \beta_p g_p$, and its maximum-likelihood estimate can be written as $\hat{\theta}(\cdot) = \hat{\beta}_1 g_1 + \cdots + \hat{\beta}_p g_p$. We refer to $\hat{\beta}_1, \ldots, \hat{\beta}_p$ as the maximum-likelihood estimates of $\beta_1, \ldots, \beta_p$, respectively, and to $\hat{\beta} = [\, \hat{\beta}_1, \ldots, \hat{\beta}_p\,]^T$ as the maximum-likelihood estimate of the coefficient vector $\beta = [\, \beta_1, \ldots, \beta_p\,]^T$. The corresponding maximum-likelihood estimates of the regression function and variance function are given by

$$\hat{\mu}(\mathbf{x}) = C'(\hat{\beta}_1 g_1(\mathbf{x}) + \cdots + \hat{\beta}_p g_p(\mathbf{x})), \qquad \mathbf{x} \in \mathcal{X},$$

and

$$\hat{\sigma}^2(\mathbf{x}) = C''(\hat{\beta}_1 g_1(\mathbf{x}) + \cdots + \hat{\beta}_p g_p(\mathbf{x})), \qquad \mathbf{x} \in \mathcal{X}.$$

In the context of logistic regression,

$$\hat{\pi}(\mathbf{x}) = \hat{\mu}(\mathbf{x}) = \frac{\exp(\hat{\beta}_1 g_1(\mathbf{x}) + \cdots + \hat{\beta}_p g_p(\mathbf{x}))}{1 + \exp(\hat{\beta}_1 g_1(\mathbf{x}) + \cdots + \hat{\beta}_p g_p(\mathbf{x}))}, \qquad \mathbf{x} \in \mathcal{X};$$

in the context of Poisson regression,

$$\hat{\lambda}(\mathbf{x}) = \hat{\mu}(\mathbf{x}) = \exp(\hat{\beta}_1 g_1(\mathbf{x}) + \cdots + \hat{\beta}_p g_p(\mathbf{x})), \qquad \mathbf{x} \in \mathcal{X}.$$

Saturated Models

Suppose now that G is identifiable and saturated. Given $\theta(\cdot) \in G$, set $\theta_k = \theta(\mathbf{x}_k)$ for $1 \le k \le d$. Then

$$l(\theta(\cdot)) = \sum_k n_k [\, \theta_k \bar{Y}_k - C(\theta_k)\,] + \log r(\mathbf{Y}; n). \qquad (4)$$

Since G is saturated, $(\theta_1, \ldots, \theta_d)$ ranges over all ordered d-tuples of real numbers as $\theta(\cdot)$ ranges over G. To maximize $l(\theta(\cdot))$, $\theta(\cdot) \in G$, we need separately to maximize $\theta_k \bar{Y}_k - C(\theta_k)$ over $\theta_k \in \mathbb{R}$ for $1 \le k \le d$.

THEOREM 13.2 Consider a linear logistic regression model in which G is identifiable and saturated. If $\bar{Y}_k$ equals 0 or 1 for some k, then there is no maximum-likelihood estimate in G of the logistic regression function. Otherwise, there is a unique maximum-likelihood estimate $\hat{\theta}(\cdot)$ in G of this function, which is the unique function $\hat{\theta}(\cdot)$ in G such that

$$C'(\hat{\theta}(\mathbf{x}_k)) = \bar{Y}_k, \qquad 1 \le k \le d. \tag{5}$$

Proof Suppose $\bar{Y}_k$ equals 0 or 1 for some k. Then, by Corollary 13.1, there is no value of $\theta_k \in \mathbb{R}$ that maximizes $\theta_k \bar{Y}_k - C(\theta_k)$. Thus, by (4), there is no maximum-likelihood estimate in G of the logistic regression function.

Suppose, conversely, that $0 < \bar{Y}_k < 1$ for all k. It then follows from Corollary 13.1 that for each k there is a unique $\hat{\theta}_k \in \mathbb{R}$ that maximizes $\theta_k \bar{Y}_k - C(\theta_k)$, which is the unique $\hat{\theta}_k \in \mathbb{R}$ such that $C'(\hat{\theta}_k) = \bar{Y}_k$. Since G is identifiable and saturated, there is a unique function $\hat{\theta}(\cdot) \in G$ such that $\hat{\theta}(\mathbf{x}_k) = \hat{\theta}_k$ for $1 \le k \le d$, which is the unique function in G that satisfies (5). We conclude from (4) that $\hat{\theta}(\cdot)$ is the unique maximum-likelihood estimate in G of the logistic regression function. ∎

The proof of the next result is left as a problem.

THEOREM 13.3 Consider a linear Poisson regression model in which G is identifiable and saturated. If $\bar{Y}_k = 0$ for some k, then there is no maximum-likelihood estimate in G of the Poisson regression function. Otherwise, there is a unique maximum-likelihood estimate $\hat{\theta}(\cdot)$ in G of this function, which is the unique function $\hat{\theta}(\cdot)$ in G that satisfies (5).

When G is identifiable and saturated and there is a maximum-likelihood estimate $\hat{\theta}(\cdot)$ in G of the canonical regression function, we regard $\hat{\theta}_k = \hat{\theta}(\mathbf{x}_k) = (C')^{-1}(\bar{Y}_k)$ as the maximum-likelihood estimate of $\theta_k = \theta(\mathbf{x}_k) = (C')^{-1}(\mu_k)$ and $\hat{\mu}_k = \hat{\mu}(\mathbf{x}_k) = C'(\hat{\theta}(\mathbf{x}_k)) = \bar{Y}_k$ as the maximum-likelihood estimate of $\mu_k = \mu(\mathbf{x}_k)$. Even when there is no maximum-likelihood estimate in G of the canonical regression function, we regard $\hat{\mu}_k = \bar{Y}_k$ as the maximum-likelihood estimate of μ_k. Then $\hat{\mu}_k$ is an unbiased estimate of μ_k.

Let $\mathbf{X}$ be the $d \times d$ invertible design matrix corresponding to a basis $g_1, \ldots, g_d$ of the identifiable and saturated linear space G. Suppose there is a maximum-likelihood estimate $\hat{\theta}(\cdot)$ in G of the canonical regression function. By Theorems 13.2 and 13.3, this function satisfies (5); that is,

$$C'(\hat{\theta}_k) = \bar{Y}_k, \qquad 1 \le k \le d, \tag{6}$$

where $\hat{\theta}_k = \hat{\theta}(\mathbf{x}_k)$ for $1 \le k \le d$. It follows from (6) that $C'(\hat{\boldsymbol{\theta}}) = \bar{\mathbf{Y}}$, where $\hat{\boldsymbol{\theta}} = [\hat{\theta}_1, \ldots, \hat{\theta}_d]^T$ and $\bar{\mathbf{Y}} = [\bar{Y}_1, \ldots, \bar{Y}_d]^T$. Consequently $\hat{\boldsymbol{\theta}} = (C')^{-1}(\bar{\mathbf{Y}})$.

Let $\hat{\boldsymbol{\beta}}$ be the coefficient vector of $\hat{\theta}(\cdot)$ relative to the indicated basis. Then $\hat{\boldsymbol{\theta}} = \mathbf{X}\hat{\boldsymbol{\beta}}$, so $\hat{\boldsymbol{\beta}} = \mathbf{X}^{-1}\hat{\boldsymbol{\theta}} = \mathbf{X}^{-1}[(C')^{-1}(\bar{\mathbf{Y}})]$. In particular, in the context of logistic regression, $\hat{\boldsymbol{\beta}} = \mathbf{X}^{-1}\text{logit}\bar{\mathbf{Y}}$; in the context of Poisson regression, $\hat{\boldsymbol{\beta}} = \mathbf{X}^{-1}\log\bar{\mathbf{Y}}$.

EXAMPLE 13.4　Consider the setup of Example 13.3 in Section 13.1, in which the logistic regression function has the form $\theta(x_1, x_2) = \beta_0 + \beta_1 \operatorname{ind}(x_1 = 1) + \beta_2 \operatorname{ind}(x_1 = 2) + \beta_3 x_2$ and consider the experimental data shown in Table 13.5. Determine

(a) the design matrix and its inverse;

(b) the maximum-likelihood estimates of $\theta_1, \theta_2, \theta_3$, and θ_4;

(c) the maximum-likelihood estimates of the regression coefficients;

(d) the treatment that is apparently best and the one that is apparently worst, vis-à-vis the probability of cure.

TABLE　13.5

x_1	x_2	n	Y	$\bar{Y}$
0	−3	200	176	.88
1	−1	100	77	.77
2	1	100	60	.60
0	3	200	34	.17

Solution　Observe that the space of additive functions of x_1 and x_2 that are linear in x_2 is identifiable and saturated.

(a) The design matrix is given by

$$\mathbf{X} = \begin{bmatrix} 1 & 0 & 0 & -3 \\ 1 & 1 & 0 & -1 \\ 1 & 0 & 1 & 1 \\ 1 & 0 & 0 & 3 \end{bmatrix},$$

whose determinant equals 6. Thus,

$$\mathbf{X}^{-1} = \frac{1}{6} \begin{bmatrix} 3 & 0 & 0 & 3 \\ -4 & 6 & 0 & -2 \\ -2 & 0 & 6 & -4 \\ -1 & 0 & 0 & 1 \end{bmatrix}.$$

(b) The maximum-likelihood estimates of $\theta_1, \theta_2, \theta_3$, and θ_4 are given by

$$\hat{\theta}_1 = \log \frac{176}{24} \doteq 1.992,$$

$$\hat{\theta}_2 = \log \frac{77}{23} \doteq 1.208,$$

$$\hat{\theta}_3 = \log \frac{60}{40} \doteq 0.405,$$

and

$$\hat{\theta}_4 = \log \frac{34}{166} \doteq -1.586.$$

(c) The maximum-likelihood estimates of the vector of regression coefficients is given by $\hat{\beta} = \mathbf{X}^{-1}\hat{\theta}$. Thus,

$$\hat{\beta}_0 = \frac{\hat{\theta}_1 + \hat{\theta}_4}{2} \doteq 0.203,$$

$$\hat{\beta}_1 = \frac{-2\hat{\theta}_1 + 3\hat{\theta}_2 - \hat{\theta}_4}{3} \doteq 0.409,$$

$$\hat{\beta}_2 = \frac{-\hat{\theta}_1 + 3\hat{\theta}_3 - 2\hat{\theta}_4}{3} \doteq 0.798,$$

and

$$\hat{\beta}_3 = \frac{-\hat{\theta}_1 + \hat{\theta}_4}{6} \doteq -0.596.$$

(d) Since $0 < \hat{\beta}_1 < \hat{\beta}_2$, treatment 2 ($x_1 = 2$) is apparently the best treatment, and the standard treatment ($x_1 = 0$) is apparently the worst treatment. ∎

EXAMPLE **13.5**   An industrial experiment is designed to improve the yield of high-quality products. There are three qualitative factors in the experiment, each of which takes on the two levels -1 and 1; here -1 corresponds to the level currently in use, and 1 corresponds to a new level that may possibly improve the yield of high-quality products. The response at any factor-level combination is the number of high-quality products Y out of the n products manufactured according to that factor-level combination. Consider the logistic regression model corresponding to the space of linear functions in the levels of the three factors and consider the basis 1, x_1, x_2, x_3 of this space. Then the logistic regression function is given by $\theta(x_1, x_2, x_3) = \beta_0 + \beta_1 x_1 + \beta_2 x_2 + \beta_3 x_3$. Consider the experimental data shown in Table 13.6. Determine

(a) the design matrix and its inverse;

(b) the maximum-likelihood estimates of θ_1, θ_2, θ_3, and θ_4;

(c) the maximum-likelihood estimates of the regression coefficients;

(d) the apparently optimal factor-level combination and the corresponding maximum-likelihood estimate of the probability of getting a high-quality product.

TABLE **13.6**

x_1	x_2	x_3	n	Y	$\bar{Y}$
-1	-1	1	200	176	.88
1	-1	-1	100	77	.77
-1	1	-1	100	60	.60
1	1	1	200	34	.17

Solution   Observe that the design has the form of an orthogonal array (note that $x_3 = x_1 x_2$ at all design points). Thus, the basis functions $1, x_1, x_2, x_3$ are orthogonal relative to the inner product corresponding to unit weights. Consequently, the space of linear functions in the levels of the three factors is identifiable and saturated.

(a) The design matrix is given by

$$\mathbf{X} = \begin{bmatrix} 1 & -1 & -1 & 1 \\ 1 & 1 & -1 & -1 \\ 1 & -1 & 1 & -1 \\ 1 & 1 & 1 & 1 \end{bmatrix}.$$

The Gram matrix corresponding to the indicated basis and inner product is given by $\mathbf{X}^T\mathbf{X} = 4I$, where I is the 4×4 identity matrix. Consequently, $\mathbf{X}^{-1} = \frac{1}{4}\mathbf{X}^T$.

(b) The maximum-likelihood estimates of $\theta_1, \theta_2, \theta_3$, and θ_4 are given in the solution to Example 13.4(b).

(c) The maximum-likelihood estimate of the vector of regression coefficients is given by $\hat{\beta} = \mathbf{X}^{-1}\hat{\theta} = \frac{1}{4}\mathbf{X}^T\hat{\theta}$. Thus,

$$\hat{\beta}_0 = \frac{\hat{\theta}_1 + \hat{\theta}_2 + \hat{\theta}_3 + \hat{\theta}_4}{4} \doteq 0.505,$$

$$\hat{\beta}_1 = \frac{-\hat{\theta}_1 + \hat{\theta}_2 - \hat{\theta}_3 + \hat{\theta}_4}{4} \doteq -0.694,$$

$$\hat{\beta}_2 = \frac{-\hat{\theta}_1 - \hat{\theta}_2 + \hat{\theta}_3 + \hat{\theta}_4}{4} \doteq -1.095,$$

and

$$\hat{\beta}_3 = \frac{\hat{\theta}_1 - \hat{\theta}_2 - \hat{\theta}_3 + \hat{\theta}_4}{4} = -0.301.$$

These computations are conveniently summarized in Table 13.7.

(d) Since $\hat{\beta}_1, \hat{\beta}_2$ and $\hat{\beta}_3$ are all negative, the apparently optimal factor-level combination is $(-1, -1, -1)$; that is, we should keep the three factors at their current levels. The corresponding maximum-likelihood estimate of the probability of

TABLE   13.7

1	x_1	x_2	x_3	$\hat{\theta}$
1	−1	−1	1	1.992
1	1	−1	−1	1.208
1	−1	1	−1	−0.405
1	1	1	1	−1.586
0.505	−0.694	−1.095	−0.301	

getting a high-quality product is given by

$$\hat{\pi}_{\text{opt}} = \frac{\exp(\hat{\beta}_0 + |\hat{\beta}_1| + |\hat{\beta}_2| + |\hat{\beta}_3|)}{1 + \exp(\hat{\beta}_0 + |\hat{\beta}_1| + |\hat{\beta}_2| + |\hat{\beta}_3|)} \doteq \frac{e^{2.596}}{1 + e^{2.596}} \doteq .931. \quad \blacksquare$$

Problems

13.13 Let Y have the binomial distribution with parameters $n \geq 2$ and π, let $y \in \{1, \ldots, n-1\}$ be the observed value of Y, let $L(y) = \binom{n}{y}\pi^y(1-\pi)^{n-y}$ be the corresponding likelihood function, and set $\hat{\pi} = y/n$. **(a)** Determine $L'(\pi)$ and show that $L'(\pi) = 0$ if and only if $\pi = \hat{\pi}$. **(b)** Determine $L''(\hat{\pi})$ and show that $L''(\hat{\pi}) < 0$. **(c)** Explain why it follows from the results of parts (a) and (b) that $L(\pi)$ is uniquely maximized at $\hat{\pi}$.

13.14 Let $Y_1, \ldots, Y_n$ be independent random variables, each having the Bernoulli distribution with parameter $\pi \in (0, 1)$. Then the probability function of $\mathbf{Y} = [Y_1, \ldots, Y_n]^T$ is given by

$$f(\mathbf{y}; \pi) = \pi^{y_1 + \cdots + y_n}(1-\pi)^{n - y_1 - \cdots - y_n}, \qquad \mathbf{y} = [y_1, \ldots, y_n]^T \in \mathcal{Y},$$

where $\mathcal{Y}$ is the collection of ordered n-tuples of 0's and 1's. The corresponding log-likelihood function can be written as $l(\pi) = n[\bar{Y}\log\pi + (1-\bar{Y})\log(1-\pi)]$, $0 < \pi < 1$, where $\bar{Y} = (Y_1 + \cdots + Y_n)/n$. Show that **(a)** if $0 < \bar{Y} < 1$, then the unique maximum-likelihood estimate of π is given by $\hat{\pi} = \bar{Y}$; **(b)** if $\bar{Y} = 0$, then the function $l(\pi) = n\log(1-\pi)$, $0 \leq \pi < 1$, is uniquely maximized at $\hat{\pi} = 0$; **(c)** if $\bar{Y} = 1$, then the function $l(\pi) = n\log\pi$, $0 < \pi \leq 1$, is uniquely maximized at $\hat{\pi} = 1$.

13.15 Let Y have the Poisson distribution with mean $n\lambda$, where n is a known positive number and λ is an unknown positive number, and set $\bar{Y} = Y/n$. **(a)** Determine the likelihood function (as a function of λ). **(b)** Determine the log-likelihood function. **(c)** Show that if $\bar{Y} > 0$, then the unique maximum-likelihood estimate of λ is given by $\hat{\lambda} = \bar{Y}$. **(d)** Show that if $\bar{Y} = 0$, then the log-likelihood function, viewed as a function on $[0, \infty)$, is uniquely maximized at $\hat{\lambda} = 0$.

13.16 Let $Y_1, \ldots, Y_n$ be independent random variables, each having the Poisson distribution with parameter $\lambda > 0$. Then the probability function of $\mathbf{Y} = [Y_1, \ldots, Y_n]^T$ is given by

$$f(\mathbf{y}; \lambda) = \frac{\lambda^{y_1 + \cdots + y_n}}{y_1! \cdots y_n!} e^{-n\lambda}, \qquad \mathbf{y} \in \mathcal{Y},$$

where $\mathcal{Y}$ is the collection of ordered n-tuples of nonnegative integers. The corresponding log-likelihood function can be written as

$$l(\lambda) = n[\bar{Y}\log\lambda - \lambda] - \log(Y_1! \cdots Y_n!), \qquad \lambda > 0,$$

where $\bar{Y} = (Y_1 + \cdots + Y_n)/n$. Show that **(a)** if $\bar{Y} > 0$, then the unique maximum-likelihood estimate of λ is given by $\hat{\lambda} = \bar{Y}$; **(b)** if $\bar{Y} = 0$, then the function $l(\lambda) = -n\lambda$, $\lambda \geq 0$, is uniquely maximized at $\hat{\lambda} = 0$.

13.17   Let $Y_1, \ldots, Y_n$ be independent random variables, each having the geometric distribution with parameter $\pi \in (0, 1)$. Then the probability function of $\mathbf{Y} = [Y_1, \ldots, Y_n]^T$ is given by

$$f(\mathbf{y}; \pi) = \pi^{y_1 + \cdots + y_n}(1 - \pi)^n, \qquad \mathbf{y} = [y_1, \ldots, y_n]^T \in \mathcal{Y},$$

where $\mathcal{Y}$ is the collection of ordered n-tuples of nonnegative integers. The corresponding log-likelihood function can be written as

$$l(\pi) = n[\bar{Y}\log\pi + \log(1 - \pi)], \qquad 0 < \pi < 1,$$

where $\bar{Y} = (Y_1 + \cdots + Y_n)/n$. Show that **(a)** if $\bar{Y} > 0$, then the unique maximum-likelihood estimate of π is given by $\hat{\pi} = \bar{Y}/(\bar{Y} + 1)$; **(b)** if $\bar{Y} = 0$, then the function $l(\pi) = n\log(1 - \pi), 0 \le \pi < 1$, is uniquely maximized at $\hat{\pi} = 0$.

13.18   Let $Y_1, \ldots, Y_n$ be independent random variables, each having the exponential distribution with mean β, and set $\bar{Y} = (Y_1 + \cdots + Y_n)/n$. Show that **(a)** the log-likelihood function is given by $l(\beta) = -n(\log\beta + \bar{Y}/\beta), \beta > 0$; **(b)** the maximum-likelihood estimate of β is given by $\hat{\beta} = \bar{Y}$.

13.19   Let $Y_1, \ldots, Y_n$ be independent random variables, each normally distributed with unknown mean θ and known variance σ^2. Show that **(a)** the log-likelihood function is given by $l(\theta) = -\frac{n}{2}\log(2\pi\sigma^2) - \frac{1}{2\sigma^2}\sum_i(Y_i - \theta)^2$; **(b)** the maximum-likelihood estimate of θ is given by $\hat{\theta} = \bar{Y}$.

13.20   Let $Y_1, \ldots, Y_n$ be independent random variables, each normally distributed with known mean μ and unknown variance θ. Show that **(a)** the log-likelihood function is given by $l(\theta) = -\frac{n}{2}\log(2\pi\theta) - \frac{1}{2\theta}\sum_i(Y_i - \mu)^2, \theta > 0$; **(b)** the maximum-likelihood estimate of θ is given by $\hat{\theta} = \frac{1}{n}\sum_i(Y_i - \mu)^2$.

13.21   Let $Y_1, \ldots, Y_n$ be independent random variables such that Y_i is normally distributed with mean $\mu(\mathbf{x}_i)$ and variance σ^2 for $1 \le i \le n$, where $\mu(\cdot)$ is an unknown member of an identifiable p-dimensional linear space G and σ^2 is an unknown positive number. **(a)** Determine the likelihood function $L(\mu(\cdot), \sigma^2), \mu(\cdot) \in G$ and $\sigma^2 > 0$. **(b)** Determine the log-likelihood function $l(\mu(\cdot), \sigma^2), \mu(\cdot) \in G$ and $\sigma^2 > 0$. **(c)** Show that for given $\sigma^2 > 0$, the likelihood and log-likelihood functions are uniquely maximized at the least-squares estimate $\hat{\mu}(\cdot)$ in G of $\mu(\cdot)$. **(d)** Choose σ^2 to maximize $L(\hat{\mu}(\cdot), \sigma^2)$.

13.22   Verify Corollary 13.2.

13.23   Verify Theorem 13.3.

13.24   Suppose G is identifiable and saturated and the maximum-likelihood estimate $\hat{\theta}(\cdot)$ in G of $\theta(\cdot)$ exists. Let $\hat{\mu}(\cdot) = C'(\hat{\theta}(\cdot))$ be the corresponding maximum-likelihood estimate of $\mu(\cdot)$. Also, let $\hat{\mu}_{\mathrm{LS}}(\cdot) = P_G[\bar{Y}(\cdot)]$ denote the least-squares estimate in G of $\mu(\cdot)$. **(a)** Does $\hat{\mu}_{\mathrm{LS}}(\cdot) = \hat{\mu}(\cdot)$ on the design set? **(b)** Does $\hat{\mu}_{\mathrm{LS}}(\cdot) = \hat{\mu}(\cdot)$ on all of $\mathcal{X}$?

13.25   In the context of Problem 13.10 in Section 13.1, consider the maximum-likelihood estimates $\hat{\beta}, \hat{\theta}$, and $\hat{\mu}$ of β, θ, and μ, respectively. Show that **(a)** $\hat{\beta} = \mathbf{D}^{-1}\mathbf{X}^T\hat{\theta} = \mathbf{D}^{-1}\mathbf{X}^T[(C')^{-1}(\hat{\mu})]$; **(b)** if $\mathbf{D} = r\mathbf{I}$, then $\hat{\beta} = r^{-1}\mathbf{X}^T\hat{\theta}$.

13.26   Consider a response variable Y that has a Poisson distribution the log of whose mean depends linearly on the level of some factor: $\log EY = \beta_0 + \beta_1 x$. An experiment is conducted to estimate the unknown coefficients β_0 and β_1 that involves 50 runs at the level $x = -1$ and 100 runs at the level $x = 1$. The observed sum of the responses on the runs at level -1 is 50 and the observed sum on the runs at level 1 is 150. Determine **(a)** the design matrix; **(b)** the maximum-likelihood estimates of β_0 and β_1.

13.27 Rework parts (a)–(c) of Example 13.4 with "logistic regression" replaced by "Poisson regression."

13.28 Rework parts (a)–(c) of Example 13.5 with "logistic regression" replaced by "Poisson regression."

13.3
Existence and Uniqueness of the Maximum-Likelihood Estimate

Consider a canonical regression model corresponding to a p-dimensional linear space G. In this section, we will investigate the existence and uniqueness of the maximum-likelihood estimate in G of the canonical regression function.

Uniqueness

Recall that the log-likelihood function is given by

$$l(\theta(\cdot)) = \sum_k n_k [\theta(\mathbf{x}_k)\bar{Y}_k - C(\theta(\mathbf{x}_k))] + \log r(\mathbf{Y}; \mathbf{n}), \qquad \theta(\cdot) \in G.$$

Let $\hat{\theta}(\cdot)$ and g be members of G and let $s \in \mathbb{R}$. Then

$$l(\hat{\theta}(\cdot) + sg) = \sum_k n_k \left\{ [\hat{\theta}(\mathbf{x}_k) + sg(\mathbf{x}_k)]\bar{Y}_k - C(\hat{\theta}(\mathbf{x}_k) + sg(\mathbf{x}_k)) \right\} + \log r(\mathbf{Y}; \mathbf{n})$$

for $s \in \mathbb{R}$, so

$$\frac{dl(\hat{\theta}(\cdot) + sg)}{ds} = \sum_k n_k g(\mathbf{x}_k)[\bar{Y}_k - C'(\hat{\theta}(\mathbf{x}_k) + sg(\mathbf{x}_k))], \quad g \in G \text{ and } s \in \mathbb{R}. \quad \textbf{(7)}$$

In particular,

$$\frac{dl(\hat{\theta}(\cdot) + sg)}{ds}\bigg|_{s=0} = \sum_k n_k g(\mathbf{x}_k)[\bar{Y}_k - C'(\hat{\theta}(\mathbf{x}_k))], \qquad g \in G. \quad \textbf{(8)}$$

Suppose $\hat{\theta}(\cdot)$ is a maximum-likelihood estimate in G of the canonical regression function and let $g \in G$. Then $l(\hat{\theta}(\cdot)) \geq l(\hat{\theta}(\cdot) + sg)$ for $s \in \mathbb{R}$; that is, the function $l(\hat{\theta}(\cdot) + sg)$, $s \in \mathbb{R}$, is maximized at $s = 0$. Consequently, the derivative of this function equals zero at $s = 0$, so we conclude from (8) that

$$\sum_k n_k g(\mathbf{x}_k)[\bar{Y}_k - C'(\hat{\theta}(\mathbf{x}_k))] = 0, \qquad g \in G, \quad \textbf{(9)}$$

which can be rewritten as

$$\sum_k n_k g(\mathbf{x}_k)C'(\hat{\theta}(\mathbf{x}_k)) = \sum_k n_k g(\mathbf{x}_k)\bar{Y}_k, \qquad g \in G. \quad \textbf{(10)}$$

We refer to (9) or (10) as the *maximum-likelihood equation* for $\hat{\theta}(\cdot)$.

Suppose there is a maximum-likelihood estimate $\hat{\theta}(\cdot)$ in G of the canonical regression function and let $\hat{\mu}(\cdot) = C'(\hat{\theta}(\cdot))$ be the corresponding maximum-likelihood estimate of the regression function. Then, by (9),

$$\sum_k n_k g(\mathbf{x}_k)[\,\bar{Y}_k - \hat{\mu}(\mathbf{x}_k)\,] = 0, \qquad g \in G. \tag{11}$$

If G contains the constant functions, then it follows from (11) with $g = 1$ that

$$\sum_k n_k[\,\bar{Y}_k - \hat{\mu}(\mathbf{x}_k)\,] = 0. \tag{12}$$

Differentiating both sides of (7) with respect to s, we get

$$\frac{d^2 l(\hat{\theta}(\cdot) + sg)}{ds^2} = -\sum_k n_k g^2(\mathbf{x}_k) C''(\hat{\theta}(\mathbf{x}_k) + sg(\mathbf{x}_k)), \qquad g \in G \text{ and } s \in \mathbb{R}. \tag{13}$$

In particular,

$$\left.\frac{d^2 l(\hat{\theta}(\cdot) + sg)}{ds^2}\right|_{s=0} = -\sum_k n_k g^2(\mathbf{x}_k) C''(\hat{\theta}(\mathbf{x}_k)), \qquad g \in G. \tag{14}$$

Since $C''(\cdot) > 0$, we conclude from (13) that

$$\frac{d^2 l(\hat{\theta}(\cdot) + sg)}{ds^2} \leq 0, \qquad g \in G \text{ and } s \in \mathbb{R}; \tag{15}$$

moreover, if G is identifiable, then

$$\frac{d^2 l(\hat{\theta}(\cdot) + sg)}{ds^2} < 0, \qquad g \in G \text{ with } g \neq 0 \text{ and } s \in \mathbb{R}. \tag{16}$$

THEOREM 13.4   Suppose the function $\hat{\theta}(\cdot) \in G$ satisfies the maximum-likelihood equation. Then $\hat{\theta}(\cdot)$ is a maximum-likelihood estimate in G of the canonical regression function. If G is identifiable, then $\hat{\theta}(\cdot)$ is the unique maximum-likelihood estimate in G of this function; but if G is nonidentifiable, there are infinitely many solutions to the maximum-likelihood equation, each of which is a maximum-likelihood estimate.

Proof   Choose $\theta(\cdot) \in G$ and set $g = \theta(\cdot) - \hat{\theta}(\cdot)$. Then, by (8) and (9),

$$\left.\frac{dl(\hat{\theta}(\cdot) + sg)}{ds}\right|_{s=0} = 0. \tag{17}$$

Also, by (13),

$$\frac{d^2 l(\hat{\theta}(\cdot) + sg)}{ds^2} \leq 0, \qquad s \in \mathbb{R}. \tag{18}$$

It follows from (17) and (18) that the function $l(\hat{\theta}(\cdot) + sg)$, $s \in \mathbb{R}$, is maximized at $s = 0$. Consequently, $l(\hat{\theta}(\cdot)) \geq l(\hat{\theta}(\cdot) + sg)$ for $s \in \mathbb{R}$. In particular,

$$l(\hat{\theta}(\cdot)) \geq l(\hat{\theta}(\cdot) + g) = l(\hat{\theta}(\cdot) + \theta(\cdot) - \hat{\theta}(\cdot)) = l(\theta(\cdot)).$$

In summary, $l(\hat{\theta}(\cdot)) \geq l(\theta(\cdot))$ for $\theta(\cdot) \in G$. Therefore, $\hat{\theta}(\cdot)$ is a maximum-likelihood estimate in G of the canonical regression function.

Suppose G is identifiable. Choose $\theta(\cdot) \in G$ with $\theta(\cdot) \neq \hat{\theta}(\cdot)$, and set $g = \theta(\cdot) - \hat{\theta}(\cdot)$. Then, by (16),

$$\frac{d^2 l(\hat{\theta}(\cdot) + sg)}{ds^2} < 0, \qquad s \in \mathbb{R}. \tag{19}$$

It follows from (17) and (19) that the function $l(\hat{\theta}(\cdot) + sg)$, $s \in \mathbb{R}$, is uniquely maximized at $s = 0$. Arguing as above, we conclude that $l(\hat{\theta}(\cdot)) > l(\theta(\cdot))$ and hence that $\hat{\theta}(\cdot)$ is the unique maximum-likelihood estimate in G of the canonical regression function.

Suppose, instead, that G is nonidentifiable and let g_0 be a nonzero function in G such that $g_0(x_k) = 0$ for $1 \leq k \leq d$. Then, by (9),

$$\sum_k n_k g(\mathbf{x}_k) [\, \bar{Y}_k - C'(\hat{\theta}(\mathbf{x}_k) + s g_0(\mathbf{x}_k)) \,]$$

$$= \sum_k n_k g(\mathbf{x}_k) [\, \bar{Y}_k - C'(\hat{\theta}(\mathbf{x}_k)) \,] = 0, \qquad s \in \mathbb{R}.$$

Consequently, for each $s \in \mathbb{R}$, the function $\hat{\theta}(\cdot) + s g_0 \in G$ satisfies the maximum-likelihood equation (9) and hence a maximum-likelihood estimate in G of the canonical regression. As s ranges over $\mathbb{R}$, we get infinitely many maximum-likelihood estimates in this manner.   ∎

Existence

Next, after stating a preliminary result, we obtain a necessary and sufficient condition for the existence of the maximum-likelihood estimate of the canonical regression function.

THEOREM 13.5   Let $h(\cdot)$ be a twice continuously differentiable function on $\mathbb{R}^p$ such that

$$\frac{d^2 h(s\mathbf{b})}{ds^2} \leq 0, \qquad \mathbf{b} \in \mathbb{R}^p \text{ and } s \in \mathbb{R},$$

and

$$\lim_{|s| \to \infty} h(s\mathbf{b}) = -\infty \qquad \text{for all } \mathbf{b} \in \mathbb{R}^p \text{ with } \mathbf{b} \neq \mathbf{0}.$$

Then $\lim_{|\mathbf{b}| \to \infty} h(\mathbf{b}) = -\infty$, and hence h has a global maximum.

In Theorem 13.5, which is required for the proof of the next two theorems, $|\mathbf{b}|^2 = (b_1^2 + \cdots + b_p^2)$ for $\mathbf{b} = [b_1, \ldots, b_p]^T \in \mathbb{R}^p$. The proof of Theorem 13.5 is omitted since it belongs in an introductory text on real analysis as an application of properties of convex functions and compact (closed and bounded) sets.

THEOREM 13.6   Consider a linear logistic regression model in which G is identifiable. There is a maximum-likelihood estimate in G of the logistic regression function if and only if there is no nonzero function $g \in G$ that satisfies the following condition: For $1 \leq k \leq d$, if $g(\mathbf{x}_k) < 0$, then $\bar{Y}_k = 0$; and if $g(\mathbf{x}_k) > 0$, then $\bar{Y}_k = 1$.

Proof   Let $\theta(\cdot)$ be a member of G and let g be a nonzero member of G. Then [see (7) and (8)]

$$\frac{dl(\theta(\cdot) + sg)}{ds} = \sum_k n_k g(\mathbf{x}_k)[\, \bar{Y}_k - C'(\theta(\mathbf{x}_k) + sg(\mathbf{x}_k))\,], \qquad s \in \mathbb{R}, \qquad \textbf{(20)}$$

and

$$\frac{dl(\theta(\cdot) + sg)}{ds}\bigg|_{s=0} = \sum_k n_k g(\mathbf{x}_k)[\, \bar{Y}_k - C'(\theta(\mathbf{x}_k))\,]. \qquad \textbf{(21)}$$

Since $C'(\theta) = e^\theta/(1 + e^\theta)$ ranges over $(0, 1)$ as θ ranges over $\mathbb{R}$,

$$0 < C'(\theta(\mathbf{x}_k)) < 1, \qquad 1 \leq k \leq d. \qquad \textbf{(22)}$$

Suppose g satisfies the condition in the theorem. It then follows from (22) that g satisfies the following condition: For $1 \leq k \leq d$, if $g(\mathbf{x}_k) \neq 0$, then $g(\mathbf{x}_k)[\, \bar{Y}_k - C'(\theta(\mathbf{x}_k))\,] > 0$. Since G is identifiable, there is at least one value of k such that $g(\mathbf{x}_k) \neq 0$; hence,

$$\sum_k n_k g(\mathbf{x}_k)[\, \bar{Y}_k - C'(\theta(\mathbf{x}_k))\,] > 0. \qquad \textbf{(23)}$$

We conclude from (21) and (23) that

$$\frac{dl(\theta(\cdot) + sg)}{ds}\bigg|_{s=0} > 0$$

and hence that $\theta(\cdot)$ does not maximize the log-likelihood function. Consequently, the log-likelihood function does not have a global maximum. Thus, if there is a nonzero function $g \in G$ satisfying the indicated condition, then there is no maximum-likelihood estimate in G of the logistic regression function.

Now $C'(\theta) \to 0$ as $\theta \to -\infty$ and $C'(\theta) \to 1$ as $\theta \to \infty$, so it follows from (20) that

$$\lim_{s \to \infty} \frac{dl(\theta(\cdot) + sg)}{ds} = \sum_{g(\mathbf{x}_k) < 0} n_k g(\mathbf{x}_k) \bar{Y}_k + \sum_{g(\mathbf{x}_k) > 0} n_k g(\mathbf{x}_k)[\, \bar{Y}_k - 1\,]. \qquad \textbf{(24)}$$

Since $0 \leq \bar{Y}_k \leq 1$ for $1 \leq k \leq n$, every term on the right side of (24) is nonpositive. Suppose g is a nonzero function in G that does not satisfy the condition in the theorem. Then at least one term on the right side of (24) is negative, so

$$\lim_{s \to \infty} \frac{dl(\theta(\cdot) + sg)}{ds} < 0$$

and hence $\lim_{s \to \infty} l(\theta(\cdot) + sg) = -\infty$. By a similar argument, $\lim_{s \to -\infty} l(\theta(\cdot) + sg) = -\infty$, and hence $\lim_{|s| \to \infty} l(\theta(\cdot) + sg) = -\infty$. In particular,

$$\lim_{|s| \to \infty} l(sg) = -\infty. \qquad \textbf{(25)}$$

Suppose there is no nonzero function $g \in G$ satisfying the indicated condition. Then (25) holds for all nonzero functions $g \in G$; that is,

$$\lim_{|s| \to \infty} l(sg) = -\infty, \qquad g \in G \text{ with } g \neq 0. \tag{26}$$

Let $g_1, \ldots, g_p$ be a basis of G and set $h(\mathbf{b}) = l(b_1 g_1 + \cdots + b_p g_p)$, where $\mathbf{b} = [b_1, \ldots, b_p]^T \in \mathbb{R}^p$. Then [see (15) and (26)] h satisfies the conditions of Theorem 13.5, so it has a global maximum and hence there is a maximum-likelihood estimate in G of the logistic regression function. ■

COROLLARY **13.3**   Consider a linear logistic regression model. If G is identifiable relative to the subset $\{ \mathbf{x}_k : 1 \leq k \leq d \text{ and } 0 < \bar{Y}_k < 1 \}$ of the design set, then there is a maximum-likelihood estimate in G of the logistic regression function.

Proof   Let g be a function in G that satisfies the following condition: for $1 \leq k \leq d$, if $g(\mathbf{x}_k) < 0$, then $\bar{Y}_k = 0$ and if $g(\mathbf{x}_k) > 0$, then $\bar{Y}_k = 1$. Then g equals zero on the indicated subset of the design set, so g is the zero function. Thus, there is no nonzero function $g \in G$ satisfying the indicated condition. We conclude from Theorem 13.6 that there is a maximum-likelihood estimate in G of the logistic regression function. ■

COROLLARY **13.4**   Consider the linear logistic regression model. If $0 < \bar{Y}_k < 1$ for $1 \leq k \leq d$, then there is a maximum-likelihood estimate in G of the logistic regression function.

Proof   By ignoring points in $\mathcal{X}$ that are not design points, we can assume that G is identifiable. The desired result now follows from Corollary 13.3. ■

EXAMPLE **13.6**   Consider the linear logistic regression model with G being the space of linear functions on $\mathbb{R}$ and consider the data shown in Table 13.8. Observe that G is identifiable. Show that there is no maximum-likelihood estimate in G of the logistic regression function.

TABLE **13.8**

x	n	Y	$\bar{Y}$
100	40	0	.0
110	25	0	.0
120	50	20	.4
130	30	30	1.0

Solution   Let g be the nonzero function in G given by $g(x) = x - 120$, which is shown in Figure 13.3. Then $g(100) < 0$ and $\bar{Y}_1 = 0$; $g(110) < 0$ and $\bar{Y}_2 = 0$; $g(120) = 0$;

$g(130) > 0$ and $\bar{Y}_4 = 1$. Thus, g satisfies the condition in Theorem 13.6, so there is no maximum-likelihood estimate in G of the logistic regression function. ∎

FIGURE   13.3

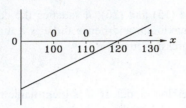

EXAMPLE 13.7   Consider the linear logistic regression model with $\mathcal{X} = \mathbb{R}$ and the data shown in Table 13.9. Investigate the existence of the maximum-likelihood estimate in G of the logistic regression function when G is the space of

(a) linear functions;   **(b)** quadratic polynomials;

(c) cubic polynomials.

TABLE   13.9

x	n	Y	$\bar{Y}$
100	40	0	.0
110	25	5	.2
120	50	20	.4
130	30	30	1.0

Solution   Since there are four design points, the spaces of linear functions, quadratic polynomials, and cubic polynomials are all identifiable.

(a) The space of linear functions is identifiable relative to $\{110, 120\}$, so it follows from Corollary 13.3 that there is a maximum-likelihood linear estimate of the logistic regression function.

(b) Let g be a quadratic polynomial satisfying the following condition: For $1 \le k \le 4$, if $g(x_k) < 0$, then $\bar{Y}_k = 0$; and if $g(x_k) > 0$, then $\bar{Y}_k = 1$. Since $\bar{Y}_1 = 0$, it follows from this condition that $g(100) \le 0$; since $\bar{Y}_4 = 1$, it follows that $g(130) \ge 0$. Moreover, since $0 < \bar{Y}_2 < 1$ and $0 < \bar{Y}_3 < 1$, it follows that $g(110) = g(120) = 0$ and hence that $g(x) = c(x - 110)(x - 120)$ for $x \in \mathbb{R}$, where c is some constant; thus, $g(100) = 200c$ and $g(130) = 200c$, so $g(100) = g(130)$. Since $g(100) \le 0$ and $g(130) \ge 0$, we conclude that

$g(100) = g(130) = 0$. Thus, $c = 0$ and hence g is the zero function. Therefore, by Theorem 13.6, there is a maximum-likelihood quadratic polynomial estimate of the logistic regression function.

(c) Consider the nonzero cubic polynomial g given by

$$g(x) = (x - 110)(x - 120)(x - 125),$$

which is shown in Figure 13.4. Now $g(100) < 0$ and $\bar{Y}_1 = 0$; $g(110) = 0$; $g(120) = 0$; $g(130) > 0$ and $\bar{Y}_4 = 1$. Thus, g is a nonzero function that satisfies the condition in Theorem 13.6, so there is no maximum-likelihood cubic polynomial estimate of the logistic regression function. [Here we could also have chosen $g(x) = (x - 110)(x - 120)(x - 130)$, so that $g(130) = 0$.] ■

F I G U R E   13.4

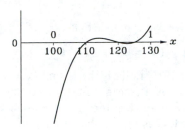

The proof of the next result is a straightforward modification of the proof of Theorem 13.6, the details being left as a problem. (In this and the next two results, $\bar{Y}_k$ can be replaced by Y_k.)

THEOREM 13.7   Consider a linear Poisson regression model in which G is identifiable. There is a maximum-likelihood estimate in G of the Poisson regression function if and only if there is no nonzero function $g \in G$ that satisfies the following condition: For $1 \le k \le d$, $g(\mathbf{x}_k) \le 0$; and if $g(\mathbf{x}_k) < 0$, then $\bar{Y}_k = 0$.

The proofs of the following two consequences of Theorem 13.7 are also left as a problem.

COROLLARY 13.5   Consider a linear Poisson regression model. If G is identifiable relative to the subset $\{\mathbf{x}_k : 1 \le k \le d \text{ and } \bar{Y}_k > 0\}$ of the design set, then there is a maximum-likelihood estimate in G of the Poisson regression function.

COROLLARY 13.6   Consider the linear Poisson regression model. If $\bar{Y}_k > 0$ for $1 \le k \le d$, then there is a maximum-likelihood estimate in G of the Poisson regression function.

EXAMPLE **13.8**   Consider the lung cancer study, which was described in Section 13.1. Let us assign the coded values of $1, \ldots, 9$, respectively, to $15 \leq T \leq 19, \ldots, 55 \leq T \leq 59$ and the coded values $1, \ldots, 7$, respectively, to $C = 0, \ldots, C \geq 35$. Then the design set is the Cartesian product of $\{1, \ldots, 9\}$ and $\{1, \ldots, 7\}$. Consider the linear Poisson regression model corresponding to the given design and the space G of all additive functions on the design set. Show that there is a maximum-likelihood estimate in G of the Poisson regression function.

Solution   Let $Y_{k_1 k_2}$ be the response variable and $n_{k_1 k_2}$ the time parameter corresponding to the design point (k_1, k_2) and set $\bar{Y}_{k_1 k_2} = Y_{k_1 k_2}/n_{k_1 k_2}$. Now let $g(x_1, x_2) = g_1(x_1) + g_2(x_2) \in G$ and suppose $g = 0$ on the subset

$$\mathcal{X}'' = \{ (k_1, k_2) : 1 \leq k_1 \leq 9, \ 1 \leq k_2 \leq 7 \text{ and } \bar{Y}_{k_1 k_2} > 0 \}$$

of the design set; that is,

$$g_1(k_1) + g_2(k_2) = 0 \qquad \text{if } 1 \leq k_1 \leq 9, 1 \leq k_2 \leq 7 \text{ and } \bar{Y}_{k_1 k_2} > 0.$$

Since $\bar{Y}_{26}, \ldots, \bar{Y}_{96}$ are all positive, we conclude that

$$g_1(k_1) + g_2(6) = 0, \qquad 2 \leq k_1 \leq 9,$$

and hence that $g_1(2) = \cdots = g_1(9)$. Similarly, by letting $k_2 = 1$, we conclude that $g_1(1) = g_1(9)$ and hence that g_1 is a constant function. In the same manner, we conclude that g_2 is a constant function. Therefore, g is a constant function and hence $g = 0$. This argument shows that G is identifiable relative to $\mathcal{X}''$. Thus, by Corollary 13.5, there is a maximum-likelihood estimate in G of the Poisson regression function.   ∎

THEOREM **13.8**   Consider the linear canonical regression model and let G_0 be a subspace of G. If there is a maximum-likelihood estimate in G of the canonical regression function, then there is a maximum-likelihood estimate in G_0 of this function.

Proof   By ignoring points in $\mathcal{X}$ that are not design points, we can assume that G and G_0 are identifiable. Suppose there is a maximum-likelihood estimate in G of the canonical regression function.

   In the context of logistic regression, we conclude from Theorem 13.6 that there is no nonzero function $g \in G$ satisfying the condition stated in that theorem. Thus, there is no nonzero function $g \in G_0$ satisfying the stated condition. Applying Theorem 13.6 with G replaced by G_0, we conclude that there is a maximum-likelihood estimate in G_0 of the logistic regression function. Similarly, in the context of Poisson regression, we conclude from Theorem 13.7 that there is a maximum-likelihood estimate in G_0 of the Poisson regression function.   ∎

THEOREM **13.9**   Consider the linear canonical regression model with $\mathbf{x}_1, \ldots, \mathbf{x}_d$ and G fixed. The probability that there is a maximum-likelihood estimate in G of the canonical regression model converges to one as $n_1, \ldots, n_d \to \infty$.

Proof Consider logistic regression. Since $0 < \pi(\mathbf{x}_k) < 1$ for $1 \leq k \leq d$, it follows from the weak law of large numbers that $P(0 < \bar{Y}_k < 1 \text{ for } 1 \leq k \leq d) \to 1$ as $n_1, \ldots, n_d \to \infty$. The desired conclusion now follows from Corollary 13.4.

Consider next Poisson regression. Since $\lambda(\mathbf{x}_k) > 0$ for $1 \leq k \leq d$, it follows from the weak law of large numbers that $P(\bar{Y}_k > 0 \text{ for } 1 \leq k \leq d) \to 1$ as $n_1, \ldots, n_d \to \infty$. The desired result now follows from Corollary 13.6. ∎

Problems

13.29 Let Y_1 and Y_2 be independent random variables such that Y_1 has the Poisson distribution with time parameter n_1 and canonical parameter θ and Y_2 has the Poisson distribution with time parameter n_2 and canonical parameter 2θ, where the positive parameter θ is unknown. **(a)** Determine the log-likelihood function corresponding to the observation of Y_1 and Y_2. **(b)** Determine $l'(\theta)$ and show that it is a strictly decreasing function of θ. **(c)** Determine the limiting value of $l'(\theta)$ as $\theta \to -\infty$ and as $\theta \to \infty$. **(d)** Use parts (b) and (c) to determine a necessary and sufficient condition on Y_1 and Y_2 for the maximum-likelihood estimate $\hat{\theta}$ of θ to fail to exist. **(e)** Find an explicit formula for $\hat{\theta}$ when it exists. [*Hint:* First solve a quadratic equation to determine $\hat{\lambda} = \exp \hat{\theta}$.]

13.30 Consider the setup of Problem 13.29, with $n_1 = 20$ and $n_2 = 30$. Use the answer to Problem 13.29(e) to determine the maximum-likelihood estimate $\hat{\theta}$ when $Y_1 = 15$ and $Y_2 = 20$.

13.31 Explain how the setup of Problem 13.29 can be viewed as a Poisson regression model.

13.32 Use Theorem 13.7 and Problem 13.3 to solve Problem 13.29(d).

13.33 Let G_0 be a subspace of a p-dimensional linear space G. Suppose there is a maximum-likelihood estimate $\hat{\theta}(\cdot)$ in G of the canonical regression function and hence a maximum-likelihood estimate $\hat{\theta}_0(\cdot)$ in G_0 of this function. Let $\hat{\mu}(\cdot) = C'(\hat{\theta}(\cdot))$ and $\hat{\mu}_0(\cdot) = C'(\hat{\theta}_0(\cdot))$ be the corresponding maximum-likelihood estimates of the regression function. Verify that **(a)** $\sum_k n_k g(\mathbf{x}_k)[\hat{\mu}(\mathbf{x}_k) - \hat{\mu}_0(\mathbf{x}_k)] = 0$ for $g \in G_0$; **(b)** if G_0 contains the constant functions, then $\sum_k n_k [\hat{\mu}(\mathbf{x}_k) - \hat{\mu}_0(\mathbf{x}_k)] = 0$.

13.34 Let $g_1, \ldots, g_p$ be a basis of G. Show that the function $\hat{\theta}(\cdot) \in G$ satisfies the maximum-likelihood equation (10) if and only if

$$\sum_k n_k g_j(\mathbf{x}_k) C'(\hat{\theta}(\mathbf{x}_k)) = \sum_k n_k g_j(\mathbf{x}_k) \bar{Y}_k, \qquad 1 \leq j \leq p. \tag{27}$$

13.35 Let $g_1, \ldots, g_p$ be a basis of G, let $\hat{\theta}(\cdot) = \hat{\beta}_1 g_1 + \cdots + \hat{\beta}_p g_p \in G$, and set $\hat{\beta} = [\hat{\beta}_1, \ldots, \hat{\beta}_p]^T$. Show that (27) can be rewritten in matrix form as

$$\mathbf{X}^T \mathbf{N} C'(\mathbf{X}\beta) = \mathbf{X}^T \mathbf{N}\bar{\mathbf{Y}},$$

where $\mathbf{N} = \mathrm{diag}(n_1, \ldots, n_d)$.

13.36 Consider the logistic regression model with $\mathcal{X} = \mathbb{R}$ and the data shown in Table 13.10. Investigate the existence of the maximum-likelihood estimate in G of the logistic regression function when G is the space of (a) linear functions; (b) quadratic polynomials; (c) cubic polynomials.

TABLE 13.10

x	n	Y	$\bar{Y}$
100	40	0	.0
110	25	5	.2
120	50	20	.4
130	30	0	.0

13.37 Consider a logistic regression model with G being the space of linear functions on $\mathbb{R}^2$ and suppose G is identifiable. Give a geometric necessary and sufficient condition for there to be *no* maximum-likelihood estimate in G of the logistic regression function.

13.38 Verify Theorem 13.7.

13.39 Verify Corollaries 13.5 and 13.6.

13.40 Consider the Poisson regression model with G being the space of linear functions on $\mathbb{R}$, and consider the data shown in Table 13.11. Observe that G is identifiable. Use Theorem 13.7 to show that there is *no* maximum-likelihood estimate in G of the Poisson regression function.

TABLE 13.11

x	n	Y
100	40	0
110	25	0
120	50	20

13.41 Consider a Poisson regression model with G being the space of linear functions on $\mathbb{R}^2$ and suppose G is identifiable. Give a geometric necessary and sufficient condition for there to be *no* maximum-likelihood estimate in G of the Poisson regression function.

13.42 Consider the Poisson regression model with $\mathcal{X} = \mathbb{R}$ and the data shown in Table 13.12. Investigate the existence of the maximum-likelihood estimate in G of the Poisson regression function when G is the space of (a) linear functions; (b) quadratic polynomials; (c) cubic polynomials.

| | **TABLE** | **13.12** |

x	n	Y
100	40	5
110	25	0
120	50	20
130	30	30

13.43 Consider the logistic regression model corresponding to a given set of data and a given identifiable linear space G. Show that if there is a maximum-likelihood estimate in G of the logistic regression function, then there is a maximum-likelihood estimate in G of the Poisson regression function in the context of the corresponding Poisson regression model for the same data.

13.4
Iteratively Reweighted Least-Squares Method

Consider a linear canonical regression model in which G is identifiable and there is a (necessarily unique) maximum-likelihood estimate $\hat{\theta}(\cdot)$ in G of the canonical regression function $\theta(\cdot)$, which is the unique solution to the maximum-likelihood equation. Let $g_1, \ldots, g_p$ be a basis of G. Then $\hat{\theta}(\cdot) = \hat{\beta}_1 g_1 + \cdots + \hat{\beta}_p g_p$. Set $\hat{\beta} = [\hat{\beta}_1, \ldots, \hat{\beta}_p]^T$. When G is unsaturated, the maximum-likelihood estimate $\hat{\theta}(\cdot)$ and its coefficient vector $\hat{\beta}$ generally cannot be found algebraically, so some numerical method of solution is required. In this section, the method that is commonly used in practice will be described.

Recall that the log-likelihood function is given by

$$l(\theta(\cdot)) = \sum_k n_k [\theta(\mathbf{x}_k)\bar{Y}_k - C(\theta(\mathbf{x}_k))] + \log r(\mathbf{Y}; \mathbf{n}), \qquad \theta(\cdot) \in G.$$

Let $\hat{\theta}_0(\cdot) \in G$ be an initial approximation to $\hat{\theta}(\cdot)$. Then

$$l(\hat{\theta}_0(\cdot) + g) = \sum_k n_k \left\{ [\hat{\theta}_0(\mathbf{x}_k) + g(\mathbf{x}_k)]\bar{Y}_k - C(\hat{\theta}_0(\mathbf{x}_k) + g(\mathbf{x}_k)) \right\}$$
$$+ \log r(\mathbf{Y}; \mathbf{n}), \qquad g \in G.$$

Consider the quadratic Taylor approximation

$$C(\hat{\theta}_0(\mathbf{x}_k) + g(\mathbf{x}_k)) \approx C(\hat{\theta}_0(\mathbf{x}_k)) + C'(\hat{\theta}_0(\mathbf{x}_k))g(\mathbf{x}_k) + \tfrac{1}{2}C''(\hat{\theta}_0(\mathbf{x}_k))g^2(\mathbf{x}_k), \quad g \in G,$$

to $C(\hat{\theta}_0(\mathbf{x}_k) + g(\mathbf{x}_k))$ for $1 \le k \le d$, which is accurate when g is suitably small. We can rewrite this approximation as

$$C(\hat{\theta}_0(\mathbf{x}_k) + g(\mathbf{x}_k)) \approx C(\hat{\theta}_0(\mathbf{x}_k)) + \hat{\mu}_0(\mathbf{x}_k)g(\mathbf{x}_k) + \tfrac{1}{2}\hat{\sigma}_0^2(\mathbf{x}_k)g^2(\mathbf{x}_k), \qquad g \in G,$$

where $\hat{\mu}_0(\cdot) = C'(\hat{\theta}_0(\cdot))$ is an approximation to $\hat{\mu}(\cdot) = C'(\hat{\theta}(\cdot))$ and $\hat{\sigma}_0^2(\cdot) = C''(\hat{\theta}_0(\cdot))$ is an approximation to $\hat{\sigma}^2(\cdot) = C''(\hat{\theta}(\cdot))$. The corresponding quadratic approximation to the log-likelihood function is given by

$$l(\hat{\theta}_0(\cdot) + g) \approx \sum_k n_k \Big\{ [\hat{\theta}_0(\mathbf{x}_k) + g(\mathbf{x}_k)]\bar{Y}_k$$

$$- [C(\hat{\theta}_0(\mathbf{x}_k)) + \hat{\mu}_0(\mathbf{x}_k)g(\mathbf{x}_k) + \tfrac{1}{2}\hat{\sigma}_0^2(\mathbf{x}_k)g^2(\mathbf{x}_k)]\Big\} + \log r(\mathbf{Y}; \mathbf{n})$$

$$= \sum_k n_k \Big\{ g(\mathbf{x}_k)[\bar{Y}_k - \hat{\mu}_0(\mathbf{x}_k)] - \tfrac{1}{2}\hat{\sigma}_0^2(\mathbf{x}_k)g^2(\mathbf{x}_k)\Big\} + \text{remainder term}$$

$$= -\frac{1}{2}\sum_k n_k\hat{\sigma}_0^2(\mathbf{x}_k)\Big(\frac{\bar{Y}_k - \hat{\mu}_0(\mathbf{x}_k)}{\hat{\sigma}_0^2(\mathbf{x}_k)} - g(\mathbf{x}_k)\Big)^2 + \text{remainder term},$$

where the two remainder terms do not involve g. Observe that maximizing this quadratic approximation to the log-likelihood function is equivalent to minimizing the weighted residual sum of squares

$$\mathrm{RSS}_0(g) = \sum_k n_k\hat{\sigma}_0^2(\mathbf{x}_k)\Big(\frac{\bar{Y}_k - \hat{\mu}_0(\mathbf{x}_k)}{\hat{\sigma}_0^2(\mathbf{x}_k)} - g(\mathbf{x}_k)\Big)^2, \qquad g \in G.$$

Consider the artificial weights $\hat{w}_{k0} = n_k\hat{\sigma}_0^2(\mathbf{x}_k)$, $1 \le k \le d$, and the artificial response data $Y_{k0} = (\bar{Y}_k - \hat{\mu}_0(\mathbf{x}_k))/\hat{\sigma}_0^2(\mathbf{x}_k)$, $1 \le k \le d$. In terms of these quantities, we can rewrite the above weighted residual sum of squares as $\mathrm{RSS}_0(g) = \sum_k \hat{w}_{k0}(Y_{k0} - g(\mathbf{x}_k))^2$ for $g \in G$.

Let $\hat{g}_0 \in G$ be the corresponding weighted least-squares estimate—that is, the function in G that minimizes this weighted residual sum of squares. To determine $\hat{g}_0$ explicitly, set $\mathbf{W}_0 = \mathrm{diag}(\hat{w}_{10}, \ldots, \hat{w}_{d0})$ and $\mathbf{Y}_0 = [Y_{10}, \ldots, Y_{d0}]^T$, choose a basis $g_1, \ldots, g_p$ of G, let $\mathbf{X}$ be the corresponding $d \times p$ design matrix, write

$$\hat{\theta}_0(\cdot) = \hat{\beta}_{10}g_1 + \cdots + \hat{\beta}_{p0}g_p \quad \text{and} \quad \hat{g}_0 = Z_{10}g_1 + \cdots + Z_{p0}g_p,$$

and set $\hat{\beta}_0 = [\hat{\beta}_{10}, \ldots, \hat{\beta}_{p0}]^T$ and $\mathbf{Z}_0 = [Z_{10}, \ldots, Z_{p0}]^T$. Then the normal equation for $\mathbf{Z}_0$ [see Equation (65) in Section 9.6] is given by $\mathbf{X}^T\hat{\mathbf{W}}_0\mathbf{X}\mathbf{Z}_0 = \mathbf{X}^T\hat{\mathbf{W}}_0\mathbf{Y}_0$, whose unique solution is given by $\mathbf{Z}_0 = (\mathbf{X}^T\hat{\mathbf{W}}_0\mathbf{X})^{-1}\mathbf{X}^T\hat{\mathbf{W}}_0\mathbf{Y}_0$. The corresponding approximation $\hat{\theta}_1(\cdot) = \hat{\theta}_0(\cdot) + \hat{g}_0$ to $\hat{\theta}(\cdot)$ is given by

$$\hat{\theta}_1(\cdot) = \hat{\beta}_{11}g_1 + \cdots + \hat{\beta}_{p1}g_p, \qquad \text{where } \hat{\beta}_1 = [\hat{\beta}_{11}, \ldots, \hat{\beta}_{p1}]^T = \hat{\beta}_0 + \mathbf{Z}_0.$$

In the same manner, starting with the approximation $\hat{\theta}_1(\cdot)$ to $\hat{\theta}(\cdot)$, we get the new approximation $\hat{\theta}_2(\cdot)$; starting with $\hat{\theta}_2(\cdot)$, we get $\hat{\theta}_3(\cdot)$; and so forth. The $(\nu + 1)$th such approximation $\hat{\theta}_{\nu+1}(\cdot)$, $\nu \ge 0$, is given in terms of the previous approximation by $\hat{\theta}_{\nu+1}(\cdot) = \hat{\theta}_\nu(\cdot) + \hat{g}_\nu$, where $\hat{g}_\nu \in G$ minimizes the residual sum of squares

$$\mathrm{RSS}_\nu(g) = \sum_k n_k\hat{\sigma}_\nu^2(\mathbf{x}_k)\Big(\frac{\bar{Y}_k - \hat{\mu}_\nu(\mathbf{x}_k)}{\hat{\sigma}_\nu^2(\mathbf{x}_k)} - g(\mathbf{x}_k)\Big)^2, \qquad g \in G,$$

with $\hat{\mu}_\nu(\cdot) = C'(\hat{\theta}_\nu(\cdot))$ and $\hat{\sigma}_\nu^2(\cdot) = C''(\hat{\theta}_\nu(\cdot))$. We can rewrite this residual sum of squares as $\mathrm{RSS}_\nu(g) = \sum_k \hat{w}_{k\nu}(Y_{k\nu} - g(\mathbf{x}_k))^2$ for $g \in G$, where $\hat{w}_{k\nu} = n_k\hat{\sigma}_\nu^2(\mathbf{x}_k)$ and

$Y_{kv} = (\bar{Y}_k - \hat{\mu}_v(\mathbf{x}_k))/\hat{\sigma}_v^2(\mathbf{x}_k)$ for $1 \le k \le d$. Set

$$\hat{\mathbf{W}}_v = \mathrm{diag}(\hat{w}_{1v}, \ldots, \hat{w}_{dv}) \quad \text{and} \quad \mathbf{Y}_v = [\, Y_{1v}, \ldots, Y_{dv} \,]^T$$

and write

$$\hat{\theta}_v(\cdot) = \hat{\beta}_{1v}g_1 + \cdots + \hat{\beta}_{pv}g_p \quad \text{and} \quad \hat{g}_v = Z_{1v}g_1 + \cdots + Z_{pv}g_p.$$

Then $\hat{\beta}_{v+1} = \hat{\beta}_v + \mathbf{Z}_v$, where $\hat{\beta}_v = [\, \hat{\beta}_{1v}, \ldots, \hat{\beta}_{pv} \,]^T$ and $\mathbf{Z}_v = [\, Z_{1v}, \ldots, Z_{pv} \,]^T$. The normal equation for $\mathbf{Z}_v$ is given by $\mathbf{X}^T\hat{\mathbf{W}}_v\mathbf{X}\mathbf{Z}_v = \mathbf{X}^T\hat{\mathbf{W}}_v\mathbf{Y}_v$, whose unique solution is given by $\mathbf{Z}_v = (\mathbf{X}^T\hat{\mathbf{W}}_v\mathbf{X})^{-1}\mathbf{X}^T\hat{\mathbf{W}}_v\mathbf{Y}_v$.

Set

$$\hat{\boldsymbol{\theta}}_v = [\, \hat{\theta}_{1v}, \ldots, \hat{\theta}_{dv} \,]^T = [\, \hat{\theta}_v(\mathbf{x}_1), \ldots, \hat{\theta}_v(\mathbf{x}_d) \,]^T = \mathbf{X}\hat{\beta}_v,$$

$$\hat{\boldsymbol{\mu}}_v = [\, \hat{\mu}_{1v}, \ldots, \hat{\mu}_{dv} \,]^T = [\, \hat{\mu}_v(\mathbf{x}_1), \ldots, \hat{\mu}_v(\mathbf{x}_d) \,]^T = C'(\hat{\boldsymbol{\theta}}_v),$$

and

$$\hat{\boldsymbol{\sigma}}_v^2 = [\, \hat{\sigma}_{1v}^2, \ldots, \hat{\sigma}_{dv}^2 \,]^T = [\, \hat{\sigma}_v^2(\mathbf{x}_1), \ldots, \hat{\sigma}_v^2(\mathbf{x}_d) \,]^T = C''(\hat{\boldsymbol{\theta}}_v).$$

Then $\hat{\mathbf{W}}_v = \mathbf{N}\hat{\boldsymbol{\sigma}}_v^2$, where $\mathbf{N} = \mathrm{diag}(n_1, \ldots, n_d)$. Observe that

$$\hat{\mathbf{W}}_v\mathbf{Y}_v = \left[\, n_1(\bar{Y}_1 - \hat{\mu}_v(\mathbf{x}_1)), \ldots, n_d(\bar{Y}_d - \hat{\mu}_v(\mathbf{x}_d)) \,\right]^T = \mathbf{N}(\bar{\mathbf{Y}} - \hat{\boldsymbol{\mu}}_v).$$

Thus, the normal equation for $\mathbf{Z}_v$ can be written as $\mathbf{X}^T\hat{\mathbf{W}}_v\mathbf{X}\mathbf{Z}_v = \mathbf{X}^T\mathbf{N}(\bar{\mathbf{Y}} - \hat{\boldsymbol{\mu}}_v)$, whose unique solution is given by $\mathbf{Z}_v = \hat{\mathbf{V}}_v\mathbf{X}^T\mathbf{N}(\bar{\mathbf{Y}} - \hat{\boldsymbol{\mu}}_v)$, where $\hat{\mathbf{V}}_v = (\mathbf{X}^T\hat{\mathbf{W}}_v\mathbf{X})^{-1}$. In summary, $\hat{\beta}_{v+1}$ is obtained from $\hat{\beta}_v$ by the following steps:

$$\hat{\boldsymbol{\theta}}_v = \mathbf{X}\hat{\beta}_v,$$

$$\hat{\boldsymbol{\mu}}_v = C'(\hat{\boldsymbol{\theta}}_v),$$

$$\hat{\boldsymbol{\sigma}}_v^2 = C''(\hat{\boldsymbol{\theta}}_v),$$

$$\hat{\mathbf{W}}_v = \mathrm{diag}(\mathbf{N}\hat{\boldsymbol{\sigma}}_v^2),$$

$$\hat{\mathbf{V}}_v = (\mathbf{X}^T\hat{\mathbf{W}}_v\mathbf{X})^{-1},$$

$$\mathbf{Z}_v = \hat{\mathbf{V}}_v\mathbf{X}^T\mathbf{N}(\bar{\mathbf{Y}} - \hat{\boldsymbol{\mu}}_v),$$

$$\hat{\beta}_{v+1} = \hat{\beta}_v + \mathbf{Z}_v.$$

If $\hat{\beta}_0$ is sufficiently close to $\hat{\beta}$, then the successive approximations $\hat{\beta}_v$ converge quadratically fast to $\hat{\beta}$; that is, the number of correct digits in the vth approximation approximately doubles at each iteration until floating-point errors in the calculations come into consideration. In automatic computer calculations, a rule must be employed for stopping the iteration process at some v and approximating $\hat{\beta}$ by $\hat{\beta}_v$. One reasonable rule is to stop when $l(\hat{\beta}_v) - l(\hat{\beta}_{v-1}) \le \epsilon$ for some small positive number ϵ (such as 10^{-5}).

The numerical method we have just described for determining $\hat{\beta}$ and hence $\hat{\theta}(\cdot)$ is referred to as the *iteratively reweighted least-squares method* since the weights $\hat{w}_{kv} = n_k\hat{\sigma}_v^2(\mathbf{x}_k) = n_kC''(\hat{\theta}_v(\mathbf{x}_k))$, $1 \le k \le d$, are recalculated during each iteration.

The general method of maximizing a function by successively maximizing quadratic Taylor approximations to this function is known as the *Newton–Raphson method*.

EXAMPLE 13.9   Obtain a more explicit form for the steps in a single iteration of the iteratively reweighted least-squares method that is applicable to logistic regression.

Solution   Recall that if $\pi = C'(\theta)$, then $\pi(1 - \pi) = C''(\theta)$. Given $c \in \mathbb{R}$ and $\mathbf{u} = [u_1, \ldots, u_d]^T$ and $\mathbf{v} = [v_1, \ldots, v_d]^T \in \mathbb{R}^d$, write $[c + u_1, \ldots, c + u_d]^T$ as $c + \mathbf{u}$, $[u_1 v_1, \ldots, u_d v_d]^T$ as $\mathbf{u} * \mathbf{v}$, and $[u_1/v_1, \ldots, u_d/v_d]^T$ as $\mathbf{u}/\mathbf{v}$. Then the steps in a single iteration can be written as follows:

$$\hat{\boldsymbol{\theta}}_\nu = \mathbf{X}\hat{\boldsymbol{\beta}}_\nu,$$

$$\hat{\boldsymbol{\pi}}_\nu = (\exp \hat{\boldsymbol{\theta}}_\nu)/(1 + \exp \hat{\boldsymbol{\theta}}_\nu),$$

$$\hat{\mathbf{W}}_\nu = \mathrm{diag}(\mathbf{n} * \hat{\boldsymbol{\pi}}_\nu * (1 - \hat{\boldsymbol{\pi}}_\nu)),$$

$$\hat{\mathbf{V}}_\nu = (\mathbf{X}^T \hat{\mathbf{W}}_\nu \mathbf{X})^{-1},$$

$$\mathbf{Z}_\nu = \hat{\mathbf{V}}_\nu \mathbf{X}^T \mathbf{N}(\bar{\mathbf{Y}} - \hat{\boldsymbol{\pi}}_\nu),$$

$$\hat{\boldsymbol{\beta}}_{\nu+1} = \hat{\boldsymbol{\beta}}_\nu + \mathbf{Z}_\nu.$$

Here $\hat{\boldsymbol{\pi}}_\nu = [\hat{\pi}_{1\nu}, \ldots, \hat{\pi}_{d\nu}]^T$, where $\hat{\pi}_{k\nu} = (\exp \hat{\theta}_{k\nu})/(1 + \exp \hat{\theta}_{k\nu})$ for $1 \leq k \leq d$. In applying the iteratively reweighted least-squares method, it is necessary to choose an initial approximation $\hat{\boldsymbol{\beta}}_0$ to $\hat{\boldsymbol{\beta}}$. Ideally, we would like to use an accurate starting value based on prior information, say, from the maximum-likelihood estimate in a previous experiment. If no such information is available, however, it is conventional to use $\hat{\boldsymbol{\beta}}_0 = \mathbf{0}$, which corresponds to $\hat{\boldsymbol{\pi}}_0 = [\frac{1}{2}, \ldots, \frac{1}{2}]^T$.   ∎

If the starting value $\hat{\boldsymbol{\beta}}_0$ in the iteratively reweighted least-squares method is too far from $\hat{\boldsymbol{\beta}}$, then the successive iterates $\hat{\boldsymbol{\beta}}_\nu$ need not converge to $\hat{\boldsymbol{\beta}}$ as $\nu \to \infty$. In fact, it may be that the corresponding values $l(\hat{\boldsymbol{\beta}}_\nu)$ of the log-likelihood function decrease rather than increase as ν increases. In the present context, there is a simple modification to the individual iterations, known as *step-halving*, that solves the problem: Given $\hat{\boldsymbol{\beta}}_\nu$ and $\mathbf{Z}_\nu$, choose m to be the smallest nonnegative integer such that

$$l(\hat{\boldsymbol{\beta}}_\nu + 2^{-m}\mathbf{Z}_\nu) \geq l(\hat{\boldsymbol{\beta}}_\nu + 2^{-(m+1)}\mathbf{Z}_\nu)$$

and set $\hat{\boldsymbol{\beta}}_{\nu+1} = \hat{\boldsymbol{\beta}}_\nu + 2^{-m}\mathbf{Z}_\nu$. With this modification, $\hat{\boldsymbol{\beta}}_\nu$ will converge to $\hat{\boldsymbol{\beta}}$ as $\nu \to \infty$ for any starting value $\hat{\boldsymbol{\beta}}_0$.

EXAMPLE 13.10   Consider the linear logistic regression model in which $\theta(x) = \beta_0 + \beta_1 x$ for $x \in \mathbb{R}$ and consider the experimental data shown in Table 13.13. Let $\hat{\beta}_0$ and $\hat{\beta}_1$ denote the maximum-likelihood estimates of β_0 and β_1, respectively.

TABLE **13.13**

x	n	Y	$\bar{Y}$
−1	60	12	.2
0	40	16	.4
1	60	36	.6

(a) Determine explicitly the maximum-likelihood equations for $\hat{\beta}_0$ and $\hat{\beta}_1$.

(b) Starting with $\hat{\beta}_{00} = 0$ and $\hat{\beta}_{10} = 0$, determine the approximations $\hat{\beta}_{01}$ and $\hat{\beta}_{11}$ from the first iteration in the iteratively reweighted least-squares method.

(c) The maximum-likelihood estimates of the regression coefficients are given by $\hat{\beta}_0 \doteq -0.4656$ and $\hat{\beta}_1 \doteq 0.8911$. Determine the corresponding maximum-likelihood estimates of θ_1, θ_2, and θ_3 and those of π_1, π_2, and π_3.

(d) Check that the numerical values given in part (c) for $\hat{\beta}_0$ and $\hat{\beta}_1$ satisfy the maximum-likelihood equations obtained in part (a).

Solution **(a)** The maximum-likelihood equations for $\hat{\beta}_0$ and $\hat{\beta}_1$ are given by

$$\sum_k n_k \frac{\exp(\hat{\beta}_0 + \hat{\beta}_1 x_k)}{1 + \exp(\hat{\beta}_0 + \hat{\beta}_1 x_k)} = \sum_k n_k \bar{Y}_k$$

and

$$\sum_k n_k x_k \frac{\exp(\hat{\beta}_0 + \hat{\beta}_1 x_k)}{1 + \exp(\hat{\beta}_0 + \hat{\beta}_1 x_k)} = \sum_k n_k x_k \bar{Y}_k$$

(see Problem 13.34 in Section 13.3). Substituting the experimental data into these equations, we get

$$60 \frac{\exp(\hat{\beta}_0 - \hat{\beta}_1)}{1 + \exp(\hat{\beta}_0 - \hat{\beta}_1)} + 40 \frac{\exp(\hat{\beta}_0)}{1 + \exp(\hat{\beta}_0)} + 60 \frac{\exp(\hat{\beta}_0 + \hat{\beta}_1)}{1 + \exp(\hat{\beta}_0 + \hat{\beta}_1)} = 64$$

and

$$-60 \frac{\exp(\hat{\beta}_0 - \hat{\beta}_1)}{1 + \exp(\hat{\beta}_0 - \hat{\beta}_1)} + 60 \frac{\exp(\hat{\beta}_0 + \hat{\beta}_1)}{1 + \exp(\hat{\beta}_0 + \hat{\beta}_1)} = 24.$$

(b) The design matrix is given by

$$\mathbf{X} = \begin{bmatrix} 1 & -1 \\ 1 & 0 \\ 1 & 1 \end{bmatrix}.$$

Also,

$$\mathbf{N} = \begin{bmatrix} 60 & 0 & 0 \\ 0 & 40 & 0 \\ 0 & 0 & 60 \end{bmatrix}.$$

Starting with $\hat{\beta}_{00} = 0$ and $\hat{\beta}_{10} = 0$, we get $\hat{\theta}_{10} = 0$, $\hat{\theta}_{20} = 0$, and $\hat{\theta}_{30} = 0$ and hence that $\hat{\pi}_{10} = .5$, $\hat{\pi}_{20} = .5$, and $\hat{\pi}_{30} = .5$. Thus, $\hat{\mathbf{W}}_0$ is given by

$$
\begin{bmatrix} n_1 \hat{\pi}_{10}(1 - \hat{\pi}_{10}) & 0 & 0 \\ 0 & n_2 \hat{\pi}_{20}(1 - \hat{\pi}_{20}) & 0 \\ 0 & 0 & n_3 \hat{\pi}_{30}(1 - \hat{\pi}_{30}) \end{bmatrix} = \begin{bmatrix} 15 & 0 & 0 \\ 0 & 10 & 0 \\ 0 & 0 & 15 \end{bmatrix},
$$

so

$$
\mathbf{X}^T \hat{\mathbf{W}}_0 \mathbf{X} = \begin{bmatrix} 1 & 1 & 1 \\ -1 & 0 & 1 \end{bmatrix} \begin{bmatrix} 15 & 0 & 0 \\ 0 & 10 & 0 \\ 0 & 0 & 15 \end{bmatrix} \begin{bmatrix} 1 & -1 \\ 1 & 0 \\ 1 & 1 \end{bmatrix} = \begin{bmatrix} 40 & 0 \\ 0 & 30 \end{bmatrix}
$$

and hence

$$
\hat{\mathbf{V}}_0 = \begin{bmatrix} 1/40 & 0 \\ 0 & 1/30 \end{bmatrix}.
$$

Consequently, $\hat{\beta}_1 = \hat{\beta}_0 + \mathbf{Z}_0 = \mathbf{Z}_0$; also, $\mathbf{Z}_0$ is given by

$$
\begin{bmatrix} 1/40 & 0 \\ 0 & 1/30 \end{bmatrix} \begin{bmatrix} 1 & 1 & 1 \\ -1 & 0 & 1 \end{bmatrix} \begin{bmatrix} 60 & 0 & 0 \\ 0 & 40 & 0 \\ 0 & 0 & 60 \end{bmatrix} \begin{bmatrix} .2 - .5 \\ .4 - .5 \\ .6 - .5 \end{bmatrix} = \begin{bmatrix} -0.4 \\ 0.8 \end{bmatrix},
$$

so $\hat{\beta}_{01} = -0.4$ and $\hat{\beta}_{11} = 0.8$.

(c) The maximum-likelihood estimates of θ_1, θ_2, and θ_3 are given by

$$
\hat{\theta}_1 = \hat{\beta}_0 + \hat{\beta}_1(-1) = \hat{\beta}_0 - \hat{\beta}_1 \doteq -1.3567,
$$

$$
\hat{\theta}_2 = \hat{\beta}_0 + \hat{\beta}_1(0) = \hat{\beta}_0 \doteq -0.4656,
$$

and

$$
\hat{\theta}_3 = \hat{\beta}_0 + \hat{\beta}_1(1) = \hat{\beta}_0 + \hat{\beta}_1 \doteq 0.4254.
$$

Thus, the maximum-likelihood estimates of π_1, π_2, and π_3 are given by

$$
\hat{\pi}_1 = \frac{\exp(\hat{\theta}_1)}{1 + \exp(\hat{\theta}_1)} \doteq .2048,
$$

$$
\hat{\pi}_2 = \frac{\exp(\hat{\theta}_2)}{1 + \exp(\hat{\theta}_2)} \doteq .3857,
$$

and

$$
\hat{\pi}_3 = \frac{\exp(\hat{\theta}_3)}{1 + \exp(\hat{\theta}_3)} \doteq .6048.
$$

(d) Checking that the given numerical values of $\hat{\beta}_0$ and $\hat{\beta}_1$ satisfy the maximum-likelihood equations obtained in part (a) amounts to checking that $60\hat{\pi}_1 + 40\hat{\pi}_2 + 60\hat{\pi}_3 \doteq 64$ and $-60\hat{\pi}_1 + 60\hat{\pi}_3 \doteq 24$. ∎

Beetle Experiment

EXAMPLE **13.11** Consider the beetle experiment, which was described in Section 13.1. Let the input variable be defined as $x = c$, where c is the concentration of CS_2. (It is more common in practice to use $x = \log_{10} c$.) Consider the logistic regression model with G being the space of quadratic functions of x and with 1, $(x - 62.8)/10$, and $[(x - 62.8)/10]^2$ being the basis of G. (Note that 62.8 is the average concentration.) Write the canonical regression function as $\theta(x) = \beta_0 + \beta_1(x - 62.8)/10 + \beta_2[(x - 62.8)/10]^2$ and its maximum-likelihood estimate as $\hat{\theta}(x) = \hat{\beta}_0 + \hat{\beta}_1(x - 62.8)/10 + \hat{\beta}_2[(x - 62.8)/10]^2$.

(a) Determine the design matrix.

(b) Use the iteratively reweighted least-squares method to determine $\hat{\beta}_0, \hat{\beta}_1, \hat{\beta}_2$.

(c) Describe an advantage of using the given basis of G rather than the usual basis $1, x, x^2$.

(d) Show the graph of $\hat{\theta}(\cdot)$ and indicate the experimental results.

(e) Show the graph of $\hat{\pi}(\cdot)$ and indicate the experimental results.

Solution (a) The design matrix is given by

$$\mathbf{X} = \begin{bmatrix} 1 & -1.374 & 1.887876 \\ 1 & -0.981 & 0.962361 \\ 1 & -0.589 & 0.346921 \\ 1 & -0.196 & 0.038416 \\ 1 & 0.196 & 0.038416 \\ 1 & 0.589 & 0.346921 \\ 1 & 0.981 & 0.962361 \\ 1 & 1.374 & 1.887876 \end{bmatrix}.$$

(It is now clear that the levels of concentration in the experiment were symmetrically distributed about 62.8.)

(b) Using the design matrix as determined in part (a), $\mathbf{n} = [59, 60, \ldots, 60]^T$ and $\mathbf{Y} = [6, 13, \ldots, 60]^T$; starting with $\hat{\beta}_0 = [0, 0, 0]^T$, we get the sequence of approximations to $\hat{\beta}$ as shown in Table 13.14. Clearly, the method has effectively converged after six iterations.

(c) An advantage of the given basis [or the basis $1, x - 62.8, (x - 62.8)^2$] over the usual basis $1, x, x^2$ is that the matrices $\mathbf{X}^T \hat{\mathbf{W}}_v \mathbf{X}$ arising during the various iterations are less nearly singular when the given basis is used than when the usual basis is used. Near singularity of these matrices leads to low relative accuracy in the floating-point computations required to carry out the method and can even prevent the method from converging (the successive iterates $\hat{\beta}_v$ can just wander around).

TABLE   13.14

ν	$\hat{\beta}_0$	$\hat{\beta}_1$	$\hat{\beta}_2$
0	0.00000	0.00000	0.00000
1	0.57610	1.49155	-0.20455
2	0.66006	2.14511	0.01296
3	0.65291	2.55360	0.37820
4	0.65406	2.78456	0.59233
5	0.65686	2.83564	0.63572
6	0.65700	2.83746	0.63721
7	0.65700	2.83746	0.63721

(d) The maximum-likelihood estimate of $\theta(\cdot)$ is given by

$$\hat{\theta}(x) \doteq 0.657 + 2.837\left(\frac{x - 62.8}{10}\right) + 0.637\left(\frac{x - 62.8}{10}\right)^2.$$

Figure 13.5 shows the graph of $\hat{\theta}(\cdot)$ and the points $(x_k, \text{logit } \bar{Y}_k)$, $1 \le k \le 7$ (note that logit $\bar{Y}_8 = \text{logit } 1 = \infty$).

FIGURE   13.5
Fitted logistic regression function

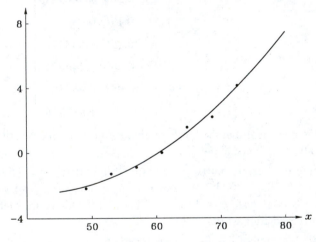

(e) The maximum-likelihood estimate of $\pi(\cdot)$ is given by

$$\hat{\pi}(x) = \frac{\exp \hat{\theta}(x)}{1 + \exp \hat{\theta}(x)},$$

where $\hat{\theta}(\cdot)$ is given in the solution to part (d). Figure 13.6 shows the graph of $\hat{\pi}(\cdot)$ and the points $(x_k, \bar{Y}_k)$, $1 \le k \le 8$.   ∎

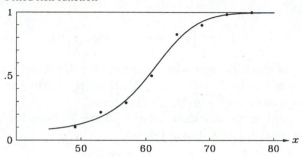

FIGURE 13.6

Fitted risk function

Binomial and Poisson Distributions

(The following discussion is mainly of illustrative value since a numerical method is not needed to obtain the maximum-likelihood estimate of the canonical parameter of the binomial or Poisson distribution.)

The Newton–Raphson method applies, in particular, when G is the space of constant functions and 1 is the basis of this space; equivalently, it applies to maximum-likelihood estimation of the canonical parameter θ of a binomial or Poisson distribution. Suppose, for a given value of $\bar{Y}$, that the maximum-likelihood estimate $\hat{\theta}$ of θ exists (that is, $0 < \bar{Y} < 1$ in the context of the binomial distribution, and $\bar{Y} > 0$ in the context of the Poisson distribution).

The form of the individual iterations in the Newton–Raphson method for determining the numerical value of $\hat{\theta}$ will now be derived starting from scratch. To this end, recall that the log-likelihood function is given by

$$l(\theta) = n[\,\theta\bar{Y} - C(\theta)\,] + \log r(Y; n), \qquad \theta \in \mathbb{R}.$$

Let $\hat{\theta}_0$ be an initial approximation to $\hat{\theta}$. Then

$$l(\hat{\theta}_0 + t) = n[\,(\hat{\theta}_0 + t)\bar{Y} - C(\hat{\theta}_0 + t)\,] + \log r(Y; n), \qquad t \in \mathbb{R}.$$

Consider the quadratic Taylor approximation to the normalizing function about $\hat{\theta}_0$, which is given by

$$C(\hat{\theta}_0 + t) \approx C(\hat{\theta}_0) + C'(\hat{\theta}_0)t + \tfrac{1}{2}C''(\hat{\theta}_0)t^2, \qquad t \in \mathbb{R}.$$

We can write this approximation as

$$C(\hat{\theta}_0 + t) \approx C(\hat{\theta}_0) + \hat{\mu}_0 t + \tfrac{1}{2}\hat{\sigma}_0^2 t^2, \qquad t \in \mathbb{R},$$

where $\hat{\mu}_0 = C'(\hat{\theta}_0)$ is an approximation to $\hat{\mu} = C'(\hat{\theta})$ and $\hat{\sigma}_0^2 = C''(\hat{\theta}_0)$ is an approximation to $\hat{\sigma}^2 = C''(\hat{\theta})$. The corresponding quadratic approximation to the log-likelihood function is given by

$$l(\hat{\theta}_0 + t) \approx n\left\{(\hat{\theta}_0 + t)\bar{Y} - [\,C(\hat{\theta}_0) + \hat{\mu}_0 t + \tfrac{1}{2}\hat{\sigma}_0^2 t^2\,]\right\} + \log r(Y; n)$$

$$= n[\,t(\bar{Y} - \hat{\mu}_0) - \tfrac{1}{2}\hat{\sigma}_0^2 t^2\,] + n[\,\hat{\theta}_0\bar{Y} - C(\hat{\theta}_0)\,] + \log r(Y; n)$$

$$= -\tfrac{1}{2}n\hat{\sigma}_0^2\Big(\frac{\bar{Y} - \hat{\mu}_0}{\hat{\sigma}_0^2} - t\Big)^2 + n[\hat{\theta}_0\bar{Y} - C(\hat{\theta}_0)]$$

$$+ \tfrac{1}{2}n\frac{(\bar{Y} - \hat{\mu}_0)^2}{\hat{\sigma}_0^2} + \log r(Y; n).$$

Clearly this quadratic approximation to the log-likelihood function is maximized by choosing t to be $\hat{\theta}_1 = (\bar{Y} - \hat{\mu}_0)/\hat{\sigma}_0^2$. Similarly, starting at $\hat{\theta}_1$, we get the next approximation $\hat{\theta}_2 = (\bar{Y} - \hat{\mu}_1)/\hat{\sigma}_1$, and so forth.

The steps in an individual iteration of the Newton–Raphson method in the present context can be written as follows:

$$\hat{\mu}_\nu = C'(\hat{\theta}_\nu),$$

$$\hat{\sigma}_\nu^2 = C''(\hat{\theta}_\nu),$$

$$Z_\nu = \frac{\bar{Y} - \hat{\mu}_\nu}{\hat{\sigma}_\nu^2},$$

$$\hat{\theta}_{\nu+1} = \hat{\theta}_\nu + Z_\nu.$$

If $\hat{\theta}_0$ is sufficiently close to $\hat{\theta}$, then $\hat{\theta}_\nu$ converges quadratically fast to $\hat{\theta}$ as $\nu \to \infty$. Otherwise, we can use step-halving: At the $(\nu + 1)$th iteration, we choose m to be the smallest nonnegative integer such that

$$l(\hat{\theta}_\nu + 2^{-m}Z_\nu) \geq l(\hat{\theta}_\nu + 2^{-(m+1)}Z_\nu)$$

and set $\hat{\theta}_{\nu+1} = \hat{\theta}_\nu + 2^{-m}Z_\nu$.

EXAMPLE **13.12**   Consider maximum-likelihood estimation of the canonical parameter θ of the binomial distribution.

(a) Write the steps for a single iteration in the Newton–Raphson method in more explicit form.

(b) Use the method to determine $\hat{\theta}$ and $\hat{\pi}$ when $\bar{Y} = .1$ and $\hat{\theta}_0 = 0$.

(c) Determine explicitly the log-likelihood function and the quadratic approximation to this function about $\hat{\theta}_0$ and show the graphs of these functions when $n = 10$.

Solution   **(a)** Here

$$C(\theta) = \log(1 + \exp[\theta]),$$

$$C'(\theta) = \frac{\exp\theta}{1 + \exp\theta},$$

and

$$C''(\theta) = \frac{\exp\theta}{1 + \exp\theta}\Big(1 - \frac{\exp\theta}{1 + \exp\theta}\Big).$$

Thus, the steps in a single iteration can be written as follows:

$$\hat{\pi}_\nu = (\exp\hat{\theta}_\nu)/(1 + \exp\hat{\theta}_\nu),$$

$$\hat{\sigma}_v^2 = \hat{\pi}_v(1 - \hat{\pi}_v),$$

$$Z_v = (\bar{Y} - \hat{\pi}_v)/\hat{\sigma}_v^2,$$

$$\hat{\theta}_{v+1} = \hat{\theta}_v + Z_v.$$

(b) The starting value is given by $\hat{\theta}_0 = 0$; correspondingly,

$$\hat{\pi}_0 = \frac{\exp 0}{1 + \exp 0} = \frac{1}{1+1} = .5,$$

$$\hat{\sigma}_0^2 = (.5)(.5) = .25,$$

and

$$Z_0 = \frac{.1 - .5}{.25} = -1.6,$$

so

$$\hat{\theta}_1 = 0 + (-1.6) = -1.6.$$

On the next iteration, we get

$$\hat{\pi}_1 = \frac{\exp(-1.6)}{1 + \exp(-1.6)} \doteq .16798,$$

$$\hat{\sigma}_1^2 \doteq (.16798)(.83202) \doteq .13976,$$

and

$$Z_1 \doteq \frac{.1 - .16798}{.13976} \doteq -0.48640,$$

so

$$\hat{\theta}_2 \doteq -1.6 + (-0.48640) \doteq -2.08640.$$

Table 13.15 lists the starting values and results of the first four iterations. Since the actual maximum-likelihood estimate of π is given by $\hat{\pi} = .1$, it is clear that there will be no practically significant changes on later iterations.

(c) The log-likelihood function is given by

$$l(\theta) = n[\theta\bar{Y} - C(\theta)] + \log\binom{n}{Y} = \theta - 10C(\theta) + \log 10, \qquad \theta \in \mathbb{R},$$

where $C(\theta) = \log(1 + e^\theta)$. The quadratic Taylor approximation to the normalizing function about $\hat{\theta}_0 = 0$ is given by

TABLE 13.15

v	$\hat{\theta}$	$\hat{\pi}$
0	0.00000	.50000
1	−1.60000	.16798
2	−2.08640	.11043
3	−2.19253	.10042
4	−2.19722	.10000

$$C(\theta) \approx C(\hat{\theta}_0) + C'(\hat{\theta}_0)(\theta - \hat{\theta}_0) + \frac{1}{2}C''(\hat{\theta}_0)(\theta - \hat{\theta}_0)^2$$

$$= \log(1 + \exp\hat{\theta}_0) + \hat{\pi}_0(\theta - \hat{\theta}_0) + \frac{1}{2}\hat{\pi}_0(1 - \hat{\pi}_0)(\theta - \hat{\theta}_0)^2$$

$$= \log 2 + \frac{\theta}{2} + \frac{\theta^2}{8}, \qquad \theta \in \mathbb{R}.$$

The corresponding quadratic approximation to the log-likelihood function about $\hat{\theta}_0$ is given by

$$l(\theta) \approx \theta - 10\left(\log 2 + \frac{\theta}{2} + \frac{\theta^2}{8}\right) + \log 10 = \log 5 - 9\log 2 - 4\theta - \frac{5}{4}\theta^2.$$

Figure 13.7 shows the graphs of the log-likelihood function and the quadratic approximation to this function about $\hat{\theta}_0$ (dashed curve).   ∎

FIGURE   13.7

Quadratic approximation to the log-likelihood function

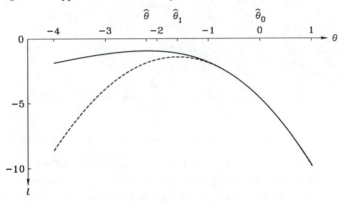

EXAMPLE 13.13   (Continued.)

(a) Investigate the behavior of the Newton–Raphson method without step-halving when $\bar{Y} = .1$ and $\hat{\theta}_0 = 2$.

(b) Use the Newton–Raphson method with step-halving to determine $\hat{\theta}$ and $\hat{\pi}$ when $n = 10$, $\bar{Y} = .1$, and $\hat{\theta}_0 = 2$.

Solution   **(a)** Starting with $\hat{\theta}_0 = 2$, on the first iteration we get

$$\hat{\pi}_0 = \frac{\exp 2}{1 + \exp 2} \doteq .881,$$

$$\hat{\sigma}_0^2 \doteq (.881)(.119) \doteq .105,$$

and

$$Z_0 \doteq \frac{.1 - .881}{.105} \doteq -7.437,$$

so $\hat{\theta}_1 \doteq 2 - 7.437 \doteq -5.437$. On the second iteration, we get

$$\hat{\pi}_1 \doteq \frac{\exp(-5.437)}{1 + \exp(-5.437)} \doteq .00434,$$

$$\hat{\sigma}_1^2 \doteq (.00434)(.996) \doteq .00432,$$

and

$$Z_1 \doteq \frac{.1 - .00434}{.00432} \doteq 22.162,$$

so $\hat{\theta}_2 \doteq -5.437 + 22.162 \doteq 16.726$. On the third iteration, we get

$$\hat{\pi}_2 \doteq \frac{\exp 16.726}{1 + \exp 16.726} \doteq 1 - 5.446 \times 10^{-8},$$

and so forth. Clearly, the Newton–Raphson method without step-halving diverges when we use the starting value $\hat{\theta}_0 = 2$.

(b) Starting with $\hat{\theta}_0 = 2$, we get $Z_0 \doteq -7.437$ as in the solution to part (a). The values of the log-likelihood function at $\hat{\theta}_0 + Z_0$, $\hat{\theta}_0 + \frac{1}{2}Z_0$ and $\hat{\theta}_0 + \frac{1}{4}Z_0$ are given by

$$l(\hat{\theta}_0 + Z_0) = \hat{\theta}_0 + Z_0 - 10\log(1 + \exp(\hat{\theta}_0 + Z_0)) + \log 10 \doteq -3.177,$$

$$l(\hat{\theta}_0 + \tfrac{1}{2}Z_0) = \hat{\theta}_0 + \tfrac{1}{2}Z_0 - 10\log(1 + \exp(\hat{\theta}_0 + \tfrac{1}{2}Z_0)) + \log 10 \doteq -1.066,$$

and

$$l(\hat{\theta}_0 + \tfrac{1}{4}Z_0) = \hat{\theta}_0 + \tfrac{1}{4}Z_0 - 10\log(1 + \exp(\hat{\theta}_0 + \tfrac{1}{4}Z_0)) + \log 10 \doteq -5.217.$$

Observe that $l(\hat{\theta}_0 + Z_0) < l(\hat{\theta}_0 + \frac{1}{2}Z_0)$, but $l(\hat{\theta}_0 + \frac{1}{2}Z_0) > l(\hat{\theta}_0 + \frac{1}{4}Z_0)$. Thus, we set $\hat{\theta}_1 = \hat{\theta}_0 + \frac{1}{2}Z_0 \doteq -1.718$. On the second iteration, we get

$$\hat{\pi}_1 \doteq \frac{\exp(-1.718)}{1 + \exp(-1.718)} \doteq .152,$$

$$\hat{\sigma}_1^2 \doteq (.152)(.848) \doteq .129,$$

and

$$Z_1 \doteq \frac{.1 - .152}{.129} \doteq -.404.$$

Since $l(\hat{\theta}_1 + Z_1) > l(\hat{\theta}_1 + \frac{1}{2}Z_1)$, we set $\hat{\theta}_2 = \hat{\theta}_1 + Z_1 \doteq -2.112$. In the same manner, we get $\hat{\theta}_3 = \hat{\theta}_2 + Z_2 \doteq -2.195$ and $\hat{\theta}_4 = \hat{\theta}_3 + Z_3 \doteq -2.197$. The starting values and results of the first four iterations (after which the Newton–Raphson method with step-halving has effectively converged) are as shown in Table 13.16. ∎

Problems

13.44 Consider the setup of Problem 13.29 in Section 13.3. Determine **(a)** the quadratic approximation to the log-likelihood function about $\hat{\theta}_0$; **(b)** the value $\hat{\theta}_1$ of θ that maximizes this quadratic approximation.

TABLE   13.16

ν	$\hat{\theta}$	$\hat{\pi}$
0	2.00000	.88080
1	−1.71831	.15209
2	−2.12223	.10695
3	−2.19504	.10020
4	−2.19722	.10000

13.45   In the context of Problem 13.30 in Section 13.3, use iteration based on Problem 13.29 and the starting value $\hat{\theta}_0 = 0$ to determine the maximum-likelihood estimate $\hat{\theta}$.

13.46   Obtain a more explicit form for the steps in a single iteration of the iteratively reweighted least-squares method that is applicable to Poisson regression.

13.47   Consider the setup of Problem 13.29 in Section 13.3. According to Problem 13.31 in that section, this setup can be viewed as a Poisson regression model. Use the result of Problem 13.46 to obtain an explicit form for the steps in a single iteration of the iteratively reweighted least-squares method that is applicable to this model and check that the result is in agreement with the answer to Problem 13.44(b).

13.48   Consider the linear Poisson regression model in which $\theta(x) = \beta_0 + \beta_1 x$ for $x \in \mathbb{R}$ and consider the experimental data (which was also used in Example 13.10) shown in Table 13.17. Let $\hat{\beta}_0$ and $\hat{\beta}_1$ denote the maximum-likelihood estimates of β_0 and β_1, respectively. **(a)** Determine explicitly the maximum-likelihood equations for $\hat{\beta}_0$ and $\hat{\beta}_1$. **(b)** Starting with $\hat{\beta}_{00} = 0$ and $\hat{\beta}_{10} = 0$, determine the approximations $\hat{\beta}_{01}$ and $\hat{\beta}_{11}$ from the first iteration in the iteratively reweighted least-squares method. **(c)** The maximum-likelihood estimates of the regression coefficients are given by $\hat{\beta}_0 \doteq -1.018$ and $\hat{\beta}_1 \doteq 0.529$. Determine the corresponding maximum-likelihood estimates of θ_1, θ_2, and θ_3 and those of λ_1, λ_2, and λ_3. **(d)** Check that the numerical values given in part (c) for $\hat{\beta}_0$ and $\hat{\beta}_1$ satisfy the maximum-likelihood equations obtained in part (a).

13.49   Consider the maximum-likelihood estimation of the canonical parameter $\theta = \log \lambda$ of the Poisson distribution with mean $n\lambda$, where n is known and λ is unknown. **(a)** Write the steps for a single iteration in the Newton–Raphson method in more explicit form. **(b)** Use the method to determine $\hat{\theta}$ and $\hat{\lambda}$ when $\bar{Y} = 0.1$ and $\hat{\theta}_0 = 0$. **(c)** Determine explicitly the log-likelihood function and the quadratic approximation to this function about $\hat{\theta}_0$ and show the graphs of these functions when $n = 10$.

TABLE   13.17

x	n	Y	$\bar{Y}$
−1	60	12	.2
0	40	16	.4
1	60	36	.6

13.50 In the context of Problem 13.49, but with the starting value of $\hat{\theta}_0 = -5$, determine the behavior of **(a)** the Newton–Raphson method without step-halving; **(b)** the Newton–Raphson method with step-halving.

13.51 Let Y have the Poisson distribution with mean λ and set $\theta = \log \lambda$. Suppose the observed value of Y equals 1. Determine **(a)** the corresponding log-likelihood function as a function of θ; **(b)** the maximum-likelihood estimate $\hat{\theta}$ of θ; **(c)** the formula for a single iteration of the Newton–Raphson method (without step-halving) for finding $\hat{\theta}$; **(d)** the approximation $\hat{\theta}_1$ to $\hat{\theta}$ based on one step of the Newton–Raphson method starting with $\hat{\theta}_0 = 100$; **(e)** an explanation for the answer to part (d); **(f)** an improvement suggested by the answer to part (e); **(g)** the number of steps required for the Newton–Raphson method to converge starting with $\hat{\theta}_0 = -10$; **(h)** an explanation for the answer to part (f); **(i)** an improvement suggested by the answer to part (h).

13.5
Normal Approximation

In this section, we derive a multivariate normal approximation to the joint distribution of the maximum-likelihood estimates of the regression coefficients and use this result to obtain nominal confidence intervals and an extension of the Wald test to logistic regression and Poisson regression.

Consider a linear canonical regression model in which G is an identifiable linear space having basis $g_1, \ldots, g_p$. The canonical regression function and its maximum-likelihood estimate can be written, respectively, as $\theta(\cdot) = \beta_1 g_1 + \cdots + \beta_p g_p$ and $\hat{\theta}(\cdot) = \hat{\beta}_1 g_1 + \cdots + \hat{\beta}_p g_p$. Set $\beta = [\beta_1, \ldots, \beta_p]^T$ and $\hat{\beta} = [\hat{\beta}_1, \ldots, \hat{\beta}_p]^T$. Recall that $n = n_1 + \cdots + n_d$. According to the next result, which is made precise and proven in advanced statistical theory, under suitable conditions, $\hat{\beta}$ and $\hat{\theta}(\cdot)$ are consistent estimates of β and $\theta(\cdot)$, respectively. (In the context of logistic regression, one of the required conditions for this result is that $n \gg 1$.)

THEOREM **13.10** Under suitable conditions, $P(\hat{\beta} \text{ exists}) = P(\hat{\theta}(\cdot) \text{ exists}) \approx 1$, $P(\hat{\beta} \approx \beta) \approx 1$, and $P(\hat{\theta}(\cdot) \approx \theta(\cdot)) \approx 1$.

Let $\mathbf{X}$ be the design matrix corresponding to the given basis and set

$$\mathbf{W} = \text{diag}(n_1 \sigma_1^2, \ldots, n_d \sigma_d^2) = \text{diag}(n_1 \sigma^2(x_1), \ldots, n_d \sigma^2(x_d)).$$

Since $\mathbf{W}$ has positive diagonal entries, it is symmetric and positive definite. The $p \times p$ matrix $\mathbf{X}^T \mathbf{W} \mathbf{X}$ is referred to as the *information matrix*; the entry in row j and column m of this matrix is given by $\sum_k n_k \sigma_k^2 g_j(\mathbf{x}_k) g_m(\mathbf{x}_k) = \sum_k n_k \sigma^2(\mathbf{x}_k) g_j(\mathbf{x}_k) g_m(\mathbf{x}_k)$. The information matrix is symmetric and positive definite, so it is invertible and its inverse $\mathbf{V} = (\mathbf{X}^T \mathbf{W} \mathbf{X})^{-1} = [v_{jm}]$ is symmetric and positive definite. If $n \gg 1$ and other conditions are satisfied, then $\mathbf{V} \approx \mathbf{0}$.

The binomial or Poisson d-sample model can be viewed as the special case of the linear canonical regression model in which $\mathcal{X} = \{1, \ldots, d\}$ and G is the space of all functions on $\mathcal{X}$. Consider the basis $I_1 = \text{ind}(x = 1), \ldots, I_d = \text{ind}(x = d)$ and label the design points as $x_k = k$ for $1 \le k \le d$. Then $\mathbf{X}$ is the $d \times d$ identity matrix, so

$$\mathbf{V} = \mathbf{W}^{-1} = \text{diag}\left(\frac{1}{n_1 \sigma_1^2}, \ldots, \frac{1}{n_d \sigma_d^2}\right).$$

THEOREM 13.11   Under suitable conditions, $\hat{\beta} - \beta$ has approximately the multivariate normal distribution with mean vector $\mathbf{0}$ and variance-covariance matrix $\mathbf{V}$.

Derivation   Recall the derivation of the quadratic approximation to the log-likelihood function about an arbitrary function $\hat{\theta}_0(\cdot)$ given in Section 13.4. In particular, the quadratic approximation to the log-likelihood function about the true canonical regression function $\theta(\cdot)$ is given by

$$l(\theta(\cdot) + g) \approx -\frac{1}{2} \sum_k n_k \sigma^2(\mathbf{x}_k)\left(\frac{\bar{Y}_k - \mu(\mathbf{x}_k)}{\sigma^2(\mathbf{x}_k)} - g(\mathbf{x}_k)\right)^2 + \text{remainder term},$$

where $\mu(\cdot) = C'(\theta(\cdot))$, $\sigma^2(\cdot) = C''(\theta(\cdot))$, the remainder term does not depend on g, and the approximation is accurate when g is suitably small. Observe that maximizing this quadratic approximation is equivalent to minimizing the weighted residual sum of squares

$$\text{RSS}(g) = \sum_k n_k \sigma^2(\mathbf{x}_k)\left(\frac{\bar{Y}_k - \mu(\mathbf{x}_k)}{\sigma^2(\mathbf{x}_k)} - g(\mathbf{x}_k)\right)^2, \qquad g \in G.$$

Consider the weights $w_k = n_k \sigma^2(\mathbf{x}_k)$, $1 \le k \le d$, and the transformed response variables $\tilde{Y}_k = (\bar{Y}_k - \mu(\mathbf{x}_k))/\sigma^2(\mathbf{x}_k)$, $1 \le k \le d$, and set $\tilde{\mathbf{Y}} = [\tilde{Y}_1, \ldots, \tilde{Y}_d]^T$. We can rewrite the weighted residual sum of squares as

$$\text{RSS}(g) = \sum_k w_k(\tilde{Y}_k - g(\mathbf{x}_k))^2, \qquad g \in G.$$

Let $\hat{g} \in G$ be the corresponding weighted least-squares estimate—that is, the function in G that minimizes this weighted residual sum of squares. Write $\hat{g} = Z_1 g_1 + \cdots + Z_p g_p$ and set $\mathbf{Z} = [Z_1, \ldots, Z_p]^T$. Then the normal equation for $\mathbf{Z}$ [see Equation (65) in Section 9.6] is given by by $\mathbf{X}^T \mathbf{W} \mathbf{X} \mathbf{Z} = \mathbf{X}^T \mathbf{W} \tilde{\mathbf{Y}}$, whose unique solution is given by $\mathbf{Z} = \mathbf{V} \mathbf{X}^T \mathbf{W} \tilde{\mathbf{Y}}$. Now

$$\begin{aligned}
\mathbf{W}\tilde{\mathbf{Y}} &= [n_1(\bar{Y}_1 - \mu(\mathbf{x}_1)), \ldots, n_d(\bar{Y}_d - \mu(\mathbf{x}_d))]^T \\
&= [n_1(\bar{Y}_1 - \mu_1), \ldots, n_d(\bar{Y}_d - \mu_d)]^T \\
&= \mathbf{N}(\bar{\mathbf{Y}} - \boldsymbol{\mu}),
\end{aligned}$$

where $\mathbf{N} = \text{diag}(n_1, \ldots, n_d)$, $\boldsymbol{\mu} = [\mu_1, \ldots, \mu_d]^T$ and $\bar{\mathbf{Y}} = [\bar{Y}_1, \ldots, \bar{Y}_d]^T$. Therefore, $\mathbf{Z} = \mathbf{V}\mathbf{X}^T\mathbf{N}(\bar{\mathbf{Y}} - \boldsymbol{\mu})$.

Since $\bar{Y}_1, \ldots, \bar{Y}_d$ are independent random variables having means $\mu_1, \ldots, \mu_d$ and variances $\sigma_1^2/n_1, \ldots, \sigma_d^2/n_d$, respectively, we see that $n_1(\bar{Y}_1 - \mu_1), \ldots, n_d(\bar{Y}_d - \mu_d)$ are independent random variables having mean 0 and variances $n_1 \sigma_1^2, \ldots, n_d \sigma_d^2$, respectively. Thus, $\mathbf{N}(\bar{\mathbf{Y}} - \boldsymbol{\mu})$ has mean vector $\mathbf{0}$ and variance–

covariance matrix $\mathbf{W}$. Consequently, $\mathbf{Z}$ has mean vector $\mathbf{0}$ and variance-covariance matrix

$$\mathbf{VX}^T\mathbf{WXV} = \mathbf{VV}^{-1}\mathbf{V} = \mathbf{V}$$

[see Theorem 5.8(f) in Section 5.3]. It follows from a multivariate form of the central limit theorem that, under suitable conditions, $\mathbf{Z}$ has approximately the multivariate normal distribution with mean vector $\mathbf{0}$ and variance–covariance matrix $\mathbf{V}$.

Now $\hat{\theta}(\cdot) = \theta(\cdot) + \hat{\theta}(\cdot) - \theta(\cdot)$ is the function in G that maximizes the log-likelihood function, so $\hat{\theta}(\cdot) - \theta(\cdot)$ is the function g in G that maximizes $l(\theta(\cdot) + g)$. If $n \gg 1$ and other conditions are satisfied, then it is probably true that $\hat{\theta}(\cdot) - \theta(\cdot)$ is close to zero (see Theorem 13.10) and hence close to the function $\hat{g}$ that minimizes the quadratic approximation $\text{RSS}(g)$ to $-2l(\theta(\cdot) + g)$; that is, $P(\hat{\theta}(\cdot) - \theta(\cdot) \approx \hat{g}) \approx 1$ and hence

$$P(\hat{\boldsymbol{\beta}} - \boldsymbol{\beta} \approx \mathbf{Z}) \approx 1. \tag{28}$$

It follows from (28) that the distribution of $\hat{\boldsymbol{\beta}} - \boldsymbol{\beta}$ is close to the distribution of $\mathbf{Z}$ and hence to the multivariate normal distribution with mean vector $\mathbf{0}$ and variance–covariance matrix $\mathbf{V}$.   ∎

In light of Theorem 13.11, we refer to $\mathbf{V}$ as the *asymptotic variance–covariance matrix* of $\hat{\boldsymbol{\beta}}$, to v_{jj} as the asymptotic variance of $\hat{\beta}_j$, to v_{jm} as the *asymptotic covariance* between $\hat{\beta}_j$ and $\hat{\beta}_m$, and to $v_{jm}/\sqrt{v_{jj}v_{mm}}$ as the *asymptotic correlation* between $\hat{\beta}_j$ and $\hat{\beta}_m$. (Even under the conditions of the theorem, v_{jj} is not necessarily a good approximation to the actual variance of $\hat{\beta}_j$; in fact, it is not clear what we should mean by the variance of $\hat{\beta}_j$ when this estimate is undefined with positive probability.)

Consider a parameter $\tau = h(\boldsymbol{\beta}) = h(\beta_1, \ldots, \beta_p)$, where $h(\boldsymbol{\beta})$ is a possibly non-linear function of $\boldsymbol{\beta}$ that is defined on an open set in $\mathbb{R}^p$ containing the true value of $\boldsymbol{\beta}$. The maximum-likelihood estimate of τ is defined by $\hat{\tau} = h(\hat{\boldsymbol{\beta}})$, with $\hat{\tau}$ being undefined if $\hat{\boldsymbol{\beta}}$ is either undefined or not in the domain of h. It is also assumed that $h(\boldsymbol{\beta})$ is a continuous function of $\boldsymbol{\beta}$ near its true value. According to Theorem 13.10, under suitable conditions, $P(\hat{\tau} \text{ exists}) \approx 1$ and $P(\hat{\tau} \approx \tau) \approx 1$.

Suppose $h(\boldsymbol{\beta})$ is a continuously differentiable function of $\boldsymbol{\beta}$ near its true value. Using (28) and applying the linear Taylor approximation to h about the true value of $\boldsymbol{\beta}$, we get

$$\hat{\tau} - \tau \approx \sum_k \frac{\partial h}{\partial \beta_k}(\boldsymbol{\beta})(\hat{\beta}_k - \beta_k)$$

$$= [\nabla h(\boldsymbol{\beta})]^T(\hat{\boldsymbol{\beta}} - \boldsymbol{\beta})$$

with probability close to 1, where

$$\nabla h(\boldsymbol{\beta}) = \left[\frac{\partial h}{\partial \beta_1}(\boldsymbol{\beta}), \ldots, \frac{\partial h}{\partial \beta_p}(\boldsymbol{\beta})\right]^T$$

denotes the gradient of h at the true value of $\boldsymbol{\beta}$. Consequently, by Theorem 13.11 [see Theorem 5.8(g) in Section 5.3 and Theorem 5.26 in Section 5.7], if $\nabla h(\boldsymbol{\beta}) \neq 0$ and other suitable conditions are satisfied, then $\hat{\tau} - \tau$ has approximately the normal distribution with mean 0 and variance $[\nabla h(\beta)]^T \mathbf{V} \nabla h(\beta) \approx 0$. In light of this result, we refer to

$$\mathrm{AV}(\hat{\tau}) = [\nabla h(\beta)]^T \mathbf{V} \nabla h(\beta) = \sum_j \sum_m v_{jm} \frac{\partial h}{\partial \beta_j}(\beta) \frac{\partial h}{\partial \beta_m}(\beta)$$

as the *asymptotic variance* of $\hat{\tau}$ and to the nonnegative square root of $\mathrm{AV}(\hat{\tau})$ as the *asymptotic standard deviation* (ASD) of $\hat{\tau}$. If $\nabla h(\beta) \neq \mathbf{0}$, then the asymptotic variance and asymptotic standard deviation of τ are both positive; otherwise, these quantities are both zero. In summary, we have derived the following result as a consequence of Theorems 13.10 and 13.11.

THEOREM 13.12 If $\nabla h(\beta) \neq \mathbf{0}$ and other suitable conditions are satisfied, then $P(\hat{\tau} \text{ exists}) \approx 1$, $P(\hat{\tau} \approx \tau) \approx 1$, $\mathrm{ASD}(\hat{\tau}) \approx 0$, and $(\hat{\tau} - \tau)/\mathrm{ASD}(\hat{\tau})$ has approximately the standard normal distribution.

The maximum-likelihood estimate of $\mathbf{W}$ is given by

$$\hat{\mathbf{W}} = \mathrm{diag}(n_1 \hat{\sigma}^2(\mathbf{x}_1), \ldots, n_d \hat{\sigma}^2(\mathbf{x}_d)).$$

The maximum-likelihood estimate of the information matrix $\mathbf{X}^T \mathbf{W} \mathbf{X}$ equals $\mathbf{X}^T \hat{\mathbf{W}} \mathbf{X}$, the entry in row j and column m of which equals $\sum_k n_k \hat{\sigma}^2(\mathbf{x}_k) g_j(\mathbf{x}_k) g_m(\mathbf{x}_k)$. The maximum-likelihood estimate of $\mathbf{V} = (\mathbf{X}^T \mathbf{W} \mathbf{X})^{-1}$ is given by $\hat{\mathbf{V}} = (\mathbf{X}^T \hat{\mathbf{W}} \mathbf{X})^{-1}$. For $1 \leq j, m \leq p$, let $\hat{v}_{jm}$ denote the entry in row j and column m of $\hat{\mathbf{V}}$, which we refer to as the maximum-likelihood estimate of v_{jm}. The standard error $\mathrm{SE}(\hat{\tau})$ of $\hat{\tau}$ is defined as the value obtained by replacing β by $\hat{\beta}$ in the formula for $\mathrm{ASD}(\hat{\tau})$; thus, $\mathrm{SE}(\hat{\tau})$ is the nonnegative square root of $[\nabla h(\beta)]^T \hat{\mathbf{V}}[\nabla h(\beta)]$. In summation form,

$$\mathrm{SE}(\hat{\tau}) = \sqrt{\sum_j \sum_m \hat{v}_{jm} \frac{\partial h}{\partial \beta_j}(\hat{\beta}) \frac{\partial h}{\partial \beta_m}(\hat{\beta})}.$$

Consider, in particular, the linear parameter $\tau = c_1 \beta_1 + \cdots + c_p \beta_p = \mathbf{c}^T \beta$ with $\mathbf{c} = [c_1, \ldots, c_p]^T$. The maximum-likelihood estimate of this parameter is given by $\hat{\tau} = c_1 \hat{\beta}_1 + \cdots + c_p \hat{\beta}_p = \mathbf{c}^T \hat{\beta}$, with $\hat{\tau}$ being undefined if $\hat{\beta}$ is undefined; the asymptotic variance of the estimate $\hat{\tau}$ is given by $\mathrm{AV}(\hat{\tau}) = \mathbf{c}^T \mathbf{V} \mathbf{c} = \sum_j \sum_m c_j c_m v_{jm}$; and its standard error is given by

$$\mathrm{SE}(\hat{\tau}) = \sqrt{\mathbf{c}^T \hat{\mathbf{V}} \mathbf{c}} = \sqrt{\sum_j \sum_m c_j c_m \hat{v}_{jm}}.$$

As a special case, $\mathrm{AV}(\hat{\beta}_j) = v_{jj}$ and $\mathrm{SE}(\hat{\beta}_j) = \sqrt{\hat{v}_{jj}}$ for $1 \leq j \leq p$.

The next result follows from Theorems 13.10 and 13.12.

THEOREM 13.13 If $\nabla h(\beta) \neq \mathbf{0}$ and other suitable conditions are satisfied, then

$$P\left(\frac{\mathrm{SE}(\hat{\tau})}{\mathrm{ASD}(\hat{\tau})} \approx 1\right) \approx 1$$

and $(\hat{\tau} - \tau)/\operatorname{SE}(\hat{\tau})$ has approximately the standard normal distribution.

[Theorems 13.12 and 13.13 are also valid in the context of the homoskedastic linear regression model, where $\mathbf{V}$ and $\hat{\mathbf{V}}$ are defined as in Section 10.3. Here τ can be a nonlinear parameter and the assumption of normality is not required, but it is required that $n \gg 1$ and that the design (that is, the design set and the positive integers $n_1, \ldots, n_d$) not be too irregular, and the results are only approximately valid. In Sections 10.3, by comparison, τ was required to be a linear parameter and the assumption of normality was also required, but the results were exactly valid and it was not required that $n \gg 1$ or that the design satisfy any regularity condition beyond the identifiability of G.]

COROLLARY 13.7 Let $1 \le j \le p$. Under suitable conditions, $P(\hat{\beta}_j \approx \beta_j) \approx 1$, $\operatorname{ASD}(\hat{\beta}_j) \approx 0$,

$$P\left(\frac{\operatorname{SE}(\hat{\beta}_j)}{\operatorname{ASD}(\hat{\beta}_j)} \approx 1\right) \approx 1,$$

and each of the random variables $(\hat{\beta}_j - \beta_j)/\operatorname{ASD}(\hat{\beta}_j)$ and $(\hat{\beta}_j - \beta_j)/\operatorname{SE}(\hat{\beta}_j)$ has approximately the standard normal distribution.

Given $\mathbf{x} \in \mathcal{X}$, set $\mathbf{g}(\mathbf{x}) = [g_1(\mathbf{x}), \ldots, g_p(\mathbf{x})]^T$. Consider the unknown parameter $\tau = \theta(\mathbf{x}) = \sum_j \beta_j g_j(\mathbf{x})$ and its maximum-likelihood estimate $\hat{\tau} = \hat{\theta}(\mathbf{x}) = \sum_j \hat{\beta}_j g_j(\mathbf{x})$. Observe that $\tau = \mathbf{c}^T \boldsymbol{\beta}$ and $\hat{\tau} = \mathbf{c}^T \hat{\boldsymbol{\beta}}$, where $\mathbf{c} = \mathbf{g}(\mathbf{x})$. Consequently, $\operatorname{AV}(\hat{\tau}) = \mathbf{c}^T \mathbf{V} \mathbf{c}$ and $\operatorname{SE}(\hat{\tau}) = \sqrt{\mathbf{c}^T \hat{\mathbf{V}} \mathbf{c}}$. Equivalently,

$$\operatorname{AV}(\hat{\theta}(\mathbf{x})) = [\mathbf{g}(\mathbf{x})]^T \mathbf{V} \mathbf{g}(\mathbf{x}) = \sum_j \sum_m v_{jm} g_j(\mathbf{x}) g_m(\mathbf{x}),$$

and

$$\operatorname{SE}(\hat{\theta}(\mathbf{x})) = \sqrt{[\mathbf{g}(\mathbf{x})]^T \hat{\mathbf{V}} \mathbf{g}(\mathbf{x})} = \sqrt{\sum_j \sum_m \hat{v}_{jm} g_j(\mathbf{x}) g_m(\mathbf{x})}.$$

Now $\partial\theta(\mathbf{x})/\partial\beta_j = g_j(\mathbf{x})$ for $1 \le j \le p$, so the gradient of $\theta(\mathbf{x})$ (vis-à-vis $\boldsymbol{\beta}$) equals $\mathbf{g}(\mathbf{x})$. Recall that the maximum-likelihood estimate of $\mu(\mathbf{x}) = C'(\theta(\mathbf{x}))$ is given by $\hat{\mu}(\mathbf{x}) = C'(\hat{\theta}(\mathbf{x}))$ and that the maximum-likelihood estimate of $\sigma^2(\mathbf{x}) = C''(\theta(\mathbf{x}))$ is given by $\hat{\sigma}^2(\mathbf{x}) = C''(\hat{\theta}(\mathbf{x}))$. From the formula $\mu(\mathbf{x}) = C'(\theta(\mathbf{x}))$ and the chain rule, we see that

$$\frac{\partial\mu(\mathbf{x})}{\partial\beta_j} = C''(\theta(\mathbf{x}))\frac{\partial\theta(\mathbf{x})}{\partial\beta_j} = \sigma^2(\mathbf{x}) g_j(\mathbf{x}), \qquad 1 \le j \le p.$$

Consequently, the gradient of $\mu(\mathbf{x})$ equals $\sigma^2(\mathbf{x})\mathbf{g}(\mathbf{x})$, which is $\sigma^2(\mathbf{x})$ times the gradient of $\theta(\mathbf{x})$. Therefore,

$$\operatorname{AV}(\hat{\mu}(\mathbf{x})) = \sigma^4(\mathbf{x})[\mathbf{g}(\mathbf{x})]^T \mathbf{V} \mathbf{g}(\mathbf{x}) = \sigma^4(\mathbf{x})\operatorname{AV}(\hat{\theta}(\mathbf{x})),$$

and

$$\operatorname{SE}(\hat{\mu}(\mathbf{x})) = \hat{\sigma}^2(\mathbf{x})\sqrt{[\mathbf{g}(\mathbf{x})]^T \hat{\mathbf{V}} \mathbf{g}(\mathbf{x})} = \hat{\sigma}^2(\mathbf{x})\operatorname{SE}(\hat{\theta}(\mathbf{x})), \qquad \mathbf{x} \in \mathcal{X}.$$

EXAMPLE **13.14**   In the context of logistic regression, obtain more explicit formulas for $\mathbf{W}$, $\hat{\mathbf{W}}$, $\mathrm{AV}(\hat{\pi}(\mathbf{x}))$, and $\mathrm{SE}(\hat{\pi}(\mathbf{x}))$ for $\mathbf{x} \in \mathfrak{X}$.

Solution   Here,

$$\mathbf{W} = \mathrm{diag}\left(n_1\pi_1(1-\pi_1), \ldots, n_d\pi_d(1-\pi_d)\right)$$

$$= \mathrm{diag}\left(n_1\pi(\mathbf{x}_1)[\,1-\pi(\mathbf{x}_1)\,], \ldots, n_d\pi(\mathbf{x}_d)[\,1-\pi(\mathbf{x}_d)\,]\right),$$

$$\hat{\mathbf{W}} = \mathrm{diag}\left(n_1\hat{\pi}(\mathbf{x}_1)[\,1-\hat{\pi}(\mathbf{x}_1)\,], \ldots, n_d\hat{\pi}(\mathbf{x}_d)[\,1-\hat{\pi}(\mathbf{x}_d)\,]\right),$$

$$\mathrm{AV}(\hat{\pi}(\mathbf{x})) = \{\pi(\mathbf{x})[\,1-\pi(\mathbf{x})\,]\}^2\,\mathrm{AV}(\theta(\mathbf{x})) = \{\pi(\mathbf{x})[\,1-\pi(\mathbf{x})\,]\}^2[\,\mathbf{g}(\mathbf{x})\,]^T\mathbf{V}\mathbf{g}(\mathbf{x}),$$

and

$$\mathrm{SE}(\hat{\pi}(\mathbf{x})) = \hat{\pi}(\mathbf{x})[\,1-\hat{\pi}(\mathbf{x})\,]\,\mathrm{SE}(\hat{\theta}(\mathbf{x})) = \hat{\pi}(\mathbf{x})[\,1-\hat{\pi}(\mathbf{x})\,]\sqrt{[\,\mathbf{g}(\mathbf{x})\,]^T\hat{\mathbf{V}}\mathbf{g}(\mathbf{x})}. \quad \blacksquare$$

EXAMPLE **13.15**   Determine the asymptotic variance and standard error of $\hat{\mu}(\mathbf{x}_2)/\hat{\mu}(\mathbf{x}_1)$ for $\mathbf{x}_1, \mathbf{x}_2 \in \mathfrak{X}$.

Solution   Observe that

$$\frac{\mu(\mathbf{x}_2)}{\mu(\mathbf{x}_1)} = \frac{C'(\theta(\mathbf{x}_2))}{C'(\theta(\mathbf{x}_1))}.$$

Thus, by the chain rule,

$$\frac{\partial}{\partial\beta_j}\frac{\mu(\mathbf{x}_2)}{\mu(\mathbf{x}_1)} = \frac{C''(\theta(\mathbf{x}_2))}{C'(\theta(\mathbf{x}_1))}g_j(\mathbf{x}_2) - \frac{C''(\theta(\mathbf{x}_1))C'(\theta(\mathbf{x}_2))}{[\,C'(\theta(\mathbf{x}_1))\,]^2}g_j(\mathbf{x}_1)$$

$$= \frac{\sigma^2(\mathbf{x}_2)}{\mu(\mathbf{x}_1)}g_j(\mathbf{x}_2) - \frac{\sigma^2(\mathbf{x}_1)\mu(\mathbf{x}_2)}{[\,\mu(\mathbf{x}_1)\,]^2}g_j(\mathbf{x}_1)$$

$$= \frac{\mu(\mathbf{x}_2)}{\mu(\mathbf{x}_1)}\left(\frac{\sigma^2(\mathbf{x}_2)}{\mu(\mathbf{x}_2)}g_j(\mathbf{x}_2) - \frac{\sigma^2(\mathbf{x}_1)}{\mu(\mathbf{x}_1)}g_j(\mathbf{x}_1)\right).$$

Consequently, $\mathrm{AV}(\hat{\mu}(\mathbf{x}_2)/\hat{\mu}(\mathbf{x}_1)) = \mathbf{c}^T\mathbf{V}\mathbf{c}$, where

$$\mathbf{c} = \frac{\mu(\mathbf{x}_2)}{\mu(\mathbf{x}_1)}\left(\frac{\sigma^2(\mathbf{x}_2)}{\mu(\mathbf{x}_2)}\mathbf{g}(\mathbf{x}_2) - \frac{\sigma^2(\mathbf{x}_1)}{\mu(\mathbf{x}_1)}\mathbf{g}(\mathbf{x}_1)\right).$$

Therefore, $\mathrm{SE}(\hat{\mu}(\mathbf{x}_2)/\hat{\mu}(\mathbf{x}_1)) = \sqrt{\hat{\mathbf{c}}^T\hat{\mathbf{V}}\hat{\mathbf{c}}}$, where

$$\hat{\mathbf{c}} = \frac{\hat{\mu}(\mathbf{x}_2)}{\hat{\mu}(\mathbf{x}_1)}\left(\frac{\hat{\sigma}^2(\mathbf{x}_2)}{\hat{\mu}(\mathbf{x}_2)}\mathbf{g}(\mathbf{x}_2) - \frac{\hat{\sigma}^2(\mathbf{x}_1)}{\hat{\mu}(\mathbf{x}_1)}\mathbf{g}(\mathbf{x}_1)\right). \quad \blacksquare$$

EXAMPLE **13.16**   In the context of logistic regression, determine the asymptotic variance and standard error of $\hat{\pi}(\mathbf{x}_2)/\hat{\pi}(\mathbf{x}_1)$ for $\mathbf{x}_1, \mathbf{x}_2 \in \mathfrak{X}$.

Solution   Applying the solution to Example 13.15, with $\mu(\mathbf{x}) = \pi(\mathbf{x})$ and $\sigma^2(\mathbf{x}) = \pi(\mathbf{x})[1 - \pi(\mathbf{x})]$, we get $\mathrm{AV}(\hat{\pi}(\mathbf{x}_2)/\hat{\pi}(\mathbf{x}_1)) = \mathbf{c}^T \mathbf{V} \mathbf{c}$, where

$$\mathbf{c} = \frac{\pi(\mathbf{x}_2)}{\pi(\mathbf{x}_1)} \left\{ [1 - \pi(\mathbf{x}_2)] \mathbf{g}(\mathbf{x}_2) - [1 - \pi(\mathbf{x}_1)] \mathbf{g}(\mathbf{x}_1) \right\}.$$

Consequently, $\mathrm{SE}(\hat{\pi}(\mathbf{x}_2)/\hat{\pi}(\mathbf{x}_1)) = \sqrt{\hat{\mathbf{c}}^T \hat{\mathbf{V}} \hat{\mathbf{c}}}$, where

$$\hat{\mathbf{c}} = \frac{\hat{\pi}(\mathbf{x}_2)}{\hat{\pi}(\mathbf{x}_1)} \left\{ [1 - \hat{\pi}(\mathbf{x}_2)] \mathbf{g}(\mathbf{x}_2) - [1 - \hat{\pi}(\mathbf{x}_1)] \mathbf{g}(\mathbf{x}_1) \right\}. \quad \blacksquare$$

Nominal Confidence Bounds and Confidence Intervals

The derivation of Theorem 12.4 in Section 12.1 is applicable in the context of the linear canonical regression model, where it yields the following more general result.

THEOREM 13.14   Suppose $\nabla h(\boldsymbol{\beta}) \neq \mathbf{0}$ and other suitable conditions are satisfied and let $0 < \alpha < 1$. Then

$$P\left(\hat{\tau} - z_{1-\alpha} \, \mathrm{SE}(\hat{\tau}) < \tau \right) \approx 1 - \alpha,$$

$$P\left(\tau < \hat{\tau} + z_{1-\alpha} \, \mathrm{SE}(\hat{\tau}) \right) \approx 1 - \alpha,$$

and

$$P\left(\hat{\tau} - z_{1-\alpha/2} \, \mathrm{SE}(\hat{\tau}) < \tau < \hat{\tau} + z_{1-\alpha/2} \, \mathrm{SE}(\hat{\tau}) \right) \approx 1 - \alpha.$$

In light of Theorem 13.14, we continue to refer to $\hat{\tau} - z_{1-\alpha} \mathrm{SE}(\hat{\tau})$ as the nominal $100(1 - \alpha)\%$ lower confidence bound for τ, to $\hat{\tau} + z_{1-\alpha} \mathrm{SE}(\hat{\tau})$ as the nominal $100(1 - \alpha)\%$ upper confidence bound for this parameter, and to

$$\hat{\tau} \pm z_{1-\alpha/2} \mathrm{SE}(\hat{\tau}) = \left(\hat{\tau} - z_{1-\alpha/2} \mathrm{SE}(\hat{\tau}), \ \hat{\tau} + z_{1-\alpha/2} \mathrm{SE}(\hat{\tau}) \right)$$

as its nominal $100(1 - \alpha)\%$ confidence interval.

EXAMPLE 13.17   Consider the beetle data as treated in Example 13.11 in Section 13.4, in which G is the space of quadratic polynomials and

$$\theta(x) = \beta_0 + \beta_1 \left(\frac{x - 62.8}{10} \right) + \beta_2 \left(\frac{x - 62.8}{10} \right)^2.$$

Determine the nominal 95% confidence interval for β_2.

Solution   The maximum-likelihood estimates of β_0, β_1, β_2, and $\mathbf{V}$, obtained during the last iteration of the iteratively reweighted least-squares method in the solution to Example 13.11(b) in Section 13.4, are given by $\hat{\beta}_0 \doteq 0.657$, $\hat{\beta}_1 \doteq 2.837$, $\hat{\beta}_2 \doteq 0.637$,

and

$$\hat{\mathbf{V}} \doteq \begin{bmatrix} 0.0291 & 0.0032 & -0.0222 \\ 0.0032 & 0.0973 & 0.0685 \\ -0.0222 & 0.0685 & 0.0988 \end{bmatrix}.$$

Since $\text{SE}(\hat{\beta}_2) \doteq \sqrt{0.0988} \doteq 0.314$ and hence $(1.96)\,\text{SE}(\hat{\beta}_2) \doteq 0.616$, the nominal 95% confidence interval for β_2 is given by $0.637 \pm 0.616 = (0.021, 1.253)$. ∎

The Wald Test

The Wald statistic corresponding to the possible value τ_0 of τ is given by $W = (\hat{\tau} - \tau_0)/\text{SE}(\hat{\tau})$. The next result follows from Theorems 13.12 and 13.13 (see the derivation of Theorem 12.5 in Section 12.1):

THEOREM 13.15   Suppose $\nabla h(\beta) \neq \mathbf{0}$ and other suitable conditions are satisfied. Then

$$P(W \leq w) \approx \Phi\left(w - \frac{\tau - \tau_0}{\text{ASD}(\hat{\tau})}\right), \qquad w \in \mathbb{R}.$$

The discussion of the Wald test in Section 12.1 is also applicable in the present context (with the preceeding theorem being used instead of Theorem 12.5 in Section 12.1).

EXAMPLE 13.18   (Continued.) Determine the P-value for the Wald test of the hypothesis that the logistic regression function is linear or, equivalently, that $\beta_2 = 0$.

Solution   The maximum-likelihood estimate of β_2 is given by $\hat{\beta}_2 \doteq 0.637$, and its standard error is given by $\text{SE}(\hat{\beta}_2) \doteq 0.314$. Thus, $W \doteq 0.637/0.314 \doteq 2.027$, so the P-value is given by $2[\,1 - \Phi(2.027)\,] \doteq .0426$. Hence, there is weak evidence against the hypothesis of linearity. ∎

Coronary Heart Disease Study

Consider now the coronary heart disease study as described in Section 13.1. Let x_1 be the coded value for age ($x_1 = 0$ for those initially in the 30–49 age group and $x_1 = 1$ for those initially in the 50–62 age group), let x_2 be the coded value for sex ($x_2 = 0$ for females and $x_2 = 1$ for males), and let x_3 be the coded value for cholesterol level ($x_3 = -3, -1, 1, 3$ according to the initial level: less than 190, between 190 and 219, between 220 and 249, and greater than or equal to 250).

Consider first the model

$$\theta(x_1, x_2, x_3) = \beta_0 + \beta_1 x_1 + \beta_2 x_2 + \beta_3 x_1 x_2 + \beta_4 x_3$$
$$+ \beta_5 x_1 x_3 + \beta_6 x_2 x_3 + \beta_7 (x_3^2 - 5)$$

for the logistic regression function. Table 13.18 lists the results obtained by computer. Based on these results, we delete the $x_3^2 - 5$ term. On the next pass, there are no surprises, so we delete the $x_2 x_3$ term. This leads to the model

$$\theta(x_1, x_2, x_3) = \beta_0 + \beta_1 x_1 + \beta_2 x_2 + \beta_3 x_1 x_2 + \beta_4 x_3 + \beta_5 x_1 x_3 \tag{29}$$

and the results shown in Table 13.19. Thus, we are not led to delete any more terms from the model. The asymptotic variance–covariance matrix of the maximum-likelihood estimates $\hat{\beta}_0, \ldots, \hat{\beta}_5$ of the regression coefficients $\beta_0, \ldots, \beta_5$ is estimated by

$$\hat{V} = \begin{bmatrix} 0.02797 & -0.02797 & -0.02615 & 0.02615 & -0.00161 & 0.00161 \\ -0.02797 & 0.04428 & 0.02615 & -0.04117 & 0.00161 & -0.00391 \\ -0.02615 & 0.02615 & 0.03525 & -0.03525 & -0.00029 & 0.00029 \\ 0.02615 & -0.04117 & -0.03525 & 0.05909 & 0.00029 & 0.00122 \\ -0.00161 & 0.00161 & -0.00029 & 0.00029 & 0.00169 & -0.00169 \\ 0.00161 & -0.00391 & 0.00029 & 0.00122 & -0.00169 & 0.00308 \end{bmatrix}.$$

To estimate the apparent effect of differing levels of cholesterol on CHD, let x_3 and $x_3 + 2$ be the coded values for consecutive levels of cholesterol. According

TABLE **13.18**

Term	Est. Coef.	SE	W	P-value
1	−3.880	0.167	−23.212	.000
x_1	1.736	0.208	8.329	.000
x_2	1.291	0.195	6.630	.000
$x_1 x_2$	−0.597	0.244	−2.452	.014
x_3	0.227	0.059	3.855	.000
$x_1 x_3$	0.185	0.055	−3.392	.001
$x_2 x_3$	0.081	0.058	1.403	.161
$x_3^2 - 5$	0.015	0.015	1.034	.301

TABLE **13.19**

Term	Est. Coef.	SE	W	P-value
1	−3.935	0.167	−23.533	.000
x_1	1.704	0.210	8.098	.000
x_2	1.361	0.188	7.247	.000
$x_1 x_2$	−0.571	0.243	−2.348	.019
x_3	0.297	0.041	7.240	.000
$x_1 x_3$	−0.192	0.056	−3.461	.001

to (29),

$$\theta(x_1, x_2, x_3 + 2) - \theta(x_1, x_2, x_3) = 2(\beta_4 + \beta_5 x_1).$$

Thus, the corresponding odds ratio depends on the age group, but not on sex.

In particular, for those initially in the 30–49 age group ($x_1 = 0$), we get

$$\theta(0, x_2, x_3 + 2) - \theta(0, x_2, x_3) = 2\beta_4,$$

and hence the corresponding odds ratio is given by

$$\frac{\pi(0, x_2, x_3 + 2)/[1 - \pi(0, x_2, x_3 + 2)]}{\pi(0, x_2, x_3)/[1 - \pi(0, x_2, x_3)]} = \exp(2\beta_4).$$

The nominal 95% confidence interval for $2\beta_4$ is given by

$$2[0.297 \pm (1.96)(0.041)] \doteq 0.594 \pm .161 = (0.433, 0.755),$$

and the 95% confidence interval for the indicated odds ratio is given by $(e^{0.433}, e^{0.755}) \doteq (1.543, 2.128)$. Thus, roughly speaking, within the younger age group and for a given sex, the odds on getting CHD double as we go from one level of cholesterol to the next higher level.

Similarly, for those initially in the 50–62 age group ($x_1 = 1$), we get

$$\theta(1, x_2, x_3 + 2) - \theta(1, x_2, x_3) = 2(\beta_4 + \beta_5),$$

and hence the corresponding odds ratio is given by

$$\frac{\pi(1, x_2, x_3 + 2)/[1 - \pi(1, x_2, x_3 + 2)]}{\pi(1, x_2, x_3)/[1 - \pi(1, x_2, x_3)]} = \exp(2(\beta_4 + \beta_5)).$$

The standard error of $\beta_4 + \beta_5$ is given by

$$SE(\beta_4 + \beta_5) = \sqrt{0.00169 - 2(0.00169) + 0.00308} \doteq 0.0374.$$

Thus, the nominal 95% confidence interval for $2(\beta_4 + \beta_5)$ is given by

$$2[0.105 \pm (1.96)(0.0374)] \doteq 0.210 \pm .147 = (0.063, 0.357),$$

and the 95% confidence interval for the indicated odds ratio is given by $(e^{0.063}, e^{0.357}) \doteq (1.065, 1.428)$. Thus, roughly speaking, within the older age group and for a given sex, the odds on getting CHD increase, but by less than 50%, as we go from one level of cholesterol to the next higher level.

It should be noted that when a model is selected based on observed data and then the same data are used to estimate the unknown parameters of the model (as we have just done in the context of the CHD study), then the resulting confidence intervals and P–values are even more suspect than usual.

Lung Cancer Study

Consider now the lung cancer study as described in Section 13.1. Let T' denote the midpoint of the corresponding range of the number of years of smoking: $T' = 17$ if $15 \le T \le 19$, $T' = 22$ if $20 \le T \le 24$ and so forth. Let C' denote the mean number of cigarettes consumed each day for the corresponding range of C: $C' =$

0 if $C = 0$, $C' = 5.2$ if $1 \le C \le 9$, and so forth. Also, set $x_1 = \log(T'/40)$ and $x_2 = \log((C' + 1)/20)$. Let the design points be ordered from left to right within rows. Then, corresponding to the first design point, we get $x_{11} = \log(17/40) \doteq -0.856$ and $x_{12} = \log(1/20) \doteq -2.996$; also, there were $n_1 = 10366$ man-years at risk and $Y_1 = 1$ cases of lung cancer. Similarly, $x_{21} = \log(17/40) \doteq -0.856$, $x_{22} = \log(6.2/20) \doteq -1.171$, $n_2 = 3121$, $Y_2 = 0$, and so forth.

Consider the corresponding Poisson regression model. Modeling the Poisson regression function $\theta(\cdot)$ as a quadratic function of x_1 and x_2, we get the results shown in Table 13.20. Thus, we are led to dropping the interaction term $x_1 x_2$ from the model for the Poisson regression. Similarly, at the next step we are led to dropping the x_1^2 term and at the third step to dropping the x_2^2 term. Upon doing so, we get the results shown in Table 13.21. This leads to the maximum-likelihood estimate

$$\hat{\theta}(x_1, x_2) \doteq -6.020 + 4.389 x_1 + 1.122 x_2$$

of the Poisson regression function, which suggests that number of years of smoking and number of cigarettes consumed each day do not interact in their effect on the rate of getting lung cancer.

TABLE 13.20

Term	Est. Coef.	SE	W	P-value
1	−6.029	0.115	−52.606	.000
x_1	4.318	0.355	12.173	.000
x_1^2	−0.578	1.039	−0.556	.578
x_2	1.219	0.179	6.812	.000
x_2^2	0.101	0.091	1.108	.268
$x_1 x_2$	0.135	0.519	0.261	.794

TABLE 13.21

Term	Est. Coef.	SE	W	P-value
1	−6.020	0.180	−74.975	.000
x_1	4.389	0.325	13.512	.000
x_2	1.122	0.142	7.896	.000

On the other hand, the corresponding maximum-likelihood estimate of the rate function is given by

$$\hat{\lambda}(x_1, x_2) = \exp(\hat{\theta}(x_1, x_2)) \doteq \exp(-6.020 + 4.389 x_1 + 1.122 x_2).$$

In terms of the variables T' and C', the maximum-likelihood estimate of the rate function is given by

$$\hat{\lambda}(T', C') \doteq \exp(-6.020)\left(\frac{T'}{40}\right)^{4.389}\left(\frac{C' + 1}{20}\right)^{1.122}$$

$$\doteq (7.855 \times 10^{-12})(T')^{4.389}(C' + 1)^{1.122}.$$

Rewriting the variables T' and C' as T and C, respectively, we get

$$\hat{\lambda}(T, C) \doteq \exp(-6.020)\left(\frac{T}{40}\right)^{4.389}\left(\frac{C+1}{20}\right)^{1.122}$$

$$\doteq (7.855 \times 10^{-12})T^{4.389}(C + 1)^{1.122}.$$

Accordingly, the number of years of smoking and the number of cigarettes consumed each day do indeed interact in their effect on the rate of getting lung cancer. Figure 13.8 shows the graphs of $\hat{\lambda}(T, C)$, $20 \leq T \leq 60$, for $C = 0, 20, 40$.

FIGURE 13.8

Dependence of fitted-rate function on number of cigarettes consumed

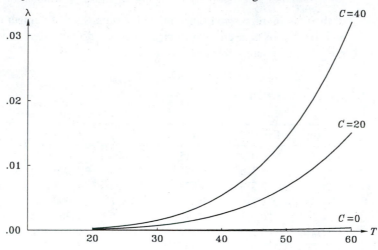

Problems

13.52 Consider the setup of Problem 13.29 in Section 13.3. Use the corresponding Poisson regression model in the answer to Problem 13.31 in that section together with Theorem 13.11 in this section to determine the approximate distribution of $\hat{\theta} - \theta$, where $\hat{\theta}$ is the maximum-likelihood estimate of θ.

13.53 Consider the setup of Problem 13.29 in Section 13.3 and the numerical values in Problem 13.30 in that section. Use the answer to Problem 13.52 in this section to determine **(a)** the nominal 95% confidence interval for θ; **(b)** the P-value for the Wald test of the hypothesis that $\theta = 0$.

13.54 Consider the data on the 2496 individuals in the CHD study whose initial serum cholesterol level was in the 190–249 range—that is, in one of the two intermediate cholesterol groups. As in the analysis in the based on all 4856 individuals in the study, we are led to the model for the logistic regression function given by (29), but we now get the results shown in Table 13.22. **(a)** Determine the nominal 95% confidence interval for $\theta(0, x_2, 1) - \theta(0, x_2, -1)$ and for the corresponding odds ratio and interpret the latter result.

The asymptotic variance–covariance matrix of $\hat{\beta}_4$ and $\hat{\beta}_5$ (which is simply the lower-right 2×2 submatrix of the asymptotic variance–covariance matrix of $\hat{\beta}_0, \ldots, \hat{\beta}_5$) equals

$$\begin{bmatrix} 0.01675 & -0.01675 \\ -0.01675 & 0.02887 \end{bmatrix}.$$

(b) Determine the nominal 95% confidence intervals for $\theta(1, x_2, 1) - \theta(1, x_2, -1)$ and the corresponding odds ratio. **(c)** Determine the P-value for the Wald test of the hypothesis that, within the older age group and for a given sex, the odds on getting CHD are the same for those whose initial serum cholesterol level is in the 190–219 range and those whose initial level is in the 220–249 range.

TABLE 13.22

Term	Est. Coef.	SE	W	P-value
1	−4.174	0.263	−15.880	.000
x_1	1.891	0.322	5.881	.000
x_2	1.502	0.295	5.096	.000
$x_1 x_2$	−0.642	0.373	−1.719	.086
x_3	0.431	0.129	3.331	.001
$x_1 x_3$	−0.387	0.170	−2.279	.023

13.55 In the context of the lung cancer study, as analyzed in this section, determine the nominal 95% confidence interval for $\theta(x_1, x_2 + 0.5) - \theta(x_1, x_2)$ and for the corresponding relative risk and interpret the latter result.

13.56 In the context of Poisson regression, determine more explicit formulas for **(a)** $\mathbf{W}$; **(b)** $\hat{\mathbf{W}}$; **(c)** $\mathrm{AV}(\hat{\lambda}(\mathbf{x}))$; **(d)** $\mathrm{SE}(\hat{\lambda}(\mathbf{x}))$.

13.57 In the context of Poisson regression, determine the asymptotic variance and standard error of $\hat{\lambda}(\mathbf{x}_2)/\hat{\lambda}(\mathbf{x}_1)$ for $\mathbf{x}_1, \mathbf{x}_2 \in \mathcal{X}$.

13.58 Determine the asymptotic variance of $\log(\hat{\mu}(\mathbf{x}_2)/\hat{\mu}(\mathbf{x}_1))$ for $\mathbf{x}_1, \mathbf{x}_2 \in \mathcal{X}$ **(a)** in the general context of this section; **(b)** for logistic regression; **(c)** for Poisson regression.

13.59 Verify that the answers to Problem 13.58 are compatible with the corresponding answers to Problem 12.13 in Section 12.1 in the context of the d-sample model. (Assume, for simplicity, that $d = 2$.)

13.60 In the context of Problem 13.10 in Section 13.1, show that **(a)** $\mathbf{V} = \mathbf{D}^{-1}\mathbf{X}^T\mathbf{W}^{-1}\mathbf{X}\mathbf{D}^{-1}$; **(b)** if $\mathbf{D} = r\mathbf{I}$, then $\mathbf{V} = r^{-2}\mathbf{X}^T\mathbf{W}^{-1}\mathbf{X}$.

13.61 Consider a parameter $\tau_1 = h_1(\boldsymbol{\beta}) = h_1(\beta_1, \ldots, \beta_p)$, where h_1 is a continuously differentiable function of $\boldsymbol{\beta}$ that is defined on an open set in $\mathbb{R}^p$ containing the true value of β, and consider a parameter $\tau_2 = h_2(\tau_1)$, where h_2 is a continuously differentiable function of τ_1 that is defined on an open interval in $\mathbb{R}$ containing the range of h_1 and hence the true value of τ_1. Then $\tau_2 = h(\beta)$, where $h(\beta) = h_2(h_1(\beta))$. The maximum-likelihood estimate of τ_2 is given by $\hat{\tau}_2 = h(\hat{\beta}) = h_2(h_1(\hat{\beta})) = h_2(\hat{\tau}_1)$, where $\hat{\tau}_1 = h_1(\hat{\beta})$ is the maximum-likelihood estimate of τ_1. Show that **(a)** $\mathrm{AV}(\hat{\tau}_2) = [\, h_2'(\tau_1)\,]^2 \, \mathrm{AV}(\hat{\tau}_1)$; **(b)** $\mathrm{SE}(\hat{\tau}_2) = |h_2'(\hat{\tau}_1)| \, \mathrm{SE}(\hat{\tau}_1)$.

13.62 (Continued.) Suppose h_2 is a strictly increasing function on $\mathbb{R}$. Let (L_1, U_1) be a nominal $100(1 - \alpha)\%$ confidence interval for τ_1 [so that L_1 is the nominal $100(1 - \alpha/2)\%$ lower confidence bound for τ_1 and U_1 is its nominal $100(1 - \alpha/2)\%$ upper confidence bound], and consider the nominal $100(1 - \alpha)\%$ confidence interval (L_2, U_2) for τ_2, where $L_2 = h_2(L_1)$ and $U_2 = h_2(U_1)$. Show that the actual coverage probability of the confidence interval (L_2, U_2) for τ_2 equals the actual coverage probability of the confidence interval (L_1, U_1) for τ_1; that is, $P(L_2 < \tau_2 < U_2) = P(L_1 < \tau_1 < U_1)$.

13.63 Let $\tilde{g}_1, \ldots, \tilde{g}_p$ be a basis of G having coefficient matrix $\mathbf{A} = [\, a_{jl}\,]$ relative to the basis $g_1, \ldots, g_p$, let $\tilde{\mathbf{X}}$ be the design matrix relative to the basis $\tilde{g}_1, \ldots, \tilde{g}_p$, and set $\tilde{\mathbf{V}} = (\tilde{\mathbf{X}}^T \mathbf{W} \tilde{\mathbf{X}})^{-1}$. Use Theorem 9.8 in Section 9.2 to show that $\tilde{\mathbf{V}} = (\mathbf{A}^T)^{-1} \mathbf{V} \mathbf{A}^{-1}$.

13.64 (Continued.) Let $\tilde{\beta} = [\, \tilde{\beta}_1, \ldots, \tilde{\beta}_p\,]^T$ denote the coefficient vector of $\theta(\cdot)$ relative to the basis $\tilde{g}_1, \ldots, \tilde{g}_p$. Use the formula $\beta = \mathbf{A}^T \tilde{\beta}$ from Equation (5) in Section 9.1 to show that **(a)** $\beta_m = \sum_j a_{jm} \tilde{\beta}_j$ for $1 \le m \le p$; **(b)** $\partial \beta_m / \partial \tilde{\beta}_j = a_{jm}$ for $1 \le j, m \le p$.

13.65 (Continued.) Consider a parameter $\tau = h(\beta) = h(\mathbf{A}^T \tilde{\beta}) = \tilde{h}(\tilde{\beta})$, where $\tilde{h}(\tilde{\beta}) = h(\mathbf{A}^T \tilde{\beta})$ for $\tilde{\beta} \in \mathbb{R}^p$. Suppose $h(\beta)$ is a continuously differentiable function of β near its true value and hence that $\tilde{h}(\tilde{\beta})$ is a continuously differentiable function of $\tilde{\beta}$ near its true value. Then, by the chain rule from multivariate calculus,

$$\frac{\partial \tilde{h}(\tilde{\beta})}{\partial \tilde{\beta}_j} = \frac{\partial h(\mathbf{A}^T \tilde{\beta})}{\partial \tilde{\beta}_j} = \sum_m \frac{\partial h}{\partial \beta_m}(\mathbf{A}^T \tilde{\beta}) \frac{\partial \beta_m}{\partial \tilde{\beta}_j} = \sum_m \frac{\partial h}{\partial \beta_m}(\beta) \frac{\partial \beta_m}{\partial \tilde{\beta}_j}.$$

(a) Use this result together with that of Problem 13.64(b) to verify that $\nabla \tilde{h}(\tilde{\beta}) = \mathbf{A}[\, \nabla h(\beta)\,]$. **(b)** Use the results of part (a) and Problem 13.63 to verify that $[\, \nabla \tilde{h}(\tilde{\beta})\,]^T \tilde{\mathbf{V}}[\, \nabla \tilde{h}(\tilde{\beta})\,] = [\, \nabla h(\beta)\,]^T \mathbf{V}[\, \nabla h(\beta)\,]$ and hence that the asymptotic variance of the maximum-likelihood estimate of τ does *not* depend on the choice of the basis of G. (Similarly, the standard error of the maximum-likelihood estimate of τ does *not* depend on the choice of the basis of G.)

13.6
The Likelihood-Ratio Test

Consider a discrete $\mathcal{Y}$-valued random variable $\mathbf{Y}$ having positive probability function $f(\mathbf{y}; \theta)$, $\mathbf{y} \in \mathcal{Y}$, where θ is an unknown member of a specified set Θ. Let θ_0 and θ_1 be distinct members of Θ and consider the simple hypothesis that $\theta = \theta_0$ and the simple alternative that $\theta = \theta_1$ (where "simple" means having a single possibility). A likelihood-ratio test of the hypothesis against the alternative rejects the hypothesis if and only if the likelihood ratio $f(\mathbf{Y}; \boldsymbol{\theta}_1)/f(\mathbf{Y}; \boldsymbol{\theta}_0)$ is sufficiently large. The Neyman–Pearson lemma in Section 12.2 provides theoretical justification for such a test. The likelihood ratio can also be written as $L(\boldsymbol{\theta}_1)/L(\boldsymbol{\theta}_0)$, where $L(\boldsymbol{\theta}) = f(\mathbf{Y}; \boldsymbol{\theta})$, $\boldsymbol{\theta} \in \Theta$, is the likelihood function.

Let Θ_0 be a nonempty subset of Θ such that $\Theta_1 = \Theta \backslash \Theta_0$ is also nonempty. Consider the compound hypothesis that $\boldsymbol{\theta} \in \Theta_0$ and the compound alternative that $\boldsymbol{\theta} \in \Theta_1$ (where "compound" means having a number of possibilities). Since the likelihood function is bounded above by 1, the function $L(\boldsymbol{\theta})$, $\boldsymbol{\theta} \in \Theta_0$, has a supremum (least upper bound)

$$L(\Theta_0) = \sup_{\boldsymbol{\theta} \in \Theta_0} L(\boldsymbol{\theta}) \leq 1.$$

Similarly, the function $L(\boldsymbol{\theta})$, $\boldsymbol{\theta} \in \Theta_1$, has a supremum $L(\Theta_1)$. A likelihood-ratio test of the compound hypothesis against the compound alternative rejects the hypothesis if and only if the likelihood-ratio $L(\Theta_1)/L(\Theta_0)$ is sufficiently large. Suppose the likelihood function is a continuous function on Θ and that every point in Θ_0 is the limit of points in Θ_1 (which is true, for example, if $\Theta = \mathbb{R}^2$ and Θ_0 is a line in $\mathbb{R}^2$). Then the supremum of the likelihood function over Θ equals its supremum over Θ_1, so the likelihood ratio can be written as $L(\Theta)/L(\Theta_0)$. Note that $1 \leq L(\Theta)/L(\Theta_0) < \infty$. Suppose further that there is a maximum-likelihood estimate $\hat{\boldsymbol{\theta}} \in \Theta$ of $\boldsymbol{\theta}$ and also a maximum-likelihood estimate $\hat{\boldsymbol{\theta}}_0 \in \Theta_0$ of $\boldsymbol{\theta}$. Then

$$L(\Theta) = \max_{\boldsymbol{\theta} \in \Theta} L(\boldsymbol{\theta}) = L(\hat{\boldsymbol{\theta}}) \quad \text{and} \quad L(\Theta_0) = \max_{\boldsymbol{\theta} \in \Theta_0} L(\boldsymbol{\theta}) = L(\hat{\boldsymbol{\theta}}_0),$$

so the likelihood ratio can be written as $L(\hat{\boldsymbol{\theta}})/L(\hat{\boldsymbol{\theta}}_0)$. Set $l(\Theta) = \log L(\Theta) \leq 0$ and $l(\Theta_0) = \log L(\Theta_0) \leq l(\Theta)$. The likelihood-ratio test rejects the hypothesis if and only if the quantity

$$D = 2[\, l(\Theta) - l(\Theta_0)\,] = 2 \log \frac{L(\Theta)}{L(\Theta_0)},$$

referred to as the *likelihood-ratio statistic*, is sufficiently large. Observe that $0 \leq D < \infty$.

Let Θ be a p-dimensional space and let Θ_0 be a p_0-dimensional subspace of Θ, where $p_0 < p$. Under suitable conditions, if the hypothesis is valid, then D has approximately the chi-square distribution with $p - p_0$ degrees of freedom. Let $0 < \alpha < 1$. The inequality $D \geq \chi^2_{1-\alpha, p-p_0}$ determines the critical region having nominal size α for the likelihood-ratio test of the hypothesis that $\boldsymbol{\theta} \in \Theta_0$; that is, if the hypothesis holds and the suitable conditions are satisfied, then $P(D \geq \chi^2_{1-\alpha, p-p_0}) \approx \alpha$. The corresponding acceptance region is given by $D < \chi^2_{1-\alpha, p-p_0}$. The P-value for the test, given by $P\text{-value} = 1 - \chi^2_{p-p_0}(D)$, has the usual interpretation: D lies in the critical region for the test of size α if α is greater than or equal to the P-value and in the acceptance region if α is less than the P-value.

The likelihood-ratio test will now be applied to logistic and Poisson regression. Let G be an identifiable p-dimensional linear space. Recall that the log-likelihood function is given by

$$l(\theta(\cdot)) = \sum_k n_k[\theta(\mathbf{x}_k)\bar{Y}_k - C(\theta(\mathbf{x}_k))] + \log r(\mathbf{Y}; \mathbf{n}), \qquad g \in G.$$

Let G_0 be a p_0-dimensional space of G, where $p_0 < p$. According to Theorem 13.8 in Section 13.3, if there is a maximum-likelihood estimate $\hat{\theta}(\cdot)$ in G of the canonical regression function, then there is a maximum-likelihood estimate $\hat{\theta}_0(\cdot)$ in G_0 of this function. Thus, it follows from Theorem 13.9 in Section 13.3 that if suitable conditions are satisfied, then $\hat{\theta}(\cdot)$ and hence $\hat{\theta}_0(\cdot)$ exist with probability close to 1.

Set

$$L(G) = \sup_{g \in G} L(g) \quad \text{and} \quad l(G) = \log L(G) = \sup_{g \in G} l(g)$$

and let $L(G_0)$ and $l(G_0)$ correspond to G_0 in a similar manner. If $\hat{\theta}(\cdot)$ exists, then

$$l(G) = l(\hat{\theta}(\cdot)) = \sum_k n_k[\hat{\theta}(\mathbf{x}_k)\bar{Y}_k - C(\hat{\theta}(\mathbf{x}_k))] + \log r(\mathbf{Y}; \mathbf{n});$$

if $\hat{\theta}_0(\cdot)$ exists, then

$$l(G_0) = l(\hat{\theta}_0(\cdot)) = \sum_k n_k[\hat{\theta}_0(\mathbf{x}_k)\bar{Y}_k - C(\hat{\theta}_0(\mathbf{x}_k))] + \log r(\mathbf{Y}; \mathbf{n}).$$

Consider the hypothesis that the canonical regression function lies in G_0 and the alternative that it lies in G, but not in G_0. If $\hat{\theta}(\cdot)$ exists, then $\hat{\theta}_0(\cdot)$ exists and the likelihood-ratio statistic is given by

$$D = 2[l(\hat{\theta}(\cdot)) - l(\hat{\theta}_0(\cdot))]$$

$$= 2\sum_k n_k\left\{[\hat{\theta}(\mathbf{x}_k) - \hat{\theta}_0(\mathbf{x}_k)]\bar{Y}_k - [C(\hat{\theta}(\mathbf{x}_k)) - C(\hat{\theta}_0(\mathbf{x}_k))]\right\}. \qquad \textbf{(30)}$$

THEOREM 13.16   If $p_0 < p$, $\theta(\cdot) \in G_0$, and other suitable conditions are satisfied, then D has approximately the chi-square distribution with $p - p_0$ degrees of freedom.

Derivation   Except on an event having small probability, which we can ignore, $\hat{\theta}(\cdot)$ and hence $\hat{\theta}_0(\cdot)$ exist.

Recall from the derivation of Theorem 13.11 in Section 13.5 that the quadratic approximation to the log-likelihood function about $\theta(\cdot)$ is given by

$$l(\theta(\cdot) + g) \approx -\tfrac{1}{2}\,\mathrm{RSS}(g) + \text{remainder term},$$

where the remainder term does not depend on g and the approximation is accurate when g is suitably small; here

$$\mathrm{RSS}(g) = \sum_k n_k\sigma^2(\mathbf{x}_k)\left(\frac{\bar{Y}_k - \mu(\mathbf{x}_k)}{\sigma^2(\mathbf{x}_k)} - g\right)^2, \qquad g \in G.$$

Let $\hat{g}$ be the function in G that minimizes $\mathrm{RSS}(g)$, $g \in G$. Since $\hat{\theta}(\cdot) = \theta(\cdot) + \hat{\theta}(\cdot) - \theta(\cdot)$ and $\hat{\theta}(\cdot) - \theta(\cdot)$ is the function $g \in G$ that maximizes $l(\theta(\cdot) + g)$, $g \in G$, we conclude that

$$l(\hat{\theta}(\cdot)) \approx -\tfrac{1}{2} \operatorname{RSS}(\hat{g}) + \text{remainder term.} \qquad \textbf{(31)}$$

Similarly, let $\hat{g}_0$ be the function g in G_0 that minimizes $\operatorname{RSS}(g)$, $g \in G_0$. Then

$$l(\hat{\theta}_0(\cdot)) \approx -\tfrac{1}{2} \operatorname{RSS}(\hat{g}_0) + \text{remainder term.} \qquad \textbf{(32)}$$

It follows from (31) and (32) that

$$D \approx \operatorname{RSS}(\hat{g}_0) - \operatorname{RSS}(\hat{g}). \qquad \textbf{(33)}$$

Recall that H is the space of all functions defined on the design set. Consider the inner product on H corresponding to the weights $w_k = n_k \sigma^2(\mathbf{x}_k)$, $1 \le k \le d$, and let h be the function on H given by

$$h(\mathbf{x}_k) = \frac{\bar{Y}_k - \mu(\mathbf{x}_k)}{\sigma^2(\mathbf{x}_k)}, \qquad 1 \le k \le d.$$

Then $\hat{g}$ is the orthogonal projection $P_G h$ of h onto G, and $\operatorname{RSS}(\hat{g}) = \|h - \hat{g}\|^2$. Similarly, $\hat{g}_0 = P_{G_0} h$ and $\operatorname{RSS}(\hat{g}_0) = \|h - \hat{g}_0\|^2$. Now $h - \hat{g} = h - P_G h$ is orthogonal to every function in G. In particular, it is orthogonal to $\hat{g} - \hat{g}_0$, and hence

$$\begin{aligned}
\operatorname{RSS}(\hat{g}_0) &= \|h - \hat{g}_0\|^2 \\
&= \|h - \hat{g} + \hat{g} - \hat{g}_0\|^2 \\
&= \|h - \hat{g}\|^2 + \|\hat{g} - \hat{g}_0\|^2 \\
&= \operatorname{RSS}(\hat{g}) + \|\hat{g} - \hat{g}_0\|^2.
\end{aligned}$$

Thus, by (33),

$$D \approx \|\hat{g} - \hat{g}_0\|^2. \qquad \textbf{(34)}$$

Let $g_1, \ldots, g_p$ be an orthonormal basis of G such that $g_1, \ldots, g_{p_0}$ is a basis of G_0 and let $\mathbf{Z} = [Z_1, \ldots, Z_p]^T$ be the coefficient vector of $\hat{g}$ relative to $g_1, \ldots, g_p$. Then $\hat{g} = Z_1 g_1 + \cdots + Z_p g_p$ and $\hat{g}_0 = Z_1 g_1 + \cdots + Z_{p_0} g_{p_0}$, so

$$\hat{g} - \hat{g}_0 = Z_{p_0+1} g_{p_0+1} + \cdots + Z_p g_p$$

and hence $\|\hat{g} - \hat{g}_0\|^2 = Z_{p_0+1}^2 + \cdots + Z_p^2$. Therefore, by (34),

$$D \approx \sum_{k=p_0+1}^{p} Z_k^2. \qquad \textbf{(35)}$$

Since $g_1, \ldots, g_p$ are orthonormal, their Gram matrix is given by $\mathbf{X}^T \mathbf{W} \mathbf{X} = \mathbf{I}$ (the $p \times p$ identity matrix), where $\mathbf{X}$ is the $d \times p$ design matrix corresponding to $g_1, \ldots, g_p$ and $\mathbf{W} = \operatorname{diag}(n_1 \sigma^2(\mathbf{x}_1), \ldots, n_k \sigma^2(\mathbf{x}_k))$. Consequently (see the derivation of Theorem 13.11 in Section 13.5), $\mathbf{Z}$ has mean vector $\mathbf{0}$ and variance–covariance matrix $\mathbf{I}$. Moreover, $\mathbf{Z}$ has approximately the multivariate normal distribution with the indicated mean vector and variance–covariance matrix; that is, the joint distribution of $Z_1, \ldots, Z_p$ is approximately that of p independent standard normal random variables. Therefore, by (35), D has approximately the chi-square distribution with $p - p_0$ degrees of freedom.   ∎

Suppose $p > 1$ and that G contains the constant functions. The maximum-likelihood constant estimate of $\theta(\cdot)$ is given by $\hat{\theta}_0(\cdot) = \hat{\theta}$, where $\hat{\theta} = (C')^{-1}(\bar{Y})$. Equation (30) for the likelihood-ratio statistic for testing the hypothesis that $\theta(\cdot)$ is a constant function simplifies to

$$D = 2\sum_k n_k \left\{ [\hat{\theta}(\mathbf{x}_k) - \hat{\theta}]\bar{Y}_k - [C(\hat{\theta}(\mathbf{x}_k)) - C(\hat{\theta})] \right\}. \tag{36}$$

If $\theta(\cdot)$ is a constant function and other suitable conditions are satisfied, then D has approximately the chi-square distribution with $p - 1$ degrees of freedom.

Suppose G is saturated and $d \geq 2$. Consider the *hypothesis of homogeneity*; that is, $\theta_1 = \cdots = \theta_d$ or, equivalently, $\mu_1 = \cdots = \mu_d$. Here G_0 is the space of constant functions, and $p_0 = 1$. Equation (36) for the likelihood-ratio statistic for testing this hypothesis simplifies to

$$D = 2\sum_k n_k \left\{ [\hat{\theta}_k - \hat{\theta}]\bar{Y}_k - [C(\hat{\theta}_k) - C(\hat{\theta})] \right\}, \tag{37}$$

where $\hat{\theta}_k = (C')^{-1}(\bar{Y}_k)$ for $1 \leq k \leq d$ and $\hat{\theta} = (C')^{-1}(\bar{Y})$. If the hypothesis and other suitable conditions are satisfied, then D has approximately the chi-square distribution with $d - 1$ degrees of freedom.

EXAMPLE 13.19   In the context of the logistic regression model, determine a formula for the likelihood-ratio statistic for testing

(a) the hypothesis that $\theta(\cdot) \in G_0$ under the assumption that $\theta(\cdot) \in G$, where G_0 is a proper subspace of G and the maximum-likelihood estimate of $\theta(\cdot)$ in G exists;

(b) the hypothesis that $\theta(\cdot)$ is a constant function [or, equivalently, that $\pi(\cdot)$ is a constant function] under the assumption that $\theta(\cdot) \in G$, where G contains the space of constant functions as a proper subspace and the maximum-likelihood estimate of $\theta(\cdot)$ in G exists;

(c) the hypothesis $\pi_1 = \cdots = \pi_d$ of homogeneity under the assumptions of a saturated model, where $d \geq 2$.

Solution   In the context of the logistic regression model,

$$\hat{\theta}(\mathbf{x}) = \log \frac{\hat{\pi}(\mathbf{x})}{1 - \hat{\pi}(\mathbf{x})}, \qquad \mathbf{x} \in \mathcal{X},$$

$$C(\hat{\theta}(\mathbf{x})) = \log \frac{1}{1 - \hat{\pi}(\mathbf{x})}, \qquad \mathbf{x} \in \mathcal{X},$$

$$\hat{\theta}_0(\mathbf{x}) = \log \frac{\hat{\pi}_0(\mathbf{x})}{1 - \hat{\pi}_0(\mathbf{x})}, \qquad \mathbf{x} \in \mathcal{X},$$

and

$$C(\hat{\theta}_0(\mathbf{x})) = \log \frac{1}{1 - \hat{\pi}_0(\mathbf{x})}, \qquad \mathbf{x} \in \mathcal{X}.$$

Set $\hat{\pi}_k = \bar{Y}_k$ for $1 \le k \le d$, $Y = \sum_k Y_k = n\bar{Y}$, and $\hat{\pi} = \bar{Y} = Y/n$.

(a) By (30) and elementary algebra, the likelihood-ratio statistic for testing the hypothesis that $\theta(\cdot) \in G_0$ under the assumption that $\theta(\cdot) \in G$ is given by

$$D = 2 \sum_k \left(Y_k \log \frac{\hat{\pi}(\mathbf{x}_k)}{\hat{\pi}_0(\mathbf{x}_k)} + (n_k - Y_k) \log \frac{1 - \hat{\pi}(\mathbf{x}_k)}{1 - \hat{\pi}_0(\mathbf{x}_k)} \right). \tag{38}$$

(b) By (38), the likelihood-ratio statistic for testing the hypothesis of homogeneity [that is, $\theta(\cdot)$ is a constant function or, equivalently, $\pi(\cdot)$ is a constant function] is given by

$$D = 2 \sum_k \left(Y_k \log \frac{\hat{\pi}(\mathbf{x}_k)}{\hat{\pi}} + (n_k - Y_k) \log \frac{1 - \hat{\pi}(\mathbf{x}_k)}{1 - \hat{\pi}} \right). \tag{39}$$

(c) By (39), the likelihood-ratio statistic for testing the hypothesis of homogeneity under the assumptions of a saturated model is given by

$$D = 2 \sum_k \left(Y_k \log \frac{\hat{\pi}_k}{\hat{\pi}} + (n_k - Y_k) \log \frac{1 - \hat{\pi}_k}{1 - \hat{\pi}} \right),$$

which can be written as

$$D = 2 \sum_k \left(Y_k \log \frac{Y_k/n_k}{Y/n} + (n_k - Y_k) \log \frac{(n_k - Y_k)/n_k}{(n - Y)/n} \right). \quad \blacksquare \tag{40}$$

Approximate Equivalence of Wald Tests and Likelihood-Ratio Tests

The canonical regression function is given in terms of a basis $g_1, \ldots, g_p$ of G by $\theta(\cdot) = \sum_j \beta_j g_j$. Observe that this function is in the subspace G_0 of G spanned by $g_1, \ldots, g_{p-1}$ if and only if $\beta_p = 0$. When there is a maximum-likelihood estimate $\hat{\theta}(\cdot)$ in G of the canonical regression function, it is given by $\hat{\theta}(\cdot) = \sum_j \hat{\beta}_j g_j$. The next result is analogous to Theorem 10.16 in Section 10.8.

THEOREM 13.17 Let G_0 be the subspace of G consisting of all functions in G whose coefficient β_p of g_p equals zero, let D be the likelihood-ratio statistic for testing the hypothesis that $\theta(\cdot) \in G_0$ under the assumption that $\theta(\cdot) \in G$, and let $W = \hat{\beta}_p / \text{SE}(\hat{\beta}_p)$ be the Wald statistic for testing the equivalent hypothesis that $\beta_p = 0$. If the hypothesis and other suitable conditions are satisfied, then $P(D \approx W^2) \approx 1$.

Derivation   Consider the inner product introduced in the derivation of Theorem 13.16. If g_p is replaced by $g_p - P_{G_0} g_p$, then β_p, $\hat{\beta}_p$ and $\text{SE}(\hat{\beta}_p)$ are unaltered [see Problem 13.65(b) in Section 13.5], so W is unaltered; also, G and G_0 are unaltered, so D is unaltered. Thus, without loss of generality, we can assume that g_p is orthogonal to G_0 and hence to each of the functions $g_1, \ldots, g_{p-1}$. If g_p is replaced by g_p/a, where $a > 0$, then G and G_0 and hence D are unaltered; also, $\hat{\beta}_p$ is replaced by $a\hat{\beta}_p$, and $\text{SE}(\hat{\beta}_p)$ is replaced by $a\,\text{SE}(\hat{\beta}_p)$ [see Problem 13.64(b) in Section 13.5], so W is unaltered. Thus, without loss of generality, we can assume that $\|g_p\| = 1$. Replacing $g_1, \ldots, g_{p-1}$ by an orthonormal basis of G_0, which again does not alter D or W, we can assume that $g_1, \ldots, g_p$ is an orthonormal basis of G and hence that its Gram matrix $\mathbf{X}^T\mathbf{WX}$ (see Theorem 9.20 in Section 9.4) is the $p \times p$ identity matrix $\mathbf{I}$. Thus, the asymptotic variance–covariance matrix $\mathbf{V}$ of $\hat{\beta}$ is given by $\mathbf{V} = (\mathbf{X}^T\mathbf{WX})^{-1} = \mathbf{I}$, so $\text{AV}(\hat{\beta}_p) = 1$ and hence $\text{ASD}(\hat{\beta}_p) = 1$. We now conclude from Theorem 13.13 in Section 13.5 that, with probability close to 1, $\text{SE}(\hat{\beta}_p) \approx 1$. Under the indicated hypothesis, $\beta_p = 0$ and hence $\hat{\beta}_p = \hat{\beta}_p - \beta_p$. Thus, with probability close to 1,

$$W^2 = \left(\frac{\hat{\beta}_p}{\text{SE}(\hat{\beta}_p)} \right)^2 \approx \hat{\beta}_p^2 = (\hat{\beta}_p - \beta_p)^2. \tag{41}$$

It follows from (28) in Section 13.5 that $P(\hat{\beta}_p - \beta_p \approx Z_p) \approx 1$. Thus, by (41),

$$P(W^2 \approx Z_p^2) \approx 1. \tag{42}$$

According to (35) with $p_0 = p - 1$,

$$P(D \approx Z_p^2) \approx 1. \tag{43}$$

We conclude from (42) and (43) that $P(D \approx W^2) \approx 1$.   ∎

It follows from Theorems 13.16 and 13.17 that if $\theta(\cdot) \in G_0$ and the other conditions of these theorems are satisfied, then the P-value for the likelihood-ratio test of the hypothesis that $\theta(\cdot) \in G_0$ is approximately equal to the P-value for the Wald test of the equivalent hypothesis that $\beta_p = 0$.

Goodness-of-Fit Test

The likelihood-ratio test can be used to test the assumption that the canonical regression function is in G. Suppose the assumptions of the canonical regression model are satisfied and G is an identifiable p-dimensional linear space of functions on $\mathcal{X}$ with $p < d$, but consider the possibility that the canonical regression function may not be in G. By ignoring points of $\mathcal{X}$ that are not design points, we can think of G as a proper subspace of the d-dimensional space H of all functions on the design set. Then the canonical regression function is in H. The likelihood-ratio test of the assumption that the canonical regression function is in G is referred to as the goodness-of-fit test of this assumption. The corresponding likelihood-ratio statistic $2[\,l(H) - l(G)\,]$, referred to as the *deviance* of G and denoted by $D(G)$, can be viewed as a measure of the lack-of-fit of the maximum-likelihood estimate in G of the canonical regression function. Under suitable conditions, if the canonical regression function is in G, then $D(G)$ has approximately the chi-square distribution with $d - p$ degrees of freedom. If G is saturated, then its deviance equals zero.

Suppose the maximum-likelihood estimate $\hat{\theta}_k = (C')^{-1}(\bar{Y}_k)$, of θ_k exists for $1 \leq k \leq d$ or, equivalently, that the maximum-likelihood estimate in H of $\theta(\cdot)$ exists. Then, by Theorem 13.8 in Section 13.3, the maximum-likelihood estimate $\hat{\theta}(\cdot)$ in G of $\theta(\cdot)$ exists. Thus, by (30),

$$D(G) = 2 \sum_k n_k \Big\{ [\hat{\theta}_k - \hat{\theta}(\mathbf{x}_k)] \bar{Y}_k - [C(\hat{\theta}_k) - C(\hat{\theta}(\mathbf{x}_k))] \Big\}. \tag{44}$$

In particular, the deviance of the space of constant functions is given by

$$D(G_{\text{con}}) = 2 \sum_k n_k \Big\{ [\hat{\theta}_k - \hat{\theta}] \bar{Y}_k - [C(\hat{\theta}_k) - C(\hat{\theta})] \Big\}, \tag{45}$$

where $\hat{\theta} = (C')^{-1}(\bar{Y})$. This formula is in agreement with (37); that is, the goodness-of-fit test of the assumption that the canonical regression function is constant is equivalent to the likelihood-ratio test of the hypothesis that $\theta_1 = \cdots = \theta_d$ under the assumption of a saturated model.

EXAMPLE 13.20 In the context of the logistic regression model, determine a formula for the deviance of

(a) G; **(b)** the space of constant functions.

Solution **(a)** By (38), the deviance of G is given by

$$D(G) = 2 \sum_k \left(Y_k \log \frac{\hat{\pi}_k}{\hat{\pi}(\mathbf{x}_k)} + (n_k - Y_k) \log \frac{1 - \hat{\pi}_k}{1 - \hat{\pi}(x_k)} \right), \tag{46}$$

where $\hat{\pi}_k = \bar{Y}_k$ for $1 \leq k \leq d$.

(b) By (46), the deviance of the space of constant functions is given by

$$D(G_{\text{con}}) = 2 \sum_k \left(Y_k \log \frac{\hat{\pi}_k}{\hat{\pi}} + (n_k - Y_k) \log \frac{1 - \hat{\pi}_k}{1 - \hat{\pi}} \right),$$

where $\hat{\pi} = \bar{Y}$, so it can be written as

$$D(G_{\text{con}}) = 2 \sum_k \left(Y_k \log \frac{Y_k/n_k}{Y/n} + (n_k - Y_k) \log \frac{(n_k - Y_k)/n_k}{(n - Y)/n} \right), \tag{47}$$

where $Y = \sum_k Y_k = n\bar{Y}$. This formula is in agreement with (40); that is, the goodness-of-fit test of the assumption that $\pi(\cdot)$ is constant is equivalent to the likelihood-ratio test of the hypothesis that $\pi_1 = \cdots = \pi_d$ under the assumption of a saturated model. Under suitable conditions, if $\theta(\cdot)$ is a constant function, then $D(G_{\text{con}})$ has approximately the chi-square distribution with $d - 1$ degrees of freedom. ∎

Let G_0 be a p_0-dimensional subspace of G. The likelihood-ratio statistic for testing the hypothesis that $\theta(\cdot) \in G_0$ under the assumption that $\theta(\cdot) \in G$ can be written in terms of the deviances of G_0 and G as

$$D = D(G_0) - D(G), \tag{48}$$

the proof being left as a problem. If the indicated hypothesis and the other conditions of Theorem 13.16 are satisfied, then D has approximately the chi-square distribution with $p - p_0$ degrees of freedom.

(The deviance of G is analogous to the lack-of-fit sum of squares for G in the context of the normal linear regression model. Conversely, in the context of the normal linear regression model, the F test can be derived as a likelihood-ratio test. Moreover, the F statistic can be written as

$$F = \frac{[\,\mathrm{LSS}(G_0) - \mathrm{LSS}(G)\,]/(p - p_0)}{S^2}.$$

Let $n \gg 1$. Then $P(S^2/\sigma^2 \approx 1) \approx 1$ and hence, with probability close to 1,

$$(p - p_0)F \approx \frac{\mathrm{LSS}(G_0) - \mathrm{LSS}(G)}{\sigma^2}.$$

Suppose the regression function is in G_0. Then, by Theorem 10.14(b) in Section 10.8, $[\,\mathrm{LSS}(G_0) - \mathrm{LSS}(G)\,]/\sigma^2$ has the chi-square distribution with $p - p_0$ degrees of freedom, and hence $(p - p_0)F$ has approximately the chi-square distribution with $p - p_0$ degrees of freedom.)

EXAMPLE **13.21**   Consider the beetle experiment, which was described in Section 13.1, and the corresponding linear logistic regression model in which G is the space G_p of polynomials on $\mathbb{R}$ of degree p.

(a) For $0 \leq p \leq 3$, determine the P-value for the goodness-of-fit test of the assumption that the logistic regression function is in G_p.

(b) Determine the P-value for the likelihood-ratio test of the hypothesis that the logistic regression function is linear under the assumption that it is a quadratic polynomial.

(c) Explain the discrepancy between the solution to part (a) for $p = 1$ and the solution to part (b).

Solution   (a) Using (44) or a statistical package, we obtain the table of deviances shown in Table 13.23. In particular, the deviance for the cubic fit is given by $D(G_3) \doteq 2.943$, and the corresponding degrees of freedom is given by $d - p = 8 - 4 = 4$. The statistic

$$\mathrm{MD} = \frac{D(G_3)}{\mathrm{DF}} \doteq \frac{2.943}{4} \doteq 0.736$$

is shown as a useful descriptive statistic. The P-value for the goodness-of-fit test of the cubic polynomial assumption is given by $1 - \chi_4^2(2.943) \doteq .567$. The regression function clearly is not constant, but the goodness-of-fit tests do not provide any evidence against the linear, quadratic, or cubic polynomial model.

TABLE   13.23

Model	D	DF	MD	P-value
Constant	284.202	7	40.600	.000
Linear	7.566	6	1.261	.272
Quadratic	2.988	5	0.598	.702
Cubic	2.943	4	0.736	.567

(b) The likelihood-ratio statistic for testing the hypothesis of linearity under the assumption that the logistic regression function is a quadratic polynomial is given by $D = D(G_1) - D(G_2) \doteq 7.566 - 2.988 = 4.578$. The corresponding P-value is given by $1 - \chi_1^2(4.578) \doteq .0324$. Thus, there is evidence against the hypothesis of linearity. By comparison (and as an illustration of Theorem 13.17), in the solution to Example 13.18 in Section 13.5, the P-value for the Wald test of the hypothesis of linearity under the same assumption was found to be .0426.

(c) According to the solution to part (a), the P-value for the goodness-of-fit test of the assumption that the logistic regression function is linear is given by $1 - \chi_6^2(D(G_1)) \doteq 1 - \chi_6^2(7.566) \doteq .272$. On the other hand, according to the solution to part (b), the P-value of the likelihood-ratio test of the hypothesis of linearity under the assumption that the logistic regression function is a quadratic polynomial is given by $1 - \chi_1^2(D(G_1) - D(G_2)) \doteq 1 - \chi_1^2(4.578) \doteq .0324$. To explain this discrepancy, assume the logistic regression function is a quadratic polynomial. The goodness-of-fit statistic $D(G_2)$ corresponding to this assumption has approximately the chi-square distribution with 5 degrees of freedom. The likelihood-ratio statistic $D(G_1) - D(G_2)$ for testing the hypothesis of linearity under the assumption that the logistic regression function is a quadratic polynomial has 1 degree of freedom. On the other hand, the deviance $D(G_1) = D(G_1) - D(G_2) + D(G_2)$ corresponding to the assumption of linearity has 6 degrees of freedom. The randomness due to the 5 degrees of freedom in $D(G_2)$ tends to swamp the evidence of nonlinearity in $D(G_1) - D(G_2)$. Thus, the power of the goodness-of-fit test of the assumption of linearity is much less than the power of the likelihood-ratio test of the hypothesis of linearity under the assumption that the logistic regression function is a quadratic polynomial when that assumption is valid.

A portion of the discrepancy between the two P-values is due to the unusually low value of the deviance of the quadratic fit relative to the mean of the chi-square distribution with 5 degrees of freedom (2.988 versus 5). Suppose the experimental data had been somewhat different and, specifically, that $D(G_1) - D(G_2) \doteq 4.578$ as in the actual experiment, but $D(G_2) = 5$. Then we would have gotten that $D(G_1) = D(G_1) - D(G_2) + D(G_2) \doteq 9.578$, and the corresponding P-value for the goodness-of-fit test of the assumption of linearity would have been given by $1 - \chi_6^2(D(G_1)) \doteq 1 - \chi_6^2(9.578) \doteq .144$, so the discrepancy between the two P-values would not have been so large. ∎

Lung Cancer Study

EXAMPLE 13.22   Consider the lung cancer study and recall the discussion of this study in Section 13.5. Determine the P-value for

(a) the goodness-of-fit test of the assumption that the Poisson regression function is a linear function of x_1 and x_2;

(b) the goodness-of-fit test of the assumption that the Poisson regression function is a quadratic polynomial in x_1 and x_2;

(c) the likelihood-ratio test of the hypothesis that the Poisson regression function is a linear function of x_1 and x_2 under the assumption that it is a quadratic polynomial in these variables.

Solution   Here $d = 63, p = 6$, and $p_0 = 3$. The three desired P-values are conveniently shown in a table of deviances (Table 13.24). These results support modeling the Poisson regression function for the lung cancer study as a linear function of x_1 and x_2.   ∎

TABLE   13.24

Model	D	DF	MD	P-value
Linear	61.032	60	1.017	.439
Quadratic	59.457	57	1.043	.386
Difference	1.575	3	0.525	.665

EXAMPLE 13.23   (Continued.) Determine the P-value for

(a) the goodness-of-fit test of the assumption that the Poisson regression function is a linear function of x_1 and x_2;

(b) the goodness-of-fit test of the assumption that the Poisson regression is an additive function of x_1 and x_2;

(c) the likelihood-ratio test of the hypothesis that the Poisson regression function is a linear function of x_1 and x_2 under the assumption that it is an additive function of these variables.

Solution   In the lung cancer data, x_1 takes on $d_1 = 9$ levels, x_2 takes on $d_2 = 7$ levels, and all combinations of one of the nine levels of x_1 and one of the seven levels of x_2 appear. Thus, $d = d_1 d_2 = 63$. Let G_1 be an identifiable and saturated space of functions of x_1, let G_2 be an identifiable and saturated space of functions of x_2, and let $G = G_1 + G_2$ be the corresponding additive space of functions of x_1 and x_2. Then $\dim(G_1) = 9$, $\dim(G_2) = 7$, and G_1 and G_2 both contain the constant functions, so G has dimension $9 + 7 - 1 = 15$. In computing the deviance corresponding to G, we can let X equal the design set $X_1 \times X_2$, where X_1 consists of the nine levels of x_1 in the data and X_2 consists of the seven levels of x_2. Then G_1 is the space of all functions on X_1, and G_2 is the space of all functions on X_2.

Let $I_1(x_1), \ldots, I_9(x_1)$ denote the indicator functions corresponding to the nine levels of x_1 and let $J_1(x_2), \ldots, J_7(x_2)$ denote the indicator functions corresponding to the seven levels of x_2. Then $1, I_1(x_1), \ldots, I_8(x_1), J_1(x_2), \ldots, J_6(x_2)$ form a basis of G.

The desired P-values are shown in a table of deviances (Table 13.25). These results again support modeling the Poisson regression function for the lung cancer study as a linear function of x_1 and x_2. ∎

TABLE 13.25

Model	D	DF	MD	P-value
Linear	61.032	60	1.017	.439
Additive	51.471	48	1.072	.340
Difference	9.561	12	0.797	.654

Coronary Heart Disease Study

EXAMPLE **13.24** Consider the coronary heart disease study. Let G_0 be the linear space of functions spanned by $1, x_1, x_2, x_1x_2, x_3, x_1x_3$ and let G be the space spanned by $1, x_1, x_2, x_1x_2, I_1(x_3), I_2(x_3), I_3(x_3), x_1I_1(x_3), x_1I_2(x_3), x_1I_3(x_3)$, where $I_1(x_3) = \text{ind}(x_3 = -3)$, $I_2(x_3) = \text{ind}(x_3 = -1)$, and $I_3(x_3) = \text{ind}(x_3 = 1)$. (In modeling the logistic regression function as being in G rather than the subspace G_0 of G, we are avoiding the somewhat dubious procedure of treating the coded levels of cholesterol in a quantitative manner.) Determine the P-value for

(a) the goodness-of-fit test assumption that the logistic regression function is in G_0;

(b) the goodness-of-fit test of the assumption that the logistic regression function is in G;

(c) the likelihood-ratio test of the hypothesis that the logistic regression function is in G_0 under the assumption that it is in G.

Solution Here $n = 16$, $p = 10$, and $p_0 = 6$. The three desired P-values are conveniently shown in a table of deviances (Table 13.26). These results support modeling the logistic regression function as in Section 13.5. ∎

TABLE 13.26

Model Space	D	DF	MD	P-value
G_0	11.032	10	1.103	.355
G	7.585	6	1.264	.270
Difference	3.447	4	0.862	.486

EXAMPLE **13.25**   Consider the data on the 806 females in the coronary heart disease study that were initially in the 50–62 age group. Use the data to determine the *P*-value for testing the hypothesis that the risk of CHD does not vary with serum cholesterol level for females within the indicated age group.

Solution   According to (47), the appropriate deviance is given by

$$D = 2\left\{ 9 \log \frac{9/58}{90/806} + 49 \log \frac{49/58}{716/806} + 12 \log \frac{12/135}{90/806} + 123 \log \frac{123/135}{716/806} \right.$$

$$\left. + 21 \log \frac{21/218}{90/806} + 197 \log \frac{197/218}{716/806} + 48 \log \frac{48/395}{90/806} + 347 \log \frac{347/395}{716/806} \right\}$$

$$\doteq 2.671,$$

and the corresponding *P*-value is given by $1 - \chi_3^2(2.671) \doteq .445$. Thus, there is no evidence that the risk of CHD varies with the level of serum cholesterol for females within the indicated age group.   ∎

Problems

13.66   In the context of the vitamin C experiment described in Example 12.7 in Section 12.1, determine the *P*-value for the likelihood-ratio test of the hypothesis that $\pi_1 = \pi_2$.

13.67   Consider the 1462 individuals in the coronary heart disease study that are initially in the 50–62 age group and ignore the distinction between men and women. Determine the *P*-value for the likelihood-ratio test of the hypothesis of homogeneity; that is, the risk of getting CHD does *not* depend on the level of cholesterol among the older age group. Interpret the result.

13.68   Consider the 1462 individuals in the coronary heart disease study that are initially in the 50–62 age group and consider the model $\theta(x_2, x_3) = \beta_0 + \beta_2 x_2 + \beta_3 x_3$ for the logistic regression function. The maximum-likelihood estimate of β_3 is given by $\hat{\beta}_3 \doteq 0.105$, and the standard error of this estimate is given by $\mathrm{SE}(\hat{\beta}_3) \doteq 0.0374$. Also, the deviance for the model equals 7.511, and the deviance for the submodel having the form $\theta(x_2, x_3) = \beta_0 + \beta_2 x_2$ equals 15.642. **(a)** Determine the *P*-value for the Wald test of the hypothesis that the risk of getting CHD does not depend on the level of cholesterol; that is, $\beta_3 = 0$. **(b)** Determine the *P*-value for the likelihood-ratio test of the hypothesis in part (a).

13.69   Consider the linear Poisson regression model, where G is an identifiable p-dimensional linear space and the maximum-likelihood estimate $\hat{\theta}(\cdot)$ in G of the Poisson regression function $\theta(\cdot)$ exists. Let $\hat{\lambda}(\cdot) = \exp \hat{\theta}(\cdot)$ denote the corresponding maximum-likelihood estimate of the rate function $\lambda(\cdot) = \mu(\cdot) = \exp \theta(\cdot)$. Verify that

$$l(G) = \sum_k n_k [\,\bar{Y}_k \log \hat{\lambda}(x_k) - \hat{\lambda}(x_k)\,] + \sum_k (Y_k \log n_k - \log Y_k!).$$

13.70 (Continued.) Let G_0 be a proper subspace of G, let $\hat{\theta}_0(\cdot)$ denote the maximum-likelihood estimate in G_0 of $\theta(\cdot)$, and let $\hat{\lambda}_0(\cdot) = \exp\hat{\theta}_0(\cdot)$ denote the corresponding maximum-likelihood estimate of $\lambda(\cdot)$. Let D denote the likelihood-ratio statistic for testing the hypothesis that $\theta(\cdot) \in G_0$. **(a)** Verify that

$$D = 2\sum_k n_k\left(\bar{Y}_k \log\frac{\hat{\lambda}(\mathbf{x}_k)}{\hat{\lambda}_0(\mathbf{x}_k)} - [\hat{\lambda}(\mathbf{x}_k) - \hat{\lambda}_0(\mathbf{x}_k)]\right).$$

(b) Verify that if G_0 contains the constant functions, then

$$D = 2\sum_k Y_k \log\frac{\hat{\lambda}(x_k)}{\hat{\lambda}_0(\mathbf{x}_k)}.$$

(See Problem 13.33 in Section 13.3.)

13.71 Consider a linear Poisson regression model in which G is saturated. Verify that

$$l(G) = \sum_k n_k(\bar{Y}_k \log\hat{\lambda}_k - \hat{\lambda}_k) + \sum_k(Y_k \log n_k - \log Y_k!),$$

where $\hat{\lambda}_k = \bar{Y}_k$ for $1 \le k \le d$.

13.72 (Continued.) Verify that the likelihood-ratio statistic for testing the hypothesis $\lambda_1 = \cdots = \lambda_d$ of homogeneity can be written as

$$D = 2\sum_k Y_k \log\frac{Y_k/n_k}{Y/n}, \qquad \text{where } Y = \sum_k Y_k = n\bar{Y}.$$

13.73 In the context of Problem 12.6 in Section 12.1, which involves the Poisson two-sample model, determine the P-value for the likelihood-ratio test of the hypothesis that $\lambda_1 = \lambda_2$ (see Problem 13.72).

13.74 Consider a linear Poisson regression model in which G is identifiable and the maximum-likelihood estimate $\hat{\theta}(\cdot)$ in G of $\theta(\cdot)$ exists. **(a)** Verify that the deviance of G is given by

$$D(G) = 2\sum_k n_k\left(\bar{Y}_k \log\frac{\hat{\lambda}_k}{\hat{\lambda}(\mathbf{x}_k)} - [\hat{\lambda}_k - \hat{\lambda}(\mathbf{x}_k)]\right),$$

where $\hat{\lambda}_k = \bar{Y}_k$ and $\hat{\lambda}(\mathbf{x}_k) = \exp\hat{\theta}(\mathbf{x}_k)$ for $1 \le k \le d$. **(b)** Verify that if G contains the constant functions, then

$$D(G) = 2\sum_k Y_k \log\frac{\hat{\lambda}_k}{\hat{\lambda}(\mathbf{x}_k)}.$$

[*Hint:* See Problem 13.33 in Section 13.3.] **(c)** Verify that the deviance of the space of constant functions can be written as

$$D(G_{\text{con}}) = 2\sum_k Y_k \log\frac{Y_k/n_k}{Y/n}, \qquad \text{where } Y = \sum_k Y_k = n\bar{Y}.$$

13.75 Verify Equation (48).

13.76 Consider the setup of Example 12.9 in Section 12.1 and the hypothesis in part (c) of that example. **(a)** Explain how the indicated setup and hypothesis can be viewed in terms of logistic regression. **(b)** Determine the P-value for the likelihood-ratio test of the hypothesis.

13.77   Consider the setup of Problem 13.29 in Section 13.3. Use the corresponding Poisson regression model in the answer to Problem 13.31 in that section to determine the likelihood-ratio statistic for testing the hypothesis that $\theta = 0$.

13.78   Use the answer to Problem 13.77 to determine the P-value for the likelihood-ratio test of the hypothesis in Problem 13.53(b) in Section 13.5.

Properties of Vectors and Matrices

The properties of vectors and matrices that are relevant to this textbook are briefly summarized here, with additional material from linear algebra being treated in Chapter 9.

Let m and n be positive integers. Consider the $m \times n$ matrix

$$\mathbf{A} = [\, a_{ij} \,] = \begin{bmatrix} a_{11} & \cdots & a_{1n} \\ \vdots & & \vdots \\ a_{m1} & \cdots & a_{mn} \end{bmatrix},$$

which has m rows and n columns. We refer to a_{ij} as the entry in row i and column j of $\mathbf{A}$. The matrix $\mathbf{A}$ is said to be a square matrix if $m = n$. The collection of $m \times n$ matrices is denoted by $\mathbb{R}^{m \times n}$.

A matrix is said to be a zero matrix, written as $\mathbf{0}$, if all of its entries equal zero. The matrix equation $\mathbf{A} = \mathbf{B}$ holds if and only if $\mathbf{A}$ and $\mathbf{B}$ have the same numbers of rows and columns and every entry of $\mathbf{A}$ equals the corresponding entry of $\mathbf{B}$.

The transpose of $\mathbf{A} = [\, a_{ij} \,]$ is the $n \times m$ matrix

$$\mathbf{A}^T = \begin{bmatrix} a_{11} & \cdots & a_{m1} \\ \vdots & & \vdots \\ a_{1n} & \cdots & a_{mn} \end{bmatrix},$$

the entry in row i and column j of which is a_{ji}. The matrix $\mathbf{A}$ is said to be symmetric if $\mathbf{A}^T = \mathbf{A}$; if so, then $\mathbf{A}$ is a square matrix. An $n \times n$ matrix $\mathbf{A} = [\, a_{ij} \,]$ is symmetric if and only if $a_{ij} = a_{ji}$ for $1 \leq i, j \leq n$.

Let $\mathbf{A}$ be an $n \times n$ matrix. The entries $a_{11}, \ldots, a_{nn}$ are referred to as the diagonal entries of $\mathbf{A}$ (a_{ii} is its ith diagonal entry), and the remaining entries (a_{ij} with $i \neq j$) are referred to as its off-diagonal entries. If the off-diagonal entries of $\mathbf{A}$ are all zero, then **A** is called a diagonal matrix and it is given by

$$\mathbf{A} = \begin{bmatrix} a_{11} & & 0 \\ & \ddots & \\ 0 & & a_{nn} \end{bmatrix}.$$

Every diagonal matrix is symmetric. The diagonal matrix

$$\begin{bmatrix} a_1 & & 0 \\ & \ddots & \\ 0 & & a_n \end{bmatrix}$$

is also written as diag($a_1, \ldots, a_n$) or as diag(**a**), where $\mathbf{a} = [\, a_1, \ldots, a_n \,]^T$.

Let $\mathbf{A} = (a_{ij})$ be an $m \times n$ matrix and let $b \in \mathbb{R}$. The scalar multiplication of **A** by b is given by

$$b\mathbf{A} = \begin{bmatrix} ba_{11} & \cdots & ba_{1n} \\ \vdots & & \vdots \\ ba_{m1} & \cdots & ba_{mn} \end{bmatrix};$$

the entry in row i and column j of $b\mathbf{A}$ is ba_{ij}. The matrix $(-1)\mathbf{A}$ is written as $-\mathbf{A}$.

Let $\mathbf{A} = [\, a_{ij} \,]$ and $\mathbf{B} = [\, b_{ij} \,]$ be $m \times n$ matrices. The sum of **A** and **B** is the $m \times n$ matrix given by $\mathbf{A} + \mathbf{B} = [\, a_{ij} + b_{ij} \,]$; that is, the entry in row i and column j of $\mathbf{A} + \mathbf{B}$ equals $a_{ij} + b_{ij}$. The difference of **A** and **B** is given by $\mathbf{A} - \mathbf{B} = \mathbf{A} + (-\mathbf{B}) = [\, a_{ij} - b_{ij} \,]$.

Let $\mathbf{A} = [\, a_{ij} \,]$ be an $l \times m$ matrix and let $\mathbf{B} = [\, b_{ij} \,]$ be an $m \times n$ matrix. Then the product

$$\mathbf{AB} = \begin{bmatrix} a_{11} & \cdots & a_{1m} \\ \vdots & & \vdots \\ a_{l1} & \cdots & a_{lm} \end{bmatrix} \begin{bmatrix} b_{11} & \cdots & b_{1n} \\ \vdots & & \vdots \\ b_{mn} & \cdots & b_{nn} \end{bmatrix}$$

of **A** and **B** is defined as the $l \times n$ matrix the entry in row i and column j of which equals

$$\sum_{k=1}^{m} a_{ik} b_{kj};$$

the number of rows of **AB** is the same as the number of rows of **A**, and the number of columns of **AB** is the same as the number of columns of **B**. Matrix multiplication is generally not commutative; that is, there are matrices **A** and **B** such that **AB** and **BA** are defined but $\mathbf{AB} \neq \mathbf{BA}$. If $\tilde{\mathbf{A}}$ is obtained from **A** by applying a permutation to the rows of **A**, then $\tilde{\mathbf{A}}\mathbf{B}$ is obtained from **AB** by applying the same permutation to the rows of **AB**; if $\tilde{\mathbf{B}}$ is obtained from **B** by applying a permutation to the columns of **B**, then $\mathbf{A}\tilde{\mathbf{B}}$ is obtained from **AB** by applying the same permutation to the columns of **AB**; if $\tilde{\mathbf{A}}$ is obtained from **A** by applying a permutation to the columns of **A** and $\tilde{\mathbf{B}}$ is obtained from **B** by applying the same permutation to the rows of **B**, then $\tilde{\mathbf{A}}\tilde{\mathbf{B}} = \mathbf{AB}$.

Let $\mathbf{A} = [\, a_{ij} \,]$, $\mathbf{B} = [\, b_{ij} \,]$, and $\mathbf{C} = [\, c_{ij} \,]$ be matrices such that **AB** and **BC** are defined. Then (**AB**)**C** and **A**(**BC**) are defined, and (**AB**)**C** = **A**(**BC**); that is, matrix multiplication is associative. Thus, without ambiguity, we can write (**AB**)**C** and

$A(BC)$ as ABC. The entry in row i and column j of ABC is given by

$$(ABC)_{ij} = \sum_k \sum_l a_{ik}b_{kl}c_{lj}.$$

If B is a diagonal matrix, then

$$(ABC)_{ij} = \sum_k a_{ik}b_{kk}c_{kj}.$$

The identity matrix $I = I_n$ is the $n \times n$ diagonal matrix all of whose diagonal entries equal 1; that is,

$$I = \begin{bmatrix} 1 & & 0 \\ & \ddots & \\ 0 & & 1 \end{bmatrix}.$$

If A is a matrix having n columns, then $AI = A$; and if A has n rows, then $IA = A$. In particular, if A is an n by n matrix, then $AI = IA = A$.

Let A and B be $n \times n$ matrices. Then $AB = I$ if and only if $BA = I$; if so, then A is said to be invertible, and B is called the inverse of A, which is written as A^{-1} (an invertible matrix has a unique inverse); otherwise, A is said to be noninvertible. The matrix A^T is invertible if and only if A is invertible, in which case $(A^T)^{-1} = (A^{-1})^T$. If A is symmetric and invertible, then A^{-1} is symmetric. The diagonal matrix $A = \text{diag}(a_1, \ldots, a_n)$ is invertible if and only if $a_1, \ldots, a_n$ are nonzero, in which case $A^{-1} = \text{diag}(a_1^{-1}, \ldots, a_n^{-1})$. If A is invertible and $\tilde{A}$ is obtained from A by applying a permutation to the rows (columns) of A, then $(\tilde{A})^{-1}$ is obtained from A^{-1} by applying the same permutation to the columns (rows) of A^{-1}. Let A and B be $n \times n$ matrices. Then AB is invertible if and only if A and B are invertible, in which case $(AB)^{-1} = B^{-1}A^{-1}$.

An $n \times n$ matrix $A = [a_{ij}]$ has a determinant

$$\det(A) = \begin{vmatrix} a_{11} & \cdots & a_{1n} \\ \vdots & & \vdots \\ a_{n1} & \cdots & a_{nn} \end{vmatrix},$$

which is a real number. The matrix A is invertible if and only if $\det(A) \neq 0$. The determinant of a diagonal matrix is the product of its diagonal elements. The determinants of 2×2 and 3×3 matrices are given explicitly by

$$\begin{vmatrix} a_{11} & a_{12} \\ a_{21} & a_{22} \end{vmatrix} = a_{11}a_{22} - a_{12}a_{21}$$

and

$$\begin{vmatrix} a_{11} & a_{12} & a_{13} \\ a_{21} & a_{22} & a_{23} \\ a_{31} & a_{32} & a_{33} \end{vmatrix} = \begin{matrix} a_{11}a_{22}a_{33} + a_{12}a_{23}a_{31} + a_{13}a_{21}a_{32} \\ - a_{11}a_{23}a_{32} - a_{12}a_{21}a_{33} - a_{13}a_{22}a_{31}. \end{matrix}$$

For $1 \leq i, j \leq n$, the determinant M_{ij} of the $(n-1) \times (n-1)$ submatrix of A obtained by deleting its ith row and jth column is referred to as the minor of a_{ij}. If A is a symmetric matrix, so is the matrix $[M_{ij}]$. Suppose A is invertible. Then its inverse can be obtained by multiplying the transpose of the matrix $[(-1)^{i+j}M_{ij}]$ by

$1/\det(\mathbf{A})$. In particular, if $n = 2$, then

$$\mathbf{A}^{-1} = \frac{1}{\det(\mathbf{A})} \begin{bmatrix} a_{22} & -a_{12} \\ -a_{21} & a_{11} \end{bmatrix}.$$

If $n = 3$, then

$$\mathbf{A}^{-1} = \frac{1}{\det(\mathbf{A})} \begin{bmatrix} M_{11} & -M_{21} & M_{31} \\ -M_{12} & M_{22} & -M_{32} \\ M_{13} & -M_{23} & M_{33} \end{bmatrix},$$

where

$$M_{11} = \begin{vmatrix} a_{22} & a_{23} \\ a_{32} & a_{33} \end{vmatrix}, \quad M_{12} = \begin{vmatrix} a_{21} & a_{23} \\ a_{31} & a_{33} \end{vmatrix}; \quad M_{13} = \begin{vmatrix} a_{21} & a_{22} \\ a_{31} & a_{32,} \end{vmatrix},$$

$$M_{21} = \begin{vmatrix} a_{12} & a_{13} \\ a_{32} & a_{33} \end{vmatrix}, \quad M_{22} = \begin{vmatrix} a_{11} & a_{13} \\ a_{31} & a_{33} \end{vmatrix}, \quad M_{23} = \begin{vmatrix} a_{11} & a_{12} \\ a_{31} & a_{32,} \end{vmatrix},$$

$$M_{31} = \begin{vmatrix} a_{12} & a_{13} \\ a_{22} & a_{23} \end{vmatrix}, \quad M_{32} = \begin{vmatrix} a_{11} & a_{13} \\ a_{21} & a_{23} \end{vmatrix}, \quad M_{33} = \begin{vmatrix} a_{11} & a_{12} \\ a_{21} & a_{22,} \end{vmatrix}.$$

(This approach to matrix inversion is generally not computationally useful when $n \geq 4$.)

Some properties involving matrices (in which the indicated matrix operations are assumed to be defined) are as follows:

$$\mathbf{A} + \mathbf{B} = \mathbf{B} + \mathbf{A};$$

$$(\mathbf{A} + \mathbf{B}) + \mathbf{C} = \mathbf{A} + (\mathbf{B} + \mathbf{C});$$

$$\mathbf{A} + \mathbf{0} = \mathbf{A};$$

$$\mathbf{A} - \mathbf{A} = \mathbf{0};$$

$$a(\mathbf{B} + \mathbf{C}) = a\mathbf{B} + a\mathbf{C};$$

$$(a + b)\mathbf{C} = a\mathbf{C} + b\mathbf{C};$$

$$a(b\mathbf{C}) = (ab)\mathbf{C};$$

$$1\mathbf{A} = \mathbf{A};$$

$$0\mathbf{A} = \mathbf{0};$$

$$(\mathbf{A}\mathbf{B})\mathbf{C} = \mathbf{A}(\mathbf{B}\mathbf{C});$$

$$\mathbf{A}\mathbf{0} = \mathbf{0};$$

$$\mathbf{0}\mathbf{A} = \mathbf{0};$$

$$\mathbf{A}(\mathbf{B} + \mathbf{C}) = \mathbf{A}\mathbf{B} + \mathbf{A}\mathbf{C};$$

$$(\mathbf{A} + \mathbf{B})\mathbf{C} = \mathbf{A}\mathbf{C} + \mathbf{B}\mathbf{C};$$

$$(a\mathbf{B})\mathbf{C} = a(\mathbf{B}\mathbf{C});$$

$$\mathbf{A}(b\mathbf{C}) = b\mathbf{A}\mathbf{C};$$

$$(\mathbf{A}^T)^T = \mathbf{A};$$

$$(a\mathbf{B})^T = a\mathbf{B}^T;$$

$$(\mathbf{A} + \mathbf{B})^T = \mathbf{A}^T + \mathbf{B}^T;$$

$$(\mathbf{AB})^T = \mathbf{B}^T\mathbf{A}^T;$$

$$\mathbf{AA}^{-1} = \mathbf{I};$$

$$\mathbf{A}^{-1}\mathbf{A} = \mathbf{I};$$

$$(\mathbf{A}^{-1})^{-1} = \mathbf{A};$$

$$(\mathbf{AB})^{-1} = \mathbf{B}^{-1}\mathbf{A}^{-1};$$

$$(a\mathbf{B})^{-1} = a^{-1}\mathbf{B}^{-1};$$

$$(\mathbf{A}^T)^{-1} = (\mathbf{A}^{-1})^T.$$

$$\det(\mathbf{AB}) = [\det(\mathbf{A})][\det(\mathbf{B})];$$

$$\det(\mathbf{A}^T) = \det(\mathbf{A});$$

$$\det(\mathbf{A}^{-1}) = [\det(\mathbf{A})]^{-1};$$

$$\det(a\mathbf{B}) = a^n \det(\mathbf{B}), \text{ where } \mathbf{B} \text{ is an } n \times n \text{ matrix.}$$

A matrix having one row is called a row vector, and a matrix having one column is called a column vector. Given $a_1, \ldots, a_n \in \mathbb{R}$, set

$$\mathbf{a} = [a_1, \ldots, a_n]^T = \begin{bmatrix} a_1 \\ \vdots \\ a_n \end{bmatrix}.$$

Then $\mathbf{a} \in \mathbb{R}^n$. We refer to a_i as the ith entry of $\mathbf{a}$ and of $\mathbf{a}^T$. If $n = 1$, then $\mathbf{a} = \mathbf{a}^T = [a_1]$ can be identified with the real number a_1.

Let the matrix $\mathbf{A} = [a_{jk}]$ have d columns, let $\mathbf{W} = \text{diag}(w_1, \ldots, w_d)$ be a $d \times d$ diagonal matrix, and let $\mathbf{b} = [b_k]$ be a d-dimensional column vector. Then the kth entry of $\mathbf{Wb}$ is $w_k b_k$, and the jth entry of $\mathbf{AWb}$ is $\sum_k w_k a_{jk} b_k$.

Let $\mathbf{b} = [b_1, \ldots, b_n]^T \in \mathbb{R}^n$. Then

$$\mathbf{b}^T\mathbf{a} = \mathbf{a}^T\mathbf{b} = [a_1, \ldots, a_n] \begin{bmatrix} b_1 \\ \vdots \\ b_n \end{bmatrix} = a_1 b_1 + \cdots + a_n b_n = \sum_{i=1}^{n} a_i b_i.$$

In particular,

$$\mathbf{a}^T\mathbf{a} = a_1^2 + \cdots + a_n^2 = \sum_{i=1}^{n} a_i^2.$$

Let $\mathbf{a} = [a_1, \ldots, a_m]^T \in \mathbb{R}^m$ and $\mathbf{b} = [b_1, \ldots, b_n]^T \in \mathbb{R}^n$. Then

$$\mathbf{ab}^T = \begin{bmatrix} a_1 \\ \vdots \\ a_m \end{bmatrix} [b_1, \ldots, b_n] = \begin{bmatrix} a_1 b_1 & \cdots & a_1 b_n \\ \vdots & & \vdots \\ a_m b_1 & \cdots & a_m b_n \end{bmatrix}.$$

Let $\mathbf{A} = [a_{ij}]$ be an $n \times n$ matrix. Then $\mathbf{A}$ is noninvertible if and only if there is a nonzero column vector $\mathbf{b}$ such that $\mathbf{Ab} = \mathbf{0}$; it is also noninvertible if and only if there is a nonzero row vector $\mathbf{b}^T$ such that $\mathbf{b}^T\mathbf{A} = \mathbf{0}$. Observe that

$$\mathbf{b}^T\mathbf{Ac} = \sum_{ij} b_i c_j a_{ij}, \qquad \mathbf{b} = [\,b_1, \ldots, b_n\,]^T \text{ and } \mathbf{c} = [\,c_1, \ldots, c_n\,]^T \in \mathbb{R}^n.$$

In particular,

$$\mathbf{b}^T\mathbf{Ab} = \sum_i \sum_j b_i b_j a_{ij}, \qquad \mathbf{b} = [\,b_1, \ldots, b_n\,]^T \in \mathbb{R}^n.$$

The matrix $\mathbf{A}$ is said to be positive semidefinite if

$$\mathbf{b}^T\mathbf{Ab} \geq 0, \qquad \mathbf{b} \in \mathbb{R}^n;$$

it is said to be positive definite if

$$\mathbf{b}^T\mathbf{Ab} > 0, \qquad \mathbf{b} \in \mathbb{R}^n \text{ and } \mathbf{b} \neq \mathbf{0}.$$

The diagonal entries of a positive semidefinite matrix are nonnegative, and the diagonal entries of a positive definite matrix are positive. A diagonal matrix is positive semidefinite if and only if its diagonal entries are nonnegative, and it is positive definite if and only if its diagonal entries are positive.

Suppose $\mathbf{A}$ is positive definite. Then it is positive semidefinite and invertible and its inverse is positive definite. Suppose $\mathbf{A}$ is positive semidefinite and symmetric. Then it is positive definite if and only if it is invertible. If $\mathbf{A}$ is positive definite and symmetric, so is $\mathbf{A}^{-1}$.

For any matrix $\mathbf{B}$, $\mathbf{BB}^T$ and $\mathbf{B}^T\mathbf{B}$ are positive semidefinite, symmetric matrices. Let $\mathbf{B}$ be a square matrix. If $\mathbf{B}$ is invertible, then $\mathbf{BB}^T$ and $\mathbf{B}^T\mathbf{B}$ are positive definite; conversely, if either $\mathbf{BB}^T$ or $\mathbf{B}^T\mathbf{B}$ is positive definite, then $\mathbf{B}$ is invertible.

An $n \times n$ matrix $\mathbf{A} = [\,a_{ij}\,]$ is said to be a lower-triangular matrix if $a_{ij} = 0$ for $j > i$—that is, if

$$\mathbf{A} = \begin{bmatrix} a_{11} & & 0 \\ \vdots & \ddots & \\ a_{n1} & \cdots & a_{nn} \end{bmatrix}.$$

The matrix $\mathbf{A}$ is said to be an upper-triangular matrix if $a_{ij} = 0$ for $i > j$—that is, if

$$\mathbf{A} = \begin{bmatrix} a_{11} & \cdots & a_{1n} \\ & \ddots & \vdots \\ 0 & & a_{nn} \end{bmatrix}.$$

The transpose of a lower-triangular matrix is an upper-triangular matrix, and the transpose of an upper-triangular matrix is a lower-triangular matrix. The determinant of a lower- or upper-triangular matrix is the product of its diagonal entries. A lower- or upper-triangular matrix is invertible if and only if its diagonal entries are all nonzero. The inverse of an invertible lower-triangular matrix is a lower-triangular matrix, and the inverse of an invertible upper-triangular matrix is an upper-triangular matrix.

Let **A** be a positive semidefinite, symmetric $n \times n$ matrix. Then there is a lower-triangular $n \times n$ matrix **L** such that $\mathbf{LL}^T = \mathbf{A}$; we refer to $\mathbf{LL}^T$ as a *Cholesky decomposition* of **A**. Now $\det(\mathbf{A}) = \det(\mathbf{LL}^T) = \det(\mathbf{L})\det(\mathbf{L}^T) = [\det(\mathbf{L})]^2$, so the determinant of **A** is nonnegative. The matrix **A** is positive definite if and only if its determinant is positive. If **A** is positive definite, then it has a unique Cholesky decomposition $\mathbf{LL}^T$ such that **L** has positive diagonal entries.

Let $\mathbf{A} = (a_{ij})$ be a positive definite, symmetric $n \times n$ matrix and let $\mathbf{LL}^T$ be the unique Cholesky decomposition of **A** such that

$$\mathbf{L} = \begin{bmatrix} l_{11} & & 0 \\ \vdots & \ddots & \\ l_{n1} & \cdots & l_{nn} \end{bmatrix}$$

has positive diagonal entries. The entries of **L** can be found successively in the order $l_{11}, \ldots, l_{n1}, l_{22}, \ldots, l_{n2}, \ldots, l_{nn}$ by equating the corresponding entries of $\mathbf{LL}^T$ to those of **A**.

Suppose, in particular, that $n = 2$. Then the equation $\mathbf{LL}^T = \mathbf{A}$ can be written as

$$\begin{bmatrix} l_{11} & 0 \\ l_{21} & l_{22} \end{bmatrix} \begin{bmatrix} l_{11} & l_{21} \\ 0 & l_{22} \end{bmatrix} = \begin{bmatrix} a_{11} & a_{12} \\ a_{21} & a_{22} \end{bmatrix}.$$

We successively solve the equations

$$l_{11}^2 = a_{11},$$

$$l_{21}l_{11} = a_{21},$$

$$l_{21}^2 + l_{22}^2 = a_{22},$$

for $l_{11} > 0$, l_{21}, and $l_{22} > 0$.

Suppose next, that $n = 3$. Then the equation $\mathbf{LL}^T = \mathbf{A}$ can be written as

$$\begin{bmatrix} l_{11} & 0 & 0 \\ l_{21} & l_{22} & 0 \\ l_{31} & l_{32} & l_{33} \end{bmatrix} \begin{bmatrix} l_{11} & l_{21} & l_{31} \\ 0 & l_{22} & l_{32} \\ 0 & 0 & l_{33} \end{bmatrix} = \begin{bmatrix} a_{11} & a_{12} & a_{13} \\ a_{21} & a_{22} & a_{23} \\ a_{31} & a_{32} & a_{33} \end{bmatrix}.$$

We successively solve the equations

$$l_{11}^2 = a_{11},$$

$$l_{21}l_{11} = a_{21},$$

$$l_{31}l_{11} = a_{31},$$

$$l_{21}^2 + l_{22}^2 = a_{22},$$

$$l_{31}l_{21} + l_{32}l_{22} = a_{32},$$

$$l_{31}^2 + l_{32}^2 + l_{33}^2 = a_{33},$$

for $l_{11} > 0$, $l_{21}, l_{31}, l_{22} > 0$, l_{32}, and $l_{33} > 0$.

Partitioned Matrices

Matrices can be *partitioned* into submatrices. For example, we can partition the 3×3 matrix

$$\mathbf{A} = \begin{bmatrix} a_{11} & a_{12} & a_{13} \\ a_{21} & a_{22} & a_{23} \\ a_{31} & a_{32} & a_{33} \end{bmatrix}$$

as

$$\mathbf{A} = \begin{bmatrix} \mathbf{A}_{11} & \mathbf{A}_{12} \\ \mathbf{A}_{21} & \mathbf{A}_{22} \end{bmatrix},$$

where

$$\mathbf{A}_{11} = \begin{bmatrix} a_{11} & a_{12} \\ a_{21} & a_{22} \end{bmatrix} \quad \mathbf{A}_{12} = \begin{bmatrix} a_{13} \\ a_{23} \end{bmatrix}$$

$$\mathbf{A}_{21} = [a_{31}, a_{32}] \quad \mathbf{A}_{23} = [a_{33}].$$

Consider partitioned matrices

$$\mathbf{A} = \begin{bmatrix} \mathbf{A}_{11} & \mathbf{A}_{12} \\ \mathbf{A}_{21} & \mathbf{A}_{22} \end{bmatrix} \text{ and } \mathbf{B} = \begin{bmatrix} \mathbf{B}_{11} & \mathbf{B}_{12} \\ \mathbf{B}_{21} & \mathbf{B}_{22} \end{bmatrix}.$$

The formula

$$\mathbf{A} + \mathbf{B} = \begin{bmatrix} \mathbf{A}_{11} + \mathbf{B}_{11} & \mathbf{A}_{12} + \mathbf{B}_{12} \\ \mathbf{A}_{21} + \mathbf{B}_{21} & \mathbf{A}_{22} + \mathbf{B}_{22} \end{bmatrix}$$

is valid provided that the indicated matrix additions are well defined; that is, $\mathbf{A}_{11}$ and $\mathbf{B}_{11}$ have the same size, and so forth. The formula

$$\mathbf{AB} = \begin{bmatrix} \mathbf{A}_{11}\mathbf{B}_{11} + \mathbf{A}_{12}\mathbf{B}_{21} & \mathbf{A}_{11}\mathbf{B}_{12} + \mathbf{A}_{12}\mathbf{B}_{22} \\ \mathbf{A}_{21}\mathbf{B}_{11} + \mathbf{A}_{22}\mathbf{B}_{21} & \mathbf{A}_{21}\mathbf{B}_{12} + \mathbf{A}_{22}\mathbf{B}_{22} \end{bmatrix}$$

is valid provided that the indicated matrix multiplications are well defined; that is, $\mathbf{A}_{11}$ has the same number of columns as $\mathbf{B}_{11}$ has rows, and so forth.

Consider a partitioned square matrix

$$\mathbf{A} = \begin{bmatrix} \mathbf{A}_{11} & \mathbf{0} \\ \mathbf{A}_{21} & \mathbf{A}_{22} \end{bmatrix},$$

where $\mathbf{A}_{11}$ and hence $\mathbf{A}_{22}$ are also square matrices. Then

$$\det(\mathbf{A}) = \det(\mathbf{A}_{11}) \det(\mathbf{A}_{22}).$$

Thus, $\mathbf{A}$ is invertible if and only if $\mathbf{A}_{11}$ and $\mathbf{A}_{22}$ are invertible, in which case

$$\mathbf{A}^{-1} = \begin{bmatrix} \mathbf{A}_{11}^{-1} & \mathbf{0} \\ -\mathbf{A}_{22}^{-1}\mathbf{A}_{21}\mathbf{A}_{11}^{-1} & \mathbf{A}_{22}^{-1} \end{bmatrix},$$

as may be checked by matrix multiplication.

In particular,

$$\mathbf{A} = \begin{bmatrix} \mathbf{A}_{11} & \mathbf{0} \\ \mathbf{0} & \mathbf{A}_{22} \end{bmatrix}$$

is invertible if and only if A_{11} and A_{22} are invertible, in which case

$$A^{-1} = \begin{bmatrix} A_{11}^{-1} & 0 \\ 0 & A_{22}^{-1} \end{bmatrix}.$$

We refer to A as a block diagonal matrix having two diagonal blocks. The last result can be extended in the obvious manner to handle block diagonal matrices having any number of diagonal blocks.

Let A be a symmetric, positive semidefinite $n \times n$ matrix, let $1 \le m < n$, and let A_{11} be an $m \times m$ submatrix of A that is obtained by deleting $n - m$ rows and the corresponding $n - m$ columns from A. Then A_{11} is symmetric and positive semidefinite; and if A is positive definite, so is A_{11}.

B

Summary of Probability

B.1
Random Variables and Their Distributions
Distributions

Let $\mathcal{Y}$ be a nonempty set and let P assign a real number $P(B)$ to subsets B of $\mathcal{Y}$. We say that P is a distribution on $\mathcal{Y}$ if it satisfies the following properties: $P(B) \geq 0$ for $B \subset \mathcal{Y}$, $P(\mathcal{Y}) = 1$, and $P(B_1 \cup B_2 \cup \cdots) = P(B_1) + P(B_2) + \cdots$ for $B_1, B_2, \ldots$ disjoint subsets of $\mathcal{Y}$.

Let P be a distribution on $\mathcal{Y}$ and let $B \subset \mathcal{Y}$. Then $P(B)$ is referred to as the probability of B. The following properties are consequences of the definition of a distribution:

$$P(\emptyset) = 0,$$

$$P(B^c) = 1 - P(B) \text{ for } B \subset \mathcal{Y},$$

$$P(B) \leq 1 \text{ for } B \subset \mathcal{Y},$$

$$P(B_2 \backslash B_1) = P(B_2) - P(B_1) \text{ for } B_1 \subset B_2 \subset \mathcal{Y},$$

$$P(B_1) \leq P(B_2) \text{ for } B_1 \subset B_2 \subset \mathcal{Y}.$$

Let $\mathcal{Y}_1 \subset \mathcal{Y}$. If P is a distribution on $\mathcal{Y}$ such that $P(\mathcal{Y}_1) = 1$, we can think of P as a distribution on $\mathcal{Y}_1$. Conversely, if P is a distribution on $\mathcal{Y}_1$, we can think of it as a distribution on any $\mathcal{Y}$ such that $P(\mathcal{Y}_1) = 1$.

Associated with a $\mathcal{Y}$-valued random variable $\mathbf{Y}$ is a distribution P on $\mathcal{Y}$. We refer to $P(B) = P(\mathbf{Y} \in B)$ as the probability of the event that $\mathbf{Y} \in B$.

Let $\mathbf{W}$ be a $\mathcal{W}$-valued random variable and let $\mathbf{g}$ be a $\mathcal{Y}$-valued function on $\mathcal{W}$. We refer to the $\mathcal{Y}$-valued random variable $\mathbf{Y} = \mathbf{g}(\mathbf{W})$ as a transform of $\mathbf{W}$. To express the distribution of $\mathbf{Y}$ in terms of $\mathbf{g}$ and the distribution of $\mathbf{W}$, let $B \subset \mathcal{Y}$ and set $A = \{w \in \mathcal{W} : \mathbf{g}(\mathbf{w}) \in B\}$. Then $P(\mathbf{Y} \in B) = P(\mathbf{W} \in A)$.

Probability Functions

A probability function on $\mathcal{Y}$ is a nonnegative function on $\mathcal{Y}$ whose sum over $\mathcal{Y}$ equals 1. Let f be a probability function on $\mathcal{Y}$. The distribution corresponding to f is given by $P(B) = \sum_{\mathbf{y} \in B} f(\mathbf{y})$ for $B \subset \mathcal{Y}$. Every distribution on a finite or countably infinite set has a probability function.

If the distribution of the random variable $\mathbf{Y}$ has a probability function f, then f is referred to as the probability function of $\mathbf{Y}$, and $\mathbf{Y}$ is said to be a discrete random variable. Let $\mathbf{Y}$ be a discrete random variable having probability function $f_{\mathbf{Y}}$. Then $f_{\mathbf{Y}}(\mathbf{y}) = P(\mathbf{Y} = \mathbf{y})$ for $\mathbf{y} \in \mathcal{Y}$ and $P(\mathbf{Y} \in B) = \sum_{\mathbf{y} \in B} f_{\mathbf{Y}}(\mathbf{y})$ for $B \in \mathcal{Y}$.

A $\mathcal{Y}$-valued random variable $\mathbf{Y}$ is said to be a constant random variable if $P(\mathbf{Y} = \mathbf{a}) = 1$ for some $\mathbf{a} \in \mathcal{Y}$. The probability function of such a random variable is given by $f(\mathbf{a}) = 1$ and $f(\mathbf{y}) = 0$ for $\mathbf{y} \neq \mathbf{a}$.

Let $\mathbf{Y} = \mathbf{g}(\mathbf{W})$ be a transform of a discrete $\mathcal{W}$-valued random variable $\mathbf{W}$. Then the probability function of $\mathbf{Y}$ can be expressed in terms of $\mathbf{g}$ and the probability function of $\mathbf{W}$ as follows: For $\mathbf{y} \in \mathcal{Y}$,

$$f_{\mathbf{Y}}(\mathbf{y}) = \sum_{w \in A} f_{\mathbf{W}}(\mathbf{w}), \qquad \text{where } A = \{\mathbf{w} \in \mathcal{W} : \mathbf{g}(\mathbf{w}) = \mathbf{y}\}.$$

Density Functions

A density function on $\mathbb{R}$ is a nonnegative function on $\mathbb{R}$ whose integral over $\mathbb{R}$ equals 1. Let f be a density function on $\mathbb{R}$. The distribution corresponding to f is given by $P(B) = \int_B f(y)\, dy$ for $B \subset \mathbb{R}$. The density function f is said to be symmetric about zero if $f(-y) = f(y)$ for $y \in \mathbb{R}$.

Let Y be a random variable having density function f—that is, whose distribution has density function f. Then $P(Y = y) = 0$ for $y \in \mathbb{R}$ and

$$P(a \leq Y \leq b) = P(a \leq Y < b) = P(a < Y \leq b) = P(a < Y < b) = \int_a^b f(y)\, dy$$

for $a \leq b$.

Distribution Functions

Let Y be a real-valued random variable and let P be its distribution. The distribution function of Y (or of P) is the function F on $\mathbb{R}$ given by

$$F(y) = P(Y \leq y), \qquad y \in \mathbb{R}.$$

Observe that

$$P(Y > y) = 1 - F(y), \qquad y \in \mathbb{R},$$

and

$$P(a < Y \le b) = F(b) - F(a), \qquad a \le b.$$

A distribution function is nondecreasing in y and it has limits 0 and 1 at $-\infty$ and ∞, respectively. A distribution on $\mathbb{R}$ is uniquely determined by its distribution function; that is, if P_1 and P_2 are distributions on $\mathbb{R}$ having the same distribution function, then $P_1(B) = P_2(B)$ for $B \subset \mathbb{R}$.

Let Y be a real-valued random variable having probability function f. The distribution function F of Y is also referred to as the distribution function corresponding to f; it is given explicitly in terms of f by

$$F(y) = \sum_{z \le y} f(z), \qquad y \in \mathbb{R}.$$

Let Y be a constant real-valued random variable, so that $P(Y = a) = 1$ for some $a \in \mathbb{R}$. Then its distribution function is given by $F(y) = 0$ for $y < a$ and $F(y) = 1$ for $y \ge a$.

Let Y be a nonnegative integer-valued random variable having probability function f and distribution function F. Then $F(0) = f(0)$, whereas

$$F(y) = f(0) + \cdots + f(y) \quad \text{and} \quad f(y) = F(y) - F(y-1), \qquad y = 1, 2, \ldots .$$

Let F be the distribution function corresponding to a distribution having density function f on $\mathbb{R}$. Then

$$F(y) = \int_{-\infty}^{y} f(z) \, dz, \qquad y \in \mathbb{R}.$$

This distribution function is continuous on $\mathbb{R}$ and $F'(y) = f(y)$ at every continuity point y of f. Also,

$$1 - F(y) = \int_{y}^{\infty} f(z) \, dz, \qquad y \in \mathbb{R},$$

and

$$F(b) - F(a) = \int_{a}^{b} f(y) \, dy, \qquad a \le b.$$

If f is symmetric about zero, then $F(-y) = 1 - F(y)$ for $y \in \mathbb{R}$.

Consider a random variable Y whose distribution function F is continuous on $\mathbb{R}$ and continuously differentiable except possibly at finitely many points. Define the function f on $\mathbb{R}$ by $f(y) = F'(y)$ at a nonexceptional point and $f(y) = 0$ otherwise. Then f is a density function of Y.

Quantiles

Let $-\infty \le c < d \le \infty$. Consider a random variable Y having a density function f that is positive on (c, d) and equals zero outside this interval. Its distribution function

F is continuous and strictly increasing on (c, d) and has limits 0 and 1 at c and d, respectively. The restriction of F to (c, d) has an inverse function F^{-1}, which is a continuous and strictly increasing function from $(0, 1)$ onto (c, d) and has limits c and d at 0 and 1, respectively. Let $0 < p < 1$. The quantity $y_p = F^{-1}(p)$, which is the unique solution to the equation $F(y_p) = p$, is referred to as the pth quantile of Y (or of its distribution, distribution function, or density function). Observe that

$$F(y_p) = \int_c^{y_p} f(y)\, dy = p \quad \text{and} \quad 1 - F(y_p) = \int_{y_p}^d f(y)\, dy = 1 - p.$$

Moreover, $F(y) < F(y_p) = p$ for $y < y_p$, and $F(y) > F(y_p) = p$ for $y > y_p$. If f is positive on $\mathbb{R}$ and symmetric about zero, then $y_{1-p} = -y_p$ for $0 < p < 1$.

Univariate Transformations

Let W be a real-valued random variable having a density function f_W that is continuous and positive on some interval and equals zero outside this interval, let F_W denote the distribution function of W, and let w_p denote its pth quantile. Let Y be a transform of W having a density function f_Y and let F_Y denote the distribution function of Y and y_p its pth quantile.

Set $Y = a + bW$, where $b > 0$. Then

$$f_Y(y) = \frac{1}{b} f_W\left(\frac{y - a}{b}\right), \qquad y \in \mathbb{R},$$

$$F_Y(w) = F_W\left(\frac{y - a}{b}\right), \qquad y \in \mathbb{R},$$

and $y_p = a + bw_p$.

Suppose W is a positive random variable and set $Y = 1/W$. Then $f_Y(y) = 0$ for $y \le 0$, and $f_Y(y) = y^{-2} f_W(1/y)$ for $y > 0$; $F_Y(y) = 0$ for $y \le 0$, and $F_Y(y) = 1 - F_W(1/y)$ for $y > 0$; and $y_p = 1/w_{1-p}$.

Suppose the density function of W is symmetric about zero and set $Y = W^2$. Then $f_Y(y) = 0$ for $y \le 0$ and

$$f_Y(y) = \frac{1}{\sqrt{y}} f_W(\sqrt{y}), \qquad y > 0;$$

$F_Y(y) = 0$ for $y \le 0$ and $F_Y(y) = 2F_W(\sqrt{y}) - 1$ for $y > 0$; and $y_p = w_{(1+p)/2}^2$.

Independence

Let $\mathbf{Y}_i$ be a $\mathcal{Y}_i$-valued random variable for $1 \le i \le n$. The random variables $\mathbf{Y}_1, \ldots, \mathbf{Y}_n$ are said to be independent if

$$P(\mathbf{Y}_1 \in B_1, \ldots, \mathbf{Y}_n \in B_n) = P(\mathbf{Y}_1 \in B_1) \cdots P(\mathbf{Y}_n \in B_n)$$

for $B_1 \subset \mathcal{Y}_1, \ldots, B_n \subset \mathcal{Y}_n$. Random variables are said to be dependent if they are not independent.

Suppose $\mathbf{Y}_1, \ldots, \mathbf{Y}_n$ are discrete random variables. Then they are independent if and only if

$$P(\mathbf{Y}_1 = \mathbf{y}_1, \ldots, \mathbf{Y}_n = \mathbf{y}_n) = P(\mathbf{Y}_1 = \mathbf{y}_1) \cdots P(\mathbf{Y}_n = \mathbf{y}_n)$$

for $\mathbf{y}_1 \in \mathcal{Y}_1, \ldots, y_n \in \mathcal{Y}_n$.

Let $\mathbf{Y}_1, \ldots, \mathbf{Y}_n$ be independent. If we permute these random variables, combine two or more of them or delete one or more of them, we get a new sequence of independent random variables; also, the transforms $\mathbf{g}_1(\mathbf{Y}_1), \ldots, \mathbf{g}_n(\mathbf{Y}_n)$ are independent, where $\mathbf{g}_1, \ldots, \mathbf{g}_n$ are functions on $\mathcal{Y}_1, \ldots, \mathcal{Y}_n$, respectively. If $\mathbf{Y}_1, \ldots, \mathbf{Y}_n, (\mathbf{Y}_{n+1}, \mathbf{Y}_{n+2})$ are independent and $\mathbf{Y}_{n+1}$ and $\mathbf{Y}_{n+2}$ are independent, then $\mathbf{Y}_1, \ldots, \mathbf{Y}_{n+2}$ are independent.

Expectation

Every nonnegative random variable Y has a mean (expected value), which is either a nonnegative number or ∞. Let Y be an arbitrary real-valued random variable. Then Y can be written as the difference of two nonnegative random variables: $Y = Y_+ - Y_-$. The mean of Y is given by $EY = EY_+ - EY_-$, which can be finite, ∞, $-\infty$, or of the indeterminate form $\infty - \infty$.

The mean of a (real-valued) random variable is determined by its distribution; that is, if two random variables have the same distribution, then they have the same mean. The mean of a random variable is also referred to as the mean of its distribution (or distribution function, probability function, or density function).

Let $\mathbf{Y}$ be a $\mathcal{Y}$-valued random variable having probability function $f_{\mathbf{Y}}$ on $\mathcal{Y}$ and let g be a real-valued function on $\mathcal{Y}$. Then $Eg(\mathbf{Y}) = \sum_{\mathbf{y}} g(\mathbf{y}) f_{\mathbf{Y}}(\mathbf{y})$.

Let Y be a constant real-valued random variable, so that $P(Y = a) = 1$ for some $a \in \mathbb{R}$. Then $EY = a$.

Let Y be a random variable having finite mean. Then $E(a + Y) = a + EY$ for $a \in \mathbb{R}$ and $E(bY) = bE(Y)$ for $b \in \mathbb{R}$.

Let Y be a random variable having a density function f_Y that is zero outside an interval (c, d) and let g be a real-valued function on (c, d). Then

$$Eg(Y) = \int_c^d g(y) f_Y(y) \, dy.$$

Let $Y_1, \ldots, Y_n$ be random variables having finite mean. Then

$$E(b_1 Y_1 + \cdots + b_n Y_n) = b_1 EY_1 + \cdots + b_n EY_n, \qquad b_1, \ldots, b_n \in \mathbb{R}.$$

If $Y_1, \ldots, Y_n$ are independent, then $E(Y_1 \cdots Y_n) = E(Y_1) \ldots E(Y_n)$.

Let Y be a nonnegative random variable such that $EY = 0$. Then $P(Y = 0) = 1$.

A (real-valued) random variable Y has finite mean if and only if $|Y|$ has finite mean, in which case $|EY| \le E|Y|$.

Let Y_1 and Y_2 be random variables such that $P(|Y_1| \le Y_2) = 1$ and Y_2 has finite mean. Then Y_1 has finite mean and $|EY_1| \le EY_2$. In particular, if $|Y|$ has finite mean, then Y has finite mean and $|EY| \le E|Y|$. Also, if $P(|Y| \le M) = 1$, where M is a finite nonnegative constant, then Y has finite mean and $|EY| \le M$.

Suppose Y_1 and Y_2 each have finite mean and $P(Y_1 \le Y_2) = 1$. Then $EY_1 \le EY_2$.

The quantity $E(Y^2)$ is called the second moment of Y. If Y has finite second moment, then its mean is also finite.

Let Y_1 and Y_2 each have finite second moment. Then $Y_1 + Y_2$ has finite second moment, and $Y_1 Y_2$ has finite mean.

Variance

Let Y have finite second moment. Then the mean of Y is finite. The variance of Y, given by $\mathrm{var}(Y) = E[(Y - EY)^2] = E(Y^2) - (EY)^2$, is also finite. The standard deviation $\mathrm{SD}(Y)$ of Y is the (nonnegative) square root of the variance of Y. Suppose, instead, that $E(Y^2) = \infty$. Then the variance and standard deviation of Y are defined to be ∞.

The variance of Y is referred to as the variance of its distribution (or distribution function, probability function, or density function). Similar terminology is used for the standard deviation of Y.

Either the second moment, variance, and standard deviation of a random variable either are all equal to ∞ or they are all finite nonnegative numbers.

Let Y be a random variable having finite variance. Then

$$\mathrm{var}(a + Y) = \mathrm{var}(Y) \quad \text{and} \quad \mathrm{SD}(a + Y) = \mathrm{SD}(Y), \qquad a \in \mathbb{R}.$$

Also,

$$\mathrm{var}(bY) = b^2 \mathrm{var}(Y) \quad \text{and} \quad \mathrm{SD}(bY) = |b|\mathrm{SD}(Y), \qquad b \in \mathbb{R}.$$

If $\mathrm{SD}(Y) > 0$, then $(Y - EY)/\mathrm{SD}(Y)$ has mean 0 and variance 1.

Let $Y_1, \ldots, Y_n$ have finite variance and let $b_1, \ldots, b_n \in \mathbb{R}$. Then $b_1 Y_1 + \cdots + b_n Y_n$ has finite variance; and if $Y_1, \ldots, Y_n$ are independent, then

$$\mathrm{var}(b_1 Y_1 + \cdots + b_n Y_n) = b_1^2 \mathrm{var}(Y_1) + \cdots + b_n^2 \mathrm{var}(Y_n).$$

If Y is a constant random variable, then $\mathrm{var}(Y) = 0$ and $\mathrm{SD}(Y) = 0$. Conversely, if $\mathrm{var}(Y) = 0$ or $\mathrm{SD}(Y) = 0$, then both quantities equal zero and Y is a constant random variable.

Let Y have finite, positive variance. According to Chebyshev's inequality,

$$P(|Y - EY| \geq c) \leq \frac{\mathrm{var}(Y)}{c^2}, \qquad c > 0.$$

An informal consequence of this inequality is that if $\mathrm{SD}(Y) \approx 0$, then $P(Y \approx EY) \approx 1$.

Random Samples

Let P be a distribution on $\mathbb{R}$ having mean μ and finite standard deviation σ. Let $Y_1, \ldots, Y_n$ be independent random variables having distribution P and hence mean μ and standard deviation σ. We refer to such random variables as forming a random sample of size n from P.

The mean, variance, and standard deviation of the sum of these random variables are given by $E(Y_1 + \cdots + Y_n) = n\mu$, $\mathrm{var}(Y_1 + \cdots + Y_n) = n\sigma^2$, and $\mathrm{SD}(Y_1 + \cdots + Y_n) = \sigma\sqrt{n}$.

The random variable $\bar{Y} = (Y_1 + \cdots + Y_n)/n$ is called the sample mean. Its mean, variance, and standard deviation are given by $E(\bar{Y}) = \mu$, $\mathrm{var}(\bar{Y}) = \sigma^2/n$, and $\mathrm{SD}(\bar{Y}) = \sigma/\sqrt{n}$. If $\sigma > 0$, then

$$\frac{\bar{Y} - \mu}{\mathrm{SD}(\bar{Y})} = \frac{\bar{Y} - \mu}{\sigma/\sqrt{n}} = \frac{Y_1 + \cdots + Y_n - n\mu}{\sigma\sqrt{n}}$$

has mean 0 and variance 1.

According to the weak law of large numbers,

$$\lim_{n \to \infty} P\left(\left| \frac{Y_1 + \cdots + Y_n}{n} - \mu \right| \leq c \right) = 1, \qquad c > 0.$$

An informal statement of this result is that $P(\bar{Y} \approx \mu) \approx 1$ for $n \gg 1$.

Let $\hat{\tau}$ be a random variable that is defined in terms of $Y_1, \ldots, Y_n$, and let τ be a real parameter such as μ, σ, or σ^2. Regarded as an estimate of τ, $\hat{\tau}$ is said to be unbiased if $E\hat{\tau} = \tau$ and to be consistent if $P(\hat{\tau} \approx \tau) \approx 1$ for $n \gg 1$. As an estimate of μ, $\bar{Y}$ is unbiased and consistent.

Let $n \geq 2$. The sample variance is given by

$$S^2 = \frac{1}{n-1} \sum_{i=1}^{n} (Y_i - \bar{Y})^2;$$

and the sample standard deviation S is the square root of the sample variance. As an estimator of σ^2, S^2 is unbiased and consistent. As an estimator of σ, S is consistent but not quite unbiased.

Normal Distribution

The standard normal distribution has mean 0 and variance 1. Its density function, which is symmetric about zero, is given by

$$\phi(z) = \frac{1}{\sqrt{2\pi}} \exp\left(-\frac{z^2}{2} \right), \qquad z \in \mathbb{R};$$

its distribution function, denoted by Φ, satisfies the formula $\Phi(-z) = 1 - \Phi(z)$ for $z \in \mathbb{R}$; and its quantiles z_p satisfy the formula $z_{1-p} = -z_p$ for $0 < p < 1$. A random variable having the standard normal distribution is referred to as a standard normal random variable.

Let $\mu \in \mathbb{R}$ and $\sigma > 0$. The normal distribution with mean μ and variance σ^2 has the density function given by

$$f(y) = \frac{1}{\sigma}\phi\left(\frac{y-\mu}{\sigma}\right) = \frac{1}{\sigma\sqrt{2\pi}} \exp\left(-\frac{(y-\mu)^2}{2\sigma^2} \right), \qquad y \in \mathbb{R};$$

its distribution function is given by

$$F(y) = \Phi\left(\frac{y-\mu}{\sigma}\right), \qquad y \in \mathbb{R};$$

and its quantiles are given by $y_p = \mu + \sigma z_p$ for $0 < p < 1$.

Let Y be normally distributed. Then $a + bY$ is normally distributed for $a, b \in \mathbb{R}$ with $b \neq 0$. Also, $Y = EY + (SD(Y))Z$, where $Z = (Y - EY)/SD(Y)$ has the standard normal distribution.

Let $Y_1, \ldots, Y_n$ be independent, normally distributed random variables and let $a, b_1, \ldots, b_n$ be constants such that $b_1, \ldots, b_n$ are not all zero. Then $a + b_1 Y_1 + \cdots + b_n Y_n$ is normally distributed.

Normal Approximation

Let Y be a random variable having finite mean μ, finite positive variance σ^2, and pth quantile y_p. The normal approximation to the distribution of Y is the normal distribution with mean μ and variance σ^2. When this approximation is accurate,

$$P(Y \leq y) \approx \Phi\left(\frac{y - \mu}{\sigma}\right), \qquad y \in \mathbb{R},$$

and $y_p \approx \mu + \sigma z_p$ for $0 < p < 1$.

Let $Y_1, \ldots, Y_n$ be a random sample of size n from a distribution having mean μ and finite positive variance σ^2. According to the central limit theorem,

$$\lim_{n \to \infty} P\left(\frac{Y_1 + \cdots + Y_n - n\mu}{\sigma\sqrt{n}} \leq z\right) = \Phi(z), \qquad z \in \mathbb{R},$$

and hence

$$\frac{\bar{Y} - \mu}{\sigma/\sqrt{n}}$$

has approximately the standard normal distribution for $n \gg 1$.

Gamma Distribution

Let $\alpha > 0$ and $\beta > 0$. The gamma distribution with shape parameter α and scale parameter β has the density function given by $f(y) = 0$ for $y \leq 0$ and

$$f(y) = \frac{y^{\alpha-1}}{\beta^\alpha \Gamma(\alpha)} \exp\left(-\frac{y}{\beta}\right), \qquad y > 0.$$

The distribution has mean $\alpha\beta$ and variance $\alpha\beta^2$. Its distribution function F equals zero on $(-\infty, 0]$. When α is a positive integer, F is given explicitly on $[0, \infty)$ by

$$F(y) = 1 - \left(\sum_{m=0}^{\alpha-1} \frac{y^m}{\beta^m m!}\right) \exp\left(-\frac{y}{\beta}\right), \qquad y \geq 0.$$

Let Y have the gamma distribution with shape parameter α and scale parameter β and let $c > 0$. Then cY has the gamma distribution with shape parameter α and scale parameter $c\beta$.

Let $Y_1, \ldots, Y_n$ be independent random variables such that Y_i has the gamma distribution with shape parameter α_i and scale parameter β for $1 \leq i \leq n$. Then

$Y_1 + \cdots + Y_n$ has the gamma distribution with shape parameter $\alpha_1 + \cdots + \alpha_n$ and scale parameter β.

Normal approximation to the gamma distribution with shape parameter α is accurate when $\alpha \gg 1$.

The exponential distribution with scale parameter β coincides with gamma distribution with shape parameter 1 and scale parameter β. Its density function is given by $f(y) = 0$ for $y \leq 0$ and

$$f(y) = \frac{1}{\beta} \exp\left(-\frac{y}{\beta}\right), \qquad y > 0;$$

its distribution function is given by $F(y) = 0$ for $y \leq 0$ and

$$F(y) = 1 - \exp\left(-\frac{y}{\beta}\right), \qquad y \geq 0;$$

its quantiles are given by

$$y_p = \beta \log \frac{1}{1-p}, \qquad 0 < p < 1;$$

and it has mean β and variance β^2.

Let Y have the exponential distribution with mean β and let $c > 0$. Then cY has the exponential distribution with mean $c\beta$.

Let $Y_1, \ldots, Y_n$ be independent random variables such that Y_i has the exponential distribution with mean β for $1 \leq i \leq n$. Then $Y_1 + \cdots + Y_n$ has the gamma distribution with shape parameter n and scale parameter β.

The gamma distribution can also be analyzed in terms of the shape parameter α and inverse-scale parameter $\lambda = 1/\beta$.

Let Y be normally distributed with mean 0 and variance σ^2. Then Y^2 has the gamma distribution with shape parameter $1/2$ and scale parameter $2\sigma^2$. Let $Y_1, \ldots, Y_n$ be independent random variables each having the normal distribution with mean 0 and variance σ^2. Then $Y_1^2 + \cdots + Y_n^2$ has the gamma distribution with shape parameter $n/2$ and scale parameter $2\sigma^2$.

Binomial Distribution

Let n be a positive integer and let $0 < \pi < 1$. The binomial distribution with parameters n and π has the probability function given by

$$f(y) = \binom{n}{y} \pi^y (1-\pi)^{n-y}, \qquad y \in \{0, \ldots, n\}$$

and $f(y) = 0$ for other values of y; here

$$\binom{n}{y} = \frac{n!}{y!\,(n-y)!}.$$

The distribution has mean $n\pi$ and variance $n\pi(1-\pi)$.

Let $\mathbf{W}_1, \ldots, \mathbf{W}_n$ be a random sample of size n from a distribution P on $\mathcal{W}$ and let B be a subset of $\mathcal{W}$ such that $0 < P(B) < 1$. Then the frequency of B among $\mathbf{W}_1, \ldots, \mathbf{W}_n$ has the binomial distribution with parameters n and $\pi = P(B)$.

Let $Y_1, \ldots, Y_N$ be independent random variables such that Y_i has the binomial distribution with parameters n_i and π for $1 \le i \le N$. Then $Y_1 + \cdots + Y_N$ has the binomial distribution with parameters $n_1 + \cdots + n_N$ and π.

Normal approximation to the binomial distribution is accurate when $n\pi(1 - \pi) \gg 1$.

The Bernoulli distribution with parameter π coincides with the binomial distribution with parameters $n = 1$ and π. Its probability function is given by $f(0) = 1 - \pi, f(1) = \pi$, and $f(y) = 0$ for other values of y. The distribution has mean π and variance $\pi(1 - \pi)$.

Let $\mathbf{W}$ be a $\mathcal{W}$-valued random variable, let B be a subset of $\mathcal{W}$ such that $0 < P(B) < 1$, and let I be the indicator random variable corresponding to the event that $\mathbf{W} \in B$, which is given by $I = 1$ if $\mathbf{W} \in B$ and $I = 0$ if $\mathbf{W} \notin B$. Then I has the Bernoulli distribution with parameter $\pi = P(\mathbf{W} \in B)$.

Let $Y_1, \ldots, Y_n$ be independent random variables, each having the Bernoulli distribution with parameter π. Then $Y_1 + \cdots + Y_n$ has the binomial distribution with parameters n and π.

Poisson Distribution

Let $\lambda > 0$. The Poisson distribution with parameter λ has the probability function given by

$$f(y) = \frac{\lambda^y}{y!} e^{-\lambda}, \qquad y \in \{0, 1, 2, \ldots\},$$

and $f(y) = 0$ for other values of y. The distribution has mean λ and variance λ.

Let $Y_1, \ldots, Y_n$ be independent random variables such that Y_i has the Poisson distribution with mean λ_i for $1 \le i \le n$. Then $Y_1 + \cdots + Y_n$ has the Poisson distribution with mean $\lambda_1 + \cdots + \lambda_n$.

The Poisson distribution with mean $\lambda = n\pi$ is an accurate approximation to the binomial distribution with parameters n and π when $n \gg 1$ and $\pi \approx 0$.

Normal approximation to the Poisson distribution is accurate when $\lambda \gg 1$.

B.2
Random Vectors

Covariance

Let X and Y have finite variance. The covariance between X and Y is given by

$$\text{cov}(X, Y) = E[(X - EX)(Y - EY)] = E(XY) - (EX)(EY).$$

In particular, $\text{cov}(Y, Y) = \text{var}(Y)$. The variance of a linear combination of X and Y is given by

$$\text{var}(aX + bY) = a^2 \text{var}(X) + b^2 \text{var}(Y) + 2ab\,\text{cov}(X, Y), \qquad a, b \in \mathbb{R}.$$

If X or Y is a constant random variable, then $\text{cov}(X, Y) = 0$. If X and Y are independent, then $\text{cov}(X, Y) = 0$. The converse to the last result is not true, however; the covariance between X and Y can equal zero even if X and Y are dependent.

Some additional properties of covariances are as follows:

$$\text{cov}(Y, X) = \text{cov}(X, Y),$$

$$C\text{cov}(a + X, b + Y) = \text{cov}(X, Y) \text{ for } a, b \in \mathbb{R},$$

$$\text{cov}(aX, bY) = ab\,\text{cov}(X, Y) \text{ for } a, b \in \mathbb{R},$$

$$|\text{cov}(X, Y)| \le \text{SD}(X)\,\text{SD}(Y).$$

Correlation

Let X and Y have finite, positive variance. Their correlation, given by

$$\rho = \text{cor}(X, Y) = \frac{\text{cov}(X, Y)}{\text{SD}(X)\,\text{SD}(Y)},$$

satisfies the inequality $-1 \le \rho \le 1$. The random variables are said to be uncorrelated if their correlation equals zero. If X and Y are independent, then they are uncorrelated, but they can be uncorrelated without being independent.

Multivariate Expectation

Let m and n be positive integers and let Y_{ij}, $1 \le i \le n$ and $1 \le j \le m$, be real-valued random variables. The random $n \times m$ matrix

$$\mathbf{Y} = \begin{bmatrix} Y_{11} & \cdots & Y_{1m} \\ \vdots & & \vdots \\ Y_{n1} & \cdots & Y_{nm} \end{bmatrix}$$

is said to have finite expectation (mean) if each of its entries has finite expectation. Let this be the case. Then the mean of $\mathbf{Y}$ is the $n \times m$ matrix given by

$$E\mathbf{Y} = \begin{bmatrix} EY_{11} & \cdots & EY_{1m} \\ \vdots & & \vdots \\ EY_{n1} & \cdots & EY_{nm} \end{bmatrix}.$$

Some properties of expectation are as follows:

$$E\mathbf{Y} = \mathbf{A} \text{ if } P(\mathbf{Y} = \mathbf{A}) = 1,$$

$$E(\mathbf{Y}^T) = (E\mathbf{Y})^T,$$

$$E(b\mathbf{Y}) = b(E\mathbf{Y}) \text{ for } b \in \mathbb{R},$$

$$E(\mathbf{A} + \mathbf{Y}) = \mathbf{A} + E\mathbf{Y}, \text{ where } \mathbf{A} \text{ is an } n \times m \text{ matrix,}$$

$$E(\mathbf{B}\mathbf{Y}) = \mathbf{B}(E\mathbf{Y}), \text{ where } \mathbf{B} \text{ has } n \text{ columns,}$$

$$E(\mathbf{YB}) = (E\mathbf{Y})\mathbf{B}, \text{ where } \mathbf{B} \text{ has } m \text{ rows.}$$

Suppose $\mathbf{Y} = [\,Y_1, \ldots, Y_n\,]^T$. Then

$$E(\mathbf{a} + \mathbf{Y}) = \mathbf{a} + E\mathbf{Y}, \qquad \mathbf{a} \in \mathbb{R}^n,$$

and

$$E(\mathbf{b}^T\mathbf{Y}) = \mathbf{b}^T E\mathbf{Y}, \qquad \mathbf{b} \in \mathbb{R}^n.$$

Covariance Matrix

Let $X_1, \ldots, X_m, Y_1, \ldots, Y_n$ have finite variance and set $\mathbf{X} = [\,X_1, \ldots, X_m\,]^T$ and $\mathbf{Y} = [\,Y_1, \ldots, Y_n\,]^T$. The covariance matrix between $\mathbf{X}$ and $\mathbf{Y}$ is the $m \times n$ matrix having entry $\mathrm{cov}(X_i, Y_j)$ in row i and column j. If $\mathbf{X}$ or $\mathbf{Y}$ is a constant random vector or if $\mathbf{X}$ and $\mathbf{Y}$ are independent, then $\mathrm{cov}(\mathbf{X}, \mathbf{Y}) = \mathbf{0}$. Some additional properties involving the covariance matrix are as follows:

$$\mathrm{cov}(\mathbf{X}, \mathbf{Y}) = E[\,(\mathbf{X} - E\mathbf{X})(\mathbf{Y} - E\mathbf{Y})^T\,],$$

$$\mathrm{cov}(\mathbf{Y}, \mathbf{X}) = [\,\mathrm{cov}(\mathbf{X}, \mathbf{Y})\,]^T,$$

$$\mathrm{cov}(\mathbf{a} + \mathbf{X}, \mathbf{b} + \mathbf{Y}) = \mathrm{cov}(\mathbf{X}, \mathbf{Y}) \text{ for } \mathbf{a} \in \mathbb{R}^m \text{ and } \mathbf{b} \in \mathbb{R}^n,$$

$$\mathrm{cov}(\mathbf{AX}, \mathbf{BY}) = \mathbf{A}\,\mathrm{cov}(\mathbf{X}, \mathbf{Y})\mathbf{B}^T \text{ if } \mathbf{A} \text{ has } m \text{ columns and } \mathbf{B} \text{ has } n \text{ columns,}$$

$$\mathrm{cov}(\mathbf{a}^T\mathbf{X}, \mathbf{b}^T\mathbf{Y}) = \mathbf{a}^T\mathrm{cov}(\mathbf{X}, \mathbf{Y})\mathbf{b} \text{ for } \mathbf{a} \in \mathbb{R}^m \text{ and } \mathbf{b} \in \mathbb{R}^n,$$

$$\mathrm{cov}(a\mathbf{X}, b\mathbf{Y}) = ab\,\mathrm{cov}(\mathbf{X}, \mathbf{Y}) \text{ for } a, b \in \mathbb{R}.$$

Variance–Covariance matrix

Let $Y_1, \ldots, Y_n$ have finite variance and set $\mathbf{Y} = [\,Y_1, \ldots, Y_n\,]^T$. The variance–covariance matrix of $\mathbf{Y}$ (or of $Y_1, \ldots, Y_n$) is the $n \times n$ matrix having entry $\mathrm{cov}(Y_i, Y_j)$ in row i and column j; in particular, the ith diagonal entry of $\mathrm{VC}(\mathbf{Y})$ is $\mathrm{var}(Y_i)$. If $Y_1, \ldots, Y_n$ are independent, then $\mathrm{VC}(\mathbf{Y}) = \mathrm{diag}(\mathrm{var}(Y_1), \ldots, \mathrm{var}(Y_n))$.

Some additional properties involving variance–covariance matrices are as follows:

$$\mathrm{VC}(\mathbf{Y}) = \mathrm{cov}(\mathbf{Y}, \mathbf{Y}),$$

$$\mathrm{VC}(\mathbf{Y}) = E[\,(\mathbf{Y} - E\mathbf{Y})(\mathbf{Y} - E\mathbf{Y})^T\,],$$

$$\mathrm{VC}(\mathbf{Y}) = \mathbf{0} \text{ if } \mathbf{Y} \text{ is a constant random vector,}$$

$$\mathrm{VC}(\mathbf{a} + \mathbf{Y}) = \mathrm{VC}(\mathbf{Y}) \text{ for } \mathbf{a} \in \mathbb{R}^n,$$

$$\mathrm{VC}(b\mathbf{Y}) = b^2\mathrm{VC}(\mathbf{Y}) \text{ for } b \in \mathbb{R},$$

$$\mathrm{var}(\mathbf{b}^T\mathbf{Y}) = \mathbf{b}^T[\,\mathrm{VC}(\mathbf{Y})\,]\mathbf{b} \text{ for } \mathbf{b} \in \mathbb{R}^n,$$

$$\mathrm{cov}(\mathbf{a}^T\mathbf{Y}, \mathbf{b}^T\mathbf{Y}) = \mathbf{a}^T[\,\mathrm{VC}(\mathbf{Y})\,]\mathbf{b} \text{ for } \mathbf{a}, \mathbf{b} \in \mathbb{R}^n,$$

$$\mathrm{VC}(\mathbf{BY}) = \mathbf{B}[\,\mathrm{VC}(\mathbf{Y})\,]\mathbf{B}^T, \text{ where } \mathbf{B} \text{ has } n \text{ columns.}$$

A variance–covariance matrix is symmetric and positive semidefinite, and it is invertible if and only if it is positive definite. The variance–covariance matrix of $\mathbf{Y}$ is noninvertible if and only if there is a vector $\mathbf{b} \in \mathbb{R}^n$ such that $\mathbf{b} \neq 0$ and $\mathbf{b}^T\mathbf{Y}$ is a constant random variable.

Multivariate Density Functions

Let n denote a positive integer. A function f on $\mathbb{R}^n$ is called a density function if f is nonnegative and its integral equals 1. Let f be a density function on $\mathbb{R}^n$. Then

$$P(B) = \int_B f(y)\, dy, \qquad B \subset \mathbb{R}^n,$$

defines a distribution on $\mathbb{R}^n$, and f is referred to as the density function of P.

Let $\mathbf{Y}$ be a random vector whose distribution has density function $f_{\mathbf{Y}}$ on $\mathbb{R}^n$. Then

$$P(\mathbf{Y} \in B) = \int_B f_{\mathbf{Y}}(\mathbf{y})\, dy, \qquad B \subset \mathbb{R}^n,$$

and $f_{\mathbf{Y}}$ is referred to as the density function of Y. If $B \in \mathbb{R}^n$ and $\int_B dy = 0$, then $P(\mathbf{Y} \in B) = 0$. In particular, $P(\mathbf{Y} = \mathbf{y}) = 0$ for $\mathbf{y} \in \mathbb{R}^n$. Let $\mathcal{Y}$ be a subset of $\mathbb{R}^n$ such that $f(\mathbf{y}) = 0$ for $\mathbf{y} \notin \mathcal{Y}$ and let g be a real-valued function on $\mathcal{Y}$. Then

$$Eg(\mathbf{Y}) = \int_{\mathcal{Y}} g(\mathbf{y}) f_{\mathbf{Y}}(\mathbf{y})\, dy.$$

Let $Y_1, \ldots, Y_n$ have joint density function $f_{Y_1, \ldots, Y_n}$. Then, for $1 \leq i \leq n$, Y_i has a density function f_{Y_i}, which is referred to as its marginal density function and which can be obtained from $f_{Y_1, \ldots, Y_n}$ by integration over the other variables: $f_{Y_i}(y_i)$ is the $(n-1)$-dimensional integral of $f_{Y_1, \ldots, Y_n}(y_1, \ldots, y_n)$ over $y_j \in \mathbb{R}$ for $j \neq i$. If

$$f_{Y_1, \ldots, Y_n}(y_1, \ldots, y_n) = f_{Y_1}(y_1) \cdots f_{Y_n}(y_n), \qquad y_1, \ldots, y_n \in \mathbb{R}, \tag{1}$$

then $Y_1, \ldots, Y_n$ are independent. Conversely, if $Y_1, \ldots, Y_n$ are independent, then the joint density function of $Y_1, \ldots, Y_n$ can be chosen so that (1) holds. If the individual density functions of $Y_1, \ldots, Y_n$ and their joint density function are continuous functions, then (1) is necessary and sufficient for $Y_1, \ldots, Y_n$ to be independent. If $Y_1, \ldots, Y_n$ have finite variance, then their variance–covariance matrix is positive definite.

Multivariate Normal Distribution

Let n be a positive integer, let $\mu \in \mathbb{R}^n$, and let Σ be an $n \times n$ positive definite symmetric matrix. The multivariate normal distribution with mean vector μ and variance–covariance matrix Σ has the density function

$$f(\mathbf{y}) = \frac{1}{(2\pi)^{n/2}\sqrt{(\det(\Sigma))}} \exp\left(-\frac{1}{2}(\mathbf{y} - \mu)^T \Sigma^{-1}(\mathbf{y} - \mu)\right), \qquad \mathbf{y} \in \mathbb{R}^n.$$

Let $Y_1, \ldots, Y_n$ be independent, normally distributed random variables having the respective means $\mu_1, \ldots, \mu_n$ and variances $\sigma_1^2, \ldots, \sigma_n^2$. Then the random vector $\mathbf{Y} = [Y_1, \ldots, Y_n]^T$ has the multivariate normal distribution with mean vector $[\mu_1, \ldots, \mu_n]^T$ and variance–covariance matrix $\mathrm{diag}(\sigma_1^2, \ldots, \sigma_n^2)$. In particular, let $Z_1, \ldots, Z_n$ be independent, standard normal random variables. Then $\mathbf{Z} = [Z_1, \ldots, Z_n]^T$ has the multivariate normal distribution whose mean vector is the zero vector, whose variance–covariance matrix is the $n \times n$ identity matrix, and whose density function is given by

$$f(\mathbf{z}) = \frac{1}{(2\pi)^{n/2}} \exp\left(-\frac{1}{2}\mathbf{z}^T\mathbf{z}\right), \qquad \mathbf{z} \in \mathbb{R}^n.$$

Let $\mathbf{Y} = [Y_1, \ldots, Y_n]^T$ have a multivariate normal distribution. The random vector obtained by permuting $Y_1, \ldots, Y_n$ or by deleting some but not all of these random variables has a multivariate normal distribution. Every linear combination $a + \mathbf{b}^T\mathbf{Y}$, where $a \in \mathbb{R}$, $\mathbf{b} \in \mathbb{R}^n$ and $\mathbf{b} \neq \mathbf{0}$, is normally distributed. Let $\mathbf{a} \in \mathbb{R}^n$ and let $\mathbf{B}$ be an invertible $n \times n$ matrix. Then the random vector $\mathbf{a} + \mathbf{BY}$ has a multivariate normal distribution. The random variables $Y_1, \ldots, Y_n$ are independent if and only if they are uncorrelated.

Let $X_1, \ldots, X_m, Y_1, \ldots, Y_n$ have a multivariate normal joint distribution and set $\mathbf{X} = [X_1, \ldots, X_m]^T$ and $\mathbf{Y} = [Y_1, \ldots, Y_n]^T$. Then $\mathbf{X}$ and $\mathbf{Y}$ are independent if and only if $\mathrm{cov}(\mathbf{X}, \mathbf{Y}) = \mathbf{0}$.

Let $\mathbf{X}$ and $\mathbf{Y}$ be independent random vectors, each having a multivariate normal distribution. Then they have a multivariate normal joint distribution.

C

Summary of Statistics

C.1

Normal Models

The Chi-Square Distribution

Let ν be a positive integer. The chi-square distribution with ν degrees of freedom coincides with the gamma distribution with shape parameter $\nu/2$ and scale parameter 2. The distribution has mean ν and variance 2ν. Its distribution function is written as $\chi^2_\nu(y)$, $y \in \mathbb{R}$, and its pth quantile is written as $\chi^2_{p,\nu}$.

If Z has the standard normal distribution, then Z^2 has the chi-square distribution with 1 degree of freedom. More generally, if $Z_1, \ldots, Z_\nu$ are independent, standard normal random variables, then $Z_1^2 + \cdots + Z_\nu^2$ has the chi-square distribution with ν degrees of freedom. Still more generally, if $\chi_1^2, \ldots, \chi_N^2$ are independent and χ_i^2 has the chi-square distribution with ν_i degrees of freedom for $1 \leq i \leq N$, then $\chi_1^2 + \cdots + \chi_N^2$ has the chi-square distribution with $\nu_1 + \cdots + \nu_N$ degrees of freedom.

Normal approximation to the chi-square distribution with ν degrees of freedom is accurate when $\nu \gg 1$.

The t Distribution

Let v denote a positive integer and let Z and χ^2 be independent random variables such that Z has the standard normal distribution and χ^2 has the chi-square distribution with v degrees of freedom. Then the random variable

$$\frac{Z}{\sqrt{\chi^2/v}}$$

has the t distribution with v degrees of freedom. Its density function is symmetric about zero; its distribution function $t_v(y)$, $y \in \mathbb{R}$, satisfies the formula

$$t_v(-y) = 1 - t_v(y), \qquad y \in \mathbb{R};$$

and its quantiles $t_{p,v}$ satisfy the formula

$$t_{1-p,v} = -t_{p,v}, \qquad 0 < p < 1.$$

If $v \gg 1$, then the t distribution with v degrees of freedom is approximately equal to the standard normal distribution. In particular if $y \in \mathbb{R}$, then $t_v(y) \approx \Phi(y)$ for $v \gg 1$; and if $0 < p < 1$, then $t_{p,v} \approx z_p$ for $v \gg 1$.

The F Distribution

Let v_1 and v_2 denote positive integers and let χ_1^2 and χ_2^2 be independent random variables such that χ_1^2 has the chi-square distribution with v_1 degrees of freedom and χ_2^2 has the chi-square distribution with v_2 degrees of freedom. Then the positive random variable

$$\frac{\chi_1^2/v_1}{\chi_2^2/v_2}$$

has the F distribution with v_1 degrees of freedom in the numerator and v_2 degrees of freedom in the denominator; its distribution function $F_{v_1,v_2}(y)$, $y \in \mathbb{R}$, satisfies the formula

$$F_{v_2,v_1}(y) = 1 - F_{v_1,v_2}(1/y), \qquad y > 0;$$

and its quantiles F_{p,v_1,v_2} satisfy the formula

$$F_{p,v_2,v_1} = \frac{1}{F_{1-p,v_1,v_2}}, \qquad 0 < p < 1.$$

Let t be a random variable having the t distribution with v degrees of freedom. Then t^2 has the F distribution with 1 degree of freedom in the numerator and v degrees of freedom in the denominator. In particular,

$$1 - F_{1,v}(y^2) = 2[\, 1 - t_v(|y|)\,], \qquad y \in \mathbb{R},$$

and

$$F_{p,1,v} = t^2_{(p+1)/2,v}, \qquad 0 < p < 1.$$

Let F have the F distribution with ν_1 degrees of freedom in the numerator and ν_2 degrees of freedom in the denominator, where ν_1 is fixed. When $\nu_2 \gg 1$, the random variable $\nu_1 F$ has approximately the chi-square distribution with ν_1 degrees of freedom. In particular, if $y > 0$, then

$$F_{\nu_1, \nu_2}(y) \approx \chi^2_{\nu_1}(\nu_1 y), \qquad \nu_2 \gg 1;$$

and if $0 < p < 1$, then

$$F_{p, \nu_1, \nu_2} \approx \frac{1}{\nu_1} \chi^2_{p, \nu_1}, \qquad \nu_2 \gg 1.$$

The Normal One-Sample Model

Let $Y_1, \ldots, Y_n$ be independent random variables having a common normal distribution with mean μ and variance σ^2. This setup is referred to as the normal one-sample model. Under this model, the sample mean $\bar{Y} = (Y_1 + \cdots + Y_n)/n$ is normally distributed with mean μ and variance σ^2/n.

Suppose $n \geq 2$. Then the sample variance

$$S^2 = \frac{1}{n-1} \sum_i (Y_i - \bar{Y})^2$$

(with $S \geq 0$) is an unbiased estimate of σ^2 [that is, $E(S^2) = \sigma^2$], $(n-1)S^2/\sigma^2$ has the chi-square distribution with $n-1$ degrees of freedom, and $\bar{Y}$ and S^2 are independent.

The estimate $\hat{\mu} = \bar{Y}$ of μ is unbiased, and S^2/n is an unbiased estimate of the variance σ^2/n of this estimate. The standard error $\mathrm{SE}(\hat{\mu}) = S/\sqrt{n}$ of the estimate is a reasonable estimate of its standard deviation $\sigma/\sqrt{n}$.

The Normal Multisample Model

Let Y_{ki}, $1 \leq k \leq d$ and $1 \leq i \leq n_k$, be independent, normally distributed random variables having common variance σ^2 and such that, for $1 \leq k \leq d$, the random variables Y_{ki}, $1 \leq i \leq n_k$, have common mean μ_k; here $d, n_1, \ldots, n_d$ are positive integers. This setup is referred to as the normal d-sample model. When $d = 1$, it is equivalent to the setup of the normal one-sample model. We refer to Y_{ki}, $1 \leq i \leq n_k$, as forming the kth random sample. This random sample has size n_k, and its sample mean is given by

$$\bar{Y}_k = \frac{1}{n_k} \sum_{i=1}^{n_k} Y_{ki},$$

which is normally distributed with mean μ_k and variance σ_k^2/n_k. Set $n = n_1 + \cdots + n_d$ and

$$\bar{Y} = \frac{1}{n} \sum_{k=1}^{d} n_k \bar{Y}_k = \frac{1}{n} \sum_{k=1}^{d} \sum_{i=1}^{n_k} Y_{ki}.$$

The between sum of squares BSS and within sum of squares WSS are given by

$$\text{BSS} = \sum_{k=1}^{d} n_k (\bar{Y}_k - \bar{Y})^2 \quad \text{and} \quad \text{WSS} = \sum_{k=1}^{d} \sum_{i=1}^{n_k} (Y_{ki} - \bar{Y})^2.$$

Suppose $n > d$ or, equivalently, $n_k > 1$ for some k. Then the pooled sample variance $S^2 = \text{WSS}/(n-d)$ (with $S \geq 0$) is an unbiased estimate of σ^2, $(n-d)S^2/\sigma^2$ has the chi-square distribution with $n - d$ degrees of freedom, and the random variables $\bar{Y}_1, \ldots, \bar{Y}_d, S^2$ are independent.

Given $c_1, \ldots, c_d \in \mathbb{R}$, consider the linear parameter $\tau = c_1 \mu_1 + \cdots + c_d \mu_d$. The estimate $\hat{\tau} = c_1 \hat{\mu}_1 + \cdots + c_d \hat{\mu}_d$ of τ is unbiased, its variance is given by

$$\text{var}(\hat{\tau}) = \sigma^2 \left(\frac{c_1^2}{n_1} + \cdots + \frac{c_d^2}{n_d} \right),$$

and

$$S^2 \left(\frac{c_1^2}{n_1} + \cdots + \frac{c_d^2}{n_d} \right)$$

is an unbiased estimate of this variance. The standard error

$$\text{SE}(\hat{\tau}) = S \sqrt{\frac{c_1^2}{n_1} + \cdots + \frac{c_d^2}{n_d}}$$

of $\hat{\tau}$ is a natural estimate of its standard deviation. In particular, let $1 \leq k \leq d$. Then $\hat{\mu}_k = \bar{Y}_k$ is an unbiased estimate of μ_k, whose standard error is given by $\text{SE}(\hat{\mu}_k) = S/\sqrt{n_k}$. Suppose next that $d \geq 2$ and let $1 \leq j, k \leq d$ with $j \neq k$. Then $\hat{\mu}_k - \hat{\mu}_j = \bar{Y}_k - \bar{Y}_j$ is an unbiased estimate of $\mu_k - \mu_j$, whose standard error is given by

$$\text{SE}(\hat{\mu}_k - \hat{\mu}_j) = S \sqrt{\frac{1}{n_j} + \frac{1}{n_k}}.$$

Confidence Bounds and Confidence Intervals

Suppose τ is a nontrivial linear parameter; that is, at least one of the numbers $c_1, \ldots, c_d$ is nonzero or, equivalently that $\hat{\tau}$ has positive variance. Then $\hat{\tau}$ is normally distributed, $(\hat{\tau} - \tau)/\text{SD}(\hat{\tau})$ has the standard normal distribution and $(\hat{\tau} - \tau)/\text{SE}(\hat{\tau})$ has the t distribution with $n - d$ degrees of freedom. Let $0 < \alpha < 1$. Then $\hat{\tau} - t_{1-\alpha, n-d} \text{SE}(\hat{\tau})$ is the $100(1 - \alpha)\%$ lower confidence bound for τ; that is, $P(\hat{\tau} - t_{1-\alpha, n-d} \text{SE}(\hat{\tau}) < \tau) = 1 - \alpha$. Also, $\hat{\tau} + t_{1-\alpha, n-d} \text{SE}(\hat{\tau})$ is the $100(1 - \alpha)\%$ upper confidence bound for τ; that is, $P(\tau < \hat{\tau} + t_{1-\alpha, n-d} \text{SE}(\hat{\tau})) = 1 - \alpha$. Finally,

$$\hat{\tau} \pm t_{1-\alpha/2, n-d} \text{SE}(\hat{\tau}) = \left(\hat{\tau} - t_{1-\alpha/2, n-d} \text{SE}(\hat{\tau}), \ \hat{\tau} + t_{1-\alpha/2, n-d} \text{SE}(\hat{\tau}) \right)$$

is the $100(1 - \alpha)\%$ confidence interval for τ; that is,

$$P\left(\hat{\tau} - t_{1-\alpha/2, n-d} \text{SE}(\hat{\tau}) < \tau < \hat{\tau} + t_{1-\alpha/2, n-d} \text{SE}(\hat{\tau}) \right) = 1 - \alpha.$$

The t Test

Given $\tau_0 \in \mathbb{R}$, consider the statistic $t = (\hat{\tau} - \tau_0)/\operatorname{SE}(\hat{\tau})$. If $\tau = \tau_0$, then t has the t distribution with $n - d$ degrees of freedom.

Let $0 < \alpha < 1$. The inequality $t \geq t_{1-\alpha, n-d}$ determines the critical region having size α for the t test of the hypothesis that $\tau \leq \tau_0$; that is, α is the maximum value of $P(t \geq t_{1-\alpha, n-d})$ that is compatible with the hypothesis. The t statistic lies in the acceptance region (complement of the critical region) if and only if τ_0 is greater than the $100(1 - \alpha)\%$ lower confidence bound $\hat{\tau} - t_{1-\alpha, n-d}\operatorname{SE}(\hat{\tau})$ for τ. The P-value for the test equals $1 - t_{n-d}(t)$. (The t statistic lies in the critical region for the t test of size α of the hypothesis if α is greater than or equal to the P-value, but not if α is less than the P-value.)

The inequality $t \leq -t_{1-\alpha, n-d}$ determines the critical region having size α for the t test of the hypothesis that $\tau \geq \tau_0$; that is, α is the maximum value of $P(t \leq -t_{1-\alpha, n-d})$ that is compatible with the hypothesis. The t statistic lies in the acceptance region if and only if τ_0 is less than the $100(1 - \alpha)\%$ upper confidence bound $\hat{\tau} + t_{1-\alpha, n-d}\operatorname{SE}(\hat{\tau})$ for τ. The P-value for the test equals $t_{n-d}(t)$.

The inequality $|t| \geq t_{1-\alpha/2, n-d}$ determines the critical region having size α for the t test of the hypothesis that $\tau = \tau_0$; that is, if the hypothesis holds, then $P(|t| \geq t_{1-\alpha/2, n-d}) = \alpha$. The t statistic lies in the acceptance region if and only if τ_0 lies in the $100(1 - \alpha)\%$ confidence interval $\hat{\tau} \pm t_{1-\alpha/2, n-d}\operatorname{SE}(\hat{\tau})$ for τ. The P-value for the test equals $2[\,1 - t_{n-d}(|t|)\,]$.

The F Test

The hypothesis of homogeneity is that $\mu_1 = \cdots = \mu_d$. The F statistic for the F test of this hypothesis is given by

$$F = \frac{\operatorname{BSS}/(d - 1)}{\operatorname{WSS}/(n - d)}.$$

If the hypothesis of homogeneity is valid, then F has the F distribution with $d - 1$ degrees of freedom in the numerator and $n - d$ of degrees of freedom in the denominator. The inequality $F \geq F_{1-\alpha, d-1, n-d}$ determines the critical region having size α for the test; that is, if the hypothesis holds, then $P(F \geq F_{1-\alpha, d-1, n-d}) = \alpha$. The P-value for the test equals $1 - F_{d-1, n-d}(F)$.

Suppose $d = 2$. Then the hypothesis of homogeneity is that $\mu_1 = \mu_2$, and the F statistic for testing this hypothesis is given by

$$F = \frac{\operatorname{BSS}}{\operatorname{WSS}/(n-2)} = \frac{n_1(\bar{Y}_1 - \bar{Y})^2 + n_2(\bar{Y}_2 - \bar{Y})^2}{S^2}.$$

The t statistic for testing the equivalent hypothesis that $\tau = 0$, where $\tau = \mu_2 - \mu_1$, is given by $t = \hat{\tau}/\operatorname{SE}(\hat{\tau})$, where $\hat{\tau} = \hat{\mu}_2 - \hat{\mu}_1 = \bar{Y}_2 - \bar{Y}_1$ and

$$\operatorname{SE}(\hat{\tau}) = S\sqrt{\frac{1}{n_1} + \frac{1}{n_2}}.$$

These two statistics are related by the formula $F = t^2$. Moreover, the P-value for the F test of the hypothesis of homogeneity coincides with the P-value for the t test of the hypothesis that $\tau = 0$.

C.2
Linear Regression
The Input-Response Model

Consider an input vector (or factor-level combination) $\mathbf{x}$ that ranges over a nonempty set $\mathcal{X}$ and a random variable Y that has finite variance and whose distribution depends on $\mathbf{x}$. We refer to Y as the response variable. Let $\mu(\mathbf{x})$ and $\sigma^2(\mathbf{x})$ denote the dependence of the mean and variance of Y on $\mathbf{x}$. The function $\mu(\mathbf{x})$, $\mathbf{x} \in \mathcal{X}$, is called the regression function. We refer to this setup as the input-response version of the regression model.

The input-response version of the homoskedastic regression model consists of the input-response version of the regression model together with the assumption that the variance of Y does not depend on $\mathbf{x}$, so that $\sigma^2(\mathbf{x}) = \sigma^2$ for $\mathbf{x} \in \mathcal{X}$. The input-response version of the normal regression model consists of the input-response version of the homoskedastic regression model together with the assumption that Y is normally distributed; accordingly, when $\mathbf{x}$ is the input vector, Y is normally distributed with mean $\mu(\mathbf{x})$ and variance $\sigma^2 > 0$.

Let G be a p-dimensional linear space of functions on $\mathcal{X}$. The input-response version of the corresponding normal linear regression model consists of the input-response version of the normal regression model together with the assumption that the regression function is in G. Let $g_1, \ldots, g_p$ be a basis of G. Under the assumption that $\mu(\cdot) \in G$,

$$\mu(\mathbf{x}) = \beta_1 g_1(\mathbf{x}) + \cdots + \beta_p g_p(\mathbf{x}), \qquad \mathbf{x} \in \mathcal{X}.$$

The uniquely determined regression coefficients $\beta_1, \ldots, \beta_p$ form the coefficient vector $\boldsymbol{\beta} = [\beta_1, \ldots, \beta_p]^T$.

The Experimental Model

Consider an experiment involving n trials, in which $\mathbf{x}_i \in \mathcal{X}$ is the input and Y_i is the random response variable on the ith trial for $1 \le i \le n$. It is assumed that the distribution of Y_i coincides with the distribution of Y in the input-response model when the input is $\mathbf{x}_i$, so that Y_i has mean $\mu(\mathbf{x}_i)$ and variance $\sigma^2(\mathbf{x}_i)$. It is also assumed that the random variables Y_i, $1 \le i \le n$, are independent. This setup, referred to as the experimental version of the regression model, is in effect throughout the remainder of this section.

According to the (experimental version of the) homoskedastic regression model, $\sigma^2(\mathbf{x}_i) = \sigma^2$ for $1 \le i \le n$; according to the normal regression model, Y_i is normally distributed with mean $\mu(\mathbf{x}_i)$ and variance $\sigma^2(\mathbf{x}_i) = \sigma^2$ for $1 \leq i \leq n$. Throughout the remainder of this section, the assumptions of the normal regression model are in effect. Set $\mathbf{Y} = [Y_1, \ldots, Y_n]^T$. Then $\mathbf{Y}$ has the multivariate normal distribution with mean vector $[\mu(\mathbf{x}_1), \ldots, \mu(\mathbf{x}_n)]^T$ and variance–covariance matrix $\sigma^2 \mathbf{I}$, where $\mathbf{I}$ is the $n \times n$ identity matrix.

Let d denote the number of distinct inputs among $\mathbf{x}_1, \ldots, \mathbf{x}_n$, so that $1 \leq d \leq n$. The distinct inputs are denoted by $\mathbf{x}'_1, \ldots, \mathbf{x}'_d$; we refer to them as the design points and to $\{\mathbf{x}'_1, \ldots, \mathbf{x}'_d\}$ as the design set. For $1 \leq k \leq d$, let $\mathcal{J}_k = \{i : \mathbf{x}_i = \mathbf{x}'_k\}$ denote the set of trials on which the input is $\mathbf{x}'_k$ and let $n_k = \#(\mathcal{J}_k)$ denote number of such trials. Then n_k is the number of repetitions of $\mathbf{x}'_k$ among $\mathbf{x}_1, \ldots, \mathbf{x}_n$. Now $\mathcal{J}_1, \ldots, \mathcal{J}_K$ are disjoint and their union is $\{1, \ldots, n\}$, so $n = \sum_k n_k$. Set $\mu_k = \mu(\mathbf{x}'_k)$ for $1 \leq k \leq d$.

For $1 \leq k \leq d$, let

$$\bar{Y}_k = \frac{1}{n_k} \sum_{i \in \mathcal{J}_k} Y_i$$

denote the sample mean of the response variables on the n_k trials having input $\mathbf{x}'_k$. Then $\bar{Y}_1, \ldots, \bar{Y}_d$ are independent random variables, and, for $1 \leq k \leq d$, $\bar{Y}_k$ is normally distributed with mean μ_k and variance σ^2/n_k. The sample mean

$$\bar{Y} = \frac{1}{n} \sum_i Y_i = \frac{1}{n} \sum_k n_k \bar{Y}_k$$

of all n response variables is normally distributed with mean

$$\frac{1}{n} \sum_i \mu(\mathbf{x}_i) = \frac{1}{n} \sum_k n_k \mu_k$$

and variance σ^2/n.

The total sum of squares $\text{TSS} = \sum_i (Y_i - \bar{Y})^2$ is a measure of the variation among the response variables. The between sum of squares $\text{BSS} = \sum_k n_k (\bar{Y}_k - \bar{Y})^2$ is a measure of the variation among $\bar{Y}_1, \ldots, \bar{Y}_d$. The within sum of squares

$$\text{WSS} = \sum_k \sum_{i \in \mathcal{J}_k} (Y_i - \bar{Y}_k)^2$$

is a measure of the variation among the response variables on trials having a common input. These sums of squares are related by the formula $\text{TSS} = \text{BSS} + \text{WSS}$. The random variables $\bar{Y}_1, \ldots, \bar{Y}_K$, WSS are independent.

Suppose $\mathbf{x}_1, \ldots, \mathbf{x}_n$ are distinct. Then $d = n$, $n_1 = \cdots = n_d = 1$, and the design set equals $\{\mathbf{x}_1, \ldots, \mathbf{x}_n\}$. Here we choose $\mathbf{x}'_k = \mathbf{x}_k$ for $1 \leq k \leq n$. Then $\mathcal{J}_k = \{k\}$, $\mu_k = \mu(\mathbf{x}'_k)$, and $\bar{Y}_k = Y_k$ for $1 \leq k \leq n$; also, $\text{WSS} = 0$.

Suppose, instead, that $\mathbf{x}_1, \ldots, \mathbf{x}_n$ are not distinct. Then $d < n$, $n_k \geq 2$ for at least one value of k, and WSS/σ^2 has the chi-square distribution with $n - d$ degrees of freedom.

Until the discussion of the goodness-of-fit test at the end of this section, the assumptions of the normal linear regression model are in effect.

Let $g_1, \ldots, g_p$ be a basis of G. The corresponding $n \times p$ design matrix is given by

$$\mathbf{X} = \begin{bmatrix} g_1(\mathbf{x}_1) & \cdots & g_p(\mathbf{x}_1) \\ \vdots & & \vdots \\ g_1(\mathbf{x}_n) & \cdots & g_p(\mathbf{x}_n) \end{bmatrix}.$$

Let $\beta_1, \ldots, \beta_p$ be the regression coefficients corresponding to the given basis and set $\beta = [\beta_1, \ldots, \beta_p]^T$. Then $\mu(\cdot) = \beta_1 g_1 + \cdots + \beta_p g_p$, so $\mu(\mathbf{x}_i) = \sum_j \beta_j g_j(\mathbf{x}_i) = (\mathbf{X}\beta)_i$ for $1 \le i \le n$ and $[\mu(\mathbf{x}_1), \ldots, \mu(\mathbf{x}_n)]^T = \mathbf{X}\beta$.

Throughout the rest of this section, it is assumed that G is identifiable; that is, there is no nonzero function $g \in G$ such that $g(\mathbf{x}_1) = \cdots = g(\mathbf{x}_n) = 0$ or, equivalently, $g(\mathbf{x}_1') = \cdots = g(\mathbf{x}_d') = 0$. Then $p \le d$ and the column vectors of $\mathbf{X}$ are linearly independent.

Least-Squares Estimation

Let g be a function on $\mathcal{X}$. For $1 \le i \le n$, the quantity $Y_i - g(\mathbf{x}_i)$ is referred to as the residual of Y_i about g. The residual sum of squares about g is given by $\text{RSS}(g) = \sum_i [Y_i - g(\mathbf{x}_i)]^2$, and the lack-of-fit sum of squares about g is given by $\text{LSS}(g) = \sum_i n_k [\bar{Y}_k - g(\mathbf{x}_k')]^2$. These two sums of squares are related by the formula $\text{RSS}(g) = \text{LSS}(g) + \text{WSS}$.

The least-squares estimate in G of the regression function is the unique function $\hat{\mu}(\cdot)$ in G such that

$$\text{RSS}(\hat{\mu}(\cdot)) = \min_{g \in G} \text{RSS}(g);$$

it is the unique function in G such that

$$\text{LSS}(\hat{\mu}(\cdot)) = \min_{g \in G} \text{LSS}(g);$$

and it is also the unique function in G that satisfies the equation

$$\sum_i g(\mathbf{x}_i)\hat{\mu}(\mathbf{x}_i) = \sum_i g(\mathbf{x}_i)Y_i, \qquad g \in G.$$

Let $g_1, \ldots, g_p$ be a basis of G. Then $\hat{\mu}(\cdot) \in G$ satisfies the previous equation if and only if

$$\sum_i g_j(\mathbf{x}_i)\hat{\mu}(\mathbf{x}_i) = \sum_i g_j(\mathbf{x}_i)Y_i, \qquad 1 \le j \le p.$$

Now $\hat{\mu}(\cdot) = \hat{\beta}_1 g_1 + \cdots + \hat{\beta}_p g_p$, where $\hat{\beta}_1, \ldots, \hat{\beta}_p$ are uniquely determined. We refer to $\hat{\beta}_1, \ldots, \hat{\beta}_p$ as the least-squares estimates of the regression coefficients $\beta_1, \ldots, \beta_p$, respectively, and to $\hat{\beta} = [\hat{\beta}_1, \ldots, \hat{\beta}_p]^T$ as the least-squares estimate of β. The matrix $\mathbf{X}^T\mathbf{X}$ is positive definite and symmetric. The system of normal equations for $\hat{\beta}_1, \ldots, \hat{\beta}_p$ can be written in matrix form as $\mathbf{X}^T\mathbf{X}\hat{\beta} = \mathbf{X}^T\mathbf{Y}$, whose unique solution is given by $\hat{\beta} = (\mathbf{X}^T\mathbf{X})^{-1}\mathbf{X}^T\mathbf{Y}$.

Sums of Squares

Let $\hat{\mu}(\cdot)$ be the least-squares estimate in G of the regression function. The lack-of-fit sum of squares for G is given by $\text{LSS}(G) = \text{LSS}(\hat{\mu}(\cdot)) = \sum_k n_k [\bar{Y}_k - \hat{\mu}(x'_k)]^2$, and the residual sum of squares for G is given by

$$\text{RSS}(G) = \text{RSS}(\hat{\mu}(\cdot)) = \sum_i [Y_i - \hat{\mu}(x_i)]^2 = \text{LSS}(G) + \text{WSS}.$$

The identifiable p-dimensional space G is said to be saturated if $p = d$. If G is saturated, then $\text{LSS}(G) = 0$ and $\text{RSS}(G) = \text{WSS}$. If G is saturated and $d = n$ and hence $p = d = n$, then $\text{LSS}(G)$, WSS, and $\text{RSS}(G)$ all equal zero.

The least-squares estimate of the regression function in the space G_{con} of constant functions is given by $\hat{\mu}(\cdot) = \bar{Y}$; correspondingly, $\text{LSS}(G_{\text{con}}) = \text{BSS}$ and $\text{RSS}(G_{\text{con}}) = \text{TSS}$.

Suppose G contains the constant functions. Then the fitted sum of squares for G is given by $\text{FSS}(G) = \sum_i [\hat{\mu}(x_i) - \bar{Y}]^2$. Moreover, $\text{FSS}(G) + \text{LSS}(G) = \text{BSS}$ and $\text{FSS}(G) + \text{RSS}(G) = \text{TSS}$. Suppose also that $\text{TSS} > 0$. Then the squared multiple correlation coefficient of G is given by

$$R^2 = \frac{\text{FSS}(G)}{\text{TSS}} = 1 - \frac{\text{RSS}(G)}{\text{TSS}}.$$

Distribution Theory

The least-squares estimate $\hat{\beta}$ of the coefficient vector β relative to the given basis has the multivariate normal distribution with mean vector β and variance–covariance matrix $\mathbf{V} = [v_{jk}] = \sigma^2 (\mathbf{X}^T\mathbf{X})^{-1}$. For $1 \le j \le p$, $\hat{\beta}_j$ is normally distributed with mean β_j and variance v_{jj}. For $1 \le j, k \le p$, $\hat{\beta}_j$ and $\hat{\beta}_k$ have covariance $v_{jk} = \sigma^2 [(\mathbf{X}^T\mathbf{X})^{-1}]_{jk}$.

The random quantities $\hat{\beta}$, $\text{LSS}(G)$, and WSS are independent. If $p < d$, then $\text{LSS}(G)/\sigma^2$ has the chi-square distribution with $d - p$ degrees of freedom.

In the remainder of this section, it is supposed that $p < n$. Set $S^2 = \text{RSS}(G)/(n - p)$, where $S \ge 0$. Then S^2 is an unbiased estimate of σ^2, $(n - p)S^2/\sigma^2$ has the chi-square distribution with $n - p$ degrees of freedom, and $\hat{\beta}$ and S^2 are independent.

The variance–covariance matrix $\mathbf{V} = \sigma^2 (\mathbf{X}^T\mathbf{X})^{-1}$ of $\hat{\beta}$ is estimated by

$$\hat{\mathbf{V}} = S^2 (\mathbf{X}^T\mathbf{X})^{-1} = \left(\frac{S}{\sigma}\right)^2 \mathbf{V};$$

in particular, $v_{jk} = \text{cov}(\hat{\beta}_j, \hat{\beta}_k)$ is estimated by

$$\hat{v}_{jk} = S^2 [(\mathbf{X}^T\mathbf{X})^{-1}]_{jk} = \left(\frac{S}{\sigma}\right)^2 v_{jk}.$$

Given $c_1, \ldots, c_p \in \mathbb{R}$, set $\mathbf{c} = [c_1, \ldots, c_p]^T$ and consider the linear parameter

$$\tau = c_1 \beta_1 + \cdots + c_p \beta_p = \mathbf{c}^T \beta.$$

The least-squares estimate $\hat{\tau} = c_1 \hat{\beta}_1 + \cdots + c_p \hat{\beta}_p = \mathbf{c}^T \hat{\beta}$ of this parameter is unbiased, its variance is given by $\mathrm{var}(\hat{\tau}) = \mathbf{c}^T \mathbf{V} \mathbf{c}$, and $\mathbf{c}^T \hat{\mathbf{V}} \mathbf{c}$ is an unbiased estimate of this variance. The standard error

$$\mathrm{SE}(\hat{\tau}) = \sqrt{\mathbf{c}^T \hat{\mathbf{V}} \mathbf{c}} = S\sqrt{\mathbf{c}^T (\mathbf{X}^T \mathbf{X})^{-1} \mathbf{c}} = \frac{S}{\sigma} \mathrm{SD}(\hat{\tau})$$

of $\hat{\tau}$ is a natural estimate of

$$\mathrm{SD}(\hat{\tau}) = \sqrt{\mathbf{c}^T \mathbf{V} \mathbf{c}} = \sigma \sqrt{\mathbf{c}^T (\mathbf{X}^T \mathbf{X})^{-1} \mathbf{c}}.$$

The standard errors of the least-squares estimates of the coefficients of the regression function are given by

$$\mathrm{SE}(\hat{\beta}_j) = \sqrt{\hat{v}_{jj}} = S\sqrt{[(\mathbf{X}^T \mathbf{X})^{-1}]_{jj}}, \qquad 1 \leq j \leq p.$$

Confidence Bounds and Confidence Intervals

The linear parameter τ is said to be trivial if $\mathbf{c} = \mathbf{0}$ or, equivalently, if $\mathrm{var}(\hat{\tau}) = 0$; otherwise, τ is said to be nontrivial, and $(\hat{\tau} - \tau)/\mathrm{SD}(\hat{\tau})$ has the standard normal distribution.

Let τ be a nontrivial linear parameter. Then $(\hat{\tau} - \tau)/\mathrm{SE}(\hat{\tau})$ has the t distribution with $n - p$ degrees of freedom. Let $0 < \alpha < 1$. Then $\hat{\tau} - t_{1-\alpha,n-p} \mathrm{SE}(\hat{\tau})$ is the $100(1 - \alpha)\%$ lower confidence bound for τ; that is, $P(\hat{\tau} - t_{1-\alpha,n-p} \mathrm{SE}(\hat{\tau}) < \tau) = 1 - \alpha$. Also, $\hat{\tau} + t_{1-\alpha,n-p} \mathrm{SE}(\hat{\tau})$ is the $100(1 - \alpha)\%$ upper confidence bound for τ; that is, $P(\tau < \hat{\tau} + t_{1-\alpha,n-p} \mathrm{SE}(\hat{\tau})) = 1 - \alpha$. Finally,

$$\hat{\tau} \pm t_{1-\alpha/2,n-p} \mathrm{SE}(\hat{\tau}) = \left(\hat{\tau} - t_{1-\alpha/2,n-p} \mathrm{SE}(\hat{\tau}),\ \hat{\tau} + t_{1-\alpha/2,n-p} \mathrm{SE}(\hat{\tau})\right)$$

is the $100(1 - \alpha)\%$ confidence interval for τ; that is,

$$P\left(\hat{\tau} - t_{1-\alpha/2,n-p} \mathrm{SE}(\hat{\tau}) < \tau < \hat{\tau} + t_{1-\alpha/2,n-p} \mathrm{SE}(\hat{\tau})\right) = 1 - \alpha.$$

The t Test

Let τ be a nonzero linear parameter and let $\hat{\tau}$ be its least-squares estimate. Given $\tau_0 \in \mathbb{R}$, consider the statistic $t = (\hat{\tau} - \tau_0)/\mathrm{SD}(\hat{\tau})$. If $\tau = \tau_0$, then t has the t distribution with $n - p$ degrees of freedom.

Let $0 < \alpha < 1$. The inequality $t \geq t_{1-\alpha,n-p}$ determines the critical region having size α for the t test of the hypothesis that $\tau \leq \tau_0$; that is, α is the maximum value of $P(t \geq t_{1-\alpha,n-p})$ that is compatible with the hypothesis. The t statistic lies in the acceptance region (complement of the critical region) if and only if τ_0 is greater than the $100(1 - \alpha)\%$ lower confidence bound $\hat{\tau} - t_{1-\alpha,n-p} \mathrm{SE}(\hat{\tau})$ for τ. The P-value for the test equals $1 - t_{n-p}(t)$. (The t statistic lies in the critical region for the t test of size α of the hypothesis if α is greater than or equal to the P-value, but not if α is less than the P-value.)

The inequality $t \leq -t_{1-\alpha,n-p}$ determines the critical region having size α for the t test of the hypothesis that $\tau \geq \tau_0$; that is, α is the maximum value of $P(t \leq -t_{1-\alpha,n-p})$ that is compatible with the hypothesis. The t statistic lies in the acceptance region if and only if τ_0 is less than the $100(1 - \alpha)\%$ upper confidence bound $\hat{\tau} + t_{1-\alpha,n-p} \, \mathrm{SE}(\hat{\tau})$ for τ. The P-value for the test equals $t_{n-p}(t)$.

The inequality $|t| \geq t_{1-\alpha/2,n-p}$ determines the critical region having size α for the t test of the hypothesis that $\tau = \tau_0$; that is, if the hypothesis holds, then $P(|t| \geq t_{1-\alpha/2,n-p}) = \alpha$. The t statistic lies in the acceptance region if and only if τ_0 lies in the $100(1 - \alpha)\%$ confidence interval $\hat{\tau} \pm t_{1-\alpha/2,n-p} \, \mathrm{SE}(\hat{\tau})$ for τ. The P-value for the test equals $2[\, 1 - t_{n-p}(|t|)\,]$.

The F Test

Let G_0 be a subspace of G having dimension $p_0 < p$, let $\hat{\mu}_0(\cdot)$ be the least-squares estimate in G_0 of the regression function, and let $\mathrm{LSS}(G_0)$ and $\mathrm{RSS}(G_0) = \mathrm{LSS}(G_0) + \mathrm{WSS}$ denote the corresponding lack-of-fit and residual sums of squares. Then

$$\mathrm{RSS}(G_0) - \mathrm{RSS}(G) = \mathrm{LSS}(G_0) - \mathrm{LSS}(G) = \sum_k n_k [\, \hat{\mu}(\mathbf{x}_k') - \hat{\mu}_0(\mathbf{x}_k')\,]^2.$$

The F statistic for testing the hypothesis that $\mu(\cdot) \in G_0$ is given by

$$F = \frac{[\, \mathrm{RSS}(G_0) - \mathrm{RSS}(G)\,]/(p - p_0)}{\mathrm{RSS}(G)/(n - p)}.$$

If the hypothesis is valid, then this statistic has the F distribution with $p - p_0$ degrees of freedom in the numerator and $n - p$ of degrees of freedom in the denominator. The inequality $F \geq F_{1-\alpha,p-p_0,n-p}$ determines the critical region having size α for the test; that is, if the hypothesis holds, then $P(F \geq F_{1-\alpha,p-p_0,n-p}) = \alpha$. The P-value for the test equals $1 - F_{p-p_0,n-p}(F)$.

Suppose $p_0 = p - 1$ and let $g_1, \ldots, g_p$ be a basis of G such that $g_1, \ldots, g_{p-1}$ is a basis of G_0. Then $\mu(\cdot) = \sum_j \beta_j g_j$, so $\mu(\cdot) \in G_0$ if and only if $\beta_p = 0$. The t statistic for testing the hypothesis that $\beta_p = 0$ is given by $t = \hat{\beta}_p / \mathrm{SE}(\hat{\beta}_p)$, and the F statistic for testing the equivalent hypothesis that $\mu(\cdot) \in G_0$ is given by

$$F = \frac{\mathrm{RSS}(G_0) - \mathrm{RSS}(G)}{\mathrm{RSS}(G)/(n - p)}.$$

These two statistics are related by the formula $t^2 = F$. Moreover, the P-value for the F test of the hypothesis that the regression function is in G_0 coincides with the P-value for the t test of the equivalent hypothesis that $\beta_p = 0$.

Saturated Models

Suppose G is saturated; that is, for every choice of $c_1, \ldots, c_d \in \mathbb{R}$, there is a $g \in G$ such that $g(\mathbf{x}_k') = c_k$ for $1 \leq k \leq d$. Since G is assumed to be identifiable, there is a unique such g; moreover, G has dimension $p = d$. Also, $\hat{\mu}(\mathbf{x}_k') = \bar{Y}_k$ for $1 \leq k \leq d$, and the lack-of-fit and residual sums of squares for G are given by $\mathrm{LSS}(G) = 0$ and $\mathrm{RSS}(G) = \mathrm{WSS}$. If G_0 is a subspace of G, then $\mathrm{RSS}(G_0) - \mathrm{RSS}(G) = \mathrm{LSS}(G_0)$.

Since G is identifiable and saturated, the $d \times d$ form

$$
\mathbf{X} = \begin{bmatrix} g_1(\mathbf{x}_1') & \cdots & g_d(\mathbf{x}_1') \\ \vdots & & \vdots \\ g_1(\mathbf{x}_d') & \cdots & g_d(\mathbf{x}_d') \end{bmatrix}
$$

of the design matrix is invertible. Set $\mathbf{N} = \mathrm{diag}(n_1, \ldots, n_d)$ and $\bar{\mathbf{Y}} = [\bar{Y}_1, \ldots, \bar{Y}_d]^T$. Then the normal equation for the least-squares estimate of β can be written as $\mathbf{X}\hat{\beta} = \bar{\mathbf{Y}}$, whose unique solution is given by $\hat{\beta} = \mathbf{X}^{-1}\bar{\mathbf{Y}}$. The variance–covariance matrix of this estimate is given by $\mathrm{VC}(\hat{\beta}) = \sigma^2 \mathbf{X}^{-1}\mathbf{N}^{-1}(\mathbf{X}^{-1})^T$.

Goodness-of-Fit Test

Suppose $p < d < n$ and the assumptions of the normal regression model are satisfied, but consider the possibility that $\mu(\cdot) \notin G$. The F statistic for testing the assumption that $\mu(\cdot) \in G$ is given by

$$
F = \frac{\mathrm{LSS}(G)/(d-p)}{\mathrm{WSS}/(n-d)}.
$$

C.3
Binomial and Poisson Models
Binomial Models

Let $Y_1, \ldots, Y_d$ be independent random variables such that, for $1 \le k \le d$, Y_k has the binomial distribution with known time parameters n_k and unknown risk π_k. We refer to this setup as the binomial d-sample model. Set $n = n_1 + \cdots + n_d$ and $Y = Y_1 + \cdots + Y_d$.

Let $1 \le k \le d$. Then Y_k has mean $n_k \pi_k$ and variance $n_k \pi_k (1 - \pi_k)$, and the random variable $\bar{Y}_k = Y_k/n_k$ has mean π_k and variance $\pi_k(1 - \pi_k)/n_k$. Thus, $\hat{\pi}_k = \bar{Y}_k$ is an unbiased estimate of π_k. Moreover,

$$
\mathrm{var}(\hat{\pi}_k) = \pi_k(1 - \pi_k)/n_k,
$$

$$
\mathrm{SD}(\hat{\pi}_k) = \sqrt{\pi_k(1 - \pi_k)/n_k},
$$

and

$$
\mathrm{SE}(\hat{\pi}_k) = \sqrt{\hat{\pi}_k(1 - \hat{\pi}_k)/n_k}.
$$

Suppose $d \ge 2$ and let $1 \le j, k \le d$ with $j \ne k$. Then

$$
E(\hat{\pi}_k - \hat{\pi}_j) = \pi_k - \pi_j,
$$

$$\mathrm{var}(\hat{\pi}_k - \hat{\pi}_j) = \frac{\pi_j(1 - \pi_j)}{n_j} + \frac{\pi_k(1 - \pi_k)}{n_k},$$

$$\mathrm{SD}(\hat{\pi}_k - \hat{\pi}_j) = \sqrt{\frac{\pi_j(1 - \pi_j)}{n_j} + \frac{\pi_k(1 - \pi_k)}{n_k}},$$

and

$$\mathrm{SE}(\hat{\pi}_k - \hat{\pi}_j) = \sqrt{\frac{\hat{\pi}_j(1 - \hat{\pi}_j)}{n_j} + \frac{\hat{\pi}_k(1 - \hat{\pi}_k)}{n_k}}.$$

The hypothesis of homogeneity is that $\pi_1 = \cdots = \pi_d$. The likelihood-ratio statistic for testing this hypothesis can be written as

$$D = 2 \sum_k \left(Y_k \log \frac{Y_k/n_k}{Y/n} + (n_k - Y_k) \log \frac{(n_k - Y_k)/n_k}{(n - Y)/n} \right).$$

If the hypothesis is valid, then D has approximately the chi-square distribution with $d - 1$ degrees of freedom.

Poisson Models

Let $Y_1, \ldots, Y_d$ be independent random variables such that, for $1 \le k \le d$, Y_k has the Poisson distribution with known time parameters n_k and unknown rate λ_k. We refer to this setup as the Poisson d-sample model. Set $n = n_1 + \cdots + n_d$ and $Y = Y_1 + \cdots + Y_d$.

Let $1 \le k \le d$. Then Y_k has mean $n_k \lambda_k$ and variance $n_k \lambda_k$, and the random variable $\bar{Y}_k = Y_k/n_k$ has mean λ_k and variance λ_k/n_k. Thus, $\hat{\lambda}_k = \bar{Y}_k$ is an unbiased estimate of λ_k. Moreover,

$$\mathrm{var}(\hat{\lambda}_k) = \lambda_k n_k, \quad \mathrm{SD}(\hat{\lambda}_k) = \sqrt{\lambda_k/n_k} \quad \text{and} \quad \mathrm{SE}(\hat{\lambda}_k) = \sqrt{\hat{\lambda}_k/n_k}.$$

Suppose $d \ge 2$ and let $1 \le j, k \le d$ with $j \ne k$. Then

$$E(\hat{\lambda}_k - \hat{\lambda}_j) = \lambda_k - \lambda_j,$$

$$\mathrm{var}(\hat{\lambda}_k - \hat{\lambda}_j) = \frac{\lambda_j}{n_j} + \frac{\lambda_k}{n_k},$$

$$\mathrm{SD}(\hat{\lambda}_k - \hat{\lambda}_j) = \sqrt{\frac{\lambda_j}{n_j} + \frac{\lambda_k}{n_k}},$$

and

$$\mathrm{SE}(\hat{\lambda}_k - \hat{\lambda}_j) = \sqrt{\frac{\hat{\lambda}_j}{n_j} + \frac{\hat{\lambda}_k}{n_k}}.$$

The hypothesis of homogeneity is that $\lambda_1 = \cdots = \lambda_d$. The likelihood-ratio statistic for testing this hypothesis can be written as

$$D = 2 \sum_k Y_k \log \frac{Y_k/n_k}{Y/n}.$$

If the hypothesis is valid, then D has approximately the chi-square distribution with $d - 1$ degrees of freedom.

C.4
Logistic Regression and Poisson Regression

Exponential Families

Consider the binomial distribution with known parameter $n \in \{1, 2, \ldots\}$ and unknown parameter $\pi \in (0, 1)$. The probability function of this distribution can be written in the exponential family form

$$f(y; n, \theta) = e^{\theta y - nC(\theta)} r(y; n), \qquad y \in \mathcal{Y}_n,$$

where $\mathcal{Y}_n = \{0, \ldots, n\}$, $r(y; n) = \binom{n}{y}$ for $y \in \mathcal{Y}_n$, $\theta = \mathrm{logit}(\pi) = \log \frac{\pi}{1-\pi} \in \mathbb{R}$, and $C(\theta) = \log(1 + e^\theta)$ for $\theta \in \mathbb{R}$.

Consider next the Poisson distribution with parameter $n\lambda$, where $n \in (0, \infty)$ is known and $\lambda \in (0, \infty)$ is unknown. The probability function of this distribution can be written in the above exponential family form with $\mathcal{Y}_n = \{0, 1, 2, \ldots\}$, $r(y; n) = n^y/y!$ for $y \in \mathcal{Y}_n$, $\theta = \log \lambda \in \mathbb{R}$, and $C(\theta) = e^\theta$ for $\theta \in \mathbb{R}$.

In either case, $\#(\mathcal{Y}_n) \geq 2$. We refer to the known parameter n as the time parameter of the distribution and to θ as the canonical parameter. The normalizing function $C(\cdot)$ can be expressed as

$$C(\theta) = \log \frac{1}{n} \sum_{y \in \mathcal{Y}_n} e^{\theta y} r(y; n), \qquad \theta \in \mathbb{R}.$$

In particular,

$$C(\theta) = \log \sum_{y \in \mathcal{Y}_1} e^{\theta y} r(y; 1), \qquad \theta \in \mathbb{R}.$$

The distribution has mean $n\mu$ and variance $n\sigma^2$, where $\mu = C'(\theta)$ and $\sigma^2 = C''(\theta)$. The function $C''(\cdot)$ is strictly positive, $C'(\cdot)$ is strictly increasing, and $C(\cdot)$ is strictly convex. The function $(C')^{-1}(\cdot)$, referred to as the link function, is used to express θ in terms of μ: $\theta = (C')^{-1}(\mu)$.

The Input-Response Model

According to the input-response version of the canonical regression model, there is a function $\theta(\cdot)$ on $\mathcal{X}$ such that, when the input vector equals $\mathbf{x} \in \mathcal{X}$ and the time parameter equals n, the response variable Y has probability function $f(\cdot; n, \theta(\mathbf{x}))$ on $\mathcal{Y}_n$. The function $\theta(\cdot)$ is called the canonical regression function. The random variable Y has mean $n\mu(\mathbf{x})$ and variance $n\sigma^2(\mathbf{x})$, where $\mu(\mathbf{x}) = C'(\theta(\mathbf{x}))$ and

$\sigma^2(\mathbf{x}) = C''(\theta(\mathbf{x}))$. We refer to $\mu(\cdot)$ as the regression function and to $\sigma^2(\cdot)$ as the variance function.

Let G be a p-dimensional linear space of functions on $\mathcal{X}$. The input-response version of the corresponding linear canonical regression model consists of the input-response version of the canonical regression model together with the assumption that the canonical regression function is in G. Let $g_1, \ldots, g_p$ be a basis of G. Under the indicated assumption, we can write the canonical regression function as

$$\theta(\mathbf{x}) = \beta_1 g_1(\mathbf{x}) + \cdots + \beta_p g_p(\mathbf{x}), \qquad \mathbf{x} \in \mathcal{X}.$$

The uniquely determined regression coefficients $\beta_1, \ldots, \beta_p$ form the coefficient vector $\boldsymbol{\beta} = [\beta_1, \ldots, \beta_p]^T$.

The Experimental Model

According to the experimental version of the canonical regression model, which is in effect throughout the remainder of this section, there are d distinct input vectors $\mathbf{x}_1, \ldots, \mathbf{x}_d$ and corresponding time parameters $n_1, \ldots, n_d$ and response variables $Y_1, \ldots, Y_d$. The response variables are independent, and, for $1 \leq k \leq d$, Y_k has probability function $f(\cdot; n_k, \theta(\mathbf{x}_k))$ on $\mathcal{Y}_{n_k}$. We refer to $\{\mathbf{x}_1, \ldots, \mathbf{x}_d\}$ as the design set.

Let $1 \leq k \leq d$ and set $\theta_k = \theta(\mathbf{x}_k)$, $\mu_k = \mu(\mathbf{x}_k) = C'(\theta(\mathbf{x}_k)) = C'(\theta_k)$, and $\sigma_k^2 = \sigma^2(\mathbf{x}_k) = C''(\theta(\mathbf{x}_k)) = C''(\theta_k)$. Then $\theta_k = (C')^{-1}(\mu_k)$. The random variable Y_k has probability function $f(\cdot; n_k, \theta_k)$, mean $n_k \mu_k$, and variance $n_k \sigma_k^2$, and the random variable $\bar{Y}_k = Y_k / n_k$ has mean μ_k and variance σ_k^2 / n_k. The random variables $\bar{Y}_1, \ldots, \bar{Y}_d$ are independent, and the random variable

$$\bar{Y} = \frac{1}{n} \sum_k Y_k = \frac{1}{n} \sum_k n_k \bar{Y}_k$$

has mean

$$\frac{1}{n} \sum_k n_k \mu_k$$

and variance

$$\frac{1}{n^2} \sum_k n_k \sigma_k^2;$$

here $n = \sum_k n_k$.

Set

$$\mathcal{Y} = \mathcal{Y}_{n_1} \times \cdots \times \mathcal{Y}_{n_d}$$

and $r(\mathbf{y}; \mathbf{n}) = \prod_k r(y_k; n_k) > 0$ for $\mathbf{y} = [y_1, \ldots y_d]^T \in \mathcal{Y}$, where $\mathbf{n} = [n_1, \ldots, n_d]^T$. Think of $\bar{y}_k = y_k / n_k$, $1 \leq k \leq d$, and

$$\bar{y} = \frac{1}{n} \sum_k y_k = \frac{1}{n} \sum_k n_k \bar{y}_k$$

as being defined in terms of $\mathbf{y} \in \mathcal{Y}$. Then the probability function of the $\mathcal{Y}$-valued random vector $\mathbf{Y} = [Y_1, \ldots, Y_d]^T$ can be written as

$$f(\mathbf{y}; \mathbf{n}, \theta(\cdot)) = \exp\left(\sum_k n_k[\theta(\mathbf{x}_k)\bar{y}_k - C(\theta(\mathbf{x}_k))]\right) r(\mathbf{y}; \mathbf{n}), \qquad \mathbf{y} \in \mathcal{Y}.$$

Observe that $\#(\mathcal{Y}) \geq 2$ and $0 < f(\mathbf{y}; \mathbf{n}, \theta(\cdot)) < 1$ for $\mathbf{y} \in \mathcal{Y}$.

The assumptions of the linear canonical regression model are in effect throughout the remainder of this section, except in the context of the goodness-of-fit test of the assumption that the canonical regression function is in G.

Let $g_1, \ldots, g_p$ be a basis of G. The corresponding $d \times p$ design matrix is given by

$$\mathbf{X} = \begin{bmatrix} g_1(\mathbf{x}_1) & \cdots & g_p(\mathbf{x}_1) \\ \vdots & & \vdots \\ g_1(\mathbf{x}_d) & \cdots & g_p(\mathbf{x}_d) \end{bmatrix}.$$

The canonical regression function $\theta(\cdot)$ and the regression function $\mu(\cdot)$ can be written in terms of the basis $g_1, \ldots, g_p$ as $\theta(\cdot) = \beta_1 g_1 + \cdots + \beta_p g_p$ and

$$\mu(\mathbf{x}) = C'(\beta_1 g_1(\mathbf{x}) + \cdots + \beta_p g_p(\mathbf{x})), \qquad \mathbf{x} \in \mathcal{X}.$$

Throughout the rest of this section, it is assumed that G is identifiable and hence that $p \leq d$.

Maximum-Likelihood Estimation

The quantity $f(\mathbf{Y}; \mathbf{n}, \theta(\cdot))$, viewed as a function of $\theta(\cdot) \in G$, is called the likelihood function and denoted by $L(\theta(\cdot))$, $\theta(\cdot) \in G$; thus,

$$L(\theta(\cdot)) = \exp\left(\sum_k n_k[\theta(\mathbf{x}_k)\bar{Y}_k - C(\theta(\mathbf{x}_k))]\right) r(\mathbf{Y}; \mathbf{n}), \qquad \theta(\cdot) \in G.$$

The function $l(\theta(\cdot)) = \log L(\theta(\cdot))$, $\theta(\cdot) \in G$, is called the log-likelihood function. Thus,

$$l(\theta(\cdot)) = \sum_k n_k[\theta(\mathbf{x}_k)\bar{Y}_k - C(\theta(\mathbf{x}_k))] + \log r(\mathbf{Y}; \mathbf{n}), \qquad \theta(\cdot) \in G.$$

Observe that $0 < L(\theta(\cdot)) < 1$ and $-\infty < l(\theta(\cdot)) < 0$ for $\theta(\cdot) \in G$.

A maximum-likelihood estimate in G of the canonical regression function is a function $\hat{\theta}(\cdot) \in G$ that maximizes $l(\theta(\cdot))$, $\theta(\cdot) \in G$. There is at most one such function. If there is such a function, the maximum-likelihood estimate in G of the canonical regression function is said to exist; otherwise, it is said not to exist or to be undefined.

Suppose $\hat{\theta}(\cdot)$ exists. Then the maximum-likelihood estimates of $\mu(\cdot)$ and $\sigma^2(\cdot)$ are given by $\hat{\mu}(\cdot) = C'(\hat{\theta}(\cdot))$ and $\hat{\sigma}^2(\cdot) = C''(\hat{\theta}(\cdot))$; also, the maximum-likelihood estimate in G_0 of the canonical regression function exists for every subspace G_0 of G.

The Maximum-Likelihood Equation

If there is a maximum-likelihood estimate in G of the canonical regression function, then it is the unique function $\hat{\theta}(\cdot)$ in G that satisfies the maximum-likelihood equation

$$\sum_k n_k g(\mathbf{x}_k) C'(\hat{\theta}(\mathbf{x}_k)) = \sum_k n_k g(\mathbf{x}_k) \bar{Y}_k, \qquad g \in G;$$

otherwise, there is no function in G that satisfies the maximum-likelihood equation. Let $g_1, \ldots, g_p$ be a basis of G. Then $\hat{\theta}(\cdot) \in G$ satisfies the maximum-likelihood equation if and only if

$$\sum_k n_k g_j(\mathbf{x}_k) C'(\hat{\theta}(\mathbf{x}_k)) = \sum_k n_k g_j(\mathbf{x}_k) \bar{Y}_k, \qquad 1 \le j \le p.$$

Let $\hat{\beta}_1, \ldots, \hat{\beta}_p$ be the coefficients of $\hat{\theta}(\cdot)$ relative to the indicated basis, so that $\hat{\theta}(\cdot) = \hat{\beta}_1 g_1 + \cdots + \hat{\beta}_p g_p$, and set $\hat{\boldsymbol{\beta}} = [\hat{\beta}_1, \ldots, \hat{\beta}_p]^T$. If $\hat{\theta}(\cdot)$ is the maximum-likelihood estimate in G of the canonical regression function, then $\hat{\beta}_1, \ldots, \hat{\beta}_p$ are referred to as the maximum-likelihood estimates of $\beta_1, \ldots, \beta_p$ and $\hat{\boldsymbol{\beta}}$ is referred to as the maximum-likelihood estimate of $\boldsymbol{\beta}$. The maximum-likelihood equation for $\hat{\boldsymbol{\beta}}$ can be written in matrix form as $\mathbf{X}^T \mathbf{N} C'(\mathbf{X}\hat{\boldsymbol{\beta}}) = \mathbf{X}^T \mathbf{N} \bar{\mathbf{Y}}$, where $\mathbf{N} = \mathrm{diag}(n_1, \ldots, n_d)$, $\bar{\mathbf{Y}} = [\bar{Y}_1, \ldots, \bar{Y}_d]^T$ and $C'(\mathbf{X}\hat{\boldsymbol{\beta}})$ is the p-dimensional column vector obtained by applying the function $C'(\cdot)$ to each of the entries of $\mathbf{X}\hat{\boldsymbol{\beta}}$.

Distribution of the Maximum-Likelihood Estimate

Set $\mathbf{W} = \mathrm{diag}(n_1 \sigma^2(\mathbf{x}_1), \ldots, n_d \sigma^2(\mathbf{x}_d))$. The $p \times p$ matrix $\mathbf{X}^T \mathbf{W} \mathbf{X}$, referred to as the information matrix, is symmetric and positive definite, as is $\mathbf{V} = [v_{jk}] = (\mathbf{X}^T \mathbf{W} \mathbf{X})^{-1}$.

In the rest of this section, it is supposed that certain suitable conditions are satisfied (including the requirement that $n \gg 1$ in the context of logistic regression). Then $P(\hat{\boldsymbol{\beta}} \text{ exists}) = P(\hat{\theta}(\cdot) \text{ exists}) \approx 1$. Moreover, $\hat{\boldsymbol{\beta}} - \boldsymbol{\beta}$ has approximately the multivariate normal distribution with mean vector $\mathbf{0}$ and variance–covariance matrix $\mathbf{V}$. In light of this result, we refer to $\mathbf{V}$ as the asymptotic variance–covariance matrix of $\hat{\boldsymbol{\beta}}$.

Consider a parameter $\tau = h(\boldsymbol{\beta})$, where h is a (possibly nonlinear) continuously differentiable function. The maximum-likelihood estimate of τ is given by $\hat{\tau} = h(\hat{\boldsymbol{\beta}})$. Let

$$\nabla h(\boldsymbol{\beta}) = \left[\frac{\partial h}{\partial \beta_1}(\boldsymbol{\beta}), \ldots, \frac{\partial h}{\partial \beta_p}(\boldsymbol{\beta}) \right]^T$$

denote the gradient of h at $\boldsymbol{\beta}$. The asymptotic variance and asymptotic standard deviation of $\hat{\tau}$ are given by $\mathrm{AV}(\hat{\tau}) = [\nabla h(\boldsymbol{\beta})]^T \mathbf{V} [\nabla h(\boldsymbol{\beta})]$ and $\mathrm{ASD}(\hat{\tau}) = \sqrt{\mathrm{AV}(\hat{\tau})}$. In particular, if $\tau = c_1 \beta_1 + \cdots + c_p \beta_p = \mathbf{c}^T \boldsymbol{\beta}$ with $\mathbf{c} = [c_1, \ldots, c_p]^T$, then $\mathrm{AV}(\hat{\tau}) = \mathbf{c}^T \mathbf{V} \mathbf{c}$. As a special case, $\mathrm{AV}(\hat{\beta}_j) = v_{jj}$ and $\mathrm{ASD}(\hat{\beta}_j) = \sqrt{v_{jj}}$ for $1 \le j \le p$. If $\nabla h(\boldsymbol{\beta}) \ne 0$, then $(\hat{\tau} - \tau)/\mathrm{ASD}(\hat{\tau})$ has approximately the standard normal distribution.

Confidence Bounds and Confidence Intervals

Set $\hat{\mathbf{W}} = \mathrm{diag}(n_1 \hat{\sigma}^2(\mathbf{x}_1), \ldots, n_d \hat{\sigma}^2(\mathbf{x}_d))$ and $\hat{\mathbf{V}} = [\hat{v}_{jk}] = (\mathbf{X}^T \hat{\mathbf{W}} \mathbf{X})^{-1}$, where $\hat{\sigma}^2(\mathbf{x}_k) = C''(\hat{\theta}(\mathbf{x}_k))$ is the maximum-likelihood estimate of $\sigma^2(\mathbf{x}_k) = C''(\theta(\mathbf{x}_k))$ for $1 \leq k \leq d$. The standard error of the maximum-likelihood estimate $\hat{\tau} = h(\hat{\beta})$ of $\tau = h(\beta)$ is given by $\mathrm{SE}(\hat{\tau}) = \sqrt{[\nabla h(\hat{\beta})]^T \hat{\mathbf{V}} [\nabla h(\hat{\beta})]}$, where

$$\nabla h(\hat{\beta}) = \left[\frac{\partial h}{\partial \beta_1}(\hat{\beta}), \ldots, \frac{\partial h}{\partial \beta_d}(\hat{\beta}) \right]^T.$$

In particular, the standard errors of the maximum-likelihood estimate $\hat{\tau} = \mathbf{c}^T \hat{\beta}$ of $\tau = \mathbf{c}^T \beta$ is given by $\mathrm{SE}(\hat{\tau}) = \sqrt{\mathbf{c}^T \hat{\mathbf{V}} \mathbf{c}}$. As a special case, the standard errors of $\hat{\beta}_1, \ldots, \hat{\beta}_p$ are given by $\mathrm{SE}(\hat{\beta}_j) = \sqrt{\hat{v}_{jj}}$ for $1 \leq j \leq p$. The standard error of $\hat{\tau}$ can be obtained from an expression for $\mathrm{ASD}(\hat{\tau})$ in terms of various parameters by replacing these parameters by their maximum-likelihood estimates.

Suppose $\nabla h(\hat{\beta}) \neq 0$. Then $(\hat{\tau} - \tau)/\mathrm{SE}(\hat{\tau})$ has approximately the standard normal distribution. Let $0 < \alpha < 1$. Then $\hat{\tau} - z_{1-\alpha} \mathrm{SE}(\hat{\tau})$ is the nominal $100(1 - \alpha)\%$ lower confidence bound for τ; that is, $P(\hat{\tau} - z_{1-\alpha} \mathrm{SE}(\hat{\tau}) < \tau) \approx 1 - \alpha$. Also, $\hat{\tau} + z_{1-\alpha} \mathrm{SE}(\hat{\tau})$ is the nominal $100(1 - \alpha)\%$ upper confidence bound for τ; that is,

$$P\left(\tau < \hat{\tau} + z_{1-\alpha} \mathrm{SE}(\hat{\tau}) \right) \approx 1 - \alpha.$$

Finally, $\hat{\tau} \pm z_{1-\alpha/2} \mathrm{SE}(\hat{\tau})$ is the nominal $100(1 - \alpha)\%$ confidence interval for τ; that is,

$$P\left(\hat{\tau} - z_{1-\alpha/2} \mathrm{SE}(\hat{\tau}) < \tau < \hat{\tau} + z_{1-\alpha/2} \mathrm{SE}(\hat{\tau}) \right) \approx 1 - \alpha.$$

Tests Involving a Single Parameter

Let $\tau_0 \in \mathbb{R}$ and set $W = (\hat{\tau} - \tau_0)/\mathrm{SE}(\hat{\tau})$, which is referred to as the Wald statistic corresponding to the possible value τ_0 of τ. If $\tau = \tau_0$, then W has approximately the standard normal distribution.

Let $0 < \alpha < 1$. The inequality $W \geq z_{1-\alpha}$ determines the critical region having nominal size α for the Wald test of the hypothesis that $\tau \leq \tau_0$: (roughly speaking), α is approximately equal to the maximum value of $P(W \geq z_{1-\alpha})$ that is compatible with the hypothesis. The Wald statistic lies in the acceptance region if and only if τ_0 is greater than the nominal $100(1 - \alpha)\%$ lower confidence bound $\hat{\tau} - z_{1-\alpha} \mathrm{SE}(\hat{\tau})$ for τ. The (nominal) P-value for the test equals $1 - \Phi(W)$.

The inequality $W \leq -z_{1-\alpha}$ determines the critical region having approximate size α for the Wald test of the hypothesis that $\tau \geq \tau_0$. The Wald statistic lies in the acceptance region if and only if τ_0 is less than the nominal $100(1 - \alpha)\%$ upper confidence bound $\hat{\tau} + z_{1-\alpha} \mathrm{SE}(\hat{\tau})$ for τ. The P-value for the test equals $\Phi(W)$.

The inequality $|W| \geq z_{1-\alpha/2}$ determines the critical region having approximate size α for the Wald test of the hypothesis that $\tau = \tau_0$: if the hypothesis holds, then $P(|W| \geq z_{1-\alpha/2}) \approx \alpha$. The Wald statistic lies in the acceptance region if and only if

τ_0 lies in the approximate $100(1 - \alpha)\%$ confidence interval $\hat{\tau} \pm z_{1-\alpha/2} \, \mathrm{SE}(\hat{\tau})$ for τ. The P-value for the test equals $2[\, 1 - \Phi(|W|) \,]$.

The Likelihood-Ratio Test

Since the likelihood function $L(\theta(\cdot))$, $\theta(\cdot) \in G$, is less than 1, it has a supremum (least upper bound)

$$L(G) = \sup_{\theta(\cdot) \in G} L(\theta(\cdot)) \le 1.$$

Similarly, since the log-likelihood function $l(\theta(\cdot))$, $\theta(\cdot) \in G$, is less than 0, it has a supremum

$$l(G) = \log L(G) = \sup_{\theta(\cdot) \in G} l(\theta(\cdot)) \le 0.$$

If the maximum-likelihood estimate $\hat{\theta}(\cdot)$ in G of $\theta(\cdot)$ exists, then

$$L(G) = \max_{\theta(\cdot) \in G} L(\theta(\cdot)) \quad \text{and} \quad l(G) = \max_{\theta(\cdot) \in G} l(\theta(\cdot)).$$

Thus,

$$l(G) = \sum_k n_k [\, \hat{\theta}(\mathbf{x}_k) \bar{Y}_k - C(\hat{\theta}(\mathbf{x}_k)) \,] + \log r(\mathbf{Y}; \mathbf{n}).$$

Let G_0 be a subspace of G having dimension p_0, where $0 \le p_0 < p$, and let $L(G_0)$ and $l(G_0)$ be defined as above. The likelihood-ratio statistic corresponding to the likelihood ratio $L(G)/L(G_0)$ is given by

$$D = 2[\, l(G) - l(G_0) \,] = 2 \log \frac{L(G)}{L(G_0)}.$$

If $\hat{\theta}(\cdot)$ exists, then $\hat{\theta}_0(\cdot)$ exists, and

$$D = 2 \sum_k n_k \left\{ [\, \hat{\theta}(\mathbf{x}_k) - \hat{\theta}_0(\mathbf{x}_k) \,] \bar{Y}_k - [\, C(\hat{\theta}(\mathbf{x}_k)) - C(\hat{\theta}_0(\mathbf{x}_k)) \,] \right\}.$$

The inequality $D \ge \chi^2_{1-\alpha, p-p_0}$ determines the critical region having nominal size α for the likelihood-ratio test of the hypothesis that the canonical regression function is in G_0: If the hypothesis is valid, then D has approximately the chi-square distribution with $p - p_0$ degrees of freedom, and hence $P(D \ge \chi^2_{1-\alpha, p-p_0}) \approx \alpha$. The P-value for the test equals $1 - \chi^2_{p-p_0}(D)$.

Suppose $p_0 = p - 1$ and let $g_1, \ldots, g_p$ be a basis of G such that $g_1, \ldots, g_{p-1}$ is a basis of G_0. Since the canonical regression function can be written as $\theta(\cdot) = \sum_j \beta_j g_j$, we see that this function is in G_0 if and only if $\beta_p = 0$. Consider the Wald statistic $W = \hat{\beta}_p / \mathrm{SE}(\hat{\beta}_p)$ and consider the hypothesis that the canonical regression function is in G_0. Under this hypothesis, $P(D \approx W) \approx 1$; thus, the P-value for the likelihood-ratio test of the hypothesis is approximately equal to the P-value for the Wald test of the equivalent hypothesis that $\beta_p = 0$.

Saturated Models

Suppose G is saturated. Suppose also that $0 < \bar{Y}_k < 1$ for $1 \le k \le d$ in the context of a binomial model and that $\bar{Y}_k > 0$ for $1 \le k \le d$ in the context of a Poisson model. Then, for $1 \le k \le d$, $\hat{\mu}_k = \bar{Y}_k$ is the maximum-likelihood estimate of μ_k, $\hat{\theta}_k = (C')^{-1}(\bar{Y}_k)$ is the maximum-likelihood estimate of θ_k, $\hat{\sigma}_k^2 = C''(\hat{\theta}_k)$ is the maximum-likelihood estimate of σ_k^2, and $\mathrm{SE}(\hat{\mu}_k) = \hat{\sigma}_k/\sqrt{n_k}$. Moreover,

$$E(\hat{\mu}_k) = \mu_k \quad \text{and} \quad \mathrm{var}(\hat{\mu}_k) = \sigma_k^2/n_k.$$

(We consider $\hat{\mu}_k$ as being defined as $\bar{Y}_k$ even if $\bar{Y}_k$ equals 0 or 1 in the context of a binomial model or $\bar{Y}_k$ equals 0 in the context of a Poisson model.) Furthermore,

$$\frac{\hat{\mu}_k - \mu_k}{\sigma_k/\sqrt{n_k}} \quad \text{and} \quad \frac{\hat{\mu}_k - \mu_k}{\hat{\sigma}_k/\sqrt{n_k}}$$

have approximately the standard normal distribution.

Suppose $d \ge 2$ and let $1 \le j, k \le d$ with $j \ne k$. The maximum-likelihood estimate of $\mu_k - \mu_j$ is $\hat{\mu}_k - \hat{\mu}_j$. Also,

$$E(\hat{\mu}_k - \hat{\mu}_j) = \mu_k - \mu_j,$$

$$\mathrm{var}(\hat{\mu}_k - \hat{\mu}_j) = \frac{\sigma_j^2}{n_j} + \frac{\sigma_k^2}{n_k},$$

and

$$\mathrm{SE}(\hat{\mu}_k - \hat{\mu}_j) = \sqrt{\frac{\hat{\sigma}_j^2}{n_j} + \frac{\hat{\sigma}_k^2}{n_k}}.$$

Moreover,

$$\frac{\hat{\mu}_k - \hat{\mu}_j - (\mu_k - \mu_j)}{\mathrm{SD}(\hat{\mu}_k - \hat{\mu}_j)} \quad \text{and} \quad \frac{\hat{\mu}_k - \hat{\mu}_j - (\mu_k - \mu_j)}{\mathrm{SE}(\hat{\mu}_k - \hat{\mu}_j)}$$

have approximately the standard normal distribution.

Since G is identifiable and saturated, the design matrix is invertible. The system of maximum-likelihood estimate equations simplifies to $C'(\mathbf{X}\hat{\beta}) = \bar{\mathbf{Y}}$, whose unique solution is given by $\hat{\beta} = \mathbf{X}^{-1}[(C')^{-1}(\bar{\mathbf{Y}})]$. Moreover, the asymptotic variance–covariance matrix of $\hat{\beta}$ is given by $\mathbf{V} = \mathbf{X}^{-1}\mathbf{W}^{-1}(\mathbf{X}^{-1})^T$.

The hypothesis of homogeneity is that $\theta_1 = \cdots = \theta_d = \theta$ for some $\theta \in \mathbb{R}$ or, equivalently, that $\mu_1 = \cdots = \mu_d = \mu$ for some $\mu \in \mathbb{R}$. Under this hypothesis, $\mu = C'(\theta)$ and $\sigma_1^2 = \cdots = \sigma_d^2 = \sigma^2$, where $\sigma^2 = C''(\theta)$. The maximum-likelihood estimates of θ, μ, and σ^2 under this hypothesis are given by $\hat{\theta} = (C')^{-1}(\bar{Y})$, $\hat{\mu} = C'(\hat{\theta}) = \bar{Y}$, and $\hat{\sigma}^2 = C''(\hat{\theta})$; and the likelihood-ratio statistic for testing the hypothesis of homogeneity can be written as

$$D = 2 \sum_k n_k \Big\{ (\hat{\theta}_k - \hat{\theta})\bar{Y}_k - [C(\hat{\theta}_k) - C(\hat{\theta})] \Big\}.$$

If this hypothesis is valid, then D has approximately the chi-square distribution with $d - 1$ degrees of freedom.

Deviance

Let $1 \leq k \leq d$. If $0 < \bar{Y}_k < 1$ in the context of logistic regression and $\bar{Y}_k > 0$ in the context of Poisson regression, we set $\hat{\theta}_k = (C')^{-1}(\bar{Y}_k)$ and say that $\hat{\theta}_k$ exists; otherwise, we say that $\hat{\theta}_k$ does not exist or that it is undefined.

Let G_{sat} be an identifiable and saturated linear space. If $\hat{\theta}_1, \ldots, \hat{\theta}_d$ exist, then

$$l(G_{\mathrm{sat}}) = \sum_k n_k[\,\hat{\theta}_k \bar{Y}_k - C(\hat{\theta}_k)\,] + \log r(\mathbf{Y}; \mathbf{n}).$$

The deviance of G is defined by $D(G) = 2[\,l(G_{\mathrm{sat}}) - l(G)\,]$. If $\hat{\theta}_1, \ldots, \hat{\theta}_d$ exist, then the maximum-likelihood estimate $\hat{\theta}(\cdot)$ in G of $\theta(\cdot)$ exists and

$$D(G) = 2 \sum_k n_k \Big\{ [\,\hat{\theta}_k - \hat{\theta}(\mathbf{x}_k)\,]\bar{Y}_k - [\,C(\hat{\theta}_k) - C(\hat{\theta}(\mathbf{x}_k))\,] \Big\}.$$

The likelihood-ratio statistic D for testing the hypothesis that the canonical regression function is in a given subspace G_0 of G can be written as $D = D(G_0) - D(G)$.

The deviance of G, which is a nonnegative statistic, is analogous to the lack-of-fit sum of squares for G in the context of the normal regression model. Suppose $p < d$. Consider the canonical regression model and let $\hat{\theta}(\cdot)$ denote the maximum-likelihood estimate of the canonical regression function under the assumption that this function is in G, but consider the possibility that this assumption is not valid. The inequality $D(G) \geq \chi^2_{1-\alpha, d-p}$ determines the critical region having nominal size α for the goodness-of-fit test of this assumption; that is, if the assumption is valid, then $D(G)$ has approximately the chi-square distribution with $d - p$ degrees of freedom and hence $P(D(G) \geq \chi^2_{1-\alpha, d-p}) \approx \alpha$. The P-value for the test equals $1 - \chi^2_{d-p}(D(G))$.

In particular, the assumption that the distribution of the response variable does not vary from one design point to another is equivalent to the assumption that the canonical regression function is constant on the design set. The maximum-likelihood constant estimate of the canonical regression function is given by $\hat{\theta} = \hat{\theta}(\cdot) = (C')^{-1}(\bar{Y})$. The corresponding deviance can be written as

$$D(G_{\mathrm{con}}) = 2 \sum_k n_k \Big\{ (\hat{\theta}_k - \hat{\theta})\bar{Y}_k - [\,C(\hat{\theta}_k) - C(\hat{\theta})\,] \Big\},$$

which coincides with the likelihood-ratio statistic for testing the hypothesis of homogeneity. If the assumption that the canonical regression function is constant is valid, then $D(G_{\mathrm{con}})$ has approximately the chi-square distribution with $d - 1$ degrees of freedom.

Logistic Regression

The probability function of the binomial distribution with known time parameter $n \in \{1, 2, \ldots\}$ and unknown parameter $\pi \in (0, 1)$ is given by

$$f(y; n, \pi) = \binom{n}{y}\pi^y(1 - \pi)^{n-y}, \qquad y \in \mathcal{Y}_n,$$

where $\mathcal{Y}_n = \{0, \ldots, n\}$. This probability function can also be written in the exponential family form

$$f(y; n, \theta) = e^{\theta y - nC(\theta)} r(y; n), \qquad y \in \mathcal{Y}_n,$$

where $r(y; n) = \binom{n}{y}$ for $y \in \mathcal{Y}_n$, $C(\theta) = \log(1 + e^\theta)$ for $\theta \in \mathbb{R}$, and $\theta = (C')^{-1}(\pi) = \text{logit}\,\pi = \log \frac{\pi}{1-\pi} \in \mathbb{R}$. The first two derivatives of $C(\cdot)$ are given by

$$C'(\theta) = \frac{e^\theta}{1 + e^\theta} \quad \text{and} \quad C''(\theta) = \frac{e^\theta}{1 + e^\theta}\left(1 - \frac{e^\theta}{1 + e^\theta}\right), \qquad \theta \in \mathbb{R}.$$

The parameter π is given in terms of θ by

$$\pi = C'(\theta) = \frac{e^\theta}{1 + e^\theta}.$$

The distribution has mean

$$n\mu = n\pi = nC'(\theta) = n\frac{e^\theta}{1 + e^\theta}$$

and variance

$$n\sigma^2 = n\pi(1 - \pi) = n\frac{e^\theta}{1 + e^\theta}\left(1 - \frac{e^\theta}{1 + e^\theta}\right).$$

Under the input-response form of the logistic regression model, when the input vector equals $\mathbf{x}$ and the time parameter equals n, the response variable has the binomial distribution with parameters n and

$$\pi(\mathbf{x}) = \frac{\exp\theta(\mathbf{x})}{1 + \exp\theta(\mathbf{x})}$$

and hence mean $n\mu(\mathbf{x}) = n\pi(\mathbf{x})$ and variance $n\sigma^2(\mathbf{x}) = n\pi(\mathbf{x})[1 - \pi(\mathbf{x})]$. We refer to $\pi(\cdot) = \mu(\cdot)$ as the risk function and to $\theta(\cdot) = \text{logit}\,\pi(\cdot)$ as the logistic regression function. Let $g_1, \ldots, g_p$ be a basis of G. Then

$$\pi(\mathbf{x}) = \frac{\exp(\beta_1 g_1(\mathbf{x}) + \cdots + \beta_p g_p(\mathbf{x}))}{1 + \exp(\beta_1 g_1(\mathbf{x}) + \cdots + \beta_p g_p(\mathbf{x}))}, \qquad \mathbf{x} \in \mathcal{X}.$$

According to the corresponding experimental model, for $1 \le k \le d$, Y_k has the binomial distribution with parameters n_k and $\pi_k = C'(\theta_k) = (\exp\theta_k)/(1 + \exp\theta_k)$. The log-likelihood function corresponding to $\mathbf{Y} = [Y_1, \ldots, Y_d]^T$ can be written as

$$l(\theta(\cdot)) = \sum_k n_k\left\{\theta(\mathbf{x}_k)\bar{Y}_k - \log(1 + \exp(\theta(\mathbf{x}_k)))\right\} + \sum_k \log\binom{n_k}{y_k}, \qquad \theta(\cdot) \in G.$$

Suppose $\hat{\theta}(\cdot)$ exists. Then the maximum-likelihood estimate of $\pi(\cdot)$ is given by

$$\hat{\pi}(\mathbf{x}) = \frac{\exp\hat{\theta}(\mathbf{x})}{1 + \exp\hat{\theta}(\mathbf{x})}, \qquad \mathbf{x} \in \mathcal{X},$$

and the maximum-likelihood estimate of

$$\mathbf{W} = \text{diag}\left(n_1\pi(\mathbf{x}_1)[1 - \pi(\mathbf{x}_1)], \ldots, n_d\pi(\mathbf{x}_d)[1 - \pi(\mathbf{x}_d)]\right)$$

is given by

$$\hat{\mathbf{W}} = \text{diag}\left(n_1 \hat{\pi}(\mathbf{x}_1)[\,1 - \hat{\pi}(\mathbf{x}_1)\,], \ldots, n_d \hat{\pi}(\mathbf{x}_d)[\,1 - \hat{\pi}(\mathbf{x}_d)\,]\right).$$

The standard error of $\hat{\pi}(\mathbf{x})$ is given in terms of the standard error of $\hat{\theta}(x)$ by

$$\text{SE}(\hat{\pi}(\mathbf{x})) = \hat{\pi}(\mathbf{x})[\,1 - \hat{\pi}(\mathbf{x})\,]\,\text{SE}(\hat{\theta}(\mathbf{x})), \qquad \mathbf{x} \in \mathcal{X}.$$

The deviance corresponding to G can be written as

$$D(G) = 2 \sum_k \left(Y_k \log \frac{\hat{\pi}_k}{\hat{\pi}(\mathbf{x}_k)} + (n_k - Y_k) \log \frac{1 - \hat{\pi}_k}{1 - \hat{\pi}(\mathbf{x}_k)} \right),$$

where $\hat{\pi}_k = \bar{Y}_k$ for $1 \le k \le d$ [and $0 \cdot \log(0) = 0$].

Let G_0 be a proper subspace of G. If $\hat{\theta}(\cdot)$ exists, then the likelihood-ratio statistic for testing the hypothesis that the logistic regression function is in G_0 can be written as

$$D = D(G_0) - D(G) = 2 \sum_k \left(Y_k \log \frac{\hat{\pi}(\mathbf{x}_k)}{\hat{\pi}_0(\mathbf{x}_k)} + (n_k - Y_k) \log \frac{1 - \hat{\pi}(\mathbf{x}_k)}{1 - \hat{\pi}_0(\mathbf{x}_k)} \right),$$

where $\pi_0(\mathbf{x}) = (\exp \hat{\theta}_0(\mathbf{x}))/(1 + \exp \hat{\theta}_0(\mathbf{x}))$. If G is saturated, then $\hat{\pi}(\mathbf{x}_k) = \bar{Y}_k$ for $1 \le k \le d$ and $\hat{\boldsymbol{\beta}} = \mathbf{X}^{-1}(\text{logit}\bar{\mathbf{Y}})$.

Poisson Regression

The probability function of the Poisson distribution with known time parameter $n > 0$ and unknown rate parameter $\lambda > 0$ is given by

$$f(y; n, \lambda) = \frac{(n\lambda)^y}{y!} e^{-n\lambda}, \qquad y \in \{0, 1, 2, \ldots\}.$$

This probability function can also be written in the exponential family form

$$f(y; n, \theta) = e^{\theta y - nC(\theta)} r(y; n), \qquad y \in \mathcal{Y}_n,$$

where $r(y; n) = n^y/y!$ for $y \in \mathcal{Y}_n = \{0, 1, 2, \ldots\}$, $C(\theta) = e^\theta$ for $\theta \in \mathbb{R}$, and $\theta = (C')^{-1}(\lambda) = \log \lambda \in \mathbb{R}$. The parameter λ is given in terms of θ by $\lambda = C'(\theta) = e^\theta$. The distribution has mean $n\mu = n\lambda = nC'(\theta) = ne^\theta$ and variance $n\sigma^2 = n\lambda = nC''(\theta) = ne^\theta$.

Under the input-response form of the Poisson regression model, when the input vector equals $\mathbf{x}$ and the time parameter equals n, the response variable has the Poisson distribution with mean $n\mu(\mathbf{x}) = n\lambda(\mathbf{x}) = n \exp \theta(\mathbf{x})$ and hence variance $n\sigma^2(\mathbf{x}) = n\lambda(\mathbf{x})$. We refer to $\lambda(\cdot) = \mu(\cdot)$ as the rate function and to $\theta(\cdot) = \log \lambda(\cdot)$ as the Poisson regression function. Let $g_1, \ldots, g_p$ be a basis of G. Then

$$\lambda(\mathbf{x}) = \exp(\beta_1 g_1(\mathbf{x}) + \cdots + \beta_p g_p(\mathbf{x})), \qquad \mathbf{x} \in \mathcal{X}.$$

According to the corresponding experimental model, for $1 \le k \le d$, Y_k has the Poisson distribution with mean $n_k \lambda_k$, where $\lambda_k = C'(\theta_k) = \exp \theta_k$, so it has variance $n_k \lambda_k$. The log-likelihood function corresponding to $\mathbf{Y} = [\,Y_1, \ldots, Y_d\,]^T$ can be written as

$$l(\theta(\cdot)) = \sum_k n_k [\,\theta(\mathbf{x}_k)\bar{Y}_k - \exp\theta(\mathbf{x}_k)\,] + \sum_k [\,y_k \log n_k - \log y_k!\,], \qquad \theta(\cdot) \in G.$$

Suppose $\hat{\theta}(\cdot)$ exists. Then the maximum-likelihood estimate of $\lambda(\cdot)$ is given by $\hat{\lambda}(\mathbf{x}) = \exp\hat{\theta}(\mathbf{x})$ for $\mathbf{x} \in \mathcal{X}$, and the maximum-likelihood estimate of

$$\mathbf{W} = \mathrm{diag}(n_1\lambda(\mathbf{x}_1), \ldots, n_d\lambda(\mathbf{x}_d))$$

is given by

$$\hat{\mathbf{W}} = \mathrm{diag}(n_1\hat{\lambda}(\mathbf{x}_1), \ldots, n_d\hat{\lambda}(\mathbf{x}_d)).$$

The standard error of $\hat{\lambda}(\mathbf{x})$ is given in terms of the standard error of $\hat{\theta}(\mathbf{x})$ by $\mathrm{SE}(\hat{\lambda}(\mathbf{x})) = \hat{\lambda}(\mathbf{x})\,\mathrm{SE}(\hat{\theta}(\mathbf{x}))$ for $\mathbf{x} \in \mathcal{X}$. The deviance corresponding to G can be written as

$$D(G) = 2\sum_k n_k \left(\bar{Y}_k \log \frac{\hat{\lambda}_k}{\hat{\lambda}(\mathbf{x}_k)} - [\,\hat{\lambda}_k - \hat{\lambda}(\mathbf{x}_k)\,] \right),$$

where $\hat{\lambda}_k = \bar{Y}_k$ for $1 \leq k \leq d$ [and $0 \cdot \log 0 = 0$]; if G contains the constant functions, then

$$D(G) = 2\sum_k Y_k \log \frac{\hat{\lambda}_k}{\hat{\lambda}(\mathbf{x}_k)}.$$

Let G_0 be a proper subspace of G. If $\hat{\theta}(\cdot)$ exists, then the likelihood-ratio statistic for testing the hypothesis that the Poisson regression function is in G_0 can be written as

$$D = D(G_0) - D(G) = 2\sum_k n_k \left(\bar{Y}_k \log \frac{\hat{\lambda}(x_k)}{\hat{\lambda}_0(x_k)} - [\,\hat{\lambda}(x_k) - \hat{\lambda}_0(x_k)\,] \right),$$

where $\lambda_0(\mathbf{x}) = \exp\hat{\theta}_0(\mathbf{x})$; if G_0 contains the contant functions, then

$$D = D(G_0) - D(G) = 2\sum_k Y_k \log \frac{\hat{\lambda}(x_k)}{\hat{\lambda}_0(x_k)}.$$

If G is saturated, then $\hat{\lambda}(\mathbf{x}'_k) = \bar{Y}_k$ for $1 \leq k \leq d$ and $\hat{\beta} = \mathbf{X}^{-1}(\log \bar{\mathbf{Y}})$.

Hints and Answers

Chapter 1

Section 1.1

1.1 1/3 **1.2 a** 1/6 **b** 1/2 **1.3** 1/6 **1.4** 1/10 **1.5** 4/9 **1.6** 1/3 **1.7** 2/3 **1.8** .67 **1.9** 1/3 **1.10** 7/15

Section 1.2

1.12 a 6, 4 **b** .6, .4 **1.13 a** 12, .6 **b** 2, .1 **c** 14, .7 **1.14 a** 8, .4 **b** 3, .15 **c** 16, .8 **d** 0, 0 **1.15 a** $\hat{P}(B_1) = .2$, $\hat{P}(B_2) = .1$ and $\hat{P}(B_1 \cup B_2) = .3$ **b** $B_1 \cap B_2$ is the collection of points in the plane whose distance from the origin is greater than eight but at most ten. **c** None of the outcomes $y_1, \ldots, y_{10}$ are in $B_1 \cap B_2$. **1.16 c** See the discussions leading up to Eq. (20) and Eq. (30) in Section 1.3. **d** Use parts (a), (b), and (c). **1.21** (c) **1.22 a** 1/6 **b** 1/4 **1.23 a** Apply de Morgan's law with B_1 and B_2 replaced by B_1^c and B_2^c, respectively. **b** Apply part (a). **c** Apply part (b). **d** Apply the commutative and distributive properties. **e** Apply parts (c) and (d). **f** Apply part (e) or argue directly from a Venn diagram.

Section 1.3

1.24 a 3/19, 16/19 **b** 10, 2 **1.26 a** 1/4 **b** 9/16 **c** 5/16 **d** 5/16 **1.27 a** 13/49 **b** 29/49 **c** 16/49 **d** 20/49 **1.28** .95 **1.29 a** 1/2 **b** 2/5 **1.30 a** 7/18 **b** 4/15 **1.31 a** 1/6 **b** 1/18 **c** 2/9 **1.32 a** 5/16 **b** 5/16 **1.33 a** 3/10 **b** 3/10 **c** 7/10 **1.34 a** 35/100 **b** 16/45 **1.35** 1 **1.36 a** $\#(\{10, 11, 12\})/38 = 3/38$ **b** $\#(\{1, \ldots, 9, 13, 14, 15\})/38 = 6/19$ **c** $\#(\{16, \ldots, 36, 0, 00\})/38 = 23/38$ **d** 0

Section 1.4

1.43 1/2 **1.44 a** 2/5 **b** 0 **c** 1 **1.45 a** 0 **b** .1 **c** .1 **d** .95 **1.46** $2/\pi$ **1.47 a** 1/6 **b** 1/3 **c** 1/3 **d** 2/3 **1.48 a** 1/18 **b** 1/36 **c** 5/36 **d** 5/18 **1.49 a** 1/10 **b** 0 **c** 3/10 **1.50 a** .001 **b** .101 **c** .099 **1.52** $(\sqrt{3} + 2\pi/3)/4$ **1.53 a** 1/8 **b** 1/8 **c** 0 **d** 1/4 **e** 1/4 **f** $\pi/4$

Section 1.5

1.55 $f(y) = 1/8$ for $y \in \{1, \ldots, 8\}$; and $f(y) = 0$ otherwise. **1.56 a** $f(0) = 1/8, f(1) = 3/8,$ $f(2) = 3/8, f(3) = 1/8$ **b** Bernoulli distribution with parameter $\pi = 3/4$ **1.57 a** Bernoulli distribution with parameter $\pi = 2/9$ **b** $f(0) = 3/18, f(1) = 5/18, f(2) = 4/18, f(3) = 3/18, f(4) = 2/18, f(5) = 1/18$ **c** $f(1) = 1/36, f(2) = 3/36, f(3) = 5/36, f(4) = 7/36,$ $f(5) = 9/36, f(6) = 11/36$ **1.58 a** Note that $f(y) = \frac{1}{y} - \frac{1}{y+1}$ **b** 1/8 **1.59** 4/5 **1.60 a** $a = 1/12$ **b** 5/16 **1.61 a** $a = 1/\alpha$ **1.62** .95 (observe that $\frac{Y}{\log 40} \leq 1 \leq \frac{Y}{\log(40/39)}$ if and only if $\frac{1}{\log(40/39)} \leq Y \leq \log 40$) **1.63** .95 (see answer to Problem 1.62) **1.64 a** $\frac{1}{3}$ **b** 0 **1.65** $g(u) = 0$ for $0 \leq u < \frac{1}{4}$, $g(u) = 1$ for $\frac{1}{4} \leq u < \frac{3}{4}$ and $g(u) = 2$ for $\frac{3}{4} \leq u \leq 1$ **1.66 a** Yes **b** No **1.67 a** Yes **b** No **1.68 a** e^{-3} **b** 1/8 **c** $(e^{-3} + 1/8)/2$ **1.69** $P(1.95 \leq Y \leq 2.05) = 1/1.95 - 1/2.05 \doteq .025016$; by Eq. (40), $P(1.95 \leq Y \leq 2.05) \approx 0.1/1.95^2 \doteq .0263$; by Eq. (41), $P(1.95 \leq Y \leq 2.05) \approx 0.1/2^2 = .025$

Section 1.6

1.70 a $F(y) = -1/y$ for $y < -1$ and $F(y) = 1$ for $y \geq -1$ **b** -2 **1.71** $c = 62.5$ and $d = 137.5$ **1.72 a** $F(y) = 0$ for $y \leq 0$, $F(y) = y^\beta$ for $0 < y < 1$, and $F(y) = 1$ for $y \geq 1$ **b** $y_p = p^{1/\beta}$ **1.73** $a = 100 \log(40/39) \doteq 2.532$ and $b = 100 \log 40 \doteq 368.89$ **1.74 a** $F(y) = (1/2)e^{y/\beta}$ for $y \leq 0$ and $F(y) = 1 - (1/2)e^{-y/\beta}$ for $y > 0$ **b** $y_p = -\beta \log \frac{1}{2p}$ for $0 < p < 1/2$ and $y_p = \beta \log \frac{1}{2(1-p)}$ for $1/2 \leq p < 1$ **1.75 a** $f(y) = (2y/\beta^2) \exp(-(y/\beta)^2)$ for $y > 0$ and $f(y) = 0$ for $y \leq 0$ **b** $y_p = \beta[\log 1/(1-p)]^{1/2}$ **1.76 a** $f(y) = 1/y^2$ for $y > 1$ and $f(y) = 0$ for $y \leq 1$ **b** $y_p = 1/(1-p)$ **1.77** $f(0) = e^{-1} \doteq .368, f(500) = e^{-.5} - e^{-1} \doteq .239, f(1000) = 1 - e^{-.5} \doteq .393$ **1.78 a** $F(y) = y/8$ for $y \in \{0, \ldots, 8\}$ **b** $y_{.25} = 2.5, y_{.5} = 4.5, y_{.75} = 6.5$ **c** $y_{.1} = 1, y_{.2} = 2, y_{.3} = 3, y_{.4} = 4, y_{.5} = 4.5, y_{.6} = 5, y_{.7} = 6, y_{.8} = 7, y_{.9} = 8$ **1.79 a** $F(0) = 1/8, F(1) = 1/2, F(2) = 7/8, F(3) = 1$ **b** $y_{.25} = 1, y_{.5} = 1.5, y_{.75} = 2$ **1.80 a** 1/32 **b** 1/32 **c** $P(2 \leq Y \leq 3) = F_Y(3) - F_Y(1) = 3/16$ **1.81 a** $F(0) = 1/6, F(1) = 4/9, F(2) = 2/3, F(3) = 5/6, F(4) = 17/18, F(5) = 1$ **b** $y_{.25} = 1, y_{.5} = 2, y_{.75} = 3$ **1.82 a** $F(1) = 1/36, F(2) = 1/9, F(3) = 1/4, F(4) = 4/9, F(5) = 25/36, F(6) = 1$ **b** $y_{.25} = 3.5, y_{.5} = 5, y_{.75} = 6$ **1.83 a** $[M, \infty)$ **b** $e^{-M/\beta}$ **c** $[0, y]$ **d** $F(y) = 0$ for $y \leq 0$, $F(y) = 1 - \exp(-y/\beta)$ for $0 < y < M$, and $F(y) = 1$ for $y \geq M$ **e** No **f** No **1.84** $\frac{d^2}{dy^2} \log f(y) = -2 \frac{e^y}{1+e^y} \left(1 - \frac{e^y}{1+e^y}\right) < 0$ **1.86 a** $(\alpha/\beta)(t/\beta)^{\alpha-1} \exp(-(t/\beta)^\alpha)$ for $t > 0$ **b** $\beta \left(\log \frac{1}{1-p}\right)^{1/\alpha}$ **c** $S(t) = \exp(-(t/\beta)^\alpha)$, $\lambda(t) = (\alpha/\beta)(t/\beta)^{\alpha-1}$; $\Lambda(t) = (t/\beta)^\alpha$ for $t > 0$ **1.87** Weibull with shape parameter α and scale parameter $\beta/2^{1/\alpha}$

Section 1.7

1.88 $F_Y(y) = y/(y + 2)$ for $y > 0$ and $F_Y(y) = 0$ for $y \leq 0$ [note that $F_W(w) = w/(w + 1)$ for $w > 0$ and $F_W(w) = 0$ for $w \leq 0$], $f_Y(y) = 2/(y + 2)^2$ for $y > 0$, $f_Y(y) = 0$ for $y \leq 0$, $y_p = 2p/(1 - p)$ for $0 < p < 1$ **1.89** $F(y) = (1/\pi) \arctan[(y - \alpha)/\beta] + 1/2$ for $y \in \mathbb{R}$; $f(y) = 1/\left(\pi\beta\{1 + [(y - \alpha)/\beta]^2\}\right)$ for $y \in \mathbb{R}$; $y_p = \alpha + \beta \tan[\pi(p - 1/2)]$ **1.90 a** Let $y > 0$. Since β is negative, we see that $\alpha + \beta W \leq y$ if and only if $W \geq (y - \alpha)/\beta$. **1.91** $y_p = a + bx_p$, where $b = b_2/b_1$ and $a = a_2 - ba_1$ **1.93** Uniform on $(a + bd, a + bc)$ **1.95 a** $100 + 10 \log[p/(1 - p)]$ **b** (66.4, 136.6) **1.96** $F(y) = 0$ for $y \leq 0$ and $F(y) = 1 - e^{-\sqrt{y}}$ for $y \geq 0$; $f(y) = [1/(2\sqrt{y})]e^{-\sqrt{y}}$ for $y > 0$ and $f(y) = 0$ for $y \leq 0$; $y_p = \left(\log \frac{1}{1-p}\right)^2$ **1.97** $F(y) = 0$ for $y \leq 1$ and $F(y) = 1 - 1/y$ for $y > 1$; $f(y) = 0$ for $y \leq 1$ and $f(y) = 1/y^2$ for $y > 1$;

$y_p = 1/(1-p)$ **1.98** $F(y) = 0$ for $y \le 0$ and $F(y) = y/(1+y)$ for $y > 0$; $f(y) = (1 + y)^{-2}$ for $y > 0$ and $f(y) = 0$ for $y \le 0$; $y_p = p/(1-p)$ **1.99** Standard logistic distribution   **1.100** Uniform on $[0, 1]$ **1.101** $F(y) = 0$ for $y \le 0$, $F(y) = \sqrt{y}$ for $0 < y < 1$, and $F(y) = 1$ for $y \ge 1$; $f(y) = 1/(2\sqrt{y})$ for $0 < y < 1$ and $f(y) = 0$ elsewhere; $y_p = p^2$

Section 1.8

1.103 a $5/36$ **b** $5/12$ [note that $P(Y_1 < Y_2) = P(Y_1 = 1, Y_2 > 1) + P(Y_1 = 2, Y_2 > 2) + \cdots + P(Y_1 = 6, Y_2 > 6)$]   **1.105 a** (I_1, S) is uniformly distributed on $\{(0, 0), (0, 1), (1, 1), (1, 2)\}$ **1.106 a** $F_{Y_1}(y_1) = 0$ for $y_1 \le 0$, $F_{Y_1}(y_1) = y_1 - y_1^2/4$ for $0 < y_1 < 2$, and $F_{Y_1}(y_1) = 1$ for $y_1 \ge 2$; $f_{Y_1}(y_1) = 1 - y_1/2$ for $0 < y_1 < 2$ and $f_{Y_1}(y_1) = 0$ elsewhere **b** $F_{Y_2}(y_2) = 0$ for $y_2 \le 0$, $F_{Y_2}(y_2) = 2y_2 - y_2^2$ for $0 < y_2 < 1$, and $F_{Y_2}(y_2) = 1$ for $y_2 \ge 1$; $f_{Y_2}(y_2) = 2(1 - y_2)$ for $0 < y_2 < 1$ and $f_{Y_2}(y_2) = 0$ elsewhere **c** $P(Y_1 \ge 1) = 1/4$, $P(Y_2 \ge 1/2) = 1/4$, $P(Y_1 \ge 1, Y_2 \ge 1/2) = 0$ **1.114 a** $f_{Y_1, Y_2}(y_1, y_2) = 1/36$ for $y_1 = y_2 \in \{1, \ldots, 6\}$ and $f_{Y_1, Y_2}(y_1, y_2) = 1/18$ for $y_1, y_2 \in \{1, \ldots, 6\}$ with $y_1 < y_2$ **b** No   **1.116** $(7/9)^4(2/9) \doteq .0813$ **1.117** Geometric with parameter $\pi = 7/9$ **1.118 a** $(5/6)^n$ **b** $1 - (5/6)^n$ **c** 4   **1.119** Geometric with parameter $\pi = 143/144$ **1.120** Replace $\#(B)$ by length(B) in Eqs. (51)–(53) and then extend the argument from $n = 2$ to $n = 3$.

Chapter 2

Section 2.1

2.1 5   **2.2** 3   **2.3** 35/18   **2.4 a** 91/36 **b** 161/36 **c** 7 (the indicated sum equals the total number of spots showing on the two dice)   **2.5** $1 - (20/19)^3 \doteq -\$0.17$ **2.6** 30%   **2.7** $-\infty$ **2.9** 1/2   **2.10 a** $\alpha/(\alpha + 1)$ **b** $\alpha/(\alpha - 1)$ for $\alpha > 1$ and ∞ for $0 < \alpha \le 1$ **2.11** $-\infty$ **2.12** $1/(1 - \beta)$ for $0 < \beta < 1$ and ∞ for $\beta \ge 1$ **2.13** $2/(3\pi)$ **2.14** 2/3 [use the solution to Example 1.43(b) in Section 1.6]

Section 2.2

2.15 -35 **2.16** 35   **2.17 a** $1 - (5/6)^{10}$ **b** $6[1 - (5/6)^{10}]$ **2.19** See Problem 1.99 in Section 1.7 or observe that $\int_0^1 \log(1 - y)\,dy = \int_0^1 \log(y)\,dy$.   **2.20 a** $1 - (10/19)^4$ **b** $1 - (18/19)^4$ **c** $3 - 2(10/19)^4 - (18/19)^4$ **2.21 a** $W_1 = 2(2I_1 - 1)$ and $W_2 = 3I_2 - 1$ **b** $EI_1 = 9/19$ and $EI_2 = 6/19$ **c** $EW_1 = -2/19$ and $EW_2 = -1/19$ **d** $W = 4I_1 + 3I_2 - 3$ **e** $-3/19$ **2.22** Y is a bounded random variable.   **2.24 a** $F(y) = 1 - y^{-2}$ for $y > 1$ and $F(y) = 0$ for $y \le 1$ **b** .01 **c** 2 **d** .2   **2.25 a** $e^{-10} \doteq 4.54 \times 10^{-5}$ **b** 10^{-1} **2.26** $1/(1 - \beta t)$ for $t < 1/\beta$ and ∞ for $t \ge 1/\beta$ **2.27** Observe that $P(Y \ge c\beta) = P(W \ge e^{c\beta t})$.   **2.28 c** $10e^{-9} \doteq 1.234 \times 10^{-3}$ (compare with the answers to Problem 2.25)   **2.29** $\mu(\mu + 1)$ **2.30** To show that the stated condition implies that $P(X \in A, Y \in B) = P(X \in A)P(Y \in B)$, choose $g(X) = \text{ind}(X \in A)$ and $h(Y) = \text{ind}(Y \in B)$ and observe that $g(X)h(Y) = \text{ind}(X \in A, Y \in B)$.

Section 2.3

2.31 2   **2.32** $P(Y \in B)[1 - P(Y \in B)]$ **2.33 a** 1 **b** 2/3   **2.34 a** 1/3 **b** 1/5 **c** 4/45   **2.35 a** 2/3 **b** 2/3 **c** 2/9   **2.36** $-1, 2, 1$ (use integration by parts)   **2.37** 6, 30   **2.38** ∞ **2.39 a** Introduce $1/w$ as a new variable of integration. **b** Make the change of variables

$w = e^y$.   **2.42** 1 square mile, $\sqrt{3}$ square miles   **2.43** 12, 120   **2.45 a** 2555/1296 **b** 2555/1296 **c** 35/6   **2.46** Y is a constant random variable.   **2.47** Y is a constant random variable.   **2.48** If $Z = Y - c$ had finite second moment for some $c \in \mathbb{R}$, then $Y = Z + c$ would have finite second moment.   **2.49 a** $\sqrt{(1 - \pi)/\pi}$ **b** $1/\sqrt{\pi}$ **c** 1   **2.52 a** $F_T(t) = 0$ for $t \le 0$, $F_T(t) = (t/\theta)^n$ for $0 < t < \theta$, and $F_T(t) = 1$ for $t \ge \theta$; $f_T(t) = n\theta^{-n}t^{n-1}$ for $0 < t < \theta$ and $f_T(t) = 0$ elsewhere **b** $ET = n(n+1)^{-1}\theta$, $E(T^2) = n(n+2)^{-1}\theta^2$ and $\text{var}(T) = n(n+1)^{-2}(n+2)^{-1}\theta^2$

Section 2.4

2.53 7/24 [use Eq. (42) and the solution to Example 2.36 in Section 2.3]   **2.54** $1/81 \doteq 1.235 \times 10^{-2}$ (which should be compared to the answers to Problems 2.25 and 2.28 in Section 2.2)   **2.55** .08   **2.56** Apply Markov's inequality to the random variable Y^2.   **2.58** 5556 **2.59** $\mu^2 + \sigma^2/n$ **2.61 a** Apply Eq. (52), with $c = 0$ **b** Use part (a) **c** Observe that $Y_i^2 = Y_i$ for $1 \le i \le n$. **d** Observe that $\bar{Y} = (Y_1 + Y_2)/2$.   **2.62 a** Observe that $E(S^2) = (ES)^2 + \text{var}(S)$.   **2.63 b** Note that $Y_i - \bar{Y} = (1 - 1/n)Y_i - (1/n)\sum_{j \ne i} Y_j$. **c** Note that the mean of $Y_i - \bar{Y}$ is $\mu_i - \bar{\mu}$.   **2.64 a** $\mu_1 - \mu_2$ **b** $\sigma_1^2/n_1 + \sigma_2^2/n_2$ **2.65** $\theta^2/[n(n+2)]$ **2.66 a** $\text{bias}(\hat{\theta}) = [n(n+1)^{-1}c - 1]\theta$, $\text{var}(\hat{\theta}) = n(n+1)^{-2}(n+2)^{-1}c^2\theta^2$, and $\text{MSE}(\hat{\theta}) = \{n(n+1)^{-2}(n+2)^{-1}c^2 + [n(n+1)^{-1}c - 1]^2\}\theta^2$ **b** $(n+2)(n+1)^{-1}$ **c** $(n+1)^{-2}\theta^2$

Section 2.5

2.68 $1 - \mu^2$ **2.69 a** $c + (d - c)U$ **b** $\beta \log \frac{1}{1-U}$ **2.70 b** $\hat{\mu} \doteq 0.567$ ($\mu = 1$) and $\text{SE}(\hat{\mu}) \doteq 0.163$ **c** $\hat{\pi} = .2$ ($\pi \doteq .368$) and $\text{SE}(\hat{\pi}) = .126$ [using Eq. (61) in Section 2.4]   **2.71** Since the quantiles of the indicated density function are given by $y_p = \sqrt{p}$ for $0 < p < 1$, we take square roots of the pseudorandom numbers.   **2.72 a** $Y = \sqrt{U}$ **b** $Y = \tan(\pi(U - 1/2))$ **2.73** This result should be obvious, because, if U is uniformly distributed on $(0, 1)$ and $Y = F^{-1}(U)$, then Y has distribution function F, and it is true, more generally, provided only that the distribution function F be continuous; but you should provide a more direct proof under the stated condition, which implies that F is a continuous and strictly increasing function on (c, d).   **2.74 a** $U = 1 - e^{-W}$ **b** $Y = \log(U/(1 - U)) = \log(e^W - 1)$ **2.75** Use $y_i = \log(\exp(w_i) - 1)$ for $1 \le i \le n$.   **2.76 a** Use $y_i = 0$ if $u_i < 1 - \pi$ and $y_i = 1$ otherwise. **b** Use $y_i = 1$, 2, or 3 according as $u_i \le 1/3$, $1/3 < u_i \le 2/3$, or $u_i > 2/3$.   **2.77** $Y = \left[\frac{1}{c} \log \frac{1}{1-U}\right]$, where $c = \log \frac{1}{\pi}$ **2.78** Use $y_i = \left[\frac{1}{c} \log \frac{1}{1-u_i}\right]$ for $1 \le i \le n$, where $c = \log \frac{1}{\pi}$.   **2.79 a** 7, 1, 0, 5, 0, 2, 1, 1, 3, 0 **b** $\hat{\mu} = 2$ ($\mu = 4$) and $\text{SE}(\hat{\mu}) \doteq 0.745$ **c** $\hat{\pi} = .2$ ($\pi \doteq .328$) and $\text{SE}(\hat{\pi}) \doteq .126$ [use Eq. (61) in Section 2.4]   **2.80** Use $[w_1/c], \ldots, [w_M/c]$, where $c = \log(1/\pi)$.

Chapter 3

Section 3.1

3.1 $\alpha = (\mu/\sigma)^2$ and $\beta = \sigma^2/\mu$ **3.2** $\frac{17}{2}e^{-3} \doteq .423$ **3.4 a** $(1/2)\sqrt{\beta\pi}$ **b** $\beta(1 - \pi/4)$ **3.5 a** $\sqrt{\beta}\Gamma(\alpha + 1/2)/\Gamma(\alpha)$ **b** $\beta\{\alpha - [\Gamma(\alpha + 1/2)/\Gamma(\alpha)]^2\}$ **3.6** $\frac{1}{\beta\sqrt{2\pi(\alpha-1)}} \approx \frac{1}{\beta\sqrt{2\pi\alpha}}$ **3.9** $f_Y(y) = \frac{\Gamma(\alpha_1 + \alpha_2)}{c^{\alpha_1 + \alpha_2 - 1}\Gamma(\alpha_1)\Gamma(\alpha_2)}y^{\alpha_1 - 1}(c - y)^{\alpha_2 - 1}$ for $0 < y < c$ and $f_Y(y) = 0$ elsewhere **3.11** $\alpha_1 = \mu[\mu(1 - \mu)\sigma^{-2} - 1]$ and $\alpha_2 = (1 - \mu)[\mu(1 - \mu)\sigma^{-2} - 1]$

Section 3.2

3.14 $\Phi(1/2) \doteq .6915$ **3.15** **a** 1.6449 **b** 1.6449 **c** 0.6745 **d** 2.5758 **3.16** $5 +$
$\Phi(-\sqrt{2}) \doteq .579$ **3.17** **a** Normal with mean 900 and standard deviation $20\sqrt{6} \doteq 49.0$
b $1 - \Phi(5/\sqrt{6}) \doteq .0206$ **3.18** $\Phi(1/\sqrt{5}) \doteq .673$ **3.19** $\Phi(\sqrt{30}/5) \doteq .863$ **3.20 a** 0 **b** 3
(integrate by parts) **c** 2 **d** $\mu^2 + \sigma^2$ **e** $E[(\mu + \sigma Z)^3] = \mu^3 + 3\mu\sigma^2$ **f** $E[(\mu + \sigma Z)^4] =$
$\mu^4 + 6\mu^2\sigma^2 + 3\sigma^4$ **g** $4\mu^2\sigma^2 + 2\sigma^4$ **3.21** **a** Gamma with shape parameter $3/2$
and scale parameter $2\sigma^2$ **b** $\frac{s^2}{\sigma^3 \sqrt{\pi/2}} \exp(-s^2/2\sigma^2)$ (see Theorem 3.3 in Section 3.1)

c $2\sigma\sqrt{2/\pi}$ (see Problem 3.5 in Section 3.1) **d** $\sigma^2(3 - 8/\pi)$ **3.22** **a** $\Phi\left(\frac{\log(y)-\mu}{\sigma}\right)$

b $\frac{1}{\sigma y}\phi\left(\frac{\log(y)-\mu}{\sigma}\right) = \frac{1}{\sigma y \sqrt{2\pi}} \exp\{-[\log(y) - \mu]^2/2\sigma^2\}$ **c** $\exp(\mu + \sigma z_p)$ **3.25** $\sigma\sqrt{2\log 2}$
3.28 $3.14 \times 10^{-5} < 1 - \Phi(4) < 3.35 \times 10^{-5}$

Section 3.3

3.29 $\Phi(-\sqrt{10}/2) \doteq .0569$ (the actual probability equals $.0318$)   **3.30** $\Phi(\sqrt{5}) \doteq .9873$. (the
actual probability is $.9893$)   **3.31** Normal approximation $= 100z_{.9}^2 \doteq 164.2$; actual value $\doteq$
168.2 **3.32 a** Gamma with shape parameter 25 and scale parameter 2 **b** $50, 100$ **c** $\Phi(1) -$
$\Phi(-0.5) \doteq .533$ (actual value $\doteq .517$)   **3.33** $2\Phi(1.5) - 1 \doteq .866$ **3.34** $2\Phi(\sqrt{3}/2) -$
$1 \doteq .614$ (see the solution to Example 2.40 in Section 2.3)   **3.35** **a** 2000 **b** 385 [note
that $P(|\bar{Y} - \mu| \geq \sigma/10) = P\left(\left|\frac{\bar{Y}-\mu}{\sigma/\sqrt{n}}\right| \geq \frac{\sqrt{n}}{10}\right)$]   **3.36** $1 - \Phi(2/\sqrt{3}) \doteq .124$ **3.37** $\Phi(2) \doteq$
$.977$ **3.38** $\Phi\left(\sqrt{\alpha} \frac{c\beta_1 - \beta_2}{\sqrt{c^2\beta_1^2 + \beta_2^2}}\right)$ **3.39** **a** $2\Phi(3/\sqrt{2.69}) - 1 \doteq .933$ (see Problem 3.38 in this
section and Problem 1.115 in Section 1.8) **b** $\Phi(-2/\sqrt{3.94}) + \Phi(9.5/\sqrt{4.8025}) - 1 \doteq .157$
3.40 Look at $\log(U_1 \cdots U_n) = \log U_1 + \cdots + \log U_n$ and use the answer to Problem 2.36 in
Section 2.3.

Chapter 4

Section 4.1

4.1 $\binom{10}{3} = 120$ **4.2** **a** $F(l) = \frac{l(l-1)}{N(N-1)}$ for $l = 2, \ldots, N$ **b** $f(l) = \frac{2(l-1)}{N(N-1)}$ for $l = 2, \ldots, N$
4.3 $1/3$ **4.4** $6!/6^6 \doteq .0154$ **4.5** $a^5 + 5a^4b + 10a^3b^2 + 10a^2b^3 + 5ab^4 + b^5$ **4.6** Write
out in terms of factorials and simplify.   **4.7** **a** Apply the binomial formula with $a = b = 1$.
b Apply the binomial formula with $a = -1$ and $b = 1$.   **4.8** **a** $\frac{6!}{2!2!2!} = 90$ **b** $\frac{90}{6^6} \doteq .00193$
c $\binom{6}{3}\frac{90}{6^6} \doteq .0386$ **4.9** $\frac{6!}{3!2!1!} = 60$ **4.10** $\sqrt{3}\, 3^{3n}/(2\pi n)$ **4.11** **a** $12,600$ **b** $302,400$
4.12 **a** 1 **b** 5 **c** 10 **d** 20 **e** 30 **4.13** Apply the multinomial formula with $a_1 = \ldots =$
$a_M = 1$.

Section 4.2

4.14 $1 - \frac{11}{6}(\frac{5}{6})^5 \doteq .263$ **4.15** **a** The function $\pi(1 - \pi)$, $0 < \pi < 1$ is uniquely maximized at $\pi = 1/2$.   **4.16** $5/16$ **4.17** $80/243$ **4.18** $135/512$ **4.19** $\binom{19}{3}(2/9)^4(7/9)^{16}$
4.20 $P(Y = y) = \binom{y+r-1}{y}\pi^y(1 - \pi)^{r-1}(1 - \pi) = \binom{r+y-1}{y}\pi^y(1 - \pi)^r$ for y a nonnegative integer   **4.21** See the solution to Example 4.8 in Section 4.1.   **4.22** **a** $.0161$ **b** $P(Y \leq 2) +$
$P(Y \geq 10) \doteq .0386$ **4.23** **a** Binomial with parameters $n = 10$ and $\pi \doteq .8186$ **b** 8.186,
$1.485, 1.219$ **4.26** $i(n+1)^{-1}, i(n+1-i)(n+1)^{-2}(n+2)^{-1}$ **4.27** $99z_{.99}^2 \doteq 536$ (Sampling with replacement and the binomial distribution are appropriate since 536 is presumably

a small proportion of the number of eligible voters. Various practical difficulties such as refusal to participate in the survey are ignored.)   **4.28 a** 194 **b** .682 **c** $n = 2908$ and $y = 531$ **4.29 a** 2100, 1750 **b** .0084   **4.30 a** .0745 (the number that survive 150 hours or more has the binomial distribution with parameters $n = 500$ and $p = e^{-3/2} \doteq .223$) **b** .0824 (the actual probability is .0837)   **4.31** $2[1 - \Phi(\sqrt{10}/2)] \doteq .114$

Section 4.3

4.32 $24(.4)(.3)(.2)(.1) = .0576$ **4.33** $\frac{13!}{5!4!3!1!}(1/4)^{13} \doteq .00537$ **4.34** $\frac{10!}{5!2!3!}(1/2)^5(1/6)^2$ $(1/3)^3 \doteq .081$ **4.35 a** $\frac{(6n)!}{(n!)^6}(1/6)^{6n}$ **b** $\frac{6^{1/2}}{(2\pi n)^{5/2}}$ **4.36** $\frac{10!}{2!5!3!}[\Phi(-1.5)]^2[\Phi(1) - \Phi(-1.5)]^5[1 - \Phi(1)]^3 \doteq .0125$ **4.37** $\frac{9!}{3!4!2!}(1/2)^3(1/4)^4(1/4)^2 = 315/2^{13}$ **4.38** .1905   **4.39** $P(Y_1 = y_1, Y_2 = y_2) = \frac{(y_1+y_2+r-1)!}{y_1!y_2!(r-1)!}\pi_1^{y_1}\pi_2^{y_2}\pi_3^{r-1}$, $\pi_3 = \frac{(y_1+y_2+r-1)!}{y_1!y_2!(r-1)!}\pi_1^{y_1}\pi_2^{y_2}\pi_3^r$ for y_1 and y_2 nonnegative integers   **4.40** $\frac{(6r)!}{(r!)^6 6^{r+1}}$ **4.41** 5/9   **4.42** 5/18   **4.43** $\frac{5!}{3!2!}(.5)^3(.3)^2 \times .2 \times \frac{9!}{5!4!}(.5)^5(.3)^4 \times .2 = \frac{5!9!}{3!2!5!4!}(.5)^8(.3)^6(.2)^2$

Section 4.4

4.45 $\frac{5^4}{4!}e^{-5} \doteq .175$ **4.46** λ/n **4.47** $2e^{-1} - 3e^{-2} \doteq .330$ **4.48 a** $(1+e^{-2\lambda})/2$ **b** $(1-e^{-2\lambda})/2$ **4.49** $f(0) + \cdots + f(4) + f(17) + f(18) + \cdots \doteq .056$ **4.50 a** .00366 (Poisson model) **b** .0527 (Poisson model) or .0524 (binomial model)   **4.51** $\frac{52!}{30!16!6!}(e^{-1/2})^{30}$ $[(1/2)e^{-1/2}]^{16}[1 - (3/2)e^{-1/2}]^6 \doteq .017$ **4.54 a** .6938 **b** .6873   **4.55 a** The normal approximation to $f(25) \doteq .07952$ is $\frac{1}{5\sqrt{2\pi}} \doteq .07979$. **b** The normal approximation to $P(Y \le 30) \doteq .863$ is $\Phi(1) \doteq .841$, and the normal approximation with the half-integer correction is $\Phi(1.1) \doteq .864$. **c** The normal approximation to the upper decile 32 of Y is $25 + 5z_{.9} \doteq 31.4$.

Section 4.5

4.56 $1 - 61e^{-6} \doteq .849$ **4.57** $\frac{125}{12}e^{-5.5} \doteq .0426$ **4.58** .0778   **4.59** $\lambda^3 e^{-5\lambda}(8\lambda^2 + 12\lambda + 3)/3$ (there must be 0, 1, or 2 typos during the third minute)   **4.60** $P(W_2 \le t) = 1 - e^{-\lambda t}$ **4.61** Exponential with inverse-scale parameter λ [note that $P(N(t+s) - N(t) \ge 1) = 1 - P(N(t+s) - N(t) = 0) = 1 - e^{-\lambda s}$].   **4.62** λ/t **4.63** $\lambda^2/(n-2)$ if $n \ge 3$ and ∞ if $n = 2$ (see Problem 3.3 in Section 3.1).   **4.64** $\left(z_{.9}^2 + 2000 + z_{.9}\sqrt{z_{.9}^2 + 4000}\right)/(2\lambda) \doteq 1041/\lambda$

[recall that the solutions to the quadratic equation $ax^2 + bx + c = 0$ are given by $x = (-b \pm \sqrt{b^2 - 4ac})/(2a)$]   **4.66 a** $\Phi\left(\frac{\lambda t - n}{\sqrt{\lambda t}}\right)$ **b** $\Phi\left(\frac{\lambda t - n}{\sqrt{n}}\right)$ **c** Yes   **4.66 a** , **b** $[1 + \lambda(t_2 - t_1)]\exp(-\lambda t_2)$ **4.67** $F_R(r) = 0$ for $r \le 0$ and $F_R(r) = 1 - \exp(-\lambda\pi r^2)$ for $r > 0$, $f_R(r) = 0$ for $r \le 0$ and $f_R(r) = 2\lambda\pi r\exp(-\lambda\pi r^2)$ for $r > 0$, $r_p = \left(\frac{1}{\lambda\pi}\log\frac{1}{1-p}\right)^{1/2}$, $ER = \frac{1}{2\sqrt{\lambda}}$ (make the change of variables $t = r^2$), $E(R^2) = \frac{1}{\lambda\pi}$, and $\text{var}(R) = \frac{4-\pi}{4\lambda\pi}$

Chapter 5

Section 5.1

5.2 1225/1296   **5.3 a** $-300/19 \doteq -15.79$, $221400/361 \doteq 613.3$, 24.76 **b** .2619 **c** .2553   **5.4** $1/\sqrt{780} \doteq .0358$ **5.5** -0.48, $-\sqrt{2/7} \doteq -.5345$ **5.10** $799/800 = .99875$

5.11 **a** $-9/10$ **b** $(9/10)(1 - I_1)$ **c** $9/190$ **5.12** $\mu_2 = \alpha + \beta\mu_1$, $\sigma_2 = \sqrt{\sigma^2 + \beta^2\sigma_1^2}$, and $\rho = \beta\sigma_1/\sqrt{\sigma^2 + \beta^2\sigma_1^2}$ **5.13** $\hat{X} = \mu_1 + [\beta\sigma_1^2/(\sigma^2 + \beta^2\sigma_1^2)](Y - \alpha - \beta\mu_1)$; $MSE(\hat{X}) = \sigma_1^2[1 - \beta^2\sigma_1^2/(\sigma^2 + \beta^2\sigma_1^2)]$ **5.14** **a** Use Theorem 5.2. **b** Use part (a). **c** Use part (b) and Eq. (16).   **5.16** **a** $\sigma_2^2 - 2b\rho\sigma_1\sigma_2 + b^2\sigma_1^2$ **b** $\rho\sigma_2/\sigma_1$ [differentiate the answer to part (a) with respect to b] **c** $\sigma_2^2(1 - \rho^2)$ **5.17** **a** Note that $ET = (b_1 + \cdots + b_n)\mu$. **b** Note that $\text{var}(\hat{\mu}) = \beta_1^2\sigma_1^2 + \cdots + \beta_n^2\sigma_n^2$. **c** Note that $\text{cov}(\hat{\mu}, T - \hat{\mu}) = \beta_1(b_1 - \beta_1)\sigma_1^2 + \cdots + \beta_n(b_n - \beta_n)\sigma_n^2$. **d** Use part (c). **e** Note that $\text{var}(T - \hat{\mu}) = (b_1 - \beta_1)^2\sigma_1^2 + \cdots + (b_n - \beta_n)^2\sigma_n^2$. **f** Use parts (d) and (e).

Section 5.2

5.18 $[1.6, 1.2, 0.8]^T$ **5.19** $[\mu_1, \alpha_1 + \beta_1\mu_1, \alpha_2 + \beta_2\mu_1 + \beta_3\alpha_1 + \beta_1\beta_3\mu_1]^T$ **5.20** **a** $3\sigma^2$ **b** $\sigma^2 I_3$ **5.21** **a** $6\sigma^2$ **b** $\sigma^2 \begin{bmatrix} 1 & 1 & 1 \\ 1 & 2 & 2 \\ 1 & 2 & 3 \end{bmatrix}$ **5.22** **a** $4/3$ **b** $\begin{bmatrix} 1 & 1/2 \\ 1/2 & 1/3 \end{bmatrix}$ **5.23** **a** 5 **b** $\begin{bmatrix} 1 & 0 & 1 \\ 0 & 1 & 0 \\ 1 & 0 & 3 \end{bmatrix}$

[use integration by parts to show that $E(Z^3) = 0$ and $E(Z^4) = 3$]   **5.25** **a** $\begin{bmatrix} \mu \\ \mu \end{bmatrix}$ **b** $\begin{bmatrix} 2\mu \\ 0 \end{bmatrix}$ **c** $\begin{bmatrix} 1 & 1 \\ 1 & -1 \end{bmatrix}$ **d** $W_1 = (Y_1 + Y_2)/2$ and $W_2 = (Y_1 - Y_2)/2$ **e** $\begin{bmatrix} 1/2 & 1/2 \\ 1/2 & -1/2 \end{bmatrix}$

5.26 **a** $\begin{bmatrix} \mu \\ \mu \\ \mu \end{bmatrix}$ **b** $\begin{bmatrix} \mu \\ 2\mu \\ 3\mu \end{bmatrix}$ **c** $\begin{bmatrix} 1 & 0 & 0 \\ 1 & 1 & 0 \\ 1 & 1 & 1 \end{bmatrix}$ **d** $W_1 = Y_1$, $W_2 = Y_2 - Y_1$ and $W_3 = Y_3 - Y_2$

e $\begin{bmatrix} 1 & 0 & 0 \\ -1 & 1 & 0 \\ 0 & -1 & 1 \end{bmatrix}$ **5.27** **a** $\mathbf{a} = \begin{bmatrix} a_1 \\ a_2 \\ a_3 \end{bmatrix}$ and $\mathbf{B} = \begin{bmatrix} b_1 & 0 & 0 \\ 0 & b_2 & 0 \\ 0 & 0 & b_3 \end{bmatrix}$

Section 5.3

5.29 $\begin{bmatrix} 0.96 & -0.48 & -0.32 \\ -0.48 & 0.84 & -0.24 \\ -0.32 & -0.24 & 0.64 \end{bmatrix}$ **5.30** **a** $\begin{bmatrix} 10 & -11 & -2 \\ -21 & 45 & 20 \\ 9 & -16 & -5 \end{bmatrix}$ **b** $\begin{bmatrix} 10 & -11 & -2 \\ -11 & 34 & 18 \\ -2 & 18 & 13 \end{bmatrix}$

5.31 **a** $\begin{bmatrix} 2 & 1 & 1 & 6 \\ 1 & 2 & 1 & 3 \\ 1 & 1 & 2 & 7 \\ 6 & 3 & 7 & 30 \end{bmatrix}$ **b** Noninvertible because (Y_1, Y_2, Y_3, Y_4) lies in a three-dimensional hyperplane in $\mathbb{R}^4$ since Y_1, Y_2, Y_3, and Y_4 are each linear functions of W_1, W_2, and W_3 (alternatively, noninvertible because its determinant equals zero) **c** $2, -1, 3$ **5.32** $\begin{bmatrix} 1 & 1 & 2 \\ 1 & 2 & 3 \\ 2 & 3 & 6 \end{bmatrix}$

(which coincides with the variance–covariance matrix found in the solution to Example 5.11)   **5.34** Observe that $b_1 Y_1 + \cdots + b_m Y_m = b_1 Y_1 + \cdots + b_m Y_m + 0 \cdot Y_{m+1} + \cdots + 0 \cdot Y_n$.   **5.36** False, since $\text{cov}(Y_1, Y_2) > 0$, $\text{cov}(Y_2, Y_3) > 0$, and $\text{cov}(Y_1, Y_3) < 0$

5.37   **a** $\sigma^2(\rho + \frac{1-\rho}{n})$ **b** $\sigma^2\rho$ **5.38**   **a** $\begin{bmatrix} \mu_1 \\ \alpha + \beta\mu_1 \end{bmatrix}$ **b** $\begin{bmatrix} \sigma_1^2 & \beta\sigma_1^2 \\ \beta\sigma_1^2 & \sigma^2 + \beta^2\sigma_1^2 \end{bmatrix}$

5.39   **b** $L = \begin{bmatrix} \sigma_1 & 0 \\ \rho\sigma_2 & \sqrt{1-\rho^2}\sigma_2 \end{bmatrix}$ **5.40** $\begin{bmatrix} 1 & \rho & \rho \\ \rho & 1 & \rho \\ \rho & \rho & 1 \end{bmatrix}$

Section 5.4

5.43 $\frac{2}{3}(4 - X - Y)$, $\frac{4}{15}$ **5.44** $\frac{\pi_3}{1-\pi_1-\pi_2}(n - X - Y)$, $\frac{n\pi_3(1-\pi_1-\pi_2-\pi_3)}{1-\pi_1-\pi_2}$ **5.45**   **a** $\alpha + \beta^T E\mathbf{X}$
b $VC(\mathbf{X})\beta$ **c** $\sigma^2 + \beta^T VC(\mathbf{X})\beta$ **5.47** Note that $Y - \hat{Y}_0 = Y - \hat{Y} + \beta_m(X_m - \hat{X}_m)$
and $\text{cov}(X_m - \hat{X}_m, Y - \hat{Y}) = 0$.   **5.48** Apply Theorem 5.17 with $m = 2$, and use
the formula $\hat{X}_2 = EX_2 + \frac{\text{cov}(X_1, X_2)}{\text{var}(X_1)}(X_1 - EX_1)$.   **5.49** Note that $\text{cov}(\hat{Y}, Y - \hat{Y}) = 0$

by Corollary 5.7.   **5.51** $117/121 \doteq .967$ **5.54**   **a** $\begin{bmatrix} 0.16 & 32.00 & 0.11 \\ 32.00 & 10,000.00 & 30.00 \\ 0.11 & 30.00 & 0.25 \end{bmatrix}$

b $1.111 + 0.243X_1 + 0.00222X_2$, 0.157 **c** $0.8 + 0.6875X_1$, 0.174 **d** $1.5 + 0.003X_2$,

0.16 **5.55**   **a** $\begin{bmatrix} 0.160 & 32.000 & 0.112 \\ 32.000 & 10,000.000 & 29.600 \\ 0.112 & 29.600 & 0.250 \end{bmatrix}$ **b** 0.157 **c** $0.76 + 0.7X_1$, 0.172

d $1.52 + 0.00296X_2$, 0.162

Section 5.5

5.56   **a** 1 **b** $1/3$ **5.57**   **a** $f_Y(y) = (2 + y)/4$ for $-1 < y < 1$ and $f_Y(y) = 0$ elsewhere
5.58   **a** $3/8$ **b** $1/6$ **5.59** $EY = 1/6$, $E(Y^2) = 1/3$, $\text{var}(Y) = 11/36$ **5.60**   **a** 0 **b** $-1/36$
c $-1/11$ **d** $(2 - X)/11$ **e** $10/33$ **5.61**   **a** $f_X(x) = (8 - x)/32$ for $0 < x < 8$ and
$f_X(x) = 0$ otherwise; $f_Y(y) = (4 - y)/8$ for $0 < y < 4$ and $f_Y(y) = 0$ otherwise   **b** $EX = 8/3$,
$\text{var}(X) = 32/9$, $EY = 4/3$, $\text{var}(Y) = 8/9$ **c** $8/3$ **d** $-8/9$ **e** $-1/2$ **f** $(8 - X)/4$ **g** $2/3$
5.62 Since $\{(y_1, y_2) \in \mathbb{R}^2 : y_1 = y_2\}$ is a line in the plane, its area is zero.   **5.63** To show
that the determinant $Y_1Y_4 - Y_2Y_3$ of the indicated matrix is nonzero with probability 1,
observe that the four-dimensional volume of $\{(y_1, y_2, y_3, y_4) \in \mathbb{R}^4 : y_1y_4 - y_2y_3 = 0\}$ is
zero.   **5.68** $\frac{\alpha_2\alpha_3}{(\alpha_2+\alpha_3)(\alpha_1+\alpha_2+\alpha_3)(\alpha_1+\alpha_2+\alpha_3+1)}$

Section 5.6

5.69 Standard Cauchy distribution   **5.70**   **a** $f_{W,Z}(w, z) = w^2/(1 + z)^2$ for $z > 0$ and
$0 < w < \min(1 + z, (1 + z)/z)$ and it equals zero elsewhere   **b** No   **5.71** $f_{Y_1, Y_2}(y_1, y_2) =$
$\frac{1}{2y_2}f_{W_1, W_2}(\sqrt{y_1 y_2}, \sqrt{y_1/y_2})$ for $y_1, y_2 > 0$ **5.72** $f_{Y_1, Y_2}(y_1, y_2) = \frac{y_1}{(1+y_2)^2}f_{W_1, W_2}\left(\frac{y_1 y_2}{1+y_2}, \frac{y_1}{1+y_2}\right)$
for $y_1, y_2 > 0$ **5.74** $f_{Y_1, Y_2}(y_1, y_2) = y_2 f_{W_1, W_2}(y_1 y_2, y_2(1 - y_1))$ for $0 < y_1 < 1$ and $y_2 > 0$
and $f_{Y_1, Y_2}(y_1, y_2) = 0$ elsewhere   **5.77** $f_{Y_1, Y_2, Y_3}(y_1, y_2, y_3) = y_3^2 f_{W_1, W_2, W_3}(y_1 y_3, y_2 y_3, y_3(1 - y_1 - y_2))$ for $y_1 > 0, y_2 > 0, y_1 + y_2 < 1$ and $y_3 > 0$ and $f_{Y_1, Y_2, Y_3}(y_1, y_2, y_3) = 0$ elsewhere
5.79 $f_{Y_1, Y_2, Y_3}(y_1, y_2, y_3) = \frac{1}{\beta^3}e^{-y_3/\beta}$ for $0 < y_1 < y_2 < y_3$ and $f_{Y_1, Y_2, Y_3}(y_1, y_2, y_3) = 0$
elsewhere   **5.80** $f_{V_1, V_2}(v_1, v_2) = 2$ for $0 < v_1 < v_2 < 1$ and $f_{V_1, V_2}(v_1, v_2) = 0$ elsewhere
5.81 $f_{V_1, V_2, V_3}(v_1, v_2, v_3) = 3!$ for $0 < v_1 < v_2 < v_3 < 1$ and $f_{V_1, V_2, V_3}(v_1, v_2, v_3) = 0$ elsewhere   **5.82**   **a** $f_{W_1, W_2}(w_1, y) = 9w_1^{-1}y^{-4}$ for $1 < w_1 < y$ **b** $f_{W_1 W_2}(y) = 9(\log y)/y^4$ for
$y > 1$ **c** $9/4$, $63/16$ [use the independence of W_1 and W_2 and show that $EW_1 = EW_2 = 3/2$
and $E(W_1^2) = E(W_2^2) = 3$]

Section 5.7

5.85 Bivariate normal with $EW = 7$, $EZ = 37$, $\text{var}(W) = 108$, $\text{var}(Z) = 604$, and $\text{cov}(W, Z)$
$= -162$ **5.87 a** $\begin{bmatrix} 1 \\ 2 \\ 3 \end{bmatrix}$, $\begin{bmatrix} 1 & 1 & 1 \\ 1 & 2 & 2 \\ 1 & 2 & 3 \end{bmatrix}$ **b** Trivariate normal with mean vector μ and

variance–covariance matrix Σ **c** $\begin{bmatrix} 1 & 0 & 0 \\ 1 & 1 & 0 \\ 1 & 1 & 1 \end{bmatrix}$ **d** Normal with mean 6 and variance 14

f $1 - \Phi(-6/\sqrt{14}) \doteq .9456$ **5.89** Trivariate normal with mean vector $\mathbf{0}$ and variance–

covariance matrix $\begin{bmatrix} 1 & \rho & \rho \\ \rho & 1 & \rho \\ \rho & \rho & 1 \end{bmatrix}$ **5.91** Observe that $P((X - EX)(Y - EY) > 0) = 2P(X \geq$

$EX, Y \geq EY) = 1/2 + (1/\pi)\arcsin\rho$ **5.92** Let X and Y have a bivariate normal joint
distribution and set $\beta = \text{cov}(X, Y)/\text{var}(X)$, $\alpha = EY - \beta EX$, and $W = Y - \alpha - \beta X$. Then
a X and W are independent. **b** W has the normal distribution with mean 0 and vari-
ance $\text{var}(Y) - [\text{cov}(X, Y)]^2/\text{var}(X)$. **5.93** Let X and Y have a bivariate normal joint
distribution and set $\beta = \text{cov}(X, Y)/\text{var}(X)$ and $\alpha = EY - \beta EX$, so that $\hat{Y} = \alpha + \beta X$ is the
best linear predictor of Y based on X. Also, set $W = Y - \hat{Y} = Y - \alpha - \beta X$. Then $Y =$
$\hat{Y} + W = \alpha + \beta X + W$, where X and W are independent and W is normally distributed with
mean 0 and variance $\text{var}(Y) - [\text{cov}(X, Y)]^2/\text{var}(X)$ **5.94** Bivariate normal with mean vector
$[EX, \alpha + \beta EX]^T$ and variance–covariance matrix

$$\begin{bmatrix} \text{var}(X) & \beta\,\text{var}(X) \\ \beta\,\text{var}(X) & \beta^2\text{var}(X) + \text{var}(W) \end{bmatrix}$$

5.95 1/2, 2, 1/2 **5.96** Trivariate normal with mean vector $\mathbf{0}$ and the variance–covariance
matrix given in the solution to Example 5.11 in Section 5.3 **5.97** Trivariate normal with
mean vector $\mathbf{0}$ and the variance–covariance matrix given in the answer to Problem 5.32 in
Section 5.3 (thus, the random variables X_1, X_2, and Y have the same joint distribution in
Problem 5.97 as they do in Problem 5.96.)

Chapter 6

Section 6.1

6.1 36/125 **6.2** Uniform on $\{2, 3, 4, 5, 6\}$ **6.3** 2/3 **6.6 a** Observe that $P(N - Y = m, Y = y) = P(N = y + m, Y = y) = P(N = y + m)P(Y = y|N = y + m)$. **6.7** 2/9 **6.8 a** 1/2
b 2/3 (if the selected coin is gold, then the other coin in the chosen drawer is gold if and only
if the first drawer is chosen) **6.9 a** 23/36 **b** 6/23 **6.10** 15/23 **6.11** $2/(2x - 1)$ for $1 \leq$
$y \leq x - 1$ and $1/(2x - 1)$ for $y = x$ **6.12** Observe that $P(\mathbf{X} = \mathbf{x}, \mathbf{Y} \in B) = \sum_{\mathbf{y} \in B} P(\mathbf{X} = \mathbf{x}, \mathbf{Y} = \mathbf{y})$.

Section 6.2

6.16 $L_1, \ldots, L_N$ is a permutation of $1, \ldots, N$, so $Y_1, \ldots, Y_N$ is a reordering of $v_1, \ldots, v_N$ and
hence $Y_1 + \ldots + Y_N = v_1 + \ldots + v_N = N\mu$. **6.17 a** $-1/(N - 1)$ **b** $(N\mu - Y_1)/(N - 1)$
c $N(N - 2)(N - 1)^{-2}\sigma^2$. **6.18 a** $10/13 \doteq 0.769$, $290/169 \doteq 1.716$, $\sqrt{290/69} \doteq 1.310$
b 10, $290/17 \doteq 17.059$, $\sqrt{290/17} \doteq 4.130$ **6.20** $N \geq 10n - 9$ **6.21** $n = N/2$ if N is
even and $n = [N/2]$ or $n = [N/2] + 1$ if N is odd. **6.23** $f(0) = \frac{1}{2} \cdot \frac{1}{4} = \frac{1}{8}, f(2) = \frac{1}{2} \cdot \frac{3}{4} +$
$\frac{1}{2} \cdot \frac{3}{4} = \frac{3}{4}, f(4) = \frac{1}{2} \cdot \frac{1}{4} = \frac{1}{8}$ **6.24** $6 \cdot \frac{3}{9} \cdot \frac{4}{10} \cdot \frac{5}{11} = \frac{4}{11}$ **6.25** $3\left(\frac{3}{9} \cdot \frac{4}{9} \cdot \frac{4}{9} + \frac{3}{9} \cdot \frac{3}{9} \cdot \frac{4}{9}\right) =$

$\frac{28}{81}$ **6.26** $\frac{3}{6} \cdot \frac{3}{5} \cdot \frac{2}{4} \cdot \frac{2}{3} \cdot \frac{1}{2} \cdot \frac{1}{1} = \frac{1}{20}$ **6.27** $\frac{3}{6} \cdot \frac{2}{5} \cdot \frac{1}{4} \cdot \frac{3}{3} \cdot \frac{2}{2} \cdot \frac{1}{1} = \frac{1}{20}$ **6.28** $\frac{1}{2}\left(\frac{1}{3}\right)^2\left(\frac{1}{6}\right)^2 =$

$\frac{1}{648}$ **6.29** $2\{[\,P(\text{RRR})\,]^2 + 3[\,P(\text{RRB})\,]^2\} = 2\left[\left(\frac{3}{6} \cdot \frac{2}{5} \cdot \frac{1}{4}\right)^2 + 3 \cdot \left(\frac{3}{6} \cdot \frac{2}{5} \cdot \frac{3}{4}\right)^2\right] = \frac{7}{50}$ (the

answer for sampling with replacement is $\frac{1}{8}$)

Section 6.3

6.30 5/11   **6.31** $\frac{4\binom{13}{5}}{\binom{52}{5}} \doteq .0020$ **6.32** $\frac{\binom{80}{5}}{\binom{100}{5}} \doteq .3193$ **6.33** $\frac{N_1 N_2 + 1}{\binom{N}{N_1}}$ **6.34** $\frac{\binom{7}{6}\binom{3}{1}}{\binom{10}{7}} \cdot \frac{2}{3} = \frac{7}{60}$

6.35 $\frac{\binom{8}{4}\binom{4}{2}}{\binom{14}{8}}$ **6.37** $\frac{\binom{20}{4}\binom{80}{16}}{\binom{100}{20}} \doteq .2437$ **6.38** 5, 6   **6.39** a $\frac{\binom{13}{4}\binom{13}{3}\binom{13}{3}\binom{13}{3}}{\binom{52}{13}} \doteq .0263$

b $\frac{\binom{13}{4}\binom{13}{3}\binom{13}{3}\binom{13}{3}}{\binom{26}{6}\binom{26}{7}} \doteq .1104$ **6.40** a $\frac{\binom{4}{3}\binom{4}{2}}{\binom{52}{5}}$ b $13 \cdot 12 \cdot \frac{\binom{4}{3}\binom{4}{2}}{\binom{52}{5}} \doteq .00144$ **6.41** a $\frac{\binom{4}{2}\binom{4}{2}\binom{44}{1}}{\binom{52}{5}}$

b $\binom{13}{2}\frac{\binom{4}{2}\binom{4}{2}\binom{44}{1}}{\binom{52}{5}} \doteq .0475$ **6.42** a $\frac{4^5}{\binom{52}{5}}$ b $\binom{13}{5}\frac{4^5}{\binom{52}{5}} \doteq .507$ **6.43** a $\frac{\binom{4}{2}\binom{4}{3}\binom{4}{6}\binom{4}{1}}{\binom{10}{4}}$

b $\frac{\binom{4}{2}\binom{4}{3}\binom{4}{6}\binom{4}{1} + \binom{4}{2}\binom{4}{0}\binom{4}{6}\binom{4}{6} + \binom{4}{1}\binom{4}{2}\binom{4}{1}\binom{4}{0}}{\binom{10}{4}}$ **6.44** $\frac{\binom{8}{5}\binom{7}{5}\binom{6}{1}}{\binom{21}{11}} \cdot \frac{2}{5}$

Section 6.4

6.47 $(x + y)/(x + \frac{1}{2})$ for $0 < y < 1$ and zero elsewhere   **6.49** a The uniform distribution on $(0, 8 - 2y)$ b $\mu(y) = 4 - y$ and $\sigma^2(y) = (4 - y)^2/3$ **6.50** a The gamma distribution with shape parameter 4 and inverse-scale parameter $1 + y$ b $\mu(y) = 4/(1 + y)$ and $\sigma^2(y) = 4/(1 + y)^2$ **6.51** a $f_X(x) = \frac{\Gamma(\alpha+\gamma)\tau^\gamma}{\Gamma(\alpha)\Gamma(\gamma)} \frac{x^{\alpha-1}}{(\tau+x)^{\alpha+\gamma}}$ for $x > 0$ b Gamma with shape parameter $\alpha + \gamma$ and inverse-scale parameter $\tau + x$ **6.52** a Gamma with shape parameter γ and inverse-scale parameter τ b Gamma with shape parameter α and inverse-scale parameter y **6.53** a Note that $(1 - e^{-x})(1 - e^{-y}) > 0$ for $x, y > 0$.   b Exponential with mean 1   c $4e^{-x-2y} - 2e^{-x-y} - 2e^{-2y} + 2e^{-y}$ for $y > 0$ d $\mu(x) = 3/2 - e^{-x}$

and $\sigma^2(x) = 5/4 - e^{-2x}$ for $x > 0$ **6.55** $\begin{bmatrix} 1 & 0 & 0 \\ 0 & 1 & 0 \\ -\beta_1 & -\beta_2 & 1 \end{bmatrix}$ **6.56** Normal with mean

$\beta_0 + \beta_1 x + \beta_2 x^2$ and variance σ^2 **6.58** a Gamma with parameters α and $\beta g(\mathbf{x})$ b Normal with mean 0 and variance $\sigma^2 g^2(\mathbf{x})$

Section 6.5

6.59 $(3x + 2)/(6x + 3)$ **6.60** $x^2/(2x - 1)$ (see Example 2.3 in Section 2.1 and Problem 6.11 in Section 6.1)   **6.64** a $\mu\,\text{var}(N)$ b μN c $\sigma^2 EN$ **6.65** Use the formula $\mu(x) = \mu + \rho(x - \mu)$, where $\rho = \text{cor}(X, Y)$.   **6.66** $-1/2$ **6.67** a $\frac{\alpha_2}{\alpha_2+\alpha_3}(1 - y_1)$ b $\frac{\alpha_2\alpha_3}{(\alpha_2+\alpha_3)^2(\alpha_2+\alpha_3+1)}(1 - y_1)^2$

c See Problem 5.67(b) in Section 5.5.   **6.68** $\frac{\mu\pi}{1-\pi}$, $\frac{\sigma^2\pi}{1-\pi} + \frac{\mu^2\pi}{(1-\pi)^2}$ **6.69** $\frac{\beta(1-\pi)}{\pi}$, $\frac{\beta^2(1-\pi^2)}{\pi^2}$ (apply Problem 6.68 with π replaced by $1 - \pi$)   **6.70** a λET b $\lambda ET + \lambda^2\text{var}(T)$ c $\lambda\text{var}(T)$ d λT e λET

Section 6.6

6.71 a $\hat{Y} = 2, 10/3, 24/5$ according as $X = 1, 2, 3$ b $52/135$ **6.72** a $2(5 - Y)/3$ b $8/3$ **6.73** a $4/(1 + Y)$ b $12/5$ **6.75** a $(\alpha + \gamma)/(\tau + X)$ b $\frac{\gamma(\gamma+1)}{(\alpha+\gamma+1)\tau^2}$ **6.76** a α/Y

b $\frac{\alpha\tau^2}{(\gamma-2)(\gamma-1)}$ for $\gamma > 2$ and infinity for $\gamma \le 2$ **6.77** a $5/4, 1/4, 1/4$ b $(X + 3)/4$, 15/16   c $3/2 - e^{-X}, 11/12$ **6.78** $(42 - X^2)/(13 - 2X)$ **6.79** See the answers to parts (b) and (c) of Problem 6.17 in Section 6.2.   **6.80 a** $\mu(X)$, where $\mu(1) = 4$, $\mu(2) = 5$, and $\mu(3) = 7$ **b** 11/3   **6.81 a** Observe that $Y - g(\mathbf{X}) = [\, Y - h(\mathbf{X})\,] + [\, h(\mathbf{X}) - g(\mathbf{X})\,]$ and that $Y - h(\mathbf{X}) = h(\mathbf{X})(W - 1)$.

Section 6.7

6.82 Normal with mean 4 and variance 3   **6.83 a** Normal with mean $\mu_2 + \rho\frac{\sigma_2}{\sigma_1}(x - \mu_1)$ and variance $\sigma_2^2(1 - \rho^2)$ **b** Normal with mean $\mu_1 + \rho\frac{\sigma_1}{\sigma_2}(y - \mu_2)$ and variance $\sigma_1^2(1 - \rho^2)$
6.84 a Normal with mean $3x_1 + 100$ and variance 25 **b** $\hat{Y} = 3X_1 + 100$ and $\text{MSE}(\hat{Y}) = 25$
c Normal with mean $2X_1 + X_2 + 50$ and variance 16 **d** $\hat{Y} = 2X_1 + X_2 + 50$ and $\text{MSE}(\hat{Y}) = 16$ **6.85 b** Normal with mean $\beta_1 x_1 + \beta_2 x_2$ and variance γ_3^2 **d** Normal with mean $(\beta_1 + \beta_2\lambda)x_1$ and variance $\beta_2^2\gamma_2^2 + \gamma_3^2$ **6.86 a** $\alpha + \beta_1\mu_1 + \beta_2\mu_2$ **b** $\text{cov}(X_1, Y) = \beta_1\sigma_1^2 + \beta_2\sigma_{12}$ and $\text{cov}(X_2, Y) = \beta_1\sigma_{12} + \beta_2\sigma_2^2$ **c** $\sigma^2 + \beta_1^2\sigma_1^2 + 2\beta_1\beta_2\sigma_{12} + \beta_2^2\sigma_2^2$ **6.87 a** 100, 500, 400, $\sqrt{4/5}$ **b** Bivariate normal with mean vector $[\,100, 100\,]^T$ and variance-covariance matrix

$$\begin{bmatrix} 400 & 400 \\ 400 & 500 \end{bmatrix}$$

c Normal with mean $20 + 4y/5$ and variance 80 **d** $\Phi(2/\sqrt{5}) \doteq .814$ **6.88 a** Bivariate normal with mean $[\,\mu, \beta\mu\,]^T$ and variance–covariance matrix

$$\begin{bmatrix} \sigma_1^2 & \beta\sigma_1^2 \\ \beta\sigma_1^2 & \sigma^2 + \beta^2\sigma_1^2 \end{bmatrix}$$

b Normal with mean $\frac{\sigma^2\mu + \beta\sigma_1^2 y}{\sigma^2 + \beta^2\sigma_1^2}$ and variance $\frac{\sigma^2\sigma_1^2}{\sigma^2 + \beta^2\sigma_1^2}$ **6.89** 2, 4, 1, 9   **6.90** See the answers to Problem 6.88(b).   **6.91 a** Trivariate normal with mean vector 0 and variance–covariance matrix

$$\begin{bmatrix} 1 & 1 & 2 \\ 1 & 3 & 4 \\ 2 & 4 & 9 \end{bmatrix}$$

b Normal with mean $x_1 + x_2$ and variance 3 (use Theorem 6.2 in Section 6.4) **c** Normal with mean $x_2/11 + 2y/11$ and variance 6/11 (use Theorem 6.6 with $\mathbf{X}$ replaced by $[\,X_2, Y\,]^T$ and Y replaced by X_1)   **6.92** Money spent on health care for a given person may increase that person's life expectancy, but people in poor health may have both unusually large expenses for health care and below normal life expectancy.   **6.93** The additional government transfer payments could have an effect on the social health of such families or on their formation.

Section 6.8

6.94 a 49/2, 49/2 **b** .164   **6.95** $(n + 1)/(n + 2)$ **6.96 a** Gamma with shape parameter 1 and inverse-scale parameter 1 **b** Gamma with shape parameter $1 + Y$ and inverse-scale parameter 2 **c** $(1 + Y)/2$ **6.97 a** Normal with mean 113.85 and variance 69.23 **b** 113.85   **6.98** 1/4 [here Eq. (47) gets replaced by $\hat{\lambda} = (\alpha + Y)/(\tau + 2)$]   **6.99 a** Normal with mean $20 + 4Y/5$ and variance 80 **b** 108   **6.100** Normal with mean $(n^{-1}\sigma^2\mu + \gamma^2\bar{Y})/(n^{-1}\sigma^2 + \gamma^2)$ and variance $n^{-1}\sigma^2\gamma^2/(n^{-1}\sigma^2 + \gamma^2)$ ($\bar{Y}$ has mean θ and variance $n^{-1}\sigma^2$)   **6.101 a** Beta with parameters $\alpha_1 + Y$ and $\alpha_2 + 1$ **b** $(\alpha_1 + Y)/(\alpha_1 + \alpha_2 + 1 + Y)$ **6.102 a** Gamma with shape parameter $\gamma + \alpha$ and inverse-scale parameter $\tau + Y$
b $(\gamma + \alpha)/(\tau + Y)$ **6.103 a** Gamma with shape parameter $\alpha + 1/2$ and inverse-scale parameter $\tau + Y^2/2$ **b** $(\alpha + 1/2)/(\tau + Y^2/2)$ **6.104 a** Uniform distribution on $\{0, \ldots, n\}$
b $n\alpha_1/(\alpha_1 + \alpha_2)$ (use the fact that the probability function of the beta-binomial distribution

with parameters $\alpha_1 + 1$, α_2 and $n - 1$ has sum 1) **6.105** Use the fact that the probability function of the negative binomial distribution with parameters $\alpha + 2$ and π has sum 1 **6.106** $\pi = 1 - \mu/\sigma^2$, $\alpha = \mu^2/(\sigma^2 - \mu)$ **6.107 a** Beta distribution with parameters $\alpha_1 + Y$ and $\alpha_2 + \alpha$ **b** $(\alpha_1 + Y)/(\alpha_1 + \alpha_2 + \alpha + Y)$

Chapter 7

Section 7.1

7.1 a 0.1 **b** 0.0447 **c** $(0.012, 0.188)$ **d** The hypothesis is rejected. **e** $2[1 - \Phi(2.236)] \doteq .0253$ **7.2 a** 31.623 **b** $(378.02, 501.98)$ **c** $Z \doteq 1.897$, so $|Z| < 1.96$ and hence the hypothesis is accepted. **d** $2[1 - \Phi(1.897)] \doteq .058$ **7.3** $\Phi(-3.541) + \Phi(-0.379) \doteq .353$ **7.4** $4(z_{.9} + z_{.975})^2 = 42$ **7.6** 387.985   **7.8** $Z \doteq 1.897 > z_{.95} \doteq 1.645$, so the hypothesis is rejected.   **7.9** .029   **7.11** $\Phi(-0.732) \doteq .232$ **7.12** $8(z_{.9} + z_{.95})^2 \doteq 69$ **7.13** Differentiate with respect to δ and use the formula $\Phi'(z) = \phi(z) = (1/\sqrt{2\pi})\exp(-z^2/2)$.

Section 7.2

7.14 b $1 - e^{-y/2}$ **c** $2\log\frac{1}{1-p}$ **7.15** $1 - 5e^{-4} \doteq .908$ [use Eq. (8) in Section 3.1] **7.16 a** Use Theorem 3.2 in Section 3.1. **b** Use Corollary 1.2 in Section 1.7. **7.17 a** Gamma distribution with shape parameter n and scale parameter β/c **b** If $c = \beta/2$, then $(Y_1 + \cdots + Y_n)/c$ has the chi-square distribution with $2n$ degrees of freedom.   **7.18 a** $f(\chi) = \chi^{\nu-1}e^{-\chi^2/2}/[2^{\nu/2-1}]\Gamma(\nu/2)]$ for $\chi > 0$ (see Theorem 3.3 in Section 3.1) **b** $\sqrt{2}\Gamma((\nu + 1)/2)/\Gamma(\nu/2)$ **c** $\nu - 2[\Gamma((\nu + 1)/2)/\Gamma(\nu/2)]^2$ **7.19** $18z_{.975}^2 \doteq 69$ (actually, $\chi_{.975,65}^2/\chi_{.025,65}^2 \doteq 1.999$) **7.20** $\nu \geq 3$ **7.21** See Problem 3.13 in Section 3.1. **7.23 a** $\alpha_1 F_{p,2\alpha_1,2\alpha_2}/(\alpha_2 + \alpha_1 F_{p,2\alpha_1,2\alpha_2})$ **b** $\alpha_2 y_p/[\alpha_1(1 - y_p)]$

Section 7.3

7.24 $(-0.030, 0.230)$ **7.26 a** $2\sigma^4/(n - d)$ **b** $\sigma\sqrt{2/(n-d)}\,\Gamma((n - d + 1)/2)/\Gamma((n - d)/2)$ **7.27 a** $\sigma^2\left(\frac{1}{n_1} + \cdots + \frac{1}{n_d}\right)$, $\sigma\sqrt{\frac{1}{n_1} + \cdots + \frac{1}{n_d}}$, $S\sqrt{\frac{1}{n_1} + \cdots \frac{1}{n_d}}$ **b** $\sigma^2 n$, $\sigma\sqrt{n}$, $S\sqrt{n}$, where $n = n_1 + \cdots + n_d$ **7.28 a** $(17.31, 17.85)$ **b** $(0.359, 1.121)$ **c** $(17.89, 18.16)$ **d** $(-0.124, 0.414)$ **7.29 a** 5.017, 0.179, 0.423 **b** $(17.842, 18.170)$ **7.30 a** 2, 3, 5, 2 **b** 24, 19 **c** 8, 42, 50 **d** 8, 21, 16.667 **e** 2.828, 4.583, 4.082 **f** $(14.813, 33.187)$ **g** $(-6.860, 16.860)$ **7.31 a** 24.944, 0.358 **b** 24.737, 25.151, $(24.688, 25.200)$ **7.32 a** 25.26, 0.476 **b** 0.177, 0.421 **c** $(24.66, 25.22)$, $(24.98, 25.54)$, $(-0.08, 0.71)$

Section 7.4

7.33 a $1 - t_8(1.808) \doteq .054$ **b** $1 - t_6(1.357) \doteq .112$ **c** $1 - t_{15}(2.258) \doteq .020$ **7.35 a** Accept [set $\tau = \mu_1 - (\mu_2 + \mu_3 + \mu_4 + \mu_5)/4$, write the hypothesis as $\tau \leq -0.1$, and note that $SE(\hat{\tau}) \doteq 0.423\sqrt{\frac{1}{11} + \frac{1}{40}}$]. **b** -0.128 **c** Note that $-0.128 < -0.1$. **d** $1 - t_{46}(1.484) \doteq .0723$ **7.36** The P-value for the t test of the hypothesis that $\mu_2 \geq \mu_4$ is given by P-value $\doteq t_{18}(2.731) \doteq .007$.   **7.37 a** $1 - t_9(2.158) \doteq .030$ **b** Yes, since the P-value for testing the hypothesis that $\mu \geq 25.3$ is given by P-value $\doteq t_9(-3.149) \doteq .006$ **7.38** No, since the P-value for testing the hypothesis that $\mu_2 \leq \mu_1$ is given by P-value $= 1 - t_{18}(1.690) \doteq .054$ **7.39** $8z_{.95}^2 \doteq 22$ **7.40** $8(z_{.9} + z_{.95})^2 \doteq 69$ **7.41** $\frac{3}{2}(z_{.9} + z_{.95})^2 \doteq 13$

Section 7.5

7.43 a The hypothesis is accepted. **b** $2[1 - t_4(2.142)] \doteq .099$ **7.44 a** Accept **b** $2t_{14}(-0.710) \doteq .489$ **7.45 a** $|t| \geq 2.262$ **b** $t \doteq 2.278$ lies in the critical region. **c** $(18.002, 18.638)$ **e** $.049$ **f** It contains 18.   **7.46** P-value $\doteq 2[1 - t_{46}(2.587)] \doteq .0129$ [The t test of size .05 rejects the hypothesis, while that of size .01 accepts it. The hypothesis itself is of dubious interest.]   **7.47 a** $2[1 - t_9(3.680)] \doteq .005$ **b** P-value $< .05$ and 95% confidence interval for μ_2 does not contain zero.   **7.50 a** 1.244, 0.220 **b** $(0.737, 1.752)$ **c** $2[1 - t_8(5.659)] \doteq .0005$ (these results are somewhat suspect because the data from patient 9 was removed after seeing this data and the data on the other nine patients)   **7.51 a** 0.833, 1.877, 2.078, 1.948 **b** 3.441, .941 **c** 3.658, 1.913 **d** 1.244, 0.902 **e** $(-0.667, 3.156)$ **f** $2[1 - t_{16}(1.380)] \doteq .187$ **g** Evidently invalid because of the large sample correlation given in the answer to part (b) (also suspect because of the reason given at the end of the answer to Problem 7.50)   **7.52 a** 0.41, 0.122 **b** 0.186 **c** $1 - t_9(3.349) \doteq .0043$ (switching from A to B would cause increased wear) **7.53 a** $S\sqrt{\frac{1}{5} + \frac{1}{5}} \doteq 2.478\sqrt{\frac{2}{5}} \doteq 1.567$, where S is the sample standard deviation of the ten averages (over the two shoes for each boy) based on the data in Problem 7.52 **b** Not substantially greater than .05   **7.54 a** $1.6(z_{.9} + z_{.975})^2 \doteq 17$ **b** $8(z_{.9} + z_{.975})^2 \doteq 84$ **7.55 a** $\text{cov}(Y_1, Y_2) = 399.5$, $\text{cor}(Y_1, Y_2) = .99875$ **b** 100 **c** $100(z_{.9} + z_{.975})^2 \doteq 1051$ **d** 80,000 **e** $80,000(z_{.9} + z_{.975})^2 \doteq 840,594$ **7.56** .952 **7.57 a** Bivariate normal with mean vector $[\mu_1, \mu_2]^T$ and variance–covariance matrix $\begin{bmatrix} \gamma + \delta & \gamma \\ \gamma & \gamma + \delta \end{bmatrix}$ **b** $\gamma/(\gamma + \delta)$

Section 7.6

7.58 a Accept **b** $1 - F_{1,14}(0.505) \doteq .489$, which agrees with the answer to Problem 7.44(b) in Section 7.5.   **7.59** Observe that $\bar{Y}_1, \ldots, \bar{Y}_d$ are independent and normally distributed with mean μ and variance σ^2/r and that $(\bar{Y}_1 + \cdots + \bar{Y}_d)/d = \bar{Y}$.   **7.60**

Source	SS	DF	MS	F	P-value
Between	0.371	3	0.124	0.640	.594
Within	7.146	37	0.193		
Total	7.518	40			

There is no evidence that the mean percentage of sugar content differs among varieties 1, 3, 4, and 5.   **7.61** P-value $\doteq 1 - F_{4,37}(1.856) \doteq .139$, so the data is compatible with the hypothesis.   **7.63** $1 - F_{1,9}(16.501) \doteq .003$, which agrees with the P-value for the t test that was found in the solution to Example 7.20 in Section 7.5.   **7.64** $1 - F_{1,18}(3.463) \doteq .079$, which agrees with the P-value for the t test that was found in the solution to Example 7.21 in Section 7.5   **7.65** $1 - F_{1,9}(11.215) \doteq .0085$, which is twice the P-value for the t test of the hypothesis that $\mu_2 \leq \mu_1$ given in the answer to Problem 7.52(c) in Section 7.5 **7.66 a** $2[1 - t_{18}(1.690)] \doteq .108$, which is twice the P-value given in the answer to Problem 7.38 in Section 7.4. **b** $1 - F_{1,18}(2.856) \doteq .108$ **7.67 a** $n(\bar{Y} - \mu_0)^2/S^2$ **b** $(\bar{Y} - \mu_0)/(S/\sqrt{n})$ **7.68** Note that $(\bar{Y}_k - \bar{Y})^2 = \bar{Y}_k^2 - 2\bar{Y}_k\bar{Y} + \bar{Y}^2$ and $(Y_{ki} - \bar{Y})^2 = Y_{ki}^2 - 2Y_{ki}\bar{Y} + \bar{Y}^2$.

Chapter 8

Section 8.1

8.1 $\frac{11}{5}x - 1$, 2.8   **8.2** $\frac{3}{7}x^2 - \frac{13}{35}x + 2$, 0.229   **8.3** **a** $a = x_1 - \frac{x_2 - x_1}{t_2 - t_1}t_1$ and $b = \frac{x_2 - x_1}{t_2 - t_1}$
b $x = x_1 + (x_2 - x_1)\frac{t - t_1}{t_2 - t_1}$ **c** $t = t_1 + \frac{x - x_1}{x_2 - x_1}(t_2 - t_1)$ **8.4** **a** $n\hat{\beta}_0 = \sum_i Y_i$ and $\hat{\beta}_1 \sum_i (x_i - \bar{x})^2 = \sum_i (x_i - \bar{x})Y_i$ **b** $\hat{\beta}_0 = \bar{Y} = (Y_1 + \cdots + Y_n)/n$ and $\hat{\beta}_1 = \sum_i (x_i - \bar{x})Y_i / \sum_i (x_i - \bar{x})^2$
8.5 **a** $\hat{\beta}_{10} = 69.5$, $\hat{\beta}_{11} = 4.775$, and $\hat{\mu}(x) = 69.5 + 4.775(x + 3) = 83.825 + 4.775x$ for $x < \hat{c}$ **b** $\hat{\beta}_{20} \doteq 84.267$, $\hat{\beta}_{21} = 0.9$, and $\hat{\mu}(x) \doteq 84.267 + 0.9(x - 3) = 81.567 + 0.9x$ for $x \geq \hat{c}$ **c** -0.583 **d** 11.122   **8.6** **a** $6\hat{\beta}_0 = \sum_i Y_i$, $70\hat{\beta}_1 = \sum_i x_i Y_i$, $1792\hat{\beta}_2/3 = \sum_i (x_i^2 - 35/3)Y_i$ **b** $\hat{\beta}_0 = \sum_i Y_i/6$, $\hat{\beta}_1 = \sum_i x_i Y_i/70$, $\hat{\beta}_2 = 3\sum_i (x_i^2 - 35/3)Y_i/1792$
c $\hat{\beta}_0 \doteq 76.883$, $\hat{\beta}_1 = 2.547$, $\hat{\beta}_2 = -0.337$, $\hat{\mu}(x) \doteq 76.883 + 2.547x - 0.337(x^2 - 35/3) \doteq 80.816 + 2.547x - 0.337x^2$ (which agrees with the result in the text)   **8.7** **a** 3.779
b 85.628   **8.8** **a** $245 + 5\hat{x}_{max} \doteq 263.893$ **b** 85.628   **8.9** $n\hat{\beta}_0 + \hat{\beta}_1 \sum_i e^{\hat{\beta}_2 x_i} = \sum_i Y_i$,
$\hat{\beta}_0 \sum_i e^{\hat{\beta}_2 x_i} + \hat{\beta}_1 \sum_i e^{2\hat{\beta}_2 x_i} = \sum_i e^{\hat{\beta}_2 x_i} Y_i$, $\hat{\beta}_0 \hat{\beta}_1 \sum_i x_i e^{\hat{\beta}_2 x_i} + \hat{\beta}_1^2 \sum_i x_i e^{2\hat{\beta}_2 x_i} = \hat{\beta}_1 \sum_i x_i e^{\hat{\beta}_2 x_i} Y_i$
(which is a nonlinear system of equations in $\hat{\beta}_0$, $\hat{\beta}_1$ and $\hat{\beta}_2$)   **8.10** **a** $58.9\frac{(x+1)(x-5)}{40} - 78.0\frac{(x+5)(x-5)}{24} + 85.2\frac{(x+5)(x+1)}{60} \doteq 80.987 + 2.630x - 0.357x^2$ **b** 3.678, 85.824 **c** 263.392
8.11 **a** $(x+3)(x+1)(x-1)(x-3)/9$ **b** 2.236 **c** -1.778 **8.12** **a** $(x+4)(x+2)x(x-2)(x-4)/45$ **b** -3.289 **c** 2.582   **8.13** **a** $p \leq 5$ **b** $p \geq 5$ **c** $p = 5$

Section 8.2

8.14 **a** 162 **b** 54   **8.15** **a** 2048 **b** 512   **8.16** 3   **8.17** **a** 72 **b** 36 runs at each level of A, 36 at each level of B, 24 at each level of C, and 24 at each level of D **8.18** 48 runs at the low level of E and 24 at the high level of E **8.19** $x_1 + x_2 + x_3 + x_4 \equiv 0 \pmod 3$ **8.20**

x_1	x_2	x_3	x_4
0	0	0	0
0	1	2	1
0	2	1	2
1	0	2	2
1	1	1	0
1	2	0	1
2	0	1	1
2	1	0	2
2	2	2	0

8.21 $x_1 + x_2 + x_3 \equiv 0 \pmod 3$ and $x_1 + 2x_2 + x_4 \equiv 2 \pmod 3$ **8.22**

x_1	x_2	x_3	x_4
0	0	2	1
0	1	1	2
0	2	0	0
1	0	1	0
1	1	0	1
1	2	2	2
2	0	0	2
2	1	2	0
2	2	1	1

8.23

x_1	x_2	x_3	x_4
−1	−1	−1	−1
1	−1	−1	1
−1	1	−1	1
1	1	−1	−1
−1	−1	1	1
1	−1	1	−1
−1	1	1	−1
1	1	1	1

Section 8.3

8.25 $\beta_0 = \mu(0)$, $\beta_1 = [\mu(1) - \mu(-1)]/2$, $\beta_2 = [\mu(-1) + \mu(1) - 2\mu(0)]/2$ **8.26 a** $x_{min} = -1$ and $x_{max} = 1$ **b** Mean response increases by 1.   **8.27 a** $\beta_5 c_1 c_2$ **b** $\beta_5 c_1 c_2$
c β_5 **8.28 a** $\beta_2 > 0$, $\beta_4 > 0$, $\beta_5^2 < 4\beta_2\beta_4$ **b** $x_1 = (\beta_3\beta_5 - 2\beta_1\beta_4)/(4\beta_2\beta_4 - \beta_5^2)$ and
$x_2 = (\beta_1\beta_5 - 2\beta_2\beta_3)/(4\beta_2\beta_4 - \beta_5^2)$ **8.29 a** $\beta_1 + 3\beta_2$ **b** $\beta_1 + 3\beta_2$ **c** $\beta_0 + \frac{3}{2}\beta_1 + \frac{7}{3}\beta_2$
d $\beta_1^2 + 6\beta_1\beta_2 + \frac{28}{3}\beta_2^2$ **e** β_2 **8.30** β_1 is the change in the mean response caused by adding
1 unit to the level of the first factor, with the second and third factors held fixed; β_2 is the
change in the mean response caused by adding 1 unit to the level of the second factor, with
the first and third factors held fixed; and β_3 is the change in the mean response caused by
adding 1 unit to the level of the third factor, with the first and second factors held fixed.
8.31 a $\beta_1 + 3\beta_2 + 7\beta_3$ **b** No, because (i) it is impossible to add 1 unit to $x_1 = x$ while
holding $x_3 = x^3$ fixed and (ii) the suggested interpretation contradicts the answer to part (a)
8.33 a $\beta_0 = \mu_3$, $\beta_1 = \mu_1 - \mu_3$, $\beta_2 = \mu_2 - \mu_3$ **b** $\mu_1 = \beta_0 + \beta_1$, $\mu_2 = \beta_0 + \beta_2$, $\mu_3 = \beta_0$
8.34 β_1 is the change in the mean response caused by changing the level of the first factor
from 3 to 1, with the second factor held fixed; β_2 is the change in the mean response caused
by changing the level of the first factor from 3 to 2, with the second factor held fixed; $\beta_2 - \beta_1$
is the change in the mean response caused by changing the level of the first factor from 1 to 2,
with the second factor held fixed; and β_3 is the change in the mean response caused by adding
1 unit to the level of the second factor with the first factor held fixed.   **8.36** $\sigma^2 \mathbf{I}$, where $\mathbf{I}$ is
the $d \times d$ identity matrix   **8.37 a** Observe that $(a - \frac{a+b}{2})^2 + (b - \frac{a+b}{2})^2 = \frac{(b-a)^2}{2}$. **b** See
the solution to Example 8.6(b)   **8.38** 26

Chapter 9

Section 9.1

9.1 linearly independent   **9.4** $[9, 6, 1, -12, 4, -4]^T$ **9.6 a** 6 **b** 15 **c** $\frac{M^2}{2} + \frac{3M}{2} + 1$

9.7 a 10 **b** 20 **c** $\frac{M^3}{6} + M^2 + \frac{11M}{6} + 1$ **9.11**   **a** $\begin{bmatrix} 1 & 0 & 0 \\ 0 & 1 & 0 \\ -c & 0 & 1 \end{bmatrix}$ **b** $\begin{bmatrix} 1 & 0 & 0 \\ 0 & 1 & 0 \\ c & 0 & 1 \end{bmatrix}$

9.12 a $\begin{bmatrix} 1 & 1 & 1 \\ 0 & 1 & 2 \\ 0 & 1 & 4 \end{bmatrix}$ **b** $\begin{bmatrix} 1 & -\frac{3}{2} & \frac{1}{2} \\ 0 & 2 & -1 \\ 0 & -\frac{1}{2} & \frac{1}{2} \end{bmatrix}$

Section 9.2

9.14 Nonidentifiable since $x_1 + x_2 + x_3 - 3 = 0$ on the design set   **9.15** Let g be a quadratic polynomial that has three or more distinct real roots. Then, by Rolle's theorem, the first derivative of g has two or more distinct real roots. Since the first derivative of g is a linear function, it must be the zero function. Thus, g must be the constant function and hence the zero function.   **9.16** Use Problem 9.15 and its answer.   **9.18** See Problems 9.15 and 9.16 and their answers and use induction.   **9.22 a** Nonidentifiable since the function $x_3 - 2x_1 - 3x_2$ is a nonzero function in G that equals zero on the design set **b** Identifiable, since, if $b_0 + b_1x_1 + b_2x_2$ equals zero at the first three design points, then $b_0 = b_1 = b_2 = 0$ **9.23** Yes, because if we apply the corresponding permutation to the entries of **b**, $\mathbf{Xb}$ is unchanged

9.24 a $\begin{bmatrix} 1 & 1 & 0 & 0 \\ 1 & 0 & 1 & 0 \\ 1 & 0 & 0 & 1 \end{bmatrix}$ **b** $[1, -1, -1, -1]^T$ **9.25 a** $\begin{bmatrix} 1 & 0 & 0 \\ 0 & 1 & 0 \\ 0 & 0 & 1 \end{bmatrix}$ **b** $\begin{bmatrix} 1 & -1 & 0 \\ 1 & 0 & 0 \\ 1 & 1 & 1 \end{bmatrix}$

Section 9.3

9.26 Unsaturated   **9.29**   **a** $-(x^3 + 9x^2 - x - 105)/48 = -(x - 3)(x + 5)(x + 7)/48$
9.30 We can choose $\mathcal{X}'$ to consist of any four noncoplanar points in $\mathbb{R}^3$. One such choice is given by $\mathcal{X}' = \{(0, 0, 0), (1, 0, 0), (0, 1, 0), (0, 0, 1)\}$.   **9.31**   **a** $\begin{bmatrix} 1 & -3 & 9 & -27 \\ 1 & -1 & 1 & -1 \\ 1 & 1 & 1 & 1 \\ 1 & 3 & 9 & 27 \end{bmatrix}$

9.32 The three points do not lie on a straight line.   **9.33** The four points do not lie on a plane.   **9.34 a** The $d \times d$ identity matrix **b** $(\mathbf{X}^{-1})^T = (\mathbf{X}^T)^{-1}$ **9.35 a** $\begin{bmatrix} 1 & 1 & 0 \\ 1 & 0 & 1 \\ 1 & -1 & 0 \end{bmatrix}$

b $(1 + x_1 - x_2)/2$, x_2, $(1 - x_1 - x_2)/2$.   **9.36**   **a** $\begin{bmatrix} 1 & 0 & 0 & 0 \\ 1 & 1 & 0 & 0 \\ 1 & 0 & 1 & 0 \\ 1 & 1 & 1 & 1 \end{bmatrix}$ **9.37** Unsaturated

by Theorem 9.10   **9.38** $g_1(x_1, x_2) = (x_2 - 2)(x_2 - 3)/2$, $g_2(x_1, x_2) = (x_1 - 1)(x_1 - 2)/2$, $g_3(x_1, x_2) = (x_2 - 1)(x_2 - 2)/2$, $g_4(x_1, x_2) = (x_1 - 2)(x_1 - 3)/2$

Section 9.4

9.39 Yes　**9.41 b** $\|g_k\|^2 = w_k$ for $1 \le k \le d$　**9.42 a** $1, x, x^2 - 5, x^3 - \frac{41}{5}x$　**b** $4r, 20r, 64r,$
$576r/5$　**9.43 a** $1, x, x^2 - 2, x^3 - \frac{17}{5}x$　**b** $5r, 10r, 14r, 72r/5$　**9.44 b** $8r$　**d**

$$\begin{bmatrix} 1 & -1 & -1 & -1 & 1 & 1 & 1 & -1 \\ 1 & 1 & -1 & -1 & -1 & -1 & 1 & 1 \\ 1 & -1 & 1 & -1 & -1 & 1 & -1 & 1 \\ 1 & 1 & 1 & -1 & 1 & -1 & -1 & -1 \\ 1 & -1 & -1 & 1 & 1 & -1 & -1 & 1 \\ 1 & 1 & -1 & 1 & -1 & 1 & -1 & -1 \\ 1 & -1 & 1 & 1 & -1 & -1 & 1 & -1 \\ 1 & 1 & 1 & 1 & 1 & 1 & 1 & 1 \end{bmatrix}$$

e $8r\mathbf{I}_8$.　**9.45 a** Same as the answer to Problem 9.44(d)　**b** $8r\mathbf{I}_8$　**d** $8r$　**9.46　a** Same
as the answer to Problem 9.44(d)　**b** $8r\mathbf{I}_8$　**d** $8r$　**9.47 a** $\begin{bmatrix} 4 & 0 & 20 \\ 0 & 20 & 0 \\ 20 & 0 & 164 \end{bmatrix}$　**b** $\mathbf{L} =$
$\begin{bmatrix} 2 & 0 & 0 \\ 0 & 2\sqrt{5} & 0 \\ 10 & 0 & 8 \end{bmatrix}$　**c** $1/2, x/(2\sqrt{5}), (x^2 - 5)/8$　**9.49** Let $b_1, \ldots, b_p \in \mathbb{R}$, at least one of
which is nonzero; set $g = b_1 g_1 + \cdots + b_p g_p$; and use Eq. (23) to conclude that $\|g\|^2 > 0$
and hence that $g \ne 0$.　**9.52** $\mathbf{b} = \mathbf{M}^{-1}\mathbf{c}$　**9.53 a** The basis functions are not orthogonal
since $\langle x_1, x_3 \rangle = 14$, G is identifiable by Theorem 9.18, and it is saturated by Theorem 9.10 in
Section 9.3.　**b** The basis functions are orthogonal, and G is identifiable but unsaturated.

Section 9.5

9.54 $\frac{17}{5}x$　**9.55 b** Both sides equal $\langle P_G h_1, P_G h_2 \rangle$.　**9.57 a** $-2/5 = -0.4$　**b** $887/400 =$
2.2175　**9.58 a** $(-18 + \sqrt{336})/6 \doteq 0.055$　**b** 2.188　**9.59 a** $1 + \sqrt{2}, 0, -2 - \sqrt{2}$
b $-\frac{2+\sqrt{2}}{14}x^2 + \frac{17+12\sqrt{2}}{35}$　**c** $2, \frac{12}{35}(3 + 2\sqrt{2}), \frac{34}{35} - \frac{24}{35}\sqrt{2}$　**9.66 a** $\sum_k h(x_k')g_k$　**9.68** $g_0 =$
$P_{G_0} g$　**9.69 a** $7x_1/5$　**b** $1, x_1, x_2, x_3 - 7x_1/5$

Section 9.6

9.70 $4x - \frac{10}{3}$　**9.71 a** $-\sqrt{2}/2, \sqrt{2}/2, \sqrt{2}/2, -\sqrt{2}/2$　**b** $0, 0, -8\sqrt{2}$　**c** $4b_0^* + 20b_2^* =$
$0, 20b_1^* = 0, 20b_0^* + 164b_2^* = -8\sqrt{2}$　**d** $5\sqrt{2}/8, 0, -\sqrt{2}/8$　**e** 2　**f** 0　**9.72 a** $-1,$
$-\sqrt{2}/2, 0, \sqrt{2}/2, 1$　**b** $0, 4 + \sqrt{2}, 0, 16 + \sqrt{2}$　**c** $5b_0^* + 10b_2^* = 0, 10b_1^* + 34b_2^* = 4 + \sqrt{2},$
$10b_0^* + 34b_0^* = 0, 34b_1^* + 130b_2^* = 16 + \sqrt{2}$　**d** $0, (-1 + 4\sqrt{2})/6, 0, (1 - \sqrt{2})/6$　**e** 3　**f** 0
9.74 b $b_1^*\langle g_1 - \bar{g}_1, g_1 - \bar{g}_1 \rangle + b_2^*\langle g_1 - \bar{g}_1, g_2 - \bar{g}_2 \rangle = \langle g_1 - \bar{g}_1, h \rangle, b_1^*\langle g_2 - \bar{g}_2, g_1 - \bar{g}_1 \rangle +$
$b_2^*\langle g_2 - \bar{g}_2, g_2 - \bar{g}_2 \rangle = \langle g_2 - \bar{g}_2, h \rangle$　**9.76** $h(x) - h(0)$ is an odd function.　**9.77** Write
$P_G h = g_1 + g_2$, where g_1 and g_2 are the orthogonal projections of $P_G h$ onto G_1 and G_2,
respectively.　**9.79** The space of polynomials of degree $d - 1$ is identifiable and saturated.
9.81 $5 + 4x_2$

Chapter 10

Section 10.1

10.1 $n\hat{\beta}_0 = \sum_i Y_i$, $\hat{\beta}_1 \sum_i (x_{i1} - \bar{x}_1)^2 + \hat{\beta}_2 \sum_i (x_{i1} - \bar{x}_1)(x_{i2} - \bar{x}_2) = \sum_i (x_{i1} - \bar{x}_1)Y_i$, $\hat{\beta}_1 \sum_i (x_{i1} - \bar{x}_1)(x_{i2} - \bar{x}_2) + \hat{\beta}_2 \sum_i (x_{i2} - \bar{x}_2)^2 = \sum_i (x_{i2} - \bar{x}_2)Y_i$ **10.2 b** Apply Equation (4) with $g = 1$ **c** Use parts (a) and (b). **10.3** Use Equation (44) in Section 9.5. **10.4** Use Equation (8) with $g = \hat{\mu}(\cdot)$. **10.7 a** $\|I_k\|^2 = n_k$ for $1 \le k \le d$ (see Corollary 9.11 in Section 9.4)

b $\hat{\mu} = \bar{Y}_k$ for $1 \le k \le d$ **10.8 b** $\begin{bmatrix} 6 & 0 \\ 0 & 70 \end{bmatrix}$ **c** $\begin{bmatrix} 0.167 & 0 \\ 0 & 0.0143 \end{bmatrix}$ **d** $[461.3, 178.3]^T$

e $76.833 + 2.547x$ **10.9 b** $\begin{bmatrix} 6 & 0 & 0 \\ 0 & 70 & 0 \\ 0 & 0 & 597.333 \end{bmatrix}$ **c** $\begin{bmatrix} 0.167 & 0 & 0 \\ 0 & 0.0143 & 0 \\ 0 & 0 & 0.00167 \end{bmatrix}$

d $[461.3, 178.3, -201.333]^T$ **e** $76.883 + 2.547x - 0.337(x^2 - \frac{35}{3})$ **10.11 b** Use Eq. (13) **10.13** Use Theorem 9.28 in Section 9.5.

Section 10.2

10.15 a 527.088, 72.933, 454.156 **b** .862 **10.16 a** See Problem 10.10 in Section 10.1. **10.17 a** Apply Eq. (8) in Section 10.1 with $g = 1$. **10.19 a** Apply Eq. (8) in Section 10.1 with $g = \hat{\mu}(\cdot) - \bar{Y}$. **b** Use part (a). **10.21 a** Apply Eq. (8) in Section 10.1 with $g = \hat{\mu}(\cdot)$. **10.24** 0, .862, .990, .993, .999, 1 **10.25** Observe that $\hat{\mu}(\mathbf{x}'_k) = \bar{Y}(\mathbf{x}'_k) = \bar{Y}_k$ for $1 \le k \le d$; alternatively, use Theorem 10.5 and observe that $\mu^*(\cdot) = \mu(\cdot)$ on the design set. **10.26** Use Eq. (22) **10.27 a** Note that $\text{cov}(\hat{\tau}_1, \hat{\tau}_2) = \text{cov}(\sum_i h_1(\mathbf{x}_i)Y_i, \sum_i h_2(\mathbf{x}_i)Y_i)$.

Section 10.3

10.28 Normal with mean β_1 and variance $\sigma^2 / \sum_i (x_i - \bar{x})^2$ **10.33 a** $\sigma^2/2r$ **b** 8 **10.34 a** $\sigma^2/8r$ **b** 2 **10.38** Use Eq. (6) in Section 9.1 or Theorem 10.3 in Section 10.1.

10.39 b $\begin{bmatrix} n & n_1 & \cdots & n_{d-1} \\ n_1 & n_1 & \cdots & 0 \\ \vdots & \vdots & & \vdots \\ n_{d-1} & 0 & \cdots & n_{d-1} \end{bmatrix}$ (the $d-1 \times d-1$ matrix consisting of the last $d-1$ rows and last $d-1$ columns of $\mathbf{X}^T\mathbf{X}$ is a diagonal matrix) **f** Note that G is saturated. **10.40 a** (72.043, 81.723), (1.130, 3.964) **b** $76.883 + 2.547x \pm 2.776\sqrt{3.039 + 0.260x^2}$ **c** (6.781, 23.785) **d** (4.521, 15.857) **e** (1.130, 3.964) **10.41 a** (75.194, 78.573), (2.053, 3.042), (-0.506, -0.168) **b-e** [See the solution to parts (b)-(e) of Example 10.11.]

Section 10.4

10.42 $\hat{\beta}_5$, $\hat{\beta}_1 - \hat{\beta}_5$, $\hat{\beta}_2 - \hat{\beta}_5$, $\hat{\beta}_3 - \hat{\beta}_5$, $\hat{\beta}_4 - \hat{\beta}_5$, $\hat{\beta}_6/25$, $\hat{\beta}_7/625$, $\hat{\beta}_8/15625$, $\hat{\beta}_9/390625$ **10.43 a** -0.277 **b** 0.0781 **c** (-0.434, -0.119) **10.44 a** -0.271 **b** 0.0705 **c** (-0.413, -0.129) **10.45 a** The change in the mean yield caused by switching from the standard variety of wheat to the new variety **b** $\begin{bmatrix} 1 & 0 \\ 1 & 1 \\ 1 & 0 \\ 1 & 1 \\ 1 & 0 \end{bmatrix}$, $\begin{bmatrix} 5 & 2 \\ 2 & 2 \end{bmatrix}$, $\begin{bmatrix} 1/3 & -1/3 \\ -1/3 & 5/6 \end{bmatrix}$

c $\begin{bmatrix} 1/3 & 0 & 1/3 & 0 & 1/3 \\ -1/3 & 1/2 & -1/3 & 1/2 & -1/3 \end{bmatrix}$ **d** $\frac{Y_1+Y_3+Y_5}{3} = 19, \frac{Y_2+Y_4}{2} - \frac{Y_1+Y_3+Y_5}{3} = 5$ [see answer to part (c)] **e** 50 **f** The plot is unrevealing. **g** For a given variety, the mean yield is an increasing function of the plot variable. **10.46 a** The change in the mean yield caused by switching from the standard variety of wheat to the new variety on a given plot **b**

$$\begin{bmatrix} 1 & 0 & -2 \\ 1 & 1 & -1 \\ 1 & 0 & 0 \\ 1 & 1 & 1 \\ 1 & 0 & 2 \end{bmatrix}, \quad \begin{bmatrix} 5 & 2 & 0 \\ 2 & 2 & 0 \\ 0 & 0 & 10 \end{bmatrix}, \quad \begin{bmatrix} 1/3 & -1/3 & 0 \\ -1/3 & 5/6 & 0 \\ 0 & 0 & 1/10 \end{bmatrix}$$

c

$$\begin{bmatrix} 1/3 & 0 & 1/3 & 0 & 1/3 \\ -1/3 & 1/2 & -1/3 & 1/2 & -1/3 \\ -1/5 & -1/10 & 0 & 1/10 & 1/5 \end{bmatrix}$$

d $\frac{Y_1+Y_3+Y_5}{3} = 19, \frac{Y_2+Y_4}{2} - \frac{Y_1+Y_3+Y_5}{3} = 5, \frac{-2Y_1-Y_2+Y_4+2Y_5}{10} = 2.2$ [see answer to part (c)] **e**

Source	SS
Fit	78.4
Residuals	1.6
Total	80.0

f .98 **g** 0.8, 0.894 **h** $\sqrt{2/3} \doteq 0.816$, (1.487, 8.513) **i** For a given variety, the mean yield appears to be a quadratic rather than a linear function of plot. **10.47 a** The change in the mean yield caused by switching from the new variety to the standard variety on a given plot **b** 5, 1.2, 10 **c** $\bar{Y} = 21, \frac{Y_1+Y_3+Y_5}{3} - \frac{Y_2+Y_4}{2} = -5, \frac{-2Y_1-Y_2+Y_4+2Y_5}{10} = 2.2$ **10.48 a** See the answer to Problem 10.47(a). **b**

$$\begin{bmatrix} 1 & 2/5 & -2 & 2 \\ 1 & -3/5 & -1 & -1 \\ 1 & 2/5 & 0 & -2 \\ 1 & -3/5 & 1 & -1 \\ 1 & 2/5 & 2 & 2 \end{bmatrix}, \quad \begin{bmatrix} 5 & 0 & 0 & 0 \\ 0 & 6/5 & 0 & 2 \\ 0 & 0 & 10 & 0 \\ 0 & 2 & 0 & 14 \end{bmatrix}, \quad \frac{1}{160}\begin{bmatrix} 32 & 0 & 0 & 0 \\ 0 & 175 & 0 & -25 \\ 0 & 0 & 16 & 0 \\ 0 & -25 & 0 & 15 \end{bmatrix}$$

c

$$\frac{1}{40}\begin{bmatrix} 8 & 8 & 8 & 8 & 8 \\ 5 & -20 & 30 & -20 & 5 \\ -8 & -4 & 0 & 4 & 8 \\ 5 & 0 & -10 & 0 & 5 \end{bmatrix}$$

d $\bar{Y} = 21, \frac{Y_1-4Y_2+6Y_3-4Y_4+Y_5}{8} = -4.375, \frac{-2Y_1-Y_2+Y_4+2Y_5}{10} = 2.2, \frac{Y_1-2Y_3+Y_5}{8} = -0.375$ [see answer to part (c)] **e**

Source	SS
Fit	79.9
Residuals	0.1
Total	80.0

f .999   **g** 0.1, 0.316   **h** 0.331, $(-8.577, -0.173)$ **i** The plot is unrevealing.

Section 10.5

10.49 The coefficient $\hat{\beta}_0$ gets replaced by $\hat{\beta}_0 - \frac{2}{3}(\hat{\beta}_2 + \hat{\beta}_4 + \hat{\beta}_6)$, and the other coefficients are unaltered. **10.50** $2(\hat{\beta}_1 - \hat{\beta}_9 - \hat{\beta}_{11} - \hat{\beta}_{13}) \doteq -5.756, (-18.036, 6.525)$ **10.51** Note that $\mathrm{cov}(\hat{\beta}_9, \hat{\beta}_{10}) = \sigma^2 \langle \frac{4}{45} x_1 x_2 + \frac{1}{45} x_3 x_4, \frac{1}{45} x_1 x_2 + \frac{4}{45} x_3 x_4 \rangle$. **10.52 a** 29.981, 15.633, 12.128, 17.506, 3.444 **b** $18(15.633^2 + 12.128^2 + 17.506^2 + 3.444^2) \doteq 12776.27$ **c**

Source	SS
Fit	12776.27
Residuals	2768.19
Total	15544.46

d $\mathrm{SE}(\hat{\beta}_0) \doteq 2.159, \mathrm{SE}(\hat{\beta}_1) = \cdots = \mathrm{SE}(\hat{\beta}_4) \doteq 2.644$ **e** $(25.504, 34.458), (10.150, 21.117), (6.645, 17.611), (12.022, 22.989), (-2.039, 8.928)$ **f** $31.267, (20.300, 42.233)$ **10.53** $(b-a)[\beta_1 + \beta_3(a+b) + \beta_5 c]$ **10.54**   **a** $\|1\|^2 = 9, \|x_1\|^2 = \|x_2\|^2 = 6, \|x_1^2 - \frac{2}{3}\|^2 = \|x_2^2 - \frac{2}{3}\|^2 = 2, \|x_1 x_2\|^2 = 4$ **b** $56, 12, 6, -11, 9, -3$ **c** $6(\hat{\beta}_1^2 + \hat{\beta}_3^2) + 2(\hat{\beta}_2^2 + \hat{\beta}_4^2) + 4\hat{\beta}_5^2 = 1860, 636$ **d** $212, 14.560$ **e** $\mathrm{var}(\hat{\beta}_0) = \sigma^2/9, \mathrm{var}(\hat{\beta}_1) = \mathrm{var}(\hat{\beta}_3) = \sigma^2/6, \mathrm{var}(\hat{\beta}_2) = \mathrm{var}(\hat{\beta}_4) = \sigma^2/2, \mathrm{var}(\hat{\beta}_5) = \sigma^2/4$ **f** $(40.554, 71.446), (-6.917, 30.917), (-26.765, 38.765), (-29.917, 7.917), (-23.765, 41.765), (-26.169, 20.169)$ **g** $2(\beta_1 + \beta_5)$ **h** $5\sigma^2/3, 18, 18.797, (-41.821, 77.821)$ **10.55**   **a** $56, 12, 6, -11, 9, 10, -6$ **b** $6(\hat{\beta}_1^2 + \hat{\beta}_3^2 + \hat{\beta}_5^2) + 2(\hat{\beta}_2^2 + \hat{\beta}_4^2 + \hat{\beta}_6^2) = 2496, 0$ **10.56**   **a** $\|1\|^2 = 9, \|x\|^2 = \|x_2\|^2 = \|x_3\|^2 = 6, \|x_1 x_2\|^2 = 4$ **b** $56, 12, -11, 10, -3$ **c** $6(\hat{\beta}_1^2 + \hat{\beta}_2^2 + \hat{\beta}_3^2) + 4\hat{\beta}_4^2 = 2226, 270$ **d** $67.5, 8.216$ **e** $\mathrm{var}(\hat{\beta}_0) = \sigma^2/9, \mathrm{var}(\hat{\beta}_1) = \mathrm{var}(\hat{\beta}_2) = \mathrm{var}(\hat{\beta}_3) = \sigma^2/6, \mathrm{var}(\hat{\beta}_4) = \sigma^2/4$ **f** $(48.396, 63.604), (2.688, 21.312), (-20.312, -1.688), (0.688, 19.312), (-14.405, 8.405)$; **g** $2(\beta_1 + \beta_4)$; **h** $5\sigma^2/3, 18, 10.607, (-11.449, 47.449)$ **10.57 a**

$$\begin{bmatrix} 9 & 0 & 0 & 0 & 0 & 0 \\ 0 & 6 & 0 & 0 & 0 & 0 \\ 0 & 0 & 6 & 0 & 0 & 0 \\ 0 & 0 & 0 & 6 & 0 & 0 \\ 0 & 0 & 0 & 0 & 4 & 2 \\ 0 & 0 & 0 & 0 & 2 & 4 \end{bmatrix}$$

c No   **d** $56, 12, -11, 10, -7, 8$ **e** $6(\hat{\beta}_1^2 + \hat{\beta}_2^2 + \hat{\beta}_3^2) + 4(\hat{\beta}_4^2 + \hat{\beta}_5^2 + \hat{\beta}_4\hat{\beta}_5) = 2418, 78$ **f** $26, 5.099$ **g** $\mathrm{var}(\hat{\beta}_0) = \sigma^2/9, \mathrm{var}(\hat{\beta}_1) = \mathrm{var}(\hat{\beta}_2) = \mathrm{var}(\hat{\beta}_3) = \sigma^2/6, \mathrm{var}(\hat{\beta}_4) = \mathrm{var}(\hat{\beta}_5) = \sigma^2/3$ **h** $(50.591, 61.409), (5.375, 18.625), (-17.625, -4.375), (3.375, 16.625), (-16.369, 2.369), (-1.369, 17.369)$ **i** $2(\beta_1 + \beta_4 + \beta_5)$ **j** $2\sigma^2, 26, 7.211, (3.051, 48.949)$ **10.58 a**

$$\begin{bmatrix} 9 & 0 & 0 & 0 & 0 & 0 & 0 \\ 0 & 6 & 0 & 0 & 0 & 0 & 0 \\ 0 & 0 & 6 & 0 & 0 & 0 & 0 \\ 0 & 0 & 0 & 6 & 0 & 0 & 0 \\ 0 & 0 & 0 & 0 & 4 & 2 & 2 \\ 0 & 0 & 0 & 0 & 2 & 4 & 2 \\ 0 & 0 & 0 & 0 & 2 & 2 & 4 \end{bmatrix}$$

c No **d** 56, 12, −11, 10, −8.25, 6.75. 3.75 **e** $6(\hat{\beta}_1^2 + \hat{\beta}_2^2 + \hat{\beta}_3^2) + 4(\hat{\beta}_4^2 + \hat{\beta}_5^2 + \hat{\beta}_6^2 + \hat{\beta}_4\hat{\beta}_5 + \hat{\beta}_4\hat{\beta}_6 + \hat{\beta}_5\hat{\beta}_6) = 2455.5$, 40.5 **f** 20.25, 4.5 **g** $\text{var}(\hat{\beta}_0) = \sigma^2/9$, $\text{var}(\hat{\beta}_1) = \text{var}(\hat{\beta}_2) = \text{var}(\hat{\beta}_3) = \sigma^2/6$, $\text{var}(\hat{\beta}_4) = \text{var}(\hat{\beta}_5) = \text{var}(\hat{\beta}_6) = 3\sigma^2/8$ **h** (49.546, 62.454), (4.096, 19.904), (−18.904, −3.096), (2.096, 17.904), (−20.107, 3.607), (−5.107, 18.607), (−8.107, 15.607) **i** $2(\beta_1 + \beta_4 + \beta_5)$ **j** $8\sigma^2/3$, 21, 7.348, (−10.618, 52.618).

Section 10.6

10.59 a $\hat{\beta}_1\sqrt{\sum_i(x_i - \bar{x})^2}/S$, where $S = \sqrt{\text{RSS}/(n-2)}$ **b** $2[1 - t_{n-2}(|t|)]$ **10.60 a** $2r/3$; $\frac{\bar{Y}_1 + \bar{Y}_3}{2} - \bar{Y}_2$, where $\bar{Y}_1$, $\bar{Y}_2$, and $\bar{Y}_3$ are the sample means of the responses at the levels −1, 0, and 1, respectively; $3\sigma^2/2r$ **b** $\frac{3}{2}(z_{.9} + z_{.975})^2 \doteq 16$ **10.61 a** −0.160 **b** 0.0695 **c** −0.043 **d** .013 **e** (−0.300, −0.020) **f** .026 **10.62** *P*-value $\doteq 2[1 - t_{22}(1.303)] \doteq$ .206, so we find no evidence that changes in the level of *D* have an effect on the quality of lube oil. **10.63 a** (−0.549, 20.549) **b** The *t* statistic lies in the acceptance region. **c** (0.451, 21.549) **d** The *t* statistic lies in the critical region. **e** No **10.64** .0256 **10.65** .048 **10.66 a** $2[1 - t_3(6.335)] \doteq .008$ **b** $2[1 - t_3(1.486)] \doteq .234$ **10.67** See the answers to Problem 10.66. **10.68 a** 6, 70, 597.333, 4147.2 **b** 76.883, 2.547, −0.337, 0.0189 **c** $(70)(2.547)^2 + (597.333)(-0.337)^2 + (4147.2)(0.0189)^2 \doteq 523.492$ **d**

Source	SS
Fit	523.492
Residuals	3.597
Total	527.088

e 0.0208, (−0.071, 0.108) **f** *P*-value $\doteq 2[1 - t_2(0.906)] \doteq .461$, so we accept the hypothesis. **10.69** The *P*-value for the *t* test of the equivalent hypothesis that the coefficient β_5 of $x_1 x_3$ equals zero is given by *P*-value $\doteq 2[1 - t_3(2.717)] \doteq .073$.

Section 10.7

10.70 $r(10\hat{\beta}_1^2 + 14\hat{\beta}_2^2)$ **10.72 a** $\hat{\mu}_0(x_1, x_2, x_3) \doteq 29.981 + 15.633x_1 + 12.128x_2 + 17.506x_3$ **b** $18(15.633^2 + 12.128^2 + 17.506^2) \doteq 12562.71$ **c**

Source	SS
Fit	12562.71
Residuals	2981.75
Total	15544.46

d $18(3.444^2) \doteq 213.56$ **e**

Source	SS
A, B, C (linear)	12562.72
D after A, B, C	213.56
Residuals	2768.19
Total	15544.46

Using a linear fit in the levels of the four factors, we conclude that changes in the level of D have a negligible effect on quality. **10.73 a** 10.049 **b**

Source	SS
Fit	10.049
Residuals	1.278
Total	11.327

c 0.261 **d**

Source	SS
G_0	10.049
G after G_0	0.261
Residuals	1.017
Total	11.327

Adding the quintic term explains another 2% of the variation in the response variable. **10.74 b**

Source	SS
Linear	454.156
Quadratic after linear	67.860
Cubic after quadratic	1.476
Quartic after cubic	3.223
Quintic after quartic	0.373
Total	527.088

86.16% of the variation in the yield of tetramer is explained by the linear fit, another 12.87% is explained by adding the quadratic term, another 0.28% is explained by adding the cubic term, another .61% is explained by adding the quintic term, and the remaining .07% is explained by adding the quintic term. **10.75 a** $4(\hat{\beta}_4^2 - \hat{\beta}_{40}^2 + \hat{\beta}_5^2 + \hat{\beta}_4\hat{\beta}_5)$ **b**

Source	SS
G_0	2226
G after G_0	192
Residuals	78
Total	2496

10.76 b Use the result of part (a) both directly and with G replaced by G_0.

Section 10.8

10.78 a $\hat{\beta}_1^2 \sum_i (x_i - \bar{x})^2 / S^2$, where $S^2 = \text{RSS}/(n-2)$ **b** $1 - F_{1,n-2}(F)$

10.79

Source	SS	DF	MS	F	P-value
A, B, C (linear)	12562.72	3	4187.57		
D after A, B, C	213.56	1	213.56	1.697	.206
Residuals	2768.19	22	125.83		
Total	15544.46	26			

Using a linear fit in the levels of the four factors, we find no evidence that changes in the level of D have an effect on the quality of lube oil. (Since $1.697 \doteq 1.303^2$, the results here are compatible with those in the answer to Problem 10.62 in Section 10.6.)   **10.80**

Source	SS	DF	MS	F	P-value
Quadratic fit	522.016	2	261.008		
Cubic term	1.476	1	1.476	0.821	.461
Residuals	3.597	2	1.798		
Total	527.088	5			

We accept the hypothesis. [Since $0.821 \doteq 0.906^2$, the F statistic is compatible with the t statistic shown in the answer to Problem 10.68(f) in Section 10.6.]   **10.81**

Source	SS	DF	MS	F	P-value
Quadratic fit	522.015	2	261.008		
Quartic after quadratic	4.699	2	2.350	6.293	.271
Residuals	0.373	1	0.373		
Total	527.088	5			

The hypothesis is acceptable.   **10.82 b** $18(\hat{\beta}_1^2 + \hat{\beta}_3^2) + 6(\hat{\beta}_2^2 + \hat{\beta}_4^2)$ **10.83**

Source	SS	DF	MS	F	P-value
Linear fit	7046.71	2	3523.36		
Lack of fit	529.07	6	88.18	0.199	.973
Within	7968.67	18	442.70		
Total	15544.46	26			

The linear fit appears to be quite satisfactory when we ignore factors C and D.   **10.84**

Source	SS	DF	MS	F	P-value
Plots	6.439	5	1.288		
Varieties after plots	0.423	3	0.141	6.878	.001
Residuals	0.655	32	0.020		
Total	7.518	40			

When we include the plot effect, it becomes clear that the mean percentage of sugar content does differ among among varieties 1, 3, 4, and 5. (See the answer to Problem 7.60 in Section 7.6.)   **10.85**

Source	SS	DF	MS	F	P-value
G_0	2226	4	556.5		
G after G_0	192	1	192.0	7.385	.073
Residuals	78	3	26.0		
Total	2496	8			

which is in agreement with the answer to Problem 10.69 in Section 10.6   **10.86 a** Use the solution to Example 10.25(a).   **10.87** $\frac{n-p}{p-p_0}\frac{R^2-R_0^2}{1-R^2}$

Chapter 11

Section 11.1

11.1 $n_1 n_4 = n_2 n_3$ **11.5** no   **11.6** No, because there are more runs with factor E at its low level than at its high level   **11.7 a** $2r$ **b** $(\bar{Y}_2 - \bar{Y}_1)/2$ **c** $2r\hat{\beta}^2$ **11.8 a** $\frac{r}{2}$ **b** $\bar{Y}_1 - \bar{Y}_2$ **c** $\frac{r}{2}\hat{\beta}^2$ **11.9 a** $\frac{2r}{3}, \frac{r}{2}$ **b** $(2\bar{Y}_1 - \bar{Y}_2 - \bar{Y}_3)/2, \bar{Y}_2 - \bar{Y}_3$ **c** $r(\frac{2}{3}\hat{\beta}_1^2 + \frac{1}{2}\hat{\beta}_2^2)$ **11.10 a** $6, \frac{9}{2}$ **b** $(2\bar{Y}_{1...} - \bar{Y}_{2...} - \bar{Y}_{3...})/2 \doteq -21.439, \bar{Y}_{2...} - \bar{Y}_{3...} \doteq -19.656$ **c** $6\hat{\beta}_1^2 + \frac{9}{2}\hat{\beta}_2^2$ **11.11 a** 345, 426.5, 427.833, 405.167, 520.167, 601.833 **b** 462.75   **11.12 a**

Source	SS	DF	MS	F	P-value
Rows	54199	5	10840	4.926	.003
Treatments	248180	5	49636	22.558	.000
Residuals	55008	25	2200		
Total	357387	35			

b 256.8333, 27.082 **c** (201.057, 312.610)   **11.13 a**

Source	SS	DF	MS	F	P-value
Treatments	248180	5	49636	13.635	.000
Residuals	109207	30	3640		
Total	357387	35			

b 256.8333, 34.834 **c** (185.693, 327.974)   **11.14 a** $(1, -1, 1, 1, 1, 1, 1)$ **b** $\hat{\beta}_0 + |\hat{\beta}_1| + \cdots + |\hat{\beta}_7| = 85$

Section 11.2

11.15 $1, x_1, x_2, x_1x_2$ **11.16** $1, x_1, x_2, x_1^2, x_1x_2, x_2^2, x_1^2x_2, x_1x_2^2, x_1^2x_2^2$ **11.17** Note that the number of design points is not a multiple of 27.   **11.18** (3, 6), (3, 7), (4, 5), (4, 8), (5, 8), (6, 7)   **11.20** $\langle x_1x_2, x_3x_4\rangle = \langle x_1x_3, x_2x_4\rangle = \langle x_1x_4, x_2x_3\rangle = 3$

10.21 b

Source	SS	DF
A	SS_A	1
B	SS_B	2
AB	SS_{AB}	2
C	SS_C	2
D	SS_D	2
E	SS_E	2
F	SS_F	2
G	SS_G	2
H	SS_H	2
Total	TSS	17

10.22 $\langle x_1 x_3, x_4 \rangle = -6$ **10.24 b**

Source	SS	DF
A	SS_A	1
B	SS_B	1
AB	SS_{AB}	1
Within	WSS	$n - 4$
Total	TSS	$n - 1$

11.28 a Note that $G_l = G_{con} \oplus G_l^0$ and $G_m = G_{con} \oplus G_m^0$. **b** Note that $G_{lm} = G_{con} \oplus G_l^0 \oplus G_m^0 \oplus G_{lm}^0$. **c** Use parts (a) and (b).

Section 11.3

11.29 $(x_1 - 1)(x_2 - 1)$, $(x_1 - 1)[(x_2 - 1)^2 - \frac{2}{3}]$, $[(x_1 - 1)^2 - \frac{2}{3}](x_2 - 1)$, $[(x_1 - 1)^2 - \frac{2}{3}][(x_2 - 1)^2 - \frac{2}{3}]$ **11.31** Yes, because (i) in computing the various sums of squares we can ignore levels of the factors that are not actually taken on in the experiment and (ii) the space of quadratic polynomials on $\{-1, 0, 1\}$ coincides with the space of all functions on this set **11.32**

Source	SS	DF	MS	F	P-value
A	4496.29	2	2248.14	25.313	.000
B	2768.69	2	1384.35	15.587	.000
C	5519.79	2	2759.89	31.075	.000
D	283.37	2	141.68	1.595	.238
AC	1232.92	4	308.23	3.470	.036
Residuals	1243.40	14	88.81		
Total	15544.46	26			

11.33 $1, x_1, x_1^2 - \frac{2}{3}, x_2, x_2^2 - \frac{2}{3}, x_1 x_2, x_1(x_2^2 - \frac{2}{3}), (x_1^2 - \frac{2}{3})x_2, (x_1^2 - \frac{2}{3})(x_2^2 - \frac{2}{3})$ **11.34 a** X_1 and X_2 are independent random variables, X_1 is uniformly distributed on $\{-1, 1\}$, and X_2 is uniformly distributed on $\{-1, 0, 1\}$. **b** x_1 has squared norm $6r$. **c** x_2 and $x_2^2 - \frac{2}{3}$ have squared norms $4r$ and $\frac{4}{3}r$, respectively. **d** x_1x_2 and $x_1(x_2^2 - \frac{2}{3})$ have squared norms $4r$ and $\frac{4}{3}r$, respectively. **e** $1, x_1, x_2, x_2^2 - \frac{2}{3}, x_1x_2, x_1(x_2^2 - \frac{2}{3})$ **11.35 a** X_1 and X_2 are independent random variables, X_1 takes on the values -1 and 1 with probabilities $1/3$ and $2/3$, respectively, and X_2 is uniformly distributed on $\{-1, 0, 1\}$. **b** $x_1 - \frac{1}{3}$ has squared norm $8r$. **c** x_2 and $x_2^2 - \frac{2}{3}$ have squared norms $6r$ and $2r$, respectively. **d** $(x_1 - \frac{1}{3})x_2$ and $(x_1 - \frac{1}{3})(x_2^2 - \frac{2}{3})$ have squared norms $\frac{16}{3}r$ and $\frac{16}{9}r$, respectively. **e** $1, x_1 - \frac{1}{3}, x_2, x_2^2 - \frac{2}{3}, (x_1 - \frac{1}{3})x_2, (x_1 - \frac{1}{3})(x_2^2 - \frac{2}{3})$ **11.36 a** Yes **b** Observe that $1, x_1 - 1, (x_1 - 1)^2 - \frac{2}{3}, x_2 - 1, (x_2 - 1)^2 - \frac{2}{3}, x_3 - 1, (x_3 - 1)^2 - \frac{2}{3}, x_4 - 1, (x_4 - 1)^2 - \frac{2}{3}$ is an orthogonal basis of G consisting of functions with positive norms.   **11.37 a**

Source	SS	DF	MS	F	P-value
A	3,528	1	3,528.0	10.438	.032
B	19,396	2	9,698.0	28.692	.004
C	8,419	2	4,209.5	12.454	.019
D	1,684	2	842.0	2.491	.198
AB	1,344	2	672.0	1.988	.251
AC	837	2	418.5	1.238	.381
AD	724	2	362.0	1.071	.424
Residuals	1,352	4	338.0		
Total	37,284	17			

b

Source	SS	DF	MS	F	P-value
A	3,528	1	3,528.0	8.288	.016
B	19,396	2	9,698.0	22.781	.000
C	8,419	2	4,209.5	9.888	.004
D	1,684	2	842.0	1.978	.189
Residuals	4,257	10	425.7		
Total	37,284	17			

11.39 $\frac{478}{3} + 14x_1 + 40x_2 + \frac{103}{4}x_3$

Section 11.4

11.41 a $SS_{blocks} = 3\sum_{i=1}^{4}(\bar{Y}_{i\cdot} - \bar{Y})^2$, $SS_{treatments} = 4\sum_{j=1}^{3}(\bar{Y}_{\cdot j} - \bar{Y})^2$ **b**

Source	SS	DF	MS	F	P-value
Blocks	25,800	3	8,600		
Treatments	7,200	2	3,600	6.767	.029
Residuals	3,192	6	532		
Total	36,192	11			

11.42 Use the fact that $I_1 + \ldots + I_5 = 1$.   **11.43 a** $-\frac{\sigma_1^2}{N-1}$ **b** $\frac{\sigma_1^2}{n_1}\frac{N-n_1}{N-1} + \frac{\sigma_1^2}{n_2}\frac{N-n_2}{N-1} + \frac{2\sigma_1^2}{N-1}$

11.44 a $\frac{\sigma_1^2}{n_1}\frac{N-n_1}{N-1} + \frac{\sigma^2}{n_1}$, $\frac{\sigma_1^2}{n_2}\frac{N-n_2}{N-1} + \frac{\sigma^2}{n_2}$ **b** $-\frac{\sigma_1^2}{N-1}$ **c** $\frac{\sigma_1^2}{n_1}\frac{N-n_1}{N-1} + \frac{\sigma_1^2}{n_2}\frac{N-n_2}{N-1} + \frac{2\sigma_1^2}{N-1} + \frac{\sigma^2}{n_1} + \frac{\sigma^2}{n_2}$

11.46 Observe that $\text{var}(g(X_{il}) - g(X_{ij})) = \text{var}(g(X_{ij})) + \text{var}(g(X_{il})) - 2\text{cov}(g(X_{ij}), g(X_{il}))$.

11.47 a Use Eq. (52) in Section 2.4.   **11.48 a** $(-1.057, -0.303)$ **b** yes $(\beta_2 - \beta_1 \doteq -0.608)$ **11.49 a** $(-0.830, -0.450)$ **b** Yes   **11.50 a**

Source	SS	DF	MS	F	P-value
Blocks	31,913	11	2,901	17.000	.000
Varieties	5,310	4	1,327	7.779	.000
Residuals	7,509	44	171		
Total	44,732	59			

b 13.064 **c** 11.408 **d** 5.333 **e** $(0.660, 22.157)$ **f** $2[1 - t_{44}(2.139)] \doteq .038$ **11.51 a**

Source	SS	DF	MS	F	P-value
Varieties	5,310	4	1,327	1.852	.132
Residuals	39,422	55	717		
Total	44,732	59			

b 26.773 **c** 11.408 **d** 10.930 **e** $(-10.496, 33.312)$ **f** $2[1 - t_{55}(1.044)] \doteq .301$

Chapter 12

Section 12.1

12.1 a $(.722, .878)$ **b** .734 **c** $1 - \Phi(1.25) \doteq .106$ **12.2 a** $-.15$, $(-.310, .010)$, .066 **b** 0.8, $(0.598, 1.002)$, .053 **c** 0.5, $(0.138, 0.862)$, .007   **12.3 a** $-.016$, .033 **b** 0.970, .026 **c** 0.804, .0034   **12.4 a** $\frac{\lambda_k}{n_k}$, $\sqrt{\hat{\lambda}_k/n_k}$ **b** $\frac{\lambda_j}{n_j} + \frac{\lambda_k}{n_k}$, $\sqrt{\hat{\lambda}_j/n_j + \hat{\lambda}_k/n_k}$

c $\tau^2\left(\frac{1}{n_j\lambda_j} + \frac{1}{n_k\lambda_k}\right)$ with $\tau = \frac{\lambda_k}{\lambda_j}$, $\hat{\tau}\sqrt{1/(n_j\hat{\lambda}_j) + 1/(n_k\hat{\lambda}_k)}$ with $\hat{\tau} = \hat{\lambda}_k/\hat{\lambda}_j$ **12.5 a** 10^{-5}, $(0.804 \times 10^{-5}, 1.196 \times 10^{-5})$ [use the answer to Problem 12.4(a)] **b** 100, $(80.4, 119.6)$

12.6 a -0.2×10^{-5}, $(-0.432 \times 10^{-5}, 0.032 \times 10^{-5})$ [use the answer to Problem 12.4(b)], .091 **b** 0.8, $(0.600, 1.000)$ [use the answer to Problem 12.4(c)], 0.050   **12.7 a** -0.005×10^{-5}, .045 **b** .968, .025   **12.8 a** β_k^2/n_k, $\hat{\beta}_k/\sqrt{n_k}$ **b** $\frac{\beta_j^2}{n_j} + \frac{\beta_k^2}{n_k}$, $\sqrt{\hat{\beta}_j^2/n_j + \hat{\beta}_k^2/n_k}$

c $\tau^2\left(\frac{1}{n_j} + \frac{1}{n_k}\right)$, $\hat{\tau}\sqrt{1/n_j + 1/n_k}$ **12.9 a** 50, $(-2.7, 102.7)$, .063 **b** 1.417, $(0.936, 1.898)$, .089   **12.10 a** 5.803, .031 **b** 1.013, .045   **12.11** $1/(4n)$, which does not depend on π **12.12** $1/(4n)$, which does not depend on λ **12.13 a** $\frac{\sigma_j^2}{n_j\mu_j^2} + \frac{\sigma_k^2}{n_k\mu_k^2}$ **b** $\frac{1-\pi_j}{n_j\pi_j} + \frac{1-\pi_k}{n_k\pi_k}$

c $\frac{1}{n_j\lambda_j} + \frac{1}{n_k\lambda_k}$

Section 12.2

12.14 .149 **12.15 a** $\frac{1}{2^9}\sum_{z=0}^{2}\binom{9}{z} \doteq .090$ **b** $\frac{1}{2^7}\sum_{z=0}^{2}\binom{7}{z} \doteq .227$ **c** $\frac{1}{2^{16}}\sum_{z=0}^{4}\binom{16}{z} \doteq .038$
12.16 a $f(0) + f(8) + \cdots + f(24) \doteq .0480$ **b** $F(0) + 1 - F(7) \doteq .0480$ **12.17 a** $F(65) +$
$1 - F(74) \doteq .088, .090$ **b** $1 - F(74) \doteq .046, .045$ **12.18** $\sum_{z=0}^{c_1}f(z; n, \lambda_0) + \sum_{z=c_2}^{\infty}f(z;$
$n, \lambda_0) = F(c_1; n, \lambda_0) + 1 - F(c_2 - 1; n, \lambda_0)$, where the integers c_1 and c_2 are defined so that
$f(z; n, \lambda_0) \leq f(y; n, \lambda_0)$ if and only if $z \leq c_1$ or $z \geq c_2$ with y being the observed value of
Y **12.19 a** $\sum_{z=0}^{c_1}f(z) + \sum_{z=c_2}^{\infty}f(z) = F(c_1) + 1 - F(c_2 - 1)$, where y_1 is the observed
value of Y_1, y is the observed value of Y, f and F denote the respective probability function
and distribution function of the binomial distribution with parameters y and $n_1/(n_1 + n_2)$,
and the integers c_1 and c_2 are defined so that $f(z) \leq f(y_1)$ if and only if $z \leq c_1$ or $z \geq c_2$
b $\sum_{z=y_1}^{y}f(z) = 1 - F(y_1 - 1)$ **12.20 a** .0870 **b** .0468 **12.21** $\left(\frac{z_{1-\alpha}\sqrt{\lambda_0} + z_{1-\beta}\sqrt{\lambda}}{\lambda - \lambda_0}\right)^2$
12.26 a $[c, \infty)$ or (c, ∞) for some $c \in (0, \infty)$ **b** $[0, c]$ or $[0, c)$ for some $c \in (0, \infty)$
12.27 $\{\mathbf{y} \in \mathbb{R}^n : (\boldsymbol{\mu}_1 - \boldsymbol{\mu}_0)^T\Sigma^{-1}\mathbf{y} \geq c\}$ or $\{\mathbf{y} \in \mathbb{R}^n : (\boldsymbol{\mu}_1 - \boldsymbol{\mu}_0)^T\Sigma^{-1}\mathbf{y} > c\}$ for some $c \in$
$\mathbb{R}$

Section 12.3

12.28 a $(.896, 1.876)$, $(.710, .867)$ **b** .975, .726 **c** $1 - \Phi(1.151) \doteq .125$ **12.31** $2\frac{N}{N+1}$
12.32 a $\frac{1}{n_k\hat{\lambda}_k}$, $\sqrt{1/(n_k\hat{\lambda}_k)}$ **b** $\frac{1}{n_j\hat{\lambda}_j} + \frac{1}{n_k\hat{\lambda}_k}$, $\sqrt{1/(n_j\hat{\lambda}_j) + 1/(n_k\hat{\lambda}_k)}$ **12.34 a** -0.172 **b** 0.842
c .015 **12.35 a** $(-0.473, 0.027)$ **b** $(0.623, 1.027)$ **c** .080 **12.36 a** $\theta = -\lambda \in (-\infty, 0)$
b $\lambda = -\theta$ **c** $C(\theta) = -\log(-\theta)$ **d** $C'(\theta) = -1/\theta$ ranges over $(0, \infty)$ as θ ranges over
$(-\infty, 0)$. **e** $C''(\theta) = 1/\theta^2$ **f** $(C')^{-1}(t) = -1/t$ for $0 < t < \infty$ **g** Mean n/λ and variance
n/λ^2 **12.37 a** $\theta = \mu/\sigma^2 \in \mathbb{R}$ **b** $\mu = \sigma^2\theta$ **c** $C(\theta) = \sigma^2\theta^2/2$ **d** $C'(\theta) = \sigma^2\theta$ ranges over
$\mathbb{R}$ as θ ranges over $\mathbb{R}$. **e** $C''(\theta) = \sigma^2$ **f** $(C')^{-1}(t) = t/\sigma^2, t \in \mathbb{R}$

Chapter 13

Section 13.1

13.1 μ, $\mu + 1.282\sigma$ **13.2 a** $\beta_0 = -\log 9 \doteq -2.197$ and $\beta_1 = \frac{1}{10}\log 9 \doteq 0.220$
b $20 + 10\frac{\log 11}{\log 9} \doteq 30.913$ **c** $27/28 \doteq .964$ **13.3** $\beta_0 + \beta_1 z_p$ **13.4** $\exp(\beta_m c) \approx 1$ or
$\exp(\theta)$ and $\exp(\theta + \beta_m c)$ are both close to zero, where $\theta = \beta_0 + \beta_1 x_1 + \cdots + \beta_m x_m$
13.5 2.489 **13.6 a** $r(\mathbf{y}; \mathbf{n}) = \prod_k \binom{n_k}{y_k}$, where $y_k \in \{0, \ldots, n_k\}$ for $1 \leq k \leq d$ **b** $r(\mathbf{y}; \mathbf{n}) =$
$\prod_k n_k^{y_k}/y_k!$, where $y_k \in \{0, 1, 2, \ldots\}$ for $1 \leq k \leq d$ **13.7** It is multiplied by the
amount $\exp(\beta_m c)$. **13.8** It is a multiplicative function of the levels of the various fac-
tors. **13.9 a** $1/5$, $e^{1/5}/(1 + e^{1/5})$ **b** $2/5$ **c** $e^{2/5}$ **d** $e^{2/5}\frac{1 + e^{(1-3x_2)/5}}{1 + e^{3(1-x_2)/5}}$ **e** $-3/5$ **f** $e^{-3/5}$
g $e^{-3/5}\frac{1 + \exp(\frac{1}{5} + \frac{2}{5}\text{ind}(x_1=1) + \frac{4}{5}\text{ind}(x_1=2) - \frac{3}{5}x_2)}{1 + \exp(\frac{1}{5} + \frac{2}{5}\text{ind}(x_1=1) + \frac{4}{5}\text{ind}(x_1=2) - \frac{3}{5}x_2 - \frac{3}{5})}$ **h** $\frac{2}{5}$ **13.11** $(\theta_1 + \theta_2)/2$, $(\theta_2 - \theta_1)/2$
13.12 $(\theta_1 + \theta_2 + \theta_3 + \theta_4)/4$, $(-\theta_1 + \theta_2 - \theta_3 + \theta_4)/4$, $(-\theta_1 - \theta_2 + \theta_3 + \theta_4)/4$, $(\theta_1 - \theta_2 -$
$\theta_3 + \theta_4)/4$

Section 13.2

13.13 a $L'(\pi) = \binom{n}{y}\pi^{y-1}(1 - \pi)^{n-y-1}(y - n\pi)$ **b** $L''(\hat{\pi}) = -n\binom{n}{y}\hat{\pi}^{y-1}(1 - \hat{\pi})^{n-y-1}$
c $L'(y) > 0$ for $\pi < \hat{\pi}$ and $L'(\pi) < 0$ for $\pi > \hat{\pi}$ **13.15 a** $\lambda^Y e^{-n\lambda}n^Y/Y!$ **b** $Y\log$

$\lambda - n\lambda + \log(n^Y/Y!)$ **13.21** **a** $\frac{1}{\sigma^n(2\pi)^{n/2}} \exp\left(-\frac{1}{2\sigma^2}\sum_i[Y_i - \mu(\mathbf{x}_i)]^2\right)$ **b** $-\frac{1}{2\sigma^2}\sum_i[Y_i - \mu(\mathbf{x}_i)]^2 - n\log\sigma - \frac{n}{2}\log 2\pi$ **d** RSS $/n$, which differs somewhat from the unbiased estimate RSS $/(n-p)$ that is used in practice   **13.24 a** Yes **b** Not necessarily   **13.26 a** $\begin{bmatrix} 1 & -1 \\ 1 & 1 \end{bmatrix}$
b $\frac{1}{2}\log\frac{3}{2} \doteq 0.203$, $\frac{1}{2}\log\frac{3}{2} \doteq 0.203$ **13.27 a** See the solution to Example 13.4(a)
b $\log 0.88 \doteq -0.128$, $\log 0.77 \doteq -0.261$, $\log 0.6 \doteq -0.511$, $\log 0.17 \doteq -1.772$ **c** -0.950, $0.415, 0.713, -0.274$ **13.28 a** See the solution to Example 13.5(a). **b** See the answer to Problem 13.27(b). **c** $-0.668, -0.349, -0.473, -0.282$

Section 13.3

13.29   **a** $\theta Y_1 + 2\theta Y_2 - n_1 e^\theta - n_2 e^{2\theta} + Y_1 \log n_1 + Y_2 \log n_2 - \log(Y_1! Y_2!)$ **b** $Y_1 + 2Y_2 - n_1 e^\theta - 2n_2 e^{2\theta}$ **c** $Y_1 + 2Y_2$, $-\infty$ **d** $Y_1 = Y_2 = 0$ **e** $\log\frac{-n_1 + \sqrt{n_1^2 + 8n_2(Y_1 + 2Y_2)}}{4n_2}$
13.30 -0.217 **13.31** Use $\mathcal{X} = \{1, 2\}$, G the one-dimensional space of functions on $\mathcal{X}$ of the form $g(x) = cx$ for $c \in \mathbb{R}$, 1 the basis of G, and θ the regression coefficient. Then the design matrix is given by $\mathbf{X} = [1, 2]^T$.   **13.33 a** Apply Eq. (11) twice. **b** Apply part (a).
13.36 a Exists. **b** Does not exist. **c** Does not exist.   **13.37** There is a line in the plane such that $\bar{Y}_k = 0$ for all $\mathbf{x}_k$'s on one side of the line and $\bar{Y}_k = 1$ for all $\mathbf{x}_k$'s on the other side ($\bar{Y}_k$ is arbitrary for $\mathbf{x}_k$'s on the line).   **13.41** There is a line in the plane such that there are no $\mathbf{x}_k$'s on one side of the line and $\bar{Y}_k = 0$ for all $\mathbf{x}_k$'s on the other side ($\bar{Y}_k$ is arbitrary for $\mathbf{x}_k$'s on the line).   **13.42 a** Exists. **b** Exists. **c** Does not exist.

Section 13.4

13.44   **a** $\theta Y_1 + 2\theta Y_2 - n_1 \exp(\hat{\theta}_0)[1 + \theta - \hat{\theta}_0 + \frac{(\theta-\hat{\theta}_0)^2}{2}] - n_2 \exp(2\hat{\theta}_0)[1 + 2(\theta - \hat{\theta}_0) + 2(\theta - \hat{\theta}_0)^2] + Y_1 \log n_1 + Y_2 \log n_2 - \log(Y_1! Y_2!)$ **b** $\hat{\theta}_0 + [Y_1 + 2Y_2 - n_1 \exp(\hat{\theta}_0) - 2n_2 \exp(2\hat{\theta}_0)]/[n_1 \exp(\hat{\theta}_0) + 4n_2 \exp(2\hat{\theta}_0)]$ **13.46** $\hat{\theta}_\nu = \mathbf{X}\hat{\boldsymbol{\beta}}_\nu$, $\hat{\boldsymbol{\lambda}}_\nu = \exp\hat{\theta}_\nu$, $\hat{\mathbf{W}}_\nu = \mathrm{diag}(\mathbf{n} * \hat{\boldsymbol{\lambda}}_\nu)$, $\hat{\mathbf{V}}_\nu = (\mathbf{X}^T\hat{\mathbf{W}}_\nu\mathbf{X})^{-1}$, $\mathbf{Z}_\nu = \hat{\mathbf{V}}_\nu\mathbf{X}^T\mathbf{N}(\bar{\mathbf{Y}} - \hat{\boldsymbol{\lambda}}_\nu)$, $\hat{\boldsymbol{\beta}}_{\nu+1} = \hat{\boldsymbol{\beta}}_\nu + \mathbf{Z}_\nu$ **13.48 a** $60\exp(\hat{\beta}_0 - \hat{\beta}_1) + 40\exp(\hat{\beta}_0) + 60\exp(\hat{\beta}_0 + \hat{\beta}_1) = 64$, $-60\exp(\hat{\beta}_0 - \hat{\beta}_1) + 60\exp(\hat{\beta}_0 + \hat{\beta}_1) = 24$
b -0.6, 0.2 **c** -1.547, -1.018, -0.490, 0.213, 0.361, 0.613 **13.49** **a** $\hat{\lambda}_\nu = \exp\hat{\theta}_\nu$, $Z_\nu = (\bar{Y} - \hat{\lambda}_\nu)/\hat{\lambda}_\nu$, $\hat{\theta}_{\nu+1} = \hat{\theta}_\nu + Z_\nu$ **b** -2.3026, 0.1 **c** $\theta - 10\exp\theta + \log 10$, $-10 + \log 10 - 9\theta - 5\theta^2$ **13.51** **a** $\theta - \exp\theta$ **b** 0 **c** $\hat{\theta}_{\nu+1} = \hat{\theta}_\nu - 1 + \exp(-\hat{\theta}_\nu)$
d $-11 + \exp 10 \doteq 22016.5$ **e** $l''(\theta) \to 0$ rapidly as $\theta \to -\infty$ **f** Step-halving **g** 104
h $l''(\theta) \to -\infty$ rapidly as $\theta \to \infty$ **i** Step-doubling (suitably defined)

Section 13.5

13.52 Normal with mean 0 and variance $1/[n_1 \exp(\theta) + 4n_2 \exp(2\theta)]$ **13.53 a** $(-0.419, -0.014)$ **b** $2\Phi(-2.100) \doteq .0357$ **13.54 a** $(0.355, 1.369)$, $(1.426, 3.933)$ **b** $(-0.344, 0.519)$, $(0.709, 1.681)$ **c** $2[1 - \Phi(0.398)] \doteq .690$ **13.55** $(0.422, 0.700)$, $(1.525, 2.014)$
13.56 a $\mathrm{diag}(n_1\lambda_1, \ldots, n_d\lambda_d) = \mathrm{diag}(n_1\lambda(\mathbf{x}_1), \ldots, n_d\lambda(\mathbf{x}_d))$ **b** $\mathrm{diag}(n_1\hat{\lambda}(\mathbf{x}_1), \ldots, n_d\hat{\lambda}(\mathbf{x}_d))$
c $\lambda^2(\mathbf{x}) \mathrm{AV}(\hat{\theta}(\mathbf{x}))$ **d** $\hat{\lambda}(\mathbf{x}) \mathrm{SE}(\hat{\theta}(\mathbf{x}))$ **13.57** $\mathrm{AV}(\hat{\lambda}(\mathbf{x}_2)/\hat{\lambda}(\mathbf{x}_1)) = \mathbf{c}^T\mathbf{V}\mathbf{c}$ and $\mathrm{SE}(\hat{\lambda}(\mathbf{x}_2)/\hat{\lambda}(\mathbf{x}_1))$
$= \sqrt{\hat{\mathbf{c}}^T\hat{\mathbf{V}}\hat{\mathbf{c}}}$, where $\mathbf{c} = \frac{\lambda(\mathbf{x}_2)}{\lambda(\mathbf{x}_1)}[\mathbf{g}(\mathbf{x}_2) - \mathbf{g}(\mathbf{x}_1)]$ and $\hat{\mathbf{c}} = \frac{\hat{\lambda}(\mathbf{x}_2)}{\hat{\lambda}(\mathbf{x}_1)}[\mathbf{g}(\mathbf{x}_2) - \mathbf{g}(\mathbf{x}_1)]$ **13.58** $\mathbf{c}^T\mathbf{V}\mathbf{c}$,
where **c** equals **a** $\frac{\sigma^2(\mathbf{x}_2)}{\mu(\mathbf{x}_2)}\mathbf{g}(\mathbf{x}_2) - \frac{\sigma^2(\mathbf{x}_1)}{\mu(\mathbf{x}_1)}\mathbf{g}(\mathbf{x}_1)$ **b** $[1 - \pi(\mathbf{x}_2)]\mathbf{g}(\mathbf{x}_2) - [1 - \pi(\mathbf{x}_1)]\mathbf{g}(\mathbf{x}_1)$
c $\mathbf{g}(\mathbf{x}_2) - \mathbf{g}(\mathbf{x}_1)$ **13.61** It follows from the chain rule that $\frac{\partial h}{\partial \beta_j}(\boldsymbol{\beta}) = h_2'(h_1(\boldsymbol{\beta}))\frac{\partial h_1}{\partial \beta_j}(\boldsymbol{\beta})$ for

$1 \le j \le p$.   **13.62** Since h_2 is strictly increasing, $h_2(L_1) < h_2(\tau_1) < h_2(U_1)$ if and only if $L_1 < \tau_1 < U_1$.

Section 13.6

13.66 $1 - \chi_1^2(4.872) \doteq .0273$, which is close to the *P*-value for the Wald test of the hypothesis that $\theta_1 = \theta_2$ [see the solution to Example 12.18(c) in Section 12.3] **13.67** $1 - \chi_2^2(2.837) \doteq .417$, so there is no evidence that the risk of getting CHD depends on the level of cholesterol among the older age group when the distinction between men and women is ignored.   **13.68 a** $2[1 - \Phi(2.806)] \doteq .0050$ **b** $1 - \chi_1^2(8.131) \doteq .0044$ (Thus, by either approach, there is strong evidence that the risk of getting CHD depends on the level of cholesterol among the older age group. This differs from the conclusion in the answer to Problem 13.67.)   **13.73** $1 - \chi_1^2(3.007) \doteq .083$, which is close to the *P*-value for the Wald test of the hypothesis that $\theta_1 = \theta_2$ [see the answer to Problem 12.35(c) in Section 12.3]   **13.76 a** Use $\mathcal{X} = \{1, 2, 3\}$, G the three-dimensional space of all functions on $\mathcal{X}$, G_0 the one-dimensional subspace of G consisting of all functions g on $\mathcal{X}$ such that $g(1) = 0$ and $g(2) = g(3)$ **b** Here $\hat{\pi}(1) = .6$, $\hat{\pi}(2) = .5$, $\hat{\pi}(3) = .3$, $\hat{\pi}_0(1) = .5$, $\hat{\pi}_0(2) = .4$, and $\hat{\pi}_0(3) = .4$, so $D = 2[30 \log \frac{.6}{.5} + 20 \log \frac{.4}{.5} + 25 \log \frac{.5}{.4} + 25 \log \frac{.5}{.6} + 15 \log \frac{.3}{.4} + 35 \log \frac{.7}{.6}] \doteq 6.215$ and hence $P - \text{value} \doteq 1 - \chi_2^2(6.215) \doteq .0447$. **13.77** $2\{\hat{\theta}Y_1 + 2\hat{\theta}Y_2 + n_1[1 - \exp(\hat{\theta})] + n_2[1 - \exp(2\hat{\theta})]\}$ **13.78** $1 - \chi_1^2(5.058) \doteq .0245$

Tables

Standard Normal Distribution Function

z	.00	.01	.02	.03	.04	.05	.06	.07	.08	.09
0.0	.5000	.5040	.5080	.5120	.5160	.5199	.5239	.5279	.5319	.5359
0.1	.5398	.5438	.5478	.5517	.5557	.5596	.5636	.5675	.5714	.5753
0.2	.5793	.5832	.5871	.5910	.5948	.5987	.6026	.6064	.6103	.6141
0.3	.6179	.6217	.6255	.6293	.6331	.6368	.6406	.6443	.6480	.6517
0.4	.6554	.6591	.6628	.6664	.6700	.6736	.6772	.6808	.6844	.6879
0.5	.6915	.6950	.6985	.7019	.7054	.7088	.7123	.7157	.7190	.7224
0.6	.7257	.7291	.7324	.7357	.7389	.7422	.7454	.7486	.7517	.7549
0.7	.7580	.7611	.7642	.7673	.7703	.7734	.7764	.7794	.7823	.7852
0.8	.7881	.7910	.7939	.7967	.7995	.8023	.8051	.8078	.8106	.8133
0.0	.8159	.8186	.8212	.8238	.8264	.8289	.8315	.8340	.8365	.8389
1.0	.8413	.8438	.8461	.8485	.8508	.8531	.8554	.8577	.8599	.8621
1.1	.8643	.8665	.8686	.8708	.8729	.8749	.8770	.8790	.8810	.8830
1.2	.8849	.8869	.8888	.8907	.8925	.8944	.8962	.8980	.8997	.9015
1.3	.9032	.9049	.9066	.9082	.9099	.9115	.9131	.9147	.9162	.9177
1.4	.9192	.9207	.9222	.9236	.9251	.9265	.9279	.9292	.9306	.9319
1.5	.9332	.9345	.9357	.9370	.9382	.9394	.9406	.9418	.9429	.9441
1.6	.9452	.9463	.9474	.9484	.9495	.9505	.9515	.9525	.9535	.9545
1.7	.9554	.9564	.9573	.9582	.9591	.9599	.9608	.9616	.9625	.9633
1.8	.9641	.9649	.9656	.9664	.9671	.9678	.9686	.9693	.9699	.9706
1.9	.9713	.9719	.9726	.9732	.9738	.9744	.9750	.9756	.9761	.9767
2.0	.9772	.9778	.9783	.9788	.9793	.9798	.9803	.9808	.9812	.9817
2.1	.9821	.9826	.9830	.9834	.9838	.9842	.9846	.9850	.9854	.9857
2.2	.9861	.9864	.9868	.9871	.9875	.9878	.9881	.9884	.9887	.9890
2.3	.9893	.9896	.9898	.9901	.9904	.9906	.9909	.9911	.9913	.9916
2.4	.9918	.9920	.9922	.9925	.9927	.9929	.9931	.9932	.9934	.9936
2.5	.9938	.9940	.9941	.9943	.9945	.9946	.9948	.9949	.9951	.9952
2.6	.9953	.9955	.9956	.9957	.9959	.9960	.9961	.9962	.9963	.9964
2.7	.9965	.9966	.9967	.9968	.9969	.9970	.9971	.9972	.9973	.9974
2.8	.9974	.9975	.9976	.9977	.9977	.9978	.9979	.9979	.9980	.9981
2.9	.9981	.9982	.9982	.9983	.9984	.9984	.9985	.9985	.9986	.9986
3.0	.9987	.9987	.9987	.9988	.9988	.9989	.9989	.9989	.9990	.9990
3.1	.9990	.9991	.9991	.9991	.9992	.9992	.9992	.9992	.9993	.9993
3.2	.9993	.9993	.9994	.9994	.9994	.9994	.9994	.9995	.9995	.9995
3.3	.9995	.9995	.9995	.9996	.9996	.9996	.9996	.9996	.9996	.9997
3.4	.9997	.9997	.9997	.9997	.9997	.9997	.9997	.9997	.9997	.9998
3.5	.9998	.9998	.9998	.9998	.9998	.9998	.9998	.9998	.9998	.9998
3.6	.9998	.9998	.9999	.9999	.9999	.9999	.9999	.9999	.9999	.9999

Quantiles of the Standard Normal Distribution

p	.000	.001	.002	.003	.004	.005	.006	.007	.008	.009
.50	0.000	0.003	0.005	0.008	0.010	0.013	0.015	0.018	0.020	0.023
.51	0.025	0.028	0.030	0.033	0.035	0.038	0.040	0.043	0.045	0.048
.52	0.050	0.053	0.055	0.058	0.060	0.063	0.065	0.068	0.070	0.073
.53	0.075	0.078	0.080	0.083	0.085	0.088	0.090	0.093	0.095	0.098
.54	0.100	0.103	0.105	0.108	0.111	0.113	0.116	0.118	0.121	0.123
.55	0.126	0.128	0.131	0.133	0.136	0.138	0.141	0.143	0.146	0.148
.56	0.151	0.154	0.156	0.159	0.161	0.164	0.166	0.169	0.171	0.174
.57	0.176	0.179	0.181	0.184	0.187	0.189	0.192	0.194	0.197	0.199
.58	0.202	0.204	0.207	0.210	0.212	0.215	0.217	0.220	0.222	0.225
.59	0.228	0.230	0.233	0.235	0.238	0.240	0.243	0.246	0.248	0.251
.60	0.253	0.256	0.259	0.261	0.264	0.266	0.269	0.272	0.274	0.277
.61	0.279	0.282	0.285	0.287	0.290	0.292	0.295	0.298	0.300	0.303
.62	0.305	0.308	0.311	0.313	0.316	0.319	0.321	0.324	0.327	0.329
.63	0.332	0.335	0.337	0.340	0.342	0.345	0.348	0.350	0.353	0.356
.64	0.358	0.361	0.364	0.366	0.369	0.372	0.375	0.377	0.380	0.383
.65	0.385	0.388	0.391	0.393	0.396	0.399	0.402	0.404	0.407	0.410
.66	0.412	0.415	0.418	0.421	0.423	0.426	0.429	0.432	0.434	0.437
.67	0.440	0.443	0.445	0.448	0.451	0.454	0.457	0.459	0.462	0.465
.68	0.468	0.470	0.473	0.476	0.479	0.482	0.485	0.487	0.490	0.493
.69	0.496	0.499	0.502	0.504	0.507	0.510	0.513	0.516	0.519	0.522
.70	0.524	0.527	0.530	0.533	0.536	0.539	0.542	0.545	0.548	0.550
.71	0.553	0.556	0.559	0.562	0.565	0.568	0.571	0.574	0.577	0.580
.72	0.583	0.586	0.589	0.592	0.595	0.598	0.601	0.604	0.607	0.610
.73	0.613	0.616	0.619	0.622	0.625	0.628	0.631	0.634	0.637	0.640
.74	0.643	0.646	0.650	0.653	0.656	0.659	0.662	0.665	0.668	0.671
.75	0.674	0.678	0.681	0.684	0.687	0.690	0.693	0.697	0.700	0.703
.76	0.706	0.710	0.713	0.716	0.719	0.722	0.726	0.729	0.732	0.736
.77	0.739	0.742	0.745	0.749	0.752	0.755	0.759	0.762	0.765	0.769
.78	0.772	0.776	0.779	0.782	0.786	0.789	0.793	0.796	0.800	0.803
.79	0.806	0.810	0.813	0.817	0.820	0.824	0.827	0.831	0.834	0.838
.80	0.842	0.845	0.849	0.852	0.856	0.860	0.863	0.867	0.871	0.874
.81	0.878	0.882	0.885	0.889	0.893	0.896	0.900	0.904	0.908	0.912
.82	0.915	0.919	0.923	0.927	0.931	0.935	0.938	0.942	0.946	0.950
.83	0.954	0.958	0.962	0.966	0.970	0.974	0.978	0.982	0.986	0.990
.84	0.994	0.999	1.003	1.007	1.011	1.015	1.019	1.024	1.028	1.032
.85	1.036	1.041	1.045	1.049	1.054	1.058	1.063	1.067	1.071	1.076
.86	1.080	1.085	1.089	1.094	1.098	1.103	1.108	1.112	1.117	1.122
.87	1.126	1.131	1.136	1.141	1.146	1.150	1.155	1.160	1.165	1.170
.88	1.175	1.180	1.185	1.190	1.195	1.200	1.206	1.211	1.216	1.221
.89	1.227	1.232	1.237	1.243	1.248	1.254	1.259	1.265	1.270	1.276
.90	1.282	1.287	1.293	1.299	1.305	1.311	1.317	1.323	1.329	1.335
.91	1.341	1.347	1.353	1.359	1.366	1.372	1.379	1.385	1.392	1.398
.92	1.405	1.412	1.419	1.426	1.433	1.440	1.447	1.454	1.461	1.468
.93	1.476	1.483	1.491	1.499	1.506	1.514	1.522	1.530	1.538	1.546
.94	1.555	1.563	1.572	1.580	1.589	1.598	1.607	1.616	1.626	1.635
.95	1.645	1.655	1.665	1.675	1.685	1.695	1.706	1.717	1.728	1.739
.96	1.751	1.762	1.774	1.787	1.799	1.812	1.825	1.838	1.852	1.866
.97	1.881	1.896	1.911	1.927	1.943	1.960	1.977	1.995	2.014	2.034
.98	2.054	2.075	2.097	2.120	2.144	2.170	2.197	2.226	2.257	2.290
.99	2.326	2.366	2.409	2.457	2.512	2.576	2.652	2.748	2.878	3.090

Quantiles of the *t* Distribution with ν Degrees of Freedom

	.6	.7	.8	.9	.95	.975	.99	.995	.999
1	0.325	0.727	1.376	3.078	6.314	12.706	31.821	63.657	318.309
2	0.289	0.617	1.061	1.886	2.920	4.303	6.965	9.925	22.327
3	0.277	0.584	0.978	1.638	2.353	3.182	4.541	5.841	10.215
4	0.271	0.569	0.941	1.533	2.132	2.776	3.747	4.604	7.173
5	0.267	0.559	0.920	1.476	2.015	2.571	3.365	4.032	5.893
6	0.265	0.553	0.906	1.440	1.943	2.447	3.143	3.707	5.208
7	0.263	0.549	0.896	1.415	1.895	2.365	2.998	3.499	4.785
8	0.262	0.546	0.889	1.397	1.860	2.306	2.896	3.355	4.501
9	0.261	0.543	0.883	1.383	1.833	2.262	2.821	3.250	4.297
10	0.260	0.542	0.879	1.372	1.812	2.228	2.764	3.169	4.144
11	0.260	0.540	0.876	1.363	1.796	2.201	2.718	3.106	4.025
12	0.259	0.539	0.873	1.356	1.782	2.179	2.681	3.055	3.930
13	0.259	0.538	0.870	1.350	1.771	2.160	2.650	3.012	3.852
14	0.258	0.537	0.868	1.345	1.761	2.145	2.624	2.977	3.787
15	0.258	0.536	0.866	1.341	1.753	2.131	2.602	2.947	3.733
16	0.258	0.535	0.865	1.337	1.746	2.120	2.583	2.921	3.686
17	0.257	0.534	0.863	1.333	1.740	2.110	2.567	2.898	3.646
18	0.257	0.534	0.862	1.330	1.734	2.101	2.552	2.878	3.610
19	0.257	0.533	0.861	1.328	1.729	2.093	2.539	2.861	3.579
20	0.257	0.533	0.860	1.325	1.725	2.086	2.528	2.845	3.552
21	0.257	0.532	0.859	1.323	1.721	2.080	2.518	2.831	3.527
22	0.256	0.532	0.858	1.321	1.717	2.074	2.508	2.819	3.505
23	0.256	0.532	0.858	1.319	1.714	2.069	2.500	2.807	3.485
24	0.256	0.531	0.857	1.318	1.711	2.064	2.492	2.797	3.467
25	0.256	0.531	0.856	1.316	1.708	2.060	2.485	2.787	3.450
26	0.256	0.531	0.856	1.315	1.706	2.056	2.479	2.779	3.435
27	0.256	0.531	0.855	1.314	1.703	2.052	2.473	2.771	3.421
28	0.256	0.530	0.855	1.313	1.701	2.048	2.467	2.763	3.408
29	0.256	0.530	0.854	1.311	1.699	2.045	2.462	2.756	3.396
30	0.256	0.530	0.854	1.310	1.697	2.042	2.457	2.750	3.385
35	0.255	0.529	0.852	1.306	1.690	2.030	2.438	2.724	3.340
40	0.255	0.529	0.851	1.303	1.684	2.021	2.423	2.704	3.307
45	0.255	0.528	0.850	1.301	1.679	2.014	2.412	2.690	3.281
50	0.255	0.528	0.849	1.299	1.676	2.009	2.403	2.678	3.261
55	0.255	0.527	0.848	1.297	1.673	2.004	2.396	2.668	3.245
60	0.254	0.527	0.848	1.296	1.671	2.000	2.390	2.660	3.232
65	0.254	0.527	0.847	1.295	1.669	1.997	2.385	2.654	3.220
70	0.254	0.527	0.847	1.294	1.667	1.994	2.381	2.648	3.211
75	0.254	0.527	0.846	1.293	1.665	1.992	2.377	2.643	3.202
80	0.254	0.526	0.846	1.292	1.664	1.990	2.374	2.639	3.195
85	0.254	0.526	0.846	1.292	1.663	1.988	2.371	2.635	3.189
90	0.254	0.526	0.846	1.291	1.662	1.987	2.368	2.632	3.183
95	0.254	0.526	0.845	1.291	1.661	1.985	2.366	2.629	3.178
100	0.254	0.526	0.845	1.290	1.660	1.984	2.364	2.626	3.174
105	0.254	0.526	0.845	1.290	1.659	1.983	2.362	2.623	3.170
110	0.254	0.526	0.845	1.289	1.659	1.982	2.361	2.621	3.166
115	0.254	0.526	0.845	1.289	1.658	1.981	2.359	2.619	3.163
120	0.254	0.526	0.845	1.289	1.658	1.980	2.358	2.617	3.160
∞	0.253	0.524	0.842	1.282	1.645	1.960	2.326	2.676	3.090

ν

Quantiles of the Chi-Square Distribution with ν Degrees of Freedom

	p							
	.005	.01	.025	.05	.95	.975	.99	.995
1	0.000	0.000	0.001	0.004	3.841	5.024	6.635	7.879
2	0.010	0.020	0.051	0.103	5.991	7.378	9.210	10.597
3	0.072	0.115	0.216	0.352	7.815	9.348	11.345	12.838
4	0.207	0.297	0.484	0.711	9.488	11.143	13.277	14.860
5	0.412	0.554	0.831	1.145	11.070	12.833	15.086	16.750
6	0.676	0.872	1.237	1.635	12.592	14.449	16.812	18.548
7	0.989	1.239	1.690	2.167	14.067	16.013	18.475	20.278
8	1.344	1.646	2.180	2.733	15.507	17.535	20.090	21.955
9	1.735	2.088	2.700	3.325	16.919	19.023	21.666	23.589
10	2.156	2.558	3.247	3.940	18.307	20.483	23.209	25.188
11	2.603	3.053	3.816	4.575	19.675	21.920	24.725	26.757
12	3.074	3.571	4.404	5.226	21.026	23.337	26.217	28.300
13	3.565	4.107	5.009	5.892	22.362	24.736	27.688	29.819
14	4.075	4.660	5.629	6.571	23.685	26.119	29.141	31.319
15	4.601	5.229	6.262	7.261	24.996	27.488	30.578	32.801
16	5.142	5.812	6.908	7.962	26.296	28.845	32.000	34.267
17	5.697	6.408	7.564	8.672	27.587	30.191	33.409	35.718
18	6.265	7.015	8.231	9.390	28.869	31.526	34.805	37.156
19	6.844	7.633	8.907	10.117	30.144	32.852	36.191	38.582
20	7.434	8.260	9.591	10.851	31.410	34.170	37.566	39.997
21	8.034	8.897	10.283	11.591	32.671	35.479	38.932	41.401
22	8.643	9.542	10.982	12.338	33.924	36.781	40.289	42.796
23	9.260	10.196	11.689	13.091	35.172	38.076	41.638	44.181
24	9.886	10.856	12.401	13.848	36.415	39.364	42.980	45.559
25	10.520	11.524	13.120	14.611	37.652	40.646	44.314	46.928
26	11.160	12.198	13.844	15.379	38.885	41.923	45.642	48.290
27	11.808	12.879	14.573	16.151	40.113	43.195	46.963	49.645
28	12.461	13.565	15.308	16.928	41.337	44.461	48.278	50.993
29	13.121	14.256	16.047	17.708	42.557	45.722	49.588	52.336
30	13.787	14.953	16.791	18.493	43.773	46.979	50.892	53.672
35	17.192	18.509	20.569	22.465	49.802	53.203	57.342	60.275
40	20.707	22.164	24.433	26.509	55.758	59.342	63.691	66.766
45	24.311	25.901	28.366	30.612	61.656	65.410	69.957	73.166
50	27.991	29.707	32.357	34.764	67.505	71.420	76.154	79.490
55	31.735	33.570	36.398	38.958	73.311	77.380	82.292	85.749
60	35.534	37.485	40.482	43.188	79.082	83.298	88.379	91.952
65	39.383	41.444	44.603	47.450	84.821	89.177	94.422	98.105
70	43.275	45.442	48.758	51.739	90.531	95.023	100.425	104.215
75	47.206	49.475	52.942	56.054	96.217	100.839	106.393	110.286
80	51.172	53.540	57.153	60.391	101.879	106.629	112.329	116.321
85	55.170	57.634	61.389	64.749	107.522	112.393	118.236	122.325
90	59.196	61.754	65.647	69.126	113.145	118.136	124.116	128.299
95	63.250	65.898	69.925	73.520	118.752	123.858	129.973	134.247
100	67.328	70.065	74.222	77.929	124.342	129.561	135.807	140.169
105	71.428	74.252	78.536	82.354	129.918	135.247	141.620	146.070
110	75.550	78.458	82.867	86.792	135.480	140.917	147.414	151.948
115	79.692	82.682	87.213	91.242	141.030	146.571	153.191	157.808
120	83.852	86.923	91.573	95.705	146.567	152.211	158.950	163.648

ν

.95, .975, and .99 Quantiles of the *F* Distribution with ν_1 Degrees of Freedom in the Numerator and ν_2 Degrees of Freedom in the Denominator

ν_1

ν_2	1	2	3	4	5	6	7	8	9	10	12	15	20	30
1	161	199	216	225	230	234	237	239	241	242	244	246	248	250
	648	800	864	900	922	937	948	957	963	969	977	985	993	1001
	4052	5000	5403	5625	5764	5859	5929	5981	6023	6056	6107	6158	6209	6261
2	18.51	19.00	19.16	19.25	19.30	19.33	19.35	19.37	19.38	19.40	19.41	19.43	19.45	19.46
	38.51	39.00	39.17	39.25	39.30	39.33	39.36	39.37	39.39	39.40	39.41	39.43	39.45	39.46
	98.50	99.00	99.17	99.25	99.30	99.33	99.36	99.37	99.39	99.40	99.42	99.43	99.45	99.47
3	10.13	9.55	9.28	9.12	9.01	8.94	8.89	8.85	8.81	8.79	8.74	8.70	8.66	8.62
	17.44	16.04	15.44	15.10	14.88	14.73	14.62	14.54	14.47	14.42	14.34	14.25	14.17	14.08
	34.12	30.82	29.46	28.71	28.24	27.91	27.67	27.49	27.35	27.23	27.05	26.87	26.69	26.50
4	7.71	6.94	6.59	6.39	6.26	6.16	6.09	6.04	6.00	5.96	5.91	5.86	5.80	5.75
	12.22	10.65	9.98	9.60	9.36	9.20	9.07	8.98	8.90	8.84	8.75	8.66	8.56	8.46
	21.20	18.00	16.69	15.98	15.52	15.21	14.98	14.80	14.66	14.55	14.37	14.20	14.02	13.84
5	6.61	5.79	5.41	5.19	5.05	4.95	4.88	4.82	4.77	4.74	4.68	4.62	4.56	4.50
	10.01	8.43	7.76	7.39	7.15	6.98	6.85	6.76	6.68	6.62	6.52	6.43	6.33	6.23
	16.26	13.27	12.06	11.39	10.97	10.67	10.46	10.29	10.16	10.05	9.89	9.72	9.55	9.38
6	5.99	5.14	4.76	4.53	4.39	4.28	4.21	4.15	4.10	4.06	4.00	3.94	3.87	3.81
	8.81	7.26	6.60	6.23	5.99	5.82	5.70	5.60	5.52	5.46	5.37	5.27	5.17	5.07
	13.75	10.92	9.78	9.15	8.75	8.47	8.26	8.10	7.98	7.87	7.72	7.56	7.40	7.23
7	5.59	4.74	4.35	4.12	3.97	3.87	3.79	3.73	3.68	3.64	3.57	3.51	3.44	3.38
	8.07	6.54	5.89	5.52	5.29	5.12	4.99	4.90	4.82	4.76	4.67	4.57	4.47	4.36
	12.25	9.55	8.45	7.85	7.46	7.19	6.99	6.84	6.72	6.62	6.47	6.31	6.16	5.99
8	5.32	4.46	4.07	3.84	3.69	3.58	3.50	3.44	3.39	3.35	3.28	3.22	3.15	3.08
	7.57	6.06	5.42	5.05	4.82	4.65	4.53	4.43	4.36	4.30	4.20	4.10	4.00	3.89
	11.26	8.65	7.59	7.01	6.63	6.37	6.18	6.03	5.91	5.81	5.67	5.52	5.36	5.20
9	5.12	4.26	3.86	3.63	3.48	3.37	3.29	3.23	3.18	3.14	3.07	3.01	2.94	2.86
	7.21	5.71	5.08	4.72	4.48	4.32	4.20	4.10	4.03	3.96	3.87	3.77	3.67	3.56
	10.56	8.02	6.99	6.42	6.06	5.80	5.61	5.47	5.35	5.26	5.11	4.96	4.81	4.65
10	4.96	4.10	3.71	3.48	3.33	3.22	3.14	3.07	3.02	2.98	2.91	2.85	2.77	2.70
	6.94	5.46	4.83	4.47	4.24	4.07	3.95	3.85	3.78	3.72	3.62	3.52	3.42	3.31
	10.04	7.56	6.55	5.99	5.64	5.39	5.20	5.06	4.94	4.85	4.71	4.56	4.41	4.25
12	4.75	3.89	3.49	3.26	3.11	3.00	2.91	2.85	2.80	2.75	2.69	2.62	2.54	2.47
	6.55	5.10	4.47	4.12	3.89	3.73	3.61	3.51	3.44	3.37	3.28	3.18	3.07	2.96
	9.33	6.93	5.95	5.41	5.06	4.82	4.64	4.50	4.39	4.30	4.16	4.01	3.86	3.70
15	4.54	3.68	3.29	3.06	2.90	2.79	2.71	2.64	2.59	2.54	2.48	2.40	2.33	2.25
	6.20	4.77	4.15	3.80	3.58	3.41	3.29	3.20	3.12	3.06	2.96	2.86	2.76	2.64
	8.68	6.36	5.42	4.89	4.56	4.32	4.14	4.00	3.89	3.80	3.67	3.52	3.37	3.21
20	4.35	3.49	3.10	2.87	2.71	2.60	2.51	2.45	2.39	2.35	2.28	2.20	2.12	2.04
	5.87	4.46	3.86	3.51	3.29	3.13	3.01	2.91	2.84	2.77	2.68	2.57	2.46	2.35
	8.10	5.85	4.94	4.43	4.10	3.87	3.70	3.56	3.46	3.37	3.23	3.09	2.94	2.78
30	4.17	3.32	2.92	2.69	2.53	2.42	2.33	2.27	2.21	2.16	2.09	2.01	1.93	1.84
	5.57	4.18	3.59	3.25	3.03	2.87	2.75	2.65	2.57	2.51	2.41	2.31	2.20	2.07
	7.56	5.39	4.51	4.02	3.70	3.47	3.30	3.17	3.07	2.98	2.84	2.70	2.55	2.39
40	4.08	3.23	2.84	2.61	2.45	2.34	2.25	2.18	2.12	2.08	2.00	1.92	1.84	1.74
	5.42	4.05	3.46	3.13	2.90	2.74	2.62	2.53	2.45	2.39	2.29	2.18	2.07	1.94
	7.31	5.18	4.31	3.83	3.51	3.29	3.12	2.99	2.89	2.80	2.66	2.52	2.37	2.20
60	4.00	3.15	2.76	2.53	2.37	2.25	2.17	2.10	2.04	1.99	1.92	1.84	1.75	1.65
	5.29	3.93	3.34	3.01	2.79	2.63	2.51	2.41	2.33	2.27	2.17	2.06	1.94	1.82
	7.08	4.98	4.13	3.65	3.34	3.12	2.95	2.82	2.72	2.63	2.50	2.35	2.20	2.03
80	3.96	3.11	2.72	2.49	2.33	2.21	2.13	2.06	2.00	1.95	1.88	1.79	1.70	1.60
	5.22	3.86	3.28	2.95	2.73	2.57	2.45	2.35	2.28	2.21	2.11	2.00	1.88	1.75
	6.96	4.88	4.04	3.56	3.26	3.04	2.87	2.74	2.64	2.55	2.42	2.27	2.12	1.94
100	3.94	3.09	2.70	2.46	2.31	2.19	2.10	2.03	1.97	1.93	1.85	1.77	1.68	1.57
	5.18	3.83	3.25	2.92	2.70	2.54	2.42	2.32	2.24	2.18	2.08	1.97	1.85	1.71
	6.90	4.82	3.98	3.51	3.21	2.99	2.82	2.69	2.59	2.50	2.37	2.22	2.07	1.89
120	3.92	3.07	2.68	2.45	2.29	2.18	2.09	2.02	1.96	1.91	1.83	1.75	1.66	1.55
	5.15	3.80	3.23	2.89	2.67	2.52	2.39	2.30	2.22	2.16	2.05	1.94	1.82	1.69
	6.85	4.79	3.95	3.48	3.17	2.96	2.79	2.66	2.56	2.47	2.34	2.19	2.03	1.86
∞	3.84	3.00	2.60	2.37	2.21	2.10	2.01	1.94	1.88	1.83	1.75	1.67	1.57	1.46
	5.02	3.69	3.12	2.79	2.57	2.41	2.29	2.19	2.11	2.05	1.94	1.83	1.71	1.57
	6.63	4.61	3.78	3.32	3.02	2.80	2.64	2.51	2.41	2.32	2.18	2.04	1.88	1.70

Index